34th EUROPEAN SYMPOSIUM ON COMPUTER AIDED PROCESS ENGINEERING / 15th INTERNATIONAL SYMPOSIUM ON PROCESS SYSTEMS ENGINEERING

VOLUME 1

COMPUTER-AIDED CHEMICAL ENGINEERING, 53

34th EUROPEAN SYMPOSIUM ON COMPUTER AIDED PROCESS ENGINEERING / 15th INTERNATIONAL SYMPOSIUM ON PROCESS SYSTEMS ENGINEERING

VOLUME 1

Edited by

Flavio Manenti
"Giulio Natta" Department of Chemistry, Materials and Chemical Engineering, Polytechnic University of Milan, Milan, Italy

Gintaras V. Reklaitis
Davidson School of Chemical Engineering, Purdue University, West Lafayette, Indiana, United States

ELSEVIER

Amsterdam – Boston – Heidelberg – London – New York – Oxford
Paris – San Diego – San Francisco – Singapore – Sydney – Tokyo

Elsevier
Radarweg 29, PO Box 211, 1000 AE Amsterdam, Netherlands
The Boulevard, Langford Lane, Kidlington, Oxford OX5 1GB, UK
50 Hampshire Street, 5th Floor, Cambridge, MA 02139, USA

Notices
Knowledge and best practice in this field are constantly changing. As new research and experience broaden our understanding, changes in research methods, professional practices, or medical treatment may become necessary.

Practitioners and researchers must always rely on their own experience and knowledge in evaluating and using any information, methods, compounds, or experiments described herein. In using such information or methods they should be mindful of their own safety and the safety of others, including parties for whom they have a professional responsibility.

To the fullest extent of the law, neither the Publisher nor the authors, contributors, or editors, assume any liability for any injury and/or damage to persons or property as a matter of products liability, negligence or otherwise, or from any use or operation of any methods, products, instructions, or ideas contained in the material herein.

British Library Cataloguing in Publication Data
A catalogue record for this book is available from the British Library

Library of Congress Cataloging-in-Publication Data
A catalog record for this book is available from the Library of Congress

ISBN (Volume 1): 978-0-443-33897-7
ISBN (Set) : 978-0-443-28824-1
ISSN: 1570-7946

For information on all Elsevier publications visit our website at https://www.elsevier.com/

Working together
to grow libraries in
developing countries

www.elsevier.com • www.bookaid.org

Publisher: Candice Janco
Acquisition Editor: Anita Koch
Editorial Project Manager: Lena Sparks
Production Project Manager: Paul Prasad Chandramohan
Designer: Mark Rogers

Typeset by STRAIVE

Content

Preface

This proceedings volume contains the papers presented at the joint conference that combined the 34th European Symposium on Computer-Aided Process Engineering (ESCAPE) and the 15th International Symposium of Process Systems Engineering (PSE), which was held in Florence, Italy on June 2-6, 2024. While the ESCAPE conferences have been convened annually since 1991, the PSE series was initiated in 1982 and has been occurring every three years, with venues rotating between the Americas, Asia and Europe. This is the fourth instance in which the two conferences have been jointly held in Europe (Trondheim 1997, Garmisch-Partenkirchen (2006) and Copenhagen (2015)).

The internationally publicized call for papers for the joint symposium drew a total of 926 abstracts, which upon review, resulted in 791 submissions of conference papers. Of the total of 668 papers which were accepted upon review, the final number of papers appearing in this proceedings volume is 590.

The scientific contents of the joint conference have been organized into eight important topics within PSE. The topic coordinators and distribution of papers within these topics are as follows:

Modeling and Simulation: 200
        Co-ordinators: Profs Ana Barbosa-Povoa, David Bogle, Fani Boukouvala

Synthesis and Design: 58
        Co-ordinators: Profs Antonio Espuna, Zdravko Kravanja, Arturo Jiminez-Gutierrz

Process Control and Operations: 77
        Co-ordinators: Profs Michael Baldea, Sebastian Engell, Sigurd Skogestad

CAPE in Sustainable Energy Applications: 70
        Co-ordinators: Profs Fabrizio Bezzo, Dominic Foo, Francois Marechal

Bioresources, Bioprocesses and Biomedical systems: 41
        Co-ordinators: Profs Antonis Kokossis, Stavros Papadokostantakis, Maria Soledad Diaz

Digitalization and Machine Learning: 85
        Co-ordinators: Profs Filip Logist, Flavio Manenti, Jinsong Zhao

Concepts, Methods and Tools: 51
       Co-ordinators: Profs Ludovic Montastruc, Manabu Kano, Heinz A Preisig

Education. In CAPE and Knowledge Transfer: 8
       Co-ordinators: Profs Gregoire Leonard, Il Moon, Patrizia Perego

The session coordinators have organized the contributions within each of these eight topics into an exciting program of parallel sessions spanning the four days of the conference. Additionally, the program features seven plenary lectures, given by outstanding researchers from the three geographic zones in which the PSE Community is active.

The collective body of knowledge captured in this proceedings volume fully represents the state of the art of developments in PSE methodology and application and is evidence of the high level of accomplishment and innovation of the global PSE community. We are grateful for the collegial efforts of the authors, members of the Scientific Committee and reviewers, topical coordinators, plenary speakers and staff of AIDIC which have resulted in an outstanding joint ESCAP 34 / PSE 2024 Symposium.

Prof. Flavio Manenti and Prof. Gintaras V Reklaitis

Scientific Committee

Chairboard

MANENTI Flavio, WP Chairman, Politecnico di Milano, Italy
REKLAITIS Gintaras V. (Rex), PSE Executive Board, Purdue University, USA

Members

ADAMS Thomas, Canada
AGRAWAL Rakesh, USA
ALMEIDA RIVERA Cristhian, The Netherlands
ARSENYEVA Olga, Ukraine
ASPRION Norbert, Germany
AVISO Kathleen, Philippines
BALDEA Michael, USA
BANDYOPADHYAY Santanu, India
BARATTI Roberto, Italy
BARBOSA POVOA Ana Paola, Portugal
BEZZO Fabrizio, Italy
BIEGLER Lorenz, USA
BOGLE David, UK
BOLLAS George, USA
BOUKOUVALA Fani, USA
BOZZANO Giulia, Italy
BROECKER Soenke, Germany
CAFARO Diego, Argentina
CAO Yankai, Canada
CHANG Chuei-Tin, Taiwan
CHEMMANGATTUVALAPPI Nishanth, Malaysia
CHENG-LIANG Chen, Taiwan
CHIEN I Lung, Taiwan
CHIU Min Sen, Singapore
CISTERNAS Luis, Chile
CREMASCHI Selen, USA
CRUZ Nicolas, Germany
DENG Chun, China
DIAZ Felipe, Chile
DOWLING Alex, USA
DURAND Helen, USA
EDEN Mario Richard, USA
EL-HALWAGI Mahmoud, USA
ENGELL Sebastian, Germany
ESPUNA Antonio, Spain
FAIRWEATHER Michael, UK
FIKAR Miroslav, Slovakia
FLORES TLACAUHUAC Antonio, Mexico
FOO Dominic C.Y., Malaysia
FORD-VERSYPT Ashlee, USA
FRIEDLER Ferenc, Hungary
FURLAN Felipe Fernando, Brazil
GANI Rafiqul, Denmark
GEORGIADIS Michail, Greece
GOMEZ Jorge Mario, Colombia
GOUNARIS Chrysanthos, USA

GROSSMANN Ignacio, USA
HANDOGO Renanto, Indonesia
HASAN Farouqe, USA
HE Yijun, China
HIRAO Masahiko, Japan
HOADLEY Andrew, Australia
HOFMANN Rene, Austria
HUSSAIN Mohd Azlan, Malaysia
IERAPETRITOU Marianthi, USA
JIMINEZ Arturo, Mexico
KANO Manabu, Japan
KAPUSTENKO Petro, Ukraine
KARIMI Iftekhar, Singapore
KAWAJIRI Yoshiaki, Japan
KELLER Andreas, Switzerland
KHEAWHOM Soorathep, Thailand
KIM Jin-Kuk, South Korea
KITTISUPAKORN Paisan, Thailand
KOKOSSIS Antonis, Greece
KONDILI Aimilia, Greece
KRASLAWSKI Andrzej, Finland
KRAVANJA Zdravko, Slovenia
KUBICEK Milan, Czech Republic
LAIRD Carl, USA
LATIFI Abderrazak, France
LE ROUX Galo Cariiillo, Brazil
LEE Jongmin, South Korea
LEE Jui-Yuan, Taiwan
LEONARD Gregoire, Belgium
LIAO Zuwei, China
LOGIST Filip, Belgium
MACIEL FILHO Rubens , Brazil
MAHALEC Vladimir, Canada
MAJOZI Thokozani, South Africa
MANCA Davide, Italy
MARECHAL Francois, Switzerland
MARTELLI Emanuele, Italy
MATOS Henrique, Portugal
MATSUMOTO Hideyuki, Japan
MCAULEY Kim, Canada
MENDEZ Carlos, Argentina
MENSHUTINA Natalia V., Russia
MHAMDI Adel, Germany
MIRCEA CRISTEA Vasile, Romania
MITSOS Alexander, Germany
MIZSEY Peter, Hungary
MOIOLI Stefania, Italy

MONTASTRUC Ludovic, France
MOON Il, South Korea
NAGY Zoltan, USA
NG Denny, Malaysia
NODA Masaru, Japan
PAPADOKONSTANTAKIS Stavros, Austria
PEREGO Patrizia, Italy
PETERS Bernhard, Luxembourg
PIERUCCI Sauro, Italy
PISTIKOPOULOS Stratos, USA
PLAZL Igor, Slovenia
PLESU Valentin, Romania
PONCE ORTEGA Jose Maria, Mexico
PREISIG Heinz, Norway
PUIGJANER Luis, Spain
REALFF Matthew, USA
RENGASWAMI Raghunathan, India
RICARDEZ-SANDOVAL Luis, Canada
RODRIGUEZ Analia, Argentina
SAMAVEDHAM Lakshminarayanan, Singapore
SECCHI Argimiro, Brazil
SEGOVIA Juan Gabriel, Mexico
SHAN-HILL WONG David, Taiwan
SIN Gurkan, Denmark

SKOGESTAD Sigurd, Norway
SMITH Robin, UK
SOHELL MANSOURI Seyed, Denmark
SOLEDAD DIAZ Maria, Argentina
SRINIVASAN Rajagopalan, India
STUBER Matt, USA
TADE Moses, Australia
TAN Raymond, Philippines
TASIC Marija, Serbia
TORRES Ana, USA
TRUSOVA Olga, Russia
TURKAY Metin, Turkey
VARBANOV Petar, Czech Republic
VECCHIETTI Aldo, Argentina
WALMSLEY Michael, New Zealand
YOO ChangKyoo, South Korea
YOSHIYUKI Yamashita, Japan
YUAN Xigang, China
YUAN Zhihong, China
ZHANG Lei, China
ZHANG Qi, USA
ZHAO Jinsong, China
ZONDERVAN Edwin, The Netherlands

Organizing Committee

Chairboard

Sauro PIERUCCI (AIDIC Servizi srl President)
Giorgio VERONESI (AIDIC General Secretary, Techint, EFCE President)

Members

Raffaella DAMERIO (AIDIC, Accounting dept.)
Manuela LICCIARDELLO (AIDIC, Events dept.)
Caterina STUCCHI (AIDIC, Events dept.)
Gaia TORNONE (AIDIC, Events dept.)

Flavio Manenti, Gintaras V. Reklaitis (Eds.), Proceedings of the 34th European Symposium on Computer Aided Process Engineering / 15th International Symposium on Process Systems Engineering (ESCAPE34/PSE24), June 2-6, 2024, Florence, Italy

Towards an Integrated Wide Approach for Sustainable Upstream Field Recovery

Shakeel Ramjanee,[a] Nilay Shah[a]

[a] Department of Chemical Engineering, Sargent Centre of Process Systems Engineering, Imperial College London, South Kensington, London, SW7 2AZ, UK.
n.shah@imperial.ac.uk

Abstract

Integrated asset modelling (also known as integrated production modelling) is the modelling and simulation of an entire production facility consisting of both subsurface and surface elements. The intent is to holistically capture the complex interactions between the individual components of the system. The full benefits of these models however are not realised for several reasons, including extensive simulation time of complex models, siloed disciplinary approaches resulting in separate disciplines owning separate elements of the framework. This research study focuses on the development of surrogate models of an integrated field set up for both a gas producing and oil producing supplemented by water injection field. Surrogate models were developed via several techniques (traditional methods such as response surface models and regression and via more complex methods such as Artificial Neural Networks, SobolGSA Random Sampling High Dimensional Modelling Representation and ALAMO) and tested to determine the accuracy versus the fully integrated asset industry tool. Global sensitivity analyses were performed on a selected number of input (manipulated) field variables on the overall outputs from the surrogate models to validate the surrogate relationship postulations prior to attempting optimisation of the developed surrogates via several techniques with positive prediction results observed at a fraction of the simulation time of the industry state of the art tool. The developed surrogate models can be deployed to optimise and forecast production in a fraction of the simulation time; given the forecasted demand of hydrocarbons within the energy mix to meet the global energy demand in the short to medium term; optimising recovery from existing fields is critical especially given the reduced capex investments into new i.e. greenfield fields being developed.

Keywords: Integrated Asset Modelling; Optimisation; Surrogate model development; Reservoir modelling; Sustainable recovery

1. Methodology

Upon setting out on this research post review of the available literature which concluded that prime focus is on reservoir rather than integrated model simplification, attempting to construct an integrated asset model test case to use a basis and develop further understanding of these tools was deemed to be the first step. The following section outlines the industry tools assessed, the subsequent test cases built and the outputs from the simulation forecasts. These models would then be used as a basis for investigating surrogate model development methods. the most widely applied integrated asset modelling tool within the upstream industry was concluded to be the Petroleum Experts (PETEX ®) software suite; as such, this was selected as the basis for the research following application and receipt of education licenses at Imperial College London. An Intel i7 core CPU of 3.4GHZ with 8GB RAM workstation was used in this research study.

1.1. Development of a Case Study Model within PETEX® Framework

Initially, a fully integrated field model (based on a central North Sea field) was developed within the industry state of the art tool, PETEX® integrated asset modelling suite to be utilised as a starting case for development of a proxy model representation. The system consisted of 3 black oil reservoir tanks within MBAL (PETEX® Material Balance zero-dimensional reservoir modelling tool), 8 producing wells across 2 production manifold, 3 production flow lines commingling at a landing host facility separator operating at 20bar. Following the construction of the test case within PETEX®, a simplified approach was adopted as a starting point for evaluating an integrated asset model proxy model proposal; 2 cases were proposed, a single gas well case coupled to an integrated network and an oil producing field supplemented by water injection drive.

1.2. Global Sensitivity Analysis within SobolGSA and Generation of RS-HDMR Proxy Meta Model Representations

Utilising the SobolGSA [4] Proxy meta modelling software (developed internally by Imperial College London specifically for global sensitivity evaluation), a global sensitivity analysis was carried out utilising the input and output data extracted for 164 single solve simulation runs within GAP. The aim was to identify which of the selected 3 manipulated variables had the highest impact on the output (gas production). The input and output data structure was set up as follows:

- Input parameter 1: Separator landing pressure (barg)
- Input parameter 2: Roughness ratio of flowline (dimensionless)
- Input parameter 3: Gas well choke differential pressure (barg)
- Output parameter 1: Optimised gas production (MMScfd)

Within SobolGSA, the input distribution was generated by specifying a lower and upper bound with the distribution type for each parameter (selected as uniform for each of the three input parameters). This generated a random data set which was then used as specified inputs within the GAP model to generate an optimised (single solve) gas production forecast (MMscfd). Utilising this input and subsequent output generated data set, a sensitivity analysis was then run within SobolGSA utilising the RS-HDMR method (Li et al, 2002) of analysis with a Sobol sequence sampling strategy.

1.3 Development of a Gas Well Integrated Asset Model Incorporating a Coupled Numerical Reservoir Model (Eclipse® and Mbal)

Following the successful development of a test integrated field single well gas within GAP and Resolve (PETEX® integrator module to couple reservoir models and surface/well models), the subsequent step of the research was to attempt to develop a numerical reservoir gas model to couple to the integrated asset model via Resolve (PETEX® integrator tool) The aim was to firstly incorporate a detailed reservoir representation to compare simulation time and results versus the simplified integrated tank reservoir model. Secondly, the aim was to extend the research to a more representative reservoir model that is typically employed by subsurface teams within the upstream industry. Two full physics gas only reservoir models were developed in Eclipse® (3D reservoir modelling tool), a single well gas reservoir model without aquifer support and a single well gas reservoir model with numerical aquifer support incorporating water breakthrough after 'X' years of production. The developed 3D reservoir (full physics) models were subsequently coupled to a network GAP model consisting of a single flowline and arrival separator via Resolve. The coupled network was then used as a basis for further proxy modelling development and sensitivity analyses. In addition, a simplified tank reservoir model with a single gas well within Mbal (PETEX®) was developed and coupled to the integrated asset model via Resolve; this was used as a comparison to compare the output and performance of the full physics reservoir model against the simplified tank model.

1.4 Development of an Oil and Water Injection Integrated Asset Model Incorporating a Coupled Numerical 3D Reservoir Model

The Norne reservoir model obtained from NTNU [3] representing a segment of the Norne field was loaded into Eclipse®. The Eclipse® based Norne reservoir model was successfully run as a standalone simulation via the Schlumberger simulation platform. The subsequent step was to couple the reservoir model to an integrated asset framework utilising the PETEX® simulation suite. Within GAP, a network model was developed comprising of 3 producing wells, a common production flowline and a production separator set an arrival pressure of 22 barg. The water injection network comprising of 2 water injection wells and a common water injection manifold was configured in a separate GAP network model and coupled to the Norne Eclipse® Reservoir simulation deck via the PETEX® Resolve platform.

1.5 Sensitivity Analysis via SobolGSA and Subsequent Development of RS-HDMR Proxy Model Representations Incorporating Time Dependency

The SobolGSA sensitivity analysis software applied to the gas well case was deployed for the oil & water injection model case and used to analyse the key manipulated variables to determine which input parameter displays the most impact on the output parameters. The input and output data sets were based on the single solve optimisation results in GAP where the simulator forecasts the optimum oil and gas production respectively for the user specified inputs and hence is time independent. The input and output parameters for the sensitivity analysis was configured as follows:

- Input Parameter 1: Production choke setting, X_o (%)
- Input Parameter 2: Water injection choke setting, X_w (%)
- Input Parameter 3: Separator landing pressure, P (barg)
- Output Parameter 1: Oil production (bbl/d)
- Output Parameter 2: Gas production (MMscfd)

164 single solve simulation runs were executed in a batch manner in GAP to generate the input and output data sets. In the absence of any other information and in line with the maximum entropy principle, the input parameter distribution was specified as uniform, extracted, and run within the integrated asset model to generate the corresponding gas and oil production outputs. Following the generation of the time independent RS-HDMR meta model within SobolGSA, an extension to apply this to a time dependent meta model was investigated. The driver for this was to factor in that the production at different time steps varies particularly when comparing early field life to late field life. The simulation runs required at least 10 minutes per forecast run within PETEX® hence required considerable simulation time to generate the full forecasted data set. SobolGSA generates an output for each individual time step i.e. for each of the 215 time steps, a sensitivity per output parameter is reported for each of the three input parameters.

1.6 Generation of Artificial Neural Network (ANN) Proxy Model Representation

Following the successful coupling of the reservoir models (both material balance within Mbal and numerical models developed within Eclipse®), to the PETEX® GAP network via Resolve, these were then utilised as a template to develop further surrogate representations of the two selected key parameters, the change in reservoir pressure with respect to time and the gas production output as a function of key specified input parameters was postulated per Eq. (1) and Eq. (2):

$$dP/dt \ f \ (G(t)) \tag{1}$$

dP/dt references the change in field pressure with respect to time and G(t) is the gas production at time t.

$$G(t) \ f \ (dP, Xg) \tag{2}$$

dP represents the differential between the field pressure and landing i.e. separator pressure (barg) and Xg represents the choke setpoint on the producing gas well (%). An Artificial Neural Network (ANN) surrogate model was generated utilising the machine learning toolbox within Matlab utilising the data set for both the integrated tank reservoir model and outputs obtained from the coupled material balance reservoir model to the integrated asset model within PETEX®. This workflow was applied to both dP/dT and G(t) functions and both tank and Eclipse® reservoir model cases with the Baysian Regularization training algorithm selected due to its applicability for noisy and difficult data sets.

1.7 Refining of Differential Pressure Profile Utilising Smoothing Spline Fit (Gas Field and Oil Field supplemented by Water Injection Cases)

Following the development of the surrogate representation of both G(t) and dP/dT functions, the results obtained for the dP/dT profile were suboptimal in comparison given the R^2 values obtained for the fit of G(t). On analysis of the differential pressure profile extracted from the PETEX® simulation forecast runs within Resolve, it was evident that there were significant oscillations within the data set that may have contributed to the sub optimal fit of the polynomial. Based on this, a smoothing spline approach was attempted to smooth out the oscillations. This data set was extracted and loaded in Matlab to generate a response fit as well as utilising the third-party surrogate modelling tool, ALAMO [3] to generate a surrogate representation between dPf/dt and G(t). Based on the revised spline differential estimate, ALAMO was used to generate a functional relationship between dPf/dt and G(t). Utilising the generated spline fit differential estimates and the extracted data profiles, an Artificial Neural Network (ANN) model was also derived for both functional relationships applied to the oil and water injection case per Eq. (3) and Eq. (4):

$$O(t) \ f \ (Xw, dP) \tag{3}$$

O(t) represents the oil production at time t, dP represents the change in field pressure less landing pressure (barg), and Xw the choke setpoint on the water injection well (%)

$$dP/dt \ f \ (Xw, O(t)) \tag{4}$$

dP/dt is the change in field pressure with respect to time and Xw represents the choke setpoint on the water injection well (%).

2.0 Results

The following section aims to summarise the key results from the conducted research at Imperial College London. The sensivity analysis carried out with SobolGSA (Imperial College London) for the gas well integrated asset model generated the following sensitivity analysis output per Figure 1 which displays Si, the first order sensitivity of each input parameter towards the output parameter; the higher the Si value i.e. closer to 1, the higher sensitivity towards the output. The separator landing pressure i.e. input parameter 1 demonstrated the largest sensitivity towards the gas production output.

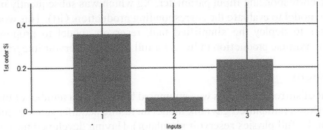

Figure 1: Sensitivity Analysis of input parameters within SobolGSA; Si (Sensitivity Index) vs Input Parameters

The subsequent RS-HDMR meta model generated within Matlab was tested and this generated a positive R^2 correlations at 0.90. The RS-HDMR model is deployed via Matlab to provide an accurate prediction of the production and field pressure profile; incorporation of time dependency demonstrated positive R^2 correlations at early field life conditions but deteriorated at later field life. Figures 2-3 depicts the production and pressure profile predictions from the most optimal surrogate models developed (Response Surface Method, ALAMO, ANN, RS-HDMR methods were deployed in this research).

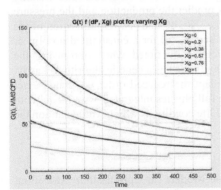

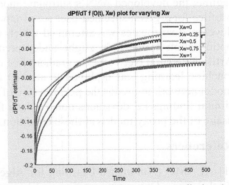

Figure 2: Gas Production prediction via surrogate model at varying water injection choke set points, Xw

Figure 3: Field differential pressure prediction (oil field) via surrogate model at varying water injection choke setpoints, Xw

A constrained non-linear optimisation framework (fmincon) based on the developed surrogate relationship for gas production and differential pressure profiles was set up utilising a custom ODE solver (adapted for stiff differential equations). The resultant cumulative gas production output compared favourably with the corresponding cumulative production output from the integrated asset simulator model. Further optimisation of the developed surrogate models (in addition to optimising the differential defined as epsilon, ε, between the output of the full physics reservoir coupled integrated asset model and the simplified tank reservoir integrated asset model) was investigated. Optimising the RS-HDMR and ANN models utilising metaheuristic methods was evaluated; genetic algorithms were employed to optimise the developed ANN's. Sobol Opt (SobolGSA global optimiser which is designed to solve global non-linear optimisation problems) was deployed to optimise the RS-HDMR models which resulted in slight overpredictions in cumulative production and revenue at optimised conditions. An optimisation framework to minimise the differential term ε was deployed in Matlab to extract the corresponding input parameter, Xg which was subsequently played back to the surrogate model to evaluate the corresponding production, G(t). This workflow would enable to user to deploy the simplified tank reservoir model to propose operational decisions to maximise production in lieu of a full physics reservoir integrated model.

3.0 Conclusions

Development of surrogate models was attempted by utilising a number of integrated asset model frameworks (simplified gas tank model including acquifer support, full physics gas tank model and a full physics reservoir simulator). Having developed the integrated asset model test cases within the industry state of the art PETEX® suite and successfully run several global sensitivity assessments, several approaches were investigated to test fit a relationship for G(t), O(t) and the field/bottom hole pressure profiles, dPf/dt and dPb/dt. Response surface fitting of polynomials demonstrated good correlation for the production functions but less so for the field pressure profiles. Imperial College London's proprietary meta modelling software, SobolGSA was deployed to determine whether extracted input and output datasets could be utilised to generate an RS-HDMR metamodel for predicting absolute production and pressure profiles for a given specified set of input parameters. Development and deploying Artificial Neural Networks (ANN) demonstrated a positive response when training the data sets utilising the Bayesian Regularization training regime; the developed ANN's surrogate models predict the production and the pressure profiles for a given specified user set of inputs i.e. Xg, dP and Xw. Optimisation of the developed surrogates was attempted which was successful for a constrained non-linear optimisation framework; slight under predictions were noted when optimising the Artificial Neural Networks via gradient free methods such as a genetic algorithm. Lastly, machine learning was deployed to attempt to adapt the reduced order integrated asset model network to predict the performance of the full physics integrated asset model with similar findings i.e. deviations noted in later field life conditions which would caution the usage of the proxy models at these conditions.

References

[1] ALAMO (Automated Learning for Algebraic Models for Optimisation) Modelling Tool, The Optimization Firm, LLC, https://minlp.com/ALAMO-modelling-tool (Software); downloaded at Imperial College London in December 2020

[2] Li, G., Wang, S., Rabitz, H., (2002), Practical Approaches To Construct RS-HDMR Component Functions, Journal of Physical Chemistry A 106, 8721-8733

[3] NTNU, Norne Reservoir Database, Department of Geoscience and Petroleum, Norne, http://www.ipt.ntnu.no/~norne/wiki1/doku.php?id=english:license,accessed at Imperial College London in September 2018

[4] SobolGSA, Imperial College London, Department of Chemical Engineering, Centre for Process Systems Engineering, London, SW72AZ, UK (Software accessed internally at Imperial College London), Downloaded January 2018

Flavio Manenti, Gintaras V. Reklaitis (Eds.), Proceedings of the 34th European Symposium on Computer Aided Process Engineering / 15th International Symposium on Process Systems Engineering (ESCAPE34/PSE24), June 2-6, 2024, Florence, Italy

Surrogate modeling application for process system emissions assessment: improving computational performances for plantwide estimations

Giulio Carnio[b], Alessandro Di Pretoro[a*], Fabrizio Bezzo[b], Ludovic Montastruc[a]

[a]*Laboratoire de Génie Chimique, Université de Toulouse, CNRS/INP, Toulouse, France*
[b]*CAPE-Lab, Dipartimento di Principi e Impianti di Ingegneria Chimica, Università di Padova, via Marzolo 9, I-35131 Padova, Italy*

alessandro.dipretoro@ensiacet.fr

Abstract

During the last decade, data driven modeling has gained a role of major interest all over the engineering fields mainly due to the need of higher computational power or, inversely, less computational demanding models for applications such as optimization, simulation, scheduling and control. Relevant contributions of process systems surrogate modeling as a support for operation optimization were already proved in literature with a reduction of the overall computational time by two orders of magnitude with respect to conventional simulations. In this research work a biogas-to-methanol plant case study is used to assess the total energy consumption and estimate the related emissions. Therefore, a modeling phase carried out via Response Surface Methodology is set up in order to obtain the analytical function that allows to estimate the equivalent CO_2 emissions over an extended range of operating conditions representing a wide interval of biomass feed composition. The study has been performed over a wide independent variables domain as well as for different sample sizes in order to compare the computational performances and the accuracy of the obtained models accordingly. The computational time was reduced by two orders of magnitude with a mean relative error lower than 1%. Given the quality of the results, this approach could be further exploited for other system variables and processes including highly non-ideal behaviour of mixtures to be treated. Furthermore, more complex sampling and different surrogate modeling strategies could be tested in order to check if even higher computational effectiveness and model accuracy could be obtained in the process systems domain.

Keywords: Response Surface Methodology, sampling, uncertainty, computational efficiency, biogas to methanol.

1. Introduction

In recent years, the digitalization of the chemical industry on a massive scale has oriented the research efforts in process systems engineering domain towards more computational effective solutions. In this perspective, several modeling alternatives, applicable to other engineering fields as well, that were proposed during the last decades in order to ease the computational criticalities have seen a renewed interest as discussed in Alizadeh et al. (2020). In particular, when dealing with complex process systems, the conventional simulations based on phenomenological models result to be the most highly time demanding step within the optimization loops. A significant example regarding the gain in terms of effectiveness obtained when replacing simulations by data-driven models can be found in Di Pretoro et al. (2022). Their research proved indeed the capability of derivative-based surrogate models to reduce the required computational time for scheduling optimization by a 2,400 factor with 99.66% solution accuracy.

Along with the digital transition, sustainability can be defined as the second current target of major concern in the industrial domain. The most established indicator to assess the environmental impact of a chemical process is the Global Warming Potential whose unit of measure is the amount (kg) of equivalent CO_2 emissions. Due to the uncertain nature of sustainable raw materials, a quick estimation of the expected emissions of the process under variable operating conditions could be a valuable decisional tool during the design and the scheduling phase of a process system. In particular, in a chemical process, the total energy consumption in terms of electricity and heat duties can be the identified as main emission item (Gadalla et al., 2006).

For this reason, a Response Surface Methodology approach, whose details are provided in the dedicated section, is exploited in this study to correlate the emissions of a biogas-to-methanol plant with the feed stream properties over a wide composition range. The obtained model was then tested to assess computational performance indicators such as computational time and model accuracy. The presentation of the case study, its process diagram as well as the operating parameters are provided in detail the next section followed by a thorough description of the modeling methodology.

2. Case study

The modeling strategy presented in this work was applied to a biogas-to-methanol case study. The choice of this system had two main purposes that are namely testing the performances of a modeling procedure on a plantwide scale, i.e. correlating the inlet and the outlet of a process involving several units in between, and assessing the accuracy of a model whose inlet parameters could vary over a wide range, since biogas properties depend on the biomass nature and the fermentation conditions. The process under analysis was modeled and simulated by means of Aspen HYSYS v12 process simulator and its simplified flow diagram is reported in Figure 1. Three main process sections can be identified: (*i*) steam reforming, (*ii*) methanol synthesis and (*iii*) product purification.

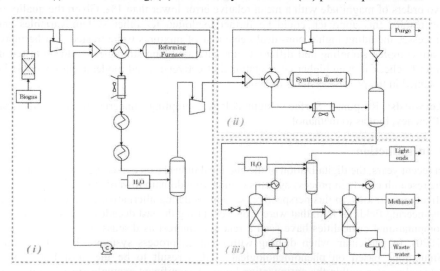

Figure 1 – Simplified Process Flow Diagram of the biogas-to-methanol plant

In the first section, the biogas fed to the process is preliminarily purified from the H_2S, compressed and mixed with water. Then, the mixture is preheated thanks to the heat recovered from the reformer product stream. In the reforming furnace the biogas is then converted into bio-syngas by supplying heat to the system. The reactor is modeled as a fixed bed plug flow reactor with a given size and its products are cooled down by means of an air cooler and a series of heat exchangers in order to condense and separate the syngas from water. The latter is then recycled back mixed with a make up stream. In the second section, the syngas is recompressed to the reactor operating pressure and preheated as in the first one. In the reactor syngas is converted into methanol with a globally exothermic reaction scheme. This second reactor as well was modeled as a fixed bed plug flow reactor. After being purified from unconverted syngas, which is recycled back to the reactor, the methanol is laminated and sent to the third section consisting of a series of distillation steps. In the first column, light ends are removed from the methanol stream and the entrained part is washed with water in a second column in order to further enhance its recovery. Finally, the last column distills methanol from water to grade AA purity.

As previously introduced, the purpose of this study is to model the plantwide emissions for variable inlet feed composition. Therefore, Table 1 reports, along with the operating parameters and the process specifications required to fulfill the final product quality, the selected domain in terms of methane and carbon dioxide partial flowrates over which the Design of Experiment will be performed.

Table 1 - Operating parameters, specifications and modeling domain

System parameters	Value	Unit
Inlet methane flowrate	10.95 ± 4.38	kmol/h
Inlet carbon dioxide flowrate	6.71 ± 2.68	kmol/h
Process specifications		
Water/methane ratio @ reformer	3.075	mol/mol
Outlet methanol purity	0.9985	kg/kg

The assessment of the environmental impact requires a deep understanding of the process. The selected indicator is the Global Warming Potential expressed as CO_2 equivalent emissions. These emissions are primarily related to heating utilities and electricity usage while the impact due to the equipment production and dismission can be neglected as already proved by the study of Gadalla et al. (2006). To be more precise, the emissions related to electricity represents only the 7.9% of the entire process energy demand while the most energy-intensive units are the reforming furnace and the reboilers of the two distillation columns. As concerns the cooling duties, since they work at ambient temperature, cooling water can be selected as utility stream, and their contribution to the overall GWP can be thus considered negligible.

So, finally, the plantwide CO_2 equivalent emissions of the biogas to methanol process were calculated according to the correlations provided in the same research article in order to generate a dataset of simulated points corresponding to different methane and carbon dioxide partial flowrate points. These values could then be employed in the following modeling step to build a surrogate model of the plant emissions according to the methodology whose details are discussed in the next section.

3. Methodology

3.1. Data generation and sampling

All literature studies dealing with metamodels agree about the fact that data sampling is the most critical step of the entire modeling procedure (McBride & Sundmacher, 2019). The quality, the nature and the distribution of the employed dataset indeed strongly affects the accuracy of the obtained model. For this reason, the data generation should follow dedicated Design of Experiment (hereafter DoE) strategies. A preliminary DoE distinction can be made between stationary and sequential modes: in the former case the entire dataset is generated at once and used in a single modeling step while, in the latter, the DoE and the modeling steps are reiterated by updating the data sampling criteria according to the model results of the previous loop with a higher computational effort. In this research study, the once-through approach is used with the goal of having an optimal space-filling over the modeling domain. Several methods for the sampling generation exist such as Monte Carlo sampling, Quasi-Monte Carlo (QMC) Sobol' design or again QMC Halton. For the biogas-to-methanol emissions modeling, the Latin Hypercube Sampling (LHS) (McKay et al., 1979) will be used since it provides good space filling and no particular criticalities in the resulting function are expected. An example of the space partition according to the LHS is provided in Figure 2 (left).

Finally, as concerns data acquisition, two main types of data are mainly exploited for surrogate modeling, namely experimental and generated ones. In this study, as already mentioned in the previous section, data were generated over the selected domain by means of phenomenological models implemented in the process simulation modules.

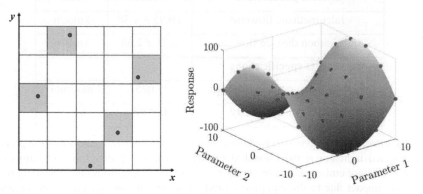

Figure 2 – left) Example of LHS sample distribution - right) Response Surface example

3.2. Modeling

The modeling strategy exploited in this research study is the Response Surface Methodology introduced by Box and Wilson (1951). The principle behind this approach is the use of a Response Surface, usually polynomial, in a n-D space to describe the relationship between n-1 explanatory variables and one or more response variables. From a mathematical point of view, the objective function to minimize is the mismatch between the F function values corresponding to the explanatory variables $\tilde{x}$ and the dataset provided by the DoE step. On the other hand, the output variables of the optimization algorithm are the coefficients K_i and k_i that respectively multiply each term and argument of the Response Surface according to the expression reported in Equation 1.

$$F(\tilde{x}) = \sum_{i=1}^{n} K_i \cdot f_i(\tilde{x}, \tilde{k}) \qquad (1)$$

For a better understanding, a graphical example is provided in Figure 2 (right).
In this study, the optimization is performed on Matlab® by means of the dedicated Curve Fitting Toolbox® that allows to select among polynomial, logarithmic or statistical functions as basis functions of the RSM. As a result, the software provides the coefficients of each function term as well as their confidence interval. Moreover, the toolbox enables the comparison between the obtained surface and a validation dataset provided by the user in order to assess model quality parameters, e.g. the sum of squared errors, while the mean relative error is calculated separately and used as reference indicator for this study. The detailed results for the biogas-to-methanol plant emissions over the selected domain are presented in the next section for different sample sizes.

4. Results

The procedure was tested for a variable sample size in order to assess the impact of the dataset dimension on the results quality. Figure 3 (left) shows the comparison between the obtained data-driven model and the simulation points for CO_2 and methane partial flowrate variations in the feed. In particular, in order to allow the distinction between the Response Surface and the validation data (green surface), the plot reports the outcome that does not correspond to the highest obtained accuracy (i.e. only 10 samples). In fact, according to the curve fitter optimizer, the best option for the approximation of the biogas-to-methanol plant emissions for any sample with 50 points or more resulted to be a 5th grade polynomial function expressed as:

$$\begin{aligned}
RS(X_1, X_2) = {} & k_1 + k_2 \cdot X_1 + k_3 \cdot X_2 + k_4 \cdot X_1 \cdot X_2 + k_5 \cdot X_1^2 + k_6 \cdot X_2^2 \\
& + k_7 \cdot X_1^3 + k_8 \cdot X_2^3 + k_9 \cdot X_1 \cdot X_2^2 + k_{10} \cdot X_2 \cdot X_1^2 \\
& + k_{11} \cdot X_1^4 + k_{12} \cdot X_1^3 \cdot X_2 + k_{13} \cdot X_1^2 \cdot X_2^2 + k_{14} \cdot X_1 \\
& \cdot X_2^3 + k_{15} \cdot X_2^4 + k_{16} \cdot X_1^5 + k_{17} \cdot X_1^4 \cdot X_2 + k_{18} \cdot X_1^3 \\
& \cdot X_2^2 + k_{19} \cdot X_1^2 \cdot X_2^3 + k_{20} \cdot X_1 \cdot X_2^4 + k_{21} \cdot X_2^5
\end{aligned} \qquad (2)$$

where X_1 and X_2 are respectively the methane and carbon dioxide flowrates and k_i are the model parameters provided by the optimizer. It is worth remarking that this model is just the result of an optimal data correlation and the obtained parameters have no particular meaning from a physical point of view.

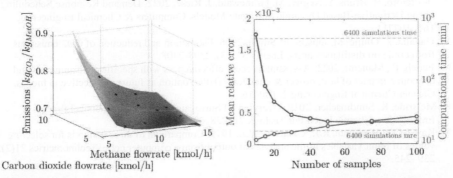

Figure 3 – left) RS vs. validation dataset - right) Computational time and mean relative error

As concerns the dataset size related performances, results are provided in Figure 3 (right). As expected, the overall computational time, given by the sum of the DoE and the function generation over the entire domain, increases with the sample size as well as the model accuracy. However, beyond 40 experimental values, the mean relative error decrease is almost negligible while the computational time keeps increasing with an almost linear trend. Therefore, beside providing a polynomial function able to replace the plantwide simulation for process environmental impact prediction, this methodology allows also to define the order of magnitude of the sample size suitable for the model generation. In practice, in case one of the operating parameters should vary, e.g. feed inlet temperature or reactor conversion, without a considerable change of the process thermodynamics or components, a new model could be easily obtained by using a dataset of about 50 samples.

5. Conclusions

The proposed methodology, based on RSM for surrogate modeling generation, provided both positive and reliable outcome for the specific case study and promising results of general validity. In particular, the research showed that building a plantwide surrogate model that allows to correlate input variables with the overall process energy consumption by keeping good accuracy is a viable solution. In terms of computational effort, the obtained data-driven model allows to perform reliable process emissions estimations over a wide range of operating conditions with a computational time reduced by orders of magnitude with respect to the conventional process simulation approach. Moreover, an optimal sample size in terms of the best compromise between accuracy and computational effort can be easily identified.

In terms of future perspectives, the obtained data-driven model could be used for optimal demand response scheduling in order to test the performances in terms of optimal solution accuracy as well as to reduce the computational time when optimal process operating conditions need to be identified during the design phase. Furthermore, different sampling strategies could be compared to detect the most suitable one for the specific system under analysis.

References

R. Alizadeh, J.K. Allen, F. Mistree, 2020. Managing computational complexity using surrogate models: a critical review. Research in Engineering Design 31, 275–298.

G. E. P. Box, K.B. Wilson, 1951. On the experimental attainment of optimum conditions. Journal of Royal Statistical Society - Series B (Methodological) 13(1), 1–45.

A. Di Pretoro, B. Bruns, S. Negny, M. Grünewald, J. Riese, 2022. Demand Response Scheduling Using Derivative-Based Dynamic Surrogate Models. Computers & Chemical Engineering 160, 107711.

M. Gadalla, Ž. Olujić, M. Jobson, R. Smith, 2006. Estimation and reduction of CO2 emissions from crude oil distillation units, Energy 31(13), 2398-2408.

M. Fedeli, F., Manenti, 2022. Assessing process effectiveness with specific environmental and economic impact of heat, power & chemicals (HPC) option as future perspective in biogas. Cleaner Chemical Engineering 2, 100016.

K. McBride, K. Sundmacher, 2019. Overview of Surrogate Modeling in Chemical Process Engineering. Chemie Ingenieur Technik 91, 228–239.

McKay, M.D., Beckman, R.J., Conover, W.J., 1979. Comparison of three methods for selecting values of input variables in the analysis of output from a computer code. Technometrics 21(2), 239–245.

Flavio Manenti, Gintaras V. Reklaitis (Eds.), Proceedings of the 34th European Symposium on Computer Aided Process Engineering / 15th International Symposium on Process Systems Engineering (ESCAPE34/PSE24), June 2-6, 2024, Florence, Italy

Enhancing Solar Photovoltaic Panel Production: A Novel Machine Learning Approach for Optimizing Mexico's Potential

Francisco Javier López-Flores,[a] César Ramírez-Márquez,[a] and José Maria Ponce-Ortega,[a*]

[a] *Chemical Engineering Department, Universidad Michoacana de San Nicolás de Hidalgo, Av. Francisco J. Múgica, S/N, Ciudad Universitaria, Edificio V1, Morelia, Mich., 58060, México*
Corresponding: jose.ponce@umich.mx

Abstract

A groundbreaking study explores the potential for large-scale solar Photovoltaic (PV) panel production in Mexico, utilizing an advanced Artificial Neural Network (ANN) model in TensorFlow to optimize various production parameters. The research underscores Mexico's strategic advantages, including its location, foreign investments, and available resources for PV panel production. The study uncovers the underrepresentation of the Americas in the global PV production market and indicates the potential for significant economic, social, and environmental benefits through scaling up production in Mexico, with module production being the most profitable segment (35.7%) and polycrystalline silicon production yielding lower earnings (13.0%). The study also identifies cell production as the highest energy consumer (39.7%). Furthermore, this research highlights the use of Machine Learning to simplify complex optimization models, offering valuable insights for policymakers, investors, and environmentalists, with the potential to enhance sustainability in the renewable energy sector. It showcases Mexico's suitability for PV panel production and paves the way for a greener future.

Keywords: PV Panel Production, Mexico's Potential, Machine Learning Integration, Economic Viability, Sustainability Benefits.

1. Introduction

The growing global need for sustainable energy in light of climate change has propelled renewable energy sources to the forefront, with solar energy emerging as a vital solution. Silicon-based solar cells have become a promising option, and currently, solar energy ranks as the third most widely used renewable energy source globally, following hydroelectric and wind power. Solar panels significantly reduce CO_2 emissions compared to fossil fuel-based electricity, with emissions ranging from 98.3 to 149.3 g CO_2 eq/kWh, making them a favorable choice for environmentally responsible energy production (Fuso Nerini et al., 2018). In recent years, solar PV technology has seen substantial growth, and it is projected to become the primary global energy source in the 21st century (Banik and Sengupta, 2021). Chinese firms played a pivotal role in reducing the production costs of solar panels, which has had a significant impact on the expansion of solar energy. Notably, a competitive bidding process initiated by Saudi Arabia in early 2021 aimed to achieve historically low solar energy production costs of 0.025 USD/kWh, which can be attributed to swift advancements in solar technology (Imam et al., 2022).

Regions like Latin America, including Mexico, have shown great potential for solar energy due to abundant resources and favorable legislative environments. Mexico, in particular, stands out due to its significant solar irradiation levels and a promising market atmosphere, capturing the attention of potential investors. Similarly, Brazil has successfully integrated PV solar energy into energy auctions, resulting in a notable surge in the number of PV solar installations interconnected with the power grid (Zhang et al., 2021). However, the challenge lies in achieving large-scale and sustainable solar panel production. Mexico currently boasts ten local solar PV manufacturers with an annual production capacity of 1.5 GW, emphasizing the growing momentum in Mexico's solar sector (Jäger-Waldau, 2019). To address this, the study introduces a comprehensive methodology that combines machine learning and multi-objective optimization models for PV panel manufacturing in Mexico.

In the realm of renewable energy, there is a noticeable shortage of reported models employing both Artificial Neural Networks (ANN) and Machine Learning. Existing research, such as the work by Rahman et al. (2021), reviews the use of machine learning, particularly ANN, for predicting renewable energy generation, offering valuable insights into sustainable energy planning and management. Similarly, Sharma et al. emphasize the importance of nanofluids in renewable energy systems and propose the use of data-driven machine learning techniques to forecast thermophysical features and assess system efficiency (Sharma et al., 2022). However, the synergy of advanced technologies remains largely uncharted in the context of renewable energy efficiency, highlighting the innovative nature of the study. The true innovation of the research lies in seamlessly incorporating the ANN model into the optimization process using the Pyomo and Optimization and Machine Learning Toolkit (OMLT) libraries, enhancing precision and applicability. This groundbreaking methodology has the potential to transform the renewable energy industry and pave the way for a sustainable and greener future.

The main text can start here.

2. Problem statement

The global demand for renewable energy and the urgency of addressing climate change have spotlighted solar photovoltaic (PV) technology as a promising solution. To illustrate the methodology's applicability, we turn to Mexico, a country with significant potential due to its strategic location, foreign investments, and available raw materials. Mexico's large land area and population, positive relations with neighboring countries, and developing nation status make it ideal for large-scale solar PV panel production. Currently, the Americas, including Mexico, have limited participation in the global solar panel market, signaling an opportunity for economic, social, and environmental benefits through increased production.

The study also addresses Mexico's solar panel demand between 2040-2047, considering the replacement of panels installed from 2015-2022. Mexico's National Electric System primarily relies on conventional technologies for energy production, with renewable sources contributing only a fraction of the installed capacity. To meet these needs, the research recommends establishing production facilities for polycrystalline silicon, wafers, cells, and modules. Conversion factors and production parameters are provided, and the energy consumption for these production plants is assumed to come from conventional technologies. Notably, the quantities of raw materials and production parameters for polycrystalline silicon are determined using an ANN model developed as part of this study.

3. Methodology

The methodology presented in Figure 1 outlines the development process for machine learning models and multi-objective optimization for solar PV panel production. It begins with creating an ANN to predict various parameters, e.g. polycrystalline silicon production. This involves data collection, preprocessing, normalization, and hyperparameter optimization to ensure efficiency. The best ANN model is then used to generate a surrogate model with the OMLT library. This surrogate model is integrated into the mathematical model, which considers different aspects of panel production, aiming to maximize Total Profit (*TOP*) and minimize emissions and water consumption. Multi-objective optimization is achieved using the ε-constraint method to find compromise solutions across the objectives.

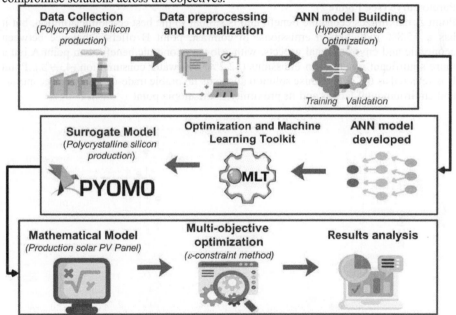

Figure 1. Schematic overview of the methodology.

4. Results

The project has been developed in the Python programming language, with the ANN model developed using TensorFlow and the Non-Linear Programming (NLP) optimization model implemented through the Pyomo library. The computations were carried out on a computer with a 2.3 GHz AMD Ryzen 7 processor and 16 GB of RAM. For ANN model development, the dataset was split into a training set (80%) and a validation set (20%), and an ANN model was trained using ReLU activation, Adam optimizer, and Mean Squared Error (MSE) loss function. Hyperparameter optimization was performed using the Hiperband algorithm, and the minimum and maximum bounds and step sizes for each hyperparameter were determined by trial-and-error experiments. The objective to be minimized was the validation loss (MSE), to mitigate overfitting. A total of 6,159 experiments were executed in 7.6 hours during the hyperparameter optimization. The best model had 12 inputs, 5 outputs, and three hidden layers with 20, 20, and 10 neurons, respectively, and a learning rate of 0.0005. The model demonstrated excellent accuracy and stability, with MSE and MAE below 0.0087 and 0.0557, and R^2

above 0.99 for all output variables, indicating a well-learned database behavior. Water consumption had the lowest error values, but differences from other variables were minimal. The surrogate model created with OMLT is equivalent to an NLP optimization model.

The optimization process considers four polycrystalline silicon production plants (ANN models), leading to a mathematical model (model for PV panel production) with 7,746 variables and 8,812 constraints when accounting for production years and multiple plants. The multi-objective approach using the ε-constraint method analyzes trade-offs between economic and environmental factors (TOP, emissions, and TWC). The Pareto curve demonstrates that as TOP increases, emissions and water consumption also rise, forming an inverse relationship. Notably, two key solutions, points A and B, are identified on the Pareto curve (see Figure 2).

Point A maximizes economic benefit and is similar to the best economic solution, but it has a 65.8% reduction in emissions. In contrast, point B offers a balance between economic and environmental aspects, with a lower economic benefit than point A but a more significant reduction in emissions (82.2%) and water consumption (43.9%). Point B is selected as the compromise solution due to its favorable trade-offs between economic and environmental factors and its proximity to the utopia point.

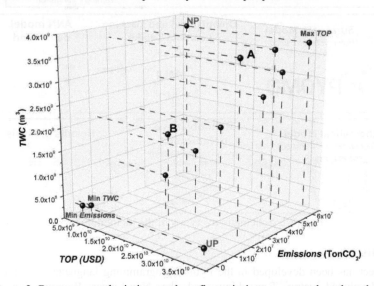

Figure 2. Pareto curve depicting total profit, emissions generated, and total water consumption.

Emissions generated by installed capacity and power produced, and water consumption by installed capacity and power generated are essential considerations. Emissions per kW of installed solar PV panels and per kWh of solar PV energy produced are significantly lower than those of fossil fuels. Water consumption per installed capacity and total water consumption per power generated are noteworthy, with values being much more environmentally friendly than other power generation methods. Notably, solution B demonstrates better efficiency in emissions generation and water use compared to solution A, reinforcing its position as the preferred compromise solution.

Figure 3 provides a breakdown of profit, energy consumption, and water consumption percentages for different production plants. Module production plants yield the highest profits (35.7%), while polycrystalline silicon production plants have the lowest earnings

(13.0%). Cell production plants consume the most energy (39.7%), while module production plants have the lowest energy consumption (15.5%). Surprisingly, wafer production plants use 89.2% of the total water, with high water consumption due to their cleaning processes. Silicon and module production plants have relatively small water consumption, accounting for 0.25% and 0.10%, respectively.

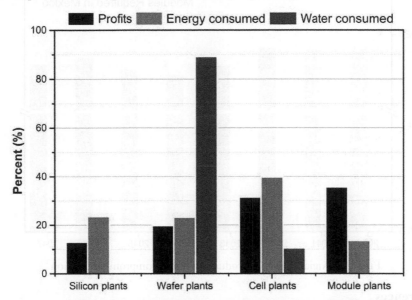

Figure 3. Profit, energy consumption, and water consumption distribution among various production plants in the compromise solution.

Throughout all years, approximately 98,261,000 modules have been manufactured (refer to Figure 4), with the majority earmarked for export and the remainder serving Mexico's solar photovoltaic energy needs. The optimal operating parameters for each polycrystalline silicon production plant, as determined through optimization using the developed ANN model. It's worth noting that each plant operates under distinct conditions, with Plant 1 boasting the highest installed capacity (9,472,000 kg) and Plant 2 possessing the lowest capacity (3,887,000 kg).

5. Conclusions

This study introduces an innovative optimization approach that quantitatively combines ANN models and mathematical models to optimize the production of solar PV panels. A comprehensive mathematical model is employed to evaluate the economic profitability, water consumption, and energy consumption of the production plants for polycrystalline silicon, wafers, cells, and PV modules. Using Mexico as a case study, the study optimizes the ANN model for polycrystalline silicon production, achieving impressive results with MSE and MAE both below 0.0087 and an R^2 value exceeding 0.99 for the validation set. Multi-objective optimization through the ε-constraint method reveals trade-offs between various conflicting objectives, shedding light on the environmental and economic aspects of different production plants. It is found that the water and energy consumption per unit of power produced during the average panel lifetime is lower than conventional technologies, but improvements in environmental terms are needed. Furthermore, the study forecasts the amount of photovoltaic waste generated upon panel disposal, with an estimated total of 812,219.8 tons, peaking in 2044 due to the high demand for solar panels. This adaptable approach provides a quantitative framework for assessing and forecasting PV solar panel production in diverse regions.

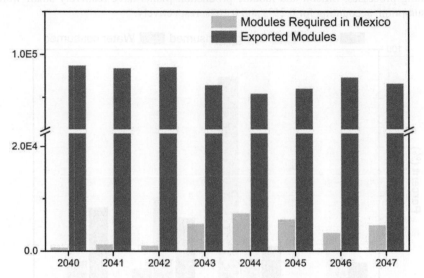

Figure 4. Annual module production in the compromise solution.

References

F. Fuso Nerini, J. Tomei, L.S. To, I. Bisaga, P. Parikh, M. Black, A. Borrion, C. Spataru, V. Castán Broto, G. Anandarajah, B. Milligan, Y. Mulugetta, 2018, Mapping synergies and trade-offs between energy and the sustainable development goals, Nat. Energy, 3, 10–15.

A. Banik, A. Sengupta, 2021, Scope, challenges, opportunities and future goal assessment of floating solar park, in: 2021 Innov. Energy Manag. Renew. Resour., 1–5.

A.A. Imam, J.O. Oladigbolu, S. Mustafa M.A., T.A. Imam, 2022, The effect of solar photovoltaic technologies cost drop on the economic viability of residential grid-tied solar PV systems in Saudi Arabia, in: 2022 Int. Conf. Electr. Comput. Energy Technol. ICECET, 1–5.

Y. Zhang, W. Jin, M. Xu, 2021, Total factor efficiency and convergence analysis of renewable energy in Latin American countries, Renew. Energy, 170, 785–795.

A. Jäger-Waldau, 2019, Joint Research Centre (European Commission) "PV status report 2019", Publications Office of the European Union, LU.

M.M. Rahman, M. Shakeri, S.K. Tiong, F. Khatun, N. Amin, J. Pasupuleti, M.K. Hasan, 2021, Prospective methodologies in hybrid renewable energy systems for energy prediction using artificial neural networks, Sustainability, 13, 2393.

P. Sharma, Z. Said, A. Kumar, S. Nižetić, A. Pandey, A.T. Hoang, Z. Huang, A. Afzal, C. Li, A.T. Le, X.P. Nguyen, V.D. Tran, 2022, Recent advances in machine learning research for nanofluid-based heat transfer in renewable energy system, Energy Fuels, 36, 6626–6658.

Flavio Manenti, Gintaras V. Reklaitis (Eds.), Proceedings of the 34th European Symposium on Computer Aided Process Engineering / 15th International Symposium on Process Systems Engineering (ESCAPE34/PSE24), June 2-6, 2024, Florence, Italy

Improving styrene polymerization through a commercial simulation software

Jeorge L. S. Amaral,[a] Rodrigo Battisti,[b*] Ricardo A. F. Machado,[a] Cintia Marangoni[a]

[a]*Department of Chemical Engineering, Federal University of Santa Catarina (UFSC), Florianópolis, 88040-900, Brazil*
[b]*Federal Institute of Education, Science and Technology of Santa Catarina (IFSC), Criciúma, 88813-600, Brazil*
rodrigo.battisti@ifsc.edu.br

Abstract

The polymerization and impregnation of Expandable Polystyrene (EPS) requires very long batch cycles when compared to other polymerization processes. Understanding the polymerization processes could be facilitated using mathematical modeling and simulation. However, the use of commercial simulators is still underexplored because of some limitations presented since these types of software tend to be more a generalist tool. Thus, the objective of this work is improving the simulation of styrene polymerization using the Aspen Plus® commercial software modified with user-implemented correlations. The free-radical polymerization kinetics was implemented, and experimental data from the literature were used to evaluate the performance of the simulation, using the standard correlation for the gel effect and with the proposed modified correlation, which was implemented by user-model routines, considering the incorporation of the impregnating agent (*n*-pentane) in the polymeric matrix. The implemented user models for the gel effect enabled a better representation of the experimental data, with minor errors from the literature. The prediction average errors of the modified model for the gel effect are five times less than for the other models. Besides the mean-square deviation is of the order of 10 times smaller in relation to the models without the gel effect and the standard. Thus, the user-model correlations implemented in the Aspen Plus® commercial software can enable the investigation of process variables that would hardly be investigated experimentally, being a versatile tool for the study of expandable polystyrene synthesis process.

Keywords: Suspension polymerization, polystyrene, blowing agent, user-model routines.

1. Introduction

Polystyrene is a thermoplastic widely used worldwide and has been extensively studied in recent decades. Achieving improvements in polymerization processes is necessary and, in this sense, the study of new initiators and suspending agents – often expensive and difficult to obtain – is recurrent (Ambrogi et al., 2016). With the same aim, the manipulation of process variables and parameters is an economical alternative for optimizing such polymeric systems, bringing considerable reductions in batch cycles without harm or changes to the properties of the produced polymer, and without the need to introduce new reagents to the original formulation (Scorah et al., 2006). Experimental studies have brought great advances in the field of suspension polymerization with the advent of multifunctional initiators and new reactor designs and configurations specific to the process. Recent literature has also presented certain alternatives in implementing the continuous production of polymers via suspension polymerization, a process traditionally conducted in batch cycles, thus providing the technical and economic advantages inherent to continuous processes (Lobry et al., 2015).

Despite the published advances regarding suspension polymerization, there are few studies that specifically deal with the impregnation of the polymer matrix using a blowing agent. Furthermore, these few existing works do not present the investigation of parameters regarding the kinetics and properties of the final product in the presence of the expanding agent. The vast majority of works concerning the impregnation of polystyrene are experimental, however, simulation is of paramount importance in the field of polymerization processes, providing low-cost support in decision-making and in the development of new production routes (Almeida et al., 2008).

The particle size distribution in the production of polystyrene via suspension polymerization was studied by Machado et al. (2000). From the population balance equation, a model was proposed to describe the evolution of the particle size distribution, and the model was validated with experimental data. The modeling and simulation of the styrene free-radical polymerization in semi-batch was investigated by Curteanu (2003). The study led to conclusions regarding the intermediate addition of initiator as a method of controlling polymer properties. Almeida et al. (2008) developed a model for predicting the stationary and dynamic behavior of a continuous styrene polymerization process. The main parameters were estimated from data of an industrial plant. To study the physical properties of polystyrene, Srivastava and Ghosh (2010) developed a molecular dynamics model for the bulk polymerization of styrene. In this work, studies were carried out regarding the glass transition of polystyrene.

More recently, Vieira and Lona (2016) used new kinetic modeling and simulations to analyze the effect of temperature on the properties of polystyrene obtained by atom transfer radical polymerization, providing a tool for analysis and optimization of this process. Currently, there are several commercial simulators that already have implemented mathematical models validated and consolidated in the literature, such as Aspen Polymers®, which provides kinetic polymerization models associated with a vast database of common chemical species in polymerization processes of greatest commercial interest. However, the mathematical models implemented in these commercial applications are comprehensive and generic, and to overcome this, many of them allow modifications to be added to study specific cases, using user routines.

The investigation of polymerization processes using commercial simulators as tools has recently gained prominence. Lesage et al. (2012) identified reaction kinetic parameters in the Aspen Custom Modeler® with the dynamic simulation of a CSTR for propylene polymerization in the gas phase to produce polypropylene, an extremely versatile thermoplastic polymer (Drummond et al., 2019). Funai et al. (2014) used Aspen Polymers® to validate a model of the Nylon® hydrolytic polymerization process in a semi-batch reactor. Kusolsongtawee and Bumroongsri (2017) presented a mathematical model of a fluidized bed polymerization reactor applied in the production of low-density polyethylene using the Aspen Custom Modeler® software to obtain polymer characteristics and investigate the flow behavior in the reactor.

In this context, the present work aims to improve styrene polymerization using Aspen Polymers® commercial software. For this purpose, the standard model of the software was modified with innovative user-model routines to consider the effect of the blowing agent in the polymerization process. After implementing the EPS impregnation model so that the effect of the blowing agent is considered, the simulation will be validated using experimental data published in the literature.

2. Methodology

2.1. Simulation design

The styrene suspension polymerization process was modeled in the Aspen Polymers® software using the batch reactor unit, as shown in the Figure 1. Benzoyl peroxide (BPO) (initiator), styrene (dispersed phase, monomer), water (continuous phase), and *n*-pentane (blowing agent) were added. The polymerization reactions occur in the organic liquid phase (drops of monomer/polymer) (Machado et al., 2007). The software's standard kinetic mechanism is implemented based on free-radical polymerization. The set of reactions included chain initiation, propagation reaction, at least one termination step and chain transfer or inhibition reaction to produce the dead polymer, with these reactions occurring simultaneously during polymerization. The parameters required for calculating the reaction rate constants, the pre-exponential factors and the activation energies were taken from Villalobos et al. (1991), Choi et al. (1988), and Mahabadi and O'Driscoll (1977). For predicting physical properties, the thermodynamic model of liquid activity coefficient of polymeric system (POLYNRTL) was used. The physicochemical properties of the polymer were calculated using Van Krevelen's group contribution method.

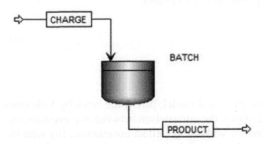

Figure 1. Block diagram of the simulation assembled in Aspen Polymers®.

2.1.1. Modifications implemented to address the gel and glass effects

The standard gel effect correlations of Aspen Polymers® are empirical models adjusted from experimental data. These refer only to the gel effect in polymerization reactions without impregnation and do not represent the glass effect. However, *n*-pentane (*n*P5) is the EPS solvent and rubberizes the polymer matrix. It is known that the addition of a solvent to the system alters the global free volume, and this alteration has a direct influence on the polymerization kinetics leading to different behaviors regarding the gel and glass effects. Therefore, in the present work, the correlations expressed in Equation (1) and Equation (2) were incorporated in Aspen Polymers® through user-model routines written in programming language, as these are based on the theory of free volume. By using user correlations, the model captures the influence of the impregnating agent on the polymeric matrix. The Equation (1) was specified to modify the termination rate constant (k_{tc}) due to the gel effect, and Equation (2) to adjust the propagation rate constant (k_p) due to the glass effect.

$$k_{tc}(X) = k_{tc0} \cdot (1 + \delta \cdot [P](X)) \cdot \left(\frac{M_{wcr}}{M_w(X)}\right)^N \cdot \exp\left[-A \cdot \left(\frac{1}{VF(X)} - \frac{1}{VF_{cr1}}\right)\right] \quad (1)$$

$$k_p(X) = k_{p0} \cdot \exp\left[-B \cdot \left(\frac{1}{VF(X)} - \frac{1}{VF_{cr2}}\right)\right] \quad (2)$$

2.2. Experimental validation

To validate the mathematical model and the modifications proposed in the simulator, part of the work presented by Villalobos et al. (1993) was replicated. The authors presented a study on the effect of adding the impregnating agent in several stages of the experimental reaction. Suspension polymerization and impregnation of polystyrene with *n*-pentane were carried out in a 3.2 L reactor with Rushton stirrer type at 350 rpm at 90 °C and an initiator concentration (BPO) of 0.01 mol·L⁻¹ of styrene. The suspending agent was tricalcium phosphate (TCP) at a concentration of 7.5 g·L⁻¹.

To validate the proposed modified model, the mean-squared error (MSE) was calculated to measure the agreement between simulated data and experimental values obtained from Villalobos et al. (1991), as shown by Equation (3).

$$MSE = \frac{1}{n}\sum_{i=1}^{n}(Y_i - \hat{Y}_i)^2 \tag{3}$$

The mean-absolute percentage error (MAPE) was also calculated, according to Equation (4), as the average between the errors of the simulated points.

$$MAPE = \frac{1}{n}\sum_{i=1}^{n}\left(\frac{Y_i - \hat{Y}_i}{Y_i}\right)\cdot 100 \tag{4}$$

3. Results and discussions

To evaluate the simulation built with the modified model, part of the work by Villalobos et al. (1993) was reproduced. Figure 2 shows the comparison between the experimental data for 7.5% *n*P5 with the simulations (i) without gel effect correlation, (ii) with the software's default correlation, and (iii) with the model's implemented user correlation.

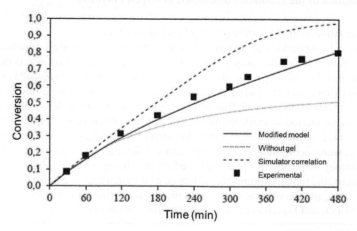

Figure 2. Comparison between the simulated and the literature data obtained.

Table 1 shows the statistical treatment data for the simulations produced without correlation for the gel effect and with the correlation implemented in the Aspen Polymers® software. In the work by Tefera et al. (1997), which deals with the selection of mathematical models, four models for polymerization via free-radicals were evaluated, and all presented errors of up to 10% in the simulations used to estimate the parameters. Therefore, according to this criterion, the model with the implemented user correlation was the only one that presented MAPE lower than 10%, considered acceptable.

Table 1. Comparative statistical analysis between the simulated models.

Time (min)	(i) Without gel		(ii) Standard model		(iii) Modified model	
	MSE	*MAPE*	*MSE*	*MAPE*	*MSE*	*MAPE*
60	0.000006	2.8%	0.000002	1.6%	0.0000061	2.8%
120	0.000359	10.3%	0.000002	0.8%	0.0003187	9.7%
180	0.001454	12.1%	0.001020	10.1%	0.0003959	6.3%
240	0.004892	16.4%	0.006790	19.4%	0.0003887	4.6%
270	0.016587	24.0%	0.014713	22.6%	0.0016061	7.5%
300	0.022769	25.3%	0.039578	33.3%	0.0003719	3.2%
330	0.038509	29.9%	0.037961	29.7%	0.0018003	6.5%
390	0.069667	35.3%	0.032371	24.0%	0.0042173	8.7%
420	0.072480	35.3%	0.035132	24.6%	0.0022415	6.2%
480	0.084570	36.4%	0.030752	21.9%	0.0005177	2.8%
	0.031129	**22.8%**	**0.019832**	**18.8%**	**0.0010786**	**5.8%**

Therefore, it can be observed that the implemented user-models for the gel effect in the commercial simulator enable a good representation of the experimental data, where the modified model with the user correlation proposed being that with the minor error about the experimental data. The average error of the predictions of the modified model for the gel effect is four times less than for the other models; besides the mean-square deviation is of the order of 10 times smaller in relation to the models without gel effect and standard. Thus, the user-model correlations implemented in the commercial software enables to investigate process variables that would hardly be investigated experimentally, being a versatile tool for the study of expandable polystyrene synthesis process.

4. Conclusions

The simulation of the EPS polymerization and impregnation process was built using the Aspen Polymers® software. The software's standard model was modified with user-model routines, so the simulation was able to represent the effect of the blowing agent on the process dynamics. The simulation with the modified model was validated using experimental data obtained from the literature. Errors of up to 10% proved the representativeness of the model. A comparison was made between the modified model, the software standard and those without gel effect, demonstrating that the simulation was an efficient tool. Therefore, this commercial simulator with modified correlation can be a versatile and efficient tool for studying EPS synthesis processes. The application allowed the mathematical model to be adapted to this specific application, which makes future investigations and manipulations of process variables less costly.

References

Almeida, A.S., Wada, K., Secchi, A.R., 2008, Simulation of styrene polymerization reactors: kinetic and thermodynamic modeling, Brazilian J. Chem. Eng., 25, 337–349, https://doi.org/10.1590/S0104-66322008000200012

Ambrogi, P.M.N., Colmán, M.M.E., Giudici, R., 2016, Modelling, simulation and experimental validation for styrene miniemulsion polymerization process, in: Kravanja, Z., Bogataj, M. (Eds.), 26th European Symposium on Computer Aided Process Engineering, Computer

Aided Chemical Engineering, Elsevier, pp. 2211–2216, https://doi.org/10.1016/B978-0-444-63428-3.50373-8

Choi, K.Y., Liang, W.R., Lei, G.D., 1988, Kinetics of bulk styrene polymerization catalyzed by symmetrical bifunctional initiators, J. Appl. Polym. Sci., 35, 1547–1562, https://doi.org/10.1002/app.1988.070350612

Curteanu, S., 2003, Modeling and simulation of free radical polymerization of styrene under semibatch reactor conditions, Cent. Eur. J. Chem, 1, 69–90, https://doi.org/10.2478/BF02479259

Drummond, C.F., Damas, M.S.P., Merlini, C., Battisti, R., 2019, Influence of clarifying agent on the properties of polypropylene copolymer industrially injected cups for Brazilian cream cheese, Int. J. Plast. Technol., 23, 170–176, https://doi.org/10.1007/s12588-019-09245-4

Funai, V.I., Melo, D.N.C., Lima, N.M.N., Pattaro, A.F., Liñan, L.Z., Bonon, A.J., Filho, R.M., 2014, Attainment of Kinetic Parameters and Model Validation for Nylon-6 Process, in: Klemeš, J.J., Varbanov, P.S., Liew, P.Y. (Eds.), 24th European Symposium on Computer Aided Process Engineering, Computer Aided Chemical Engineering, Elsevier, pp. 1483–1488, https://doi.org/10.1016/B978-0-444-63455-9.50082-9

Kusolsongtawee, T., Bumroongsri, P., 2017, Optimization of Energy Consumption in Gas-Phase Polymerization Process for Linear Low Density Polyethylene Production, Energy Procedia 138, 772–777, https://doi.org/10.1016/j.egypro.2017.10.055

Lesage, F., Nedelec, D., Descales, B., Stephens, W.D., 2012, Dynamic Modelling of a Polypropylene Production Plant, in: Bogle, I.D.L., Fairweather, M. (Eds.), 22nd European Symposium on Computer Aided Process Engineering, Computer Aided Chemical Engineering, Elsevier, p. 1128, https://doi.org/10.1016/B978-0-444-59520-1.50084-1

Lobry, E., Lasuye, T., Gourdon, C., Xuereb, C., 2015, Liquid–liquid dispersion in a continuous oscillatory baffled reactor – Application to suspension polymerization, Chem. Eng. J., 259, 505–518, https://doi.org/10.1016/j.cej.2014.08.014

Machado, F., Lima, E.L., Pinto, J.C., 2007, Uma revisão sobre os processos de polimerização em suspensão, Polímeros, 17, 166–179, https://doi.org/10.1590/S0104-14282007000200016

Machado, R.A.F., Pinto, J.C., Araújo, P.H.H., Bolzan, A., 2000, Mathematical modeling of polystyrene particle size distribution produced by suspension polymerization, Brazilian J. Chem. Eng., 17, 395–407, https://doi.org/10.1590/S0104-66322000000400004

Mahabadi, H.K., O'Driscoll, K.F., 1977, Concentration Dependence of the Termination Rate Constant During the Initial Stages of Free Radical Polymerization, Macromolecules, 10, 55–58, https://doi.org/10.1021/ma60055a009

Scorah, M.J., Dhib, R., Penlidis, A., 2006, Modelling of free radical polymerization of styrene and methyl methacrylate by a tetrafunctional initiator, Chem. Eng. Sci., 61, 4827–4859, https://doi.org/10.1016/j.ces.2006.03.018

Srivastava, A., Ghosh, S., 2010, Evaluating the Glass Transition Temperature of Polystyrene by an Experimentally Validated Molecular Dynamics Model, Int. J. Multiscale Comput. Eng., 8, https://doi.org/10.1615/IntJMultCompEng.v8.i5.80

Tefera, N., Weickert, G., Westerterp, K.R., 1997, Modeling of free radical polymerization up to high conversion. I. A method for the selection of models by simultaneous parameter estimation, J. Appl. Polym. Sci., 63, 1649–1661, https://doi.org/10.1002/(SICI)1097-4628(19970321)63:12<1649::AID-APP16>3.0.CO;2-V

Vieira, R.P., Lona, L.M.F., 2016, Simulation of temperature effect on the structure control of polystyrene obtained by atom-transfer radical polymerization, Polímeros, 26, 313–319, https://doi.org/10.1590/0104-1428.2376

Villalobos, M.A., Hamielec, A.E., Wood, P.E., 1993, Bulk and suspension polymerization of styrene in the presence of n-pentane. An evaluation of monofunctional and bifunctional initiation, J. Appl. Polym. Sci., 50, 327–343, https://doi.org/10.1002/app.1993.070500214

Villalobos, M.A., Hamielec, A.E., Wood, P.E., 1991, Kinetic model for short-cycle bulk styrene polymerization through bifunctional initiators, J. Appl. Polym. Sci, 42, 629–641, https://doi.org/10.1002/app.1991.070420309

Flavio Manenti, Gintaras V. Reklaitis (Eds.), Proceedings of the 34th European Symposium on Computer Aided Process Engineering / 15th International Symposium on Process Systems Engineering (ESCAPE34/PSE24), June 2-6, 2024, Florence, Italy

CFD-based investigation of the efficiency enhancement due to microstructure reorientation in structured packing

Christopher Dechert[a]*, Iris M. Baumhögger[a], Eugeny Y. Kenig [a]

[a] *Chair of Fluid Process Engineering, Paderborn University. Pohlweg 55, D-33098 Paderborn, Germany*
christopher.dechert@upb.de

Abstract

Wetting quality is one of the deciding factors for the efficiency of separation columns. In this work the computational fluid dynamics (CFD) technique is applied to examine the impact of liquid-phase physical properties and surface structuring on wetting in structured packings. The simulation results demonstrate a substantial impact of both factors on the interfacial area. Under certain conditions, a smooth packing surface can perform better than a microstructured surface. The orientation of the microstructure has a strong impact on its performance. Furthermore, the average liquid flow angle is not solely dependent on the packing macrostructure.

Keywords: structured packings, microstructure, wetting, CFD.

1. Introduction

Structured packings are column internals widely used in gas-liquid separation operations. Their specific geometry largely determines the fluid flow and hence the overall performance of separation columns. This is why it has been a focus of numerous studies toward enhanced mass transfer and reduced pressure drop. Computational fluid dynamics (CFD) is an ideal tool to investigate local flow phenomena within structured packings. However, simulations of a complete column or even a packing stack remain numerically unfeasible. Petre et al. (2003) were among the first to suggest a small representative elementary unit (REU) of the packing geometry that allowed application of CFD for the investigation of local flow phenomena. This idea was later used by several groups to simulate two-phase flow dynamics (e.g., Olenberg and Kenig, 2017; Hill et al., 2019). The results of these studies revealed that the wetted area of structured packing is often lower than the surface area of the packing material and that the gas-liquid interfacial area strongly depends on the liquid load and contact angle. Up to now, REUs with smooth surfaces have been analyzed, while surface structuring (used in real packings) has only been accounted for by modifying contact angles (Bertling et al., 2023). Our recent study (Dechert and Kenig, 2022) revealed that this way cannot guarantee sufficient accuracy in representing the microstructure impact on liquid flow. For this reason, in the present study, REUs of single packing sheets with explicitly geometrically represented microstructures are investigated to determine the influence of microstructures on liquid flow. Different properties of the liquid phase, e.g., liquid load, density, viscosity, surface tension, and contact angle are varied independently and the influence on the gas-liquid interfacial area, liquid flow morphology, liquid holdup, and effective flow angle in the packing is analyzed. The results provide a comprehensive view on the wetting in structured packings as well as an evaluation of the impact of different flow parameters.

2. Microstructure

This investigation is based on the structured packing Mellapak 250Y. Fabricated by Sulzer Chemtech, this packing is available with different surface structuring, namely, without microstructures, with wavy microstructures, and with hilly microstructures. In this work, only the smooth and wavy surface structures were investigated. Our analysis of the shape of the wavy microstructure (l-grooved) was performed using optical and laser microscopes. From the image obtained with the optical microscope (Figure 1), the dimensions of the wavy shape of the microstructure were estimated, namely, a height of approximately 0.0002 m and a wavelength of 0.0015 m. The structure was represented with a sinusoidal shape for CAD modeling.

Figure 1: Sulzer packing microstructure captured by an optical microscope. Left: top view, right: side view.

3. Modeling

3.1. Governing Equations

This work focuses on the liquid wetting and follows the approach of Olenberg and Kenig (2017) in which a quasi-stagnant gas phase is assumed. The movement of the gas phase is thus solely influenced by its interaction with the liquid phase. Further assumptions include incompressible and isothermal flow with Newtonian behavior. The continuity and momentum equations describing the two-phase fluid system are as follows:

$$\nabla \cdot \boldsymbol{u} = 0 \tag{1}$$

$$\rho \frac{\partial \boldsymbol{u}}{\partial t} + \rho \nabla \cdot (\boldsymbol{u}\boldsymbol{u}) = -\nabla p + \rho \boldsymbol{g} + \eta \nabla^2 \boldsymbol{u} + \boldsymbol{f}_\sigma \tag{2}$$

where ρ is density, t is time, $\boldsymbol{u}$ is velocity vector, p is pressure, $\boldsymbol{g}$ is gravity, η is dynamic viscosity and $\boldsymbol{f}_\sigma$ is a source term, which acts as a volume force and accounts for the surface tension effects at the free phase interface (Brackbill et al., 1992). The volume-of-fluid method (VOF) proposed by Hirt and Nichols (1981) was selected to resolve the free surface movement. In the original VOF formulation, the interface is resolved with a finite thickness and its exact location is not known. This leads to inaccurate calculations of interfacial normal vectors and interfacial fluxes. To determine the exact location and to increase the accuracy of the simulations, the piece-wise linear interface calculation (PLIC) method originally proposed by Youngs (1982) was applied. To govern turbulence within the Reynolds-averaged Navier-Stokes (RANS) modeling, the Reynolds stress term is implemented in Equation (2) and the k-omega SST model is employed.

3.2. Modification for Fully Periodic Flow

For a fully periodic simulation, all values and gradients at the faces of the periodic boundaries must have the same value. For the velocity, this is straightforward because the fluids are incompressible. However, pressure is not periodic. It consists of two parts, the dynamic pressure p_{dyn} and the static pressure p_{stat}:

$$p = p_{dyn} + p_{stat} \tag{3}$$

The dynamic pressure is, again, periodic due to incompressibility. The static pressure is affected by the driving forces, in this case, gravity. Therefore, the pressure is only periodic at the boundaries orthogonal to gravity. To achieve a periodic pressure at the other boundaries, the static pressure difference caused by gravity is excluded from the overall pressure field. This is realized by substituting Equation (3) in Equation (2):

$$\rho \frac{\partial u}{\partial t} + \rho \nabla \cdot (uu) = -\nabla p_{dyn} + \rho g + \eta \nabla^2 u + f_\sigma + f_{stat} \tag{4}$$

For a stagnant gas phase, f_{stat} reads as follows:

$$f_{stat} = -\nabla p_{stat} = -\rho_g \cdot g \tag{5}$$

3.3. Computational Domain and Boundary Conditions

In this work, the liquid flow over single packing sheets of a Mellapak 250Y packing was investigated. The REU was designed following Olenberg and Kenig (2017). The computational domain (Figure 2) has a quadratic shape with side lengths of 0.032 m and a height of 0.004 m.

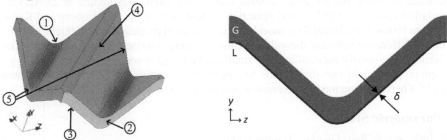

Figure 2: Computational domain for the flow over the packing sheet. Left: complete region with numbered boundaries, right: liquid film for the idealized liquid inlet (boundary number 1)

Two different flow conditions were investigated in this work. The first one is the idealized inlet condition, i.e., a liquid film covering the packing at the inlet and having a constant velocity and a constant film thickness. This corresponds to an ideal liquid-phase distribution. The second flow condition is the fully periodic flow, representing the liquid distribution for a perfectly developed flow inside a packing element. For the ideal flow condition, the boundary conditions including volume fraction α and contact angle θ are given in Table 1. For the fully periodic flow, boundaries number 1 and 2 are periodically coupled, the other boundaries remain unchanged. The velocity u and the film thickness δ of the liquid at the idealized inlet are determined using Nusselt film theory (Nusselt, 1916) and hence depend on the liquid load. As initial condition for the idealized inlet, the whole domain is filled with stagnant gas (u=0, α=0). Since there is no real inlet in case of fully periodic flow, the liquid load is set using a corresponding amount of liquid in the packing foldings.

Table 1: Boundary conditions of the idealized inlet (numbers according to Figure 2

boundary number	physical meaning	mathematical formulation
1	inlet	u=f(y), ∇p=0, α=f(y)
2	outlet	∇u=0, ∇p=0, $\nabla \alpha$=0
3	bottom wall/packing sheet	u=0, ∇p=0, α=f(θ)
4	gas phase	∇u=0, p=0, $\nabla \alpha$=0
5	sides open to neighbor REUs	periodic

3.4. Implementation

For the solution of the model equations, the interFoam solver from the open-source software package OpenFOAM 10 was utilized. This tool is based on the finite volume discretization. The equations and methods presented in Section 3.1 as well as the k-omega SST turbulence model are parts of the software package used. We extended the package by implementing the pressure adjustment described in Section 3.2. Furthermore, a calculation method to determine the size of the interfacial area based on the PLIC method and a method to evaluate the average liquid flow angle were implemented. In all simulations, the time step was adjusted automatically, to maintain the Courant number below 1 and its value near the interface below 0.5.

4. Model Validation

For the validation of our setup, the results obtained by Olenberg and Kenig (2017) with the commercial software tool STAR-CCM+ were used. They studied the Montz B1-250 packing, with dimensions similar to those of Mellapak 250Y, and varied the contact angle between 10 and 70° and liquid load between 10 and 90 m³/(m²·h). The system studied consisted of water and air with constant fluid characteristics estimated at 25 °C. Our simulation results revealed a good optical agreement with those of Olenberg and Kenig (2017) in terms of the liquid flow morphology, number of rivulets and film thickness. The average difference between the values of the resulting interfacial area was 10 %. The deviations are mostly encountered for small contact angles and high liquid loads, e.g., complete wetting in our simulations vs. incomplete wetting by Olenberg and Kenig (2017). Therefore, the model presented in our work can be judged as validated.

5. Parametric Studies

5.1. Influence of liquid load and contact angle

Bertling et al. (2023) found out that inlet effects in a Mellapak 250Y packing are negligible after the flow length of just one REU. Comparing the height of one REU of 0.03182 m to the height of one packing element, which is typically between 0.2 and 0.3 m (Kister, 1992), it is evident that the fully periodic flow is more appropriate for describing the flow dynamics within structured packings then the idealized inlet. Therefore, for the parametric studies, the fully periodic flow was chosen. The liquid morphologies shown in Figure 3, were obtained by performing the parametric study within the limits given in Section 4 using the REU with fully periodic flow.

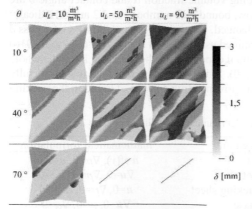

Figure 3: Liquid flow morphologies for the fully periodic flow

The study discriminates three distinct flow regimes below the loading point. At low liquid loads and contact angles, the liquid strictly follows the macrostructure of the packing sheet. The average liquid flow angle aligns with the corrugation angle of the packing. Increasing either the liquid load or the contact angle results in thicker films and some liquid spilling over the macrostructure, shifting the liquid flow angle towards the vertical. For contact angles or liquid loads exceeding certain threshold values, some liquid drains off the packing surface. In this arrangement, the drained liquid trickles out of the computational domain (out of the boundary number 4, Figure 2), making the intended liquid load unattainable. This drainage point was reached at higher flow rates of the 70 ° contact angle.

5.2. Variation of density, viscosity, and surface tension

Along with liquid load and contact angle, physical properties of the liquid phase may also significantly affect wetting behavior. We varied these properties individually within the limits determined according to the typical values encountered in separation processes. Density was varied between 500 and 2000 kg/m³, viscosity between $1 \cdot 10^{-4}$ and 1 Pa·s, and surface tension was varied between 0.015 and 0.072 N/m. With increasing density, decreasing viscosity or decreasing surface tension, the liquid flow pattern more and more deviates from the path dictated by the packing macrostructure toward overflow and drainage. Similarly, these trends in physical properties result in a higher average liquid flow angle, reduced liquid hold-up and an increase in interfacial area. Varying density, viscosity, and surface tension within the limits mentioned above changes the flow pattern significantly, from strictly following the macrostructure to liquid drainage. Such large physical system property variations can result from significant concentration changes over the column height (typical in e.g. distillation processes). Taking this effect into account can thus be beneficial for the choice of packing enhancing the column efficiency.

5.3. Surface Structuring

The l-grooved microstructure described in Section 2 is horizontally oriented in Sulzer packings (see Figure 4). For a comparison, the microstructure was rotated by 90 ° to investigate the effect of its orientation on liquid wetting. The parametric study shown in Section 5.1 was now performed for the l-grooved and the rotated l-grooved microstructures. Figure 5 demonstrates that the preferred surface structure for achieving the highest interfacial area depends on the liquid load. At low liquid loads, the rotated l-grooved microstructure resulted in the highest interfacial area, but reached the drainage point at nearly 40 m³/(m²·h). Thus, in contrast to a smooth surface, this microstructure destabilizes liquid flow. On the contrary, the drainage point of the l-grooved microstructure is reached at a liquid load of 140 m³/(m²·h), what indicates a stabilizing effect of this microstructure. For intermediate liquid loads, surface structuring seems to have little reason, as the wet pressure drop is increased by draining liquid.

Figure 4: Top view of l-grooved microstructure. Left: horizontal orientation, right: 90 ° rotated

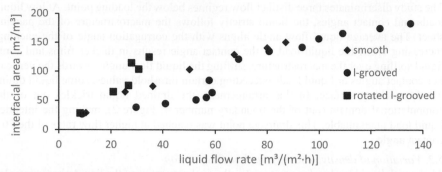

Figure 5: Interfacial area vs. liquid load for a contact angle of 40° in REUs with smooth, l-grooved, and rotated l-grooved surface

6. Conclusions

The influence of liquid-phase physical properties and surface structuring on the wetting of single structured packing sheets was investigated using CFD methods. The flow phenomena were analyzed based on a single REU. Three flow regimes below the loading point with different liquid morphologies were identified. Each of the physical properties of the liquid-phase studied has an impact on the morphology. Large variations can even lead to a change in the flow regime. Surface structuring also strongly influences the wetting conditions and existence of the three flow regimes. It was found that the l-grooved microstructure shifts the drainage point toward higher liquid loads, potentially reducing the pressure drop compared to smooth packing surface. In contrast, reorienting the l-grooved microstructure can enhance interfacial area under low liquid loads, but also shifts the drainage point to low liquid loads. Under some conditions, smooth packing surface can perform even better than the microstructured surface. The obtained results demonstrate the strong influence of the liquid phase physical properties and the surface structuring on the wetting in structured packing and consequently on the efficiency of the whole column. Future work will be focused on the integration of these properties into the CFD-based development of correlations for liquid hold-up and interfacial area.

References

R. Bertling et al., 2023, Simulation of liquid flow in structured packings using CFD-methods, Chemical Engineering Science, 269, 118405

J. U. Brackbill et al., 1992, A continuum method for modeling surface tension, Journal of Computational Physics, 100, 335-354

C. Dechert, E. Y. Kenig, 2022, CFD-based investigation of the packing microstructure influence on droplet behavior and film flow, Proc. 12th Conf. Distillation & Absorption, Toulouse, 2022

S. Hill et. al., 2019, Quantifizierung der Trenneffizienz einer strukturierten Packung mittels numerischer Simulation, Chemie Ingenieur Technik, 91, 12, 1833-1841

C. W. Hirt, B. D. Nichols, 1981, Volume of fluid (VOF) method for the dynamics of free boundaries, Journal of Computational Physics, 39, 201-225

H. Z. Kister, 1992, Distillation Design, Mc Graw Hill, Boston

W. Nusselt,1916, Die Oberflächenkondensation des Wasserdampfes, VDI-Zeitschr., 60, 541-546

A. Olenberg, E. Y. Kenig, 2017, Numerical simulation of two-phase flow in representative elements of structured packings, Computer Aided Chemical Engineering, 40, 2089-2094

C. F. Petre et. al., 2003, Pressure drop through structured packings: Breakdown into the contributing mechanisms by CFD modeling, Chemical Engineering Science, 58, 163-177

D. L. Youngs, 1982, Time-Dependent Multi-material Flow with Large Fluid Distortion, in Numerical Methods for Fluid Dynamics, Academic Press, 273-285

Flavio Manenti, Gintaras V. Reklaitis (Eds.), Proceedings of the 34th European Symposium on Computer Aided Process Engineering / 15th International Symposium on Process Systems Engineering (ESCAPE34/PSE24), June 2-6, 2024, Florence, Italy

A lumped parameter approach for determining the pressure gradient in gas-liquid annular flows

Nicolò Varallo[a,*], Giorgio Besagni[a], Riccardo Mereu[a]

Politecnico di Milano, Department of Energy, Via Lambruschini 4a, 20156 Milano, Italy
Corresponding author: *nicolo.varallo@polimi.it*

Abstract

This study defines a statistic-derived lumped parameter approach to determine the pressure gradient in two-phase annular flows. The statistical model was defined by coupling: (1) the ordinary least squares method (OLS) to determine the relationship between the variables, (2) the variance inflation factor (VIF) to check for multicollinearity issues, and (3) the least absolute shrinkage and selector operator (LASSO) to select the relevant predictors. Finally, a lumped parameter approach is derived based on the classification and regression tree (CART) approach. The model identifies the liquid and gas Reynolds numbers, the liquid phase properties, the pipe diameter, and the surface tension as significant variables influencing the two-phase pressure gradient.

Keywords: Downwards annular flow, pressure gradient, statistical analysis, lumped parameter approach.

1. Introduction

Two-phase gas-liquid annular flows are observed in a broad range of industrial processes, such as production and pipeline systems for oil and gas distribution, steam generators, boiling water reactors, and emergency core cooling facilities for the protection of nuclear reactors (Zhu et al., 2021).

Annular flow occurs in a vertical pipe at a very high gas flow rate and low liquid velocity when the gas pushes the liquid closer to the wall, allowing the maintenance of the liquid film. It is hence characterised by a central gas core surrounded by a thin liquid film flowing along the pipe wall. Interfacial waves appear along the streamwise direction due to the shear generated by the gas. Additionally, gas bubbles are entrained in the liquid film and liquid droplets in the gas core.

One of the most relevant flow properties at the global scale is the pressure drop along the pipe. To this end, this study proposes a statistic-derived lumped parameter approach for determining the pressure gradient in two-phase gas-liquid annular flows.

A dataset has been created from literature correlations, considering different operating conditions, pipe diameters, and fluid properties.

The paper is structured as follows. Section 2 presents the variables used to conduct the statistical analysis, Section 3 describes the statistical approach, and the results are discussed in Section 4. Finally, conclusions are drawn, and future studies are proposed.

2. Variables

2.1. Pressure gradient correlations

Two approaches can be used to calculate the pressure gradient. The former deems the two-phase flow as a pseudo-fluid characterised by suitably averaged properties of the liquid and gas phases. The latter, named the separated flow approach, considers the two-phase flow as artificially divided into two streams, each moving through its dedicated pipe, with the assumption that the velocity of each phase remains constant within the zone occupied by the phase (Xu et al., 2012).

The separated flow approach is used in this study, and it can be classified into two categories: the Φ_l^2, Φ_g^2 based method and the Φ_{lo}^2, Φ_{go}^2 based method. ϕ_l^2, Φ_g^2, Φ_{lo}^2, and Φ_{go}^2 are two-phase friction multipliers. Consequently, all correlations based on the separated flow approach obtain the two-phase frictional pressure gradient from one of the following expressions.

$$\left.\frac{\Delta p}{\Delta L}\right|_{tp,f} = \Phi_l^2 \left.\frac{\Delta p}{\Delta L}\right|_l = \Phi_g^2 \left.\frac{\Delta p}{\Delta L}\right|_g = \Phi_{lo}^2 \left.\frac{\Delta p}{\Delta L}\right|_{lo} = \Phi_{go}^2 \left.\frac{\Delta p}{\Delta L}\right|_{go} \tag{1}$$

In Equation (1), $\Delta p/\Delta L|_{tp,f}$ is the two-phase frictional pressure gradient, $\Delta p/\Delta L|_l$ is the frictional pressure gradient which would exist if the liquid phase is assumed to flow alone, $\Delta p/\Delta L|_g$ is the frictional pressure gradient which would exist if the gas phase is assumed to flow alone, $\Delta p/\Delta L|_{lo}$ is the frictional pressure gradient which would exist if the total mixture is assumed to be liquid, and $\Delta p/\Delta L|_{go}$ is the frictional pressure gradient which would exist if the total mixture is assumed to be gas.

The first and most important empirical model based on the separated flow approach was proposed by Lockhart and Martinelli in 1949. They proposed empirical curves that link the two-phase friction multiplier, Φ_l, to the Lockhart and Martinelli parameter, defined as

$$X = \left(\frac{\Delta p/\Delta L|_l}{\Delta p/\Delta L|_g}\right)^{1/2} \tag{2}$$

Chisholm (1967) developed a mathematical expression for the Lockhart and Martinelli 1949 empirical curves, which has been widely employed for the calculation of the pressure gradient:

$$\Phi_l = 1 + \frac{C}{X} + \frac{1}{X^2} \tag{3}$$

In Equation (3) C is a constant that depends on the characteristics of the flow under investigation. Based on the previous expression, many mathematical derivations have been developed.

Some years later, Chisholm (1972) provided a mathematical representation of the Baroczy graphical procedure, which predicts the pressure gradient for steam-gas mixtures in smooth tubes for evaporating flows. The obtained equations are also applicable to predict the pressure gradient for different two-phase fluids and link the two-phase friction multiplier, Φ_{lo}^2, to a physical coefficient defined as

$$\Gamma = \left(\frac{\Delta p/\Delta L|_{go}}{\Delta p/\Delta L|_{lo}}\right)^{1/2} \tag{4}$$

As in the case of the Lockhart-Martinelli equation, many researchers have based their correlations on the Chisholm expression. Finally, Friedel (1979) proposed a separated flow model-based correlation for the two-phase friction multiplier, considering surface tension effects through the Weber number, We. The Froude number, Fr, was also included to account for the importance of flow inertia against the external force.

The correlations for the two-phase friction multiplier used in this study are reported in Table 1, Table 2, and Table 3.

2.2. Variables selection

Starting from the correlations presented in Table 1, Table 2, and Table 3, the pressure gradient dataset has been defined by varying the operating conditions, phase properties, and pipe diameter. In particular, the variables applied to the correlations and used as predictors for the statistical analysis are listed below.

- Gas Reynolds number: from 10 000 to 40 000.
- Liquid Reynolds number: from 2 000 to 5 500.
- Pipe diameter: from 10 mm to 35 mm.
- Fluids: the correlations are usually applied and validated with different gas-liquid pairs. These include the flow of air-water and the condensation of vapor-water and halogenated refrigerants. The independent variables that characterize the fluids are the gas and liquid densities and dynamic viscosities ($\rho_g, \rho_l, \mu_g, \mu_l$), along with surface tension σ for the interaction between the fluids. The fluids considered are: (I) air-water at 22 °C and 50 °C, (II) vapor-water at saturation temperature and atmospheric pressure, (III) refrigerants at saturation temperature and atmospheric pressure (R22, R314a, R125, R32, R236ea, R114, R152a, R12).

The average value obtained from the different correlations is used to build the dataset, resulting in 3 696 data points with 8 independent variables. As these variables have a left-skewed distribution, they are used in the statistical method as log-transformed.

3. Methods

The statistical method was defined by coupling the ordinary least squares method (OLS) to determine the relationship between dependent and independent variables, the variance inflation factor (VIF) to check for multicollinearity issues, and the least absolute shrinkage and selector operator (LASSO) to select the significative predictors. Finally, based on the regression results, the classification and regression tree approach (CART) is used to segment the dataset and to define a lumped parameter approach for determining the pressure gradient. The statistical procedure is shown in Figure 1.

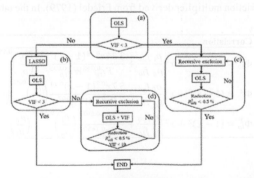

Figure 1. OLS-VIF-LASSO procedure.

Table 1. Two-phase friction multiplier derived from Lockhart and Martinelli (1949). In the table x is the gas volume fraction, D_h is the pipe diameter, and G is the mass flux.

Author	Correlation	Notes
Chisholm, 1967	$\Phi_l^2 = 1 + \dfrac{C}{X} + \dfrac{1}{X^2}$	$C = 5$ for viscous-viscous flow, $C = 10$ for turbulent-viscous flow $C = 12$ for viscous-turbulent flow, $C = 20$ for turbulent-turbulent flow
Hasan and Rhodes, 1967	$\Phi_l^2 = 1 + \dfrac{C}{X} + \dfrac{1}{X^2}$	$C = 3.218\left(\dfrac{2000}{G}\right)^{0.3602}\left(\dfrac{\rho_l}{\rho_g}\right)^{0.262}$
Mishima and Hibiki, 1996	$\Phi_l^2 = 1 + \dfrac{C}{X} + \dfrac{1}{X^2}$	$C = 21(1 - e^{-3190\,D_h})$
Sun and Mishima, 2009	$\Phi_l^2 = 1 + \dfrac{C}{X} + \dfrac{1}{X^2}$	$C = 1.79\left(\dfrac{Re_g}{Re_l}\right)^{0.4}\left(\dfrac{1-x}{x}\right)^{0.5}$
Awad and Muzychka, 2004	$\Phi_l^2 = \left[1 + \left(\dfrac{1}{X^2}\right)^{0.307}\right]^{1/0.307}$	
Muzychka and Awad, 2010	$\Phi_l^2 = 1 + \dfrac{C}{X} + \dfrac{1}{X^2}$	$C = 18.02; m = 1.014$ for turbulent-turbulent flow

Table 2. Two-phase friction multiplier derived from Chisholm (1972). In the table x is the gas volume fraction, L_a is the Laplace constant, and G is the mass flux.

Author	Correlation	Notes
Chisholm, 1972	$\Phi_{lo}^2 = 1 + (\Gamma^2 - 1)[B_{CH}x^{0.875} + x^{1.75}]$	$B_{CH} = \begin{cases} 55/\sqrt{G}, & 0 < \Gamma \leq 9.5 \\ 520/\Gamma\sqrt{G}, & 9.5 < \Gamma \leq 28 \\ 15000/\Gamma^2\sqrt{G}, & \Gamma > 28 \end{cases}$
Tran et al., 1999	$\Phi_{lo}^2 = 1 + (4.3\Gamma^2 - 1)[Nx^{0.875}(1 - x)^{0.875} + x^{1.75}]$	$N = \dfrac{[\sigma/g(\rho_l - \rho_g)]^{0.5}}{D}$
Uller-Steinhagen and Heck, 1986	$\Phi_{lo}^2 = \Gamma^2 x^3 + [1 + 2x(\Gamma^2 - 1)](1 - x)^{1/3}$	
Xu and Fang, 2012	$\Phi_{lo}^2 = \Gamma^2 x^3 + [1 + 2x(\Gamma^2 - 1)(1 - x)^{1/3}][1 + 1.54(1 - x)^{0.5}La^{1.47}]$	

Table 3. Two-phase friction multiplier derived from Friedel (1979). In the table x is the gas volume fraction.

Author	Correlation
Friedel, 1979	$\Phi_{lo}^2 = (1 - x)^2 + x^2\dfrac{\rho_l f_{go}}{\rho_g f_{lo}} + \dfrac{3.24x^{0.78}(1 - x)^{0.224}}{Fr_{tp}^{0.045}We_{tp}^{0.035}}\left(\dfrac{\rho_l}{\rho_g}\right)^{0.91}\left(\dfrac{\mu_g}{\mu_l}\right)^{0.19}\left(1 - \dfrac{\mu_g}{\mu_l}\right)^{0.7}$
Friedel, 1980	$\Phi_{lo}^2 = (1 - x)^2 + x^2\dfrac{\rho_l f_{go}}{\rho_g f_{lo}} + \dfrac{5.7x^{0.7}(1 - x)^{0.14}}{Fr_{tp}^{0.09}We_{tp}^{0.007}}\left(\dfrac{\rho_l}{\rho_g}\right)^{0.85}\left(\dfrac{\mu_g}{\mu_l}\right)^{0.36}\left(1 - \dfrac{\mu_g}{\mu_l}\right)^{0.2}$
Cavallini et al., 2002	$\Phi_{lo}^2 = (1 - x)^2 + x^2\dfrac{\rho_l f_{go}}{\rho_g f_{lo}} + \dfrac{1.26x^{0.6978}}{We_{go}^{0.1458}}\left(\dfrac{\rho_l}{\rho_g}\right)^{0.3278}\left(\dfrac{\mu_g}{\mu_l}\right)^{-1.181}\left(1 - \dfrac{\mu_g}{\mu_l}\right)^{3.477}$

4. Results

The overall fit model is based on the adjusted coefficient of determination, which is equal to 97.33%. Table 4 shows the regression results given by the OLS-VIF-LASSO procedure shown in Figure 1.

Table 4. Regression model results for the log-transformed pressure gradient.

	Coefficient (β)	Standard error	t-value	Significance	VIF
Intercept	5.527	5.943×10^{-2}	93	***	-
Re_g [-]	4.976×10^{-5}	4.695×10^{-2}	105.985	***	1
Re_l [-]	2.235×10^{-4}	4.098×10^{-6}	54.543	***	1
D_h [mm]	-1.534×10^2	5.499×10^{-1}	-278.932	***	1
ρ_l [kg/m³]	-2.872×10^{-4}	3.937×10^{-5}	-7.293	***	3.62
μ_l [kg/m³]	1.053×10^{-3}	3.986×10^{-1}	26.414	***	2.05
σ [N/m]	35.56	4.762×10^{-1}	74.670	***	5.27

By definition of linear regression, a change of a variable x_j of Δx_j, when the rest of the model variables are kept constant leads to a change in the log-transformed variable $\Delta y/y = \beta_j \Delta x_j$. The log-transformation applied to the pressure gradient, $\Delta p/L$, is $y = ln(\Delta p/L)$, so it can be derived that the relative change in the pressure gradient due to the change in x_j is $\Delta(\Delta p/L)/(\Delta P/L) = e^{\beta_j \Delta x_j} - 1$.

The influence of the phases velocities is studied varying the gas and liquid Reynolds numbers. Increasing the gas Reynolds number by 1000 increases the pressure gradient of +5.1 %. The same was found for the liquids Reynolds number since an increase of 100 in its value, results in an increase in the pressure gradient of +2.3%. Increasing the pipe diameter by 1 mm implies a reduction in the pressure gradient of -14.2%. Another relevant predictor identified by the model is the surface tension. An increase of 1 mN/m in the surface tension leads to an increase in the pressure gradient of +3.6%.

Finally, the model excluded the gas properties, but both the density and viscosity of the liquid are significant predictors. In particular, the increase in the liquid density decreases the pressure gradient (-2.9% for $\Delta\rho_l = 100$ kg/m³), while the increase in the liquid viscosity increases it (+10.5% for $\Delta\mu_l = 0.1$ mPas).

The graphical representation of the results, provided by the regression tree (Figure 2), can be immediately used to determine the pressure gradient, once the operating conditions, phase properties, and pipe diameter are known.

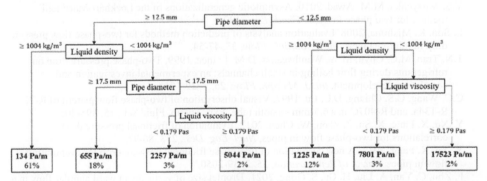

Figure 2. Regression tree.

5. Conclusions

This study proposed a statistics-derived lumped parameter model to determine the pressure gradient in two-phase annular flows.

The statistical model implemented identifies the pipe diameter, the gas and liquid Reynolds numbers, the liquid phase properties, and the gas-liquid surface tension as significative predictors. An increase in the phases Reynolds numbers increases the pressure gradient. Conversely, an increase in the liquid density and a decrease in the liquid viscosity reduces the pressure gradient, and the same was found for an increase in the pipe diameter.

The proposed approach could be applied to other operating conditions and flow regimes as a further development.

References

A. M. Aliyu, L. Lao, A. A. Almabrok, H. Yeung, 2016, Interfacial shear in adiabatic downward gas/liquid co-current annular flow in pipes, *Exp. Therm. Fluid Sci.*, 72, 75-87.

M. M. Awad, Y. S. Muzychka, 2004, A simple asymptotic compact model for two-phase frictional pressure gradient in horizontal pipes, *ASME International Mechanical Engineering Congress and Exposition*, 47098.

A. Cavallini, G. Censi, D.D. Col, G.A. Longo, L.Rossetto, 2002, Condensation of halogenated refrigerants inside smooth tubes, *HVAC&R Res.*, 8, 4.

D. Chisholm, 1967, A theoretical basis for the Lockhart-Martinelli correlation for t wo-phase flow, *Int. J. Heat Mass Transf.*, 10, 1767-1778.

D. Chisholm,1972, Pressure gradients due to friction during the flow of evaporating two-phase mixtures in smooth tubes and channels. *Int. J. of Heat Mass Transfer*, 16, 347-358.

L. Friedel, 1979, Improved friction pressure drop correlations for horizontal and vertical two-phase pipe flow, *European two-phase group meeting, Ispra, Italy*.

L. Friedel, 1980, Pressure drop during gas/vapour-liquid flow in pipes, *Int. Chem. Eng*, 20, 352-367.

A.R. Hasan, E. Rhodes, 1983, Effects of mass flux and system pressure on two-phase friction multiplier, Chem. Eng. Commun, 17, 209-229.

R.W. Lockhart, R.C. Martinelli, 1949, Proposed correlation of data for isothermal two-phase, two-component flow in pipes, *Chem. Eng. Prog*, 1, 39-48.

K. Mishinma, T. Hibiki, 1996, Some characteristics of air-water two-phase flow in small diameter vertical tubes, *Int. J. Multiph. Flow*, 22, 703-712.

H. Müller-Steinhagen, K. Heck, 1986, A simple friction pressure drop correlation for two-phase flow in pipes, *Chem. Eng. Process*, 20, 297-308.

Y.S. Muzychka, M.M. Awad, 2010, Asymptotic generalizations of the Lockhart-Martinelli method for two phase flows, *J. Fluids Eng.*, 132

L. Sun, K. Mishima, 2008, Evaluation analysis of prediction methods for two-phase flow pressure drop in mini-channels, *Int. J. Multiph. Flow*, 35, 47-54.

T.N. Tran, M.C. Chyu, M.W. Wambsganss, D.M. France, 1999, Two-phase pressure drop on refrigerants during flow boiling in small channels: an experimental investigation and correlation development, *Int. J. Multiph. Flow, 26, 1739-1754* .

C.C. Wang, C.S. Chiang, D.C. Lu, 1997, Visual observation of two-phase flow pattern of R-22, R-134a, and R-407C in a 6.5-mm smooth tube, Exp. Therm. Fluid Sci., 15, 395-405.

Y. Xu, X. Fang, X. Su, Z. Zhou, W. Chen, 2012, Evaluation of frictional pressure drop correlations for two-phase flow in pipes, *Nucl. Eng. Des.*, 253, 86-97.

Y. Xu, X. Fang, 2012, A new correlation of two-phase frictional pressure drop for evaporating flow in pipes, *Int. J. of Refrigeration*, 35, 2039-2050.

F. Zhu, C. Yan, A. Liu, H. Gu, S. Gong, 2021, Droplet size of vertically upward annular flow in a narrow rectangular channe, *Chem. Eng. Res. Des.*, 174, 107-115.

Flavio Manenti, Gintaras V. Reklaitis (Eds.), Proceedings of the 34th European Symposium on Computer Aided Process Engineering / 15th International Symposium on Process Systems Engineering (ESCAPE34/PSE24), June 2-6, 2024, Florence, Italy

Foliar Uptake Models for Biocides: Testing Structural and Practical Identifiability

Enrico Sangoi[a], Federica Cattani[b], Faheem Padia[b], Federico Galvanin[a*]

[a]*Department of Chemical Engineering, University College London (UCL), Torrington Place, WC1E 7JE London, United Kingdom*
[b]*Process Studies Group, Global Sourcing and Production, Syngenta, Jealott's Hill International Research Centre, Berkshire, Bracknell, RG42 6EY, United Kingdom*
f.galvanin@ucl.ac.uk

Abstract

As the global population grows and resources become scarcer, ensuring adequate food production is an urgent challenge. Developing safer biocides is crucial for optimizing crop yields and meeting rising food demands. Mathematical models play a pivotal role in understanding complex biological systems. This work focuses on developing a reliable model for characterizing the process of foliar biocide uptake in plants. A systematic modelling strategy involves the following steps: 1) formulating candidate models for foliar uptake; 2) conducting identifiability tests to verify that candidate model parameters can be determined from observations; 3) selecting the best model based on a-posteriori statistics from observations; 4) design experiments for achieving a precise estimation of model parameters from data. This paper presents a study on structural and practical identifiability of compartmental models for foliar biocide uptake (steps 1 and 2), considering experimental data limitations and initial condition variability. These findings will guide further model-based experimentation.

Keywords: model identification, foliar uptake, identifiability analysis, compartmental models.

1. Introduction

As the world's population continues to grow and the planet's resources remain limited, ensuring sufficient food production becomes a crucial challenge both in the present and for the coming decades. In tackling this issue, the development of improved and safer biocides will be essential to optimize crop yields and meet the increasing demand for food (Sharma et al., 2019). Mathematical models prove to be valuable tools to better understand the phenomena at play and formulate innovative solutions for the creation of new agricultural products. Several studies have been conducted on the foliar uptake of pesticides to develop mathematical models capable of describing the system behaviour. Among the models in the literature there are phenomena-specific empirical correlations (Forster et al., 2004), compartmental models (Bridges and Farrington, 1974), and diffusion-based models (Tredenick et al., 2019).

However previous studies did not assess systematically whether the model parameters can be accurately and precisely estimated, a key aspect to consider to ensure reliability of model predictions. This work aims at developing a mathematical model that can be effectively applied to describe the system and predict the dynamic uptake profile on different combinations of plant and product, while ensuring identifiability of the model parameters, i.e. that they can be uniquely determined from the system inputs and outputs

(Miao et al., 2011), including in the model building strategy also the practical experimental limitations that will act as constraints when calibrating the model.

2. Methodology

The general model identification procedure that is used in this project to model the foliar uptake of pesticides, based on Franceschini and Macchietto (2008), is presented in Figure 1. The whole modelling procedure is divided in 5 key steps: i) the formulation of candidate models, ii) the conduction of model identifiability tests, iii) model-based design of experiments (MBDoE) for model discrimination, iv) MBDoE for precise parameter estimation, and v) the model validation. This paper focuses on the step (2) "model identifiability tests", highlighted in Figure 1.

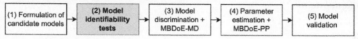

Figure 1 Model building strategy applied in the project, based on Franceschini and Macchietto (2008).

The general form considered for a dynamic model formulated as a set of differential and algebraic equations is:

$$\begin{cases} f(\dot{x}, x, u, \theta, t) = 0 \\ \hat{y} = g(x, u, \theta, t) \end{cases} \tag{1}$$

In Eq. (1), $x \in \mathbb{R}^{N_x}$ is a vector of state variables, $y \in \mathbb{R}^{N_y}$ the vector of predicted model outputs, $u \in \mathbb{R}^{N_u}$ the vector of known system inputs, $\theta \in \mathbb{R}^{N_\theta}$ the model parameters vector. The foliar uptake models that will be presented in this work are expressed as systems of ODEs, therefore can be expressed as in Eq. (1).

The question that is under assessment when testing the identifiability of a mathematical model is if the model parameters θ can be uniquely determined from the given system input $u(t)$ and the measurable system output $y(t)$ (Miao et al., 2011). Identifiability can be assessed in different ways and identifiability tests can be classified in two main categories: *a-priori* structural identifiability (not requiring preliminary experimental data) and *a-posteriori* practical identifiability tests. An a-priori structural analysis test is conducted in ideal conditions, assuming total observability of the system, i.e. all the states *y* can be observed at any time. On the other hand, a-posteriori practical identifiability test considers limitations on the quantity and quality of the data that can be practically retrieved from the system and that will be used for the estimation of the model parameters.

2.1. A-priori structural identifiability

Several methods can be applied for testing structural identifiability, among which Taylor's series, generating series, similarity transformation, direct test, differential algebra approaches. Miao et al. (2011) and Chis et al. (2011) give an extended overview of the different methods, discussing also advantages and disadvantages. Chis et al. (2011) showed that most of the methods scale badly with the number of states and parameters, and in this work the differential algebra approach is chosen because it was shown to have less limitations than the other methods. To test a priori structural identifiability with the differential algebra method the software DAISY (Bellu et al., 2007) is used.

2.2. A-posteriori practical identifiability

Practical identifiability has been tested with the correlation matrix method. The correlation matrix approach looks at the correlation between pairs of parameters in the model by means of the correlation coefficient R_{ij} between parameters θ_i and θ_j. To build the correlation matrix $\mathbf{R}$ $[N_\theta \times N_\theta]$, the ij-th entry R_{ij} is obtained from the variance-covariance matrix of the model parameters $\mathbf{V_\theta} = \{V_{\theta,ij}\}$ as in Eq. (2).

$$R_{ij} = \frac{V_{\theta ij}}{\sqrt{V_{\theta ii} V_{\theta jj}}}, \quad i,j = 1, \dots N_\theta \tag{2}$$

This identifiability technique is applied a-posteriori since it requires experimental data. In fact, the matrix $\mathbf{V_\theta}$ is obtained starting from the dynamic sensitivity matrix $\mathbf{S}$, which brings the information about the model structure and the distribution of the sampling points, and the variance-covariance matrix of the measurements $\mathbf{\Sigma_y}$, characterizing the variability in the experimental data. The sensitivity $s_{ij}(t_k)$ of the i-th response y_i to the j-th parameter θ_j at the k-th sampling time t_k is calculated as in Equation (3).

$$s_{ij}(t_k) = \frac{\partial y_i(t_k)}{\partial \theta_j} \tag{3}$$

The matrices $\mathbf{S}$ and $\mathbf{\Sigma_y}$ are combined to evaluate the Fisher Information Matrix (**FIM**), Eq.(4). The matrix $\mathbf{V_\theta} = \{V_{\theta,ij}\}$ is then obtained as the inverse of the FIM (Eq. 5).

$$\mathbf{FIM} = \mathbf{S}^T \cdot \mathbf{\Sigma_y^{-1}} \cdot \mathbf{S} \tag{4}$$

$$\mathbf{V_\theta} = \mathbf{FIM}^{-1} \tag{5}$$

The correlation coefficient R_{ij} describes how closely linked the parameters θ_i and θ_j are. From the definition in Equation (2), it can be seen that high values of R_{ij} are associated to parameter pairs having a covariance comparable with their variances. Values of the correlation coefficient R_{ij} higher than 0.99 lead to singular FIM, therefore are a symptom of practical non-identifiability (Rodriguez-Fernandez et al. 2006).

3. Model description

A compartmental model for foliar uptake of pesticides is formulated starting from a description of the system structure, as shown in Figure 2. The system is divided into the following compartments: droplet, store, cuticle, and leaf tissue. Three model structures (A, B, C) will be considered in this study. In case C, an additional compartment "environment" is included in the model, representing the AI lost form the droplet due to volatility, while other loss terms preventing the AI penetration such as photo- or chemical-instability of the molecule are represented with the transfer rate $k_{drop-loss}$. The store compartment refers to AI crystallized on the leaf surface. The model equations are a system of ODEs describing the dynamic change in mass of AI in the different compartments, as expressed in Equation (6).

$$\frac{dm_i}{dt} = \sum_j (k_{ji}m_j - k_{ij}m_i) \tag{6}$$

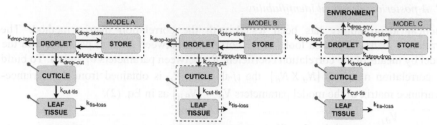

Figure 2 Graphical representation of the three compartmental model structures A, B and C. The pins departing from the compartments indicate the observed states in the system.

In (7) m_i [μg] indicates the mass of AI in compartment i, and k_{ij} [μg/min] the transfer rate of AI from compartment i to compartment j. With respect to the generic formulation presented in Eq. (1), the vector of state variables is $\mathbf{x} := \{m_i\}$, and the model parameters are $\boldsymbol{\theta} := \{k_{ij}\}$. No inputs $\mathbf{u}$ are present in the system. The vector of the observables output $\mathbf{y}$ changes among the three models: in the case A and C three states are observed, i.e. $m_{deposit}$ (being $m_{droplet+}m_{store}$), $m_{cuticle}$ and m_{tissue}. In case B only the total amount of AI inside the leaf is measured, not decoupling the cuticle and the cellular tissue, so the observed states are $m_{deposit}$ and the sum $m_{cuticle} + m_{tissue}$.

4. Results

The results of the identifiability tests are here presented along with a discussion of the findings and a comparison between the structural and practical identifiability results.

4.1. A-priori structural identifiability

For the structural identifiability analysis let indicate with $\boldsymbol{\theta}^*$ the true value of the parameters $\boldsymbol{\theta}$. In the context of structural identifiability a model is defined globally identifiable if the parameters $\boldsymbol{\theta}$ are uniquely distinguished, locally identifiable if a finite number of solutions larger than one is obtained for $\boldsymbol{\theta}$, and nonidentifiable if the system input-outputs lead to an infinite number of parameters values (Bellu et al., 2007). The results of the structural identifiability test conducted with the software DAISY are summarised in Table 1. Model parameters result to be non-identifiable in all the three scenarios considered if no information about initial conditions is available. From this result it can be concluded that knowing the initial conditions of the system is crucial to ensure model identifiability, otherwise parameters would not be identifiable even in the ideal case of noise-free measurements and total observability of the system.

When initial conditions of the system are known, the model results to be globally identifiable in scenario A, locally identifiable in scenario B and non-identifiable in the scenario C. In scenario C the model is non-identifiable even with known initial conditions

Table 1 Summary of the a-priori structural identifiability results obtained with DAISY for the three scenarios A, B and C considering known and unknown initial conditions in the system.

Structural identifiability	Scenario A	Scenario B	Scenario C
Unknown initial conditions	Model non-identifiable	Model non-identifiable	Model non-identifiable
Known initial conditions	Model globally identifiable	Model locally identifiable	Model non-identifiable

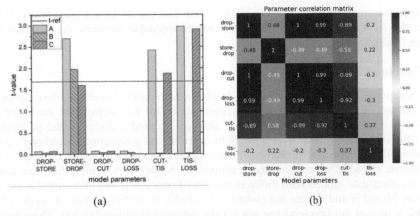

(a)   (b)

Figure 3 (a) The t-values of the parameter estimates confidence intervals, compared to the reference t-test at 0.95 significance. (b) Heatmap showing the parameters correlation matrix for scenario A.

because the parameters $k_{drop-env}$ and $k_{drop-loss}$ cannot be distinguished and infinite solutions in the form of Eq. (7) are obtained

$$k_{drop-env} + k_{drop-loss} = k^*_{drop-env} + k^*_{drop-loss} \tag{7}$$

The structural analysis results described above indicate that these two parameters cannot be uniquely determined from the system outputs considered.

4.2. A-posteriori practical identifiability

The correlation matrix method presented in the methodology section is a local test performed around a nominal value of the parameters θ^*. The values of θ^* used in this study are obtained from a preliminary parameter estimation on foliar uptake data provided by Syngenta and a *t*-test with 95% significance is conducted to assess the precision of parameter estimates. The *t*-test results are represented in Figure 3a, where low *t*-values compared to a reference are associated to parameters with very large variance. In all the scenarios the three parameters $k_{drop-store}$, $k_{drop-cut}$ and $k_{drop-loss}$ are difficult to estimate with a good precision from data, while the other parameters have acceptable *t*-values i.e. larger than *t*-reference. In scenario B, differently form A and C, also parameters $k_{cut-tis}$ and $k_{tis-loss}$ fail the *t*-test, and this result is in good agreement with the a-priori structural analysis presented in section 4.1. In fact, the result obtained for model B suggests that the transport-barrier role of the cuticle is hard to describe with this compartmental model if experimental information about the AI distribution within the leaf structure is not available.

The parameter correlation matrix is reported with the heatmap in Figure 3b only for scenario A for sake of conciseness. The model passed the a-priori structural identifiability test under these conditions, however when looking at the practical identifiability results, strong correlations between parameters emerge. The correlation coefficient among all the pairs of parameters in the set $\{k_{drop-store}, k_{drop-cut}, k_{drop-loss}\}$ result to be 0.99 or higher, indicating practical non-identifiability of these parameters (Rodriguez-Fernandez et al. 2006). From the *t*-test (Fig. 3a) high variance for these parameters was observed, while the analysis of correlation coefficient R_{ij} (Eq. 2) in the correlation matrix (Fig. 3b) lead to conclude that also their covariance is extremely large. From these results it emerges that higher model complexity must be sustained by more experimental information to ensure the identifiability of the full set of model parameters. Other strategies such as model reparameterization can be applied to overcome the identifiability issues (Quaglio et al., 2020).

5. Conclusions

In this paper, a systematic framework to develop a mathematical model for the description of foliar uptake of biocides is presented, focusing on the first steps of the methodology shown in Figure 1. The model building strategy adopted includes model parameters identifiability testing, which is done both a-priori and a-posteriori. Verifying model identifiability is an essential prerequisite to apply the model in the following steps of the procedure, i.e. MBDoE and parameter estimation. Model identifiability has been tested in three candidate models with different number of parameters and observed states in the system. Model B and C were not globally identifiable from the structural analysis, while model A passed the structural test but not the practical identifiability test. These results indicate that the model complexity achievable is strongly limited by the observability on the system. Future work will include the formulation of alternative model structures, also considering diffusion-based models, to ensure parameter identifiability as a requirement for the application of model-based design of experiments techniques in foliar uptake.

Acknowledgements

The authors gratefully acknowledge the support of the Department of Chemical Engineering, University College London, UK and Syngenta.

References

Bellu, G., Saccomani, M.P., Audoly, S., D'Angiò, L., 2007. DAISY: A new software tool to test global identifiability of biological and physiological systems. Computer Methods and Programs in Biomedicine 88, 52–61.

Bridges, R.C., Farrington, J.A., 1974. A compartmental computer model of foliar uptake of pesticides. Pestic. Sci. 5, 365–381

Chis, O.-T., Banga, J.R., Balsa-Canto, E., 2011. Structural identifiability of systems biology models: A critical comparison of methods. PLoS ONE 6.

Forster, W.A., Zabkiewicz, J.A., Riederer, M., 2004. Mechanisms of cuticular uptake of xenobiotics into living plants: 1. Influence of xenobiotic dose on the uptake of three model compounds applied in the absence and presence of surfactants into Chenopodium album, Hedera helix and Stephanotis floribunda leaves. Pest Manag. Sci. 60, 1105–1113

Franceschini, G., Macchietto, S., 2008. Model-based design of experiments for parameter precision: State of the art. Chemical Engineering Science, Model-Based Experimental Analysis 63, 4846–4872.

Miao, H., Xia, X., Perelson, A.S., Wu, H., 2011. On Identifiability of Nonlinear ODE Models and Applications in Viral Dynamics. SIAM Review 53, 3–39

Quaglio, M., Fraga, E.S., Galvanin, F., 2020. A diagnostic procedure for improving the structure of approximated kinetic models. Comput. Chem. Eng. 133, 106659.

Rodriguez-Fernandez, M., Egea, J.A., Banga, J.R., 2006. Novel metaheuristic for parameter estimation in nonlinear dynamic biological systems. BMC Bioinformatics 7, 483

Sharma, A., Kumar, V., Shahzad, B., Tanveer, M., Sidhu, G.P.S., Handa, N., Kohli, S.K., Yadav, P., Bali, A.S., Parihar, R.D., Dar, O.I., Singh, K., Jasrotia, S., Bakshi, P., Ramakrishnan, M., Kumar, S., Bhardwaj, R., Thukral, A.K., 2019. Worldwide pesticide usage and its impacts on ecosystem. SN Appl. Sci. 1, 1446

Tredenick, E.C., Farrell, T.W., Forster, W.A., 2019. Mathematical Modelling of Hydrophilic Ionic Fertiliser Diffusion in Plant Cuticles: Lipophilic Surfactant Effects. Plants 8, 202.

Flavio Manenti, Gintaras V. Reklaitis (Eds.), Proceedings of the 34th European Symposium on Computer Aided Process Engineering / 15th International Symposium on Process Systems Engineering (ESCAPE34/PSE24), June 2-6, 2024, Florence, Italy

Dynamic Multiscale Hybrid Modelling of a CHO cell system for Recombinant Protein Production

Oliver Pennington,[a]* Sebastián Espinel Ríos,[b] Mauro Torres Sebastian,[a] Alan Dickson,[a] Dongda Zhang,[a]*

[a]*University of Manchester, Manchester, Oxford Road, M1 3AL, UK*
[b]*Princeton University, New Jersey, NJ 08544, USA*
oliver.pennington-3@postgrad.manchester.ac.uk; dongda.zhang@manchester.ac.uk

Abstract

Multiscale hybrid modelling of biosystems utilises advantageous aspects of several modelling approaches, from the physical interpretations of kinetic modelling to the power of a data-driven Artificial Neural Network (ANN). This study implements multiscale modelling to gain insight into the production of Trastuzumab (Herceptin) from Chinese Hamster Ovary (CHO) cells under challenging dynamics. A reduced metabolic network is subject to enzyme constraints with a Dynamic Metabolic Flux Analysis (*ec*DMFA) approach and integrated within a macro-scale hybrid kinetic model. The model can simulate batch processes with defined initial conditions and provide insight into the systems behaviour as a response to changes in the cell culture medium. On the intracellular level, the influence from extracellular perturbations can be observed, in addition to giving an estimated production rate of unmeasured by-products. Overall, this model can be potentially used as a reliable digital twin to estimate the underlying fed-batch process dynamics for future model predictive control and process optimisation.

Keywords: Multiscale modelling, Machine learning, Kinetic modelling, Metabolic flux analysis.

1. Introduction

Bioprocesses link masses of global research interests, including the production of renewable fuels, plastics, and many other high value bioproducts. From an economic standpoint, the UK bioeconomy alone is worth roughly £220 billion (as of 2018), and is set to double by 2030, providing a platform for developing sustainable products and processes (Harrington, 2018). Such growth requires overcoming many challenges, including deficient metabolic and secretory phenotypes for protein production, low yields in reactor scale-up, by-product accumulation (leading to heightened separation costs and product loss), and significant batch-to-batch variation, generating quality-control challenges. Mammalian cell lines alone account for around 70% of therapeutic recombinant protein production (O'Flaherty et al., 2020). Utilising machine learning can play a key role in accurately simulating, controlling, and optimising these complex systems, where mechanistic models can struggle to find a balance between oversimplification and over parameterization. With the fourth industrial revolution becoming ever prevalent, the transition to bioprocess digitalisation is underway.

This study aims to unite the macro-scale and micro-scale aspects of a batch cell culture, while incorporating the advantages of an ANN to overcome dynamics that are challenging for physically derived bio-kinetic models to capture. Another challenge to overcome is the infinitely possible solution space in constraint-based modelling – it is essential to add constraints and define a cell's objective function that narrows the solution space into a region of biologically plausible scenarios. Our modelling methodology aspires to be simple in application, with minimal data requirement and computational expense, while maximising simulation accuracy, process insight, and extrapolation potential for process control and optimisation. This enables the modelling methodology described in this work best applicable to complex bioprocesses, such as mammalian cell lines, and new bioprocesses, where knowledge is limited but rapid scale-up is desired. With 20-30 new mammalian made products gaining FDA approval each year (O'Flaherty et al., 2020), there is certainly demand for modelling processes based on limited knowledge.

The study at hand looks at the growth of CHO cells from glucose across a period of 10 days, with measurements of twenty-nine medium component concentrations (including biomass, glucose, glutamine, lactate, ammonia, Trastuzumab and other amino acids) taken every 24 hours, provided by a previous study (Torres et al., 2019).

2. Methodology

2.1. Macro-scale kinetic modelling

To include as much physical information as possible, a Monod-inspired kinetic model was applied to the dynamic extracellular medium. The extracellular medium concentrations biomass (X), glucose (G), glutamine (Gln), Trastuzumab (P), lactate (Lac), ammonia (Amm), and 9 essential amino acids were simulated. The system of Ordinary Differential Equations (ODEs) for the first six states are described as in

$$\frac{dX}{dt} = X \, \mu_{\max_G} \frac{G}{K_G + G} \quad (1) \qquad \frac{dG}{dt} = - X \, v_{\max_G} \frac{G}{K_G + G} \quad (2)$$

$$\frac{dGln}{dt} = - X \, v_{\max_{Gln}} \frac{Gln}{K_{Gln} + Gln} \quad (3) \qquad \frac{dP}{dt} = X \, Y_{PX} \quad (4)$$

$$\frac{dLac}{dt} = X \, Y'_{LacG} \frac{G}{K_G + G} \quad (5) \qquad \frac{dAmm}{dt} = X \, Y_{AmmX} \quad (6)$$

where $\mu_{\max_G}$ refers to the maximum specific growth rate of biomass, with $v_{\max \, i}$ being the maximum substrate uptake flux of substrate i, K_i being the affinity constant for a given substrate i, and Y_{ij} being the yield coefficient between products i and j. For the remaining extracellular components (9 essential amino acids), an empirical correlation is used, where $\mathbf{A}$ is the vector of 9 essential amino acids, with $\boldsymbol{\alpha}$ being a vector of coefficients, as in

$$\nabla \mathbf{A} = \frac{dG}{dt} \boldsymbol{\alpha} \quad (7)$$

Parameters were found using a stochastic optimisation algorithm, Particle Swarm Optimisation (PSO), that minimised a mean squared error objective function across n datapoints, as in

$$\min \frac{1}{n} \sum_{t,meas} \left(\left(\frac{X_t - X_{meas_t}}{\sigma_X} \right)^2 + \sum_i \left(\frac{C_{i_t} - C_{i_{meas_t}}}{\sigma_i} \right)^2 \right) \quad (8)$$

The objective function minimises the difference between the simulated extracellular concentration profiles and the measured averages at each timepoint; each term is weighted by the experimental standard deviation of component i; σ_i.

2.2. Hybrid modelling

Challenging growth dynamics lead to the requirement of time-varying parameters for efficiently capturing transitions between growth phases. However, parameters allowed to vary over time are not defined as functions of time, but as functions of time-varying state variables. Such functions are defined by an Artificial Neural Network (ANN), and it is the combination of the ANN with the physical model described in Equations 1-7 that forms the hybrid macro-scale model. The first parameter allowed to vary is the maximum permitted glucose flux; $v_{\max_G}$. This parameter was deemed dynamic due to the excessive initial glucose consumption rates alongside low cell counts. Physically, it can be inferred that initial uptake fluxes may be high due to the initial challenge that the cell faces of adapting to the medium, reallocating resources to allow for optimal growth. Deviations in $v_{\max_G}$ were penalised to avoid overfitting of the experimental data. The second time varying parameter was the lactate yield coefficient; Y'_{LacG}. A similar observation was made with extremely high initial lactate production fluxes; often referred to as the Warburg effect (O'Brien et al., 2020). The two aforementioned parameters, $v_{\max_G}$ and Y'_{LacG}, are the two outputs of the ANN, which can then be used to simulate the model. This is shown in Figure 1. An appropriate ANN structure was determined using the Akaike Information Criterion with correction for small sample sizes ($AICc$), alongside hyperparameter optimisation, to avoid overfitting.

2.3. ecDMFA

A reduced metabolic network of CHO cells obtained from a genome-scale model (Jiménez del Val et al., 2023) was used in the frame of Metabolic Flux Analysis (MFA). Reduced metabolic networks have the benefit of greatly lowering computational expense for dynamic systems, while still being able to capture key intracellular behaviour. Dynamic Metabolic Flux Analysis (DMFA) is effectively a series of consecutive MFA problems solved simultaneously with different extracellular inputs and outputs. The reduced network contains 144 reactions linked by 101 metabolites and forms the mass balance defined by a stoichiometric matrix (**S**), that accounts for dilution from biomass growth, and the vector of fluxes ($\mathbf{v}_t$), as in

$$\mathbf{S}\,\mathbf{v}_t = \mathbf{0} \qquad (9)$$

where the time subscript (t) refers to the constraint being valid at every time-step the constraint is applied to, despite changes in the values of intracellular flux. The network is

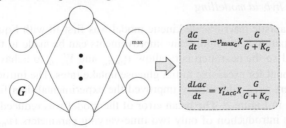

Figure 1 - ANN representation for the simulation of time-varying parameters based on system components and dynamics.

further constrained the vector of measured transport (*trans*) fluxes (a subset of $\mathbf{v}_t$), in accordance with the mass balances of the macro-scale hybrid kinetic model. The measured transport flux vector is defined by changes in the vector of measured extracellular component concentrations (**C**), as in

$$\mathbf{v}_{trans_t} = -\frac{1}{X}\frac{d\mathbf{C}}{dt}_t \qquad (10)$$

where biomass concentration (X) normalises transport fluxes per number of cells. Biomass specific growth rate (μ) is effectively a transport flux for the biomass component and is accounted for in the stoichiometric matrix in Equation 9.

To help narrow down the solution space of the flux-based modelling, further constraints can be employed. In this study, enzyme-constraints are utilised to generate biologically favourable flux distributions. Each reaction j has an associated flux and enzyme, with a corresponding molar mass (M_{r_j}) and turnover rate (k_{cat_j}). These parameters can be used alongside a reaction flux to estimate the mass of each enzyme required for the predicted flux. The total enzyme mass is constrained to a known maximum (M_E), as in

$$\sum_j \frac{M_{r_j} v_{j_t}}{k_{cat_j}} \leq M_E \qquad (11)$$

where parameters for Equation 11 were obtained from an existing enzyme-constrained modelling study on CHO cells (Yeo et al., 2020). To simulate the most realistic flux profiles possible is to employ an objective function that renders observable cell phenotypes. One example is efficient operation, where a cell aims to minimise the magnitude of all reaction rates within it to minimise wasted energy sources, such as ATP. Furthermore, one should consider cell stability – drastically inconsistent fluxes are unlikely to be correct so substantial changes in the same reactions flux should be penalised over each iteration time gap Δt. Therefore, minimum flux objective and flux inconsistency penalty are combined into one quadratic (convex) objective function, as in

$$\min \sum_{t=t_0}^{T} \mathbf{v}_t + \lambda \sum_{t=t_0+\Delta t}^{T} (\mathbf{v}_t - \mathbf{v}_{t-\Delta t})^T (\mathbf{v}_t - \mathbf{v}_{t-\Delta t}) \qquad (12)$$

where λ is the penalty weight – a tuneable parameter. The *ec*DMFA is thus given by the optimisation of Equation 12, subject to the constraints given in Equations 9-11.

3. Results and Discussion

3.1. Macro-scale hybrid modelling

The purely mechanistic macro-scale kinetic model was first fit with Equations 1-6 with no time-varying parameters. Significant improvements can be made to the 26.8% mean error present, and so the next step is to allow v_{max_G} and Y'_{LacG} to behave as a dynamic parameter to account for excessive initial glucose uptake rates. The introduction of time-varying parameters v_{max_G} drastically improved the experimental data fits, examples of which are shown in Figure 2. The mean error of the model was reduced from 26.8% to 6.9% through the introduction of only two time-varying parameters (v_{max_G} and Y'_{LacG}). The parameters show a decreasing trend that stabilise in the latter stages of the batch process. This aligns with the idea of the cell culture beginning to stabilise, supporting the theory that glucose is not initially directly driving cell growth like it is in the latter stages of the batch process, likely related to resource allocation and adaptation phenomena.

3.2. ecDMFA

The *ec*DMFA was successfully applied to the macro-scale hybrid model concentration profiles with samples taken 6 hours apart. Gradient-based optimisation was chosen since Equations 9-12 describe a convex optimisation problem, allowing the global optimum to be found. The flux results for the metabolic network are assessed to further validate *ec*DMFA simulation results throughout the experimental timeframe, while giving insight into the dynamic intracellular reaction network; a sample of which is shown in Figure 2. In the stable cell growth phase, the flux distribution stemming from glucose uptake is shown to be channelled into the TCA cycle, with less carbon being wasted in the form of lactate production. This observation aligns with current cellular understanding that lactate inhibits its own production, so central carbon flux will be directed away from its

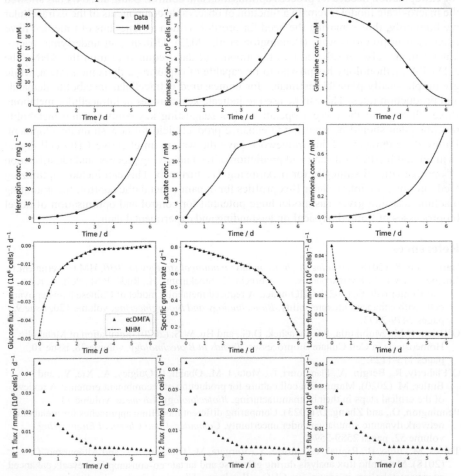

Figure 2 – Macro-scale Hybrid model (MHM) plots of Equations 1-6 and fit to *ec*DMFA with Equation 10, and samples of three intracellular reaction (IR) fluxes from the central carbon metabolism.

production beyond the initial high-production phase. Enzyme-constraints were not limiting, reflecting that not all enzymes are activated at once. Individual enzyme requirements can also be scrutinised, with noticeable contributions coming from the central carbon metabolism where large fluxes are directed; not all large fluxes require significant quantities of active enzymes since some enzymes have high turnover rates.

4. Conclusions

In this study a multi-scale hybrid model has been successfully constructed and scrutinised to gain insight into a complex CHO cell batch process for the production of Trastuzumab. The novelty of the work lies in the combination of a macro-scale hybrid kinetic model with a micro-scale *ec*DMFA modelling methodology. The introduction of time-varying parameters allowed complex macro-scale dynamics to be captured while sticking as closely as possible to the original kinetic model structure, thus maximising the capacity for extrapolation needed for process optimisation and control. Using an ANN has allowed the macro-scale model to remain a function of observable components of the extracellular medium only, thus removing the need for profiles written as functions of time that have been seen in previous work (Pennington et al., 2023). Utilizing an underlying macro-kinetic model also greatly reduces the amount of data required to train the ANN. The *ec*DMFA methodology has proven to be capable of capturing a dynamic system, while giving physically plausible estimates for unmeasured extracellular metabolite demands and production rates. This can be used to indicate key changes of intracellular metabolic fluxes throughout the process operation, thus suggesting key pathways and metabolite reactions that should be modified to enhance product yield for new strain development. Future developments of this framework pave the way to an effective CHO cell digital twin, including the modelling and prediction of the underlying process, and identification of optimal control strategies for maximising Trastuzumab. This can include optimizing feed rates or even intracellular flux profiles for dynamic metabolic control. Incorporating machine learning gives this model huge potential for control and optimisation of novel bioprocesses of limited physical understanding and experimental data.

References

Harrington, R. (2018). *Growing the bioeconomy: a national strategy to 2030*, HM Government.H.

Jiménez del Val, I., Kyriakopoulos, S., Albrecht, S., Stockmann, H., Rudd, P.M., Polizzi, K.M., and Kontoravdi, C. (2023). CHOmpact: A reduced metabolic model of Chinese hamster ovary cells with enhanced interpretability. *Biotechnology and Bioengineering*, volume 120, issue 9, pages 2479-2493.

O'Brien, C.M., Mulukutla, B.C., Mashek, D.G., and Hu, W.S. (2020). Regulation of Metabolic Homeostasis in Cell Culture Bioprocesses. *Trends in Biotechnology*, volume 38, issue 10, pages 113-1127.

O'Flaherty, R., Bergin, A., Flampouri, E., Mota, L.M., Obaidi, I., Quigley, A., Xie, Y., and Butler, M. (2020). Mammalian cell culture for production of recombinant proteins: A review of the critical steps in their biomanufacturing. *Biotechnology Advances*, volume 43.

Pennington, O., and Zhang, D. (2023). Comparing different modelling approaches for metabolic network dynamic simulation under uncertainty. *Computer Aided Chemical Engineering*, volume 52, pages 2589-2594.

Torres, M., Julio, B., Rigual, Y., Latorre, Y., Vergara, M., Dickson, A.J., and Altamirano, C. (2018). Metabolic flux analysis during galactose and lactate co-consumption reveals enhanced energy metabolism in continuous CHO cell cultures. *Chemical Engineering Science*, volume 205, pages 201-211.

Yeo, H.C., Hong, J., Lakshmanan, M., and Lee, D.Y. (2020). Enzyme capacity-based genome scale modelling of CHO cells. *Metabolic Engineering*, volume 60, pages 138-147.

Flavio Manenti, Gintaras V. Reklaitis (Eds.), Proceedings of the 34th European Symposium on Computer Aided Process Engineering / 15th International Symposium on Process Systems Engineering (ESCAPE34/PSE24), June 2-6, 2024, Florence, Italy

Dynamic Modelling Approaches in Life Cycle Assessment: A Case Study for the Evaluation of Power-to-Hydrogen

Laura R. Piñeiro*, Silvia Moreno, Ángel Luis V. Perales, Bernabé A. Fariñas, Pedro Haro

Department of Chemical and Environmental Engineering.
Escuela Técnica Superior de Ingeniería, Universidad de Sevilla.
Camino de los Descubrimientos. Seville, Spain
*lromero4@us.es

Abstract

This study introduces a model developed in MATLAB that uses different analytical approaches to investigate the impact of incorporating temporal considerations in Life Cycle Assessment (LCA), encompassing dynamic Life Cycle Inventory (LCI), the analysis of varied time horizons (TH), and the application of an absolute metric to illustrate the temporal evolution of climate impacts. Results demonstrate that selecting a shorter TH can lead to higher CO_2 equivalent emissions due to the increased impact of methane. However, the variability of the impact of the production of hydrogen from water electrolysis is minimal across different time horizons due to the dominant role of CO_2 emissions.

Keywords: Time-dependent life cycle assessment, Hydrogen, Electrolysis, Greenhouse gas emissions, Global Warming Potential

1. Introduction

LCAs are a widely used methodology for the systematic analysis of the potential environmental impacts of products or services during their entire life cycle. However, in conventional LCA the effect of Green House Gas (GHG) emissions at a certain time in the future is distorted, as emissions released in different times are added, and then, a static characterization factor is applied, usually the Global Warming Potential with a 100-year time horizon (GWP_{100}), modelling emissions as if they occur at the start of the analytical time horizon (TH), which fails to bring accurate evaluation values (Zieger, 2020).

Dynamic LCA (DLCA) research is increasingly generating interest, as it provides a more precise method of assessing impacts. Nonetheless, there is currently a lack of methodological guidelines to time incorporation in LCAs. However, the way in which dynamism is applied can considerably affect the environmental impact results.

This work delves into different methodologies for incorporating time in LCA, using as a case study the production of hydrogen via water electrolysis. The electrolyzer technology selected for the case study was an Alkaline electrolyzer, which is well-suited for large-scale hydrogen production in centralized facilities due to their lower capital costs and high production capacity and remain an attractive choice despite having a lower efficiency than Proton Exchange Membrane (PEM) electrolyzers (Senza, n.d.).

2. Methodology

2.1. Goal and scope

The system boundary was set as cradle to gate and the selected functional unit was 1 kg of hydrogen. All environmental burdens were allocated to hydrogen production, as the only by-product of electrolysis is oxygen, and this gas is usually vented (Bareiß et al., 2019). The impact category considered is climate change and it was based on the major GHG (i.e., CO_2, CH_4 and N_2O). The attributional LCA was implemented in MATLAB, using data of the Ecoinvent 3.8 database from the SigmaPro 9.5 software. The inventory data used was that of reference (Koj et al., 2017). Nevertheless, as the previous study proved that the impact of the construction phase was negligible for the impact category under consideration, only emissions from the operation phase were included in this study.

The Life Cycle Inventory (LCI) of this work integrates factors that vary over the temporal span of the LCA, thereby mirroring more accurately the real-world conditions of the evaluated system. The two variations included were related to technological advances:

- Electricity mix changes were implemented in the inventory data, following the Global Ambition predictions for Spain (TYNDP, 2022), which provides dispatch estimations from 2030 until 2050.
- The increasing efficiency of electrolysers over time was taken into account. In 2030 the efficiency was considered to be 49 kWh/kg H_2 (Deloitte, 2021). On the other hand, as the Oxford Institute for Energy Studies (2022) indicates that the efficiency aimed for Alkaline electrolysers is less than 45 kWh/kg, that was the efficiency implemented for the last year of the study (2050). For the in-between years, a linear evolution was assumed.

2.2. Static metric

The selected metric for the static approach is the Global Warming Potential, as it is the most frequently used metric (Keller, 2022). It is a normalized metric, meaning that it compares the changes resulting from a substance to those resulting from an equivalent amount of a reference gas, typically CO_2 (Peters et al., 2021), which allows to express emissions as CO_2-equivalents. For the metric GWP, the change measured is the integral of the radiative forcing of a substance over a given TH (IPCC, 2021).

This work firstly conducts a static approach to illustrate the influence of selecting a certain TH in conventional LCA. Two TH were chosen for this case study: the GWP_{100} because this is the TH most commonly used (even if there is no particular reason for selecting this particular time frame (The Guardian, 2011)), as well as 20 years, which is occasionally employed as a substitute for the GWP_{100} (EPA, n.d.). The static model calculates CO_2 equivalent emissions yearly with Eq. (1), by multiplying of the inventory data of the given year by a constant normalized metric (i.e., the GWP_{20} or the GWP_{100}).

$$kg\ CO_{2-eq} = Inventory \times Normalized\ metric \tag{1}$$

2.3. Dynamic metric

Two dynamic modelling approaches were conducted: one of them using a normalized metric in order to compare the results with the static approach, and the other one applying an absolute metric, with the aim of reflecting the evolution of impacts over time.

2.3.1. Normalized metrics of the dynamic approach

The normalized metric used for the dynamic modelling was the GWP with a flexible TH. The use of a changing TH allows to be consistent with the chosen time frame for the evaluation of the results. In this work, the chosen time frame for the evaluation of impacts was 25 years with the purpose of emphasizing the effect of CH_4, as the shorter the TH, the larger becomes the impact of CH_4 because its radiative forcing is much larger than that of CO_2 over a short period of time. Moreover, an evaluation period of a 100 years was also included with the aim of comparing the result with its static counterpart.

In this section, CO_2 equivalent emissions were also calculated with Eq. (1), but the value of the normalized metric changes every year, as its TH varies over time, and it is calculated as the difference between the evaluation period and the year in which emissions are released. For example, for an evaluation period of 25 years, emissions that occur in the first year will be evaluated for 25 years using the GWP_{25}, while emissions released during the second year will be evaluated for 24 years and so forth. This approach contrasts with the common practice applied in static LCA, which considers a fixed TH of 100 years and results in an inconsistency between the chosen TH and the time frame in which impacts are evaluated (e.g., as an emission released in year t when using a fixed TH of 100 years will be evaluated until the year t+100 (Levasseur et al., 2010)).

2.3.2. Absolute metrics of the dynamic approach

The absolute metric selected for the dynamic modelling approach was the Absolute Global Temperature change Potential (AGTP). Absolute metrics offer more comprehensive information of climate impacts (Peters et al., 2021) as they elucidate the connection between the moment in which emissions occur and their impact. This impact is expressed in a shared physical property (Ericsson et al., 2013), such as K per kg emitted in the case of the AGTP. Furthermore, absolute metrics allows to differentiate among diverse trajectories that lead to comparable climate impacts, aiding in avoiding adverse consequences, such as surpassing irreversible climate tipping points (Breton et al., 2018).

The AGTP per kg of each GHG was calculated according to the latest version of the IPCC to date (IPCC, 2021). The AGTP in K per kg of hydrogen is obtained by adding the emission of each GHG gas by their respective AGTP.

3. Results and discussion

3.1. *Normalized metric*

In this section the effect of changing the TH when using the GWP statically is presented. Moreover, static and dynamic approaches in LCAs are compared to analyze the influence of considering the temporal effects of GHG emissions.

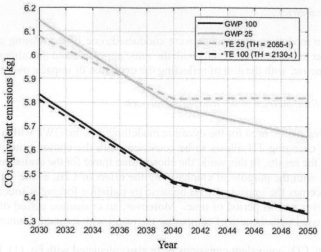

Figure 1. CO_2 equivalent emissions for a changing LCI and static (GWP_{100} and GWP_{25}) and flexible TH (TE_{25} and TE_{100}).

Figure 1 shows that for the same static metric (i.e., the GWP_{100} or the GWP_{20}), CO_2 equivalent emissions decrease with time. This is due to the two technological advances that were considered in this work, which results in a decreased amount of emissions.

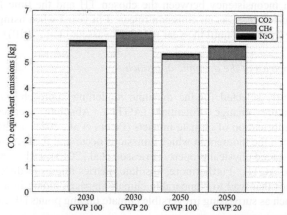

Figure 2. CO_2 equivalent emissions broken down into CO_2, CH_4 and N_2O.

Figure 2 shows that the increased CO_2 emissions each year shown in Figure 1 when using the GWP_{20} compared to the GWP_{100} is due to the increased impact of methane. As the GWP is calculated dividing the integral of the radiative forcing of a gas by the integral of the radiative forcing of CO_2 (IPCC, 2021) over the selected TH, choosing a short time frame will give a greater weight to short lived emissions like methane, as this gas has a much higher radiative efficiency over a short period of time than CO_2 (IPCC, 2021). This phenomenon also explains why the use of a flexible TH with an evaluation time of 25 years results in higher CO_2 emissions than the GWP_{20} past 2036 (as the TH is smaller than that of the GWP_{20} past this point in time). It is also noteworthy that in Figure 1 CO_2 equivalent emissions when using an evaluation time of 25 years are apparently constant past 2040. This is due to two contrasting actions: on the one hand, emissions are lower due to the incorporated technical advances, and on the other hand, the weighting of CH_4 emissions increases the later the emission is released because the TH gets smaller. Notwithstanding, this latter trend is not shown when selecting an evaluation time of 100 years, as CO_2 equivalent emissions remain almost identical to those obtained with the GWP_{100} since the TH is sufficiently large (even in the last year of the evaluation), and thus, the impact of CH_4 does not show much variability.

Figures 1 and 2 show the potential differences in the impacts obtained when conducting a LCA by selecting a certain TH. These changes could be significantly amplified in a system in which CH_4 emissions are more predominant than in this system. Therefore, it can be concluded that the production of hydrogen via water electrolysis is not that sensitive to the chosen TH (as its impacts are mainly due to CO_2 emissions that are not affected by the selected time frame).

3.2. Absolute metric

The AGTP is an instantaneous metric, meaning that it provides an estimation of the indicator's value at a specific moment in time following an emission (Breton et al., 2018). Two evaluation periods were selected: 25 and 100 years. Since the activity is considered to start in 2030, the AGTP in Figure 3 shows the temperature change in 2055 and 2130 due to the emissions that were released each year for the production of 1 kg of hydrogen and demonstrates that all emissions occurring from 2030 until 2050 have a greater impact on 2055 than on 2130.

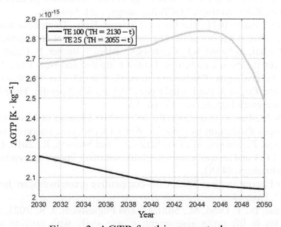

Figure 3. AGTP for this case study.

4. Conclusion

In this study, a variety of analytical approaches were developed in MATLAB for shedding light on the influence of the incorporation of time in LCA. These ranged from the use of a dynamic LCI to the analysis of the effects of selecting different TH. Furthermore, by implementing an absolute metric (i.e., the AGTP), this model illustrated the temporal evolution of climate impacts, which introduces a different approach to incorporating time in LCA practices, instead of opting for expressing emissions as CO_2-equivalents. It was also proven even if a shorter TH results in greater CO_2 equivalent emissions due to the greater impact of methane, the production of hydrogen from water electrolysis does not show a great variability with the selected TH, as CO_2 emissions are predominant.

It can be concluded that this work enhances our understanding of the implications of the inclusion of time into LCAs, which can aid practitioners in delivering more comprehensive LCAs that may well result in more informed conclusions and enhanced decision-making for the development of more effective environmental policies.

Acknowledgements

This study has been supported by grant PID2020-114725RA-I00 of the project GH2T funded by MCIN/AEI/ 10.13039/501100011033 and by the "European Union".

References

Bareiß K., de la Rua C., Möckl M. and Hamacher T., 2019, Life cycle assessment of hydrogen from proton exchange membrane water electrolysis in future energy systems, Applied Energy, vol 237, Pages 862-872.

Breton, C., Blanchet, P., Amor, B., Beauregard, R., and Chang, W. S., 2018, Assessing the climate change impacts of biogenic carbon in buildings: A critical review of two main dynamic approaches. Sustainability, 10(6).

Deloitte, 2021, Fueling the future of mobility: hydrogen electrolyzers, Hydrogen Articles Collections, Page 5.

EPA, n.d., Understanding Global Warming Potentials, https://www.epa.gov/ghgemissions/understanding-global-warming-potentials.

Ericsson, N., Porsö, C., Ahlgren, S., Nordberg, Å., Sundberg, C., & Hansson, P. A. (2013). "Time-dependent climate impact of a bioenergy system–methodology development and application to Swedish conditions." *GCB Bioenergy, 5*(5), 580-590.

IPCC, 2021, Climate Change 2021: The Physical Science Basis. Contribution of Working Group I to the Sixth Assessment Report of the Intergovernmental Panel on Climate Change [Masson-Delmotte, V., P. Zhai, A. Pirani, S.L. Connors, C. Péan, S. Berger, N. Caud, Y. Chen, L (eds.)]. Cambridge University Press, Cambridge, United Kingdom and New York, NY, USA

Keller M., 2022, Is Hydrogen Production through Natural Gas Pyrolysis Compatible with Ambitious Climate Targets in the United States? A Location-Specific, Time-Resolved Analysis, Environmental Research Letters, vol. 17, no. 12.

Koj J., Wulf C., Schreiber A. and Zapp P., 2017, Site-Dependent Environmental Impacts of Industrial Hydrogen Production by Alkaline Water Electrolysis, Energies, vol. 10, Issue 7.

Levasseur, A., Lesage, P., Margni, M., Deschênes, L., and Samson, R., 2010, Considering time in LCA: Dynamic LCA and its application to global warming impact assessments, Environmental Science and Technology, 44 (8).

Oxford Institute for Energy Studies, 2022, Cost-competitive green hydrogen: how to lower the cost of electrolysers?, Page 15.

Peters, G. P., Aamaas, B., T. Lund, M., Solli, C., and Fuglestvedt, J. S., 2021, Alternative "global warming" metrics in life cycle assessment: a case study with existing transportation data. Environmental Science and Technology, 45(20), 8633-8641.

SenzaHydrogen, n.d., PEM Hydrogen Generator VS Alkaline Hydrogen Generator. https://senzahydrogen.com/pem-hydrogen-generator-vs-alkaline-hydrogen-generator.html.

The Guardian, What are CO2e and global warming potential (GWP)?, 2011, https://www.theguardian.com/environment/2011/apr/27/co2e-global-warming-potential.

TYNDP, 2022, Scenario Report – additional Downloads. https://2022.entsos-tyndp-scenarios.eu/download/ Accesed 30 Nov. 2023.

Zieger, V., Stenger, A., Breton, C., Blanchet, P., Amor, B., 2020, "Impact of GHGs temporal dynamics on the GWP assessment of building materials: A case study on bio-based and non-bio-based walls," Building and Environment, 107210, DOI: 10.1016/j.buildenv.2020.107210

Flavio Manenti, Gintaras V. Reklaitis (Eds.), Proceedings of the 34th European Symposium on Computer Aided Process Engineering / 15th International Symposium on Process Systems Engineering (ESCAPE34/PSE24), June 2-6, 2024, Florence, Italy

Transient Flow-Assisted Kinetic Modelling and Reaction Network Identification for Pyrazole Synthesis

Fernando Vega-Ramon,[a] Linden Schrecker,[b] Miguel Angel de Carvalho Servia,[c] King Kuok (Mimi) Hii,[b] Dongda Zhang [a,*]

[a] *Department of Chemical Engineering, the University of Manchester, Engineering Building, Manchester M13 9PL, United Kingdom*
[b] *Department of Chemistry, Imperial College London, 82, Wood Lane, London W12 0BZ, United Kingdom*
[c] *Department of Chemical Engineering, Imperial College London, Exhibition Road, London SW7 2AZ, United Kingdom*
[*] *Corresponding Author. Email: dongda.zhang@manchester.ac.uk*

Abstract

Due to their structural versatility and wide range of biological activities, pyrazoles serve as important motifs in a variety of organic chemicals such as pharmaceuticals and pesticides. An established and economically favourable route for the synthesis of pyrazoles is the Knorr reaction, which involves the condensation of di-carbonyl and hydrazine reactants. Despite being researched extensively, previous kinetic studies on the Knorr reaction have been mainly qualitative and based on empirical observations. Because of this, the development of a comprehensive kinetic model capable of predicting product distribution and regioselectivity under different operating conditions remains a challenge.

In this study, a microkinetic analysis framework was combined with transient PFR experimentation techniques to perform kinetic modelling and reaction network identification. On the basis of literature observations and gathered spectroscopic and time-series kinetic data, two plausible reaction mechanisms were proposed for the condensation of symmetrical 1,3-diketones with phenyl hydrazine. Sparse-regression approaches were then adopted to identify kinetically negligible pathways within the proposed reaction networks, reducing the complexity of their corresponding microkinetic models while retaining their original fitting accuracy. The robustness of the reduced models was further confirmed through uncertainty analysis; the results indicate that most kinetic parameters are identifiable and statistically significant, suggesting the advantage of high throughput data generation via transient experiments.

Finally, a model-based design of experiments (MbDoE) methodology was adopted to identify the optimal set of experimental conditions for discriminating among the two proposed models. This work therefore provides a new avenue for the systematic analysis of complex organic reactions assisted by transient flow experimentation.

Keywords: Transient flow reactor, pyrazoles, kinetic modelling, model discrimination.

W. & S. Manca (Eds.), Proceedings of the 34th European Symposium on Computer Aided Process Engineering / 15th International Symposium on Process Systems Engineering (ESCAPE34/PSE24), June 2–6, 2024, Florence, Italy © 2024 Elsevier B.V. All rights reserved. http://dx.doi.org/10.1016/B978-0-443-28824-1.50013-4

1. Introduction

Owing to the wide availability of 1,3-diketone and hydrazine precursors, the Knorr reaction is a popular pathway for the synthesis of substituted pyrazoles. Previous literature studies (Sloop et al., 2008; Norris et al., 2005) have qualitatively demonstrated the influence of reaction parameters (such as feed ratios, reactant substituent groups, or pH) on the reaction rate and product regioselectivity. The construction of a detailed kinetic model capable of describing such effects is therefore a vital requirement for process optimisation, reaction network identification, and scale-up considerations.

A common challenge in the development of comprehensive kinetic models for complex chemical systems is the need for reliable reaction data over a diverse range of conditions. An emerging technique for the efficient generation of time-series data is transient flow experimentation, whereby a flow reactor is operated under unsteady-state conditions to emulate a collection of several batch experiments. In the present work, this high throughput experimentation tool was combined with kinetic model construction and MbDoE methodologies to elucidate potential reaction mechanisms for the addition of phenyl hydrazine to substituted diketones, namely acetyl acetone and heptane-3,5-dione.

2. Experimental Methodology

2.1. Transient Flow Setup

Ethanolic reagent solutions were kept in bottles; inlet tubing lines connected these to two Gilson 305 HPLC pumps controlled via a master/slave system. The pump outlets fed into a Valco T-piece stainless steel mixer submerged in a silicon oil bath. The mixer outlet fed into a 5.10 m (4.13 mL) stainless steel tubular reactor, which was also submerged in the oil bath. The operating temperature was maintained around the desired set point (70 °C) by means of PID control. The reactor effluent was then passed through a custom cooling system consisting of a Peltier thermoelectric cooler connected to an aluminum block heat sink. In-line IR spectroscopic analysis was then performed using a Mettler Toledo ReactIR™ 15 equipped with a micro flow cell. A back pressure regulator was set to 8.1 bar to ensure all solvents remain in solution for the tested temperatures. A more detailed account of the experimental setup can be found on previous work (Schrecker et al., 2023).

2.2. Time-series Kinetic Experiments

Transient ramps were performed by introducing step changes or linear ramps to the volumetric flowrates of the hydrazine and diketone reactants, while simultaneously monitoring the product concentration at the reactor outlet. Namely, three different kinds of transient experiments were performed:

- Residence time ramps, whereby the total volumetric flowrate is decreased while keeping the hydrazine:diketone ratios unchanged. This is equivalent to running a collection of batch experiments with the same initial concentrations but varying reaction times.
- Stoichiometry ramps, whereby the flowrates of reactants are ramped linearly at a fixed cumulative flowrate. This yields a collection of experiments with the same residence time but varying initial concentrations.
- Bivariate ramps, whereby a stoichiometry ramp is performed up to a final diketone:hydrazine ratio, after which a residence time ramp is carried out over a complete reactor volume.

3. Modelling Methodology

3.1. Reaction Mechanism Proposal

Two reaction schemes were constructed on the basis of gathered spectroscopic data and literature observations; these are presented in Fig. 1 below. Equations for the rate of change of concentration of all species were derived by applying the law of mass-action.

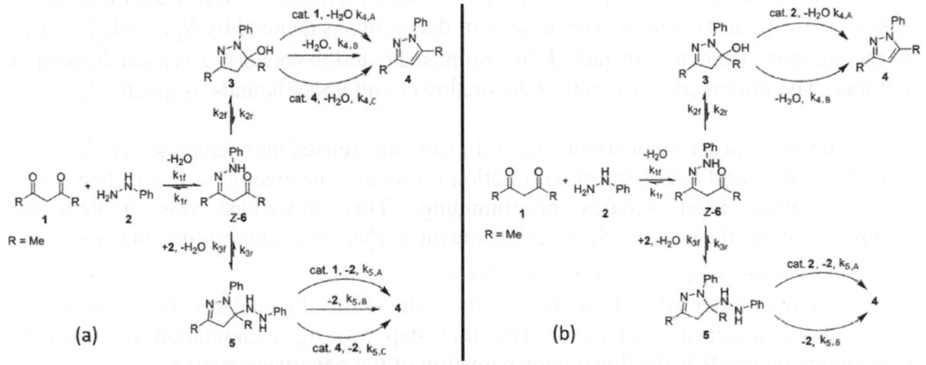

Figure 1: Proposed reaction networks for the reaction between 1,3-diketones and phenyl hydrazine. (a) Model 1 adapted from (Schrecker et al., 2022); (b) Model 2.

Model 1 in Fig. 1(a) corresponds to the reaction network proposed in our previous work (Schrecker et al., 2022). The first step in the mechanism involves the formation of a hydrazone intermediate, which has been isolated and identified in previous literature studies (Sloop et al., 2008). The general consensus in literature is that the hydrazone species undergoes ring-closing (Step 2) to form a hydroxypyrazolidine intermediate, which then aromatises to the final pyrazole product (Step 4). In addition, we also hypothesised an alternative reaction pathway involving the formation (Step 3) and aromatization (Step 5) of a di-addition intermediate, which was detected via mass spectroscopy analysis. The key difference between the two networks concerns the catalysis of the aromatization reactions; whereas in Model 1 these are product- and diketone-catalysed pathways, in Model 2 they are assumed to be hydrazine-catalysed. The rationale behind this modification is that higher hydrazine:diketone ratios favoured product yields on reactant stoichiometry ramp experiments.

3.2. Sparse Regression-based Parameter Estimation

Kinetic parameter estimation was formulated as a nonlinear, constrained optimisation problem presented in Eq. (2a) – Eq. (2c). Inspired by sparse regression approaches, the objective function to be minimised is a least-squares expression which incorporates an additional sparsity regularization term. The rationale behind this term is that model fitting and model reduction can be accomplished simultaneously by penalizing the number of active reaction pathways: For kinetically significant steps their corresponding constants take some non-zero value $k_j \gg \varepsilon$ and the penalty term approaches unity, whereas for kinetically negligible steps the penalty term approaches 0 as $k_j \approx 0$.

$$\min_k \sum_{i=1}^{N_p} \left(C_{i,E} - C_{i,S}\right)^T \Lambda \left(C_{i,E} - C_{i,S}\right) + \omega_s \sum_{j=1}^{N_k} \frac{k_j}{k_j + \varepsilon} \qquad (2a)$$

$$\text{st.} \quad \frac{d\boldsymbol{C}}{d\tau} = f(\boldsymbol{C}, \boldsymbol{k}) \tag{2b}$$

$$\boldsymbol{k_{lb}} < \boldsymbol{k} < \boldsymbol{k_{ub}} \tag{2c}$$

where $\boldsymbol{C_{i,E}}$ and $\boldsymbol{C_{i,S}}$ are vectors for the experimentally measured and simulated concentrations of species, respectively. $\boldsymbol{\Lambda}$ is a weighting matrix used to scale the residuals and ω_s is the penalty weight. The number of datapoints is denoted by N_p, while N_k refers to the number of kinetic constants $\boldsymbol{k}$ in a proposed reaction network. ε is a small, positive number. The underscripts lb and ub denote lower and upper bounds, respectively.

The differential process constraints Eq. (2b) were discretised into algebraic profiles using Adams method and backward differentiation formulas. The arising NLP was then solved via sequential least squares programming. This procedure was implemented computationally through the S*cipy* environment in *Python* programming language.

3.3. Kinetic Parameter Uncertainty Analysis
Following parameter estimation, the fidelity of the reduced model structures was further assessed via uncertainty analysis. The first step towards computation of parameter confidence intervals is the linear approximation of the covariance matrix:

$$\boldsymbol{V} = \boldsymbol{H^{-1}} \cdot \frac{S_r(\boldsymbol{k^*})}{(N_p - N_k)} \tag{3}$$

where $\boldsymbol{V}$ is the variance-covariance matrix, $\boldsymbol{H}$ is the Hessian matrix of the objective function at the optimum (approximated through central differences) and $S_r(\boldsymbol{k^*})$ refers to the minimised objective function value.

Joint confidence intervals for the estimated constants were computed from the covariance matrix following the procedure described by Franceschini et al. (2008), while their marginal variances were directly taken from the diagonal elements of the matrix.

3.4. Model Discrimination via Design of Experiments
In this work, the two proposed reaction networks incorporate different assumptions regarding the product-formation steps. A recurring challenge in kinetic model discrimination is that reaction networks comprising different elementary steps and degrees of detail can yield equally satisfactory fitting of experimental data. An MbDoE methodology was therefore adopted to design a discriminatory transient experiment; this was formulated as an optimisation problem Eq. (4a)-Eq. (4d), whereby the discrepancy between the two candidate models over a residence time ramp was maximised with respect to the feed concentrations:

$$\max_{\boldsymbol{C_0}} \quad \Delta\boldsymbol{C_P}(\boldsymbol{C_0}, \boldsymbol{k_1}, \boldsymbol{k_2}, \tau)^T \cdot \Delta\boldsymbol{C_P}(\boldsymbol{C_0}, \boldsymbol{k_1}, \boldsymbol{k_2}, \tau) \tag{4a}$$

$$\text{st.} \quad \Delta\boldsymbol{C_P} = \boldsymbol{C_{P,1}}(\boldsymbol{C_0}, \boldsymbol{k_1}, \tau) - \boldsymbol{C_{P,2}}(\boldsymbol{C_0}, \boldsymbol{k_2}, \tau) \tag{4b}$$

$$\frac{d\boldsymbol{C_1}}{d\tau} = f(\boldsymbol{k_1}, \boldsymbol{C_1}), \quad \frac{d\boldsymbol{C_2}}{d\tau} = g(\boldsymbol{k_2}, \boldsymbol{C_2}) \tag{4c}$$

$$\boldsymbol{C_{0,lb}} < \boldsymbol{C_0} < \boldsymbol{C_{0,ub}} \tag{4d}$$

where C_0 is a vector of initial conditions (in this case optimisation variables), and ΔC_P is a vector of discrepancies between the product concentration profiles $C_{P,1}$ and $C_{P,2}$ predicted by Model 1 and Model 2, respectively.

4. Results

4.1. Kinetic Model Fitting and Uncertainty Results

The reduced kinetic models exhibited good fitting performance, with an overall mean percentage error of 3.27% and 3.63% for Model 1 and Model 2, respectively. The simulated product profiles for a residence time ramp experiment with the acetyl acetone reactant are presented in Fig. 2 below.

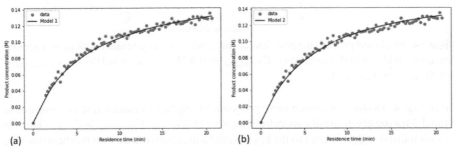

Figure 2: Model fitting results for residence time ramp experiments with acetyl acetone. (a) Model 1; (b) Model 2.

By carefully tuning the sparsity penalty weight, the trade-off between model fitting accuracy and model complexity was controlled. In this fashion, kinetically negligible reaction pathways in the proposed mechanisms were automatically identified and removed from their corresponding model structures. For both models and diketone reactants, the initial nucleophilic attack (Step 1), ring-closure (Step 2) and formation of di-addition intermediate (Step 3) were deemed to be approximately irreversible reactions. Moreover, the computed confidence intervals and correlation matrix for the non-zero kinetic constants indicate that most parameters are statistically identifiable and uncorrelated, as exemplified on Fig. 3 for Model 1 with the acetyl acetone reactant.

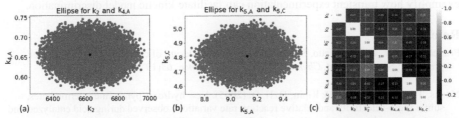

Figure 3: (a) and (b) Joint confidence ellipsoids of some kinetic parameters in Model 1 (acetyl acetone case); (c) Correlation matrix plot for kinetic parameters in Model 1.

Comparison of the estimated constants for the acetyl acetone and heptane-3,5-dione datasets can also provide insight on steric influences on the reaction rate. For example, the formation of the di-addition intermediate was found to be faster for the shorter-chained substrate, whereas the opposite is true for the initial nucleophilic attack.

4.2. Model Discrimination Results

The optimal initial conditions for a discriminatory residence time ramp experiment were determined following the procedure described in Section 3.4.

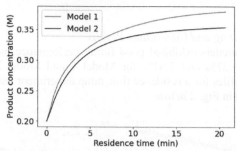

Figure 4: Predicted process trajectories for discriminatory experiment with acetyl acetone reactant, with initial conditions $C_{Diketone} = 0.4$ M, $C_{PhHy} = 0.186$ M, $C_{Pyrazole} = 0.2$ M, $C_{H_2O} = 0.014$ M.

From Fig. 4, Model 1 exhibits a more pronounced product formation rate as compared to Model 2 for the given initial concentrations where $C_{Diketone}, C_{Pyrazole} > C_{PhHy}$. This can be rationalised with reference to the key dissimilarities between the two mechanisms: The final aromatization steps are product- and diketone-catalysed for Model 1, as opposed to hydrazine-catalysed pathways in Model 2. Conducting this transient experiment will thus guide the modification and selection of an appropriate model structure for future studies.

5. Conclusions

In this work, state-of-the-art transient flow experimentation techniques were coupled with model construction and reduction methodologies to perform kinetics investigations on the Knorr pyrazole reaction. The model fitting accuracy and uncertainty results indicate that the proposed reaction networks are feasible representations of the process. Notably, analysis of the estimated constants in the reduced model structures allow for the identification of kinetically negligible reactions and the evaluation of steric effects on the reaction rate. A preliminary model-based design of experiments was also adopted to exemplify how transient experimentation can facilitate kinetic model discrimination.

References

Franceschini, G., & Macchietto, S. (2008). Model-based design of experiments for parameter precision: State of the art. *Chemical Engineering Science, 63*(19), 4846–4872.

Norris, T., Colon-Cruz, R., & Ripin, D. H. B. (2005). New hydroxy-pyrazoline intermediates, subtle regio-selectivity and relative reaction rate variations observed during acid catalyzed and neutral pyrazole cyclization. *Organic & Biomolecular Chemistry, 3*(10), 1844.

Schrecker, L., Dickhaut, J., Holtze, C., Staehle, P., Vranceanu, M., Hellgardt, K., & Hii, K. K. (Mimi). (2023). Discovery of unexpectedly complex reaction pathways for the Knorr pyrazole synthesis *via* transient flow. *Reaction Chemistry & Engineering, 8*(1), 41–46.

Sloop, J. C., Lechner, B., Washington, G., Bumgardner, C. L., Loehle, W. D., & Creasy, W. (2008). Pyrazole formation: Examination of kinetics, substituent effects, and mechanistic pathways. International Journal of Chemical Kinetics, 40(7), 370–383.

Flavio Manenti, Gintaras V. Reklaitis (Eds.), Proceedings of the 34th European Symposium on Computer Aided Process Engineering / 15th International Symposium on Process Systems Engineering (ESCAPE34/PSE24), June 2-6, 2024, Florence, Italy

Application of Appropriate Kinetic Models in Developing Pharmaceutical Drug Substance Manufacturing Processes

Maitraye Sen* and Alonso J. Arguelles

Synthetic Molecule Design and Development, Eli Lilly and Company, IN, USA
Corresponding author: sen_maitraye@lilly.com

Abstract

The active pharmaceutical ingredient (API) is usually synthesized through a series of chemical reactions. Depending on the complexity of the reaction network, it is common to form multiple impurities. The API synthesis process should be designed to minimize the impurity formation. A kinetic model can provide better process insights for doing so. However, developing a kinetic model in the early stages of process development is challenging due to scarcity of data. This work presents a framework for developing fit-for-purpose kinetic models from limited information or data. Step-by-step guidance on model structure determination and parameter estimation have been discussed.

Keywords: Active pharmaceutical active ingredient, Kinetic model, Design space.

1. Introduction

Meeting the quality specification (i.e., purity) of the API is paramount. The API synthesis process must be designed to meet the quality specifications consistently. Hence, a detailed process understanding should be built from the outset. Kinetic models combined with experimentation can provide important insight into the process. Since the API synthesis often involves complex reaction network, developing the kinetic model is not trivial. Moreover, the chemistry is not fully known, and the analytical method for profiling the different chemical species is not matured during the early phases of process development. Hence, a comprehensive data and information is not available at this stage. This often hinders identifying appropriate kinetic models, since most practitioners aim for an optimal model structure and parameter estimates. An optimal kinetic model is not necessary to guide process development (Sen et al., 2021). This work presents a framework for developing fit-for-purpose kinetic models. A fit-for-purpose model is relevant only within a desired operating space, where the synthesis process is designed. As such, a simple practice-based kinetic model development workflow has been demonstrated with the help of a complex telescoped reaction network described below.

2. Process chemistry

The simplified reaction scheme is depicted in Figure 1. It is a telescoped reaction with three sub-steps: Step 1a, 1b, 1c. In step 1a, the deprotection of hydrazide **1** is carried out in a system of aqueous HCl and m-xylene to produce the hydrazine hydrochloride **2.HCl**. The end of reaction (EOR) mixture undergoes a wash and TEA addition to neutralize the HCl. In step 1b, enol **3** is methylated using trimethyl orthoformate (TMOF) to produce methyl enol ether **4**. The EOR mixture from step 1b is concentrated via distillation. The EOR mixtures of steps 1a and 1b are combined in step 1c, where the enol ether **4** reacts with hydrazine **2** to form pyrazole **5** that is isolated via a crystallization step.

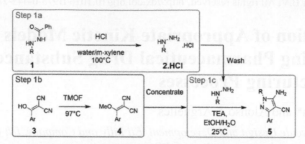

Figure 1: Simplified step 1a reaction scheme

Step1a: It is an acid-catalyzed multi-phasic reaction (slurry of compound **1** particles in water+m-xylene). Two clear liquid layers are produced at the end of reaction. The reaction completes in ~24 h at the target conditions. Traces of other hydrazide or di-hydrazide impurities in **1** forms hydrazine that may react in step 1c to form impurity **5-IM1** (Figure 3C).

Step 1b (Figure 2): Enol **3** is O-methylated by treatment with neat TMOF to produce methyl enol ether **4**. The reaction starts as a suspension and turns clear in short order. Three compounds related to enol **3** are formed during step 1b: i) the desired product **4** (~90%), ii) N-methylated impurity **4-IM1** (~7%), and iii) methyl ester **4-IM2** (~1%). The amount of each compound formed at reaction completion at target conditions is provided in brackets (as %area from HPLC). This reaction takes ~12 h to complete at the target conditions. The methyl formate is removed by distillation following the 1b reaction.

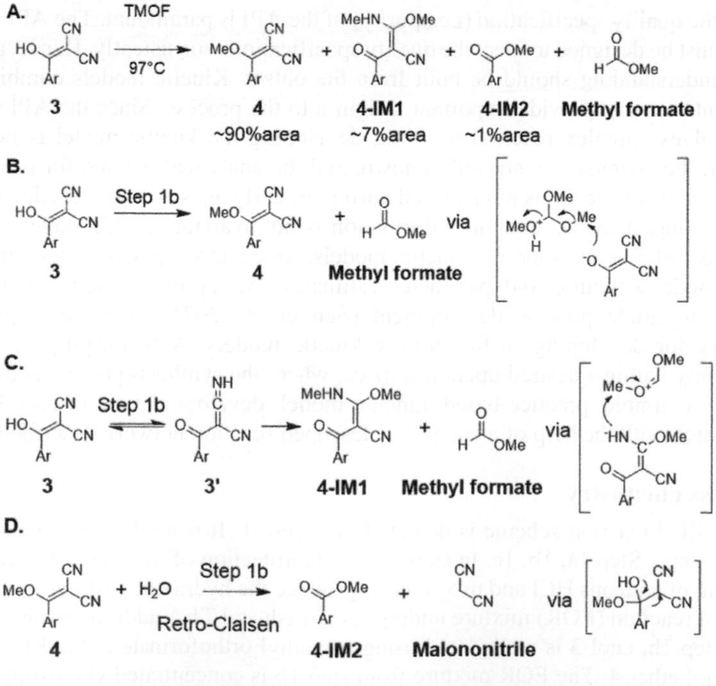

Figure 2: Mechanism of formation of 4, 4-IM1, 4-IM2 in step 1b

Step 1c (Figure 3): Enol ether **4** and alkyl hydrazine **2** react to form pyrazole **5**. This reaction takes ~4 h to complete at target conditions. Many impurities are formed in step 1c (Figure 3B). Figure 3C depicts the major impurities and their amount (%area from HPLC) formed at target condition at the end of the overall reaction. The crystallization step must reject the impurities to the desired levels during the isolation of **5**.

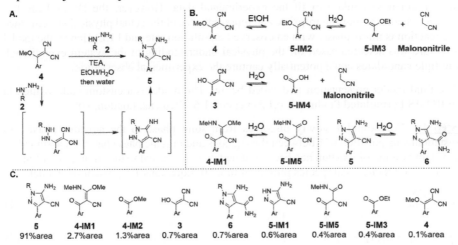

Figure 3: Schematic of step 1c

3. Kinetic model development

Step 1a has an aqueous and organic (m-xylene) phase. The solid particles of **1** is dispersed in the aqueous phase and assumed as a lumped solid phase. m-xylene is on the top of the aqueous phase. A schematic of the physical system is shown in Figure 4A. The reaction scheme was simplified for the model (Figure 4A). Sublimation of the benzoic acid (BzOH) is prevented by m-xylene.

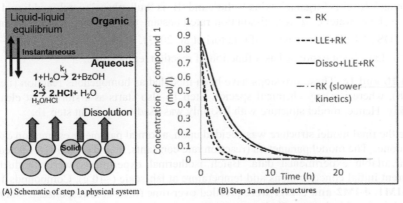

(A) Schematic of step 1a physical system   (B) Step 1a model structures

Figure 4: Schematic and model structure of step 1a

The structure and estimated parameter values determine a models' behavior. Three model structures incorporating the following physical phenomena were considered: i) reaction kinetics (RK) only, ii) liquid-liquid equilibrium (LLE) and RK and iii) dissolution rate (Disso), LLE and RK. Figure 4B shows a preliminary prediction from the three model structures when the kinetic parameters (k_1, k_2) are kept constant across them. The estimates of the kinetic parameters were obtained from historical data of similar reactions. The RK and LLE+RK structures predict reaction completion at less than 10 hours at target condition. The closest prediction to the experimental observation is the Disso+LLE+RK structure. RK model structure with different values of the kinetic parameters (i.e., with slower kinetics) could also fit the experimental data. However, the Disso+LLE+RK model structure is more appropriate and representative of the actual physical system since the reaction is multi-phasic with an observed dissolution rate and LLE. Hence, the model structure was selected based off the physical understanding of the system even though multiple candidates could potentially capture the experimental observation.

The final model presentation is detailed below. The model is a custom code written in gPROMS Formulated Products® (gFP) version 1.6 (Siemens, London, UK).

Step 1a: Compound **1** slowly dissolves from solid phase to the aqueous phase. An instantaneous partition of **1** between the aqueous and organic phase has been assumed. A constant volume of reaction has been assumed, which is equal to the summation of initial volumes of all chemical species present at time=0. Equations 1-5 present the model.

$$K_i = \frac{x_i^{aq}}{x_i^{org}} \tag{1}$$

$$\frac{dx_i}{dt} = -r_i + \alpha_i(x_i^s - x_i^{aq}) \tag{2}$$

$$x_i^{aq} = \frac{x_i^B K_i}{K_i + 1} \tag{3}$$

$$x_i^B = x_i^{org} + x_i^{aq} \tag{4}$$

$$r_i = f(x_i^{aq}) \tag{5}$$

Here, i= compound **1**, hydrazide impurity or di-hydrazide impurity (if present in **1**). 'K' is the partition coefficient. 'x' is the concentration (mol/l). Superscripts 'aq' stands for 'aqueous', 'org' is 'organic', 's' is 'solid', and 'B' is 'bulk phase' (combined liquid phase of m-xylene+water). 'α' is a dissolution rate constant with units of (s^{-1}) that lumps the term (DS$_w$/Vh) in equation (1) of (Hattori et al. 2013). 'r' is the reaction rate ($\frac{mol}{l\,s}$). 'K' and 'α' have been expressed as a function of reaction temperature.

Step 1b and 1c: These two steps have been modeled as homogenous reaction (i.e., -r$_j$= dx$_j$/dt), where j is any chemical species. Step 1b also starts as a slurry that clears out quickly. Hence, model structure with RK only was used to represent step 1b.

Once the final model structure was decided upon, a formal parameter estimation exercise was done. The model parameters (reaction rate constants, α, and K) were estimated from the available experimental data. Batch isothermal experiments were conducted at different initial concentrations and temperature at lab-scale (~10 g). Compounds **1, 3, 4, 5, 4-IM1, 4-IM2 and 5-IM1** were profiled over time (measured responses). Only three experiments with time-profiled concentration were available, hence limited data.

The sensitivity of each model parameter to the measured responses was calculated. Only the model parameters with reasonable sensitivities towards the measured responses were estimated. Including the less sensitive parameters in the parameter estimation deteriorates the overall performance of the algorithm (Sen et al., 2021). Hence, they were fixed at a value determined from previous experience with similar reactions. For example, Table 1 presents the sensitivity indices of the step 1a model parameters towards the concentration of **1** during a reaction at a given temperature and initial concentration. The sensitivity analysis was run by producing 80000 scenarios sampled using built-in Sobol sampling (Sobol, 1993) in gFP. The least sensitive parameter is k_2, hence excluded from parameter estimation. It was fixed at a reasonable value determined from other similar reactions.

Table 1: Sensitivity indices of step 1a model parameters

Model parameter	Sensitivity indices
k_1	0.768
k_2	4.02E-8
α	0.236
K	0.034

4. Results and discussion

The kinetic model was used to obtain important process insights as detailed below.

Step 1a: Figure 5A presents the fractional yield of **2.HCl** (with respect to the initial concentration of **1**) at the end of 24 h, as a function of initial concentration of **1** (x-axis) and temperature (color coded). Total of 40000 simulations were run at various combinations of initial concentration of **1** and reaction temperature (sampled via Sobol, 1993 method). The reaction slows down significantly at lower temperatures. The desired conversion of **1** cannot be achieved at lower temperatures within the target completion time of 24 h. Since cycle time is important, maintaining a reasonable reaction temperature is critical. A temperature of 94 °C is affordable without slowing down the reaction significantly. It results in 98% conversion at the end of 24 h. Hence, reaction temperature should be always $\geq$ 94 °C. Moreover, at higher temperatures, slight variability in the initial concentration of **1** from target has no significant impact on the reaction rate.

Step 1b: Figure 5B shows **4-IM1**, a major impurity of step 1b. Total of 47000 simulations sampled via Sobol, 1993 method were run. The amount of **4-IM1** (y-axis) at the end of 12 h is plotted as a function of temperature (x-axis) and initial concentration of **3** (color coded). The spread of the colored dots representing the initial concentration of **3** is narrow compared to the change observed with variation in temperature. Clearly, the amount of **4-IM1** formed is more sensitive to the temperature than the initial concentration of **3**. A similar relationship of **4** (product of 1b) with temperature and initial concentration of **3** is seen. Since, both **4** and **4-IM1** form from **3** in two parallel and kinetically competitive reactions, there is no way to slow down one without slowing down the other. Hence, some amount of **4-IM1** will be formed that should be rejected during crystallization.

Step 1c: Total of 40000 simulations sampled via Sobol, 1993 method were run. Figure 5C presents the amount of **5** at the end of 4 h (y-axis) as a function of the ratio between initial concentration of **2** and **4** (x-axis) and reaction temperature (color coded). If the ratio is <1 the yield of **5** decreases significantly because the excess of **4** reacts to form impurities. **4** has a higher affinity to react with **2**, but in absence of **2**, it partakes in other side reaction. Figure 5C also shows that rate of formation of **5** is insensitive at temperature<30 °C (scattered dots). However, a distinct region appears at temperature>30 °C that shows that the rate decreases (amount of **5** formed decreases). This is because the impurity formation rate increases. Hence, it is important to maintain a ratio $\geq$1 and temperature<30 °C.

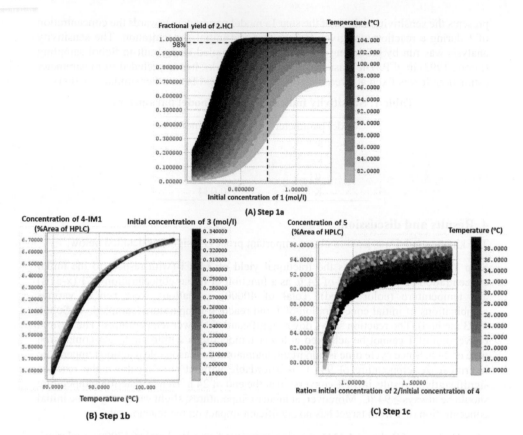

Figure 5: Contour plots of step 1a, step 1b and step 1c

5. Conclusion

Kinetic models enhance process understanding, bridging any knowledge gap resulting from limited data. A framework for developing fit-for-purpose kinetic model has been presented. The current framework is much simpler than the other alternatives of model structure determination and parametrization discussed by Sen et al., 2021. We have applied this framework to minimize impurity formation in a complex telescoped reaction.

References

- Y. Hattori, Y. Haruna, M. Otsuka, 2013. Dissolution process analysis using model-free Noyes-Whitney integral equation, Colloids and Surfaces B: Biointerfaces,102, 227-231.
- M. Sen, A.J. Arguelles, S.D. Stamatis, S. Garcia-Munoz, S. Kolis, 2021. An optimization-based model discrimination framework for selecting an appropriate reaction kinetic model structure during early phase pharmaceutical process development, React. Chem. Eng., 6 (11), 2092-2103.
- I.M. Sobol, 1993. Sensitivity Estimates for Nonlinear Mathematical Models, Mathematical Modelling and Computational Experiments, 4, 407-414.

Flavio Manenti, Gintaras V. Reklaitis (Eds.), Proceedings of the 34th European Symposium on Computer Aided Process Engineering / 15th International Symposium on Process Systems Engineering (ESCAPE34/PSE24), June 2-6, 2024, Florence, Italy

DAF Turbidity Removal Dynamics Model Employing LSTM Networks with Monte Carlo Dropout

Felipe M. M. Sousa,[a] Ivo J. Cunha,[a] Vinícius M. Muller,[a] Flávio V. Silva[a]

[a]*University of Campinas, Cidade Universitária Zeferino Vaz Av. Albert Einstein, 500, Zipcode 13083-852, Campinas-SP, Brazil*
flaviovs@unicamp.br

Abstract

Dissolved Air Flotation (DAF), a well-established technology, is still not fully comprehended despite its longstanding use. The complex nature of DAF limits deterministic model development, impeding optimization endeavors. This study explores an alternative approach by leveraging deep learning algorithms, specifically Long Short-Term Memory networks (LSTMs). LSTMs, proficient at discerning intricate time-series patterns from extensive datasets, are employed to construct empirical models for turbidity removal in a laboratory-scale DAF prototype. Additionally, a Bayesian approximation using Monte Carlo Dropout (MCD) is introduced to gauge uncertainty derived from information gaps in the database. The results showed that MCD-LSTM can predict the overall dynamics with errors within ±0.7 and ±1.3 Nephelometric Turbidity Units (NTU) for training and test datasets, respectively. The prediction errors for unseen data hinted at the potential for inference enhancement by incorporating new input variables.

Keywords: dissolved air flotation, long short-term memory, Monte Carlo Dropout.

1. Introduction

Water, crucial for human development, exists worldwide in diverse forms. Despite its apparent abundance, only about 3% of Earth's water has low dissolved solids, suitable for consumption and industry. Additionally, only 13% of fresh water is readily available in aquifers, rivers, and lakes, while most are locked in glaciers. This scarcity amplifies the significance of accessible water resources (Glantz, 2018).

Regulatory agencies in each country propose diverse regulations to maintain potability standards and mitigate potential adverse effects from water consumption. These regulations specify the responsibilities of handling the water supply and set acceptable standards for biological, chemical, and physical parameters.

In terms of physical aspects, the primary variable under analysis is turbidity, characterized as an indirect measure of the concentration of colloidal or suspended particles in the fluid. Turbidity can influence the optical and organoleptic aspects of water, the development of gastrointestinal diseases (Morris et al., 1996; Schwartz, 2000), setbacks in the final quality of products in industrial processes, and potential equipment deterioration. Thus, efficient separation techniques are essential to mitigate these possible problems.

One promising separation method is dissolved air flotation (DAF), a technology known for several decades with records dating back to the 1960s in Finnish clarification units. Although it has been employed for a long time, a comprehensive description of the phenomena involved in DAF is still not entirely understood. Various theories have been proposed (Edzwald, 1995; Fukushi et al., 1995), but the complexity of the system hinders the attainment of deterministic models that are easily applicable. Simulations based on computational fluid dynamics (CFD) have been proposed with great results (Rodrigues and Béttega, 2018; Rybachuk and Jodłowski, 2019). However, they demand high computational effort and time, hindering potential optimization projects and implementation of model-based automatic process control strategies.

Given this impasse, empirical modeling employing deep learning emerges as a potential alternative for constructing a mathematical model capable of representing the FAD process due to the greater ease of obtaining experimental data on important operational parameters. However, few studies have been developed within this paradigm (Souza et al., 2021), indicating a promising research area.

In this work, we aim to investigate the performance of Long Short-Term Memory networks (LSTMs), architectures developed to learn intricate time-series patterns from a large volume of data, to model DAF dynamics. Furthermore, a Bayesian approximation using Monte Carlo Dropout (MCD) was introduced to quantify uncertainty from the lack of information from the whole state space. These algorithms were utilized to determine empirical models for turbidity removal in a DAF prototype at a laboratory scale.

2. Materials and Methods

2.1. DAF prototype

Figure 1 depicts the DAF prototype developed in the Laboratory of Chemical Systems Engineering (LESQ) at the Chemical Engineering Faculty, University of Campinas, Brazil. This pilot plant was utilized in this work to gather data on the dynamics of turbidity removal under different inlet flows, inlet turbidity, saturation pressure, and recycle flow. The operational range of these variables is detailed in Table 1.

The raw effluent is initially introduced into the coagulation/flocculation tank for pretreatment. It is then transferred to the flotation tank, where it combines with the recycle flow containing microbubbles generated by the needle valve. This stream is regulated by an automated stepper motor, adjusting the valve according to the desired flow rate. Part of the clarified water is diverted to a sand filter and redirected to the saturation tank. This vessel promotes high-pressure air saturation of the water used in the recycle flow.

Table 1 - Input variables value investigation space.

Variable	Range	Unit
Inlet flow	1.1 - 1.6	L/min
Inlet turbidity	15.2 - 40.4	NTU
Saturation pressure	3.5 - 6.0	bar
Recycle flow	0.25 - 0.36	L/min

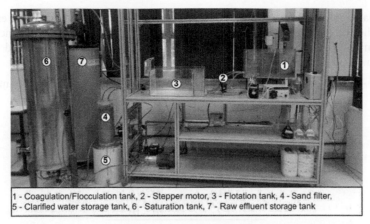

1 - Coagulation/Flocculation tank, 2 - Stepper motor, 3 - Flotation tank, 4 - Sand filter,
5 - Clarified water storage tank, 6 - Saturation tank, 7 - Raw effluent storage tank

Figure 1 - DAF prototype

The raw effluent was synthesized using red latosol soil, a prevalent type of clay found in São Paulo/Brazil, replicating typical conditions in water treatment facilities. In the physical-chemical stage, sodium aluminate 1% (v/v) and tannin SG 5% (v/v) solutions were employed to neutralize the charged suspended particles and to promote the formation of larger flocs for subsequent separation in the flotation tank. Jar tests determined the flow rates of these solutions for a 21 NTU sample, the mean value of most experimental runs.

The data acquisition rate of the SCADA system was configured at 1 s. However, due to the sluggish system dynamics, the data employed in the models were discretized into 1-minute intervals, using an average of 60 samples to determine the values of all variables. The entire database spanned 38 hours, containing a total of 2,280 data points.

2.2. Development of the MCD-LSTM networks

Python was used to build the MCD-LSTM models using Tensorflow and Keras functional API. The latter enables the inclusion of the dropout layer during the inference phase. The selected input variables for the models include inlet flow, inlet turbidity, recycle flow, pressure, temperature of the recycle stream, water level in the saturation tank, and outlet turbidity in previous timesteps. The target output was the outlet turbidity in future steps.

The database was normalized between -1 and 1 to avoid biased models, and it was sequentially split into training and test datasets using a 4/1 proportion, respectively. The loss function employed for the optimization stage was selected as the mean squared error.

In this study, the periods of the input variables ranged from 1 to 10 min, while the output time series ranged from 1 to 5 min. For each temporal pair, 250 models were trained using Optuna, an easy-to-use hyperparameter optimization framework for automating the search process. The successive halving pruning method was employed to suppress suboptimal configurations and early-stopping with zero-improvement threshold and 10-fails tolerance to halt the training stage when overfitting was detected. The selected search space for the hyperparameters is summarized in Table 2.

According to Gal and Ghahramani (2015), using dropout regularization in deep neural networks during both training and inference steps allows for an interpretation resembling a Bayesian approximation termed MCD. This method involves deriving mean estimates and predictive uncertainty through k stochastic forward passes, conducted concurrently to minimize computational time effectively. This study employed a hundred forward passes to calculate the quantities mentioned above.

Table 2 - Search space for hyperparameters using the Optuna optimization framework.

Variable	Range
Neurons	7-700
Dropout rate	0.1-0.5
Learning rate	10^{-7}-10^{-2}
L1	10^{-5}-10^{-1}
L2	10^{-5}-10^{-1}

3. Results and Discussions

The best-performing MCD-LSTM network explored in this study was comprised of 165 neurons, a 45.8% dropout rate, L1 and L2 norms set at $1.1 \cdot 10^{-5}$ and $9.2 \cdot 10^{-5}$ respectively, an initial learning rate of $8.5 \cdot 10^{-3}$, and utilized two previous data samples to predict a single future step. The results highlighted increasing errors as more extended frame data was used. Moreover, similar trends were observed for models forecasting over longer prediction horizons.

The dropout rate of 45.8% potentially indicates over-parametrization, a concern that can impact the model's generalization capabilities. To address this issue, we implemented the optimization framework Optuna as a measure to mitigate the potential effects of over-parametrization. However, it is important to note that despite employing Optuna, complete immunity from the impact of over-parametrization cannot be guaranteed.

The determination of an appropriate dropout rate lacks a universal guideline, but pertinent literature often recommends values ranging from 0.3 to 0.5 for complex tasks. Srivastava et al (2014) highlighted that a dropout rate of 0.5 appears to approach optimality across a diverse spectrum of networks and tasks.

Figure 2 illustrates the predictions using the training dataset. It is noticed that there is a high correlation between the predicted and the real data, and this is within the uncertainty boundaries for most of the prediction horizon. The most significant errors and uncertainty boundaries occurred in transition points between experimental runs, corresponding to drastic changes in the input conditions that can carry uncorrelated information within the recurrent layers. Excluding these points, the trained data error was approximately ±0.7 NTU. Thus, the trained model could predict the outlet turbidity within reasonable errors, indicating a good training phase.

Figure 3 shows the generalization performance using the test dataset. The predictions for unseen data captured the overall dynamics trend, but they were slightly off for most of the runs, even when the uncertainty boundaries were considered. Similar to the training stage, the most significant errors were calculated during the changes between experimental runs and reached an absolute maximum of 2.8 NTU. Excluding the transitions between runs, the test errors were within ±1.3 NTU, considerably higher than the training errors. A similar trend was observed by Souza et al (2021) in which the generalization test for their DAF prototype model using sliding window and feedforward neural networks was 2.5 times greater than the training errors.

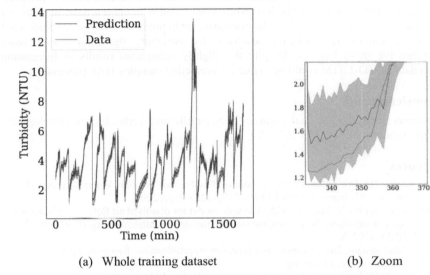

(a) Whole training dataset   (b) Zoom

Figure 2 - Predictions of MCD-LSTM using the training dataset.

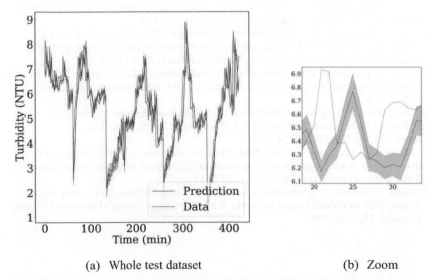

(a) Whole test dataset   (b) Zoom

Figure 3 - Predictions of the MCD-LSTM model using the test dataset.

Although the results were worse than expected, it illustrates that MCD-LSTM forecasts the data dynamics. This observation could indicate that some important input variables for the process, such as post-flocculation pH and treatment-solutions flow rates, could be introduced into the model to aggregate more information about the previous stage, as well as optical information concerning the bubble concentration and size in the flotation tank.

4. Conclusions

The MCD-LSTM networks to predict turbidity removal in a DAF system revealed intriguing potential and limitations. The results showcased promising outcomes in capturing the overall dynamics of turbidity removal within the training dataset. However, the model exhibited higher prediction errors when faced with unseen data, indicating a need for additional input variables. The discrepancies in predictions suggested a potential avenue for enhancing predictive accuracy by enriching the model with more comprehensive input features. Despite the slightly suboptimal results in forecasting unseen data, MCD-LSTM could be viable for predicting complex DAF processes.

Acknowledgments

The authors thank the National Council for Scientific and Technological Development (CNPq) grant number 141165/2021-9.

References

J.K. Edzwald, 1995. Principles and applications of dissolved air flotation. Water Science and Technology 31. https://doi.org/10.1016/0273-1223(95)00200-7

K. Fukushi, N. Tambo, Y. Matsui, 1995. A kinetic model for dissolved air flotation in water and wastewater treatment. Water Science and Technology 31. https://doi.org/10.1016/0273-1223(95)00202-X

Y. Gal, Z. Ghahramani, 2015. Dropout as a Bayesian Approximation: Representing Model Uncertainty in Deep Learning.

M.H. Glantz, 2018. Water security in a changing climate, World Meteorological Organization.

R.D. Morris, E.N. Naumova, R. Levin, R.L. Munasinghe, 1996. Temporal variation in drinking water turbidity and diagnosed gastroenteritis in Milwaukee. American Journal of Public Health 86, 237–239. https://doi.org/10.2105/AJPH.86.2.237

J.P. Rodrigues, R. Béttega, 2018. Evaluation of multiphase CFD models for Dissolved Air Flotation (DAF) process. Colloids and Surfaces A: Physicochemical and Engineering Aspects 539, 116–123. https://doi.org/10.1016/j.colsurfa.2017.12.015

Y. Rybachuk, A. Jodłowski, 2019. Mathematical model of dissolved air flotation (DAF) based on impulse conservation law. SN Applied Sciences 1, 541. https://doi.org/10.1007/s42452-019-0560-y

J. Schwartz, 2000. Drinking water turbidity and gastrointestinal illness in the elderly of Philadelphia. Journal of Epidemiology & Community Health 54, 45–51. https://doi.org/10.1136/jech.54.1.45

A.C.O. Souza, N.L. Ferreira, F.V. Silva, 2021. Empirical modeling of turbidity removal in a dissolved air flotation system: application of artificial neural networks. Water Supply 21, 3946–3959. https://doi.org/10.2166/ws.2021.152

N. Srivastava, G. Hinton, A. Krizhevsky, I. Sutskever, R. Salakhutdinov, 2014. Dropout: A Simple Way to Prevent Neural Networks from Overfitting. Journal of Machine Learning Research 15, 1929–1958.

Flavio Manenti, Gintaras V. Reklaitis (Eds.), Proceedings of the 34th European Symposium on Computer Aided Process Engineering / 15th International Symposium on Process Systems Engineering (ESCAPE34/PSE24), June 2-6, 2024, Florence, Italy

Design and Optimization of a sustainable process for the transformation of glucose into high added value products

Carlos Rodrigo Caceres[a] , Eduardo Sánchez-Ramirez[a] , Juan Gabriel Segovia-Hernández[a,*]

[a]Department of Chemical Engineering, Universidad de Guanajuato, Campus Guanajuato, Guanajuato, 36050, Mexico

* gsegovia@ugto.mx

Abstract

This study unveils the design of a sustainable biorefinery for the simultaneous production of levulinic acid (LA), gamma-valerolactone (GVL), furfural (FF), and hydroxymethylfurfural (HMF) from corn stover. The utilization of this raw material enables the revalorization of agricultural residue. Employing Aspen Plus simulation, three distinct scenarios were assessed to achieve a purity exceeding 98% for the four compounds. These scenarios prioritize the preferential production of specific compounds. A sustainable process design was accomplished through multi-objective optimization using the MODE-TL algorithm, incorporating environmental and economic objectives. The optimal design for each scenario was successfully determined. Notably, Scenario 2 demonstrated the most sustainable outcome, with minimal values for annualized total cost ($3.157 * 10^7$ dollars/year), environmental impact ($7.710 * 10^6$ points/year), and energy consumption ($1.756 * 10^9$ MJ/year).

Keywords: biorefinery, sustainability, DETL method.

1. Introduction

Annually, the global accumulation of agricultural waste surpasses 998 million tons, with Mexico contributing a substantial 76 million tons to this staggering figure. Traditionally, these residues have been disposed of through environmentally detrimental burning, exacerbating global $CO2$ emissions to an alarming 8.68 billion tons. Recognizing the latent potential within lignocellulosic materials, a paradigm shift emerges—biorefineries present a transformative opportunity to convert these residues into high-value-added products. This proactive transition aligns seamlessly with the United Nations' principles of sustainable development, specifically targeting Sustainable Development Goals (SDGs) related to affordable and clean energy, as well as responsible consumption and production. Lignocellulosic biomass, a versatile resource, opens doors to a myriad of possibilities. The US National Renewable Energy Laboratory has identified 30 promising products from lignocellulosic biomass, including levulinic acid (LA), gamma-valerolactone (GVL), furfural (FF), and hydroxymethylfurfural (HMF). These compounds find applications in diverse industries such as pharmaceuticals, polymers, and solvents, serving as intermediates to produce other valuable compounds. Notably, the markets for GVL and FF exceed 700 and 500 million USD/year, respectively, while LA and HMF markets reach 28 and 61 million USD/year. This study leverages Aspen Plus to design a biorefinery for the simultaneous production of these compounds (>98%

purity) from corn stover through acid dilute pretreatment. Three scenarios, delineating the division of the LA stream into either purification or transformation into GVL, are explored at proportions of 50/50, 75/25, and 25/75. The optimization of each scenario involves the implementation of Differential Evolution with Tabu List (DETL). The sustainability assessment comprehensively considers total annualized cost (TAC), Eco-99, and energy consumption metrics. This research pioneers the optimization of agricultural waste utilization, placing a strong emphasis on the intersection of economic viability and environmental responsibility, thereby marking a significant stride toward sustainable practices.

2. Methodology

2.1. Biorefinery design: Within an extensive biorefinery framework, the process is methodically segmented into the sections: pretreatment, hydrolysis, reaction, and purification. The objective of pretreatment is to disintegrate the lignocellulosic matter, rendering it more amenable for subsequent processing. In the hydrolysis stage, either enzymes or acidic solutions are deployed to breaking down the biomass into its constituent sugars. The ensuing reaction stage is pivotal, where sugars undergo a transformative process, evolving into chemicals and other valuable products. Finally in the purification section, the desired end-products are isolated. Notably, the conceptualization and design of this process were meticulously formulated based on insights garnered from an exhaustive literature review. Aspen Plus software was used, providing a sophisticated platform for process simulation and optimization. The resolution of the thermodynamic complexities was achieved through the utilization of the NRTL-HOC model. This model accurately forecasts the equilibrium of mixtures of polar compounds, providing a study framework for process design.

The biomass employed is corn stover, a highly abundant residue in Mexico (48.2×10^9 kg/year, Contreras-Zarazúa et al., 2022). The composition of the biomass is (%w/w):44.38 % cellulose, 33.63% hemicellulose and 22.37 % lignin. Calculations are based on a feed rate of 15,000 kg/h, equivalent to 127,500 tons/year. This represents 10% of the corn stover in Guanajuato state in 2015 (SAGARPA ,2015). The biomass undergoes pretreatment in reactor R1, with a water-to-biomass ratio of 2:1. Dilute acid pretreatment was selected because allows shorter residence times and high hemicellulose conversions (Conde-Mejía et al., 2012). The acid-to-water ratio is 0.77 kg of sulfuric acid per 1000 kg of water. The pretreatment conditions are a T=158°C, and P=6.76 bar (Contreras-Zarazúa et al., 2022).

In Figure 1, we present a simplified diagram of the plant. The effluent emerging from pretreatment reactor R1 undergoes neutralization in reactor R2. Through a filtration process, two streams are obtained: a liquid stream rich in hemicellulose (containing 10% xylose), which is directed towards FF production, and a solid stream comprising cellulose and lignin. In reactor R3, cellulose undergoes conversion into glucose, while lignin is separated through filtration from the reactor effluent. Column 1 is used for recovery of sulfuric acid. The bottoms from Column C1, primarily composed of glucose, are divided into two equal parts. One part is utilized to produce levulinic acid and formic acid in reactor R4, while the other is employed in HMF production in reactor R6, the same reactor where xylose is transformed into FF. Most of the sulfuric acid recovery is achieved in bottoms of column 2. The distillate from column C2, which contains levulinic acid as the primary compound of interest is split using splitter 2. Using the splitter 2 this stream can be directed either to the purification stage in column C6 or to reactor R5 to produce GVL, followed by subsequent purification in column C3. Different scenarios were considered for the proportion in which this stream is divided to give precedence to manufacturing a specific compound. Three scenarios resulting from dividing this stream in the proportions: 50/50, 25/75, and 75/25 as shown in Table 1 and Figure 1. Subsequently, the design and optimization were carried out for the three scenarios. Moving towards the lower section of the diagram, following reactor R6, columns C4 and C5 are situated, specifically dedicated to obtaining HMF and FF respectively as the bottom products.

Table 1. Different scenarios according to division of distillate stream of C2

Scenario	Percentage sends to column C6	Percentage sends to reactor R5
1 (50/50)	50	50
2 (25/75)	25	75
3 (75/25)	75	25

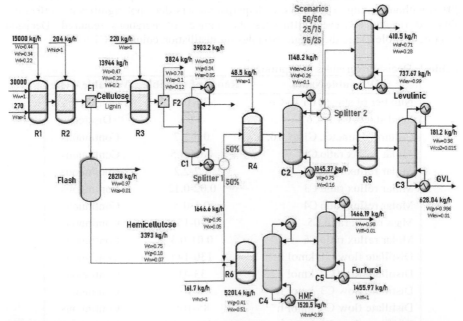

Figure 1. Simplified diagram and mass balance of the scenario 1 for proposed biorefinery.

2.2. Optimization: The objective functions are the economics and environmental indices: total annualized cost (TAC), the Eco-indicator-99 (Eco-99) and the total requirement of energy for the process. This guarantee obtains a sustainable design for the biorefinery. The design was carried out by minimizing the multi-objective function:

$$Min(TAC, EI99, \text{Energía}) = f(N, N_f, RR, D, d) \tag{1}$$

Subject to: $y_i P_c \geq x_i P_c, w_i F_c \geq u_i F_c$ .Where N is the number of stages for the distillation columns, Nf is the feed stage, RR is the reflux ratio, D is the distillate flow rate and d is the column diameter. The total number of decision variables was 30. The optimization problem is constrained by ensuring that the purities ($y_i P_c$) are at least as high as $x_i P_c$ and that the recovery flows of the products ($w_i F_c$) are greater than or equal to $u_i F_c$. To execute the optimization, Aspen Plus was connected to Microsoft Excel via COM technology. Regarding objective functions, the total annualized cost is calculated using the modular cost technique (Turton, 2003). The Eco-indicator 99 is calculated using equation 2 (Contreras et al, 2019). Where:ω is the weighting factor for damage, δ_i is the impact value for category i, αs, αsl and αel are the steam, steel and electricity required. The energy requirements are the sum of the cooling and heating and are obtained from the simulation.

$$Eco\text{-}99 = \Sigma \omega \cdot \delta_i \cdot \alpha s_i + \omega \cdot \delta_i \cdot \alpha sl_i + \Sigma \omega \cdot \delta_i \cdot \alpha el_i \tag{2}$$

The EDTL method was implemented using Microsoft Excel and Aspen Plus. Excel sends the decision variable vector to Aspen, which evaluates process variables. After simulation, Aspen sends back the resulting vector to Excel, where objective function values are assessed, and new decision variable values are proposed. Parameters included 120 initial individuals, 400 generations, a tabu list of 60 individuals, a tabu radius of 0.0001, and crossover and mutation factors of 0.9 and 0.3, respectively (Alcocer et al., 2019) without tuning. The choice of hyperparameters does not significantly affect the obtained results, however, influences the number of iterations required. Decision variables (Table 2) are the design variables for distillation columns (C1 to C6).

Table 2. Decision variables for optimization

Variable Name	Range of variable	Variable Type
Number of stages, C1 to C6	20-100	Discrete
Feed stage, C1 to C6	3-99	Discrete
Column diameter, C1 to C6, m	0.5-1.7	Continuous
Molar reflux ratio C1	0.02-0.5	Continuous
Molar reflux ratio C2	1.2-2.2	Continuous
Molar reflux ratio C3	0.02-0.12	Continuous
Molar reflux ratio C4	0.1-0.5	Continuous
Mass reflux ratio C5	10-15	Continuous
Molar reflux ratio C6	0.01-0.3	Continuous
Distillate flow C1, kmol h^{-1}	130-144	Continuous
Distillate flow C2, kmol h^{-1}	33-41	Continuous
Distillate flow C3, kmol h^{-1}	8-10	Continuous
Distillate flow C4, kmol h^{-1}	85-105	Continuous
Distillate flow C5, kg h^{-1}	1400-1700	Continuous
Distillate flow C6 kmol h^{-1}	10-14	Continuous

3.Results

This section unveils the primary outcomes of the multiobjective optimization process, satisfying all prescribed purity and recovery constraints. The graphical representations depict the Pareto fronts delineating trade-offs between pairs of objective functions. To distill a singular optimal solution the utopia, point methodology was deployed. This method strategically positions an ideal solution at the Pareto front's extremity, where enhancing one objective leads to a compromise in the other. This approach facilitates the selection of an optimal design that strikes a balance between competing objectives.

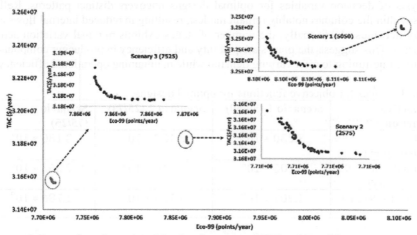

Figure 2. Pareto front between objective functions: TAC and Eco-99

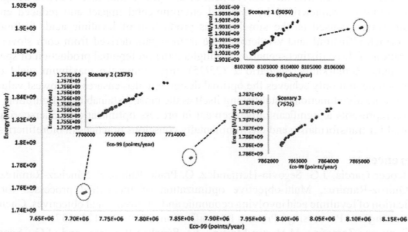

Figure 3. Pareto front between objective functions: Energy and Eco-99

Figure 2 illustrates the TAC vs Eco-99 graph, showcasing a competitive relationship between these functions. This mean, the most cost-effective process tends to have a more pronounced environmental impact. This due to the mutual influence of both objective functions by the reboiler duty. Moving to Figure 3, the plot depicts the relationship between energy and Eco-99, revealing a linear correlation. In distillation columns, the steam flow within the reboiler impacts both Eco-99 values and energy consumption.

Table 3 illustrates objective function values for optimal designs across various scenarios, marked by red points in Figures 2 and 3. The optimal solution within scenario 2 showcases superior values across all objective functions. This scenario prioritizes the highest quantity of GVL. In terms of the eco-indicator, Scenarios 3 and 1 display increments of 2% and 5%, respectively, regarding scenario 2. Moreover, the energy requirement in Scenario 2 was observed at its lowest, with increases of 1.7% and 8% in Scenarios 3 and 1, respectively, emphasizing Scenario 2 as the most sustainable design. Analysis of decision variables for optimal designs uncovers distinct patterns. Reflux ratios within the columns notably remain modest, resulting in reduced internal flows and energy demands. Additionally, the number of stages exhibits minimal variation across scenarios. This suggests the plant's adaptability and efficiency in producing compounds, enabling fine-tuning to meet varying demand while maintaining operational efficiency."

Table 3. Values for objective functions for optimal designs

Objective Function	Scenario 1 (50/50)	Scenario 2 (25/75)	Scenario 3 (75/25)
TAC (dollars/year)	$3.250 * 10^7$	$3.157 * 10^7$	$3.180 * 10^7$
Eco-indicator 99 (points/year)	$8.105 * 10^6$	$7.710 * 10^6$	$7.863 * 10^6$
Energy (MJ/year)	$1.901 * 10^9$	$1.756 * 10^9$	$1.786 * 10^9$

4.Conclusions

This study concludes with the achievement of an optimal and cost-effective design for the biorefinery, characterized by minimal environmental impact and reduced energy consumption. Focused on the simultaneous production of levulinic acid, furfural, 5-hydroxymethylfurfural, and γ-valerolactone from sugars derived from corn stover, the exploration of three distinct scenarios highlighted the preferential production of specific compounds. Remarkably, Scenario 2 (25/75), prioritizing the maximum yield of γ-valerolactone, not only achieves the optimal design but also ensures the lowest values in all three objective functions. establishing itself as the most sustainable configuration. This success represents a significant advancement in process optimization, underscoring the potential for transformative and environmentally conscious practices in biorefineries.

References

H. Alcocer-García, J.G. Segovia-Hernández, O. Prado-Rubio, E. Sánchez-Ramírez and J.J. Quiroz-Ramírez, Multi-objective optimization of intensified processes for the purification of levulinic acid involving economic and environmental objectives, Chemical Engineering and Processing - Process Intensification, 2019,136, 123–137.

G. Contreras-Zarazúa, M.Martin-Martin, E. Sánchez-Ramirez and J.G. Segovia-Hernández, Furfural production from agricultural residues using different intensified separation and pretreatment alternatives. Economic and environmental assessment, Chemical Engineering and Processing - Process Intensification,2021, 171.

R. Turton, R. C. Bailie, W. B. Whiting, and J. A. Shaeiwitz, Analysis, Synthesis, and Design of Chemical Processes, Second Edition, PrenticeHall, 2003.

Flavio Manenti, Gintaras V. Reklaitis (Eds.), Proceedings of the 34th European Symposium on Computer Aided Process Engineering / 15th International Symposium on Process Systems Engineering (ESCAPE34/PSE24), June 2-6, 2024, Florence, Italy

Formulation-independent pharmaceutical dry granulation model via gray box approach

Kanta Sato[a,b*], Shuichi Tanabe[a], Susumu Hasegawa[a], Manabu Kano[b]

[a] *Formulation Technology Research Laboratories, Daiichi Sankyo Co., Ltd., 1-12-1 Shinomiya, Hiratsuka, Kanagawa 254-0014, Japan*
[b] *Department of Systems Science, Kyoto University, Yoshida-Honmachi, Sakyo-ku, Kyoto 606-8501, Japan*
sato.kanta.m6@daiichisankyo.co.jp

Abstract

In the roller compaction process, powder blends are compacted into ribbons by opposite-direction rotating rolls and subsequently milled to obtain granules. Since ribbon density is an important material attribute affecting product quality, controlling it is essential. We developed a novel mechanistic model describing the impact of process parameters and material attributes on ribbon density. By incorporating the pressure of a preconsolidation powder blend into the model, the compression coefficient of the powder blend in the small-scale uniaxial compression tests can be applied to the large-scale roller compactor. This allows us to understand the relationship between process parameters and ribbon density from small-scale experiments with few materials. A gray box model was developed to predict ribbon density, combining a first principle model based on the Johanson model and a statistical model. The accuracy of the gray box model was validated through manufacturing data, achieving precise predictions with a Root Mean Square Error of Cross Validation of 0.032 in relative ribbon density. This approach is expected to significantly reduce material consumption and enhance comprehension of the roller compaction process. The validation results demonstrate its capacity to predict ribbon density using only a minimal amount of experimental material, provided that a compressibility coefficient is within the range of the training data.

Keywords: Roller compaction, Gray box modeling, Pharmaceutical manufacturing, Process modeling, Scale up

1. Introduction

To achieve the desired product quality, it is essential to accurately understand the relationship between process parameters and product quality but understanding the relationship requires significant resources and costs. Predictive modeling is crucial in understanding these relationships. First principle modeling requires deep insight into the process. In some cases, parameters may be required that are not easily accessible during normal operation. The serial gray box model supplements certain parameters of a first principle model with statistical models (Ahmad et al., 2020). According to Von Stosch et al. (2014), statistical models can be applied with limited process knowledge, but they require large amounts of data compared to a first principle model, and the quality of the model can only be trusted in the vicinity of the data from which the model was derived. Gray box models balance the advantages and disadvantages of first principle models and statistical models.

The granulation process is a crucial step that improves the flowability of the powder and increases the manufacturing efficiency of subsequent processes (Kleinebudde, 2004). The roller compaction (RC) process is a continuous process in which powder blends are compacted into ribbons by rolls and subsequently milled into granules. In the RC process, ribbon density control is critical to ensure the quality of the final products. Johanson (1965) proposed a first principle model for RC, the rolling theory of granular solids that correlates materials attributes and process parameters with the ribbon density in a simplified two-dimensional space as shown in Figure 1(a).

To obtain the model parameters, RC manufacturing must be performed over a wide range of ribbon densities on the target manufacturing machine as material attributes (Raynolds et al., 2010). When data sets based on specific materials are limited, there may be a discrepancy between predicted and experimental ribbon densities. Amini and Akseli (2020) succeeded in predicting ribbon density utilizing the Johanson model by substituting the RC manufacturing with uniaxial compression tests, as shown in Figure 1(b). This was achieved under limited conditions such as the roll speeds ranging from 2 to 6 min^{-1} where the preconsolidation density would remain constant regardless of the process parameters. Considering the actual process development scenario requiring a comprehensive evaluation of the process parameters to identify the design space, a robust first principle modeling approach that can take the variation of preconsolidation density into account is required. To accommodate varying production speeds, manufacturing at higher roll speeds may be required.

This study aims to build a model predicting ribbon density under a wide range of process parameters and to set process parameters that achieve the desired ribbon density from a small amount of material. By regarding preconsolidation powder conditions as variable parameters, we resolved the discrepancy between RC and uniaxial compression tests. A gray box model was built to predict ribbon density. A statistical model was developed to estimate the preconsolidation parameters, which were difficult to measure, and the ribbon density was calculated using a first principle model with the estimated parameters.

2. Material and Methods

2.1. Materials

This study used four formulations A to D. Formulation A consisted of 72 % PEARLITOL 50 C as mannitol (Roquette, France), 18 % CEOLUS PH-101 as microcrystalline cellulose (Asahi Kasei, Japan), and 8 % NISSO HPC-SL as hydroxypropyl cellulose (HPC) (Nippon Soda, Japan), to provide the ideal relative ribbon density range (Zinchuk et al., 2004). Formulations B to D consisted of different ratios of plastic material such as CEOLUS KG-1000, microcrystalline cellulose (MCCKG) (Asahi Kasei, Japan), and brittle material such as dibasic calcium phosphate anhydrous (DCPA) (Fuji Chemical Industries, Japan) to provide different compaction behavior. Compositions of MCCKG

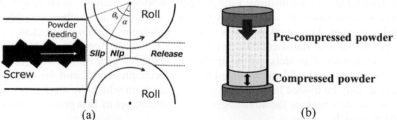

Figure 1 Schematic diagram of (a) roller compaction process, (b) uniaxial compression test

and DCPA are 78 % and 20 % for formulation B, 49 % each for formulation C, and 20 % and 78 % for formulation D. 2 % HyQual 5712, magnesium stearate (SpecGx, USA) was included as a lubricant in all formulations.

RC manufacturing was conducted using a roller compactor FP90×30 (Freund Turbo, Japan) over a wide range of process parameters, i.e., roll pressure of 8–27 MPa, roll speed of 6–20 min[-1], and roll gap of 0.75–1.04 mm.

The compressibility coefficient (K) for each formulation was determined through uniaxial compression tests. 300 mg of the powder blends were placed in a 10 mm diameter die and compacted with a flat-faced punch at 10 mm/min, and then the density was measured. The density of the compacts obtained in the uniaxial compression test was measured in the range of 2.5 MPa to 100 MPa to cover the ribbon density obtained in RC manufacturing.

2.2. First principle model

The powder blend fed by the screw experiences the screw feed pressure (P_0) in the slip region, resulting in the preconsolidation relative density (γ_0). The relative density is the ratio of the true density of the material to the density of the ribbon. When the roll angle (θ) becomes less than the nip angle (α), compaction of the powder begins. This area is called the nip region, and the compaction of the powder increases as θ decreases. At $\theta=0$, the distance between the rolls is the shortest, and the powder experiences the maximum pressure (P_{max}) from the rolls. For a detailed derivation of P_{max}, please refer to Johanson's original paper (Johanson, 1965). The powder has K, and the relative density of the compacted ribbon (γ_R) is expressed by Eq. (1).

$$\gamma_R = \gamma_0 \left(P_{max}/P_0 \right)^{1/K} \tag{1}$$

Eq. (2) is given by rearranging the logarithm of Eq. (1).

$$\ln \gamma_R = \ln \gamma_0 + 1/K \ln P_{max} - 1/K \ln P_0 \tag{2}$$

Raynolds et al. (2010) set $P_0=1$, regardless of the material attributes and process parameters, and set the slope of $\ln \gamma_R$ and $\ln P_{max}$ in Eq. (2) as $1/K$ with an intercept of $\ln \gamma_0$. They found differences in K and γ_0 between the uniaxial compression test and RC. In RC, a velocity gradient occurs in the nip region, and the powder blend near the roll surface moves faster than the powder blend farther away (Muliadi et al., 2012), which deviates from Johanson's assumption that the powder blend moves at a uniform velocity. We assumed that K remains constant regardless of the machines. Under the assumption, the discrepancy in K between RC and uniaxial compression test calibration was due to the different assumed values of P_0 and γ_0. If there is a velocity gradient in the preconsolidation powder blend around the rolls in RC due to process parameters, both P_0 and γ_0 expected to vary accordingly. It is usually difficult to measure P_0 and γ_0 during normal operation because the preconsolidated powder blend in RC exists in the closed space. Since the slip region remains constant within the same machines, its volume can also be considered constant, given a sufficient feed of powder blend. Once P_0 is determined, γ_0 can be deduced consequently. We introduced an unmeasurable parameter ζ to represent the relationship between P_0 and γ_0 in Eq. (3).

$$\zeta = \gamma_0 \left(1/P_0 \right)^{1/K} \tag{3}$$

γ_R is expressed by Eq. (4) based on Eq. (1) and Eq. (3).

$$\gamma_R = \zeta P_{max}^{1/K} \tag{4}$$

Using a statistical model, ζ was calculated from measured manufacturing results and predicted from the powder blend attributes and process parameters. It was used to calculate γ_R in Eq. (4). Eq. (2) can be transformed into Eq. (5).

$$\ln P_{max} = K(\ln \gamma_R - \ln \gamma_0) + \ln P_0 \tag{5}$$

K was obtained from the uniaxial compression test and used for RC. Since the only pressure applied to the preconsolidated powder blend in the uniaxial compression test is atmospheric, the preconsolidation relative density can be considered constant for the same material. Consequently, K can be calculated from the slope of $\ln P_{max}$ and $(\ln \gamma_R - \ln \gamma_0)$.

2.3. Statistic model

Gaussian Process Regression (GPR) was used to build the statistical model with K-Fold Cross Validation (KFCV; K=10). In KFCV, the dataset is divided into K equal subsets, with one subset used as test data and the remaining K−1 subsets used as training data. This process is repeated K times and the results are averaged to evaluate the performance of the model. Prediction performance was evaluated using the coefficient of determination R^2 and the Root Mean Square Error Cross Validation (RMSECV) or Root Mean Square Error Prediction (RMSEP). The input parameters used for the GPR model were the same as those for the first principle model, i.e., true density, bulk density, wall friction angle, effective angle of internal friction, roll force, roll gap, roll speed, and K. The importance of the input parameters was evaluated using Permutation Importance (PI). ζ was calculated from Eq. (4) using the dataset for each run.

2.4. Case Study

All formulations were used in the model building and cross validation confirmed the accuracy of the ribbon density predictions. Then, to predict RC manufacturing results using only small-scale experiments, a model was built using RC manufacturing data for three of the four formulations to predict the relative ribbon density of the untrained formulation, and the relative ribbon density of the remaining formulation was predicted.

3. Results

Figure 2 shows the relationship between K, roll force, roll gap, roll speed, and ζ, where ζ is derived from manufacturing data. K increased with higher DCPA content in formulations B to D. Formulation A showed a significantly higher value than the other three formulations. One factor that can be inferred is the existence of the binder HPC in Formulation A, which is not included in Formulations B to D. ζ was found to vary not only with formulation but also with manufacturing conditions. Therefore, it is suggested

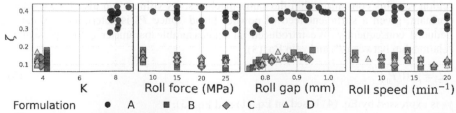

Figure 2 Scatter plots matrix of process parameters and ζ.

that ζ should not be treated as a fixed parameter based solely on material attributes, but rather as a variable parameter that also depends on process parameters.

Table 1 shows the prediction accuracy of ζ using the statistical model. All formulations achieved excellent prediction accuracy for ζ, with RMSECV of 0.010 and R^2 of 0.993. The prediction of ζ was significantly affected by K, as indicated by the PI. Figure 3(a) shows the scatter plots of observed versus predicted values of γ_R using ζ obtained in cross validation. An accurate prediction of the RMSECV of 0.032 and R^2 of 0.962 indicates that γ_R can be accurately predicted using the gray box model approach.

Table 1 displays predicted results for the untrained formulation. The RMSEP for ζ from formulations A to D were 0.244, 0.013, 0.006, and 0.009, respectively. Figure 3(b) shows the results for γ_R calculated using the predicted ζ, with corresponding RMSEP values of 0.518, 0.044, 0.023, and 0.039. These results indicate practical prediction accuracy, except for formulation A. In formulation A, a significant difference in the distribution of the largest PI in the statistical model compared to the other formulations was observed.

Table 1 Prediction accuracy of ζ in case studies and PI for modeling

Target formulation			All data	A	B	C	D
ζ prediction	Cross validation	RMSECV	0.010	0.009	0.010	0.011	0.011
		R^2	0.993	0.860	0.993	0.992	0.993
	Prediction of the untrained formulation	RMSEP	–	0.244	0.013	0.006	0.009
		R^2	–	−40.266	0.838	0.910	0.801
Permutation Importance in cross validation	True density (g / mL)		0.263	0.095	0.087	0.295	0.320
	Bulk density (g / mL)		0.081	0.045	0.078	0.044	0.040
	Wall friction angle (°)		0.001	0.210	0.001	0.001	0.000
	Effective angle of internal friction (°)		0.023	0.309	0.027	0.030	0.012
	Roll force (MPa)		0.037	0.710	0.036	0.035	0.042
	Roll gap (mm)		0.084	1.331	0.076	0.076	0.091
	Roll speed (min^{-1})		0.003	0.057	0.003	0.003	0.002
	K (-)		0.348	0.039	0.376	0.393	0.290

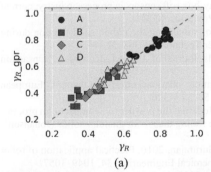 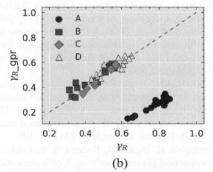

(a)   (b)

Figure 3 (a) Cross validation of γ_R for all formulations: experimental vs predicted. Data fitted with RMSECV=0.032, R^2=0.962. (b) Predictive validation of γ_R on untrained formulation. Data fitted with (A) RMSECV=0.518, R^2=-61.411, (B) RMSECV=0.044, R^2=0.776, (C) RMSECV=0.023, R^2=0.897, (D) RMSECV=0.039, R^2=0.720.

This difference can be attributed to the different distribution of the manufacturing data in Figure 2. Specifically, K values, which is the largest PI in the statistical model, were significantly different for formulation A from the other three formulations, resulting in an extrapolation in the GPR model built with the other three formulations.

While the dataset used in this study was limited to specific formulations, future research should use a wider range of datasets to improve the generalizability and robustness of the model. This study showed the potential variability in manufacturing parameters for ζ and the effect of K on ζ, demonstrating the complexity of the interaction between formulation and manufacturing processes.

The gray box modeling used in this study captured the differences between formulations with different material attributes while incorporating the variability in manufacturing parameters to achieve high predictive accuracy. By combining a statistical approach based on experimental data with mechanistic modeling based on phenomena, this method offers a new way to predict the behavior of formulations under different manufacturing conditions. This advancement is expected to enable innovative formulation design that significantly reduces development time and cost. In addition, by increasing the predictability of manufacturing processes, the gray box model is expected to contribute to improved robustness and quality in pharmaceutical formulation development.

4. Conclusion

This study presents a gray box modeling approach to predict ribbon density in pharmaceutical manufacturing, which is essential for product quality in the roller compaction process. This gray box model achieved accurate predictions with RMSECV of 0.010 and 0.032 for ζ and relative ribbon density. We have established a predictive relationship between preconsolidation pressure and relative density, which is critical for optimizing the manufacturing process. Model validation results show that the ribbon density can be predicted with only a small amount of experimental material as long as K is within the range of the training data. Our research contributes a validated, adaptable model to the pharmaceutical industry that promises improvements in product quality and streamlined the manufacturing process. The model's implications include resource reduction and an improved understanding of the manufacturing process.

References

I. Ahmad, A. Ayub, M. Kano, I.I. Cheema, 2020, Gray-box soft sensors in process industry: current practice, and future prospects in era of big data, Processes, 8, 243.

H. Amini, I. Akseli, 2020, A first principle model for simulating the ribbon solid fraction during pharmaceutical roller compaction process, Powder Technology, 368, 32–44.

J. Johanson, 1965, A rolling theory for granular solids, Journal of Applied Mechanics, Transaction of ASME, 32, 842–848.

P. Kleinebudde, 2004, Roll compaction/dry granulation: pharmaceutical applications, European Journal of Pharmaceutics and Biopharmaceutics, 58, 317–326.

A. R. Muliadi, J. D. Litster, C. R. Wassgren, 2012, Modeling the powder roll compactionn process: Comparison of 2-D finite element method and the rolling theory for granular solids (Johanson's model), Powder Technology, 221, 90–100.

G. Reynolds, R. Ingale, R. Roberts, S. Kothari, B. Gururajan, 2010, Practical application of roller compaction process modeling, Computers and Chemical Engineering, 34, 1049–1057.

M. V Stosch, R. Oliveira, J. Peres, S. F.de Azevedo, 2014, Hybrid semi-parametric modeling in process systems engineering: Past, present and future, Computer and Chemical Engineering, 60, 86-101.

A. V. Zinchuk, M. P. Mullarney, B. C. Hancock, 2004, Simulation of roller compaction using a laboratory scale compaction simulator, International Journal of Pharmaceutics, 269, 403–415.

Flavio Manenti, Gintaras V. Reklaitis (Eds.), Proceedings of the 34th European Symposium on Computer Aided Process Engineering / 15th International Symposium on Process Systems Engineering (ESCAPE34/PSE24), June 2-6, 2024, Florence, Italy

Bayesian Hybrid Models for Simulation of Microbial Biohydrogen Photo-Production Processes

Shraman Pal,[a] Bovinille Anye Cho,[b] Antonio Del Rio Chanona,[c] Dongda Zhang,[b] Max Mowbray [c*]

[a]Department of Chemical Engineering, IIT Kharagpur, Kharagpur, West Bengal, India
[b]The University of Manchester, Engineering Building A, Manchester, M13 9PL, UK
[c]Imperial College London, South Kensington Campus, London, SW7 2AZ, UK
*email of the corresponding author: mmowbray@ic.ac.uk

Abstract

Biohydrogen is widely regarded as the renewable energy of the future due to the qualities of zero emissions under combustion, high calorific value, and ambient synthesis. However, biohydrogen microbial photo-production systems are complex, providing challenge to simulate with mechanistic models and necessitating hybrid modelling. In this work, we present an effective methodology that leverages adjoint sensitivity analysis to identify a Bayesian neural network (BNN) hybrid model for prediction of a microbial photo-production process. The use of the Bayesian paradigm provides natural quantification of parametric uncertainties. We demonstrate that the scheme identifies an accurate model (3-11% mean percentage error) with well calibrated uncertainty estimates without mechanistic assumption on kinetics. In future, we will develop methods to enable the model transfer across scales.

Keywords: Bayesian Hybrid Modelling, Variational Inference, Biohydrogen.

1. Introduction

1.1 Dynamic modelling of biohydrogen photo-production process

The large-scale combustion of fossil fuels to meet the world's energy is of increasing concern due to global climate change and energy security challenges arising from dwindling fossil fuel supplies. For these reasons, biohydrogen has been coined as "the energy of the future" due to its high calorific value and other appealing properties (Teke et al, 2023). Although biosynthesised by a plethora of microorganisms, *Rhodopseudomonas palustris* (*R. palustris*) has demonstrated the continuous production of biohydrogen during all growth phases, including the stationary phase, lasting significantly longer observed with other microbial species (Anye Cho et al, 2021). As a result, *R. palustris's* biohydrogen production has been extensively researched with first principles mechanistic models to facilitate experimental design and bioprocess upscaling to industrial scale from the usual laboratory scales and pilot plant scales (Anye Cho et al, 2021). However, the complexity of bioprocess dynamics results in high investment burden for development and validation of mechanistic model structures. Yet, reported mechanistic model predictions observe high uncertainty (up to 20% variation), often attributed to the difficulty in describing thousands of intracellular metabolite transformations with a few lumped biokinetics parameters. This has led to interest in the development of data-driven approaches.

1.2 Data-driven approaches and uncertainty estimation

The use of data driven approaches has been reviewed with use of Artificial Neural Network (ANN) reported widely (Teke *et al.*, 2023). However, most of these studies predicted biohydrogen yield and/or biohydrogen productivity at the final fermentation time step, as a function of fixed process variables providing time invariant predictions. Such simplifications fall short of the underlying bioprocess description involving thousands of metabolite reactions, typically time varying over the course of fermentation. Thus, these approaches are suboptimal when utilized for biohydrogen bioprocess optimization. The few dynamic models reviewed in Teke *et al.* (2023) reported satisfactory overall biohydrogen production prediction (i.e., $R^2 = 0.939$), but poor prediction performance in the lag phase, thus requiring improved training approaches. For instance, algorithms predicting multiple steps ahead when provided with only the initial conditions, could alleviate prediction errors for the various growth phases (i.e., lag phase) (Del Rio Chanona et al, 2017).

However, none of the reported algorithms considered modelling uncertainties. Additionally, hybridising these algorithms with mechanistic models could benefit both predictive extrapolation and interpolation. However, this has not been addressed in the literature for biohydrogen production to the best of our knowledge. Therefore, this work aims to identify a Bayesian neural network (BNN) hybrid model for prediction of *R. palustris's* biohydrogen production process. The BNN's posterior predictive distributions enable the expression of parameter uncertainty, and direct quantification of the hybrid model uncertainty as a result. However, no literature investigation has reported this so far. Additionally, we highlight the versatility of adjoint sensitivity analysis for estimating the gradients of models with relatively high (i.e., >50) numbers of data driven parameters. Previously, hybrid models have typically been estimated using either (i) two-steps consisting of nonlinear programming and then supervised learning (Rogers et al, 2023), or (ii) via the forward sensitivities approach (von Stosch et al, 2014). Hence, we interrogate the proposed hybrid BNN's performance and provide in-depth discussions.

2. Methodology

2.1 Hybrid models for dynamic modelling of biohydrogen photo-production systems

In this work, we assume the availability of a dataset $D = \{\tau_n, u_n\}_{n=1:N}$, where $\tau_n = (x_{t_0}^n, x_{t_1}^n, \dots, x_{t_T}^n)$ describes the evolution of the system state, $x \in \mathbb{R}^{n_x}$, observed at discrete points over a time horizon, $t_k \in \{t_0, \dots, t_T\}$, and $u_n \in \mathbb{R}^{n_u}$ describes operational conditions that are fixed for a trajectory $n \in \{1, \dots, N\}$, over the course of the horizon. Specifically, in this work, we assume that we have measurements of the state $x = [c_X, c_S, c_H]^T$ representative of the biomass concentration $(g\ L^{-1})$, c_X, glycerol concentration (mM), c_S, and biohydrogen collection (mL), c_H, within a small scale photobioreactor with volume, $V = 500\ mL$. The operational conditions, u_n, denote the incident light intensity, $I\ (Wm^{-2})$. We assume that the photobioreactor (PBR) is a Schott bottle based PBR and is well-mixed by a magnetic stirrer. We wish to identify an approximating systems model as:

$$\dot{x} = f(x, u, \theta) = \mu(x, u\ \theta)x \tag{1}$$

where $f: \mathbb{R}^{n_x} \times \mathbb{R}^{n_u} \times \mathbb{R}^{n_\theta} \to \mathbb{R}^{n_x}$ represents a general description of the continuous time state dynamics; and we assume $\theta \in \mathbb{R}^{n_\theta}$ represent parameters of a data-driven model, which is embedded within the model structure provided by $f(\cdot)$ to predict time-varying parameters as a function of state. In this work, we specify this data-driven element as $\mu: \mathbb{R}^{n_x} \times \mathbb{R}^{n_u} \times \mathbb{R}^{n_\theta} \rightarrow \mathbb{R}^{n_x}$, which defines specific production or consumption rates of the state components via a neural network model structure. This means we simply leverage structure from the mass balance. We assume that estimation of the parameters, θ, is characterised by uncertainty, and so are interested in the use of Bayesian estimation practice for the identification of θ. This leads to discussion of BNN modelling.

2.2 Bayesian Neural Networks

Bayesian estimation practice is underpinned through use of Bayes' rule for the inference of a posterior parameter distribution, $p(\theta \mid \mathcal{D})$, given definition of the likelihood of the data under the parameters, $p(\mathcal{D} \mid \theta)$, and choice of a prior distribution, $p(\theta)$. Estimation of the posterior for the general case of nonlinear models is challenging and usually dependent on the use of sampling schemes. However, sampling observes an exponential increase in expense with parameter dimensionality, n_θ, which leaves its application dependent on relatively low dimensional parameter spaces (i.e. $n_\theta < 20$).

Variational inference is an approach that poses Bayesian estimation as optimisation. The aim is to identify the best approximation of the posterior, $q(\theta)$, from a variational family, $\mathcal{Q}$. The optimum $q^*(\theta)$ minimises the Kullback-Liebler (KL) divergence to the true posterior (Mowbray et al, 2022). Typical parameterisation of the variational family is the joint of independent Gaussian distributions as inherited through the mean-field approximation, which describes the variational distribution as $q(\theta \mid \mu, \Sigma)$, where $\mu \in \mathbb{R}^{n_\theta}$ represents the mean, and $\Sigma = \text{diag}([\sigma_1^2, \dots, \sigma_N^2]) \in \mathbb{R}^{n_\theta \times n_\theta}$ represents a diagonal covariance matrix. The predictive distribution of the model is then expressed by sampling the parameters from the variational distribution. It is worth noting that this means that BNNs typically have twice the number of parameters than their frequentist counterparts have for a given structure. Estimation of $q^*(\theta \mid \mu, \Sigma)$ can be approached by maximising a surrogate objective, known as the evidence lower bound (ELBO), which when evaluated empirically with K samples, $\theta^k \; \forall k \in \{1, \dots, K\}$, from $q(\theta \mid \mu, \Sigma)$ takes the form:

$$\max_{\mu,\Sigma} \frac{1}{K} \sum_{k=1}^{K} \log p(\mathcal{D} \mid \theta^k) + \log p(\theta^k) - \log q(\theta^k \mid \mu, \Sigma) \tag{2}$$

The ELBO identifies the parameters of $q(\cdot)$ that maximise the expected sum of the likelihood of the data under the parameters, and a regularisation term that penalises deviations of the variational posterior approximation from the prior defined. Here, we estimate the moments of the approximation by use of the reparameterization trick, and by making assumption such that the likelihood reduces to the negative sum of square errors.

2.3 Integrating Bayesian neural networks and hybrid models

Within the scope of our hybrid model, we seek means to integrate BNNs for modelling biohydrogen production in continuous time. We formalise the problem as follows:

$$\min_{\mu,\Sigma} \frac{1}{K} \sum_{k=1}^{K} \left[\sum_{n=1}^{N} [\phi^n(x^k(t_n), \theta^k)] - \log p(\theta^k) + \log q(\theta^k \mid \mu, \Sigma) \right]$$

$$s.t. \quad \dot{x}^k - f(x^k, u^k, \theta^k) = 0, \qquad \forall k \tag{3}$$

$$\theta^k = \mu + \Sigma \varepsilon^k, \quad \forall k$$

$$\varepsilon^k \in \mathrm{E}, \quad \forall k$$

where $\phi^n: \mathbb{R}^{n_x} \times \mathbb{R}^{n_\theta} \rightarrow \mathbb{R}$ evaluates the squared error to the data at discrete points in time, $\mathrm{E} = \{\varepsilon^1, \dots, \varepsilon^K\}$ describes a set of samples with $\varepsilon \sim \mathcal{N}(\mathbf{0}, \mathbf{I})$, and $\mathbf{I} \in \mathbb{R}^{n_\theta \times n_\theta}$ is the identity matrix. The second set of constraints describe the reparameterization trick, which provides description of the parameter samples from the variational approximation deterministically in terms of its moments. We can estimate the moments of the variational approximation as usual through use of approximate second order optimisation schemes, which leverage first-order gradient information only. We can define the gradient of our ELBO objective with respect to parameters, $\nabla_{\mu,\Sigma} J$, as follows:

$$\nabla_{\mu,\Sigma} J = \sum_{k=1}^{K} [\nabla_{\mu,\Sigma} J_{LSQ}^k + \nabla_{\mu,\Sigma} J_{REG}^k] \tag{4}$$

where $\nabla_{\mu,\Sigma} J_{LSQ}^k$ is the gradient of our least squares objective with respect to the variational parameters, and $\nabla_{\mu,\Sigma} J_{REG}^k$ is the gradient of the Bayesian regularization term. Estimation of $\nabla_{\mu,\Sigma} J_{REG}^k$ leverages automatic differentiation. Estimation of $\nabla_{\mu,\Sigma} J_{LSQ}^k$ is proceeds via adjoint sensitivity analysis.

2.4 Adjoint sensitivity analysis

Estimation of $\nabla_{\mu,\Sigma} J_{LSQ}^k$ via adjoint sensitivity analysis is appealing because it scales according to the number of BNN parameters only (i.e., $\mathcal{O}(2n_\theta)$ rather than $\mathcal{O}(n_x 2n_\theta)$ as is the case with the forward sensitivities). The procedure follows. Firstly, given a sample of the network parameters, the state dynamics are integrated forward from an initial condition, $x^k(t_0) = x_0^k$. The solution, $x^k(t)$, is stored in memory at checkpoints over the horizon. The cost is also evaluated at the discrete points in the time horizon for which one has measurements, $\phi^n(x^k(t_n), \theta^k) \, \forall n$, and stored in memory. This enables definition of the adjoint ODE system, the solution of which, $\lambda^k(t) \in \mathbb{R}^{n_x}$, is identified by solving a terminal value problem, starting from the end of the horizon, t_N, and proceeding backwards towards the initial time, t_0. The terminal condition is defined by the partial derivatives of the terminal cost with respect to the state, $\lambda^k(t_N) = \phi_x^N(x^k(t_N), \theta^k)$. Due to the discrete nature of the objective, discontinuities exist in the adjoint solution, which is enforced via a jump condition. With the availability of both the state and adjoint solutions, as well as evaluations of the cost function, we can calculate the partial derivative of the least squares objective with respect to the jth variational parameter as (Chachuat, 2007):

$$\nabla_{[\mu,\Sigma]_j} J_{LSQ}^k = \sum_{n=1}^{N} \left[\phi_{[\mu,\Sigma]_j}^n(x^k(t_n), \theta^k) + \int_{t_{n-1}}^{t_n} \lambda^k(t)^T f_{[\mu,\Sigma]_j}(x^k, u^k, \theta^k) dt \right] \\ + \lambda^k(t_0)^T x_{[\mu,\Sigma]_j}^k(t_0) \tag{5}$$

where $\phi_{[\mu,\Sigma]_j}^n : \mathbb{R}^{n_x} \times \mathbb{R}^{n_\theta} \to \mathbb{R}$ is the partial derivative of the least squares cost; $f_{[\mu,\Sigma]_j} : \mathbb{R}^{n_x} \times \mathbb{R}^{n_u} \times \mathbb{R}^{n_\theta} \to \mathbb{R}^{n_x}$ is the partial derivatives of the state dynamics; and finally $x_{[\mu,\Sigma]_j}^k \in \mathbb{R}^{n_x}$ defines the partial derivatives of the initial state components, all of which are defined with respect to the jth variational parameter, and may be estimated through automatic differentiation. Essentially, we see that the partial derivative is comprised of three terms describing the influence of the parameters on the cost, the dynamics, and the initial conditions, respectively (Chachuat, 2007). In practice, several implementations exist to perform equivalent procedure to adjoint sensitivity analysis. In this work, we used the adjoint procedure that is defined within the JAX-based library, DIFFRAX (Kidger, 2022). We estimate gradients over mini batches of samples drawn from the variational distribution, with updates of the variational parameters at each iteration provided by the ADAM optimizer.

3. Case Study

We assume that we have a small dataset generated from a ground truth nonlinear kinetic model descriptive of our biohydrogen photo-production system, as presented in [2]. We assume isothermal operation, such that we drop terms expressing temperature dependence from the original work:

$$\dot{c}_X = \mu_m . \mu(I). c_X \qquad \dot{c}_S = -Y_{XS}.\dot{c}_X - m. c_X \qquad \dot{c}_H = \alpha_m \alpha_{H2}(I)c_X \qquad (6)$$

where $\mu_m = 0.147$, $Y_{XS} = 9.656$, $m = 0.0139$, $\alpha_m = 51.332$, and $\mu(I)$ and $\alpha_{H2}(I)$ describe the effects of incident light intensity on the system. These terms approximate the spatial heterogeneity of the effects of light intensity, by using the average of light intensity within the reactor under assumption on its geometry and the use of the lambert beer law to describe its spatial decay from the PBR surface. Please see the original paper [2] for more information. The ground truth model is sampled under different incident light intensities, $I \in \{100, 150, 175, 200\}$ (Wm^{-2}), to generate a dataset of $k = 4$ different trajectories under the model dynamics, starting from the same initial condition of $x(t_0) = [0.061, 51.517, 0.100]^T$. Each trajectory is associated with $(N + 1 =)$ 24 measurements over a horizon of $t \in [0, 240]$ hours. The trajectory corresponding to the 3rd condition of $I = 175$ (Wm^{-2}) was used for validation, the others for training. Through the framework proposed in Section 2, we aim to identify a Bayesian hybrid model using the training data available. We define a neural network structure consisting of 4 hidden layers with rectified linear unit activation functions, and a hyperbolic tangent over the final hidden layer. This enables us to scale the predictions of the data-driven element to a pre-defined range. The input data were standardised to have a variance equal to 1, based on the mean and standard deviation of each state component. Dormand Prince's 8/7 method was utilized for forward numerical integration of the system state, as well as the backward adjoint integration. The prior on the parameters was defined as $\theta_i \sim \mathcal{N}_i(0.0, 0.01)$ for all parameters in the network.

The model is assessed via the mean absolute percentage error (MAPE), and the coverage probability (CP). The CP defines the probability with which a certain number, c, of standard deviations of the model's prediction, will cover the residual between the data, and the mean prediction. Please see (Mowbray et al, 2022) for definition of these metrics. Table 1 details the model's performance on the training and validation datasets.

Table 1: Summary of results for BNN hybrid model in training and validation

Predictive task	State component	MAPE %	CP ($c = 3$)	CP ($c = 2$)
	c_X	5.91	1.00	1.00
Training	c_S	7.63	1.00	1.00
	c_H	10.68	1.00	0.95
	c_X	5.76	1.00	1.00
Validation	c_S	3.23	1.00	1.00
	c_H	10.61	1.00	1.00

The results presented in Table 1 demonstrate a highly accurate predictive model in the mean, both in training and validation. This is highlighted by MAPE of 3-11% across both training and validation tasks for all state components. Additionally, the model performance is consistent across predictive tasks. Analysis of the CP indicates that uncertainty may not be perfectly calibrated, given that we should expect a reduction in the CP when $c = 2$. Ideally, in this case CP should be 0.95, instead of 0.997 as is the case for $c = 3$. This is only observed for the c_H state component in training. The validation performance is visualized by Fig. 1, which shows the model's prediction from the initial state over the horizon relative to the data available. Analysis of the figure highlights the predictive accuracy in the mean and emphasizes the calibration of the model uncertainty.

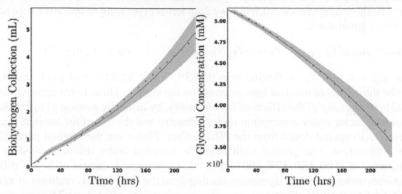

Figure 1: Visualization of model predictive performance. Left - biohydrogen; right - substrate. Blue scatter points represent measurements; orange line plot represents mean prediction; blue shaded region represents the 75th and 25th percentiles of the model's predictive distribution.

4. Conclusions

In this paper, we demonstrated the construction of a Bayesian hybrid model using mean-field variational inference and adjoint sensitivity analysis. The model identified made limited mechanistic assumptions beyond the use of mass balance. However, it was able to identify accurate predictive performance in the region of 3-11% mean percentage error. As a result, this framework, provides avenue to mitigate dependence on mechanistic model construction practice, which can be highly time consuming. In future, we will investigate the effect of integrating our approach with additional mechanistic assumptions on the kinetics of the system. Furthermore, we will explore use of the model in the context of transfer learning to enable adaptation to different scales and geometries of biohydrogen photo-production processes.

References

Teke GM, Anye Cho B, Bosman CE, Mapholi Z, Zhang D, Pott RWM. Towards industrial biological hydrogen production: a review. World Journal of Microbiology 2023.

Anye Cho B, Ross BS, du Toit JP, Pott RWMC, del Río Chanona EA, Zhang D. Dynamic modelling of Rhodopseudomonas palustris biohydrogen production: Perturbation analysis and photobioreactor upscaling. Int J Hydrogen Energy 2021;46:36696–708.

Del Rio-Chanona EA, Fiorelli F, Zhang D, Ahmed NR, Jing K, Shah N. An Efficient model Construction Strategy to Simulate Microalgal Lutein Photo-Production Dynamic Process. Biotechnol Bioeng 2017;114:2518–27.

Rogers AW, Song Z, Ramon FV, Jing K, Zhang D. Investigating 'greyness' of hybrid model for bioprocess predictive modelling. Biochem Eng J 2023;190.

von Stosch M, Oliveira R, Peres J, Feyo de Azevedo S. Hybrid semi-parametric modeling in process systems engineering: Past, present and future. Comput Chem Eng 2014;60:86–101.

Mowbray, M., Kay, H., Kay, S., Caetano, P. C., Hicks, A., Mendoza, C., ... & Zhang, D. (2022). Probabilistic machine learning based soft-sensors for product quality prediction in batch processes. Chemometrics and Intelligent Laboratory Systems, 228, 104616.

Chachuat, B. (2007). Nonlinear and dynamic optimization: From theory to practice.

Kidger, P. (2022). On neural differential equations. arXiv preprint arXiv:2202.02435.

Flavio Manenti, Gintaras V. Reklaitis (Eds.), Proceedings of the 34th European Symposium on Computer Aided Process Engineering / 15th International Symposium on Process Systems Engineering (ESCAPE34/PSE24), June 2-6, 2024, Florence, Italy

Loss-in-Weight feeder performance prediction using Machine Learning

Hikaru G. Jolliffe[a], Carlota Mendez Torrecillas[a], Gavin Reynolds[b], Richard Elkes[c], Hugh Verrier[d], Michael Devlin[a], Bastiaan Dickhoff[e], and John Robertson[a].

[a]CMAC, 99 George Street, Glasgow, G1 1RD, UK.
[b]Oral Product Development, PT&D, Operations, AstraZeneca UK Limited, Charter Way, Macclesfield SK10 2NA, UK.
[c]GSK Ware R&D, Harris's Lane, Ware, Hertfordshire, SG12 0GX, UK.
[d]Pfizer Research and Development UK Ltd, Ramsgate Road, Sandwich CT13 9NJ, UK.
[e]DFE Pharma GmbH & Co. KG, Kleverstrasse 187, 47568 Goch, Germany.
hikaru.jolliffe@strath.ac.uk

Abstract

There has been significant drive in recent years to gain greater understanding of unit operations useful for continuous direct compaction, and to leverage data-driven approaches such as Machine Learning to extract trends from large and complex datasets. In this work, an approach using three Machine Learning models to predict the parameters in an equation for Loss-in-Weight feeder performance are presented. Industrially-relevant feeders with multiple screws per feeder are studied, and the approach allows feed factor decay to be predicted using material properties and equipment choice as inputs. Using a wide range of excipients and Active Pharmaceutical Ingredients (APIs) for testing shows good performance against industrially-relevant targets. The approach presented here would be useful for equipment pre-selection activities prior to experimental work.

Keywords: feeder, modelling, Machine Learning, powders.

1. Introduction

Driven by a need to improve efficiency and ever-increasing R&D costs in the pharmaceutical sector, the use of alternative production methods such as Continuous Manufacturing is gaining ground (Lee et al., 2015; Plumb, 2005; Schaber et al., 2011). Towards this, there is a need to develop mathematical models of key unit operations, as this allows simulations and digital Design-of-Experiments (DoE) to be rapidly carried out (Lee et al., 2020). One such key unit operation is the Loss-in-Weight (LIW) feeder.

LIW feeders are key unit operation in continuous, semi-continuous and batch manufacturing systems. Frequently the first unit operation in secondary processing, in essence they are hoppers of powder sitting above a screw or screws, which mechanically convey powder in a controlled manner as they turn; LIW feeders supply or 'feed' material to subsequent unit operations. Differences between designs of feeder can range from hopper size and shape; design of internal agitator; and screw number (e.g. twin-screw as opposed to single screw), design, size, and speed. In the literature there is an extensive body of work characterising feeders and evaluating their mass flow accuracy, reliability, and variability (Bascone et al., 2020; Bostijn et al., 2019; Engisch and Muzzio, 2012; Fernandez et al., 2011; Gao et al., 2011; Yadav et al., 2019).

Literature models for feeder performance typically require regression with experimental data (of feed factor with hopper contents) to determine model parameters, after which the equation can be used for that given material with that given equipment configuration *i.e.* quantities of material are required for testing, and model transferability is limited. In the present work, a Machine Learning-based approach developed, where the inputs are material properties and equipment configuration, and the outputs are model parameters that would otherwise require regression. The dataset used for training and testing spans two types of feeders (GEA Compact Feeder and Gericke GZD200.22) with multiple screws each. Multiple materials are included: 19 excipients and two generic Active Pharmaceutical Ingredients (APIs), which have been further supplemented by the inclusion of literature data (25 APIs/grades, 3 excipients) from Pfizer (Shier et al., 2022).

2. Materials, equipment, experiments

2.1. Feeders
The GEA Compact Feeder comprises a flat bottom cylindrical hopper with interchangeable twin screws slightly offset from the centre. The operating volume of the hopper is 2 L (total volume is 2.5 L) and has three gearboxes available 455:1 (1-64 rpm), 235: 1 (1-124 rpm) and 63:1 (1-460 rpm) giving a speed range 1-460 rpm for both screws sets: 20 mm pitch double concave (20C) and 10 mm pitch double concave; both screws are 20 mm diameter. The internal hopper agitator rotates horizontally about the hopper axis. The GZD200.22 is a flat-bottomed feeder of 10 L hopper volume, with twin double-concave screws of 22 mm diameter. There are three screws that are represented in the database: the 11 mm pitch 11C screws, the 22 mm pitch 22C screws, and the 33mm pitch 33C screws. The internal hopper agitator rotates horizontally about the hopper axis.

2.2. Materials
The materials used in the used in the present work include those in the base dataset (19 excipients and a generic API) supplemented by literature data from Pfizer (Shier et al., 2022). The additional literature data is predominantly proprietary substances, many for which there is data (both material characterisation and feeder performance data) for various grades or batches (3 excipients, 25 API grades and batches representing 16 APIs).

2.3. Material characterisation
Bulk and tapped densities were measured with an AutoTap (Quantachrome). Shear tests were done with a Powder Flow Tester (Brookfield). Stability and Variable Flow Rate were done in a FT4 Powder Rheometer (Freeman Technology), as were compressibility and permeability. A MasterSizer 3000 (Malvern) was used for particle size.

2.4. Volumetric discharge experiments
Base dataset of GEA Compact Feeder and Gericke GZD200.22 experiments are volumetric discharge runs conducted from an initially full hopper. Screw speeds used depend on feeder: 30, 60, 90 or 230 RPM for the Compact Feeder, and 170, 340, or 510 RPM for the GZD200.22. Runs have been started with screws initially uncharged (empty). Feeders have been run until mass flow stops, and no refills have been conducted. Literature data (Shier et al., 2022) are from gravimetric runs, and whilst this is different to the base dataset (volumetric runs), as subsequent modeling did not consider screw speed as a factor (and treated runs at different screw speeds as replicates) the two datasets are compatible. Some runs in the literature data used mesh grids at the end of the screws to break up any clumps of powder and improve flow, and where there was indication this impacted feeder performance those materials were excluded (not a common occurrence).

3. Data processing and Machine Learning approach structure

A modified feeder performance equation from literature has been used when regressing parameters using raw data (Wang et al., 2017):

$$ff = ff_{max} - ff_{max}e^{-\beta F} \tag{1}$$

Where *ff* is feed factor, the mass flow per turn of the screw(s) (g/rev), and *F* is hopper fill fraction. There are two fitting parameters: maximum feed factor ff_{max} (g/rev), the initial (typically highest) feed factor when the hopper is full; and rate constant β (dimensionless), a measure of how quickly the feed factor decays from ff_{max}. High β means feed factor is stable for longer and then precipitously decays, while low β means a more gradual but earlier decay (Figure 1A).

Raw feeder performance data of feed factor against hopper fill fraction mass (calculated from hopper mass, volume, and material bulk density). This allows a comparison across feeders of different hopper volumes, and allows value ranges of the β parameter to be defined: once β is at or above 15, the feed factor decay trajectories are functionally the same for the 30%–80% hopper fill region of interest (Figure 1A). This region of interest is used as the modeling efforts are not concerned with the initial rise from zero feed factor (Figure 1B), and the fact that the final decay as the hopper is emptying is likely governed by substantially different powder flow phenomena and moreover would not be experienced in practice as refill would have occurred before this point is reached. As β is very sensitive to flat regions of data, a maximum of 15 has been set for this parameter.

3.1. Machine learning model structure, predictors, algorithms

The goal of the Machine Learning was to predict model parameters ff_{max} and β. Three Machine Learning models (developed and implemented in Matlab) have been used: one to predict ff_{max} (ensemble of trees with LSBoost algorithm), one to make an initial coarse estimate of β (the *class* of β; ensemble of trees with AdaBoost algorithm), and a third model to use this *class* as an additional input predictor and make a refined prediction of β (ensemble of trees with LSBoost algorithm). The class of β is a rough first estimate, and there are four classes: class 15 (values of β at 15), class 10 (values between 10 and 15),

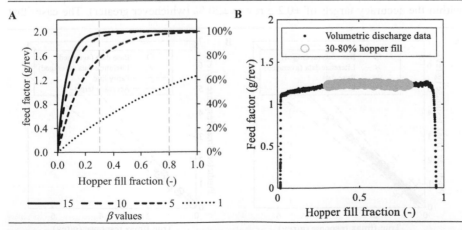

Figure 1. Decay of feed factor. A: example Eq. (1) decay defined with ff_{max} = 2 g/rev, ff_{min} = 0 g/rev. B: example volumetric discharge data with key 30–80 % hopper fill region highlighted.

class 5 (values between 5 and 10), and class 1 (values between 1 and 5). Class 15 materials have a steady flow with precipitous decay, and class 1 materials have a gradual decay that starts much sooner. Aside from material properties used as predictors to the Machine Learning model, the following equipment properties are also predictors: feeder model (categorical), hopper volume, screw choice (categorical), screw diameter, and screw pitch. There is also ideal feed factor, the product of bulk density and screw free volume; the latter is calculated from geometry assuming the clearance volume around the screw plays no part in material transport (Bascone et al., 2020; Yu, 1997).

Training and test datasets have been split with an approximately 80/20 ratio. Conditions (feeder-screw-material combinations) often have multiple runs, either true replicate runs, or runs conducted at different RPMs. This results in some variability between parameters regressed for that condition, particularly for the β parameter. For the training dataset, the average value of regressed parameters has been used, but duplicated equivalent to the number of replicates as weighting. For the testing dataset, averages have been used but they have not been duplicated to not overestimate model accuracy. There are 215 runs in the training set, and 49 in the test set (21 after run averaging). However, there are some material property gaps in the literature data *i.e.* measurements that were not needed or performed for the purposes of that literature data. Excluding these from analysis (missing predictor impact is beyond present work scope) leaves 36 runs (13 after averaging).

4. Results and discussion

Accuracy targets represent practical considerations for feeder operation: for ff_{max} the target was ±0.2 g/rev up to 1.0 g/rev and ±20 % thereafter. For β the target was accuracy such that the location of a 10 % drop from ff_{max} is within ± 50 g in terms of hopper contents (this represents different thresholds depending on hopper volume and material density).

The Machine Learning model for ff_{max} has been trained such that 90.2 % of the training data points are within the accuracy targets (Figure 2). In general it was easier to train materials in the base dataset (primarily excipients) than it is for literature data materials (primarily proprietary APIs). The most important predictor is ideal feed factor, the product of screw capacity and bulk density. In testing, a majority (11/13) data points were within the accuracy targets of ±0.2 g/rev or ±20 % (whichever greater). The cases that

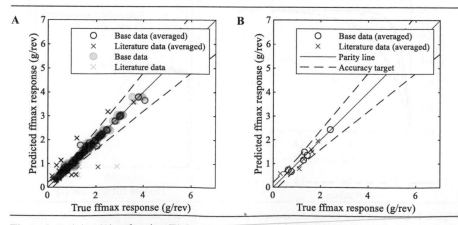

Figure 2. Training (A) and testing (B) for ff_{max}. Unaveraged data points are shown in a lighter shade.

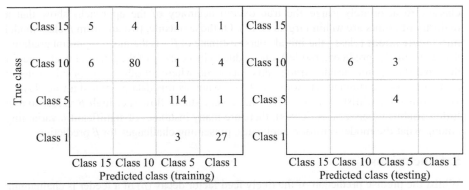

	Class 15	Class 10	Class 5	Class 1		Class 15	Class 10	Class 5	Class 1
Class 15	5	4	1	1	Class 15				
Class 10	6	80	1	4	Class 10		6	3	
Class 5			114	1	Class 5			4	
Class 1			3	27	Class 1				

Figure 3. Training (A) and testing (B) for classification of β.

fell outside the accuracy targets (although only barely outside) were from the literature data, mirroring the more challenging training agreement of this dataset.

While classification of β does not have accuracy targets as such (inaccuracies are passed on to the refined prediction) the classification accuracy can still be assessed (Figure 3). Most data points are either class 5 or 10 *i.e.* few are at the extremes. Training is straightforward (27/30 class 1 correctly classified, 114/115 class 5 correctly classified, 80/91 class 10 correctly classified) although it is slightly challenging for class 15 (5/11 correctly classified). In testing, a majority are correctly classified (6/9 class 10 cases, all four class 5 cases) with no misclassifications greater than one class difference.

Finding trends between the predictors and β is more challenging than for ff_{max}. Leaving aside the possibility that there is some unmeasured material property or material properties that are required to get a clear picture of the link, the parameter is also simply more variable, even among replicates (Figure 4). Nevertheless, 88 % of training data points (averaged replicates) fall within the specified accuracy threshold for beta (±50 g of material in the hopper to accurately predict location of 10 % drop from initial feed factor). Testing β model accuracy using predicted classes as an input (instead of known

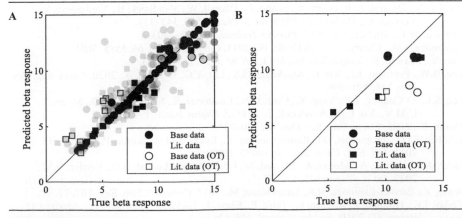

Figure 4. Training (A) and testing (B) for prediction of β. Data points outside of accuracy thresholds are marked 'OT'. Unaveraged data points are shown in a lighter shade.

classes, to accurately gauge the true overall accuracy of the approach) show that a majority of points are within targets (9/13). Of those outside, two are from β class model misclassifications (class – the initial coarse estimate of β – is the most important predictor for the refined β prediction model), one is due to inaccuracy of the β model itself, and for the final data point outside targets, this was a run where all other runs of this material (which are in the training set) were of class 5 whereas this data point had a true class of 10. Ultimately, whilst many materials have very stable flow (*i.e.* high β values and/or little variability between replicates), there are many instances of significant β variability among what the model considers replicates, presenting challenges for β prediction.

5. Conclusions

Whilst the precise prediction of the likely feed factor decay (β) of a feeder is challenging – unsurprising given the variability of this in real data even among replicates – the prediction of the magnitudes of likely feed factor (ff_{max}) is accurate. Taken together, the approaches presented here offer a way to gain insight into likely equipment performance with reduced need for experimentation with the feeders.

Acknowledgements

This work has been funded by the Medicines Manufacturing Innovation Centre project (MMIC), UK (project ownership: Centre for Process Innovation, CPI). Funding has come from Innovate UK and Scottish Enterprise. Founding industry partners with significant financial and technical support are AstraZeneca and GSK. The University of Strathclyde (via CMAC) is the founding academic partner. Pfizer are project partners and have provided key technical input and data. Project partners DFE Pharma have provided materials and technical input, and project partners Gericke AG have provided technical equipment support and advice. Project partners Siemens and Applied Materials have provided key software and software/IT expertise. Andrew Shier (now of GSK) significantly contributed to the present work via the literature data (Shier et al., 2022).

References

Bascone, D., Galvanin, F., Shah, N., Garcia-Munoz, S., 2020. *Ind. Eng. Chem. Res.* 59, 6650–6661.
Bostijn, N., Dhondt, J., Ryckaert, A., Szabó, E., Dhondt, W., Van Snick, B., Vanhoorne, V., Vervaet, C., De Beer, T., 2019. *Int. J. Pharm.* 557, 342–353.
Engisch, W.E., Muzzio, F.J., 2012. *Powder Technol.* 228, 395–403.
Fernandez, J.W., Cleary, P.W., McBride, W., 2011. *Chem. Eng. Sci.* 66, 5585–5601.
Gao, Y., Muzzio, F., Ierapetritou, M., 2011. *AIChE J.* 57, 1144–1153.
Lee, B.W., Peterson, J.J., Yin, K., Stockdale, G.S., Liu, Y.C., O'Brien, A., 2020. *Chem. Eng. Res. Des.* 156, 495–506.
Lee, S.L., O'Connor, T.F., Yang, X., Cruz, C.N., Chatterjee, S., Madurawe, R.D., Moore, C.M.V., Yu, L.X., Woodcock, J., 2015. *J. Pharm. Innov.* 1–9.
Plumb, K., 2005. Chem. Eng. Res. Des. 83, 730–738.
Schaber, S.D., Gerogiorgis, D.I., Ramachandran, R., Evans, J.M.B., Barton, P.I., Trout, B.L., 2011. *Ind. Eng. Chem. Res.* 50, 10083–10092.
Shier, A.P., Kumar, A., Mercer, A., Majeed, N., Doshi, P., Blackwood, D.O., Verrier, H.M., 2022. *Int. J. Pharm.* 625, 122071.
Wang, Z., Escotet-Espinoza, M.S., Ierapetritou, M., 2017. *Comput. Chem. Eng.* 107, 77–91.
Yadav, I.K., Holman, J., Meehan, E., Tahir, F., Khoo, J., Taylor, J., Benedetti, A., Aderinto, O., Bajwa, G., 2019. *Powder Technol.* 348, 126–137.
Yu, Y., 1997. (Doctoral Thesis). University of Wollongong.

Flavio Manenti, Gintaras V. Reklaitis (Eds.), Proceedings of the 34th European Symposium on Computer Aided Process Engineering / 15th International Symposium on Process Systems Engineering (ESCAPE34/PSE24), June 2-6, 2024, Florence, Italy

Process Design of Carbon-Neutral Routes for Methanol Synthesis

Omar Almaraz, Srinivas Palanki*, Jianli Hu

Dept. of Chemical and Biomedical Engineering, West Virginia University, WV 26506
*srinivas.palanki@mail.wvu.edu

Abstract

Two different microwave-assisted carbon-neutral routes are designed for the manufacture of methanol. The first process involves the dry reforming of methane with carbon dioxide for producing synthesis gas in a microwave reactor, which is subsequently converted to methanol. The second process involves the direct hydrogenation of carbon dioxide in a novel microwave reactor. Simulation results show that both processes can produce 206.5 t/h of methanol at a purity of 99.9%, which is the current production rate of methanol at Natgasoline LLC using the traditional steam reforming process. Furthermore, both novel processes have significant decarbonization potential. Heat integration tools are employed to demonstrate that both processes have energy savings potential.

Keywords: methanol, microwave-assisted, decarbonization.

1. Introduction

Methanol, a versatile and cost-effective alcohol, finds a multitude of industrial applications across various sectors. One primary use of methanol is as a feedstock in the production of chemicals, such as formaldehyde and acetic acid, essential building blocks for the manufacturing of plastics, resins, and coatings. Additionally, methanol serves as a crucial solvent in industries like pharmaceuticals, where it facilitates the extraction and synthesis of various pharmaceutical products. In the energy sector, methanol plays a pivotal role as a clean-burning fuel, both in its pure form and as a component of alternative fuels like biodiesel. Its role as a fuel extends further with the emerging trend of methanol-powered fuel cells for electricity generation. Moreover, methanol is employed in the production of renewable energy sources like biodiesel, contributing to sustainable energy solutions. With its diverse applications, methanol stands as a key player in fostering innovation and sustainability across a spectrum of industrial processes. The global methanol market size was valued at $28.78 billion in 2020 and is projected to reach $41.91 billion by 2026 (Market research, 2023). Methanol has traditionally been produced from natural gas by first converting methane to syngas and then converting syngas to methanol (Aimiuwu et al., 2020). However, this is a very energy intensive process and produces a significant amount of the greenhouse gas carbon dioxide. Hence, is necessary to look for more efficient alternative routes to manufacture methanol.

The search for carbon-neutral routes for the manufacture of chemicals in industrial scale has led to interest in utilizing novel microwave reactors instead of conventional heat sources that utilize the combustion of hydrocarbon fuels which produce greenhouse gases. Microwave-assisted synthesis also offers numerous advantages, including accelerated reaction rates, enhanced selectivity, and improved yields, compared to traditional heating methods. The controlled and uniform heating achieved with microwave reactors has been applied to various chemical processes, such as organic synthesis, catalysis, and polymerization. Notable studies include the work of Kappe (2004) on continuous-flow microwave reactors for organic synthesis, highlighting the potential for scale-up applications. More recently, Yan et al. (2023) demonstrated the use of a novel reactor utilizing microwave/electromagnetic irradiation to produce clean hydrogen.

In this research, two different microwave-assisted carbon-neutral routes are designed for the manufacture of methanol. The first route involves the dry reforming of methane in a novel microwave reactor to produce synthesis gas (syngas), which is subsequently catalytically converted to methanol. The second route involves the reaction of carbon dioxide captured from the atmosphere with green hydrogen using a novel $CuO/ZnO/ZrO2$ structured catalyst developed at Oakridge National Laboratory to directly form methanol in a microwave reactor. The objective is to produce 206.5 t/h of methanol at a purity of 99.9%, which is the current production rate of methanol at Natgasoline LLC in Beaumont, Texas (USA) using the traditional steam reforming process (Haque et al, 2021).

2. Process Modeling

2.1. Methodological Framework
The first step in the methodology involves the development of a comprehensive steady-state simulation model for the two different routes mentioned in the previous section. This simulation model is developed using the process simulation software ASPEN Plus by integrating relevant process information, feed specifications, and operating conditions obtained from the literature. Next, a heat exchanger network (HEN) is designed to reduce utility costs using the ASPEN Energy Analyzer (AEA) v11.0 software. The AEA software identifies the optimal configuration of the HEN that maximizes energy efficiency while meeting the targeted utility consumption, capital investment, and surface area requirements by formulating and solving a mixed-integer linear programming problem. By exchanging heat between streams at different temperature levels, the HEN aims to minimize the use of external utilities and enhance the overall process efficiency.

2.2. Indirect Conversion of Carbon Dioxide to Methanol via Dry Reforming
Figure 1 shows the process flow diagram of the dry reforming process to convert methane to syngas and then to methanol. The objective is to produce 206.5 t/h of methanol at a purity of 99.9%, of methanol. The process is initiated by introducing a natural gas stream sourced from a pipeline. This incoming gas is then split into two separate streams. The initial portion, which is 313.6 kmol/h of natural gas, is utilized as fuel for the separation processes in the form of reboiler duties. Concurrently, the second portion, consisting of 2,339.3 kmol/h of natural gas, undergoes a preliminary separation treatment to remove any free liquids and particles present in the gas stream. The preheated gas stream is directed to reactor 1, where a crucial desulfurization step occurs. Reactor 1 operates at a temperature of 375 °C and a pressure of 100 kPa, ensuring optimal conditions for converting any organic sulphur compounds into H_2S if present. After the desulfurization step, the stream from reactor 1 is combined with CO_2 in a specific ratio of 0.5 CH_4/CO_2, promoting the equilibrium conversion of CH_4 and enhancing overall process efficiency. This mixture is then heated to a temperature of approximately 800 °C, preparing it for the subsequent step in the novel microwave syngas reactor. This reactor operates under specific conditions, at a temperature of 800 °C and a pressure of 100 kPa. Within the microwave syngas reactor, the pretreated natural gas stream undergoes reactions resulting in the formation of syngas—a mixture of carbon monoxide (CO) and hydrogen (H_2). Importantly, these reactions occur in an equilibrium state at low pressure, preventing

carbon formation and ensuring the absence of higher hydrocarbons in the resulting gas stream (Pham et al., 2020). Higher pressures (e.g. above 500 kPa) lead to significant coke formation and it is important to avoid such conditions.

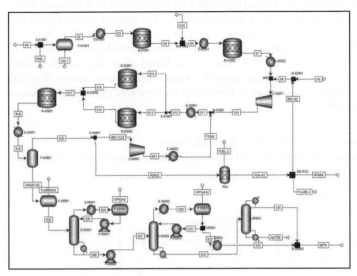

Figure 1: Flowsheet for Indirect Conversion of Carbon Dioxide to Methanol via Dry Reforming

To achieve the desired syngas composition, H_2 is added to the product gas stream, maintaining a stoichiometric number around 2 during operation. The resulting gas mixture is then compressed to a pressure of 7.6 MPa using a dedicated compressor. Following compression, the syngas stream is further heated to a temperature of approximately 188 °C, preparing it for subsequent methanol production stages. The heated and pressurized syngas is evenly distributed to twin methanol reactors, operating at a temperature of 254 °C and a pressure of 7.6 MPa. The hydrogenation of CO_2 and CO, facilitated by Cu- and Zn-based catalysts (CuO and Al_2O_3 or ZnO), drives the conversion process, leading to methanol production. Upon exiting the methanol reactors, the gas stream, comprising methanol, water, unreacted H_2, CO, CO_2, CH_4, and N_2, undergoes heat exchange with the reactor's feed gas. The gas stream is subsequently directed to a third reactor for the final conversion stage. Operating at a temperature of approximately 215 °C and a pressure of 7.6 MPa, this third reactor enables the completion of the conversion process, ensuring optimal methanol production. The outlet gas from this reactor consists of methanol, water, and unreacted gases. To separate crude methanol from the gas stream, a separator is employed. This separation step occurs at a temperature of approximately 40 °C and a pressure of 6.7 MPa. Following separation, a portion of the unreacted gas is repressurized and recycled back into the twin methanol reactors, optimizing resource utilization. Another portion of the unreacted gas is purged into the flare, minimizing the buildup of unwanted gases in the system.

The purification of crude methanol involves an additional separator and three distillation columns. Initially, the crude methanol undergoes a depressurization process in the separator, resulting in the venting of an off-gas stream primarily composed of CO_2, H_2, and CH_4. The bottom stream from the separator is then directed to the first distillation column, where further separation occurs. This column effectively separates the remaining CO_2 and trace amounts of other gases, which are subsequently flared. The bottom product from the first column is then pumped to the second distillation column for further separation. Within this column, the desired top product, characterized by a required purity of 99.9 mol% methanol, is collected as a liquid distillate. The bottom product from this column constitutes a mixture of methanol and water, which undergoes the final purification step in the third distillation column. This column enables the collection of methanol with a required purity of 99.9 mol% as a liquid distillate, which is then combined with the top products from the second column. The bottom stream from the third column mainly consists of pure water. The combined flow rate of purified methanol gives a purity of 99.9 mol%.

2.3. Direct Hydrogenation of Carbon Dioxide to Methanol

Figure 2 shows the process flow diagram of the novel process to directly convert carbon dioxide and methane to methanol.

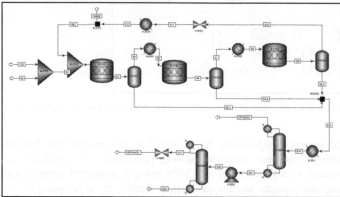

Figure 2: Direct Hydrogenation of Carbon Dioxide and Methane to Methanol

The stream of green H_2 is mixed with CO_2 and utilized for the conversion to methanol directly using a novel $CuO/ZnO/ZrO2$ structured catalyst (Balyan et al., 2023). Since the conversion is low (13%), it is necessary to utilize three reactors in series with intermediate removal of crude methanol product along with recycle of unreacted reactants as shown in Figure 2. All the reactors are heated using microwave radiation and operate at 160-220 °C and 140-970 kPa. The crude methanol stream is sent to a series of two distillation columns. The first distillation column operates at a pressure of 200 kPa and is utilized to separate the carbon dioxide and other non-condensable gases from the methanol stream. The second distillation column operates at a pressure of 812 kPa and produces a top stream that is 206.5 t/h of 99.9% pure methanol. The bottom stream is mostly water and used elsewhere in the plant as a produced water.

3. Simulation Results

ASPEN Plus software was employed to simulate the process of converting methane to syngas and syngas to methanol. For methane conversion, the RKSoave thermodynamic property method was used, while the NRTL method was employed for syngas conversion to methanol. The microwave reactors were modeled as yield reactors and the experimentally reported results of yield from the literature (Yan et al., 2023; Balyan et al., 2023) were used in the simulations. In the base case simulation of dry reforming without heat integration, the feed stream consisting of 2,325 kmol/h of methane is utilized, resulting in a final product stream of 206.5 t/h of methanol with a purity of 99.9%. In the base case simulation of direct hydrogenation of carbon dioxide to methanol, 20,562 kmol/h of hydrogen is reacted with 6,123 kmol/h of CO_2, resulting in a final product stream of 206.5 t/h of methanol with a purity of 99.9%. This successful conversion showcases the effectiveness of the proposed indirect dry reforming process as well as the direct hydrogenation of CO_2 for methanol production. Table 1 provides a comparison between dry reforming, direct hydrogenation of carbon dioxide and traditional steam reforming. The dry reforming process exhibits remarkable efficiency by utilizing only 36% of the methane compared to the traditional steam reforming process, while the direct hydrogenation of carbon dioxide utilizes no methane. This substantial reduction/elimination of methane consumption is significant from both economic and environmental perspectives, as it allows for the preservation of valuable resources and minimizes the cost and environmental impact of methane extraction and utilization. It is important to note that while the conventional process *produces* 287.5 kmol/h of CO_2, the dry reforming process *consumes* 4,079 kmol/h of CO_2 and the direct hydrogenation of CO_2 process *consumes* 5,287.7 kmol/h of CO_2. While this shows a significant decarbonization potential, it should be noted that the conventional steam reforming process does not utilize any hydrogen whereas both the dry reforming process as well as the direct hydrogenation process utilize significant amounts of hydrogen, which must be produced from green methods to extract the decarbonization potential shown in these simulations. One green method that has shown promise is the use of chemical looping techniques for producing hydrogen with sequestration-ready CO_2 (Riley et al., 2023).

Table 1: Decarbonization Potential of Novel Processes as Compared to Steam Reforming

	Steam Reforming (kmol/h)	Microwave Dry Reforming (kmol/h)	CO_2 Hydrogenation (kmol/h)
Methane to Syngas	6,430.0	2,325.0	0.0
Hydrogen Added	0.0	10,642.7	20,562.3
Water Added to Process	13,506.8	0.0	0.0
Methanol Produced	6,455.0	6,445.0	6,445.0
Total CO_2 Produced	261.4	-4,079.1	-5,287.7

Overall, the base case simulation results highlight the significant potential of the microwave-assisted processes in utilizing carbon dioxide as a valuable feedstock. To achieve optimal operation and enhance energy utilization, a heat exchanger network (HEN) is designed to facilitate efficient heat exchange and minimize utility costs. The heat integration tool within the ASPEN Energy Analyzer software enables the generation of multiple feasible solutions for the HEN optimization. The integration of the HEN within the microwave-assisted process demonstrates its potential in significantly reducing the reliance on hot utility and cold utility consumption. Table 2 shows the comparison

Table 2: Comparison of Heat Duties

	Steam Reforming (MW)	Microwave Dry Reforming (MW)	CO_2 Hydrogenation (MW)
Base Case Hot Utility	808.3	387.7	299.8
Base Case Cold Utility	1,106.0	650.0	545.7
Heat Integrated Hot Utility	165.0	137.2	121.0
Heat Integrated Cold Utility	378.1	387.4	250.9

between the base case and the heat integrated (HI) scenario and it is observed that there are significant improvements in energy utilization. For both novel microwave processes, there are energy savings via heat integration. After heat integration, the CO_2 hydrogenation process utilizes the least amount of hot and cold utilities.

4. Conclusions

The results of this simulation study provide compelling evidence for the significant potential of microwave-assisted dry reforming as well as microwave-assisted direct carbon hydrogenation as promising methods to produce methanol. The novel processes offer numerous advantages over traditional steam reforming techniques, making them attractive alternatives in the field of catalytic methanol production with significant decarbonization potential. One notable advantage of microwave-assisted dry reforming over conventional steam reforming is its lower methane consumption. The consumption of methane can be eliminated in the microwave-assisted direct hydrogenation of CO_2 if hydrogen is produced from green methods and the energy utilized in the separation train is hydrocarbon-free. Both novel processes utilize significant amounts of CO_2, which help in decarbonization. The simulation studies also demonstrate the significant potential in reducing utility costs via the application of heat-integration. Overall, these results highlight the importance of continued research in the field of microwave-assisted reactions for developing carbon-neutral processes.

5. Acknowledgment

This study was supported by the United States Department of Energy.

References

G. Aimiuwu, E. Osagie and O. Omoregbe, 2020, "Process Simulation for the Production of Methanol via CO_2 Reforming of Methane Route," Chemical Product and Process Modeling, 17:69-79.

S. Balyan, K. Tewari, B. Robinson, C. Jiang, Y. Wang, J. Hu, 2023, "CO_2 Hydrogenation to Olefins in Microwave-Thermal Hybrid Heating Reactor," Journal of Catalysis Science & Technology (submitted)

M. E. Haque, N. Tripathi, and S. Palanki, 2021, "Development of an Integrated Process Plant for the Conversion of Shale Gas to Propylene Glycol," Industrial & Engineering Chemistry Research, 60 (1), 399-441.

C.O. Kappe, 2004, "Controlled Microwave Heating in Modern Organic Synthesis," Angew. Chem. Int. Ed., 43, 6250-6284.

Market Research (2023). Methanol Market by Feedstock (Natural Gas, Coal, Biomass), https://www.marketresearch.com/MarketsandMarkets-v3719/Methanol-Feedstock-Natural-Gas-Coal-30408866/ (accessed 04.04.23)

T. Pham, K.S. Ro, L. Chen, D. Mahajan, T.J. Siang, U.P.M. Ashik, J.I. Hayashi, D.P. Mihn, D.V.N. Vo, 2020, "Microwave-Assisted Dry Reforming of Methane for Syngas Production: A Review," Environ Chem. Lett., 18, 1987–2019.

J. Riley, M. Bobek, C. Atallah, R. Siriwardane, and S. Bayham, 2023, "Syngas and H_2 Production from Natural Gas Using $CaFe_2O_4$ - Looping: Experimental and Thermodynamic Integrated Process Assessment," J. Hydrogen Energy, 48, 29898-29915.

K. Yan, X. Jie, X. Li, J. Horita, J. Stephens, J. Hu, Q. Yuan, 2023, "Microwave-Enhanced Methane Cracking for Clean Hydrogen Production in Shale Rocks," International Journal of Hydrogen Energy, 48, 41, 15421-15432.

Flavio Manenti, Gintaras V. Reklaitis (Eds.), Proceedings of the 34th European Symposium on Computer Aided Process Engineering / 15th International Symposium on Process Systems Engineering (ESCAPE34/PSE24), June 2-6, 2024, Florence, Italy

Efficient use of energy in distillation: Advancing towards the electrification of the chemical industry

Zinet Mekidiche-Martínez,[a*] José A. Caballero,[a] Juan A. Labarta[a]

[a]Inttitute of Chemical Process Engineering, University of Alicante. Carretera de S. Vicente s.n. 03690, Alicante, Spain
caballer@ua.es

Abstract

In the pursuit of sustainable development and reduced CO2 emissions, industries are transitioning from fuel-based to electrified processes. However, challenges in energy efficiency arise, particularly in sectors like chemicals and pharmaceuticals, which contribute significantly to energy consumption and emissions. The study focuses on improving the low thermodynamic efficiency of distillation processes, crucial for purification in these sectors.

Exploring advanced distillation configurations such as thermally coupled distillation, the study aims to enhance energy efficiency, decrease global energy demands in high-energy distillation processes, and advance electrification. It acknowledges the potential increase in operating costs when substituting combustion-generated heat with electrical sources. The analysis adopts a dual perspective, examining total annualized costs economically and considering environmental impacts through life cycle analysis, utilizing the ReCiPe metric at midpoint and endpoint levels.

1. Main Text

There has been considerable interest in Heat Pump Assisted Distillation (HPAD); however, there are limited contributions that simultaneously consider the synthesis of distillation sequences, heat pumping, heat integration, and thermal coupling. Noteworthy works in this area include the study by Kazemi et al. (2018), which explores configurations for a three-component separation. Kiss and Smith (2020) addressed the separation of a five-component alcohol mixture, considering common alternatives along with heat pumping. Miao et al. (2023) optimized the separation of light hydrocarbons with divided wall columns and heat pumping, although the column sequence was not explicitly optimized. Yuan et al. (2022) provided a comprehensive analysis, systematically considering all basic configurations, energy integration, mechanical recompression, and bottom flashing. According to Yuan et al. (2022), excluding HPAD from the synthesis of distillation sequences, traditional thermal energy is more cost-effective than electricity, and compressors rank among the most expensive equipment in the chemical industry. However, considering the imperative to reduce reliance on fossil fuels and transition to green electricity, it becomes crucial to incorporate heat pump-assisted distillation as a viable option during the synthesis of distillation sequences. Regardless of energy integration considerations, the initial step in designing a distillation-based separation sequence involves generating (explicitly or implicitly) the complete space of alternatives. Caballero and Grossmann (2006) and Giridar and Agrawal (2010) demonstrated that, for zeotropic mixtures with a single feed, the search space comprises 'regular configurations': configurations consisting of exactly N-1 columns (N being the number of components to be separated). This set of regular configurations can be systematically generated from structurally different sequences of separation tasks, or the equivalent Basic configurations, as per Agrawal's definition. While starting from a given configuration may yield several intensified alternatives, there is currently no definitive methodology, except for generating Divided Wall Columns (DWCs) and possibly some other intensification alternatives, that allows for the complete generation of the intensified configurations' search space. An excellent review on advances in distillation intensification can be found in Jiang and Agrawal (2019).

In this paper, we generate and evaluate, using the chemical process simulator Aspen-HYSYS, the complete search space for the separation of a three-component mixture. We simultaneously consider thermally coupled, heat integration, single compressor Vapor Recompression (VRC), Bottom Flashing, Heat-Integrated Columns with Internal Heat-Coupled Distillation (HICiD) following the suggestion of Harwardt and Marquardt (2012), and DWC. This approach allows us to identify relatively simple new arrangements that, to the best of our knowledge, have not been previously published. We can select a set of configurations that can be further improved using arrangements and ideas mentioned earlier (e.g., multi-staged compression, intermediate boilers/condensers integrated with VRC, etc.). The remainder of the paper is structured as follows. First, we describe the motivating example used as a case study. Then, we define the methodology used, demonstrating how to systematically generate all HPAD sequences for separating a three-component mixture, including intensified alternatives, and discuss how this strategy could be extended to mixtures with more components. In this section, we also introduce the total cost analysis and life cycle methodology used. Subsequently, we use the case study, involving a Benzene-Toluene-Xylene (BTX) mixture, to compare resulting arrangements considering full electrification, partial electrification, or no electrification at all, from both an economic and environmental perspective. Finally, we conclude with some remarks.

2. Example

To showcase the effectiveness of the algorithm, we will explore the separation of a Benzene, Toluene, and p-Xylene mixture. Our goal is to achieve a minimum purity level of 0.995 for each individual product. The specific data for this example can be found in Table 1. Data for the case study.

Table 1. Data for the case study.

Components	Composition (mol fraction)		
Benzene	0.3	Feed Flow	200 kmol/h (18430 kg/h)
Toluene	0.4	Pressure	101.3 kPa
P-xylene	0.3		
Cold Utilities	Cost ($/kW·y)	Hot Utilities	Cost ($/kW·y)
water (20-15 ℃)	11.4	LP Steam (~2 bar 120 ℃)	277.5
Thermodynamics Peng Robingson (default Aspen-HYSYS parameters) Cost estimation based on correlations by (Turton et al, 2013) Electricity 0.1 $/kWh interest = 10% in 10 years		HP Steam (~10 bar 180 ℃)	292.18

Starting from the base case (direct sequence), simulations of two additional configurations were performed: indirect sequence, and divided wall distillation column,. All these three initial configurations were further modified using energy integration techniques. To do that, several energy integration strategies or techniques were implemented, including boiler-condenser integration (HI), thermal coupling (TC), vapor recompression (MVR), bottom flashing (MVR (Botton flashing)), and internal heat integration (HDiC). Additionally, these different techniques were also combined and customized to achieve the maximum energy savings and process efficiency. With the combination of all these options, a total of 77 configurations are obtained, which will be analyzed below.

3. Analysis

3.1 Total Cost Analysis

This study focuses on evaluating the economic aspects of different configurations, considering Annual Operating Costs (including labour, maintenance, energy, and supplies), Capital Costs (initial investments in equipment, infrastructure, licenses, etc.), and the Total Annualized Cost (TAC), which combines both operating and capital costs over a specific period. The TAC is calculated as the sum of capital and operational costs, with equations provided for each. The capital investment cost considers factors like equipment unit cost, adjusted to the relevant year based on the CEPCI index. The operational cost is a product of an annualization factor and electricity consumption. The annualization factor for capital investment is computed using an equation involving the annual interest rate and amortization period. Detailed formulas and correction factors are provided for estimating equipment unit cost. The study emphasizes the importance of this economic analysis for making informed decisions about different design alternatives. A figure in the paper breaks down the total costs, capital costs, and operating costs for each configuration, providing crucial data for a comprehensive economic assessment and comparative analysis.

3.2 Life Cycle Assessment Analysis

The Life Cycle Assessment (LCA) methodology, widely used for environmental impact evaluation, involves several key phases: Goal and Scope Definition, life Cycle Inventory (LCI), Life Cycle Impact, Assessment (LCIA) and Interpretation.
In this study, the ReCiPe 2008 method within the Ecoinvent Database was chosen for LCIA, assessing eighteen impact subcategories transformed into three endpoint categories: ecosystem quality, human health, and resource depletion. This structured LCA methodology provides comprehensive insights into the environmental impacts of analysed systems. The 18 midpoint categories evaluate impacts on various environmental aspects, including resource depletion, climate change, acidification, eutrophication, toxicity, ozone layer depletion, water warming, land use, and water consumption. The outcomes are measured through three endpoint categories: loss of biodiversity, loss of habitat quality, and impact on human health.
In conclusion, the employed LCA methodology, exemplified by the ReCiPe 2008 method, offers a robust framework for assessing and understanding the environmental

impacts of complex systems, guiding the formulation of targeted objectives for sustainability enhancement.

4. Results

4.1 Cost Analysis Results

The graph now displays only the configurations that are within the 50% range of the total annual cost compared to the best obtained configuration.

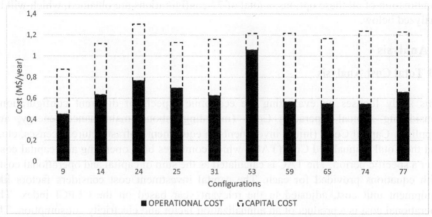

Figure 1. Analysis of Optimal Configurations: Economic Evaluation of Studied Setups.

Upon examining the figure 1, clear patterns and significant differences between various configurations can be identified. Some configurations stand out for having low operating costs, making them attractive options for long-term operations. On the other hand, certain configurations may require higher initial investments (capital costs) but could result in lower operating costs over time. In the base case, representing the direct separation sequence, the annual utility cost is 2.28 million dollars, with a manufacturing cost (COM) of 898,720 dollars per year, resulting in a Total Annualized Cost (TAC) of 2.370 million dollars per year.

Simulation results indicate that the analysed energy integration methods effectively reduce utility costs, enhancing the overall cost-effectiveness of distillation processes. After analysing energy, environmental impacts, and TAC, configuration 9 emerges as optimal, employing a direct sequence with bottom flashing in reboiler BC and condenser B, along with bottom flashing in reboiler C and condenser A. This configuration significantly reduces energy consumption and operational costs by substituting steam with electricity, resulting in a 63% reduction in TAC. The next favourable option is configuration 14, a direct sequence with vapor recompression in reboiler BC and condenser B, along with bottom flashing in reboiler C. It achieves a 53.7% decrease in TAC compared to the base case. Configuration 25, a direct sequence with thermal coupling and vapor recompression in condenser B and reboiler C, is also promising, with a 52.38% reduction in TAC. Another noteworthy alternative is configuration 31, incorporating a direct sequence with internal heat integration in the BC enriching section and vapor recompression in reboiler A and condenser BC. This results in a 51% reduction in TAC. Internal energy integration involves dividing the distillation column into enrichment and stripping sections, enhancing overall energy efficiency through internal heat transfer between process streams. In any case, by adopting these more efficient configurations, industries can gain a competitive advantage while advancing their commitment to sustainable practices. It is evident that investing in energy-efficient solutions is not only an economic necessity but also crucial for building a greener and more resilient industry for the future of our planet.

4.2 LCA Results

The Figure 2 represents the damage impact categories within the ReCiPe 2008 methodology at the endpoint level for all the studied configurations. Each impact category is labelled on the x-axis, while the magnitude of the impact is shown on the y-axis.
The interesting aspect of this graph is that it shows how different configurations or scenarios can have different environmental impacts. To provide a more detailed analysis, we present a refined version of the graph, exclusively showcasing the most cost-effective configurations.

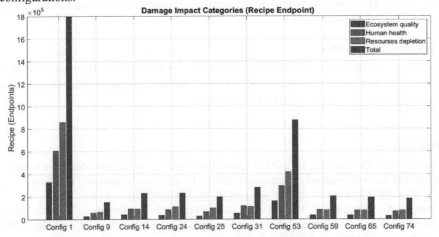

Figure 2. Measurement of Ultimate Environmental Impact Through Three Endpoint Categories in the Life Cycle Assessment (LCA) Process.
In the Figure 2 the first bar for each configuration corresponds to 'Ecosystem Quality,' the second to 'Human Health,' the third to 'Resources Depletion,' and the last bar represents the 'Total' assessment.
The analysis yields a crucial insight: while optimizing costs is essential, an equal emphasis must be placed on considering environmental impact when evaluating configurations. Fortunately, many cost-efficient configurations also demonstrate low environmental impact. This discovery underscores a promising synergy between profitability and sustainability, emphasizing the importance of prioritizing solutions that offer both financial and ecological benefits in industrial and production environments.

5. Conclusions

The study focused on enhancing distillation column efficiency for increased electrification and decarbonization, incorporating simultaneous heat integration alternatives. While individually not groundbreaking, these strategies significantly improved energy efficiency, reducing external energy needs, costs, fossil fuel dependence, and environmental impacts. Pressure swing distillation for benzene, toluene, and p-xylene separation proved effective, with Configuration 9 emerging as optimal. This configuration transitioned from steam to electricity, resulting in a 63% decrease in Total Annualized Cost (TAC). Other options with vapor recompression and thermal coupling demonstrated substantial cost reductions (53.7% to 52.38%) and environmental benefits. The positive correlation between lower energy requirements, costs, and environmental impacts suggests potential win-win scenarios. Life Cycle Assessment identified hotspots, emphasizing the need for eco-friendly solutions. Achieving emission reduction goals requires complementary approaches like hydrogen technologies and CCUS. Energy integration played a crucial role, with Configuration 9 standing out for sustained process improvement. The findings underscore the urgency of addressing challenges in alignment with the Paris Agreement goals.

6. Acknowledgements

The authors acknowledge financial support to the Spanish «Ministerio de Ciencia e Innovación», Spain under project PID2021-124139NB-C21 and to the «Generalitat Valenciana» under project PROMETEO/2020/064.

7. References

Caballero, J.A., Grossmann, I.E., 2006. Structural Considerations and Modeling in the Synthesis of Heat-Integrated−Thermally Coupled Distillation Sequences. Ind. Eng. Chem. Res. 45, 8454–8474. https://doi.org/10.1021/ie060030w

Harwardt, A., Marquardt, W. (2012). Heat-integrated distillation columns: Vapor recompression or internal heat integration? AIChE J. 58, 3740–3750. https://doi.org/10.1002/aic.13775.

Jiang, Z., & Agrawal, R. (2019). Process intensification in multicomponent distillation: A review of recent advancements. Chem Eng Res and Des. 147, 122-145. https://doi.org/10.1016/j.cherd.2019.04.023

Kazemi, A., Mehrabani-Zeinabad, A., Beheshti, M. (2018). Evaluation of various heat pump assisted direct, indirect, Petlyuk and side stripper sequences for three-product separations. Chem. Eng. Sci. 181, 19–35. https://doi.org/10.1016/j.ces.2018.02.007.

Kiss, A.A., Smith, R. (2020). Rethinking energy use in distillation processes for a more sustainable chemical industry. Energy. 203, 117788. https://doi.org/10.1016/j.energy.2020.117788.

Miao, G., Ma, Y., Yang, C., Tong, B., Li, G., Xiao, J. (2023). The effective synthesis of heat-pump assisted distillation process with multiple columns for light hydrocarbon separation.Chem. Eng. Sci. 269, 118449. https://doi.org/10.1016/j.ces.2023.118449

Shah, V.H., Agrawal, R., 2010. A matrix method for multicomponent distillation sequences. AIChE J. 56, 1759–1775. https://doi.org/10.1002/aic.12118

Yuan, H., Luo, Y., Yuan, X., 2022. Synthesis of heat-integrated distillation sequences with mechanical vapor recompression by stochastic optimization. Comput. Chem. Eng. 165, 107922. https://doi.org/10.1016/j.compchemeng.2022.107922

Flavio Manenti, Gintaras V. Reklaitis (Eds.), Proceedings of the 34th European Symposium on Computer Aided Process Engineering / 15th International Symposium on Process Systems Engineering (ESCAPE34/PSE24), June 2-6, 2024, Florence, Italy

Modeling Fed-batch Cultures of Yeast for the Production of Heterologous Proteins - an Industrial Experimental Study

Micaela Benavides[a,b], Pascal Gerkens[b], Gaël de Lannoy[b], Laurent Dewasme[a] and Alain Vande Wouwer[a,*]

aSystems, Estimation, Control and Optimization (SECO), University of Mons, 7000 Mons, Belgium bGlaxoSmithKline, 1330 Rixensart, Belgium
*alain.vandewouwer@umons.ac.be

Abstract

This work reports on the development of a dynamic model of yeast cultures based on a few industrial vaccine production data sets, hypothesizing on the structure of the kinetics, and testing various parametrizations. The proposed model describes the catabolism of a genetically modified strain of the yeast *Saccharomyces cerevisiae*. The main metabolic mechanisms are translated into simple multiplicative activation and inhibition kinetic factors, avoiding the use of nonlinear switching functions, and involving only a few measurable concentrations, i.e., biomass, glucose, and ethanol. A parameter identification study is carried out, based on the minimization of a non- linear least squares criterion, and taking measurement noise into account. A sensitivity analysis is performed, where the resulting Fisher Information Matrix is used to characterize the precision of the parameter estimates. A model extension, including dissolved oxygen, is also proposed as a promising alternative.

Keywords: Mathematical Modeling, Parameter identification, Estimation, Biotechnology, *Saccharomyces cerevisiae*.

1. Introduction

Antigen production is an essential step for vaccine development, usually achieved with the help of a vector such as a genetically modified yeast strain. The latter is cultivated in scaled-up bioreactors, where the prediction of critical compound concentration profiles can be achieved by dynamic process models. Following advanced genetic studies, the yeast *S. cerevisiae* has been successfully modified for the production of recombinant vaccines (Silva et al., 2022).

During culture scale-up, the bioreactor environmental conditions and the culture medium must meet some quality attributes, such as a high yield of recombinant protein expression and, in turn, vaccine efficiency (Vieira Gomes et al., 2018). One way to meet these conditions consists in controlling the input flow rate, based on a model-based feedback loop using the available measurements, i.e. biomass, substrate, ethanol, ammonium, and dissolved oxygen concentrations. However, all these state variables should be measurable online and, therefore, require expensive equipment. Software sensors are an interesting alternative, providing unmeasurable state estimates using the available mathematical models and measurement device configuration (Bogaerts and Vande Wouwer, 2003), (Dewasme et al., 2009).

The metabolism of *S. cerevisiae* is ruled by its respiratory capacity (Sonnleitner and Käppeli, 1986). When operated at low glucose levels, the yeast culture oxidizes all the available glucose into biomass and carbon dioxide (CO2). This metabolic regime is called respirative and only a part of the respiratory capacity is used. When the respiratory capacity becomes insufficient to oxidize all the available glucose, and acts as a bottleneck, the yeast metabolism switches to the fermentation pathway, where the glucose excess fermentates into ethanol. This regime is also known as the Crabtree effect or overflow metabolism (Deken, 1966). It should be noticed that, in presence of ethanol, a switch back to the respirative regime leads to ethanol consumption, thanks to the remaining part of the respiratory capacity (Wills, 1990). Fed-batch applications often aim to avoid this "short-term Crabtree effect" by using exponential feeding to maintain a low glucose concentration during yeast cultivation (Patel and Thibault, 2009; Dewasme et al., 2011). Several other models have been suggested to expand the unstructured model of Sonnleitner and Käppeli (1986). For instance, Richelle et al. (2014) have considered the consumption of nitrogen as an evident limiting factor of biomass growth and a responsible factor for the production of trehalose. However, most of these models consider short-term cultures, which usually do not allow triggering several metabolic switches, for instance leading to a second ethanol production phase as depicted in (Grosfils, 2007).

The main motivation of this work is to present a mechanistic model of yeast culture experiments operated in fed-batch mode including several metabolic switches. In the next section, the experimental conditions are briefly described. Section 3 develops the model based on macroscopic mass balance equations and the selection of kinetic structures combining different modulation factors. Section 4 compares the most promising models and conclusions are drawn in section 5.

2. Materials and Methods

In this study, a recombinant yeast strain of *S. cerevisiae* is cultivated in fed-batch mode. A total of three cultures have been conducted with an initial volume of 5.5 L, and a stirrer speed set to 260 rpm. The temperature was maintained at 30°C throughout all the experimental sessions, while the pH was regulated at 5, using a base solution. The bioreactor was equipped with an in-line pO2 sensor delivering dissolved oxygen in percents, a stirrer motor controlled to maintain this pO2 above 60%, and a peristaltic pump controlling the feed flow rate. The culture duration was set to 95 hours, and offline measurements of optical density, glucose, ethanol, and ammonium were taken every 2 hours. The optical density provides the concentration of biomass based on a dry weight calibration. The cultures were conducted in GSK laboratories, and any other detail remains confidential.

3. Model development

Two candidate models are proposed. The first model (Model 1), considers three reactions with variables including biomass (X), glucose (S), and ethanol (E). This model counts 13 parameters to be estimated. Model 2 simplifies the process description by using only two reactions and incorporates oxygen information.

3.1. Model 1

Model 1 focuses on the macroscopic description of both respirative and fermentative regimes, as shown in Table 1. The first reaction considers glucose oxidation, the second reaction describes the respiro-fermentative pathway while the third reaction considers the possible ethanol consumption. The first and third reactions occur in the respirative regime while the first and second ones correspond to the fermentative regime.

The first reaction rate r_1 considers Monod kinetics to describe glucose uptake, limited by the available respiratory capacity either inhibited by the biomass density (second factor) or the presence of ethanol (third factor). The second reaction rate r_2 is also ruled by a Monod factor related to glucose uptake and limited by the respiratory capacity depending on the biomass density. The third reaction rate r_3 is driven by the presence of ethanol using Monod kinetics, a respiratory capacity limitation factor comparable to reactions 1 and 2, and an inhibition factor by substrate, explaining the preferential selection, as main substrate, of glucose over ethanol.

Mass balance is applied to each macro-reaction, yielding a differential equation system eqs. (4) to (6). $D = F_{in}/V$ is the dilution rate where F_{in} is the inlet feed flow rate and V the bioreactor volume.

Table 1: Model 1

Stoichiometric equations	Reactions and differential equations	
$S \xrightarrow{r_1 X} k_1 X$	$r_1 = \mu_{m1} \cdot \frac{S}{K_{S1}+S} \cdot \frac{1}{1+\frac{X}{K_{IX}}} \cdot \frac{1}{1+\frac{E}{K_{IE}}}$	(1)
$S \xrightarrow{r_2 X} k_2 E + k_4 X$	$r_2 = \mu_{m2} \cdot \frac{S}{K_{S2}+S} \cdot \frac{1}{1+\frac{X}{K_{IX}}}$	(2)
$E \xrightarrow{r_3 X} k_3 X$	$r_3 = \mu_{m3} \cdot \frac{E}{K_E+E} \cdot \frac{1}{1+\frac{X}{K_{IX}}} \cdot \frac{1}{1+\frac{S}{K_{IS}}}$	(3)
	$\frac{dX}{dt} = (k_1 r_1 + k_3 r_3 + k_4 r_2) \cdot X - D \cdot X$	(4)
	$\frac{dS}{dt} = -(r_1 + r_2) \cdot X - D \cdot S + D \cdot S_{in}$	(5)
	$\frac{dE}{dt} = (k_2 r_2 - r_3) \cdot X - D \cdot E$	(6)
	$\frac{dV}{dt} = D \cdot V = F_{in}$	(7)

X:Biomass concentration (optical density)[g/L]; S:Substrate concentration [g/L]; E:Ethanol concentration [g/L]; S_{in}:Inlet feed substrate concentration [g/L]; F_{in}:Feed flow rate[L/h]; V:Culture volume[L]; k_i:Yield coefficients [g/g]; K_{S1}, K_{S2}:Substrate half-saturation constants[g/L]; K_E:Ethanol half-saturation constant[g/L]; K_{IX}, K_{IE} and K_{IS}: Biomass, ethanol and substrate inhibition constants [g/L]; μ_{mi}: Maximum specific rate constants [g/gX/h].

3.2. Model 2

Model 2 merges the first two reactions of Model 1 into one reaction, assuming that a clear separation of glucose oxidation and fermentation mechanisms is difficult to observe. The second reaction of Model 2 still describes ethanol oxidation but also considers the additional dissolved oxygen variable O, assumed to drive ethanol consumption. Applying mass balance to all the involved macroscopic species leads to the new differential equations eqs. (10) to (14).

Table 2: Model 2

Stoichiometric equations	Reactions and differential equations	
$S \xrightarrow{r_1 X} k_1 X + k_2 E$	$r_1 = \mu_{m1} \cdot \frac{S}{K_{S1}+S} \cdot \frac{1}{1+\frac{X}{K_{IX}}} \cdot \frac{1}{1+\frac{E}{K_{IE}}}$	(8)
$k_4 E + O \xrightarrow{r_2 X} k_3 X$	$r_2 = \mu_{m2} \cdot \frac{O}{K_O+O} \cdot \frac{E}{K_E+E} \cdot \frac{1}{1+\frac{X}{K_{IX}}} \cdot \frac{1}{1+\frac{S}{K_{IS}}}$	(9)
	$\frac{dX}{dt} = (k_1 r_1 + k_3 r_2) \cdot X - D \cdot X$	(10)
	$\frac{dS}{dt} = -r_1 \cdot X - D \cdot S + D \cdot S_{in}$	(11)
	$\frac{dE}{dt} = (k_2 r_1 - k_4 r_2) \cdot X - D \cdot E$	(12)
	$\frac{dO}{dt} = -r_2 \cdot X - D \cdot O + OTR$	(13)
	$\frac{dV}{dt} = D \cdot V = F_{in}$	(14)

O : concentration of dissolved oxygen [g/L]; OTR: Oxygen transfer rate [g/Lh]; K_o: Oxygen half-saturation constant[g/L].

OTR is the oxygen transfer rate from the gas to the liquid phase (eq. 13) and it is a variable that is challenging to measure, requiring a gas analyzer. An interesting measurement approach is proposed in (Rocha, 2003) where both oxygen and nitrogen gas fractions and flow rates are measured at process input and output. An alternative way to approximate OTR is also to use the volumetric transfer coefficient of oxygen, $k_L a$, as follows (Papapostolou et al., 2019): $OTR = k_L a(O_{sat} - O)$, where the difference between the dissolved oxygen concentration at saturation (O_{sat}) and the actual concentration (O) in the liquid phase is assumed to be proportional to the OTR in a specific bioreactor environment modeled by $k_L a$ and assumed constant. However, even approximating the constant $k_L a$ value is not trivial and the latter depends on several factors, such as the agitation speed, the viscosity of the culture medium, and the stirrer/tank geometry.

To overcome these difficulties, and since the dissolved oxygen tension pO2 measurements are available, the OTR is replaced by the derivative of pO2, assuming that this signal tracks the OTR trajectory. Consequently, Equation (13) is modified as follows:

$$\frac{dm}{dt} = -r_2 \cdot X - D \cdot m + \frac{dpO2}{dt} \tag{15}$$

where m represents a metabolic variable that is related to the concentration of dissolved oxygen. Since the concentration of pO2 is given in percentage, the value of m is also expressed in percentage. Additionally, O is replaced by m in Equation (9), and the parameter to be identified, K_O, is also expressed in percentage.

4. Identification of the proposed models

Two experimental data sets were considered for parameter identification, while a third data set was used for model cross-validation. A weighted least-squares criterion representing the weighted distances between the data and the model predictions is minimized and reads as follows:

$$J(\theta) = \sum_{i=1}^{N} \left[\left(y_i(\theta) - y_{i,meas} \right)^T \cdot W_i^{-1} \cdot \left(y_i(\theta) - y_{i,meas} \right) \right] \tag{16}$$

where $y_{i,meas}$ is the measurement vector, y_i is the model prediction vector, θ is the parameter vector, N is the number of measurement samples, and W_i^{-1} is the weighting matrix. The minimization of J is considered in the following nonlinear programming problem:

$$\hat{\theta} = \arg \min_{\theta} J(\theta) \tag{17}$$

Among the several available optimizers present in the MATLAB libraries, the Nelder-Mead function, achieving the optimization by application of the simplex algorithm, and used in the routine 'fminsearch', is selected. To assess the accuracy of the parameter estimates, the Fisher Information Matrix (FIM) (Walter and Pronzato, 1997) is built:

$$F = \sum_{i=1}^{N} \left(\frac{\partial y}{\partial \theta}(t_i) \right)^T \Sigma_i^{-1} \left(\frac{\partial y}{\partial \theta}(t_i) \right) \tag{18}$$

The FIM contains information about the measurement noise (via the inverse of the measurement error covariance matrix Σ_i^{-1}), and the output sensitivity function $\partial y / \partial \theta$ at each point i in time. The inverse of the FIM provides an optimistic estimate (i.e., a lower bound) of the parameter estimation error covariance matrix $\widehat{Cov}$. In practice, the parameter estimation error standard deviations can be inferred from the square roots of the diagonal elements of $\widehat{Cov}$, as follows:

$$\hat{\sigma}_i(\hat{\theta}_i) = \sqrt{F_{ii}^{-1}} \tag{19}$$

The coefficients of variations (CV) are therefore calculated as $CV_i = \frac{\hat{\sigma}_i}{\hat{\theta}_i}$. The parameter values along with their respective CV for the two models previously described are presented in Figure 1. Model 1 has the best curve fitting as highlighted by the root-mean-square error (RMSE) values in Figure 1, which can be explained by the higher number of parameters providing more degrees of freedom to fit the data.

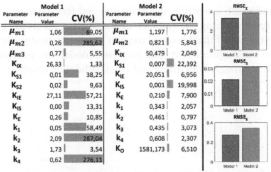

Parameter Name	Model 1 Parameter Value	CV(%)	Parameter Name	Model 2 Parameter Value	CV(%)
μ_{m1}	1,06	69,05	μ_{m1}	1,197	1,776
μ_{m2}	0,26	285,62	μ_{m2}	0,821	5,843
μ_{m3}	0,77	5,55	K_{IX}	50,479	2,049
K_{IX}	26,33	1,33	K_{S1}	0,007	22,392
K_{S1}	0,01	38,25	K_{IE}	20,051	6,956
K_{S2}	0,02	9,63	K_{IS}	0,001	19,998
K_{IE}	27,11	57,21	K_E	0,210	7,900
K_{IS}	0,00	13,31	k_1	0,343	2,057
K_E	0,26	10,85	k_2	0,461	0,797
k_1	0,05	58,49	k_3	0,435	3,073
k_2	2,09	287,04	k_4	0,608	2,307
k_3	1,73	3,54	K_O	1581,173	6,510
k_4	0,62	276,11			

Figure 1: Parameter estimate values with their respective CVs and RMSEs of the biomass (X), glucose (S), and ethanol (E) predictions of the 2 models.

However, despite higher RMSEs, it can also be observed that the maximum CV values of Model 2 are lower than those of Model 1 which presents high CVs for all parameters involved in the second reaction. This suggests that the simplification done in Model 1, where reactions 1 and 2 are merged, is consistent with the inclusion of dissolved oxygen. The curve fitting of both models is shown in Figures 2 and 3 in direct validation using one of the two available data sets. For the sake of confidentiality, all values are normalized. It is worth mentioning that the quality of the model fitting on the other data set, used in direct validation, is similar. Based on these results, both models seem to properly reflect the process behavior. The results indicate that despite a higher RMSE value, Model 2 presents the advantage of better reproducing the data trends, more specifically when ethanol enters its second production phase at the end of the culture. Additionally, the cross-validation of Model 1 and Model 2 using the third experimental data set is presented in Figure 4, suggesting the good prediction capacity of both models. The same conclusions from the direct validations apply to the cross-validation since it can be observed that Model 2 provides a better prediction of the end of the experiment while Model 1 still presents a slightly better precision (the corresponding RMSE value is ±6% less). It is worth noting that the ethanol measurement device has a limited sensitivity assimilating too small non-zero values to a small constant. All initial conditions are however identified during the cross-validation, allowing the model to start below the device sensitivity threshold.

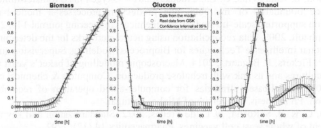

Figure 2: Estimation of biomass, glucose and ethanol using the proposed Model 1.

5. Conclusions

In this study, a mathematical model has been proposed, that predicts the time evolution of biomass, glucose, and ethanol concentrations in cultures of the yeast *S. cerevisiae*. The model takes into account the reactions of glucose oxidation, glucose fermentation, and ethanol oxidation. Additionally, the model is based on the structuration of the kinetic laws, which are based on the product of activation and inhibition factors. A second model has been developed which simplifies the oxidation and fermentation of the glucose into one reaction, thereby reducing the number of parameters involved. In this model, the inclusion of the dynamics of the dissolved oxygen has shown to significantly improve the model formulation and its prediction. The proposed models are confirmed to have good prediction capability through both direct and cross-validation results.

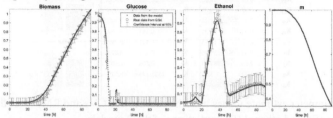

Figure 3: Estimation of biomass, glucose, ethanol and O^* using the proposed Model 2.

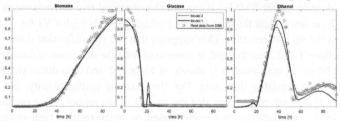

Figure 4: Cross-validation of the proposed Model 1 and 2, with normalized RMSE values of 0.8725 and 0.9304, respectively.

References

P. Bogaerts, A. Vande Wouwer, 2003. Software sensors for bioprocesses. ISA Transactions 42.

R. H. D. Deken, 1966. The crabtree effect: A regulatory system in yeast. J . gen. Microbiol. 44.

L. Dewasme, P. Bogaerts, A. Vande Wouwer, 2009. Monitoring of bioprocesses: mechanistic and data-driven approaches. Studies in Computational Intelligence, (Computational Intelligent Techniques for Bioprocess Modelling, Supervision and Control, Maria do Carmo Nicoletti, Lakhmi C. Jain, eds.). Springer Verlag, pp. 57–97.

L. Dewasme, B. Srinivasan, M. Perrier, A. V. Wouwer, 2011. Extremum-seeking algorithm design for fed-batch cultures of microorganisms with overflow metabolism. Journal of Process Control.

A. Grosfils, 2007. First principles and black box modelling of biological systems. Ph.D. thesis, Université libre de Bruxelles.

A. Papapostolou, E. Karasavvas, C. Chatzidoukas, 2019. Oxygen mass transfer limitations set the performance boundaries of microbial pha production processes–a model-based problem investigation supporting scale-up studies. Biochemical engineering journal 148, 224–238.

N. Patel, J. Thibault, 2009. Data reconciliation using neural networks for the determination of kLa. Computational Intelligence Techniques for Bioprocess Modelling, Supervision and Control.

A. Richelle, P. Fickers, P. Bogaerts, 2014. Macroscopic modelling of baker's yeast production in fed-batch cultures and its link with trehalose production. Computers & chemical engineering 6.

I. Rocha, 2003. Model-based strategies for computer-aided operation of recombinant e. coli fermentation. Ph.D. thesis, University of Minho, Portugal.

A. J. D. Silva, C. K. d. S. Rocha, A. C. de Freitas, 2022. Standardization and key aspects of the development of whole yeast cell vaccines. Pharmaceutics 14 (12), 2792.

B. Sonnleitner, O. Käppeli, 1986. Growth of saccharomyces cerevisiae is controlled by its limited respiratory capacity: formulation and verification of a hypothesis. Biotechnology and bioengineering 28 (6), 927–937.

A. M. Vieira Gomes, T. Souza Carmo, L. Silva Carvalho, F. Mendonça Bahia, N. S. Parachin, 2018. Comparison of yeasts as hosts for recombinant protein production. Microorganisms 6 (2), 38.

E. Walter, L. Pronzato, 1997. Identification of parametric models. Springer Verlag New-York.

C. Wills, 1990. Regulation of sugar and ethanol metabolism in saccharomyces cerevisiae. Critical reviews in biochemistry and molecular biology 25 (4), 245–280.

Flavio Manenti, Gintaras V. Reklaitis (Eds.), Proceedings of the 34th European Symposium on Computer Aided Process Engineering / 15th International Symposium on Process Systems Engineering (ESCAPE34/PSE24), June 2-6, 2024, Florence, Italy

Performance of Piecewise Linear Models in MILP Unit Commitment: Difference of Convex vs. J1 Approximation

Felix Birkelbach[a,*], Victor São Paulo Ruela[a,b], René Hofmann[a]

[a]TU Wien, Institute of Energy Systems and Thermodynamics, Vienna, Austria
[b] Tata Steel Ijmuiden, Ceramics Research Centre, Velsen-Noord, The Netherlands
felix.birkelbach@tuwien.ac.at

Abstract

For unit commitment problems, mixed integer linear programming (MILP) is the most widely used solution approach, mainly due to the availability of fast reliable solvers that can handle large problem instances. One challenge with this approach is to incorporate complex non-linear operating behavior in the model. Typically, this is done by employing piecewise-linear models to approximate the non-linear functions and then use specialized MILP formulations to incorporate them in the MILP model.

The most efficient way to incorporate piecewise linear functions with more than two variables in MILP is the logarithmic formulation by Vielma et al. It has been shown to outperform other formulations by a large margin. Even so, it is not routinely used in literature – probably because it puts stringent requirements on the approximation. Specifically, the triangulation has to be compatible with the J1 triangulation.

Recently an algorithm for finding piecewise-linear approximations has been proposed: The difference of convex (DoC) method. It has been shown to achieve high accuracies with only a few linear pieces, thus promising fast performance in MILP. However, DoC approximations are not compatible with the fast Log-formulation.

In this paper, we compare the MILP performance of DoC approximation with J1 approximation. As a use case, we consider a unit commitment problem of the energy system with a thermal storage unit. The two approximation methods are used to model non-linear operating behavior of a thermal storage unit. The solving times for different problem sizes and approximation accuracies are compared and guidelines for achieving the best performance for unit commitment are proposed.

Keywords: energy system, unit commitment, MILP, piecewise linear approximation.

1. Introduction

Increasingly strict emission targets put a lot of pressure on industry to operate their energy supply systems more efficiently. To make sure that energy demand is covered at all times, the operation of energy system has to be planned accurately. This planning task is usually formulated as a unit commitment (UC) problem. One of the most popular methods to formulate and solve UC problems is mixed integer linear programming (MILP). The huge advances in MILP algorithms in recent years allows it to solve UC problems very quickly (Koch et al., 2022). MILP has the advantage (compared to evolutionary algorithms and heuristic approaches) that it reliably finds the optimal operation trajectory because, as a deterministic optimization method, it provides global optimality guarantees. UC can be used to organize energy supply in a cost-optimal manner and to ensure that energy demand is always met.

To compute reliable operation trajectories, the technical limitations of the units in the energy systems have to be represented accurately. Since the functions that describe the operating behavior are typically non-linear, implementing them in MILP is not straightforward. The operating behavior has to be approximated with a piecewise linear function. Unfortunately, such piecewise linear functions can impact the solving time of the UC problem severely. The reason is that in order to represent a piecewise linear function in MILP, auxiliary variables have to be introduced to distinguish the pieces. Various MILP formulations for piecewise linear functions, which differ in the number of auxiliary variables and constraints, have been proposed (Vielma et al., 2010) and their performance was assessed (Brito et al., 2020; Silva and Camponogara, 2014). Even though all three cited studies found that logarithmic coding outperforms the other formulations by a large margin, it is not routinely used – probably because it puts stringent requirements on the approximation. Specifically, that the triangulation has to be compatible with the J1 triangulation.

Finding an accurate piecewise linear approximation with the fewest number of linear pieces is a challenge in itself, on par with solving the UC problem. The challenge is to divide the domain of the function into pieces and to assign a linear function to each of them so that the difference between the approximation and the function is minimal. Basic linear elements are referred to as simplices (triangles in 2 dimensions). The naïve approach to compute a piecewise linear approximation is to cover the domain of the function with a grid, which is in turn used to construct a triangulation. Unfortunately, this approach generally requires a lot of linear pieces to reach the desired accuracy and thus the MILP performance will be poor. For that reason, specialized algorithms have been proposed. One approach revolves around recursively subdividing triangles into smaller triangles until the target accuracy is reached (Rebennack and Kallrath, 2015), another uses iterative mesh refinement (Obermeier et al., 2021). Even though both approaches can find good piecewise linear approximations, neither produces approximations that are compatible with logarithmic coding. Recently, we compared the J1 approach with the one by Rebennack & Kallrath in a UC problem and found that, even if approximation with J1 needs more linear pieces, the solving time of the UC problem is only a fraction thanks to compatibility with logarithmic coding (Birkelbach et al., 2023). For that reason, we developed a method to compute J1 compatible approximations and thus allows for logarithmic coding (Birkelbach, 2024).

Recently, another interesting approach to achieve faster solving times has been proposed: the difference of convex (DoC) approach (Kazda and Li, 2021). They use polyhedra (polygons) instead of simplices (triangles) to significantly reduce the number of linear pieces to reach the target accuracy. They demonstrated their approach by comparing it with the one by Rebennack & Kallrath. Because of the impressive results, also faster solving times can be expected compared to general simplex models. The key question now is: Does approximation with polyhedra or a J1 triangulation yield the fastest solving time in a UC problem?

In this paper, we compare the DoC approach with the J1 approach by applying them to a use case from one of our projects. We briefly introduce the relevant background on MILP formulations and the approximation methods. Then we discuss the effect that the various combinations of approximation and MILP formulation have on the UC problem. Finally, we assess the solving time of the UC problem and propose some recommendations to achieve the fastest solving time with piece-wise linear models.

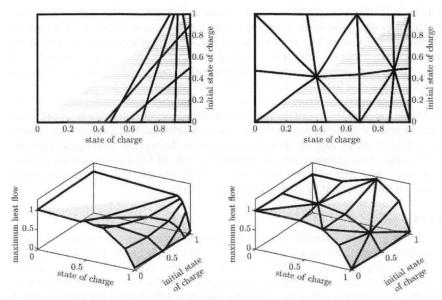

Figure 1 Illustration of the piecewise linear approximation of the function that describes the maximum charging rate of the heat storage with a target RMSE of 0.02 (axes are normalized). Difference of convex (left), J1 triangulation (right). Top view (top), isometric view (bottom).

2. Methods

In this section, we will briefly introduce the methods that were used in this study. Where a detailed discussion is not possible due to space restrictions, we will provide references.

2.1. MILP formulations for piecewise linear functions

In their seminal paper "Mixed-Integer Models for Nonseparable Piecewise-Linear Optimization" (Vielma et al., 2010) discuss and analyze six different formulations for representing piecewise linear functions in MILP. For the present study, three of these formulations are relevant: Multiple Choice (MC), disaggregated convex combination with logarithmic coding (DLog) and the logarithmic branching convex combination (Log). The three key differences are: 1) the number of auxiliary binary variables grows linearly for MC and logarithmically for DLog and Log. 2) the number of auxiliary continuous variables increases considerably with the number of linear pieces for DLog than for MC and Log. 3) MC and DLog can be used with any type of convex polytope, while Log requires a J1 triangulation.

Performance studies showed that for functions with only a few linear pieces, MC yields the best performance, while DLog and Log have an advantage if the function is comprised of many linear pieces. In general, Log outperformed DLog. (Vielma et al., 2010)

2.2. Piecewise linear approximation

The goal of piecewise linear approximation for MILP is to find a piecewise linear function that fulfills the accuracy requirements, typically in terms of the root mean square error (RMSE) or the maximum absolute error (MAE), and that will yield the best possible solving time for the MILP problem. In general this means that the approximation should require as few linear pieces as possible, but compatibility with efficient MILP formulation does also play a role.

The difference of convex (DoC) method (Kazda and Li, 2021) finds a piecewise linear approximation with an heuristic using the difference of two convex functions to represent any non-convex function. The resulting linear pieces are polyhedra, i.e. polygons in two dimensions (see Figure 1, left). They showed that DoC produces approximations with significantly fewer linear pieces than the method by Kallrath and Rebennack. To represent these approximations in MILP, either the MC or DLog can be used.

To compute piecewise linear approximations that are compatible with the Log formulation, we used the algorithm that we published recently (Birkelbach, 2024). It uses a combination of a gradient based optimizer with a heuristic to find a good approximation. Figure 1 (right) shows such an approximation where the typical "union jack" pattern is clearly visible. This approximation is compatible with all MILP formulations proposed in (Vielma et al., 2010). In this study we consider MC and Log. DLog was omitted because it is outperformed by Log (Vielma et al., 2010).

3. Use Case

For evaluating the performance of each modeling approach and the effect of the MILP formulation, we used the same UC model as (Koller et al., 2019). In this UC model, a very simple energy system (see sketch in Figure 2) consisting of a generating unit (GU) and a packed bed regenerator (PBR) as heat storage unit. The GU produces both heat and electricity. Electricity is sold at the electricity market at a fluctuating but known price. To allow the GU to shut down during times of low electricity prices, the PBR is used to store heat and supply it later. The UC model has a prediction horizon of 8 days with time steps of 1 hour. This results in 192 time steps. More details on the model in (Koller et al., 2019). The typical output of the UC problem is shown in Figure 2. The diagram on the top shows how the heat demand is covered by either a heat flow from the GU or from the PBR. The fluctuating electricity price that makes operation of the GU uneconomic at some times is also shown. In the bottom diagram, the operating trajectory of the PBR is shown. The bars illustrate heat flow to and from the storage. The line shows the state of charge.

To model the behavior of the PBR in the UC model comprehensively, the dependence of the maximum charging and discharging power on the state of charge has to be considered. The maximum power diminishes, when the storage is almost fully charged or discharged, because the power depends on the position of the thermocline inside the packed bed. When the storage operates in partial charging and discharging cycles (which will usually

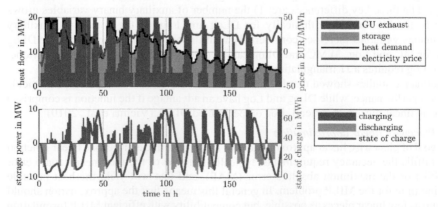

Figure 2 Illustration of the output of the unit commitment problem.

be the case), the state of charge at the time of switching also has to be considered. (Koller et al., 2019)

4. Results

The UC problem and the DoC algorithm were formulated using YALMIP R20210331 (Löfberg, 2004) in Matlab R2023b. The problems were solved using Gurobi 10.0.0 on a 128-core system (AMD EPYC 7702P) with 256 GB RAM. The J1 approximations were computed using Matlab's optimization toolbox.

For our use case, we used a simulation model to generate a data set that gives the maximum charging/discharging rate of the storage for various initial states of charge (colored dots in Figure 1). We then approximated this dataset with both the DoC and the J1 method. As accuracy metric we chose the root mean square error (RMSE) and performed the approximation with three target accuracies 0.08, 0.04 and 0.02. Table 1 shows number of linear pieces that each method requires to approximate both the charging and discharging behavior with the target accuracy. The results show, that the DoC method indeed requires fewer linear pieces than the triangulation approach (see Figure 4, top).

Table 1 also shows the number of auxiliary variables that are required to implement each of the approximations in MILP depending on the MILP formulation. The number of binary variables can be reduced significantly with both logarithmic methods especially for the detailed approximations. The Log method stands out, because it reduces both the number of binary and continuous variables.

In the next step, we used each approximation in the UC problem and measured the time it took to solve the model to an optimality gap of 10^{-2}. The results are shown in the last column in Table 1. For RMSE 0.08, DoC performs significantly better than J1 because it can reach the target accuracy with very few linear pieces. (So few actually that some of the auxiliary variables can be eliminated during pre-solve). Starting at moderate accuracies, the J1 method starts to outperform the DoC method, even though it uses more linear pieces. At an RMSE of 0.02 the UC problem with the J1 approximation is three times faster than the one with DoC.

5. Conclusions

We studied the performance of two methods to compute piecewise linear approximations of non-linear functions for MILP. While the difference of convex (DoC) approach is

Table 1 Performance metrics of the two approximation methods. Best performing method for each accuracy is highlighted bold.

	Target RMSE	Number of linear pieces	MILP formulation	Number of binary variables	Number of continuous variables	Solving time of UC in s
DoC	0.08	4	MC	4	8	**0.7283**
			DLog	2	16	2.5547
	0.04	12	MC	12	24	13.0714
			DLog	6	48	35.2150
	0.02	26	MC	26	52	103.0004
			DLog	8	104	416.1805
J1	0.08	10	MC	10	20	12.6132
			Log	5	14	4.4169
	0.04	20	MC	20	40	131.5463
			Log	7	22	**5.7963**
	0.02	32	MC	32	64	551.9540
			Log	8	30	**36.6738**

capable of finding approximations with fewer linear pieces than the J1 approach, this advantage is in large parts offset by effect of the MILP formulation. While the DoC yielded the best performance with the crude approximation, the J1 approach performed significantly better for the more detailed ones.

The results also show that –at least in the accuracy range that we studied– DoC does not experience speedup with logarithmic coding (DLog); MC was the best choice throughout. The reason is that MC requires much fewer auxiliary continuous variables than DLog. For the J1 approach, on the other hand, the Log formulations always performed better.

To summarize, our results suggest as a general rule of thumb that DoC is the best choice for problem instances where a few linear regions suffice to reach the target accuracy. DoC models should be paired with the MC formulation. In case more than a few linear pieces are required, J1 approximation is the method of choice. Even if it needs a few more linear pieces, this is outweighed by its compatibility with the Log formulation.

References

Birkelbach F. Piecewise linear approximation for MILP using J1 compatible triangulations [in preparation] 2024.

Birkelbach F, Kasper L, Schwarzmayr P, Hofmann R. Operation Planning with Thermal Storage Units Using MILP: Comparison of Heuristics for Approximating Non-Linear Operating Behavior. Proc. ECOS 2023, Las Palmas De Gran Canaria, Spain: 2023, p. 1345–50. https://doi.org/10.52202/069564-0122.

Brito BH, Finardi EC, Takigawa FYK. Mixed-integer nonseparable piecewise linear models for the hydropower production function in the Unit Commitment problem. Electr Power Syst Res 2020;182:106234. https://doi.org/10.1016/j.epsr.2020.106234.

Kazda K, Li X. Nonconvex multivariate piecewise-linear fitting using the difference-of-convex representation. Comput Chem Eng 2021;150:107310. https://doi.org/10.1016/j.compchemeng.2021.107310.

Koch T, Berthold T, Pedersen J, Vanaret C. Progress in mathematical programming solvers from 2001 to 2020. EURO J Comput Optim 2022;10:100031. https://doi.org/10.1016/j.ejco.2022.100031.

Koller M, Hofmann R, Walter H. MILP model for a packed bed sensible thermal energy storage. Comput Chem Eng 2019;125:40–53. https://doi.org/10.1016/j.compchemeng.2019.03.007.

Löfberg J. YALMIP : a toolbox for modeling and optimization in MATLAB. 2004 IEEE Int. Conf. Robot. Autom. IEEE Cat No04CH37508, Taipei, Taiwan: IEEE; 2004, p. 284–9. https://doi.org/10.1109/CACSD.2004.1393890.

Obermeier A, Vollmer N, Windmeier C, Esche E, Repke J-U. Generation of linear-based surrogate models from non-linear functional relationships for use in scheduling formulation. Comput Chem Eng 2021;146:107203. https://doi.org/10.1016/j.compchemeng.2020.107203.

Rebennack S, Kallrath J. Continuous Piecewise Linear Delta-Approximations for Bivariate and Multivariate Functions. J Optim Theory Appl 2015;167:102–17. https://doi.org/10.1007/s10957-014-0688-2.

Silva TL, Camponogara E. A computational analysis of multidimensional piecewise-linear models with applications to oil production optimization. Eur J Oper Res 2014;232:630–42. https://doi.org/10.1016/j.ejor.2013.07.040.

Vielma JP, Ahmed S, Nemhauser G. Mixed-Integer Models for Nonseparable Piecewise-Linear Optimization: Unifying Framework and Extensions. Oper Res 2010;58:303–15. https://doi.org/10.1287/opre.1090.0721.

Flavio Manenti, Gintaras V. Reklaitis (Eds.), Proceedings of the 34th European Symposium on Computer Aided Process Engineering / 15th International Symposium on Process Systems Engineering (ESCAPE34/PSE24), June 2-6, 2024, Florence, Italy

Performance evaluation of several reverse osmosis process configurations for the removal of N-nitrosomethylethylamine (NMEA) from wastewater

Mudhar Al-Obaidi[a,b], Alanood Alsarayreh[c], Iqbal M. Mujtaba[d]

[a]*Technical Institute of Baquba, Middle Technical University, Baquba 32001, Iraq*
[b]*Technical Instructor Training Institute, Middle Technical University, Baghdad 10074, Iraq*
[c]*Chemical Engineering Department, Mu'tah University, Al Karak 61710, Jordan*
[d]*Chemical Engineering Department, University of Bradford, Bradford BD7 1DP, UK.*

I.M.Mujtaba@bradford.ac.uk

Abstract

Reverse Osmosis (RO) process has been extensively used for the elimination of toxic contaminants from wastewater. Development of an appropriate RO process in terms of process configuration (design) and operating conditions is essential for the effective removal of contaminants. For the first time, this study investigates different RO configurations for effective elimination of NMEA from wastewater using model based techniques. These configurations include series, parallel, and tapered RO process. Using a complete process model, simulation of different RO configurations is carried out for a given set of operating conditions and the best RO configuration is selected. The tapered configuration of retentate-permeate reprocessing design of RO process has been found to be more effective in terms of removal of NMEA from the wastewater despite its high energy consumption.

Keywords: Reverse Osmosis, Wastewater treatment, N-nitrosomethylethylamine (NMEA), Modelling, Simulation, Performance Analysis.

1. Introduction

Reverse osmosis (RO) is a water purification technique utilised in wastewater treatment and water desalination. Harmful particles are eliminated by RO using a semipermeable membrane (Osorio et al., 2022). N-nitrosomethylethylamine (NEMA), is a member of the N-nitrosamine family, was regarded as a human carcinogenic chemical based on the US EPA and IARC classifications (IARC, 1987). According to statistics, 250×10^{-6} ppm of NMEA is the cancer risk limit as stipulated by the US EPA, IRIS (USEPA, 1993). As a result, several water authorities worldwide are subjected to regulations on the maximum permissible level of NEMA concentration in reclaimed water and drinking water for human consumption.

Given their high hydrophilic properties and lower molecular weight, the N-nitrosamine family members NDMA (N-nitrosodimethylamine), NMEA, and NPYR (N-nitrosopyrrolidine) are thought to be resistant to total elimination through the RO process (Alaba et al., 2017). There has been a great deal of studies done in the open literature on the removal of the most-stubborn NDMA molecules utilising the RO process (Fujioka et al., 2014; Al-Obaidi et al., 2018a, 2018b; Takeuchi et al., 2018; Szczuka et al., 2020). In this regard, only Fujioka et al. (2014) conducted the related research on the removal of NEMA (molecular weight 88.06 g/mol) from wastewater using the spiral wound RO process. In this context, Fujioka et al. (2014) employed a single stage RO treatment system consisting of three membrane modules of ESPA2-4040 spiral wound elements linked in a series. The results showed that the NEMA rejections are 72%, 82%, and 87.2%, respectively, for 4, 6.51, 10.1 atm of feed pressure. Fujioka et al. (2014) considered only a lab-scale RO process and did not consider different RO configurations, which is the focus of this study. Accordingly, four distinct RO process configurations of six membrane modules are proposed. Using a simulation-based model, the optimal design will be selected while considering a fixed set of the inflow conditions (temperature, flow rate, concentration, and feed pressure). The performance metrics of the selected configuration will be compared against the results of Fujioka et al. (2014) to demonstrate the significance of this study.

2. Modelling of a spiral wound module

Al-Obaidi et al. (2018d) developed a mathematical model to assess the impact of inlet conditions on the performance metrics of RO process for desalination. Compared to the previous models developed by Al-Obaidi et al. (2018a,b,c), Al-Obaidi et al. (2018d) established an inclusive model that characterises the influence of water temperature on the water and solute transport parameters of the membrane. In the current study, the same model was calibrated to systematically envisage the performance of different configurations of RO process towards the removal of NEMA from wastewater. Table 1 presents the comprehensive model that was coded and solved using gPROMS software.

3. Model Validation

The model estimations are compared against those collected data of Fujioka et al. (2014) of three spiral wound modules connected in a series. Table 2 and 3 show the marginal errors between the model estimations and experimental data of the studied performance indicators. This in turn assures the robustness of the model. Thus, it is fair to utilise the model to measure the performance of different configurations of RO process towards the removal of NMEA from wastewater. However, it should be noted that there is a considerable discrepancy between the model prediction and experimental data particularly at low operating pressure (Table 3). This can be attributed to the assumption made during the development of this model. This is the validity of Da Costa et al. (1994) who provided relationships between the spacer characteristics and pressure drop in the feed channel. Accordingly, different values of feed spacer characteristic A* and n (dimensionless) (Eq. 6 of Table 1) were investigated for each feed spacer. Seemingly, these values do not accurately fit the conditions of very low pressure of RO process.

4. Proposed configurations of multistage RO process

Fig. 1(A, B, C, and D) depicts the proposed configurations of RO system under investigation for the elimination of NMEA from wastewater. Six spiral wound modules of RO process are considered to propose four different configurations (configuration A: series, configuration B: parallel, configuration C: tapered 3:2:1 of retentate-retentate reprocessing design (RR), and finally configuration D: tapered 3:2:1 of retentate-permeate reprocessing design (RP)). Note, configuration D (Fig. 1) is characterised by the existence of an energy recovery device (ERD) that is installed between the second and third stages to ensure an effective treatment of the permeate in the third stage. This is utilised to increase the reprocessing permeate pressure (1 atm) to a higher value in order to sustain sufficient driving power within the third stage's membranes and throughout the stages that follow. The design features of the spiral wound membrane element (made by Hydranautics, Oceanside, CA., USA) and the maximum limits of inlet conditions are provided in Table 4. Table 5 presents the water transport parameter and NEMA solute transport parameter of the membrane used.

Table 1. Model of RO system (Al-Obaidi et al., 2018d)

No.	Model equation	Description
1	$Q_p = NDP_{fb}\, A_{w(T)}$	Permeate flow rate (Q_p)
2	$A_{w(T)} = A_{w(25\,°C)}\, TCF_p\, F_f$	Water transport parameter of membrane ($A_{w(T)}$)
3	$TCF_p = \exp[0.0343\,(T-25)] \quad < 25\,°C$ $TCF_p = \exp[0.0307\,(T-25)] \quad > 25\,°C$	Temperature correction factor of permeate
4	$NDP_{fb} = P_{fb} - P_p - \pi_b + \pi_p$	Net driving pressure of feed and brine (NDP_{fb})
5	$P_{fb} = P_f - \frac{\Delta P_{drop,E}}{2}$	Feed-brine pressure (P_{fb})
6	$\Delta P_{drop,E} = \frac{9.8692 \times 10^{-6}\, A^*\rho_b\, U_b^2\, L}{2d_h\, Re^n}$	Pressure drop along membrane length ($\Delta P_{drop,E}$)
7	$Re = \frac{\rho_b\, d_h\, Q_b}{t_f\, W\, \mu_b}$	Reynolds number (Re)
8	$Q_b = \frac{Q_f + Q_r}{2}$	Bulk flow rate in the feed channel (Q_b)
9	$Q_f = Q_r + Q_p$	Mass balance
10	$Q_f\, C_f - Q_r\, C_r = Q_p\, C_p$	Solute balance
11	$\pi_b = 0.7994\, C_b\, [1 + 0.003\,(T-25)]$	Osmotic pressure in the brine (π_b)
12	$\pi_p = 0.7994\, C_p\, [1 + 0.003\,(T-25)]$	Osmotic pressure in the permeate (π_p)
13	$C_b = \frac{C_f + C_r}{2}$	Bulk concentration (C_b)
14	$C_r = C_f\, [1 - Rec]^{-Rej}$	Retentate concentration (C_r)
15	$C_p = \frac{C_f}{Rec}\, [1 - (1 - Rec)]^{(1-Rej)}$	Permeate concentration (C_p)
16	$C_w = C_p + \left(\frac{C_f + C_r}{2} - C_p\right) exp\left(\frac{Q_p/A_m}{k}\right)$	Solute concentration on the membrane wall (C_w)
17	$J_w = \frac{B_s\, Rej}{(1-Rej)}$ and $Q_s = B_{s(T)}\,(C_w - C_p)$	Water flux (J_w) and solute flux (Q_s)
18	$B_{s(T)} = B_{s(25\,°C)}\, TCF_s$	Solute transport parameter of membrane ($B_{s(T)}$)
19	$TCF_s = 1 + 0.05\,(T-25) \quad < 25\,°C$ $TCF_s = 1 + 0.08\,(T-25) \quad > 25\,°C$	Temperature correction factor of solute TCF_s
20	$Rej = \frac{C_f - C_p}{C_f}$	Solute rejection (Rej)
21	$k = 0.664\, k_{dc}\, Re_b^{0.5}\, Sc^{0.33} \left(\frac{D_b}{d_h}\right) \left(\frac{2d_h}{L_f}\right)^{0.5}$	Mass transfer coefficient (k)
22	$Sc = \frac{\mu_b}{\rho_b\, D_b}$	Schmidt number (Sc)
23	$Rec = \frac{Q_p}{Q_f} = \frac{(C_r - C_f)}{(C_r - C_p)}$	Water recovery (Rec)
24	$SEC_{RO} = \frac{P_f\, Q_f}{Q_p\, \varepsilon pump}$	Specific energy consumption (SEC_{RO})
25	$\rho_b = 498.4\, m_f +$ $\sqrt{[248400\, m_f^2 + 752.4\, m_f\, C_b]}$ and $m_f = 1.0069 - 2.757 \times 10^{-4}\, T$	The physical property: Density (ρ_b)
26	$D_b =$ $6.725 \times 10^{-6}\, \exp(0.1546 \times 10^{-3}\, C_b - \frac{2513}{T+273.15})$	The physical property: Diffusivity (D_b)
27	$\mu_b = 1.234 \times 10^{-6}\, \exp(0.0212\, C_b + \frac{1965}{T+273.15})$	The physical property: Viscosity (μ_b)

$F_f\ (-)$:fouling factor=1 for a new membrane); P_f and P_p (atm): feed and permeate pressure; U_b (m/s): bulk velocity; d_h (m): hydraulic diameter of feed spacer channel; t_f (m): feed channel height; W (m): membrane width; k_{dc} (-): constant; L_f (m): length of filament in the spacer mesh; $\varepsilon pump$ (-) pump efficiency;

Table 2. Comparison of experimental (Fujioka et al., 2014) and model predictions of retentate flowrate and product concentration (operating conditions: 2.43x10^{-3} m³/s, 250x10^{-6} ppm, 10.1 atm, and 20 °C)

Experimental $Q_r^{Exp.}$x10³ (m³/s)	Theoretical $Q_r^{The.}$x10³ (m³/s)	Relative error%	Experimental $C_p^{Exp.}$x10⁸ (ppm)	Theoretical $C_p^{The.}$x10⁸ (ppm)	Relative error%
2.23	2.27	1.79	3.19	3.00	5.95

Table 3. Experimental and modelled rejections of NEMA at three inlet pump pressures and fixed 2.43x10^{-3} m³/s, 250x10^{-6} ppm, and 20 °C

Pressure (atm)	Experimental rejection%	Theoretical rejection%	Relative error%
4	72.2	64.52	10.62
6.51	82.1	79.73	2.87
10.1	87.2	87.99	-0.90

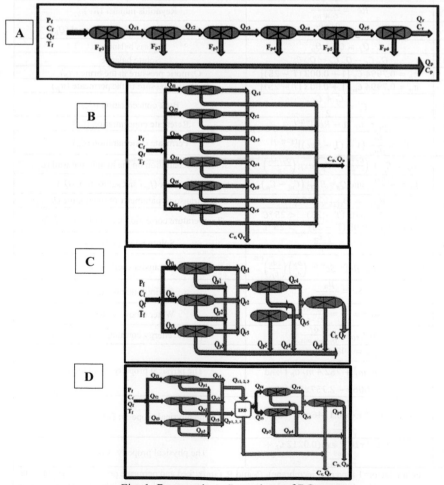

Fig. 1. Proposed configurations of RO system

Table 4. Design features of the spiral wound membrane module (Fujioka, 2014)

Feature	Value	Feature	Value
Type	ESPA2-4040	Diameter (m)	0.1003
Thickness of feed spacer t_f (m)	6.6×10^{-4}	Pump and ERD efficiencies	80%, 80%
Active area A (m^2)	7.9	Max. pump pressure (atm)	41.05
Length L (m)	0.9	Max. and Min. feed flowrate (m^3/s)	0.001, 0.005
Width W (m)	8.7778	Max. temperature (°C)	45
Length of filament in the spacer mesh L_f (m)			0.006

Table 5. Membrane transport parameters of NMEA (Fujioka et al., 2014; Al-Obaidi et al., 2018c)

Compound	Water transport parameter $A_w \times 10^6$ (m/s atm) at 20 °C	Solute transport parameter $B_s \times 10^6$ (m/s) at 20 °C
NMEA	1.0730	1.14

5. Assessment of the proposed configurations of RO system

The current study focuses on evaluating the performance indicators of four proposed configurations of RO process (Figure 2) working at a unique set of inlet conditions of 250×10^{-6} ppm (feed concentration), 10 atm (feed pressure), 2.43×10^{-3} m^3/s (feed flow rate), and 20 °C (feed temperature). The proposed configurations are specifically designed within six membrane modules and therefore they represent the twice size of the design of Fujioka et al. (2014). Please note that Fujioka et al. (2014) utilised three modules (same membrane area and design parameters) in a series configuration. Table 5 presents the associated results of the performance metrics of each proposed configuration for the selected inlet conditions. Table 6 ascertains the prosperity of configuration D (321 Tapered RP) as it gains the highest removal of NEMA of 98.79% and the lowest product concentration compared to other tested configurations, including the design of Fujioka et al., 2014. Interestingly, this is a clear improvement of NMEA removal compared to that of 87.2% obtained by Fujioka et al., 2014. However, configuration D has got the lowest productivity and the highest specific energy consumption. The lowest product concentration of configuration D is obtained due to feeding low concentration streams into stage 3, which improves the RO system efficiency. However, the low quantity of feed water of stage 3 can explain the reason of low productivity of the configuration D. This clarification explains the demerit of configuration D as it has got a highest specific energy consumption than the other configurations. In this context, it should be noted that parallel configuration has been introduced as the compromise configuration of having a high removal of NMEA of 89.13%, the maximum productivity, and the lowest specific energy consumption.

Table 6. Summary of the results of the proposed configurations (A, B, C, and D) for the removal of NMEA from wastewater

Configuration	Cp plant (ppm)	Cr plant (ppm)	Rej (-)	Productivity (m^3/day)	SEC (kWh/ m^3)
A: Series	3.75E-08	2.74E-07	84.99	21.423	3.447
B: Parallel	2.72E-08	2.92E-07	89.13	33.220	2.223
C: 321 Tapered RR	2.78E-08	2.90E-07	88.89	32.181	2.295
D: 321 Tapered RP	3.00E-09	2.57E-07	**98.79**	5.542	6.257
Fujioka et al. (2014): Series of three elements	3.19E-8	2.68E-7	87.2	16.819	4.435

6. Conclusions

In this study, the removal of NMEA organic compound from wastewater was evaluated considering four different proposed configurations of RO process. This is the first attempt to appraise the performance metrics of six spiral wound membrane modules of RO process configured as series, parallel, tapered of retentate-retentate design and tapered of retentate-permeate design, towards the removal of NMEA. Also, in terms of size of the process, it is one-fold times bigger than the lab scale RO process of Fujioka et al. (2014) Systematically, the comparison was made using a mathematical model of the process with considering a fixed set of inlet conditions. The tapered configuration retentate-permeate design was found to be the best due to its remarkable removal efficiency of 98.79% compared to 87.2% that can be found in open literature.

References

A. Szczuka, N. Huang, J.A. MacDonald, A. Nayak, Z. Zhang, W.A. Mitch, 2020. N-Nitrosodimethylamine formation during UV/hydrogen peroxide and UV/chlorine advanced oxidation process treatment following reverse osmosis for potable reuse. Environmental Science & Technology, 54(23), pp.15465-15475.

H. Takeuchi, N. Yamashita, N. Nakada, H. Tanaka, 2018. Removal characteristics of N-nitrosamines and their precursors by pilot-scale integrated membrane systems for water reuse. International journal of environmental research and public health, 15(9), p.1960.

IARC, IARC monographs on the evaluation of carcinogenic risks to humans: Overall evaluations off carcinogenicity: An updating of IARC monographs volumes 1-42: Supplement 7, International agency for research on cancer, 1987.

M.A. Al-Obaidi, C. Kara-Zaïtri, and I.M. Mujtaba, 2018a. Simulation of full-scale reverse osmosis filtration system for the removal of N-nitrosodimethylamine from wastewater. Asia-Pacific Journal of Chemical Engineering, 13(2), p.e2167.

M.A. Al-Obaidi, J.P. Li, S. Alsadaie, C. Kara-Zaïtri, and I.M. Mujtaba, 2018b. Modelling and optimisation of a multistage reverse osmosis processes with permeate reprocessing and recycling for the removal of N-nitrosodimethylamine from wastewater using species conserving genetic algorithms. Chemical Engineering Journal, 350, pp.824-834.

M.A. Al-Obaidi, C. Kara-Zaïtri, and I.M. Mujtaba, 2018c. Simulation and optimisation of spiral-wound reverse osmosis process for the removal of N-nitrosamine from wastewater. Chemical Engineering Research and Design, 133, pp.168-182.

M.A. Al-Obaidi, A.A. Alsarayreh, A.M. Al-Hroub, S. Alsadaie, and I.M. Mujtaba, 2018d. Performance analysis of a medium-sized industrial reverse osmosis brackish water desalination plant. Desalination, 443, pp.272-284.

P.A. Alaba, Y.M. Sani, S.F. Olupinla, W.M.W. Daud, I.Y. Mohammed, C.C. Enweremadu, O.O. Ayodele, 2017. Toward N-nitrosamines free water: Formation, prevention, and removal. Critical reviews in environmental science and technology, 47(24), pp.2448-2489.

S.C. Osorio, P.M. Biesheuvel, E. Spruijt, J.E. Dykstra, A. van der Wal, 2022. Modeling micropollutant removal by nanofiltration and reverse osmosis membranes: considerations and challenges. Water Research, p.119130.

T. Fujioka, S.J. Khan, J.A. Mcdonald, A. Roux, Y. Poussade, J.E. Drewes, L.D. Nghiem, 2014. Modelling the rejection of N-nitrosamines by a spiral-wound reverse osmosis system: Mathematical model development and validation. Journal of Membrane Science, 454, 212–219.

T. Fujioka, 2014. Assessment and optimisation of N-nitrosamine rejection by reverse osmosis for planned potable water recycling applications.

USEPA, Integrated Risk Information System (IRIS), http://www.epa.gov/iris/ (1993).

Flavio Manenti, Gintaras V. Reklaitis (Eds.), Proceedings of the 34th European Symposium on Computer Aided Process Engineering / 15th International Symposium on Process Systems Engineering (ESCAPE34/PSE24), June 2-6, 2024, Florence, Italy

Optimizing Lactic Acid Recovery from Vinasse: Comparing Traditional and Intensified Process Configurations

Lucas de Oliveira Carneiro,[a] Romildo Pereira Brito,[a] Karoline Dantas Brito,[a]

[a]*Federal University of Campina Grande, R. Aprígio Veloso 882, Campina Grande 58429-900, Brazil*
karoline.dantas@ufcg.edu.br

Abstract

Biological processes are the major source of lactic acid, one of the chemical commodities of great interest to the industry, because of the optical purity obtained in such processes. One of these processes can be the ethanol fermentation, which produces a considerable amount of lactic acid as by-product. However, fermentation results in high diluted streams with a considerable amount of other organic acids and impurities that bring more complexity to the purification process. Therefore, the objective of the present work is to develop the optimal design of the lactic acid recovery process from vinasse using intensification technologies such as mechanic vapor recompression and reactive divided wall column (RDWC). Two configurations were evaluated for the process, one conventional with reactor and column distillation, and an intensified with a single RDWC. The Total Annual Cost (TAC) was optimized, and the results compared in the bases of operational, economic, and environmental indicators. The intensified configuration presents the lowest TAC, but the capital cost of this process is a little higher than conventional configuration. In the other side, the conventional configuration, even with the lowest CAPEX, present high operational costs, and carbon emissions, beside more methanol consumption and water use.

Keywords: Reactive Distillation, Process Intensification, Lactic Acid Recovery.

1. Introduction

Lactic acid (LAc) is one of the most important chemical commodities due to its large commercial application in the food and pharmaceutical industries. The LAc molecule has a chiral carbon that gives it optical activity, and, therefore, can be found in two optical isomers: D(-)-LAc and L(+)-LAc, the latter being the most desired in the industry as it does not pose risks to human health. The request for optical purity implies in the fact that, currently, 90% of the world's lactic acid production occurs via biotechnological routes.

The use of biotechnological routes, however, presents some disadvantages, such as high dilution, presence of other organic acids and impurities, and the strong interaction between lactic and water (GONZÁLEZ-NAVARRETE et al., 2022). Therefore, several purification methods have been investigated such as precipitation, extraction, adsorption, short-path evaporation, and membrane separation, but the most used in the literature is the reactive distillation (RD) technology. In this process, lactic acid reacts with an alcohol, especially methanol, to form an ester. As the ester formed has greater thermal stability than the lactic acid, it is recovered in a distillation column and is subsequently hydrolyzed to form lactic acid again in a reactive distillation column (PAZMIÑO-MAYORGA et al., 2021).

Given the importance of lactic acid to the industry and the need to reduce or reuse industrial waste, the objective of this work is to develop a process for recovering the lactic acid present in vinasse produced during ethanol fermentation. The main contributions of this work to the existing literature consist of considering multiple species in the feed stream, which directly impacts methanol consumption and energy supply, and evaluating the recovery of lactic acid at lower concentration levels, with 2% wt.

2. Process Design

The process to produce lactic acid from vinasse consists of three areas, as highlighted in Figure 1. The process flowsheet was modeled using Aspen Plus V12 software. UNIQUAC-HOC was the thermodynamic model used, which uses the UNIQUAC equation to predict the non-ideality of the liquid phase and the Hayden O'Connell equation of state to predict the dimerization of organic acids in the vapor phase.

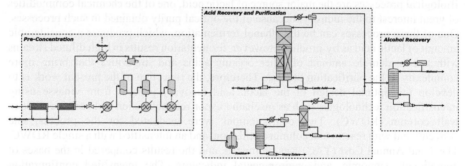

Figure 1. Process flowsheet to produce lactic acid from vinasse.

The first area of the process is pre-concentration, in which the vinasse feeds a triple-effect evaporation so that most of the water is removed. This removal is necessary because esterification reactions are limited by chemical equilibrium, even in the presence of catalysts (PÖPKEN *et al.,* 2000), and the presence of water directs the reaction toward the reactants. In this paper, the vapor produced in the last stage of evaporation is used as a heating fluid at some points in the process after mechanical vapor recompression (MVR) (CARNEIRO *et al.,* 2022).

After pre-concentration, the concentrated vinasse is sent to the acid recovery and purification area, where lactic acid (LAc) is esterified with methanol (MeOH), forming methyl lactate (MetLAc). Next, methyl lactate is separated from the remaining esters and organic acids and hydrolyzed into lactic acid.

To increase the esterification conversion, the Amberlyst 15 ion exchange was used as a catalyst (PÖPKEN *et al.,* 2000; PAZMIÑO-MAYORGA *et al.,* 2021). The reactions considered in the process modeling are shown below in Eq. (1) to Eq. (6).

$$C_3H_6O_3\ (LAc) + CH_3OH(MeOH) \leftrightarrow C_4H_8O_3(MetLAc) + H_2O \tag{1}$$

$$C_2H_4O_2\ (HAc) + CH_3OH(MeOH) \leftrightarrow C_3H_6O_2(MetHAc) + H_2O \tag{2}$$

$$C_4H_6O_4\ (SUc) + CH_3OH(MeOH) \leftrightarrow C_5H_8O_4(MMS) + H_2O \tag{3}$$

$$C_5H_8O_4\ (MMS) + CH_3OH(MeOH) \leftrightarrow C_6H_{10}O_4(DMS) + H_2O \tag{4}$$

$$2C_3H_6O_3 \ (LAc) \leftrightarrow C_6H_{10}O_5(DLAc) + H_2O \qquad\qquad (5)$$

$$C_3H_6O_3 \ (LAc) + C_6H_{10}O_5(DLAc) \leftrightarrow C_9H_{14}O_7(TLAc) + H_2O \qquad (6)$$

Two configurations were evaluated for the acid recovery and purification area. Figure 2 compares the two configurations.

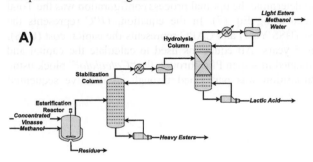

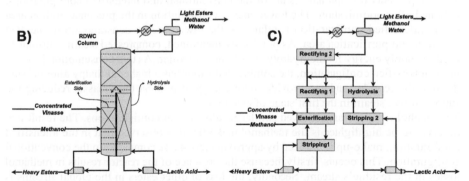

Figure 2. A) Flowsheet of the conventional configuration; B) Flowsheet of the intensified configuration (RDWC); and C) Sections and interconnections of the RDWC column.

The conventional configuration consists of 1) esterification reactor, which operates under conditions of temperature and pressure sufficient to vaporize the water and esters formed; 2) stabilization column, which aims to remove heavy esters and glycerol that were vaporized in the reactor; and 3) hydrolysis column, which have the objective to convert the methyl lactate formed in the reactor into lactic acid.

The intensified configuration consists of coupling the esterification reactor, the stabilization column, and the hydrolysis column in a single reactive divided-wall column (RDWC). The concentrated vinasse feeds the left side of the wall, where the esterification reactions occur. The vaporized esters, as well as the excess of methanol, and the water produced, rise to rectification section 1. The vapor leaving rectification section 1 contains light esters, water, and excess methanol. This vapor is directed to the rectification section 2, where all methanol, light esters produced, and excess water will be removed as top product. The liquid leaving the rectification section 2 - that consists of methyl lactate and water - is divided between the two sides of the RDWC. In the hydrolysis section – the right side of the wall - methyl lactate is converted into lactic acid; the vapor stream from this section goes directly to the rectification section 2, while the liquid stream goes to the stripping section 2, where the lactic acid is obtained as a bottom product.

The last area of the process is the alcohol recovery area, which have the function to separate the methanol from water and any residues of methyl lactate.

3. Process Optimization

The objective function used to determine the optimal process configuration was the Total Annual Cost (TAC), calculated using Eq. (7). In the equation, *OPC* represents the operational cost of the process (US\$/Year) and CAPC represents the capital cost (US\$). The payback considered was 5 years. All equations used to calculate the capital and operational costs were implemented in Aspen Plus through the "*Calculator*" block using Fortran language. TAC optimization was performed through an iterative sequential procedure.

$$TAC = OPC + \frac{CAPC}{Payback} \tag{7}$$

4. Results and Discussion

Table 1 presents the optimal results of the conventional and intensified configurations, obtained after optimization. The lower energy consumption in the preconcentration area of the intensified configuration is related to the lower energy consumption in the acid recovery and purification area. As previously mentioned, compressed steam is primarily used to supply energy for the hydrolysis column's reboiler. As this consumption is lower in the intensified configuration, the compressed vapor has a higher enthalpy after passing through the reboiler and can provide more energy to preheat the vinasse, reducing the supply of live steam in the first stage of evaporation.

The alcohol recovery area showed similar results in both configurations. The result that deserves to be highlighted is the methanol make-up. It is observed that in the intensified configuration, make-up is reduced by approximately 44 % compared to the conventional configuration. This occurs, firstly, because the presence of the reactor results in methanol losses in the residue's stream. Secondly, the loss of other esters in the global streams is smaller in the intensified configuration, especially of dimethyl succinate (DMS). DMS losses in the conventional configuration are 2.54 kmol/h, while in the intensified configuration these losses are equal to 0.09 kmol/h. Considering that, for each kmol of DMS formed, 2 kmol of methanol is consumed, the formation and loss of this ester ultimately represents a loss of methanol.

The amount of cooling water decreased significantly in the intensified configuration, going from 2,524.5 m³/h to 2,201.1 m³/h. This result improves the environmental indicators of the process, considering that the reduction in water consumption is being sought after by industries around the world.

When analyzing Table 1, it is possible to observe that the intensified configuration presented a reduction of 9.5 % in the total operational cost when compared to conventional configuration. This result highlights how the use of intensification techniques is advantageous for separation processes, especially distillation. However, it is important to highlight that this result would be even better if the steam used in all reboilers in the distillation columns were of the same pressure.

The capital cost, however, negatively impacts the TAC of the intensified configuration, as it presents an increase of 3.32 % in relation to the conventional configuration. The

main equipment that contributes to a higher capital cost are the vinasse preheaters, the
compressor and the RDWC.

Table 1. Optimal results for the conventional and intensified configurations.

	Conventional	*Intensified*
Pre-Concentration Area		
Heat Duty (MW)	11.33	9.02
Compressor Horsepower (MW)	6.07	6.34
Capital Cost (MMU$)	10.19	10.40
Operational Cost (MMU$/Y)	5.49	5.11
Acid Recovery and Purification Area		
Heat Duty (MW)	14.09	12.27
Condenser Duty (MW)	-10.55	-8.57
Capital Cost (MMU$)	1.87	2.11
Operational Cost (MMU$/Y)	1.28	1.50
Alcohol Recovery Area		
Heat Duty (MW)	0.69	0.70
Cooling Duty (MW)	-4.04	-4.15
Methanol Make-Up (t/h)	0.32	0.18
Capital Cost (MMU$)	0.56	0.53
Operational Cost (MMU$/Y)	1.64	1.01
Heat Duty (MW)	16.34	14.26
Cooling Duty (MW)	-14.59	-12.72
Electricity (MW)	6.07	6.34
Operational Cost (MMU$/Y)	8.42	7.62
Capital Cost (MMU$)	12.62	13.04
Total Annual Cost (MMU$/Y)	10.94	10.27
Total Carbon Emissions (kt/Y)	47.78	44.61

A parameter that can define the choice for one or another configuration is the payback.
In the calculations of the economic analysis, a period of 5 years was considered for the
payback; however, if this time is extended to 10 years, the TAC of the intensified
configuration will be even more attractive (8.92 MMU$/year) than that of the
conventional configuration (9.68 MMU$/year). Therefore, higher paybacks favor the
implementation of the intensified configuration.

When the environmental results are observed, it is possible to conclude that, even
requiring more high-pressure steam, the intensified configuration reduces CO_2 emissions
by 6.63 % when compared to the conventional configuration. This result is mainly a
function of the lower low-pressure steam consumption in the hydrolysis reboiler of the
intensified configuration. Figure 3 and Figure 4 show the optimal flowcharts obtained.

5. Conclusions

This work evaluated two process configurations for the recovery of lactic acid present in
vinasse. The process flowsheets were simulated and optimized in the Aspen Plus. With
the optimized results, it was possible to observe that the use of a reactive divided-wall
column (RDWC) showed improvements in the main operational indicators, specific
consumption and operational cost, and environmental indicators (CO_2 emissions). The
intensification configuration with RDWC presented a reduction in the TAC of the
process, but its capital cost is higher than the conventional configuration. However, in

higher payback periods, the advantage of RDWC over the conventional configuration becomes even more significant. Finally, proposals for process intensification are very attractive, primarily in a scenario of search for more sustainable processes.

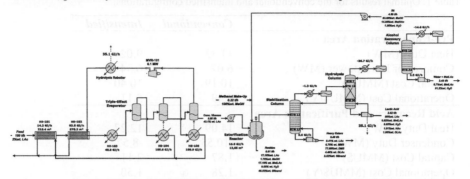

Figure 3. Optimal results for conventional configuration.

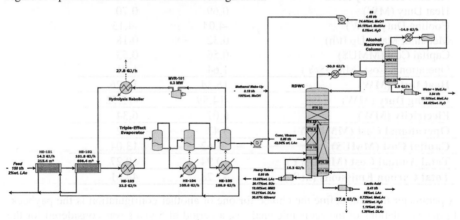

Figure 4. Optimal results for intensified configuration.

6. References

L. D. O. CARNEIRO, R. P. S. MATOS, W. B. RAMOS, R. P. BRITO, K. D. BRITO, 2022, Sustainable design and optimization of the recovery process of acetic acid from vinasse: Operational, economic, and environmental analysis, Chemical Engineering and Processing - Process Intensification, v. 181, p. 109176.

C. GONZÁLEZ-NAVARRETE, E. SÁNCHEZ-RAMÍREZ, C. RAMÍREZ-MÁRQUES, S. HERNÁNDEZ, E. COSSÍO-VARGAS, J. G. SEGOVIA-HERNÁNDEZ, 2022, Innovative Reactive Distillation Process for the Sustainable Purification of Lactic Acid, Industrial & Engineering Chemistry Research, v. 61, p. 621-637.

I. PAZMIÑO-MAYORGA, M. JOBSON, A. A. KISS, 2021, Conceptual design of a dual reactive dividing wall column for downstream processing of lactic acid, Chemical Engineering and Processing - Process Intensification, v. 164, p. 108402.

T. PÖPKEN, L. GÖTZE, J. GMEHLING, 2000, Reaction Kinetics and Chemical Equilibrium of Homogeneously and Heterogeneously Catalyzed Acetic Acid Esterification with Methanol and Methyl Acetate Hydrolysis, Industrial & Engineering Chemistry Research, v. 39, p. 2601-2611.

Flavio Manenti, Gintaras V. Reklaitis (Eds.), Proceedings of the 34th European Symposium on Computer Aided Process Engineering / 15th International Symposium on Process Systems Engineering (ESCAPE34/PSE24), June 2-6, 2024, Florence, Italy

Comparison of machine learning based hybrid modelling methodologies for dynamic simulation of chemical reaction networks

Harry Kay , Fernando Vega-Ramon, Dongda Zhang*

Department of Chemical Engineering, University of Manchester, Oxford Road, Manchester, M1 3AL, UK.
Corresponding author, email: dongda.zhang@manchester.ac.uk

Abstract

Accurate and reliable prediction of state variables and performance indicators within chemical reaction networks is essential for the control and optimisation of chemical process kinetics. Traditionally, kinetic modelling is employed to derive a set of ordinary differential equations. However, simplifications are often introduced via assumptions throughout the modelling process. These assumptions may reduce the model integrity and generalisability to unique conditions. Data driven modelling can overcome some of the aforementioned challenges but is characterised by further problems such as the lack of interpretability and data quality issues. A solution to overcome these challenges is the integration of kinetic and data-driven modelling, named hybrid modelling. Here, the kinetic component retains the underlying structure of the reaction network, and the data-driven component accommodates for any missing information within the kinetic model.

In order to identify the most promising hybrid model structure and construction strategy, different techniques were adopted and benchmarked against a traditional pseudo steady state kinetic model for a C-16 hydroisomerisation reaction network. It was found that hybrid models often outperform alternative methods, offering high accuracy, and overcoming a range of challenges associated with pure kinetic and data-driven modelling, highlighting their potential to be used for industrial chemical reaction applications.

Keywords: Machine learning, dynamic simulation, reaction networks, hybrid modelling.

1. Introduction

Mathematical models were historically derived to express complex relationships between predictor and response matrices in effort to describe any given physical phenomena of interest. Such models were developed for many reasons including predictive capabilities of current and future states, control, optimisation, reducing experimental burdens, accelerating system design (e.g., faster development of industrial processes, and rapid scale-up transitions) and investigations of underlying system behaviour (e.g., variable dependencies). In many fields, physical insight and knowledge can be introduced to derive equations through first-principles and experimental observations used to estimate model parameters or add empirical correlations to increase the predictive performance. It is, however, typical that certain parameters and variables within industrial processes possess too complex of interactions to adequately model via first principles. In such cases, the non-linearities can be better represented via the paradigm of machine learning, thus enabling the modelling of complicated dynamic systems. Machine learning lends itself naturally to Industry 4.0, with the large quantity of process information available through sensor recordings and experiments.

Recent works have incorporated black-box models within many fields including soft sensing (Kay et al., 2022), image classification (Rawat & Wang, 2017), model-based design of experiment (Greenhill et al., 2020), and chemical reaction engineering (Yang et al., 2020). Although capable of formulating accurate predictive models, black-box machine learning suffers from lack of interpretability and usually lacks the ability to generalise to other systems or sufficiently different conditions. Ideally, a middle ground exists that preserves the advantages of both physical and data-driven modelling yet does not retain any associated disadvantages. A solution that provides this is known as hybrid modelling where both mechanistic and data-driven information is incorporated into a single model.

In the literature, two major forms exist to characterise hybrid modelling:

- Using machine learning to estimate dynamic changes within key parameters of a defined kinetic model. Commonly, this method is applied to systems with complex phenomena where parameters may deviate through time such as in bioprocess modelling (Rogers et al., 2023; Zhang et al., 2019). This approach offers increased interpretability as no change to the structure is made. The complex non-linear dynamics of the parameters are instead evaluated by data-driven methods.

- Extending a defined kinetic structure with an unknown data driven error parameter to account for missing information (Quaghebeur et al., 2022). This is commonly referred to as a hybrid discrepancy model and is implemented to rectify an error with the model structure itself.

2. Problem Statement

This study focuses on an in-depth analysis and comparison of the major forms of hybrid modelling and delves further into different methodologies of solving each system. More specifically, we will focus on simulating a dynamic C16 hydroisomerisation reaction. A rigorous microkinetic model (Figure 1) was derived in (Vega-Ramon et al., 2023) and reduced using optimisation to remove comparatively insignificant reactions. From this, the pseudo steady state hypothesis (PSSH) was applied to simplify the kinetic structure (Figure 1) and remove any dependencies on non-observable states (surface coverages). The model parameters were then lumped to ensure identifiability for better quality solutions upon optimisation and the resulting model used as a benchmark for traditional modelling techniques. The microkinetic model was exploited to generate 8 training datasets with varying initial mass fractions of $C_{16}(x_{c16})$, mono branch product x_{mo}, multi branch product x_{mul}, cracking byproducts x_{cn}, and catalyst concentrations C. The training datasets were then used to determine the kinetic constants of the PSSH model. Three testing dataset were chosen to mimic the experimental data with mass fractions of $x_{c16} = 1.00$, $x_{mo} = 0.00$, $x_{mul} = 0.00$, $x_{cn} = 0.00$, and catalyst concentrations of 0.1 wt.%, 0.3 wt.% and 0.5 wt.%.

$$\frac{dx_{c16}}{dt} = c_{\theta_T} \cdot (-k_1 \cdot x_{c16} \cdot \theta_V + k_1^- \cdot \theta_{c16})$$

$$\frac{d\theta_{c16}}{dt} = \frac{c_{r_0}}{\rho_b} \cdot [k_1 \cdot x_{cn} \cdot \theta_V + k_2^- \cdot \theta_{mo} - (k_1^- + k_2) \cdot \theta_{c16}]$$

$$\frac{dx_{mo}}{dt} = \ldots$$

$$\frac{d\theta_{mo}}{dt} = \ldots$$

Pseudo-steady state hypothesis

$$\frac{dx_{mul}}{dt} = \ldots$$

+

$$\frac{d\theta_{mul}}{dt} = \ldots$$

Parameter lumping

$$\frac{dx_{cn}}{dt} = c_{\theta_T} \cdot (k_7 \cdot \theta_{cn} - k_7^- \cdot x_{cn} \cdot \theta_V)$$

$$\frac{d\theta_{cn}}{dt} = \frac{c_{r_0}}{\rho_b} \cdot [k_7^- \cdot x_{cn} \cdot \theta_V + k_5 \cdot \theta_{mul} + k_8 \cdot \theta_{mo} - (k_5^- + k_7 + k_8^-) \cdot \theta_{cn}]$$

$$\frac{dx_{c16}}{dt} = \frac{-(\overline{k_1} + \overline{k_2} + \overline{k_3}) \cdot x_{c16}}{1 + \overline{k_7} \cdot x_{c16} + \overline{k_8} \cdot x_{mo} + \overline{k_9} \cdot x_{cn}}$$

$$\frac{dx_{mo}}{dt} = \frac{\overline{k_1} \cdot x_{c16} - (\overline{k_4} + \overline{k_5}) \cdot x_{mo}}{1 + \overline{k_7} \cdot x_{c16} + \overline{k_8} \cdot x_{mo} + \overline{k_9} \cdot x_{cn}}$$

$$\frac{dx_{mul}}{dt} = \frac{\overline{k_2} \cdot x_{c16} + \overline{k_4} \cdot x_{mo} + \overline{k_6} \cdot x_{cn}}{1 + \overline{k_7} \cdot x_{c16} + \overline{k_8} \cdot x_{mo} + \overline{k_9} \cdot x_{cn}}$$

$$\frac{dx_{cn}}{dt} = \frac{\overline{k_3} \cdot x_{c16} + \overline{k_5} \cdot x_{mo} - \overline{k_6} \cdot x_{cn}}{1 + \overline{k_7} \cdot x_{c16} + \overline{k_8} \cdot x_{mo} + k_9 \cdot x_{cn}}$$

Figure 1 - Schematic showing the microkinetic model and the resulting reduced model after the pseudo steady state hypothesis was applied and parameters were lumped

3. Methodology

Here we will introduce two strategies for solving the hybrid modelling frameworks namely the one-step and two-step approaches. However, we must define an appropriate metric to ascertain parameters within the kinetic structure which are not adequately described through being limited to a constant value. Ideally, this is achieved through some determination of uncertainty, as the parameters holding the largest uncertainty are ultimately less confident in their assigned deterministic value. In this work, a measure of uncertainty is made through a leave-two-out optimisation of the pseudo steady state model, where two of the training datasets are withheld from the optimiser through each iteration and the variance of the optimal parameters taken. In this manner, a form of parametric uncertainty is calculated, and the highest variances can be associated with kinetic constants that require to be estimated at each individual timestep.

Firstly, we will formally define the two-step methodology as shown in Figure 2 where parameter $\overline{k_1}$ has been identified as best suited to be time varying.

$$\left. \frac{\Delta x_{c16}}{\Delta t} \right|_{t=0} = \frac{-(\overline{k}_{1,t=0} + \overline{k}_2 + \overline{k}_3) \cdot x_{c16}}{1 + \overline{k}_7 \cdot x_{c16} + \overline{k}_8 \cdot x_{mo} + \overline{k}_9 \cdot x_{mul}}$$

$$\vdots$$

$$\left. \frac{\Delta x_{c16}}{\Delta t} \right|_{t=n} = \frac{-(\overline{k}_{1,t=n} + \overline{k}_2 + \overline{k}_3) \cdot x_{c16}}{1 + \overline{k}_7 \cdot x_{c16} + \overline{k}_8 \cdot x_{mo} + \overline{k}_9 \cdot x_{mul}}$$

$$\frac{dx_{c16}}{dt} = \frac{-(\overline{k_1} + \overline{k_2} + \overline{k_3}) \cdot x_{c16}}{1 + \overline{k_7} \cdot x_{c16} + \overline{k_8} \cdot x_{mo} + \overline{k_9} \cdot x_{cn}}$$

$$\frac{dx_{mo}}{dt} = \frac{\overline{k_1} \cdot x_{c16} - (\overline{k_4} + \overline{k_5}) \cdot x_{mo}}{1 + \overline{k_7} \cdot x_{c16} + \overline{k_8} \cdot x_{mo} + \overline{k_9} \cdot x_{cn}}$$

$$\frac{dx_{mul}}{dt} = \frac{\overline{k_2} \cdot x_{c16} + \overline{k_4} \cdot x_{mo} + \overline{k_6} \cdot x_{cn}}{1 + \overline{k_7} \cdot x_{c16} + \overline{k_8} \cdot x_{mo} + \overline{k_9} \cdot x_{cn}}$$

$$\frac{dx_{cn}}{dt} = \frac{\overline{k_3} \cdot x_{c16} + \overline{k_5} \cdot x_{mo} - \overline{k_6} \cdot x_{cn}}{1 + \overline{k_7} \cdot x_{c16} + \overline{k_8} \cdot x_{mo} + \overline{k_9} \cdot x_{cn}}$$

Figure 2 - Schematic of the two-step hybrid modelling approach for a single time varying kinetic constant k_1.

This methodology is defined by initially discretising the set of time varying parameters as constants over set time intervals between each measured datapoint. This optimization was performed using interior point optimisation (IPOPT) in Pyomo. Once the optimal set of constants were obtained, a neural network can be constructed and employed to learn the correlations between the set of state parameters and Pd catalyst loading weight percentage, and the relevant set of time varying kinetic constants describing the system of differential equations between the current and next future states. This strategy was also applied to the development of the discrepancy model.

The second hybrid modelling strategy is the one-step approach where the network and physical parameters are optimised and evaluated simultaneously. This can be addressed in multiple manners such as in (Psichogios & Ungar, 1992) where the sensitivity approach is taken. Using the forward sensitivities, however, requires the solving of excessive numbers of ODE's scaling linearly with respect to the numbers of parameters. The adjoint sensitivity method can be used to alleviate this burden which scales constantly with number of parameters and is therefore far less computationally expensive. This approach was implemented for solving neural ODE's in (Chen et al., 2018). In this work, the time varying and constant parameters are solved simultaneously through a combination of automatic differentiation and gradient descent. To evaluate and assess the performance of all strategies and models, two metrics were defined being the mean absolute percentage error (MAPE) (1), and the mean absolute error (MAE) (2).

$$MAPE = \frac{1}{N} \sum_{i=1}^{N} \frac{|y_i - f(x_i|\theta)|}{y_i} \cdot 100 \tag{1}$$

$$MAE = \frac{1}{N} \sum_{i=1}^{N} |y_i - f(x_i|\theta)| \tag{2}$$

4. Results and Discussion

Through the analysis as described in the methodology, three parameters were determined to hold the largest uncertainty in the order of k_1, k_9, and k_6. A multistep ahead approach is taken to obtain the predictions of the neural networks on the training sets by inputting the network outputs through the set of ODE's that define the chemical reaction network and using these as inputs to ascertain the time varying constants that represent the kinetics until the next future state. The prediction results for the PSSH, discrepancy, one-step and two-step models are presented in Table 1 below where the discrepancy model has four time varying parameters (one per ODE) and the others have three.

Table 1 – Average MAPE and MAE testing results (over the 3 testing datasets) for the pseudo steady state benchmark model, discrepancy model, one-step time varying parameter model, and two-step time varying parameter model.

Model	Mean average percentage error / %				Mean absolute error			
	x_{c16}	x_{mo}	x_{mul}	x_{cn}	x_{c16}	x_{mo}	x_{mul}	x_{cn}
PSSH	10.3	6.0	71.1	18.9	0.030	0.023	0.030	0.022
Discrepancy (two-step)	15.8	12.4	36.8	14.0	0.048	0.049	0.008	0.018
One-step	4.6	4.5	65.0	10.3	0.017	0.016	0.010	0.009
Two-step	7.9	7.2	16.7	10.0	0.023	0.030	0.006	0.010

From Table 1, it is evident that the discrepancy model performed the worst out of the hybrid configurations, yielding the largest mean absolute errors often being nearly 100% larger than the rest. This is reciprocated with the mean average percentage errors with the exception of the multi branch product predicting similarly to the other configurations. This can be reasoned due to the inherent structure of the ODE system not missing information but the parameters being misinformed instead. The PSSH benchmark model seemed to accurately represent the reactant and mono branch product mass fractions with

high accuracy, however evidently lacked knowledge regarding the multi branch product
and cracking byproduct mass fractions.

Both of the hybrid model frameworks that simulated time varying parameters improved
significantly over the benchmark both in terms of MAPE and MAE. The simultaneous
estimation of the network and physical parameters (i.e., one step approach) through
automatic differentiation offered an average absolute improvement in MAPE and MAE
of 5.5 % and 0.013 respectively. The large percentage errors associated with the
prediction of the multi branch product result from mass fractions close to zero. Using the
two-step approach provided average absolute improvements in MAPE and MAE of 11.7
% and 0.009. The mass fraction profiles estimated by the models for a single testing
dataset can be seen in Figure 3.

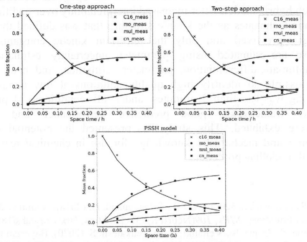

*Figure 3 - Predicted mass fraction profiles for initial conditions: $x_{c16} = 1$, $x_{mo} = 0$, $x_{mul} =$
0, $x_{cn} = 0$, $C = 0.3$ wt. % using the one and two-step hybrid models and the PSSH model
(markers represent measured data and the solid curves represent the models predictions).*

From Figure 3, it is clear that the PSSH model lacks necessary information regarding the
trend of the multi branch product profile, and has an observable discrepancy for the
cracking by-products, mono branch products and for the C16 profile. Both of the time
varying parameter approaches provided the required missing information for the multi
branch product profile and amended the predictions to exhibit the correct trend, reducing
the MAE by 67 % and 80 % for the one and two step approaches respectively. In
addition, the overshoot at the terminal of the C16 reaction is mitigated and the mean
absolute errors corresponding to the cracking by-products reduced by 55 % and 59 %.
In general, the models focused on minimising the MAE/MAPE accompanying the
profiles closer to zero as these would generate the largest standardised errors upon a
small change in predictions. The two-step approach is limited by the parameter
estimation in the first stage of its methodology as, the neural network is purposely
trained to correlate the predictor matrix to these values, whereas the one-step approach is
not. However, as the physical and data-driven parameters are simultaneously estimated,
the network does not learn any soft constraints imposed through parameter estimation,
leading to rapid changes in gradients in the mass fraction profiles. Hence, as shown in
Figure 3, the curves are less smooth and tend to oscillate around the measured values,
which is not physically meaningful.

Furthermore, additional penalties had to be imposed within the one-step method to prevent any kinetic constants becoming less than zero. If such penalties were not introduced, this constraint would often be broken, whereas this was never found in the two-step approach as the constraints were preserved from parameter estimation.

5. Conclusion

In conclusion, three hybrid models were constructed to simulate a dynamic chemical reaction network, proving to be useful for developing accurate, interpretable models for predicting reaction profiles. The discrepancy model performed the worst failing to improve upon the benchmark PSSH model, however, the one and two-step methodologies for time varying parameters improved upon the benchmark MAE by over 34%. Such hybrid models offer increased interpretability over traditional black box machine learning techniques as the kinetic structure that was defined through analysis and modelling is preserved and only the uncertain kinetic constants manipulated (typically these are estimated using parameter uncertainty estimation). Through the use of automatic differentiation, parameter estimation, and neural networks, it was shown possible to identify high quality solutions at low computational cost (network training was less than 5 minutes under all conditions). Upon testing of the optimised frameworks, good predictive performance and generalisation capabilities were exhibited. This study has presented the potential of integrating machine learning and mechanistic knowledge for use in chemical reaction networks and as a general modelling procedure.

References

Chen, R. T. Q., Rubanova, Y., Bettencourt, J., & Duvenaud, D. (2018). Neural Ordinary Differential Equations. *NIPs*, *109*(NeurIPS), 31–60. https://arxiv.org/abs/1806.07366v5

Greenhill, S., Rana, S., Gupta, S., Vellanki, P., & Venkatesh, S. (2020). Bayesian Optimization for Adaptive Experimental Design: A Review. *IEEE Access*, *8*, 13937–13948.

Kay, S., Kay, H., Mowbray, M., Lane, A., Mendoza, C., Martin, P., & Zhang, D. (2022). Integrating Autoencoder and Heteroscedastic Noise Neural Networks for the Batch Process Soft-Sensor Design. *Industrial and Engineering Chemistry Research*, *61*(36), 13559–13569.

Psichogios, D. C., & Ungar, L. H. (1992). A hybrid neural network-first principles approach to process modeling. *AIChE Journal*, *38*(10), 1499–1511.

Quaghebeur, W., Torfs, E., De Baets, B., & Nopens, I. (2022). Hybrid differential equations: Integrating mechanistic and data-driven techniques for modelling of water systems. *Water Research*, *213*, 118166.

Rawat, W., & Wang, Z. (2017). Deep convolutional neural networks for image classification: A comprehensive review. *Neural Computation*, *29*(9), 2352–2449.

Rogers, A. W., Song, Z., Ramon, F. V., Jing, K., & Zhang, D. (2023). Investigating 'greyness' of hybrid model for bioprocess predictive modelling. *Biochemical Engineering Journal*, *190*, 108761.

Vega-Ramon, F., Wang, W., Wu, W., & Zhang, D. (2023). Developing a rigorous chemical reaction network reduction strategy for n-hexadecane hydroisomerisation. *Computer Aided Chemical Engineering*, *52*, 157–162.

Yang, F., Dai, C., Tang, J., Xuan, J., & Cao, J. (2020). A hybrid deep learning and mechanistic kinetics model for the prediction of fluid catalytic cracking performance. *Chemical Engineering Research and Design*, *155*, 202–210.

Zhang, D., Del Rio-Chanona, E. A., Petsagkourakis, P., & Wagner, J. (2019). Hybrid physics-based and data-driven modeling for bioprocess online simulation and optimization. *Biotechnology and Bioengineering*, *116*(11), 2919–2930.

Flavio Manenti, Gintaras V. Reklaitis (Eds.), Proceedings of the 34th European Symposium on Computer Aided Process Engineering / 15th International Symposium on Process Systems Engineering (ESCAPE34/PSE24), June 2-6, 2024, Florence, Italy

A Mathematical Framework to Study the Impact of Technological Learning on Portfolio Planning

Pooja Zen Santhamoorthy,[a] Selen Cremaschi[a,*]

[a]*Department of Chemical Engineering, Auburn University, AL 36849, USA*
selen-cremaschi@auburn.edu

Abstract

Technological learning leads to cost reduction in investment and operation. Different learning factors have different impacts on portfolio planning. This work develops a mathematical framework based on optimization for studying the impact of technological learning on portfolio planning. The integer programming (IP) model minimizes the investment and operation costs of the project portfolio by optimizing the selection of processes and technologies based on the nature of the learning factors considered. A case study of planning carbon capture (CC) to meet the annual capture targets while accounting for technological learning due to shared knowledge reveals that more capture facilities are deployed in the early planning periods to accelerate technological learning. The total cost of capture was reduced by $170 million when a log-linear curve with a learning rate of 0.30 was used to model the technological learning.
Keywords: Learning effect, Integer programming (IP), Carbon Capture (CC)

1. Introduction

Economic competitiveness enables a technology's deployment, while its stable commercialization also depends on technology cost reduction (Elia et al., 2021). Multiple factors drive technology cost reductions, such as investment in research and development (learning-by-researching), shared/spill-over knowledge (learning-by-copying), functionality improvement (learning-by-using), and economies-of-scale (learning-by-doing) (Upstill and Hall, 2018). Different learning curves are used to mathematically represent technology cost reductions (Anzanello and Fogliatto, 2011). Each learning factor can affect different aspects of portfolio planning, and the extent depends on the nature of the learning curve selected. For example, learning-by-using can reduce operational costs, while learning-by-copying can reduce the investment cost of a technology. The technology should be implemented as early as possible to get the utmost benefit from the former learning factor. But for the latter, deployment should be delayed to gain experience from as many other deployed projects. Thus, a strategic deployment of technologies is necessary to maximize the cost reductions from different learning factors over the planning period.

This paper introduces a multiperiod optimization model to study the effect of technological learning on the selection of processes and technologies to produce a set of products over a planning period. The model's objective is to minimize the total cost of technology deployment and operation to meet the product demands.

2. Mathematical Model Formulation

An optimization-based framework is developed to integrate the processes ($i \in I$) and technologies ($j \in J$) under a single network for planning technology deployment to produce products ($k \in K$) with the available raw materials ($r \in R$) over a planning horizon

($t \in T$). Figure 1 illustrates the overall superstructure. Based on the demand for the products k and the availability of materials (R_{rt}) over time t, the processes P_i are selected to produce products X_{kt} using one of the compatible and cost-effective technologies TG_j to minimize the total cost of technology deployment and operation, while considering the cost reduction due to different learning factors along the planning horizon.

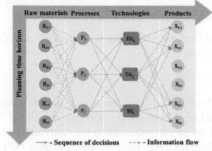

Figure 1. Superstructure of processes and technologies selection to produce products.

2.1. Binary Variables

The sets $i \in I$, $j \in J$, $k \in K$, and $t \in T$ represent the processes, technologies, products, and planning periods. The sets $m1 \in M1$, $m2 \in M2$, .., $mm \in MM$ and $n1 \in N1$, $n2 \in N2$, .., $nn \in NN$ represent different learning effects on investment and operating costs. A binary variable $y_{i,j,k,t}$ is defined in Eq. (1) to represent the selection of a process i and a compatible technology j to produce product k at time t. Two sets of binary variables, $bm_{i,j,k,t,m1,...,mm}$ and $bn_{i,j,k,t,n1,...,nn}$, are introduced in Eqs. (2) and (3) to track learning due to different learning effects in investment and operation of the technology used to produce product k.

$$y_{i,j,k,t} = \begin{cases} 1, \text{if process } i \text{ uses tech } j \text{ to produce product } k \text{ at time } t \\ 0, \text{otherwise} \end{cases} \quad (1)$$

$$bm_{i,j,t,k,m1,...,mm} = \begin{cases} 1, \text{if process } i \text{ using tech } j \text{ to produce product } k \text{ has} \\ \quad \text{reduced investment cost due to learnings } m1,..,mm \text{ at time } t \\ 0, \text{otherwise} \end{cases} \quad (2)$$

$$bn_{i,j,k,t,n1,...,nn} = \begin{cases} 1, \text{if process } i \text{ using tech } j \text{ to produce product } k \text{ has} \\ \quad \text{reduced operational cost due to learnings } n1,..,nn \text{ at time } t \\ 0, \text{otherwise} \end{cases} \quad (3)$$

2.2. Parameters

The amount of product k that can be produced by process i with technology j at time t is represented by the parameter $D'_{i,j,k,t}$, which depends on the availability of resources. The annualized investment to implement a facility with a capacity that produces $D'_{i,j,k,t}$ amount of product k using technology j is represented by the parameter $IC_{i,j,k,t}$. The annual operating cost is represented by the parameter $OC_{i,j,k,t}$. The cumulative extent of cost reduction in the investment and operating costs due to different learning effects $m1,..,mm$ and $n1,..,nn$ at time t are represented by two sets of parameters, $\alpha_{i,j,k,t,m1,...,mm}$ and $\beta_{i,j,k,t,n1,...,nn}$, respectively. The demand for product k at time t is represented by $D_{k,t}$.

2.3. Constraints

Equations (4) and (5) ensure process i uses at most one technology j to produce product k, and the same technology is used in the subsequent periods. For a technology deployed in a process (i.e., $y_{i,j,k,t}=1$), one binary variable corresponding to each considered learning effect $M1,..,MM$ and $N1,..,NN$ on investment and operating costs takes a value of one, by

Eqs. (6) and (7). Otherwise, the binary variables $bm_{i,j,k,t,m1,...,mm}$ and $bn_{i,j,k,t,n1,...,nn}$ take a value of zero. Equation (8) ensures that the annual demand for each product ($D_{k,t}$) is met by producing $D'_{i,j,k,t}$ amount using the selected processes and technologies.

$$\sum_{j\in J}{}^{'}y_{i,j,k,t} \leq 1 \qquad \forall i \in I, k \in K, t \in T \tag{4}$$

$$y_{i,j,k,t} \geq y_{i,j,k,t-1} \qquad \forall i \in I, j \in J, k \in K, t \in T \tag{5}$$

$$\sum_{mm\in MM} bm_{i,j,k,t,m1,...,mm} = y_{i,j,k,t} \quad \forall i \in I, j \in J, k \in K, t \in T, M1,..,mm \in MM \tag{6}$$

$$\sum_{nn\in NN} bn_{i,j,k,t,n1,...,nn} = y_{i,j,k,t} \quad \forall i \in I, j \in J, k \in K, t \in T, n1 \in N1,.., nn \in NN \tag{7}$$

$$\sum_{i\in I, j\in J}{}^{'} D'_{i,j,k,t}\, y_{i,j,k,t} \geq D_{k,t} \qquad \forall k \in K, t \in T \tag{8}$$

2.4. Objective

The objective is to minimize the total cost (*TC*) of producing a set of products over the planning horizon while meeting annual demands. The cost consists of annualized investment and annual operating costs for the deployed technologies, which are computed accounting for the cost reduction due to technological learning, as given in Eq. (9).

$$Min\ TC = \sum_{i,j,k,t} \left(IC_{i,j,k,t} \left(y_{i,j,k,t} - \sum_{m1,..,mm} \alpha_{i,j,k,t,m1,..,mm}\, bm_{i,j,k,t,m1,..,mm} \right) \right)$$
$$+ \sum_{i,j,k,t} \left(OC_{i,j,k,t} \left(y_{i,j,k,t} - \sum_{n1,..,nn} \beta_{i,j,k,t,n1,..,nn}\, bn_{i,j,k,t,n1,..,nn} \right) \right) \tag{9}$$
$$i \in I, j \in J, k \in K, t \in T, m1 \in M1,.., mm \in MM, n1 \in N1,.., nn \in NN$$

Using technology j for process i to produce product k at time t leads to an investment ($IC_{i,j,k,t}$) and an operation cost ($OC_{i,j,k,t}$) which are added to the total project cost with the variable, $y_{i,j,k,t}$. The parameters $\alpha_{i,j,k,t,m1,..,mm}$ and $\beta_{i,j,k,t,n1,..,nn}$ represent the cumulative extent of cost reduction in the investment and operating costs due to different learning effects $m1,..,mm$ and $n1,..,nn$ at time t. The reduction extents are accounted for in the total cost with the binary variables tracking the learning, $bm_{i,j,k,t,m1,..,mm}$ and $bn_{i,j,k,t,n1,..,nn}$.

2.5. Implementation of technological learning models

Equations (1) to (9) yield an integer programming model (IP) that optimizes the selection of processes and technologies based on technological learning and product demands over a planning period. Incorporating technological learning models into the developed multiperiod portfolio planning model is demonstrated with an example of implementing learning from shared/spill-over knowledge (learning-by-copying). The extent of this learning-by-copying on a technology to be deployed at time t depends on the number of processes that already use the same technology to produce the same product k. Let the set $m1 \in M1 = \{1, 2, ..., I'\}$ represent the learning effects due to shared knowledge, where I' represents the total number of processes in the network. The binary variable $bm_{i,j,k,t,m1,...,mm}$ tracks the technological learning extent for a process based on the number of processes already using the technology (computed using $y_{i,j,k,t}$) with parameter, $L1_{m1} = \{1, 2, ..., I'\}$. For example, for a fourth-of-a-kind process, the binary variable $bm_{i,j,k,t,4,...,mm}$ takes a value of 1 as $L1_4 = 4$, and the summation of the variable $y_{i,j,k,t-1}$ is 3. Equation (6) ensures that only one binary variable is selected per source using this technology for this learning effect. Equation (11) enforces that the same binary variable is selected at subsequent periods so that the learning extent does not change over time.

$$\sum_{ml \in M1} (L1_{ml} - 1) bm_{i,j,k,t,ml,..,mm} \leq \sum_{i \in I} y_{i,j,k,t-1} \quad \forall i \in I, j \in J, k \in K, t \in T \quad (10)$$

$$bm_{i,j,k,t,ml,..,mm} \geq bm_{i,j,k,t-1,ml,..,mm} \quad \forall i \in I, j \in J, k \in K, t \in T, mm \in MM \quad (11)$$

The implementation of the learning curve to the developed shared/spill-over knowledge-based learning model is demonstrated using two different learning curves, log-linear and S-curve (Anzanello and Fogliatto, 2011). For the log-linear curve, the cost reduction extent ($LC1_{j,k,ml}$) for an ml^{th} process that uses the technology j to produce k depends on the learning rate for the technology $LR1_{j,k}$, as shown in Eq. (12), and it can take a value between 0 and 1. To avoid complete cost reduction due to infinite learning, the maximum cost reduction extent possible for the technology ($MR_{j,k}$) is introduced in the model by defining the cost reduction as a piecewise function such that $LC1_{j,k,ml}=MR_{j,k}$. The S-curve initially has slow progress in learning and then a rapid ascent. The cost reduction extent for the ml^{th} process depends on the learning rate, $LR2_{j,k}$, which takes a value greater than zero, and the inflection point in the curve, $TR2_{j,k}$, as given in Eq. (13).

$$LC1_{j,k,ml} = \begin{cases} 1 - ml^{-LR1_{j,k}}, \text{if } 1 - ml^{-LR1_{j,k}} \leq MR_{j,k} \\ MR_{j,k}, \quad \text{if } 1 - ml^{-LR1_{j,k}} \geq MR_{j,k} \end{cases} \quad \forall j \in J, k \in K, t \in T, ml \in M1 \quad (12)$$

$$LC2_{j,k,ml} = MR_{j,k}(1 + e^{-LR2_{j,k}(ml-TR2_{j,k})})^{-1} \quad \forall j \in J, k \in K, t \in T, ml \in M1 \quad (13)$$

3. Results and Discussion

The capabilities of the developed model are demonstrated with a case study of planning carbon capture (CC) from 20 different emission sources over a period of 20 years (discretized into 20 equal time periods). The carbon composition in the flue gas varies from 4 to 47%. The annual capture target increases biannually by 5% (from 30% to 70%). Two post-combustion capture technologies, absorption using aqueous monoethanolamine (ABS-MEA) and pressure swing adsorption using methyl viologen exchanged zeolite Y (PSA-MVY), are considered. The emission sources data and the cost models from Hasan et al. (2015) are used to construct model parameters.

Under no technological learning, the model optimizes the selection of emission sources to minimize the total capture cost while meeting the annual capture targets. ABS-MEA is preferred for sources with low carbon composition in flue gas, while PSA-MVY is cost-effective for sources with high carbon composition. The effect of cost reduction in investment on the portfolio planning decisions is analyzed by considering the technological learning from shared knowledge for PSA-MVY. Two different learning curves, log-linear and S-curve (Eqs. 12 and 13), are considered. Figure 2 shows the extent of cost reduction for the Nth-of-a-kind facility due to shared knowledge when the learning curves' parameters are varied. The maximum cost reduction extent possible is taken as 20%. The curves 1 to 3 represent the log-linear curves having learning rates (LR1) of 0.05, 0.10, and 0.30. At higher rates, the cost reduction extent reaches the maximum value from learning from fewer facilities using the technology. Curves 4 to 6 represent the S-curves having learning rates (LR2) of 1.5, 3, and 1.5 and inflection points (TR2) of 4, 4, and 6. The curves with higher learning rates and lower inflection points require learning from less number of facilities to reach the maximum cost reduction.

The models are formulated in Python V3.9.12 using PYOMO V6.4.1 and solved to an optimality gap of 1% using CPLEX V20.10 on an AMD Ryzen Threadripper PRO 3955WX 3.89 GHz processor with 16 cores and utilizing a maximum of 64 GB RAM.

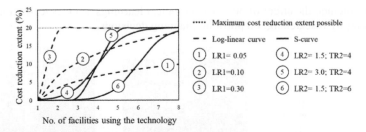

Figure 2. The extent of cost reduction due to technology learning from shared knowledge for different parameters of the log-linear curve and the S-curve.

Figure 3 shows the portfolio planning decisions for the cases when technological learning for PSA-MVY is considered and modeled using a log-linear curve with a learning rate of 0.10 and for the case not considering learning. Figures 3(a) and 3(b) show the CO_2 composition in flue gas and the total CO_2 emitted by the sources. The technology deployment decisions for the sources to meet annual capture targets over the planning period are shown in Figures 3(d) and 3(c) for the cases with and without learning, respectively. In the early planning periods, sources with high emissions are selected for capture. Other sources are selected to meet the higher capture targets over time. Comparing the results for the two cases in Figures 3(c) and 3(d), it can be seen that when technological learning is considered, some facilities using PSA-MVY are implemented in earlier periods to accelerate the learning for the later deployed facilities. On the other hand, some facilities are deployed in later periods to make the capture more cost-effective with technological learning. Source 2 uses ABS-MEA for capture in the case without technological learning. However, when the PSA-MVY learning is considered, PSA-MVY becomes more cost-effective than ABS-MEA for source 2 by the time of its capture facility deployment. The total capture cost decreases from \$25.98B to \$25.88B when the PSA-MVY investment cost decreases by the log-linear curve with a rate of 0.10.

The effect of modeling the learning with different learning curves and using different learning parameters on the planning decisions is shown in Figure 4. For the log-linear learning curve with a learning rate (LR1) of 0.05, the number of facilities using PSA-MVY at different planning periods does not change from that in the case with no learning. But as the learning rate is increased (LR1= 0.10 and LR1= 0.30), more facilities using PSA-MVY are deployed in the early planning periods to accelerate learning and to reduce the total capture cost further to \$25.88B and \$25.81B. For the S-curve with a high learning rate (LR2= 3) and a higher inflection point (TR2= 6), though the capture cost decreases due to technological learning, the deployment decisions remain unaffected. On the other hand, when a lower learning rate and inflection point (LR2= 1.5l; TR2= 4) is used, the deployment decisions are similar to the log-linear case with LR1=0.30, but its capture cost is still high due to its longer slow learning phase. The solution time (wall clock time) for these cases varies from 120 seconds to 15,600 seconds.

4. Conclusions and Future Directions

This work developed a multi-period optimization model that assists the selection of processes and technologies to produce a set of products while meeting product demands and accounting for the effect of technological learning on investment and operating costs. A case study of CC demonstrated the capability of the model to minimize the total cost of capture planning by accelerating technological learning due to shared knowledge by implementing more facilities in the early planning periods. Future work will extend the analysis to include the effect of other technological learnings, such as learning-by-doing and learning-by-researching, on portfolio planning decisions.

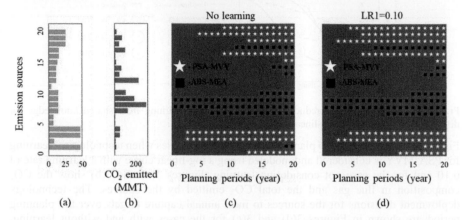

Figure 3. (a) CO₂ composition of the emission sources. (b) Total CO₂ emitted by the sources over the planning period. Deployment of capture technologies to the sources over the planning periods to meet the biannually increasing capture targets: (c) when technological learning is not considered, and (d) when learning due to shared knowledge for PSA-MVY is considered.

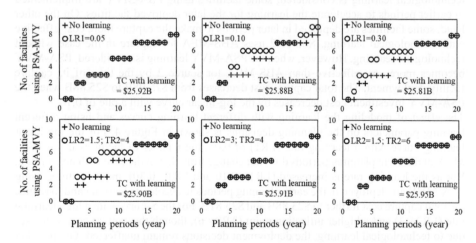

Figure 4. No. of facilities using PSA-MVY at different planning periods when technological learning from shared knowledge is considered and modeled with log-linear (LR1) and S-curves (LR2 & TR2) for different learning parameters.

References

M. J. Anzanello & F.S. Fogliatto (2011). Learning curve models and applications: Literature review and research directions. International Journal of Industrial Ergonomics, 41(5), 573-583.

A. Elia, M. Kamidelivand, F. Rogan, & B. Ó. Gallachóir (2021). Impacts of innovation on renewable energy technology cost reductions. Renew. Sustain. Energy Rev., 138, 110488.

M.F. Hasan, E. L. First, F. Boukouvala, & C. A. Floudas (2015). A multi-scale framework for CO₂ capture, utilization, and sequestration: CCUS and CCU. Comput. Chem. Eng., 81, 2-21.

G. Upstill & P. Hall (2018). Estimating the learning rate of a technology with multiple variants: The case of carbon storage. Energy Policy, 121, 498-505.

Flavio Manenti, Gintaras V. Reklaitis (Eds.), Proceedings of the 34th European Symposium on Computer Aided Process Engineering / 15th International Symposium on Process Systems Engineering (ESCAPE34/PSE24), June 2-6, 2024, Florence, Italy

Dynamic Hybrid Model for Nanobody-based Antivenom Production (scorpion antivenon) with *E. coli* CH10-12 and *E. coli* NbF12-10.

David Camilo Corrales[a], Susana María Alonso Villela[b], Balkiss Bouhaouala-Zahar[c,d], Julien Cescut[a], Fayza Daboussi[a,b], Michael O'Donohue[b], Luc Fillaudeau[b], César Arturo Aceves-Lara[b*]

[a] *INRAE, UMS (1337) TWB, 135 Avenue de Rangueil, 31077 Toulouse, France*
[b] *TBI, Université de Toulouse, CNRS, INRAE, INSA, 135 Avenue de Rangueil, 31077 Toulouse, France*
[c] *Laboratoire des Biomolécules, Venins et Applications Théranostiques (LBVAT), Institut Pasteur de Tunis, 13 Place Pasteur, BP-74, 1002 Le Belvédère, Tunis, Tunisia*
[d] *Faculté de Médecine de Tunis, Université Tunis El Manar, 15 rue Djebel Lakhdhar, 10007 Bab Saadoun Tunis, Tunisia*
Corresponding author: aceves@insa-toulouse.fr

Abstract

Immunotherapy is a specific treatment for scorpion stings, with antibody fragments being used to neutralize scorpion neurotoxins. Nanobodies (VHH), fragments of camelid antibodies, were successfully produced intracellularly in *Escherichia coli WK6* in fed-batch cultures. Their production was further enhanced by dynamic modelling. Two dynamic modelling approaches to describe nanobody CH12-10 production as a function of the induction temperature are proposed in this work. The first one is a kinetic model and the second is a hybrid approach, coupling a mass balance with support vector machine (SVM). Both models were calibrated and validated with independent data sets. Results reveal that the hybrid model procures better predictions than the kinetic model. Finally, the hybrid model was improved by retraining the SVM Model, resulting in a Normed Root Mean Square Error (NRMSE) values between 0.1148 and 0.8523.

Keywords: Nanobodies, *Escherichia coli*, Hybrid Model, SVM, Dynamic Models

1. Introduction

Scorpion stings pose a serious health problem in tropical and subtropical countries (Chippaux and Goyffon, 2008). Serotherapy targets the neurotoxins in the scorpion venom by using fragments of equine antibodies (F(ab)'$_2$). Engineered, toxin-specific Nanobodies (VHH), which are fragments of camelid antibodies, have been found to have better tissue penetration and lower molecular weight (15 kDa vs 100 kDa) than their equine counterparts (Bouhaouala-Zahar *et al.* 2011). These nanobodies are the best candidates to produce bispecific antibodies (Deffar *et al.* 2009). Their production in microbial hosts, such as *Escherichia coli*, is a frequent practice in the pharmaceutical industry to enhance antiserum production (Alonso Villela *et al.* 2023). However, defining optimal production conditions (induction, temperature, etc.) is a trial-and-error exercise. The production of recombinant proteins is usually modeled using Monod or Luedeking-Piret kinetics (Zheng *et al.* 2005; Hua *et al.* 2006), and rarely using dynamic models (Alonso Villela *et al.* 2021). Hybrid modeling approaches have recently gained importance (Badr *et al.* 2023, Kaya *et al.* 2023). Accordingly, in this paper we have compared the use of a kinetic model and a hybrid kinetic model integrating a Support Vector Machine (SVM).

2. Materials and Methods

2.1. Microbial strains and culture medium

Cultures were carried out with ampicillin-resistant recombinant strains *E. coli* CH10-12 (Alonso Villela *et al.* 2021) and *E. coli* NbF12-10 (Hmila *et al.* 2010). The nanobodies were produced after induction with a 1 mM IPTG pulse. Cultures were initiated by spreading a small aliquot of a glycerol stock on LB (Lysogeny Broth) agar plates. After incubation, a single well-isolated colony was used to inoculate 15 mL LB medium. The bioreactor inoculum was prepared by diluting 1 mL of the LB culture in 100 mL of Minimal Medium (MM, 5 g/L glucose) and incubated at 37 °C. MM was also used for the bioreactor batch. However, for fed-batch and production phases, a modified MM with 300 g/L glucose was used, both described elsewhere (Alonso Villela *et al.* 2021).

2.2. Experimental setup

Cultures were conducted in a 5 L bioreactor (Biostat® B-DCU, Sartorius) with a working volume of 2 L. The bioreactor was equipped with dissolved oxygen (pO_2), pH, temperature (T), and pressure (P) sensors, and two six-flat-blade Rushton turbines a three-blade marine impeller. Temperature was set to 37 °C for the batch and fed-batch modes, pO_2 kept over 15 %, and the pH was regulated at 7 for the entire culture. Induction of *E. coli* CH10-12 was carried at temperatures 28 °C, 29 °C, 30 °C, 32 °C, 33 °C, and 37 °C, for a duration of 6 h, and up to 35 h. *E. coli* NbF12-10 was carried at 29 °C for up to 35 h. The batch phase was started with 100 mL of inoculum in 1.5 L MM. Upon glucose depletion, the feed was started (μ of 0.38 h^{-1}) by the BioPAT® MFCS (Sartorius) software, and during the production phase the feed was set to 4 g/h of glucose.

2.3. Biomass quantification

The optical density (OD) of the cell culture was measured by spectroscopy at 600 nm. Biomass cell dry weight (cdw) was quantified by gravimetry. During the batch phase an aliquot (5 to 10 mL) of culture was removed and filtered in pre-weighted polyamide filters. During the fed-batch and production phases 1.5 mL of culture were centrifuged in a pre-weighted Eppendorf tube.

2.4. Glucose quantification

Glucose was quantified both by enzymatic analysis and high-performance liquid chromatography (HPLC) coupled with a photodiode array (UV)

2.5. Nanobody quantification

Periplasmic proteins were extracted by osmotic shock and purified by IMAC. The eluates were separated using SDS-PAGE in reducing conditions and later quantified by image densitometry according to (Alonso Villela *et al.*, 2020).

2.6. Gas analysis

Inlet and outlet gas were analyzed for carbon dioxide (CO_2) and oxygen (O_2) using photoacoustic spectroscopy and magneto-acoustic spectroscopy, respectively.

3. Model description

The bioreactor is assumed to be an infinitely mixed culture without any transfer limitation. The mass balances for biomass, glucose, proteins and volume are given by Eq. (1) – (4).

$$\frac{dX}{dt} = r_X - X * \frac{F_{in}}{V} \tag{1}$$

$$\frac{dS}{dt} = \frac{F_{in}}{V}(S_{in} - S) - r_S - S * \frac{F_{in}}{V} \tag{2}$$

$$\frac{dP}{dt} = r_P - P * \frac{F_{in}}{V} \tag{3}$$

$$\frac{dV}{dt} = F_{in} \tag{4}$$

X and S are the concentrations of biomass and glucose, in g/L, and P is the concentration of nanobody proteins, in mg/L. F_{in} is the glucose flowrate, in L/h, V is the volume of the liquid phase in the bioreactor, in L, S_{in} is the concentration of glucose in the feed, in g/L, and r_X is the production rate of biomass, in g/h, and defined by Eq. (5). r_S is the consumption rate of glucose, in g/h, as defined by Eq. (6).

$$r_X = \mu * X \tag{5}$$

$$r_S = -\frac{1}{Y_{SX}} * \mu * X \tag{6}$$

With μ as the specific growth rate, in h^{-1}, as defined by a Monod kinetics in Eq. (7), and Y_{SX} as the yield coefficient of substrate, in g cdw/g S.

$$\mu = \frac{\mu_{max} * S}{K_S + S} \tag{7}$$

$$\mu_{max} = \mu'_{max} * T + b \tag{8}$$

μ_{max} is the maximum specific growth rate in h^{-1} and K_S is the substrate saturation coefficient, in g/L. μ'_{max} is a temperature-dependent factor in h^{-1}, T is the temperature of the bioreactor, in °C, and b is a temperature-independent factor, in h^{-1}. The production rate of nanobody proteins, r_P, is in mg/h, and defined as follows:

$$r_P = \begin{cases} 0 & \textit{if there is not induction} \\ q_P * X & \textit{if there is an induction} \end{cases} \tag{9}$$

Where q_p is the specific productivity of the nanobody production (mg_{protein}/g cdw/h). Two models were proposed to predict the specific productivity of the nanobody production. In the first one, q_P is modelled using a temperature-dependent equation, and is described in section *3.1. Dynamic kinetic Model*. In the second one, q_P is modelled with a hybrid model using a Support Vector Machine (SVM), described in section *3.2. Hybrid Model approach*.

3.1. Dynamic kinetic model

The dynamic kinetic model uses a kinetic for q_P that is function of temperature as presented in Eq. (11) and established by (Alonso Villela *et al.*, 2021).

$$q_P = -\alpha * \mu + \gamma_1 * \exp(-Ap_1/T) - \gamma_2 * \exp(-Ap_2/T) + \beta \tag{11}$$

With α as the growth-dependent factor, in mg/g cdw, γ_1 and γ_2 as the pre-exponential coefficients, in mg/g cdw/h, Ap_1 and Ap_2 as the activation and inactivation coefficient, respectively in °C, and β as the temperature-independent coefficient in mg/g cdw/h.

3.2. Dynamic hybrid model approach

The hybrid model integrates a SVM learner to compute q_P. The SVM was trained with experimental data provided by (Alonso Villela *et al.*, 2021).

$$q_P = f(T, \mu) \tag{12}$$

Four kernel functions (linear, quadratic, cubic and Gaussian) were tested using Matlab2020a. The Gaussian kernel obtained the best performance using 5 k-fold cross validation.

3.3. Statistical analysis

Models prediction quality for protein dynamic production were analyzed using the normed root mean square error (NMRSE) defined as:

$$NRMSE = \sqrt{\frac{\sum_{i=1}^{T}(Y_{exp} - \hat{Y})^2}{n}} / (Y_{exp,\,max} - Y_{exp,\,min}) \tag{13}$$

With n the number of experimental data, Y_{exp} the experimental data corresponding to time t_i, $\hat{Y}$ the data simulated at t_i, Y_{min}, and Y_{max} the minimum and maximum data values.

4. Results

4.1. Model calibration

The models were calibrated using *E. coli* CH10-12 fed-batch cultures induced at 28, 32, 33 and 37 °C using the *fmincon* function Matlab2020a. Table 1 shows model calibrated parameters. μ_{max} and Y_{sx} have similar values ($0.67 < \mu_{max} < 0.87$ and $0.29 < Y_{sx} < 0.47$ respectively) to those reported in literature (Alonso Villela et al., 2021).

Table 1. Model Parameters

Parameters	μ_{max}	K_s	Y_{sx}	μ'_{max}	b
Value	0.7853	0.0001	0.2361	0.3538	0.0726

Figure 1 shows both kinetic model and hybrid model simulations for two data sets (28 °C and 32 °C). Both models correctly estimate the final production of biomass and substrate.

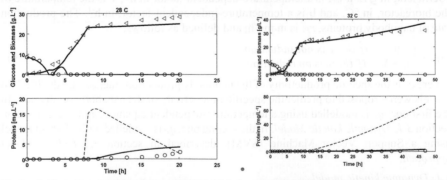

Figure 1. Experimental data of *E. coli* CH10-12 induced at 28 °C (left) and 32 °C (right) used for calibration of glucose (o), biomass (<) and proteins (o), kinetic model (- -), and hybrid model (–).

The novel hybrid model provides a better estimation of the dynamics of protein production than the kinetic model. Models were validated with three independent data sets of *E. coli* CH10-12 fed-batch cultures induced at temperatures 29 and 30 °C, and *E. coli* NbF12-10 induced at 29 °C. Figure 2 shows both model simulations and experimental data for two data sets of *E. coli* CH10-12 (29 °C and 30 °C). Once again, the hybrid dynamic model has a better prediction for proteins than the kinetic model.

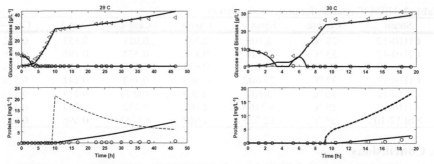

Figure 2. Experimental data of *E. coli* CH10-12 induced at 29 °C (left) and 30 °C (right) used for validation of glucose (o), biomass (<) and proteins (o), kinetic model (- -), and hybrid model (–).

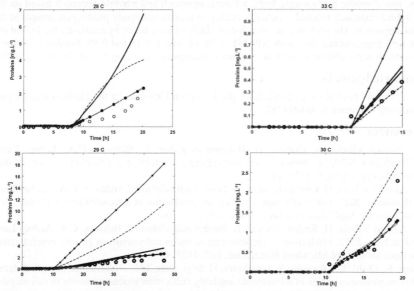

Figure 3. Experimental data for protein concentration (o) in *E. coli* CH10-12 induced at 28 °C (upper-left), 33 °C (upper-right), 29 °C (lower-left), and 30 °C (lower-right), hybrid model for calibration (- -), Case 1 (*-), Case 2 (·-) and Case 3 (–).

The hybrid model, SVM, was retrained using experimental data other than calibration to improve protein simulations. Four cases were compared:
- Case 1: CH10-12 culture data used for calibration.
- Case 2: CH10-12 culture data used for calibration and induced at 30 °C.
- Case 3: CH10-12 culture data used for calibration and induced at 29 °C, and 30 °C.
- Case 4: CH10-12 culture data used for calibration and induced at 29 °C, 30 °C, and NbF12-10 data induced at 29 °C.

Table 2 reveals that the hybrid approach achieved better accuracy than the kinetic approach. Case 2 and Case 3 have similar NRMSE values (Table 2 and Figure 3). These results show that the hybrid model could be improved by retraining the experimental data. Finally, the hybrid model is able to simulate the production of the protein correctly for two different *E. coli* strains (Table 2).

Table 2. NRMSE analysis for proteins simulation

Strain	Culture	Kinetic	Case 1	Case 2	Case 3	Case 4
CH10-12	28 °C	3.109	0.577	0.911	0.153	0.152
CH10-12	33 °C	7.427	0.118	0.534	0.195	0.170
CH10-12	37 °C	18.011	0.635	1.049	0.453	0.429
CH10-12	32 °C	29.802	0.860	0.168	0.141	0.538
CH10-12	30 °C	3.107	0.209	0.0879	0.115	0.125
CH10-12	29 °C	3.2463	2.102	4.132	0.254	0.421
NbF12-10	29 °C	12.775	4.010	8.6452	0.696	0.852

5. Conclusions

The data from recombinant protein production produced in *E. coli* CH10-12 fed-batch cultures were used to build two dynamic models to optimize the nanobody production for future cultures. In this study, two dynamic models were built, a kinetic approach and a hybrid approach based on SVM. The hybrid approach reached a better accuracy to predict nanobody production compared to the kinetic approach. The protein concentration predictions were better by retraining the hybrid model using new experimental data with NRMSE values between 0.11 and 0.85. Finally, this approach shows that it is possible to accurately simulate protein production from two different *E. coli* strains.

Acknowledgments

This work was funded (or co-funded) by the European Union under the Horizon Europe project Bioindustry 4.0, grant n. 101094287

References

S.M. Alonso Villela, H. Kraïem, B. Bouhaouala-Zahar, C. Bideaux, C.A. Aceves Lara, L. Fillaudeau, 2020. A protocol for recombinant protein quantification by densitometry. MicrobiologyOpen, 9, 1175–1182.

S.M. Alonso Villela, H. Ghezal-Kraïem, B. Bouhaouala-Zahar, C. Bideaux, C.A. Aceves Lara, L. Fillaudeau, 2021. Effect of temperature on the production of a recombinant antivenom in fed-batch mode. Appl Microbiol Biotechnol, 105, 1017–1030.

S.M. Alonso Villela, H. Kraïem-Ghezal, B. Bouhaouala-Zahar, C. Bideaux, C.A. Aceves Lara, L. Fillaudeau, 2023. Production of recombinant scorpion antivenoms in *E. coli*: current state and perspectives. Appl Microbiol Biotechnol, 107, 4133–4152

S. Badr, K. Oishi, K. Okamura, S. Murakami, H. Sugiyama, 2023. Hybrid modelling and data-driven parameterization of monoclonal antibody cultivation processes: Shifts in cell metabolic behavior, Comput Aided Chem Eng, 52, 985-990.

B. Bouhaouala-Zahar, R. Ben Abderrazek, I. Hmila, N. Abidi, S. Muyldermans, M. El Ayeb, 2011. Immunological aspects of scorpion toxins: current status and perspectives. Inflamm Allergy Drug Targets 10, 358–368

J.P. Chippaux, M. Goyffon, 2008. Epidemiology of scorpionism: A global appraisal. Acta Trop, 107, 71–79.

K. Deffar, H. Shi, L. Li, X. Wang, X. Zhu, 2009. Nanobodies - the new concept in antibody engineering. Afr J Biotechnol, 8, 2645–2652.

I. Hmila, D. Saerens, R.B. Abderrazek, C. Vincke, N. Abidi, Z. Benlasfar, J. Govaert, M.E. Ayeb, B. Bouhaouala-Zahar, S. Muyldermans, 2010. A Bispecific Nanobody to Provide Full Protection against Lethal Scorpion Envenoming. FASEB j, 24 (9), 3479–3489

X. Hua, D. Fan, Y. Luo, X. Zhang, H. Shi, Y. Mi, X. Ma, L. Shang, G. Zhao, 2006. Kinetics of High Cell Density Fed-batch Culture of Recombinant *Escherichia coli* Producing Human-like Collagen. Chin J Chem Eng, 14, 242–247.

S. Kaya, H. Kaya, A.W. Rogersa, D. Zhang, 2023. Integrating hybrid modelling and transfer learning for new bioprocess predictive modelling, Comput Aided Chem Eng, 52, 985-990.

Z.Y. Zheng, S.J. Yao, D.Q. Lin, 2005. Using a kinetic model that considers cell segregation to optimize hEGF expression in fed-batch cultures of recombinant *Escherichia coli*. Bioprocess Biosyst Eng, 27, 143–152.

Flavio Manenti, Gintaras V. Reklaitis (Eds.), Proceedings of the 34th European Symposium on Computer Aided Process Engineering / 15th International Symposium on Process Systems Engineering (ESCAPE34/PSE24), June 2-6, 2024, Florence, Italy

Hybrid Modeling of PEM Fuel Cell: Machine Learning of Equilibrium Water Content within a 1D Cell Model

Ievgen Golovin [a], Kilian Knoppe [a], Christian Kunde [b], Achim Kienle[a,b]

[a]*Otto von Guericke University, Universitätsplatz 2, 39106 Magdeburg, Gemany*
[b]*Max Planck Institute for Dynamics of Complex Technical Systems, Sandtorstraße 1, 39106 Magdeburg, Germany*
ievgen.golovin@ovgu.de

Abstract

Low-temperature proton-exchange membrane fuel cells (PEMFCs) require careful water management in order to keep the membrane hydrated for high proton conductivity and to keep the gas channels from flooding with liquid water. Accurate dynamical models are needed for model-based approaches for controller and soft sensor design. In the case that some properties of the fuel cell components are not well known or physical models for them are not available, a hybrid model, combining a first-principles-based model with data-based submodels, may be applied. This work explores how a specific submodel of a PEMFC, namely the equilibrium water content of the membrane, can be determined from measurement data that would be readily available from a fully assembled fuel cell. Synthetic data generated from a reference PEMFC model are used to train the hybrid model. Results for data with different sampling rates and levels of added noise are compared to study the required data quality for successful training.

Keywords: PEMFC, hybrid modeling, machine learning

1. Introduction

The pivotal component of a PEMFC is the membrane. It provides the separation of gases between the anode and cathode gas channels, as well as the transport of protons. Ensuring sufficient membrane hydration is essential to maintain high proton conductivity and, consequently, achieve efficient power generation (Singh et al., 2022). Model-based water management requires the establishment of reliable dynamical models of PEMFC. Special attention should be directed to the membrane submodel, as it often includes empirical dependencies derived from specific experiments conducted in the earlier phases of PEMFC development (Liso et al., 2016). One challenge here is the accurate description of the water entering the membrane. The equilibrium water content of the membrane depends on the water vapor activity and the temperature in the gas phase. The reliability of empirical relations determined for specific membrane materials or operating ranges is a concern if they are applied outside of these conditions. Determining this relation directly from measurement data of a fully assembled fuel cell instead might increase model fidelity with limited additional effort.

In this paper, a hybrid PEMFC model is proposed (Martensen et al., 2023), offering the advantage of preserving the physical states within the system and the possibility to learn the equilibrium water content from data. Here, it is assumed that the equilibrium water content function can be approximated using an artificial neural network (ANN). To determine the parameters of the ANN, synthetic data sampled from the output signals of a reference model are used. Output signals that can typically be measured on a fully assembled PEMFC are selected for this purpose. The influence of data sampling rates and measurement noise on the training process and resulting function is studied.

2. Materials and Methods

2.1. Reference model of a PEMFC

In this case study, a reference model is used to generate synthetic data in place of actual measurements. The PEMFC reference model is based on (Mangold et al., 2010), while the details of the membrane model can be found in (Neubrand, 1999). This is a 1D along-the-channel, first-principle, single-phase, non-isothermal model that is divided into the following parts: anode and cathode gas channels, gas diffusion layers, catalyst layers, membrane and electrode. The resulting model consists of a set of partial differential equations (PDEs) and algebraic equations.

For numerical simulations, the spatial coordinate was discretized using the finite volume method with N_{cv} control volumes. The resulting system of differential and algebraic equations (DAEs) has the following structure:

$$\begin{bmatrix} I & 0 \\ 0 & 0 \end{bmatrix} \begin{bmatrix} \dot{x}_d \\ \dot{x}_a \end{bmatrix} = \begin{bmatrix} f_d(t, x_d, x_a, u) \\ f_a(t, x_d, x_a, u) \end{bmatrix} \tag{1}$$

with initial conditions for $t_0 = 0$

$$\begin{aligned} x_d(t_0) &= x_{d,0}, \\ f_a(t, x_{d,0}, x_{a,0}, u) &= 0, \end{aligned} \tag{2}$$

where $x_d \in \mathbb{R}^{11N_{cv}}$ is the vector of dynamic states that consists of the concentrations of species in the anode (H_2, H_2O) and cathode (O_2, N_2, H_2O) gas channels, water content in the membrane, electrical potential difference, internal energies in gas channels, and total energy of the solid, $x_a \in \mathbb{R}^{12N_{cv}+1}$ is the vector of algebraic states that consists of the mole fractions of the species in the catalyst layers, electrical potential and current density of the membrane, temperatures of gases in the channels and solid, pressures in the gas channels, and voltage of the fuel cell, $u \in \mathbb{R}^{11}$ is the vector of inputs that consists of input molar flows of reactants and water, temperature of gases at the inlets, coolant temperature, output pressures of the channels, and electrical current.

Combining the dynamic and algebraic states together and considering the output equation, the reference model of PEMFC can be rewritten in compact form as

$$\begin{aligned} M\dot{x} &= f(t, x, u), \\ y &= h(t, x, u), \end{aligned} \tag{3}$$

where $y \in \mathbb{R}^3$ is the vector of outputs that consists of the voltage and output flows of water vapor from the gas channels.

The water content at the membrane surface is assumed to be in equilibrium with the gas phase of the neighboring catalyst layer. There is a number of different equilibrium water content functions or so-called sorption isotherms that can be found in the literature (Dickinson et al., 2020). For the reference model, two polynomial relations are used that have been fitted to experiments for the temperature points 30 °C (Springer et al., 1991) and 80 °C (Hinatsu et al., 1994):

$$\Lambda_{30} = 36\, a_{H_2O}^3 - 39.85\, a_{H_2O}^2 + 17.81\, a_{H_2O} + 0.043,$$
$$\Lambda_{80} = 14.1\, a_{H_2O}^3 - 16\, a_{H_2O}^2 + 10.8\, a_{H_2O} + 0.3, \tag{4}$$

where a_{H_2O} is the relative humidity in the gas phase.

For temperatures between these values, linear interpolation is applied, resulting in the following function of two variables, see Fig. 1,

$$\Lambda = f_\Lambda(a_{H_2O}, T) \tag{5}$$

that is used in the anode and cathode catalyst layer for each control volume.

2.2. Hybrid model of a PEMFC

The hybrid model is trained on data in order to reproduce the physical behavior of a PEMFC, represented in this case study by the reference model.

Each instance of f_Λ in the reference model is replaced in the hybrid model by

$$\hat{f}_\Lambda = a_{H_2O}\, f_{ANN}(v, \theta), \tag{7}$$

where f_{ANN} is a feedforward ANN with a single hidden layer with hyperbolic tangent activation function and a linear output layer, $v = [a_{H_2O} \quad T]^\mathsf{T}$ is the vector of the ANN inputs and $\theta \in \mathbb{R}^{N_\theta}$ is a parameter vector that consists of the weights and biases of the ANN layers.

In the resulting hybrid model structure

$$M\dot{x} = \tilde{f}(t, x, \tilde{x}, u),$$
$$\tilde{x} = f_{\tilde{x}}(x), \tag{6}$$
$$y = h(t, x, u),$$

$\tilde{x}$ are auxiliary variables in the right-hand-side function that are calculated by the submodel $f_{\tilde{x}}(x)$. The function $f_{\tilde{x}}$ comprises all appearances of the equilibrium water content function $\Lambda = f_\Lambda(a_{H_2O}, T)$ in the discretized model.

2.3. Parametrization problem

In this paper, reconstruction of the function from the PEMFC output vector is considered. The objective function based on the mean squared error (MSE) is used as follows:

$$J(\theta) = \frac{1}{N_{exp}} \sum_{i=1}^{N_{exp}} (y_i^s - \hat{y}_i^s(\theta))^2, \tag{8}$$

where, $y^s \in \mathbb{R}^{3 \times N_{exp}}$ is the min-max scaled output vector from the collected data, N_{exp} is number of sample points, and $\hat{y}^s \in \mathbb{R}^{3 \times N_{exp}}$ is the min-max scaled output vector of the hybrid model calculated for the parameter set of ANN θ.

3. Numerical setup

The reference and hybrid models are implemented in MATLAB 2022b (The MathWorks, Inc., Natick, MA, USA). For the discretization $N_{cv} = 20$ is used, which results in 220 dynamic and 241 algebraic states. To solve these models efficiently, the ode15s solver with sparse mass matrix M and additional Jacobian pattern options are used. In addition, for the right-hand-side functions of the state equations in (3) and (6), MEX-functions are generated and used for the simulations.

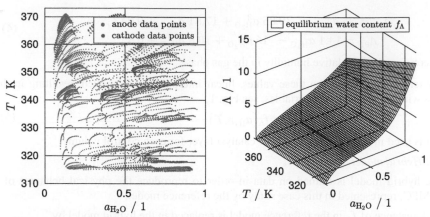

Figure 1 – Locations of data points for internal variables (left, every 20th point is
plotted) and shape of reference equilibrium water content f_Λ (right).

The reference model is used to generate output data of the PEMFC for the parametrization
of the hybrid model. The function $\hat{f}_\Lambda$ is supposed to approximate the f_Λ well over its entire
domain, which is restricted to values $0.05 \leq a_{H_2O} \leq 0.95$ and $313 \leq T \leq 373$ K. For
this purpose, an input trajectory is manually designed such that, according to the reference
model, the domain of f_Λ is well covered. The resulting input trajectory contains 14 set
points with 2 min of transition time and 8 min of steady-state time. To smoothen the
trajectories during the transition, shape-preserving piecewise cubic interpolation was
used. In Fig. 1, coverage of the domain for f_Λ is depicted. The sample data $d = (y, t_k)$
for input trajectory u contains the output matrix $y \in \mathbb{R}^{3 \times N_{exp}}$ and time vector $t_k \in \mathbb{R}^{N_{exp}}$.
The sampling time $t_s = 1$ s is chosen for the nominal case 1. To study how the quality of
data influences training, three additional data sets with a different sampling time and
added normally distributed noise are generated:

- case 2: $t_s = 5$ s, no added noise;

- case 3: $t_s = 1$ s, added normally distributed noise ($\sigma = 0.5\%$ for voltage, $\sigma = 2\%$ for output flows);

- case 4: $t_s = 5$ s, added normally distributed noise ($\sigma = 0.5\%$ for voltage, $\sigma = 2\%$ for output flows).

The ANN f_{ANN} includes 10 neurons in the hidden layer resulting in $N_\theta = 41$ parameters
that need to be estimated from training. For the ANN inputs and outputs min-max scaling
is used. The initialization of the ANN parameters needs to be approached with caution to
ensure that a numerical solution for the overall model can be calculated. Here, the
parameters of the hidden layer are initialized randomly and the parameters of the output
layer are initialized such that $\hat{f}_\Lambda$ is approximately linear in a_{H_2O} and constant in T,
yielding the initial set of ANN parameters θ_{init}.

The training of the ANN is based on minimizing the objective in (8). Here, MATLAB's
fminunc is used with default options except step tolerance set to 10^{-16} and specified
objective gradient. The objective gradient is calculated using the complex-step method.

4. Results

The model was trained separately for cases 1-4, starting from the same initial parameters θ_{init}. The results are visualized in Fig. 2 and Fig. 3. Fig. 2 demonstrates the convincing agreement between the output of the trained hybrid model and the sample data with added noise, i.e. case 2. Fig. 3 shows the errors between equilibrium water content in the trained hybrid model and the reference model. The mean values of the errors $\left| \hat{f}_{\Lambda} - f_{\Lambda} \right|$ are summarized in Table 1. Despite the decrease in data quality in the considered cases, only insignificant errors are observed in the domain of interest, except near its boundaries, where the density of data points is lower.

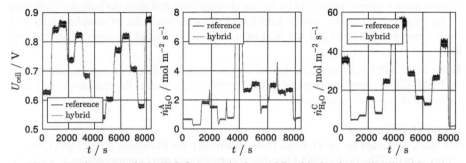

Figure 2 – Outputs of PEMFC for sample data with added noise and trained hybrid model. Cell voltage (left), water vapor flow at anode (middle) and cathode (right).

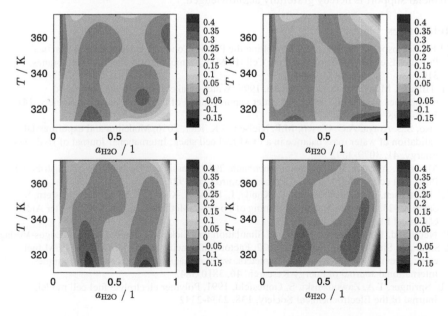

Figure 3 – Error $\hat{f}_{\Lambda} - f_{\Lambda}$ between equilibrium water content in the trained hybrid model and the reference model. Training on datasets for the nominal case 1 (top left), case 2 (top right), case 3 (bottom left), and case 4 (bottom right).

Table 1 – Mean values of the absolute error $|\hat{f}_\Lambda - f_\Lambda|$

Cases	1	2	3	4
Mean error	2.31e-2	2.26e-2	3.89e-2	4.14e-2

5. Conclusions

In this work, a hybrid model for a PEMFC containing a data-based submodel for the equilibrium water content of the membrane has been presented. It has been shown that the equilibrium water content can be accurately approximated from the output signals that can be measured on a fully assembled PEMFC. In addition, reduced sampling rates and added measurement noise did not lead to significantly reduced accuracy or overfitting issues.

Future work will be dedicated to the training of the hybrid model on real measurements of a PEMFC and corresponding design of experiments, as well as the extension to a larger number of data-based submodels.

Acknowledgements

This work is funded by the German Federal Ministry for Economic Affairs and Climate Action as part of the project "KI-Grundlagenentwicklung mit Leitanwendungen Virtuelle Sensorik und Brennstoffzellenregelung (KI-Embedded)" (FKZ: 19I21043E). The financial support is hereby gratefully acknowledged.

References

E.J.F. Dickinson, G. Smith, 2020, Modelling the Proton-Conductive Membrane in Practical Polymer Electrolyte Membrane Fuel Cell (PEMFC) Simulation: A Review, Membranes, 10, 310

J.T. Hinatsu, M. Mizuhata, H. Takenaka, 1994, Water uptake of perfluorosulfonic acid membranes from liquid water and water vapor, Journal of the Electrochemical Society, 141, 1493-1498

V. Liso, S.S. Araya, A.C. Olesen, M.P. Nielsen, S.K. Kær, 2016, Modeling and experimental validation of water mass balance in a PEM fuel cell stack, International Journal of Hydrogen Energy, 41, 3079-3092

M. Mangold, A. Bück, R. Hanke-Rauschenbach, 2010, Passivity based control of a distributed PEM fuel cell model, Journal of Process Control, 20, 292-313

C.J. Martensen, C. Plate, T. Keßler, C. Kunde, L. Kaps, A. Kienle, A. Seidel-Morgenstern, S. Sager, 2023, Towards Machine Learning of Power-2-Methanol Processes, Computer Aided Chemical Engineering, 52, 561-568

W. Neubrand, 1999, Modellbildung und Simulation von Elektromembranverfahren, Logos-Verlag

R. Singh, A. Singh Oberoi, T. Singh, 2022, Factors influencing the performance of PEM fuel cells: A review on performance parameters, water management, and cooling techniques, International Journal of Energy Research, 46, 3810-3842

T.E. Springer, T.A. Zawodzinski, S. Gottesfeld, 1991, Polymer electrolyte fuel cell model, Journal of the Electrochemical Society, 138, 2334-2342

Flavio Manenti, Gintaras V. Reklaitis (Eds.), Proceedings of the 34th European Symposium on Computer Aided Process Engineering / 15th International Symposium on Process Systems Engineering (ESCAPE34/PSE24), June 2-6, 2024, Florence, Italy

On the development of hybrid models to describe delivery time in autoinjectors

Andrea Friso[a], Mark Palmer[b], Gabriele Bano[c], Federico Galvanin[a*]

[a] *Department of Chemical Engineering, University College London, Gower Street, London WC1E6BT, United Kingdom*
[b] *GlaxoSmithKline, Park Road, Ware, SG120DP, United Kingdom*
[c] *GlaxoSmithKline, 1250S Collegeville (PA) 19426, United States*
Correspoding author e-mail: f.galvanin@ucl.ac.uk

Abstract

Autoinjector (AJ) devices are medical devices that enable subcutaneous self-administration of medicines. A key challenge for AJs is the evaluation of injection time because in can impact on the usability of the autoinjector. Several models have been presented in literature to describe the injection time in AJs with different grade of complexity and accuracy. The limitation of the existing models is that they depend only on geometrical properties of the AJs, rheology of the fluid injected and the speed of the plunger. However, these models do not consider the dependence of friction on the distance travelled by the piston. In this paper a new hybrid model is developed to describe the injection time in AJs, integrating physics-based elements from the literature with data driven components. Results show that the proposed model is able to predict in an accurate way the experimentally measured friction force acting on the plunger.

Keywords: mathematical modelling, hybrid models, Gaussian processes.

1. Introduction

The use of autoinjectors (AJs) has grown in recent years due to the ease of use of these devices and the fact that they can be used without the intervention of medical personnel. AJs are used to deliver the drug necessary to treat various diseases such as diabetes and arthrosis, and for each treatment the characteristics of the AJ change (Vijayaraghavan 2012). The key parameter of these devices is the injection time, i.e., the time required to inject all the drug. This parameter is crucial for the use of autoinjectors. In fact, very long delivery times have the potential to usability because the patient may not be able to hold the injector in place for the required period of time. This could lead to early removal of the autoinjector which would result in an incomplete dose. Very short delivery times may also be undesirable if this leads to an increase in injection pain. For this reason, the accurate design, i.e., geometry of AJs, is pivotal. One tool that can be useful in the design of AJs are mathematical models to describe the forces acting on the AJs. The use of mathematical models, in the form of differential and algebraic equations (DAEs), plays an important role in the design and optimization of operating conditions of injection devices. Different models to describe the forces involved in AJs operation have been proposed in the literature (Rathore et al., 2011, Zhong et al. 2021). These models describe the movement and the behavior of the drug as it flows through the various components of the device such as the syringe barrel and the needle. These forces are:

- Driving force: this force is the force that pushes the plunger of the syringe making the injection process possible;

- Hydrodynamic force: this force is classified as resistance force because it is opposed to the driving force. This force refers to the fluid dynamic effects of the medication inside the barrel;
- Friction force: this force, as the hydrodynamic one, is classified as resistance force because it opposes the movement of the piston. This force is generated by friction due to the contact between the barrel and the piston sliding in the syringe.

In Fig. (1) a schematic representation of an AJ with the forces acting on the system is reported.

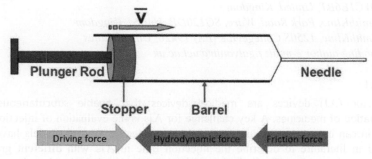

Figure 1: Schematic representation of the structure and forces acting on the autoinjector (Rathore et al. 2011)

Models proposed in literature manage to describe in an accurate way the friction force and the driving force, but so far there are no adequate models to describe the dynamics of the friction force. According to the literature the friction force should only depend on the velocity of the plunger. Instead, from the measured friction force, it is observed that the friction force depends also on the travelled distance. This project aims to propose a hybrid model capable of predicting the values of the friction force obtained from experimental measurements at different conditions of plunger speed. The hybrid model will combine a mechanistic model with a Gaussian process (Rasmussen and Williams, 2006) to represent the model mismatch to obtain a model that can predict the variable value of the friction forces in AJs at different operating conditions.

2. Proposed framework and methodology

2.1. Mathematical models – Mechanistic models

In this section the proposed framework to develop the hybris model is presented and explained. Mechanistic models of AJs are usually described by a system of differential and algebraic equations (DAEs) expressed in the following general form:

$$\begin{cases} \mathbf{f}(\dot{\mathbf{x}}(t), \mathbf{x}(t), \mathbf{u}(t), \boldsymbol{\theta}, t) = 0 \\ \hat{\mathbf{y}} = g(\mathbf{x}(t)) \end{cases} \tag{1}$$

In Eq. (1) $\mathbf{x}$ is the array of the N_x state variables, $\dot{\mathbf{x}}$ is the array of the state variables derivatives, $\mathbf{u}$ is the $N_u \times 1$ array of the experimental inputs, $\boldsymbol{\theta}$ is the N_θ-dimension array of the model parameters, t is the time, $\hat{\mathbf{y}}$ is the $N_{\hat{y}}$-dimensional vector of predicted outputs for the measured variables in the system under analysis. Every system that is governed by physics laws theoretically has a model that can describe accurately its behavior, but this "true" AJ model involves phenomena in the definition of friction forces that are too complex to be identified and for this reason it is necessary to use an approximated model. When an approximate model described by Eq. (1) is used to obtain output predictions these will be different from the output of an ideal "true" model. This difference is called model mismatch and can be expressed in the following form:

$$\boldsymbol{\varepsilon}^M = \mathbf{y}^{true} - \hat{\mathbf{y}} \tag{2}$$

In Eq. (2), $\boldsymbol{\varepsilon}^M$ is the model mismatch, $\mathbf{y}^{true}$ is the true model prediction and $\hat{\mathbf{y}}$ is the approximate model prediction described by Eq. (1). Given this expression the set of experimental measurements $\mathbf{y}$ can expressed as sum of the model prediction and the prediction mismatch:

$$\mathbf{y} = \hat{\mathbf{y}} + \boldsymbol{\varepsilon}^M + \boldsymbol{\varepsilon}_y{}^2 \tag{3}$$

where $\boldsymbol{\varepsilon}_y{}^2$ is the vector of measurement errors, here assumed to be normally distributed with zero mean and variance matrix $\boldsymbol{\sigma}^2$. If the model mismatch is high, this means that the adopted mechanistic model is unable to describe the data adequately and therefore its fitting must be improved. The method proposed in this case is not to directly improve the fitting of the model by acting on the approximate model itself and refining its structure but to use a data-driven model to directly model the mismatch between approximate model prediction and measurement. Combining the data driven model with the mechanistic model will result in a hybrid model able to predict the output adequately.

2.2. Mathematical models – Gaussian processes

Hybrid models refer to a category of models that combine different models or techniques to describe a system. In this case the two used models are a mechanistic model, capable of describing the physics of the system and an empirical model, used to describe the model mismatch. The selected data-driven model selected to model the mismatch is the Gaussian process (GP). GPs are a particular class of non-parametric models that can be used both for classification and regression. GPs extend the concept of a multivariate Gaussian distribution to create a distribution over an infinite-dimensional vector of functions, such that every finite sample of function values is Gaussian distributed. GPs aim to model an unknown set of functions, f_i, given a set of measurements. In this case study the set of functions will be used to describe the model mismatch. These models are fully specified with the prior mean function, $\mu(\cdot)$, and the kernel function, $k(\cdot,\cdot)$:

$$f_i \sim GP\big(\mu_i(\cdot), k_i(\cdot,\cdot)\big) \tag{4}$$

The mean function gives information about the mean of the test points prior to observing data and the kernel function returns the covariance between points. Given a dataset, the posterior distribution of a point x^* is determined using the available noisy data and their distribution: $D_i := \{X, y_i\}$, where $X = [x_1, .., x_N]$. At this point the posterior mean (m_i) and variance (Σ_i) of x^* are determined:

$$m_i(x^*) = \mathbb{E}(f_i^*|x^*, D) = k_i(x^*, X)[K_i(X, X) + \sigma_i^2]^{-1}\big(y_i - \mu_i(x^*)\big) + \mu_i(x^*) \tag{5}$$

$$\Sigma_i(x^*) = \mathbb{V}(f_i^*|x^*, D) = k_i(x^*, x^*)k_i(x^*, X)[K_i(X, X) + \sigma_i^2]^{-1}K_i(X, x^*) \tag{6}$$

Where $\mathbb{E}$ is the expectation, $\mathbb{V}$ is the variance, K is the covariance matrix and σ_i^2 is the noise. At this point the posterior mean and the variance can be used:

$$f(x^*) \sim \mathcal{N}\big(m(x^*), \Sigma_i(x^*)\big) \tag{7}$$

There are several options for the function to use to describe the kernel. For this application the radial basis function (RBF) kernel has been selected. This kernel is parameterized using a length scale, $l > 0$, that can either be scalar or a vector with the same dimensions of the input. This kernel is given by the expression in Eq. (8).

$$k(x_i, x_j) = \sigma^2 exp\left(-\frac{d(x_i, x_j)^2}{2l^2}\right) \tag{8}$$

where l is the length scale of the kernel, $d(\cdot,\cdot)$ is the Euclidian distance and σ^2 the output variance. Once the GP is defined it is possible to define the output using Eq. (3) where ε^M is obtained using the GP

3. Case study

The proposed framework has been tested on the prediction of the friction force in the AJs devices using a dataset obtained running real experiments.

3.1. Dataset

The available data were obtained from experiments conducted with a constant plunger speed such that the total balance of forces was zero. To maintain a constant piston speed and to measure the force necessary for movement, a dynamometer was used. To obtain the value of the friction force throughout the experiment, the fact that the summation of the acting forces was zero was exploited (the acceleration of the system is equal to zero):

$$F_{driving} = F_{hydrodynamic} + F_{friction} \tag{10}$$

This was followed by the assumption that the hydrodynamic model used to calculate the hydrodynamic force, Eq. (11) (Rathore et al. 2011), was correct and thus using Eq. (10) it was possible to estimate the friction force.

$$F_{hydrodynamic} = \frac{8\pi d_s^4 \mu L}{d_n^4} \bar{v}_s + \frac{3}{2}\rho \frac{\pi d_s^6}{4d_n^4} \bar{v}_s^2 \tag{11}$$

In the following table the nomenclature and the values of all the parameters are reported.

Table 1: Variables nomenclature – Hydrodynamic model

Parameters name	Physical meaning
d_s	Syringe diameter
d_n	Needle diameter
μ	Formulation viscosity
ρ	Formulation density
L	Needle length
$\bar{v}_s$	Mean velocity of the plunger

The experiments, run with the same type of AJ but at different velocities of the piston, are repeated four times to evaluate the variance of the total force, the measured output of the system.

3.2. Mechanistic model

The mechanistic model selected to describe the physics law behind the friction force is the model proposed by Rathore et al. (2011):

$$F_{friction} = \left(\frac{2\pi \mu_{oil} r_s l_{stop}}{d_{oil}}\right) \bar{v}_s \tag{12}$$

In this model there are parameters related to the AJ geometry and parameters used to describe the rheology of the lubrication layer in the inner wall of the barrel.

Table 2: Variable nomenclature - Friction model

Parameters name	Physical meaning
r_s	Syringe radius
l_{stop}	Stopper length
d_{oil}	Thickness of the lubrication layer
μ_{oil}	Oil viscosity

3.3. Gaussian process

A GP on the form described in *Section 2.2.1* has been implemented and trained using the data of 8 experiments. Once the data-driven model had been trained, the predictive capabilities of the empirical model were validated using the experimental data of the last experiment.

4. Results

In this section the results obtained using the proposed framework are shown. In Fig. (1) the profile of the measured friction force is reported.

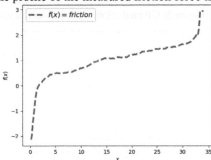
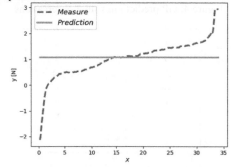

Figure 2: Profile of the measured friction

Figure 3: Comparison of the profiles of the predicted friction and of the measured friction

From Fig. (3) it is possible to notice that the mechanistic model is not capable to describe the profile of the friction in an adequate way. Fig. (2) shows that the friction force depends on the position of the piston (x) and this variable is not included in the model. At this point using Eq. (2) ε^M was evaluated. The prediction mismatch obtained was used to train the GP, Fig. (4). Fig. (4) shows the profile of predictions mismatch the function obtained using the GP and the observation used to derive the function. In Fig. (4) the predicted mean of the GP (μ) is obtained using only two observations. The predicted function in this case is not accurate, in fact the 95% confidence interval is very high, so it is necessary to increase the number of observations used. In Fig. (5) the function obtained using 8 experimental observations is reported. From this figure it can be seen that using more observation the GP managed to create a function able to describe in an adequate way the profile of the prediction mismatch. At this point it is necessary to assess whether the GP is also capable to predict the value of ε^M. In Fig. (6) the predicted ε^M and the measured mismatch comparison is reported. From this figure it can be noticed that the GP can model the prediction mismatch in an accurate way when the system is in steady state conditions (between the two red dotted lines). Fig. (7) shows the friction force measured from the experiment and the friction force predicted using the hybrid model.

5. Conclusion and future work

In this project an alternative approach to combine mechanistic models and data-driven models (GPs) for the description of friction forces in AJ devices has been presented. The novelty of the approach is that the role of the data-driven part is not to improve the performance of the mechanistic model but is to model the difference between the experimental data and the model predictions.

A. Friso et al.

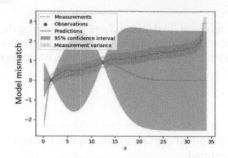

Figure 4: GP predictions with 2 observations

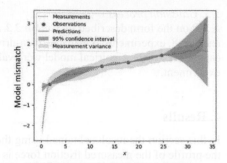

Figure 5: GP predictions with 4 observations

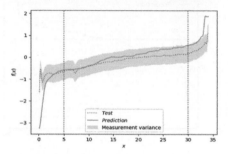

Figure 6: Comparison between the predicted ε^M and the measured ε^M

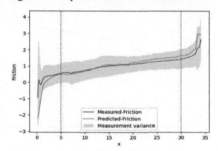

Figure 7: Comparison between the measured friction and that predicted by the hybrid model.

The results show that this hybrid model manages to predict in an accurate way the measured friction in case of homoscedastic variance. The future evolution of this work will include i) the comparison of the proposed approach with state estimation methods based on extended Kalman filters in order to be able to consider the variability in the hybrid model prediction; ii) the analysis of results considering different variance models to describe the distribution of measurement errors. Repeated experiments, in fact, showed that in some cases the variance from repeated runs is not constant in time but heteroscedastic.

Acknowledgements

The authors gratefully acknowledge the support of the Department of Chemical Engineering, University College London, UK, and GSK

References

Rathore N., Pranay P., Eu B., Ji W., Walls E. "Variability in syringe components and its impact on functionality of delivery systems." PDA Journal of Pharmaceutical Science Technology, Volume 65, 2011.

Zhong X., Guo T., Vlachos P., Veilleux J.C., Huaiqiu S. G., Collins D.S., Ardekani A.M., "An experimentally validated dynamic model for spring-driven autoinjectors.", International Journal of Pharmaceutics, Volume 594, 2021.

Rasmussen C.E., Williams C.K.I., "Gaussian processes for machine Learning", The MIT Press, 2006.

Vijayaraghavan, R., Autoinjector Device for Rapid Administration of Life Saving Drugs in Emergency (Review Paper). *Defence Science Journal*, *62*(5), 307-314, 2012.

Flavio Manenti, Gintaras V. Reklaitis (Eds.), Proceedings of the 34th European Symposium on Computer Aided Process Engineering / 15th International Symposium on Process Systems Engineering (ESCAPE34/PSE24), June 2-6, 2024, Florence, Italy

Dynamic modeling of particle size and porosity distribution in fluidized bed spray agglomeration

Eric Otto,[a,*] Robert Dürr,[b] Achim Kienle,[a,c] Andreas Bück,[d] Evangelos Tsotsas[a]

[a]*Otto von Guericke University Magdeburg, Universitätsplatz 2, 39106 Magdeburg, Germany*
[b]*Magdeburg-Stendal University of Applied Sciences, Breitscheidstraße 2, 39104 Magdeburg, Germany*
[c]*Max Planck Institute for Dynamics of Complex Technical Systems, Sandtorstraße 1, 39106 Magdeburg, Germany*
[d]*Friedrich-Alexander-University Erlangen-Nuremberg, Cauerstraße 4, 91058 Erlangen, Germany*
eric.otto@ovgu.de

Abstract

Population balance modeling is a powerful to tool for the simulation of particle formation processes such as fluidized bed spray agglomeration (FBSA), where agglomerates are formed from primary particles by binary aggregation. In addition to the agglomerate volume, agglomerate porosity is important for the characterization of product particles since it affects various physical properties. In this contribution a new method is proposed to incorporate porosity into a population balance model. To this end an empirical relationship between the volume of solids within the agglomerate and its total volume including voids is utilized. Since this relationship is different for primary particles and agglomerates, respectively, the evolution of the number density distribution for both types of particles is modeled. The proposed population balance equations are validated by fitting three kinetic parameters to experimental data and comparing measured particle size distributions with model predictions, showing good agreement.

Keywords: population balance modeling, agglomerate porosity.

1. Introduction

Continuous fluidized bed spray agglomeration (FBSA) is a size enlargement process for the production of solid particles. Here, a liquid binding agent is sprayed onto the surface of a fluidized particle bed. After particle collision and binder drying, new agglomerates are formed. In the continuous process configuration (Fig. 1), subject of this contribution, there is a constant primary particle feed and product particle withdrawal.

Physical particles properties, such as the volume and porosity of the resulting agglomerates are important, since they determine agglomerate characteristics, such as solubility, mechanical strength and flowability, and thereby the economic value of the product. For process monitoring,

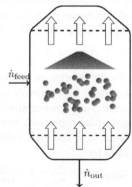

Figure 1: Process scheme with particle feed, agglomeration in the process chamber and withdrawal

control and optimization, it is therefore crucial to have accurate models describing their evolution. Typically, multi-dimensional population balance equations (PBEs) are used as macroscale models (Iveson, 2002), describing the evolution of particle distributions with internal properties by means of partial differential equations. Alternatively, stochastic Monte-Carlo methods can be applied to model certain subprocesses such as particle wetting, collision and drying in detail at the expense of increased computational effort (Singh and Tsotsas, 2019).

In previous contributions, the authors developed one-dimensional population balance models (PBMs) for continuous FBSA processes and fitted parameters to experimental particle size distributions (Otto et al., 2021, 2022). In these contributions however, the porous nature of agglomerates has only been incorporated by assuming an average reduced agglomerate density, resulting in models that do not accurately capture the influence of porosity on the measured particle size distributions. In this contribution a new approach to population balance modeling for the FBSA production of agglomerates described by their volume and porosity is presented. To this end, an empirical relationship between the two descriptors presented in Singh and Tsotsas (2019) and Strenzke et al. (2022) is utilized. In contrast to models in the literature (Iveson, 2002; Poon et al., 2008), the approach presented here allows for the use of only a one-dimensional PBE. Since primary particles and agglomerate have different characteristic porosity distributions, primary particle and agglomerate populations are considered separately, resulting in two modeled distributions. The proposed system of PBEs in validated by fitting its kinetic parameters to experimental particle size distributions. Additionally, model implications for application in process control are discussed.

2. Process model

2.1. Agglomerate porosity distribution

In the following a model for the relation between agglomerate volume and porosity is presented. It is assumed that an agglomerate consists of a solid and a gaseous (air/void) fraction. Liquid components are neglected, due to their low volume fraction. The porosity of an agglomerate is defined as the ratio between gas volume v_g and volume packed by solid material v_s. Since the dried binder volume is small compared to the total volume of all primary particles it is neglected and v_s is given by

$$v_s = N_p v_p = N_p \frac{\pi}{6} d_p^{\ 3}, \tag{1}$$

with the number of primary particles N_p and the primary particle diameter d_p. Here it is assumed, that primary particles are monodispersed and spherical. Although this is a simplifying assumption, the influence of polydispersity of primary particles on agglomerate porosity can be considered as negligibly small (Singh and Tsotsas, 2022). Using v_{tot}, the total volume of the agglomerate (including voids), the porosity ε_{agg} is given by

$$\varepsilon_{agg} = 1 - \frac{v_s}{v_{tot}} = 1 - \frac{v_s}{v_s + v_g}, \tag{2}$$

From the above equation, it follows that ε_{agg}, v_g or v_{tot} can be used equivalently to fully characterize an agglomerate if v_s is given. Singh and Tsotsas (2019) provide the following empirical relationship between the porosity and the number of primary particles

$$\varepsilon_{agg} = 1 - 0.465 N_p \left(\frac{k}{N_p}\right)^{\frac{3}{D_f}}, \tag{3}$$

with parameters D_f, the so-called fractal dimension, and k, the prefactor. In order to obtain this relationship, v_{tot} is defined as the volume of an equivalent sphere with radius R_e based on the gyration radius of the agglomerates. The morphological parameters D_f and k are determined empirically by correlating the number of primary particles N_p in an agglomerate to the measured gyration radius for a number of agglomerates. Strenzke et al. (2022) determined $D_f = 2.89$ and $k = 0.82$ for a continuous FBSA experiment with constant process conditions. Additionally, Singh and Tsotsas (2019) showed that it is possible to linearly correlated the morphological parameters to process conditions, namely the gas inlet temperature T_g and the binder concentration x_b. By rearranging Eqs. (1), (2) and (3), the following power law relationship between v_s and v_{tot} is obtained

$$v_s(v_{tot}) = 0.465^{D_f/3} k v_p^{1-D_f/3} v_{tot}^{D_f/3}, \tag{4}$$

Note that the above equation assumes that v_s and v_{tot} can take continuous values, whereas Eq. (1) implies discrete values due to the discrete nature of N_p. Thus, Eq. 4 can be seen as a continuous extension of the underlying discrete map. Furthermore, it is important to remember that Eq. (4) only hold for agglomerates. The primary particle porosity ε_p is constant in general, for the FBSA experiments with non-porous glass beads considered here $\varepsilon_p = 0$.

2.2 Population balance equation

In Otto et al. (2021) a population balance model for the evolution of the particle size distribution during a continuous FBSA process has been presented. Therein, the particle population is represented by the number density distribution (NDD) $n(t, v_s)$ with solid volume as internal coordinate. In order to describe the agglomeration kinetics, parameters of an empirical agglomeration kernel have been fitted to measurements of particle size distributions (Strenzke et al., 2020), where the particle size distribution is measured by means of image analysis (Camsizer X2, Microtrac Retsch). The measured particle size is represented by a diameter d, which can be transformed to volume using the standard assumption of spherical particles. This measured volume, however, includes voids and therefore is not the volume v_s but v_{tot}. To account for this gap between measured NDD $n(t, v_{tot})$ and modeled NDD $n(t, v_s)$, the literature provides more detailed population balance models. Usually, the NDD is augmented by an additional internal variable, e.g., the agglomerate porosity ε_{agg} (Iveson, 2002) or the void/gas volume v_g within the agglomerate (Poon et al., 2008), i.e., $n = n(t, v_s, v_g)$. According to the previous subsection, any of the variables ε_{agg}, v_g or v_{tot} can be used as second internal variable, since one of these determines the others. In both Iveson (2002) and Poon et al. (2008) it is assumed that the solid volume, the volume of gas and therefore the total agglomerate volume v_{tot} is conserved during binary aggregation which directly determines the resulting agglomerate porosity. However, for the FBSA process considered in this contribution, the nonlinearity of Eq. (4), shows that since v_s is conserved during agglomeration, v_{tot} is not. Furthermore, Eq. (4) shows that v_s and v_{tot} are not independent variables, eliminating the need for a multidimensional PBE. In fact, the derivative of Eq. (4) given by

$$\left.\frac{dv_s(v_{tot})}{dv_{tot}}\right|_{agg} = \frac{0.465 k D_f}{3} v_p^{1-D_f/3} v_{tot}^{D_f/3-1}, \tag{5}$$

defines a transformation between $n(t, v_{tot})$ and $n(t, v_s)$ by

$$n(t, v_{tot}) = n(t, v_s(v_{tot})) \left.\frac{dv_s(v_{tot})}{dv_{tot}}\right|_{agg}, \tag{6}$$

which is the desired equation connecting the measured and the modeled NDD for a population of agglomerates.

From the discussion at the end of the previous subsection it is clear that Eq. (5) and thus Eq. (6) only holds for actual agglomerates. For primary particles, on the other hand, the solid volume is equal to the total particle volume, i.e.

$$\frac{dv_s(v_{tot})}{dv_{tot}}\bigg|_p = 1. \tag{7}$$

Thus, for a general particle population, it is necessary to distinguish the NDDs for primary particles $n_p(t, v_s)$ and agglomerates $n_a(t, v_s)$ with

$$n(t, v_s) = n_p(t, v_s) + n_a(t, v_s). \tag{8}$$

When computing the measured NDD $n(t, v_t)$ corresponding to Eq. (8), the different transformations have to be considered, resulting in

$$n(t, v_{tot}) = n_p\big(t, v_s(v_{tot})\big)\frac{dv_s(v_{tot})}{dv_{tot}}\bigg|_p + n_a\big(t, v_s(v_{tot})\big)\frac{dv_s(v_{tot})}{dv_{tot}}\bigg|_{agg}. \tag{9}$$

The different transformations and their effect on the total particle size distribution are showcased in Fig. 2, where example NDDs of a primary particle and an agglomerate population are shown depending on v_s and v_{tot} respectively. While the primary particle NDD is the same in both coordinates, the agglomerate NDD appears flattened and shifted towards higher volumes in the v_{tot}-coordinate.

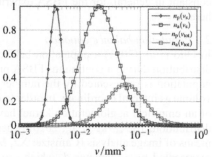

Figure 2: Modelled NDDs with respect to v_s (blue) and measured NDDs with rescpect to v_{tot}.

In order to obtain PBEs for the evolution of $n_p(t, v_s)$ and $n_a(t, v_s)$, aggregation between primary particles and primary particles (pp), agglomerates and primary particles (pa) as well as agglomerates and agglomerates (aa) have to be modeled separately, as presented schematically in Fig. 3. The respective population balances are obtained by balancing the particle accumulation at volumes v_s with the respective fluxes

$$\frac{dn_p(t, v_s)}{dt} = \dot{n}_{p,feed}(t, v_s) - D_{pp}(t, v_s) - D_{pa}(t, v_s) - \dot{n}_{p,out}(t, v_s) \tag{10}$$

$$\frac{dn_a(t, v_s)}{dt} = B_{pp}(t, v_s) + B_{pa}(t, v_s) + B_{ap}(t, v_s) + B_{aa}(t, v_s) - D_{pa}(t, v_s) \\ - D_{aa}(t, v_s) - \dot{n}_{a,out}(t, v_s) \tag{11}$$

Here, $\dot{n}_{p,feed}$ is the primary particle feed and $\dot{n}_{p,out}$ as well as $\dot{n}_{a,out}$ are modeling the particle withdrawal. For the specific modeling of these terms, we refer the reader to Otto et al. (2021). The variables B and D denote particle birth and death respectively, given by the following equations (Hulburt and Katz, 1964)

$$B_{xy}(t, v_s) = \frac{1}{2}\int_0^{v_s} \beta_{xy}(t, u_s, v_s - u_s)n_x(t, u_s)n(t, v_s - u_s)dv_s \quad \forall(x,y) \in \{p, a\}^2 \tag{12}$$

and

$$D_{xy}(t, v_s) = \int_0^\infty \beta_{xy}(t, u_s, v_s)n_x(t, u_s)n(t, v_s)dv_s \quad \forall(x,y) \in \{p, a\}^2 \tag{13}$$

The agglomeration kinetics are described by the agglomeration kernel $\beta(t, u_s, v_s)$, representing the volume dependent rate of agglomeration. In this standard form of the agglomeration term, it is assumed, that the kernel only depends on the solid volume part of the agglomerate and therefore on the agglomerate mass. This is a simplifying assumption, since the rate of agglomeration depends on multiple factors such as the binder coverage of the particle surface which itself may be increased by higher total agglomerate volume.

Various kernel functions, both mechanistic and empirical, with varying complexity are presented in the literature. Kernel identification results from Otto et al. (2022) suggest that the agglomeration behavior is different for the three types of particle interaction described above. Therefore, in this contribution we distinguish between the kernel β_{pp}, $\beta_{pa} = \beta_{ap}$ and β_{aa}. The simplest kernel model is the constant, i.e., size and time independent kernel function which will be adopted here.

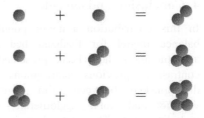

Figure 3: Agglomeration between a) primary particle and primary particle, b) primary particle and agglomerate and c) agglomerate and agglomerate.

2.3 Model discussion

The presented correlations of number of primary particles and agglomerate porosity by fractal dimension and prefactor have been applied to FBSA in batch configuration (Singh and Tsotsas, 2019) and one continuous experiment (Strenzke et al., 2022). In both settings process conditions have been kept constant. For the general case of non-constant process conditions, the morphological descriptors D_f and k will change and the power law of Eq. (3) and thus Eq. (4) will generally not be valid. In this case there will be no unique, i.e., time-independent, relationship between v_{tot} and v_s and a population balance equation with two internal coordinates may be required to capture the dynamics of the process. This could be circumvented by assuming a slow and uniform change of D_f and k depending on a process condition $u(t)$, e.g.

$$T \frac{dD_f(t)}{dt} = -D_f(t) + Ku(t). \tag{14}$$

In order to apply the PBEs in a model-based process control context, the state, i.e., primary particle and agglomerate distributions have to be measured. Current measurement technology, however, does not distinguish between primary particles and agglomerates. Therefore, either advanced measurement technology has to be developed, e.g., image-based distinction or the distributions have to be estimated using model-based observer or soft-sensor techniques.

3. Model validation

In order to validate the proposed model, the three yet unidentified kernel parameters are fitted to the continuous FBSA experiment presented in Strenzke et al. (2022). In accordance with Otto et al. (2021) the focus is on the steady state particle size distribution measured after 120 minutes. For the parametrization of the feed and withdrawal terms in the population balance as well as the numerical discretization of the PDEs, we refer the reader to Otto et al. (2021). Using the Matlab implemented surrogate optimization method (Gutmann, 2001) to minimize the L_2-error between experimental and calculated distribution, the parameters $\beta_{pp} = 0.2327 \cdot 10^{-9}$, $\beta_{pa} = 0.1537 \cdot 10^{-9}$ and $\beta_{aa} =$

0.0158 · 10^{-9} are found. In Fig. 5 the measured and calculated normalized volume distributions

$$q_3(d) = \frac{d^3 n(d)}{\int_0^\infty d^3 n(d) dd}, \quad n(d) = n\big(v_{tot}(d)\big)\frac{dv_{tot}}{dd}, \quad v_{tot}(d) = \frac{\pi}{6}d^3 \qquad (15)$$

in steady state are presented, showing good agreement.

4. Conclusion and outlook

In this contribution a novel population balance model for fluidized bed spray agglomeration has been presented. In contrast to previous contributions, it is distinguished between solid agglomerate volume and total agglomerate volume including voids. The solid volume part of the

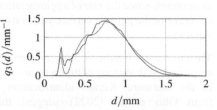

Figure 4: Comparison between experimental (blue) and calculated (red) normalized volume distribution in steady state.

agglomerate is utilized as internal coordinate of the modeled number density distribution, resulting in a classical one-dimensional population balance equation. In order to derive the measurable particle size distribution which is based on the total agglomerate volume, an empirical correlation between solid and total volume is incorporated. Since this correlation, based on the particle fractal dimension, is different for primary particles and agglomerates, both populations are modeled separately. The presented model is validated by fitting three kinetic parameters and comparing the particle size distribution predictions with the measurements of a continuous experiment. The results are in good agreement for the steady state distributions.

Future research directions are twofold. By using correlations of different process conditions to the fractal dimension, their influence on the agglomerate porosity will be modeled and validated with additional experiments. Furthermore, the presented model will be used as basis for model-based process control of agglomerate size and porosity.

Acknowledgements

This work is funded by the DFG SPP 2364 "Autonomous processes in particle technology" (project ID: 504524147). The financial support is hereby gratefully acknowledged.

References

H. Gutmann, 2001. Journal of Global Optimization 19, 201–227.

H. Hulburt, S. Katz, 1964. Chemical Engineering Science 19 (8), 555–574.

S. M. Iveson, 2002 Powder Technology 124 (3), 219–229.

E. Otto, R. Dürr, G. Strenzke, S. Palis, A. Bück, E. Tsotsas, A. Kienle, 2021. Advanced Powder Technology 32 (7), 2517–2529.

E. Otto, A. Maksakov, R. Dürr, S. Palis, A. Kienle, 2022. IFAC-PapersOnLine 55 (7), 260–265.

J. M.-H. Poon, C. D. Immanuel, F. J. Doyle, III, J. D. Litster, 2008. Chemical Engineering Science 63 (5), 1315–1329.

A. K. Singh, E. Tsotsas, 2019 Powder Technology 355, 449–460.

A. K. Singh, E. Tsotsas, 2022. Chemical Engineering Science 247, 117–022.

G. Strenzke, R. Dürr, A. Bück, E. Tsotsas, 2020. Powder Technology 375, 210–220.

G. Strenzke, M. Janocha, A. Bück, E. Tsotsas, 2022. Powder Technology 397, 117–111.

Flavio Manenti, Gintaras V. Reklaitis (Eds.), Proceedings of the 34th European Symposium on Computer Aided Process Engineering / 15th International Symposium on Process Systems Engineering (ESCAPE34/PSE24), June 2-6, 2024, Florence, Italy

DoE-integrated Sparse Identification of Nonlinear Dynamics for Automated Model Generation and Parameter Estimation in Kinetic Studies

Wenyao Lyu[a], Federico Galvanin[a]

[a]Department of Chemical Engineering, University College London (UCL), Torrington Place, London, WC1E 7JE, United Kingdom
f.galvanin@ucl.ac.uk

Abstract

Digital twins have revolutionized manufacturing by utilising robust kinetic models to predict the behaviour of biochemical reaction systems under variable operating conditions. Identifying accurate expressions for reaction system models formulated as sets of differential and algebraic equations (DAEs) is challenging due to numerous state variables and kinetic parameters, compounded by limited observations and experimental errors. To enhance reliability, model structure confirmation should precede parameter estimation and validation. Traditional model-building approaches require prior knowledge of candidate models, whereas model generation methods like Sparse Identification of Nonlinear Dynamics (SINDy) only require the definition of a library of function terms. In this paper, we propose a new model identification framework, named Design of Experiment-integrated Sparse Identification of Nonlinear Dynamics (DoE-SINDy) to iteratively generate, evaluate, and select candidate models to represent systems with minimal training data. Tested on a simulated case study, DoE-SINDy succeeds in identifying an assumed true model efficiently with a limited experimental budget and noisy datasets.

Keywords: model identification, model generation, kinetic studies, sparse regression, data-driven modelling

1. Introduction

Digitisation is driving a profound transformation in manufacturing sectors by implementing digital twins based on robust kinetic models for representing the dynamic behaviour of reaction systems modelled as a set of differential and algebraic equations. Identifying these kinetic models from limited observations is very challenging due to their complex model structure, numerous variables, and parameters (Quaglio et al., 2020). Constructing a suitable model for digital twins often requires costly experimentation and a substantial investment of time and analytical resources. Furthermore, the limited observations and inevitable experimental errors further hinder the adequate identification. The reliability of identified models is determined by the adequacy of model structure (i.e., set of equations) and parameter precision. Model structure adequacy is a statistical measure of how well a model fits the data, while parameter precision is related to the uncertainty of parameter estimates. Thus, it is necessary to identify a model with an accurate model structure and reduce the uncertainty in the parameters as much as possible. Traditional model-building approaches involve model selection for identifying the most accurate model structure from a set of known model structures (Asprey and Macchietto, 2000), and model modification for correcting the unsuitable structure in the selected model. Challenges may arise when there is insufficient theoretical understanding of the system to generate candidate models or when the 'true' (i.e., most suitable) model is not among the candidate models, a common scenario in chemical and biochemical processes. Model generation approaches such as SINDy (Brunton et al., 2016) and AI-DARWIN (Chakraborty et al., 2021) are affected by several limitations: 1) potential generation of unidentifiable models; 2) absence of a rigorous model evaluation stage; 3) model generation is extremely sensitive to experimental conditions explored in the training set; 4) lack of a model selection stage. Motivated by these challenges, a new systematic model identification framework named DoE-SINDy is proposed in this paper for simultaneously generating model structures and estimating parameters from a small dataset to represent a dynamic system. To realise it, this framework aims to integrate model generation techniques and design of experiments (DoE) to expedite the identification of kinetic models.

2. Methodology

A dynamic chemical reaction model can be formulated as a set of differential and algebraic equations (DAEs) in the form

$$\mathbf{f}(\dot{\mathbf{x}}, \mathbf{x}, \mathbf{u}, t, \boldsymbol{\theta}) = 0$$
$$\hat{\mathbf{x}} = \mathbf{h}(\mathbf{u}, t, \boldsymbol{\theta}) \tag{1}$$

where $\mathbf{f}$ and $\mathbf{h}$ are respectively an N_f and N_x vector of model equations, $\mathbf{x}$ is an N_x vector of state variables, $\mathbf{u} \in U$ is an N_u vector of control input variables, t is time and $\boldsymbol{\theta}$ is an N_θ vector of model parameters, $\hat{\mathbf{x}}$ represents an N_x vector of model predictions for a measurable set of system state variables $\mathbf{x}$. This research focuses on simultaneously identifying the model structure $\mathbf{f}$, and parameters $\boldsymbol{\theta}$.

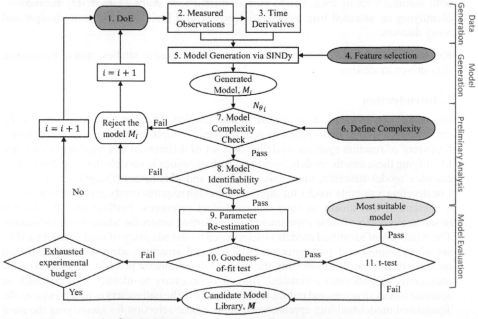

Figure 1 Framework of DoE-SINDy for identifying the most suitable models from experimental data.

SINDy is a main sparse regression-based approach that is capable of deriving models with minimal prior knowledge of physical mechanisms. As demonstrated in the literature by Brunton et al. (2016), SINDy operates under the assumption that a process model can be efficiently represented by only a few important terms that govern the dynamics. It adeptly integrates the most suitable terms from a candidate term library $\mathbf{F(X)}$ and estimates parameters $\boldsymbol{\theta}$ to formulate model expressions, which strikes an equilibrium between model accuracy and complexity. Thus, it averts over-fitting while preserving the underlying physics of the discovered model. DoE-SINDy begins with a Design of Experiments (DoE), where operating conditions are specified based on underlying physical constraints (block 1 in Figure 1). Subsequently, time-varying measurements, such as concentrations of specific components, are collected along the trajectory, and the time derivatives of these measurements are numerically approximated due to their continuity in time (blocks 2 and 3 in Figure 1). As an extension of the original SINDy, DoE-SINDy utilises the original SINDy for model generation (blocks 4, 5 in Figure 1), employing an open-source library called *PySINDy* (de Silva et al., 2020; Kaptanoglu et al., 2022). In contrast to the original SINDy, DoE-SINDy is a closed-loop framework, iteratively generating multiple models from increasing-size datasets to ensure the identification of an assumed true model as shown in Figure 1. Additionally, DoE-SINDy incorporates a preliminary analysis stage (blocks 6, 7, 8 in Figure 1) to reject unidentifiable models and those with excessive terms. Identifiability, as defined in Miao et al., (2011), refers to the property of a dynamic system where the parameters $\boldsymbol{\theta}$ can be uniquely determined from the given system input $\mathbf{u}$ and the measurable system output $\mathbf{x}$. DoE-SINDy implements sensitivity-based practical identifiability analysis based on the calculation of Fisher Information Matrix (FIM), denoted as $\boldsymbol{H}_\theta$. When analysing the identifiability over a trajectory, according to Waldron et al. (2019), $\boldsymbol{H}_\theta$ is computed as the summation of $\boldsymbol{H}_\theta(t_m)$ at each sampling point

$$[\boldsymbol{H}_\theta]_{kl} \approx \sum_{m=1}^{N_m} \frac{1}{\sigma_n^2} q_{nk}(t_m^i) q_{nl}(t_m^i), \text{ for } k, l = 1, \dots, N_\theta \tag{2}$$

In (2), $q_{nk}(t_m)$ is the *nk*-th element of the sensitivity $\boldsymbol{Q}$ representing the sensitivity of the response x_n to the parameter θ_k at time sampling point t_m, calculated according to Franceschini et al. (2008). When $\boldsymbol{H}_\theta$ is full rank, the model is considered locally identifiable. Conversely, when $\boldsymbol{H}_\theta$ is singular, this indicates that the model is not identifiable (Miao et al., 2011). Since the assumption of noise distribution is not considered when implementing Sequentially Thresholded Least-squares algorithm (STLSQ) to estimate the parameters in the model generation section. DoE-SINDy incorporates a Maximum Likelihood (ML) estimator to robustly re-estimate model parameters (block 9 in Figure 1) under the assumption of measurements corrupted by Gaussian noise with zero mean and constant standard deviation σ.

Subsequentially, DoE-SINDy includes a model evaluation stage for selecting accurate model(s) from the generated models via a χ^2 test (block 10 in Figure 1) and assessing parameter precision via a *t*-test (block 11 in Figure 1).

3. Model description

We test the performance of DoE-SINDy to the same case study used in Quaglio et al. (2020). The goal is to identify an assumed true kinetic model of a batch reaction system described in equation (3).

$$\frac{dC_A}{dt} = -k_1 C_A^2$$

$$\frac{dC_B}{dt} = k_1 C_A^2 - k_2 C_B \qquad\qquad\qquad (3)$$

$$\frac{dC_C}{dt} = k_2 C_B$$

In equation (3), the reaction rate constants are denoted by k_1 and k_2. To collect experimental data in silico, we specified various initial concentrations of A within the range from 40 to 250 mol m^3 s^{-1} to simulate the process multiple times. The duration of each experiment is 350 s, and measurements are taken from 100 to 300 s every 10 s. 20 experiments are simulated for data collection, and the initial concentration of species A is designed based on a Latin Hypercube Sampling (LHS) experimental design. Gaussian distributed noise with zero mean and constant standard deviation $\sigma = 0.1$ mol m^{-3} is added to the simulated data to generate in-silico measurements.

4. Results

The performance of the model generation of DoE-SINDy is presented here along with a discussion of the findings and a comparison between the efficiency of identifying the true model between DoE-SINDy and the original SINDy.

4.1 Algorithm efficiency using a specific LHS dataset

The structure and parameters of the model generated through iterations are significantly influenced by the training data. Even when using identical measurements from the same 20 experiments, the generated models may exhibit diversity.

To illustrate how the order of data loading impacts the model structure in each iteration, we examine four different data loading sequences α, β, γ, and δ. It is worth noting that, SINDy estimates the parameters of every term in a model, as there are 4 terms in the true model, the number of parameters N_θ of the SINDy generated model having the identical model structure as the true model is 4 ($\theta_1, \theta_2, \theta_3$ and θ_4) instead of 2 (k_1 and k_2). To simply visualise the models generated over the iterative process, N_θ is used to represent the model structure.

As illustrated in Figure 2, in general, with the increase of number of iterations, N_θ decreases, eventually converging to 5 parameters. The true model with 4 parameters is successfully identified at one of the iterations during the model identification process, but only with sequences α and γ. These observations show that the training data used is crucial in determining whether the true model is identified.

Identifiability results show that the generated models are unidentifiable, as DoE-SINDy generated models have anti-correlated parameters, i.e., $\theta_1 = k_1$ and $\theta_2 = -k_1$, $\theta_3 = -k_2$ and $\theta_4 = k_2$. Because of this, the parameters cannot be precisely estimated.

The results from χ^2 test show that a model is falsified for under-fitting even when the true model structure is identified. For example, the χ^2 of the model with 4 parameters identified with sequence γ is $\chi^2 = 1.81 \times 10^4$ which is larger than the reference $\chi^2_{0.95} = 8.16 \times 10^2$. However, there are negligible deviations between the measurements and the predictions. In that case, the true model may not be selected out of the generated models according to the results of χ^2 test.

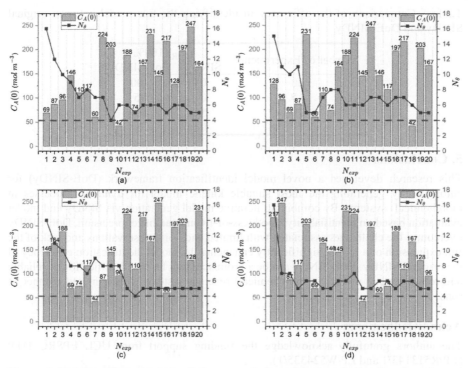

Figure 2 Number of parameters of the generated model at each iteration based on 4 different experimental data loading sequences (a) α, (b) β, (c) γ, and (d) δ. The red line refers to the true number of parameters N_θ, which is 4, bars refer to the defined initial concentration of A in each LHS experiment. The line illustrates N_θ for the model generated at each iteration.

4.2 Capability in general conditions and comparison with SINDy

To compare original SINDy and DoE-SINDY we investigate four different data sizes (5, 10, 15 and 20 experiments). For each data size, 20 datasets are generated. Within a single dataset, LHS is employed to design the initial concentrations of A for each experiment, and the code DoE-SINDy is executed 100 times with shuffled sequences to mitigate the impact of sequence on model generation. The Overall Success Rate, denoted as η_{OSR}, and defined as the ratio between 'successful datasets' out of 20 datasets at this data size, is introduced as a metric to quantify the efficiency in identifying the true model structure.

Table 1 shows that the original SINDy only succeeds in identifying the true model within a dataset obtained from 10 experiments, while fails in all datasets obtained from 5, 15, and 20 experiments. Conversely, DoE-SINDy consistently succeeds with nearly all datasets at the data sizes of 10, 15 and 20 experiments. This observation reveals that by identifying models iteratively can significantly enhance the efficiency of identifying the true model. Additionally, it is beneficial to explore more sequences before conducting additional experiments.

Table 1 Comparison of the efficiency in identifying the true model between original SINDy and DoE-SINDy.

Number of Experiments	$\eta_{OSR}/100\%$	
	SINDy	DoE-SINDy
20	0	100
15	0	95
10	5	95
5	0	0

5. Conclusions

This research developed a novel model identification framework (DoE-SINDy) for automatically generating the most suitable kinetic model for dynamic processes in bio(chemical) systems. By combining iterative model generation via SINDy and model evaluation based on identifiability, goodness-of-fit and parameter precision, DoE-SINDy overcomes the challenges associated with limited and noisy observations and lack of the full expression of a model, significantly enhancing the model identification efficiency of the original SINDy. DoE-SINDy can be a promising tool for the identification of reliable and robust kinetic models in bio(chemical) reaction systems with a limited number of runs.

Acknowledgements

The authors gratefully acknowledge the funding support from UCL EPSRC DTP (EP/R513143/1 and EP/W524335/1).

References

S.P. Asprey and S. Macchietto, 2000. Statistical tools for optimal dynamic model building. Computers & Chemical Engineering, 24(2-7), pp.1261-1267.

S.L. Brunton, J.L. Proctor, and J.N. Kutz, 2016. Discovering governing equations from data by sparse identification of nonlinear dynamical systems. Proceedings of the national academy of sciences, 113(15), pp.3932-3937.

A. Chakraborty, A. Sivaram, and V. Venkatasubramanian, 2021. AI-DARWIN: A first principles-based model discovery engine using machine learning. Computers & Chemical Engineering, 154, p.107470.

G. Franceschini and S., Macchietto, 2008. Model-based design of experiments for parameter precision: State of the art. Chemical Engineering Science, 63(19), pp.4846-4872.

A.A. Kaptanoglu, B.M. de Silva, U. Fasel, K. Kaheman, A.J. Goldschmidt, J.L. Callaham, C.B. Delahunt, Z.G. Nicolaou, K. Champion, J.C. Loiseau, and J.N. Kutz, 2021. PySINDy: A comprehensive Python package for robust sparse system identification. arXiv preprint arXiv:2111.08481.

H. Miao, X. Xia, A.S. Perelson and H. Wu, 2011. On identifiability of nonlinear ODE models and applications in viral dynamics. SIAM review, 53(1), pp.3-39.

M. Quaglio, L. Roberts, M.S.B. Jaapar, E.S. Fraga, V. Dua, and F. Galvanin, 2020. An artificial neural network approach to recognise kinetic models from experimental data. Computers & Chemical Engineering, 135, p.106759.

C. Waldron, A. Pankajakshan, M. Quaglio, E. Cao, F. Galvanin, and A. Gavriilidis, 2019. Closed-loop model-based design of experiments for kinetic model discrimination and parameter estimation: benzoic acid esterification on a heterogeneous catalyst. Industrial & Engineering Chemistry Research, 58(49), pp.22165-22177.

Flavio Manenti, Gintaras V. Reklaitis (Eds.), Proceedings of the 34th European Symposium on
Computer Aided Process Engineering / 15th International Symposium on Process Systems
Engineering (ESCAPE34/PSE24), June 2-6, 2024, Florence, Italy

Computational Fluid Dynamics and Trust-Region Methods to Optimize Carbon Capture Plants with Membrane Contactors

Hector A. Pedrozo[a], Grigorios Panagakos[a], and Lorenz T. Biegler[a,*]

*[a]Department of Chemical Engineering, Carnegie Mellon University, 5000 Forbes
Avenue, Pittsburgh, PA 15213, USA*
lb01@andrew.cmu.edu

Abstract

In this study, we extend the trust-region filter method to optimization problems including external functions from Computational Fluid Dynamics (CFD) simulations and equation-oriented Aspen Plus rigorous models. To show the optimization framework, we address the optimal design of an MEA-based carbon capture plant employing hollow fiber membrane contactors, where we formulate an optimal design problem, by considering the minimization of the CO_2 avoided cost of the carbon capture plant. The optimization problem is formulated in Pyomo and solved with IPOPT; the CFD model is implemented in Comsol Multiphysics. We create a framework to process the algorithmic data and run the CFD simulations and Aspen Plus via a Python code. This trust-region filter application leads to an effective strategy to optimize integrated process models, including rigorous CFD models. The implementation is effective in handling data, setting up the rigorous simulators, and automatically running these programs.

Keywords: CFD; trust-region method; optimal design; membrane contactors; EO model

1. Introduction

Current challenges in Process Systems Engineering include the formulation of multi scale optimization problems to shed light on the truth potential of promising technologies for relevant problems, such as the optimal design of carbon capture technologies. Performing multi-scale optimization usually involves the formulation of hybrid optimization problems, where explicit equations along with external functions (also known as black-box functions) are embedded in the mathematical formulation simultaneously (Pedrozo et al., 2022, 2021a). The external functions are usually associated with rigorous models developed in specific programs, so their analytical form is not available.

In order to guarantee finding an optimal solution that satisfies first-order optimality conditions, special solution strategies are required that should be tailored for these hybrid problems. In the present study, we present an optimization framework to formulate mathematical problems including rigorous Computational Fluid Simulations (CFD) in Comsol Multiphysics® and complex Aspen Plus models in the formulation of a nonlinear programming (NLP) mathematical problem, showing that the method works across different platforms. We develop a customized implementation of the trust-region filter method described by Yoshio and Biegler (2021). This method not only ensures optimality conditions but also allows us to tackle integrated process models, which encompass rigorous CFD models. A special feature of this solution strategy is the capability to profit from derivative information to reduce the number of rigorous model calls, and consequently, improving the performance of the algorithm (Biegler et al., 2014).

As case study, we address the optimal design of a relevant carbon capture technology using hollow fiber membrane (HFM) contactors. With growing concerns regarding CO_2 emissions from human and industrial activities, immediate decarbonization solutions are increasingly demanded. Retrofitting existing power and industrial plants with post-combustion carbon capture technology is an attractive short-term strategy in the current context. Among the available technologies, absorption-based carbon capture stands out due to its maturity. In order to enhance the performance of the absorber unit, HFM contactors have been proposed due to their operational flexibility, scalability, modularity, and large interfacial areas (Rivero et al., 2020). We focus on the optimization of an MEA-based carbon capture plant using HFM contactors, where the objective function to be minimized is the CO_2 avoided cost. This problem requires a comprehensive approach and incorporates a rigorous CFD model for partially wetted membrane contactors.

2. Methodology

The general form of this hybrid problem can be represented as follows:

$$\min_{v,w} f(v,w,y) \tag{1.1}$$
$$s.t. \quad g(v,w,y) \leq 0 \tag{1.2}$$
$$y = t(w) \tag{1.3}$$

where $f(v,w,y)$ is the objective function, $g(v,w,y)$ are the inequality constraints (including the variable bounds); $t(w)$ represents the rigorous (or 'truth') model; y are the response variables of these truth models, w are the matching input variables, and v are the remaining variables for the problem.

Trust region methods tackle problem (1) by formulating a trust-region subproblem that includes reduced models (RMs), instead of the rigorous ones, and solving it in a search domain where the RMs are good enough to improve either the objective function or the infeasibility. The trust-region subproblem at iteration k can be expressed as follows:

$$\min_{v,w} f(v,w,y) \tag{2.1}$$
$$s.t. \quad g(v,w,y) \leq 0 \tag{2.2}$$
$$y = r_k(w) \tag{2.3}$$
$$\|u - u_k\| \leq \Delta_k \tag{2.4}$$

where the truth model $t(w)$ has been replaced by a reduced model $r_k(w)$, and trust region constraints are formulated using the decision variables u, the current base point u_k, and the trust region size Δ_k. When derivative information is available from the truth model, we introduce zero and first order corrections to the RMs to update it. In this way, the RMs are exact at the base point used for the trust region constraint and we could use any reduced model function $\tilde{r}(w)$ to build them, as shown in Eq. (3). Although these methods provide convergence guarantees for any $\tilde{r}(w)$, the performance of the algorithm improves significantly when the accuracy of $\tilde{r}(w)$ improves. Here, we build $\tilde{r}(w)$ by using the ALAMO approach (Cozad et al. 2014), as follows:

$$r_k(w) = \tilde{r}(w) + \left(t(w_k) - \tilde{r}(w_k)\right) + \left(\nabla t(w_k) - \nabla \tilde{r}(w_k)\right)^T (w - w_k) \tag{3}$$

3. Process modeling

We apply the trust-region method to an optimal design problem of a carbon capture technology using MEA-based absorption with membrane contactors. The process flowsheet considered for this process is the conventional CO_2 absorption process using only one amine heat exchanger for heat recovery (Pedrozo et al., 2023), where the packing column for absorption has been replaced by membrane contactor modules. This includes absorption unit that operates at counter-current, providing mass transfer area to promote the CO_2 absorption in the liquid solvent. The input streams are a lean solvent stream and flue gas (CO_2-rich gas), while the output streams are the amine CO_2-rich stream and the clean gas. The rich solvent stream is pumped to a heat exchanger to preheat before CO_2 desorption. The regeneration process is carried out in a stripping column, using conventional structured packing. This process includes a reboiler, which represents the main energy consumption of the carbon capture plant. The regenerated solvent is obtained at the bottom of the column, and is fed to the amine heat exchanger for energy recovery. Then, it is mixed with makeup streams and conditioned with a cooler to be recycled at the absorber unit.

Regarding the modeling approaches, the flowsheet is implemented in Pyomo (Hart et al. 2017) and solved with IPOPT (Wächter and Biegler, 2006). The models for the heat exchanger, the pump, and the mixer are formulated using explicit equations, based on the literature (Pedrozo et al., 2021b). On the other hand, the key process units are modeled as external function and calculated using specialized programs. In particular, the truth model for the absorption unit is a rigorous CFD model developed in Comsol Multiphysics®, while the truth model for the regenerator process is implemented in Aspen Plus.

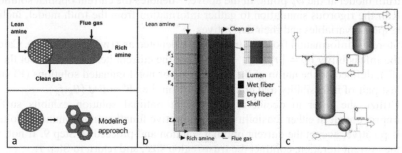

Figure 1: Rigorous models for the absorber and the regenerator. a) Membrane module and modeling approach. b) 2D, partially wetted fiber representation, considering four domains: lumen, wet fiber, dry fiber, and shell. c) Aspen Plus model for the regeneration process

We show a representation of the membrane module in Fig. 1a, where the solvent is fed in the lumen of hollow fiber and the flue gas is in the shell. For the absorber model, we consider homogeneous flow distribution, i.e., same operating condition for each hollow fiber. In this way, we determine a hypothetical shell radius (r_4) for modeling purposes. Therefore, we can model the transport phenomena for one fiber, which is considered to be representative of the system. To build such a model, we use a 2-D axisymmetric approach that allows a detailed description of the transport phenomena within reasonable CPU times (Rivero et al., 2020). The geometric representation ensures accuracy when angular variations are insignificant, and it offers 3D representations of the fiber without the computational overhead of a full 3D CFD simulation. The CFD approach used for membrane contactors enables the incorporation of mass transport processes involving convection, diffusion, and chemical reactions for individual fibers. The mathematical model of the absorption using membrane contactors includes four sections (see Fig. 1b): i) lumen side, ii) wetted membrane, iii) dry membrane, and iv) shell side.

To tackle the numerical problem, we use a mapped meshing technique, including a fine distribution of cells at the gas-liquid interphase, as shown in Fig. 1b. Regarding the derivative information, we leverage the optimization sensitivity module in Comsol Multiphysics® to provide the corresponding data.

The rigorous model for the regeneration process is implemented in Aspen Plus and shown in Fig. 1c. A "RadFrac" block is used to model the column, as discussed in Pedrozo et al., (2023). In order to obtain accurate sensitivity information, the model is run with the equation-oriented (EO) mode in Aspen Plus.

4. Algorithm

We have employed a trust region filter method based on Yoshio and Biegler (2021). In contrast to penalty functions, filter methods construct the constrained optimization problem as a biobjective problem, aiming at minimizing both the objective function and the constraint violation independently. The main algorithm includes the following steps:

1. Create reduced models based on Eq. (3), by utilizing information from the rigorous models (i.e., simulations in Comsol Multiphysics® and Aspen EO), including response variable values and their derivatives. The bounds of optimization variables, u, are modified based on the trust region size Δ_k, and the base point u_k.
2. Solve the trust region subproblem in Pyomo (problem (2)).
3. Store the current optimal solution for input variables w_k^*, decision variables u_k^*, and the objective function f_k^*. Then, configure the rigorous simulation to evaluate the truth model at the w_k^* point. In the above, * denotes the current optimal solution.
4. Run the rigorous simulation to gather information from the truth model, including response variables and their derivatives.
5. Store the information from the truth model, denoted as $t(w_k^*)$ and $\nabla t(w_k^*)$. Calculate the infeasibility $\theta_k^* = \|r_k(w_k^*) - t(w_k^*)\|$ at the current iteration k and set the filter (F'), defined as the union of the existing set of nondominated solutions (F) and the last pair of infeasibility and objective function, i.e., $F' = F \cup (\theta_k, f_k)$.
6. Utilize the filter to decide if the current optimal solution exhibits sufficient improvement in either feasibility or the objective function (nondominated solution). If positive, accept the current optimal solution and proceed to Step 9. If not, reject the current iteration, contract the trust-region size, and return to Step 1.
7. Verify the convergence criterion ($\theta_k^* < \epsilon$) to assess if the infeasibility is sufficiently low and if an optimal solution has been reached. If the infeasibility exceeds the target tolerance (ϵ), proceed to Step 8.
8. Check the switching criterion (see Yoshio and Biegler (2021)) to classify the current optimal solution as an f-step (move to Step 9) or a θ-step (move to Step 10). f-steps are constrained to relatively small infeasibilities.
9. For f-step, we have an improvement of the objective function over the infeasibility. The trust region size (Δ_k) is increased and the filter remains unaltered to prevent missing potential optimal solutions in subsequent iterations. Proceed to Step 11.
10. For θ-steps, the acceptance of the current optimal solution depends on a reduction in infeasibility, and this information is used to update the filter to promote convergence and enhance accuracy for future iterations. The trust region size (Δ_k) may increase, decrease, or remain unchanged based on the updating criteria (Yoshio and Biegler, 2021).
11. Update the base point, and the pair (θ_k, f_k). Then, return to Step 1.

5. Case study

As a case study, we consider the optimization of a capture plant designed to purify the flue gas generated by a coal-fired power plant. The flue gas has a CO_2 concentration of

15 mol% and a mass flowrate of 288 t/h. As specification design, we consider a CO_2 recovery higher than 90 %. The objective function is the minimization of the CO_2 avoided cost, as calculated in the literature (Pedrozo et al., 2023).

Regarding the optimization variables, we consider the following decision variables for the capture plant: the regenerator height, the output temperature of the amine heat exchanger to preheat the CO_2-rich streams, the mass flow rate of the lean amine stream, the operating temperature of the absorber (isothermal operation), the MEA concentration of the lean amine stream (w_{MEA}), the CO_2 molar fraction of the lean stream (x_{CO_2}), the inner radius of fibers (r_1), the membrane thickness (r_3 -r_1), hypothetical shell radius (r_4), the fiber length (l_f), and the porosity (ε). We note that the rigorous CFD model allows inclusion of geometric features as optimization variables in the optimal design problem.

6. Results

When solving the hybrid optimization problem, which includes a CFD and Aspen EO simulations, through the trust region framework, the problem successfully converges within 33 iterations, as shown in Fig. 2. It is observed that at the beginning of the optimization process, we have an important reduction in the objective function and f-steps in the algorithm, while we mainly obtain important reduction in the infeasibility in the final iterations (θ-steps), until the convergence criterion is achieved ($\theta<10^{-4}$).

Regarding the CPU time of the solution procedure, it demands 1260 s. The breakdown of this time reveals that running the rigorous model consumes 98% of the total time, while the CPU time associated with solving the optimization subproblems is negligible.

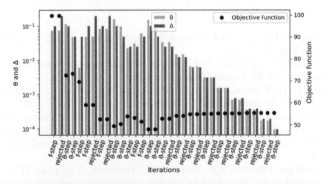

Figure 2: Iterations of the trust-region filter method. θ: infeasibility metric, Λ: trust region size

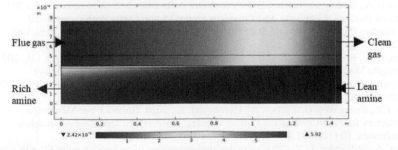

Figure 3: CO_2 molar concentration (mol/m^3) profile for a single hollow fiber. Re_G=16.56; x_{CO_2}=0.15; Re_L=4.55; w_{CO_2}=5.31 wt%; w_{MEA}=40 wt%; T_S=320 K; r_1=0.375 mm; r_3-r_1=0.130 mm; r_4-r_3=0.359 mm; l_f=1.46 m; ε=0.55

The optimal design for the carbon capture plant using HFM contactors exhibits favorable performance metrics, including a CO_2 avoided cost (objective function) of 55.33 \$/t-CO_2, and a specific energy demand of 3.46 GJ/t-CO_2 when utilizing MEA solvent. These outcomes indicate that this technology can serve as a viable competitor to traditional absorbers employing structured packing, particularly in certain application when modularity and flexibility are desirable.

Regarding the results associated with the CFD model, Fig. 3 shows the optimal CO_2 concentration profiles in a hollow fiber, including the optimal values for the decision variables associated with the design of the absorber module. Here, we report the Reynolds numbers for the gas (Re_G) and liquid (Re_L) phase.

7. Conclusions

The trust-region filter method is a potent tool to build optimization frameworks to address hybrid problems that include external functions from specialized codes (e.g., CFD simulations). Our framework shows proficiency in data management, Python-based setup of Comsol Multiphysics® and Aspen Plus, and automated program execution. As these platforms are widely used across engineering science, the solution strategy enables the integration of existing rigorous models in NLP formulations. Moreover, our solution strategy is generalizable beyond the tools used in the current demonstration.

We show the efficiency of the method by solving an optimal design problem for a CO_2 capture plant using hollow fiber membrane (HFM) contactors. Numerical results indicate that the CO_2 avoided cost associated with this technology is 55.33 \$/t-CO_2.

References

Biegler, L.T., Lang, Y., Lin, W., 2014. Multi-scale optimization for process systems engineering. Comput. Chem. Eng. 60, 17–30.

Cozad, A., Sahinidis, N. V., & Miller, D. C. (2014). Learning surrogate models for simulation-based optimization. AIChE Journal, 60(6), 2211-2227.

Hart, W. E., Laird, C. D., Watson, J. P., Woodruff, D. L., Hackebeil, G. A., Nicholson, B. L., & Siirola, J. D. (2017). Pyomo-optimization modeling in python (Vol. 67, p. 277). Springer.

Pedrozo, A., Valderrama-Ríos, C.M., Zamarripa, M., Morgan, J., Osorio-Suárez, J.P., Uribe-Rodríguez, A., Diaz, M.S., Biegler, L.T., 2023. Equation-Oriented Optimization Applied to the Optimal Design of Carbon Capture Plants Using Rigorous Models. Ind. Eng. Chem. Res. 62, 7539–7553.

Pedrozo, H.A., Rodriguez Reartes, S.B., Bernal, D.E., Vecchietti, A.R., Díaz, M.S., Grossmann, I.E., 2021a. Hybrid model generation for superstructure optimization with Generalized Disjunctive Programming. Comput. Chem. Eng. 154, 107473.

Pedrozo, H.A., Rodriguez Reartes, S.B., Vecchietti, A.R., Diaz, M.S., Grossmann, I.E., 2021b. Optimal Design Of Ethylene And Propylene Coproduction Plants With Generalized Disjunctive Programming And State Equipment Network Models. Comput. Chem. Eng. 107295.

Pedrozo, H.A., Rodriguez Reartes, S.B., Vecchiettic, A.R., Diaz, M.S., Grossmann, I.E., 2022. Surrogate Modeling for Superstructure Optimization with Generalized Disjunctive Programming. 14th International Symposium on Process Systems Engineering – PSE 2021+, Kyoto, Japan.

Rivero, J.R., Panagakos, G., Lieber, A., Hornbostel, K., 2020. Hollow Fiber Membrane Contactors for Post-Combustion Carbon Capture: A Review of Modeling Approaches. Membranes, 10 (12), 382.

Wächter, A., & Biegler, L. T. (2006). On the implementation of an interior-point filter line-search algorithm for large-scale nonlinear programming. Mathematical programming, 106, 25-57.

Yoshio, N., Biegler, L.T., 2021. Demand-based optimization of a chlorobenzene process with high-fidelity and surrogate reactor models under trust region strategies. AIChE J. 67, e17054.

Flavio Manenti, Gintaras V. Reklaitis (Eds.), Proceedings of the 34th European Symposium on Computer Aided Process Engineering / 15th International Symposium on Process Systems Engineering (ESCAPE34/PSE24), June 2-6, 2024, Florence, Italy

Dynamic Modelling of Electrodialysis with Bipolar Membranes using NARX Recurrent Neural Networks

Giovanni Virruso,[a,b] Calogero Cassaro,[a] Waqar Muhammad Ashraf,[b] Alessandro Tamburini,[a] Vivek Dua,[b] I. David L. Bogle,[b,*] Andrea Cipollina,[a] Giorgio Micale[a]

[a]Dipartimento di Ingegneria, Università degli studi di Palermo, Viale delle scienze Ed 6, Palermo, 90128, Italia
[b]Centre for Process Systems Engineering, Department of Chemical Engineering, University College London, Torrington Place, London, WC1E 7JE, United Kingdom
d.bogle@ucl.ac.uk

Abstract

Electrodialysis with bipolar membranes (EDBM) is an innovative and effective process for the simultaneous production of acid and base solutions from salty streams. It has been proven to play a key role in several circular economy approaches to valorize waste industrial brines, but it can also be used for in situ generation of chemicals, especially in remote areas. The adoption of such technology at industrial scale requires reliable modelling tools capable of predicting both dynamic and stationary operations as process conditions vary, such as energy supplied to the system and the target concentration of chemicals. In this study, nonlinear autoregressive models with exogenous inputs (NARX) were applied for the first time to EDBM to predict the behaviour of this complex and non-linear process. Thus, an effective and low computational demanding neural-based modelling tool was developed. As a preliminary step, the network was trained with three different datasets, generated by a fully validated model. The best architecture was chosen to give good performance, testing the network with a new dataset. The NARX network accurately predicts the different behaviour of EDBM outputs (i.e. voltage and solutions conductivities) showing low average discrepancies between predicted and true values (lower than 0.5 %). These results suggest the possibility of using neural network-based models to effectively optimize and control EDBM process. Next step will focus on the training and validation of a network obtained with a set of data from a real EDBM plant.

Keywords: EDBM, BMED, Circular economy, Brine valorization, Artificial neural networks.

1. Introduction

Industrial processes are responsible for generation of waste streams which need to be treated to reduce their impact and reach environmental standards. In several cases, especially in the field of water treatment (Badruzzaman et al., 2009) and salt production (Turek et al., 2005), mixed salt solutions are produced. These streams represent a valuable unconventional source of minerals which can be recovered through circular economy approaches, improving the sustainability of industrial processes. Recently, the Horizon 2020 Water Mining project (*Grant Agreement*, 2020) has proved a minimum liquid discharge approach to valorize desalination brine, recovering chemicals, high value salts and fresh water. This is accomplished through the adoption of different units based on membrane, thermal and reactive technologies. A crucial role in the treatment chain is played by ElectroDialysis with Bipolar Membranes (EDBM) unit, a membrane-based technology capable of producing acid and base solutions (Herrero-Gonzalez et al., 2023). These chemicals can be reused inside the chain as reagents for the reactive unit, but also to perform pH adjustment, regeneration of ion-exchange resins and cleaning. EDBM can also be suitable for *in-situ* generation of chemicals ready to use in all the industrial fields where medium-low concentration of acid and base are need (i.e., 1-5 wt.%). This motivates the importance of studying and optimizing this technology to promote its adoption on an industrial scale.

EDBM employs a salt solution and water as feed streams, and through the use of electrical energy is able to simultaneous produce acid and base streams and reduce the salinity of the salt solution. In this process ion-exchange membranes are employed, both monopolar (i.e., anionic and cationic) and bipolar. These membranes allow the selective passage of ions depending on their electrical charge. The bipolar system also promotes water dissociation into protons and hydroxide ions. The membranes are stacked together placing polymeric spacers between them to form the solution channels: acid, base and salt channels. The three solution channels and the ion-exchange membranes constitute the repetitive unit of an EDBM stack, also known as a triplet. Placing several triplets between a couple of electrodes and supplying a direct current allows the process to generate the desired amount of chemicals.

The EDBM process has been modelled adopting a first principles approach based on mass balances and on electro-neutrality requirements (Mier et al., 2008). These models show good performance compared to experimental results, although they require some calibration parameters to be estimated (Culcasi et al., 2022). A major drawback is that they require a significant computational effort, especially when dynamic behaviors are simulated, reducing the attractiveness of these modeling tools, especially for advanced control strategies.

To overcome these limits, AI-based models can be used to speed up predictive analysis and optimization (Ashraf & Dua, 2023). Among them, artificial neural networks have shown good performance in predicting the behavior of membrane-based processes (Asghari et al., 2020). Neural Networks (NN) are black box models capable of reproducing nonlinear relationships between inputs and outputs. The NN structure is composed of processing elements (called nodes or neurons) interconnected using weights and grouped into layers (Jing et al., 2012). The information flow between neurons of different layers can be directed from input to output (feedforward network), or from the output of a layer to the input of a preceding layer (recurrent or feedback network) or also between neurons of the same layer (lateral connections) (Porrazzo et al., 2013). Nonlinear autoregressive models with exogenous inputs (NARX) neural networks have been successfully implemented in many dynamic systems with complex nonlinearities, such as hysteresis phenomena (Wang & Song, 2014).

This study applies for the first time a NN to the EDBM process. Specifically, NARX neural networks were used to develop an effective modelling tool of a commercial EDBM unit. Data to train the network were obtained through a modified version of a first-principles model (Virruso et al., 2024). A semi-industrial scale EDBM unit (Cassaro et al., 2023) was simulated under transient operating conditions, investigating the effect of varying current density and solutions flowrates. The neural network model was tested using an additional dataset (not used during the training phase) showing an average error lower than 0.5 % for multi-step predictions.

2. Methodology

2.1. Data generation

The data used to develop the NARX model were obtained from a dynamic version of a multi-scale model (Virruso ct al., 2024), used to simulate a commercial EDBM unit, FT-ED 1600-3 (FuMA-Tech GmbH). The EDBM unit operates adopting the feed and bleed process configuration for all the lines (i.e., acid, base, and salt), Figure 1. Four different datasets were generated by varying the operating conditions as a function of time and observing the dynamic response of the system. Fixed inlet concentrations of the solutions were used in all the datasets. A concentration of 1.5 mol L^{-1} of NaCl was employed for the salt steam, while, for acid and base line, water with a low salt content (1 mmol L^{-1} of NaCl) was utilized. Step changes were given to the outlet flowrates of the solutions and to the current density to investigate several operating conditions of interest for industrial applications. For each dataset consecutive changes were simulated with alternating periods of stable conditions (about 0.9 h) to observe the system's behavior, adopting a discretization time of 100 s. Details about datasets are reported in Table 1.

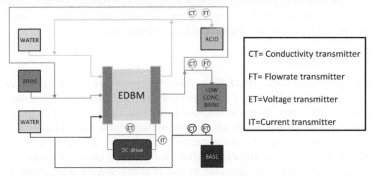

Figure 1. Schematic representation of an EDBM unit operated in the feed and bleed mode. Transmitters are depicted to show the variables used as inputs and outputs in the NARX model.

The variables subjected to step changes represent the inputs of the network and they were chosen to include all the major operating variables that are commonly manipulated in EDBM applications (Virruso et al., 2023). As output of the neural network model, physically meaningful and easy to measure online variables were selected. In particular, the outlet conductivity of the three solutions and the voltage applied to the stack were chosen. The former provides a partial indication of the outlet concentration reached (more detailed analysis must be performed offline), while the latter provides an indication of the energy consumed by the EDBM.

Table 1. Datasets obtained with the mechanistic model. The variation range for each variable, the operation time simulated and the number of step changes are reported for each dataset.

Dataset	$Q_{acid,out}$ (L min^{-1})	$Q_{base,out}$ (L min^{-1})	$Q_{salt,out}$ (L min^{-1})	I (A m^{-2})	$\dfrac{Q_{base,out}}{Q_{base,out}}$	Duration (h)	N. step changes
1	0.5-1.5	0.5-1.5	1-2.25	200-500	1.5-2	10.8	12
2	1-3	1-3	1.5-4.5	200-500	1.5	16.7	19
3	1.8-2.8	1.8-2.8	2.6-4.1	200-400	1.5	10.6	10
4	1.3-2.3	1.3-2.3	1.9-3.4	200-500	1.5	5.5	6

2.2. NARX neural network structure

The network structure presents four inputs (i.e., solutions outlet flowrates and current density) and four outputs (i.e, voltage applied to the stack and solutions outlet conductivities). A single hidden layer was used as this has been demonstrated to provide satisfactory results in different cases, both for feedforward (Jing et al., 2012) and recurrent networks (Cadenas et al., 2016). In addition to the number of nodes in the hidden layer, the order of the Tapped Delay Line (TDL) of input and output must be selected for NARX neural network. The value of TDL represents how many past values of inputs and outputs are needed by the network to estimate the current output (Wang & Song, 2014).

The Mean Squared Error (MSE), as shown in Eq (1), was used as cost function to train the network.

$$MSE = \frac{1}{N} \sum_{k=1}^{N} (y(k) - d(k))^2 \tag{1}$$

Where $y(k)$ represents the predicted value at time step k, while $d(k)$ is the desired value. As activation functions, the Elliot symmetric sigmoid and the linear function were adopted for the hidden layer and the output layer, respectively.

3. Results and discussion

To train the NARX network, *MATLAB's Neural net time series* toolbox was employed. The structure described above was implemented and the Levenberg-Marquardt training algorithm was used. Three different datasets (1 to 3 in Table 1) were employed for the training phase, while dataset 4 was adopted for subsequent testing. The NARX network was trained in series-parallel configuration (also known as open loop), while its performance was evaluated using the parallel configuration (also known as closed loop) for multi-step predictions. Initialization of the network weight was adopted to avoid instability issues in closed loop. The training data were randomly divided into three subsets: training, validation and testing subsets. In particular, 80 % of the data was used for training, 10% for validation and the remaining 10% for verification. It is important to note that this division of results is useful to stop the training phase. The training phase stops at the number of epochs at which the error for the validation set starts to increase and uses the weights and biases determined at that point (Porrazzo et al., 2013).

To define the number of nodes in the hidden layer and the order of TDL of inputs and outputs, different networks were trained by changing the number of nodes between 1 and 15 as well as the number of TDL of the outputs between 1 and 5. The MSE obtained in dataset 4 was calculated using the network in parallel configuration and predicting the entire dataset. The average value of MSE between all the outputs was adopted as the evaluation criterion of the network performance. The best performance was obtained for a NARX network with 12 nodes and a TDL of 3 for the output. Finally, it was checked whether an increase in performance could be obtained using a non-zero value for the TDL of the input, and values between 1 and 3 were investigated. No improvements were observed, consequently a zero order TDL was used for the input.

The generalization performance of the NARX network in predicting dataset 4 are shown in Figure 2. As it can be observed the NARX network, in parallel configuration and doing multi-step predictions, is able to reliably predict the entire dataset 4. The dynamics of all the EDBM outputs were accurately predicted despite the much faster behavior of the voltage compared to the conductivities. The local error between the predicted value and the true value was evaluated. Average values lower than 0.5 % were found for all the outputs demonstrating the high prediction capability of the NARX networks.

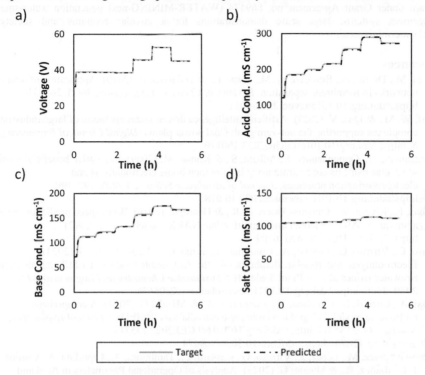

Figure 2. Comparison between the NARX results and the true value for the fourth dataset. The NARX neural network was employed in closed loop mode for multi-step prediction. The time-dependent profiles are shown of: a) voltage applied to the stack, b) outlet conductivity of the acid line, c) outlet conductivity of the base line and d) outlet conductivity of the salt line.

4. Conclusions

This work applies a neural network to electrodialysis with bipolar membranes for the first time. A recurrent dynamic network, NARX, was used to study the EDBM unit dynamic behaviour in feed and bleed operation mode. A NARX neural network with four inputs and four outputs and one hidden layer was employed. The best number of nodes and the order of tapped delay line (TDL) of input and output was determined evaluating different architectures. As criterion to select the best architecture the MSE obtained on the entire dataset 4 was used, adopting a parallel configuration and doing multi-step predictions. The selected network presents 12 nodes in a hidden layer as well as a 3rd order TDS for the output and no TDL for the input. The generalization performance was shown demonstrating that the different dynamic responses of the EDBM outputs can be predicted with high precision (average error <0.5 %). These results show the effectiveness of NARX networks in describing complex nonlinear electro-membrane systems, suggesting that they can be used for advanced control systems and to optimize EDBM process. Future research will also investigate the use of real data, obtained from an operating pilot scale EDBM unit, to assess whether the effect of non-ideal phenomena and noisy signals can affect the network performance.

Acknowledgements

This project has received funding from the European Union's Horizon 2020 research and innovation program under Grant Agreement no. 869474 (WATER-MINING-next generation water-smart management systems: large scale demonstrations for a circular economy and society). https://watermining.eu/.

References

Asghari, M., Dashti, A., Rezakazemi, M., Jokar, E., & Halakoei, H. (2020). Application of neural networks in membrane separation. *Reviews in Chemical Engineering, 36*(2), 265–310. https://doi.org/10.1515/revce-2018-0011

Ashraf, W. M., & Dua, V. (2023). Artificial intelligence driven smart operation of large industrial complexes supporting the net-zero goal: Coal power plants. *Digital Chemical Engineering, 8*. https://doi.org/10.1016/j.dche.2023.100119

Badruzzaman, M., Oppenheimer, J., Adham, S., & Kumar, M. (2009). Innovative beneficial reuse of reverse osmosis concentrate using bipolar membrane electrodialysis and electrochlorination processes. *Journal of Membrane Science, 326*(2), 392–399. https://doi.org/10.1016/j.memsci.2008.10.018

Cadenas, E., Rivera, W., Campos-Amezcua, R., & Heard, C. (2016). Wind speed prediction using a univariate ARIMA model and a multivariate NARX model. *Energies, 9*(2). https://doi.org/10.3390/en9020109

Cassaro, C., Virruso, G., Culcasi, A., Cipollina, A., Tamburini, A., & Micale, G. (2023). Electrodialysis with Bipolar Membranes for the Sustainable Production of Chemicals from Seawater Brines at Pilot Plant Scale. *ACS Sustainable Chemistry and Engineering, 11*(7), 2989–3000. https://doi.org/10.1021/acssuschemeng.2c06636

Culcasi, A., Gurreri, L., Cipollina, A., Tamburini, A., & Micale, G. (2022). A comprehensive multi-scale model for bipolar membrane electrodialysis (BMED). *Chemical Engineering Journal, 437*, 135317. https://doi.org/10.1016/J.CEJ.2022.135317

Grant agreement. (2020). https://doi.org/10.3030/869474

Herrero-Gonzalez, M., López, J., Virruso, G., Cassaro, C., Tamburini, A., Cipollina, A., Cortina, J. L., Ibañez, R., & Micale, G. (2023). Analysis of Operational Parameters in Acid and Base Production Using an Electrodialysis with Bipolar Membranes Pilot Plant. *Membranes, 13*(2). https://doi.org/10.3390/membranes13020200

Jing, G., Du, W., & Guo, Y. (2012). Studies on prediction of separation percent in electrodialysis process via BP neural networks and improved BP algorithms. *Desalination, 291*, 78–93. https://doi.org/10.1016/j.desal.2012.02.002

Mier, M. P., Ibañez, R., & Ortiz, I. (2008). Influence of ion concentration on the kinetics of electrodialysis with bipolar membranes. *Separation and Purification Technology, 59*(2), 197–205. https://doi.org/10.1016/j.seppur.2007.06.015

Porrazzo, R., Cipollina, A., Galluzzo, M., & Micale, G. (2013). A neural network-based optimizing control system for a seawater-desalination solar-powered membrane distillation unit. *Computers and Chemical Engineering, 54*, 79–96. https://doi.org/10.1016/j.compchemeng.2013.03.015

Turek, M., Dydo, P., & Klimek, R. (2005). Salt production from coal-mine brine in ED–evaporation–crystallization system. *Desalination, 184*(1–3), 439–446. https://doi.org/10.1016/J.DESAL.2005.03.047

Virruso G, Cassaro C, Culcasi A, Cipollina A, Tamburini A, Bogle I, & Micale G. (2024). Multi-scale modelling of an Electrodialysis with Bipolar Membranes pilot plant and economic evaluation of its potential (submitted). *Desalination.*

Virruso, G., Cassaro, C., Tamburini, A., Cipollina, A., & Micale, G. D. M. (2023). *Performance Evaluation of an Electrodialysis with Bipolar Membranes Pilot Plant Operated in Feed & Bleed Mode.* https://doi.org/10.3303/CET23105013

Wang, H., & Song, G. (2014). Innovative NARX recurrent neural network model for ultra-thin shape memory alloy wire. *Neurocomputing, 134*, 289–295. https://doi.org/10.1016/j.neucom.2013.09.050

Flavio Manenti, Gintaras V. Reklaitis (Eds.), Proceedings of the 34th European Symposium on Computer Aided Process Engineering / 15th International Symposium on Process Systems Engineering (ESCAPE34/PSE24), June 2-6, 2024, Florence, Italy

Modelling a Geothermal Vaporiser: A First Step Towards a Digital Twin

Theo Brehmer-Hine, Wei Yu, Brent Young*

Chemical and Materials Engineering Department, The University of Auckland, New Zealand
*b.young@auckland.ac.nz

Abstract

'Digitalisation' is a current mega-trend for assisting industry decision-making. The geothermal industry is not immune to this trend, given the strong incentives to improve efficiency whilst maintaining product quality, safety and productivity. The 'digital twin' can be considered as one of the core concepts of digitalisation, and yet it is still at the conceptual stage for many manufacturing operations. The first step in digital twin implementation is to build digital models since they are the essential components of digital twins. However, how to select suitable models from many different types of models of different fidelity for the geothermal industry is an open question which has not yet been addressed. In this paper, we will compare two modelling approaches: first principles (using Python) and hybrid process simulation (using Aspen HYSYS) for simulating industrial-scale geothermal vaporizer dynamics. The development of these one-way connection models is useful as a first step towards a digital twin and allows a plant model with more flexibility for vaporiser modelling.

A comparison was made between models and the tools used in their development based on attributes of performance, availability, ease of use and model fidelity. The models from both approaches showed adequate predictions at multiple steady-state conditions. However, the Aspen HYSYS model performed slightly better. There were only slight differences in prediction performance and processing speeds. For the availability and ease of use of these tools, a trade-off would need to be made in deciding which modelling approach to use, weighing a monetary investment in using Aspen HYSYS against the time investment in building the first principles model.

Keywords: Digital twin, Model Fidelity, Vaporiser

1. Introduction

To meet New Zealand's (NZ's) goal of net-zero greenhouse gas (GHG) emissions by 2050, development into disruptive decarbonisation technologies in the process heat sector will be required. Digital twin (DT) technology shows the potential to improve energy efficiency and reduce greenhouse gas emissions. As evaporators and vaporisers are essential to NZ industries such as the dairy processing industry and the geothermal power industry, implementing DT technology on evaporators or vaporisers should be considered by all industries using these units.

A Digital Twin (DT) in a manufacturing context consists of a virtual representation of a production system 'twinned' with the physical system. A DT can run simultaneously with the physical system, whereby, an automatic and bi-directional connection of data flows between the physical and virtual systems. The role of DTs in manufacturing systems is to predict and optimise the behaviour of the systems in real-time. Sub-categories of digital shadows and digital models exist at lower levels of integration between model and system. A digital shadow is where the automatic flow of data is unidirectional from the physical system to the virtual system, while a digital model has no automated flow of data between the systems (Kritzinger *et al.*, 2018). A subtle adjustment to this classification was made by Yu *et al.* (2022), where the term digital manager is used to describe DTs with a two-way connection. This distinction is made to encompass DTs used in plant control systems rather than in plant design or plant retrofits.

The first step in DT development is in building a digital model of a system. In this work, an investigation was conducted into various models of evaporator and Organic Rankine Cycle (ORC) systems. The modelling approaches can be broadly categorised into first principles, 'black box' and 'grey box' modelling. A significant portion of the models seen in the literature have been developed using commercial process simulators such as Aspen Plus, Aspen HYSYS, VMG Symmetry, gProms and Modelica. Models developed using these software programs display a process flow diagram of this system itself and are capable of modelling both steady-state and dynamic systems. This is similar to first principles (FP) based models developed in software such as MATLAB, however, in FP models no visual representation of the system is typically shown. A significant number of empirical ('black box') models have also been developed for these systems in steady state and dynamics, however, these also lack visual representations. Of these models, only Soares *et al.* (2019) developed a model with a connection between the model and a system.

This paper is organised as follows. In Section 2, the vaporiser's basic information is provided. In Section 3, two modelling approaches: first principles and simulation (Aspen HYSY) are discussed. In Section 4, the performance of these models will be compared based on their dynamic response performance as while as attributes of performance: performance, availability, ease of use and model fidelity. Finally, the conclusions of this paper are presented.

2. System description

The vaporiser belonging to one NZ geothermal binary plant is investigated in this paper. The plant can generate a gross net power of 35 MW during normal operation. The system uses normal pentane as the working fluid. The vaporiser has multiple tube sets, for brine and geothermal vapour. The vaporiser configuration is shown in Fig. 1. The brine tubes

have two passes, which carry out the sub-heating and a small amount of vaporisation of the working fluid, while the geothermal vapour carries out the remainder of the vaporisation and a degree of superheating of the working fluid in a single pass. The working fluid flows perpendicular to the flow of the geothermal fluids.

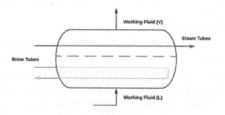

Figure 1: The vaporiser shell and tube configuration

The composition of a geothermal fluid is strongly related to its thermodynamic properties and therefore requires a 'fit for purpose' equation of state to be modelled most effectively. With access to the geothermal plant's compositional data, it was noted that the vapour is largely comprised of water with a small amount of carbon dioxide. Minuscule amounts of methane, ammonia, hydrogen sulfide, hydrogen and nitrogen were also shown in this analysis, however, the mole fractions were less than 0.0001 and it was therefore deemed unnecessary to include these components. The brine is assumed to be pure water as the changes to the thermodynamic properties of the fluid are relatively small when compared with vapour. The simplified composition (mole fraction) of the separated vapour used in the following models is shown in Table 1.

Table 1: Simplified composition of the separated geothermal vapour

Component	Composition (Mole fraction)
Water	0.985
Carbon Dioxide	0.015

3. Vaporiser Modelling

A vaporizer and associated models of the plant were developed in Aspen HYSYS, and a separate vaporizer model was also developed in Python. All models were developed and tuned against snapshots of the plant's steady-state conditions. The vaporiser models were then transitioned to dynamic simulations to study their dynamic responses.

3.1. Aspen HYSYS model

Aspen HYSYS is widely used in both industry and academia in analysing process engineering applications in steady state and dynamics. As such, Aspen HYSYS was selected for the development of a dynamic model of the vaporiser. In addition to this, a plant model of the working fluid cycle was developed in a steady state.

There is no unit operation provided by Aspen HYSYS that replicates a multiple tube-set heat exchanger. The approach taken is based on the method developed by Proctor *et al.* (2016), where two coolers were used to model the heat transfer of each tube-set in addition to a heater modelling the heating of the shell side. The LMTD was calculated for each section and these were multiplied by the static UA values calculated based on plant data. The resultant heat flows were added and used for specifying the shell-side heater. The Aspen HYSYS schematic is shown in Fig. 2.

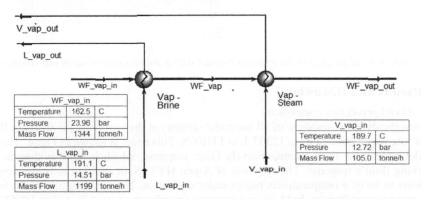

Figure 2: The Aspen HYSYS schematic of the vaporiser model. The 'Vap-Brine' and 'Vap-Steam' refer to the brine and geothermal steam of the vaporiser.

Careful consideration must be taken in selecting the Equations of State (EOS) as this is necessary for an accurate model as it handles the calculation of state variables under given conditions. Modelling a vapour mixture of water and carbon dioxide can be difficult due to the complex interactions between polar water and non-polar carbon dioxide molecules.

After testing different EOS packages, the Cubic-Plus-Association (CPA) EOS was selected based on model prediction accuracy.

3.2. First principles (FP) model

The FP model was developed in Python using the solve_ivp integrator from SciPy integrate, where the explicit Runge-Kutta method of order 3 (2) was used in solving the ODE. The model utilised Coolprop for the thermodynamic package. The governing equations of the vaporiser are the working fluid mass and energy balance, the geothermal fluid mass and energy balance, and the energy balance of the wall. The following assumptions were made for simplification of the solution of the differential equations:

- o   No axial heat conduction, only heat transport in the working fluid, wall or geothermal fluid.
- o   Spatially constant fluid pressure $p(t)$.
- o   The three tube passes are each discretised into a number (n) of lumped sub-volumes, each of which corresponds to a sub-volume of the shell.
- o   Static geothermal fluid and working fluid temperature profiles for each section.
- o   Inlet and outlet enthalpy are averaged to approximate the temperature of the section.
- o   The mass flow rate for each fluid is equal in all sections.
- o   The geothermal vapour composition is assumed to contain CO_2 and water.

In developing this model, the working fluid was segmented into three vertical sections. These corresponded to the three passes of geothermal fluid through the vaporiser. A diagram describing this configuration is shown in Fig. 3.

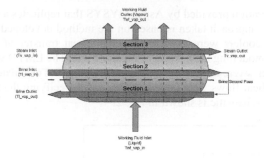

Figure 3: Configuration of the Python FP model with a discretisation of seven sub-volumes

4. Results and Discussion

4.1. Model prediction comparison

To test the dynamic response of all the outlet streams of the vaporiser, the working fluid flow rate was increased from 1230T/h to 1310T/h. This result is shown in Fig. 4. The FP model appears to show a comparatively faster response, which is most evident in the working fluid's response. The response of Aspen HYSYS, in addition to being slower, appears to be of a comparatively higher-order derivative. This is thought to result from the more comprehensive hold-up and pressure specifications used in Aspen HYSYS's heat exchanger model which mimics physical systems. This could be addressed by implementing more rigorous mass and momentum balances into the ODE or by tuning the model dynamically (e.g., Pili *et al.*, 2017; Huster *et al.*, 2018). When comparing the FP model against the outlet temperatures produced in Aspen HYSYS, a Root Mean Squared Error (RMSE) of 0.124∘C, 0.401∘C and 0.250∘C was shown for the brine, vapour and working fluid respectively. The mean average percentage error (MAPE) was 0.039%, 0.211% and 0.093% for brine, vapour and working fluid, respectively.

4.2. Model comparison

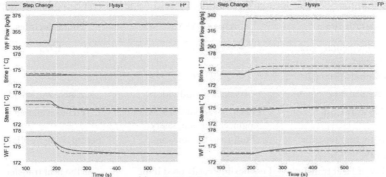

Figure 4: Step change of the working fluid (A) and brine (B) flowrates and the temperature response of the vaporiser for the Aspen HYSYS model and the Python FP model

The comparison criteria are based on Alford *et al.*, (2022). This includes the model performance availability and ease of use of the software. The comparison criteria are also extended to include the aforementioned model fidelity framework outlined by Yu *et al.*, (2022). The summary of the comparisons is listed in Table 2.

<u>Performance:</u> The model performance considers prediction accuracy and the processing time required.

<u>Availability:</u> The availability considers the accessibility and the monetary investment of these software tools. Python and CoolProp are both available worldwide at no cost, while an Aspen HYSYS dynamic license, though also globally available, costs approximately 85% of a process engineering graduate's annual salary in New Zealand.

<u>Ease of use:</u> This considers the difference in time, effort and skill requirements of developing each model. For the entire FP model, a relatively small percentage of the written code is used for the calculation of heat transfer, whereas the remaining code is used in setting up the thermodynamic properties and state variables. This setup includes the developed empirical equations and CoolProp. This differs from the Aspen HYSYS model which has a wide range of easily accessible thermodynamic property packages.

<u>Model fidelity:</u> The model fidelity framework proposed by Yu *et al.*, (2022) was used which includes three distinct attributes of a Digital Twin: looks-like, behaves-like and connects to. Both models developed have been validated against multiple steady-state data points and are capable of modelling dynamically. These models therefore would be classified as 3-D in the behaves-like attribute.

5. Conclusions

Two models: a plant model of the ORC cycle developed in Aspen HYSYS and a first principles-based model of the vaporiser developed in Python using a discretisation approach were developed. These models were used to compare the advantages and disadvantages of different modelling approaches for DT implementation. This was used to further understand the key strengths and limitations of the software tools used, and the model development process. We conclude that there were slight differences in prediction performance and processing speeds. For the availability and ease of use of these tools, a trade-off would need to be made in deciding which modelling software to use, weighing a monetary investment in using Aspen HYSYS against the time investment in using Python.

Table 2: Summary of the comparisons of the FP and Aspen HYSYS model attributes

Model Attribute	Aspen HYSYS	First Principles
Average Prediction Error of Steady State	0.065	1.773
Simulation Speed (s)	93.09	80.09
Availability	Available worldwide	Available worldwide
Software Cost	85% of a graduate process engineer's annual salary	Freely available
Set-up (thermodynamic properties)	Built-in to the software	Set-up using CoolProp
Flexibility of Heat transfer	Less flexible, easily used if the unit operation is available	As the heat transfer equation is formulated the high degree of flexibility
Development Time (Estimate)	1 week	4 months
Quality Assurance/Robustness	Tested commercially by a professional team of people	Tested by one person
Behaves-like	3D	3D
Looks-like	2D	2D
Connects-to	1D	1D

Acknowledgments

This research has been supported by the programme "Ahuora: Delivering sustainable industry through smart process heat decarbonisation", an Advanced Energy Technology Platform, funded by the New Zealand Ministry of Business, Innovation and Employment.

References

M. Alford, I. Udugama, W. Yu & B.R. Young, B. R., 2022, Flexible digital twins from commercial off-the-shelf software solutions: A driver for energy efficiency and decarbonisation in process industries? *Chemical Product and Process Modeling*, 17(4), 395-407.

W. Kritzinger, M. Karner, G. Traar, J. Henjes & W. Sihn, W., 2018, Digital Twin in manufacturing: A categorical literature review and classification. *In IFAC-PapersOnLine*, 51(11), 1016–1022.

M. J. Proctor, W. Yu, R. D. Kirkpatrick & B. R. Young, 2016, Dynamic modelling and validation of a commercial scale geothermal organic Rankine cycle power plant. *Geothermics*, 61, 63–74.

R. M. Soares, M. M.Câmara, T. Feital & J. C. Pinto, 2019, August. Digital Twin for monitoring of industrial multi-effect evaporation. *Processes*, 7(8), 537.

W. Yu, P. Patros, B. R.Young, E. Klinac & T. G. Walmsley, 2022, Energy digital twin technology for industrial energy management: Classification, challenges and future. *Renewable and Sustainable Energy Reviews*, 161, 112407.

Flavio Manenti, Gintaras V. Reklaitis (Eds.), Proceedings of the 34th European Symposium on Computer Aided Process Engineering / 15th International Symposium on Process Systems Engineering (ESCAPE34/PSE24), June 2-6, 2024, Florence, Italy

Development and Application of a Simplified Non-Ideal Mixing Model for Semi-Batch Crystallization

Jan Trnka[a], Giovanni Maria Maggioni[b], František Štěpánek[a],*

[a]*Department of Chemical Engineering, UCT Prague, Technická 3, 16628 Prague, Czech Republic*
[b]*Engineering & Technology – PTD, Bayer AG, 51368 Leverkusen, Germany*
**Frantisek.Stepanek@vscht.cz*

Abstract

In this work, we investigate the effect of mixing intensity on the crystal size distribution (CSD) in a semi-batch crystallizer via population balance modelling. We have coupled balance equations with a simplified model of turbulent mixing developed by Baldyga et al. (1997). This approach was previously applied to isothermal reactive crystallization by Ståhl and Rasmuson (2009), but its potential was not fully explored. We have extended the model in two directions. First, we have developed an adaptive time step strategy for increased stability of the numerical solution and faster computation. Second, we have implemented the non-isothermal case, thus enabling the study of coupled cooling-antisolvent crystallization. Based on our simulations, we explain how the non-ideality of mixing affects nucleation and why the average crystal size reaches a maximum with increasing mixing intensity.

Keywords: crystallization, modelling, turbulent mixing

1. Introduction

Mathematical modelling plays a crucial role in facilitating effective design and control of crystallization processes. However, most models currently used in industry and academia assume typically perfect mixing, failing to capture the system dependency on mixing dynamics and limiting our understanding of the process. Although the use of computational fluid dynamics (CFD) is possible in principle, this approach is generally too demanding for extensive parametric studies. To mitigate this issue, we present an efficient implementation of a simplified mixing model for semi-batch crystallization.

2. Modelling

2.1. Turbulent mixing modelling

The model developed and described by Baldyga et al. is based on the description of meso- and micro-mixing phenomena, governed by their corresponding characteristic times t_{meso} and t_{micro} (Eq. (1) and Eq. (2), respectively). The energy dissipation rate in the feed region (ε_{eff}) is calculated as a multiple of the specific power input (Eq. (3)). The model describes mixing as an expansion of a discrete feed volume ('drop') by engulfment of the bulk volume. The evolution of the volume fraction of mesomixed (X_{meso}) and micromixed (X_{micro}) regions is then described by Eq. (4) and Eq. (5). It is assumed that a possible reaction occurs in the micromixed volume. The change of molar concentration in the droplet of volume fraction X_{micro} for the species i (c_i) is modelled by Eq. (6), where

$\langle c_i \rangle$ is the bulk concentration. All bulk variables are further denoted in the paper by angle brackets. Bulk concentrations do not change during the drop expansion. This model can be used for the description of mixing in a semi-batch process by discretizing the feed stream into a series of droplets, each described individually.

$$t_{meso} = 2 \left(\frac{d^2}{\varepsilon_{eff}} \right)^{1/3} \tag{1}$$

$$t_{micro} = \frac{1}{E} = 12.7 \left(\frac{\nu}{\varepsilon_{eff}} \right)^{1/2} \tag{2}$$

$$\varepsilon_{eff} = k_\varepsilon \frac{P}{m} = k_\varepsilon \frac{P_0 N_s^3 D^5}{V} \tag{3}$$

$$\frac{dX_{meso}}{dt} = \frac{1}{t_{meso}} X_{meso} (1 - X_{meso}) \tag{4}$$

$$\frac{dX_{micro}}{dt} = E X_{micro} \left(1 - \frac{X_{micro}}{X_{meso}} \right) \tag{5}$$

$$\frac{dc_i}{dt} = E \left(1 - \frac{X_{micro}}{X_{meso}} \right) (\langle c_i \rangle - c_i) + r_i \tag{6}$$

2.2. Crystallization model equations

During the droplet expansion, each of the two regions (drop and bulk) is described by its own set of equations, coupled via mixing. Crystallization is modelled by a population balance equation (PBE). It is further assumed here that breakage and agglomeration are negligible. However, they could be easily included in the modelling framework. Note that crystal nuclei are assumed to form at negligible size.

2.2.1. Expanding drop equations

In the droplet, crystallization is modelled by Eq. (7) (completed by its boundary condition). The concentration of solute product (c_{Pr}) changes during the droplet expansion according to Eq. (8), while other mixture components according to Eq. (6). With respect to the model of Ståhl and Rasmuson, we have included the energy balance equation, Eq. (9), omitting here reaction, crystallization, and mixing heat for simplicity.

$$\frac{\partial f}{\partial t} + G \frac{\partial f}{\partial L} = E \left(1 - \frac{X_{micro}}{X_{meso}} \right) (\langle f \rangle - f), \qquad f(0, t) = \frac{J}{G} \tag{7}$$

$$\frac{dc_{Pr}}{dt} = E \left(1 - \frac{X_{micro}}{X_{meso}} \right) (\langle c_{Pr} \rangle - c_{Pr}) + r_{Pr} - \frac{3 k_v \rho_{cr}}{M_{Pr}} G \int_0^\infty L^2 f dL \tag{8}$$

$$\frac{dT}{dt} = E \left(1 - \frac{X_{micro}}{X_{meso}} \right) \frac{\rho_{bulk}}{\rho} \frac{c_{p,bulk}}{c_p} (\langle T \rangle - T) \tag{9}$$

2.2.2. Equations for bulk

The equations describing the bulk variables are similar to those of the droplet. However, we assume that the intensive variables are not affected by the droplet expansion (i.e., the bulk acts as a reservoir). Thus, no mixing term is present. Note that, for non-isothermal processes, the temperature of the bulk changes due to heat exchange with the jacket at temperature T_j.

$$\frac{\partial \langle f \rangle}{\partial t} + G_{bulk} \frac{\partial \langle f \rangle}{\partial L} = 0, \qquad \langle f \rangle (0, t) = \frac{J_{bulk}}{G_{bulk}} \tag{10}$$

$$\frac{d \langle c_{Pr} \rangle}{dt} = r_{Pr,bulk} - \frac{3 k_v \rho_{cr}}{M_{Pr}} G_{bulk} \int_0^\infty L^2 \langle f \rangle dL \tag{11}$$

$$\frac{d \langle T \rangle}{dt} = \frac{UA(T_j - \langle T \rangle)}{\rho_{bulk} c_{p,bulk} V_{bulk}} \tag{12}$$

3. Model implementation and initial conditions

As mentioned, the feed stream is discretized into droplets of volume $\Delta V = Q \Delta t$, where Δt is the chosen feed time step. Clearly, we assume here that the feed flow rate (Q) is constant. Note that Δt must be greater than the time of drop expansion. The initial state of droplet is equal to pure feed. The model equations are then solved for each droplet individually. The initial value of the volume fractions is defined by Eq. (13). The initial state of the bulk varies for each droplet and is equal to the state of the droplet at the end of previous step as described by Eq. (14-16). The population balance equations are solved via high-resolution finite volume method with Koren flux limiter. The resulting system of ODEs is numerically integrated using an explicit Runge–Kutta method (RK45 in the Python solver implementation). Only primary nucleation and size-independent growth were assumed in the simulations.

$$X_{micro}(0) = X_{meso}(0) = \Delta V / V \tag{13}$$

$$\langle c_i \rangle(0)_0 = c_{i0}, \qquad \langle c_i \rangle(0)_{k+1} = c_i(\Delta t)_k \tag{14}$$

$$\langle f \rangle(L, 0)_0 = f_0, \qquad \langle f \rangle(L, 0)_{k+1} = f(L, \Delta t)_k \tag{15}$$

$$\langle T \rangle(0)_0 = T_0, \qquad \langle T \rangle(0)_{k+1} = T(\Delta t)_k \tag{16}$$

We have studied two systems representing a common use of a semi-batch crystallizer: the reactive isothermal precipitation of benzoic acid (BA) (Ståhl and Rasmuson, 2009) and the non-isothermal antisolvent crystallization of aspirin (Lindenberg et al., 2009). For the reactive crystallization, following the original work of Ståhl and Rasmuson, we have assumed a fast, mass transport-limited reaction. The kinetic and thermodynamic data for each system are provided in the respective references.

4. Results

4.1. General simulation results – benzoic acid

Figure 1A shows the supersaturation in the bulk (S_b) and the maximal supersaturation reached in each droplet ($S_{d,max}$) over the course of the feeding time. The existence of two "parallel" supersaturation profiles significantly influences the crystallization process. We have identified three distinct nucleation phases, marked on both plots in Figure 1. In the first phase, nucleation is driven by the high supersaturation achieved in the droplet, forming a sharp peak in CSD. In the second phase, the bulk supersaturation is high enough to initiate nucleation in the bulk as well, leading to the formation of second peak. In the third phase the nucleation in the bulk is again negligible. However, the receding drop supersaturation is still high enough to support formation of new nuclei (Figure 1B).

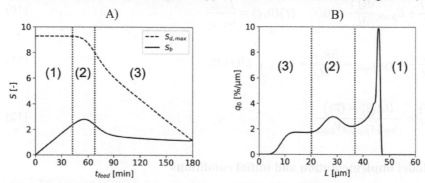

Figure 1: Typical simulation results for BA: A) supersaturation profile, B) population density

4.2. Numerical stability

Our simulations revealed that an arbitrary choice of both feed and crystal size discretization can result in an oscillating behavior in CSD. Due to discretization of the feed, the final CSD is comprised of crystal "subpopulations" formed in each droplet, typically resulting in a peak in the distribution. However, these peaks lead to oscillations only if their width (W_i) covers multiple lengths of size interval (ΔL). To quantify this mismatch, we define number K as the number of intervals covered by each droplet peak (Eq. (17), note the identity with the Courant Number), and postulate that $K_i \leq 1, \forall i$ is a sufficient condition to eliminate the oscillation. As a result, the oscillation fade with an increasing number of droplets (shortening of Δt) for chosen ΔL, as we show in Figure 2A.

$$K_i = \frac{W_i}{\Delta L} = \frac{\overline{G}_i \Delta t}{\Delta L} \tag{17}$$

By setting the value of K as a constant for every droplet, we can compute the length of an adaptive feed time step from Eq. (17). Since this simple strategy leads to unreasonably long steps at growth rates close to zero (i.e., low supersaturation), we remedy by introducing the heuristic parameter θ (Eq. (18)), which serves as a "limiter" for time step increase. Our simulations indicate that a suitable choice seems $\theta = 1.1$. Additionally, we assume that the mean growth rate can be well approximated by the value from the previous step. As demonstrated in Figure 2B, this approach presents an optimal use of computational time as better precision is achieved compared to the case of constant time step at discretization of feed into the same number of droplets. We have concluded that for $K = 1$, this strategy provides a satisfactory balance between eliminating oscillations, minimizing computational time, and ensuring high accuracy.

$$\Delta t_{i+1} = \min\left(\frac{K\Delta L}{\overline{G}_i}, \theta \Delta t_i\right) \tag{18}$$

4.3. Effect of mixing intensity

Our simulations indicate that increasing the mixing intensity leads to higher bulk supersaturation, as shown in Figure 3A. This causes two effects: increased overall growth and increased bulk nucleation. The influence of mixing on the final CSD is depicted in

Figure 3B. At low ε_{eff}, nucleation in the drop being dominant, intensified mixing leads to bigger particles as the effect of increased growth prevails over raise in bulk nucleation. However, once the bulk nucleation gets high enough to form a second peak, further increase of ε_{eff} leads to a shift of the mass of distribution to the second peak at smaller sizes. This transition is reflected in the dependency of volume-based median particle size ($L_{3,50}$) on ε_{eff} (Figure 3C) and explains the existence of maximum in this curve. Note that the maximum is reached when the distribution shifts from unimodal to bimodal.

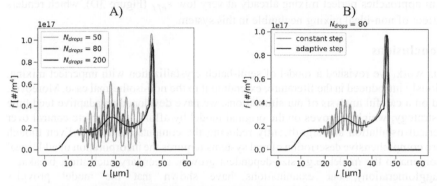

Figure 2: Oscillations in CSD of BA: A) final CSD for various feed discretization levels, B) comparison of constant and adaptive feed time step results at the same number of droplets

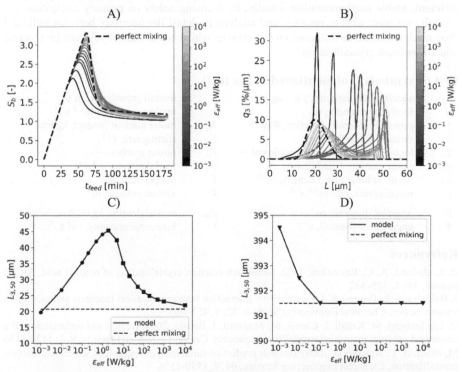

Figure 3: System dependency on mixing intensity: A) bulk supersaturation (BA), B) final CSD (BA), C) median particle size (BA), D) median particle size (aspirin)
● – only drop nucleation peak, ■ – bimodal distribution, ▼ – only bulk nucleation peak

The maximum in this dependency has been previously experimentally measured, e.g., by Åslund and Rasmuson (1992) for benzoic acid and modelled by Zauner and Jones (2000) for the case of calcium oxalate. However, both studies assumed the crystal size drops at high ε_{eff} solely due to secondary nucleation and breakage caused by high shear forces. We have shown that this behaviour can be qualitatively described only accounting for primary nucleation and size-independent growth. Notably, the model converges to ideal mixing for high ε_{eff}, as one would reasonably expect. Note that in the case of aspirin the system approaches perfect mixing already at very low ε_{eff} (Figure 3D), which renders the effect of non-ideal mixing negligible in this system.

5. Conclusions

In this work, we revisited a model of semi-batch crystallization with imperfect mixing, previously introduced in the literature, extending it to the non-isothermal case. Moreover, based on a careful analysis of our simulations, we have developed an adaptive feed time step strategy, which improves on the original model by affording complete control over numerical oscillation and significantly reducing the computational time. Even though a more comprehensive description of real systems requires the incorporation of additional phenomena to the model (e.g., size-dependent growth, secondary nucleation, breakage, or agglomeration), our examinations have shown that the model provides a good qualitative description of real phenomena, in agreement with existing literature. It thus forms the basis for new, more detailed developments, while proposing a new, more efficient, stable implementation. Finally, by focusing solely on primary nucleation and growth, we were able to uncover and analyze in detail the interplay between bulk and drop nucleation and its influence on crystal size, which is particularly relevant for reactive and antisolvent crystallization.

List of symbols (not mentioned in the text)

c_p	specific heat capacity, J K^{-1} kg^{-1}		G	overall growth rate, m s^{-1}
d	feed pipe diameter, m		J	nucleation rate, # m^{-3} s^{-1}
f	population density function, # m^{-4}		M_{Pr}	molar mass of product, kg mol^{-1}
k_v	volume shape factor, −		N_s	stirring rate, s^{-1}
k_ε	local energy dissipation coefficient, −		P_0	power number, −
q_0	norm. number-based density, m^{-1}		Q	feed flow rate, m^3 s^{-1}
q_3	norm. volume-based density, m^{-1}		T	temperature, K
r	reaction rate, mol m^{-3} s^{-1}		V	system volume, m^3
D	impeller diameter, m		ρ_{cr}	crystal density, kg m^{-3}
E	engulfment constant, s^{-1}		ν	kinematic viscosity, m^2 s^{-1}

References

B. L. Åslund, Å. C. Rasmuson, 1992, Semibatch reaction crystallization of benzoic acid, AIChE journal, 38, 3, 328-342

J. Bałdyga, J. R. Bourne, S. J. Hearn, 1997, Interaction between chemical reactions and mixing on various scales, Chemical Engineering Science, 52, 4, 457-466

C. Lindenberg, M. Krättli, J. Cornel, M. Mazzotti, J. Brozio, 2009, Design and optimization of a combined cooling/antisolvent crystallization process, Crystal Growth and Design, 9, 2, 1124-1136

M. Ståhl, Å. C. Rasmuson, 2009, Towards predictive simulation of single feed semibatch reaction crystallization, Chemical Engineering Science, 64, 7, 1559-1576

R. Zauner, A. G. Jones, 2000, Scale-up of continuous and semibatch precipitation processes, Industrial & engineering chemistry research, 39, 7, 2392-2403

Flavio Manenti, Gintaras V. Reklaitis (Eds.), Proceedings of the 34th European Symposium on Computer Aided Process Engineering / 15th International Symposium on Process Systems Engineering (ESCAPE34/PSE24), June 2-6, 2024, Florence, Italy

Analysis of the preferred ethylene production route from carbon dioxide at a supply chain level: results of mathematical modelling for a Teesside case study

Grazia Leonzio,[a,b]* Nilay Shah[b]

[a]*Sargent Centre for Process Systems Engineering, Department of Chemical Engineering, Imperial College London, London SW7 2AZ, UK*
[b]*Department of Mechanical, Chemical and Materials Engineering, University of Cagliari, via Marengo 2, 09123 Cagliari, Italy*
**grazia.leonzio@unica.it*

Abstract

Currently, new routes for producing chemical building blocks are required with the aim to support the energy and feedstock transition. Considering both global demand and production capacity, ethylene is the most important organic chemical and for this reason alternative production routes (based on carbon dioxide and water) have been investigated and screened in terms of costs and emissions in one of our previous works. In this research, the best alternative ethylene production technology is suggested at a supply chain level for the Teesside cluster (UK) through the development of two different mathematical models for the supply chain. Results show that the best ethylene production route is based on methanol-to-olefin plant where methanol is produced by syngas obtained from carbon dioxide-water co-electrolysis. Through a global sensitivity analysis based on a surrogate model, it is found that the carbon dioxide utilization cost has the highest impact on the supply chain total cost. The optimization of the electrolytic cell could help with cost reduction.

Keywords: carbon dioxide, ethylene, supply chain, mathematical modelling, global sensitivity analysis.

1. Introduction

Nowadays there is a mismatch between the amount of carbon dioxide (CO_2) that is emitted and the amount that is used, forcing researchers and companies to suggest new and alternative CO_2 conversion routes to bridge this gap. Among different CO_2-based products, ethylene is desirable due to its large market and has high value compared to other reduction products. For these reasons, new production systems have been proposed and investigated in the literature as in Ioannou et al. (2020) and Berkelaar et al. (2022). On the other hand, in one of our previous works we modelled and compared the cost and global warming potential of different ethylene production routes starting from CO_2 and water (H_2O) (Leonzio et al., 2023). Here, the investigated technologies are the following: the electrochemical process (tandem and direct CO_2 electrochemical reduction to ethylene processes), methanol-to-olefin (MTO) plant with methanol obtained from CO_2 hydrogenation with blue and green hydrogen and from CO_2 electrochemical reduction, and MTO process with methanol obtained from syngas produced in a solid oxide electrolytic cell (SOEC) for the CO_2-H_2O co-electrolysis. We found that the electrochemical tandem process is the most promising from an economic point of view while the MTO process using methanol produced by syngas from CO_2-H_2O co-electrolysis in a solid oxide electrolytic cell (SOEC) is the most preferred from an environmental point of view.

After this first screening, in this research, we aim to explore the best route for ethylene production (between the two suggested above) at a supply chain level, by developing two different mixed integer linear programming (MILP) models for the supply chain located in the Teesside cluster (UK). Two different models are used to verify the independence of results by the used method. Through the development of a surrogate model, a Global Sensitivity Analysis (GSA) is conducted to determine the most significant factors for total cost and ethylene production cost.

2. Methodology

2.1. Mathematical model of supply chain

In the considered supply chain, CO_2 is captured from flue gas, transported via pipeline and stored/used for ethylene production through the two routes suggested in our previous work (Leonzio et al., 2023). Therefore, a carbon capture, utilization and storage (CCUS) supply chain is proposed here. In particular, the framework is located in the Teesside cluster (UK) where the Endurance reservoir is the CO_2 storage site with a supposed storage capacity of 20 $MtCO_2$. Companies like Sabic, Ineos, CF Fertiliser, Tioxide Europe and BOC Linde are the CO_2 source sites, while the CO_2 conversion site is set in the Sabic plant (Geels, 2022). Capture technologies like absorption with mono-ethanolamine, pressure and vacuum swing adsorption, membrane are taken into account. In that designed cluster, total CO_2 emissions were 351.5 $MtCO_2$ in 2019 and must be reduced by 54.8 % to achieve the target fixed for the future (2050) (Geels, 2022).

Two MILP models are developed to optimize the framework: one is based on a node structure as reported by Elahi et al. (2014) while the other one is non-node based as proposed by Leonzio et al. (2019). In both models, CO_2 is flowing from sources to storage/utilization sites but in the first case a material balance is added as a constraint for each node.

The optimal topology is provided by the minimization of the total cost (e.g. the sum of CO_2 capture and compression, CO_2 transportation, CO_2 storage and CO_2 utilization costs) solved by using the AIMMS software tool with CPLEX 12.7.1 as the solver.

2.2. Surrogate model

A surrogate model method used in the literature is the random sampling-high dimensional model representations (RS-HDMR) (Lambert et al., 2016), which is implemented here to express the supply chain total cost and ethylene production cost as a function of CO_2 capture and compression cost, CO_2 transportation cost, CO_2 storage cost and CO_2 utilization cost. The following equation (with 1 as the maximum grade for parameters β and α) is considered to be regressed (see Eq. 1):

$$F(x) = fo + \alpha_1^{cc} \cdot \varphi_1(x_{cc}) + \alpha_1^{ct} \cdot \varphi_1(x_{cT}) + \alpha_1^{cs} \cdot \varphi_1(x_{cs}) + \alpha_1^{cu} \cdot \varphi_1(x_{cU}) +$$

$$+\beta_{1,1}^{cc,ct} \cdot \varphi_1(x_{cc}) \cdot \varphi_1(x_{cT}) + \beta_{1,1}^{cc,cs} \cdot \varphi_1(x_{cc}) \cdot \varphi_1(x_{cs}) + \beta_{1,1}^{cc,cu} \cdot \varphi_1(x_{cc}) \cdot \varphi_1(x_{cu}) +$$

$$+\beta_{1,1}^{ct,cs} \cdot \varphi_1(x_{ct}) \cdot \varphi_1(x_{cs}) + \beta_{1,1}^{ct,cu} \cdot \varphi_1(x_{ct}) \cdot \varphi_1(x_{cu}) + \beta_{1,1}^{cs,cu} \cdot \varphi_1(x_{cs}) \cdot \varphi_1(x_{cu}) \quad (1)$$

with *F(x)* the supply chain total cost/ethylene production cost, *fo* a constant to be evaluated, x_{cc}, x_{ct}, x_{cs} x_{cu} the uncertain variables respectively for CO_2 capture and compression cost (cc), CO_2 transportation cost (ct), CO_2 storage cost (cs) and CO_2

utilization cost (cu), α and β parameters to be found and $\varphi_1(x_i)$ the Legendre orthogonal polynomial, as in Eq. (2) (Lambert et al., 2016):

$$\varphi_1(x_i) = \sqrt{3} \cdot (2 \cdot x_i - 1) \qquad (2)$$

with i the each factor. In order to determine the value of each parameter, Sobol sampling (to generate 40 sampled points (Kleijnen, 2017)) is used for each factor which is varied between +/- 20 % of the base value according to a triangular probability distribution. For each input, the total cost and ethylene production cost are found through the supply chain optimization. The generated inputs and founded outputs are used to find the values of parameters through the software SobolGSA.

2.3. Global sensitivity analysis

A GSA is carried out to quantify the overall uncertainty in a key performance indicator (KPI) by varying simultaneously all factors inside their uncertainty range. In this work, the supply chain total cost and ethylene production cost are the KPIs, while the CO_2 capture and compression cost, CO_2 transportation cost, CO_2 storage cost and CO_2 utilization cost are the considered factors. KPIs are evaluated through the developed surrogate models while each factor is taken into account through the respective Legendre orthogonal polynomial function for which 1,000 Sobol samples are considered for the uncertainty according to a triangular probability distribution (a range between +/- 20 % of the basic value is assumed for each factor) (Sobol, 2001). The GSA is developed by using the software tool SobolGSA, estimating total-order Sobol sensitivity indices.

3. Results and discussions

3.1. Results of supply chain optimization

The optimization results show that, for both mathematical models, the best production route is based on the MTO plant with methanol from syngas because this is the only one selected for ethylene production. The supply chain cost is 5.9 \$/kg$_{Ethylene}$ and 5.7 \$/kg$_{Ethylene}$ for the models based on Leonzio et al. (2019) and Elahi et al. (2014) respectively. A sensitivity analysis is conducted changing the CO_2 reduction rate and evaluating costs as in Figure 1: in both models, at a higher amount of captured CO_2, the percentage of CO_2 utilization cost on the total cost increases because at a fixed storage capacity more CO_2 must be converted to achieve a higher reduction target.

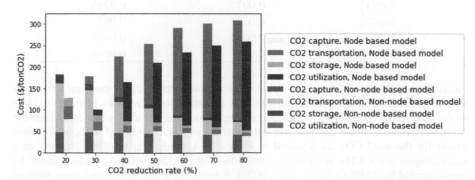

Figure 1 Cost analysis of the CCUS supply chain in both models at different CO_2 reduction rates

3.2. Results of surrogate model

Through the software SobolGSA and according to the RS-HDMR method, the values of parameters for the surrogate models related to supply chain total cost and ethylene production cost are figured out as reported in Table 1. Data are shown here for both supply chain model structures.

Table 1 Values of parameters for CCUS supply chain surrogate models

Parameter	Non-node based structure		Node based structure	
	Total cost (k\$/y)	Ethylene production cost (\$/t$_{Ethylene}$)	Total cost (k\$/y)	Ethylene production cost (\$/t$_{Ethylene}$)
fo	430,338	6,019	333,754	5,778
α_1^{cc}	4,303	60.18	4,928	85.32
α_1^{ct}	3,963	55.43	479	8.29
α_1^{cs}	-152	-2.12	161	2.78
α_1^{cu}	23,049	322	18,670	323
$\beta_{1,1}^{cc,ct}$	411	5.75	331	5.73
$\beta_{1,1}^{cc,cs}$	276	3.85	290	5.02
$\beta_{1,1}^{cc,cu}$	618	8.64	507	8.77
$\beta_{1,1}^{ct,cs}$	1,255	17.55	860	14.88
$\beta_{1,1}^{ct,cu}$	-243	-3.39	-337	-5.83
$\beta_{1,1}^{cs,cu}$	-598	-8.36	-317	-5.49

3.3. Results of global sensitivity analysis

Values of total-order Sobol sensitivity indices for each factor are reported in Table 2. A value higher than 0.05 makes the input significant (Zhang et al., 2015): only the CO_2 utilization cost is significant for total and production cost in both supply chain models.

Table 2 Total-order Sobol sensitivity index for total cost/ethylene production cost for the supply chain and for each factor (1=CO_2 capture and compression cost factor; 2=CO_2 transportation cost factor; 3=CO_2 storage cost factor; 4=CO_2 utilization cost factor)

Sensitivity index	Non-node based structure	Node based structure
Stot[1]	0.0075	0.0240
Stot[2]	0.0043	0.0029
Stot[3]	0.0059	0.0047
Stot[4]	0.9827	0.9688

According to this result, a sensitivity analysis is conducted keeping constant other cost factors and changing only the significant CO_2 utilization cost factor in a range +/- 53 % of the base value by using the developed surrogate models. The analysis is carried out for both models to evaluate the specific cost (\$/tCO_2 avoided) with and without a carbon credit for the used CO_2, as reported in Figure 2. According to the IPCC report, only technologies with CO_2 abatement costs lower than 220 \$/tCO_2 avoided could be implemented by 2030 (Ostovari et al., 2023). Results show that at the base case, without carbon credit, only the non-node based model causes a supply chain total cost (228 \$/tCO_2

avoided) higher than that suggested by IPCC. On the other hand, the supply chain based
on node modelling has a cost of 182 \$/tCO$_2$ avoided.

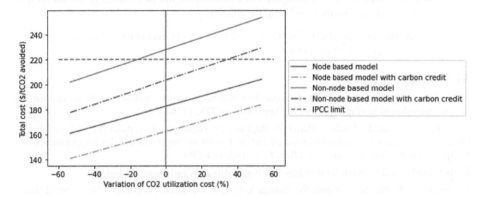

Figure 2 Total cost of CCUS supply chain changing the CO$_2$ utilization cost factor
compared to the base case in both models and the IPCC limit

In the first case, a reduction of CO$_2$ utilization cost of about 20 % allows a total cost below
the suggested limit. When a credit (45 \$/tCO$_2$ (Carbon Credit, 2023)) for only CO$_2$
utilization is provided, both mathematical models at the base case ensure a total cost lower
than that of the IPCC limit. In addition to the economic incentives, it is important to
reduce the ethylene production cost so that, some suggestions are provided. It is needed
to decrease the methanol production cost through the reduction of syngas production cost
obtained by the SOEC. The highest influence of operating cost on methanol production
through this route is reported by Adnan and Kribia (2020), while the high cost of syngas
production through a SOEC is suggested by Redissi and Bouallou (2013). In a SOEC,
operating costs have a great influence on total cost and the highest impact is due to the
electricity cost (Freire Ordonez et al., 2021). For these reasons, further studies on SOECs
should be conducted in order to improve Faradaic efficiency and current density and
minimize the current leakage so that a lower power for syngas production will be needed.
Low-cost and large-scale commercialization of SOECs requires a consistent current
density higher than 5 A/cm, which is however hard to achieve, even for a high temperature
electrolysis system (Cao et al., 2022). At this value of current density, a Faradaic
efficiency of about 100% should be ensured so that the development of new catalysts and
electrodes that are able to ensure this value is suggested. In addition to the above advice,
lowering the temperature to a range of 500-700 °C could help in the reduction of
maintenance costs by mitigating materials and maintenance issues. However, reducing
the operating temperature decreases the electrocatalytic activity of electrode materials,
requiring the development of high-performance materials.

4. Conclusions

Two different mathematical models (node and non-node based) are considered here for a
CCUS supply chain producing ethylene. Different models are used in order to provide the
independence of results from the model formulation. It is found that the best ethylene
production route is based on the MTO plant, ensuring a cost of 5.9 \$/kg$_{Ethylene}$ and 5.7
\$/kg$_{Ethylene}$ respectively for the non-node based and node based models. A surrogate model

is built with the RS-HDMR method to find through a GSA that, the most significant factor for cost is the CO_2 utilization cost. In order to have the supply chain cost lower than the limit fixed by IPCC, more research and optimization should be conducted for the SOEC by improving current density and Faradaic efficiency.

Acknowledgments: Funding by the mobility of young researcher grant of Cagliari University and Imperial College London are gratefully acknowledged.

References

M.A. Adnan, M.G. Kibria, 2020. Comparative techno-economic and life-cycle assessment of power-to-methanol synthesis pathways, Appl. Energy 278, 115614

L. Berkelaar, J. van der Linde, J. Peper, A. Rajhans, D. Tiemessen, L. van der Ham, H. van den Berg, 2022, Electrochemical conversion of carbon dioxide to ethylene: Plant design, evaluation and prospects for the future, Chem. Eng. Res. Des., 182, 194–206

Carbon credit, 2023. Available at: https://carboncredits.com/carbon-prices-today/

J. Cao, Y. Li, Y. Zheng, S. Wang, W. Zhang, X. Qin, G. Geng, B. Yu, 2022, A Novel Solid Oxide Electrolysis Cell with Micro-/Nano Channel Anode for Electrolysis at Ultra-High Current Density pover 5 Acm^{-2}, Adv. Energy Mater. 12, 2200899

D. Freire Ordonez, N. Shah, G. Guillen-Gosalbez, 2021, Economic and full environmental assessment of electrofuels via electrolysis and co-electrolysis considering externalities, Appl. Energy 286, 116488

N. Elahi, N. Shah, A. Korre, S. Durucan, 2014. Multi-period Least Cost Optimisation Model of an Integrated Carbon Dioxide Capture Transportation and Storage Infrastructure in the UK, Energy Procedia 63, 2655–2662

F.W. Geels, 2022, Conflicts between economic and low-carbon reorientation processes: Insights from a contextual analysis of evolving company strategies in the United Kingdom petrochemical industry (1970–2021), Energy Res. Soc. Sci. 91, 102729

I. Ioannou, S.C. D'Angelo, A.J. Martín, J. Pérez-Ramírez, G. Guillén-Gosálbez, 2020, Hybridization of Fossil- and CO2-Based Routes for Ethylene Production using Renewable Energy, ChemSusChem, 13, 6370–6380

J.P.C. Kleijnen, 2017, Regression and Kriging metamodels with their experimental designs in simulation: A review, Eur. J. Oper. Res., 256, 1, 1, 1-16.

R.S.C. Lamberta, F. Lemke, S.S. Kucherenko, S. Song, N. Shah, 2016, Global sensitivity analysis using sparse high dimensional model representations generated by the group method of data handling, Mat. Comp. Simul., 128, 42–54

G. Leonzio, B. Chachuat, N. Shah, 2023, Towards ethylene production from carbon dioxide: Economic and global warming potential assessment, Sustain. Prod. Consum. DOI:10.1016/j.spc.2023.10.015

G. Leonzio, F.U. Foscolo, E. Zondervan, 2019. Sustainable utilization and storage of carbon dioxide: Analysis and design of an innovative supply chain, Comput. Chem. Eng., 131, 106569

H. Ostovari, L. Kuhrmann, F. Mayer, H. Minten, A. Bardow. 2023. Towards a European supply chain for CO2 capture, utilization, and storage by mineralization: Insights from cost-optimal design, J. CO2 Util. 72, 102496

Y. Redissi, C. Bouallou, 2013, Valorization of Carbon Dioxide by Co-Electrolysis of CO2/H2O at High Temperature for Syngas Production, Energy Procedia 37, 6667 – 6678

I.M. Sobol, 2001, Global sensitivity indices for nonlinear mathematical models and their Monte Carlo estimates, Mathematics & Computers in Simulation 55, 271-280

X.Y. Zhang, M.N. Trame, L.J. Lesko, S. Schmidt, 2015. Sobol Sensitivity Analysis: A Tool to Guide the Development and Evaluation of Systems Pharmacology Models, CPT: pharmacometrics & systems pharmacology, 4, (2), 69–79.

Flavio Manenti, Gintaras V. Reklaitis (Eds.), Proceedings of the 34th European Symposium on Computer Aided Process Engineering / 15th International Symposium on Process Systems Engineering (ESCAPE34/PSE24), June 2-6, 2024, Florence, Italy

Comparative Assessment of Flexible Natural Gas Monetisation Processes to Products Under Uncertainties: Agent-Based Modelling Approach

Noor Yusuf,[a] Ahmed Al-Nouss,[a] Tareq Al-Ansari[a*]

[a]College of Science and Engineering, Hamad Bin Khalifa University, Qatar Foundation, Doha 34110, Qatar
Talansari@hbku.edu.qa

Abstract

With the increased demand for cleaner energy resources, natural gas is a bridging fuel for smoothening the transition to renewables. Power, liquified natural gas (LNG), ammonia, and urea have attracted significant attention among the various monetisation routes. The different monetisation routes differ regarding production technologies, associated energy and utilities requirements, released emissions, and operational flexibility. Despite the estimated demand growth in product demand, each natural gas monetisation process is subject to exogenous market uncertainties. This work evaluates the flexibility of natural gas monetisation processes to LNG, ammonia, urea, and power by investigating plant design configurations and natural gas production allocation to different production routes. The commercial software Aspen HYSYS is used for process modelling and simulation, followed by identifying the operational flexibility of each process. The simulation results and forecasted price and demand data are then used as input into an agent-based model to identify the optimal annual natural gas allocation to different processes subject to environmental and economic objectives. Overall, the study provides decision-makers with a systematic approach to evaluate the effectiveness of flexible natural gas allocation to different processes based on technical, economic, and environmental aspects. The results of Qatar's case study demonstrate the importance of prioritising Power and LNG production to maximise profitability and hedge against risks. On the other hand, ammonia production is maximised to offset environmental emissions and tackle CO_2 emissions.
Keywords: Natural gas, Agent-Based Modelling, Uncertainties, Operational Flexibility.

1. Introduction

Since 2020, the countries involved in the multilateral process after the Paris Agreement for climate change have been submitted national climate action plans, reflecting the timeline of actions taken to tackle emissions, including switching from coal to natural gas and increasing share for renewables. Hence, countries with abundant natural gas reserves must consider the needs of different markets when planning natural gas supply chains. Raw natural gas can be physically or chemically monetised into value-added products to enhance the economic value of natural gas when targeting international markets. Amongst the different investigated natural gas monetisation routes, liquified natural gas (LNG), compressed natural gas (CNG), synthetic fuels produced via gas-to-liquids process (GTL), methanol, ammonia, and ethylene have proved to have significant interest in the international markets, supported by industrial, transport, agricultural, and household needs. The selection of monetisation routes is a strategic decision-making problem that involves multiple parameters, including the geographic location of the natural gas field (i.e., onshore or offshore), the composition of produced natural gas, the climate of the producing country, the distance to the targeted markets, and the demand in final markets.

To meet the increased demand for natural gas, producing countries proactively responded by assessing the feasibility of expanding new natural gas projects or deploying new ones. Yet, market shifts supported by demand fluctuations due to changes in market preferences and the emergence of decarbonisation policies had jeopardised the expansion of projects, especially during the COVID-19 pandemic, when the growth of fossil-based commodities declined dramatically. From another perspective, renewables have attained attention in the last few years, supported by decarbonisation and diversification objectives. For example, European countries accelerated the efforts to switch to renewable resources for power generation after the Russian-Ukrainian conflict. This raises the importance of operational flexibility in energy management so that producing countries can proactively react to market changes. In addition to sustaining economic profitability, operational flexibility allows natural gas-producing companies to meet environmental targets and utilise resources effectively. In the literature, operational flexibility for single-product natural gas systems has been explored intensively at a process-based level by evaluating the acceptable system operating boundaries at fixed equipment sizes (Bhosekar & Ierapetritou, 2020; Verleysen et al., 2021; Yusuf et al., 2022). Yet, a gap has been identified in exploring how different flexible production systems react in an integrated multi-product natural gas supply chain. This study builds on previous work by Yusuf et al., (2023) by investigating the optimal annual natural gas capacity to four monetisation routes using agent-based modelling: power, LNG, ammonia, and urea, subject to technical, economic, and environmental aspects. Hence, this work provides decision-makers with a holistic approach to assessing the optimal natural gas allocation strategy.

2. Methodology and Data

The model built as part of this study includes an agent-based model (ABM) to undertake sustainable planning in Qatar's energy sector. The developed framework is a decision-making tool that enables the prediction of portfolio decisions for the downstream natural gas industry as a response to meet rapidly changing demand. The advantage of production flexibility allows the evaluation of natural gas allocation based on economic and environmental scenarios.

2.1 ABM simulation

An agent is a self-contained unit with a set of characteristics and behaviours. Particular rules govern its interactions with other agents and the environment in which it resides (Lopez-Jimenez et al., 2018). In this study, two major groups of agents interact to meet the demand for LNG, power, ammonia and urea while adhering to economic and environmental limitations. The proposed ABM model replicates the yearly choices of the energy sector over a period of 34 years (n=34), with the primary agent, represented by the natural gas generating system NG, interacting with D_i, $i \in D$, where D is the set of downstream industries. Both agents interact in response to the local and global economy.

Table 1: Characteristics of agents.

Agent	Attributes	Behaviours
NG system (*NG*)	- Total NG capacity C (billion cubic meters-BCM). - Yearly NG distribution capacities C_{ng} (BCM)	- Allocate NG to Power (*p*), LNG

	- Yearly NG allocation to Power Qp, LNG Qn, Ammonia Qa, and Urea Qu. - Global Warming Potential from generating each product (kg CO_2–eq)	(n), Ammonia (a), and Urea u.
Downstream Industries (Di)	- Yearly production capacities Q_{Di} (kg) - Yearly power local demand Pi (kWh)	- Determine the best production sinks (S).

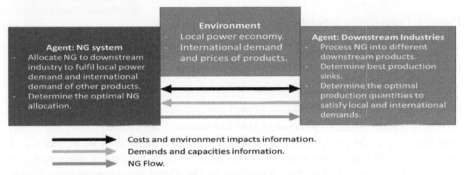

Figure 1: Representation of ABM elements.

2.1.1 Agents

Agent: NG System

The natural gas system NG is the model's central element because it is the agent that ensures the allocation of yearly extracted NG capacities Cng. To accomplish this goal, the NG grid allocates NG to 4 sinks: LNG, power, ammonia and urea, so that the total grid capacity C is allocated. The sink is selected using one of two strategies: 1) allocate NG to satisfy local power production demand (p) or 2) find the optimal allocation of NG to LNG (n), Ammonia (a), and Urea u based on international demands. Each decision is influenced by the economic and environmental factors discussed in section 3 and the operations and characteristics associated with the downstream industries Di.

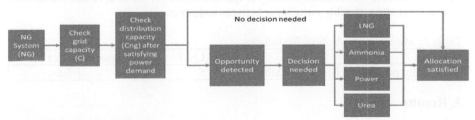

Figure 2: NG system behavior.

Agent: Downstream Industries System

This analysis considers four downstream industries as agents Bi, including LNG, ammonia, power and urea production. As part of their behaviour, power is satisfied first based on the local demand. Then, the ideal NG allocation blend to the downstream industries sink is identified. The amounts allocated from the NG grid to downstream industries are determined by the yearly production capacity of each industry Di, power local demand *Pi*, and minimum flexibility production of each industry *Dmin*. Figure 3 demonstrates the downstream industries system behaviour.

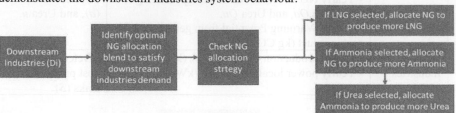

Figure 3: Downstream industries system behaviour.

To analyse the performance of *NG* and *Di*, economic *EC* and environmental *EI* indicators for each strategy are computed at each time step *n* using the following equations (1-4):

Strategy (NG):

$$EC = Q_{i,NG} \times c_{i,NG} \text{ where } i = p \, (Power), n \, (LNG), a \, (Ammonia), and \, u \, (Urea) \tag{1}$$

$$EI = Q_{i,NG} \times e_{i,NG} \text{ where } i = p \, (Power), n \, (LNG), a \, (Ammonia), and \, u \, (Urea) \tag{2}$$

such that $Q_{i,NG}$ is the yearly NG allocation to downstream products in BCM, and $c_{i,NG}$ is the net cost from this allocation in \$/BCM based on HYSYS simulation. Whereas $e_{i,NG}$ is the unit of environmental impact quantified as the global warming potential (GWP) associated with the generation of downstream products estimated through HYSYS simulation.

Strategy (Di):

$$EC = Q_{i,D} \times c_{i,D} \quad \text{where } i = p \, (Power), n \, (LNG), a \, (Ammonia), and \, u \, (Urea) \tag{3}$$

$$EI = Q_{i,D} \times e_{i,D} \quad \text{where } i = p \, (Power), n \, (LNG), a \, (Ammonia), and \, u \, (Urea) \tag{4}$$

such that $Q_{i,D}$ is the yearly generation of downstream products from natural gas in kWh for power and in MMTPA for others and $c_{i,D}$ is the net profit of generating downstream products in \$/kWh for power and in \$/MT for others. Whereas, $e_{i,D}$ is the GWP associated with generating downstream products. Both $c_{i,D}$ and $e_{i,D}$ are estimated using Process Economic Analyzer and Energy Analyzer in Aspen HYSYS. The minimum production flexibility of each downstream industry is considered, as per Table 2, to meet committed demand and allow flexible switchover between industries.

Table 2: Minimum production flexibilities.

Process	LNG	Ammonia	Urea
Minimum production flexibility (MMTPA)	5	3.8	5.6

3. Results

Two scenarios were developed to evaluate the flexibility of allocating NG to downstream industries within Qatar's energy industry. The first scenario considers the economic expenses associated with each downstream industry as a decision-making criterion. Under this scenario, the ABM model assigns percentages of NG allocation to each industry, with the cheapest technology receiving the greatest share. The second scenario makes judgments based on each industry's environmental performance, independent of the profit generated. Although the economic and environmental components are the primary selection criteria, the agents' behaviour is significantly influenced by shifting yearly capabilities and varying demands and prices.

Scenario 1: Economic restrictions

To meet local demands of power and international demands of downstream products, the NG allocation system recommends a poly allocation strategy under the economic scenario. Nonetheless, LNG is given the highest proportion after satisfying local power demand, with partial participation from ammonia in some years (figure 4-5). This distribution is owing to the cheap cost of producing electricity from natural gas.

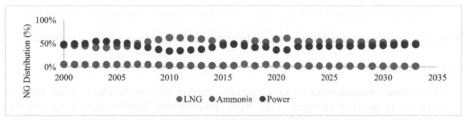

Figure 4. NG allocation mix to downstream industries under scenario 1.

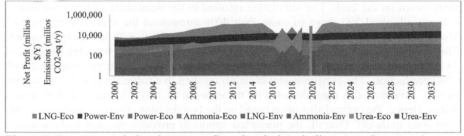

Figure 5. Downstream industries net profit and emissions indicators under scenario 1.

Scenario 2: Environmental restrictions

Under environmental constraints, the NG allocation system continues to use a poly allocation strategy for LNG and ammonia and domination of power generation (figure 6). Compared to the preceding scenario, the contribution of LNG is reduced dramatically (figure 7). This decline is primarily due to the significant emissions it generates compared to the ammonia-urea route. In this scenario, the ammonia-urea route is more dominant, allowing urea production to participate. Compared to the prior scenario, the environmental case presents an intriguing environmentally friendly option that decreases average GWP emissions by 4%. However, deploying the requires a 55% rise in prices.

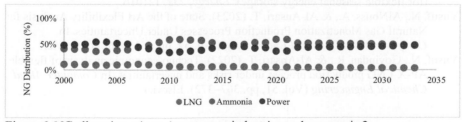

Figure 6. NG allocation mix to downstream industries under scenario 2.

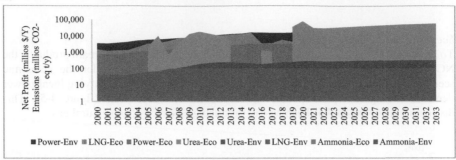

Figure 7. Downstream industries net profit and emissions indicators under scenario 2.

4. Conclusions

With increased uncertainties in final markets, natural gas supply chains must respond proactively to market changes. This work investigated flexible annual natural gas allocation to power, ammonia, LNG, and urea subject to economic and environmental objectives using ABM. In the economic scenarios, the agent decided on the allocation share based on the costs. The simulation resulted in the dominance of LNG production with an allocated NG share of more than 40% throughout the studied time horizon. Meanwhile, the environmental scenario prioritised processes with low CO_2 emissions. This resulted in increasing the share of the ammonia-urea route and reducing the share of NG allocation to LNG by 12%, influenced by the high CO_2 emissions associated with LNG production. The environmental criterion resulted in a reduction of GWP emissions by 4%. However, this route comes with the expense of a 55% increase in deployment prices. Hence, the optimal allocation based on the two decision criteria is based on a cost-emissions trade-off. Overall, the presented approach provides decision-makers with a holistic approach to decision-making during times of uncertainty. It integrates technical knowledge along with market data for prompt decision-making.

References

Bhosekar, A., & Ierapetritou, M. (2020). Modular design optimization using machine learning-based flexibility analysis. *Journal of Process Control*, *90*, 18–34.

Lopez-Jimenez, J., Quijano, N., & Wouwer, A. Vande. (2018). On the Use of Agent-Based Modeling for Smart Farming. In *2018 22nd International Conference on System Theory, Control and Computing (ICSTCC)* (pp. 348–353). IEEE.

Verleysen, K., Parente, A., & Contino, F. (2021). How sensitive is a dynamic ammonia synthesis process? Global sensitivity analysis of a dynamic Haber-Bosch process (for flexible seasonal energy storage). *Energy*, *232*, 121016.

Yusuf, N., AlNouss, A., & Al-Ansari, T. (2023). State of the Art Flexibility Analysis for Natural Gas Monetization Production Processes Under Uncertainties. In *Computer Aided Chemical Engineering* (Vol. 52, pp. 1615–1621). Elsevier.

Yusuf, N., Govindan, R., & Al-Ansari, T. (2022). Techno-economic analysis of flexible AP-X LNG production process under risks and uncertainties. In *Computer Aided Chemical Engineering* (Vol. 51, pp. 367–372). Elsevier.

Flavio Manenti, Gintaras V. Reklaitis (Eds.), Proceedings of the 34th European Symposium on Computer Aided Process Engineering / 15th International Symposium on Process Systems Engineering (ESCAPE34/PSE24), June 2-6, 2024, Florence, Italy

Production of High-Purity Methane via Sorption-Enhanced CO_2 Methanation in an Adiabatic Packed Bed Reactor

Giuseppe Piso[a*], Piero Bareschino[a], Claudio Tregambi[a], Francesco Pepe[a], Erasmo Mancusi[a]

[a]Dipartimento di Ingegneria, Università degli Studi del Sannio, Benevento 82100, Italy

* giuseppe.piso@unisannio.it

Abstract

In-situ water removal can enhance the CO_2 methanation conversion degree according to Le Chatelier's principle. Sorption-enhanced methanation (SEM) using a bifunctional solid with Ni serving as catalyst and zeolite 13X for water removal can, in principle, be carried out in an adiabatic packed-bed reactor so that the produced heat will be used for water removal over the desorption phase. Under this perspective, this study presents a model of a SEM process based on an adiabatic fixed bed reactor. The results indicate that the production of pure methane occurs for a quite long time, showing promising outlooks for this technology.

Keywords: Synthetic natural gas, sorption enhanced methanation, mathematical model, packed bed reactor, 13X zeolite.

1. Introduction

The conversion of captured CO_2 and green H_2 to chemicals represents a promising route for both leveraging excess energy from renewable power plants and long-term energy storage. On the one hand, the high exothermicity of the CO_2 methanation reaction imposes limits on the maximum conversion degree. On the other hand, it is important to note that, due to the reaction stoichiometry, even a 99 % conversion of CO_2 would ensure just a 95 % mole fraction of methane, not matching the minimum regulatory requirements for direct gas grid injection in most Countries (97.5 %, according to Erdener et al., 2023). To overcome this challenge, several reactor concepts and layouts have been proposed ranging from multiple adiabatic packed beds with inter-cooling (Bareschino et al., 2021, Mancusi et al., 2021) and optional product recycle (Bareschino et al., 2022) to micro-structured reactors consisting of multiple micro tubes, filled by catalyst, and surrounded by a coolant fluid (Brachi et al., 2023).

In-situ water removal emerges as a key strategy to enhance the efficiency of the Sabatier reaction by addressing its thermodynamic constraints. Bifunctional solids, combining a methanation catalyst with a water-removing material such as zeolite (Gómez et al., 2023, Wei et al., 2021), were explored for this purpose. The main challenge in using zeolites for in-situ water removal is the high temperature reached during CO_2 methanation. Indeed, the zeolite's water adsorption capacity diminishes as temperature rises. In commercial methane production processes, reactors operate in a cascade (Rönsch et al., 2016), with the last reactor receiving lower reactant concentrations. Consequently, the heat generated during methanation in the final reactor is significantly less than in earlier stages. This results in temperatures below 450 °C in the last reactor, making it an ideal scenario for a sorption-enhanced methanation (SEM) process utilizing zeolites (Wei et

,al., 2021). Nevertheless, effective heat management is critical in SEM process, where high temperatures significantly impact sorption capacity (Kiefer et al., 2022). At the same time, it is crucial to observe that the heat produced during the SEM process can be used to remove the adsorbed water. Indeed, to sustain a continuous product supply, a minimum of two alternating reactors are required. While in one reactor sorption-enhanced methanation occurs, the other is dried. Heat produced during methanation can be efficiently managed using shell-and-tube reactors (Bareschino et al., 2023). Alternatively, adiabatic reactors facilitate the storage of produced heat within the bed, which can be harnessed during the drying phase.

This study introduces an adiabatic reactor tailored for sorption-enhanced methanation, employing a dynamic heterogeneous model. The model considers both inter- and intra-phase gradients, acknowledging for the non-stationary nature of sorption-enhanced methanation. Model predictions encompass axial variations in temperature and concentrations of all chemical species. A sensitivity analysis was conducted to evaluate the impact of variations in operating pressure and volumetric gas flow rate on process performances, including methane purity at the reactor exit and effective time length.

2. Mathematical Model

A dynamic model for an adiabatic reactor used in sorption-enhanced methanation is discussed. This model considers temperature and concentrations changes along the axis direction, addressing inter-phase and intra-phase gradients, is dynamic, reflecting the inherently non-stationary nature of sorption-enhanced methanation.

CO_2 methanation reaction Eq. (1), i.e. the Sabatier reaction, is a combination of CO methanation and reverse water gas shift reaction (Koschany et al., 2015):

$$CO_2 + 4H_2 \leftrightarrow CH_4 + 2H_2O \tag{1}$$

In the case examined in this study, the fed mixtures contain a minimum of 80 % volume of CH_4, while the remaining 20 % is composed by H_2 and CO_2 in stochiometric ratio. As a result, the maximum temperatures attained remain strictly below 450 °C, and the presence of CO can be safely neglected (Koschany et al., 2015), so the mass balances just consider chemical species i (=CH_4, CO_2, H_2, H_2O). In detail, for the bulk gas phase axial dispersion, convection, and mass transfer between gas and solid phase are considered:

$$\varepsilon_b \frac{\partial c_{i,g}}{\partial t} = \varepsilon_b D_{ax,i} \frac{\partial^2 c_{i,g}}{\partial z^2} - \frac{\partial (v c_{i,g})}{\partial z} - k_{m,i} a_v \left(C_{i,g} - C_{s,i}\big|_{r_p} \right) \tag{2}$$

while the overall mass balance only considers convection and mass transfer between gas and solid phase:

$$\varepsilon_b \frac{\partial c_{tot,g}}{\partial t} = -\frac{\partial (v c_{tot,g})}{\partial z} - \sum_{i=1}^{Ns} k_{m,i} a_v \left(C_{i,g} - C_{s,i}\big|_{d_{p/2}} \right) \tag{3}$$

Where, z is the axial position along each reactor, v the gas velocity (m·s^{-1}) ε_b represent the bed porosity, c (mol.m^{-3}) the gas concentration for gas (g) and solid (s) phase, $D_{ax,i}$ (m^2·s^{-1}) and $k_{m,i}$ (m·s^{-1}) effective axial dispersion and gas solid mass transfer.

Dynamical mass balance in the solid phase for CH_4, CO_2, and H_2 considers the simultaneous diffusion and reaction:

$$\varepsilon_p \frac{\partial c_{i,s}}{\partial t} = \varepsilon_p D_{eff,i} \left(\frac{\partial^2 c_{i,s}}{\partial r_c^2} + \frac{1}{r_c} \frac{\partial c_{i,s}}{\partial r_c} \right) + \rho_{cat} f_{cat} v_i r_i \tag{4}$$

While mass balance in the solid phase for H_2O considers adsorption too:

$$\varepsilon_p \frac{\partial C_{H_2O,s}}{\partial t} = \varepsilon_p D_{eff,i}\left(\frac{\partial^2 C_{H_2O,s}}{\partial r_c^2} + \frac{1}{r_c}\frac{\partial C_{H_2O,s}}{\partial r_c}\right) + \rho_{cat}\nu_{H_2O}r_{H_2O} - \rho_{ads}f_{ads}\frac{\partial q}{\partial t} \qquad (5)$$

Where r_c represents the radial position along the pellet, q water load (mol·kg⁻¹), ρ_{cat} and ρ_{cat} catalyst and zeolite densities (kg.m⁻³), ε_b the catalyst porosity and $D_{eff,i}$ (m²·s⁻¹) effective diffusion of species i in solid and r_i (mol .kgcat⁻¹.s⁻¹) the rate of consumption or formation of i-species . Finally, in Eqs. (4)-(5), the reaction and adsorption terms are weighed by the volumetric fraction per unit volume f_{cat} and f_{ads}, respectively. Several authors (e.g. Gomez et al., 2023) showed that at the typical temperatures at which CO₂ methanation takes place, H₂O and CO₂ co-adsorption is negligible, thus just water adsorption is here considered.

The energy balance considers axial dispersion and convection, and heat transfer between gas and solid phase:

$$\varepsilon_b\rho_{gas}c_{p,gas}\frac{\partial T_g}{\partial t} = \lambda_{ax}\frac{\partial^2 T_g}{\partial z^2} - \rho_{gas}c_{p,gas}\nu\frac{\partial T_g}{\partial z} + h_f a_v\left(T_s - T_g\right) \qquad (6)$$

Due to the high thermal conductivity of the solid, intra-particle temperature gradients can be neglected and the energy balance for the solid phase is:

$$(1 - \varepsilon_b)(\rho_{cat}f_{cat} + \rho_{ads}f_{ads})c_{p,s}\frac{\partial T_s}{\partial t} = h_f a_v\left(T_s - T_g\right) + (1 - \varepsilon_b)\rho_{cat}f_{cat}(-\Delta H_m)r +$$
$$(1 - \varepsilon_b)\rho_{ads}f_{ads}(-\Delta H_{ads})\frac{\partial q}{\partial t} \qquad (7)$$

In Eqs. (6)-(7), T (K) represent gas (g) and solid (s) temperature, c_p (J.kg⁻¹.K⁻¹) the heat capacity, h_f gas-solid heat transfer coefficient (W·m⁻²·K⁻¹), while ΔH_m and ΔH_{ads} are the reaction and adsorption enthalpies, respectively.

To complete the gas and solid mass and enthalpy balances, classic Danckwerts boundary conditions were imposed at the reactor inlet and outlet, while symmetry and flux continuity represented the boundary conditions for the solid phase.

According to Mette et al. (2014) a linear driving force (LDF) model has been used to describe the adsorption rate:

$$\frac{\partial q}{\partial t} = K_{LDF}(q^* - q) \qquad (8)$$

and the equilibrium adsorption capacity, q^*, is described by the Dubinin- Ashtakov model (Mette et al., 2014). The kinetic model proposed by Koschany et al. (2015) for Ni-catalysed CO₂ methanation, which neglects side reactions, has been adopted. Moreover, these kinetics have been recently used to model a sorption-enhanced methane synthesis. For reader convenience, the reaction kinetic is:

$$r_m = \frac{k(P_{H_2}P_{CO_2})^{0.5}\left(1 - \frac{P_{CH_4}P_{H_2O}^2}{P_{H_2}^4 P_{CO_2}K_{eq}}\right)}{\left(1 + K_{OH}\frac{P_{H_2O}}{P_{H_2}^{0.5}} + K_{H_2}P_{H_2}^{0.5} + K_{mix}P_{CO_2}^{0.5}\right)^2} \qquad (9)$$

The mathematical model has been solved using the commercial software package COMSOL Multiphisycs®. The size of the PDE set is large, but model order reduction techniques are available for future studies (Cutillo et al., 2023).

The Shomate equation and data from the NIST Chemistry WebBook were used to calculate heat capacities and enthalpy of formation of all the gaseous species The temperature dependence of mass and heat transport coefficients, diffusivities, and gas properties were considered through the state-of-the-art correlations (see Bareschino et al.,

2023). The operating conditions, reactor volumes, and catalyst properties used in the simulations are reported in Table 1.

Table 1 – Parameters used in the simulations.

Parameter	Value	Parameter	Value
c_{ps}	1100 J·kg^{-1}·K^{-1}	M_{cat}	0.064 kg
λ_S	0.25 W·m·K^{-1}	M_{ads}	1.22 kg
L	6 m	ρ_c	2350 kg.m^{-3}
d_t	1 m	T_{in}	300 °C
d_p	2.5 mm	P	3 bar
$\Delta H°_{ads}$	-45·10^3 J·mol^{-1}	$\Delta H°_m$	-165·10^3 J·mol^{-1}

3. Results and discussion

As already discussed, the proposed model applies to the last reactor in a cascade, designed to operate at lower temperatures. Accordingly, the composition of the feed gas in this research comprises CH$_4$ (80 %$_v$), H$_2$ (16 %$_v$), and CO$_2$ (4 %$_v$), maintaining a stoichiometric ratio.

Fig. 1 shows temperature and the distribution of adsorbed water along the reactor axis for several time instants. The temperature on catalyst layers closest to the reactor inlet quickly increases at the beginning, as the fresh feed meets a dry catalyst. The cold feed lowers the temperature of these catalyst layers and thus favors the absorption of the produced water.

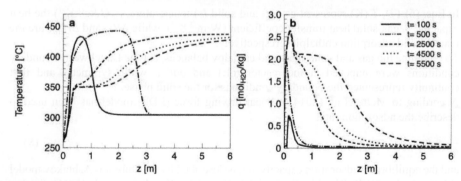

Figure-1 Temperature profiles (a) and adsorbed water (b) along the reactor axis for several time instants.

It is crucial to emphasize the bifunctional nature of the solid employed in this context. Consequently, even after reaching the maximum adsorption capacity in a specific section, the solid continues to act as a catalyst. Thus, as the catalyst layers near the inlet approach their maximum water adsorption capacity, the methanation reaction persists, albeit no longer water adsorption occurs. Over time, the innermost catalyst layers become water-saturated, impeding the methanation reaction in SEM conditions. This results in a reduction of reactant conversion, leading to a decrease in both heat generation and, consequently, overall temperature. The spatial profiles depicted in Fig.1 reveal a distinct pattern: the heat generated by the methanation reaction is concentrated in the reactor sections near the outlet, whereas water tends to be predominantly adsorbed on the sections closer to the inlet. These distributions are conducive to an optimal counter-current bed regeneration process.

Fig. 2 reports gas compositions and temperature at the reactor outlet. Methane concentration after 4200 s crosses below the 97.5 % line, as can be seen in Fig.2. (a), thus setting the maximum time length for reactor operation.

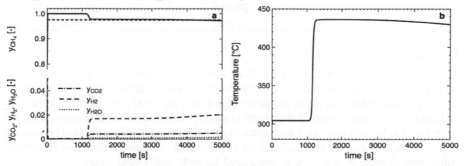

Figure-2 Gas molar fractions (a) and temperature (b) at the reactor outlet for F_{stp}=600 m³/h, dashed line representing the threshold value of y_{CH4}=0.975 for direct gas grid injection.

By analyzing the time series reported in Fig.2 (a), we can see that H_2 and CO_2 breakthroughs always precede the H_2O one, being more evident for H_2. This phenomenon is due to the limiting effect of the chemical reaction resulting in a lower conversion rate than convective mass transport thus a slip of reactants occurs (Kiefer et al., 2022). For the assigned operating conditions (Tab.1), a dry mixture is present at the reactor outlet, but for times larger than 4200 s the methane percentage drops below 97.5 %. Therefore, the time limit for this type of operation is not set by the breakthrough of the water but by the slip phenomenon of unconverted reactants. It is important to note, however, that up to this time, the reactor produces a mixture with a methane percentage greater than 97.5 %. To pursue process intensification, the effect of different pressure values and three different molar flow rates has been considered. As the pressure increases, the gas velocity decreases, which results in a longer utilization time for the same flow rate (see Fig. 3). Moreover, the pressure has a positive effect on conversion too. Indeed, CO_2 methanation is usually carried out at pressures greater than 10 bar (Rönsch et al., 2016) while for sorption enhanced methanation 100 % conversions are attained for a sufficient long-time even at pressure lower than 10 bar. Therefore, the increase in pressure ensures longer operating times for the same volumetric flow rate as summarized in Tab. 2.

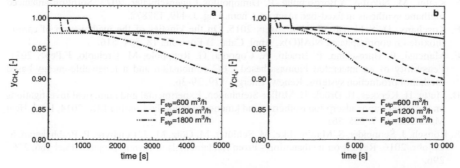

Figure-3 CH₄ molar fraction at reactor exit for different volumetric gas flow rates at P=3 bar (a) and P=5 bar (b). Dotted line represents the threshold value of y_{CH4}=0.975.

Table 2 – Operating time for several pressure and volumetric flow rates.

F_{stp} m^3/h	P=3 bar	P= 5 bar
600	4200 s	7960 s
1200	2040 s	3810 s
1800	1230 s	2450 s

4. Conclusions

This study shows that an adiabatic packed bed reactor, filled with bifunctional solids combining a methanation catalyst with a water-removing material, can be used as the last unit in a methanation cascade to produce a high-purity (y_{CH4}>97.5 %) methane stream. In particular, it has been observed that water is mainly adsorbed on the first catalyst layers, so drying will have to take place in counter-current. Process intensification can be pursued by increasing the reactor pressure, indeed even for high volumetric flow rate a methane output rate of more than 97.5 % is guaranteed for sufficiently large times.

References

P. Bareschino, G. Piso, C. Tregambi, F.Pepe, E. Mancusi, 2023. Numerical modelling of a sorption-enhanced methanation system. Chem. Eng. Sci, 277, 118876.

P. Bareschino, E. A. Cutillo, C. Tregambi, F. Pepe, G. Continillo, E. Mancusi, 2022. Periodic Oscillations in Methane Reactor: Effects of the Main Operating Parameters, Comput. Aided Chem. Eng. 51 1-6.

P. Bareschino, E. Mancusi, C. Tregambi, F. Pepe, M. Urciuolo, P. Brachi, G. Ruoppolo, 2021, Integration of biomasses gasification and renewable-energies-driven water electrolysis for methane production, Energy, 230, 120863

P. Brachi, P. Bareschino, C. Tregambi, F. Pepe, M. Urciuolo, G. Ruoppolo, E. Mancusi, 2023, Assessing the feasibility of an integrated CLC-methanation system using solar dried and torrefied biomasses as a feedstock, Fuel, 331, 125951.

E. A. Cutillo, E. Mancusi, K. Bizon, P. Bareschino, G. Continillo, 2023, A Reduced Order Model for the Prediction of the Dynamics of a Methane Reactor, Comput. Aided Chem. Eng. 52 11999–1204.

B.C. Erdener, B. Sergi,O.J. Guerra, A.Lazaro Chueca, K. Pambour, C. Brancucci, B.M. Hodge, 2023. A review of technical and regulatory limits for hydrogen blending in natural gas pipelines. Int. J. Hydrogen Energy 48, 5595–5617.

L. Gómez, I. Martínez, M.V. Navarro, R. Murillo, 2023, Selection and optimisation of a zeolite/catalyst mixture for sorption-enhanced CO_2 methanation (SEM) process, Journal of CO2 Utilization, 77, 102611.

F. Kiefer, M. Nikolic, A.Borgschulte, P. Dimopoulos Eggenschwiler, 2022. Sorption-enhanced methane synthesis in fixed-bed reactors. Chem. Eng. J. 449, 137872.

F. Koschany, D. Schlereth, O. Hinrichsen, 2015, On the kinetics of the methanation of carbon dioxide on coprecipitated NiAl(O)x, Appl. Catal. B Environ., 18, 504–516.

E. Mancusi, P. Bareschino, P. Brachi, A. Coppola, G. Rupppolo, M. Urciuolo, F.Pepe, 2021, Feasibility of an integrated biomass-based CLC combustion and a renewable-energy-based methanol production systems, Renew. Energy, 179, 29-36.

B. Mette, H. Kerskes, H. Drück, H. Müller-Steinhagen, Experimental and numerical investigations on the water vapor adsorption isotherms and kinetics of binderless zeolite 13X, 2014, Int. J. Heat Mass Transf., 7, 555–561.

S. Rönsch, J. Schneider, S. Matthischke, M. Schlüter, M. Götz, M., J. Lefebvre, P. Prabhakaran, S. Bajohr, 2016. Review on methanation - From fundamentals to current projects. Fuel 166, 276–296.

L. Wei, H. Azad, W. Haije, H. Grenman, W. de Jong, 2021. Pure methane from CO_2 hydrogenation using a sorption enhanced process with catalyst/zeolite bifunctional materials, Appl. Catal. B Environ. 297, 120399

Flavio Manenti, Gintaras V. Reklaitis (Eds.), Proceedings of the 34th European Symposium on Computer Aided Process Engineering / 15th International Symposium on Process Systems Engineering (ESCAPE34/PSE24), June 2-6, 2024, Florence, Italy

Phase-field modeling for freeze crystallization in a binary system

Xiaoqian Huang[a], Aurélie Galfré[a], Françoise Couenne[a], Claudia Cogné[a]

[a]*Universite Claude Bernard Lyon 1, LAGEPP UMR5007 CNRS, 43 boulevard du 11 Novembre 1918, Villeurbanne 69100, France*
Xiaoqian.huang@univ-lyon1.fr

Abstract

This article investigates the freeze crystallization process from water-sodium chloride solution using the Phase-field model. This process (from 273.15 K to the eutectic point 252.05 K) generates two phases: a solid phase comprising solely ice and a liquid phase consisting of a concentrated solution. The phase-field model is employed to simulate the process and depict the specific dendritic structure of the solid. The challenge is to reproduce the exclusion of the solute in the ice while respecting the thermodynamic consistency of the system. The article aims to provide a straightforward thermodynamic approach to address this complexity and reproduce the solute trapping mechanisms. Additionally, the model is extended to a more concentrated solution. The conclusion suggests that lower growth velocity is favorable for ice purification and emphasizes the importance of avoiding side dendrites.

Keywords: Phase-field model, thermodynamics, freeze crystallization, saline solution

1. Introduction

Freeze crystallization, a commonly employed separation process in the food, pharmaceutical, and water treatment industries, is known for its efficiency in preserving delicate molecules and its energy efficiency. The conventional approach for modeling this process in chemical engineering is as follows: liquid and solid phases are in two distinct regions separated by a sharp interface. Conservation equations are solved in each phase and explicit algebraic conditions are imposed at the interface (Najim and Krishnan, 2022). However, this approach encounters challenges in accurately tracking the moving interface, often relying on assumptions of a sharp and discontinuous interface.

Phase-field (PF) model addresses these issues by monitoring the ice front and its complex dendritic morphology by providing a spatially diffuse interface along a finite width. This diffuse interface is represented through an order variable φ. This model serves as a widely employed numerical tool in material sciences, particularly for the investigation of microstructure evolution. While the majority have been developed to address metal solidification, a few have reported studies on ice crystallization kinetics. One of the major challenges in ice crystallization modeling is the vacancy of the solute in the solid phase. To avoid salt trapping in ice, the system's thermodynamics has to be consistent and well-adjusted in the model equations. Yuan et al. (2020) studied the ice crystallization from salt solution without addressing this issue: solute is incorporated in the solid phase, which is not thermodynamically consistent. Van der Sman (2016) and Li and Fan (2020) tried to tackle this problem by introducing an interaction coefficient of components depending on φ to define each phase's free energy. In this article, this challenge is addressed by a straightforward thermodynamic approach. In the first part, the PF model is introduced.

Then, we focus on the introduction of a pseudo-component in the model equations to mimic thermodynamically the salt behaviour. Subsequently, the developed model will be applied to the freeze crystallization of a binary system composed of H_2O-NaCl (A-B).

2. Phase-field model

The system under consideration consists of a small volume of H_2O-NaCl (A-B) liquid mixture. It undergoes a phase transition between solid and liquid: pure ice crystal grows due to the undercooling degree whereas the solute (salt) remains in the liquid phase. The PF model deals with the interface-tracking problem by using a continuous order variable φ. The system comprises a diffuse interface, a solid phase and liquid phase. φ takes value 0 for the solid phase (S), 1 for the liquid phase (L), and intermediate values $]0, 1[$ in the diffuse interface. The PF model is based on the Ginzburg–Landau free energy functional of the biphasic system, energy, mass and the so-called Allen-Cahn partial differential equations defined over the whole system.

2.1. Governing equations

The Landau-Ginzburg Gibbs free energy functional G (J) guarantees the thermodynamic consistency of the heterogeneous system. It reads:

$$G = \int \left[g(T, \varphi, x_B) + \frac{\epsilon_\varphi^2}{2} |\nabla \varphi|^2 \right] dV \tag{1}$$

The 1st term represents the free energy density g (J/m^3) (Eq. (2)) depending on φ, molar fraction of the solute x_B (-), and temperature T (K); the 2nd term is the energy gradient in the interface with the interface gradient coefficient ϵ_φ ((J/m)$^{1/2}$):

$$g(T, \varphi, x_B) = g^S(T, x_B^S) + h(\varphi)\left(g^L(T, x_B^L) - g^S(T, x_B^S)\right) + wp(\varphi) \tag{2}$$

g comprises each bulk energy (g^S and g^L) and the energy barrier of the phase transition w(J/m^3); $h(\varphi) = \varphi^3(6\varphi^2 - 15\varphi + 10)$ is a monotonously increasing polynomial and $p(\varphi) = \varphi^2 (1 - \varphi)^2$ is a double-well function: they are chosen to ensure local minima with respect to each phase and guarantee the thermodynamic conditions (Callen, 1985). The molar fraction at the interface x_B is related to liquid phase x_B^L and solid phase x_B^S as:

$$x_B = x_B^S + h(\varphi)(x_B^L - x_B^S) \tag{3}$$

The bulk energy of liquid phase is obtained by applying a mixture law:

$$g^L(T, x_B^L) = x_A^L \mu_A^L + x_B^L \mu_B^L \tag{4}$$

Where, for an ideal phase:

$$\mu_i^L(T, x_B^L) = \mu_i^{L,o}(T) + RT \ln x_i^L \quad (i = A \text{ or } B) \tag{5}$$

o indicates the pure component. Similar equations are used for the solid phase. By assuming local equilibrium conditions, we have:

$$\frac{\partial g^L}{\partial x_B^L} = \frac{\partial g^S}{\partial x_B^S} \Rightarrow \mu_B^L(T, x_B^L) - \mu_A^L(T, x_B^L) = \mu_B^S(T, x_B^S) - \mu_A^S(T, x_B^S) \tag{6}$$

The dynamical evolution of φ is given by the Allen-Cahn equation:

$$\frac{\partial \varphi}{\partial t} = -M_\varphi \frac{\delta G}{\delta \varphi} = M_\varphi \left[\epsilon_\varphi^2 \nabla^2 \varphi - h'(\varphi)\Delta g - wp'(\varphi) \right] \tag{7}$$

With M_φ ((J.s)$^{-1}$) the interface mobility coefficient and $\frac{\delta G}{\delta \varphi}$ the variational derivative of G wrt φ. The notation $'$ denotes derivatives wrt φ.

The thermodynamic driving force Δg is:

$$\Delta g = g^L(T, x_B^L) - g^S(T, x_B^S) - (x_B^L - x_B^S)\frac{\partial g^L}{\partial x_B^L} \tag{8}$$

Only diffusive transport mechanism is considered in the model. The dynamic evolutions of molar fraction x_B and temperature T are given by:

$$\frac{\partial x_B}{v_m \partial t} = -\nabla . J_B = \nabla . M_B(\varphi)\nabla\frac{\delta G}{\delta x_B} \tag{9}$$

$$\frac{\partial e}{\partial t} = -\nabla . J_e = \nabla . \lambda(\varphi)\nabla T \tag{10}$$

e (J/m^3) is the internal energy density, J_B (mol/m^2/s) and J_e (W/m^2) are respectively the densities of solute and heat fluxes. $M_B(\varphi) = \frac{D_B x_B(1-x_B)}{v_m RT}$ (mol^2/(J.s.m)) is the mobility coefficient of B, with v_m(m^3/mol) the constant molar volume, $D_B(\varphi)$ (m^2/s) the diffusivity coefficient of B in A and $\lambda(\varphi)$ (W/(m.K)) the thermal conductivity coefficient depending on the φ evolution.

The internal energy is postulated by:

$$e(T, \varphi, x_B) = x_A e_A(T, \varphi) + x_B e_B(T, \varphi) \tag{11}$$

Where

$$e_i(T, \varphi) = e_i^S(T) + h(\varphi)(e_i^L(T) - e_i^S(T)) \quad (i = A \, or \, B) \tag{12}$$

2.2. Interface equilibrium conditions

The parameters such as ϵ_φ, w and interface thickness ξ (m) are related to physical properties as interface energy σ (J/m^2) by considering the equilibrium conditions at interface: $\Delta g = 0, \frac{\partial \varphi}{\partial t} = 0$ and at melting temperature of the solution T^m(K). The interface thickness and surface tension are determined (Kim et al., 1999): $\xi = \frac{4\sqrt{2}}{\sqrt{w}}\epsilon_\varphi$ and $\sigma = \frac{\sqrt{2w}\epsilon_\varphi}{6}$.

2.3. Pseudo-component

The PF method requires the definition of the Gibbs energy density of each phase (Eq. (4)). At equilibrium, the thermodynamic driving force Δg must be zero (Eq. (8)). Furthermore, the local equilibrium condition (Eq. (6)) needs to be respected. In the real phase diagram of the system (solid curves in Fig. 1 represent real phase diagram with T^E and x_B^E, the temperature and the concentration at eutectic point for H$_2$O-NaCl mixture), the solidus curve is along the y-axis, i.e. variable x_B^S values zero within the study conditions. To mimic this behavior, the pseudo-component is used and enables to satisfy the PF method requirements (Bayle, 2020).

Only two properties are needed for the determination of the pseudo-component: pseudo melting point T_B^m at $x_B = 1$, and pseudo melting latent heat L_B^m (J/mol). Assuming the solution is ideal and neglecting the impact of thermal capacities, the liquidus equation is obtained by equal chemical potential of each component:

$$\mu_i^L(T) - \mu_i^S(T) = L_i^m\left(1 - \frac{T}{T_i^m}\right) + RT ln\left(\frac{x_i^L}{x_i^S}\right) = 0 \quad (i = A \, or \, B) \tag{13}$$

By fixing T_B^m smaller than T_A^m but higher than absolute zero, a pseudo latent heat L_B^m constant is then estimated by minimizing the sum of the squared errors between the temperature calculated for the ideal liquidus (eq.13) with real solidus (dashed curve ① in Fig.1) and simulated results. It could be noticed that the partition coefficient $k = \frac{x_B^S}{x_B^L}$ is zero for real solidus and the real liquidus requires the activity coefficient of the solute. Only the ideal solution is considered in this paper because the system is dilute. The pseudo-component is evaluated for $x_B = [0, 0.06]$ and $T^m = [260.15\ K, 273.15\ K]$. The difference between the estimated liquidus and ideal liquidus is small, i.e. 0.002 K.

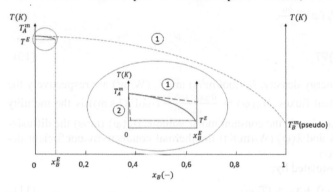

In the encircled part of Fig. 1, the pseudo solidus is represented by dashed curve ② with the pseudo properties estimated. With the known T_B^m, L_B^m and liquid phase properties, the other thermodynamic properties required in the PF model, such as internal energy and heat capacity, can be easily calculated.

Figure 1: Illustration of pseudo-component properties determination

2.4. Numerical implementation

Physical properties	Phase	Values	Numerical parameters	Values
Thermal conductivity coefficient λ (W/(m.K))	L	0.562	Mesh number	300
	S	2.2190	Grid size dx=dy (μm)	0.29
Diffusion coefficient of the solute D_B (m²/s)	L	6.10^{-9}	Time step dt (ns)	5.23
	S	$1.733.10^{-13}$	Interface thickness ξ (m)	$3.25.10^{-6}$
Surface tension σ (J/m²)		0.0773	Interfacial gradient coefficient ϵ_φ ((J/m)$^{1/2}$)	$3.73.10^{-4}$
Molar volume v_m (m³/mol)		18.10^{-6}	Magnitude of energy barrier w (J/m³)	$3.37.10^{-7}$
Melting temperature of H_2O T_A^m (K)		273.15	Strength of anisotropy δ (-)	0.08
Latent heat of H_2O L_A^m (J/mol)		6030	Number of branches m (-)	6
Pseudo melting temperature of NaCl T_B^m (K)		2	Interface mobility coefficient M_φ ((J.s)$^{-1}$)	2, 4
Pseudo latent heat of NaCl L_B^m (J/mol)		77.3		
Initial liquid temperatures T^0(K)		258.15, 263.15, 268.15		
Initial liquid concentration x_B^0 (-)		10^{-5}, 10^{-4}		

Table 1: Physical and numerical parameters

The model (Eq. (7) (9) (10)) is implemented in MATLAB with the parameters in table 1. The time integration is solved by the explicit Euler method. Discretization and divergence are respectively calculated by the finite difference method and 9-points Laplacian. The boundary conditions are fixed to be zero Neumann conditions. The initial conditions are:
liquid phase $x^2 + y^2 \geq r^2 : \varphi = 1, T = T^0,\ x_B^L = x_B^0$;
solid phase $x^2 + y^2 \leq r^2 : \varphi = 0, T = T_A^m, x_B^S = 0$.

with r the initial radius of the nuclei, and solid phase at $T_A^m = 273.15\ K$ the ice melting temperature, several initial liquid temperatures T^0 and concentrations x_B^0 are studied. The anisotropy function is included in gradient coefficient ϵ_φ:

$$\epsilon_\varphi(\theta) = \epsilon_\varphi^o \left(1 + \delta \cos(m\theta)\right) \tag{16}$$

with δ the strength of the anisotropy, m the number of branches and θ the angle between the direction of normal velocity and x-axis: $\theta = \arctan\left(\frac{\nabla\varphi_y}{\nabla\varphi_x}\right)$ (Kobayashi, 1993).

3. Results and discussion

During freeze crystallization, the solute is not incorporated in the ice and accumulates at the liquid-solid interface. If the solute's diffusion velocity is slower than the interface mobility, it becomes trapped in the solid phase, resulting in a non-zero partition coefficient by forming solute brines or pockets for experimentation. Thermodynamically, this manifests as a chemical potential jump at the interface, indicating the local solute trapping effect on crystal growth and solute distribution. Regarding the temperature field, the simulated system experiences heating due to the released latent heat, leading to a reduction in undercooling and a subsequent slowdown in growth velocity. PF model effectively reproduces these phenomena.

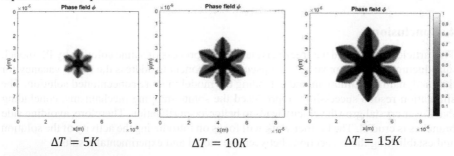

$$\Delta T = 5K \qquad\qquad \Delta T = 10K \qquad\qquad \Delta T = 15K$$

Figure 2: Dendrites growth for different undercooling with $x_B^0 = 10^{-5}$ at $t = 1.52\ ms$

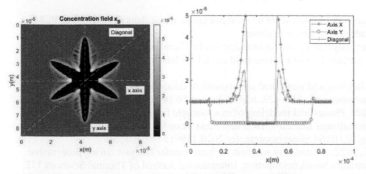

Figure 2: Concentration profile for $\Delta T = 15K$ at $t = 1.52ms$

In our study, the impacts of undercooling degree ($\Delta T = T^0 - T_A^m$) and solution concentration are investigated. Higher undercooling results in increased crystal growth velocity and more robust branches (Fig. 2). The solute distribution is depicted in cross-sections along the X, Y axis, and the diagonal section (Fig. 3). Solute accumulates between the main branches with lower growth velocity, while less amount of solute is observed on the branches' tips with faster velocity. The diagonal section shows the solute integration in side branches with the second minor jump. The X-axis cross-section reveals that a lower undercooling leads to an earlier side branches growth due to the limit of solute accumulation (Fig.4 (a)). The solute trapping effect intensifies with increasing concentration. Figure 4 (b) demonstrates that the appropriate undercooling and mobility coefficient M_φ can alleviated spurious effects by adjusting the relative velocity of crystal growth and solute diffusion. All these results support also the applicability of our pseudo-component approach to higher concentration while maintaining ice purity.

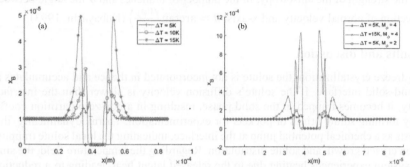

Figure 4: Concentration profile along the X axis for (a) different undercooling (5,10,15K) with $x_B^o = 10^{-5}$ and (b) more concentrated solution $x_B^o = 10^{-4}$ at $t = 1.52ms$

4. Conclusions

This article investigated the freeze crystallization process of saline solution by PF model, introducing a novel approach with a pseudo-component to address the solute vacancy for thermodynamic equilibrium and extending the model to more concentrated solution. The simulation results successfully reproduced the solute trapping mechanisms, concluding that lower crystal growth velocity leads to better ice purification. Therefore, avoiding side branches is crucial. The Further work will focus on introducing the activity of the solution and establishing the connections between the model and experimental results.

References

Bayle, R., 2020. Simulation des mécanismes de changement de phase dans des mémoires PCM avec la méthode multi-champ de phase.

Callen, H.B., 1985. Thermodynamics and an Introduction to Thermostatistics. Wiley.

Kim, S.G., Kim, W.T., Suzuki, T., 1999. Phase-field model for binary alloys. PHYSICAL REVIEW E.

Kobayashi, R., 1993. Modeling and numerical simulations of dendritic crystal growth. Physica D: Nonlinear Phenomena 63, 410–423. https://doi.org/10.1016/0167-2789(93)90120-P

Li, J.-Q., Fan, T.-H., 2020. Phase-field modeling of macroscopic freezing dynamics in a cylindrical vessel. International Journal of Heat and Mass Transfer 156, 119915. https://doi.org/10.1016/j.ijheatmasstransfer.2020.119915

Najim, A., Krishnan, S., 2022. A similarity solution for heat transfer analysis during progressive freeze-concentration based desalination. International Journal of Thermal Sciences 172, 107328. https://doi.org/10.1016/j.ijthermalsci.2021.107328

van der Sman, R.G.M., 2016. Phase field simulations of ice crystal growth in sugar solutions. International Journal of Heat and Mass Transfer 95, 153–161. https://doi.org/10.1016/j.ijheatmasstransfer.2015.11.089

Yuan, H., Sun, K., Wang, K., Zhang, J., Zhang, Z., Zhang, L., Li, S., Li, Y., 2020. Ice crystal growth in the freezing desalination process of binary water-NaCl system. Desalination 496, 114737. https://doi.org/10.1016/j.desal.2020.114737

Flavio Manenti, Gintaras V. Reklaitis (Eds.), Proceedings of the 34th European Symposium on Computer Aided Process Engineering / 15th International Symposium on Process Systems Engineering (ESCAPE34/PSE24), June 2-6, 2024, Florence, Italy

Design and Optimization of a Steam-assisted Adsorption Process for Direct Air Capture

Adam Ward [a,b], Maria M. Papathanasiou [a,b], Ronny Pini [a,*]

[a] *Department of Chemical Engineering, Imperial College London, UK*
[b] *Sargent Centre for Process Systems Engineering, Imperial College London, UK*
* *email: r.pini@imperial.ac.uk*

Abstract

Purification of CO_2 from atmospheric air via a steam-assisted temperature-vacuum swing adsorption (S-TVSA) process is a promising approach for efficiently achieving greenhouse gas removal. In this work, we present a computational framework for design and optimization of S-TVSA direct air capture processes by employing detailed numerical simulations, variance-based sensitivity analysis, and black-box optimization. We develop a numerical simulation platform for S-TVSA processes through solution of the governing dynamic material, momentum, and energy balance equations via a finite volume approach. We then use the developed simulator to conduct variance-based sensitivity analysis to quantify the influence of each process operating condition on all key process KPIs, in terms of both first and second order effects. Further, we conduct constrained multi-objective optimization to maximize the efficiency of S-TVSA direct air capture in terms of maximum productivity and minimum energy usage, while achieving high CO_2 purity. The results show that the system performance is strongly non-linear with respect to the operating decisions, and that process design by rigorous optimization is central to obtaining near-optimal performance. Further, we identify that under optimal operating conditions, the energy usage of S-TVSA direct air capture is not prohibitively large for wide-scale deployment – but the system productivity is low. This challenges the emerging view of co-locating direct air capture to low-carbon electricity and heat provision without consideration of the available land footprint in the vicinity of such resources. Results recommend that significant future research efforts should be dedicated towards enhancing the productivity of S-TVSA direct air capture processes to enable their deployment at climate-relevant scales.

Keywords: Direct air capture, process simulation, process optimization, global sensitivity analysis, temperature-vacuum swing adsorption

1. Introduction

Greenhouse gas removal (GGR) is to play a critical role in achieving net-zero carbon emissions by mid-century, in-line with international climate commitments (IEA, 2022). GGR allows handling of residual CO_2 emissions for which there are few other decarbonization options, as well as addressing historical emissions. A leading technological solution for providing GGR is the direct removal of CO_2 from atmospheric air by chemical separation in a direct air capture (DAC) process (Deutz & Bardow, 2021). Among the available options for DAC, chemical adsorption onto the surface of solid adsorbents in a steam-assisted temperature-vacuum swing adsorption (S-TVSA) process is a promising approach. An S-TVSA process operates in a complex sequence of steps where CO_2 is adsorbed from the air at ambient conditions and is subsequently recovered from the sorbent at high purity by heating the sorbent under vacuum using a steam purge.

The design of environmentally effective S-TVSA DAC processes is very challenging owing to several key factors, including the complexity of the process cycle, the presence of multiple conflicting performance targets, and the large number of operating decisions (step durations, operating pressures/temperatures, gas flow rates). To date, S-TVSA processes have been designed according to heuristic guidelines (Stampi Bombelli et al, 2020; Young et al, 2021). Owing to the highly non-linear nature of the system, we can anticipate that such an approach does not yield a near-optimal process. Further, there has been very little published work which aims to quantitatively understand the relationship between the operational decisions and the performance of the system (Young et al, 2023).

In this work, we present the development and application of a mathematical model to enhance the understanding of S-TVSA DAC via a rigorous computational operational assessment. We deploy dynamic numerical process simulation, variance-based sensitivity analysis, and black-box optimization to model, design, and optimize an S-TVSA process with the aim of better understanding the system operation and the reasonable limits of process performance for the current state-of-the-art technology.

2. Methods

2.1. Temperature-vacuum swing adsorption process

We consider a 5-step steam-assisted temperature-vacuum swing adsorption (S-TVSA) process for purification of CO_2 from humid ambient air comprising $CO_2/N_2/H_2O$ using a fixed bed of APDES-NFC adsorbent (Stampi-Bombelli et al, 2020). The cycle steps are 1) adsorption, 2) blowdown, 3) heating, 4) desorption, and 5) pressurisation and cooling (Young et al, 2023). In the adsorption step, ambient air is fed to the column and CO_2/H_2O are selectively adsorbed onto the surface of the adsorbent. In the blowdown and heating steps, the pressure in the column is reduced using a vacuum pump and the temperature is increased using a heating jacket to remove residual N_2 from the system, and to prepare optimal conditions for efficient desorption of CO_2 in the following step. In the desorption step, the column is further heated using a direct steam purge under vacuum to collect a high-purity CO_2 product. Finally, in the pressurization and cooling step, the column is returned to ambient pressure by introducing air, and the temperature is reduced by flowing cooling water in the column jacket. The sequence of cycle steps is then repeated, with CO_2 being captured and recovered in a semi-batch fashion.

2.2. Process simulation

The dynamics of the S-TVSA process have been simulated using a detailed 1D adsorption column model (Ward & Pini, 2022). The model equations are comprised of material, momentum, and energy balances. The material balance equations describe axially dispersed plug flow of a mixture of ideal gases in a packed bed of adsorbent pellets, as well as the transfer of mass between the gas-phase and the adsorbed-phase in the column. The momentum balance is Darcy's law for the pressure drop inside a packed-bed column. The energy balance equations describe several important mechanisms of heat transfer, including conduction, convection, heat released by adsorption, and heat exchanged with the heating jacket.

The resulting model equations are a system of partial differential equations (PDEs), which we provide in Table 1. The model equations are first discretized with respect to space by applying a weighted essentially non-oscillatory (WENO) finite volume scheme using N = 10 volume elements to yield a system of ordinary differential equations (ODEs) (Haghpanah et al, 2013). The system of ODEs contains 60 equations, describing the evolution over time of pressure, composition, temperature, and adsorbed amount at all locations in the column. The system of ODEs is integrated with respect to time using the variable-order stiff *ode15s* solver in MATLAB. The equations are integrated subject to cyclic boundary conditions representing the 5-step temperature-vacuum swing adsorption process, described in Section 2.1, until the attainment of cyclic steady state (CSS). Once CSS is achieved, the process performance is evaluated in terms of the following key performance indicators (KPIs): CO_2 purity, productivity, electrical energy usage, and thermal energy usage.

Table 1. System of non-dimensional material, momentum, and energy balance equations for simulation of adsorption column dynamics.

Overall material balance:	$\frac{\partial \bar{p}}{\partial \tau} - \frac{\bar{p}}{\bar{T}}\frac{\partial \bar{T}}{\partial \tau} = -\bar{T}\frac{\partial}{\partial Z}\left(\frac{\bar{p}\bar{v}}{\bar{T}}\right) - \psi\bar{T}\sum_{i=1}^{n_c}\frac{\partial x_i}{\partial \tau}$
Component material balance:	$\frac{\partial y_i}{\partial \tau} + \frac{y_i}{\bar{p}}\frac{\partial \bar{p}}{\partial \tau} - \frac{y_i}{\bar{T}}\frac{\partial \bar{T}}{\partial \tau} = \frac{1}{Pe}\frac{\bar{T}}{\bar{p}}\frac{\partial}{\partial Z}\left(\frac{\bar{p}}{\bar{T}}\frac{\partial y_i}{\partial Z}\right) - \frac{\bar{T}}{\bar{p}}\frac{\partial}{\partial Z}\left(\frac{y_i\bar{p}\bar{v}}{\bar{T}}\right) - \frac{\bar{T}}{\bar{p}}\psi\frac{\partial x_i}{\partial \tau}$
Solid-phase material balance:	$\frac{\partial x_i}{\partial \tau} = \alpha_i(x_i^* - x_i)$
Pressure drop:	$-\frac{\partial \bar{p}}{\partial Z} = \frac{150}{4r_p^2}\left(\frac{1-\epsilon}{\epsilon}\right)^2\frac{v_0 L}{p_0}\mu\bar{v}$
Column energy balance:	$\frac{\partial \bar{T}}{\partial \tau} + \Omega_2\frac{\partial \bar{p}}{\partial \tau} = \Omega_1\frac{\partial^2 \bar{T}}{\partial Z^2} - \Omega_2\frac{\partial}{\partial Z}(\bar{p}\bar{v}) + \sum_{i=1}^{n_c}\left[(\sigma_i - \Omega_3\bar{T})\frac{\partial x_i}{\partial \tau}\right] - \Omega_4(\bar{T} - \bar{T}_w)$
Wall energy balance:	$\frac{\partial \bar{T}_w}{\partial \tau} = \Pi_1\frac{\partial^2 \bar{T}_w}{\partial Z^2} + \Pi_2(\bar{T} - \bar{T}_w) - \Pi_3(\bar{T}_w - \bar{T}_a)$

2.3. Variance-based sensitivity analysis

We have conducted a variance-based sensitivity analysis of the process KPIs (CO_2 purity, productivity, electrical energy usage, and thermal energy usage) with respect to all the operational decisions of the S-TVSA process (cycle step durations, operating pressure/temperature, gas flowrates). Sensitivity analysis has been conducted in this work by coupling the process simulator described in Section 2.2 to the SobolGSA software package (Kucherenko, 2013). To apply the SobolGSA software to our system, we have supplied the software with the process KPIs corresponding to 4,096 operating points determined by quasi-random sampling via the Sobol sequence. The bounds applied to the process operating conditions for quasi-random sampling are provided in Table 2.

Table 2. Upper and lower bounds of process operating conditions used for variance-based sensitivity analysis and process optimization.

	t_{ads} [s]	t_{heat} [s]	t_{des} [s]	p_L [bar]	T_H [K]	v_F [m/s]	v_s [m/s]
Lower bound:	1,000	500	1,000	0.05	363	0.003	0.0015
Upper bound:	15,000	1,500	40,000	0.5	373	0.01	0.005

2.4. Process optimization

The performance of the S-TVSA process has been optimized to maximize the efficiency of the process in terms of maximum productivity and minimum energy usage. The design is conducted subject to the requirement that the CO_2 product gas should have a purity of at least 95%. The optimization problem is formulated as a constrained multi-objective optimization with the following form:

$$\min_{\theta}\left[-Pr, W_{eq}\right]$$
$$\text{s.t.} \quad \theta_L \leq \theta \leq \theta_U$$
$$Pu_{CO_2} \geq 95\%$$

where Pr is the productivity, W_{eq} is the specific equivalent work, Pu_{CO_2} is the CO_2 product purity, and θ is the vector of process operating conditions. θ_L and θ_U are the lower and upper bounds, respectively, applied to the operating conditions for optimization. The concept of the specific equivalent work has been adopted to combine the electrical and thermal energy usage into a single value for the purposes of conducting multi-objective optimization. The bounds used for optimization are identical to those used for variance-based sensitivity analysis (Section 2.3) and are provided in Table 2. The implicit constraint on the CO_2 purity is handled using a penalty function approach. The constrained multi-objective optimization problem has been solved using the non-dominated sorting genetic algorithm II (NSGA-II). We run the algorithm for 100 generations with a population size of 140. The NSGA-II algorithm has been widely applied to the design of adsorption-based processes and has been previously shown to be effective at identifying optimal process performance (Ward & Pini, 2022).

3. Results & Discussion

3.1. Variance-based sensitivity analysis

Variance-based sensitivity analysis has been conducted to quantify the influence of each process operating condition on each process KPI. The most significant interactions between the operating conditions and the process KPIs are presented in Table 3 ($S \geq 0.1$). We can see that the influence of the operating conditions on the process performance is predominantly controlled by first order effects, with at least $\approx 50\%$ of the oberseved variability in each KPI being attributed to significant first order effects. However, there are also significant second order effects for the CO_2 purity, electrical energy usage, and thermal energy usage. We can see that the most important operating conditions for optimizing the process performance are the duration of the adsorption (t_{ads}) and desorption (t_{des}) steps, and the desorption pressure (p_L). Particularly, the adsorption step duration and desorption pressure have a significant effect on all the conflicting process KPIs. This result underlines the need to apply rigorous optimization

Table 3. First and second order Sobol indices for the effect of operating conditions on the process KPIs. Only significant effects are presented in the table ($S \geq 0.1$).

KPI	CO₂ purity	Productivity	Electrical energy	Thermal energy
1st order effects:	t_{ads} (0.44)	t_{ads} (0.33)	t_{ads} (0.43)	t_{ads} (0.19)
	p_L (0.23)	t_{des} (0.42)	p_L (0.13)	p_L (0.24)
		p_L (0.11)		
2nd order effects:	t_{ads}, p_L (0.16)		t_{ads}, p_L (0.15)	t_{ads}, p_L (0.13)

to such a system design, as the non-linear interactions between the decisions and the process performance will yield heuristic process design extremely challenging, and likely sub-optimal.

3.2. Process optimization

The performance of the S-TVSA process has been optimized to maximize the system productivity and minimize the energy usage, while achieving a CO_2 product purity of at least 95%, as described in Section 2.4. The resulting Pareto front is presented in Figure 1(a). We can see that the minimum energy usage of the process, $W_{eq,min} = 1.66$ MJ/kg, is relatively low. Such an energy usage is comparable in magnitude to the energy usage of deployments of adsorption-based separations in other energy systems applications (*e.g.,* post-combustion CO_2 capture) (Haghpanah et al, 2013). Therefore, despite previous literature having expressed significant concerns over the high energy usage of DAC processes, we find that the process is not prohibitively costly in terms of energy usage. We find that the total energy usage of the process is strongly dominated by thermal energy usage, emphasizing the need for co-location of DAC units to a supply of low-carbon heat to enable environmental effectiveness – such as geothermal heating or industrial waste heat. However, this need for co-location to particular resources is challenged by the very low productivity of the system at optimal operating conditions. We find that the maximum productivity, $Pr_{max} = 1.63 \times 10^{-3}$ mol/m^3/s, is very low. To contextualize the magnitude of this productivity, deployments of adsorption-based separations to post-combustion carbon capture typically yield a productivity of order $\mathcal{O}(10^0)$ mol/m^3/s (Ward & Pini, 2022). The productivity of the system is a strong indication of the required land footprint at climate-relevant scales. Therefore, we can anticipate that there is a significant practical conflict between the need to co-locate to specific energy resources to enable environmental effectiveness, and the large land footprint of the system. Development of design approaches to increase the maximum productivity of the system should form a central aspect of future work on S-TVSA processes for DAC to allow for practical deployment at the required scale.

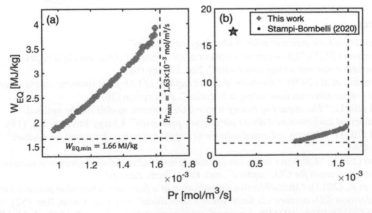

Figure 1. (a) Constrained productivity/energy usage Pareto front for the S-TVSA process. (b) Comparison of process performance by formal optimization (orange circles) and heuristic design guidelines (blue star). The dashed lines in each panel correspond to the corresponding single-objective optima for each objective.

In Figure 1(b), we present a comparison between the Pareto front obtained in this work, and the process performance obtained by applying heuristic design guidelines suggested for S-TVSA processes for DAC in previous work (Stampi-Bombelli et al, 2020). We can see that application of rigorous process optimization using the NSGA-II algorithm has resulted in significantly improved process performance, as compared to the heuristic design. We find that the maximum productivity of the system is increased by +629%, and the minimum energy usage of the process is reduced by -90.2%. As anticipated in Section 3.1, this result underlines the critical need to apply rigorous computational optimization techniques when designing such a strongly non-linear system with multiple conflicting performance targets.

4. Conclusions

In conclusion, we have developed a computational simulation, sensitivity analysis, and optimization framework for the design and optimization of steam-assisted adsorption processes applied to direct air capture. We simulate the process performance by numerical solution of the governing material, momentum, and energy balance equations using a high-resolution finite volume scheme. The process simulator is coupled to both variance-based sensitivity analysis and rigorous multi-objective process optimization to allow for a comprehensive assessment of the system design. Sensitivity analysis reveals the strongly non-linear nature of the design problem, highlighting the influence of multiple process operating conditions on several conflicting process KPIs. Through optimization of the system performance, we find that the energy usage of the process is not prohibitively high. However, we find that the productivity of the system is very low, falling several orders of magnitude below benchmark applications of adsorption-based separations in other energy systems applications. This implies that the land footprint of such processes at climate relevant scales will be a significant deployment constraint. We therefore contend that a central aspect of future work on steam-assisted adsorption processes for DAC should be optimization of the adsorbent selection and contactor design to target enhanced system productivity. Finally, future work will aim to conduct techno-economic optimization of the system performance – accounting particularly for practical factors such as pressure drop, heat transfer, and flow non-idealities at large-scales.

References

IEA (2022). *"Direct air capture: a key technology for net zero"*.
Deutz & Bardow (2021). *"Life-cycle assessment of an industrial direct air capture process based on temperature-vacuum swing adsorption"*. Nat. Energy (6).
Stampi-Bombelli et al (2020): *"Analysis of direct capture of CO_2 from ambient air via steam-assistem temperature-vacuum swing adsorption"*. Adsorption (26).
Young et al (2021): *"The impact of binary water-CO_2 isotherm models on the optimal performance of sorbent-based direct air capture processes"*. Energy Environ. Sci. (14).
Young et al (2023): *"Process informed adsorbent design guidelines for direct air capture"*. Chem. Eng. J. (456).
Ward & Pini (2022): *"Efficienct Bayesian optimization of industrial-scale pressure-vacuum swing adsorption processes for CO_2 capture"*. Ind. Eng. Chem. Res. (61).
Haghpanah et al. (2013): *"Multiobjective optimization of a four-step adsorption process for postcombustion CO_2 capture via finite volume simulation"*. Ind. Eng. Chem. Res. (52).
Kucherenko (2013): *"SobolHDMR: A general-purpose modeling software"*. Methods Mol Biol.

Flavio Manenti, Gintaras V. Reklaitis (Eds.), Proceedings of the 34th European Symposium on Computer Aided Process Engineering / 15th International Symposium on Process Systems Engineering (ESCAPE34/PSE24), June 2-6, 2024, Florence, Italy

Process Robustness Evaluation for Various Operating Configurations of Simulated Moving Bed Chromatography

Kensuke Suzuki[a], Tomoyuki Yajima[a], Yoshiaki Kawajiri[a,b]*

[a]*Department of Materials Process Engineering, Nagoya University, Furo-cho, Nagoya 464-8603, Japan*
[b]*School of Engineering Science, LUT University, Mukkulankatu 19, 15210 Lahti, Finland*
kawajiri@nagoya-u.jp

Abstract

Simulated Moving Bed (SMB) chromatography is widely used in industry as a continuous separation technique. While various operating configurations have been proposed in the past, there has been a lack of attempts to evaluate the robustness of these configurations against uncertainties such as model inaccuracies and flow rate discrepancies. In this study, we quantify the uncertainty of product purity as posterior predictive distributions. In particular, the following three configurations are considered: the conventional SMB, 3-Zone SMB, and F-shaped SMB. It was found that the F-shaped SMB was the most robust configuration concerning the purity of two products, extract and raffinate.

Keywords: Robust Design; Uncertainty Quantification; Preparative Chromatography; Simulated Moving Bed

1. Introduction

Simulated Moving Bed (SMB) chromatography is widely employed as a continuous separation technique. This technique has been used for applications in many different industries such as petrochemicals, sugars, and pharmaceuticals. Figure 1(a) illustrates the schematic diagram of the conventional SMB (hereafter referred to as CSMB), which consists of four chromatographic columns. As can be seen in this figure, desorbent and feed are injected, while extract and raffinate are withdrawn simultaneously. After a certain period of time (step time), the SMB process switches the inlet and outlet ports simultaneously in the clockwise direction. This switching mimics the counter-current operation between the liquid and solid phases. In addition to CSMB, various operating configurations have been proposed for SMB processes. For example, Figure 1(b) represents the 3-Zone SMB (hereafter referred to as 3-Zone), and Figure 1(c) represents the F-shaped SMB (hereafter referred to as F-shaped), as reported by Kawajiri and Biegler (2008). Unlike CSMB, these two operating configurations do not involve a recycle stream.

The robustness of various operating configurations in the SMB process is of utmost importance and should be carefully evaluated. During the actual operation of the SMB process, it is often observed that the desired purity and recovery are not achieved due to various uncertainties, such as systematic bias of pump flow rates and observation errors. Therefore, the robustness of the SMB process, which quantifies the deviations from the desired purity, is a critical metric, particularly in industries that demand severe purity requirements, for instance, pharmaceuticals. However, few comparative assessments of robustness across different operating configurations have been addressed (Mota et al., 2007), while other performance metrics such as productivity and desorbent consumption have been investigated.

In this study, we quantify the uncertainty of purity as posterior predictive distributions to evaluate the robustness of the three operating configurations. We compare the uncertainty in purity for the three operating configurations: CSMB, 3-Zone, and F-shaped. In this comparison, a binary separation problem is considered using four chromatographic columns. This research considers four major uncertainties: flow rate discrepancies by pumps, feed concentration errors, uncertainties in model parameters (model errors)(Suzuki et al., 2021), and observation errors. Using the posterior predictive distributions, we evaluate the uncertainty in model predictions for these four uncertainties.

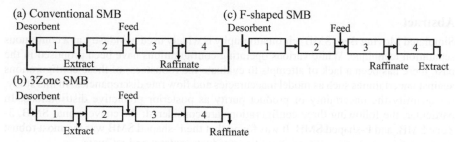

Figure 1. Schematic diagrams of (a) Conventional SMB, (b) 3-zone SMB, and (c) F-shaped SMB.

2. Methods

2.1. Mathematical Modeling of SMB

In this study, the kinetic model is employed as the governing equation for the internal dynamics of SMB columns. Detailed information on the kinetic model can be found in Grosfils et al. (2007). This model is represented by the following two partial differential equations:

$$\frac{\partial C_i^j(x,t)}{\partial t} = -\frac{1-\varepsilon_b}{\varepsilon_b}\frac{\partial q_i^j(x,t)}{\partial t} - u_j\frac{\partial C_i^j(x,t)}{\partial x}, \qquad i=1,2, \qquad j=1,2,\dots,4; \tag{1}$$

$$\frac{\partial q_i^j(x,t)}{\partial t} = k_i\left(\frac{H_i C_i^j(x,t)}{1+b_1 C_1^j(x,t)+b_2 C_2^j(x,t)} - q_i^j(x,t)\right). \tag{2}$$

In the above equations, $C_i^j(x,t)$ and $q_i^j(x,t)$ represent the liquid and solid phase concentrations of component i within the jth column, respectively; x and t denote spatial and temporal coordinates, respectively; ε_b refers to the porosity; u_j represents the internal flow rate in the jth column; k_i stands for the mass transfer coefficient; H_i signifies the Henry's constant, and b_i denotes the affinity coefficient. Note that a competitive Langmuir adsorption isotherm, which is a nonlinear isotherm model, is adopted in Eq. (2). As illustrated in Figure 1, there exist four independent flow rates in SMB: desorbent u_D, extract u_E, feed u_F, and raffinate u_R. Details about the flow balances between columns can be found in Kawajiri and Biegler (2008). In this study, model parameters $\boldsymbol{\theta} \in \mathbb{R}^6$ and operating parameters $\boldsymbol{u} \in \mathbb{R}^6$ are defined in the following vector forms: $\boldsymbol{\theta} = [H_1, H_2, k_1, k_2, b_1, b_2]$ and $\boldsymbol{u} = [u_4, u_D, u_E, u_F, C_{F,1}, C_{F,2}]$, where $C_{F,1}$ and $C_{F,2}$ are the feed concentrations for the two components.

The purity of the two products, extract and raffinate, is obtained from the average product concentrations when the SMB process reaches a cyclic steady state (CSS). The SMB process reaches CSS under sufficient operating time. The product concentration $\bar{C}_{\text{Prod},p,i}$ is an average concentration of extract (E) and raffinate (R) for each component under CSS. These four product concentrations are represented as the average product concentration vector $\overline{\boldsymbol{C}}_{\text{Prod}} \in \mathbb{R}^4$, where $\overline{\boldsymbol{C}}_{\text{Prod}} = [\bar{C}_{\text{Prod},E,1}, \bar{C}_{\text{Prod},E,2}, \bar{C}_{\text{Prod},R,1}, \bar{C}_{\text{Prod},R,2}]$. The purity of each product Pur_p, where $p = E, R$, is obtained from $\overline{\boldsymbol{C}}_{\text{Prod}}$.

2.2. Posterior Predictive Distribution of Product Purity

A statistical model accounting for observation errors in SMB experiments is formulated using the average product concentration $\overline{\boldsymbol{C}}_{\text{Prod}}$ and the parameter $\boldsymbol{\phi}$. In practice, $\overline{\boldsymbol{C}}_{\text{Prod}}$ is determined using analytical devices, for example, HPLC. These analytical measurements always contain observation errors. Assuming that the observation errors follow a Gaussian distribution, the observation (realization) $\bar{c}_{\text{Prod}}$ can be expressed as:

$$\bar{c}_{\text{Prod}} = \bar{C}_{\text{Prod}}(\boldsymbol{\phi}) + \boldsymbol{\epsilon}, \qquad \boldsymbol{\epsilon} \sim p(\boldsymbol{\epsilon}) = \prod_{m=1}^{4} N(0, \sigma_{\text{obs}}), \tag{3}$$

where $\boldsymbol{\phi} \in \Omega$ represents a set of parameters that have uncertainty in this study, and $\boldsymbol{\phi} = [\boldsymbol{\theta}, \boldsymbol{u}]$. It is assumed that $\boldsymbol{\theta}$ and $\boldsymbol{u}$ are independent of each other. $\bar{C}_{\text{Prod}}(\boldsymbol{\phi})$ is a deterministic model with $\boldsymbol{\phi}$ as input and $\overline{\boldsymbol{C}}_{\text{Prod}}$ as output, i.e., $\bar{C}_{\text{Prod}}: \mathbb{R}^{12} \rightarrow \mathbb{R}^4$. Here, $\boldsymbol{\epsilon} \in \mathbb{R}^4$ is the measurement error vector. It is assumed that each component of $\boldsymbol{\epsilon}$ follows an independent and identically distributed normal distribution $N(0, \sigma_{\text{obs}})$. Based on Eq. (3), the realization $\bar{c}_{\text{Prod}}$ through the statistical model $p(\overline{\boldsymbol{C}}_{\text{Prod}}|\boldsymbol{\phi})$ are defined as follows:

$$\bar{c}_{\text{Prod}} \sim p(\overline{\boldsymbol{C}}_{\text{Prod}}|\boldsymbol{\phi}). \tag{4}$$

The posterior predictive distribution is obtained by marginalizing the statistical model with respect to the posterior distribution of the parameters:

$$p(\overline{\boldsymbol{C}}_{\text{Prod}}|D) = \int_{\Omega} p(\overline{\boldsymbol{C}}_{\text{Prod}}|\boldsymbol{\phi}) p(\boldsymbol{\phi}|D) \, d\boldsymbol{\phi} \tag{5}$$

where, $p(\boldsymbol{\phi}|D)$ represents the posterior distribution of $\boldsymbol{\phi}$, i.e., the uncertainty obtained from the data D. Here, since $\boldsymbol{\theta}$ and $\boldsymbol{u}$ are independent, the posterior distribution $p(\boldsymbol{\phi}|D)$, which is model parameter uncertainty, can be expressed as the product of the multivariate probability density functions for $\boldsymbol{\theta}$ and $\boldsymbol{u}$ as: $p(\boldsymbol{\phi}|D) = p(\boldsymbol{\theta}|D)p(\boldsymbol{u})$. Here, $p(\boldsymbol{\theta}|D)$ is the posterior distribution of $\boldsymbol{\theta}$ which is obtained by Bayes' theorem (Yamamoto et al., 2021). In this study, operating parameter uncertainty $p(\boldsymbol{u})$ is assumed as a multivariate probability distribution based on our experimental knowledge.

The probability of obtaining a product where the purity is at most r, $P(Pur_p < r)$, is defined using a cumulative distribution function (CDF) as follows:

$$F_{Pur_p}(r) = \int_0^r p_{Pur_p}(r|D)\, dr, \qquad p = E, R \tag{6}$$

where $p_{Pur_p}(r|D)$ is a predictive distribution of purity Pur_p that is marginalized over the purity of product p using Eq. (5). Note that $F_{Pur_p}(r)$ equals 1 at $r = 100$. Using this $F_{Pur_p}(r)$, the probability that purity is greater than r, $P(Pur_p \geq r)$, is defined as follows:

$$P(Pur_p \geq r) = 1 - F_{Pur_p}(r). \tag{7}$$

2.3. Optimization of SMB Operations

To ensure a fair comparison for the three configurations, CSMB, 3-Zone, and F-shaped, the same optimization problem is solved for each operating configuration. Our optimization problem aims to maximize throughput as the objective, with four internal flow rates and the step time as decision variables. Inequality constraints are applied to the internal flow rates, specifying upper and lower bounds. Additionally, inequality constraints are imposed to set lower bounds on the purity of the two products, extract and raffinate. The optimization techniques are described in detail elsewhere (Kawajiri and Biegler, 2008).

3. Results

3.1. Posterior Distribution and Optimal Operations

The posterior distribution $p(\boldsymbol{\theta}|D)$ was estimated by Sequential Monte Carlo (SMC) (Yamamoto et al., 2021). The left side of Table 1 displays the median and 95% credible intervals for the marginal posterior distribution of each parameter in $\boldsymbol{\theta}$. Here, $p(\boldsymbol{\theta}|D)$ was estimated using SMC with computer-generated batch experimental data D that injects a small amount of feed into a single column which is used in the system of SMB (Suzuki et al., (2021)). The true values of the model parameters were adopted from Grosfils et al., (2007).

Table 1. Details of the model parameter uncertainty $p(\boldsymbol{\theta}|D)$ and the operating parameter uncertainty $p(\boldsymbol{u})$.

Model parameter uncertainty		Operating parameter uncertainty		
Parameter	Median with 95% CI[(A)]	Parameter	Mean	SD[(B)]
$H_1\ [-]$	$3.23\ ^{+1.05\times10^{-2}}_{-9.99\times10^{-3}}$	$u_4\ [m/h]^{(C)}$	u_4^*	2.00×10^{-2}
$H_2\ [-]$	$6.49\ ^{+2.43\times10^{-2}}_{-2.35\times10^{-2}}$	$u_D\ [m/h]$	u_D^*	2.00×10^{-2}
$k_1\ [s^{-1}]$	$3.21\ ^{+3.40\times10^{-2}}_{-3.42\times10^{-2}}$	$u_E\ [m/h]$	u_E^*	2.00×10^{-2}
$k_2\ [s^{-1}]$	$1.90\ ^{+2.78\times10^{-2}}_{-2.65\times10^{-2}}$	$u_F\ [m/h]$	u_F^*	2.00×10^{-2}
$b_1\ [vol\%^{-1}]$	$0.211\ ^{+8.77\times10^{-3}}_{-8.42\times10^{-3}}$	$C_{F,1}\ [vol\%]$	1.20	1.50×10^{-2}
$b_2\ [vol\%^{-1}]$	$0.414\ ^{+2.05\times10^{-2}}_{-1.99\times10^{-2}}$	$C_{F,2}\ [vol\%]$	1.20	1.50×10^{-2}

(A) 95% credible interval; (B) standard deviation; (C) u_4 does not have uncertainty in operating configurations other than CSMB, where there is no recycle flow.

The operating condition of the three operating configurations was found by the optimization where the target purity was 98%. The lower bound for the purity constraint for both extract and raffinate was set at 99.0% in which a safety margin of 1% was added to the target purity of 98%. The upper and lower limits for internal flow rate were set at 4.0 m/h and 0.1 m/h, respectively. The model parameters were assumed to be at the median of the posterior distribution, $p(\boldsymbol{\theta}|D)$.

We assumed $p(\boldsymbol{u})$ as a multivariate normal distribution, where random variables are independent of each other, as discussed in Section 2.2. The mean and standard deviation of each random variable in operating parameter uncertainty $p(\boldsymbol{u})$ are shown in Table 1. The mean values of the probability density functions for flow rates were the optimal solutions for each operating configuration. The standard deviation was set to 0.5% of the upper limit of the internal flow rate, 4.0 m/h. For the probability density functions of feed concentrations, the mean values were taken from the feed concentrations assumed in the optimization, $C_{F,i} = 1.2$ [vol%] for $i = 1,2$, and the standard deviation was set to 1.25% of the mean. The observation error, $p(\boldsymbol{\epsilon})$ in Eq. (3), was assumed to have a mean of zero and standard deviation σ_{obs} of 3.0×10^{-4} vol%.

3.2. Posterior Predictive Distribution

The posterior predictive distributions of purity for the three operating configurations under the given uncertainties in Table 1 were estimated using the Monte Carlo method. In the Monte Carlo method, 10,000 sets of parameters were sampled from $p(\boldsymbol{\phi}|D)$, and dynamic simulations of SMB were carried out until the process reached CSS each time, which was repeated 10,000 times. Additionally, 1,000 observation errors, $\boldsymbol{\epsilon}$, were sampled from $p(\boldsymbol{\epsilon})$ and added to the 10,000 simulations. As a result, $p(\overline{\boldsymbol{C}}_{\mathrm{Prod}}|D)$ was approximated with 10,000,000 samples. The joint posterior predictive distribution of purity was obtained from the resulting $p(\overline{\boldsymbol{C}}_{\mathrm{Prod}}|D)$.

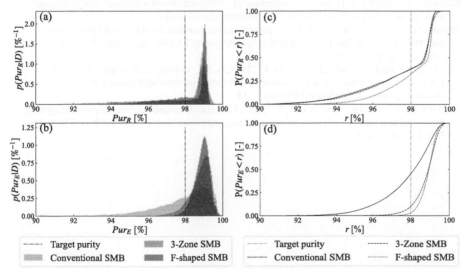

Figure 2. Marginal posterior distribution of raffinate purity (a) and extract purity (b) and cumulative distribution function of raffinate purity (c) and extract purity (d) for the three configurations: CSMB, 3-Zone, and F-shaped.

The robustness of raffinate purity is the highest in F-shaped, but differences among the three configurations are small. Figure 2 (a) and (c) show the marginal posterior predictive distributions and the CDFs, respectively, for raffinate purity of CSMB, 3-Zone, and F-shaped. The values of $P(Pur_R \geq 98.0)$ for raffinate purity are 0.620, 0.617, and 0.675 for CSMB, 3-Zone, and F-shaped, respectively, where notable differences cannot be seen.

Regarding extract purity, on the other hand, F-shaped demonstrates substantially higher robustness than the other two configurations. Figure 2 (b) and (d) show the marginal posterior predictive distributions and the CDFs of extract purity for the three configurations. The probability of obtaining purity values of 98.0% or higher, given by $P(Pur_E \geq 98.0)$ in Eq. (6), for CSMB, 3-Zone, and F-shaped yields values of 0.546, 0.899, and 0.970, respectively, where the highest value is achieved by F-shaped.

4. Conclusions

In this study, we quantified the robustness of three operating configurations: Conventional SMB, 3-Zone, and F-shaped. By comparing the posterior predictive distributions, we found that the robustness varies among the operating configurations under the same uncertainty. Regarding the purity of extract and raffinate, F-shaped demonstrated the highest robustness among the three operating configurations. This result implies that the robustness of purity can be highly influenced by operating configuration.

References

Grosfils, V., Levrie, C., Kinnaert, M., Vande Wouwer, A., 2007. A systematic approach to SMB processes model identification from batch experiments. Chem. Eng. Sci. 62, 3894–3908.

Mota, J. P. B., Araújo, J. M. M., Rodrigues, R. C. R., 2007. Optimal Design of Simulated Moving-Bed Processes under Flow Rate Uncertainty. AIChE Journal 53, 2630–2642.

Kawajiri, Y., Biegler, L.T., 2008. Comparison of configurations of a four-column simulated moving bed process by multi-objective optimization 433–442.

Suzuki, K., Yajima, T., Kawajiri, Y., "Comprehensive quantification of model prediction uncertainty for simulated moving bed chromatography", Computer Aided Chemical Engineering 49, 943-948 (2022).

Suzuki, K., Harada, H., Sato, K., Okada, K., Tsuruta, M., Yajima, T., Kawajiri, Y., 2021. Utilization of operation data for parameter estimation of simulated moving bed chromatography. J Adv Manuf Process 1–18.

Yamamoto, Y., Yajima, T., Kawajiri, Y., 2021. Uncertainty quantification for chromatography model parameters by Bayesian inference using sequential Monte Carlo method. Chem. Eng. Res. Des. 175, 223–237.

Flavio Manenti, Gintaras V. Reklaitis (Eds.), Proceedings of the 34th European Symposium on Computer Aided Process Engineering / 15th International Symposium on Process Systems Engineering (ESCAPE34/PSE24), June 2-6, 2024, Florence, Italy

State-Based Shrinkage Behavior for Waste Incineration Modeling

Lionel Sergent[a,b]*, Abderrazak M. Latifi[a], François Lesage[a], Jean-Pierre Corriou[a] and Alexandre Grizeau[b]

[a] *Université de Lorraine, CNRS, LRGP, F-54000 Nancy, France*
[b] *SUEZ, F-67590, Schweighouse-sur-Moder, France*
lionel.sergent@univ-lorraine.fr

Abstract

Waste incineration reduces the volume of incoming solid material by up to 90%. Thus, the shrinkage behavior must be addressed in the modeling of incinerators. Most common mathematical representations of shrinkage express the volume of the representative particle as a direct function of its composition. Although simple, this approach does not work well with solid mixing when included in a global model. A new approach was developed which considers two additional distributed state variables, namely contraction and internal porosity. Both follow their own partial differential equation, governed by reaction rates and mixing. This allows to keep the waste bed model fully described by a set of partial differential and algebraic equations, solvable via classical techniques. The new shrinkage sub-model is used in a complete waste bed behavior model and allows to analyze bed thickness.

1. Introduction

Within the EU, roughly 30% of household trash is incinerated, most commonly with grate kilns as depicted in figure 1. Despite their wide industrial use, their design and operation still raise several problems mainly due to the complex phenomena involved: the heterogeneity of the waste, the complex nature of pyrolysis, the stiff dynamics of combustion, the radiatiave heat exchanges, the disordered granular solid mass movement on the grate, and the lack of local measurements. Specifically, solid shrinkage cannot be neglected, as bed thickness is a crucial operational parameter. Additionally, volume reduction indirectly influences heat and mass transfers. The understanding of the phenomena involved is important to develop fine models intended to enhance energy recovery, improve pollution control and reduce maintenance costs.

The objective of this paper is to develop an approach to address the problem of solid shrinkage in the modeling of an industrial incinerator. The developed approach may be adapted to other reactive shrinking or expanding porous media.

2. Literature Review

2.1. Principle of the Most Common Approach

A first thoroughly studied example of similar solid volume reduction can be found in food dehydration (Qiu et al., 2015), or in ceramic drying. Semi-empirical equations are developed to link bulk density, internal porosity, shrinkage and moisture content.

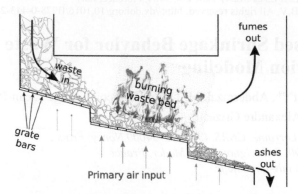

Figure 1: Schematic view of a typical grate incinerator

Especially, some equations have the general form:

$$V_p = V_{p,0} \times f(w_{\text{moisture}}) \tag{1}$$

where V_p is the volume of a representative particle, $V_{p,0}$ is its initial volume, and $f(w_{\text{moisture}})$ expresses the shrinkage as a function of the moisture content w_{moisture}.

The models used for solid combustion are often similar in nature, but generally use the fixed carbon, organic matter and moisture contents as arguments for the shrinkage function. The rationale is that moisture and organics mostly leave internal pores in the particle as they escape, leaving a structure of fixed carbon *i.e.* char. Once that fixed carbon consumes, the internal pores are liberated to extra-particular space, and the volume of the particle decreases drastically. Assuming no initial internal porosity, this process is described in figure 2.

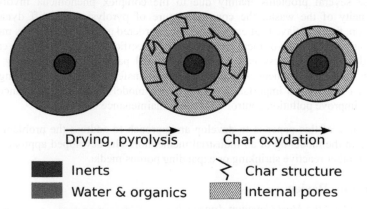

Figure 2: Model of a volume reduction of a particle undergoing drying, pyrolysis and carbon oxidation.

The expression of the shrinkage function $f(w_{\text{moisture}}, w_{\text{organics}}, w_{\text{char}})$ becomes quite complex (Menard, 2003), but follows the principle of Eq. (1).

2.2. Limitations

This formulation describes quite well the volume reduction effect at a particle level and is therefore suitable for discrete element modeling (Brosch et al., 2014). However, it does not suit continuous porous medium approaches. For instance, within a finite volumes simulation, if a cell with very low fixed carbon content receives some fixed carbon from an adjacent cell, the receiving cell will experience a non-physical sharp increase in internal porosity. This has pushed continuous medium modelers to simulate mixing by periodically exchanging (shrinking) finite volume cells (Menard, 2003) or by implementing elaborate collapse mechanisms (Hermansson and Thunman, 2011). These approaches may be justified by the discontinuous nature of mixing. However, they do not express the mixing inside the partial differential equations. The mixing is instead treated by events within the numerical integration. Hoang et al. (2022) also reported numerical instabilities.

3. Proposed Approach

3.1. Assumptions and Principle

The proposed approach does not change the physics described but is rather a slight paradigm change. We introduce two new state variables: contraction ν and internal porosity φ. The contraction represents how much local material has shrunk compared to a reference state and is calculated as:

$$\nu = \frac{V_p}{V_{\text{ref}}} \tag{2}$$

where V_{ref} is a reference volume, for instance the initial volume of the particle. For simplicity, external porosity ϵ is assumed constant and spatial variation is assumed to be 1-dimensional along the vertical axis z. For reasons explained by Shin and Choi (2000), these assumptions are reasonable for combustion happening on a grate.

Reactions may induce a change in local internal porosity. Therefore, we assume that for each reaction j there exists a coefficient, denoted β_j, called internal porosity production factor, in m^3 of internal pore created per kg of solid material reacted, such that:

$$\frac{\partial V_{c\varphi}}{\partial t} = \sum_{j \in \mathbb{T}} \beta_j V_c r_j \tag{3}$$

where V_c is a control volume, $V_{c\varphi}$ is the volume of internal pores within that control volume, r_j is the local rate of reaction j in kg m^{-3} s^{-1} and $\mathbb{T}$ is a set indexing for all considered chemical and physical transformations. The internal porosity production factors depend on the state variables. We can translate the process given in figure 2 into the following new relations:

$$\beta_{\text{drying}} = \frac{1}{\rho_{\text{moisture}}} \tag{4}$$

$$\beta_{\text{pyrolysis}} = \frac{1}{\rho_{\text{organics}}} - \frac{\psi_{\text{char, pyrolysis}}}{\rho_{\text{char}}} \tag{5}$$

$$\beta_{\text{combustion}} = -\frac{(1 - \epsilon)\varphi}{\rho_{\text{app},Cf}} \tag{6}$$

where ρ_i represents the true density of solid i while $\rho_{app,i}$ represents its apparent density. $\psi_{char,\ pyrolysis}$ is the coefficient of production of char (fixed carbon) through pyrolysis in kg of char produced per kg of organic material pyrolyzed.

3.2. Resulting Equations

Working out the partial mass balance on an elementary control volume yields:

$$\frac{\partial(\nu\rho_{app,i})}{\partial t} = \nu S_i \quad \text{where} \quad S_i = \sum_{j\in T}\psi_{ij}r_j \tag{7}$$

is the source term due to reactions. Eq. (3) reduces to:

$$(1-\epsilon)\frac{\partial(\nu\varphi)}{\partial t} = \nu\sum_{j\in T}\beta_j r_j \tag{8}$$

Since the volume of a particle is the sum of the volume of its solid constituents and internal pores, the evolution of contraction may be derived from Eq. (2):

$$\frac{\partial\nu}{\partial t} = \frac{\nu}{1-\epsilon}\sum_{j\in T}\gamma_j r_j \tag{9}$$

$$\gamma_j = \left(\beta_j + \sum_{i\in S}\frac{\psi_{ij}}{\rho_i}\right) \tag{10}$$

where S is a set indexing for the considered solid constituents. The γ_j appear naturally when deriving the contraction evolution and may be termed contraction factor of reaction j.

3.3. Mixing

Mixing may be modeled by a dispersion term. The mass flux J_i of constituent i is defined by analogy with the first Fick's law as:

$$J_i = \rho_{app}D_s\frac{\partial w_i}{\partial z} \tag{11}$$

where D_s is a dispersion coefficient (assumed constant) and w_i are the mass fractions. Working out the updated Eq. (7-9) yields the following final form:

$$\frac{\partial\rho_{app,i}}{\partial t} = D_s\frac{\partial}{\partial z}\left(\rho_{app}\frac{\partial w_i}{\partial z}\right) + \sum_{j\in T}r_j\left(\psi_{ij} - \frac{\rho_{app,i}}{1-\epsilon}\gamma_j\right) - \frac{\rho_{app,i}}{1-\epsilon}\sum_{k\in S}\frac{D_s}{\rho_k}\frac{\partial}{\partial z}\left(\rho_{app}\frac{\partial w_k}{\partial z}\right) \tag{12}$$

$$\frac{\partial\varphi}{\partial t} = \frac{1}{1-\epsilon}\sum_{j\in T}r_j(\beta_j - \varphi\gamma_j) + \frac{D_s}{1-\epsilon}\sum_{i\in S}\frac{1}{\rho_i}\frac{\partial}{\partial z}\left(\rho_{app}\frac{\partial w_i}{\partial z}\right) \tag{13}$$

$$\frac{\partial\nu}{\partial t} = \frac{l}{1-\epsilon}\sum_{j\in T}\gamma_j r_j + \frac{\nu}{1-\epsilon}D_s\sum_{i\in S}\frac{1}{\rho_i}\frac{\partial}{\partial z}\left(\rho_{app}\frac{\partial w_i}{\partial z}\right) \tag{14}$$

which have the advantage of representing both mixing and shrinkage within a system of transport partial differential equations, solvable for instance with shrinking finite volumes techniques (interface tracking methods). Alternatively, a fixed grid may be used, but it would require the use of interface capturing methods (with a solid vertical velocity).

4. Application to Solid Combustion

4.1. Fixed Packed-Bed

The developed model of volume reduction has been embedded in a general combustion model described in (Sergent et al., 2023). For a dynamic pipe-bowl combustion, the temporal profiles obtained at mid-height in the packed bed are represented in figure 3.

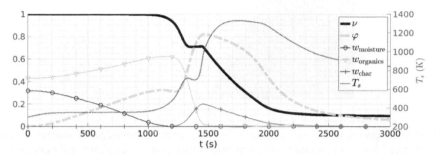

Figure 3: Temporal profiles of local variables describing the solid at mid-height during fixed pipe-bowl combustion. The scale for the solid temperature is given on the right.

After a brief initial heating phase of roughly 100 s, the temperature stabilizes at 100°C (373 K on the graph), moisture slowly dries up, internal porosity φ increases and the contraction ν remains at 1. At 1100 s, internal porosity experiences a slight decrease under the effect of mixing (which tends to homogenize the different layers of material), becoming apparent while no intense reaction is taking place. Between 1200 and 1400 s, organic materials get pyrolyzed under increasing temperature. A small portion of the produced char consumes right away while the rest accumulates. This explains why contraction starts decreasing while internal porosity keeps on increasing and is a direct consequence of the assumptions made for the internal porosity production factors β_j. At 1400 s, contraction increases slightly because of mixing. After 1500 s, the char burns away, bringing intense heat, liberating internal porosity and contracting the combustible bed. After 2500 s, only inerts remain and the system settles towards a thermal equilibrium between the overhead flame radiative heating and the convective cooling from the provided combustion air.

4.2. Grate Waste Burning

Together with the walking column approach (Shin and Choi, 2000), a model for the horizontal solid movement, and a model for the gas phase, the presented volume reduction approach may be used to study the evolution of bed thickness on graphs such as the one presented in figure 4.

The bed thickness variations produced by the simulation are in overall accordance with visual observation made on a real incinerator in France. Shortly after 5 meters from the waste input there is a hotspot associated with a low bed thickness zone. This corresponds to observed abnormal degradation of grate bars. The increases in bed thickness after 6 and 8 meters are due to a slow-down of the grate bars towards the end of the kiln to ensure complete combustion.

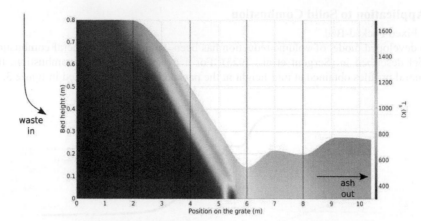

Figure 4: Side view of a burning waste bed on a grate, flattened for visualization. The vertical bars correspond to different primary air zones. See figure 1 for context.

5. Conclusion and Perspectives

A new mathematical formulation for shrinkage effects in a reacting packed-bed of granular is developed. It requires the addition of two new local dynamic state variables but allows the expression of non-trivial volume reduction combined with mixing within partial differential equations. This approach is used to model a burning waste bed on a traveling grate and study its thickness.

This formulation also opens perspectives towards the description of the continuous medium with dynamic statistical distributions, including particle diameter distributions, which would be much better suited to take into account the extremely heterogeneous nature of waste.

References

Brosch, B., Scherer, V., Wirtz, S., 2014. Simulation of municipal solid waste incineration in grate firing systems with a particle based novel Discrete Element Method. VGB Powertech 75–83.

Hermansson, S., Thunman, H., 2011. CFD modelling of bed shrinkage and channelling in fixed-bed combustion. Combustion and Flame, 158, 988–999.

Hoang, Q.N., Van Caneghem, J., Croymans, T., Pittoors, R., Vanierschot, M., 2022. A novel comprehensive CFD-based model for municipal solid waste incinerators based on the porous medium approach. Fuel, 326, 124963.

Menard, Y., 2003. Modélisation de l'incinération sur grille d'ordures ménagères et approche thermodynamique du comportement des métaux lourds. PhD, Institut National Polytechnique de Lorraine, France.

Sergent, L., Lesage, F., Latifi, A.M., Corriou, J.-P., Grizeau, A., 2023. Dynamic modeling of a waste incinerator furnace in view of supervision, Computer Aided Chemical Engineering, 52, 589–594.

Shin, D., Choi, S., 2000. The combustion of simulated waste particles in a fixed bed. Combustion and Flame, 121, 167–180

Flavio Manenti, Gintaras V. Reklaitis (Eds.), Proceedings of the 34th European Symposium on
Computer Aided Process Engineering / 15th International Symposium on Process Systems
Engineering (ESCAPE34/PSE24), June 2-6, 2024, Florence, Italy

Process Simulation of biofuel production from Waste Cooking Oil

Suresh Kumar Jayaraman

AVEVA Group Plc, Lake Forest, California 92630, United States of America
Suresh.jayaraman@aveva.com

Abstract

Increasing oil demand, climate change and depletion of fossil fuel play vital role in alternate fuel research. This study explores a design methodology to produce HEFA (Hydroprocessed Esters and Fatty Acids) biofuel from waste cooking oil (WCO). Waste cooking oil is used as a renewable resource to avoid food crops being used for fuel production and to reduce the raw material cost. This article involves developing a detailed steady state simulation of production of HEFA jet fuel & green diesel from WCO using first principles unit-operations models and rigorous thermodynamics on a next generation commercial simulation tool, AVEVA Process Simulation. The production process involves three key steps, Hydrogenation & Hydrodeoxygenation, Hydrocracking, and Separation of the products to obtain desired jet fuel and diesel.

Keywords: Waste cooking oil, biofuel, process simulation, sustainability

1   Introduction

The contribution to greenhouse gas emissions (GHG) by aviation industries is growing faster in recent decades compared to other mode of transport (Catarina I. Santos, 2018). Thus, aviation industries are actively seeking for a sustainable crude oil jet fuel substitute (Maria Fernanda Rojas Michaga, 2022). Bio-jet fuel, also known as Sustainable Aviation Fuel (SAF) is a biofuel used to power aircraft that has similar properties like traditional jet fuel with a lesser carbon footprint (Sustainable Aviation Fuels). Biofuel is made from renewable biomass that comes from plants and animals. Food crops are usually used for biofuel production, but researchers are focusing on waste resources and non-food crops to make the process profitable (Jayaraman, 2023).

Waste Cooking Oil (WCO) serves as an economical feedstock for biofuel production (Monika, 2023). WCO is usually disposed or dumped into the drainage system, and the inappropriate disposal of WCO increases environmental pollution and contaminates both terrestrial and aquatic habitats (Omojola Awogbemi, 2021). Thus, the interest in using WCO as a feedstock has increased in the recent years to make the biofuel production economical and sustainable. The utilization of WCO contributes to the concept of circular economy.

This paper focuses on utilizing WCO as a viable feedstock to produce bio jet-fuel & green diesel and simulation hydrotreatment of WCO using AVEVA Process Simulation.

2   Process Description

2.1   Properties and chemical composition of WCO

Oil goes through three different types of reaction, namely thermolytic, oxidative and hydrolytic during the process of frying (Ján Cvengroš, 2004). The physical property like surface tension, viscosity, saponification, flash point, free fatty acid, and specific heat change after frying. These chemical changes make the cooking oil undesirable for human consumption and can be used for biofuel production. Research shows that the reported physicochemical properties and fatty acid composition of WCO change with the degree of usage (Omojola Awogbemi, 2021) (Monika, 2023). The properties and free fatty acid composition of WCO play a vital role in the quality and quantity of biofuel produced. Fatty Acid composition of WCO samples is listed in Table 1.

Table 1: Fatty acid composition of WCO samples

Type of Free Fatty Acid	Carbon chain	(Wan Nur Aifa Wan Azahar, 2017)	(Jeong-Hun Kim, 2021)	(Arjun B. Chhetri, 2008)	(O. Awogbemi, 2020)
Oleic acid	C18:1	43.67	56.98	52.9	7.9
Linoleic acid	C18:2	11.39	31.21	13.5	37.35
Palmitic acid	C16:0	38.35	7.9	20.4	54.75
Stearic acid	C18:0	4.33	3.15	4.8	-
Myristic acid	C14:0	1.03	-	0.9	-
Palmitoleic acid	C16:1	-	-	4.6	-
Linolenic acid	C18:2	0.29	0.76	0.8	-
Others	-	0.94	0	2.1	0

2.2   HEFA – Process Description

HEFA biofuel is produced by the Hydrogenation of the WCO, and the various steps in the process is shown in the Figure 1. The key steps of the process is hydrogenation and Hydrodeoxygenation of WCO, then the hydrocracking of the saturated and deoxygenated products, followed by the separation process to obtain the lights, jet fuel and diesel (Bealu, 2017).

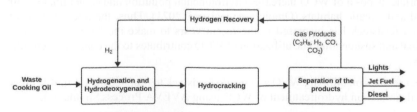

Figure 1: Key steps of Hydrotreatment of WCO.

3   Biofuel production process

3.1   Process Simulation

Process simulations are used to assess the commercial feasibilities of the proposed processes. The Next generation process simulation software, AVEVA™ Process Simulation developed by AVEVA, is used in this work.

3.2   Component & Thermodynamic selection

The composition of WCO in this paper is based on the work done by (O. Awogbemi, 2020) as shown in Table 1. Triglycerides (Triolein, Trilinolein, and Tripalmitin) and respective FFA (Oleic acid, Linoleic acid, and Palmitic acid) & hydrocarbons are added from DIPPR (Design Institute for Physical Properties) and SIMSCI data bank. The fluid contains hydrocarbons, hydrogen, carbon dioxide and light gases, so Redlich-Kwong-Soave (SRK) and Ideal Gas Law thermodynamic properties are used in this simulation. Missing thermodynamic properties are entered manually in the fluid.

3.3   Production process

The production process of biofuel from WCO is presented in the Figure 4 based on the work of (Cláudia J.S. Cavalcanti, 2022).

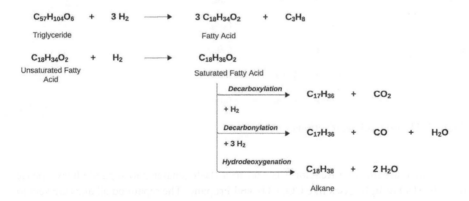

Figure 2: Hydrogenation and Hydrodeoxygenation reactions.

The WCO feed is mixed with Hydrogen stream for the Hydrogenation reaction for 100% fractional conversion of triglyceride into saturated fatty acid (SFA) at 400 C and 9.2 MPa in Reactor 1. Except for palmitic acid, Oleic acid and Linoleic acid goes through saturation to form stearic acid. So, the product stream mainly consists of palmitic acid, stearic acid, and propane. Then SFA goes through a series of reactions to produce alkanes, CO_2, CO, and H_2O at 400 C and 9.2 MPa in the presence of NiMo/Al2O3 as catalyst (Bambang Veriansyah, 2012) The Hydrogenation and Hydrodeoxygenation reactions in Reactor 1 and Reactor 2 are shown in the Figure 2.

The product stream from Reactor 2 goes through two flash separators to separate alkanes from water, gases and unused H_2. A component separator is used to separate H_2 from water and tail gases. 30% of unused H_2 is recycled back to the H_2 make up tank and the rest of H_2 along with the alkanes is sent to Reactor 3 for Hydrocracking and CO & CO_2 Methanation reactions at 50 C and 3 MPa. Reactions in Reactor 3 are shown in the Figure 3. This reactor is operated with an excess of 10% H_2 to avoid deactivation of the catalyst. Hydrocracking helps with increasing carbon range to C7 – C18. Isomerization can be considered at this step to increase the production of jet fuel compared to the diesel (Bealu, 2017). Conversion reactor is used in the simulation for all three reactions due to the lack of reaction kinetics data in the literature. All three reactions are very specific to the triglycerides present in the feedstock.

CO & CO_2 Methanation

CO	+	$3 H_2$	$\longrightarrow$	CH_4	+	H_2O
CO_2	+	$4 H_2$	$\longrightarrow$	CH_4	+	$2 H_2O$

Hydrocracking

$C_{16}H_{34}$	+	H_2	$\longrightarrow$	$C_{13}H_{28}$	+	C_3H_8
$C_{17}H_{36}$	+	H_2	$\longrightarrow$	$C_{14}H_{30}$	+	C_3H_8
$C_{18}H_{38}$	+	H_2	$\longrightarrow$	$C_{15}H_{32}$	+	C_3H_8
$C_{17}H_{36}$	+	H_2	$\longrightarrow$	C_9H_{20}	+	C_8H_{18}
$C_{18}H_{38}$	+	H_2	$\longrightarrow$	$C_{10}H_{22}$	+	C_8H_{18}
$C_{17}H_{36}$	+	H_2	$\longrightarrow$	$C_{10}H_{22}$	+	C_7H_{16}
$C_{18}H_{38}$	+	H_2	$\longrightarrow$	$C_{11}H_{24}$	+	C_7H_{16}

Figure 3: Hydrocracking reactions.

The product stream from Reactor 3 is sent to a flash separator to separate heavy phase from H_2, H_2O & light gases like CO, CO_2, and Propane. The separated alkanes are sent to a distillation column with 22 theoretical stages, a reflux ratio of 0.9, total condenser, and kettle reboiler. The feed enters at Stage 12 and the column is operated at 1 atm pressure and condenser temperature is specified to be 60 C. Sensitivity analysis was performed to find the optimal reflux ratio and the feed stage to keep the energy consumption low. The vapor product of the column mostly consists of propane, and a small fraction of C7 – C9 (lights), The distillate product of the column consists of hydrocarbons in the range of C7 – C10 (jet fuel) and the bottom product of the column consists of hydrocarbons in the range of C10 – C18 (Diesel). The simulation results of feed and product mass flowrates are shown in Table 2. The ratio of oil to H_2 presented in this study is 0.038, which is considerably less than similar work reported in the literature.

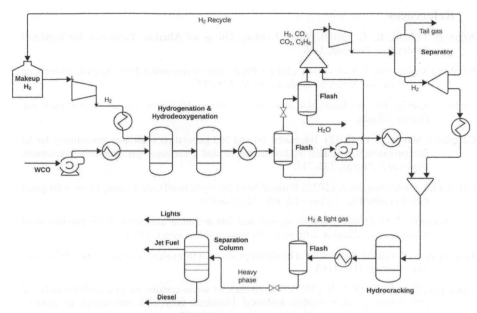

Figure 4: Process flowsheet of the hydrotreatment of WCO.

Table 2: Simulation results of feed and product mass flowrates.

		Flowrate (kg/hr)
Raw material feed	WCO	1074
	Hydrogen	41
Products	Lights	51.8
	Jet Fuel	128.2
	Diesel	710.6

4 Conclusion

The process of Hydrotreatment of waste cooking oil to obtain jet fuel and green diesel was simulated using AVEVA Process Simulation based on the studies reported in the literature. The simulation results show that 710.6 kg/hr of green diesel and 128.2 kg/hr of jet fuel can be produced from 1074 kg/hr of WCO. Green diesel is a second generation of biofuel can be used as an alternate for petroleum diesel and the jet fuel can be used in the aviation gasoline range and the byproduct propane can be used to generate heat within the process. The simulation model shows that waste cooking oil can be used as a viable feedstock to produce quality renewable fuels like green diesel, and jet fuel. This process suggests that waste cooking oil can be converted to a sustainable renewable energy and addresses the serious concerns of food crop being used for biofuel production. The usage of WCO as the feedstock considerably decreases the raw material cost, thereby reducing the operating cost of the production plant.

References

Arjun B. Chhetri, K. C. (2008). Waste Cooking Oil as an Alternate Feedstock for Biodiesel Production. *Energies*, 3-18.

Bambang Veriansyah, J. Y.-A.-W.-G. (2012,). Production of renewable diesel by hydroprocessing of soybean oil: Effect of catalysts,. *Fuel,*, 578-585,.

Bealu, Z. (2017). *Process Simulation and Optimization of Alternative Liquid Fuels Production.* Doctoral thesis.

Catarina I. Santos, C. C. (2018). Integrated 1st and 2nd generation sugarcane bio-refinery for jet fuel production in Brazil: Techno-economic and greenhouse gas emissions assessment. *Renewable Energy*, 733-747.

Cláudia J.S. Cavalcanti, M. A. (2022). Simulation of the soybean oil hydrotreating process for green diesel production,. *Cleaner Chemical Engineering.*

Ján Cvengroš, Z. C. (2004). Used frying oils and fats and their utilization in the production of methyl esters of higher fatty acids. *Biomass and Bioenergy*, 173-181.

Jayaraman, S. K. (2023). Simulation of biodiesel production from algae. *Computer Aided Chemical Engineering*, 1187-1192.

Jeong-Hun Kim, Y.-R. O.-A.-S. (2021). Valorization of waste-cooking oil into sophorolipids and application of their methyl hydroxyl branched fatty acid derivatives to produce engineering bioplastics. *Waste Management*, 195-202.

Maria Fernanda Rojas Michaga, S. M. (2022). Bioenergy with carbon capture and storage (BECCS) potential in jet fuel production from forestry residues: A combined Techno-Economic and Life Cycle Assessment approach. *Energy Conversion and Management*, 115346.

Monika, S. B. (2023). Biodiesel production from waste cooking oil: A comprehensive review on the application of heterogenous catalysts,. *Energy Nexus,*, 100209.

O. Awogbemi, F. I. (2020). Effect of usage on the fatty acid composition and properties of neat palm oil, waste palm oil, and waste palm oil methyl ester. *International Journal of Engineering and Technology*, 110-117.

Omojola Awogbemi, D. V. (2021). Advances in biotechnological applications of waste cooking oil. *Case Studies in Chemical and Environmental Engineering*, 100158.

(n.d.). *Sustainable Aviation Fuels.* Office of Energy Efficiency & Renewable Energy.

Wan Nur Aifa Wan Azahar, R. P. (2017). Mechanical performance of asphaltic concrete incorporating untreated and treated waste cooking oil. *Construction and Building Materials*, 653-663.

Flavio Manenti, Gintaras V. Reklaitis (Eds.), Proceedings of the 34th European Symposium on Computer Aided Process Engineering / 15th International Symposium on Process Systems Engineering (ESCAPE34/PSE24), June 2-6, 2024, Florence, Italy

A virtual entity of the digital twin based on deep reinforcement learning model for dynamic scheduling process

Jinglin Wang[a] , Jinsong Zhao[a,b]

[a]*Department of Chemical Engineering, Tsinghua University, Beijing 100084, China*
[b]*Beijing Key Laboratory of Industrial Big Data System and Application, Tsinghua University, Beijing, China*
jinsongzhao@tsinghua.edu.cn

Abstract

Digital twins (DT) are increasingly recognized as a transformative technology in the process industry. In complex chemical industrial systems, the management of operations is fraught with multiple steps, resource constraints, and dynamic events such as anomalies and equipment maintenance. Current scheduling methods for dynamic scenarios, though capable of handling these complexities, are primarily manual methods based on on-site information and are unable to respond to fault in time. These methods often result in slow and uneconomical outcomes and could not fit the uncertain and volatile production environment, especially process where safety is a problem of great concern. This paper introduces a DT framework that incorporates a deep reinforcement learning (DRL) model based on heterogenous graph neural networks (HGNN) as its core virtual entity, which could contract features from basic structure regardless of the constraints of size of problems. This HGNN-DRL design is also able to meet the needs of uncertainty and safety, fulfilling both of the requirements for full lifecycle adaptability and real-time interaction which is hard to effectively implement by traditional methods.

Keywords: Digital Twin, schedule, dynamic event, safety.

1. Introduction

Within chemical industrial parks, malfunctions in critical equipment can trigger cascading disruptions to production schedules, culminating in significant economic losses. Consequently, the implementation of a rational and effective maintenance plan is paramount for ensuring operational stability and profitability. For instance, disruptions in coal chemical industrial ports, such as equipment breakdowns, processing blockages, and conveyor mistracking not only induce production pauses but also incur demurrage charges and other financial penalties associated with delays.

The integration of smart manufacturing technology is propelling a new era of synergy between advanced information technology and various sectors of the manufacturing industry. Digital twin (DT) is considered one of the most promising technologies for realizing intelligent manufacturing which can achieve seamless integration and interaction between the physical and informational realms, causing increasing interest of chemical industries. There are documented instances of DTs in the realm of specific equipment and unit operations (Kender et al., 2022; Spinti et al., 2022; W. Wang et al., 2021) and on broader process scale (Koulouris et al., 2021; Perez et al., 2022), facilitating lifecycle monitoring and prediction as well as fault identification.

In contrast to the traditional approach of manually adjusting the schedule and introducing passive maintenance after failures, DT enable predictive maintenance and dynamic scheduling by monitoring equipment and diagnosing failures in advance through cyber-physical connections, and help to avoid the delays and losses caused by slow and unstable human response.

In DT implementations, employing machine learning techniques to create surrogate models has proven beneficial in reducing the duration of process simulations and decreasing the online computational load (Galeazzi et al., 2022). Among various machine learning approaches, deep reinforcement learning (DRL) is increasingly being utilized in production processes, showing distinct advantages in computation speed and optimization effectiveness as well as the ability to handle the flexibility and uncertainty in production processes (Gao et al., 2023; Panzer & Bender, 2022; X. D. Wang et al.; Yan et al., 2022). This work introduces a novel digital twin (DT) framework that comprehensively incorporates dynamic scheduling processes with integrated safety considerations. It details the development of a virtual entity encompassing a novel DRL algorithm, HGNN-DRL method, designed to simultaneously solve scheduling problems and provide real-time responses to dynamic events. The framework is applicable to diverse flexible job shop scheduling problems (FJSP), including transportation and some batch processes, as demonstrated by an example to laytime in coal chemical industrial ports.

2. Methodology

2.1. Proposed DT

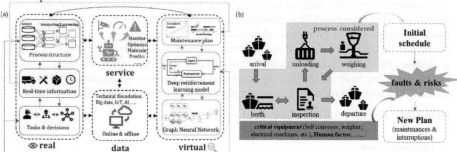

Figure 1 (a)DT framework. Bidirectional connections ensure synchronization. (b)An example of dynamic laytime schedule evolves in response to fault detection and risk assessment.

With five-dimensional modelling methodology (Fei Tao et al., 2019), this paper establishes the following framework which has 5 fundamental components: physical entity, virtual entity, connections, data, and services. The DT fitting into the category of a generic DT model in system-hierarchy (F. Tao et al., 2022). Dimensionally, our focus is on physical properties of key objects, as well as the behaviour and scheduling rules. Functionally, the emphasis is on implementing basic planning arrangements and completing maintenance plans based on timely information.

DT framework of real-time scheduling (Fig. 1a) adapts to processes by extracting base structure information. Its physical entity mirrors scheduling elements (storage, transport, etc.), and is modelled as a dynamic FJSP. The virtual counterpart, centred around a DRL model, is the core part of DT. It updates the problem information to a graph structure and extracts features to be used by an intelligent agent to generate an efficient schedule. In order to ensure safety, real-time measurement data should be fed into intelligent models to identify abnormal variables, and then develop maintenance plans by means of fault detection, diagnosis, and prediction (Bi et al., 2022).

Here, an example of laytime schedule in coal chemical industrial ports is elucidated. The workflow which can be described as FJSP is depicted in Figure 1b. During the laytime, transportation, manual inspection, unloading, weighing, and other tasks are initiated. Several factors such as berth occupation, device examination and faults on conveyors, weighers and other devices, impact the scheduling, thereby disrupting the former plans. DT monitors key variables (Figure 1b) and triggers plan updates upon detected faults. This enables proactive preventive maintenance at optimal intervals, minimizing machine downtime while the flexible model preserves schedule efficiency.

2.2. Scheduling model of the virtual entity

FJSP traditionally employs disjunctive graphs (Figure 2a) with operation nodes (O_{ij}), source, sink, conjunctive arcs (indicating job operation priorities), and disjunctive arcs (linking operations on the same machine). Solving FJSP involves selecting and directing input/output disjunctive arcs for each operation node. While various methods (mathematical programming, heuristics, metaheuristics) exist, DRL has recently demonstrated superior speed and performance (Panzer & Bender, 2022; Qin et al., 2023). DT serves as an interface linking the physical and digital worlds by representing the past, present, and predicting future states of physical objects, offers a unified framework integrating existing technologies to integrate functions, especially in adapting to dynamic scenarios and real-time changes. Although there are many solutions available to solve similar problems, we still need to select and design rapid response measures that meet the DT requirements.

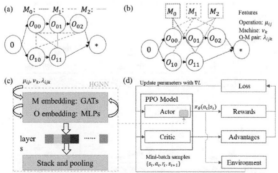

Figure 2 Information extraction and utilization with graphs. (a)Disjunctive graph. (b)Heterogenous graph. (c)HGNN. (d)PPO process

Considering the need for universality and rapid response in DT, one innovative DRL method based on heterogeneous graph neural networks (HGNN) (Song et al., 2023) from recent literature inspired us to construct a generic solution. This approach redefines the conventional disjunctive graph into a heterogeneous graph structure (Figure 2b). Extracting features by encoding operation nodes, machine nodes, and arcs via HGNN into specific dimensions help the model to accommodate samples of varying sizes. Then HGNN forms a part of the actor component of the proximal policy optimization (PPO) model which is capable of rapidly calculating FJSP overall outperforming other methods (Figure 2c). However, the model does not account for dynamic event insertion, limiting its ability to handle real-world scenarios like maintenance or unexpected incidents.

We expand upon the aforementioned method, transforming its application from static scenarios to dynamic ones. This enables rapid adaptation to fluctuations in production demands and unforeseen events, while incorporating equipment downtime and workflow interruptions. Consequently, the problem representation closely approximates real-world production situations. Figure 2d illustrates the learning process of the DRL model, showcasing the interactive loop between the PPO model agent and the environment. This environment is encoded using "OpenAI Gym 0.25.0" and the model is developed with "PyTorch 1.12.1".

3. Performance

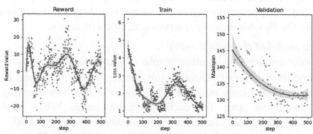

Figure 3 Model training and validation process.

The model is trained on randomly generated FJSP instances, with dynamic events applied to half of the machines. The duration of events d follows a normal distribution according to completion time CT. $\text{CT} = \max_i \sum_j p_{ij\bar{k}}, k = 0,1,\ldots m-1, d\sim N(\mu,\sigma^2)$, $\mu = 0.2\text{CT}, \sigma = 0.5\mu$. The model validation process is applied on a dataset provided by Song(Song et al., 2023).

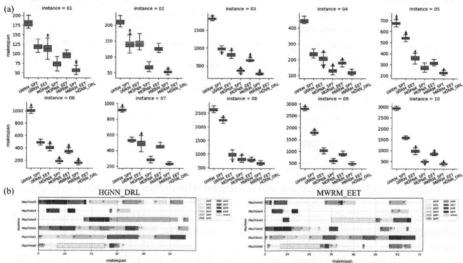

Figure 4 Tests on Brandimarte dataset. (a) Makespans calculated by different rules with insertions of dynamic events. (b) Mk01 instance with dynamic events.

To evaluate the performance of the proposed deep reinforcement learning model with samples of varying scales and characteristics, the model was assessed by comparison of other methods using FJSP instances. Meta-heuristic algorithms like the genetic algorithm and swarm intelligence algorithm are not effective for dynamic FJSP as they are designed for static manufacturing settings (Qin et al., 2023). Therefore, to evaluate the dynamic FJSP model, a comparison with heuristic rules is a viable approach. Some heuristic rules that combine machine dispatching rules like Shortest Processing Time (SPT) and Earliest End Time (EET) with job sequencing rules such as Most Operation Number Remaining (MOPNR), Most Work Remaining (MWKR) and Least Work Remaining (LWKR) are usually used in tests and show advantageous results(Lei et al., 2022).

We compared our DRL model with heuristic rules on a public dataset of 10 instances (Mk01 to Mk10) which vary in size and structure (Brandimarte, 1993). The aforementioned dynamic event generation scheme is employed to produce 100 randomly generated cases for each instance, upon which each method is applied to rigorously evaluate their performance under diverse scenarios. HGNN-DRL method demonstrated the capability to efficiently solve given problems, delivering responses within a few seconds (0.5 to 8 seconds). It outperformed heuristic rules on 9 instances but gave less efficient results than the Most Work Remaining and Earliest End Time rule on the small-scale Mk01 instance. Nonetheless, the DRL model provided robust and generally better plans for large-scale instances with more operations and longer duration (Figure 4a). Additionally, the solution of Mk01 instance with stochastic dynamic events is shown as an example (Figure 4b). Cause there is no standardized solution gap for dynamic FJSP, we compared mean makespan on each instance with MWRM-EET method, the heuristic rule with the smallest solution, and find an average improvement of 11%.

4. Conclusion

This paper establishes a generic DT framework for dynamic scheduling processes, applicable to various processes such as laytime schedule of ports and batch processes which can be described as similar structures. A novel HGNN-DRL model is proposed as the core of its virtual entity. By dynamically integrating real-time data and fault detection technology, it empowers the system to autonomously establish proactive maintenance plans and implement swift emergency responses to fault scenarios, optimizing efficiency and minimizing downtime. By means of numerical experiments, its capabilities and comprehensive performances that outperforming heuristic methods are demonstrated, which shows its potential to reduce waste and swiftly respond to equipment status as well as to design reasonable maintenance plans. The DT could be a high-efficient and inherent safety design for scheduling processes in chemical industries. To the best of the author's knowledge, this methodology has a little utilization in similar scenarios.

In future work, the DT can incorporate various constraints, such as energy consumption optimization (Mokhtari & Hasani, 2017), work pre-emption, and maintenance demands. Bi-directionally connection between the physical and virtual entities would be built. Additionally, advanced fault diagnosis approaches will be applied to more challenging real-world industrial cases, validating its flexibility and effectiveness that demonstrate application potential in chemical manufacturing field.

5. Acknowledgement

The authors gratefully acknowledge support from the National Science and Technology Innovation 2030 Major Project (2018AAA0101605) of the Ministry of Science and Technology of China and the National Natural Science Foundation of China (No.62003004).

References

X. Bi, Qin, R., Wu, D., Zheng, S., & Zhao, J. (2022). One step forward for smart chemical process fault detection and diagnosis. Computers & Chemical Engineering, 164, 107884. doi:https://doi.org/10.1016/j.compchemeng.2022.107884

P. Brandimarte. (1993). Routing and scheduling in a flexible job shop by tabu search. Annals of Operations Research, 41(3), 157-183. doi:10.1007/BF02023073

A. Galeazzi, Prifti, K., Gallo, F., & Manenti, F. (2022). A Methodology for The Optimal Surrogate Modelling of Digital Twins Using Machine Learning. In L. Montastruc & S. Negny (Eds.), Computer Aided Chemical Engineering (Vol. 51, pp. 1543-1548): Elsevier.

Y. P. Gao, Chang, D. F., & Chen, C. H. (2023). A digital twin-based approach for optimizing operation energy consumption at automated container terminals. Journal of Cleaner Production, 385. doi:10.1016/j.jclepro.2022.135782

R. Kender, Roessler, F., Wunderlich, B., Pottmann, M., Golubev, D., Rehfeldt, S., & Klein, H. (2022). Development of control strategies for an air separation unit with a divided wall column using a pressure-driven digital twin. Chemical Engineering and Processing-Process Intensification, 176. doi:10.1016/j.cep.2022.108893

A. Koulouris, Misailidis, N., & Petrides, D. (2021). Applications of process and digital twin models for production simulation and scheduling in the manufacturing of food ingredients and products. Food and Bioproducts Processing, 126, 317-333. doi:10.1016/j.fbp.2021.01.016

K. Lei, Guo, P., Zhao, W., Wang, Y., Qian, L., Meng, X., & Tang, L. (2022). A multi-action deep reinforcement learning framework for flexible Job-shop scheduling problem. Expert Systems with Applications, 205, 117796. doi:https://doi.org/10.1016/j.eswa.2022.117796

H. Mokhtari, & Hasani, A. (2017). An energy-efficient multi-objective optimization for flexible job-shop scheduling problem. Computers & Chemical Engineering, 104, 339-352. doi:https://doi.org/10.1016/j.compchemeng.2017.05.004

M. Panzer, & Bender, B. (2022). Deep reinforcement learning in production systems: a systematic literature review. International Journal of Production Research, 60(13), 4316-4341. doi:10.1080/00207543.2021.1973138

H. D. Perez, Wassick, J. M., & Grossmann, I. E. (2022). A digital twin framework for online optimization of supply chain business processes. Computers & Chemical Engineering, 166. doi:10.1016/j.compchemeng.2022.107972

Z. Qin, Johnson, D., & Lu, Y. (2023). Dynamic production scheduling towards self-organizing mass personalization: A multi-agent dueling deep reinforcement learning approach. Journal of Manufacturing Systems, 68, 242-257. doi:https://doi.org/10.1016/j.jmsy.2023.03.003

W. Song, Chen, X., Li, Q., & Cao, Z. (2023). Flexible Job-Shop Scheduling via Graph Neural Network and Deep Reinforcement Learning. IEEE Transactions on Industrial Informatics, 19(2), 1600-1610. doi:10.1109/tii.2022.3189725

J. P. Spinti, Smith, P. J., & Smith, S. T. (2022). Atikokan Digital Twin: Machine learning in a biomass energy system. Applied Energy, 310. doi:10.1016/j.apenergy.2021.118436

F. Tao, Xiao, B., Qi, Q. L., Cheng, J. F., & Ji, P. (2022). Digital twin modeling. Journal of Manufacturing Systems, 64, 372-389. doi:10.1016/j.jmsy.2022.06.015

F. Tao, Zhang, H., Liu, A., & Nee, A. Y. C. (2019). Digital Twin in Industry: State-of-the-Art. IEEE Transactions on Industrial Informatics, 15(4), 2405-2415. doi:10.1109/tii.2018.2873186

W. Wang, Wang, J., Tian, J., Lu, J., & Xiong, R. (2021). Application of Digital Twin in Smart Battery Management Systems. Chinese Journal of Mechanical Engineering, 34(1). doi:10.1186/s10033-021-00577-0

X. D. Wang, Hu, X. F., & Wan, J. F. Digital-twin based real-time resource allocation for hull parts picking and processing. Journal of Intelligent Manufacturing. doi:10.1007/s10845-022-02065-1

Q. Yan, Wang, H. F., & Wu, F. (2022). Digital twin-enabled dynamic scheduling with preventive maintenance using a double-layer Q-learning algorithm. Computers & Operations Research, 144. doi:10.1016/j.cor.2022.105823

Flavio Manenti, Gintaras V. Reklaitis (Eds.), Proceedings of the 34th European Symposium on Computer Aided Process Engineering / 15th International Symposium on Process Systems Engineering (ESCAPE34/PSE24), June 2-6, 2024, Florence, Italy

The assessment of LNG export scenarios for Qatar in the European Gas Market

Majed M. A. Munasser[a], Sabla Y. Alnouri[a] and Abdelbaki Benamor [a,b]

[a]*Gas Processing Center, College of Engineering, Qatar University, Doha, Qatar*
[b]*Department of Chemical Engineering, College of Engineering, Qatar University, P.O. Box 2713, Doha, Qatar*
sabla@qu.edu.qa

Abstract

The growing energy demand in Europe, combined with strained political relations with Russia due to the ongoing Ukrainian conflict, has prompted the European continent to seek alternative energy sources and reduce its dependence on Russian natural gas imports in the coming years. This need for diversification has become even more critical following disruptions to the pair of offshore pipelines, Nord Streams 1 and 2, which traverse the Baltic Sea and connect Russia to Germany. Therefore, it is imperative to conduct an in-depth analysis of the current situation in the EU to identify viable future strategies for diversifying its gas imports. This paper examines the current state of the LNG market in Europe and identifies the primary gas suppliers and importers in the European gas market. Based on this, a diversification plan which entails the EU establishing competitive, long-term gas import opportunities with Qatar is presented. There are multiple avenues through which Qatar can expand its LNG trade contracts with the EU, involving various routes. To determine the most optimal routes for natural gas exports, a Linear Program (LP) mathematical model was utilized. This model assists in choosing the best LNG export scenarios from Qatar to Europe, taking into account factors such as shipping logistics, European regasification terminals, and the existing network of pipelines for gas distribution throughout the European continent. The primary pipelines considered in this study for distribution are assumed to be under the operation of the European Network of Transmission System Operators for Gas (ENSTOG). Hence, multiple LNG shipping routes from Qatar to Europe in conjunction with major European regasification facilities were explored, followed by a distribution plan utilizing ENSTOG's pipeline infrastructure. The economic viability of the proposed export scenarios was assessed by incorporating several economic factors into the model's objective function. As such, several case studies for LNG export from Qatar to Europe were studied, by identifying the most efficient routes for shipping, regasification, and distribution using Europe's existing pipeline infrastructure. Additionally, introducing regasification routes through intermediary countries, such as Turkey, was also studied. The results suggest that Qatar has the capacity to meet up to a 50% disruption in Russian gas supplies to the European Union

Keywords: LNG; Europe; Qatar; Regsification; Pipeline

1. Introduction

As the global energy landscape undergoes a transformation, securing a consistent supply of natural gas has become increasingly vital. According to estimates from the International Energy Agency (IEA), natural gas currently contributes to 25% of the world's electricity generation and accounts for 22% of global energy consumption (IEA, 2022). Europe's natural gas and liquefied natural gas (LNG) consumption has reached 550 billion cubic meters per annum. In order to guarantee a continuous supply of gas resources and meet Europe's natural gas needs, strategic LNG suppliers must be chosen. Qatar plays a pivotal role in the golabl LNG export market, commanding nearly 35% of the global market share. Over the past decade, the balance of power between Russia and Qatar has been a significant factor in the global energy landscape, as noted Eser et al. (2019). The LNG sector has seen remarkable growth due to a shift toward cleaner energy, particularly natural gas. This trend is driven by increased energy demand and a reduced interest in carbon-intensive fuels. LNG's versatility has made natural gas more attractive on the international stage, enabling it to reach more locations, whereby unused gas reserves are being explored (Meza and Koc, 2021). In 2021-2022, global LNG imports reached 372.3 million tons, marking a 16.2 million-ton increase from the previous year (IEA, 2022). The annual growth rate of LNG imports has been steadily climbing at 4.5%, driven by economic recovery and the growing demand for cleaner, lower CO_2-emitting energy sources (IEA, 2022). LNG imports have the advantage of being less susceptible to political issues and control compared to traditional pipeline systems. However, effective utilization of LNG requires well-established infrastructure, including LNG terminals, regasification units, and pipeline networks for integrating regasified gas into existing transmission networks. Europe has a substantial reliance on external natural gas supplies. Recent challenges have emerged in Europe after 2020 and the beginning of 2021, fueled by increased European demand following the COVID-19 pandemic and the Russian-Ukrainian crisis. The political tensions stemming from the conflict have added pressure, particularly concerning the threat of disruption to pipeline supplies. In 2021-2022, European demand reached between 530 to 560 BCM, with a significant portion supplied by Russia, approximately 230 BCM (a combination of LNG and pipeline sources) (Meza et al., 2021). Pipelines accounted for around 155 BCM, passing through Poland, Nord Stream 1 in Ukraine, and the Turkish Stream. The majority of European gas is produced in Norway and distributed across Europe through an extensive network, covering approximately 22-24% of European demand. Regarding LNG supply to Europe, Qatar and the USA have emerged as primary suppliers, each contributing between 25-30 BCM of LNG. Russia, the US, Algeria, and Nigeria follow closely, supplying 17.8, 13.9, and 14 BCM, respectively. The USA is currently the leading supplier of LNG with 24.1 million metric tons per annum (MTPA), followed by Qatar with 15.9 MTPA (Ashkanani, and Kerbache, 2023). Replacing Russian natural gas is more challenging than substituting oil or coal due to differences in infrastructure, transportation, and storage. Moreover, Europe will continue to require significant natural gas imports, especially as most of Europe's pipeline infrastructure is designed for Russian gas imports (Dorigoni, and Portatadino, 2008). These alternative supplies will primarily need to arrive as LNG by sea. Ensuring supply security also necessitates maintaining sufficient gas storage levels so that European countries can withstand sudden interruptions in gas deliveries. In response to the Russian-Ukrainian conflict, the EU's strategy aims to diversify its natural gas sources by 2030, with a focus on importing more LNG. This presents an opportunity for Qatar, to enter the European market.

2. Methodology

This paper studies the current situation of the LNG market in Europe, and identifies the main gas suppliers and gas importers in the European gas market. Moreover, a diversification plan that involves Qatar taking this opportunity to present itself as a reliable gas supplier to Europe. There exist multiple prospects for Qatar to expand its LNG trade contracts with the EU, via different routes. In order to identify the best possible routes, a Linear Program (LP) mathematical model has been developed to help identify optimal LNG exporting scenarios from Qatar to Europe. The model accounts for shipping, regasification terminals in Europe, and existing network pipe structures that can be used for gas distribution within the European continent operated by the European Network of Transmission System Operators for Gas (ENSTOG, 2023) illustrated in Figure 1. Several LNG shipping routes from Qatar to Europe coupled with major European regasification units and the pipeline infrastructure via ENSTOG were utilized.

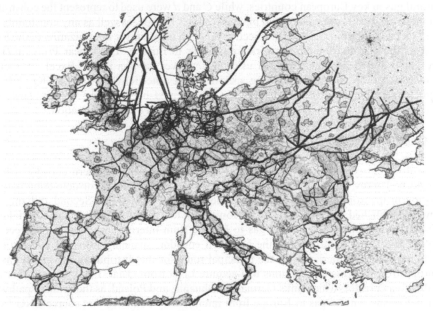

Figure 1: ENSTOG pipeline network via energypost.eu (ENSTOG, 2023)

A linear programming model was utlized in this work. It stands as one of the prevalent deterministic models employed for managing LNG supply and demand. The rationale for selecting this method for optimizing the LNG supply and demand problem is due to the simplicity and transparency that linear programming tools offer (Liberti and Kucherenko, 2005). Moreover, their computational efficiency is crucial for handling large-scale problems (Liberti and Kucherenko, 2005). The model is particularly suited for resource allocation and cost minimization objectives, representing various components of the LNG supply chain effectively. The availability of efficient solver tools, and the balanced trade-off between accuracy and complexity further support its adoption for optimizing LNG distribution networks (Liberti and Kucherenko, 2005). However, it should be noted that that linear programming may not capture all nuances of the real-world LNG market.

Within this model, the intricate interplay of LNG supply and demand is distilled into linear equations. The primary goal was to optimize the allocation of resources, specifically minimizing the overall cost while meeting specified demand constraints. This optimization model incorporates many factors, including shipping expenses,infrastructure costs, and production outlays. As such, the model was formulated according to Eq. (1-4).

$$Min. C^T x \qquad \qquad (1)$$
$$st. Ax = b \qquad \qquad (2)$$
$$st. Cx \geq d \qquad \qquad (2)$$
$$x \geq 0 \qquad \qquad (3)$$

Where x represents the LNG allocations from Qatar to Europe, as well as the pipeline allocations from regasification terminals across Europe to various destinations, C^T represents the various cost parameters (including shipping, regasification, regasification expansion, liquefaction, and storage). A and b were used to represent the demand for natural gas in key European countries, while C and d were used to represent the constrains on the regasification terminals and pipelines across Europe, as well as any constraints on the LNG supply from Qatar. The objective function involves the minimization of the total cost of the gas distribution network from Qatar to Europe. The Linear Program (LP) presented above was implemented using the "Solver" option in MS-Excel 2021 on a Windows laptop with the following specifications: Intel(R) Core i7-7820HQ CPU @ 2.90GHz 2.90 GHz, installed RAM 16.0 GB, and a 64-bit operating System. The code will be available for public access in an extended version of this publication.

3. Case Studies

Four different scenarios were assessed. First, employing the Southern corridor, which connects Turkey, Italy, Greece, and Croatia, as the primary route for delivering natural gas to Europe from intermediary nations receiving LNG from Qatar. Second, utilizing the newly established pipelines extending from Spain to Europe, such as the Midi Catalonia pipeline, for transporting natural gas to Europe from intermediary nations that receive LNG from Qatar. Third, using the Northern corridor, spanning from the UK through Belgium to the Netherlands, as the principal route for the distribution of natural gas to Europe from intermediary nations that acquire LNG from Qatar. Finally, opting for the Baltic Sea corridor, which links Germany, Lithuania, and Poland, as the primary pathway for delivering natural gas to Europe from intermediary countries that source LNG from Qatar. Each of the scenarios described above were conducted by introducing a Russian gas supply disruptions of 50%, and the respective allocations were then attained to meet the European demand based on the gas shortage identified. The capacities of the pipeline networks were acquired from ENTSOG, with a specific emphasis on countries located at cross-border points. Figure 2 below provides a visual representation. Moreover, the regasification capacities of the different countries involved in each scenario were used in the model, as summarized in Table 1.

Table 1: Summary of maximum regasification capacities used for Scenarios 1-4 (BP, 2022) & (IGU 2023)

Country	MTPA	Country	MTPA
Turkey	838	France	650
Italy & Greece	161	Belgium	161
Croatia	212	Germany	490
Spain	965	Poland	43
UK	219	Lithuania	88
Netherlands	159		

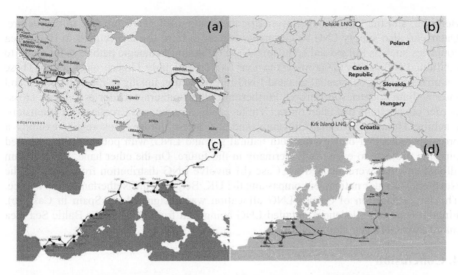

Figure 2: Visual representations of pipe corridors for Scenarios 1-4 including (a) Southern Corridor (b) Northern Corridor (c) Spain Corridor and (d) Baltic Sea Corridor

The model results have yielded preliminary LNG export strategies for Qatar. It has been determined that Qatar has the capacity to meet up to a 50% disruption in Russian gas supplies to the European Union. Figure 3 illustrates the Optimal percentage of LNG allocated per country with respect to the total LNG export preditions from Qatar to Europe for the different scenarios that were investigated.

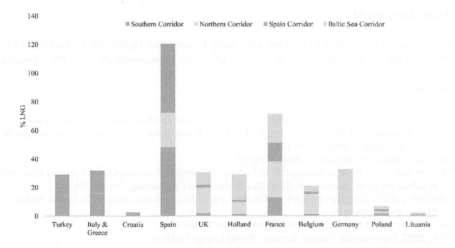

Figure 3: Optimal percentage of LNG allocated per country with respect to the total LNG export preditions from Qatar to Europe for all scenarios combined

To accommodate higher demand, Qatar would need to reconsider its existing contracts in Asian markets. Notably, Case (a) and (Case d) emerged as more cost-effective options when compared to the other scenarios studied. Despite this, Qatar can still utilize the Northern corridor to serve regions with increased demand, including the UK, France,

Belgium, and the Netherlands, leveraging the existing ENSTOG pipeline infrastructure. Qatar can also supply LNG to Germany, which, in turn, can distribute it to Baltic Sea countries such as Poland and Lithuania. Case (a) outlines strategic pathways for Qatar to facilitate LNG supply to Europe via nations like Turkey, Italy, Greece, and Croatia. These countries possess substantial gas and LNG infrastructure, linking to major pipelines and intermediate countries in Europe. Furthermore, the southern corridor establishes cross-border connections that interconnect Turkey, Italy, Greece, and Croatia, with extensions reaching Hungary, Austria, Slovenia, and the Swiss border. This region exhibits a promising demand outlook for both natural gas and LNG, with potential for proposed pipeline projects to even reach Germany in the future. On the other hand, the optimum allocations that were attained for Case (b) involve LNG distribution from Qatar to the Northern Corridor nations, encompassing the UK, Belgium, the Netherlands, and France. The largest portion of Qatar's LNG allocation was designated for Spain in Case (c). Finally, Case (d) allocations entailed LNG being sent from Qatar to the Baltic Sea area mainly via Germany and Poland.

4. Conclusion

A comprehensive mathematical model designed to optimize Qatar's natural gas exports to Europe. The model takes into account various critical factors including demand, supply, pricing, and transportation. It successfully integrates both economic and technical aspects of the natural gas industry, offering a holistic solution that minimizes the costs associated with meeting the European market demands. Four distinct case study scenarios for transporting natural gas from Qatar to Europe were explored, and significant implications   for   Qatar's   natural   gas   industry   were   discussed.

Aknowlegement

This paper was supported by Qatar university Project No. QUST-2-CENG-2023-1535. The findings achieved herein are solely the responsibility of the authors.

References

A. Meza, and M. Koç, 2021, The LNG trade between Qatar and East Asia: Potential impacts of unconventional energy resources on the LNG sector and Qatar's economic development goals. Resources Policy, 2021, 70, 101886

BP, 2022, Statistical Review of World Energy

D.A. Wood, 2016, Natural gas imports to Europe: The frontline of competition between LNG and pipeline supplies. Journal of Natural Gas Science and Engineering, 36, A1-A4

ENSTOG. 2023, European network of transmission system operators for gas. Available from: https://www.entsog.eu/

IGU, 2023, World LNG Report 2022. International Gas Union

IEA, 2022, International Energy Agency, World Energy Outlook

L. Liberti, and S. Kucherenko, 2005, Comparison of deterministic and stochastic approaches to global optimization, 12, 3, 263-285

P. Eser, N. Chokani, and R. Abhari, 2019, Impact of Nord Stream 2 and LNG on gas trade and security of supply in the European gas network of 2030. Applied Energy, 238, 816-830

S. H. Ashkanani, and L. Kerbache, 2023, Enhanced megaproject management systems in the LNG industry: A case study from Qatar. Energy Reports, 9, 1062-1076

S. Dorigoni, and S. Portatadino, 2008, LNG development across Europe: Infrastructural and regulatory analysis. Energy Policy, 36, 9, 3366-3373

Flavio Manenti, Gintaras V. Reklaitis (Eds.), Proceedings of the 34th European Symposium on Computer Aided Process Engineering / 15th International Symposium on Process Systems Engineering (ESCAPE34/PSE24), June 2-6, 2024, Florence, Italy

Hybrid deep learning model for evaluations of protein-ligand binding kinetic property

Yujing Zhao, Qilei Liu[*], Yu Zhuang, Yachao Dong, Linlin Liu, Jian Du, Qingwei Meng, Lei Zhang[*]

State Key Laboratory of Fine Chemicals, Frontiers Science Center for Smart Materials Oriented Chemical Engineering, Institute of Chemical Process Systems Engineering, School of Chemical Engineering, Dalian University of Technology, Dalian 116024, China
Qilei Liu (liuqilei@dlut.edu.cn), Lei Zhang (keleiz@dlut.edu.cn)

Abstract

A growing consensus is emerging that optimizing the binding kinetic parameters is crucial to ensure better drug efficacy. Therefore, *in silico* methods are necessary to predict the kinetic parameters of drug-protein binding. In this study, we demonstrate the application of an attention mechanism to derive a deep learning-based structure-kinetics relationship model for the dissociation rate constants (k_{off}) of inhibitors targeting drug-protein interactions. This model has the capability to provide accurate predictions for evaluating the k_{off} of protein-ligand complexes. To the best of our knowledge, the deep learning model developed in this work achieves the state-of-the-art prediction accuracy and outperforms other commonly used machine learning-based generic dissociation kinetic models (e.g., random forest models) in terms of precision and efficiency.

Keywords: drug binding kinetics, deep learning, dissociation kinetic constant, Mol2vec, attention mechanism.

1. Introduction

Drugs play a significant role in disease prevention and safeguarding public health. The process of discovering a new drug is an arduous and costly endeavour, often surpassing a duration of 10 years and requiring a budget exceeding one billion dollars. The binding/unbinding kinetic properties of protein-ligand complexes, such as the dissociation rate constant (k_{off}), are increasingly recognized as crucial factors influencing drug efficacy and safety during the drug optimization process. Although various experimental methods, molecular dynamics simulations, and surrogate modeling techniques have been developed to measure or estimate k_{off}, the current approaches, both experimental and *in silico*, suffer from limitations in providing high-throughput and accurate predictions. Therefore, this study aims to address this gap by introducing a novel data-driven dissociation kinetic model. This model employs a hybrid deep learning architecture that combines a convolutional neural network with the attention mechanism for the prediction of k_{off} values in protein-ligand complexes across multiple targets.

2. Development of the Dissociation Kinetic Model

2.1. Dissociation Kinetic Data Preparation

In this study, a comprehensive dissociation kinetic database is established by collecting dissociation kinetic data from four datasets, namely KIND (Schuetz et al., 2020), BindingDB (Gilson et al., 2016), PDBbind-koff-2020 (Liu et al., 2022), and the work by

Amangeldiuly et al. (2020). For the database construction, the dissociation kinetic data contributed by the aforementioned datasets are pre-processed based on the following criteria:

- Entries involving ligand SMILES that cannot be read by the RDKit tool (Landrum, 2006) or the ligand Morgen fragments cannot be found in the Mol2vec method (Jaeger et al., 2018) are deleted.
- Entries associated with protein-ligand complexes having a Morgen fragment number between 20 and 74 and an amino acid number between 40 and 500 are retained.
- Entries related to protein-ligand complexes with pk_{off} values ranging from 0 to 5 and without uncertain inequality signs (e.g., ">", "<") are retained.
- Entries involving protein-ligand complexes with more than one protein chain are excluded.
- Ligands conforming to Lipinski's "Rule of Five" are retained.
- Entries involving charged or ionic ligands are excluded.
- Duplicate entries are retained only once.

After applying the above criteria, a total of 2,716 unique protein-ligand pk_{off} values are obtained for our database, as well as the corresponding SMILES strings of ligands and FASTA strings of proteins. This dataset is divided into a training dataset (2,175 samples), a validation dataset (270 samples), and a test dataset (271 samples), following an 8:1:1 ratio, respectively. The pk_{off} values in our database cover a wide range from 0 to 5, representing the majority of cases in terms of pk_{off} values. A higher pk_{off} value indicates a longer duration of the inhibitor candidate binding to the protein. In **Figures 1(a-b)**, the distributions of the numbers of Morgan fragments for ligands and the amino acids for proteins in our database are presented, respectively. These distributions approximate the normal distribution, indicating that our dissociation kinetic database encompasses diverse samples of ligands and proteins.

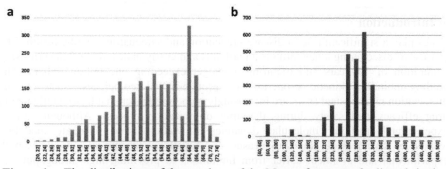

Figure 1. **a** The distributions of the numbers of the Mogan fragments for ligands in the established database. **b** The distributions of the numbers of the amino acids for proteins in the established database.

2.2. Input Representation

In our dissociation kinetic model, the protein-ligand complexes are represented using text-based strings because accessing their three-dimensional structures is time-consuming. The ligand is represented using the Mol2vec descriptor, which is a unique set of Morgan fragments with embedded feature vectors for each molecule. To obtain the Mol2vec descriptor, the SMILES string of the ligand is first transformed into fragments using the Morgan algorithm with a radius of 1. Then, a search approach is employed to embed the fragments with feature vectors, where the default dimension is 300. It is important to note that the Morgan fragments used in Mol2vec provide greater interpretability compared to plain text-based SMILES strings. Each ligand sample is represented by a two-dimensional matrix with dimensions of 74×300. The complete ligand feature is obtained by summing the vectors of individual Morgan fragments. For more details about the Mol2vec descriptor, please refer to Jaeger et al. (2018).

The protein is represented using higher-order amino acid strings with embedded feature vectors. Each higher-order amino acid consists of three adjacent amino acids in the protein sequence, which is provided in the FASTA format. This representation captures the interactions among amino acids. A total of 6,715 higher-order amino acid strings are generated from the proteins in our dissociation kinetic database. Each higher-order amino acid string is then embedded with a feature vector using the embedding method in the PyTorch library (Paszke et al., 2019). The feature size for each embedded vector is set to 30. Based on the higher-order amino acid strings, each protein sample is represented by a two-dimensional matrix with dimensions of 500×30. The complete protein feature is obtained by summing the vectors of individual higher-order amino acid strings.

2.3. Implementation of the Dissociation Kinetic Model

The architecture of the end-to-end text-based deep learning model for predicting pk_{off} is illustrated in **Figure 2**. Take a protein-ligand complex as an example. The deep learning model begins with the SMILES string of the ligand and the amino acid sequence (FASTA string) of the protein. These inputs are then transformed into the ligand matrix (Mol2vec descriptor: Morgan fragments with identifier features) and the protein matrix (higher-order amino acids with embedding features), respectively. For the ligand matrix, it undergoes feature extraction through a fully connected (FC) layer and an attention layer. On the other hand, the protein matrix is processed by a one-dimensional convolution (Conv) layer and a MaxPool layer to reduce the row size before being passed through an attention layer. This step helps reduce unnecessary computational costs in the attention layer. At last, the processed features of the ligand matrix are aggregated using a readout operation that adds up the matrix rows. The same process is applied to the protein matrix. The resulting features from both matrices are concatenated and passed through subsequent FC layers to predict the pk_{off} values for the protein-ligand complex.

3. Leverage a Hybrid Deep Learning Model for Dissociation Rate Evaluations

A novel data-driven dissociation kinetic model is developed in this work to predict pk_{off} using a hybrid deep learning architecture comprising a convolutional neural network and the attention layer. The hyperparameters (epoch, batch size, dropout rate, etc.) are determined through a trial-and-error approach based on our empirical knowledge. The loss function employed in our dissociation kinetic model is the mean squared error (*MSE*), commonly used in regression tasks. To minimize the loss function, we optimize the model parameters using the adaptive moment optimizer (Kingma and Ba, 2014), with the learning rates ranging from 0.0005 to 0.00005. A decay factor of 0.5 is applied to update the learning rates when the loss function of the validation set does not decrease with epochs. The evaluation criteria for assessing the prediction results of our dissociation kinetic model in estimating pk_{off} are the Pearson coefficient (*r*) and the root mean square error (*RMSE*). Higher values of *r* and lower values of *RMSE* indicate better predictive performance of the dissociation kinetic model. The deep learning-based dissociation kinetic model is trained 20 times to prevent fortuitously exceptional outcomes. The dissociation kinetic model is implemented using the PyTorch library (Paszke et al., 2019) and the Python language (Oliphant, 2007). The training and prediction results for pk_{off} of the deep learning model are shown in **Figure. 3**.

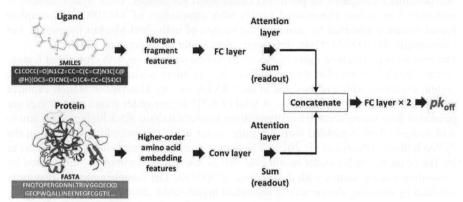

Figure 2. The architecture of the deep learning model.

Figure 3. Training and prediction results for pk_{off} of the deep learning model. **a** The loss function (*MSE*) varying with the epochs during the training process of the deep learning model. **b** The metrics function (*r*) varying with the epochs during the training process of the deep learning model. **c** The prediction results of the deep learning model for pk_{off}.

The *MSE* and *r* values of our deep learning model for the training, validation, test sets are as follows: 0.238±0.018, 0.299±0.020, 0.292±0.014 and 0.869±0.010, 0.838±0.012, 0.836±0.009, respectively. **Figures 3(a-b)** depict the variations of the loss function (*MSE*) and the metrics function (*r*) varying with the epochs during the training process of the deep learning model, respectively. These results demonstrate that our developed deep learning model does not suffer from overfitting and can effectively extrapolate to new samples. Furthermore, the incorporation of the attention mechanism accelerates the learning process of our model. The deep learning model achieves convergence at around 40th epoch when attention layers are utilized, whereas it requires more than 80 epochs to achieve similar results without attention layers. **Figure 3(c)** illustrates the prediction results of our optimal deep learning model on the test set, demonstrating an acceptable prediction ability with an *r* value of 0.840. This indicates that our deep learning model can provide qualitative evaluations of the dissociation kinetics of protein-ligand complexes. Importantly, our model is lightweight, with only 15,603 training parameters, compared to other affinity-based deep learning models such as KDEEP (Jiménez et al., 2018), which typically have around 1,000,000 training parameters. This highlights the efficiency of our model in terms of training and extrapolations.

To highlight the superior prediction accuracy of our deep learning model in estimating pk_{off} values, we compare it with other machine learning-based generic dissociation kinetic models described in the literature (**Table 1**). Amangeldiuly et al. (2020) compiled a database containing 501 protein-ligand complexes with experimental k_{off} from the public literature. They used this dataset to develop an optimal random forest (RF) algorithm, achieving an r value of 0.78 and *RMSE* of 0.82 on the validation set, and an r value of 0.75 and *RMSE* of 1.10 on the test set. Liu et al. (2022) created a database consisting of 680 experimental k_{off} values and their corresponding protein-ligand complex structures. They developed a general RF model based on this database to predict pk_{off} values. The final model was selected using the evaluation criteria of r and *RMSE* values: for the training set, $r = 0.968$ and *RMSE* = 0.474; for the validation set, $r = 0.706$ and *RMSE* = 0.986; and for the test set, $r = 0.501$ and *RMSE* = 0.891. The model developed by Liu et al. (2022) shows signs of overfitting, as indicated by its performance on the training, validation, and test datasets. On the other hand, it is unclear whether the model proposed by Amangeldiuly et al. (2020) suffers from overfitting, as they have not provided the performance on the training dataset. In comparison, our deep learning model demonstrates superior performance on the training set ($r = 0.869\pm0.010$ and *RMSE* = 0.488 ± 0.018), validation set ($r = 0.838\pm0.012$ and *RMSE* = 0.546 ± 0.018), and test set ($r = 0.836\pm0.009$ and *RMSE* = 0.540 ± 0.013). This improvement can be attributed to the advantages of our deep learning architecture and the abundance of our database. Moreover, our model enjoys faster computational speed compared to the aforementioned models, thanks to the utilization of text descriptors. In contrast, the calculations of descriptors in the previous models rely on three-dimensional structures of protein-ligand complexes. These analyses strongly support the conclusion that our deep learning model outperforms other known machine learning-based generic dissociation kinetic models in predicting pk_{off} values.

Table 1. Performance comparisons between the deep learning model developed in this work and the existing models in literature.

Model	Dataset	*r*	*RMSE*
This work	Training	0.869±0.010	0.488±0.018
	Validation	0.838±0.012	0.546±0.018
	Test	**0.836±0.009**	**0.540±0.013**
RF model (Amangeldiuly et al., 2020)	Training	Unknown	Unknown
	Validation	0.78	0.82
	Test	0.75	1.10
RF model (Liu et al., 2022)	Training	0.968	0.474
	Validation	0.706	0.986
	Test	0.501	0.891

4. Conclusions

A comprehensive dissociation kinetic database is constructed by gathering 2,716 unique k_{off} values for various ligands and targets from the literature. Utilizing this database, an end-to-end deep learning-based model is developed for rapid k_{off} predictions. Our deep learning model achieves superior performance compared with the existing machine learning-based generic dissociation kinetic models, as evidenced by the *MSE* and r values obtained on the training, validation, and test sets. Specifically, the *MSE* values are 0.238±0.018, 0.299±0.020, and 0.292±0.014, while the *r* values are 0.869±0.010, 0.838±0.012, and 0.836±0.009, respectively. These results demonstrate that our model surpasses other known machine learning-based generic dissociation kinetic models and achieves the state-of-the-art prediction accuracy in predicting pk_{off}.

Acknowledgement

The authors are grateful for financial supports of "the Fundamental Research Funds for the Central Universities" [DUT22YG218], China Postdoctoral Science Foundation [2022M710578], and National Natural Science Foundation of China [22208042].

References

T.E. Oliphant, 2007. Python for scientific computing. Computing in science & engineering, 9(3), pp.10-20.

S. Jaeger, S. Fulle, and S. Turk, 2018. Mol2vec: unsupervised machine learning approach with chemical intuition. Journal of chemical information and modeling, 58(1), pp.27-35.

N. Amangeldiuly, D. Karlov, and M.V. Fedorov, 2020. Baseline model for predicting protein–ligand unbinding kinetics through machine learning. Journal of Chemical Information and Modeling, 60(12), pp.5946-5956.

M.K. Gilson, T. Liu, M. Baitaluk, G. Nicola, L. Hwang, and J. Chong, 2016. BindingDB in 2015: a public database for medicinal chemistry, computational chemistry and systems pharmacology. Nucleic acids research, 44(D1), pp.D1045-D1053.

J. Jiménez, M. Skalic, G. Martinez-Rosell, and G. De Fabritiis, 2018. K deep: protein–ligand absolute binding affinity prediction via 3d-convolutional neural networks. Journal of chemical information and modeling, 58(2), pp.287-296.

H. Liu, M. Su, H.X. Lin, R. Wang, and Y. Li, 2022. Public data set of protein–ligand dissociation kinetic constants for quantitative structure–kinetics relationship studies. ACS omega, 7(22), pp.18985-18996.

G. Landrum, 2006. RDKit: Open-source cheminformatics. 2006. Google Scholar.

D.P. Kingma, and J. Ba, 2014. Adam: A method for stochastic optimization. arxiv preprint arxiv:1412.6980.

D.A. Schuetz, L. Richter, R. Martini, and G.F. Ecker, 2020. A structure–kinetic relationship study using matched molecular pair analysis. RSC Medicinal Chemistry, 11(11), pp.1285-1294.

A. Paszke, S. Gross, F. Massa, A. Lerer, J. Bradbury, G. Chanan, T. Killeen, Z. Lin, N., Gimelshein, L. Antiga, and A. Desmaison, 2019. Pytorch: An imperative style, high-performance deep learning library. Advances in neural information processing systems, 32.

Flavio Manenti, Gintaras V. Reklaitis (Eds.), Proceedings of the 34th European Symposium on Computer Aided Process Engineering / 15th International Symposium on Process Systems Engineering (ESCAPE34/PSE24), June 2-6, 2024, Florence, Italy

Information sharing for cost-effective risk mitigation in supply chains: A methodological framework

Shivam Vedant[a,b], Rahul Kakodkar[a,c], Natasha J. Chrisandina[a,c], Catherine Nkoutche[a,c], Eleftherios Iakovou[a,b,d], Mahmoud M. El-Halwagi[a,c], Efstratios N. Pistikopoulos[a,b,c] *

[a]*Texas A&M Energy Institute, Texas A&M University, College Station, TX 77845*
[b]*Department of Multidisciplinary Engineering, Texas A&M University, College Station 77840*
[c]*Artie McFerrin Department of Chemical Engineering, Texas A&M University, College Station 77840*
[d]*Department of Engineering Technology and Industrial Distribution, Texas A&M University, College Station, TX 77843*
stratos@tamu.edu

Abstract

Process systems and their supply chains are becoming increasingly susceptible to a plethora of exogenous factors that can induce disruptions. Information from supply chain stakeholders regarding an impending disruption can help decision-makers plan operation schedules to minimize their impact and improve supply chain resilience. Multiscale modeling and optimization frameworks often assume perfect information. Nevertheless, this assumption fails to capture the value of information sharing about potential upcoming disruptions from supply chain stakeholders who often have precious and timely advanced knowledge. Notably, there can be a lag in information flows sharing which can impede the ability of decision-makers to plan for and then respond to an impending change of state. Lack of timely and effective information sharing among stakeholders erodes supply chain resilience. To this end, we present a first effort mathematical programming-based methodological framework to quantify the impact of information sharing for supply risk mitigation. In the multiperiod modeling framework, the information set is made available to the decision-makers at different time periods in the planning horizon. The nominal scenario, wherein information is made available at the start of the temporal horizon is compared to scenarios wherein the information is made available at later stages. In doing so, the economic benefit of sharing information is quantified, and managerial insights are obtained.

Keywords: information flow, supply chain optimization, multiscale modeling, supply chain risk mitigation, resilience

1. Introduction

Supply chains play a crucial role in facilitating the production and distribution of essential (and often critical for national security) commodities and products, ranging from energy carriers and specialty chemicals to consumer goods. Chemical and energy supply chains comprise interconnected chemical process plants that encompass various unit operations, while also establishing multi-echelon connections with key stakeholders such as suppliers, distributors, third party logistics operators (3PLs), industrial customers, and consumers alike. However, geographically dispersed chains are increasingly exposed to exogenous factors which renders them vulnerable to disruptions occurring across multiple operational scales, resulting in significant human and economic consequences (El-Halwagi et al., 2020) and reduced supply chain resilience (Iakovou and White, 2020). Incidents such as the Bhopal gas leak (1984), the Deepwater Horizon oil spill (2010), and the Beirut explosion (2020), have underscored the inherent risks posed by operational discord. Additionally, global "black swan" disruptions, including the COVID-19 pandemic, natural disasters, wars, tariff wars, and labour shortages continue to have a profound impact on business continuity, leading to significant financial losses, disruptions in production, and escalated costs of critical consumer goods. New modelling approaches are needed to design cost-competitively resilient supply chains and assess the relevant trade-offs ("optionalities") between cost efficiency and resilience (Gopal et al., 2023).

Multiscale modeling and optimization can be applied to plan and manage end-to-end supply chains with awareness to phenomena occurring at disparate spatiotemporal scales. Decisions taken at different temporal levels can be categorized into strategic (long-term), tactical (mid-term), and operational (short-term) (Shapiro, 1999). Often, supply chain stakeholders (such as an upstream supplier at a faraway location) have intimate knowledge and local insights about upcoming regionalized disruptions, such as labor strikes, resource shortages, social upheaval, and manufacturing capacity reductions. While the flow of resources and information have been modeled under an integrated purview (Perez et al., 2023), the role of efficient and timely information sharing during pre, and post stages of disruptions is not well studied. To this end, we present a first effort methodology to gauge the impact of information sharing for upcoming disruption scenarios. Operational and scheduling decisions for a fixed network-design are analyzed under varying levels of access to information. The multiperiod problem is formulated as a linear program (LP) and different scenarios are compared where in the decision-maker has: 1) early and perfect information regarding upcoming disruptions; 2) delayed access to information at various temporal levels; 3) no insight until the disruption occurs. The primary goals of the framework are to: 1) quantitatively assign an economic value to information sharing; and 2) determine the optimal time epoch of sharing information. The system is modeled in the *energiapy* python package (Kakodkar and Pistikopoulos, 2023).

2. Methodology

The developed framework utilizes a rolling-horizon multiperiod methodology to model the sharing of information and the corresponding adjustments for supply chain recovery. At the beginning of each period, the decision-maker has perfect information on disruptions and operational parameters (for that individual period) and has an anticipated dataset for the following periods which may be updated through information sharing. To model the temporal flow of information, the information set is updated at discrete time epochs of the temporal horizon. For example, in the perfect information (nominal) scenario, the decision-maker has access to the entire information set about the upcoming disruption at the start of the planning period itself. In scenarios with time lag, the decision-maker gains access to updated information from other supply chain stakeholders only in later time periods. In terms of mathematical modeling, this corresponds to solving distinct problems for varying partitions of the temporal horizon: few with the expected information set (P1) and others with the updated information set (P2). Notably, the end state of the problem with expected information (P1) is used to initialize the latter (P2) to maintain continuity. For instance, in a scenario where the supply chain tiers obtain revised information at time period $K+1$ concerning a deviation occurring at time period D from the anticipated dataset, the state variables (at time period $K+1$) from problem (P1) can be employed to initialize problem (P2) with the updated information set (Figure 1). Subsequently, problem (P2) is solved for time periods $K+1$ to N with the updated information set. Furthermore, it is assumed that once the disruption occurs, the information is immediately made available across the supply chain stakeholders.

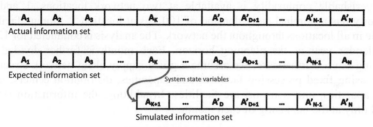

Figure 1. Modeling methodology for information flow

The system is represented using the resource task network (RTN) methodology and modeled as a multiperiod LP. Notably, to incorporate both partial and complete failures normalized capacity factors are introduced to processes and transports to capture the effects of supply chain disruptions (Eqns. 1 & 2). Notations and descriptions for parameters and variables reported in the rest of the manuscript are shown in Table 1.

Table 1. Notations and Descriptions for parameters and variables

Notation	Description
$\mathcal{R}^{trans}$	Set of resources that can be transported
$P_{l,p,t}$	Production level of p
$Cap^{f-P}_{l,p,t}$	Capacity factor for the production levels of process p
$Cap^{P}_{l,p,t}$	Installed production capacity for process p
$Exp^{F}_{f,l,l',r,t}$	Amount exported between sources and sinks for an individual resource
$Cap^{f-F}_{f,l,l',t}$	Capacity factor for export between locations
$Cap^{F}_{f,l,l',t}$	Installed export capacity of a transport mode
$Vopex^{N}_{t}$	Operating costs for the network
Φ^{N}_{t}	Penalty incurred due to unmet demand for the network
$InvCost^{N}_{t}$	Inventory storage costs for the network

$$P_{l,p,t} \leq Cap^{f-P}_{l,p,t} \cdot Cap^{P}_{l,p,t} \tag{1}$$

$$\sum_{r \in \mathcal{R}^{trans}} Exp^{F}_{f,l,l',r,t} \leq Cap^{f-F}_{f,l,l',t} \cdot Cap^{F}_{f,l,l',t} \tag{2}$$

The objective function is the minimization of total cost over the finite planning horizon. This includes the operating costs (for processing, maintaining inventory, and transport), and penalties incurred due to unmet demand. Since, this is a distribution planning problem and no value is being added to the commodity within the network, procuring and selling prices are assumed to be fixed. The cost objective is thus given by Eq 3:

$$\sum_{t \in \mathcal{T}^{network}} Vopex_t^N + InvCost_t^N + TransCost_t^N + \Phi_t^N \tag{3}$$

3. Motivating Example

Figure 2 depicts the distribution planning problem (DPP) adapted with modifications that is based on an indicative tactical problem proposed and studied by Ivanov et al. (2014). A non-perishable commodity is available at two network locations, 1 and 6. The commodity has a nominal daily demand of 100 kgs at location 5. Storage facilities are available in all locations throughout the network. The analysis is conducted for 12 months of a calendar year as the planning horizon. Each month is further divided into 30 scheduling days. Furthermore, we assume a given supply chain network structure, i.e., encompassing fixed processing facilities and capacities, distribution centers, inventory buffer capacities, and transport arc facilities. In addition, the information set can be updated only at the beginning of each month.

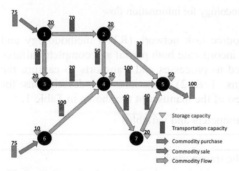

Figure 2. Case Study - Distribution Network

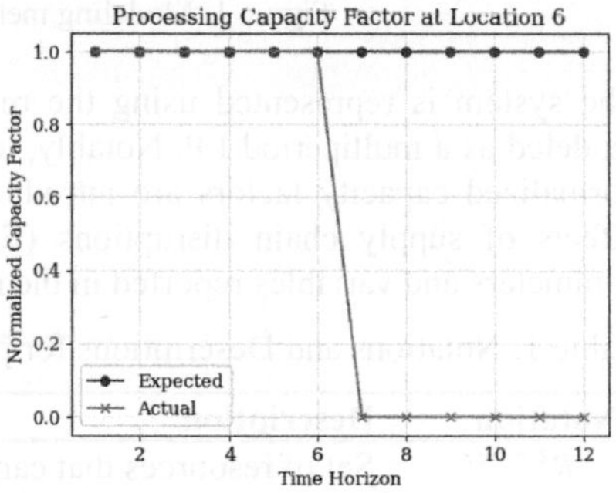

Figure 3. Case Study - Expected vs Actual resource availability

Information Sharing Scenarios

The processing facility at location 6 is going to experience a catastrophic failure at the beginning of month 7. To compare the cost of information sharing, the overall cost is obtained for three scenarios:

1) Scenario 1 - Information is shared at the beginning of month 1;
2) Scenario 2 - Information is shared at the beginning of month 6; and
3) Scenario 3 - Information is shared at the beginning of month 7 when the disruption occurs.

To assess the resilience of the supply chain network, we are using the overall demand fill rate as a key performance indicator:

$$Fill\ Rate = \frac{Total\ demand\ satisfied\ over\ the\ planning\ horizon}{Total\ demand\ over\ the\ planning\ horizon}$$

4. Results and Discussions

Initially, the DPP is solved with the expected information set (problem P1) to obtain the system state variables (inventory levels) values at the beginning of months 6, and 7. The initial inventory values for all locations were zero. This follows from the fact the under the assumption of no disruption, the system is fully capable of meeting the daily demand without storing the commodity. Using these values, instances of problem (P2) are initialized for their respective scenarios 2, and 3. The results obtained from the four scenarios are reported below:

Table 2. Cost performances for different scenarios

Scenario	Total Cost ($USD)	Fill Rate (%)
Scenario 1	657,231.25	88
Scenario 2	657,231.25	88
Scenario 3	662,274.0	87.5

Results show that Scenario 3 incurs the highest cost, as the information set is updated concurrently with the disruption event, hindering supply chain risk mitigation and response. Furthermore, Scenario 3, has a lower demand fulfilment (fill) rate indicating lower supply chain resilience due to lack of timely information. Higher costs for Scenario 3 are due to the incurred penalty resulting from unsatisfied demand. However, Scenarios 1 and 2 exhibit equitable cost performance and demand fill rates. The observed behaviour suggests a temporal threshold (at month 6) after which when information is shared, the overall cost increases. It can also be seen from the inventory build-up results that the network does not start building up inventories until after month 6 (Figure 4). This observation suggests that sharing information earlier than that, does not result in a competitive advantage. This managerial insight could provide valuable guidance for decision-makers to determine a deadline for information sharing to avoid incurring additional penalties.

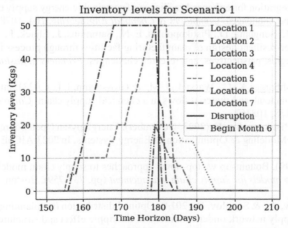

Figure 4. Inventory levels at network locations during Scenario 1

5. Conclusions and Future Directions

In this work, we have presented a first effort towards a methodological framework to model the value of information sharing about upcoming disruptions in the supply chains.

The approach involves partitioning the problem into two parts: before (P1) and after the information is received (P2). After receiving the latest information regarding an upcoming disturbance, the optimization problem (P2) is constructed by utilizing the system state variables (such as inventory, product flows, etc.) from the initial formulation (P1) for initialization. The results show that early sharing information not only has economic benefits, but also enables higher supply chain performance. Moreover, the system can accommodate some lag in information transfer without incurring a high penalty. Thus, the framework allows for the quantitative assessment of not only the cost of information sharing but also provides an ideal time of sharing critical information. In the near future, we intend to expand the presented work by including the early sharing of probability-based disruption scenarios in the analysis through stochastic modeling.

6. Acknowledgements

Research was sponsored partially by the ARM (Advanced Robotics for Manufacturing) Institute through a grant from the National Institute of Standards and Technology (NIST) and was accomplished under Agreement Number W911NF-17-3-0004. The views and conclusions contained in this document are those of the authors and should not be interpreted as representing the official policies, either expressed or implied, of the Office of the Secretary of Defense or the U.S. Government. The U.S. Government is authorized to reproduce and distribute reprints for Government purposes notwithstanding any copyright notation herein. Furthermore, the authors acknowledge the financial support from the Texas A&M Energy Institute.

7. References

Iakovou, E., & White, C. (2020). How to build more secure, resilient, next-gen US supply chains. *Brookings Institute Tech street*. https://www.brookings.edu/articles/how-to-build-more-secure-resilient-next-gen-u-s-supply-chains/

Chrisandina, N. J., Vedant, S., Iakovou, E., Pistikopoulos, E. N., & El-Halwagi, M. M. (2022). Multi-scale integration for enhanced resilience of sustainable energy supply chains: Perspectives and challenges. *Computers & Chemical Engineering*, 164, 107891.

El-Halwagi, M. M., Sengupta, D., Pistikopoulos, E. N., Sammons, J., Eljack, F., & Kazi, M. K. (2020). Disaster-resilient design of manufacturing facilities through process integration: principal strategies, perspectives, and research challenges. *Frontiers in Sustainability*, 1, 595961.

Perez, H. D., Harshbarger, K. C., Wassick, J. M., & Grossmann, I. E. (2023). Integrating information, financial, and material flows in a chemical supply chain. *Computers & Chemical Engineering*, 178, 108363.

Kakodkar, R., & Pistikopoulos, E. (2023, November). Energiapy-an Open Source Python Package for Multiscale Modeling & Optimization of Energy Systems. In 2023 AIChE Annual Meeting. AIChE.

Shapiro, J. F. (1999). Bottom-up vs. top-down approaches to supply chain modeling. In *Quantitative models for supply chain management* (pp. 737-759). Boston, MA: Springer US.

Ivanov, D., Pavlov, A., & Sokolov, B. (2014). Optimal distribution (re) planning in a centralized multi-stage supply network under conditions of the ripple effect and structure dynamics. European Journal of Operational Research, 237(2), 758-770.

Gopal, C., Tyndall, G., Partsch, W., & Iakovou, E. (2023). Breakthrough Supply Chains: How Companies and Nations Can Thrive and Prosper in an Uncertain World. McGraw Hill Professional.

Flavio Manenti, Gintaras V. Reklaitis (Eds.), Proceedings of the 34th European Symposium on Computer Aided Process Engineering / 15th International Symposium on Process Systems Engineering (ESCAPE34/PSE24), June 2-6, 2024, Florence, Italy

Optimization of Operating Conditions Using a Crystallizer Model with Local Temperature Control

Saki Iizuka[a], Kano Ishikawa[a], Sanghong Kim[a]

[a]*Tokyo University of Agriculture and Technology, 2-24-16, Naka-cho, Koganei-shi, Tokyo, 184-8588, Japan*
sanghong@go.tuat.ac.jp

Abstract

In the crystallization processes, our primary topics are the quality and the productivity of crystalline particles. In this study, local temperature control was focused to investigate how much the quality and the productivity can be improved. A two-dimensional distributed parameter system model of a batch cooling crystallizer with local temperature controllers was developed. A multi-objective optimization problem was solved using the model. It was found that local temperature control reduced the operation time and the control error of particle size up to 14.4 % and 44.2 %, compared to constant cooling at 0.30 W, respectively without worsening the other objective function.

Keywords: Modeling, Crystallization, Distributed parameter system, Local control

1. Introduction

Crystallization is a chemical process by which crystals are formed from solution. In chemical and pharmaceutical industries, crystallization has been used for separation and purification. In the crystallization processes, the quality and the productivity of crystalline particles are primary interests. The quality can be characterized by particle size distribution and monodisperse crystalline particles with desired size is preferred.

In order to increase the productivity of particles with desired particle size distribution, Ma (2021) reported pulsed ultrasound enhanced continuous reactive crystallization. When the ultrasonic probe was inserted into single and multistage mixed suspension mixed product removal (MSMPR) crystallizer, particles with narrow particle size distribution could be obtained. In order to increase the productivity without worsening the product quality, Mesbah (2011) reported a model-based control approach of a semi-industrial batch evaporative crystallizer. The heat input to the crystallizer was manipulated by the state estimation in the feedback control system. Pascual (2022) reported that the high yield of particles was produced by properly setting the residence time and the agitation rate in the continuous cooling crystallization using a MSMPR system.

Most studies mentioned that local conditions affect crystallization. However, these previous studies have not focused on manipulation of local conditions in the crystallizer. For example, Mesbah (2011) defined only a heat input for a crystallizer, and it was assumed that the crystallizer was heated uniformly with a heat input. When local conditions are manipulated, monodisperse particles with desired size can be obtained without worsening the productivity. Therefore, in this study, to find better crystallization operations, a simulation system in which local conditions could be manipulated was constructed, and local operations were optimized.

2. Crystallizer model and local operation

Two-dimensional distributed parameter system (DPS) model of a batch cooling crystallizer with imaginary local temperature controllers was constructed.

2.1. Distributed parameter system model

DPS model is useful to find better crystallization operations, because DPS model can calculate conditions in the crystallizer more accurately than lumped parameter system model. To construct DPS model, a target system was divided into multiple units.

2.2. Crystallization model

Cooling crystallization is one of typical methods of crystallization. In the developed model, heat transfer and crystal growth were considered. Governing equations of heat transfer and crystal growth shown in Eqs. (1) to (4) (Perry at al., 1997) (Kim et al., 2023) were embedded in each unit.

$$\frac{dT}{dt} = -\frac{1}{C}(Q_{out} - Q_{in}) \tag{1}$$

$$\frac{dr}{dt} = G \tag{2}$$

$$G = \begin{cases} k_g \exp\left(\frac{-E_a}{RT}\right)(c - c_s)^\gamma & (c \geq c_s) \\ 0 & (c_s > c) \end{cases} \tag{3}$$
$$\tag{4}$$

Here, T [K] is temperature, t [s] is time, C [J/K] is heat capacity, Q_{in} [W] is heat input rate, Q_{out} [W] is heat output rate, r [μm] is particle size, G [μm/s] is crystal growth rate, k_g [(μm/min) (g-solute/g-solvent)$^{-\gamma}$] is rate constants for the crystal growth, E_a [J/mol] is activation energy of the crystal growth, R [J/mol/K] is universal gas constant, c [g-solute/g-solvent] is concentration in the solution, c_s [g-solute/g-solvent] is solubility, and γ [-] is exponential parameters on supersaturation for the crystal growth.

2.3. Local operation

In this study, it was assumed that temperature of each unit can be manipulated by an imaginary controller. To represent the local temperature controller, local heat transfer rate Q_c [W] was added to Eq. (1) as shown in Eq. (5). Temperature of each unit can be controlled by Q_c. Q_c was optimized when optimization problems were solved.

$$\frac{dT}{dt} = -\frac{1}{C}(Q_{out} - Q_{in} + Q_c) \tag{5}$$

3. Case study

The proposed model was evaluated through an example. In this example, a multi-objective optimization problem was solved, because there is a trade-off between the quality and the productivity of particles. One of objective functions f_1 is standardized operation time t_f/t_{ref} [-] as shown in Eq. (7), which related to the productivity. t_f [s] is operation time. t_{ref} [s] is representative value of operation time: 60 s. The other objective function f_2 is standardized control error of particle size [μm] as shown in Eq. (8), which related to the quality of crystalline particles. r_{ref} [μm] is representative value of particle size: 1 μm. N [-] is the number of particles, and r_{set} [μm] is particle size setpoint. When the problem was solved, scalarizing was used as shown in Eq. (6). w_1 [-] and w_2 [-] are weights.

$$f(\boldsymbol{Q}_{c,all}) = w_1 f_1 + w_2 f_2 \tag{6}$$

$$f_1 = \frac{t_f}{t_{ref}} \tag{7}$$

$$f_2 = \frac{\sum_{n=1}^{N} \sqrt{\frac{1}{N}(r_n - r_{set})^2}}{r_{ref}} \tag{8}$$

3.1. Simulation conditions

In this example, simulation conditions were set referencing a previous study (Kim et al., 2023). A target system was divided into three units, and each unit is called upper, middle, and lower unit. Initial particle sizes in each unit were set as shown in Table 2. Particles in each unit were monodisperse. Quality of initial particles were low, because the initial particle sizes were different between each unit, and it meant particles in the target system were polydisperse. The range of Q_c was from 0 to 0.90 W. The value of Q_c was changed every 20 minutes. The values of parameters and other conditions were shown in Table 3.

$$c_s = -8.707 + 9.669 \times 10^{-2}T - 3.610 \times 10^{-4}T^2 + 4.590 \times 10^{-7}T^3 \tag{9}$$

Table 2 Initial particle sizes in each unit

Unit name	Particle size r [μm]	Number of particles N [-]
Upper	100	30,000
Middle	150	30,000
Lower	200	30,000

Table 3 The values of parameters and conditions

Parameter name	Value
Target system size (width, depth, height)	9 cm, 3 cm, 9 cm
Mesh size (width, depth, height)	9 cm, 3 cm, 3 cm
Time step of crystal growth and heat transfer calculation	5 s
Production volume set point	15.0 g
r_{set}	500 μm
Initial temperature	313.15 K
k_g	2.26×10^8 (μm/min) (g-solute/g-solvent)$^{-\gamma}$
E_a	3.62×10^4 J/mol
R	8.31 J/mol/K
c	0.30 g-solute/g-solvent
γ	1.14
C	53.04 J/K

3.2. Results and discussion

Values of the objective functions for feasible solutions, pareto optimal solutions, and solutions with constant cooling rates, which were 0.15, 0.30, 0.45, 0.60, 0.75, and 0.90 W, were shown in Fig. 1. It was found that objective functions became smaller by local temperature control than those with constant cooling rate in all units. Compared to constant cooling at 0.30 W, which was represented by the second dot of solution with constant cooling from the right in Fig 1, local temperature control reduced operation time and control error of particle size up to 14.4 % and 44.2 %, respectively without worsening the other objective function.

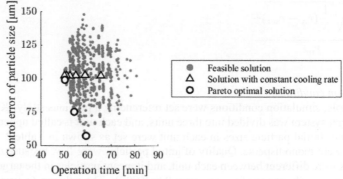

Fig. 1 Values of the objective functions

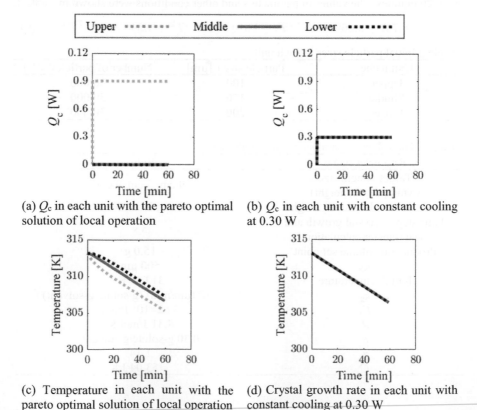

(a) Q_c in each unit with the pareto optimal solution of local operation

(b) Q_c in each unit with constant cooling at 0.30 W

(c) Temperature in each unit with the pareto optimal solution of local operation

(d) Crystal growth rate in each unit with constant cooling at 0.30 W

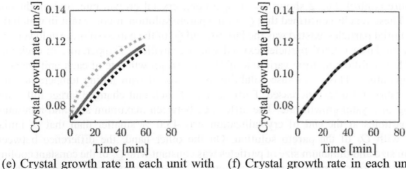

(e) Crystal growth rate in each unit with the pareto optimal solution of local operation

(f) Crystal growth rate in each unit with constant cooling at 0.30 W

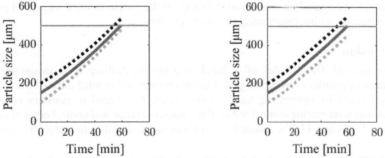

(g) Particle size in each unit with the pareto optimal solution of local operation

(h) Particle size in each unit with constant cooling at 0.30 W

Fig.2 Simulation results of the example

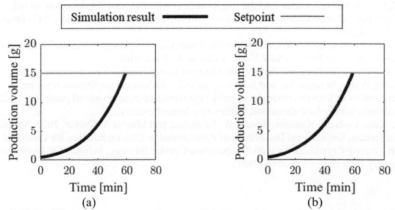

Fig. 3 Production volume (a) with a pareto optimal solution of local operation (b) with constant cooling at 0.30 W

Next, the detailed information on one of pareto optimal solutions, which is represented by the first dot of pareto optimal solutions from the right of in Fig 1, and one of solutions with constant cooling at 0.30 W were compared to show the effectiveness of local temperature control. Fig. 2 shows Q_c, temperature, crystal growth rate, and production volume. These results confirmed that Q_c of the pareto solution at upper unit in which the smallest initial particles existed was the largest, and Q_c of the pareto solution at lower unit in which the largest initial particles existed was the smallest. Temperature of each unit changed lower following time variation of Q_c. Crystal growth rate of each unit changed with temperature. This is because solubility is decreased with falling temperature and supersaturation ratio was increased. Particle size of each unit changed larger following increasing of crystal growth rate. The difference between maximum size and minimum size of particles at the end of crystallization was 43.0 % smaller than that of initial particles with Q_c of the pareto solution. On the other hand, the difference between maximum size and minimum size of particles was constant with Q_c of the constant cooling. Then, control error of particle size at the end of crystallization with Q_c of the pareto solution was 44.2 % smaller than that with Q_c of the constant cooling. Fig. 3 shows production volume, which confirmed that operation times of each simulation result were almost equal. As a result, it was showed that Q_c of the pareto solution improved particle size distribution, without worsening the operation time.

4. Conclusion

Two-dimensional DPS model of a batch cooling crystallizer with imaginary local temperature controllers was developed. Operation time and control error of particle size were minimized by optimizing local cooling rates. Compared to constant cooling at 0.30 W, local temperature control reduced the operation time and control error of particle size up to 14.4 % and 44.2 %, respectively without worsening the other objective function.

References

Ali Mesbah, Adrie E. M. Huesman, Herman J. M. Kramer, Zoltan K. Nagy, and Paul M. J. Van den Hof, 2011, Real-time control of a semi-industrial fed-batch evaporative crystallizer using different direct optimization strategies, AIChE Journal, 57, 6, 1557-1569

Gladys Kate Pascual, Philip Donnellan, Brian Glennon, Barbara Wood, and Roderick C. Jones, 2022, Design and optimization of the single-stage continuous mixed suspension-mixed product removal crystallization of 2-Chloro-N-(4-methylphenyl)propenamide, ACS Omega, 7, 16, 13676-13686

Perry, Robert H., Green, Don W., and Maloney, James O., 1997, Perry's Chemical Engineers' Handbook Seventh Edition, New York, America: McGraw-Hill.

Yiming Ma, Zhixu Li, Peng Shi, Jiawei Lin, Zhenguo Gao, Menghui Yao, Mingyang Chen, Jingkang Wang, Songgu Wu, and Junbo Gong, 2021, Enhancing continuous reactive crystallization of lithium carbonate in multistage mixed suspension mixed product removal crystallizers with pulsed ultrasound, Ultrasonics Sonochemistry, 77

Youngjo Kim, Yoshiaki Kawajiri, Ronald W. Rousseau, and Martha A. Grover, 2023, Modeling of Nucleation, Growth, and Dissolution of Paracetamol in Ethanol Solution for Unseeded Batch Cooling Crystallization with Temperature-Cycling Strategy, Industrial & Engineering Chemistry Research, 62, 6, 2866-2881

Flavio Manenti, Gintaras V. Reklaitis (Eds.), Proceedings of the 34th European Symposium on Computer Aided Process Engineering / 15th International Symposium on Process Systems Engineering (ESCAPE34/PSE24), June 2-6, 2024, Florence, Italy

System identification of an industrial cascade refrigeration system

Jun Chang[a], Wei Yu[a], James Carson[b], and Brent Young[a]

Ahuora – Centre for Smart Energy Systems
[a]Department of Chemical and Materials Engineering, University of Auckland, 5 Grafton Road, Auckland 1010, New Zealand
[b]School of Engineering, University of Waikato, Hamilton 3240, New Zealand
jcha708@aucklanduni.ac.nz

Abstract

The cascade refrigeration cycle, characterized by multiple refrigeration cycles operating at different temperature levels, poses unique challenges for control and optimization. For developing optimisation or control strategies, a robust and reliable model must be developed. This paper presents a study of the system identification of an industrial cascade refrigeration system based on the simulation data from a validated first principles model. The research employs a subspace identification method to develop a data-driven dynamic model that accurately represents the transient behaviour of the cascade system. The identified model incorporates the interactions between the high and low-temperature refrigeration cycles, accounting for variations in operating conditions and load demands. Verification of the identified model is conducted using separate simulation data, demonstrating its capability to predict system behaviour. The insights gained from the system identification process lay the foundation for the development of advanced control strategies aimed at optimizing energy consumption and maintaining operational stability in the industrial cascade refrigeration system.

Keywords: cascade refrigeration, system identification, data-driven model.

1. Introduction

Process utilities, namely refrigeration and heating, account for a significant share of the overall process industry energy usage. Vapor-compression cycles (VCCs) have been widely utilized in industrial refrigeration. This technology is increasingly employed in process heating and heat recovery as an alternative to fossil-fuel-based heating to reduce carbon emissions. For optimal performance across a wider temperature range and maximal energy efficiency, the designs of VCCs tend to become much more complicated, as demonstrated in a case study in this research where a R744 (carbon dioxide)-R717 (ammonia) cascade refrigeration system (-50 °C) is coupled with an R717 heat pump (90 °C). However, a VCC does not always operate at the design operating condition because a coupled process or manufacturing process is not continuous, meaning that there are no constant thermal loads on both the cooling and heating sides. Instead, it is often subject to cyclical on/off operation, which deteriorates energy efficiency and equipment longevity. Controller tuning is also an issue due to the nonlinearity of VCC components (Goyal *et al.*, 2019). Ineffectual control over the complex vapor compression system not only reduces efficiency but also damages the compressor. He *et al.* (1997) and Rasmussen and Alleyne (2006) pointed out that multiple single-input/single-output (SISO) control loops were less efficient due to the cross-coupling among the variables. Therefore, to maximize the benefits of a vapor-compression system, optimization and advanced control systems are needed, ideally guided by a dynamic process model. Several studies (Rasmussen and Alleyne, 2006; Jain and Alleyne, 2015; Wang *et al.*, 2021) demonstrated the benefits of a multi-input/multi-output (MIMO) control or optimization strategy for VCCs.

System identification involves the development of mathematical models that accurately represent the dynamic behaviour of a system based on input/output data. The resulting data-driven model will be used as an internal model for an adaptive model predictive control (MPC) in a future study. Whilst there is an accurate first principles model available (Chang et al., 2023), data-driven models are preferred implementation format for model predictive control. Data-driven models are favoured over first principles models due to their flexibility and adaptability (Drgoňa et al., 2020), and they also can capture multivariable non-linear dynamics from plant data that necessarily simplified first principles models may not be to.

This study presents a system identification of an industrial R744-R717 cascade refrigeration system as a part of developing a model predictive controller. In prior work (Chang *et al.*, 2023), a validated first-principles-based dynamic model of the industrial system was developed using a commercial process simulator. In this current work, based on dynamic response data from the validated simulation of the industrial process, state-space models were estimated using the System Identification Toolbox in MATLAB. The models were also compared to a validation dataset and evaluated for their prediction performance.

2. System description

The simplified R744-R717 cascade cycle shown in Fig.1 provides 300 kW of cooling at temperatures ranging from -20 to -45 °C. The system has a separate R452a refrigeration unit on the R744 receiver to avoid excessively high pressure. The secondary fluid for the R744 evaporator is a water-ethylene-glycol mixture (thermal fluid) and the flow is manipulated with a variable speed pump.

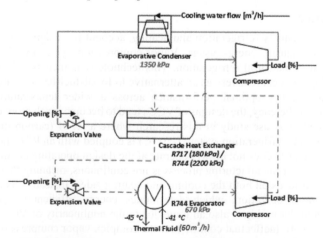

Figure 1. A schematic diagram of the R744 (bottom) - R717 (top) cascade refrigeration system. The design pressures for the heat exchangers and temperatures for the thermal fluid are shown. The manipulated variables are indicated with arrows.

Each cycle in the cascade cycle includes a flooded evaporator with a gas-liquid separator, and oil-flooded screw compressor, an oil separator, a condenser with receiver, and an electric expansion valve. The heat absorbed at the R744 evaporator plus the compressor work is released to the R717 cycle at the cascade heat exchanger. Liquid R717 refrigerant is vaporized at the cascade heat exchanger. The saturated vapor is then compressed at the compressor. The superheated discharge flows through the evaporative condenser while releasing heat to the cooling water.

3. First principles model

To generate simulation data for system identification, the cycle is modelled using dynamic process simulator, Symmetry v2020.4.0.52 (SLB, USA). The model is simple enough to tune the parameters for fitting the plant data, and the model can simulate key system dynamics. Based on the plant configuration, it is assumed that each flooded evaporator (the R744 evaporator and the R717 side of the cascade heat exchanger in Fig.1) is combined with the corresponding gas-liquid separator. Thus, they are modelled as a heat exchanger with an inlet flow of partially evaporated refrigerant and an outlet flow of saturated vapor. The level in the heat exchanger is maintained by the expansion valve. The model from the previous study is also modified in order to drive the operating conditions to desired conditions. The evaporative condenser was initially modelled as a cooler component, but now it is modelled as a heat exchanger with cooling water for more fidelity. The inlet of the cooling water temperature is assumed to be constant, and the flow rate is manipulated to adjust the duty.

4. System identification

4.1. Input/output variables selection

For output variable selection, the priority is to select the least number of variables that can show the dynamic status of the system. For the purpose outlined in the introduction, the four pressures (evaporation and condensation pressures for each refrigerant) are selected. The pressures intuitively display the status of the system.

In order to determine the input variables for the system identification procedure, the manipulatable input variable candidates were subjected to open-loop step tests to preliminarily analyse the dynamics of the system. The step tests also provide time constants for the output variables which are used for designing an appropriate pseudo random binary sequence (PRBS) signal for MIMO system identification.

The magnitudes of the step change were within a 10 % deviation from the initial values. This was due to the unstable cycle dynamic. In particular, completely drying or flooding the heat exchangers is to be avoided, which generate fluctuating data and can cause the simulation (and the system) to become unstable.

From the preliminary results (Table 1), the R744 condensation pressure exhibits the highest gain from both compressors, which could explain the existing control loop, R744 condensation pressure – R717 compressor load. The step test of the condenser cooling water shows the highest magnitude of the gain on the R717 condensation pressure.

For these valves, the process gains for all outputs are lower than other inputs, and it was impossible to estimate the time constants due to significant fluctuation in the data. The fluctuations could be attributed by numerical errors owing to the static head fluctuation within the heat exchangers and the closed cycle behaviour. It should be noted that the process gains of valve-pressure pairs are expected to be negligible in the actual case due to the system configuration with the gas-liquid separator and the flooded-evaporator. Unlike a direct expansion system where an expansion valve regulates the superheat of refrigerant vapour at the suction of the compressor, the valves in this case only work as 'liquid make-up' valves for regulating the separator level. Furthermore, for a stable simulation with a PRBS input signal, it is desired to maintain the levels in the separator.

Table 1. Process gains (Kp) and time constant (in round brackets) for input-output pairs

Kp [kPa/%] (τ [s])	R744 evaporation pressure	R744 condensation pressure	R717 evaporation pressure	R717 condensation pressure
R744 compressor	-1.6 (7)	10.7 (57)	1.4 (60)	9.2 (108)
R717 compressor	-0.3 (-)	-7.6 (51)	-1.3 (41)	0.9 (10)
Cooling water flow rate	-0.6 (109)	-9.4 (83)	-1.5 (70)	-22.4 (136)
R744 valve	0.5 (-)	0.4 (-)	0 (-)	0.1 (-)
R717 valve	0 (-)	1.4 (-)	0.2 (-)	0 (-)

The selected input variables are both compressor loadings and the cooling water flow rate. The level control loops between the valves and the separators are kept closed for a stable simulation. The valves actions can affect the pressures, but since the gains are negligible, and they are likely to open and close repeatedly, the impact can be considered a noise.

4.2. PRBS input signal

In designing a suitable PRBS signal for the particular system, the steps illustrated in Sarath Yadav and Indiran (2019) are followed. The amplitude and the bandwidth are determined based on gain, time constant, and dead time from the step tests. From the preliminary step tests, no dead times were observed. The amplitudes for compressors and cooling water flow were 10 % and 5 % respectively. The bandwidth (Ω) is estimated using the highest time constant in Table 1 as:

$$\Omega = \frac{1}{\tau} \tag{1}$$

4.3. MIMO state-space identification

The input variables were subjected to the PRBS signal independently to simulate the output dynamic responses. The simulation data were then processed to have zero mean. With multiple I/O data, subspace identification method (N4SID) was chosen to estimate MIMO state-space models of order 1 through 15. The model was then compared to the validation data set to evaluate model fit. The model fit was evaluated by the normalised root mean squared error (NRMSE).

$$fit = 100(1 - \frac{\|y - \hat{y}\|}{\|y - mean(y)\|}) \tag{2}$$

5. Results and discussion

The model fits for different model orders are summarised in Fig. 2.

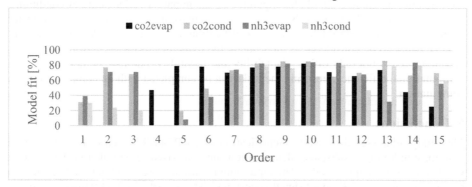

Figure 2. Model fit against validation data for models of order 1 to 15.

The models of order 1 to 6 are not sufficient to predict the output dynamics. From 7th order model, the model fit is around 70 %. The models of order 8 and 9 show adequate fits of around 80 % throughout the output variables. From the 10th order model onward, the fit for some output variables deteriorate, which can be attributed overfitting. Using the principle of parsimony, the minimum model order required for the system is 8th order (Fig. 3).

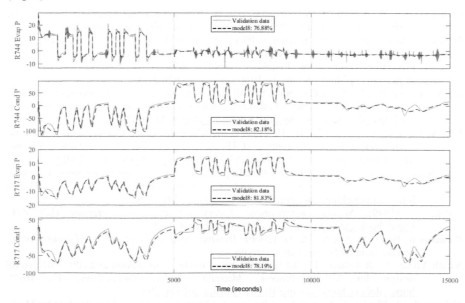

Figure 3. The 8th order model response plotted against the validation data series. The PRBS input is applied to R744 compressor from 0 to 5,000 s, R717 compressor from 5,000 to 10,000 s, and the cooling water flow rate from 10,000 to 15,000 s.

Although 80 % prediction seems promising for controller design, it should be noted that the model is derived from limited simulation data and the model is relatively higher order, compared to Rasmussen and Alleyne (2006).

considering molecular distribution. Later, Stochastic Reconstruction (SR) methods were developed, where molecules are constructed using specific methods and probability density functions (PDFs) to describe structural attributes (Hudebine and Verstraete, 2004). The molecular type-homologous series (MTHS) method further enhanced these models by introducing homologous series, thus reducing optimization variables and increasing computational efficiency (Wu and Zhang, 2010). Recent advancements include the integration of two-dimensional homologous series distributions and the cloud model (CM) for better handling molecular distribution uncertainty (Zhang et al., 2022). This has led to more precise molecular libraries and distribution fluctuation descriptions. High-Resolution Mass Spectrometry (HRMS) data now provide DBE distribution data for vacuum gas oil (VGO) and heavier oil products in reconstruction processes (Guan et al., 2023). However, there is still limited research on DBE distribution patterns of petroleum components, and current models have not fully utilized these patterns from HRMS.

This article presents a new molecular reconstruction method utilizing dual two-dimensional homologous series distributions (Dual-2D-HSD), focusing on core structure-side chain distribution and DBE-carbon number distribution. The first 2D-HSD, based on core structure-side chain distribution, creates a molecular library essential for reconstruction. The second 2D-HSD, based on the DBE distribution pattern from HRMS, uses the JSD method to measure distribution consistency, thereby quantitatively representing the proposed distribution patterns and enhancing the alignment of molecular distribution results with actual conditions. Combining these dual 2D-HSD approaches in the optimization process enhances property prediction accuracy and aligns with real molecular DBE distribution patterns. This method marks a significant advancement in molecular reconstruction, particularly in efficiently handling complex molecular structures and addressing challenges in scenarios with limited data.

2. Methods

2.1. Building a Molecular Library Using the First 2D-HSD.

The initial step of the molecular reconstruction method involves creating a molecular library using the first 2D-HSD and the structure-oriented lumping (SOL) method with 21 structural increments. This two-step process first establishes core structures, such as the A1-1 core structure in Figure 1, and then attaches side chains of varying lengths to each core, forming the complete molecular library. This library is crucial for later stages of molecular reconstruction. The first 2D-HSD mapped the molecular library and molar fractions of complex petroleum components, identifying 19 core structure types and 172 core structures. Side chains with 1 to 50 carbon atoms were added to each core structure, forming a library of 8600 molecules. The first dimension HSD detailed the distribution of core structures per type, while the second dimension HSD showed side chain length distribution per core structure, using the gamma distribution for characterization.

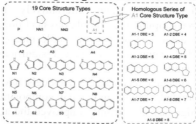

Figure 1. 19 Core structure types and homologous series of the A1 core structure type.

2.2. The Second 2D-HSD and the DBE Distribution Patterns

Due to limitations in the precision of analytical techniques, existing HRMS can only provide molecular abundance distribution data represented in terms of the DBE dimension for complex petroleum components. As a result, the distribution patterns derived from HRMS are applicable in the DBE dimension. The DBE value for each molecule can be calculated from its structural formula using the Eq. (1), where C is carbon, H is hydrogen, X is halogen atoms, and N is nitrogen.

$$DBE = rings + \pi \, bonds = C - \frac{H + X}{2} + \frac{N}{2} + 1 \tag{1}$$

Using the molecular library from the first 2D-HSD and given molar fractions, two-dimensional molecular abundance graphs in DBE and carbon number dimensions are created. Figure 2a shows the DBE dimension molecular abundance for N6 core structure type. Figure 2b, the vertical distribution from 2a, displays the relative abundance for varying DBE values, with blue points summing up molecular abundances per DBE row in 2a, and red lines indicating fitted gamma distribution curves. Figure 2c shows relative abundance for DBE = 9 in 2a, depicting carbon number distribution for molecules with identical DBE values. Figure 2d marks the maximum values in each row from 2a with red dots and a purple trend line, illustrating the change trend in maximum values per row.

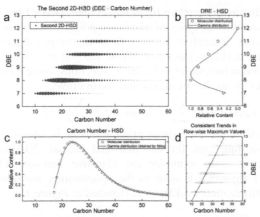

Figure 2. (a) The second 2D-HSD diagram (in core structure type N6). (b) Distribution of molecular relative abundance along the DBE axis (vertical direction in figure 2a). (c) Distribution of molecular relative abundance along the carbon number axis (horizontal direction for DBE=9 in figure 2a). (d) The trend of the maximum points of each row.

Based on HRMS data (Guan et al., 2021) on molecular distribution in petroleum components, the DBE representation reveals: (a) Summing molecules with identical DBE values shows their cumulative molecular abundance in the DBE dimension follows a gamma distribution, depicted in Figure 2b. (b) For a specific DBE value, molecular abundance distribution in the carbon number direction also follows a gamma distribution, as shown in Figure 2c. (c) In the DBE versus carbon number two-dimensional graph, each row (representing the same DBE value) has maximum abundance points that follow a consistent trend, shifting gradually towards the upper-right direction, as demonstrated in Figure 2d.

2.3. Optimization Process for Molecular Reconstruction Based on Dual-2D-HSD

To optimize molecular reconstruction in the first 2D-HSD level, the goal is to minimize discrepancies between calculated and experimental properties. This involves using a weighted average of the absolute relative deviations of physical properties, as outlined in Eq. (2), with the optimization variables being the distribution's parameters.

$$F_1 = \sum_{i=1}^{N} \mu_i \left| \frac{P_i^{cal} - P_i^{exp}}{P_i^{exp}} \right| \tag{2}$$

In the second 2D-HSD level for molecular reconstruction, the aim is to quantitatively match the molecular abundance distribution within each core structure type with existing distribution patterns. The objective function evaluates the difference between the actual molecular abundance and the proposed patterns. In specific calculations: (1) For the proposed molecular distribution patterns (a) and (b), corresponding gamma distributions are fitted based on the actual distribution data. The dissimilarity is measured using the Jensen-Shannon Divergence (JSD), derived from the Kullback-Leibler Divergence (KLD), as detailed in Eq. (3) to (5). Here, P is the gamma distribution from fitting, and Q is the actual distribution. A smaller JSD indicates a closer match to the proposed patterns.

$$KLD(P||Q) = \sum_x P(x) \log \frac{P(x)}{Q(x)} \tag{3}$$
$$M = 0.5 \times (P + Q) \tag{4}$$
$$JSD = 0.5 \times KLD(P||M) + 0.5 \times KLD(Q||M) \tag{5}$$

(2) For molecular distribution pattern (c) in the second 2D-HSD level, linear correlation (R^2) evaluates the trend consistency of maximum points in each row. Higher R^2 means more consistent trends, indicating a closer fit to the proposed pattern. The objective function, detailed in Eq. (6), includes three terms: the first two assess agreement with the molecular distribution in horizontal and vertical directions, while the third term measures the trend consistency in maximum value points across rows.

$$F_2 = \eta_1 \sum JSD_{horizontal} + \eta_2 \sum JSD_{vertical} + \eta_3 \sum -R_i^2 \tag{6}$$

The objective function for the optimization problem is obtained by summing the respective objective functions for the first and second 2D-HSD levels, and the optimization objective is to minimize the objective function. The particle swarm optimization (PSO) algorithm is used as the optimization algorithm, and the overall computation and optimization workflow are depicted in Figure 3.

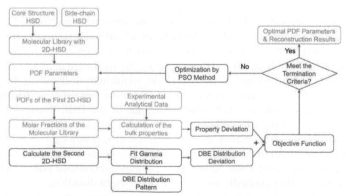

Figure 3. the optimization process and the calculation of the objective function based on the Dual 2D-HSD.

3. Results and discussion

The Dual-2D-HSD molecular reconstruction model was applied to VGO's molecular reconstruction, alongside a traditional 2D-HSD model for comparison. The optimization problem was solved using MATLAB 2023a and the PSO algorithm with a particle swarm of 100. Results from both methods are in Table 1, with the absolute relative error (ARE) measuring computational deviations for each physical property.

Table 1. Calculation results of the 2D-HSD and the Dual-2D-HSD

	2D-HSD	Dual 2D-HSD
Optimization Time	769s (100%)	316s (41.1%)
Property	ARE of the Calculation Result	
Carbon	0.23 %	0.24 %
Hydrogen	1.64 %	1.53 %
Sulfur	0.00 %	0.00 %
Nitrogen	0.00 %	0.00 %
Paraffin	0.00 %	0.00 %
Naphthene	0.00 %	0.00 %
Aromatic	0.00 %	0.00 %
10 vol% BP	0.56 %	2.13 %
30 vol% BP	1.49 %	1.89 %
50 vol% BP	0.38 %	0.96 %
70 vol% BP	0.01 %	1.71 %
90 vol% BP	1.77 %	3.09 %
95 vol% BP	2.11 %	2.74 %
MARE	0.63 %	1.10 %

The Dual-2D-HSD method showed a mean absolute relative error (MARE) of 1.10%, compared to 0.63% for the 2D-HSD method. The computation time for Dual-2D-HSD was only 41.1% of that required for 2D-HSD. This indicates that including molecular distribution knowledge reduces the optimization space and speeds up the algorithm, improving the potential of molecular reconstruction algorithms for industrial use.

To evaluate how the proposed molecular distribution patterns affect optimization, we extracted molecular abundance data for the core molecule type S4 from both 2D-HSD and Dual-2D-HSD methods. The results are shown in Figures 4 and 5, respectively. Each figure features a central molecular abundance distribution chart with DBE and carbon number coordinates. The left side shows the vertical molecular abundance distribution, and the right side displays the horizontal distribution for each DBE value.

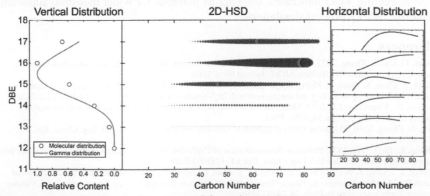

Figure 4. Molecular distribution obtained from the 2D-HSD method.

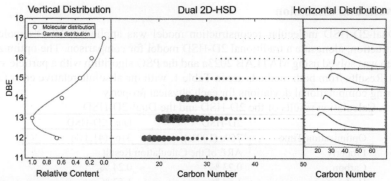

Figure 5. Molecular distribution obtained from the Dual-2D-HSD method.

Figure 4 shows that without molecular distribution pattern knowledge, the 2D-HSD method's molecular reconstruction optimization lacks clear distribution patterns in both horizontal and vertical directions. There are significant variances in molecular abundance and distribution across different rows, with a higher concentration of molecules at larger DBE values, contradicting experimental findings. Additionally, the maximum abundance points in each row are sporadically placed, not aligning with the proposed distribution pattern (c). In contrast, Figure 5 illustrates that incorporating molecular distribution pattern knowledge in the Dual-2D-HSD method results in a clearer gamma distribution both horizontally and vertically. The molecular distribution skews towards lower DBE values, matching real-world distributions more closely. Furthermore, the maximum abundance points in each row show a consistent upward-rightward trend. Thus, the Dual-2D-HSD method achieves more realistic molecular abundance distributions, aligning better with the distribution patterns than the 2D-HSD method.

4. Conclusions

This study enhances the existing 2D-HSD molecular reconstruction method by adding a second 2D-HSD representation in the DBE dimension, creating the Dual-2D-HSD method. By utilizing JSD to measure the deviation between distributions, it quantitatively incorporates molecular distribution patterns derived from HRMS data into this new representation. The Dual-2D-HSD method not only yields accurate computational results more quickly but also provides more realistic molecular distributions than the traditional 2D-HSD method. This approach's efficiency, accuracy, and rationality have been proven in reconstructing VGO molecules, showing its potential for wider industrial application to improve factory process simulations.

References

D. Guan, G. Cai, L. Zhang, 2023. Dual-objective optimization for petroleum molecular reconstruction based on property and composition similarities. AIChE Journal 69, e18108.
D. Guan, Z. Chen, X. Chen, Y. Zhang, Q. Qi, Q. Shi, S. Zhao, C. Xu, L. Zhang, 2021. Molecular-level heavy petroleum hydrotreating modeling and comparison with high-resolution mass spectrometry. Fuel 297, 120792.
Hudebine, D., Verstraete, J.J., 2004. Molecular reconstruction of LCO gasoils from overall petroleum analyses. Chemical Engineering Science 59, 4755–4763.
Y. Wu, N. Zhang, 2010. Molecular Characterization of Gasoline and Diesel Streams. Ind. Eng. Chem. Res. 49, 12773–12782.
C. Zhang, K. Bi, T. Qiu, 2022. Molecular Reconstruction of Crude Oil: Novel Structure-Oriented Homologous Series Lumping with a Cloud Model. Ind. Eng. Chem. Res. 61, 18810–18820.
Y. Zhang, L. Zhang, Z. Xu, N. Zhang, K.H. Chung, S. Zhao, C. Xu, Q. Shi, 2014. Molecular Characterization of Vacuum Resid and Its Fractions by Fourier Transform Ion Cyclotron Resonance Mass Spectrometry with Various Ionization Techniques. Energy Fuels 28, 7448–7456.

Flavio Manenti, Gintaras V. Reklaitis (Eds.), Proceedings of the 34th European Symposium on Computer Aided Process Engineering / 15th International Symposium on Process Systems Engineering (ESCAPE34/PSE24), June 2-6, 2024, Florence, Italy

Process Modelling for Photo-Iniferter RAFT with Multiple Chain Transfer Agents

Rui Liu,[a] Xiaowen Lin,[a] Xi Chen,[a, b] Antonios Armaou[c, d]

[a]*State Key Laboratory of Industrial Control Technology, College of Control Science and Engineering, Zhejiang University 310027, Hangzhou China.*
[b]*National Center for International Research on Quality-targeted Process Optimization and Control, Zhejiang University 310027, Hangzhou China. xi_chen@zju.edu.cn*
[c]*Chemical Engineering Department, University of Patras, Patras 26504, Greece.*
[d]*Chemical Engineering & Mechanical Engineering Departments, Pennsylvania State University, College Park, PA 16802 USA. armaou@upatras.gr & armaou@psu.edu*

Abstract

Model development constitutes a pivotal part in process systems engineering. Specifically, modelling polymerization processes with newly clarified mechanisms holds significance in guiding experiment/process design and optimizing polymer production. Advanced photo-induced reversible deactivation radical polymerization (RDRP) techniques, such as photo-iniferter reversible addition-fragmentation chain-transfer (PI-RAFT), leverage light irradiation to control polymer growth. The introduction of diverse chain transfer agents (CTAs) into the PI-RAFT system yields varied polymer generation rates and microstructural properties, thereby affording expanded opportunities for material discovery and applications in high-value industries. In this study, an accelerated Monte Carlo (MC) model at a microscopic resolution is established for the dynamic PI-RAFT processes. Subsequently, the impact of multiple CTAs on the final polymers of the PI-RAFT process is investigated via the developed model. The accuracy and efficiency of the established accelerated MC model are validated against the equation-oriented deterministic method. The insights derived from this study provide guidance towards the experimental design and process optimization in PI-RAFT polymerization.

Keywords: Monte Carlo simulation, PI- RAFT, microscopic scale, process modelling.

1. Introduction

Developing modelling approaches for polymerization processes, practically those involving pioneering and newly clarified mechanisms, is of great demands to advance kinetic comprehension and facilitate industrial applications. Over the last decades, reversible deactivation radical polymerization (RDRP) has shown significant promise among various polymerization techniques due to its ability for precise control over polymer growth. RDRP enables the production of polymers with desired properties through establishing dynamic equilibriums among different polymer species within the system (Zhou et al., 2020). Furthermore, several photo-induced strategies for radical (re)generation have been applied in RDRP, leveraging light irradiation as an external stimulus (Corrigan et al., 2020). Such photo-RDRP techniques contribute to environmentally sustainable polymerization processes and enhance the potential of resulting polymers for biocompatible applications.

Among diverse photo-RDRP techniques, the photo-iniferter reversible addition-fragmentation chain transfer (PI-RAFT) emerges as a powerful and versatile photo-RDRP polymerization method. PI-RAFT enables the production of polymers with intricate and well-defined structures through meticulously controlled polymerization processes. The introduction of diverse chain transfer agents (CTAs) into the PI-RAFT system further yields polymers with varied microstructural properties, including molecular weight distribution (MWD) (Lehnen et al., 2023). As microstructural polymeric properties directly influence the end-use functionalities of polymer products, the development of efficient modelling approaches for PI-RAFT with microscopic resolution holds significance in reducing the number of experiments required for process optimization.

Focusing on computational modelling methods with embedded MWD, they can be classified into two categories: deterministic equation-based methods and stochastic Monte Carlo (MC) methods (Liu et al., 2023). The former encounters challenges in accurately describing polymeric properties since direct solution of the underlying ordinary differential equation systems is computationally infeasible due to memory resource limitations and stiffness. These hurdles are surpassed via coarse graining and/or order reduction for computational efficiency at the cost of limited applicability to simpler mechanisms (Saldívar-Guerra, 2020). In contrast to deterministic methods, MC simulations enable the characterization of individual polymers based on probability theory, providing more microscopic-scale information, albeit at a computational expense due to tracking all discrete reaction events alongside time (Cole et al., 1994).

In this study, an accelerated MC model, coupled with the constant-number MC (CNMC) approach, is established for dynamic PI-RAFT processes with multiple CTAs. The evolution of MWDs from processes with varying ratios of two distinct CTAs is simulated. The accuracy and efficiency of the developed accelerated MC model are validated against the deterministic method of moments (MoM). This study provides an opportunity for the further design and optimization of polymerization processes.

2. Photo-iniferter RAFT polymerization process

The schematic representation of PI-RAFT polymerization, featuring a single CTA species is presented in Figure 1. The corresponding kinetic mechanism is detailed in reactions 1-16 in Table 1. Herein, R^*, M, TR, and T^* denote primary radicals, monomers, CTA, and thiocarbonylthio radicals, respectively. D_n, P_n^*, and TP_n^* represent dead polymers, radical polymers, and inactive polymers with chain lengths of n, respectively. Also, P_nTR and P_nTP_m signify one-arm adduct and two-arm adduct radicals, wherein the latter exhibits chain lengths of n and m on each arm. The directional arrows in the chemical equations of Table 1, labelled "hv", indicate reactions triggered by light irradiation.

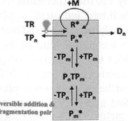

Figure 1. Reaction mechanism scheme for PI-RAFT polymerization.

Upon light irradiation, TR agents undergo photo-dissociation, yielding primary radicals (R^*) and thiocarbonylthio radicals (T^*) (reaction 13 in Table 1). Similarly, inactive polymers decompose under light, leading to the production of radical polymers and thiocarbonylthio radicals T^* (reaction 15 in Table 1). Primary radicals and radical polymers engage reversibly with T^* to form CTA and inactive polymers, respectively (reactions 14 & 16 in Table 1). Chain growth ensues through propagation reactions, where primary radicals and radical polymers react with monomers (reactions 1-2 in Table 1). Termination reactions result in the generation of dead polymers with the irreversible consumption of radicals (reactions 3-6 in Table 1). The core of the PI-RAFT processes is the dynamic equilibrium among three distinct polymer species: radical polymers, inactive polymers, and RAFT adduct radicals. This equilibrium is established through reversible addition and fragmentation reactions (reactions 7-12 in Table 1). The radical polymers partake in addition reactions with TR and inactive polymers TP_n^* to generate one-arm and two-arm adduct radicals, respectively (reactions 7, 9, and 11 in Table 1). Reversely, the adduct radicals undergo fragmentations in either direction to generate radical polymers (reactions 8, 10, and 12 in Table 1).

Table 1. Kinetic mechanism for PI-RAFT

	description	event	equation	rate constant *
1	Ini-propagation	ini-propagation	$R^*+M \rightarrow P_1^*$	$k_{i,R} = 200\ \text{M}^{-1}\text{s}^{-1}$
2	Propagation	propagation	$P_n^*+M \rightarrow P_{n+1}^*$	$k_p = 200\ \text{M}^{-1}\text{s}^{-1}$
3		termination	$R^*+R^* \rightarrow D_0$	$k_t = 2\times10^4\ \text{M}^{-1}\text{s}^{-1}$
4		termination	$P_n^*+R^* \rightarrow D_n$	$k_t = 2\times10^4\ \text{M}^{-1}\text{s}^{-1}$
5	Termination	termination by combination	$P_n^*+P_m^* \rightarrow D_{n+m}$	$k_{tc} = 10^4\ \text{M}^{-1}\text{s}^{-1}$
6		termination by disproportionation	$P_n^*+P_m^* \rightarrow D_n+D_m$	$k_{td} = 10^4\ \text{M}^{-1}\text{s}^{-1}$
reactions related to chain transfer agent (CTA) of TR				
7		addition	$TR+P_n^* \rightarrow P_nTR$	$k_{addT} = 2.857\times10^3\ \text{M}^{-1}\text{s}^{-1}$
8		fragmentation	$P_nTR \rightarrow TR+P_n^*$	$k_{-addT} = 5.714\times10^7\ \text{s}^{-1}$
9	Equilibrium	addition	$TP_n^*+R^* \rightarrow P_nTR$	$k_{\beta T} = 1.333\times10^8\ \text{M}^{-1}\text{s}^{-1}$
10	for CTA of TR	fragmentation	$P_nTR \rightarrow TP_n^*+R^*$	$k_{-\beta T} = 2.667\times10^{12}\ \text{s}^{-1}$
11		addition	$TP_n^*+P_m^* \rightarrow P_nTP_m$	$k_{addPT} = 2\times10^3\ \text{M}^{-1}\text{s}^{-1}$
12		fragmentation	$P_nTP_m \rightarrow TP_n^*+P_m^*$	$k_{-addPT} = 4\times10^7\ \text{s}^{-1}$
13		photodissociation	$TR \xrightarrow{hv} T^*+R^*$	$k_{bsT1} = 5\times10^{-9}\ \text{s}^{-1}$
14	Photo-iniferter	addition	$T^*+R^* \rightarrow TR$	$k_{-bsT1} = 5\ \text{M}^{-1}\text{s}^{-1}$
15	for CTA of TR	photodissociation	$TP_n^* \xrightarrow{hv} T^*+P_n^*$	$k_{bsT2} = 5\times10^{-10}\ \text{s}^{-1}$
16		addition	$T^*+P_n^* \rightarrow TP_n^*$	$k_{-bsT2} = 5\ \text{M}^{-1}\text{s}^{-1}$
reactions related to chain transfer agent (CTA) of XR				
17		addition	$XR+P_n^* \rightarrow P_nXR$	$k_{addX}\ 1.7\times10^2\ \text{M}^{-1}\text{s}^{-1}$
18		fragmentation	$P_nXR \rightarrow XR+P_n^*$	$k_{-addX} = 3.4\times10^{10}\ \text{s}^{-1}$
19	Equilibrium	addition	$XP_n^*+R^* \rightarrow P_nXR$	$k_{\beta X} = 2.267\times10^{10}\ \text{M}^{-1}\text{s}^{-1}$
20	for CTA of XR	fragmentation	$P_nXR \rightarrow XP_n^*+R^*$	$k_{-\beta X} = 4.533\times10^{18}\ \text{s}^{-1}$
21		addition	$XP_n^*+P_m^* \rightarrow P_nXP_m$	$k_{addPX} = 68\ \text{M}^{-1}\text{s}^{-1}$
22		fragmentation	$P_nXP_m \rightarrow XP_n^*+P_m^*$	$k_{-addPX} = 1.36\times10^{10}\ \text{s}^{-1}$
23		photodissociation	$XR \xrightarrow{hv} X^*+R^*$	$k_{bsX1} = 1.2\times10^{-4}\ \text{s}^{-1}$
24	Photo-iniferter	addition	$X^*+R^* \rightarrow XR$	$k_{-bsX1} = 1\times10^2\ \text{M}^{-1}\text{s}^{-1}$
25	for CTA of XR	photodissociation	$XP_n^* \xrightarrow{hv} X^*+P_n^*$	$k_{bsX2} = 9\times10^{-5}\ \text{s}^{-1}$
26		addition	$X^*+P_n^* \rightarrow XP_n^*$	$k_{-bsX2} = 1\times10^2\ \text{M}^{-1}\text{s}^{-1}$

* Kinetic rate constants reported in the literature (Lehnen et al., 2023) were used in this work.

In more intricate PI-RAFT systems, the inclusion of an additional CTA species denoted as XR, enhances the efficiency and control of the process. The kinetic mechanism of PI-RAFT with two CTAs is tabulated in Table 1. The relevant chemical equations pertaining to XR can be found in reactions 17-26 in Table 1. Photodissociation of XR and inactive polymers XP_n* produce primary radicals and radical polymers through light irradiation, respectively, alongside thiocarbonylthio radicals $X*$. This photodissociation undergoes reversible reactions to reform agents XR and inactive polymers XP_n*. Furthermore, the introduction of a second CTA species establishes a second equilibrium of addition-fragmentation reactions based on XR (reactions 17-22 in Table 1).

The polymerization process of PI-RAFT with two CTA species comprises 14 distinct reactants, eight of which are polymers with an additional dimension of information regarding chain lengths. To accurately model this process, it is necessary to track the concentrations of non-polymer reactants, including monomers (M), CTA with T (TR), CTA with X (XR), primary radicals ($R*$), and thiocarbonylthio radicals with T and X ($T*$ and $X*$). For a microscopic understanding of the polymeric information, the structures of individual polymers, including radical polymers (P_n*), dead polymers (D_n), inactive polymers with T and X (TP_n* and XP_n*), one-arm adduct radicals with T and X (P_nTR and P_nXR), and two-arm adduct radicals with T and X (P_nTP_m and P_nXP_m), should be recorded over time. It is worth noting that describing the individual two-arm adduct radical poses a significant challenge, as it requires recording chain lengths of two dimensions on both arms. Predicting the evolution of microscopic polymer properties with all possible chain lengths as a function of time through solving a large-scale ODE system with thousands of equations is computationally infeasible. However, MC simulation offers a mathematically straightforward tool for modelling processes characterized by complex mechanisms and provides results at the microscopic scale.

3. PI-RAFT process modelling with constant-number Monte Carlo

In a general MC simulation for polymerizations, the reaction rate R is defined as the product of the kinetic rate constant (k) and the instantaneous numbers of the reactants (n) in the system (Gillespie, 1977). For reactions involving single reactant A, the reaction rate is calculated as:

$$R = k \times n_A \tag{1}$$

For reactions involving two distinct reactants A and B, the reaction rate is calculated as:

$$R = \frac{k \times n_A \times n_B}{VNA} \tag{2}$$

For reactions with two reactants of the same species B, the reaction rate is calculated as:

$$R = \frac{k \times n_B \times (n_B - 1)}{2 \times VNA} \tag{3}$$

where, VNA denotes the product of the simulated volume size and the Avogadro constant. The instantaneous probability ($\mathbb{P}$) of each reaction taking place is calculated as the proportion of the corresponding reaction rate to the summation of the reaction rates of all reactions (R_{SUM}). Random numbers are generated to select the next reaction. Subsequently, the system states are updated based on the selected reaction, and the reaction time is evaluated by a time interval defined as:

$$\tau = -\frac{1}{R_{SUM}} ln(rnd) \tag{4}$$

This process repeats until a preset time or monomer conversion is reached. It should be noted that the simulated volume size remains constant, and the total number of particles changes forward in time. A sufficiently large volume size is necessary for accurate modeling, which results in smaller time intervals and more discrete simulation steps, as evidenced in Eqs. (2)-(4). However, simulating a dynamic PI-RAFT process with high monomer conversion requires a significant number of steps for reaction selection and system state updates, resulting in longer computational time. In the rest of this article, the general MC simulation technique is referred to as constant-volume MC (CVMC).

CNMC is a variation of MC simulation for process modelling (Khalili et al., 2010). In CNMC, a system with an initialized total number (N) of particles is considered. At each simulation step, the selected reaction leads to a net loss or increment (Δn) in the total particle number within the system. To maintain the total particle number, CNMC simulation either bootstraps Δn particles to replenish the system in case of a net loss, or randomly discards Δn particles from the system in case of a net increment. To accurately represent the concentrations of reactants in the real system, the simulated volume size is recalculated at each CNMC simulation step. In batch PI-RAFT processes, the total particle number tends to decrease as monomers are integrated into polymers, which necessitates increasing the simulated volume size to maintain the fixed total number. With comparable precision in MWDs at the later stages of the process with high monomer conversion, CNMC simulation is more computationally efficient than CVMC as the simulated volume size is relatively small in the initial stage of the CNMC simulation.

4. Results and discussions

Simulations for dynamic PI-RAFT processes were conducted. All simulations were performed on a personal desktop computer with 32 GB of RAM and an Intel i7-10700 CPU core running Windows 11 OS using Python version 3.

4.1. Simulations of PI-RAFT with single CTA for microscopic properties

A batch PI-RAFT process with one single CTA is simulated through developing a CVMC model, a CNMC model, and a deterministic MoM model. The initial concentrations of monomers and CTA are $2\ M \cdot L^{-1}$ and $0.04\ M \cdot L^{-1}$. The volume size (VNA) is fixed to 10^5 in CVMC simulation. The initial volume size (VNA) is set to 10^4 with a fixed particle number of 20,400 in CNMC simulation. The simulated results are presented in Figure 2. As shown in Figure 2(a)-(b), the macroscopic properties of polymers obtained from the CNMC simulation is in good agreement with the results from the deterministic MoM and the CVMC simulation. The MWD from the deterministic MoM is absent in Figure 2(c) due to its inability for such microscopic properties. The computational time required for CVMC and CNMC simulations to model the process until 95% monomer conversion at the microscopic scale is 2470.4 s and 266.7 s, respectively, representing an acceleration of approximately 10 times.

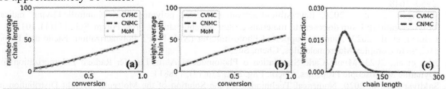

Figure 2. Simulation results using CVMC simulation, CNMC simulation, and deterministic MoM. (a) number-average and (b) weight-average chain lengths as a function of monomer conversion; (c) MWDs of polymers at 95% monomer conversion.

4.2. Simulations of PI-RAFT with varied ratios of CTAs for microscopic properties

The PI-RAFT processes were simulated using two distinct CTAs, and the resulting MWDs at 95% monomer conversion are shown in Figure 3. The simulations were performed using CNMC with a fixed particle number of 20,400. The initial sum concentration of the two CTAs was set at $0.04\ M \cdot L^{-1}$. The varying transparencies of color in the figure represent results from systems with different ratios of the two CTAs, where curves with less transparency indicate the use of less *XR* and more *TR*. Due to the relatively small number of particles used in the simulations, the MWDs displayed a jagged appearance. Furthermore, the average chain lengths obtained were consistent across different ratios of CTAs, but the MWDs exhibited broader widths with higher ratios of *XR*. The development of the CNMC models with microscopic resolution is essential for a comprehensive understanding of the kinetic characteristics in PI-RAFT.

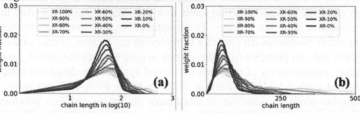

Figure 3. MWDs of polymers at 95% monomer conversion obtained from CNMC simulations, displayed with chain length axes in (a) $\log_{10}$ scale and (b) linear scale. The varying transparencies of color represent the results from systems utilizing different ratios of the two CTAs. The numbers in the legends denote for the initial proportions of *XR* in the system.

5. Conclusions

The study has presented the development of CVMC, CNMC, and deterministic MoM models for predicting and understanding dynamic PI-RAFT processes at the microscopic scale. The effects of varying CTA ratios on the MWDs of polymers were further investigated through simulations. The acceleration and accuracy of the developed CNMC model have been validated against the CVMC simulations and the MoM model. The presented CNMC simulation is efficient and flexible for integration into prospective tasks related to process design and optimization.

Acknowledgments

The financial support of the National Key Research and Development Project (No. 2023YFB3307803) is gratefully acknowledged by the first, second, and third authors. The financial support of the University of Patras, Medicus program (No. 81816) is gratefully acknowledged by the fourth author. The financial support from the China Scholarship Council is gratefully acknowledged by the first and third authors.

References

Y. Zhou, et al., 2020, Role of External Field in Polymerization: Mechanism and Kinetics, Chem. Rev., 120(5), 2950–3048.

N. Corrigan, et al., 2020, Reversible-deactivation radical polymerization (Controlled/living radical polymerization): From discovery to materials design and applications, Prog. Polym. Sci., 111(1), 101311.

A. Lehnen, et al., 2023, Xanthate-Supported Photo-Iniferter (XPI)-RAFT Polymerization: Facile and Rapid Access to Complex Macromolecules, Chem. Sci., 3.

R. Liu, et al., 2023, Monte Carlo Simulation o Photoinduced Atom-Transfer Radical Polymerization for Dynamic Microscopic Properties, Chem. Eng. Sci., 276(15), 118811.

E. Saldívar-Guerra, 2020, Numerical Techniques for the Solution of the Molecular Weight Distribution in Polymerization Mechanisms, State of the Art, Macromol. React. Eng., 14(4), 2000010.

J. Cole, et al., 1994, Monte Carlo Simulation of Radiative Heat Transfer in Rapid Thermal Processing (RTP) Systems, MRS Proceedings, 342, 425.

D. Gillespie, 1977, Exact Stochastic Simulation of Coupled Chemical Reactions, J. Phys. Chem., 81(25), 2340–2361.

S. Khalili, et al., 2010, Constant Number Monte Carlo Simulation of Population Balances with Multiple Growth Mechanisms, AIChE J., 56(12), 3137-3145.

Flavio Manenti, Gintaras V. Reklaitis (Eds.), Proceedings of the 34th European Symposium on Computer Aided Process Engineering / 15th International Symposium on Process Systems Engineering (ESCAPE34/PSE24), June 2-6, 2024, Florence, Italy

A Study on Thermodynamic Efficiency and Economic Viability of ORC-Based Geothermal Hydrogen Production Systems

Jinyue Cui,[a,b,*] Muhammad Aziz[a,b]

[a]*Department of Mechanical Engineering, The University of Tokyo, 7-3-1 Hongo, Bunkyo-Ku, Tokyo, 113-8656, Japan*
[b]*Institute of Industrial Science, The University of Tokyo, 4-6-1 Komaba, Meguro-ku, Tokyo, 153-8505, Japan*
jinyuecui@g.ecc.u-tokyo.ac.jp

Abstract

The imperative for sustainable hydrogen production is intensifying amidst the transition to cleaner energy paradigms. This study critically evaluates the thermodynamic efficiency and economic viability of hydrogen production harnessing geothermal energy, utilizing Organic Rankine Cycle (ORC) systems. Analysing three working fluids—R236fa, R245fa, and R600a—across diverse geothermal fluid temperatures, we determine their performance metrics in terms of energy efficiency and levelized cost of hydrogen (LCOH). Our findings reveal R245fa achieving a peak efficiency of approximately 0.11 in Double Flash Integrated-ORC systems, while R600a stands out with the lowest LCOH, approximately $5.22/kg at 300°C, underlining its economic efficiency across a range of geothermal conditions. These results highlight R600a as the most efficient and economically viable candidate for geothermal hydrogen production. This study underscores the pivotal role of working fluid selection and suggests that bespoke system configurations could markedly bolster the sustainability and economic feasibility of hydrogen production, contributing to the strategic optimization of geothermal energy utilization in green hydrogen production initiatives.

Keywords: geothermal, hydrogen, renewable energy, levelized cost, ORC

1. Introduction

As the global energy sector confronts the dual challenges of environmental degradation and the looming scarcity of fossil fuel reserves, hydrogen has emerged as a pivotal energy carrier for the future. Its high energy yield and clean combustion profile make it an attractive alternative in the quest for sustainable energy solutions (Rosen and Koohi-Fayegh, 2016). Nevertheless, traditional hydrogen production methods are often marred by significant environmental drawbacks, necessitating a shift towards 'green' hydrogen synthesized from renewable energy sources to alleviate ecological impacts.

In this context, geothermal energy stands out as an exemplary renewable source due to its reliability, renewability, and low environmental footprint. It represents a beacon of hope for the continuous and environmentally conscious production of hydrogen. The literature underscores the versatility of geothermal energy for hydrogen production, with methodologies ranging from direct utilization of geothermal steam to advanced hybrid processes combining heat and electricity for the steam electrolysis (Balta et al., 2010). Anifantis et al. (2017) provided insight into the integration of geothermal energy with hydrogen technologies and their comprehensive energy management strategies. Alirahmi et. al. (2022) proposed and optimized a geothermal-driven multi-generation system, showcasing significant potential in sustainable and cost-effective hydrogen production. Despite recognition of its potential, the detailed application of geothermal energy in hydrogen production through sophisticated thermodynamic systems, such as single and double flash systems integrated with Organic Rankine Cycles (ORC), is not thoroughly investigated.

This study endeavours to bridge the knowledge gap by conducting a thorough thermodynamic and economic analysis of geothermal-driven hydrogen production systems. Figure 1 serves as a foundational schematic, mapping out the energy conversion pathways within different ORC configurations. By scrutinizing these systems for their operational efficiency and economic potential, the research aims to decode the complex potential of geothermal energy in sustainable hydrogen production. We dissect the feasibility of deploying geothermal resources and probe the economic ramifications, driving the discourse forward to enhance scientific understanding and inform the development of policies fostering sustainable energy practices.

2. Process description

This study conducts an in-depth evaluation of energy efficiency and levelized cost of hydrogen (LCOH) in geothermal-driven ORC systems, as delineated by the thermodynamic processes illustrated in Figure 1. The schematic in Figure 1 depicts geothermal hot water serving as the geofluid, which is the primary energy source for the ORC configurations that generate electricity to power a Polymer Electrolyte Membrane (PEM) electrolyzer for hydrogen production. The analysis spans across a comprehensive range of geofluid inlet temperatures, with a focus on three organic working fluids—R236fa, R245fa, and R600a—each chosen for their respective thermodynamic suitability to the thermal profiles of the geothermal inputs.

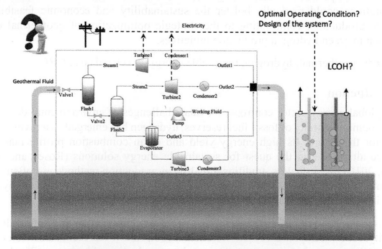

Figure 1. Schematic Overview of Thermodynamic Processes in Geothermal-Driven Hydrogen Production Using ORC Systems.

At lower geofluid temperatures, between 120°C to 140°C, the standard ORC setup is utilized, as shown in Figure 2a. Here, R236fa and R245fa are employed due to their lower thermal stability and efficiency at these temperatures, while R600a is considered for its potential at slightly higher operational temperatures within this range. Moving into the intermediate temperature bracket of 160°C to 200°C, a Single Flash ORC integration comes into play, optimizing the energy extraction from the geothermal steam and enhancing the subsequent electricity generation for the PEM electrolyzer, as captured in Figure 2b.

The methodology advances as the geofluid inlet temperatures reach higher levels, from 220°C to 300°C, where a more sophisticated Double Flash Integrated-ORC system is warranted, as illustrated in Figure 2c. This system is designed to harness the full thermal potential of high-temperature geothermal resources, utilizing dual evaporation stages to maximize energy recovery before electricity generation for hydrogen production. Throughout all these configurations, the three selected working fluids are continuously assessed for their performance in maximizing energy efficiency and minimizing the LCOH.

The study's core objective is to determine the optimal system configuration that provides the highest energy efficiency and the lowest LCOH for sustainable hydrogen production. By examining each ORC system and working fluid under varying temperature conditions, we aim to provide a robust set of data that will inform future design and implementation strategies for geothermal-based hydrogen production systems. The findings from this research will significantly contribute to the development of renewable energy technologies, driving forward the use of geothermal resources in the clean energy transition.

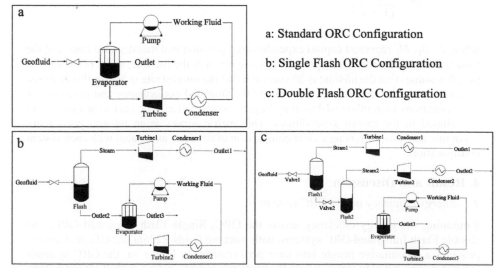

a: Standard ORC Configuration

b: Single Flash ORC Configuration

c: Double Flash ORC Configuration

Figure 2. Process Diagrams of ORC System Configurations for Hydrogen Production.

3. Methodology

3.1. Energy Efficiency of Geothermal-Driven Hydrogen Production Systems

The energy efficiency of various configurations is quantitatively defined as the ratio of power output to heat duty. This critical metric serves as an indicator of the system's proficiency in converting geothermal heat into usable forms of energy—either as electrical power or direct heat—vital for the hydrogen production process. A thorough analysis of the thermodynamic cycles inherent in the single/double flash and ORC configurations is conducted, with a keen focus on their capacity to effectively harness and transform geothermal energy. By calculating the ratio of the system's power output against the heat duty provided by the geothermal resource, the study meticulously evaluates the energy conversion efficiency of each configuration. This evaluation is not just a measure of performance but also a crucial step in refining system designs. It aims to maximize energy efficiency and opens pathways for innovative energy conservation methods in geothermal-driven hydrogen production systems, aligning with the goals of sustainable energy development.

3.2. LCOH from Geothermal Sources

To assess the economic feasibility of the geothermal-driven hydrogen production systems, we adopt the LCOH, particularly focusing on the shipping of hydrogen as a significant economic factor. The LCOH is defined by the following equation:

$$LCOH = \frac{\sum_{t=1}^{n} \frac{I_t + O_t}{(1+r)^t}}{\sum_{t=1}^{n} \frac{M_t}{(1+r)^t}} \tag{1}$$

where I_t, O_t, M_t represent capital expenditures, operation and maintenance cost, and the total amount of hydrogen transported, respectively. n is the lifetime, and r is the interest rate. It assumes that the lifetime is 20 years and that the interest rate is 5% for this analysis. By calculating the LCOH, we can evaluate the financial performance and potential for cost reductions in geothermal-based hydrogen transportation. The cost of working fluid is included in the capital expenditures. This metric serves as an integral part of our economic analysis, providing a clear comparison of the costs associated with each system configuration.

4. Result and Discussion

4.1. Energy efficiency across ORC systems

Examining the energy efficiency across the ORC, Single Flash Integrated-ORC, and Double Flash Integrated-ORC systems, with working fluids R236fa, R245fa, and R600a, provides a quantitative insight into their performance. In Figure 3a, the ORC system's efficiency was noted to increase with the evaporation temperature, with R245fa reaching an efficiency peak near 0.048 at higher temperatures, significantly outperforming R236fa and R600a.

Transitioning to the Single Flash Integrated-ORC (Figure 3b), an improvement in efficiency was observed for all working fluids, with R245fa showing a notable efficiency

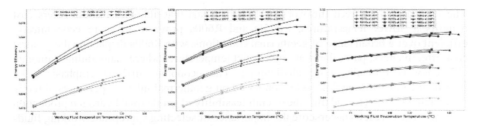

Figure 3. Energy Efficiency of ORC Systems.

value of 0.066 at elevated temperatures. Meanwhile, R600a presented a steep increase in efficiency, indicative of its effective thermal properties in this hybrid system.

The Double Flash Integrated-ORC system (Figure 3c) demonstrated the highest energy efficiencies, with R245fa achieving efficiencies close to 0.11 at the highest geothermal conditions. In comparison, R600a's efficiency closely followed, affirming its suitability for high-temperature applications within the most advanced system configuration.

These results underscore the importance of system selection in conjunction with working fluid choice to optimize energy efficiency for geothermal-driven hydrogen production. The clear trend across all figures indicates the superior performance of integrated systems, particularly the Double Flash Integrated-ORC, in harnessing geothermal energy for efficient hydrogen production.

4.2. Levelized cost of hydrogen

Our comprehensive assessment of the levelized cost of hydrogen (LCOH) from geothermal-driven systems, as depicted in Figures 4a through 4c, showcases the economic variations between the working fluids R236fa, R245fa, and R600a across a spectrum of geothermal inlet temperatures. R245fa emerges as the most thermodynamically efficient across the board, while R600a displays an unparalleled economic advantage, consistently maintaining the lowest LCOH, particularly at higher temperatures where it proves to be most cost-effective. R236fa, while starting economically at cooler temperatures with an LCOH around \$25.64/kg at 120°C, becomes less financially viable as inlet temperatures rise, with costs climbing to \$6.21/kg at 300°C. Conversely, R600a exhibits remarkable economic stability, with initial costs around \$21.57/kg at 120°C and peaking at about \$5.22/kg at 300°C. This analysis highlights R600a as the most economical option for geothermal hydrogen production across all assessed conditions, suggesting its potential to drive down costs in geothermal energy applications.

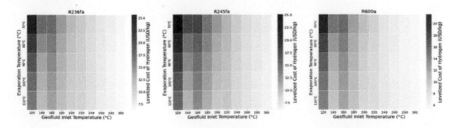

Figure 4. LCOH of Geothermal Hydrogen Production

The comparison between the working fluids reveals that although R236fa offers cost benefits at lower temperatures, R245fa and R600a prove to be more economically efficient over a broad range of geothermal conditions, with R600a displaying the greatest cost-effectiveness, particularly at higher inlet temperatures where maintaining economic efficiency is typically more challenging. The economic analysis emphasizes the significance of working fluid selection based on geothermal inlet temperatures, directly influencing the LCOH and thus the overall feasibility of geothermal hydrogen production systems. These results underscore the importance of selecting appropriate working fluids and system configurations to optimize geothermal hydrogen production, emphasizing their pivotal role in enhancing both sustainability and economic feasibility in this domain.

5. Conclusions

In this comprehensive study, we evaluated the thermodynamic efficiency and the levelized cost of hydrogen production using geothermal-driven Organic Rankine Cycle (ORC) systems. Our findings highlight R245fa as the most thermodynamically efficient working fluid across various geothermal conditions, while R600a excels in economic viability by maintaining the lowest levelized cost of hydrogen production. This contrast underlines the importance of carefully selecting working fluids, balancing thermodynamic superiority and cost-effectiveness, to optimize geothermal hydrogen production systems. The study advocates for the integration of advanced ORC configurations, emphasizing that a strategic approach encompassing both R245fa for efficiency and R600a for economic benefits can significantly enhance the sustainability and economic feasibility of hydrogen production, guiding future research towards improving hydrogen's role as a clean and sustainable energy vector.

References

A.S. Anifantis, A. Colantoni, S. Pascuzzi, 2017, "Thermal energy assessment of a small scale photovoltaic, hydrogen and geothermal stand-alone system for greenhouse heating," Renewable Energy, 103, 115-127.

M.A. Rosen, S. Koohi-Fayegh, 2016, "The prospects for hydrogen as an energy carrier: an overview of hydrogen energy and hydrogen energy systems," Energy, Ecology and Environment, 1, 10-29.

M.T. Balta, I. Dincer, A. Hepbasli, 2010, "Potential methods for geothermal-based hydrogen production," International Journal of Hydrogen Energy, 35, 10, 4949-4961.

S.M. Alirahmi et al., 2022, "Green hydrogen & electricity production via geothermal-driven multi-generation system: Thermodynamic modeling and optimization," Fuel, 308, 122049.

Flavio Manenti, Gintaras V. Reklaitis (Eds.), Proceedings of the 34th European Symposium on Computer Aided Process Engineering / 15th International Symposium on Process Systems Engineering (ESCAPE34/PSE24), June 2-6, 2024, Florence, Italy

Regional Capacity Expansion Planning of Electricity and Hydrogen Infrastructure: A Norwegian Case Study

Kang Qiu[a], Espen Flo Bødal[a], Brage Rugstad Knudsen[a]

[a]*SINTEF Energy Research, Sem Sælands vei 11, Trondheim 7034, Norway*
kang.qiu@sintef.no

Abstract

The impact of hydrogen production, transmission, and storage is examined for a regional energy system in Norway. To this end, the optimal investment decisions are determined using a combined electricity and hydrogen capacity expansion model, which is expanded to include hydropower production. A case study of a regional energy system located in the Haugaland region of southwestern Norway is presented. This study analyses the driving factors behind the investment decisions of hydrogen production, storage, and transmission. Results of the case study indicate that hydrogen storage investments are influenced by varying hydrogen demand, while the locality of hydrogen production is influenced by the location of hydrogen demand, electric generation and the transport cost of both electricity and hydrogen.

Keywords: Hydrogen, Capacity expansion planning, Energy systems optimization.

1. Introduction

Hydrogen (H_2) can play a crucial role in abating CO_2 emissions in hard-to-abate sectors, such as industry and transportation. Flexible H_2 production is often mentioned to enable more cost-efficient integration of variable renewable energy (VRES) in the energy system as it can be used to balance variations in energy availability, thus reducing energy curtailment and increasing utilization of energy system infrastructure (Lange et al., 2023). Flexible H_2 demand in hard-to-abate sectors can deliver much of the same flexibility for the power system as H_2 storage systems for large-scale electricity storage, without the low round-trip efficiency of reconverting H_2 to electricity in fuel cells or H_2 gas turbines. The causes and effects of renewable power generation, transmission, the production of H_2, and the projected end-use demands of a regional energy system are tightly coupled. This necessitates a detailed analysis of the driving factors behind investment decisions. This paper investigates the role of green H_2 in the regional energy system using a linear capacity expansion planning model of the combined H_2 and electricity system. Production, storage, and transport of both energy carriers are modelled with the same level of detail, including energy storage in H_2 tanks and hydropower reservoirs which will become increasingly important as the shares of variable renewable energy of the total energy production increases.

The case study proposed is the regional energy system in Haugalandet, Norway. Haugalandet is a region in south-western Norway with significant industrial activity in the gas and metal sectors resulting in a potential local H_2 demand (Voldsund et al., 2021). Furthermore, Haugalandet has large hydropower capacities and significant potential for both on- and offshore wind generation. The export of power out of the region is considered through both onshore and offshore interconnectors. The locality of H_2 production facilities and their driving factors are analyzed, in particular, whether and when capacity expansion close to demand or close to energy source is optimal. Additionally, the resulting needs for storage and transport investments are analyzed as a part of assessing the needed H_2 infrastructure to supply potential demand from varying renewables.

2. Method

In this work, a capacity expansion model is used to determine the cost-optimal electricity and H_2 infrastructure. The modelling framework optimizes the investment capacity of production and conversion technologies, and transport infrastructure. Investment limitations due to resource limitations are included. Balance constraints keep track of production, demand, conversion, storage, and curtailment in every node for the energy carriers.

Electricity is generated from existing hydropower and wind power plants. Furthermore, offshore, and onshore wind capacities are given as an investment option. New storage technologies for the electricity side are modelled as batteries. Power is transported via transmission lines. Proton exchange membrane (PEM) electrolyzers produce H_2 while PEM fuel cells convert H_2 to electricity. They represent the coupling between the electricity and the H_2 system. H_2 storage is modelled as an investment option in the H_2 system. The joint electricity and H_2 optimization presented in (Bødal et al., 2020) is extended to model hydropower. The storage constraint is updated to include hydropower by adding regulated (I_{tin}^{reg}) and unregulated inflow (I_{tin}^{ureg}) and allowing for spillage (e_{tin}^{spill}) of water, modelled as

$$s_{tin} = s_{(t-1)in} + \eta_i^{in} e_{tin}^{in} - \left(\frac{1}{\eta_i^{out}}\right) e_{tin}^{out} + I_{tin}^{reg} + I_{tin}^{ureg} - e_{tin}^{spill}, \quad \forall i \in S.$$

Where s_{tin} is the storage level, e_{tin}^{in} or e_{tin}^{out} is the energy added to or subtracted from the storage at efficiencies η_i^{in} and η_i^{out} respectively. Indices are t for time, i for unit and n for node. In addition, there is a minimum production limit that is only active for hydropower storage. This states that water used for energy production and spilled water is at least equal to the unregulated inflow:

$$\left(\frac{1}{\eta_i^{out}}\right) e_{tin}^{out} + e_{tin}^{spill} \geq I_{tin}^{ureg}$$

3. Case study

The case study investigates the joint, optimal electricity and H_2 planning for the Haugaland region in Norway towards 2030[1]. Only regional demand and transport are considered for H_2, while power can be exported out of the region through offshore interconnectors abroad, or through onshore transmission lines into different market regions in Norway. The transmission capacity of the electrical grid is modelled from NVE atlas (NVE, 2023) and Statnett plans for the Haugaland area (Statnett, 2015). Investment costs for electrical transmission are determined from costs estimates by Statnett for new

---

[1] The model and data are available at: https://github.com/espenfb/HEIM/tree/master/case_haugaland.

420 kV transmission line Blåfalli-Gismarvik in the area (Statnett, 2020), while for the H_2 system compressed gas truck transport is assumed (Mintz et al., 2006).

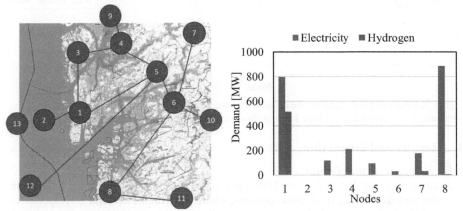

Figure 1: Left: Proposed case study of the Haugaland region. The region is modelled in eight nodes (node 1-8) and four market nodes (node 9-12). Right: Average electricity and H_2 demand per node.

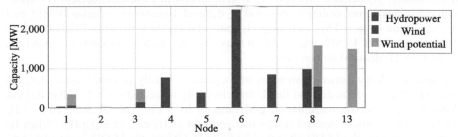

Figure 2: Hydropower, installed wind, and potential wind investment capacity per node. Node 13 represents possible offshore wind investment, while nodes 1-8 describe onshore wind installations and investment potential.

Both investment and operating decisions are determined in an integrated optimization within a planning horizon of one year. The case study has a geographical resolution modelled in thirteen nodes (Figure 1, left), whereof nine are joint electric and H_2 nodes and four are market nodes that enable electricity import and export out of the region. The joint electric and H_2 nodes (nodes 1-8, node 13) have a specified energy demand (Figure 1, right), hydropower capacities, and existing and potential wind generation (Figure 2).

The H_2 demand is based on a case study for decarbonizing local industry and transport (Voldsund et al., 2021), which includes local metal production and a ferry. While the future H_2 demand in existing industries and transport is limited, there exist opportunities for large-scale H_2 demand in node 1. This H_2 demand can arise from fuel-switching of existing offshore gas turbines, transport to Europe through existing pipelines or new industries, such as green steel production at the local industry area (Voldsund et al., 2021, p. 2). In this case, we assume that new green steel production is established and results in an H_2 demand of 90,000 (t H_2)/y.

We investigate five permutations of the case study. First, with constant and variable H_2 demand profiles in 4.1. Then, the effect of limitations in transmission investments and new wind power is analyzed in 4.2. and 4.3 respectively.

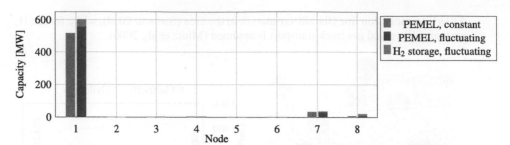

Figure 3: Effect of H₂ demand on the system. The green bar plot shows investments under constant H₂ demand while the blue bar plot shows investments under fluctuating H₂ demand.

4. Results and Discussion

4.1. Effect of H_2 demand profiles

The capacities and location of PEMEL investments are shown in Figure 3 for both constant and fluctuating H_2 demand profiles, where the average demand corresponds to Figure 1. In this case, no wind investments are allowed.

With constant H_2 demand profiles, the system invests in PEM electrolyzers (PEMEL) with capacities corresponding to the local H_2 demand. Investments in extra PEMEL and H_2 storage are triggered by fluctuating H_2 demand. In node 1, the extra PEMEL and H_2 storage capacity is 43 MW and 44 MWh respectively, which corresponds to around 8% of the local H_2 demand for one hour.

Transport of energy carriers from nodes where power is mostly generated (nodes 6,7, and 8) to where most of the demand for H_2 and electricity is located (node 1) is handled via expanding the electric transmission grid capacity by 9.1%. The levelized cost of H_2 (LCOH) with constant demand results in a unit cost of H_2 of 2.25 €/(kg H_2). When H_2 demand fluctuates the LCOH rises to 2.30 €/(kg H_2). Here, the LCOH is calculated by dividing the added investments and operational costs for the entire system due to H_2 production by the quantities of H_2 produced. The added costs are found by calculating the increased total system costs for a case with H_2 demand to a case without H_2 demand, where all other parameters are the same. Thus, the LCOH includes all costs related to the H_2 supply, incl. production, storage, and transportation in both the electricity and H_2 system.

4.2. Constrained electricity transmission system

While previous results indicate that the most cost-optimal way of providing H_2 to the system is to transport electricity by expanding the grid and producing H_2 at the demand site. Transmission grid development is becoming a bottleneck to the global energy transition due to prohibitively long construction times of 5-15 years and regulations that needs modernization (International Energy Agency, 2023). Lack of transmission capacity are restricting renewable generation and electrification projects that are important for emission reduction. This section discusses scenarios where investments in the electrical grids are not possible.

Figure 4 shows investments in the H_2 system technologies without and with constraints on grid investments, respectively, leading to expansion of the H_2 transport system. H_2 transport is modelled as compressed gaseous trucks with limited transport capacity per branch. The investments costs in the required capacity of trucks, trailers, and terminal are

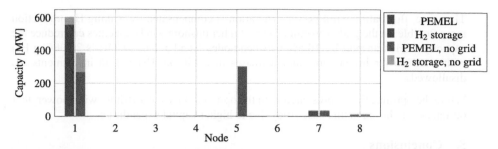

Figure 4: Investments into PEMEL and H_2 storage without and with limited grid investments.

determined with a techno-economic assessment model of the H_2 infrastructure (Mintz et al., 2006). With constrained electricity transmission, the PEMEL investments shift from node 1 to node 5. PEMEL investment in node 1 is reduced to 270 MW and only can serve 51% of the local H_2 demand, while node 5 is increased to 301 MW. Node 5 is a preferred location for H_2 conversion as it lies between nodes with high hydropower capacities (nodes 4-8) and the site of the highest H_2 demand in node 1. Subsequently, new investments are made in H_2 transport capacity between nodes 5 and 1 that are equal to the amount of PEMEL capacity in node 5. In total, only 2% (12 MW) of additional PEMEL investments are observed compared to the previous case with unconstrained grids.

Investments in H_2 storage capacity in node 1 rise significantly to 114 MW, which is 22% of the local H_2 demand for one hour and an increase of 161%. Negligible amounts of H_2 storage are seen in node 5. H_2 storage is co-located with the variable H_2 demand in node 1 on the energy-deficit side of the transmission constraint to reduce the investments in truck transport capacity. In node 5, H_2 can be produced and transported at a constant rate with a steady electricity supply from hydropower. The rest of the H_2 system remains largely the same as before because H_2 is produced locally at the demand sites. The LCOH is 2.50 €/(kg H_2) when grid investments are disallowed.

4.3. Effects on wind power

In this section, we discuss what happen to the previous cases when options to invest in wind power is added. The capacity expansion planning model invests in high capacities of onshore wind, which are located at nodes 1, 3, and 8 (shown in Figure 5). Onshore wind is particularly lucrative in this case study, as it is located close to nodes of high demand.

The new wind investments increase the capacity for electricity generation by 26.4% with and without modelled H_2 demand. As wind power is generated at locations with already high energy demands, the grid capacity does not need to be expanded to accommodate wind power investments. In the case of H_2 demand, wind investments are not shown to impact the electric transmission capacity. In the case of constrained transmission grids, adding investments in variable wind power results in 144 MW (+20.7 %) H_2 storage capacity, which is 26% of the local H_2 demand for one hour.

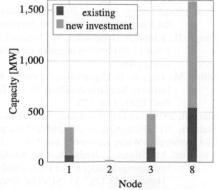

Figure 5: Existing onshore wind power capacities and optimal investments

Therefore, fluctuating wind power generation is compensated by making H_2 production more flexible. In the grid-constrained case, higher onshore wind capacities can reduce the capacity of H_2 transport by 10.5% between nodes 1 and 5. The LCOH is 2.30€/(kg H_2) with wind power investments and increases to 2.48€/(kg H_2) if grid investments are disallowed.

While the optimization is presented with the option to invest in offshore wind power, that option is not chosen in this case study due to higher capital expenditure.

5. Conclusions

The combined electricity and H_2 capacity expansion planning model with an extension to model hydropower has been presented and applied to a case study of a Norwegian regional energy system. The results indicate the impact of H_2 demand, electricity transmission, and renewable wind generation on the production, storage, and transport of H_2. Fluctuating local H_2 demand incentivizes flexibility in H_2 production and investments in H_2 storage. If further investments in electric transmission are allowed, the cost-optimal way to provide H_2 to the system is to produce at the location of demand. In this case, transmitting electricity is preferred over transporting H_2. When grid investments are prohibited, the production of H_2 moves to nodes which are closer to the hydropower capacities in the system while H_2 is transported to the demand site. Wind power does not increase the need for grid capacity investments, indicating that the transmission capacity in the case study can handle the integration of VRES. H_2 storage investments increase due to new fluctuating wind power generation, while H_2 transport capacities decrease due to the co-location of H_2 demand and wind power production. The LCOH ranges from 2.25-2.50 €/(kg H_2), where the main cost driver is electricity grid capacity.

Acknowledgement

This publication has been funded by the project Hydrogen pathways 2050, coordinated by IFE. The authors gratefully acknowledge the financial support from the Research Council of Norway (grant 326769) and the user partners Gassco, Equinor and Statkraft.

References

Bødal, E.F., Mallapragada, D., Botterud, A., Korpås, M., 2020. Decarbonization synergies from joint planning of electricity and hydrogen production: A Texas case study. Int. J. Hydrog. Energy.

International Energy Agency, 2023. Electricity Grids and Secure Energy Transitions: Enhancing the Foundations of Resilient, Sustainable and Affordable Power Systems. OECD.

Lange, H., Klose, A., Lippmann, W., Urbas, L., 2023. Technical evaluation of the flexibility of water electrolysis systems to increase energy flexibility: A review. Int. J. Hydrog. Energy 48.

Mintz, M., Gillette, J., Elgowainy, A., Paster, M., Ringer, M., Brown, D., Li, J., 2006. Hydrogen Delivery Scenario Analysis Model for Hydrogen Distribution Options. Transp. Res. Rec. 1983, 114–120.

Statnett, 2020. Konsesjonssøknad: Ny 420 kV-forbindelse Blåfalli-Gismarvik.

Statnett, 2015. Forsyning av økt kraftforbruk på Haugalandet. Statnett.

The Norwegian Water Resources and Energy Directorate, n.d. NVE Atlas

Voldsund, M., Straus, J., Jordal, A.B.K., Knudsen, B.R., 2021. CO2-fangst fra hydrogenproduksjon ved Haugaland Næringspark : rapport fra CLIMIT idéstudie "Haugaland CCS," 31. SINTEF Energi AS.

Flavio Manenti, Gintaras V. Reklaitis (Eds.), Proceedings of the 34th European Symposium on Computer Aided Process Engineering / 15th International Symposium on Process Systems Engineering (ESCAPE34/PSE24), June 2-6, 2024, Florence, Italy

Framework based on a coupled input-yield model for infrastructure-driven agriculture

Farhat Mahmood, *Tareq Al-Ansari

College of Science and Engineering, Hamad Bin Khalifa University, Qatar Foundation, Doha, Qatar
talansari@hbku.edu.qa

Abstract

As the population and demand for food increases, the need to produce food locally in a sustainable manner is imperative for a country's food security. Traditional agriculture in arid regions faces difficulties due to non-arable land and harsh climate, hindering crop growth. Controlled environment agriculture, particularly greenhouses, offers a practical solution by protecting crops from external elements and supplying necessary resources. However, controlled environment agriculture systems require significant energy and water resources to sustain optimal conditions for crop development. In greenhouse operations, the innovative approach of the potential repurposing of excessive cooling from underutilised mega infrastructure could lead to cost savings and more efficient production. Therefore, this research introduces a novel approach that utilises the infrastructure from the 2022 FIFA World Cup™ for agricultural purposes. It presents a unique methodology for the dual use of large-scale facilities, focusing on integrating agricultural systems near stadiums using existing cooling systems and space. The results indicate that at optimal conditions, such as day and night temperatures of 18 °C and 24 °C, tomatoes can be produced for \$3.50 kg^{-1}. Furthermore, the developed framework is flexible and can be applied globally for co-utilising infrastructure for agricultural production.

Keywords: Resource optimisation, Greenhouses, Energy, Crop yield, Infrastructure co-utilisation

1   Introduction

The global population is growing rapidly, leading to a significant rise in the demand for essential resources like food, energy, and water. By 2050, it's estimated that the need for these resources will surge by approximately 60% for food, 55% for water, and 80% for energy (Larsson, 2018). Therefore, it is critical to enhance the production of these resources in an efficient and environmentally friendly manner.

In terms of food production, in regions like Qatar, with extreme weather and limited water, traditional agriculture is impractical, leading to heavy reliance on food imports, which can be wasteful and costly (Mahmood et al., 2020). Furthermore, standard agricultural practices and food transportation contribute significantly to global greenhouse gas emissions, exacerbating climate change and further impacting agricultural productivity. Conversely, greenhouses provide a controlled environment for plant growth, regulating temperature and humidity, making them a sustainable solution for food production in challenging climates and helping to mitigate food security issues.

Greenhouses in arid regions like Qatar require significant energy, affecting their cost-effectiveness and sometimes leading to shutdowns during summer (Mahmood et al., 2023a). However, Qatar's innovative approach to repurposing infrastructure from the

2022 FIFA World Cup™ offers an innovative solution. The advanced cooling systems from the seven state-of-the-art stadiums initially used to maintain temperatures during the tournament can be integrated with greenhouse systems. This repurposing can substantially reduce the production costs for greenhouses, enabling efficient, year-round food production. This strategy not only optimises stadium infrastructure post-event but also supports Qatar's goals for food security and sustainable development.

2   Methodology

For optimal growth and development, plants cultivated in greenhouse settings need precise control over microclimate and irrigation (Engler and Krarti, 2021). Greenhouses provide the provision to and maintain these ideal conditions throughout different seasons, utilising various technologies and methods. Therefore, this study focuses on closed greenhouses as they are more energy-efficient in arid climates than open ones, reducing cooling needs and better maintaining conditions for crop cultivation.

2.1   Greenhouse requirement

Greenhouses in challenging environments need continuous energy, water, and CO_2 to support plant growth (Mahmood et al., 2023b). Energy is essential for regulating the internal climate, primarily for cooling. Furthermore, calculating the greenhouse's thermal load is crucial for efficiency and creating suitable growing conditions, especially in regions with extreme climates. The schematic of the closed greenhouse is demonstrated in Figure 1.

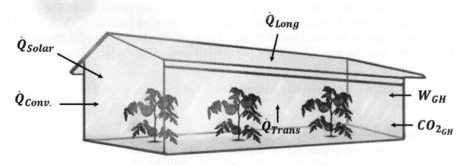

Figure 1: Closed greenhouse schematic.

2.2   Greenhouse input model

The following equation calculates the cooling requirement of the greenhouse (Van Beveren et al., 2013):

$$C_g \frac{dT_{GH}(t)}{dt_h} = \dot{Q}_{Solar}(t) + \dot{Q}_{Conv.}(t) + \dot{Q}_{Long}(t) + \dot{Q}_{Trans.}(t) - \dot{Q}_{Cooling}(t) \qquad (1)$$

Water is primarily used to irrigate crops and sustain the necessary humidity levels within the greenhouse. Greenhouses can optimise their water use by accurately assessing water requirements, ensuring that plants receive the right amount of hydration without wastage. The water requirement is determined by the following (Van Beveren et al., 2013):

$$W_{GH}(h,d) = \frac{d\chi(h,d)}{dt_h} - E(h,d) - C(h,d) \tag{2}$$

CO_2 plays a pivotal role in photosynthesis, and its addition to the greenhouse atmosphere can enhance both the quality and yield. Plants steadily utilise the available CO_2, leading to a decrease in its concentration over time. Therefore, extra CO_2 must be introduced from an external source to maintain the optimal levels necessary for plant growth. The amount of CO_2 required is given as follows (Stanghellini et al., 2012):

$$CO_{2_{GH}}(h,d) = \frac{dCO_2(h,d)}{dt_h} + CO_{2_{pa}}(h,d) \tag{3}$$

2.3 Greenhouse yield model

TOMGRO, which considers various greenhouse parameters, estimates crop yield in a greenhouse setting (Jones et al., 1999). The value of N is calculated by the following:

$$\frac{dN(h,d)}{dt_h} = N_m f_n(T_{GH}(h,d)) \tag{4}$$

The *LAI* is given by:

$$\left.\begin{array}{l} \dfrac{dLAI(h,d)}{dt_h} = \delta_{LEA} p_d \lambda_{LEA}(T_{GH}) \dfrac{e^{(\beta(N-N_b))}}{1 + e^{(\beta(N-N_b))}} \dfrac{dN(h,d)}{dt_h} \quad if: LAI(h,d) \\ \qquad\qquad \leq LAI_{max} \\[2mm] \dfrac{dLAI(h,d)}{dt_h} = 0 \quad if: LAI(h,d) > LAI_{max}. \end{array}\right\} \tag{5}$$

The following equation gives the rate of dry fruit development of the plant:

$$\frac{dW_f(d)}{dt_d} = GR_{net}(d)\alpha_f f_f(T_d(d))(1 - e^{(-v(N(d)-N_{FF}))})g(T_{GH}(d)) \quad if: N \tag{6}$$
$$> N_{FF}$$

The following equation calculates the total weight of the plant:

$$\frac{dW(d)}{dt} = \frac{dW_f(d)}{dt} + (V_{max} - p_l)p_d\frac{dN(d)}{dt} \tag{7}$$

The total mature fruit weight is given as follows:

$$\frac{W_M(d)}{dt} = D_F(T_{GH})(W_F(d) - W_M(d)) \quad when \quad N > N_{FF} + K_F \tag{8}$$

The details of the variables and parameters used in the greenhouse model can be found in Mahmood et al. (2023b).

2.4 Cost of production

Greenhouse production costs consist of the initial capital and the operational cost. The total cost is given as follows:

$$Total\ cost = Capital\ cost\left(\frac{r(1+r)^n}{(1+r)^n - 1}\right) + Operational\ cost \tag{9}$$

The total cost of greenhouse operation includes factors like the discount rate (r) and the greenhouse's expected lifespan (n) in years. The discount rate is set at 0.1, while the lifespan is assumed to be 25 years. The cooling system's cost is reduced by utilising the existing cooling system infrastructure, contributing significantly towards reducing the overall capital cost.

$$Capital\ cost = A_{GH}S_{cost} + A_{GHS}C_{cost}N_{CR} \tag{10}$$

The capital cost is given as follows:

Operational cost

$$= \sum_{h=1}^{6480} \left(E_{Elec}C_{Elec} + m_{H_2O}\ C_{H_2O} + m_{CO_2}\ C_{CO_2} \right) + F_C + L_C \tag{11}$$

The cost of different inputs is given in Table 1.

Table 1: Cost of different greenhouse inputs.

Inputs	Symbol	Price
Structure construction	S_{cost}	\$25 m^{-2}
Electricity	C_{Elec}	\$0.036 kWh^{-1}
Water	C_{H_2O}	\$1.43 m^{-3}
CO$_2$	C_{CO_2}	\$2 kg^{-1}
Fertilizer	F_C	\$3.21 m^{-2}
Labour cost	L_C	\$15.23 m^{-2}

2.5 Greenhouse optimisation

The model for the greenhouse is optimised using decision variables to identify parameters that reduce production costs. The objective function is given by:

$$min\left(\frac{Capital\ cost\ \left(\frac{r\ (1+r)^n}{(1+r)^n - 1} \right) + Operational\ cost}{Total\ yield\ produced} \right) \tag{12}$$

2.6 Co-utilisation of infrastructure

Post-World Cup, some stadiums are planned to be downsized or repurposed, reducing their cooling needs and creating surplus capacity. This excess capacity presents an opportunity to support local food production by integrating stadium cooling systems with nearby greenhouse systems. This approach could significantly lower greenhouse capital costs, as cooling is a major financial factor.

3   Results and discussion

The framework calculates the greenhouse's yield and energy, water, and CO$_2$ needs in Qatar by utilising data on solar radiation and outside temperature from a standard typical meteorological year (TMY) file. Furthermore, the optimisation determines the decision variables that produce the lowest costs associated with yield production.

3.1 Optimum decision variables

This model was simulated for a greenhouse with an area of 625 m^2 (25m x 25m), focusing on tomato cultivation. It involves optimising various factors over the harvest period, including day and night temperatures, RH, CO$_2$ levels, LAI, and the material used for the greenhouse covering. The optimal decision variables are a nighttime temperature of 18°C, while the daytime temperature is higher, at 24°C. The relative humidity level of 75%, with a CO$_2$ concentration of 1000 ppm. In terms of output, the total yield achieved is

24.96 tons, producing a unit yield of 46.84 kgm^{-2}. The total operational cost amounts to $ 46,442.34, resulting in a cost of $3.50 kg^{-1}.

3.2 Infrastructure driven greenhouses

This framework employs a generalised model with optimal decision parameters for greenhouses of various sizes adjacent to the constructed stadiums. Qatar's small, uniformly topographic area of 11,571 km^2 allows for using the same TMY data for all stadiums. Figure 2 illustrates the potential yield that could be achieved by constructing greenhouses adjacent to the stadiums using the available land and cooling. The greenhouse adjacent to Education City Stadium produces the lowest yield, at 239.19 tons, due to limited space, while the greenhouse near Lusail Stadium produces the highest yield, at 2568.59 tons, due to higher available land and cooling capacity. The proposed greenhouses can produce a total yield of 8240.91 tons by co-utilising the stadium's infrastructure. Figure 3 illustrates the variation in cooling capacity usage among the stadiums' greenhouses as it correlates with the available land and the capacity of each cooling system. Education City Stadium requires a minimal cooling capacity of just 1.64 %, in contrast to the ones near Al-Rayyan Stadium, which demand the highest cooling usage at 24.56 %. These results indicate that the proposed greenhouses only require a small part of the available cooling to operate throughout the year.

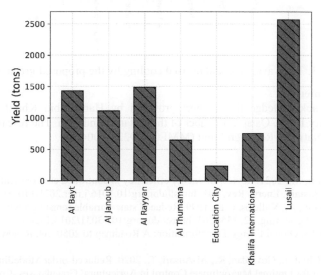

Figure 2: Potential yield produced by the proposed greenhouses.

4 Conclusion

Open-field agriculture is unfeasible in arid regions like Qatar due to harsh climates, infertile land, and water scarcity. However, closed greenhouses offer a year-round solution but with high capital costs. Therefore, this study proposes using unutilised and sophisticated infrastructure from FIFA World Cup 2022™ for food production. The framework incorporates a model for greenhouse requirements and crop yield based on parameters such as temperature, humidity, CO_2 levels, leaf area index, plant density, and covering material. Economic optimisation identifies optimal conditions for the most cost-effective yield, including specific temperature, humidity, and CO_2 settings.

The proposed framework is both general and flexible, developed to be applicable across various geographical locations with differing infrastructures. Utilizing existing infrastructure can enhance food sustainability by utilizing available resources for food production. This approach not only optimizes the use of existing infrastructure and systems but also reduces the need for additional investments and environmental impacts associated with new constructions. By adapting and enhancing current infrastructures, food can be produced sustainably and cost-efficiently.

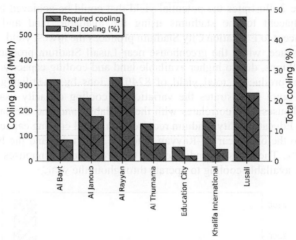

Figure 3: Total available and utilised cooling by the proposed greenhouses.

Acknowledgment

The authors acknowledge the support provided by Hamad bin Khalifa University, Education, City Doha, Qatar, a member of the Qatar Foundation. The research is funded by the Qatar National Research Fund (MME01-0922-190049).

References

Engler, N., Krarti, M., 2021. Review of energy efficiency in controlled environment agriculture. Renew. Sustain. Energy Rev. 141. https://doi.org/10.1016/j.rser.2021.110786

Jones, J.W., Kenig, A., Vallejos, C.E., 1999. Reduced state-variable tomato growth model. Trans. Am. Soc. Agric. Eng. 42, 255–265. https://doi.org/10.13031/2013.13203

Larsson, M., 2018. Global Energy Transformation: A Roadmap to 2050. Int. Renew. Energy Agency.

Mahmood, F., Ghiat, I., Govindan, R., Al-Ansari, T., 2020. Reduced-order Modelling (ROM) Approach for Optimal Microclimate Control in Agricultural Greenhouses. Comput. Aided Chem. Eng. 48, 1879–1884. https://doi.org/10.1016/B978-0-12-823377-1.50314-1

Mahmood, F., Govindan, R., Bermak, A., Yang, D., Al-Ansari, T., 2023a. Data-driven robust model predictive control for greenhouse temperature control and energy utilisation assessment. Appl. Energy 343. https://doi.org/10.1016/j.apenergy.2023.121190

Mahmood, F., Luqman, M., Al-Ansari, T., 2023b. Application of a dynamic hybrid energy-crop yield model to support the co-utilisation of mega infrastructure for food security. Environ. Technol. Innov. 31. https://doi.org/10.1016/j.eti.2023.103157

Stanghellini, C., Bontsema, J., De Koning, A., Baeza, E.J., 2012. An algorithm for optimal fertilization with pure carbon dioxide in greenhouses. Acta Hortic. 952, 119–124. https://doi.org/10.17660/ActaHortic.2012.952.13

Van Beveren, P.J.M., Bontsema, J., Van Straten, G., Van Henten, E.J., 2013. Minimal heating and cooling in a modern rose greenhouse. IFAC Proc. Vol. 4, 282–287. https://doi.org/10.3182/20130828-2-SF-3019.00026

Flavio Manenti, Gintaras V. Reklaitis (Eds.), Proceedings of the 34th European Symposium on Computer Aided Process Engineering / 15th International Symposium on Process Systems Engineering (ESCAPE34/PSE24), June 2-6, 2024, Florence, Italy

Integrating effort- and gradient-based approaches in optimal design of experimental campaigns

Marco Sandrin[a,b], Benoît Chachuat[a*], Constantinos C. Pantelides[a,b]

[a]*Sargent Centre for Process Systems Engineering, Department of Chemical Engineering, Imperial College London, London SW7 2AZ, United Kingdom*
[b]*Siemens Industry Software, London W6 7HA, United Kingdom*
b.chachuat@imperial.ac.uk

Abstract

Model-based design of optimal experimental campaigns comprising multiple parallel runs can prove computationally challenging. Effort-based methods can help in overcoming some of these challenges through discretising the experimental design space. However, the quality of the resulting approximate solutions depends heavily on this a priori discretisation. This paper presents a methodology for integrating the appealing features of effort-based methods with those of conventional gradient-based approaches, with a view to computing maximally-informative campaigns of experiments for improving parameter precision. The effectiveness of the methodology is demonstrated on a case study involving a microbial culture dynamic model.

Keywords: model-based design of experiments, optimal experiment design, parameter precision

1. Introduction

Model-based design of experiments (MBDoE) is instrumental to accelerating the development of predictive mechanistic models in Process Engineering (Franceschini and Macchietto, 2008). Much of the work in this area to date has focused on sequential MBDoE aiming to improve parameter precision, where experiments are designed and executed one-at-a-time by maximising an appropriate measure of their information content using gradient-based methods. For nonlinear models, the resulting optimisation problems are typically nonconvex, making them prone to converge to suboptimal solutions or even fail when local optimisation techniques are used.

Effort-based methods (Fedorov and Leonov, 2014; Kusumo et al., 2022) are particularly suited for designing campaigns that comprise multiple experiments to be executed in parallel. They employ a discretisation of the experimental design space into a finite set of candidate experiments, and then determine the number of replicates (the *efforts*) of each candidate, with a view to maximising the information content of the combined experiments. This leads to a (possibly mixed-integer) convex program that can be solved reliably using state-of-the-art (mixed-integer) nonlinear optimisation techniques. A caveat of this approach is that the quality of the solution is strongly affected by the extent to which the set of candidates covers the experimental design space of interest.

In this contribution, we propose a methodology for integrating the appealing features of effort-based methods with those of conventional gradient-based approaches for designing maximally-informative parallel experimental campaigns. Following an initial discretisation of the experimental design spaces, the values of the efforts are first optimised to determine which candidate experiments should be included in the experimental campaign. In a second step, the selected experiments are refined using a gradient-based search to further increase the information content. The proposed methodology is implemented as a new experiment design solver within the gPROMS modelling framework (Siemens Industry Software, 1997–2023). We illustrate its effectiveness and benefits on a case study involving a microbial culture model.

2. Methodology

2.1. Problem definition

We consider a system with n_x experimental controls $x \in \mathbb{X} \subset \mathbb{R}^{n_x}$ and n_y measured responses $y \in \mathbb{R}^{n_y}$,

$$y = \eta(\theta, x) + \epsilon \tag{1}$$

where $\theta \in \mathbb{R}^{n_\theta}$ are the uncertain parameters in the model η. For simplicity, measurements are assumed to be independent, and the measurement error $\epsilon \in \mathbb{R}^{n_y}$ is assumed to have zero mean and uncorrelated homoscedastic covariance Σ_y.

We consider a campaign comprising $N_t > 0$ experimental runs to be executed in parallel, with the objective of generating data for the estimation of the model parameters. Such campaigns often include repeated runs with identical experimental controls (*replicates*), so the experimental design ξ can be defined as (Fedorov and Leonov, 2014)

$$\xi := \begin{Bmatrix} x_1 & \dots & x_{N_c} \\ p_1 & \dots & p_{N_c} \end{Bmatrix} \tag{2}$$

where $N_c \leq N_t$ is the number of distinct experiments, and $p_i \in (0, N_t]$ the number of replications (also called *effort*) of the i-th experimental candidate with controls x_i for $i = 1, \dots, N_c$ so that $\sum_{i=1}^{N_c} p_i = N_t$. The set of experimental controls $\{x_i, \dots, x_{N_c}\}$ is the *support* of the experimental design ξ, supp(ξ).

Determining the optimal design ξ^* entails maximising some scalar information criterion ϕ over all possible numbers of supports $N_c \leq N_t$, the experimental controls x_i and the corresponding efforts p_i simultaneously.

$$\xi^* \in \arg\max_{\xi} \phi(\xi) \tag{3}$$

Since our focus is on campaigns which are optimal for model calibration, the information criterion ϕ is based on the Fisher information matrix (FIM), $M \in \mathbb{R}^{n_\theta \times n_\theta}$ computed as (Atkinson et al., 2007)

$$M(\xi, \theta) := \sum_{i}^{N_c} p_i A(x_i, \theta) \tag{4}$$

where A is the atomic matrix associated with the experiment candidate x_i, and is calculated as

$$A(x_i, \theta) := \frac{\partial \eta}{\partial \theta}(x_i, \theta)^{\mathrm{T}} \Sigma_y^{-1} \frac{\partial \eta}{\partial \theta}(x_i, \theta) \tag{5}$$

2.2. Integration of effort- and gradient-based approaches

The proposed approach couples sampling- and gradient-based designs. Following a discretisation of the experimental space, the methodology iterates between solving an effort-based design—which provides the optimal number of supports and the values of the experimental efforts—and performing a conventional, gradient-based search over the space of the experimental controls. This approach is similar to that proposed by Vanaret et al. (2021); however, it is crucial to combine the algorithms in an iterative methodology, as optimality cannot be guaranteed after performing each step only once.

One possible way to integrate the two steps is to use the solution of the sampling-based design as the initial guess for the gradient-based design. Then, the refined supports may be added to the initial set of experiment candidates, and the exact design performed again to update the values of the efforts. A variation of the above involves taking only one iteration of the gradient-based algorithm, before returning to the effort-based formulation.

2.2.1. Effort-based step

The first optimisation step builds on the continuous-effort approach (Kusumo et al., 2022; Vanaret et al., 2021), in which the experimental design space is discretised into a finite set of experiment candidates $\mathbb{X}_s := \{x_i, \dots, x_{N_s}\} \subset \mathbb{X}$ with $N_s \gg N_t$, for instance through the application of low-discrepancy sampling (Sobol', 1967). The search ξ^* in Eq. (3) is then reduced to a search over the efforts p_i associated with each experiment $x_i \in \mathbb{X}_s$,

$$\left(p_1^*, \dots, p_{N_s}^*\right) \in \arg\max_{p} \phi\left(p\right) \tag{6}$$

where the optimal efforts $\left(p_1^*, \dots, p_{N_s}^*\right)$ are allowed to assume zero values. The support of the optimal design ξ^* is obtained after the solution of (6) as

$$\text{supp}(\xi^*) := \{x_j \in \mathbb{X}_s : p_j > 0\} \tag{7}$$

The optimal exact local design is computed by solving the following integer nonlinear program (INLP), in which the D-optimal design criterion is considered:

$$\max_{p} \log \det \sum_{i=1}^{N_s} p_i \mathbf{A}\left(x_i, \boldsymbol{\theta}_0\right) \tag{8}$$

$$\text{s.t.} \sum_{i=1}^{N_s} p_i = N_t, \quad p_i \in \mathbb{Z}_+, \forall i \tag{9}$$

Since we are computing a local design, the sensitivities and atomic matrices in (5) are calculated at a nominal parameter value $\boldsymbol{\theta}_0$. Except for the integrality restrictions $p_i \in \mathbb{Z}_+$, the optimisation problem (8)–(9) is convex since the objective function is concave and the constraint is linear in the efforts p_i.

Solution of this INLP relies on an outer-approximation (OA) decomposition algorithm (Duran and Grossmann, 1986; Fletcher and Leyffer, 1994) as in Sandrin et al. (2023). The algorithm is initialised by solving a continuous nonlinear program (NLP) obtained by relaxing the integrality restrictions to $p_i \geq 0$. This NLP subproblem provides an upper bound on the exact design optimum, while a lower bound is computed by rounding the optimal efforts of the continuous design via apportionment techniques such as that proposed by Pukelsheim and Rieder (1992). Subsequently, the OA iterates between solving a mixed-integer linear (MILP) master subproblem—which updates the upper bound on the information content and provides a new exact design candidate—and evaluating the D-optimality criterion in lieu of a primal subproblem as there are no continuous decision variables. A single linear cut is added to the master subproblem at each iteration through linearising the convex objective at the solution point $\hat{p}$ from the previous iteration if the corresponding FIM in Eq. (4) is non-singular; otherwise, the integer cut $\|p - \hat{p}\|_1 \geq 1$ is appended to the master subproblem.

2.2.2. Gradient-based step

The solution of the exact design in (8)–(9) yields the set $\mathbb{X}_c^* = \{x_1^*, ..., x_{N_c}^*\} \subseteq \mathbb{X}_s$ together with the values of the associated efforts p_i^*, $i = 1, ..., N_c$. These can then serve as initial guesses for a gradient-based algorithm aiming to simultaneously design a set of N_c distinct experiments.

The gradient-based experiment design problem, based on the D-optimal criterion, makes use of the EXPDES solver in gPROMS which employs the following objective function:

$$\max_{x \in \mathbb{X}} \left(\det \sum_{i=1}^{N_c} p_i^* \mathbf{A}(x_i, \boldsymbol{\theta}_0) \right)^{1/N_\theta} \tag{10}$$

The function in (10) is highly nonconvex and therefore global optimality of the solution obtained cannot be guaranteed. However, since the initial guesses were obtained via the solution of the effort-based design problem, they should already be in highly informative regions of the experimental space; and any locally optimal solution obtained starting from these guesses should represent an improvement in the objective function.

The optimal solution x_i^{**} of the problem (10) is appended to the original set of experimental candidates i.e. $\mathbb{X}_s^* = \mathbb{X}_s \cup \{x_1^{**}, ..., x_{N_c}^{**}\}$. The next iteration of the effort-based approach is based on the augmented set $\mathbb{X}_s^*$. The overall iteration terminates when two successive executions of the effort-based step return identical optimal solutions.

3. Case study

The proposed methodology is tested on a model describing the fermentation of baker's yeast in a semi-batch reactor. Assuming Monod-type kinetics for biomass growth and substrate consumption, the system is described by the following set of ODEs:

$$\frac{dy_1}{dt} = (r - u_1 - \theta_4) y_1 \tag{11}$$

$$\frac{dy_2}{dt} = -\frac{r y_1}{\theta_3} + u_1 (u_2 - y_2) \tag{12}$$

with $r = \dfrac{\theta_1 y_2}{\theta_2 + y_2}$ $\tag{13}$

where y_1 [gL^{-1}] is the biomass concentration; y_2 [gL^{-1}] the substrate concentration; u_1 [h^{-1}] the dilution factor; u_2 [gL^{-1}] the substrate concentration in the feed; and θ_i, $i = 1, ..., 4$ the model parameters with nominal values $\boldsymbol{\theta}_0 = [0.1, 0.1, 0.1, 0.1]^T$.

A campaign comprising $N_t = 4$ parallel experiments is to be designed. We assume a fixed experiment duration of $\tau = 10$ h and $N_{sp} = 5$ equidistant sampling times for the output measurements y_1 and y_2. The initial conditions are $y_1(0) = 7$ and $y_2(0) = 0.1$. We assume time-invariant profiles for the experimental controls $u_1 \in [0.05, 0.2]$, $u_2 \in [5, 35]$.

For the effort-based design, we consider two different discretisations of the experimental design space, with $N_s = 20$ and $N_s = 100$ respectively. The effort-based optimisation is solved with gPROMS' mixed-integer nonlinear programming solver MINLPOA. The relaxed NLP is solved using gPROMS' sequential quadratic programming solver

NLPSQP with both feasibility and optimality tolerances set to 10^{-9}. Master MILP subproblems are solved using XPRESS (FICO, 1983–2023) with relative and absolute convergence tolerances of 10^{-6} and 10^{-9}, respectively. The OA iterations are terminated when the absolute gap between the master solution value and the incumbent is below 10^{-6}. The gradient-based optimisation is solved by NLPSQP with optimality tolerance set to 10^{-9}.

3.1. Discretisation of the experimental space with 20 samples

The optimal design determined by the effort-based step consists of three supports, $\mathbb{X}_c^* = \{(0.1813, 31.25), (0.0640, 19.06), (0.1391, 34.06)\}$ with corresponding efforts $\boldsymbol{p}^* = \{1, 1, 2\}$ and a D-optimality criterion based on Eq. (8) of 43.15. The improved design after the gradient-based step consists of the following updated supports, $\mathbb{X}_c^{**} = \{(0.1983, 35), (0.05, 5), (0.1277, 35)\}$. Notice how the supports in the refined campaign move to the boundary of the experimental space $\mathbb{X}$ compared to their (sampled) effort-based counterparts. The updated D-optimality criterion is 44.28, demonstrating that the gradient-based step is effective and can significantly increase the information content of the experimental campaign in the situation where the initial discretisation of the design space is rather sparse. A subsequent run of the effort-based optimisation on the augmented candidate set does not improve the results. Figure 1a shows the distribution of the samples over the experimental design space, together with the supports selected by the effort-based step and those refined through the gradient-based step.

3.2. Discretisation of the experimental space with 100 samples

The optimal design determined by the effort-based step now includes four supports, $\mathbb{X}_c^* = \{(0.1391, 34.06), (0.1191, 33.83), (0.1988, 34.77), (0.0558, 8.984)\}$ with corresponding efforts $\boldsymbol{p}^* = \{1, 1, 1, 1\}$ and a D-optimality criterion of 44.01. The subsequent gradient-based step refines the supports to $\mathbb{X}_c^{**} = \{(0.1621, 35), (0.1096, 35), (0.1897, 35), (0.05, 5)\}$ and increases the D-optimality criterion to 44.33. As in the previous case, a second effort-based step does not change the solution and therefore the algorithm terminates. Figure 1b plots the distribution of samples, supports of the optimal effort-based design, and refined experiments within the experimental space.

Comparing the results obtained with the case considered above, we note that with higher numbers of samples, the experimental design space is covered more evenly, and therefore the margin for improvement that can be achieved via the gradient-based refinement step is smaller. On the other hand, even with $N_s = 20$, refinement does succeed in obtaining a solution that is superior to that achieved by the effort-based approach applied to the $N_s = 100$ case (objective function values 44.28 and 44.01 respectively).

4. Conclusions

We proposed a methodology for combining effort- and gradient-based optimisation steps to design more informative parallel experimental campaigns. The addition of a gradient-based search over the experimental space leads to designing more informative campaigns compared to those obtained after a sole effort-based optimisation. We illustrated this methodology on a simple case, where the benefits are particularly significant when the initial discretisation of the experimental design space is sparse. This gives us confidence that the methodology could also be effective in higher-dimensional problems.

Future work will entail investigating various degrees of interaction between the effort- and gradient-based steps, with a view to providing optimality guarantees. Moreover, we wish to extend the work from local designs to more robust approaches which account for uncertainty in the values of the model parameters.

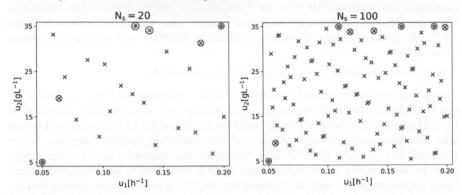

Figure 1. Distribution of the samples over the design space (blue crosses), supports of the optimal design after the effort-based step (circled blue crosses), and refined experiments after the gradient-based step (circled red points).

Acknowledgements

This project has received funding from the European Union's Horizon 2020 research and innovation programme under the Marie Skłodowska-Curie grant agreement No. 955520 (Digitalgaesation).

References

A.C. Atkinson, A.N. Donev, R. Tobias, 2007. *Optimum experimental design, with SAS*. Oxford University Press.

M.A. Duran, I.E. Grossmann, 1986. An outer-approximation algorithm for a class of mixed-integer nonlinear programs, *Mathematical Programming*, **36**, 307–339.

V.V. Fedorov, S.L. Leonov, 2014. *Optimal design for nonlinear response models*. CRC Press.

FICO, 1983–2023. Xpress Solver, https://www.fico.com/en/products/fico-xpress-optimization

R. Fletcher, S. Leyffer, 1994. Solving mixed integer nonlinear programs by outer approximation, *Mathematical Programming*, **66**, 327–349.

G. Franceschini, S. Macchietto, 2008. Model-based design of experiments for parameter precision: State of the art, *Chemical Engineering Science*, **63**, 4846–4872.

K.P. Kusumo, K. Kuriyan, S. Vaidyaraman, S. García-Muñoz, N. Shah, B. Chachuat, 2022. Risk mitigation in model-based experiment design: A continuous-effort approach to optimal campaigns, *Computers & Chemical Engineering*, **159**, 107680.

F. Pukelsheim, S. Rieder, 1992. Efficient rounding of approximate designs. *Biometrika*, **79**, 763–770.

M. Sandrin, K.P. Kusumo, C.C. Pantelides, B. Chachuat, 2023. Solving for exact designs in optimal experiment campaigns under uncertainty [Manuscript submitted for publication to ADCHEM 2024].

Siemens Industry Software, 1997–2023. gPROMS, https://www.siemens.com/global/en/products/automation/industry-software/gproms-digital-process-design-and-operations.html

I.M. Sobol', 1967. On the distribution of points in a cube and the approximate evaluation of integrals, *USSR Computational Mathematics and Mathematical Physics*, **7**, 86–112.

C. Vanaret, P. Seufert, J. Schwientek, G. Karpov, G. Ryzhakov, I. Oseledets, N. Asprion, M. Bortz, 2021. Two-phase approaches to optimal model-based design of experiments: how many experiments and which ones?, *Computers & Chemical Engineering*, **146**, 107218.

Flavio Manenti, Gintaras V. Reklaitis (Eds.), Proceedings of the 34th European Symposium on Computer Aided Process Engineering / 15th International Symposium on Process Systems Engineering (ESCAPE34/PSE24), June 2-6, 2024, Florence, Italy

Modelling and Simulation of a Forward Osmosis Process

Matej Ružička[a], Ingrid Helgeland[b] and Radoslav Paulen[a]

[a]*Faculty of Chemical and Food Technology, Slovak University of Technology in Bratislava, Bratislava, Slovakia*
[b]*Aquaporin A/S, Lyngby, Denmark*
xruzickam@stuba.sk

Abstract

Forward osmosis (FO) is an effective alternative to a more established reverse osmosis (RO) process, which is used mainly for water extraction and concentration of aqueous solutions. This work is focused on modelling of the FO process to understand its behaviour, characteristic features required for its subsequent optimisation. We develop the mathematical model of the process based on mass balances and theoretical foundations from mass transfer theory. We then fit the model using the experimental data. We study the appropriateness of using white, grey, and black box modelling principles. The results favour the grey-box strategy as it accounts for common over-simplifications of the model on mass transfer effects, yet it respects the fundamental laws.

Keywords: Membrane Processes, Forward Osmosis, Reverse Osmosis, Mathematical Models

1. Introduction

Membrane technology is well established in process industry mostly for purification and concentration of aqueous solutions. A popular membrane process, e.g., for seawater desalination is reverse osmosis (RO). Compared to RO, the forward osmosis (FO) process is less popular (Kucera, 2015). Its potential is though vast in water purification and solutions concentration. While in RO, a high external hydraulic pressure is applied to the feed solution, in FO, the separation is based on a natural osmotic pressure. This enables various benefits such as harmlessness to feed solutions and promising energy savings.

To stand up to the expectations, the FO plants need to be optimally designed and operated (Ali et al., 2021). For this purpose, mathematical modelling is needed to reveal characteristic (static and dynamic) behaviour of the process under different design and operational decisions. One of the biggest challenges in membrane process modelling is to reliably predict the membrane flux given system state (temperature, pressure, species concentrations, etc.). Although mass transfer theory provides several well understood mechanisms and theoretical foundations, it is often indicated that the experimenters encounter deviations from the ideal (theoretical) behaviour (Khan et al., 2023).

In this paper, we employ white (first-principles model derived from literature), grey (first-principles model combined with a regression model fitted with experimental data), and black box modelling (simple regression model) principles to identify a mathematical

Acknowledgements: This work is funded by the Slovak Research and Development Agency (project no. APVV-21-0019), by the Scientific Grant Agency of the Slovak Republic (grant no. 1/0691/21), and by the European Commission (grant no. 101079342, Fostering Opportunities Towards Slovak Excellence in Advanced Control for Smart Industries).

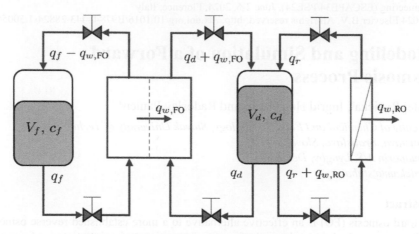

Figure 1: Scheme of the forward osmosis process.

model of an FO process. The achieved results indicate that the grey box modelling is the best alternative as the obtained model is the most reliable one.

2. Problem Description

The FO process essentially consists of feed and draw tanks and membrane modules (FO and RO). The process system is schematically depicted in Figure 1. The draw solution is prepared such that its osmotic pressure is much higher than the osmotic pressure of the feed solution. This gives a prerequisite for establishing a driving force for the separation of water from the feed, i.e., osmotic pressure difference. An FO membrane module, which involves a functional layer (made of aquaporin in our case), provides an interface between the feed and draw solutions. By design and thanks to the natural phenomena, water passes from the feed to the draw solution. As a result, the feed solution gets concentrated. Consequently, draw solution gets diluted and needs to be recovered for the effective continuation of the process. For this purpose, an RO membrane module with a similar or same type of membrane as in FO module is placed in the system. For the separation of water to occur through the RO membrane, external pressure is required that be supplied by a pump. Final products of the process are the purified water and concentrated feed. If the materials of the used membranes and composition of the draw tank are selected appropriately, the separation of water from the draw solution (regeneration of the draw solution) is more energy efficient than using the RO directly for the feed solution.

It is, however, not only the design of the process chemical side (the used membrane and draw solution materials) that contributes to the overall energy efficiency. The process should be designed (decisions should be made on types and sizes of membrane modules and used pumps) and operated in an optimal manner. The optimal operation should take into account dynamic behaviour of the system. One should decide whether the draw tank should be kept in a steady-state condition (water removal through the RO module equal to the water flow through FO module) or whether the system should operate in some form of a cyclic regime (e.g., switching the separation through the RO membrane on/off periodically). For this purpose, a mathematical model of the process is developed here.

3. Process Modelling

For the development of a mathematical model, we assume perfect mixing in the tanks, fully developed hydraulic profiles, ideal membrane rejection (perfect passage of water and no passage of other substances), and common and constant temperature and density of all streams. We use the following notation. Indexes w, f, d, FO, and RO stand for water, feed, draw, forward and reverse osmosis, respectively. V stands for the volume (in L), q for flowrate (in L/h), J for permeate flux (in LMH, i.e., L/m²/h), A for the area of a membrane (in m²), ΔP for the transmembrane pressure (in bar), and c for the concentration (in kg/L). Mass balance of the system in Figure 1 then reads as:

$$\frac{dV_f(t)}{dt} = -q_{w,\text{FO}}(t) = -A_{\text{FO}}J_{w,\text{FO}}(c_f(t), c_d(t)), \tag{1a}$$

$$\frac{dV_d(t)}{dt} = q_{w,\text{FO}}(t) - q_{w,\text{RO}}(t) = A_{\text{FO}}J_{w,\text{FO}}(c_f(t), c_d(t)) - A_{\text{RO}}J_{w,\text{RO}}(c_d(t), \Delta P), \tag{1b}$$

$$\frac{dc_f(t)}{dt} = \frac{c_f(t)}{V_f(t)}A_{\text{FO}}J_{w,\text{FO}}(c_f(t), c_d(t)), \tag{1c}$$

$$\frac{dc_d(t)}{dt} = \frac{c_d(t)}{V_d(t)}\left[A_{\text{FO}}J_{w,\text{FO}}(c_f(t), c_d(t)) - A_{\text{RO}}J_{w,\text{RO}}(c_d(t), \Delta P)\right]. \tag{1d}$$

The commonly used models for water flux through forward and reverse osmosis membranes can be obtained from the literature (Khan et al., 2023) as:

$$J_{w,\text{FO}}(c_f(t), c_d(t)) = k_w(\pi_d(c_d(t)) - \pi_f(c_f(t))), \tag{2}$$

$$J_{w,\text{RO}}(c_d(t), \Delta P) = k_w(\Delta P - \pi_d(c_d(t))), \tag{3}$$

where k_w stands for water permeability. The osmotic pressure π can be modelled as (Foley, 2013):

$$\pi(c) = iRTc, \tag{4}$$

where i is the van't Hoff factor, R is the gas constant, and T is the temperature.

The mathematical model presented in Eqs. (1)–(4) is a first-principles model. Using a white-box modelling principle, one can plug in the parameter values to obtain a final instance of the model ready for simulation, optimisation, etc. A common situation in practice is that the simplifications made by the process modelling assumptions result in inadequacies when model simulation results are compared to the experimental measurements. To counteract these, a grey box modelling approach (Cozad et al., 2015) is an alternative. This amounts to selecting an altered functional form of certain phenomena in the developed model while respecting the fundamental laws such as mass balances. In the case of FO process modelling, we opt for parameterising the permeate flux by a polynomial in $c_f(t)$ and $c_d(t)$. A black box approach is another alternative, when the mass balance cannot be constituted reliably, e.g., due to missing information on system inflows and outflows. In such a case, one could form the prediction model of the water flux through a membrane similarly to the grey box approach, yet mass balance equations would not be included in the model fitting procedure.

4. Model Fitting

Assuming an experiment conducted over a certain time period with the permeate fluxes, volumes, and concentrations measured, one can train the grey box model via:

$$\min_{p} \sum_{k=1}^{N} \sum_{m \in \{FO,RO\}} \frac{(J_{w,m}^{exp}(t_k) - J_{w,m}(t_k))^2}{2\sigma_{J_{w,m}}^2} + \sum_{s \in \{f,d\}} \frac{(V_s^{exp}(t_k) - V_s(t_k))^2}{2\sigma_{V_s}^2} + \frac{(c_s^{exp}(t_k) - c_s(t_k)^2)}{2\sigma_{c_s}^2} \quad (5a)$$

$$\text{s.t. Eqs. (1a)–(1d), } V_f(0) = V_{f,0}, \ V_d(0) = V_{d,0}, \ c_f(0) = c_{f,0}, \ c_d(0) = c_{d,0}, \quad (5b)$$

$$J_{w,FO}(c_f(t), c_d(t)) = f_{w,FO}(c_f(t), c_d(t), p), \quad (5c)$$

$$J_{w,RO}(c_d(t), \Delta P) = f_{w,RO}(c_d(t), \Delta P, p), \quad (5d)$$

where N is the number of experimental measurements, upper index exp denotes the measured values and σ is the standard deviation of the measurement noise. Conditions (5b) involve mass balance equations as well as initial conditions of the states, which can also be estimated. Equations (5c) and (5d) represent polynomials of an appropriate order, whose parameters p are to be estimated. Due to the presence of Eqs. (1a)–(1d), (5) results in a dynamic optimisation problem.

The black box approach is much simpler in our studied case. It involves solving:

$$\min_{p} \sum_{k=1}^{N} \sum_{m \in \{FO,RO\}} \frac{(J_{w,m}^{exp}(t_k) - J_{w,m}(t_k))^2}{2\sigma_{J_{w,m}}^2} \quad (6a)$$

$$\text{s.t. } J_{w,FO}(c_f(t), c_d(t)) = f_{w,FO}(c_f^{exp}(t), c_d^{exp}(t), p), \quad (6b)$$

$$J_{w,RO}(c_d(t), \Delta P) = f_{w,RO}(c_d^{exp}(t), \Delta P, p). \quad (6c)$$

Problem (6) represents a simple regression. This approach can be regarded to be of static nature as any consequence of taking measurements in subsequent time steps is neglected.

5. Results

The experimental measurements were gathered using a setup similar to Figure 1 with an FO membrane module containing hollow fibre Aquaporin RO/FO membrane and using orange juice as the feed solution and K-lac as a draw solution. K-lac is used as it offers high values of osmotic pressure. The temperatures of the streams were kept constant during the experiment. The main difference of the experimental setup, compared to Figure 1, was the absence of RO membrane, thus no draw solution regeneration took place. The experiment took 4 hours and 25 minutes, after the initial period with full recycle regime for the process start up. The starting volume in the feed tank was 250 L. The corresponding sucrose concentration of the orange juice was 11.2 Brix.

Throughout the experiment, fresh draw solution has been used and thus its concentration was kept constant. Three different concentrations were used consecutively: 20% w/w for the first hour of the experiment, 40% w/w in the following 84 minutes, and 60% w/w for the rest of the experiment duration. After 45 minutes of the experiment run, feed volume was too low. Therefore, 36 L of fresh water was added over a period of 10 minutes.

Measurements of concentration and volume in the feed tank as well as FO membrane water flux were measured every ten seconds, yielding 1,590 data points for each measured output. Volume (concentration) measurements were not recorded for one interval (two intervals) with duration of 10 minutes, which slightly reduces the dataset cardinality. The conditions of the experiment, i.e., the absence of the RO part, simplify the problem (5), making the Eqs. (1b), (1d) and (3) (and consequently Eqs. (5d) and (6c)) unnecessary.

We implemented the problems (5) and (6) in CasADi (Andersson et al., 2019) via python interface. We considered linear and quadratic forms of polynomials in Eq. (5c). We have estimated the values of standard deviations of measurement errors using the experimental data. The results of fitting the black box model are shown in Figure 2 for the best fit

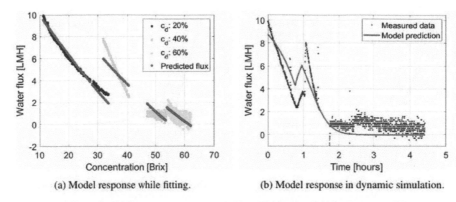

(a) Model response while fitting.   (b) Model response in dynamic simulation.

Figure 2: Performance assessment of prediction by the black box model.

obtained that is by the quadratic form of Eq. (6b), which showed the smallest lack-of-it using the statistical p-value test. We show the exact result of fitting (left hand plot, Figure 2a) and the performance of the fitted model in dynamic simulation of the experiment according to the model (1) (right hand plot, Figure 2b).

Based on the experimental data, we can conclude that higher the feed concentration, higher flux measurements variations. We thus cannot expect any of the trained models to be particularly suitable for accurate prediction in such conditions. From the fitting results of the black box model (Figure 2a), it is apparent that the fit is best for the data in the first hour of the experiment with low concentrations of feed and draw solutions. The fit gets evidently worse already for the values $c_d = 40\%$w/w. For the last part of the experiment (with high feed concentrations), one cannot reliably conclude appropriateness of the fit. Looking at the Figure 2b, one can observe that the model performs even worse and if used for further purpose, e.g., process design or optimisation, its results would be unsatisfactorily when compared with reality. This is caused by the model being fitted with the experimental values of concentrations that did not respect the mass balance (lacking data reconciliation), resulting in the unrealistic predictions.

When fitting the grey box model, the best form of the flux equation (5c) turned out to be linear, which already shows a large difference between black box and grey box models. We can highlight that the grey box modelling approach results in a simpler model. At this point, we can also assess the appropriateness of the white box model. If the white box model was appropriate, i.e., if the FO flux can be described by Eqs. (4) and (2), the grey box model would turn out to contain similar coefficients multiplying feed and draw concentrations. This is not the case that we observe here. The difference in the magnitude of the coefficients is a factor of 50. We thus conclude that the white box model is unsuitable in this application.

The results of the grey box modelling are depicted in Figure 3 that compares it simultaneously with black box model under the dynamic simulation of the conducted experiment. As we can see, the grey box model represents measured data more accurately. This is especially true when observing the values of volume in the feed tank, which is the output variable measured most accurately. We can also observe the already expected trend that the model cannot produce reliable predictions for higher values of the feed concentrations. Another experiment would be necessary to improve it.

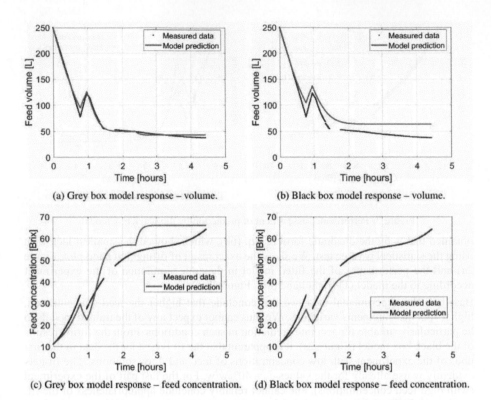

(a) Grey box model response – volume.   (b) Black box model response – volume.

(c) Grey box model response – feed concentration.   (d) Black box model response – feed concentration.

Figure 3: Comparison of model responses of grey box and black box model.

6. Conclusions

We studied modelling of a forward osmosis process as a prerequisite for process optimisation. We compared three approaches to model building: white, grey, and black box. The results show that grey box approach is superior as it can account for nonideal behaviour of the membrane (as compared to what white box model would assume based on simplified mass transfer theory) and, at the same time, it respects the mass balances during the training (compared to the black box approach that neglects such information).

References

S. M. Ali, S.-J. Im, A. Jang, S. Phuntsho, H. K. Shon, 2021. Forward osmosis system design and optimization using a commercial cellulose triacetate hollow fibre membrane module for energy efficient desalination. Desalination 510, 115075.

J. A. E. Andersson, J. Gillis, G. Horn, J. B. Rawlings, M. Diehl, 2019. CasADi – A software framework for nonlinear optimization and optimal control. Mathematical Programming Computation 11 (1), 1–36.

A. Cozad, N. V. Sahinidis, D. C. Miller, 2015. A combined first-principles and data-driven approach to model building. Computers & Chemical Engineering 73, 116–127.

G. Foley, 2013. Membrane Filtration: A Problem Solving Approach with MATLAB. Cambridge University Press.

M. A. W. Khan, M. M. Zubair, H. Saleem, A. AlHawari, S. J. Zaidi, 2023. Modeling of osmotically-driven membrane processes: An overview. Desalination, 117183.

J. Kucera, 2015. Reverse osmosis. John Wiley & Sons, Ltd.

Flavio Manenti, Gintaras V. Reklaitis (Eds.), Proceedings of the 34th European Symposium on Computer Aided Process Engineering / 15th International Symposium on Process Systems Engineering (ESCAPE34/PSE24), June 2-6, 2024, Florence, Italy

Adoption Dynamics of District Heat Networks: An Agent-Based Commercial Model

Thomas Cowley,[a] Emma Morris,[a] Timothy Hutty,[a] Solomon Brown[a]

[a]Department of Chemical & Biological Engineering, University of Sheffield, Sheffield, S1 3JD, England, UK
s.f.brown@sheffield.ac.uk

Abstract

This study uses Agent-Based Modelling to analyse the adoption of District Heating (DH) networks in the commercial sector. It examines the impact of financial incentives and policy changes, demonstrating varied responses based on organizational Social Value Orientations. The findings emphasize the need for targeted policy interventions to accelerate sustainable energy transitions.

Keywords: Agent-based model, DHN expansion, District Heat Network

1.   Introduction

The global shift towards renewable and low-carbon energy, emphasized by the 2016 Paris Agreement, aligns with the UK's goal of achieving Net-Zero emissions by 2050 (The Sixth Carbon Budget, 2020). The reliance on natural gas in the commercial sector is seen as unsustainable and policy shifts towards lower emissions, result in increasing gas costs (Gadenne et al, 2011). District heating (DH) is emerging as an alternative for commercial areas, offering efficient heat distribution (Energy Saving Trust, 2018). Its benefits include reduced reliance on natural gas and grid electricity, lower carbon emissions, and cost savings. Expanding DH networks, however, faces challenges like securing reliable heat users, uptake particularly from commercial stakeholders (Busch et al, 2014). Policy interventions can encourage commercial participation (Dowson, 2016). Agent-Based Modelling (ABM) is uniquely suited for simulating commercial adoption and forecasting DH network expansion due to its ability to model individual stakeholder behaviours and interactions in stochastic systems, other methods use a more aggregated approach may not capture emergent. (Busch et al, 2014). Policies promoting DH expansion are critical, with financial support for commercial DH infrastructure being essential (Webb et al, 2015). Empowering local authorities to develop DH strategies can improve uptake. The focus on incentivizing commercial building is vital, as they significantly influence DH network capacity utilization and regional emission reduction. This work is informed by literature on the adoption of renewable technologies like heat pumps and EVs (Vuthi et al, 2022.; Ghorbani et al, 2020) and uses ABM to analyse the commercial expansion of DH networks, assessing the effect of gas price increase, developer engagement, and installation cost subsidy.

2.   Methodology

2.1.   Model Framework

The model aims to simulate the decision-making process and outcomes related to connecting to the DH network, integrating empirical data and theoretical constructs. It focuses on decision-making among commercial organizations, defined as entities with more than five employees. These entities are characterized using the IAD framework (Shah et al, 2020), with much of the data derived from Energy Performance Certificates (EPC) (Gov.uk, n.d). The study examines 30 diverse organizations in Royston, Barnsley, capturing spatial dynamics through geographical information system (GIS). Agent interactions within the model include events, transitions, state charts, and action charts. Network extensions will be funded the developer, whilst the retrofit, maintenance and fuel costs will be associated with the organisation (Domestic Building Services Compliance Guide, 2021). Gas, electricity, and district heating costs increase with inflation at 2%. Each agent has a boiler replacement age, based on a distribution for average boiler breakdowns (CDW Engineering, 2022). Each heating system is updated by +1 for each year of simulation. Thus, updating other parameters such as efficiency which decrease with age.

2.2.   *Agent Heterogeneity and Classification*

Organizations are assumed to have the necessary knowledge for making independent decisions about utilities (Bijvoet, 2017). The challenge in modelling organizations lies in their heterogeneous behaviour, particularly how attitudes and priorities affect decisions. The Theory of Social Value Orientation (SVO) is used to understand this diversity (Ghorbani et al, 2020). SVO, which accounts for varying motives and goals among agents, categorizes them into four types (Fouladvand et al, 2022):

1.   Altruistic: Maximizes social and environmental benefits regardless of gain.
2.   Cooperative: Maximize both social/environmental benefits and their own.
3.   Individualistic: Concentrate solely on maximizing profits.
4.   Competitive: Strive to maximize their benefits while minimizing others.

All public buildings are altruistic, as they tend to follow government recommendations (Todnem et al, 2005). Commercial organizations with higher revenues can afford cooperative behaviour, while most others behave individualistically (Sharan, 2011). The value orientation of the agent determines the distribution of acceptable payback, and downtime values. It also determines the difficulty factor distribution in acquiring initial CAPEX investment. Each parameter uses a density function PERT-beta distribution. The distributions have been chosen based on standard commercial change management requirements and values defined as 'cost effective' when evaluating previous networks (Dowson, 2016).

2.3.   *Energy considerations*

Organizations are assigned an Energy Performance Certificate (EPC), rated from A (most efficient) to G. Each rating corresponds to a normal distribution of efficiency values, and maintenance costs. Organizations already using green energy sources are less likely to switch, hence, it's assumed these organizations will not opt for DH.

2.4.   *Agent Decision-Making*

The payback period for each agent takes into consideration the capital, operational cost and monetary savings (Sharan, 2011). The calculated payback needs to be less than the organisation's required payback (a distribution determined by the value orientation) to move onto the next part of the decision-making process. Eq. 1 X describes this:

$$Payback = \frac{(IC_{1,2} \cdot A)}{\left(\frac{E_d \cdot A \cdot C_{1,2}}{\eta_{1,2}} + M_{1,2}\right) - \left(\frac{E_d \cdot A \cdot C_3}{\eta_3} \cdot M_3\right)} \tag{1}$$

Where $IC_{1,2}$ ($£m^{-2}$) is the installation cost of gas and electric heating, respectively. A (m^2) is the floor area of the building, E_d ($kWhm^{-2}y^{-1}$) is the energy demand.

Installation cost has the units $£m^{-2}$. $C_{1,2,3}$ $(£kWh^{-1})$ is the cost of gas, electricity, and district heating, respectively. $M_{1,2,3}$ $(£y^{-1})$ is the maintenance cost for gas, electric, and district heating, respectively. $\eta_{1,2,3}$ is the efficiency of the boiler, electric heater, and district heating, respectively. Variables in this equation are characterised by specific statistical distributions, representing their variability; η, $installation\ cost$, $M_{1,2,3}$ ~ $Triangular(a, b, c)$, where a is min, b is max, c is mode; $C_{1,2,3}$ ~ $N(\mu_{1,2,3}\sigma^2_{1,2,3})$, where μ is the mean, σ is the standard deviation.

Another barrier to connecting is the amount of downtime required to implement the change (Todnem et al, 2005). The allowance of downtime varies with the organisation's motivation for change which is dependent on value orientation (Fouladvand et al, 2022), and the actual acceptable downtime calculated for each agent follows a PERT-beta distribution. District heating relies on a densely populated area to be effective (Dowson, 2016). Literature suggests that there needs to be a demand of $260MWh\ ha^{-1}y^{-1}$ (Fouladvand et al, 2022). Therefore, a parameter for the maximum distance from the pipeline is 100m to ensure minimum heat density is met.

There is a decision-making flowchart for each agent in the model that includes 8 transitions and various. Transition 1 is initiated by either a heating system failure, its aging affecting efficiency significantly (Balaras et al, 2005), a surge in gas prices beyond normal inflation or an EPC rating below average (Bush et al, 2017).

Transition 2 reflects the scenarios prompting consideration of district heating (DH). This transition is activated by active policies, such as, installation cost subsidies, gas taxes, or gas price increases. Transition 3 occurs upon direct interaction with the developer, where agents again assess the adequacy of their heating system, but through transition 5 bypass whether location, subsidies or gas price affect the decision. If an agent considers a connection here, surrounding agents are also prompted to consider it (Sharan, 2011).

Stages (transitions 6-8) in the state chart represent the agent's internal change management system. Two Action Charts evaluate the payback period (Transition 6) and installation time (Transition 7). Costs for DH installation vary between £15-82m^{-2} for gas centrally heated buildings and £112-141m^{-2} for electrically heated buildings (Dowson, 2016). DH costs range from £0.0551-0.1491kWh^{-1}, electricity from £0.2191-0.2299kWh^{-1}, and gas from £0.0955-0.116kWh^{-1}, based on a 2019 energy price report (Heat and Buildings Strategy, 2021). Installation times range from 7-365 days (Dowson, 2016). Despite having a management system, change often occurs reactively, discontinuously, and ad-hoc, with a 70% failure rate in change programs (Todnem et al, 2005). A percentage of agents who decide to join DH may fail due to external factors (Transition 9) (Sharan, 2011). The success rate distribution is influenced by the agent's value orientation. Agents that pass these evaluations are considered to have 'Joined' the network. If not, they either repair their existing system if it's below the boiler replacement age or purchase a new system.

3. Results

3.1. *Case study*

The model includes scenarios that simulate various policy interventions, each configuration was run only once. 'Invitation to Join' indicates the DH network developer inviting all agents within a feasible distance to connect. 'Increase in Gas Cost' simulates a government policy to raise natural gas costs by a certain percentage. 'Decrease in Installation Cost' represents a subsidy from the government or local authority, modelled as a percentage reduction in installation costs.

Table 1 – Scenario and policy configuration

Simulation Run	Invitation to Join	Increase in Gas Cost	Decrease in Installation Cost
1	No	No	No
2	Yes	No	No
3	Yes	No	20%
4	No	No	20%
5	No	No	50%
6	Yes	50%	No
7	No	50%	No

Figure 1 shows organizational decision-making over 10 years, highlighting agent stages with various interventions. Key observations include: 17-27% of agents joining in the first year and 43-57% over 10 years in all simulations; the duration in each decision stage varies; without intervention, many agents remain reluctant to replace heating systems; interventions, such as developer invitations and decreased installation costs, shift more agents to the consideration phase, potentially speeding up decisions to join DH networks; the primary barrier remains the decision to replace heating systems, with financial incentives crucial for adoption; only 4 agents were initially unable to join due to location, reducing as more connected; installation time wasn't a significant barrier in decision-making, overshadowed by policy impacts.

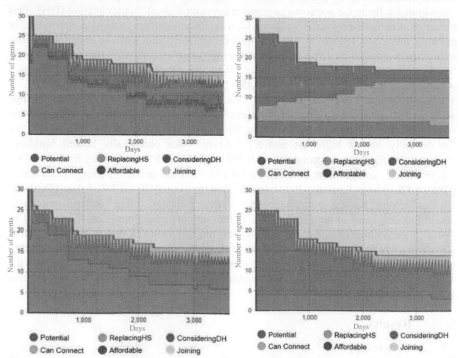

Figure 1 – Number of agents in each state over 10-year period. a; No intervention, b; Invitation from developer, c; 20% Installation subsidy, d; Invitation and 20% installation subsidy

Figure 2 rate at which agents join a district heating network over a 10-year period. Key findings are that the most effective strategy for connecting the maximum number of agents involves sending invitations to all organizations within a feasible area and offering financial incentives. A 20% discount on installation costs proves as effective as a 50% discount over a decade, although the higher subsidy results in faster agent connection, but it may not be economically viable for the developer. Interestingly, a 50% increase in energy bills due to economic or governmental factors significantly increases the rate of agents joining, similar to a 50% subsidy, but with a quicker impact. The study also reveals that a sudden rise in energy bills, could accelerate the rate of agents considering. However, this is contingent upon the availability of other heating systems like heat pumps. The research further indicates that while subsidies appeal to agents with a short-term financial focus, increasing gas prices offer broader benefits.

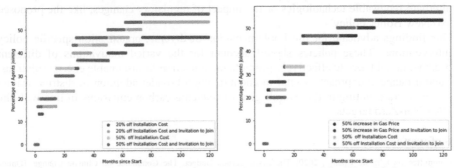

Figure 2 – shows the percentage of agents joining the network over 10-year period with x axis being months. a; Installation subsidy, b; most effective policies

Figure 3 shows the number of agents joining in their SVO category. Over 10 years, up to 63% of Altruistic and 61% of Individualistic joined the network. These participation rates reflect expected behaviours from each SVO type. Findings indicate that gas price increases impact Social and Environmental agents more significantly than Profit agents, who require lower installation costs for immediate investment return. Profit-driven agents prioritize short-term gains over long-term benefits, while socially driven agents are more responsive to economic changes than developer encouragement. Natural gas price fluctuations could lead to a notable increase in network participation, this suggests that profit-driven organizations might need specific policy interventions, especially those with lower energy demands.

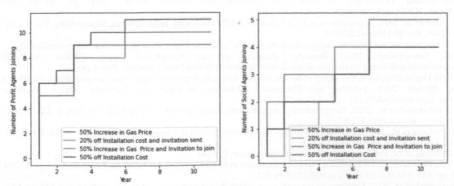

Figure 3 – The effectiveness of different policy interventions with each value orientation. a; Individualistic, b; Altruistic.

4.   Discussion & Conclusions

This research, employing an agent-based model within the Institutional Analysis and Development framework, has identified crucial factors that influence commercial entities' decisions to adopt district heating networks. The study underscores the effectiveness of financial incentives tailored to the varied SVOs of organizations. Immediate benefits such as subsidies appeal to individualistic entities, while cooperative and altruistic organizations are inclined towards long-term financial gains like reduced energy costs.

The analysis demonstrates that economic factors, including the prospects of energy price inflation and policy shifts, are pivotal in influencing the adoption of district heating networks. However, the scope of the model does not extend to the potential adoption of alternate renewable technologies or the impact of regulatory changes, like the proposed 'No New Boiler Policy'.

Our findings advocate for the implementation of diverse, organization-specific policy interventions. These policies should account for the varied motivations of different organizational types to effectively promote the transition to sustainable energy solutions. Such a nuanced approach is vital in encouraging a broader adoption of district heating systems, contributing to the overall goal of reducing carbon emissions in line with the UK's Net-Zero targets.

References

Committee on Climate Change, 2020, The Sixth Carbon Budget, The Committee on Climate Change. [Online Available: www.theccc.org.uk/publications]

D. Gadenne et al., 2011, The influence of consumers' environmental beliefs and attitudes on energy saving behaviours, Energy Policy, Vol. 39, No. 12, pp. 7684–7694.

Energy Saving Trust, 2018, What is District Heating? [Online Available: https://energysavingtrust.org.uk/what-district-heating/]

J. Busch et al., 2015, Emergence of District-Heating Networks; Barriers and Enablers in the Development Process. [Online Available: research.ncl.ac.uk/media/sites/researchwebsites/ibuild/iBUILD%20CP3.pdf]

M. Dowson, 2016, Connecting Existing Buildings to District Heating Networks. [Online Available: www.usdn.org/uploads/cms/documents/161214_-_connecting_existing_buildings_to_dhns_-_technical_report_00.pdf]

J. Webb, 2015, Improvising innovation in UK urban district heating, Energy Policy, Vol. 78, pp. 265-272, doi: 10.1016/j.enpol.2014.12.003.

P. Vuthi et al., 2022, Agent-based modeling (ABM) for urban neighborhood energy systems, Energy Inform 5, 55, doi: 10.1186/s42162-022-00247-y.

A. Ghorbani et al., 2020, Growing community energy initiatives from the bottom up, Energy Research & Social Science, Vol. 70, Dec. 2020, doi: 10.1016/j.erss.2020.101782.

I. Nikolic and A. Ghorbani, 2011, A method for developing agent-based models of socio-technical systems, International Conference on Networking, Sensing and Control, ICNSC 2011, pp. 44–49, doi: 10.1109/ICNSC.2011.5874914.

K. Shah et al., 2020, Application of an IAD-Enhanced IREPP Framework to Island States, Sustainability, Vol. 12, No. 7: 2765, doi: 10.3390/su12072765.

UK Government, n.d., Find an energy certificate. [Online Available: https://www.gov.uk/find-energy-certificate]

UK Government, 2021, Domestic Building Services Compliance Guide. [Online Available: https://assets.publishing.service.gov.uk/media/61b880b4e90e07044462d865/Domestic_Part_L.pdf]

CDW Engineering, 2022, Probability of failure. [Online Available: https://cdwengineering.com/probability-of-failure/]

X. Bijvoet, 2017, Learning and Competitive Behaviour in Open DH Networks. [Online Available: http://repository.tudelft.nl/]

J. Fouladvand et al., 2022, Energy security in community energy systems, J Clean Prod, Vol. 366, p. 132765, doi: 10.1016/j.jclepro.2022.132765.

R. Todnem et al., 2005, Organisational change management, Journal of Change Management, Vol. 5, No. 4, pp. 369–380, doi: 10.1080/14697010500359250.

V. Sharan, 2011, Fundamentals of Financial Management, Pearson.

C. A. Balaras et al., 2005, Heating energy consumption in European apartment buildings, Energy Build, Vol. 37, No. 5, pp. 429–442, doi: 10.1016/j.enbuild.2004.08.003.

R. E. Bush et al., 2017, Empowering innovations to unlock district heating in the UK, J Clean Prod, Vol. 148, pp. 137–147, doi: 10.1016/j.jclepro.2017.01.129.

The Secretary of State for Business, 2021, HM Government – Heat and Buildings Strategy.

Flavio Manenti, Gintaras V. Reklaitis (Eds.), Proceedings of the 34th European Symposium on Computer Aided Process Engineering / 15th International Symposium on Process Systems Engineering (ESCAPE34/PSE24), June 2-6, 2024, Florence, Italy

Assessment of Technical and Economic Viability in the Transformation of Refinery Vacuum Residue Waste into Cleaner Fuels

Ammr M. Khurmy [a,b] , Usama Ahmed [a,c] , Ahmad Al Harbi [d] , Taha Aziz *[e]

[a]Chemical Engineering Department, King Fahd University of Petroleum and Minerals, Dhahran 31261, Saudi Arabia
[b]Petrochemical and Conversion Industries Sector, Ministry of Investment of Saudi Arabia, Riyadh 12382, SaudiArabia
[c]Interdisciplinary Research Center for Hydrogen and Energy Storage, King Fahd University of Petroleum and Minerals, Dhahran 31261, Saudi Arabia
[d]Engineering Department, Sadara Chemical Company, Jubail 35412, Saudi Arabia
[e] Department of Mathematics, Dammam Community College, King Fahd University of Petroleum and Minerals, KFUPM Box 5084, Dhahran, 31261, Saudi Arabia
* Corresponding Author's E-mail: taha.aziz@kfupm.edu.sa

Abstract

Methanol and hydrogen, as cleaner fuel options, hold significant potential for decarbonizing the petrochemical industry (Hamid et al., 2020). This study aims to establish a process integration framework for the simultaneous production of methanol and hydrogen from vacuum residue while minimizing sulfur and carbon emissions (Gudiyella et al., 2018). This investigation involves the development of two distinct process models. In the first case, vacuum residue is subjected to gasification using oxygen and steam, with the resulting syngas being processed to yield both methanol and hydrogen. The second case follows a similar process model to the first but places a stronger emphasis on methanol production from vacuum residue. Both models are subjected to a comprehensive techno-economic comparison, considering various factors such as methanol and hydrogen production rates, specific energy requirements, carbon conversion, CO_2 emissions, overall process efficiencies, and project feasibility. The comparative analysis reveals that the second case, focused on methanol production, offers several advantages. It reduces the specific energy requirements by 86.01% compared to the first case. Additionally, CO_2 emissions are reduced by 69.76% in the second case compared to the first. Overall, the second case demonstrates superior project feasibility, showcasing enhanced process performance and reduced production costs in comparison to the first case.

Keywords: gasification; vacuum residue; carbon capture and utilization; methanol synthesis; hydrogen; process integration

1. Introduction

The global refining industry is loaded with mounting challenges, including stringent environmental regulations, thin profit margins and evolving demand patterns. The focus on reducing sulfur content in residual fuel oil for domestic use such as heavy fuel oil (HFO) is intensifying due to environmental concerns. The shift toward lower sulfur levels that are reaching 0.5 wt% requires expensive flue gas treatment for HFO combustion and prompting refineries to seek alternative conversion methods for heavy residues. Upgrading these residues to lighter products emerges as a key strategy to enhance refinery margins while addressing environmental standards (Zuideveld et al., 2000). Ships, as major consumers of HFO can contribute significantly to air pollution and greenhouse gas emissions. The shipping industry's share in global CO_2, SO_x, and NO_x emissions is substantial. Reducing carbon emissions is now a global priority. The transition to a green and low-carbon economy will require prompting initiatives for decarbonization across all sectors. The International Maritime Organization (IMO) is actively engaged in reducing greenhouse gas emissions from international shipping, fostering exploration into low-carbon and zero-carbon fuels such as LNG, methanol, biodiesel, hydrogen, and ammonia (Meng et al., 2022). Gasification is a promising technology to convert heavy residues to lighter products, and it could be integrated with various other technologies including water-gas-shift, CCUS and combined heat and power (CHP) to create a more efficient and sustainable process. Syngas, the main product of gasification, can be further processed to produce high-purity hydrogen, methanol or other Fischer Tropsch (FT) chemicals (Alibrahim et al., 2021). (Ahmed, 2020) developed a model for coal conversion to an alternative technology that integrates coal gasification and natural gas reforming. The alternative technology showed 4.28% higher efficiency, 18.3% reduction in greenhouse gases, 13% lower fuel production cost, and 34.3% lower CO_2 emissions. Fayez et al. (2021) developed a vacuum residue to methanol (VRTM) process and achieved 90 t/h methanol production with 99.9% purity. The VRTM cycle offers 49.5% energy efficiency which is 1.6% higher than the SRTM (steam reforming to methanol) process and a 14% lower unit cost compared to the conventional SRTM process. The aim of this study is to develop a process model to produce methanol and hydrogen by the gasification of vacuum residue. The second goal is to modify the process model to focus only on methanol followed by performing technical and economic analysis to evaluate the process feasibility of each case.

2. Modelling and Simulation:

In the design basis and simulation methodology, Aspen Plus V12 is employed to simulate the proposed process model for four different case studies. The Peng–Robinson with Boston–Mathias alpha function (PR-BM) property package is selected to determine the physical state of the chemical components in the process. The vacuum residue is treated as an unconventional component and is defined based on its elemental composition, including proximate and ultimate analyses. The VR gasification results were validated with the literature (Choi et al., 2007). The dual production of methanol and hydrogen involves several reaction steps, including gasification, water–gas shift (WGS), and methanol synthesis. The simulation involves different reactor models within Aspen Plus. The gasification unit is simulated using two reactor models, namely RYield and RGibbs. The RYield reactor model decomposes the vacuum residue into its elemental components based on the ultimate and proximate analysis of the feed. Subsequently, the RGibbs reactor model determines the yield of the gas product by minimizing Gibbs free energy. Mass yield is linked to the ultimate and proximate analysis using the calculator tool and Fortran code. For modeling the water–gas shift reaction (WGS), the REquil reactor model is employed, which performs phase and chemical equilibrium calculations. However, it is noted that REquil is suitable for equilibrium reactions of known stoichiometry. The methanol synthesis is simulated using the RPlug reactor model with Langmuir-

Hinshelwood–Hougen–Watson reaction kinetics. The design specification and operational parameter of major equipment models are summarized in Table 1.

Table 1: Design Specification and operational parameter of the proposed process models.

Equipment	Description	Aspen Model
Gasification Unit	Temp/Pressure = 1355 °C/ 4 MPa Steam/fuel ratio = 0.5 Oxygen to fuel ratio = 0.84 VR Flow Rate = 101 kg/sec Carbon conversion = 99%	RYield, RGibbs
Water Gas Shift (WGS)	Sour Catalyst (Co-Mo) 2 Adiabatic reactors Steam/ CO = 2 CO Conversion = 99.4%	REquil
AGR Unit	Rectisol Process (Methanol Solvent) Temp/Pressure = -45 °C/3.2 MPa H_2S Removal = 100ppbv CO_2 Removal = 99%	RadFrac, Flash
Methanol Synthesis	Temp/Pressure = 180 °C /8 MPa $CuO/Al_2O_3/ZnO$ catalyst	RPlug

3. Process Description:

Two case studies have been developed in this study dual production of H_2 and methanol as represented in Figure 1 and Figure 2. Case 1 focused on the dual production of methnaol and hydrogen, werheas, case 2 focuses mainly on the production of methanol. The ASU employs high- and low-pressure columns to separate oxygen from air at cryogenic condition to supply the gasification unit .

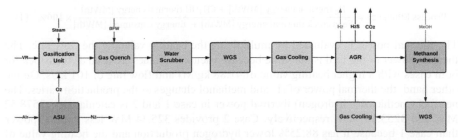

Figure 1: Vacuum residue to methanol and H_2 production with CO_2 capture (case 1).

Both the processes involve vacuum residue gasification using oxygen and steam, followed by syngas cooling using a gas quenching method. WGS unit is used to adjust the hydrogen-to-carbon ratio for methanol synthesis. The first gas cooling unit cools the shifted syngas for the Acid Gas Removal (AGR) unit utilizing methanol as a solvent to absorb H_2S and CO_2 . The methanol synthesis unit employs a single reactor with multiple injections of cold recycled streams. This process enhances equilibrium conversion by decreasing the temperature of the exothermic reaction. The process design includes a second WGS unit and a gas cooling unit to maximize CO conversion and produce H_2 and CO_2.

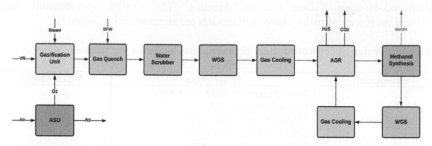

Figure 2. Vacuum residue to methanol production with CO_2 capture (case 2).

4. Results and Discussion

4.1. Hydrogen and Methanol Production:

Two case studies have been developed in this project to produce methanol and hydrogen from high sulfur content vacuum residue. The methanol production in case 1 and 2 is calculated as 57.49 kg/s and 129.92 kg/s, respectively. On the other hand, the hydrogen production in case 1 and 2 is calculated as 15.92 kg/s and 1.87 kg/s respectively. Furthermore, the net fuel production in case 1 and 2 is calculated as 73.41 kg/s and 131.79 kg/s, respectively. In comparison to the net fuel specific production energy between case 1 and 2, there is a potential of producing 79.6% more fuel when recycling 98.7% of the unreacted syngas to the methanol synthesis unit as in Case 2.

4.2. Process Performance Analysis:

Another important technical parameter indicator is the process efficiency, which evaluates the utilization of the power requirement in methanol and hydrogen production. The process efficiency is calculated using the thermal power of the feedstock and products in addition to the power requirement for methanol and hydrogen production as represented in Equation 1.

$$\text{Process Efficiency} = \frac{H_2 \text{ thermal energy [MWth]} + CH_3OH \text{ thermal energy [MWth]}}{\text{Feed stock thermal energy [MWth]} + \text{ Energy consumed [MWth]}} \times 100\% \quad (1)$$

The thermal power is evaluated by multiplying the heating value by the flow rate. The thermal power of the feed is the same because vacuum residues are used as feedstock for both cases with a higher heating value of 41.88 kg/MJ and flow rate of 101 kg/s. On the other hand, the thermal power of H_2 and methanol changes as the production varies. The net fuel (methanol + hydrogen) thermal power in case 1 and 2 is calculated as 3578.52 MW and 3253.18 MW, respectively. Case 2 provides 325.34 MW less thermal power than case 1 because it has 88.25% lower hydrogen production and the heating value of hydrogen (141.7 MJ/kg) is six times the heating value of methanol (23 MJ/kg). The calculated process efficiency for case 1 and 2 is 46.5% and 42.9%, respectively. In a comparison between case 1 and 2, the process efficiency of case 2 is 7.74% lower compared to case 1. Case 2 has a lower process efficiency because it has a 9.1% lower net fuel thermal power and only a 3.48% lower power requirement compared to case 1. Similarly, The specific energy requirement of the net fuel in case 1 and 2 is calculated as 21.23 kg/GJ and 39.49 kg/GJ, respectively, as represented in Figure 3. Moreover, The calculated CO_2 specific emission in case 1 and 2 is 2.97 and 0.8974, respectively. The comparative analysis shows that the case 2 design offers a 69.79% lower CO_2 specific emission than case 1.

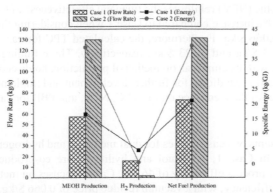

Figure 3: Methanol and hydrogen production rates with specific energy requirements

5. Project feasibility:

The economic feasibility analysis of methanol and hydrogen production from vacuum residue reveals promising results. CAPEX and OPEX were calculated to perform economic analysis and finding the total production cost (TPC) and minimum selling price (MSP) which arc crucial project feasibility indicators, with considering the fluctuation of vacuum residue price between 0.022 \$/kg and 0.11 \$/kg. CAPEX includes the cost of land, equipment, machinery, construction, and engineering while OPEX includes the cost of raw material, utility as well as the cost of labor and maintenance CAPEX has been converted to the annualized capital charge (ACC) assuming a project life of 30 years and an interest rate of 10% using the relation shown in the following equation (Eq 2).

$$ACC = \frac{[i(1+i)^n]}{[i(1+i)^n - 1]} \times CAPEX \qquad (2)$$

Case 2, emphasizing methanol production exhibits a significantly lower TPC, positioning it as a more economically viable option than Case 1. The competitive edge is further evident in a higher product selling price and a shorter payback period.

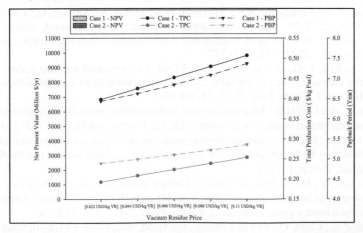

Figure 4: Project feasibility indicators of Case 1 and Case 2 at different VR feed price.

The net present value (NPV) reinforces the financial attractiveness of Case 2, showcasing a substantial 27.29% increase compared to Case 1 when considering the average vacuum residue price of 0.066 $/kg. Furthermore, the calculated TPC for case 1 and 2 has been calculated as 0.453 $/kg and 0.223 $/kg, respectively. These findings suggest that the proposed process, with its emphasis on methanol production, holds economic feasibility, making it a compelling candidate for further development and implementation. Figure 4 represent the effect of the feed price on the TPC, NPV, and PBP for case 1 and 2.

Conclusion:

This research presents two case studies for dual methanol and hydrogen production from vacuum residue. In Case 1, methanol and hydrogen are co-produced from vacuum residue, yielding a process efficiency of 46.5%. Case 2 focuses on methanol production, achieving 42.9% efficiency, with a vacuum residue price of 0.066 $/kg, Case 2 exhibits a lower total production cost (TPC) of 0.223 $/kg, compared to 0.453 $/kg in Case 1. Furthermore, Case 2 demonstrates superior technical and environmental performance, boasting a 50.6% lower production cost, 27.29% higher net present value, and a shorter payback period, indicating enhanced project feasibility over Case 1.

Acknowledgements

The authors would like to acknowledge support provided by the Deanship of Research Oversight and Coordination (DROC) at the King Fahd University of Petroleum & Minerals (KFUPM) for funding this work through Project no. INHE2308.

References

Ahmed, U. Techno-economic analysis of dual methanol and hydrogen production using energy mix systems with CO2 capture. Energy Convers. Manag. 2020, 228, 113663.

Alibrahim, H.A.; Khalafalla, S.S.; Ahmed, U.; Park, S.; Lee, C.-J.; Zahid, U. Conceptual design of syngas production by the integration of gasification and dry-reforming technologies with CO_2 capture and utilization. *Energy Convers. Manag.* 2021, *244*, 114485.

Al-Rowaili, F.N.; Khalafalla, S.S.; Al-Yami, D.S.; Jamal, A.; Ahmed, U.; Zahid, U.; Al-Mutairi, E.M. Techno-economic evaluation of methanol production via gasification of vacuum residue and conventional reforming routes. Chem. Eng. Res. Des. 2021, 177, 365–375.

Choi, Y.-C.; Lee, J.-G.; Yoon, S.-J.; Park, M.-H. Experimental and theoretical study on the characteristics of vacuum residue gasification in an entrained-flow gasifier. Korean J. Chem. Eng. 2007, 24, 60–66.

Gudiyella, S. *et al.* (2018) 'An experimental and modeling study of vacuum residue upgrading in supercritical water', *AIChE Journal*, 64(5), pp. 1732–1743.

Hamid, U. *et al.* (2020) 'Techno-economic assessment of process integration models for boosting hydrogen production potential from coal and natural gas feedstocks', *Fuel*, 266, p. 117111.

Meng, L.; Liu, K.; He, J.; Han, C.; Liu, P. Carbon emission reduction behavior strategies in the shipping industry under government regulation: A tripartite evolutionary game analysis. *J. Clean. Prod.* 2022, *378*, 134556.

Zuideveld, P.L.; Chen, Q.; van den Bosch, P.J.W.M. Integration of Gasification with Thermal Residue Conversion in Refineries. *Gasif. Technol. Conf.* 2000, 1–15

Flavio Manenti, Gintaras V. Reklaitis (Eds.), Proceedings of the 34th European Symposium on Computer Aided Process Engineering / 15th International Symposium on Process Systems Engineering (ESCAPE34/PSE24), June 2-6, 2024, Florence, Italy

Modelling of a Cartridge-based Fixed Bed Radial Flow Reactor for Methanol Synthesis

Frank Sauerhöfer-Rodrigo,[a,] Manuel Rodríguez,[a] Ismael Díaz[a]

[a]Dpto. Ingeniería Química Industrial y del Medio Ambiente, Universidad Politécnica de Madrid, C/ José Gutiérrez Abascal, 2, 28006, Madrid, Spain
frank.sauerhofer.rodrigo@alumnos.upm.es

Abstract

A novel mathematical model for a multitubular cartridge-based fixed bed radial flow reactor for methanol synthesis has been developed. This type of reactors has the advantage of having a lower effective bed thickness, resulting in a lower pressure drop and thus, improving the operational costs of the plant. The model has been evaluated using the original and a refitted Van der Graaf kinetics, showing a much better methanol yield using the latter. The sensitivity analysis results show that above 20 cartridges per tube, the methanol yield barely improves. Additionally, there is a trade-off between the pressure drop and the methanol yield. It is suggested to perform a full techno-economic and sustainability assessment including the full recycle loop to compare it with the conventional multitubular fixed bed reactor.

Keywords: methanol synthesis, radial flow reactor, reactor modelling, cartridge-based fixed bed reactor.

1. Introduction

The European Green Deal requires industries to decarbonize the industrial energy systems by achieving net-zero greenhouse gas emissions by 2050 (European Commission, 2019). This entails that conventional fossil fuels and fossil-based commodity chemicals will have to be replaced by non-fossil products. There are many alternatives to fossil fuels, amongst which synthetic fuels seem very promising at an industrial scale. Among all the non-fossil fuel alternatives to conventional fossil fuels, e-methanol from captured CO_2 and green hydrogen from electrolysis is one of the key products industries are currently looking at due to its outstanding combustion properties and the wide spectrum of methanol-derived products.

The main fixed bed reactor technologies used for methanol synthesis are the adiabatic reactor and the isothermal reactor (Bisotti et al., 2022; Bozzano and Manenti, 2016). Adiabatic systems use multiple beds in series with intercooling. Nevertheless, hot and cold spots can be found, leading to low reaction rates, catalyst deactivation or byproducts formation. On the other hand, isothermal reactors are cooled either by boiler feed water (Lurgi steam rising converter) or a cooling gas. However, there are heat transfer limitations, which increase the construction costs. The two main challenges of the fixed bed reactors used in industry are the pressure drop and the heat removal management.

New reactor geometries have arisen in order to overcome some of the challenges that traditional fixed bed reactors suffer from. In 2014, Davy Process Technology patented a novel reactor design for methanol production (Gamlin, 2014). The current assignee of the patent is Johnson Matthey Davy Technologies Limited. The technology consists of a multitubular reactor where each tube is filled with multiple cartridges (up to 200) in series and the shell side contains a cooling medium. Within each cartridge, the catalyst is placed in an annular container. The gas enters at the top of the cartridge to the inner cylindric hole, flows radially outwards and react adiabatically (see Figure 1). Then the gas is collected and cooled down in contact with the tube wall. Then the gas goes to the next cartridge to repeat the reaction-cooling process. This configuration results in a lower effective bed length, which translates in a lower pressure drop, and a high heat transfer.

In this work, a new model for the cartridge-based fixed bed reactor for methanol synthesis is presented. The model contains mass, heat and momentum balances for each cartridge. The model is used to compare the performance of this novel reactor configuration with the conventional multitubular reactor with tubes filled with catalyst particles and a cooling medium contained in the shell side. The performance is compared based on the pressure drop and overall carbon conversion. Additionally, a sensitivity analysis will be carried out to determine the effect of catalyst distribution along different number of cartridges per tube.

2. Reactor model

The reactor model is divided in two sections: the reaction section and the cooling section. The boundaries of the reaction section are limited to the catalyst container walls (see Figure 1). The cooling section boundaries are limited to the volume where the gas flows downwards and it is in direct contact with the tube wall (see Figure 1). The model has been developed in Aspen Custom Modeler V14. The thermodynamic package used is Peng-Robinson.

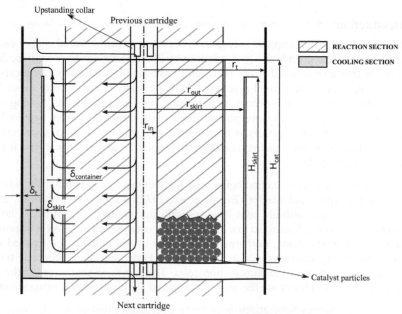

Figure 1. Longitudinal cross-section of a cartridge

2.1. Reaction section

It is assumed that the gas is well distributed along the entire catalyst bed and that there is no channeling or maldistribution. Axial dispersion is neglected, and the annular catalyst container is adiabatic. The mass balance is shown in Eq. (1), where F_i is the molar flow of species i, r is the radial dimension in the catalyst container, $r_j(s)$ is the reaction rate of reaction j at a distance s from the particle centre, $v_{i,j}$ is the stoichiometric coefficient of species i in reaction j, V_p is the particle volume and ρ_b is the bed density.

$$\frac{dF_i}{dr} = 2\pi r H_{cat}\rho_b \int_0^{\frac{V_p}{3}} \sum_{j=1}^{3} v_{i,j} r_j(s) \, dV_p \tag{1}$$

The concentration of each species within the catalyst particle is defined by the mass continuity equation and the Fick's first law shown in Eq. (2), where r_i is the total reaction rate of species i, J_i is the diffusion molar flux, C_i is the concentration of species i and A_p is the spherical superficial area. Those equations are subjected to the boundary conditions shown in Eq. (3), where ρ_g is the gas density, M_g is the molecular mass of the gas and y_i is the molar fraction of species i in the bulk gas phase.

$$r_i(s)dV_p - d\left(J_i(s) \cdot A_p(s)\right) = r_i(s)dV_p - d\left(\left(-D_{e,i}\frac{dC_i}{ds}\right) \cdot A_p(s)\right) = 0 \tag{2}$$

$$\frac{dC_i}{ds} = 0 \quad s = 0; \quad C_i = y_i\rho_g M_g \quad s = R_p \tag{3}$$

The energy balance is shown in Eq. (4), where H is the enthalpy flow. The momentum balance is shown in Eq. (5) (Froment et al., 2011), using the friction factor defined by Tallmadge (1970) (Erdim et al., 2015). d_p is the particle diameter, ε_b is the bed voidage, Re_g is the Reynolds number and u_g is the superficial velocity of the gas.

$$\frac{dH}{dr} = 0 \tag{4}$$

$$\frac{dP}{dr} = -\frac{\rho_g u_g^2}{d_p}\frac{1-\varepsilon_b}{\varepsilon_b^3}\left(4.2\left(\frac{1-\varepsilon_b}{Re_g}\right)^{\frac{1}{6}} + 150\frac{1-\varepsilon_b}{Re_g}\right) \tag{5}$$

The boundary conditions for each cartridge are as follows:

$$F_i = F_{i,0}, \, T = T_0, \, P = P_0 \quad r = r_{in} \tag{6}$$

2.2. Cooling section

It is assumed that the flow is fully developed in the cooling section. The temperature of the external surface of the tube wall (T_w) is assumed to be the same as that of the cooling media. The mass flow and compositions remain steady and equal to the inlet conditions along the entire cooling section. Additionally, the pressure drop in this section is neglected. The heat balance is shown in Eq. (7). The thermal resistance is calculated with the definition presented by McAdams and Frost (1922) and Steynberg et al. (2004) and shown in Eq.7.. The wall thermal conductivity of steel (k_{steel}) is assumed to be 60 W/ (K·m) (Philippe et al., 2009).

$$dH - \left(\frac{\delta_t}{k_{steel}2\pi(r_t + \delta_t/2)dz} + \frac{1}{h2\pi r_t dz}\right)^{-1}(T - T_w) = 0 \qquad (7)$$

2.3. Model parameters

The wall effects on the bed voidage are neglected, resulting in a uniform porosity of the catalyst bed throughout the entire annulus. The porosity is calculated with the correlation proposed by Sodré and Parise (1998) using the hydraulic diameter (Eq. 15 of the original paper, with $B = 0.387$).

The effective diffusion coefficients are calculated using the approach proposed by Loomerts et al. (2000) for a multicomponent system using a combined bulk binary diffusivity and Knudsen diffusion. The Knudsen diffusion is calculated as proposed by Vedrine (1983). A pore radius of 10 nm is taken (Lommerts et al., 2000). The binary gas-phase diffusion coefficient for each species is calculated with Fuller et al.'s method (Fuller et al., 1966). According to Graaf et al. (1990), ε/τ has a value of 0.123.

The heat transfer coefficient in the annulus region with the gas flowing downwards is calculated with the correlation proposed by Gnielinski (2009).

3. Study case

The case to be studied is adapted from the case used by Yusup et al. (2010) and used by Leonzio (2020) to study different kinetic models. The feed conditions and composition are shown in Table 4 of Yusup et al.'s work (2010). The reactor geometry is based on that of the conventional reactor presented in the aforementioned works and adapted to the cartridge-based reactor technology (see Table 1). The number of cartridges is varied in order to check how sensible the results are to the number of cartridges. The height of each cartridge is equal to the total reactor length divided by the number of cartridges per tube.

The performance indicators to be evaluated in this study are carbon conversion per pass and pressure drop. In a full methanol synthesis scheme, the recycle loop would impact the overall carbon conversion as well as the energy expenditure to compress and re-heat the recycled gas. A sensitivity analysis has been carried out to check the impact of the number of cartridges over the two performance indicators using the two kinetics proposed. As seen in Figure 2 (top), the original Van der Graaf kinetics results in a lower methanol yield than the refitted kinetics. However, in both cases there is a higher yield towards methanol than with the conventional methanol reactor, which is ~1 t/(m³-catalyst h) according to Blug et al. (2014). Additionally, there seems to be a plateau when increasing the number of cartridges keeping the same catalyst mass, approaching a quasi-isothermal behaviour at a certain point within the reactor (Figure 2 bottom). Nevertheless, the pressure drop increases when increasing the number of cartridges per tube, as the effective bed thickness through which the gas passes becomes higher. The peak methanol production using the original Van der Graaf kinetics with two cartridges is a result of the gas being heated by the shell fluid after the first cartridge, reaching a temperature sufficiently high to significantly increase the reaction extent in the next adiabatic reaction section. With one cartridge, the temperature is not high enough to have a high reaction rate and produce as much methanol as with two cartridges. With more than two cartridges, the gas cools down in between cartridges, limiting up to some extent the reaction rate.

Table 1. Reactor geometry and design parameters

Parameter	Units	Value	Parameter	Units	Value
Total catalyst loading	kg	35,778.5	δ_{skirt}	mm	1
Catalyst density	kg/m^3	1100	$\delta_{container}$	mm	0.3
Tube diameter	m	0.1	d_p	mm	0.5
δ_t	mm	3.2	Number of tubes		1521
r_{in}	m	0.0183	Reactor length	m	7.26
r_{out}	m	0.0432	H_{cat}/H_{CANS}		0.95
r_{skirt}	m	0.0675	T_w	°C	250

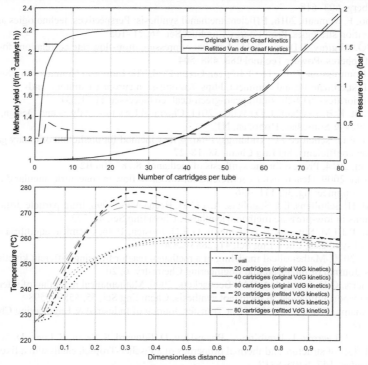

*Figure 2. Carbon conversion and pressure drop vs number of cartridges per tube (top
figure) and temperature profiles along the reactor (bottom figure)*

4. Conclusions

A novel reactor model for methanol synthesis from syngas using a multitubular cartridge-based fixed bed reactor has been presented. Two different kinetics have been used. The results of the sensitivity analysis showed that for a given catalyst loading and a certain reactor length, there is a trade-off between the pressure drop of the reactor and the methanol yield. Above 20 cartridges the pressure drop significantly increases while the carbon conversion barely increases. Additionally, the methanol yield per mass of catalyst is significantly higher than that of a conventional reactor. It is suggested to perform a full techno-economic and sustainability assessment to compare this type of reactor with the conventional reactor including the recycle loop so that the energy consumption and carbon footprint of other pieces of equipment within it are accounted for as well.

5. Acknowledgments

The authors would like to acknowledge Repsol S.A. for supporting this work.

References

F. Bisotti, M. Fedeli, K. Prifti, A. Galeazzi, A. Dell'Angelo, F. Manenti, 2022, Impact of Kinetic Models on Methanol Synthesis Reactor Predictions, In Silico Assessment and Comparison with Industrial Data. Ind Eng Chem Res, 61, 2206–2226

M. Blug, J. Leker, L. Plass, A. Günther, 2014, Methanol generation economics, In: M. Bertau, H. Offermanns, L. Plass, F. Schmidt, H.-J. Wernicke (Eds.), Methanol: The Basic Chemical and Energy Feedstock of the Future: Asinger's Vision Today. Springer Berlin Heidelberg, Berlin, Heidelberg, 603–618

G. Bozzano, F. Manenti, 2016, Efficient methanol synthesis: Perspectives, technologies and optimization strategies, Prog Energy Combust Sci, 56, 71–105

E. Erdim, Ö. Akgiray, I. Demir, 2015, A revisit of pressure drop-flow rate correlations for packed beds of spheres, Powder Technol 283, 488–504

European Commission, 2019, The European Green Deal, The European Commission - Striving to be the first climate-neutral continent, https://commission.europa.eu/strategy-and-policy/priorities-2019-2024/european-green-deal_en (accessed 11.21.23)

G.F. Froment, K.B. Bischoff, J. De Wilde, 2011, Chemical Reactor-Analysis and Design, 3rd ed., John Wiley & Sons, Inc., Hoboken, NJ

E.N. Fuller, P.D. Schettler, J.C. Giddings, 1966, A new method for prediction of binary gas-phase diffusion coefficients, Ind Eng Chem, 58, 18–27

T.D. Gamlin, 2014, Process for the synthesis of methanol, 8,785,506 B2

V. Gnielinski, 2009, Heat transfer coefficients for turbulent flow in concentric annular ducts, Heat Transfer Engineering, 30, 431–436

G.H. Graaf, H. Scholtens, E.J. Stamhuis, A.A.C.M. Beenackers, 1990, Intra-particle diffusion limitations in low-pressure methanol synthesis, Chem Eng Sci, 45, 773–783

G.H. Graaf, P.J.J.M. Sijtsema, E.J. Stamhuis, G.E.H. Joosten, 1986, Chemical equilibria in methanol synthesis, Chem Eng Sci, 41, 2883–2890

G. Leonzio, 2020, Mathematical modeling of a methanol reactor by using different kinetic models, Journal of Industrial and Engineering Chemistry, 6, 20

B.J. Lommerts, G.H. Graaf, A.A.C.M. Beenackers, 2000, Mathematical modeling of internal mass transport limitations in methanol synthesis, Chem Eng Sci, 55, 5589–5598

W.H. McAdams, T.H. Frost, 1922, Heat Transfer, Journal of Industrial & Engineering Chemistry, 14, 13–18

R. Philippe, M. Lacroix, L. Dreibine, C. Pham-Huu, D. Edouard, S. Savin, F. Luck, D. Schweich, 2009, Effect of structure and thermal properties of a Fischer-Tropsch catalyst in a fixed bed, Catal Today, 147, S305–S312

Sodré, J.R. Parise, J.A.R., 1998, Fluid flow pressure drop through an annular bed of spheres with wall effects, Exp Therm Fluid Sci 17, 265–275

A.P. Steynberg, M.E. Dry, B.H. Davis, B.B. Breman, 2004, Fischer-tropsch reactors, Stud Surf Sci Catal, 152, 64–195

J.A. Tallmadge, 1970, Packed bed pressure drop—an extension to higher Reynolds numbers, AIChE Journal

J.C. Vedrine, 1983, Mass Transport in Heterogeneous Catalysis, In: Bénière, F., Catlow, C.R.A. (Eds.), Mass Transport in Solids, NATO Advanced Science Institutes Series, Springer, Boston, MA, 505–536

S. Yusup, N.P. Anh, H. Zabiri, 2010, A Simulation Study of an Industrial Methanol Reactor Based on Simplified Steady-State Model, International Journal of Recent Research and Applied Studies, 5, 213–222

Flavio Manenti, Gintaras V. Reklaitis (Eds.), Proceedings of the 34th European Symposium on Computer Aided Process Engineering / 15th International Symposium on Process Systems Engineering (ESCAPE34/PSE24), June 2-6, 2024, Florence, Italy

Sustainability and Quality-by-Digital Design of an Integrated End-to-End Continuous Pharmaceutical Process

Timothy J. S. Campbell[a], Chris D. Rielly[a], Brahim Benyahia[a*]

[a]Department of Chemical Engineering ,Loughborough University, Leicestershire, UK
b.benyahia@lboro.ac.uk

Abstract

An end-to-end mathematical model of a continuous ibuprofen process, which integrates both upstream and downstream processing, was developed to identify optimal and sustainable design and operation options and deliver quality assurance. The main objectives include maximizing productivity, minimizing waste, and developing systematic methods to implement Quality-by-Design (QbD) at the plant-wide level. To achieve these objectives, a set of quality and environmental key performance indicators were considered along with a set of Critical Quality Attributes (CQA). A variance-based Global Sensitivity Analysis (GSA) was used to capture the impact of the uncertainties and variability amongst a large set of process parameters and material attributes to help identify the minimum set of Critical Material Attributes (CMA), and Critical Quality Attributes (CQA). Two methods were implemented and compared, namely the estimability analysis and the Analytical Hierarchy Process (AHP). The design space associated with the CQA was identified based on the minimum set of the CPP and CMA which proved critical, ensuring safe and reliable operating ranges for quality assurance. The plant model was simulated and optimized using gPROMs Formulated products 2.3.1 with data obtained from literature.

Keywords: Integrated Continuous Pharmaceutical Plant, Global Sensitivity Analysis, Estimability, Quality-by-Design, Design Space, Analytical Hierarchy Process.

1. Introduction

The increased adoption of continuous manufacturing in the pharmaceutical and biopharmaceutical sectors has grown the demand for more reliable and robust mathematical models and decision-making tools to improve technology selection, process design and optimization, scale-up and scale-out, and the development of effective process and quality control strategies. However, despite the advantages of integrated continuous pharmaceutical manufacturing, their development is still hampered by the complexity inherent to large sets of interactive units, and lack of systematic methods to deliver build-in quality assurance (Benyahia et al., 2018; Campbell et al., 2022). The implementation of QbD and Quality-by-Control at the plant-wide scale requires an in-depth understanding of the interplay between large sets of CPP, CMA, and CQA. It is crucial to identify the minimum set of CPP and CMA for the determination of the most reliable design space which captures more effectively the CQA variability and allows more robust control strategies.

Continuous manufacturing has many benefits for industry, such as reduced waste and costs, and enhanced operating flexibility and resilience. The Robust digital design tools and high-fidelity mathematical models can be used to build quality assurance that could address these critical needs and enable the other advantages to be realized. The increased adoption of advanced digital technologies such as plant-wide mathematical model and digital twins and advanced enterprise and plant-wide control and optimization strategies are anticipated to foster even further the advantages of immerging technologies in Pharma such as continuous manufacturing and autonomous plants.

However, the development of plant-wide mathematical models for pharmaceutical and biopharmaceutical plants is still very limited and hampered by many technical challenges. Moreover, most of the published plant-wide model-based investigations were limited to economic or techno-economic analysis using shortcut methodologies that require relatively less accurate mathematical models. Only a few examples discussed the optimal operation and plant-wide control of integrated continuous pharmaceutical plant (Benyahia et al., 2018, Lakerveld et al., 2015).

Over the last years, QbD has become a benchmark for quality management in the pharmaceutical and biopharmaceutical industries. However, its implementation at the plant-wide level or in integrated processes is still very limited due inherent challenges, such as the large sets or interactive CPP and CMA that significantly impact the intermediate quality attributes and CQA. The identification of the minimum set of CPP and CMA is paramount in QbD to deliver built-in quality assurance for drug safety and efficacy. Plant-wide mathematical models can play a crucial role in understanding the effect of the different process parameters and therefore identify more reliably the design space. One of the most effective methods to determine the criticality of the process parameters material attributes is based on the sensitivity analysis to help reduce the dimensionality of the design space (Bano et al, 2018). However, a method based solely on the sensitivity analysis and disregards the impact of the correlations between the CPP and CMA may lead to poor design space or/and ineffective control strategies.

In this work, we propose a methodology inspired by our recently developed Quality-by-Digital Design (QbDD) framework where the mathematical models can be fully adopted at all developments stages to address product quality and process operation requirements. Firstly, a variance-based GSA was used to capture the impact of uncertainty and variability amongst a large set of process parameters and material attributes. Two methods were implemented and compared, namely the estimability analysis and the AHP. The design space associated with the CQA was identified based on the minimum set of the CPP and CMA which proved critical, ensuring safe and reliable normal operating ranges for quality assurance.

2. Methods

2.1. Process description

The end-to-end process to produce ibuprofen tablets from synthesis to tableting is shown In Figure 1, firstly, Ibuprofen is synthesized through three reactions steps conducted in plug flow reactors in series. The first plug flow reactor involves Friedel–Crafts acylation, while the PhI(OAc)2-mediated 1,2-aryl migration occurs in the second reactor, and finally, ibuprofen product is formed by hydrolysis in the third reactor. The reaction outlet stream is then purified in a continuous liquid-liquid extractor to recover most of the ibuprofen in the organic solvent. Next, three mixed-suspension-mixed-product-removal (MSMPR) crystallizers are used to crystallize the ibuprofen, in which the supersaturation point is maintained by the manipulation on the antisolvent flowrate (Water) and the temperature (between 283 K and 313 K). Subsequently, the filtration and wash-filtration stages are applied to separate ibuprofen crystals from the mother liquor and reaction impurities and residues. The next processing step is a wet granulation where the API is combined with the excipients (i.e., glucose and microcrystalline cellulose) in presence of a liquid binder. The liquid is then removed in a spray dryer before the granules are fed to the final tablet press.

2.2. Proposed digital framework

Figure 2 shows the proposed digital framework. The plant-wide mathematical model of the integrated end-to-end continuous plant for ibuprofen manufacture was first created based on data obtained from the literature. The model was then used to optimize process design and operation options to achieve acceptable quality bounds along with optimal environmental and economic and environmental key performance indicators (KPI). The formulation of the optimization problem is not presented here for the sake of brevity.

The GSA was then performed for all process parameters with respect to all CQA and KPI. The proposed variance-based GSA was performed using a quasi-random sampling technique (Sobol). Subsequently, the effects of the process parameters and material attributes were ranked based on the estimability analysis and AHP. A criticality threshold was then established to identify the minimum set of the CPP and CMA followed by the construction of the design space.

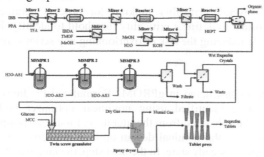

Figure 1. Process flow diagram of the continuous integrated end-to-end manufacturing of Ibuprofen. Isobutyl benzene (IBB), Propanoic acid (PPA), triflic acid (TFA), Iodobenzene diacetate (IBDA), Trimethyl orthoformate (TMOF), Methanol (MeOH), Potassium Hydroxide (KOH), Heptane (HEPT), microcrystalline cellulose.

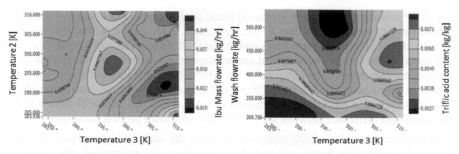

Figure 4. A sample of reduced dimensionality Design Spaces.

4. Conclusions

A plant-wide model of an integrated continuous pharmaceutical process for ibuprofen manufacture was developed as an effective decision-making tool for plant optimization, and most importantly for QbDD. The integrated continuous manufacturing plant involves a large set of interactive process parameters and material attributes that impact several CQA and KPI. Two methods were proposed to identify the minimum set of CPP and CMA, which capture more effectively the set of reduced dimensionality design spaces allowing more reliable built-in quality assurance. The proposed methods were both capable to deliver an overall ranking of the process parameters and material attributes based on their effects on the CQA and KPI. Despite some similarities in the overall ranking, the estimability analysis outperforms the AHP as it systematically excludes highly correlated CPP and CMA, besides the criticality threshold can be fine-tuned to meet more effectively the risk assessment criteria. Nevertheless, the AHP may deliver a fast and computationally cheaper way to rank process parameters and material attributes based on a quantitative approach that captures the variability of the targeted CQA and KPI. However, the method may be limited due the need to introduce a vector of the relative importance weights, which may be biased by the expert opinion. This bias may be potentially minimized with the availability of prior knowledge on the intrinsic risks and impact of deviations in each of the CQA on drug safety and efficacy.

References

Bano, G., Wang, Z., Facco, P., Barolo M. & Ierapetritou, M., 2018. A novel and systematic approach to identify the design space of pharmaceutical processes. Computers & Chemical Engineering, 115, 309-322.

Benyahia, B. (2018). Applications of a plant-wide dynamic model of an integrated continuous pharmaceutical plant: Design of the recycle in the case of multiple impurities. Computer Aided Chemical Engineering, Elsevier B.V. 41, 141-157.

Campbell, T., Rielly, C. & Benyahia, B. (2022). Digital design and optimization of an integrated reaction-extraction-crystallization-filtration continuous pharmaceutical process. Computer Aided Chemical Engineering 51, 775-780.

Fysikopoulos, D., Benyahia, B., Borsos, A., Nagy, Z. K., & Rielly, C. D. (2019). A framework for model reliability and estimability analysis of crystallization processes with multi-impurity multi-dimensional population balance models. Computers & Chemical Engineering, 122, 275-292.

R Lakerveld, B Benyahia, PL Heider, H Zhang, A Wolfe, C Testa, S Ogden, ...(2015) The application of an automated control strategy for an integrated continuous pharmaceutical pilot plant. Organic Process Research & Development 19 (9), 1088-1100

Yao, K.Z., Shaw, B.M., Kou, B., McAuley, K.B., Bacon, D.W. (2003). Modelling ethylene/butene copolymerization with multi catalyst: parameter estimability and experimental design. Polym. React. Eng. 11 (3), 563–588.

Flavio Manenti, Gintaras V. Reklaitis (Eds.), Proceedings of the 34th European Symposium on Computer Aided Process Engineering / 15th International Symposium on Process Systems Engineering (ESCAPE34/PSE24), June 2-6, 2024, Florence, Italy
© 2024 Elsevier B.V. All rights reserved. http://dx.doi.org/10.1016/B978-0-443-28824-1.50059-4

Development of a fast-charging protocol considering degradation using high-fidelity lithium-ion batteries

Chanho Kim[a], Minsu Kim[b], Junghwan Kim[a, *]

[a]*Department of Chemical and Biomolecular Engineering, Yonsei University, 50 Yonsei-ro, Seoul, 03722, Republic of Korea*
[b]*Department of Chemical Engineering, Massachusetts Institute of Technology, Cambridge, MA, 02139, USA*
kjh24@yonsei.ac.kr

Abstract

As demand for electric vehicles and grid stability increases, optimizing lithium-ion batteries (LIB) usage has become crucial. To address this, advanced battery management systems are spotlight as continuously monitor battery internal phenomena. Fast charging of LIB is one of the challenges for advanced battery management. Applying high current can reduce charging time, but it also accelerates battery degradation. Therefore, development of fast-charging protocol considering degradation condition is crucial. Porous electrode theory (PET) models can suggest high fidelity battery model. However, due to the wide design space and numerous complex degradation mechanisms, limitations exist using PET model. Using model-free approach, such as genetic algorithm (GA), can handle this problem. In this study, we applied GA for parameter identification and fast-charging protocol development. Identified parameters are applied to the PET model and we employed GA to minimize charging time while considering voltage and temperature constraints. Using the optimal charging protocol achieved by model-free approach, charging performance can be compared with experiments to identify the improvement.

Keywords: Lithium-ion battery, Parameter estimation, Fast-charging protocol, Degradation, Porous electrode theory

1. Introduction

Intercalation-based batteries are attracting attention with the advancement of industries such as electric vehicles, mobile devices, and energy grids. Among various materials, lithium-ion batteries (LIBs) are the most widely used due to their high energy density and low self-discharge. However, long charging times and reduced performance by degradation are remaining challenges. They can be overcome by charging under conditions that suppress degradation. According to battery chemistry, high temperatures and voltages accelerates battery degradation (Kumar et al., 2023). Charging under high current can reduce charging time, but it also increases temperature and voltage. Thus, designing a charging strategy reducing charging time and suppressing degradation is crucial. However, since experimenting cycling behavior of LIBs is a time-consuming and expensive task, it can be handled inexpensively by numerical battery model.

Various numerical modeling approaches exist to describe the charging/discharging behavior of LIBs. Each model is divided into equivalent circuit model (ECM) and electrochemical model (EM). EM is preferred because it can depict internal phenomena such as concentration and temperature distribution. The most widely used EM approach is the porous electrode theory (PET) model, and widely used modeling tools include PyBaMM (Sulzer et al., 2021), LIONSIMBA (Torchio et al., 2016), and PETLION (Berliner et al., 2021). However, model-based charging strategy development has limitations due to hundreds or thousands of complex degradation mechanisms. Complex degradation mechanisms make charging strategy development a large-scale optimization problem (Jiang et al., 2022). Developing an optimization-based model-free approach can solve the challenges (Ouyang et al., 2015).

In this research, we build a high-fidelity battery model and develop a charging strategy to suppress degradation. The parameters of PET model are estimated to depict battery behavior. This improves efficiency of optimization since experiments can be handled inexpensively. However, complex relationships of parameters can cause undesirable result like local optimum. To solve it, we divided 17 parameters to three sets and conducted one main and two fine optimization steps. Parameters were selected by sensitivity analysis (Li et al, 2020). First optimization with six parameters depicts main charging tendency. Then, after fixing first optimization results, secondary optimization with other six parameters were conducted for fine tuning. Likewise, last optimization was executed while preceding results are applied. Several studies have been conducted to identify current level in the multistage constant current (MCC) stage. Voltage and SOC are used as charging switching criteria. Liu et al. (2018) consistently reduce current of the MCC stage by randomly selected voltage and analyze effect on degradation. Jiang et al. (2022) divided SOC into 20 % intervals and propose a charging considering temperature and voltage constraints. We focus on the criteria for distinguishing between current levels and charging strategies in MCC stage. As a result, by SOC-based charging design and optimal current level, charging conditions don't accelerating degradation are obtained.

2. Methodology

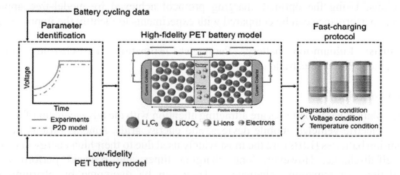

Figure 1. Schematic of fast-charging protocol development using high-fidelity LIBs model

Fig. 1 depicts a schematic diagram of generating a high-fidelity PET model by parameter estimation and designing fast-charging protocol with it. This section discusses theoretical description, parameter estimation, and fast-charging protocol problem.

2.1. Porous electrode theory (PET) model

PET is a mathematical modeling method that describes cycling behavior of LIBs and depicts internal phenomena with high accuracy based on study of Newman et al. (1975)

and Fulller et al (1994). PET model consists of partial differential equations for mass and charge conservation at cathode, separator, and anode and describes the electrochemical dynamics of solid particles and electrolytes. Many governing equations are linked through ion flux by Bulter-Volmer equation in eqn (1).

$$j(z,t) = 2\frac{i_0}{F} \sinh\left[\frac{F}{2RT(z,t)}\eta\right]$$
(1)

Where j is the ionic flux, i_0 is the exchange current density, F is the Faraday constant, η is the electrode overpotential. The diffusion of lithium ions inside each solid particle is described by eqn (2), where eqn (3) is the boundary condition.

$$\frac{\partial}{\partial t}c_s(z,t) = \frac{1}{z^2}\frac{\partial}{\partial z}\left[z^2 D_{eff}^s \frac{\partial}{\partial z}c_s(z,t)\right]$$
(2)

$$\frac{\partial}{\partial z}c_s(z,t) = 0 \ (z = 0), \qquad \frac{\partial}{\partial z}c_s(z,t) = -\frac{j(z,t)}{D_{eff}^s} \ (z = R_s)$$
(3)

Where $c_s(z,t)$ is concentration of solid particles, z is the radial direction of ion insertion to particle, t is time, R_s is radius of particles, D_{eff}^s is effective solid-phase diffusivity. SOC calculated as a function of average particle concentration is by eqn (4)

$$SOC(t) := \frac{1}{L_n c_s^{max,n}} \int_0^{L_n} c_s^{avg}(z,t)dz$$
(4)

L_n is thickness of negative electrode, $c_s^{max,n}$ is maximum concentration of lithium-ions is the negative electrode, and $c_s^{avg}(z,t)$ is average concentration of solid-phase. The movement of electrons in electrode is described by Ohm's law described by eqn (5).

$$\frac{\partial}{\partial z}\left[\sigma_{eff}\frac{\partial}{\partial z}\Phi_s(z,t)\right] = aFj(z,t)$$
(5)

Where σ_{eff} is the effective conductivity of the electrodes, $\Phi_s(z,t)$ is the solid potential and a is the ratio of particle surface area to volume. Voltage is expressed as eqn (6) by the difference in current collector solid potential of each electrode.

$$V(t) := \Phi_s(0,t) - \Phi_s(L,t)$$
(6)

The heat transfer phenomenon inside the battery, such as the generation of various heat sources, is expressed as eqn (7).

$$\rho C_p \frac{\partial}{\partial t}T(z,t) = \frac{\partial}{\partial z}\left[\lambda\frac{\partial}{\partial z}T(z,t)\right] + Q_{ohm}(z,t) + Q_{rxn}(z,t) + Q_{rev}(z,t)$$
(7)

Where ρ is the density, C_p is the specific heat, λ is the thermal conductivity, $Q_{ohm}(z,t)$ is a heat source caused by the movement of lithium ions, $Q_{rxn}(z,t)$ is caused by ionic flux and overpotential, and $Q_{rev}(z,t)$ is caused by the change in entropy of the electrode. A detailed description of the PET model is in reference (Newman et al., 1975)

2.2. Parameter identification

Since PET model has significant number of parameters, design space becomes excessively large to employ model mathematically. Thus, in this study we applied derivative-free optimization, especially genetic algorithm (GA), that doesn't use mathematical data of objective functions. GA repeats cycle of selection, crossover, mutation, and evaluation until evaluation result reaches optimization tolerance. After optimization is finished, we achieve optimal parameters. GA has been used in various field like NCC feed composition (Kim et al, 2023) and plastic recycling (Lee et al., 2022)

As an objective function we used the sum of difference between experimental dataset (Gun et al., 2015) and simulation result of MATLAB-based LIONSIMBA. Constant power-constant voltage (CP-CV) step is conducted. Lower and upper bounds are 0 and 1 since we applied reverse min-max scaling expressed as eqn (8).

$$x_i = x_{scaled}(x_{max} - x_{min}) + x_{min} \tag{8}$$

2.3. Fast-charging protocol

SOC has been used as charging stage distinction criteria, but no exact guideline what % to start and how much intervals to separate exists (Attia et al., 2020). To discuss SOC section split mechanism, we split SOC more finely with 10 % intervals from 20 % to 80 % and applied GA to find out best combination. After 80 % SOC, CV charging protocol is executed, but the result won't be considered in objective function.

We designed objective function considering charging time and degradation constraints. Upper and lower bound is 5 and 1 C-rate. Objective function is expressed as eqn (9).

$$Obj = Charge\ time + penalty * (violation\ counts) + CV\ stop\ penalty \tag{9}$$

Temperature constraint is 313.15 K and voltage constraint is 4.1 V. Number of constraint violation are counted during the simulation. If CV step stopped before 95 % SOC, it is considered as problematic charging and penalty is added. After optimization is finished, objective function result and best C-rate combination are returned.

3. Results

3.1. Parameter identification

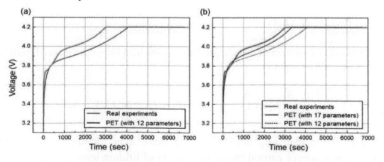

Figure 2. Parameter identification result (a) With 12 parameters (6 for main + 6 for fine fitting) (b) Final results with 5 additional parameters

In this study, we started identification process with top twelve parameters from the sensitivity analysis by Li et al (2020). From Li et al.'s study, we selected six parameters, thickness and active material volume fraction of cathode and anode, anode reaction rate coefficient, and cathode maximum ionic concentration for main curve fitting. Then, fine optimization is conducted with the other six parameters, which are particle radius and diffusion coefficient of cathode and anode, cathode reaction rate coefficient, and anode maximum ionic concentration. Result until this procedure is illustrated in Fig. 2(a).

After a few cycles of main and fine fitting process is done, we conducted analysis about relationship between parameter change and CV step beginning time to the rest of parameters from Li et al.'s study by changing one parameter while fixing the others. As a result, Bruggeman coefficient and particle surface area for positive and negative electrode, and transference number are selected for secondary fine optimization. Then, final optimization with five extracted parameters to achieve the result of Fig. 2(b).

Fig. 2(b) shows the voltage plot of both parameter identification using simulation and experimental data. As figure illustrates, simulation result using identified parameters have similar charging tendency with experimental curve. Since identified parameters can imitate the battery charge behavior, we can use simulation results instead of experiments.

3.2. Fast-charging protocol

Table 1. Fast-charging protocol result

SOC section	C rate (1C = 2.6A)	Time (s)
20 ~ 30 %	5.8569 C	65
30 ~ 40 %	3.2568 C	109
40 ~ 50 %	2.4387 C	145.5
50 ~ 60 %	2.2436 C	161
60 ~ 70 %	1.7594 C	203.5
70 ~ 80 %	1.2478 C	298.5
Total time		**982.5 s**

Optimal C rate for each SOC sections and total time required is shown in the Table 1. According to Table 1, C rate decreases per each interval during the charge MCC protocol.

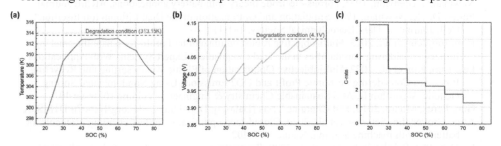

Figure 3. Battery behavior plot for optimal charge strategy. (a) X-axis: SOC, Y-axis: Temperature (K) (b) X-axis: SOC, Y-axis: Voltage (V) (c) X-axis: SOC, Y-axis: C-rate

As Fig. 3 illustrates, optimal output minimizes charging time while obeying both 4.1 V and 313.15 °C degradation constraints. MCC protocol is known for a fast-charging and difficulty in temperature control (Jiang et al, 2022). However, Fig. 3 shows that apply high current at the beginning then decrease per interval can stably control temperature. Moreover, Fig. 3(a) depicts temperature is main constraint until SOC reaches 60%. After 60%, Fig. 3(b) shows that voltage regulates the charge current of the rest part.

4. Conclusion

To minimize the time cost and degradation, optimal charging strategy is required. Repetitive experiments, however, takes immense times and expenses. Therefore, numerical simulation with optimization can be appropriate alternative. In this study, we used GA to identify battery parameters and determine the optimum MCC combination. The results suggest GA could find out parameters that imitate the charge tendency and find out best charging protocols efficiently. Thus, this research demonstrates suggested charge protocol search method can be a proper solution for the situation. We propose that cyclic optimization of parameters for better imitation and discrete optimization using MCC interval as variable can be considered as follow-up research.

References

Jiang, Berliner, Lai, Asinger, Zhao, Herring, Bazant, Braatz, 2022, Fast charging design for Lithium-ion batteries via Bayesian optimization, Applied Energy, 307

Ouyang, Chu, Lu, Li, Han, Feng, Liu, 2015, Low temperature aging mechanism identification and lithium deposition in a large format lithium iron phosphate battery for different charge profiles, Journal of Power Sources, 286

Liu, Zou, Li, Wik, 2018, Charging Pattern Optimization for Lithium-Ion Batteries With an Electrothermal-Aging Model, IEEE Transactions on Industrial Informatics, 14(12)

Kumar, Sathyamurthy Kim, Panchal, Ali. Et al, 2023, A state-of-the-art review on advancing battery thermal management systems for fast-charging, Applied Thermal Engineering, 226

Sulzer, Marquis, Timms, Robinson, Chapman, 2021, Python Battery Mathematical Modelling (PyBaMM), Journal of Open Research Software, 9

Torchio, Magni, Gopaluni, Braatz, Raimondo, 2016, LIONSIMBA: A Matlab Framework Based on a Finite Volume Model Suitable for Li-Ion Battery Design, Simulation, and Control, Journal of The Electrochemical Society, 163(7)

Berliner, Cogswell, Bazant, Braatz, 2021, Methods—PETLION: Open-Source Software for Millisecond-Scale Porous Electrode Theory-Based Lithium-Ion Battery Simulations, Journal of The Electrochemical Society, 168(9)

Newman, Tiedemann, 1975, Porous-electrode theory with battery applications, AIChE Journal

Fuller, Doyle, Newman, 1994, Simulation and Optimization of the Dual Lithium Ion Insertion Cell, Journal of The Electrochemical Society, 141(1)

Kim, Joo, Kim et al., 2023, Multi-objective robust optimization of profit for a naphtha cracking furnace considering uncertainties in the feed composition, Expert Systems With Applications

Lee, Lim, Kim et al., 2022, Multiobjective Optimization of Plastic Waste Sorting and Recycling Processes Considering Economic Profit and CO2 Emissions Using Nondominated Sorting Genetic Algorithm II, ACS Sustainable Chemistry & EngineeringVolume 10, Issue 40

Gun, Perez, Moura, 2015, Fast Charging Tests [Dataset], Dryad https://doi.org/10.6078/D1MS3X

Attia, Grover, Jin, Serverson et al., 2020, Closed-loop optimization of fast-charging protocols for batteries with machine learning, Nature, 397-402

Li, Cao, Jöst, Ringbeck, Kuipers et al., 2020, Parameter sensitivity analysis of electrochemical model-based battery management systems for lithium-ion batteries, Applied Energy, 269

Flavio Manenti, Gintaras V. Reklaitis (Eds.), Proceedings of the 34th European Symposium on Computer Aided Process Engineering / 15th International Symposium on Process Systems Engineering (ESCAPE34/PSE24), June 2-6, 2024, Florence, Italy

Applying an Efficient Approach for Modeling and Optimization of Membrane Gas Separation Processes Using Maxwell-Stefan Theory

Héctor Octavio Rubiera Landa,[a,*] Joeri F. M. Denayer[a]

[a]*Department of Chemical Engineering, Vrije Universiteit Brussel, Pleinlaan 2, B-1050 Elsene, Brussels, Belgium*
hector.octavio.rubiera.landa@vub.be

Abstract

The Maxwell-Stefan theory (M-S) provides a comprehensive framework for mass transfer that can be applied for describing multicomponent diffusion in microporous materials, e.g., zeolites and metal-organic frameworks (MOFs), see e.g., Krishna (1990). M-S theory has been applied successfully to model membrane separation processes, cf. Krishna (2014), relying on principles that can accurately estimate competitive adsorption equilibria, such as the Ideal Adsorbed Solution Theory (IAST) by Myers & Prausnitz (1965). In this work, we incorporate IAST to solve transient mass balances of membrane processes with an efficient strategy developed by Fechtner & Kienle (2018) for chromatographic separations modeling. To render accurate & efficient computations of the required M-S mass transfer fluxes, we apply analytical expressions for partial derivatives of adsorbed concentrations w.r.t. fluid concentrations developed by Rubiera Landa et al. (2013). We implement this efficient approach in three types of membrane process models of increasing detail to analyze propane/propylene separations on thin-layered membranes made of zeolitic imidazolate frameworks (ZIFs). In the first model, we consider only the thin-layered membrane. The second model considers a "batch" process, which includes mass balances around membrane, permeate & retentate compartments. The last more-detailed model considers a spatially-distributed asymmetric hollow-fiber membrane system, with flux estimations obtained from the less-detailed models. Robustness & efficiency of the proposed M-S/IAST calculation approach are tested with an optimization study using the more-detailed model, where we apply a genetic algorithm to identify process operation variables (i.e., decision variables) that yield optimal performance for this hydrocarbon separation example.

Keywords: Maxwell-Stefan approach, Ideal Adsorbed Solution Theory (IAST), membrane separation processes, Multi-Objective Optimization (MOO)

1. Introduction

Adsorption-based separation processes have gained considerable attention in the last decades due to their potential for providing suitable alternatives to energy-intensive, well-established technologies, such as distillation processes. Owing to their operational flexibility, they constitute an attractive option for electrification of the chemical industry. This is particularly important in the production of light hydrocarbon olefins, viz. ethylene & propylene, and their associated separation & purification steps. Designing & developing these kinds of processes is a challenging task, which requires fundamental understanding of mass transfer and adsorption principles. The Maxwell-Stefan approach (M-S) provides a comprehensive theoretical framework based on Irreversible

Thermodynamics that has been developed by Krishna (1990, 2014) to describe mass transfer in microporous materials, including e.g., adsorbent particles and thin-layered membranes. Adsorption equilibrium thermodynamics is included in this approach to consider the effect that local adsorbed concentrations produce in the effective transport of species inside the microporous medium. IAST is a well-established alternative to describe competitive equilibria in microporous materials using single-component adsorption isotherms. In combination with M-S theory, permeation behavior of many gas mixtures & materials can be predicted & analyzed in detail.

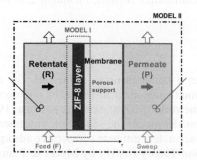
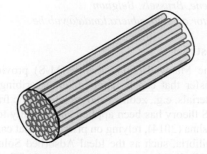

Figure 1: Graphical illustration of membrane processes that apply ZIF-8 thin-layered membrane for propylene/propane mixture separation. *Left:* Thin-layered membrane (Model I) & "Batch" process (Model II). *Right:* Spatially-distributed hollow-fiber membrane module (Model III).

Investigated membrane separation processes: Graphical illustrations of thin-layered membrane separation processes considered in this work are displayed in Fig. 1. A high-pressure feed gas (F) mixture of propylene (component 1) & propane (component 2) is continuously supplied to the process on the retentate side (R), whereas the permeate side (P) remains under low pressure. As a result, a pressure differential is established across a ZIF-8 membrane, generating concentration gradients and inducing a preferential permeation of the faster-diffusing propylene through the microporous material, which is exploited to achieve separation. A defect-free ZIF-8 thin layer is assumed; mass transfer resistance of the porous support is ignored, as well as any mechanism of membrane degradation.

2. Methods & Algorithms

Efficient implementation to solve M-S flux equations: The explicit, compact matrix form of the flux equations derived from M-S approach, used to describe multicomponent mass transfer (Krishna (1990, 2014)), is:

$$N = -\sigma \Delta \Gamma \nabla_\xi q \; ; \quad \Gamma := diag\left(\frac{q_i}{c_i}\right) J_c, \quad i = 1, \dots, N_c, \tag{1}$$

with Jacobian, J_c, of fluid-phase concentrations, c, w.r.t. loadings, q, at equilibrium. Ideal gas behavior & isothermal conditions are assumed. Diffusional frictional effects of the confined species are quantified by $\sigma\Delta$, which take M-S diffusivities, $\mathcal{D}_i^{MS}$, & adsorbed-phase molar fractions, x_i, into account. Thermodynamic factors, Γ, require knowledge of adsorption equilibria, $c = f^{-1}(q)$, which are only available analytically for a few simple adsorption equilibrium equations. Therefore, from a calculation point of view, an inversion of equilibria, $q = f(c)$, and corresponding evaluation of Jacobian elements, $\partial c_i / \partial q_j$, $i, j = 1, \dots, N_c$, are required to implement Eq. (1) in mass balance equations of the membrane processes studied herein. To overcome this difficulty, the Inverse Function Theorem is invoked, so Jacobian, J_c, in Eq. (1) can be written as

$$J_c \equiv \mathcal{J}_q^{-1}; \quad N = -\sigma \Delta \, diag\left(\frac{q_i}{c_i}\right) \mathcal{J}_q^{-1} \nabla_\xi q, \quad N = [N_1, \dots, N_{N_c}]^T, \tag{2}$$

whereby, analytical formulae by Rubiera Landa et al. (2013), applicable to IAST, are used to calculate $\mathcal{J}_q^{-1}$—see details in Rubiera Landa and Denayer (2023). This renders efficient & robust M-S flux calculations where Γ are obtained in closed form with minimal computational cost.

Applied membrane models: *Model I: Transient model for thin-layered membrane.* The applicable 1-D mass balances around the thin-layered membrane—see Fig. 1—are:

$$\frac{\partial q_i}{\partial t} = \frac{1}{\rho} \frac{\partial N_i}{\partial z} \quad i = 1, \dots, N_c, \tag{3}$$

with N_i given by Eq. (2), and implicitly-defined IAST equilibria, $\varphi(c, c^0, q, T) = 0$. This yields a partial differential-algebraic equation (PDAE) system, with applicable IC & BCs. After sufficiently long time, the process reaches steady-state, i.e., flux values become constant & separation performance is evaluated then. *Model II: "Batch" membrane process.* An additional level of complexity is introduced by considering steady-state mass balances around well-mixed retentate (R) & permeate (P) sides, including the membrane and inlet & outlet molar flow rates (see Fig. 1). This yields the following nonlinear equation in the unknown, y_1^R, for the binary mixture case, $N_c = 2$, considered here—cf. van de Graaf et al. (1999):

$$f(y_1^R) = \frac{\dot{n}_1^F - N_1^{SS} A}{[\dot{n}_1^F + \dot{n}_2^F] - [N_1^{SS} A + N_2^{SS} A]} - y_1^R = 0 \; ; \; y_1^R + y_2^R = 1 \; ; \; N_i^{SS} = f(y_1^R). \tag{4}$$

Eqs. (4) are solved iteratively by applying Algorithm 1. The computations of steady-state M-S fluxes, N_i^{SS}, with Model I, Eqs. (3), are embedded in the solution of Eq. (4). Afterward, performance metrics are evaluated.

Algorithm 1: Batch membrane mass balance (Model II)

1: Specify feed parameters, $\dot{n}^F, y_1^F$, operating pressures, p^R, p^R; & membrane area, A.
2: Assign a guess value for $y_1^P \leftarrow y_1^{P(*)}$
3: Compute y_1^R from $f(y_1^R) = 0$ with Eqs. (4), applying Eqs. (3) for steady-state flux values, N_1^{SS}, N_2^{SS}.
4: Close the steady-state mass balances with $\dot{n}^r = \dot{n}^F - \sum_{i=1}^{N_c} N_i^{SS} A = \dot{n}^F - \dot{n}^P$
5: Verify if $y_1^P - \frac{\dot{n}_1^P}{\sum_{i=1}^{N_c} \dot{n}_i^P}$, else, iterate until fulfilled.
6: Calculate performance metrics: perm-selectivity, $\Phi_{Sel,i}$; permeability, $\Phi_{Perm,i}$.

Model III: Asymmetric hollow fiber membrane. We take the following 1-D spatially-distributed, steady-state model by Gabrielli et al. (2017), to calculate retentate side (R) molar flow rate & compositions:

$$\frac{d}{dz} \dot{n}^R = \dot{n}^F m \frac{y_1^R - \frac{p^P}{p^R} y^*}{y^*} \; ; \qquad \frac{d}{dz} y_1^R = \frac{1}{\dot{n}^F} m \frac{\left(y_1^R - y^*\right)\left(y_1^R - \frac{p^P}{p^R} y^*\right)}{y^* \dot{n}^R}, \tag{5a}$$

with dimensionless parameter, $m := \frac{\Pi_1 A p^R}{\dot{n}^F}$. The total membrane area for N_f hollow fibers of length, L_f, with outer radius, r_o, is: $A = A_f N_f = 2\pi r_o L_f N_f$. Local molar fraction at position, $z = z^*$, is,

$$y^* = \frac{1+(\alpha-1)\left(\frac{p^P}{p^R}+y_1^R\right)-\left[\left[1+(\alpha-1)\left(\frac{p^P}{p^R}+y_1^R\right)\right]^2-4\frac{p^P}{p^R}\alpha(\alpha-1)y_1^R\right]^{1/2}}{2\frac{p^P}{p^R}(\alpha-1)}. \tag{5b}$$

The IC is: $\dot{n}^R(z=0) = \dot{n}^F$ $y_1^R(z=0) = y_1^F$. Perm-selectivity, defined as the ratio of steady-state permeances, $\alpha := \Pi_1/\Pi_2$, is supplied as a parameter to the model and computed with Models I or II. Mass balances around the fiber module are closed with:

$$\dot{n}^F = \dot{n}^R(z=L_f) + \dot{n}^P(z=L_f);$$
$$\dot{n}^F y_1^F = \dot{n}^R(z=L_f)y_1^R(z=L_f) + \dot{n}^P(z=L_f)y_1^P(z=L_f). \tag{5c}$$

Solution of membrane models: In Model I, the PDAE system, Eqs. (3), is solved by applying the Method-of-lines approach (Schiesser (1991)). A cell-centered finite volume method (FVM)—see e.g., Hundsdorfer and Verwer (2003)—is applied to discretize along spatial coordinate, z, yielding an index-1 DAE system to integrate numerically, until steady-state is attained. In Model II, Eq. (4) in the unknown, y_1^R, is solved with MATLAB™ function 'fsolve'. In Model III, Eqs. (5), the set of ODEs are integrated from $z = 0$ to $z = L_f$. All models use MATLAB™ function 'ode15s' for integration.

Propylene/propane separation using ZIF-8: In this study, we apply parameters at $T = 308\ K$ reported by Zhang et al. (2012) for the propylene/propane/ZIF-8 system, including single-component adsorption equilibria & zero-coverage diffusivity values used in M-S flux calculations. Competitive equilibria are calculated with IAST & Jacobian formulation, Eq. (2).

Estimation of perm-selectivity & permeabilities from Models I & II using sampling technique: Models I & II provide permeability values and performance metrics, perm-selectivity, α, & permeability, $\Phi_{Perm.,i}$, based exclusively on M-S flux calculations. These metrics are then supplied to the more-detailed Model III. Latin Hypercube Sampling design (McKay et al. (1979))—function 'lhsdesign' in MATLAB™—is used to explore these metrics. A design with $M = 100$ samples is applied for variables, $p^R \in [1.5 \times 10^5, 4.5 \times 10^5]\ Pa$ & $y_1^R \in [0.4, 0.85]$, keeping constant $p^P = 10^3\ Pa$.

Application of multi-objective optimization (MOO) to analyze process performance of Model III: We formulate the following bound-constrained MOO task:

$$maximize\ \{\Phi_{Pu,i}(u), \Phi_{Re,i}\},\ \text{s. t.}\ u^{lb} \leq u \leq u^{ub}, \tag{6}$$

with purity, $\Phi_{Pu,i}$, recovery, $\Phi_{Re,i}$ & decision variables, u. Other relevant objectives include specific membrane area, $\Phi_{A,i}$, & specific energy consumption, $\Phi_{En,i}$. Due to the robustness and fast calculation of Model III, a black-box function approach can be employed to solve task (6). Therefore, the optimal separation performance of the hollow-fiber membrane module can be explored by applying genetic algorithms (GA).

Table 1. Design variables & bounds for MOO task.

Variable	$A, [m^2]$	$p^R, [Pa]$	$p^P, [Pa]$
Lower bound	$0.47(50\ fibers)$	1.5×10^5	1×10^4
Upper bound	$3.77(400\ fibers)$	4.5×10^5	8×10^4

We use the NSGA-II (Deb et al. (2002)) variant 'gamultiobj' available in MATLAB™. Applied decision variables are listed in Table 1 with corresponding bounds. A population size of 75 function evaluations with a maximum budget of 150 generations is specified. Additional simulation parameters are: molar feed flow rate, $\dot{n}^F =$

0.05 mol/s; hollow-fiber outer diameter, $r_o = 1.5 \times 10^{-3}\ m$ (1.5 mm); fiber length, $L_f = 1\ m$; membrane layer thickness, $\delta = 2.5 \times 10^{-6}\ m$ (2.5 $microns$); and permeances, $\Pi_1 = 4.2 \times 10^{-8}\ mol/(m^2 \cdot s \cdot Pa)$ & $\Pi_1 = 3.9 \times 10^{-10}\ mol/(m^2 \cdot s \cdot Pa)$. Propylene feed compositions, $y_1^F = \{0.4, 0.6, 0.85\}$, are considered for comparison.

3. Results & Discussion

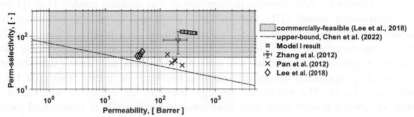

Figure 2. Robeson plot for propylene/propane pair, displaying: Model I sampling result; experimental measurements by Lee et al. (2018) & Pan et al. (2012); & estimation by Zhang et al. (2012) for comparison.

Estimation of perm-selectivity & permeability values with Model I: Fig. 2 illustrates a Robeson plot for the propylene/propane pair with the performance result obtained by applying Model I. Propylene to propane perm-selectivity has relatively close values between 115 and 125, with propylene permeabilities in the range [250,400] $Barrer$— (1 $Barrer = 3.35 \times 10^{-16}\ mol/(m \cdot s \cdot Pa)$). Similar results are obtained with Model II. Accordingly, ZIF-8 thin-layered membranes perform beyond the known upper bound for polymeric membranes reported by Chen et al. (2022) & fall inside the projected commercially-viable region discussed in Lee et al. (2018). Experimental data of some authors is also displayed in Fig. 2, showing deviations from model predictions. Factors such as membrane fabrication and processing methods, porous support & MOF synthesis contribute to the observed performance discrepancies, as discussed by Lee et al. (2018). The models herein discussed assume ideal membranes, ignoring porous support effects; these should be considered in more-detailed process models.

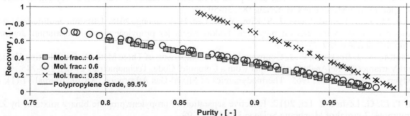

Figure 3. Computed propylene recovery vs. propylene purity Pareto frontiers for different feed compositions, $y_1^F = \{0.4, 0.6, 0.85\}$. Design variables & bounds applied are listed in Table 1.

MOO results for Model III: Fig. 3 illustrates MOO results for the asymmetric hollow-fiber membrane module (cf. Fig. 1). The optimizer identifies the trade-off between recovery & purity for each feed composition, y_1^F, considered. The best result in terms of these conflicting metrics is obtained when a stream enriched with propylene is processed by the membrane module. Pareto frontiers also show that this is a challenging separation, where purity values close to the desired propylene-grade purity (99.5%) are difficult to attain with a single hollow-fiber module, as specified. This aspect of membrane performance for propylene/propane separations has also been discussed in detail by Yamaki et al. (2022), where membrane cascades have been proposed as an alternative to meet stringent downstream propylene purity requirements at feasible product recoveries.

4. Conclusion

We have presented a detailed strategy to analyze the separation performance of membrane processes based on thin-layers of microporous adsorbents. We considered models of increasing complexity, incorporating information of less-detailed models (I & II) into a spatially-distributed model (III) that provides a more-detailed representation of a membrane separation operation. M-S flux calculations in a propylene/propane/ZIF-8 example were incorporated efficiently by applying analytical thermodynamic factor expressions. Separation performance was analyzed with a GA, confirming the robustness of our modeling approach, identifying optimal recovery vs. purity trade-offs for the example considered. The presented strategy provides an attractive alternative for materials screening studies from first principles and can assist in conceptual process design of challenging gas separation tasks.

Acknowledgement: We thank the support of *Agentschap Innoveren & Ondernemen*, Belgium, in the frame of Moonrise-MOONSHOT Project HBC.2020.2612 (3E210291): *"Hybrid membrane/sorption technology for more efficient C2 and C3 separations"*.

References

X. Y. Chen, A. Xiao, D. Rodrigue, 2022. Polymer-based Membranes for Propylene/Propane Separation. Separation & Purification Reviews 51 (1), 130-142.

K. Deb, A. Pratap, S. Agarwal, T. Meyarivan, 2002. A Fast and Elitist Multiobjective Genetic Algorithm: NSGA-II. IEEE Transactions on Evolutionary Computation 6 (2), 182-197.

M. Fechtner, A. Kienle, 2018. Efficient simulation and equilibrium theory for adsorption processes with implicit adsorption isotherms - Ideal adsorbed solution theory. Chemical Engineering Science 177, 284-292.

P. Gabrielli, M. Gazzani, M. Mazzotti, 2017. On the optimal design of membrane-based separation processes. Journal of Membrane Science 526, 118-130.

W. Hundsdorfer, J. G. Verwer, 2003. Numerical solution of time-dependent advection-diffusion-reaction equations. Springer, ISBN 3-540-03440-4.

R. Krishna, 1990. Multicomponent surface diffusion of adsorbed species: A description based on the Generalized Maxwell-Stefan equations. Chemical Engineering Science 45 (7), 1779-1791.

R. Krishna, 2014. The Maxwell-Stefan description of mixture diffusion in nanoporous crystalline materials. Microporous and Mesoporous Materials 185, 30-50.

M. J. Lee, M. R. A. Hamid, J. Lee, J. S. Kim, Y. M. Lee, H.-K. Jeong, 2018. Ultrathin zeolitic-imidazolate framework ZIF-8 membranes on polymeric hollow fibers for propylene/propane separation. Journal of Membrane Science 559, 28-34.

M. D. McKay, R. J. Beckman, W. J. Conover, 1979. A Comparison of Three Methods for Selecting Values of Input Variables in the Analysis of Output from a Computer Code. Technometrics 21 (2), 239-245.

A. L. Myers, J. M. Prausnitz, 1965. Thermodynamics of Mixed-Gas Adsorption. AIChE Journal 11 (1), 121-127.

Y. Pan, T. Li, G. Lestari. Z. Lai, 2012. Effective separation of propylene/propane binary mixtures by ZIF-8 membranes. Journal of Membrane Science 390-391,93-98.

H. O. Rubiera Landa, J. F. M. Denayer, 2023. Efficient modeling and simulation of gas separations applying Maxwell-Stefan approach and Ideal Adsorbed Solution Theory. Chemical Engineering Science (submitted).

H. O. Rubiera Landa, D. Flockerzi, A. Seidel-Morgenstern, 2013. A Method for Efficiently Solving the IAST Equations with an Application to Adsorber Dynamics. AIChE Journal 59 (4), 1263-1277.

W. E. Schiesser, 1991. The numerical method of lines: integration of partial differential equations. Academic Press, ISBN 9780128015513.

J. M. van de Graaf, F. Kapteijn, J. A. Moulijn, 1999. Modeling Permeation of Binary Mixtures Through Zeolite Membranes. AIChE Journal 45 (3), 497-511.

T. Yamaki, N. Thuy, N. Hara, S. Taniguchi, S. Kataoka, 2022. Design and Evaluation of Two-Stage Membrane-Separation Processes for Propylene-Propane Mixtures. Membranes 12 (2), 163.

C. Zhang, R. P. Lively, K. Zhang, J. R. Johnson, O. Karvan, W. J. Koros, 2012. Unexpected Molecular Sieving Properties of Zeolitic Imidazolate Framework-8. The Journal of Physical Chemistry Letters 3, 2130-2134.

Flavio Manenti, Gintaras V. Reklaitis (Eds.), Proceedings of the 34th European Symposium on Computer Aided Process Engineering / 15th International Symposium on Process Systems Engineering (ESCAPE34/PSE24), June 2-6, 2024, Florence, Italy

Modelling of the Co-gasification of Waste Plastic and Biomass using Oxygen-CO₂ Mixtures

Nomadlozi L. Khumalo[a], Bilal Patel[a]

[a]*Department of Chemical Engineering, University of South Africa, Corner of Christiaan de Wet Road and Pioneer Avenue, Florida 1709, South Africa*
patelb@unisa.ac.za

Abstract

Gasification provides a promising solution for valorizing plastic waste by using high temperatures and gasifying agents to convert the waste into versatile syngas. However, the gasification of plastic waste comes with various operational challenges. Co-gasifying plastic waste with biomass offers several advantages, including the reduction of operational challenges, decreased tar formation, and the production of syngas with improved quality and higher energy content.

The study investigates the impact of various factors on the syngas produced from the co-gasification of sawdust (SD) and plastic waste (low-density polyethylene, PE) using different blend ratios (BRs) of PE (25 %, 50 %, and 75 %) and different operating conditions, including equivalence ratio (ER) and CO_2/C ratios (0.6 and 1.4). The study utilized a non-stoichiometric equilibrium model within Aspen Plus to analyze the co-gasification process. The key findings include the production of syngas with high hydrogen (H_2) content, high Lower Heating Value (LHV) of 9.2 MJ/Nm^3, and high H_2/CO ratios. These favorable outcomes were achieved at low ER values below 0.4 and a CO_2/C ratio equal to 0.6.

Keywords: co-gasification, blend ratio, equivalence ratio; synergistic effect, plastic waste.

1. Introduction

The extensive use of plastics has resulted in significant plastic waste (PW) generation, causing environmental pollution (Midilli et al. 2022). Currently, over 79% of plastic waste is disposed in landfills or ends up in the environment (Tejaswini et al. 2022). Landfilling and incineration are both not feasible environmental solutions to waste plastics. Chemical valorization of PW through gasification has shown to be an effective conversion technology, addressing environmental concerns, and reducing reliance on fossil fuels. Gasification involves the high-temperature conversion of carbonaceous components with limited oxygen, resulting in the production of syngas, which is a valuable energy product (Rafey et al. 2023). Despite its high conversion efficiency, mono-gasification of plastic waste faces challenges such as polymer agglomeration and high tar yield. To overcome these limitations, biomass (BM) is a suitable complement to plastic waste due to it being carbon-neutral, environmentally friendly, and abundantly available (Mariyam et al. 2022). Co-gasification process, combining biomass and plastic waste, mitigates operational challenges associated with mono-gasification, reduces tar formation, and improves the quality and energy content of the syngas produced (Shahbaz et al. 2020). However, a gap in the existing literature on the impact of using a mixture of oxygen and carbon dioxide as a gasifying agent during the co-gasification of biomass (pine sawdust) and plastic waste (low-density Polyethylene) has been identified. Carbon dioxide promotes the dry reforming reaction, which increases combustible gases and, when mixed with oxygen, facilitates auto-thermal operation (Pinto et al. 2016).

Previous research, such as studies conducted by Kaydouh et al. (2022), De et al. (2021) Ruoppolo et al. (2012), and Wang et al. (2021), primarily focused on co-gasification of biomass and plastic waste, but typically utilized single gasifying agents like air, steam, oxygen, or carbon dioxide, instead of mixtures of gasifying agents. The study aims to investigate the impact of feedstock composition, which includes biomass, plastic waste, and their blended feedstocks (e.g., (25 % PE + 75 % biomass), (50 % PE + 75 % biomass) and (75 % PE + 25 % biomass), using an oxygen-carbon dioxide mixture as the gasifying agent. The focus will be on assessing how these variables affect the composition of the syngas (including the concentration H_2), the H_2/CO ratio of the syngas, and the Lower Heating Value (LHV) of the syngas. The research will also explore the potential synergistic interaction between biomass and plastic waste during the co-gasification process.

2. Modelling framework

A non- stoichiometric process model was developed in Aspen Plus for the co-gasification of biomass and plastic (Khumalo et al. 2023). The Peng-Robinson-Boston-Mathias (PR-BM) thermodynamic model was selected (Ramzan et al. 2011). The enthalpy and density functions for biomass (sawdust) and plastic waste (PE) were defined using HCOALGEN and DCOALIGT algorithms. The feedstocks considered was polyethylene (PE), biomass (sawdust) and various blend ratios; (25% PE + 75 % biomass), (50 % PE + 50% biomass) and (75% PE + 25% biomass). Biomass properties were specified based on ultimate analyses (Carbon (C): 45.5 %, Hydrogen (H): 5 %, Oxygen (O): 47.1 %, Nitrogen (N): 0.05 % and ash: 2.35 %) and proximate analyses (Fixed Carbon (FC): 18.45 %, Volatile Matter (VM): 79.2 %, ash: 2.35 % and Moisture Content (MC): 5.76 %). Similarly, the ultimate analyses of plastic waste (C: 85.81 %, H:13.86 %, O: 0 %, N: 0.12 %, S:0.06 % and ash: 0.15 %), proximate analysis of plastic waste (FC: 0 %, VM: 99.85 %, ash content: 0.15 % and MC: 0.02 %) (Al Amoodi et al. 2013) was also specified. Gasifying agents such as oxygen and carbon dioxide were specified with variations in equivalence ratio (ER) between 0.1 -1 and CO_2/C ratios equal to 0.6 and 1.4.

The gasifier was modelled utilising a RYield reactor, which was used for the decomposition of non- conventional components into elements. These elements, together with the gasifying agents, enter a RGibbs reactor, which uses Gibbs energy minimization to predict product distribution The products pass through a cyclone unit, which separates solids (ash) from the gaseous product. The effect of feedstock composition and gasifying agent flowrate, on the product gas composition, H_2/CO ratio and Lower Heating Value (LHV) was determined. The LHV was calculated using equation (1)

$$LHV_{syngas} = 10.79\ X_{H2} + X_{CO} + 35.83\ X_{CH4} \tag{1}$$

where, X is the molar fraction of the gaseous component (Tavares et al. 2018).

3. Results and Discussion

3.1 Effect of the equivalence ratio on H_2 composition in the product gas for various blend ratios when carbon dioxide – oxygen mixture is used as gasifying agent.

The data in Figure 3.1 (a) and (b), illustrates that at low equivalence ratio (ER) values below 0.4 the hydrogen (H_2) content is high for all feedstocks, however as ER increases further the H_2 content decreases. Figure 3.1 (a) show that the addition of plastic waste, (PW), (polyethylene), (PE) to the feedstock in the presence of oxygen – carbon dioxide mixture as a gasifying agent increases the H_2 composition in the product gas, with the highest H_2 composition of 30% achieved at an ER equal to 0.14 and a fixed CO_2/C ratio equal to 0.6 from the blend ratio of (75 % PE + 25 % biomass). The increase in H_2 composition can be attributed to PE, which acts as a H_2 donor to the radicals generated from biomass (BM) pyrolysis thus stabilizing those radicals (Ahmed et al. 2011). In Figure 3.1 (b), the results show that the H_2 composition produced at higher carbon dioxide flowrates i.e., CO_2/C ratio equal to 1.4, is lower than that obtained at lower carbon dioxide flowrates i.e.CO_2/C ratio equal to 0.6. This phenomenon occurs because the addition of carbon dioxide at a high flowrate promotes certain chemical reactions, such as the reverse dry reforming reaction and reverse water gas shift reaction which is responsible for a decrease in the H_2 content of the product gas (Islam 2020). The synergy between BM and PW, when an oxygen-carbon dioxide mixture is used as gasifying agent, is evident at low CO_2/C ratio of 0.6 instead of 1.4. This is observed in Figure 3.1 (a) as the highest hydrogen content was obtained from the blended feedstocks such as a blend ratio of (25 % PE + 75 % biomass) instead of a single feedstock, whereas, in Figure 3.1 (b) there was no synergistic interaction between BM and PW since the highest H_2 content was obtained from PE feedstock.

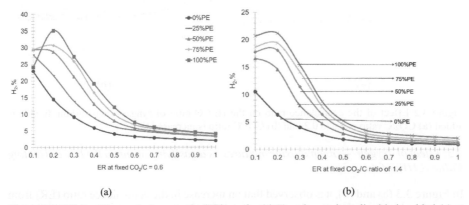

(a)   (b)

Figure 3.1 Effect of the equivalence ratio (ER) on the (a) H_2, when carbon dioxide is added (a) at low flowrate through CO_2/C ratio equals to 0.6 and (b) CO_2/C equals to 1.4.

3.2. Effect of oxygen – carbon dioxide mixtures as a gasifying agent on the H₂/CO ratio of the syngas.

Figure 3.2 (a) and (b) show that as the (ER) increases from 0.1 to 1, using an oxygen-carbon dioxide mixture as a gasifying agent, the H_2/CO ratios of the syngas decrease for various blends of feedstocks. The highest H_2/CO ratios in the syngas are achieved at low ER values below 0.4. When ER is below 0.4, certain chemical reactions like the Boudouard reaction, partial oxidation reaction, water-gas shift reaction, reverse methanation CO_2 reforming reaction, and methanation reaction are favourable, leading to an enhancement in the H_2/CO ratios of the syngas. Moreover, Figure 3.2 (a) and (b) indicate that when polyethylene is added to the feedstock mixtures, the H_2/CO ratio of the syngas increases when an oxygen-carbon dioxide mixture is used as the gasifying agent. This increase can be attributed to the high H_2 content in PE material. However, the use of an oxygen-carbon dioxide mixture as the gasifying agent does not lead to a synergistic interaction between biomass and polyethylene since the recommended H_2/CO ratio of 2 was not obtained from the blended feedstocks, but from 100 % PE. In addition, Figure 3.2 (a) reveals that the H_2/CO ratio of 2 was obtained from the PE feedstock and not from the blended feedstocks. This suggests that for the co-gasification of biomass and PW, the use of an oxygen-carbon dioxide mixture as a gasifying agent is not suitable for achieving the desired syngas quality.

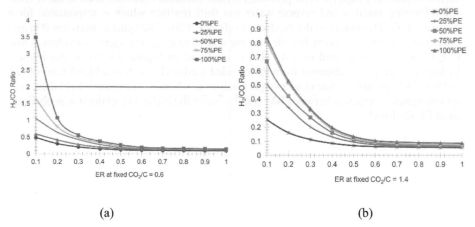

(a)   (b)

Figure 3.2 Effect of the equivalence ratio on the H_2/CO ratio of syngas when carbon dioxide is added (a) CO_2/C ratio equal to 0.2 and (b) CO_2/C ratio is increased to 1.4.

3.3. Effect of oxygen – carbon dioxide mixtures as a gasifying agent on the Lower Heating Value (LHV) of the syngas

In Figure 3.3 (a) and (b), it is observed that an increase in the equivalence ratio (ER) from 0.1 to 1, with fixed CO_2/C ratios at 0.6 and 1.4, initially causes feedstocks with higher polyethylene percentages to increase in Lower Heating Value (LHV). However, as ER continues to increase, the LHV subsequently decreases. This trend is consistent for all feedstock types, except for the LHV of biomass, which decreases as ER increases. At ER below 0.4, there is an increase in LHV due to chemical reactions, such as the Boudouard reaction, partial oxidation, reverse water gas reaction, dry reforming reaction, and methanation reaction, which favour combustible gases such as H_2, CO and CH_4. Figure

3.3 (a) and (b) also show that an increase in the PE percentage in the feedstocks, when an oxygen-carbon dioxide mixture is used as the gasifying agent, leads to an increase in the LHV of the syngas, this is attributed to light and stable hydrocarbons formed by thermal cracking of polymer chains, which contributes to the increase in combustible gases ,which consequently increases the LHV of the syngas (Parrillo et al. 2023). Furthermore, Figure 3.3 (a) and (b) demonstrate that when the CO$_2$/C ratio increases from 0.6 to 1.4, the LHV of the syngas decreases. In Figure 3.3 (a), the highest LHV of 9.2 MJ/Nm^3 is produced from when ER is 0.1 and CO$_2$/C ratio equal to 0.6 at a blend ratio of (25 % PE + 75 % biomass). Whereas, in Figure 3.3 (b), the highest LHV is obtained from a single feedstock of polyethylene. This suggests that the use of an oxygen-carbon dioxide mixture as a gasifying agent enhances the LHV of the syngas at low carbon dioxide flow rates with a CO$_2$/C ratio of 0.6. Consequently, this indicates that during the co-gasification of biomass and polyethylene in the presence of an oxygen-carbon dioxide mixture as a gasifying agent, a synergistic interaction between biomass and polyethylene occurs, leading to higher LHV values when low ER values below 0.4 and at fixed CO$_2$/C ratio equal to 0.6 are utilised.

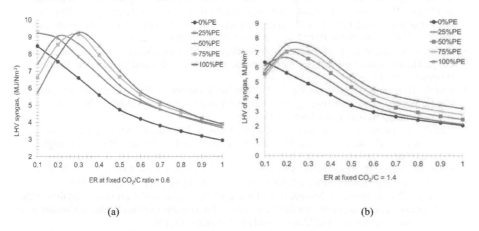

(a)   (b)

Figure 3.3 (a) and (b) Effect of the equivalence ratio (ER) on the Lower Heating Value (LHV) of syngas when carbon dioxide flowrate is varied (a) CO$_2$/C ratio equal to 0.6 and (b) CO$_2$ /C ratio equals to 1.4.

4. Conclusion

The co-gasification of biomass and plastic waste using an oxygen-carbon dioxide mixture as the gasifying agent has revealed some interesting findings. Specifically, increasing the polyethylene percentage in the blended feedstocks leads to higher levels of H$_2$ in the syngas, as well as an increase in the LHV of the syngas, which reaches a maximum of 9.2 MJ/Nm^3 at low ER values below 0.4 and low carbon dioxide flowrate through CO$_2$/C ratio of 0.4. This improvement can be attributed to the fact that PE has a high volatile matter and low oxygen content. The most significant improvements are observed in feedstock blend ratio containing (25 % PE + 75 % biomass) compared to other blended feedstock ratios. A high-quality product composition and energy content of syngas are most favourable when the equivalence ratio (ER) is kept low, below 0.4, and with a fixed CO_2/C ratio equal to 0.6. These conditions lead to a more efficient syngas production process when co-gasifying biomass and polyethylene.

References

I.I. Ahmed, N. Nipatummakul, A.K. Gupta, (2011), Characteristics of Syngas From Co-gasification of Polyethylene and Woodchips, Applied Energy, 88,165 – 174.

N. Al Amoodi, P. Kannan, A. Al Shoaibi, C. Shrinivasakannan, (2013), Aspen Plus Simulation of Polyethylene Gasification under Equilibrium Conditions, Chemical Engineering Communication, 200, 7 – 9.

S. De, S. Pati, (2021), Model Development and Thermodynamic Analysis of Biomass Co-gasification using Aspen Plus, Indian Chemical Engineer, 63,172 – 183.

M. D.W Islam, (2020), Effect of Different Gasifying Agents (Steam. H_2O_2, Oxygen, CO_2 and Air) on Gasification Parameters, International Journal of Hydrogen Energy, 45,31760 – 31774.

M. Kaydouh, N. Hassan, (2022), Thermodynamic Simulation of the Co-gasification of Biomass and Plastic Waste for Hydrogen-Rich Syngas Production, Results in Engineering, 100771.

N. L Khumalo, B. Patel, (2023), A Parametric Analysis of the Co-gasification of Biomass and Plastic Waste using Aspen Plus, Computer Aided Chemical Engineering,52,1421 – 1426.

S. Mariyam, M. Shahbaz, T. Al-Ansari, H.R, Mackey, G. McKay, (2022), A Critical Review on Co-gasification and Co-pyrolysis for Gas Production, Renewable and Sustainable, Energy Reviews, 161, 112349.

A. Midilli, H. Kucuk, M. Haciosmanoglu, U. Akbulut, I. Dincer, (2022), A Review on Converting Plastic Wastes into Clean Hydrogen via Gasification for better Sustainability, International Journal Energy Research, 46,4001 – 32.

F. Parrillo, F. Ardolino, C. Boccia, G. Cali, D. Marotto, A. Pettinau, U. Arena, (2023), Co-gasification of Plastics Waste and Biomass in a Pilot Scale Fluidied Bed Reactor, Energy, 273,127220.

F. Pinto, R. Andre, M. Miranda, D. Neves, F. Varela, J. Santos, (2016), Effect of Gasification Agent on Co-gasification of Rice Production Wastes Mixtures, Fuel,180,407 – 416.

A. Rafey, K. Pal. A. Bohre, A. Modak, K.K, Pant, (2023), A state-of-the-art review on the on the Technological Advancements for the Sustainable Management of Plant Waste in Consort with the Generation of Energy and Value – Added Chemicals, 13,420.

N. Ramzan, A. Ashraf, S. Naveed, A. Malik, (2011), Simulation of Hybrid Biomass Gasification Using Aspen Plus: A Comparative Performance Analysis for Food, Municipal Solid and Poultry Waste, Biomass and Bioenergy ,35, 3962 – 3969.

G. Ruoppolo, P. Ammendola, R. Chirone, F. Miccio, (2012), H_2-rich Syngas Production by Fluidized Bed Gasification of Biomass and Plastic Fuel, Waste Management, 32,724 – 732.

M. Shahbaz, T. Al-Ansan, M. Inayat, S. A, Sulaiman, P. Parthasarathy, G. McKay, (2020), A Critical Review on the Influence of Process Parameters in Catalytic Co-gasification: Current Performance and Challenges for a Future Prospectus, Renewable and Sustainable Energy Reviews,134,110382.

R. Tavares, E. Monteiro, F. Tabet, A. Rouboa, (2020), Numerical Investigation of Optimum Operation Conditions of Syngas and Hydrogen Production from Biomass Gasification using Aspen Plus, Renewable Energy,146,1309 – 1314.

M.S.S.R Tejaswini, P. Pathak, S. Ramkrishna, P.S. Ganesh, (2022), A Comprehensive Review on Integrative Approach for Sustainable Management of Plastic Waste and its Associated Externalities, Science of the Total Environment, 825,153973.

Z. Wang, J, Li, K. G, Burra, X, Liu, X. Li, M. Zang, T. Lei, (2021), Synergetic Effect on CO_2-Assisted Co-gasification of Biomass and Plastics, Journal of Energy Resource Technology,143,031901-8

Flavio Manenti, Gintaras V. Reklaitis (Eds.), Proceedings of the 34th European Symposium on Computer Aided Process Engineering / 15th International Symposium on Process Systems Engineering (ESCAPE34/PSE24), June 2-6, 2024, Florence, Italy

Impact assessment of CO$_2$ capture and low-carbon hydrogen technologies in Colombian oil refineries

Erik López Basto [a,b*] , Gijsbert Korevaar [a] , Andrea Ramírez Ramírez [c] ,

[a] *Department of Engineering Systems and Services, Faculty of Technology, Policy, and Management, Delft University of Technology, The Netherlands*

[b] *Cartagena Refinery. Ecopetrol S.A., Cartagena de Indias, Colombia*

[c] *Department of Chemical Engineering, Faculty of Applied Sciences, Delft University of Technology, The Netherlands*

Corresponding author email: e.lopezbasto@tudelft.nl

Abstract

This research uses system optimization to assess short, medium, and long-term scenarios to achieve the committed CO$_2$-emission goals of Ecopetrol while minimizing potential adverse impacts such as incremental operational costs and utility demand. Two Colombian refineries are used as a case study: a medium-complexity and a high-complexity refinery. The study explores whether the level of complexity plays a significant role in the results.

Potential technologies were ranked using a multi-criteria decision analysis. The system analysis and optimization were done in Linny-R, a mixed integer linear programming software package developed by TU Delft. In the short-term (2030) scenario, the selected technologies include low-carbon H$_2$ produced from Steam Methane Reformer units with carbon capture and storage and H$_2$ produced from renewable electricity sources. The medium and long-term (2050) scenario also included biomass gasification, naphtha reforming, and the cracking unit, all with carbon capture and storage. The refineries were modelled using on-site company data. The results indicate that using low-carbon H$_2$ and carbon capture and storage to flue gases would allow to reach the net zero target. Furthermore, the results show that the level of complexity in a refinery significantly impacts the decarbonization deployment pathways. The high-complexity refineries benefited from using low-carbon H$_2$ as feedstock while the medium-complexity refinery relied on a combination of carbon capture and low-carbon H$_2$ as an alternative fuel. This research highlights the potential to achieve substantial CO$_2$ emissions reductions with less impact on the total operational cost by using the amount of excess refinery gas generated when H$_2$ is used as fuel in boilers and process furnaces. A significant challenge remains in identifying suitable applications for surplus refinery fuel gas beyond its conventional use in combustion within boilers and furnaces.

Keywords: Carbon capture and storage, low-carbon hydrogen, oil refinery decarbonization, multi-criteria decision analysis, and system optimization

1. Introduction

The global commitment to reduce greenhouse gas emissions is driving a shift in our energy preference, moving away from fossil fuels towards cleaner energy sources with lower carbon footprints. This shift is expected to decrease the demand for fossil fuels in transportation. However, refineries play a crucial role beyond fuel supply; they serve as a source of raw materials for manufacturing base chemicals, speciality chemicals, and fuels for the shipping and aviation sectors. Due to the significance of these products, refineries will continue to play a pivotal role in the foreseeable future. Consequently, it is crucial to evaluate and develop a strategy for decarbonization (Oliveira & Schure, 2020).

In 2018, Ecopetrol was responsible for approximately 4% of the Colombian GHG emissions (IDEAM et al, 2022), and it has committed to reducing its GHG emissions to 75 % of the level emitted in 2019 **by 2030** (5.9 Mt CO_{2eq}/y), corresponding **to scopes 1 and 2**. In addition, the long-term strategy of Ecopetrol aims to achieve **net-zero carbon emissions for scopes 1 and 2 by 2050** (note that the target does not include scope 3 emissions i.e.,l indirect emissions not included in scope 2 that occur in the value chain of the reporting company, including both upstream and downstream emissions). Nowadays, the downstream sector is responsible for 55 % of the company's GHG emissions under scopes 1 and 2, with the refineries contributing to 98 % of those emissions (Canova, 2021). This study focuses on a case study based on two Colombian oil refineries that have different levels of complexity. An overview of the key characteristics of both refineries is shown in Table 1. Note that H_2 use differs between the refineries. In the high-complexity configuration refinery, H_2 production contributes 33% of the emissions, whereas, in the medium-complexity configuration refinery, it corresponds to 7 % of total CO_2 emissions. This study aims to a) assess the potential for decarbonization in each refinery by using low carbon H_2 as a feedstock and as a fuel, and by capturing CO_2 from flue gas streams and b) evaluate whether the level of complexity, which is often overlooked in this type of assessment, affects the results.

Table 1. Main characteristics of the refineries case study

		Cartagena	Barrancabermeja
	Unit	**Value**	**Value**
Complexity level[1]		High	Medium
Crude oil Capacity	Mt/y	11.45	11.95
Annual CO_2 emissions	Mt CO_2-eq /y	2.5	3.1
Gas fuel consumption	PJ/y	22.6	40.3
Electricity production	PJe/y	2.93	3.53
Steam production	PJth/y	3.5	28.63
Hydrogen production	kt/y	84	28.7
Total Conversion Yield	%	96.7 %	77 %
H_2 consumption index	t H_2-consumed / feedstock	0.015	0.0051

1: The refinery complexity is defined by the Nelson Complexity Index, which quantifies the type of process units in a refinery and their capacity relative to the atmospheric distillation unit by assigning a factor (Kaiser, 2017).

2. Methodology

The methodology is composed of three stages. In the first stage, promising low-carbon technologies for hydrogen production and CO_2 capture technologies were identified. The assessment considered three time periods, i.e., short-term (by 2030), medium-term (by 2040), and long-term (by 2050). The short-term period includes technologies with a TRL larger than 8 that could be deployed before 2030; the long-term period includes technologies currently at a TRL of 3 or larger. The second stage is composed of two steps: (i) selecting suitable technologies, and (ii) gathering data for case studies based on the complexity level of the refinery. Five technologies were selected for this study and are presented in Table 2 and Table 3.

In terms of CCS, post-combustion capture in flue gas from boilers, furnaces, reformers, and the FCC plant. The capture was done using MEA 30%wt with a 90% CO_2 capture rate based on work reported in (IEAGHG, 2017). For the SMR unit, it was considered a 95% CO_2 capture rate using ADIP-X solvent (45 %wt. MDEA conc. and 5 %wt. Piperazine conc.) in the out-stream from the water gas shift reactor(25 barg and 350 C) based on work done by Meerman et al. (2012). Table 4 shows the techno-economic parameters of the CO_2 capture technologies selected.

Table 2. Selected low-carbon H_2 process

	Technology	Produc.sub-method	Feedstock	Horizon
SMR+CC	Thermochemical	Steam Reforming	Natural Gas	Short term
Ren Elec + PEM El.	Electrolysis	PEM electrolysis	Water + Ren. elec	Short term
Biomass Elec + PEM El.	Electrolysis	PEM electrolysis	Water + Biomass	Short term
Biomass gasif. + CC.	Thermochemical	Gasification	Biomass	Long term
Naphtha Reforming + CC.	Thermochemical	Steam Reforming	Low-grade Naphtha	Long term

Table 3. General techno-economic parameters of low-carbon H_2 technologies.

	SMR	PEM El.	Biomass Elec	Biomass gasif.	Naphtha Reform.
Emission factor, kg CO2/kg H2	9.31	0	† (up) $_+$ 1.7 kg CO_2/ kWe	†(up) $_+$ 19.5 kg CO_2/kg H_2	-
		*66.6 kWh/ kg			
Yield	2.95 % vol. H2/NG	H_2 (2030) 60.4 (2040)	1.13 MWe / t dry biomass	0.1 t H_2 / t dry biomass (IG)	0.0032 t H_2/bl
Electricity consump., MWe/t H2	0.36	54.7 (2050)	0	1.57	0.007 MWe/ bl
Steam consump., t Steam / t H2	7.3 (export)	0	0	11.6	0.035 t/bl
Fuel gas consump., GJ/ t H2	73.8	0	0	-	0.25 GJ/bl
Capex 2022, M€/ t/d H2	1.7-2.6	3.6-5.7	4.4-8.7 M€/MWe	4.3 – 5.9	15-17 M€ /kbld
Fix Opex (Capex %)	4%	3%	3%	3%	3.5%
Econ. Lifetime, year	25	20	20	25	30

12% Discounting rate * *Including 19% additional consumption for auxiliary equipment.* † *Upstream: 126.5 kg CO2 / t dry biomass kbld: Kilo Barrel per day*

Regarding CO_2 storage, the geological reservoir used in the model follows the estimation of capacity onshore and offshore potential in Colombia reported by Younis et al., (2023). Table 5 shows the calculated CO_2 footprint for transportation and storage alternatives. Data for the case studies used confidential information from the on-site refinery processes (e.g., yields, mass and energy balance, operational cost), as well as scientific and industrial publications available in the literature, and information gathered from expert interviews. The mass, energy, and emissions balances were estimated for the annual operation of each process unit under normal conditions. As the raw data used in this study is confidential, values are reported at the block process level.

In the final stage, a system model was developed that represents the process and interactions of the technologies under evaluation. They are designed to represent the capacity, limits, and availability of the case study in interaction with the existing processes in the oil refineries. The two case studies were modeled using Linny-R (Bots, 2022) a mixed-integer linear programming (MILP) with a Gurobi MILP solver. In this software, the refinery system can be represented by a block diagram. Each block corresponds to a process and the connections between blocks represent an energy or mass stream. The model was built in layers and at a section level of detail. Figure 1 shows a screenshot of the main layer in Linny-R. Additionally, all feedstocks/products were related to a process through a linear function. Finally, the balances shown in the model were made to represent a daily basis. The model runs 7 steps, every step corresponding to 5 calendar years, starting in 2020 and ending in 2050.

The main objective of the optimization function is centered on the maximization of cash flow within distinct blocks following the scheduling of CO_2 emissions reductions, products, and availability of feedstocks and technologies. Feedstocks, products, and processes are considered variables in Linny-R. Every process and feedstock/product can be set at a low and up limit (Capacity). A list of data sets allows to establish a schedule of production or capacity or prices in time to feedstocks, products, and processes. Two separate models were built, one for each refinery. The model is composed of 486 variables for the high-complex refinery and 461 for the medium-complex refinery model.

Table 4. Carbon capture processes

Process	CO_2 partial press., bar	Chem. solvent	Electricity demand, kWh/t CO_2 cap.	Heat demand, GJ/ t CO_2 cap.	Capex, € 2022/ t CO2	CO_2 cap. cost, €/t CO2
SMR+CC	3-6	ADIP-X. 50% wt.	2.1 (Absorber= 2.7 barg)	1.97 (Desorber= 1.9 barg/ 115 C)	124.4	54
SMR+CC	0.08					
Biomass gasification	0.1					
Boilers	0.05-0.09	MEA. 30% wt.	64.55 (HC) 58.25 (MC)	3.68 (HC) 3.65 (MC)	142	56.7
NGCC-CHP	0.043					
Process furnaces	0.08-0.10					
FCC	0.1-0.17					

Source of CO2 partial pressure data: Calculated based on SIGEA (2019). IRENA, (2021)
Economic lifetime: 20 years. Discount rate: 12%. Fix opex: 4% capex.

Table 5. CO_2 footprint of transportation and storage alternatives.

Refinery	On-shore pipeline, kg CO_2 /t CO_2 stored	Off-shore pipeline, kg CO_2 /t CO_2 stored	Off-shore shipping, kg CO_2 /t CO_2 stored
Cartagena. (HC))	2.5	2.9	18.3
Barrancabermeja. (MC)	1.8	9.6	22.5

Based on Younis et al., (2023), Khoo & Tan, (2006), (Knoope et al., 2015), (Yoo et al., 2013)

The baseline for CO_2 emissions and data collection was for the year 2019. Cost and prices for CAPEX and OPEX were updated to 2022 using the Chemical Engineering Plant Cost Index, and a currency conversion rate from dollars or Colombian pesos to euros in 2022. Prices associated with feedstocks, products, and production capacities were set according to the Colombian context. Technologies that involve biomass as feedstock considered tree species of Eucalyptus available in Colombia (E. camaldulensis, E. grandis, and E. globulus). The capacity of utilities and refining process represent the actual capacity of both oil refineries. Fuel production capacity (i.e. gasoline, diesel, and jet fuel) was defined according to the Ecopetrol long-term production scheduling strategy. Colombian electricity grid connection to refineries is 70 MW capacity, 85 €/ MWh (XM, 2020), and with a Carbon footprint of 186 kg CO_2 / MWhe (128 kg CO_2 / MWhe, according to Unidad de Planeación Minero Energética (UPME) (UPME., 2019).

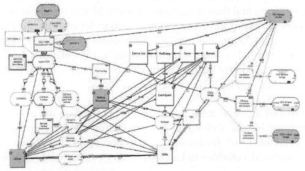

Figure 1. Linny-R Model. Screenshot of the main layer

3. Results and discussion

The impact of implementing the low-carbon H2 technologies on the energy and CO_2 balances of the two refineries is shown in Table 5. In both cases, the results show a significant reduction of CO_2 emissions in the short term (26% and 23% for HC and MC configurations, respectively) and they meet the target of CO_2 neutrality in 2050. The SMR with CC is the most cost-effective technology to produce hydrogen, with a production price between 1.0 and 1.5 €/kg lower than H_2 based on PEM electrolysis (using REN between 30-40 €/MWh). It reaches the same production price in the long-term horizon with a REN of 30 €/MWh and a forecasted incremental price of natural gas.

Hydrogen based on Biomass+ CC emerges as a viable option to achieve CO_2 emissions neutrality in the long term horizon, despite having a production cost that is 2-3 €/kg higher than the SMR-CC and REN-PEM electrolyzer alternatives. This is due to the significant advantage it provides as a "negative emissions process (NET)" when both biomass H_2 and the biomass electricity process are accompanied by CCS. Biomass processes alone (without CCS) contribute to a CO_2 reduction of 13% for both refinery configurations, with respect to the baseline. The results do not show any limitation concerning the availability of CO_2 underground reservoir. For the MC refinery, the onshore capacity (64 Mt CO_2) was enough until 2050 horizon. The HC Refinery will use 100 % onshore (12 Mt CO_2) and 32 % Colombian offshore of its CO_2 storage capacity by 2050. Similarly, the MC refinery will use 93 % of the on-shore of its CO_2 storage capacity by 2050.

Table 6. Energy and CO2 impacts in oil refineries

High Complex	Base Line	Short Term	Long Term	Medium Complex	Base Line	Short Term	Long Term
Oil refinery throughput, kbld	155	200	197	Oil refinery throughput, kbld	239	242	242
kt/y	7718	9959	9810	kt/y	11891	12034	12034
Total Hydrogen Demand, kt/d	211	287 / 136%	508 / 241%	Total Hydrogen Demand, kt/d	79	85 / 107%	358 / 454%
H2 Production processes				H2 Production processes			
SMR, kt/d	211	155	110	SMR, kt/d	79	10	30
Biomass gasification, kt/d	0	0.0	25.0	Biomass gasification, kt/d	0	0	0
Naphtha Reformer, kt/d	0	0.0	106.4	Naphtha Reformer, kt/d	0	0.0	120.0
Natural Gas Feedstock, TJd	26.2	20.7	15.9	Natural Gas Feedstock, TJd	10.6	2.6	5.2
Total Gas fuel demand, TJd	37	45.4 / 121%	49.7 / 133%	Total Gas fuel demand, TJd	78	79.8 / 102%	97.1 / 124%
Natural Gas (external source), TJd	11.7	17.2 / 147%	3.4 / 29%	Natural Gas (external source), TJd	14.9	16.8 / 113%	13.1 / 87%
Refinery off-gas (internal source)	25.8	28.3 / 100%	46.4 / 180%	Refinery off-gas (internal source)	63.1	63.0 / 100%	84.0 / 133%
Total electricity demand, MWh	96	448 / 467%	646 / 666%	Total electricity demand, MWh	93	261 / 280%	502 / 539%
Fossil fuel source, MWh	96	45 / 46%	36 / 37%	Fossil fuel source, MWh	93	42 / 45%	25 / 27%
External grid, MWh	0	67	70	External grid, MWh	0	53	69
Renewables, MWh	0	336	539	Renewables, MWh	0	166	408
Total Steam Demand, kt/d	9	10.3 / 117%	21.2 / 241%	Total Steam Demand, kt/d	8	8.8 / 115%	25.0 / 327%
Total CO2 emissions, kt CO2/y	2405	1836 / 74%	0 / 0%	Total CO2 emissions, kt CO2/y	3500.2	2704 / 77%	0 / 0%

The impact of the refinery's level of complexity on the utilization of low-carbon hydrogen is evident in the short-term scenario. In the high-complexity (HC) refinery, a substantial reduction (26%) in CO_2 emissions was achieved, with 36% attributed to bio-electricity with CC used in the electrolyzers. In contrast, the medium-complexity (MC) refinery achieved a 17% reduction in CO_2 emissions, with 74% credited to bio-electricity with CC. For the long term, MC refineries gained an edge through the implementation of CO_2 capture from flue gas produced by fossil fuel combustion, resulting in an 85% reduction in total CO_2 emissions, in comparison to the 72% reduction achieved in the HC refinery configurations. The decarbonization technologies for the high-complexity refinery are oriented towards renewable energy (REN) and electrolyzers; whereas for the medium-complexity refinery, the optimization focuses on biomass + CC electricity and gasification to produce low-carbon hydrogen. In both cases, the Naphtha reformer + CC process was shown to be the less expensive way to produce low-carbon hydrogen, playing a significant role in the decarbonization pathways in the long-term horizon. Biomass-based H_2 with CC emerges as a viable option post-2030, despite having a higher hydrogen production cost. The benefits will likely far outweigh the negative impact on the Oil refinery's cash flow.

4. Conclusions

The combination of low-carbon hydrogen production and CO_2 capture technologies provides for both types of oil refineries a pathway to achieve the CO_2 reduction target committed by Ecopetrol in the short-term (75 %) and CO_2 neutrality in the long-term.

The level of complexity in oil refineries significantly impacts the decarbonization process, with the high-complexity refinery benefiting from low-carbon H_2 as feedstock for the processes and the medium-complexity refinery relying more on CO_2 capture in combination with hydrogen as an alternative fuel. However, the CO_2 emissions reduction is limited because of the avoiding flaring of surplus fuel gas generated as a consequence of the fuel shifting process.

However, a significant challenge lies in identifying suitable applications for surplus refinery fuel gas beyond its conventional use in combustion within boilers and furnaces will be necessary in future research.

References

Bots, P. (2022). Linny-r. In Version 1.0.15. https://sysmod.tbm.tudelft.nl/linny-r/docs/?15

Canova, W. (ecopetrol). (2021). Decarbonization Strategy of Ecopetrol Downstream sector. LARTC The Latin American Refining Technology Conference, 18.

IDEAM, Fundación Natura, PNUD, MADS, DNP, C. (2022). Informe del inventario nacional de gases efecto invernadero 1990-2018 y carbono negro 2010-2018 de Colombia . www.cambioclimatico.com

IEAGHG. (2017). IEAGHG Technical Review August 2017 Understanding the Cost of Retrofitting CO2 Capture in an Integrated Oil Refinery. August. https://ieaghg.org/docs/General_Docs/Reports/2017-TR8.pdf

IRENA. (2021). Reaching zero with renewables - Capturing Carbon. In /Technical-Papers/Capturing-Carbon.

Kaiser, M. J. (2017). A review of refinery complexity applications. Petroleum Science, 14(1), 167–194. https://doi.org/10.1007/s12182-016-0137-y

Khoo, H. H., & Tan, R. B. H. (2006). Life cycle investigation of CO2 recovery and sequestration. Environmental Science and Technology, 40(12), 4016–4024. https://doi.org/10.1021/es051882a

Knoope, M. M. J., Ramírez, A., & Faaij, A. P. C. (2015). Investing in CO2 transport infrastructure under uncertainty: A comparison between ships and pipelines. International Journal of Greenhouse Gas Control, 41, 174–193. https://doi.org/10.1016/j.ijggc.2015.07.013

Meerman, J. C., Hamborg, E. S., van Keulen, T., Ramírez, A., Turkenburg, W. C., & Faaij, A. P. C. (2012). Techno-economic assessment of CO2 capture at steam methane reforming facilities using commercially available technology. International Journal of Greenhouse Gas Control, 9, 160–171. https://doi.org/10.1016/j.ijggc.2012.02.018

Oliveira, C., & Schure, K. M. (2020). Decarbonisation options for the Dutch refinery sector. Dec.

SIGEA. (2019). Ecopetrol Emissions report.

Unidad de Planeación Minero Energética (UPME). (2019). Documento de cálculo del Factor de emisión de SIN 2018.

XM. (2020). Precio promedio y energía transada. http://www.xm.com.co/Paginas/Mercado-de-energia/preciopromedio-y-energia-transada.aspx.

Yoo, B. Y., Choi, D. K., Kim, H. J., Moon, Y. S., Na, H. S., & Lee, S. G. (2013). Development of CO2 terminal and CO2 carrier for future commercialized CCS market. International Journal of Greenhouse Gas Control, 12, 323–332. https://doi.org/10.1016/j.ijggc.2012.11.008

Younis, A., Suarez, L., Lap, T., & Edgar, Y. (2023). Exploring the spatiotemporal evolution of bioenergy with carbon capture and storage and decarbonization of oil refineries with a national energy system model of Colombia. 50(July). https://doi.org/10.1016/j.esr.2023.101232

Flavio Manenti, Gintaras V. Reklaitis (Eds.), Proceedings of the 34th European Symposium on Computer Aided Process Engineering / 15th International Symposium on Process Systems Engineering (ESCAPE34/PSE24), June 2-6, 2024, Florence, Italy

A Task Graph Parallel Computing Architecture for Distillation Column Simulation on Flash Granularity

Shifeng Qu,[a] Shaoyi Yang,[a] Wenli Du,[a] Feng Qian[a]

[a]Key Laboratory of Smart Manufacturing in Energy Chemical Process, Ministry of Education, East China University of Science and Technology, Shanghai 200237, China

Abstract

Simulations of distillation columns stand as the most time-consuming unit operation calculations, profoundly affecting both the efficiency and practical feasibility of process simulation, design and optimization. The complexity of distillation columns stems from the involvement of multiple chemical components as well as extensive energy and mass transfer stages. In this paper, a novel parallel computing architecture is presented specifically for the distillation process. The architecture revolutionizes distillation calculations by decomposing them into smaller, independently and concurrently computable subtasks, i.e., flash calculations and solving tridiagonal matrices for material balance. At the core of the architecture is the creation of an executable task graph, which maps out dependencies between subtasks and incorporates conditional tasking for efficient control-flow management during iterative material and energy balance calculations. The proposed architecture was applied to distillation cases with different computation intensities to ascertain its effectiveness. The results reveal that under optimal parallel granularity, parallelization substantially improves the efficiency of distillation simulation.

Keywords: parallel computing, process simulation, distillation column, task graph computation

1. Introduction

Process simulation employs mathematical equations and physical models to emulate the behavior and interactions of various components and unit operations under specified conditions. Utilizing process simulation software for distillation processes empowers engineers to comprehend and forecast component separation efficiencies. This facilitates the design and scaling of distillation towers and associated equipment to meet specific production demands and economic objectives. Simulating distillation processes further allows for the optimization of operating conditions, offering vital theoretical and data support for production efficiency, cost control, and safety in the chemical and petroleum industries.

However, distillation columns are the most computation-intensive units in process simulations based on the sequential-modular (SM) approach, demanding extensive iterative calculations. The modular nature of the SM approach facilitates individual module solving and debugging, simplifying error identification and rectification. Despite its limitations in handling processes with nested iterations or complex topologies, the SM approach remains indispensable and is widely adopted in commercial simulation software (Pantelides and Renfro, 2013). In this study, we refined the distillation calculation architecture within the SM framework, integrating parallel computation to speed up distillation simulations. The novel architecture involves dissecting the material balance calculation of each component, bubble point calculation on each stage, and enthalpy conservation calculations into separate subtasks, with their interdependencies mapped according to the distillation algorithm. Based on the dependencies among subtasks, our parallel architecture then generates the computable task graph for distillation simulation, laying the groundwork for parallel processing of material balance equations solving and flash calculations at the components and stages level, respectively. Loop control flow is supported through the integration of conditional task nodes into the task graph, thereby facilitating the control decisions for inner and outer iterations during distillation simulation (Huang et al., 2021). In Section 3, we employed a toy column example and a practical de-butanizer column to substantiate the effectiveness of the proposed architecture.

2. Parallel computing architecture for distillation simulation

In chemical engineering, distillation stands out as a prevalent separation technique that capitalizes on the varying boiling points of mixture components to effectively separate lighter and heavier elements. The parallel computing architecture introduced in this paper is intended for parallelizing the bubble-point (BP) distillation algorithm. The BP method is based on equilibrium stages, involving the iterative solving of the MESH equations, namely, **M**aterial balance for each component, phase-**E**quilibrium relation for each component, mole-fraction **S**ummations and energy (ent**H**alpy) balance. Specifically, the vapor phase composition $y_{i,j}$ in the material balance equations is eliminated using the phase equilibrium equations, denoted as follows:

$$y_{i,j} = K_{i,j} x_{i,j} \tag{1}$$

where K signifies the phase equilibrium constant, x refers to the composition of the liquid phase, with i indicating the index of the component and j representing the index of the stage in the column. After elimination, the original material balance equations become M-K equations. Each equation within the M-K system can be uniformly expressed in the following form:

$$A_{i,j} x_{i,j-1} + B_{i,j} x_{i,j} + C_{i,j} x_{i,j+1} = D_{i,j} \tag{2}$$

The coefficients in the above equations can each be represented as follows:

$$\begin{cases} A_{i,j} = V_j + \sum_{k=1}^{j-1} \left(F_k - S_k - G_k \right) - V_1 \\ B_{i,j} = -\left[K_{i,j} \left(V_j + G_j \right) + V_{j+1} + \sum_{k=1}^{j} \left(F_k - S_k - G_k \right) - V_1 + S_j \right] \\ C_{i,j} = K_{i,j+1} V_{j+1} \\ D_{i,j} = -F_j Z_{i,j} \end{cases} \tag{3}$$

where V_j and L_j are the vapor and liquid flow rates leaving the jth tray, respectively. F represents the feed flow rate, while G and S are the sidestream flow rates for the vapor and liquid phases, respectively. Consequently, the M-K equation set for the ith component can be expressed as the following tridiagonal matrix:

$$\begin{bmatrix} B_1 & C_1 & & & & & \\ A_2 & B_2 & C_2 & & & & \\ & & \vdots & & & & \\ & A_j & B_j & C_j & & & \\ & & \vdots & & & & \\ & & & A_{N-1} & B_{N-1} & C_{N-1} \\ & & & & A_N & B_N \end{bmatrix} \begin{bmatrix} x_1 \\ x_2 \\ \vdots \\ x_j \\ \vdots \\ x_{N-1} \\ x_N \end{bmatrix} = \begin{bmatrix} D_1 \\ D_2 \\ \vdots \\ D_3 \\ \vdots \\ D_{N-1} \\ D_N \end{bmatrix} \qquad (4)$$

Initial values for V_j on each stage and $K_{i,j}$ for each component are set to linearize the M-K equations before solving the distillation system. Then, the M-K tridiagonal matrix of each component is solved to obtain its liquid phase composition x_j on each stage. After normalizing the $x_{i,j}$ on each stage, the bubble point temperature and phase equilibrium constants for the liquid phase of each stage are computed through property calculations. We adopt press-vapor fraction (PV) flash calculation to supersede the bubble point calculation. This is achieved by assigning a near-zero value to the vapor fraction to determine an approximate temperature for the bubble point. The results of PV flash calculations are used to update the temperature, component phase equilibrium constants, and vapor phase composition for each stage. After each update, the liquid composition is checked to ensure it satisfies the S equations, namely:

$$\left| \left(\sum_{i=1}^{c} x_i \right)_j - 1.0 \right| \le \varepsilon_c \quad (1 \le j \le N) \qquad (5)$$

where ε_c is the tolerance, determined following user-defined computational precision. If not all stages meet the above criteria, the material balance tridiagonal matrix is re-solved using the $K_{i,j}$ values obtained from the PV flash calculations. This control flow is referred to as the K loop. The loop is executed multiple times until the S equations are satisfied, moving to the energy balance computation in the outer loop to calculate the updated values of V_j. Convergence in the outer loop is confirmed by evaluating if the variation in V_j across successive iterations falls below a predefined threshold. If the above conditions are not met by all stages, the program will cycle back to the K loop (inner loop), where a new round of material balance and bubble point calculations are performed using the updated V_j values, continuing until the realization of energy balance.

In each iteration of the K loop, bubble point calculations are conducted based on the liquid phase composition $x_{i,j}$ and pressure P_j of each stage to update the stage temperature T_j. The bubble point calculations for stages are independent subtasks, indicating no dependency or communication overhead between them, and thus can be executed simultaneously without waiting for results of each other. Likewise, the solving of the M-K tridiagonal matrix for components can also be performed in parallel. Figure 1 illustrates the revised BP algorithmic diagram in our proposed parallel architecture. As shown in the diagram, bubble point calculations and the solving of tridiagonal matrices are parallelized across all the stages and components, respectively. In our architecture, the simulation of distillation is divided into subtasks with distinct functionalities. These subtasks, along with their dependencies as determined by the distillation algorithm shown in the diagram, are added to the executable task graph. The task scheduler sequences the tasks according to their interdependencies, ensuring that all prerequisite tasks are completed before the commencement of a new task. The workflow of the proposed architecture is shown in Figure 2.

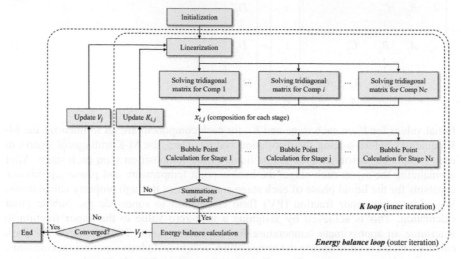

Figure 1: Algorithm diagram of the bubble point method after parallelization.

Notably, it is safe for multiple threads to simultaneously read data from the same memory, provided no thread is writing to that data item at the same time. Data races occur when two or more threads access the same memory location concurrently, and at least one of the accesses is for writing. Data races may lead to inconsistent read results, as different threads might observe various 'intermediate' states of the data item being written (Dolan et al., 2018). In our architecture, the distillation column configurations set on the user interface, which remain unchanged throughout the simulation process, are thus shared for reading by multiple threads. Variables that are updated during the inner and outer iteration, like $x_{i,j}$, V_j, $K_{i,j}$, are subject to data races due to repeated modifications by the parallelized subtasks. Therefore, private copies of these data items are created for each subtask to avoid interference. The subtasks process their own data copy and merge the results after parallelization.

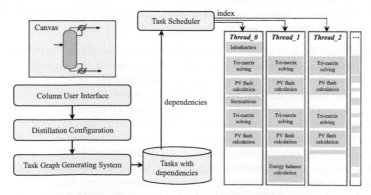

Figure 2: Architecture of distillation parallel simulation.

In addition to static tasks such as PV flash calculations and tridiagonal matrix solving, the algorithm also requires tasks consistently monitoring the convergence status of the inner and outer loops and exiting them appropriately. In our architecture, the distillation task graph is designed for dynamic execution, supporting loop control flow decisions without being torn into a flat directed acyclic graph. Specifically, conditional task nodes are introduced to the task graph to determine the subsequent static tasks for execution based on the returned value of convergence evaluation, and thus guide the executor to move between the inner and outer loops.

3. Case study

We validated the effectiveness of the proposed task graph parallel architecture on two columns with different computation intensities. One is a toy case for ethylene-ethane separation, with only 3 components and 10 stages. The other is a de-butanizer column from a practical cracked gas separation process, primarily tasked with separating butane (C4 hydrocarbons like butane and isobutane) from heavier hydrocarbons in cracked gas. The de-butanizer involves 32 components and 44 stages. Both columns were simulated using the Soave–Redlich–Kwong equation of state. Experiments were conducted on a Windows 11 x86 64-bit machine equipped with a 12th Gen Intel(R) Core(TM) i5-12490F CPU at 3000 MHz, featuring 6 cores and 12 logical processors, and 32 GB of memory. The whole project was compiled using Ninja with C++19 standards. Notably, we were unable to conduct a comparison of the simulation time for the distillation column across various commercial software. It is impossible to perform time profiling on the distillation code of these software programs due to the inability to access their source code.

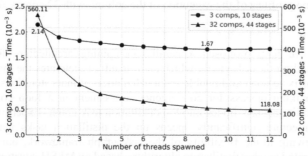

Figure 3: Execution time of columns over an increasing number of threads.

Figure 3 shows the trend of computation time for the toy case and the de-butanizer as the number of spawned threads increases. All the presented results are an average of 10 runs. It can be seen that the simulation efficiency of the toy case shows limited improvement with parallelization. Compared to the de-butanizer, the toy case features a column with much fewer stages and components, thus its PV flash calculations and tridiagonal matrix solutions entail a much smaller problem size and shorter computation time. When parallelizing inherently small tasks, the overheads - including thread management and context switching - can become a substantial part of the total computation time. As the number of threads increases, it will come to a point where the additional overhead surpasses any performance benefits gained from parallelization. Furthermore, the dependencies between subtasks also significantly impact parallelization effectiveness. For the toy case, the tridiagonal matrix solving task for each inner iteration can be distributed for parallel execution across up to three threads. When the number of spawned threads surpasses three, due to task dependencies, some threads become idle, reducing the overall parallel efficiency. When the thread number reaches ten, the PV flash calculation tasks achieve full parallelism. Spawning additional threads will no longer result in further acceleration of the simulation.

In contrast to the toy case, the de-butanizer with a significantly larger subtask problem size, experiences a remarkable simulation speed increase, approaching almost fivefold. However, it is evident from Figure 3 that the improvement becomes increasingly marginal as the number of threads is close to 12. This is because the thread count approaches the number of CPU logical cores, and simultaneously executing too many parallel tasks under limited resources results in parallel efficiency degradation. Figure 4 displays the time profiling of the de-butanizer execution, with a magnified view of the period between 50 milliseconds and 80 milliseconds. The tridiagonal matrix solving, PV flash calculations, and energy balance calculations are executed sequentially in alignment with the distillation algorithm. The time-consuming tasks of the inner loop are evenly distributed among the threads.

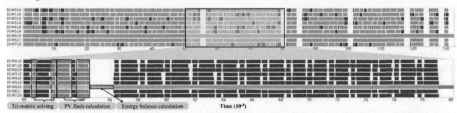

Figure 4: Time profiling of the execution of the column with 32 comps and 44 stages.

4. Conclusion

In summary, we have proposed a parallel computing architecture for distillation simulation based on the task graph, which significantly accelerates the material balance tridiagonal matrix solution and bubble point calculation of inner iterations. We facilitate the decision-making process for convergence assessment in a loop control flow by integrating conditional task nodes into the distillation task graph. Experiments show the proposed architecture significantly speeds up the distillation tasks with appropriate parallelism. Owing to the flexibility of the task graph framework, our work can be seamlessly adapted to other distillation algorithms based on MESH equations, such as the inside-out method and the sum-rate method. Given the flexibility in setting task node dependencies within our architecture, the distillation task graph can be involved in assembling any complex chemical process task graph by linking it with other unit operation tasks, realizing parallel simulation of whole process flowsheets.

References

Pantelides, C. C., Renfro, J. G., 2013. The online use of first-principles models in process operations: Review, current status and future needs. Computers & Chemical Engineering, 51, 136-148.

Huang, T. W., Lin, D. L., Lin, C. X., Lin, Y., 2021. Taskflow: A lightweight parallel and heterogeneous task graph computing system. IEEE Transactions on Parallel and Distributed Systems, 33(6), 1303-1320.

Dolan, S., Sivaramakrishnan, K. C., Madhavapeddy, A., 2018. Bounding data races in space and time. ACM SIGPLAN Notices, 53(4), 242-255.

Flavio Manenti, Gintaras V. Reklaitis (Eds.), Proceedings of the 34th European Symposium on Computer Aided Process Engineering / 15th International Symposium on Process Systems Engineering (ESCAPE34/PSE24), June 2-6, 2024, Florence, Italy

Digital Twin Technology In The Thermal Processing Industry Of Granular Material Based On The Extended Discrete Element Method (XDEM)

Bernhard Peters[a], Prasad Adhav[a], Daniel Louw[a], Navid Aminnia[a], Xavier Besseron[a]

University of Luxembourg,
bernhard.peters@uni.lu

Abstract

Granular material is ubiquitous in our day-to-day environment and is dealt with in a variety of industries such as mineral, pharmaceutical, chemical, agriculture, biochemical and food to name a few. Apart from the storage and transport of granular material, it requires often thermal treatment in conjunction with a chemical conversion. Already the dynamics of granular material are complex, and the thermal conversion of particulate material in contact with a fluid phase poses an additional level of complexity. Being aware of the fundamental mechanisms and the interactions is a key factor for producing the optimal process and product. An understanding is effectively accomplished via research on a particle scale for which the Extended Discrete Element Method (XDEM) is a perfectly suited simulation platform. It evaluates the dynamic and thermodynamic state of individual particles of granular matter being processed. Each particle exchanges momentum, mass and heat with the surrounding fluid for which the state is predicted with Computational Fluid Dynamics (CFD). In order to achieve utmost flexibility, particles are assigned specific material properties and a variety of processes may be attached to a particle covering heat-up, arbitrary chemical reactions or reaction mechanisms and phase changes e.g. drying coating or melting. Innovative and fast algorithms that are enhanced with a hybrid parallelisation strategy based on OpenMP and MPI reach industrial scales. Thus, applications are feasible on medium-sized workstations but also on high-performance computers within reasonable computational times.

Keywords: particulate matter, thermal processing, CFD-XDEM coupling, digital twin

1. Introduction

Numerous engineering challenges in processing industry, agriculture and food industry, pharmaceutical industry, construction, raw material processing and renewable energies to name a few involve a particulate phase with its thermo- dynamics and a fluid phase. Some predominant examples are the production or processing of sand, coal, fertilizer, corn, coffee, nuts, renewable fuels e.g. biomass. Simulating such industrial processes with computers allows for optimising the design of factories, production units, waste management plants, etc. and are therefore of strategic interest. Indeed, these impacts, for example, the quality of the products, the energy, the efficiency and the time required to produce and process them or the amount of waste generated. However, as stated above, such simulation involves multi-physics phenomena (particle mechanics and thermo-dynamics, fluid dynamics structural analysis). Although they can be treated by pure continuous approaches such as multi-phase flows, current computer resources and recent advances in research by Krause et al. (2017); Mahmoudi et al. (2016); Mohseni et al. (2019); Pozzetti et al. (2018); Pozzetti and Peters (2017); Peters and Pozzetti (2017) allow now describing the particulate phase by a discrete method and thus, provide a deeper insight into the underlying physics. Hence, it is now conceivable that each part of the multi-physics simulation is performed by the most suited approach, executed concurrently and then coupled. It follows almost naturally an evolution in design tools: starting from computer-aided design (CAD) and engineering (CAE) in the beginnings and boosting virtual reality (VR) to the next level of sophistication of virtual prototyping as

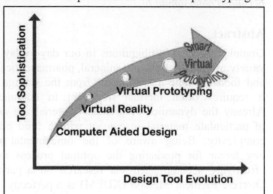

depicted in Figure 1. Adding the multi-physics aspect to virtual prototyping creates a digital twin for smart virtual prototyping as the final target in Figure 1.

In addition to engineering requirements, societal needs demand digital twin technology i.e. virtual prototyping that allows a shift from current empirical-based practice to an advanced multi-physics simulation technology. Often, engineers opt for "copy & paste technology" of already proven components and systems from previous generations. It results in a conservative design

Figure 1: Evolution of design tools to reach smart virtual prototyping

with little potential for innovative ideas. These limitations are removed by virtual prototyping that is among "Gartner's Top 10 Strategic Technology Trends for 2017" as stated by Gartner. While already single-physics software platforms for computational fluid dynamics (CFD) as mentioned by Phuc et al. (2016) and discrete element method (DEM) including thermodynamics of the particulate phase, referred to as extended discrete element method (XDEM) summarised by Peters et al. (2018) exist, an integrating framework for executing discrete and continuous single-physics modules in a highly scalable parallel mode does not exist. A deployment of a highly performing multi-physics framework is hindered by the fact that a simple coupling of two or more parallelised modules ends in a sequential application because data between individual modules is not exchanged. Rather than developing an "all-in-one" multi-physics software platform within a single code, a coupling of "best-of-the-class" single-physics software modules in a multi-physics coupling framework for engineering applications is preferred. Consequently, the modules of the XDEM software consist of a DEM module for the motion of particles and a thermodynamic module for the thermodynamic state of particles, OpenFOAM describing fluid dynamics, and Calculix representing the structural analysis. This methodology paves the path towards virtual prototyping that comprises a particulate phase in contact with a fluid phase and walls. It closes a technological gap and has the potential to revolutionise design and operation of plants and factories impacting heavily on society.

2. Simulation Platform

As above-mentioned, the methodology relies on a seamless integration of "best-of-the-class" software modules into a framework that allows a smooth interaction between the individual packages as depicted in Figure 2. Currently, the advanced multi-physics simulation platform (AMST) couples the discrete element method (DEM) to describe motion of a granular media, finite element analysis (FEM) for structural analysis, extended discrete element method (XDEM) representing the thermodynamic state of particles and computational fluid dynamics (CFD) evaluating the fluid-/thermodynamics of a single or multi-phase flow. This coupling concept has ultimate flexibility because it is easily extendible by additional modules e.g. electromagnetics and existing modules are simply replaced by alternative modules for fitting the requirements to a better

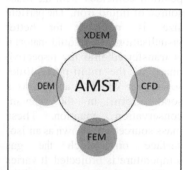

Figure 2: A framework for integrating "best-of-the-class" software modules to from the advanced multi-physics simulation technology (AMST).

degree. Each of the modules is equipped with an interface to the remaining modules of the simulation platform so that a correct exchange of relevant data between all modules is provided. In general, the data is exchanged in a bi-directional way which ensures a full two-way coupling between each of the two modules leading to an 8-way coupling for all modules in Figure 2 involved. However, a one-way coupling or less interactions between modules may also be applicable under certain conditions if the effect of one module on the other is negligible or even non-existent.

3. Results and Discussion

The versatility of the multi-physics simulation platform is represented by a variety of application addressing material processing in a blast furnaces, combustion of biomass, a tool analysis of an abrasive water cutting nozzle and additive manufacturing.

For blast furnace operation as shown for the cohesive zone in Figure 3, a hot blast enters the reactor through the tuyeres where it burns coke to produce energy for heating the incoming ore and coke and to form carbon monoxide as a reducing agent for the iron bearing particles. The carbon monoxide enriched blast flows upward through the packed bed indicated by the velocity vectors in the Figure 3. The reduction process of iron oxides namely, hematite, magnetite and wustite is described by reversible reactions including a forward and backward reaction rate. Thermodynamic equilibrium for the magnitude of forward and backward reaction rates being equal depends on the temperature and gas composition. Carbon monoxide is formed as a product of the reduction that is transferred into the gas phase. It will be in contact with the next upper coke layer where the Boudouard reaction ($C + CO_2 \leftrightarrow 2\ CO$) generates carbon monoxide. It flows into the upper ore layer where it is available to continue the reduction process. This mechanism repeats itself over the alternating ore and coke layers, however, with decreasing intensity due to falling temperatures. After iron oxides are reduced by carbon monoxide and the particle temperature reaches the melting temperature, iron melts and forms a liquid phase on the particle surface. These particles appear in the Figure 3 as ore layers of which the position coincides with the mass source of liquid iron. The particle size is reduced for better visualisation. The liquid material is transferred into the respective phase of the multi-phase Euler solver, for which it appears as a source term in the relevant conservation equation. These mass sources are shown as an iso-surface on which the gas temperature is projected. It varies over the cohesive zone by app. 900 K indicating that the gas temperature is not suited to identify the location of the cohesive zone by continuum approaches e.g. two-fluid model.

Heated up by radiation, depicted by the temperature distribution in Figure 4, the pellets for biomass combustion undergo drying, pyrolysis and gasification along distinct regions of the grate. The gaseous products of pyrolysis i.e. methane, carbon monoxide, carbon dioxide, hydrogen and vapour are transferred into the gas phase shown in Figure 4 as the distribution of tar and are transported towards the boiler indicated by the streamlines coloured by temperature. These products burn under the influence of secondary air and flue gas injection to form carbon dioxide and vapour. Char as a solid product of pyrolysis burns with primary air by a heterogeneous reaction until ash leaves the furnace.

An extended and complex 6-way-coupling in shown in Figure 5 for an abrasive water jet cutting nozzle. Sand particles entering the

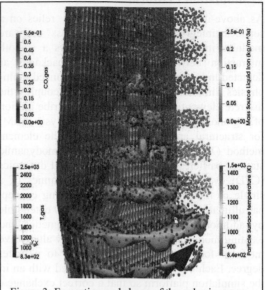

Figure 3: Formation and shape of the cohesive zone

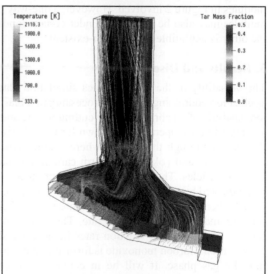

Figure 4: Distribution of tar, gas and particle temperature in a biomass furnace. Wood pellets enter the furnace and move over the forward acting grate.

mixing chamber are accelerated by a water jet of app. 200 m/s. A multi-phase mixture consisting of air, water and particle streams through the focussing tube and interacts heavily with the inner wall of the housing. The solid structure is excited by these impacts and deforms which is shown as displacements.

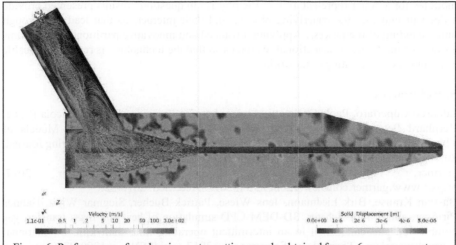

Figure 6: Performance of an abrasive water cutting nozzle obtained from a 6-way momentum coupling between fluid (CFD), particles (DEM) and nozzle housing (FEM).

The fitness and capabilities of the XDEM platform is underlined by applying it for additive manufacturing in Industry 4.0 as depicted in Figure 6, for which a high-energy laser beam scans over a layer of metal power and, thus, melting it for fusion with the solid build part. It allows investigating into the operation parameters such as laser power, speed, hatch size and powder properties and layer preparation.

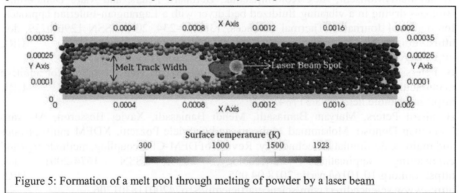

Figure 5: Formation of a melt pool through melting of powder by a laser beam

Rather than resolving the flow around particles and inside the interstitial space as done by Saraei and Peters et al. (2023), which requires enormous computational resources already for small computational domains, the current approach applies space-averaging to estimate relevant fluid dynamics variables. Consequently, transfer for heat, mass and momentum is estimated by well known correlations such as Nusselt or Schmidt numbers that has proven to be sufficiently accurate in numerous applications.

4. Conclusions

The current contribution addresses a flexible technology to build an advanced multi-physics simulation platform (AMST) by integrating "best-of-the-class" software modules into a coupled simulation framework. This technology allows for an extension depending on the multi-physics requirements and for easy replacement of one or more individual modules for a better representation of the physics in question. Results presented provide a deep insight into the underlying physics and their interactions that lead to thorough understanding of the process. Applying advanced and innovative partitioning algorithms allows for moderate computational resources so that the technology is readily applicable without super-computing frameworks.

5. References

Edoardo Copertaro, Paolo Chiariotti, Alvaro Antonio Estupinan Donoso, Nicola Paone, Bernhard Peters, and Gian Marco Revel. XDEM for Tuning Lumped Models of Thermochemical Processes Involving Materials in the Powder State. Engineering Journal, 20:187-201, 2016. ISSN 01258281. doi: 10.4186/ej.2016.20.5.187.

Gartner. Top 10 Strategic Technology Trends for 2017. https://www.gartner.com/newsroom/id/3482617. Accessed: 2018-02.

Bastian Krause, Birk Liedmann, Jens Wiese, Patrick Bucher, Siegmar Wirtz, Hannes Piringer, and Viktor Scherer. 3D-DEM-CFD simulation of heat and mass transfer, gas combustion and calcination in an intermittent operating lime shaft kiln. International Journal of Thermal Sciences, 2017. ISSN 1290-0729. doi: https://doi.org/10.1016/j.ijthermalsci.2017.03.017.

Shibo Kuang, Zhaoyang Li, and Aibing Yu. Review on modelling and simulation of blast furnace. Steel research international, 89(1):1700071n/a, 2018. ISSN 1611-3683.

A. H. Mahmoudi, X. Besseron, F. Hoffmann, M. Markovic, and B. Peters. Modelling of the biomass combustion on a forward acting grate using XDEM. Chemical Engineering Science, 142:32-41, 2016. doi: 10.1016/j.ces.2015.11.015.

Mohammad Mohseni, Alex Kolomijtschuk, Bernhard Peters, and Marc Demoulling. Biomass drying in a vibrating fluidized bed dryer with a Lagrangian-Eulerian approach. International Journal of Thermal Sciences, 138:219-234, 2019. ISSN 1290-0729. doi: https://doi.org/10.1016/j.ijthermalsci.2018.12.038.   URL http://www.sciencedirect.com/science/article/pii/S1290072917314011.

B. Peters and G. Pozzetti. Flow characteristics of metallic powder grains for additive manufacturing. EPJ Web of Conferences, 13001:140, 2017. URL http://hdl.handle.net/10993/31734.

Bernhard Peters, Maryam Baniasadi, Mehdi Baniasadi, Xavier Besseron, Al- varo Estupinan Donoso, Mohammad Mohseni, and Gabriele Pozzetti. XDEM multi-physics and multi-scale simulation technology: Review of DEM-CFD coupling, methodology and engineering applications. Particuology, 2018. ISSN 1674-2001. doi: https://doi.org/10.1016/j.partic.2018.04.005.   URL http://www.sciencedirect.com/science/article/pii/S1674200118301706.

Pham Van Phuc, Shuichi Chiba, and Kazuo Minami. Large scale transient CFD simulations for buildings using OpenFOAM on a world's top-class supercomputer. In 4th OpenFOAM User Conference 2016, 2016.

G. Pozzetti and B. Peters. Evaluating erosion patterns in an abrasive water jet cutting nozzle using XDEM. Advances in Powder Metallurgy & Particulate Materials, pages 191-205, 2017. URL http://hdl.handle.net/10993/31360.

G. Pozzetti, X. Besseron, A. Rousset, and B. Peters. A parallel multiscale DEM-VOF method for large-scale simulations of three-phase flows. Proceedings of ECCM-ECFD 2018, 2018. URL http://hdl.handle.net/10993/35831.

Sina Hassanzadeh Saraei and Bernhard Peters. Immersed boundary method for considering lubrication effects in the CFD-DEM simulations. Powder Technology, page 118603, 2023.

Flavio Manenti, Gintaras V. Reklaitis (Eds.), Proceedings of the 34th European Symposium on Computer Aided Process Engineering / 15th International Symposium on Process Systems Engineering (ESCAPE34/PSE24), June 2-6, 2024, Florence, Italy

Model-based Design of Experiments for the Identification of Kinetic Models of Amide Formation

Emmanuel Agunloye,[a] Muhammad Yusuf,[b] Thomas W. Chamberlain,[b] Frans L. Muller,[b] Richard A. Bourne,[b] Federico Galvanin[a*]

[a]*Department of Chemical Engineering, University College London, WC1E 7JE London, United Kingdom*
[b]*Institute of Process Resarch and Development, Schools of Chemistry and Chemical and Process Engineering, University of Leeds, LS2 9JT Leeds, United Kingdom*
f.galvanin@ucl.ac.uk

Abstract

Model-based design of experiments (MBDoE) techniques have been applied to various process systems in the scientific community to optimally determine a minimum number of informative experiments to enable identification of a kinetic model structure with precisely determined parameters. The effectiveness of MBDoE techniques is however deeply affected by parametric uncertainty. To evaluate the effect of parametric uncertainty on model effectiveness, this work compares and evaluates two different MBDoE approaches: 1) LHS-MBDoE, where the Latin-hypercube sampling (LHS) precedes MBDoE application; and 2) robust MBDoE, where MBDoE techniques apply ab-initio via either the expected value or worst-case approach. Using experimental and in-silico data, MBDoE methodologies were tested on a pharmaceutically relevant reaction system involving homogeneous amide formation, which can be described using reversible chemical kinetics. The performances of the two MBDoE approaches were assessed using i) the χ^2 lack-of-fit test; ii) the Student's t-test; and iii) the determinant of Fisher information matrix (FIM) as posterior scalar measure of information.

Keywords: Amide formation, Latin-hypercube sampling, kinetic models, model-based design of experiment, robust model-based design of experiment.

1. Introduction

Extensive experimentations occur across the pharmaceutical life cycle to characterise drug candidates and their processes to produce effective, safe and profitable drugs. Predominantly, the pharmaceutical industry employs statistical design of experiments (DoE) techniques, which include factorial and random designs (Destro and Barolo, 2022). DoE campaigns often require a large number of time-consuming experiments generated with limited information about the system, causing a waste of time and resources. Informative experiments can be designed using Model-based DoE (MBDoE) techniques (Franceschini and Macchietto, 2008), which are optimal experimental designs for gathering maximum information. MBDoE techniques have increasing applications in several fields, and across the entire product life cycle from development to manufacturing (Abt et al., 2018).

MBDoE application requires four key elements: 1) physics-based modelling, to account for the underlying knowledge about the process system; 2) design space characterisation, to specify the boundary values of process variables (such as input concentrations and process temperature), usually dictated by the equipment operating range; 3) definition of experimental design criteria, to state the objective to optimise from the experimental study; 4) testing criteria, to assess the performance of MBDoE (Franceschini and Macchietto, 2008). To design optimal experiments for parameter precision, MBDoE procedures can be initialised following two methodologies: conventional MBDoE and robust MBDoE (Asprey and Macchietto, 2002). The conventional methodology requires prior parameter estimates with their covariance matrix as initial guess, usually estimated from preliminary data obtained from experiments designed using statistical DoE, to compute the posterior Fisher information matrix in the design space. On the other hand, the robust methodology, which does not require prior parameter estimates, solves the experimental design problem in the uncertain parameter space to compute the expected information, which can be done via two approaches: expected value and worst-case approach. In this work, we compare and evaluate conventional and robust MBDoE methodologies using an optimal experimental design software platform called "SimBot" controlling remotely an automated smart flow reactor ("LabBot") for kinetic model identification in the synthesis of a pharmaceutically relevant homogeneous amide formation.

2. Model identification algorithm

Fig. 1 shows a segment of the SimBot algorithm highlighting five modules: 1) model structure; 2) preliminary DoE; 3) parameter estimation; 4) MBDoE for parameter precision, and 5) model validation, where the precisely identified model can be tested.

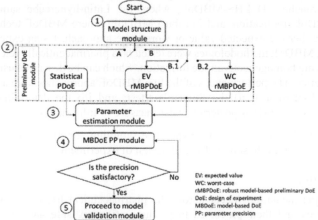

Fig. 1: The SimBot framework for model identification using MBDoE for parameter precision.

The model structure defined in Module 1 can be formulated generally as a set of differential and algebraic equations (DAEs):

$$f(\dot{x}(t), x(t), u(t), \theta, t) = 0 \tag{1}$$

$$y(t) = g(x(t)) \tag{2}$$

where t is the elapsed time in a batch system or space time in a steady-state system. $x \in \mathbb{R}^{N_x}$ is the vector of state variables representing the process system and $\dot{x}$ the first derivative of the state variables with respect to time. $u \in \mathbb{R}^{N_u}$ is the vector of control (manipulated) variables whereas $\theta \in \mathbb{R}^{N_\theta}$ is the vector of unknown model parameters to be estimated within a continuous realisable set $\boldsymbol{\Theta}$; $\boldsymbol{y} \in \mathbb{R}^{N_y}$ is the vector of measured (output) variables.

Module 2 in the SimBot can be used to design preliminary experiments following two different methodologies: Latin-hypercube sampling (LHS) and robust MBDoE. Design of experiments in the LHS technique follows 4 steps: i) the LHS partitions the design space $\boldsymbol{\Phi}$ equally into $N_{exp}{}^{N_\varphi}$ hypercubes $\boldsymbol{\Phi}_i$, ii) selects randomly N_{exp} hypercubes $\boldsymbol{\Phi}_{i,s}$ from this universal space $\boldsymbol{\Phi}$ while ensuring no two selected hypercubes ($\boldsymbol{\Phi}_{i,s}$ and $\boldsymbol{\Phi}_{j,s}$) are in the same partitioned space, iii) then generates statistically random numbers within [0,1], one for each design space dimension, and iv) lastly multiplies each random number by the dimension range to situate a designed experiment in each selected hypercube $\boldsymbol{\Phi}_{i,s}$. With the partitioning, the LHS explores the entire design space ("space filling design"). Robust MBDoE, on the other hand, directly employs the model in the design space without prior parameter estimates $\boldsymbol{\theta}_0$. Rather, this methodology employs an uncertain parameter space $\boldsymbol{\Theta}$ to design preliminary experiments. It is an ab-initio optimization technique to maximise a scalar measure, in this work the determinant, of the Fisher information matrix in the design space. The Fisher information matrix (FIM) for a sampled point in the design space can be expressed as (Quaglio et al., 2018):

$$H_\theta(\boldsymbol{\theta}, \boldsymbol{\varphi}) = \sum_i^{N_y} \frac{1}{\sigma_i^2} \boldsymbol{Q}_i^T \boldsymbol{Q}_i \tag{3}$$

where $\boldsymbol{Q}_i$ is the matrix of the sensitivity coefficients q_{ij} defined as:

$$q_{ij} = \frac{\partial \hat{y}_i}{\partial \theta_j} \qquad i = 1, \dots, N_y; i = 1, \dots, N_\theta \tag{4}$$

and σ_i the standard deviation of the measurement error associated with the measurement of the j-th response variable (here assumed to be uncorrelated and independent from the sampled experiments); $\boldsymbol{\varphi}$ is the experimental design vector containing the decision variables being manipulated in the experiments. Two alternative robust MBDoE approaches have been implemented in the SimBot: 1) expected value and 2) worst case, expressed respectively as:

$$\boldsymbol{\varphi}_{EV} = \arg\max_{\varphi \in \Phi} \; \underset{\theta \in \Theta}{E} \; \{|H_\theta(\boldsymbol{\theta}, \boldsymbol{\varphi})|\}; \quad \boldsymbol{\varphi}_{WC} = \arg\max_{\varphi \in \Phi} \; \underset{\theta \in \Theta}{min} \; \{|H_\theta(\boldsymbol{\theta}, \boldsymbol{\varphi})|\} \tag{5,6}$$

In Module 3, the SimBot calculates the prior parameter estimates from preliminary DoEs by minimizing the negative log-likelihood function (Bard, 1974):

$$\widehat{\boldsymbol{\theta}} = \arg\min_{\theta \in \Theta} \; \frac{1}{2} \sum_{i=1}^{N_{exp}} \{\log(2\pi)^{N_y} + \log\det \boldsymbol{V}_y + [\boldsymbol{y}_i - \widehat{\boldsymbol{y}}_i(\boldsymbol{\theta})]^T \boldsymbol{V}_y^{-1}[\boldsymbol{y}_i - \widehat{\boldsymbol{y}}_i(\boldsymbol{\theta})]\} \tag{7}$$

$\boldsymbol{V}_y$ is the measurement covariance matrix. Parameter estimation performance is assessed by the module by using a χ^2 lack-of-fit test and the parameter t-test given respectively as:

$$\chi^2 = \sum_{j=1}^{N_{exp}} \sum_{i=1}^{N_y} \frac{(y_{ij} - \hat{y}_{ij})^2}{\sigma_{ii}^2} \tag{8}$$

$$t_i = \frac{\theta_i}{t\left(1 - \frac{\alpha}{2}, N_{exp}N_y - N_\theta\right)\sqrt{V_{ii}}} \, ; \, \forall i = 1, \dots, N_\theta \tag{9}$$

V_{ii} is the ith diagonal entry of the inverse of the Fisher information matrix $\mathbf{H}(\widehat{\boldsymbol{\theta}}, \boldsymbol{\varphi})$. t_i must be greater than t_{ref} and χ^2 less than χ^2_{ref}, the reference values being obtained at $(1 - \alpha)$ confidence level ($\alpha = 0.05$ in this work) and $\left(N_{exp}N_y - N_\theta\right)$ degrees of freedom.

Module 4 for MBDoE is the optimal design for parameter precision algorithm that determines the impact of the SimBot software in any application. The algorithm employs the physics-based model with prior parameter estimates to search the design space $\boldsymbol{\Phi}$ for optimal design of experiment $\boldsymbol{\varphi} \in \mathbb{R}^{N_\varphi}$ that would minimise the uncertainty region of model parameters (in other words, maximise parameter precision and model robustness) using a scalar measure of the Fisher information matrix. With $\mathbf{H}_0$ denoting information from previous N_{exp} experiments, the expected marginal posterior Fisher information matrix covariance $\mathbf{H}_{N_{exp}+1}(\widehat{\boldsymbol{\theta}}, \boldsymbol{\varphi})$ at a new experimental design can be obtained as:

$$\mathbf{H}_{N_{exp}+1}(\widehat{\boldsymbol{\theta}}, \boldsymbol{\varphi}) = \mathbf{H}_0 + \sum_i^{N_y} \frac{1}{\sigma_i^2} \boldsymbol{Q}_i^T \boldsymbol{Q}_i \tag{10}$$

A metric of cumulative Fisher information (i.e., the determinant) evaluated as $|\sum \mathbf{H}|$ can be used to assess the performance of alternative MBDoE approaches.

3. Pharmaceutical case study

To evaluate the relative performance of alternative MBDoE methodologies, we consider a case study related to homogeneous amide formation ($R'CONH_2$) in a flow reactor system, where an amine (RNH_2) reacts with an ester ($R'COOR''$) reversibly as:

$$RNH_2 + R'COOR'' \underset{r_b}{\overset{r_f}{\rightleftharpoons}} R'CONH_2 + R''OR \tag{R1}$$

where $R''OR$ is an alcohol by-product.

This synthesis was conducted in a smart flow reactor situated at the University of Leeds and controlled remotely using the MBDoE-driven software operated from University College London in a cloud-based experimental platform. The process control variables are amine and ester inlet concentrations $\boldsymbol{c}(0) = \boldsymbol{c}_0$, residence time τ and the reactor temperature T while the reactor exit concentration of amide $c_3(\tau)$ is the response measured at steady state. The synthesis system operates as a plug flow reactor and can be modelled as a set of differential and algebraic equation with the design vector: $\boldsymbol{\varphi} = [T, \tau, \mathbf{c}_0^T]^T$ and the four Arrhenius parameters describing the forward (f) and backward (b) reactions in R1 to estimate the set of model parameters $\boldsymbol{\theta} = [k_{f,0}, E_{a,f}, k_{b,0}, E_{a,b}]$.

4. Results

The information contents measured using the Fisher information matrix of the conducted experiments via the three approaches, namely LHS followed by MBDoE (LHS-MBDoE), robust worst-case (rMBDoE-WC), robust expected value (rMBDOE-EV) are shown in Fig. 2. The LHS-MBDoE methodology designed the least number of experiments (5 experiments) that precisely determined kinetic parameters in the amide formation model, followed by the worst-case robust MBDoE, which required the design of 7 experiments before reaching a statistically satisfactory parameter precision. The expected value approach, however, failed to reach the target parameter precision even after 10 experiments. Experimentations were conducted for the designed experiments in the LHS-MBDoE and data used in the initial parameter estimation (first 4 experiments) and parameter recalibration with 5 experiments (one additional experiment added to the first 4), the additional experiment being designed using a D-optimal MBDoE for parameter precision. With the amide formation model correctly identified using the LHS-MBDoE methodology, the model was simulated using the designed conditions from the robust MBDoE approaches and the simulation results were corrupted with noise generated randomly following a Gaussian distribution ($\mu = 0.0$ and $\sigma = 0.0003$) to obtain in-silico data for the robust MBDoE approaches and save experimental resources in this study. Yielding reproducible results, the three approaches passed the χ^2 lack-of-fit test in describing the data as shown in Table 1.

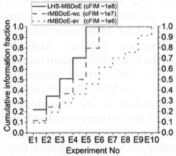

Fig. 2: Cumulative Fisher information fraction with experimental runs for the MBDoE techniques: LHS-MBDoE satisfying performance test at Exp 5 with $|FIM| = 2.6 \times 10^8$; rMBDoE (wc) for robust MBDoE with worst-case satisfying performance test at Exp 6 with $|FIM| =$

3.2×10^7; and rMBDoE (ev) for robust MBDoE with expected value fluctuating in information around $|FIM| = 4.4 \times 10^6$.

Table 1: χ^2 and t-values statistics

	LHS-MBDoE	rMBDoE (wc)	rMBDoE (ev)
χ^2 (χ^2_{ref} = 26.3)	$7.3 \cdot 10^{-9}$	22.8	16.5
$[t_1, t_2, t_3, t_4]$ (t_{ref} = 1.78)	[38.99, 5.23, 23.14, 3.42]	[7.24, 17.11, 3.07, 2.07]	[26.38, 23.79, 0.07, 2.36 $\cdot 10^{-9}$]
N_{exp}	5	6	10

The 5 experiments from the LHS-MBDoE yielded the highest cumulative experimental information, the determinant of the cumulative Fisher information matrix in Fig. 2 being 2.6×10^8, an average value of 5.2×10^7 per experiment. The experimental information of the last experiment designed using the MBDoE for parameter precision was 7.7×10^7 with all parameter t-values after 5 experiments being larger than the reference t_{ref}. On the other hand, the 6 experiments in the worst-case robust MBDoE could only attain a determinant of 3.2×10^7 in the cumulative Fisher information matrix, an average value of 5.3×10^6 per experiment. Nevertheless, this still produced a precise estimation of model parameters, with t-values being larger than t_{ref}.

The maximum cumulative information attained at the end of 10 designed experiments using rMBDoE-EV is much lower compared to the other cases, with a determinant value of 4.4×10^6. Most of the experiments designed using this approach yielded little information about the amide model and the worst experiment in LHS-MBDoE contains more information than the cumulative model information achieved in the 10 experiments sequentially designed using rMBDoE-EV. Data from the 4 preliminary experiments designed by rMBDoE-EV carried nearly zero information about the model, initially estimating poor values for the parameters. Due to the high parameter correlation in the amide formation reversible kinetics reported in Eq. (R1), subsequent MBDoE for parameter precision could not improve poor performing parameter set without destabilizing their well performing counterpart, the technique therefore remaining stuck in improving only the first 2 parameters at the expense of the last 2: after 10 experiments the parameter t-values $\{= [26.38, 23.79, 0.07, 2.36 \cdot 10^{-9}]^T\}$. As reported in Asprey and Macchietto (2002), the performance of the expected value approach in designing preliminary experiments depends on a known parameter distribution $p(\boldsymbol{\theta})$ in the parameter space $\boldsymbol{\Theta}$.

5. Conclusions

This paper evaluated and compared the performance of three MBDoE approaches in designing informative experiments in amide formation synthesis: 1) a Latin hypercube sampling for designing preliminary DoEs followed by MBDoE for parameter precision (LS-MBDoE); 2) a robust MBDoE using a worst case approach (rMBDoE-WC) employed ab-initio to design preliminary experiments and then subsequently to design further experiments for parameter precision; 3) a robust MBDoE using an expected value approach (rMBDoE-EV). The experimental information acquired from these methodologies was quantified using the determinant of the Fisher information matrix and the model parameter precision using the t-test. The LHS-MBDoE methodology combining exploration from the LHS and exploitation from the MBDoE performed most satisfactorily, attaining the highest cumulative information in the fewest runs of experimentation. None of the robust approaches matched this performance. Particularly, the expected values preliminary experiments ill-conditioned the model to design sub-optimal experiments that generated very low information on model parameters. Careful definition of prior model parameter information is therefore required before applying the expected value approach.

Acknowledgements

This project has received funding from the EPSRC (EP/R032807/1 and EP/X024016/1). The support is gratefully acknowledged.

References

F. Destro, M. Barolo, 2022. International Journal of Pharmaceutics 620

G. Franceschini, S. Macchietto, 2008, Model-based design of experiments for parameter precision: State of the art, Chemical Engineering Science, Volume 63, Issue 19, Pages 4846–4872, ISSN 0009-2509, https://doi.org/10.1016/j.ces.2007.11.034

M. Quaglio, E.S. Fraga, F. Galvanin, 2018. ChERD, Vol. 136, Pages 129 – 143, doi.org/10.1016/j.cherd.2018.04.041

S.P. Asprey, S. Macchietto, 2002. Designing robust optimal dynamic experiments, Journal of Process Control 12 (2002) 545–556

V. Abt, T. Barz, M.N. Cruz-Bournazou, C. Herwig, P. Kroll, J. Möller, R. Pörtner, R. Schenkendorf, R., 2018, Model-based tools for optimal experiments in bioprocess engineering, Current Opinion in Chemical Engineering, Volume 22, 2018, Pages 244-252, ISSN 2211-3398, https://doi.org/10.1016/j.coche.2018.11.007

Y. Bard, 1974, Nonlinear parameter estimation, Academic Press.

Flavio Manenti, Gintaras V. Reklaitis (Eds.), Proceedings of the 34th European Symposium on Computer Aided Process Engineering / 15th International Symposium on Process Systems Engineering (ESCAPE34/PSE24), June 2-6, 2024, Florence, Italy

A Holistic Approach for Model Discrimination, Multi-Objective Design of Experiment and Self-Optimization of Batch and Continuous Crystallization Processes

Xuming Yuan[a] , Brahim Benyahia[a*]

[a]*Loughborough University, Department of Chemical Engineering, Epinal Way, Loughborough, Leicestershire, LE11 3TU, UK*
B.Benyahia@lboro.ac.uk

Abstract

In this study, a systematic approach is proposed to determine the best mathematical model candidate and reduce model uncertainty at a lower experimental cost, followed by model-based self-optimization to maximize process performance and product quality. The proposed method starts with the structural identifiability analysis of all model candidates, where only the structurally identifiable models can survive. Using prior knowledge, if available, or data from initial experiments, a set of preliminary/nominal parameters are estimated for the survived models followed by model discrimination based on a F-test. If more than one model passes the test, model-discrimination design of experiment (MDDoE) will be performed to identify the best model. Next, the most estimable parameter set is revealed by the estimability analysis. Afterwards, optimal experimental designs are identified based on different seeding and temperature operating strategies in the context of multi-objective model-based design of experiment (MBDoE) formulation, where the D-optimality criterion of the estimable parameters and the experimental resources are optimized. Using the optimal experimental campaign, the model parameters of the best model are updated on the fly followed by model-based self-optimization to deliver the optimal operation and design strategies along with the design space, where the attainment of the targeted Critical Quality Attributes is guaranteed. A cooling crystallization process of paracetamol (batch/continuous) is used to demonstrate and validate the proposed approach. The results show that the best model can be systemically identified while minimizing the model uncertainties and maximizing prediction capabilities. Using the most reliable model, more robust and precise self-optimization and tracking of the design space can be at lower experimental cost, demonstrating the potential effectiveness of the proposed method in crystallization processes.

Keywords: Crystallization, Model Discrimination, Model-Based Design of Experiment, Self-Optimization, Estimability Analysis.

1. Introduction

Throughout the development stages of pharmaceutical processes, the availability of reliable mathematical models is critical for the effective exploration of the design space and the delivery of robust and cost-effective designs, operating and control strategies (Benyahia et al., 2021; Liu and Benyahia, 2022). In pharmaceutical processes, crystallization plays an important role as a separation and purification technique, which usually involves complex kinetics and mechanisms, including primary and secondary nucleation, growth, and dissolution, as well as agglomeration and breakage. Therefore, it is a common situation that a crystallization process can be described by various models, which usually involves a large set of parameters. Consequently, it is difficult to determine the most appropriate model, and it leads to the high requirement on the amount and quality of experimental data for model identification. Hence, it is challenging to build a reliable model for a crystallization process, especially at the early stage of the process development where data availability is limited.

To address such issues in the establishment of reliable crystallization models, Yuan and Benyahia (2023) proposed a methodology to investigate and identify the crystallization models comprehensively, which incorporated structural identifiability analysis, practical identifiability (estimability) analysis and model-based design of experiments (MBDoE). The proposed approach ensures that only the structurally identifiable models can survive and reduces the parameter uncertainty, while the estimability of the parameters can be assessed a priori or a posteriori. The implementation of the approach in the case of a seeded batch cooling crystallization process of paracetamol revealed the structural identifiability of the investigated model and the reduced uncertainty of the parameters through D-optimal design. However, the approach does not provide a further screening criterion for the structurally identifiable model candidates, and the estimability analysis is independent from the MBDoE. One potential complementary method is model-discrimination design of experiment (MDDoE), suggested by Sen et al. (2021) where the objective is to maximize the difference between two model candidates in the experiments to be designed. Another approach in Kilari et al.'s work (2023) applied a combination of Akaike information criterion (AIC) and F-test to determine the most appropriate model for the crystallization process. Nonetheless, the proposed approaches neglected investigating the structural and practical identifiability (estimability) of the models.

Hence, this paper provides a holistic approach for the development of cooling crystallization models, in which structural identifiability analysis, AIC-test and F-test, MDDoE, estimability analysis and MBDoE are systematically incorporated in the proposed methodology. This approach is implemented in an in-silico case study of the cooling crystallization of paracetamol to demonstrate the benefits of its application in the development of crystallization models. In addition, a following multi-objective optimization of the batch crystallization process is performed in the case study to show how the improved model is utilized to enhance the desired CQAs, where the results indicate the potential effectiveness of the proposed methodology in the establishment of reliable and predictive crystallization models.

2. Methodology

The framework of the proposed methodology is shown in Figure 1. At the start, several model candidates describing the same crystallization process are available. The first step is the investigation of the structural identifiability of the models (Test 1) based on a set of observable outputs, where only the structurally identifiable models can survive. Once the structurally identifiable models are recognized, preliminary experiments are conducted to generate the experimental data for preliminary parameter estimation, giving the nominal parameter sets. If any available prior knowledge of the nominal parameters exists, it can be introduced in the form of nominal values in which case the preliminary experiments are not required. The nominal parameters are then used to build the sensitivity matrices or covariance matrices. Next, AIC-test and F-test (Test 2) are carried out, which aim at screening the model that best fits the experimental data. If more than one model candidate passes the tests, MDDoE will be performed with the objective of maximizing the difference between the models. The resulting optimal experimental campaign will be conducted, and the parameters are re-estimated, and the best model

A Holistic Approach for Model Discrimination, Multi-Objective Design of   393
Experiment and Self-Optimization of Batch and Continuous Crystallization
Processes

candidate is determined by running Test 2 again. It should be noted that this MDDoE step can be conducted iteratively if one single optimal experiment is insufficient to discriminate the models. The estimability of the best model is analyzed later, revealing the most estimable parameters subset, which is used to compute the reduced Fisher information matrix (FIM) in the subsequent MBDoE. With the new experimental data collected from the optimized experimental profiles, the new parameter estimates are obtained from re-estimation. Finally, the uncertainties of the new estimates are evaluated (Test 3). A new MBDoE problem can be formulated and solved if the parameter uncertainty is higher than the targeted level. Otherwise, the workflow is terminated.

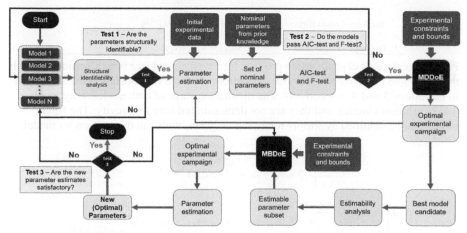

Figure 1. Proposed framework for Model Based Design of Experiments.

2.1. Mathematical Models

A cooling crystallization process of paracetamol is selected as the case study in this paper. The real process involves primary and secondary nucleation, growth, and nucleation, as well as agglomeration and breakage. The crystallization processes are commonly simulated by using population balance equations (PBEs) and mass and energy balance equations. The finite difference approach is applied in this work to discretize the crystal size L and the distribution $n(L)$ for the solution of the PBEs. The solubility of paracetamol (C^*) follows a quadratic function of the temperature.

$$C^* = p_0 + p_1 T + p_2 T^2 \tag{1}$$

The absolute supersaturation in the crystallization process is given by:

$$S = C - C^* \tag{2}$$

Two model candidates are investigated in this study, where the first model has the primary and secondary nucleation, growth, and dissolution terms, while the second model has the additional agglomeration and breakage terms. The kinetic equations and parameters of Model 1 and Model 2 are summarized in Table 1 and 2, respectively.

Table 1. Kinetic equations and parameters of Model 1.

Kinetics	Equations	Parameters
Primary Nucleation	$J_1 = k_{b1} S^{b_1}$	k_{b1}, b_1
Secondary Nucleation	$J_2 = k_{b2} S^{b_2} \mu_2^{j_2}$	k_{b2}, b_2, j_2
Growth	$G = k_g S^g$	k_g, g
Dissolution	$D_s = k_{ds}(-S)^{ds}$	k_{ds}, ds

Table 2. Kinetic equations and parameters of Model 2.

Kinetics	Equations	Parameters
Primary Nucleation	$J_1 = k_{b1}S^{b_1}$	k_{b1}, b_1
Secondary Nucleation	$J_2 = k_{b2}S^{b_2}\mu_2^{j_2}$	k_{b2}, b_2, j_2
Growth	$G = k_g S^g$	k_g, g
Dissolution	$D_s = k_{ds}(-S)^{ds}$	k_{ds}, ds
Agglomeration Caused Birth	$B_{agg,i} = \frac{1}{2}K_a L_i^5 \Delta L_i \sum_{j=1}^{i} \frac{n(L_i^3 - L_j^3)}{(L_i^3 - L_j^3)^{\frac{2}{3}}} N_j$	K_a
Agglomeration Caused Death	$D_{agg,i} = K_a n(L_i)\Delta L_i \sum_{j=1}^{N_L}(L_i^3 + L_j^3)N_j$	
Breakage Caused Birth	$B_{brk,i} = 2K_b \Delta L_i \sum_{j=i+1}^{N_L} L_j^{\gamma-1} N_j$	K_b, γ
Breakage Caused Death	$D_{brk,i} = K_b L_i^{\gamma} n(L_i)\Delta L_i$	

For the continuous crystallization process, let τ be the residence time. Considering both the PBEs and the mass balances, the crystallization stage of Models 1 and 2 are shown in Equations 3 and 4, respectively. The dissolution stage of both models only involves the dissolution kinetics, and they are not demonstrated here for brevity. The batch model can be obtained by simply removing the terms with τ in the continuous model.

$$\begin{cases} \frac{dN_1}{dt} = -\frac{G}{2\Delta L_1}N_1 + J_1 + J_2 + \frac{N_{1,in}-N_1}{\tau} \\ \frac{dN_i}{dt} = -\frac{G}{2\Delta L_i}N_i + \frac{G}{2\Delta L_i}N_{i-1} + \frac{N_{i,in}-N_i}{\tau} \\ \frac{dN_{N_L}}{dt} = -\frac{G}{2\Delta L_{N_L}}N_{N_L} + \frac{G}{2\Delta L_{N_L-1}}N_{N_L-1} + \frac{N_{N_L,in}-N_{N_L}}{\tau} \\ \frac{dC}{dt} = -3\rho k_v G\mu_2 + \frac{C_{in}-C}{\tau} \end{cases} \quad (3)$$

$$\begin{cases} \frac{dN_1}{dt} = -\frac{G}{2\Delta L_1}N_1 + J_1 + J_2 + B_{agg,1} - D_{agg,1} + B_{brk,1} - D_{brk,1} + \frac{N_{1,in}-N_1}{\tau} \\ \frac{dN_i}{dt} = -\frac{G}{2\Delta L_i}N_i + \frac{G}{2\Delta L_i}N_{i-1} + B_{agg,i} - D_{agg,i} + B_{brk,i} - D_{brk,i} + \frac{N_{i,in}-N_i}{\tau} \\ \frac{dN_{N_L}}{dt} = -\frac{G}{2\Delta L_{N_L}}N_{N_L} + \frac{G}{2\Delta L_{N_L-1}}N_{N_L-1} + B_{agg,N_L} - D_{agg,N_L} + B_{brk,N_L} - D_{brk,N_L} + \frac{N_{N_L,in}-N_{N_L}}{\tau} \\ \frac{dC}{dt} = -3\rho k_v G\mu_2 + \frac{C_{in}-C}{\tau} \end{cases} \quad (4)$$

2.2. Model Discrimination

The structural identifiability of both models is confirmed by using the Matlab toolbox GenSSI 2.0, developed by Chis, Banga and Balsa-Canto (2011), where the selected observables in this case study are the concentration, mean crystal size and crystal counts. Afterwards, the in-silico preliminary experiments as well as the corresponding parameter estimation are carried out and the models are assessed based on the AIC-test and F-test summarized in Table 3, which reveals that Model 1 is superior to Model 2 in this case study, as it has a lower AIC value, and the F-test strongly supports this model. Although the real process seems to exhibit some agglomeration and breakage, the effects are deemed not influential enough based on the preliminary experiments, thus the penalty term of the number of model parameters for Model 2 in AIC-test dominates, while Model 1 capture the main dynamics in the preliminary experiments with a simpler form. It is also worth noting that the computational effort for the parameter estimation for Model 2 is much higher than Model 1 (10 min against 4 h on a PC equipped with an 11th Gen Intel(R) Core (TM) i5-11400 CPU @ 2.60GHz Processor with 16GB RAM). Hence, Model 1 is the best model in this case study.

Table 3. AIC-test and F-test results for Model 1 and Model 2

Model	AIC	p-value from F-test
Model 1	-181.3478	1
Model 2	11.9647	

A Holistic Approach for Model Discrimination, Multi-Objective Design of
Experiment and Self-Optimization of Batch and Continuous Crystallization
Processes
395

2.3. Estimability Analysis and Multi-Objective Model-Based Design of Experiment

The estimability analysis of the parameters in Model 1 based on the preliminary experimental results in an estimability rank of the model parameters (Table 4), where all the parameters estimable except the secondary nucleation rate constant k_{b2}. Afterwards, a multi-objective MBDoE is performed, which aims at minimizing the determinant of the inverse of the FIM of the estimable parameter subset (D-optimal design) and the amount of paracetamol used in the experiment. Temperature cycling with holds and intermittent seed addition are implemented, and the decision variables include the cooling/heating rates and the durations, mean crystal size of the seed, seed addition amount and time, initial temperature, residence time and sampling times. The mathematical formulation of the optimization problem is shown in Equation 5. Figure 2 (a) and (b) show the Pareto front of the problem (the blue star represents the selected optimal profile for the experiment in this case study) and the corresponding optimal seed addition profile. The parameter estimation using both the preliminary data and the new data from the optimized experiment is performed, giving the new parameter estimates, and the reduction of parameter uncertainty is illustrated by the joint confidence regions before and after the MBDoE, where Figure 2 (c) gives an example of the parameters k_g and ds.

Table 4. Estimability ranking of the parameters in Model 2

Rank	Parameter	Description
1	ds	Dissolution power number
2	b_1	Primary nucleation power number
3	g	Growth power number
4	k_{ds}	Dissolution rate constant
5	k_{b1}	Primary nucleation rate constant
6	j_2	Secondary nucleation power number of total surface area
7	k_g	Growth rate constant
8	b_2	Secondary nucleation power number
9	k_{b2}	Secondary nucleation rate constant

$$
\min_{R,\Delta t_R,T_{ini},t_{samp},d_{seed},\mu_{0,seed},\Delta t_{seed},\tau} \det(FIM_{est}^{-1}) \ \& \ \min_{R,\Delta t_R,T_{ini},t_{samp},d_{seed},\mu_{0,seed},\Delta t_{seed},\tau} m_{PCM,seed}
$$

$$
\text{s. t.} \quad -0.5 \leq R_i \leq 0 \ (i = 1,3,5)
$$
$$
0 \leq R_i \leq 1 \ (i = 2,4,6)
$$
$$
0 \leq \Delta t_{R,j} \ (1 \leq j \leq 12)
$$
$$
\sum_{j=1}^{12} \Delta t_{R,j} = 240
$$
$$
10 \leq T_{ini} + \sum_{i=1}^{6} R_i \Delta t_{R,2i-1} \leq 45 \ (1 \leq l \leq 6)
$$
$$
0 \leq t_{samp,1}
$$
$$
t_{samp,10} \leq 240
$$
$$
t_{samp,k} - t_{samp,k+1} \leq 0 \ (1 \leq k \leq 9)
$$
$$
50 \leq d_{seed} \leq 150
$$
$$
100 \leq \mu_{0,seed,l} \leq 2000 \ (1 \leq l \leq 6)
$$
$$
0 \leq \Delta t_{seed,l} \ (1 \leq l \leq 6)
$$
$$
\sum_{l=1}^{6} \Delta t_{seed,l} = 240
$$
$$
12.5 \leq \tau \leq 100
$$

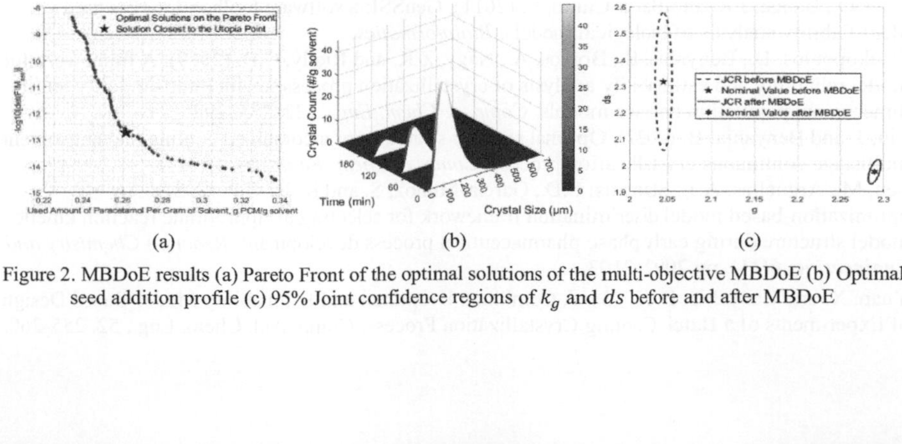

(5)

Figure 2. MBDoE results (a) Pareto Front of the optimal solutions of the multi-objective MBDoE (b) Optimal seed addition profile (c) 95% Joint confidence regions of k_g and ds before and after MBDoE

2.4. Self-Optimization of the Continuous Cooling Crystallization Process

With the updated parameters, the self-optimization of the continuous cooling crystallization process is conducted subsequently, aiming at maximizing the mean crystal size at the end of the process, where the initial condition of the self-optimization stage is the final condition of the MBDoE stage. The mathematical formulation of the bilevel optimization problem is not given for the sake of brevity. The optimal temperature profile of the whole process (MBDoE and self-optimisation) is shown in Figure 3 (a), while the resulting mean crystal size trajectory is shown in Figure 3 (b). The confidence band of the mean crystal size is also shown in Figure 3 (c), where the narrower confidence band reveals the reduction in uncertainties thanks to the proposed MBDoE.

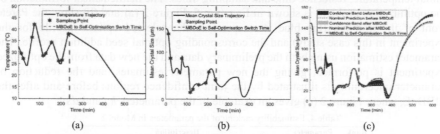

| (a) | (b) | (c) |

Figure 3. Optimization results of the continuous cooling crystallization process (a) Optimal temperature profile (b) Mean crystal size trajectory (c) 95% confidence band of the mean crystal size trajectory.

3. Conclusions

A novel holistic approach is proposed for the development of crystallization models in this study, which incorporates model discrimination through structural identifiability analysis, AIC and F-test, MDDoE, estimability analysis and MBDoE. The proposed approach was implemented to a cooling crystallization process of paracetamol, where two model candidates were investigated. The results showed that the second model has superior performance. The subsequent multi-objective MBDoE combined with the estimability analysis successfully reduced the uncertainty of the parameters through the optimized experiment at a lower cost. The last step consists in the self-optimization which is aimed at finding the optimal profile that maximizes the mean crystal size. The results demonstrate the effectiveness of the proposed method in the establishment of reliable and predictive crystallization models.

References

Benyahia, B., Anandan, P.D. and Rielly, C. (2021). Robust Model-Based Reinforcement Learning Control of a Batch Crystallization Process. In: *9th ICSC*, 89-94.

Chis, O., Banga, J.R. and Balsa-Canto, E. (2011). GenSSI: a software toolbox for structural identifiability analysis of biological models. *Bioinformatics*.

Fysikopoulos, D., Benyahia, B., Borsos, A., Nagy, Z.K. and Rielly, C.D. (2019). A framework for model reliability and estimability analysis of crystallization processes with multi-impurity multi-dimensional population balance models. *Comp. & Chem. Eng.*, 122, 275–292.

Liu, J. and Benyahia, B. (2022). Optimal start-up strategies of a combined cooling and antisolvent multistage continuous crystallization process. *Comp. & Chem. Eng.*, 159, 107671.

Sen, M., Arguelles, A.J., Stamatis, S.D., García-Muñoz, S. and Kolis, S.P. (2021). An optimization-based model discrimination framework for selecting an appropriate reaction kinetic model structure during early phase pharmaceutical process development. *Reaction Chemistry and Engineering*, 6(11), pp.2092–2103.

Yuan, X. and Benyahia, B. (2023). A Combined D-optimal and Estimability Model-Based Design of Experiments of a Batch Cooling Crystallization Process. Comp. Aid. Chem. Eng., 52, 255-260.

Flavio Manenti, Gintaras V. Reklaitis (Eds.), Proceedings of the 34th European Symposium on Computer Aided Process Engineering / 15th International Symposium on Process Systems Engineering (ESCAPE34/PSE24), June 2-6, 2024, Florence, Italy

Computer-Aided Assessment of a Droplet Absorber for CO_2 Direct Air Capture

Maria F. Gutierrez[a*], Kasimhussen Vhora[a,b], Andreas Seidel-Morgenstern[a], Gábor Janiga[b], Peter Schulze[a]

[a]*Max Planck Institute for Dynamics of Complex Technical Systems Magdeburg, Sandtorstr. 1, 39106 Magdeburg, Germany*
[b]*Otto von Guericke University Magdeburg, Universitätsplatz 2, 39106 Magdeburg, Germany*
gutirrezsnchez@mpi-magdeburg.mpg.de

Abstract

In this study, a droplet absorber capable of capturing CO_2 from air was assessed by means of CFD simulations and differential mass balances. The absorber uses a NaOH solution dispersed in micro droplets that drives the flow of air in a co-current arrangement. Three different absorber geometries were evaluated to study the influence of the height and the absorber diameter on the absorption performance. According to the results, the specific surface area is reduced along the absorber due to droplet collision and coalescence; however, after two meters the specific surface area is increased again. Analysis of simulation results showed that changes in the rate of collision, coalescence and droplet breakup generate this behaviour. The longest absorber showed the best performance in terms of capture rate (1.1 kg CO^2/(h-m^3 abs)) and energy consumption (1 kWh/kg CO_2), while changes in absorber diameter did not influence the absorber performance.

Keywords: CO_2 absorption, CFD simulation, capture rate, energy consumption.

1. Introduction

Carbon Dioxide Removal (CDR) technologies are crucial to achieve net negative CO_2 emissions to the atmosphere and climate neutrality (IPCC, 2023). The CODA (Carbon-negative sODA ash) project has been studying a CDR technology for the absorption of CO_2 from air using a sodium hydroxide (NaOH) aqueous solution to produce sodium carbonate (Na_2CO_3) in a more sustainable way (BMBF & FONA, 2023). While in the traditional Solvay process nearly 0.5 kg CO_2/kg soda are emitted, in the CODA process around 0.4 kg CO_2/ton soda are removed from the air (besides the emission avoidance). When CO_2 is captured directly from the atmosphere (Direct Air Capture, DAC), large amounts of air should be processed due to its low concentration (currently ~417 ppm). The economic feasibility of DAC systems relies on the design of suitable equipment able to provide a high volumetric capture rate (smaller capital cost) with the lowest energy demand (less operational costs), minimizing total costs for CO_2 DAC. Previous efforts to achieve this resulted in investigation of spray absorbers (Stolaroff et al., 2008), counter-current packed absorbers (Mazzotti et al., 2013) and cross flow packed absorbers (Holmes & Keith, 2012). With the aim to reduce the air pressure drop (energy demand) and to avoid the use of packing and fans (capital cost), a droplet absorber in which the liquid flow drives the gas flow in a co-current mode and could tolerate crystal formation has been studied in the CODA project. The performance of such an absorber was previously evaluated using a simple model that neglected the droplet coalescence (Gutierrez et al., 2023). However, evaluation of the technology under more realistic conditions is still required.

In the present study, the CO_2 capture rate of the droplet absorber is predicted through a combination of differential mass balances and Computational Fluid Dynamics (CFD) simulations. Three different absorber geometries were evaluated to observe the influence of the absorber diameter and length on the surface area, the capture rate and the energy consumption due to pumping. The use of CFD for the design of a droplet absorber for DAC applications is the novelty of this study.

2. Droplet absorber

A nozzle plate with holes of 180 μm diameter arranged in a triangular pattern and 0.9 mm distance (middle-middle) produced the droplets in the studied absorber. In all geometries studied the holes in the nozzle represented 3.3% of the total nozzle plate area. The liquid volumetric flow rate was set to ensure the operation in the Rayleigh breakup regime of round liquid jets in quiescent air, where the surface tension force is the predominant breakup mechanism (Huimin Liu, 1999). This ensured the vertical fall of the droplets and avoiding the loss of surface area due to collision with the absorber wall. The fall of the droplets drives the movement of the air inside the absorber by friction so that the liquid and the gas flow are arranged in a co-current mode. Three different geometries were studied: a base case, an absorber with the same diameter but with a longer height (Hx3) and an absorber with the same length but with a wider diameter (Dx2). Details on the base case was provided in a previous publication (Gutierrez et al., 2023) and specific details on the geometries used and liquid flow rate is provided in Table 1.

Table 1. Absorber geometry and liquid flow rate

Absorber variable	Units	Base case	Dx2	Hx3
Absorber diameter	mm	28.4	56.8	28.4
Absorber height	m	1.54	1.54	5
Number of nozzles	-	841	3341	841
Liquid flow	L/h	157.72	626.56	157.72

2.1. Material balance

The absorption of CO_2 from the air by droplets of a NaOH solution was described using the differential mass balance shown in Eq. (1). In this equation y_{CO_2} represents the mole fraction of CO_2 in the gas phase, z is the absorber height, K_{G,CO_2}^c is the overall mass transfer coefficient of CO_2 in the gas phase, a is the specific surface area, $\dot{G}'$ is the molar flow per cross sectional area, $y_{co_2}^*$ is the CO_2 gas mole fraction in equilibrium with the CO_2 liquid bulk concentration, P is the pressure of the system, R is the universal gas constant and T is the temperature of the system.

$$\frac{dy_{CO_2}}{dz} = -\frac{K_{G,CO_2}^c a}{\dot{G}'}\left(y_{co_2} - y_{co_2}^*\right)\frac{P}{RT} \tag{1}$$

The overall mass transfer coefficient was calculated using the two-resistance theory as described in a previous publication (Gutierrez et al., 2023). This coefficient is a function of kinetic and solubility parameters as well as hydrodynamic variables, namely the droplet diameter, the gas and the liquid velocities. In contrast with our previous publication, the hydrodynamic variables here were not kept constant along the absorber. In addition, in this case, the water loss by evaporation was neglected, the temperature was considered constant along the absorber (10°C) and the liquid entering the absorber was saturated in sodium carbonate. Other assumptions and considerations that remain the same are here shortly summarized and reported in detail in our previous publication:

- The change of $\dot{G}'$ was neglected because CO_2 concentrations are very small and water evaporation is not considered.

- Phase equilibrium concentration ($y_{CO_2}^*$) was calculated by using Henry's law and assuming ideal gas. The required CO_2 concentration in the liquid phase was calculated with differential mass balances in the liquid phase.
- The reaction is irreversible due to the high concentration of hydroxide ions (OH^-) in comparison with that of CO_2 in the liquid phase.
- The effect of the reaction in the mass transfer was considered by using the Enhancement factor (E), which is a function of the Hatta number (Ha). Due to the considered operational conditions, the droplet absorber remains in the "very fast" reaction regime, where $Ha > 3$ and $E = Ha$ (Danckwerts, 1970). Consequently, the liquid mass transfer coefficient depends on the kinetic rate constant, the concentration of CO_2 and OH^- in the liquid phase and the liquid diffusivity of CO_2.

2.2. CFD Simulation

The hydrodynamic variables, the pressure of the system and specific surface area were calculated by means of a detailed CFD simulation performed in the Simcenter STAR CCM+ software (version 2020.1, Siemens Product Lifecycle Management Software Inc., Plano, TX, USA). For that, a digital twin of the absorber was built and used as base to construct a volumetric domain, which was partitioned into discrete cells to generate a polyhedral mesh. The Reynolds Averaged Navier-Stokes (RANS) equations were discretized using the finite volume method. The turbulence model k-ω Menter SST two-layer was used for the closure of the RANS equations. The Eulerian-Lagrangian framework was used to model the flow of liquid water droplets and air inside the cylindrical absorber (the Lagrangian model for tracking droplets and the Eulerian model for the air). In the CFD simulation, the liquid droplets were represented by a reduced number of parcels. Each parcel consisted of particles assumed to be spherical, sharing identical properties such as diameter, velocity and density. Each parcel was treated as a source term integrated into the governing equations of mass and momentum at its specific location. The no-time counter (NTC) collision model was used to simulate the interaction between droplets that occurs during collisions and coalescence. For the Hx3 geometry, the KHRT droplet breakup model (based on the Kelvin-Helmholtz and the Rayleigh-Taylor theories) was also used. Two boundary conditions were set: the wall boundary condition and the bottom pressure outlet.

The results of droplet velocity and gas velocity from the CFD simulations were used to calculate mean values in segments of 10 mm height, so that their radial variation was neglected. In each segment, the total surface area was calculated and divided into the segment volume to obtain the specific surface area along the absorber. The total model (CFD simulation + differential balances) was validated with experimental data on the base case geometry. Both the induced gas velocity and the CO_2 concentration in the outlet of the absorber could be well described by the model within a 2.2% of relative error. Mesh validation and evaluation of the parcel size was done for the base case. Details on such validations will be provided in a future publication.

2.3. Capture rate and energy consumption

The droplet absorber was evaluated in terms of the capture rate (r, directly related with the capital costs) and the energy consumption (E, directly related with the operational costs). The performance variables were evaluated according with Eq. (2) and Eq. (3). In the last equation, ρ_L is the liquid density, g is the gravity, ΔP_{nozz} is the nozzle plate pressure drop, $\dot{F}_L$ is the volumetric flow rate, η_p is the pump efficiency (85%) and A is the cross sectional area. The nozzle plate pressure drop was calculated with a correlation proposed in the literature for pressure drop across micro-orifices (Ushida et al., 2014).

$$r(z) = K_{G,CO_2}^c a\left(y_{co_2} - y_{co_2}^*\right)\frac{P}{RT} \tag{2}$$

$$E = \frac{(\rho_L gH + \Delta P_{nozz})\dot{F}_L/\eta_p}{\int_0^H \left[K_{G,CO_2}^c a\left(y_{co_2} - y_{co_2}^*\right)\frac{P}{RT}\right]A dz} \tag{3}$$

3. Results and Discussion

The specific surface area along the absorber for the studied geometries is presented in Figure 1 (a) to (c). Due to droplet coalescence, the specific surface area is reduced within the first two meters of absorber. As presented in Figure 1 (d), 76% of the reduction of the number of droplets occurs in the same region. Despite the loss in surface area, the specific surface area is in the range of other type of absorbers (random and structured packing).

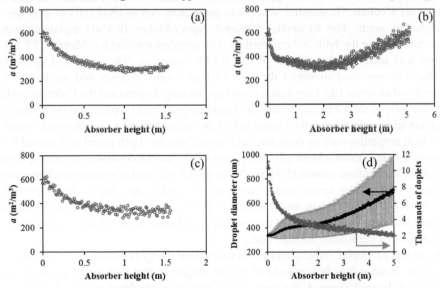

Figure 1: Specific surface area for the (a) Base case, (b) Hx3 and (c) Dx2 geometries and (d) mean droplet diameter and droplet count for the Hx3 geometry.

Interestingly, the specific surface area starts to increase after the first two meters of absorber height (Figure 1 (b)). The specific surface area in a segment of absorber can be calculated as $a_s = N\pi d^2/V_s$, where N is the number of droplets in the segment, d is the mean droplet diameter in the segment and V_s is the segment volume. In the first two meters of absorber (see Figure 1 (d)), the diameter of the droplets is increased and the droplet count is heavily reduced due to coalescence ($\downarrow N \gg \uparrow d$), therefore the specific surface area is reduced. After this height, the breakup of droplets becomes significant, which slows down the reduction of droplet number (N). Considering the standard deviation of the droplet diameter (shown as error bar in Figure 1 (d)), big droplets prone to breakage can be found in the absorber after 2.5 m height. In this way, the loss of droplets due to collision is compensated with the generation of droplets due to breakup (although the rate of collision is still higher than the rate of breakup). At the same time, the droplet diameter (d) keeps increasing along the absorber due to coalescence. After 2.5 m height, the number of droplets decrease linearly (not as heavily as before) and the droplet diameter keeps increasing linearly ($\downarrow N < \uparrow d$), and since $a_s \propto d^2$, the specific surface area starts to linearly increase despite the reduction in the number of droplets. On the other hand, the increase of the absorber diameter does not affect significantly the behavior of specific surface area along the absorber, only a slight dispersion of this variable is observed (Figure 1 (c)).

CFD results on droplet and induced gas velocities for the Hx3 geometry is presented in Figure 2 (a). Results for the base case and the Dx2 geometry follow the same trend until the respective absorber length for those geometries. While the droplet velocity always increases along the absorber, the induced gas velocity reaches a constant value after 2.5 m. The gravity force is always acting on the droplets, which causes the constant increment on the liquid velocity along the absorber. For the air, the force of the droplets downwards and the friction force with the wall in the opposite direction of the movement reach an equilibrium at 2.5 m. Consequently, the relative velocity between the droplets and the air increases continuously after 2.5 m, which also increases the probability of breakup. In contrast to other absorbers, the gas and liquid flows are tightly related in the droplet absorber. Results here obtained are useful to determine the gas to liquid ratio in the droplet absorber for different geometries, which is required for further process design.

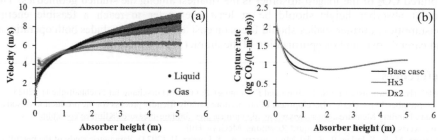

Figure 2: (a) Droplet and induced air velocity for the Hx3 geometry and (b) capture rate for the three geometries studied.

Results of capture rate along the absorber are presented in Figure 2 (b). Capture rate of the droplet absorber is also in the same range of packed absorbers (1.2-2.2 kg CO₂/(h-m³) in counter-current arrangement (Mazzotti et al., 2013) and 0.68 kg CO₂/(h-m³) in cross-flow arrangement (Holmes & Keith, 2012)). For all cases, capture rate decreases rapidly in the first meter of absorber height, mainly due to the loss in surface area. After one meter of absorber, the decrease in capture rate slows down, and for the longer absorber (Hx3) a slight increase in the capture rate is observed. In general, differences in capture rate between the three studied geometries is small and all geometries have a similar performance in terms of capture rate.

On the other hand, the energy consumed per mass of absorbed CO₂ is 1.38 kW-h/kg CO₂ for the base case, 1.03 kW-h/kg CO₂ for the Hx3 geometry and 1.15 kW-h/kg CO₂ for the Dx2 geometry. This parameter is in the same range of reported values for other droplet absorbers (1.2-2.5 kW-h/kg CO₂ (Stolaroff et al., 2008). In comparison with packed absorbers, the droplet absorber does not require a fan to generate the gas flow. However, the total energy consumption of the droplet absorber is still higher than that reported for packed absorbers (0.555 kW-h/kg CO₂ in counter-current arrangement (Zeman, 2008) and 0.082 kW-h/kg CO₂ in cross-flow arrangement (Keith et al., 2018)). The lowest energy consumption reported for packed absorbers was obtained using a special low pressure drop packing in a cross-flow arrangement and a discontinuous pumping of the liquid (Keith et al., 2018). For the droplet absorber, results on the energy consumption showed that the longest absorber has the lowest energy consumption. Moreover, the results allowed identifying that the absorber height should be at least one meter to make the energy consumption reaching a reasonable value. Below this height, the capture of CO_2 is not enough to compensate the use of energy in the absorber. Despite of the higher operational costs, the droplet absorber could still represent a feasible technology for DAC in terms of capital costs and the energetic optimization is still to be done. Besides obtaining a higher capture rate (smaller absorber) than the cross-flow absorber, packing and fan costs could be avoided by using the droplet absorber.

4. Conclusions

The combination of detailed CFD simulations and differential mass balances allowed assessing three different geometries of a droplet absorber for DAC. The CFD simulations showed that the flow of gas is driven by friction with the falling droplets in the absorber. The performance of the droplet absorber in terms of specific surface area and capture rate is comparable to that reported for packed absorbers. The specific surface area and the capture rate decrease in the first 2 meters of absorber but starts to increase due to changes in droplet diameter and number of droplets. The amount of energy consumed per mass of captured CO_2 of the longest absorber is the smallest among the studied geometries. The droplet absorber height should be at least one meter to reach a feasible energy consumption. Further studies should focus on cost analysis that includes both operational and capital costs and on optimization of the droplet absorber design.

References

BMBF (Bundesministerium für Bildung und Forschung), & FONA (Forschung für Nachhaltigkeit). (2023, September 13). CODA – Development of an environmentally friendly process for the production of soda ash. Funding Measures. https://www.fona.de/en/measures/funding-measures/KlimPro/coda.php

Danckwerts, P. V. (1970). Gas-Liquid Reactions. McGraw-Hill.

Gutierrez, M. F., Schulze, P., Seidel-Morgenstern, A., & Lorenz, H. (2023). Parametric study of the reactive absorption of CO2 for soda ash production. In A. C. Kokossis, M. C. Georgiadis, & E. Pistikopoulos (Eds.), 33rd European Symposium on Computer Aided Process Engineering (Vol. 52, pp. 2935–2940). Elsevier.

Holmes, G., & Keith, D. W. (2012). An air-liquid contactor for large-scale capture of CO2 from air. Philosophical Transactions of the Royal Society A: Mathematical, Physical and Engineering Sciences, 370(1974), 4380–4403.

Huimin Liu. (1999). Fundamental Phenomena and Principles in Droplet Processes. In Science and Engineering of Droplets (pp. 121–237). William Andrew.

IPCC. (2023). Climate change 2023: synthesis report. https://www.ipcc.ch/report/ar6/syr/

Keith, D. W., Holmes, G., Angelo, D. St., & Heidel, K. (2018). A Process for Capturing CO2 from the Atmosphere. Joule, 2(8), 1573–1594

Mazzotti, M., Baciocchi, R., Desmond, M. J., & Socolow, R. H. (2013). Direct air capture of CO2 with chemicals: Optimization of a two-loop hydroxide carbonate system using a countercurrent air-liquid contactor. Climatic Change, 118(1), 119–135.

Stolaroff, J. K., Keith, D. W., & Lowry, G. V. (2008). Carbon dioxide capture from atmospheric air using sodium hydroxide spray. Environmental Science and Technology, 42(8), 2728–2735.

Ushida, A., Hasegawa, T., & Narumi, T. (2014). Anomalous phenomena in pressure drops of water flows through micro-orifices. Microfluidics and Nanofluidics, 17(5), 863–870.

Zeman, F. (2008). Experimental results for capturing CO2 from the atmosphere. AIChE Journal, 54(5), 1396–1399.

Flavio Manenti, Gintaras V. Reklaitis (Eds.), Proceedings of the 34th European Symposium on Computer Aided Process Engineering / 15th International Symposium on Process Systems Engineering (ESCAPE34/PSE24), June 2-6, 2024, Florence, Italy

Modelling and Parametric Analysis for Improving Technical Performance of Industrial-Scale Basic Oxygen Furnace Gas Fermentation to Isopropyl Alcohol

Gijs J. A. Brouwer[a]*, Haneef Shijaz[a] and John A. Posada[a]

[a]*Delft University of Technology, Department of Biotechnology, Van der Maasweg 9, 2629HZ Delft, The Netherlands*
*g.j.a.brouwer@tudelft.nl

Abstract

The iron and steel industry is responsible for 30% of all industrial CO_2 emissions, largely emitted via hot Basic Oxygen Furnace (BOF) gas (CO, H_2, CO_2). Gas fermentation can convert the BOF gas into valuable chemicals such as the disinfectant and platform chemical isopropyl alcohol (IPA), which is currently only produced with petrochemical cracking. The goal of this research was to model the state-of-the-art industrial-scale BOF gas fermentation to IPA and identify the key process parameters affecting technical performance. The designed and modelled industrial-scale process was based on the published LanzaTech technology with *Clostridium autoethanogenum*. In the process model, the IPA is purified through extractive distillation with pure glycerol as an entrainer. During sensitivity analysis, eleven process parameters were investigated for their effect on the eighteen chosen technical Key Performance Indicators (KPIs). These process parameters are (product selectivity ($Y_{IPA/CO}$), CO volumetric mass transfer rate (VMT_{CO}), CO conversion, reactor dilution rate (D), temperature off-gas condenser, biomass separation liquid loss, extractive distillation glycerol mole fraction, extractive distillation molar reflux ratio, glycerol recycle purge, broth recycle purge and anaerobic digestion waste conversion). The sensitivity analysis identified that the key technical parameters affecting the KPIs are the gas fermentation parameters (CO conversion, VMT_{CO}, $Y_{IPA/CO}$, and D) as well as the biomass filtration liquid loss. Moreover, increasing CO conversion, VMT_{CO}, $Y_{IPA/CO}$ as well as decreasing D and the biomass filtration liquid loss all individually had the greatest positive impact on the KPIs. Thus, this study has successfully synthesised and modelled a state-of-the-art industrial-scale BOF gas fermentation to IPA process and identified the key process parameters to improve its technical process performance. These findings can be used both to optimise the BOF gas fermentation to IPA process, and perform further economic evaluations and environmental impact assessment.

Keywords: BOF gas fermentation, Isopropyl Alcohol (IPA), process modelling, sensitivity analysis, technical performance.

1. Introduction

The iron and steel industry is responsible for 30 % of all industrial CO_2 emissions (IEA, 2020). A large portion of CO_2 is emitted through flaring of the energy-rich off-gas containing CO, H_2 and CO_2 called Basic Oxygen Furnace (BOF) gas. However, BOF gas can be converted, through gas fermentation, into valuable chemicals such as the disinfectant and platform chemical isopropyl alcohol (IPA) (Liew *et al.*, 2022). Currently, there is no green alternative for the polluting petrochemical production of IPA through cracking. BOF gas fermentation to IPA could serve this growing IPA market with a sustainable alternative.

In 2020, LanzaTech commercialised industrial-scale BOF gas fermentation to ethanol using a genetically engineered and patented (Simpson *et al.*, 2012), prototrophic (Annan *et al.*, 2019), anaerobic acetogen (*Clostridium autoethanogenum*, *C. autoethanogenum*) (Köpke and Simpson, 2020). Usually, the gas-liquid mass transfer is limiting the reaction rate and volumetric productivity (q_i) of gas fermentations. However, alcohols like ethanol have gas-liquid mass transfer enhancing properties (Puiman *et al.*, 2022a), giving a CO volumetric mass transfer rate (VMT_{CO}) up to 8.5 g_{CO}/L/h for the industrial-scale external-loop gas-lift reactor used by LanzaTech (Puiman *et al.*, 2022b). The effect of IPA on the mass transfer was not investigated, but according to Keitel and Onken (1982, p. 94) the mass transfer-enhancing properties of ethanol and isopropanol (i.e., IPA) are similar. More recently, LanzaTech published promising results of a pilot-scale process to produce IPA from BOF gas (Liew *et al.*, 2022). Nonetheless, the industrial-scale IPA process and the key parameters affecting its performance are not yet known. Therefore, the goal of this research was to model the state-of-the-art industrial-scale BOF gas fermentation to IPA and identify the key process parameters affecting technical performance. First, the 46,000 $kton_{IPA}$/year (Liew *et al.*, 2022) industrial-scale production process was modelled (see section 2) and its technical performance assessed (see section 3). Then, a sensitivity analysis was performed to identify the key process parameters to improve technical process performance (see section 4). Lastly, conclusions and recommendations are given (see section 5).

2. Modelling of Industrial-Scale Basic Oxygen Furnace Gas Fermentation to Isopropyl Alcohol

This section describes the modelling of the industrial-scale state-of-the-art BOF gas fermentation to IPA, including heat integration and recycles, in Aspen Plus V12.0. The most relevant process units and conditions are schematically given (see Figure 1). The property method used is Non-Random Two-Liquid (NRTL), with Mix Non-Conventional (MIXNC) for the gas fermentation and biomass filtration. The model was based on the published LanzaTech technology (Liew *et al.*, 2022; Köpke and Simpson, 2020; Handler *et al.*, 2016) with the genetically engineered, prototrophic *C. autoethanogenum* (Simpson *et al.*, 2012; Annan *et al.*, 2019). The gas fermentation stoichiometry was approached as a black box and obtained through thermodynamics (Heijnen and Van Dijken, 1992), assuming only acetic acid and biomass as by-products. This resulted in an overall black box stoichiometry (see Equation 1) as input for the stoichiometric reactor (RStoich) in the model. Acetone and ethanol were not included as by-products due to their minor fraction produced (Liew *et. al.,* 2022), but were accounted for with the other by-products. Besides, reaction kinetics were not included, since the gas fermentation was assumed limited by the volumetric mass transfer of CO (VMT_{CO}; Liew *et al.*, 2022).

$$-8.24\,CO - 1.58\,H_2 - 2.63\,H_2O - 0.0462\,NH_3 - 0.111\,NaOH \rightarrow$$
$$0.185\,CH_{1.75}O_{0.5}N_{0.25} + 1.00\,IPA + 4.83\,CO_2 + 0.111\,Acetic\ Acid + \tag{1}$$
$$0.111\,Na^+$$

The applied industrial-scale reactor conditions and mass transfer rate were described by Puiman *et al.* (2022b). The product recovery mentioned by Liew *et al.* (2022) was split into biomass cross-flow filtration and extractive distillation (Simpson *et al.*, 2012), with pure glycerol as an entrainer (see Figure 1). The purges, to prevent accumulation of impurities, were anaerobically digested (Figure 1; Liew *et al.*, 2022). The biogas resulting from the anaerobic digestion and the fermenter off-gas is combusted with an excess of O_2, to generate heat as a utility for internal use (see Figure 1).

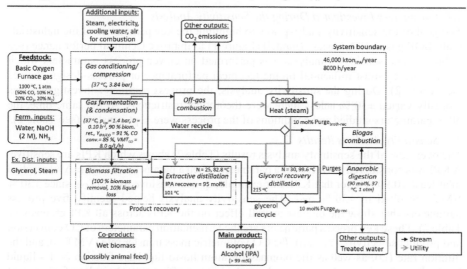

Figure 1: The schematic overview of the Basic Oxygen Furnace gas fermentation to isopropyl alcohol (IPA) adapted and extended from Liew *et al.* (2022). D: dilution rate, biom. ret.: biomass retention, $Y_{IPA/CO}$: product selectivity, CO conv.: CO conversion, VMT_{CO}: volumetric mass transfer rate of CO in the reactor, N: number of distillation stages.

3. Technical Performance Assessment of the Industrial-Scale Basic Oxygen Furnace Gas Fermentation to Isopropyl Alcohol

The key performance indicators (KPIs) were selected based on their indication of gas fermentation performance, overall substrate-to-product performance, or technical performance. Besides productivity and carbon efficiency, this leads to KPIs such as the consumption of chemicals/ substrates, the consumption of utilities, as well as the generation of wastes and process-related CO_2 emissions. Moreover, KPIs that influence economics are the required reactor volume ($V_{Reactor}$) and the IPA titer. Together, these eighteen KPIs indicate the overall technical process performance (see Table 1).

Table 1: Key Performance Indicator (KPI) results of the base case scenario of the BOF gas fermentation to IPA process model. Furthermore, the VMT_{CO}: CO volumetric mass transfer rate, R_{IPA}: IPA production rate, $V_{Reactor}$: required reactor volume, q_{IPA}: volumetric productivity, and glycerol$_{MF-ED}$: mole fraction of glycerol at the extractive distillation.

Parameters	Base Case	Units	KPI Results	Base Case	Units	KPI Results	Base Case	Units
Set parameters:			*Process KPIs:*			*Emission & waste:*		
VMT_{CO}	8.000	g_{CO}/L/h	R_{IPA}	5755	kg_{IPA}/h	CO_2 Off-gas	4.387e+4	kg/h
CO conversion	0.850	Mole frac.	Carbon Efficiency	3.022e-1	Mole frac.	CO_2 Biogas	9778	kg/h
CO Feed (BOF gas)	2.662e+4	kg/h	$V_{Reactor}$	2829	m³	Process CO_2 emission	5.364e+4	kg/h
Glycerol$_{MF-ED}$	5.131e-2	Mole frac.	q_{IPA}	2.083	g_{IPA}/L/d	Waste in Wastewater	1459	kg/h
Dilution rate	0.100	h⁻¹	IPA titer	22.30	g/L	Biomass Out	457.9	kg/h
Product selectivity	9.101e-1	Cmole frac.						
Purities:			*Feeds:*			*Utility consumption:*		
IPA purity	9.913e-1	mass%	Feed Glycerol	7440	kg/h	Electricity	8835	kW
Glycerol rec. purity	9.590e-1	mass%	Feed Water	28.35	m³/h	Cooling water	1050	-GJ/h
			Feed NaOH	435.8	kg/h	LP-Steam	118.1	GJ/h
			Feed NH₃	102.2	kg/h	HP-Steam	424.3	GJ/h

4. Sensitivity Analysis of the Industrial-Scale Basic Oxygen Furnace Gas Fermentation to Isopropyl Alcohol

4.1. Parameters Investigated During the Sensitivity Analysis

The goal of the sensitivity analysis was to highlight the key parameters of the industrial-scale BOF gas to IPA process (model) (Figure 1) to improve technical performance (see Table 1). The sensitivity analysis was performed on eleven process parameters which were expected most influential on the technical performance or process design choices (see Table 2). During the sensitivity analysis, the process parameter investigated was typically varied ±30 % unless there were theoretical restrictions. At the same time, the other parameters studied and conditions of the process were kept constant (see Table 2).

4.2. Sensitivity Analysis Results

The overview of the sensitivity analysis results (Table 2) shows which KPIs are sensitive. A KPI was considered sensitive for a parameter, when a parameter change ($\pm\Delta\%$) resulted in at least $\pm 0.90\ \Delta\%$ for that KPI. For example, q_{IPA} is sensitive to VMT_{CO}, since +30 % VMT_{co} resulted in +30 % q_{IPA}. The sensitivity analysis helped identify five process parameters that show the biggest overall effect on the KPIs across all KPI groups (see Table 2). These key process parameters are the fermentation parameters of CO conversion and product selectivity ($Y_{IPA/CO}$), the CO volumetric mass transfer rate (VMT_{CO}), and the dilution rate (D), as well as the biomass filtration liq-to-liq phase fraction (= 1 - liquid loss). These key process parameters have the biggest effect on technical performance of the industrial-scale BOF gas-to-IPA fermentation process and are a good focus point to improve technical performance. Based on the sensitivity analyses, the following trends were found:

Higher CO conversion and $Y_{IPA/CO}$ improve technical performance (see Table 2). Both CO conversion and $Y_{IPA/CO}$ influence most KPIs (Table 2) of which the most important economic indicators are R_{IPA}, carbon efficiency and $V_{Reactor}$. However, $V_{Reactor}$ increases along with R_{IPA} and results in more utility consumption. Therefore, at higher CO conversion the indicated q_{IPA} does not increase, only relatively more BOF gas is used. Besides, technical performance improves at a higher $Y_{IPA/CO}$ and less by-product is formed, consequently reducing the utility consumption, waste and process-related CO_2 emissions (Table 2). Lower dilution rate seems to improve technical performance (see Table 2). Overall, utility consumption, waste and process-related CO_2 emissions increase with increasing D, while the R_{IPA} decreases and vice versa (Table 2). However, the increase in HP-steam, water, and glycerol consumption at a lower D might provide a trade-off against the R_{IPA} and carbon efficiency. Also, at lower D an IPA titer above 25 gIPA/L (Köpke *et al.*, 2016) can, in reality, limit process performance due to product inhibition. Additionally, Wang *et al.* (2023) has shown that an alcohol titer has an optimum mass transfer enhancing effect, where a higher titer reduces the mass transfer. Thus, for this process, the IPA titer likely has an optimum around 25 g_{IPA}/L. Higher CO volumetric mass transfer rate improved technical performance (see Table 2). Overall, at a higher VMT_{CO} the KPIs indicate increased technical performance, generating less waste and relatively more product (q_{IPA}) in a more concentrated process (lower $V_{Reactor}$) (Table 2). Lower biomass filtration liquid loss seems to improve technical performance (see Table 2). A lower biomass filtration liquid loss (liq-to-liq frac., Table 2) resulted in the KPIs indicating improved technical performance. Because of the reduction of broth loss, consequently less product loss. As a result, the R_{IPA} increased and the consumption of fresh feeds reduced. However, utility consumption increased, possibly giving a trade-off if the utility consumption increases more than the R_{IPA}. Extractive distillation affects technical performance, with a lower molar reflux ratio ($RR_{mol,ED}$) indicating improved technical performance (see Table 2). The investigated parameters of the extractive distillation (Gly_{MF-ED} and $RR_{mol,ED}$) showed that the $C_{IPA,Reactor}$, glycerol feed, and CO_2 biogas are sensitive to the Gly_{MF-ED} (Table 2). Whereas, only the LP-steam consumption is sensitive to the $RR_{mol,ED}$. A 30 % lower $RR_{mol,ED}$ decreases the LP-steam 1.38 times as much, while R_{IPA} and carbon efficiency only change -1.13 % (Table 2). Combining these insights, the extractive distillation itself is considered important for process performance.

Higher temperature of the off-gas condenser and a lower purified glycerol purge fraction improved technical performance (see Table 2). The off-gas condenser showed less utility consumption and little reduction in performance at increased temperature, thus higher $T_{off-gas\ cond.}$ could improve technical performance (see Table 2). Whereas, technical performance improved for a lower glycerol purge fraction, giving less glycerol consumption, waste production, and process-related CO_2 emissions (see Table 2). Usually, there is a trade-off in KPIs for each parameter. These KPIs have opposite trends and both affect the technical process performance. This makes the overall effect of a parameter change hard to predict.

Table 2: Sensitivity analysis results. The lower boundary, base case, and upper boundary results were linearly fitted. If the adjusted $R_2 \geq$ 0.95, then the linear fit slope is given. Otherwise, the resulting Δ%. VMT_{CO}: CO volumetric mass transfer rate, the liq-to-liq frac.: 1 - liquid loss during biomass filtration, Gly_{MF-ED}: extractive distillation glycerol molefraction, $RR_{mol,ED}$: molar reflux ratio during extractive distillation. R_{IPA}: IPA production rate, q_{IPA} volumetric productivity, LP-steam: low-pressure steam, and HP-steam: high-pressure steam. A sensitivity > ±0.90 Δ% of the parameter is bold.

KPI Group	KPI	Gas Fermentation				Off-gas treatment	Microfiltration
		CO conc. (±5 %)	VMT_{co} (±30 %)	Product selectivity (±5 %)	Dilution rate (±30 %)	$T_{off-gas\ cond.}$ (R_x -1.80 %, +11.50 %)	liq-to-liq frac. (±1 %)
Process KPIs	R_{av}	0.99 (R²= 1.000)	1.02 (R²= 0.990)	1.04 (R²= 0.990)	LB -0.08 %; UB +1.78 %	**-3.56 (R²= 1.000)**	1.04 (R²= 1.000)
	Carbon efficiency	0.99 (R²= 1.000)	1.02 (R²= 0.990)	1.07 (R²= 0.988)	LB -0.08 %; UB -0.10 %	0.70 (R²= 1.000)	1.04 (R²= 1.000)
	q_{IPA}	No effect	1.00 (R²= 1.000)	No effect	No effect	0.86 (R²= 1.000)	No effect
	IPA titer	0.06 (R²= 1.000)	1.04 (R²= 1.000)	-0.02 (R²= 1.000)	-0.13 (R²= 0.979)	No effect	0.74 (R²= 0.997)
	Vinasse	1.00 (R²= 1.000)	No effect	No effect	No effect	No effect	No effect
Feeds	Water feed	0.98 (R²= 1.000)	-1.14 (R²= 0.953)	3.70 (R²= 1.000)	1.04 (R²= 1.000)	LB -0.08 %; UB +1.78 %	**-3.56 (R²= 1.000)**
	Glycerol feed	0.99 (R²= 1.000)	-1.10 (R²= 0.950)	0.13 (R²= 0.092)	1.00 (R²= 0.999)	LB 0.00 %; UB -0.10 %	0.70 (R²= 1.000)
	NaOH feed	1.00 (R²= 1.000)	-16.97 (R²= 1.000)	-16.97 (R²= 1.000)	1.00 (R²= 0.999)	LB 0.00 %; UB +0.03 %	0.86 (R²= 1.000)
	NH3 feed	1.00 (R²= 1.000)	No effect	-0.70 (R²= 1.000)	No effect	No effect	No effect
Utility consumption	Electricity	-0.18 (R²= 0.999)	-0.12 (R²= 0.061)	0.09 (R²= 0.999)	0.11 (R²= 0.997)	-0.77 (R²= 0.983)	No effect
	Cooling water	0.98 (R²= 0.998)	-0.65 (R²= 0.958)	-1.26 (R²= 0.999)	0.59 (R²= 0.997)	LB -0.05 %; UB -0.07 %	0.20 (R²= 1.000)
	LP-steam	1.28 (R²= 1.000)	LB -14.39 %; UB +4.24 %	-0.16 (R²= 0.970)	0.19 (R²= 0.987)	LB -0.05 %; UB -0.32 %	1.44 (R²= 1.000)
	HP-steam	1.85 (R²= 1.000)	-1.33 (R²= 0.968)	-3.48 (R²= 1.000)	1.21 (R²= 0.991)	LB 0.00 %; UB -0.14 %	0.77 (R²= 1.000)
Emission & waste	CO_2 off-gas	-0.34 (R²= 1.000)	No effect	0.09 (R²= 0.999)	0.11 (R²= 0.997)	-0.77 (R²= 0.983)	No effect
	CO_2 biogas	1.00 (R²= 1.000)	LB +9.17 %; UB -21.36 %	-1.26 (R²= 0.999)	0.59 (R²= 0.992)	LB -0.08 %; UB -0.07 %	1.61 (R²= 1.000)
	Process CO_2 emission	-0.09 (R²= 1.000)	LB +8.96 %; UB -3.89 %	-0.16 (R²= 0.970)	0.19 (R²= 0.987)	0.29 (R²= 0.999)	No effect
	Waste in wastewater	0.99 (R²= 1.000)	LB +46.37 %; UB -18.69 %	-3.48 (R²= 1.000)	0.33 (R²= 0.999)	1.24 (R²= 1.000)	No effect
	Biomass Out	1.00 (R²= 1.000)	No effect	-0.70 (R²= 0.999)	0.43 (R²= 0.984)	No effect	No effect

KPI Group	KPI	Extractive distillation		Purges before recycle		Anaerobic digestion
		Gly_{MF-ED} (±30 %)	$RR_{mol,ED}$ (±30 %)	Glycerol Purge (±30 %)	Fresh Purge (±30 %)	Waste conc. frac. (±10 %)
Process KPIs	R_{av}	LB -1.23 %; UB +0.24 %	LB -0.36 %; UB +0.03 %	LB -0.08 %; UB +0.27 %	**0.95 (R²= 1.000)**	LB 0.00 %; UB +0.11 %
	Carbon efficiency	**1.07 (R²= 1.000)**	LB -0.27 %; UB +0.02 %	**1.00 (R²= 1.000)**	LB -0.63 %; UB +0.35 %	LB 0.04 %; UB -0.56 %
	q_{IPA}	No effect	No effect	No effect	No effect	No effect
	IPA titer	LB +34.80 %; UB -1.81 %	LB +6.69 %; UB -0.63 %	-0.08 (R²= 0.995)	-0.09 (R²= 0.956)	-0.28 (R²= 1.000)
	Vinasse	No effect	No effect	No effect	No effect	No effect
Feeds	Water feed	LB -1.23 %; UB +0.24 %	LB -0.36 %; UB +0.03 %	-0.08 (R²= 0.995)	**0.95 (R²= 1.000)**	0.12 (R²= 1.000)
	Glycerol feed	**1.07 (R²= 1.000)**	LB -0.27 %; UB +0.02 %	0.30 (R²= 0.092)	LB -0.63 %; UB +0.35 %	0.02 (R²= 1.000)
	NaOH feed	No effect	No effect	LB -5.23 %; UB +12.68 %	No effect	-0.08 (R²= 1.000)
	NH3 feed	No effect	No effect	LB -4.83 %; UB +1.52 %	No effect	No effect
Utility consumption	Electricity	LB -0.06 %; UB 0.00 %	LB -1.13 %; UB +0.08 %	0.11 (R²= 1.000)	LB 0.00 %; UB -0.13 %	0.12 (R²= 1.000)
	Cooling water	LB -1.70 %; UB +0.69 %	LB -1.13 %; UB +0.08 %	LB -1.26 %; UB +0.05 %	LB -0.33 %; UB 0.16 %	0.02 (R²= 1.000)
	LP-steam	-0.14 (R²= 0.998)	LB -0.06 %; UB +0.03 %	LB -0.06 %; UB +0.03 %	LB +0.96 %; UB +2.55 %	-0.06 (R²= 1.000)
	HP-steam	LB -2.21 %; UB -7.10 %	-0.01 (R²= 0.950)	-0.01 (R²= 0.997)	-0.27 (R²= 1.000)	-0.28 (R²= 1.000)
Emission & waste	CO_2 off-gas	No effect	LB -0.36 %; UB +0.03 %	-0.08 (R²= 0.995)	0.95 (R²= 1.000)	0.12 (R²= 1.000)
	CO_2 biogas	0.96 (R²= 0.994)	LB +0.27 %; UB +0.02 %	-0.63 (R²= 0.999)	No effect	1.01 (R²= 1.000)
	Process CO_2 emission	0.18 (R²= 0.994)	LB +0.18 %; UB +0.01 %	0.18 (R²= 1.000)	No effect	0.29 (R²= 1.000)
	Waste in wastewater	0.54 (R²= 0.998)	LB 0.15 %; UB -0.01 %	0.51 (R²= 1.000)	0.05 (R²= 0.979)	-5.14 (R²= 1.000)
	Biomass Out	No effect	No effect	No effect	No effect	No effect

5. Conclusions

A state-of-the-art industrial-scale BOF gas fermentation to IPA was modelled based on the LanzaTech pilot reported in the literature. IPA purification using extractive distillation is an energy- and glycerol-intensive process with room for improvement. Besides, the process KPIs (R_{IPA}, carbon efficiency, $V_{Reactor}$, q_{IPA}, IPA titer), feeds, utility consumption, process-related CO_2 emission, and waste generated could be used to identify the key parameters of the BOF gas fermentation to IPA. The key parameters identified are the gas fermentation parameters (CO conversion, VMT_{CO}, $Y_{IPA/CO}$, and D) and the liquid loss during biomass filtration. The technical process performance could be improved by increasing either the CO conversion, VMT_{CO}, or $Y_{IPA/CO}$, and decreasing D or the biomass filtration liquid loss. These key parameters can be used to optimise the technical process performance. However, this is just an indication based on the process KPIs, feeds, utility consumption, process-related CO_2 emissions and waste generated. To further assess the process performance, a Techno-Economic Evaluation and Life-Cycle Assessment should be done. These insights can be combined to assess the overall (i.e., technical, economic and environmental) performance of the BOF gas fermentation to IPA process.

References

F. J. Annan, B. Al-Sinawi, C. M. Humphreys, R. Norman, K. Winzer, M. Köpke, S. D. Simpson, N. P. Minton, A. M. Henstra, 2019, Engineering of vitamin prototrophy in clostridium ljungdahlii and clostridium autoethanogenum, Applied Microbiology and Biotechnology, 103, 4633–4648

R. M. Handler, D. R. Shonnard, E. M. Griffing, A. Lai, I. Palou-Rivera, 2016, Life cycle assessments of ethanol production via gas fermentation: Anticipated greenhouse gas emissions for cellulosic and waste gas feedstocks, Industrial and Engineering Chemistry Research, 55, 3253–3261

J. Heijnen, J. Van Dijken, 1992, In search of a thermodynamic description of biomass yields for the chemotrophic growth of microorganisms, Biotechnology and Bioengineering, 39, 833–858

International Energy Agency, 2020, Iron and steel technology roadmap, https://www.iea.org/reports/iron-and-steel-technology-roadmap

G. Keitel, U. Onken, 1982, The effect of solutes on bubble size in air-water dispersions, Chemical Engineering Communications, 17, 85–98

M. Köpke, S. Simpson, F. Liew, W. Chen, 6 2016, Fermentation process for producing isopropanol using a recombinant microorganism, US Patent 9,365,868

M. Köpke, S. D. Simpson, 2020, Pollution to products: recycling of 'above ground' carbon by gas fermentation, Current Opinion in Biotechnology, 65, 180–189

F. E. Liew, R. Nogle, T. Abdalla, B. J. Rasor, C. Canter, R. O. Jensen, L. Wang, J. Strutz, P. Chirania, S. De Tissera, A. P. Mueller, Z. Ruan, A. Gao, L. Tran, N. L. Engle, J. C. Bromley, J. Daniell, R. Conrado, T. J. Tschaplinski, R. J. Giannone, R. L. Hettich, A. S. Karim, S. D. Simpson, S. D. Brown, C. Leang, M. C. Jewett, M. Köpke, 2022, Carbon-negative production of acetone and isopropanol by gas fermentation at industrial pilot scale, Nature Biotechnology, 40 (3), 335–344

L. Puiman, B. Abrahamson, R. G. van der Lans, C. Haringa, H. J. Noorman, C. Picioreanu, 2022b, Alleviating mass transfer limitations in industrial external-loop syngas-to-ethanol fermentation, Chemical Engineering Science, 259, 117770

L. Puiman, M. P. Elisiário, L. M. Crasborn, L. E. Wagenaar, A. J. Straathof, C. Haringa, 2022a, Gas mass transfer in syngas fermentation broths is enhanced by ethanol, Biochemical Engineering Journal, 185, 108505

S. D. Simpson, M. Köpke, F. Liew, W. Y. Chen, 2012, Recombinant microorganisms and uses therefor, patent Number: WO2012/115527A2

Y. Wang, X. Shen, H. Zhang, T. Wang, 2023, Marangoni effect on hydrodynamics and mass transfer behavior in an internal loop airlift reactor under elevated pressure, AIChE Journal

Flavio Manenti, Gintaras V. Reklaitis (Eds.), Proceedings of the 34th European Symposium on Computer Aided Process Engineering / 15th International Symposium on Process Systems Engineering (ESCAPE34/PSE24), June 2-6, 2024, Florence, Italy

Model-Based Determination of a Stationary Phase Gradient in Liquid Chromatography Using Optimal Control Theory

Alexander Eppink,[a] Heiko Briesen[a,*]

[a]Technical University of Munich, Process Systems Engineering, Gregor-Mendel-Straße 4, 85354 Freising, Germany
*heiko.briesen@tum.de

Abstract

Several studies have shown that a stationary phase with a gradient of the ion exchange capacity or particle diameter can improve the column efficiency in liquid chromatography operating in gradient elution mode. However, no general method exists to obtain the optimal stationary phase gradient. This contribution closes this methodological gap by using optimal control theory. We combine control parametrization with a direct-single-shooting method to obtain optimal stratified stationary phases. We demonstrate our approach with an ion exchange chromatographic column. For the chosen case study, the optimal ion exchange capacity gradient predicts an increase of the separation efficiency by 36.3 % compared to a homogeneous column.

Keywords: optimal control, liquid chromatography, stationary phase gradient, efficiency

1. Introduction

In analytical liquid chromatography, the goal is to reach the most efficient columns possible. In general, homogeneous columns with constant properties in the axial direction are used. The required resolution is achieved by regulating the mobile phase composition, temperature, column length, or flow rate.

Nevertheless, using series-coupled columns with different properties has been established in recent years (Alvarez-Segura et al., 2016). The essential disadvantage of coupled columns is the additional void volume between the columns. Therefore, the concept of continuous graded stationary phases has been developed in several articles. Codesido et al. (2019) concluded that chromatographic packings with linear particle size gradients increase the efficiency compared to homogeneous packings. Note that we here use the established terminology in chromatography. The gradient here is not to be confused with the mathematical gradient of the profile but refers to the profile itself. Horváth et al. (2021) investigated convex and concave functions in addition to linear particle size gradients and found that convex gradients improve the column performance even more. In another contribution, Horváth et al. (2023) numerically investigated a gradient of an ion exchange chromatographic column.

However, there is no generally established method to actually design an optimal stationary phase gradient. We present a model-based approach that uses optimal control theory to determine an optimal continuous property of the stationary phase as a function of the axial position in the column.

2. Case study

We use an ion exchange chromatographic column to demonstrate that the presented method is highly effective. We aim to increase the retention time difference between two components by determining an optimal function of the ion exchange capacity Q in the axial direction z. In order to describe the chromatograms of the two components at the outlet of the column, we use the equilibrium dispersive model in Eq. (1). For more information regarding the model equations, the reader is referred to Horváth et al. (2023).

$$\frac{\partial c_i}{\partial t} = -\frac{u_0}{1 + k_i}\frac{\partial c_i}{\partial z} + \frac{D_{ax}}{1 + k_i}\frac{\partial^2 c_i}{\partial z^2}, \quad i \in \{1,2\} \tag{1}$$

c_i is the concentration of the corresponding component i; t is the time variable; u_0 is the mobile phase velocity; D_{ax} is the axial dispersion coefficient. The retention factor k_i of component i is calculated by Eq. (2).

$$k_i = \frac{1 - \varepsilon}{\varepsilon} K_i \left(\frac{Q}{\varphi}\right)^{n_i}, \quad i \in \{1,2\} \tag{2}$$

ε is the internal column porosity; K_i is the ion exchange selectivity of component i; n_i is the charge of the solute ion of component i. The eluent concentration φ is due to the gradient elution mode a function of time t and axial position z (Eq. (3)).

$$\varphi = \varphi_0 + \theta \cdot \left(t - \frac{z}{u_0}\right) \tag{3}$$

φ_0 is the initial eluent concentration; θ is the parameter describing the constant slope of the gradient elution mode.

To solve Eq. (1), we need to formulate an initial condition and boundary conditions. At time $t = 0$, the column is filled with pure eluent and is flowed through by an inlet stream, which contains both components. From $t = t^{switch}$ onwards, only pure eluent flows through the column. The outlet of the column is implemented as a constant flux. This results in the initial condition in Eq. (4) and the boundary conditions in Eq. (5) - (6).

$$c_i(t = 0, z) = 0 \quad \forall \; z \in [0, L], \; i \in \{1,2\} \tag{4}$$

$$c_i(t, z = 0) = \begin{cases} c_i^{in} & \forall \; t \in [0, t^{switch}] \\ 0 & \forall \; t \in (t^{switch}, t^{end}] \end{cases}, \quad i \in \{1,2\} \tag{5}$$

$$\frac{\partial c_i(t, z = L)}{\partial z} = 0 \quad \forall \; t \in [0, t^{end}], \; i \in \{1,2\} \tag{6}$$

L is the total length of the column; c_i^{in} is the inlet concentration of component i; t^{switch} is the time point when the inlet stream changes to pure eluent; t^{end} is the end time of the simulation.

3. Optimization criterion

The case study aims to maximize the separation of two components through an optimally chosen gradient of the ion exchange capacity Q. To reach the maximal separation, the retention time $t_{r,1}$ of the earlier eluting component can be minimized, and the retention time $t_{r,2}$ of the later eluting component can be maximized. Since increasing the retention time $t_{r,2}$ of the second component also increases the total analysis time, we want to prevent this solution by the formulation of the objective functional J. Objective functional J maximizes the separation of the two components $(t_2 - t_1)$, and an additional term A becomes nonzero if the retention time $t_{r,2}$ exceeds the maximum allowable retention time t_r^{max}. Moreover, we define a range of available minimal and maximal ion exchange capacities Q^{min} and Q^{max}.

$$\min J = -(t_2 - t_1) + A \tag{7}$$

$$\text{s. t. } Eq.\,(1) - (6) \tag{8}$$

$$t_{r,2} \leq t_r^{max} \Rightarrow A = 0, \quad t_{r,2} > t_r^{max} \Rightarrow A = 10^5 \cdot t_{r,2} \tag{9}$$

$$Q \geq Q^{min} \tag{10}$$

$$Q \leq Q^{max} \tag{11}$$

t_1 is the time point when the first component is fully eluted; t_2 is the time point when the second component reaches the outlet (cf. Figure 2).

4. Numerical solution

To obtain the chromatograms of the two different components, we need to solve the partial differential Eq. (1). The method of lines transforms Eq. (1) into a system of ordinary differential equations by discretizing the spatial variable z using 200 equidistantly distributed discretization points. First-order partial derivatives are approximated using a two-point upwind scheme, and second-order partial derivatives are approximated using a three-point central scheme. The resulting system of ordinary differential equations is solved using ode15s in MATLAB (Version R2023a, supplier: The MathWorks, Natick, Massachusetts). The optimal control problem is reduced to a conventional optimization problem by combining control parametrization with a direct-single-shooting method (Goh and Teo, 1988). In the first iteration, it is assumed that the optimal function is linear. Therefore, the ion exchange capacities at the inlet and outlet are optimally determined, with all capacities linearly interpolated between these two points. In the second iteration, a third control variable is added in the middle of the first two. The three control variables are again optimally determined with continuous information linearly interpolated in between. With each iteration, one additional control variable is added between every two control variables. Using this strategy of successive refinement, any continuous function can be approximated. In this work, four iterations, and thus nine control variables are used to find the optimal function of the ion exchange capacity Q in the axial direction z. The reader is referred to our previous contribution for detailed information regarding the optimal control algorithm (Eppink et al., 2023). The numerical gradient-based solver fmincon in MATLAB solves the resulting optimization problem. Table 1 specifies the model parameters chosen in the same range as in Horváth et al. (2023). The first value of the ion exchange selectivity K_i, the charge of the dissolved ion n_i, and the inlet concentration c_i^{in} correspond to the first component, and the second value to the second component, respectively.

Table 1: Numerical parameters.

Parameter	Value
L	$0.2\ [m]$
u_0	$4.17 \cdot 10^{-3}\ [m/s]$
D_{ax}	$1 \cdot 10^{-8}\ [m/s^2]$
ε	$0.4\ [-]$
K_i	$3.36, 44.14\ [-]$
φ_0	$0.001\ [mol/L]$
θ	$1.67 \cdot 10^{-4}\ [mol/(L \cdot s)]$
t_r^{max}	$200\ [s]$
n_i	$1, 2\ [-]$
Q^{min}, Q^{max}	$1 \cdot 10^{-4}, 10 \cdot 10^{-3}\ [eqiv/L]$
c_i^{in}	$0.3, 1.0\ [g/L]$
t^{switch}	$50\ [s]$
t^{end}	$230\ [s]$

5. Results

The goal of the objective functional J is to separate the two components as far as possible with a maximum retention time t_r^{max} of the second component. Figure 1 shows the ion exchange capacity Q as a function of the axial position z for the optimal column and the corresponding homogeneous column. The ion exchange capacity Q is minimal at the inlet of the optimal column and increases with increasing slope in the axial direction z. Note that the optimal gradient of the ion exchange capacity Q is in good agreement with the results of Horváth et al. (2023).

The separation of the two components is much larger using the optimal column than the homogeneous column (Figure 2). The higher resolution of the optimal column is only due to the higher retention time difference between the two components and is not caused by minimizing physical bandwidth effects. Overall, the separation width is increased by 36.3 % using the optimal column compared to the homogeneous column. If no constraint is formulated in the optimization criterion, the solver would further increase the resolution by increasing the retention time $t_{r,2}$ of the later eluting component. However, this would lead to an overall higher analysis time and would not guarantee comparability between the optimal stationary phase gradient and the homogeneous column.

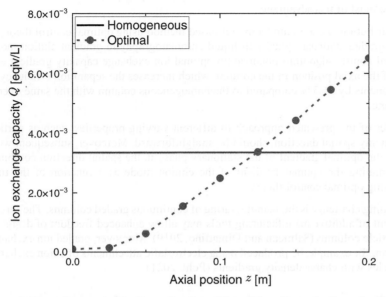

Figure 1: Ion exchange capacity Q as a function of the axial position z for the optimal column and the homogeneous column.

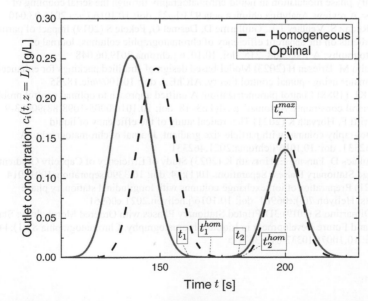

Figure 2: Outlet concentration $c_i(t, z = L)$ of both components over time t for the optimal column and the homogeneous column with the same retention time $t_{r,2}$.

6. Conclusion

Various works have already shown that a gradient of the stationary phase leads to an increase in the efficiency of a column in liquid chromatography in gradient elution mode. We are convinced that model-based methods like the one presented in this work will help to make the most of this advantage.

In this contribution, we presented a model-based method using optimal control theory to determine graded stationary phases in liquid chromatography in gradient elution mode. The optimal control algorithm obtained an optimal ion exchange capacity gradient as a function of the axial position in the column, which increases the separation efficiency of two components by 36.3 % compared to the homogeneous column with the same overall analysis time.

The transfer of the presented approach to different varying properties, such as particle diameter in any spatial direction, should be straightforward. Moreover, subsequent work could find the optimal gradient of the stationary phase in the spatial direction combined with determining the optimal gradient of the elution mode as a function of the time variable using optimal control theory.

One remaining challenge is the manufacturing of continuous graded columns. The further development of additive manufacturing tools may allow enhanced freedom of design in graded particle columns (Salmean and Dimartino, 2019). Moreover, graded ion exchange columns can, for example, be produced using electrostatic attachments of anion exchange latex particles with charge density gradients (Pohl, 2021).

References

Alvarez-Segura T, Torres-Lapasió JR, Ortiz-Bolsico C, García-Alvarez-Coque MC (2016) Stationary phase modulation in liquid chromatography through the serial coupling of columns: A review. Analytica chimica acta 923:1–23. doi: 10.1016/j.aca.2016.03.040

Codesido S, Rudaz S, Veuthey J-L, Guillarme D, Desmet G, Fekete S (2019) Impact of particle size gradients on the apparent efficiency of chromatographic columns. Journal of chromatography. A 1603:208–215. doi: 10.1016/j.chroma.2019.06.048

Eppink A, Kuhn M, Briesen H (2023) Model-based design of stratified packings for enhanced mass transfer using optimal control theory. AIChE J. doi: 10.1002/aic.18285

Goh CJ, Teo KL (1988) Control parametrization: A unified approach to optimal control problems with general constraints. Automatica 24(1):3–18. doi: 10.1016/0005-1098(88)90003-9

Horváth S, Gritti F, Horváth K (2021) Theoretical study of the efficiency of liquid chromatography columns with particle size gradient. Journal of chromatography. A 1651:462331. doi: 10.1016/j.chroma.2021.462331

Horváth S, Lukács D, Farsang E, Horváth K (2023) Study of Efficiency of Capacity Gradient Ion-Exchange Stationary Phases. Separations 10(1):14. doi: 10.3390/separations10010014

Pohl CA (2021) Preparation of ion exchange columns with longitudinal stationary phase gradients. Heliyon 7(5):e06961. doi: 10.1016/j.heliyon.2021.e06961

Salmean C, Dimartino S (2019) 3D-Printed Stationary Phases with Ordered Morphology: State of the Art and Future Development in Liquid Chromatography. Chromatographia 82(1):443–463. doi: 10.1007/s10337-018-3671-5

Flavio Manenti, Gintaras V. Reklaitis (Eds.), Proceedings of the 34th European Symposium on Computer Aided Process Engineering / 15th International Symposium on Process Systems Engineering (ESCAPE34/PSE24), June 2-6, 2024, Florence, Italy

Design and simulation of hydrate-based desalination using R-152a refrigerant.

Sief Addeen Aldroubi [a], Umer Zahid [a,b], Hassan Baaqeel [a,c] *

[a] Department of Chemical Engineering, King Fahd University of Petroleum and Minerals, Dhahran 31261, Saudi Arabia
[b] Interdisciplinary Research Center for Membranes & Water Security, King Fahd University of Petroleum & Minerals, Dhahran 31261, Saudi Arabia
[c] Interdisciplinary Research Center for Renewable Energy & Power Systems, King Fahd University of Petroleum & Minerals, Dhahran 31261, Saudi Arabia
baaqeel@kfupm.edu.sa

Abstract

Innovative desalination technologies are coming into greater focus to meet the mounting problem of securing freshwater resources. Among them, Gas Hydrate-Based Desalination is one of the intriguing possibilities, distinguished by its exceptional water recovery capabilities despite the water's salinity. In this study, R-152a is selected as a hydrate former due to its hydrate formation properties at a high temperature and lower pressures compared to conventional refrigerants. The results show that R-152a is an efficient hydrate formation refrigerant when compared to using propane and methane, resulting in a specific energy consumption of 3.72 kWh/m^3. Moreover, it possesses a low global warming potential (GWP), promoting both energy efficiency and environmental sustainability. Finally, a sensitivity analysis was performed to study the effect of brine concentration on the overall performance of the hydrate-based desalination process.
Keywords: R-152a, hydrate-based desalination, hydrates, energy

1. Introduction

Accessing fresh and clean water is becoming increasingly challenging. To address these difficulties, highlighting desalination becomes crucial. Desalination is a process that eliminates salts and impurities from seawater, offering a consistent supply of pure water and potential solutions to freshwater scarcity. Given that approximately 97% of Earth's water is saline, studying and exploring diverse desalination techniques is imperative. These techniques are broadly categorized into three groups: Thermal-based desalination, Membrane-based desalination, and emerging methods such as Gas-hydrate based desalination and Capacitive Deionization (CDI). Multistage flash (MSF) and multiple effect distillation (MED), the most widely spread thermal based technologies. Even though they are efficient in producing fresh water, their high energy requirements associated with the water phase change limit their application. On the other hand, the most common membrane technology used, accounting for around 60% of the pure water produced is RO desalination (World Bank, 2019). However, RO technology is limited to water with low total dissolved solids (TDS) and its feasibility is weakened by the fact that RO systems require a pre-treatment plant where chemical additives are added to the sea water and high operating pressure reaching 80 bar (Subramani and Jacangelo, 2015). Hence, it is crucial to develop and investigate new desalination technologies like hydrate-based desalination. Hydrate-based desalination is a process where pure water is forced to form a crystal-like structure known as hydrate around a certain guest molecule at temperatures above the water's freezing point. Melting the hydrate produces salt-free water and the hydrate former guest molecule which can be used again in the process. Both formation and dissociation temperature of hydrate are dependent on the hydrate former chosen (Babu et al., 2018). Gas hydrates formed aren't considered chemical compounds; instead, the enclosed guest interacts with water molecules through relatively weak van der Waals forces. A range of studies have explored gas hydrate-based desalination. (Sahu et al., 2018) identified propane and ethane as suitable candidates for this purpose, while (Cha and Seol, 2013) proposed the use of cyclopentane and cyclohexane as secondary hydrate guests to increase gas hydrate formation temperature. (He et al., 2018) focused on improving the efficiency of gas hydrate-based desalination processes by developing a process that utilizes LNG cold energy, significantly reducing energy consumption. It was reported that the resulting specific energy consumption of the process was reduced from 65.14 kWh/m^3 when using an external refrigeration cycle to 0.84 kWh/m^3 when using LNG cold energy to cool the process. (He et al., 2020) proposed a process where cyclopentane was used as a hydrate former to increase the hydrate formation temperature to 280.25 K. However, during the hydrate dissociation cyclopentane tends to form an emulsion requiring a secondary complicated process to separate cyclopentane and water. (McCormack and Andersen, 1995) pioneered the use of R-141b refrigerant as a hydrate former in two Hydrate-based desalination pilot plants in the US. The outcomes showed promise, enabling further exploration into utilizing refrigerants as hydrate formers. Refrigerants, when employed in this role, tend to decrease energy consumption because they typically form at higher temperatures and lower pressures. Recent studies by (Mok et al., 2022) and (Dongre et al., 2022) compared the effectiveness of various refrigerants as hydrate formers, highlighting R-152a as the top choice due to its superior kinetics, lowest Global Warming Potential (GWP) and conversion rates. According to their studies, R-152a forms hydrates at higher temperatures and lower pressures compared to propane and other refrigerants like R-22. Moreover, the kinetics of R-152a remains unaffected by seawater salt concentration, making it a strong candidate for optimizing hydrate-based desalination processes, particularly in countries without LNG imports. This study aims to simulate the hydrate-based desalination process using R-152a as a hydrate former.

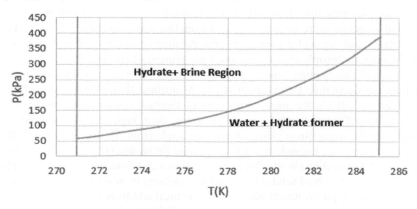

Figure 1: Phase equilibria data for R-152a at 5 wt% NaCl

2. Process design

2.1. Refrigeration cycle

In the phase diagram of R-152a hydrates shown above, it is evident that at a particular pressure, the feed needs to be cooled below the equilibrium temperature for hydrate formation. At our operating pressure of 300 kPa, the equilibrium temperature is approximately 284 K. To ensure hydrate formation, the reaction runs at 283 K. Because the hydrate formation reaction is exothermic, the temperature of the reactor should be controlled at 283 K to prevent hydrate crystal dissociation. An ethylene-glycol cycle which is constantly cooled in heat exchanger-5 is responsible for controlling the temperature; hence preventing the dissociation of hydrate crystals. For fair comparison of the process performance with previous works, the same working fluid and heat exchanger network in the refrigeration cycle is simulated in this work. (He et al., 2018)

2.2. Hydrate based desalination process

2.2.1. Process Description

Figure 2 below shows the process flow diagram of hydrate-based desalination along with the external refrigeration cycle. The simulation details of this work are shown in Table 1. R-152a is usually stored in containers up to 500 kPa at temperature ranges of 293-298 K to remain in gaseous state, this storage condition is beneficial as it reduces the energy required to compress the feed. R-152a (293 K and 310 kPa) is cooled by propane in heat exchanger 4 to reach the hydrate formation conditions. The feed sea water at ambient conditions (298.15 K and 101.3KPa) is pumped to 315 kPa to account for any pressure losses in the pipes and heat exchangers. Then, it undergoes cooling in two successive heat exchangers (1 and 2) by pure water and brine respectively, before being further cooled by propane. The cooled feed (R-152a and seawater) enters the hydrate formation reactor forming hydrate crystals. The reactor effluent containing a mixture of crystals, brine, and excess hydrate former, undergoes separation in a three-phase separator where hydrate crystals are separated from the mixture entering a dissociation vessel. Seawater is used to increase the temperature of the vessel disturbing the equilibrium conditions, producing pure water and R-152a which might be used again in the process.

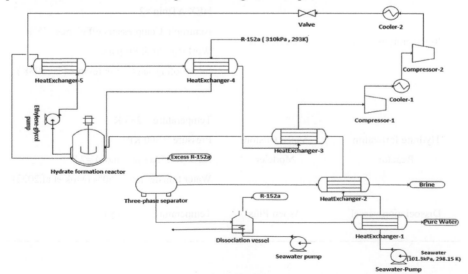

Figure 2: Process flow diagram of hydrate-based desalination

2.2.2. Hydrate formation and dissociation

The hydrate formation reactor was modelled using Aspen Custom modeler. The following reactions are hydrate formation and dissociation reactions respectively.

$$C_2H_4F_2 + 7.7H_2O -> C_2H_4F_2 \cdot (7.7H_2O) \tag{1}$$

$$C_2H_4F_2 \cdot (7.7H_2O) -> C_2H_4F_2 + 7.7H_2O \tag{2}$$

A published hydrate model (Chen and Guo, 1998) was employed to characterize the composition of the refrigerant R-152a within the hydrate structure. The resulting composition of R-152a in the hydrate phase exhibited a value of 0.996, indicating its limited solubility in water. Consequently, no flash calculation was deemed necessary for assessing gas/liquid interactions. This result aligns with existing literature and serves as validation for the model. Furthermore, both mass and energy balances were conducted to ensure the accuracy of the process simulation. Validation of the model was also achieved by comparing the calculated enthalpy of hydrate formation, which measured at -84 kJ/mol, closely resembling the value reported by (Mok et al., 2021) of -82.2 kJ/mol.

$$Mass\ balance: V_{in}y_i + L_{in}x_i = L_{out}x_i + Hx_i^H + V_{out}y_i \tag{3}$$

$$Water\ transfer\ rate: \frac{L_{in}x_i}{L_{out}x_i} \tag{4}$$

Table1. Process simulation details

Unit/ system	Aspen Model	Parameters
Refrigeration cycle	Aspen Plus	HEX ΔTmin =2 Isentropic Compressors efficiency: 75% Working Fluid: Propane Intercooling stage compressors (T= 293K)
Hydrate formation Reactor	Aspen Custom Modeler	Temperature = 283 K Pressure = 300 KPa H2O/ R-152a (mol./mol.) =7.7 Water transfer rate = 60% (Mok et al,2022)
Dissociation vessel	Aspen Plus	Temperature = 298.15 K

3. Results and discussion

3.1. Specific energy consumption

$$SEC = \frac{Total\ Work}{Volume\ of\ Pure\ water\ produced\ per\ hour} \tag{5}$$

It is usually an indicator of the energy requirement of the process. In the case where the hydrate former recycling is not considered, it was reported that when using methane as a hydrate former to purify produced water, the specific energy consumption was 99.6 kWh/m³, given that it was assumed that methane enters the process at 4000 kPa and further compressed to 8000 kPa (Babu et al., 2021). In addition, when using propane as a hydrate former the SEC was reported to be 65.14 kWh/m³ (He et al., 2018). At a water transfer rate of 60% (Mok et al., 2022), 8.32 m³/h of pure water is produced from this proposed process, the specific energy tends to be 3.72 kWh/m³. This proves that using R-152a tends to increase the possibility of commercializing this desalination technology. Figure 3 shows the energy analysis of the process. It can be clearly shown that, the highest energy consuming equipment are the ones in the refrigeration cycle, which imply that if any renewable energy source or sustainable refrigeration system to be integrated with the process instead of the traditional refrigeration cycle, it would decrease the SEC even more.

3.2. Water salinity and specific energy consumption

It is important to analyze the effect of sea water salinity in terms of NaCl concentration on specific energy consumption (SEC). The concentrations of NaCl from 10% to 70% are studied. Referring to figure 4, as the concentration of NaCl increases, SEC increases. This is because as water becomes more saline, less water is produced per hour; hence more energy is consumed per m3 of water. This study ensures that investigating new hydrate formers with high water transfer rate can promote the commercialization of this desalination technique even at high saline solutions.

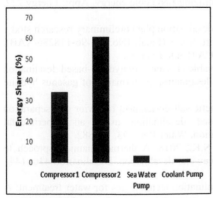

Figure 3: Energy share per equipment

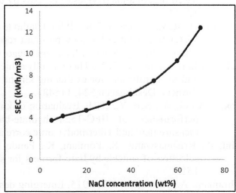

Figure 4: SEC vs Water Salinity

4. Conclusions

Utilizing R-152a as a hydrate former significantly enhances the efficiency of the hydrate-based desalination process by making the hydrate crystal forms at higher temperature and lower pressure. This results in a water transfer rate of 60% and a SEC of 3.72 kWh/m^3. Such enhancement in the performance will make this desalination technique attractive for commercialization. Even though the specific energy consumption decreased drastically when using R-152a, the compressors of the refrigeration cycle are the major contributors to energy consumption, which make exploring the possibility of renewable energy integration crucial to achieve lower energy consumption.

Acknowledgement

The author(s) would like to acknowledge the support provided by the Deanship of Scientific Research at King Fahd University of Petroleum & Minerals (KFUPM).

References

Babu, P., Bollineni, C., Daraboina, N., 2021. Energy Analysis of Methane-Hydrate-Based Produced Water Desalination. Energy Fuels 35, 2514–2519.

Babu, P., Nambiar, A., He, T., Karimi, I.A., Lee, J.D., Englezos, P., Linga, P., 2018. A Review of Clathrate Hydrate Based Desalination To Strengthen Energy–Water Nexus. ACS Sustain. Chem. Eng. 6, 8093–8107.

Cha, J.-H., Seol, Y., 2013. Increasing Gas Hydrate Formation Temperature for Desalination of High Salinity Produced Water with Secondary Guests. ACS Sustain. Chem. Eng. 1, 1218–1224.

Chen, G.-J., Guo, T.-M., 1998. A new approach to gas hydrate modelling. Chem. Eng. J. 71, 145–151.

Dongre, H.J., Deshmukh, A., Jana, A.K., 2022. A thermodynamic framework to identify apposite refrigerant former for hydrate-based applications. Sci. Rep. 12, 16688.

He, T., Chong, Z.R., Babu, P., Linga, P., 2020. Techno-Economic Evaluation of Cyclopentane Hydrate-Based Desalination with Liquefied Natural Gas Cold Energy Utilization. Energy Technol. 8, 1900212.

He, T., Nair, S.K., Babu, P., Linga, P., Karimi, I.A., 2018. A novel conceptual design of hydrate based desalination (HyDesal) process by utilizing LNG cold energy. Appl. Energy 222, 13–24.

McCormack, R.A., Andersen, R.K., 1995. Clathrate desalination plant preliminary research study. Water treatment technology program report No. 5 (Final) (No. PB-96-118286/XAB). Thermal Energy Storage, Inc., San Diego, CA (United States).

Mok, J., Choi, W., Seo, Y., 2022. Theoretically achievable efficiency of hydrate-based desalination and its significance for evaluating kinetic desalination performance of gaseous hydrate formers. Desalination 524, 115487.

Mok, J., Choi, W., Seo, Y., 2021. Evaluation of kinetic salt-enrichment behavior and separation performance of HFC-152a hydrate-based desalination using an experimental measurement and a thermodynamic correlation. Water Res. 193, 116882.

Sahu, P., Krishnaswamy, S., Ponnani, K., Pande, N.K., 2018. A thermodynamic approach to selection of suitable hydrate formers for seawater desalination. Desalination 436, 144–151.

Subramani, A., Jacangelo, J.G., 2015. Emerging desalination technologies for water treatment: A critical review. Water Res. 75, 164–187.

World Bank, 2019. The Role of Desalination in an Increasingly Water-Scarce World. World Bank, Washington, DC.

Flavio Manenti, Gintaras V. Reklaitis (Eds.), Proceedings of the 34th European Symposium on Computer Aided Process Engineering / 15th International Symposium on Process Systems Engineering (ESCAPE34/PSE24), June 2-6, 2024, Florence, Italy

Microfluidic Dynamics coupled with Populational Balance Equations to further describe Water/diesel Microemulsions

Nadia Gagliardi Khouri[*],[a] Rubens Maciel Filho [a]

[a] State University of Campinas (UNICAMP), School of Chemical Engineering, Av. Albert Einstein, 500, Campinas, 13083-852, Brazil.

*khouri.g.nadia@gmail.com

Abstract

Reducing the environmental impact of pollution emissions stemming from diesel engine combustion is a pressing concern. One promising approach is the utilization of water/diesel (W/D) microemulsions that mitigate pollutant emissions due to the microexplosion process. It also elevates combustion efficiency, improves the piston performance, and does not require adaptations on the engine. In a novel work, a phenomenological model was developed to produce (W/D) microemulsions through mechanical agitation aided by emulsifiers. This methodology effectively characterizes the distribution of water droplets through a population balance evaluation, and it allows a precise depiction of the micrometric scale through a refined model. However, a shortcoming of this approach is that the vessel geometry cannot be accounted by only using a populational balance methodology. Therefore, this study aims to apply microfluidics concepts together with the population balance equations. The computational fluid dynamic (CFD) studies were conducted in Ansys Fluent software and the partial differential equations related to the population balance were coupled in the simulation. The mathematical model approach and the software developed in Python were validated by comparing the results obtained with other computational/experimental data. It is worthwhile highlighting that the proposed simulation described in more detail the emulsification process and the impact of the vessel's geometry could be evaluated.

Keywords: Computational fluid dynamics, Population balance, Breakage functions, Fuel upgrading.

1. Introduction and objective

The main advantages of diesel engines are their high energy efficiency, allowing for their use in heavy vehicles. However, the main disadvantage of diesel combustion is the high emission of polluting gases. (Debnath et al., 2015). Especially the nitrogen monoxide (NO) reaction, which occurs at temperatures around 1300 °C (Equation 1) (Driscoll, 1997).

$$N_2 + O_2 \rightarrow 2NO \quad (T\sim1300°C) \tag{1}$$

The emission of nitrogen monoxide can be prevented by lowering the combustion temperature. However, this can affect the engine's performance. Consequently, efforts are made to employ techniques that reduce the temperature without compromising engine efficiency. Diesel upgrading by using a water microemulsion is an effective method to decrease pollutant emissions. This process involves a microexplosion, which lowers the combustion temperature due to the water vaporization's endothermic nature (Patel and Dhiman, 2021).

The water/diesel (W/D) microemulsions can be formed through various processes, but agitation coupled with emulsifiers stands out (Supriyanto et al., 2021). The manufacturing of this microemulsion depends on the operational and physical variables of the process, and these can influence the engine efficiency and pollutant emissions. Experimental tests were employed to examine the W/D microemulsions and their combustion, but there was limited exploration into the formulation of phenomenological models. Recently, an article addressed the use of modeling based on population balances to describe the formation of W/D microemulsions through mechanical homogenization using the Python programming language (Khouri et al., 2023). In general, Khouri et al. (2023) considered the following hypotheses to model the micro-droplets of water in diesel:

- Use of a simplified population balance (Figure 1)
- Test of two breakage functions (g(v)) applied to emulsions consolidated in the literature (Coulaloglou and Tavlarides, 1977; Andersson and Andersson, 2006) to determine the most suitable for the water/diesel system.
- Numerical solution by the Discrete Method.

$$\frac{\partial n(v,t)}{\partial t} = N_g - D_g$$

❖ Water distribution (n) identified by the droplets volume (v);
❖ Time-dependent process (t);
❖ Homogeneous mixture.

❖ The process is described by the birth (N_g) and death (D_g) of droplets by breakage functions (g(v));
❖ Coalescence is not considered.

Figure 1: Simplified populational balance and its considerations.

As modeling results revealed, the breakage function by Coulaloglou and Tavlarides (1977) is suitable only for emulsions at larger scales. Additionally, the breakage model by Andersson and Andersson (2006) did not represent any of the data sets (micrometric and nanometric) – this unsatisfactory result stems from limitations in the model, where only the breakage of millimetric droplets is represented. Therefore, an adjustment was made to the second model to depict the breakage of micrometric droplets. Consequently, the adjusted model yielded reasonable results.

Expanding the models of the W/D system are important, as the combustion process and pollutant emissions can be analyzed in more depth after developing a tool that further describes this formation. Advancements in phenomenological studies can be achieved by extending the modeling through computational fluid dynamics (CFD). Thus, the scope of this work focused on the application of CFD coupled with population balances (PB) in the Ansys Fluent to simulate an agitation system that produce a W/D microemulsion. By using simulation techniques with CFD, the direct influence of the system's geometry and fluid viscosity could be evaluated, addressing limitations of the methodology that solely relies on population balance equations (Becker et al., 2014).

2. Methodology

The simulation was performed in the Ansys V14.5 software, and the objective was to compare it with the results obtained by Khouri et al. (2023). The operational and physicochemical conditions of W/D emulsion homogenization used in this work are described by Table 1 (Dataset A).

Table 1: Dataset A (Rastogi et al., 2019) conditions and results of the W/D microemulsification.

Water volume fraction	5.3% (v/v)
Emulsifier composition	50% Tween© 80 + 50% Span© 80
Interfacial tension	3.8×10^{-3} N/m
Mixing time	2 min
Rotational speed	2500 rpm
Dispersing element diameter	8 mm
Temperature	25°C
Final droplet size diameter	6 ± 2 μm

Initially, the geometries were constructed in CAD (computer-aided design) format using the *Design Modeler* environment. According to the information from Dataset A, the process was conducted in a 10 mL glass vial with a mixing element of 8 mm. Therefore, a cylinder with a diameter of 20 mm and a height of 50 mm was used to represent the glass vial, along with an impeller of a diameter equal to 8 mm. Additionally, in the region around this impeller, a cylinder was added to define and characterize the rotation in the subsequent steps. In other words, the geometry was separated into two cell zones: a stationary domain (outer_fluid) and a mixing domain (inner_fluid).

Next, the system mesh was generated by discretizing the geometry in the *Meshing* environment. Micro-mixing systems can be described using tetrahedral elements using cells in the order of 6.10^5 (Shiea et al. 2022). The Advanced Size Function was set to 'On: Proximity and Curvature' to refine mesh quality. The sizing specified for this geometry is detailed in Table 2. Skewness and orthogonality criteria were considered to assess the quality of the generated mesh. Skewness quantifies the shape of the elements to determine if they possess an equilateral property. Ansys recommends that the skewness must be lower than 0.95. Orthogonality evaluates the angle of the faces/edges of the elements. In the simulator, orthogonal quality varies in values from 0-1, and it is recommended to use values greater than 0.1.

Table 2: Mesh sizing information and characteristics.

Num Cells Across Gap	3
Min Size	4×10^{-3} mm
Proximity Min Size	4×10^{-3} mm
Max Face Size	0,4 mm
Max Size	0,8 mm
Growth Rate	1,2
Minimum Edge Length	0,250 mm
Inflation	Program Controlled

After the mesh generation, the CFD simulation can be implemented in *Fluent*. In summary, the setup used was:

- **General:** transient and gravity of -9.8 m/s².
- **Models:** Multiphase (Eulerian), Viscous (Standard k-epsilon), Population Balance (Discrete).
- **Materials/Phases:** diesel (first phase) and water (second phase).
- **Cell Zone conditions:** Mesh motion = 2500 rpm in the mixing domain (inner_fluid).
- **Mesh interfaces** – contact region between the stationary (outer_fluid) and mixing domain (inner_fluid).
- **Boundary conditions** – none, due to the transient state (i.e., no inlets or outlets were defined).
- **Solution Methods** – Phase Couple SIMPLE.
- **Solution Initialization** – Gauge pressure = 1 atm, Water volume fraction = 0.053, 0-bin fraction = 1, Time step size = 1s, Number of steps = 120, Max iteration/time step = 100.

The population balance was included through the 'addon-module' present in the Fluent library. To use this module, it was necessary to activate the Multiphase (Eulerian) model. The population balance was solved using the discrete method to compare with the results obtained by the baseline study (Khouri et al., 2023). The breakage function used was the Luo model, considering the influence of interfacial tension – an important parameter for the system (Khouri et al., 2023). Flow definition and calculation of its parameters were also developed at this stage. It was observed that the turbulence in systems with homogenization could be adequately represented by the k-epsilon model (Becker et al., 2014). After defining the operational and physical variables according to Table 1, it was possible to obtain the results of the CFD simulations.

3. Results and Discussion

The mesh generated from the data in Table 2 successfully met all the necessary criteria to ensure its quality (Table 3). In other words, the generated elements were above the order of 10^5, the maximum skewness was below 0.95, and the minimum orthogonality was above 0.1. The average and standard deviation of the skewness and orthogonality were also within reasonable values for the system's application.

Table 3: Mesh statistics and quality parameters.

Nodes	303886
Elements	1402469
Min Skewness	1.182×10^{-6}
Max Skewness	0.848
Average Skewness	0.227 ± 0.122
Min Orthogonality	0.183
Max Orthogonality	0.999
Average Orthogonality	0.867 ± 0.088

The CFD simulation obtained a water droplet distribution of 3.698 ± 1550 µm(Figure 2). The residual values during the iterations had a transient behavior. In the final set, the continuity presented a residual of 3×10^{-5} and the bin-fractions residues were in the order of 10^{-4}, indicating an acceptable convergence. The performance of the Luo-model to describe the formation of W/D microemulsions has a Standardizes Mean Difference 1.312 in comparison to the only-population balance model (Khouri et al. 2023).

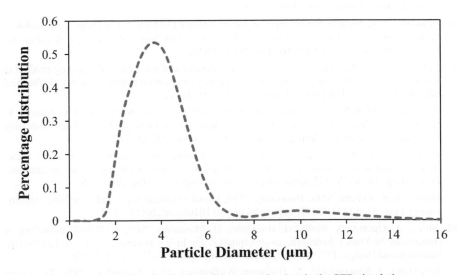

Figure 2: Population distribution of the water droplets in the CFD simulation.

Thess results reflects the potential in using Ansys to represent the W/D microemulsions, but further studies should be conducted to improve the results. A suggestion is to implement the adjusted model in the breakage kernel by using a User-Defined Function (UDF). The challenge to apply this UDF is that Ansys only accepts coding in C language. Therefore, the adjusted model must be re-structured and adapted since its original format is in Python. Other improvement that could be employed in this case, is to use the Standard Method of Moments (SMM) to achieve a convergence in a shorter processing time. The Discrete solution method can be computationally exhausting to systems that require a more refined distribution to be represented, (i.e., number of bins).

4. Conclusion

Therefore, the Ansys Fluent software is a capable approach to describe the behavior of W/D microemulsion by using the Population Balance available in the "addon-module". The proposed simulation described the emulsification process more adequately by considering the impact of the vessel's geometry and the mixture viscosity. Further studies could implement an UDF to represent the breakage and solve by the SMM to refine the results.

References

R. Andersson, B. Andersson, 2006, Modeling the breakup of fluid particles in turbulent flows, AIChE Journal, 52, 2031-2038. https://doi.org/10.1002/aic.10832.

P. J. Becker, F. Puel, A. Dubbelboer, J. Janssen, N. Sheibat-Othman, 2014, Coupled population balance–CFD simulation of droplet breakup in a high pressure homogenizer, Computers & Chemical Engineering, 68, 140-150, https://doi.org/10.1016/j.compchemeng.2014.05.014.

C.A. Coulaloglou, L.L. Tavlarides, 1977, Description of interaction processes in agitated liquid-liquid dispersions, Chemical Engineering Science, 32, 1289-1297, https://doi.org/10.1016/0009-2509(77)85023-9.

B.K. Debnath, U.K. Saha, N. Sahoo, 2015, A comprehensive review on the application of emulsions as an alternative fuel for diesel engines, Renewable and Sustainable Energy Reviews, 42, 196-211, http://dx.doi.org/10.1016/j.rser.2014.10.023.

J. A. Driscoll, 1997, Acid rain demonstration: The formation of nitrogen oxides as a by-product of high-temperature flames in connection with internal combustion engines, Journal of Chemical Education, 74, 12, 1424, https://doi.org/10.1021/ed074p1424.

N. G. Khouri, J. O. Bahú, N. T. Miranda, C. B. Batistella, M. R. W. Maciel, V. O. C. Concha, R. Maciel Filho, 2023, Diesel upgrading: A modeling of its microemulsions, Fuel Processing Technology, 239, 107545, https://doi.org/10.1016/j.fuproc.2022.107545.

K. R. Patel, V. D. Dhiman, 2021, A review on emission and performance of water diesel micro-emulsifed mixture-diesel engine, International Journal of Environmental Science and Technology, 19, 8027-8042, https://doi.org/ 10.1021/acs.energyfuels.9b02870.

P. Rastogi, N.S. Kaisare, M.G. Basavaraj, 2019, Diesel emulsion fuels with ultralong stability, Energy Fuel, 33, 12227-12235, https://doi.org/10.1016/j.fuel.2018.09.074.

M. Shiea, A. Querio, A. Buffo, G. Boccardo, D. Marchisio, 2022, CFD-PBE modelling of continuous Ni-Mn-Co hydroxide co-precipitation for Li-ion batteries, Chemical Engineering Research and Design, 177, 461-472, https://doi.org/10.1016/j.cherd.2021.11.008.

E. Supriyanto, J. Sentanuhady, A. Dwiputra, A. Permana, M.A. Muflikhun, 2021, The recent progress of natural sources and manufacturing process of biodiesel: a review, Sustainability, 13, 5599, https://doi.org/10.3390/su13105599.

Flavio Manenti, Gintaras V. Reklaitis (Eds.), Proceedings of the 34th European Symposium on
Computer Aided Process Engineering / 15th International Symposium on Process Systems
Engineering (ESCAPE34/PSE24), June 2-6, 2024, Florence, Italy

Cleaning Kinetic Model for Carbohydrates- and Protein-based Stains in Automatic Dishwasher

José E. Roldán-San Antonio,[a,*] Carlos Amador,[b] Mariano Martín,[a] Kevin Blyth,[b]
Vania C. Croce Mago[b]

[a] Department of Chemical Engineering, University of Salamanca, Plaza Caídos 1-5,
Salamanca, 37008, Spain
[b] Procter & Gamble Ltd., Newcastle Innovation Centre, Newcastle Upon Tyne, NE12
9BZ, UK.
joseenrique@usal.es

Abstract

A mechanistic model capable of describing the cleaning mechanism of protein-based and
carbohydrate-based stains in an automatic dishwasher (ADW) is proposed. The effect of
several factors such as pH, temperature, water hardness, bleaching and chelating agents,
surfactants, builder, and functional polyacrylate in different cycles of dishwashers were
studied. A three – stage procedure is used. First, a statistical analysis based on response
surface methodology (RSM) was conducted to identify the impact of the most significant
variables in the cleaning of each stain. Both stains cleaning was highly dependent by
enzymatic action. However, protein-based stain was the only where its colour was
compromised by the action of bleaching agents. Next, the kinetic mechanisms of the
significant technologies, were proposed and integrated in a unique dynamic model based
on non-linear differential equations, including the stability of enzymes and bleaching
compounds. Subsequently, the parameter estimation was done employing Levenberg-
Marquardt nonlinear least squares algorithm. Finally, a validation of the mechanistic
model for each stain was carried out, getting a determination coefficient higher than 0.8
for carbohydrate-based and protein-based stains respectively, capturing most of the
variation in stain cleanliness.

Keywords: Stain Removal, Automatic Dishwasher, Cleaning Model.

1. Introduction

Millions of people every day clean dishes, being the dishwashing a common practice
around the world. Automatic dishwasher contributes with a significant reduction of water
and energy consumption (75% and 25% respectively) compared to hand washing
(Berkholz et al., 2010). The cleaning inside an ADW involves complex physical and
chemical processes which affect the cleaning performance being highly linked between
(Pérez-Mohedano et al., 2017). The optimization of cleanliness, sustainability and
operating costs is only possible by understanding the cleaning mechanisms involved.
Several techniques have been used to analyze the stain removal behavior during cleaning
such as micromanipulation technique based on a stainless-steel probe to measure the
adhesive forces (Liu et al., 2002), millimanipulation technique to study highly adhesive
soils (Ali et al., 2015) or Fluid Dynamic Gauging (FDG) to explore the thickness
evolution of the stain (Gordon et al., 2010). The change in the thickness of the stain, as
well as the forces that condition its structure and adhesion to the surface where it is found,
are highly dependent on the chemical conditions present in the wash. In this way,

detergent's ingredients such as enzymes, bleaches, surfactants, buffers, and builders are key factors in stains cleaning. Enzymes promote a more environmentally friendly washing by the reducing the wash time, pH, and temperature. Bleaches provide a germicidal action together with a stain color reduction. Surfactants aim to reduce the surface tension, increasing the wettability of the stain and its dissolution in the washing media. On the other hand, the control of pH and water hardness is an important factor that is carried out by buffers and builders. Currently, the combination of activated bleaching systems and enzymes such as amylases and proteases are highly employed since the hydrolysis and bleaching of stains as well as disinfection during washing is possible at low temperatures without compromising the stability of the active compounds present (Bianchetti et al., 2015).

In this way, a detailed understanding of the cleaning mechanism of each of the key ingredients on the different types of stains is crucial to improve the cleaning performance and reduce the environmental impact, promoting the achievement of the objectives set by United Nations in terms of sustainability and responsible consumption and production (United Nations, 2023). For these reasons, the present work is focused on the identification of the most important factors and mechanisms involved in carbohydrate – based and protein-based stains, proposing a cleaning model for each stain where it is included the stability of the key cleaning ingredients such as enzymes and bleaching compounds. The validation of the models was carried out using experimental data.

2. Material and Experimental Procedure

2.1. Stain technical samples

Sprayed layers of starch and minced meat on a melamine base were considered as stain technical samples, acquired from Centre for Testmaterials. The dimensions of samples were 12 cm x 10 cm x 0.12 cm.

2.2. Detergent formulation

Amylase and protease enzymes with an activity of 4% and 9.71% respectively were considered. The activated bleaching system was composed by TAED as bleach activator, manganese - based bleaching catalyst and sodium percarbonate with an activity of 92%, 2% and 100% respectively. Insoluble precipitate formation and pH were controlled by chelating, builder, sodium carbonate and silicate – based compound. Additionally, polymers and non – ionic surfactants were also considered in the study.

2.3. Apparatus

A Miele GSL200 dishwasher was considered to run all the experiments. In this appliance 8 different cycles were conducted. All cycles have the same stages, a main wash, cold rinse, and hot rinse. For main wash, a time distribution of 17 – 69 min and temperature levels of 40 - 50 °C were considered. In the case of hot rinse, durations between 23 – 26 min and target temperatures range of 55 - 65 °C were performed. The duration of the cold rinse is the same for all cycles and its temperature profile is set by the temperature of the main wash stage.

2.4. Experimental procedure

Initially, the water supply tanks are filled, spiking the corresponding levels of water hardness for each experiment. Then, a pair of tiles are located inside an ADW, one on the left and the other on the right – hand, placing a blank tile at the front. To mime the soil found in the domestic ADWs, a defrosted ballast soil is added using the same level for all experiments. Next, the pH meters are located inside an ADW. Once the corresponding cycle is running, the pH and temperatures profiles are recording over time by pH meters.

Once the cycle finishes, the stain removal index (SRI) is measured as response variable. SRI is computed by the color difference between the stain before and after washing, employing the eq (1), (Copley, 2017; Neiditch et al., 1980) where IN_i and FN_i are the initial and final noticeability for the stain i. The execution of the experiments was carried out based on a design of experiments (DOE) provided by the custom design platform of JMP Pro 16. A total of 217 experiments were conducted, employing a 10 % for validation.

$$SRI_i = \frac{IN_i - FN_i}{IN_i} \tag{1}$$

3. Modelling Methodology

A three – step methodology was considered to build a non – linear differential model to describe the cleaning mechanism of protein-based and carbohydrate – based stains in ADWs. Figure 1 describes the steps of the methodology followed. Once all the experiments were conducted based on the design of experiments, a response surface methodology was employed to identify the most significant variables at their effects on the cleaning of each stain, considering a statistical significance of 0.01 as reference. Based on the statistical analysis, a kinetic mechanism for each key factor is proposed, formulating a non-linear differential model for each stain. Subsequently, an estimation of the adjustable parameters of the main cleaning mechanisms was done employing Levenberg-Marquardt nonlinear least squares algorithm. Finally, a validation of the mechanistic model for each stain was carried out by correlating the estimated and measured SRIs.

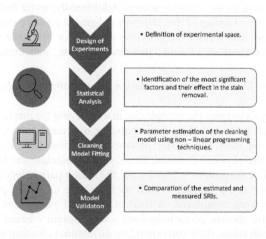

Figure 1: Cleaning modelling methodology for carbohydrate – based and protein – based stains.

3.1. Mechanistic Cleaning Modelling

During the cleaning, a combination between the physical removal and bleaching of the stain takes place. Physical removal by enzymatic action promotes a loss of color since a fraction of the components that provide the color is solubilized. Simultaneously, the bleaching compounds act on the remaining fraction of the stain removing the color. Assuming a homogeneous composition distribution of the stain, the total physical stain removal can be computed by eq (2), where is included the easy and difficult rate of removal due to the transition from a fast-cleaning stage at the beginning of the cycle to a slower one experimentally witnessed. This curvature change in the cleaning is considered by μ_i, a fitting parameter. Additionally, k_i^{Phys} is the cleaning constant which depends on the main factors involved in the physical removal. The union between the physical removal and bleaching from bleaches is showed by eq (3), where SRI_i^{Total} represents the total stain removal including physical removal and loss of color and m_i^d is the color remaining fraction of the stain. This fraction can be computed as the difference between the total reaming fraction and the bleached fraction of the stain i.

$$\frac{dSRI_i^{Phys}}{dt} = \mu_i \cdot k_i^{Phys} \cdot \left(f_i^E - SRI_i^{E,Phys}\right) + k_i^{Phys} \cdot \left(f_i^D - SRI_i^{D,Phys}\right) \tag{2}$$

$$\frac{dSRI_i^{Total}}{dt} = -\frac{dm_i^d}{dt} \tag{3}$$

4. Results and Discussion

4.1. Cleaning factors selection for carbohydrate – and protein – based stains.

From the response surface methodology, the most important variables in carbohydrate – based and protein – based stains removal are identified. Table 1 summarizes the results. The statistical analysis shows that time and enzymatic action by amylase are the main two drivers for cleaning of the carbohydrate – based stain. However, bleaching compounds provide a negative impact in the cleaning of this stain because they decompose the amylase, not impacting on the stain bleaching. Additionally, water hardness also gives a negative impact on the cleaning of this stain since deposits on stain surface by metal ions from water hardness may hinder the amylase action. For the protein-based stain, the cleaning is highly dependent of time, protease, amylase, and bleaching compounds concentrations.

4.2. Cleaning model formulation and validation for carbohydrate – based and protein – based stains.

The statistical analysis reduces the complexity in identifying the mechanisms involved in the removal of each stain and their corresponding differential model parameters. For carbohydrate – based stain, the model parameters selected were those related to the amylase effect on the physical removal, employed to calculate the kinetic constant of physical removal. For the protein – based stain, the fitting parameters related to physical removal by amylase and protease as well as bleaching effect by peracetic acid and hydrogen peroxide were considered. A total of 8 and 11 parameters coming from the cleaning mechanistic models of carbohydrate - and protein - based stain were fitted, respectively. For both stains, their corresponding differential cleaning models have a high statistical significance since p - values lower than 0.01 were obtained in the ANOVA of the regressions.

After the fitting parameter estimation, a validation of the mechanistic model for each stain was carried out comparing the estimated and measured SRIs. Figure 2 shows a scatter plot made up by the predicted stain removal index versus the experimental ones for both stains. The models can capture the most part of the variation in the SRI since determination coefficient of 0.93 and 0.89 were obtained for carbohydrate – based and protein – based stain respectively. However, the reduced amount of experimental data between 30 and 50 % of the SRI in Figure 2a provides outliers in this region.

Table 1: Selected factors for differential models of carbohydrate – and protein – based stains from statistical analysis.

	Represented Factors	p -value > F
Carbohydrate - based stain	TAED	0.0003
	Amylase	<0.0001
	Water Hardness	0.0007
	Main wash time	<0.0001
	Amylase – Amylase	<0.0001
Protein - based stain	Sodium percarbonate	<0.0001
	Protease	<0.0001
	Amylase	<0.0001
	Non - ionic surfactants mixture	0.044
	Main wash time	<0.0001
	Sodium percarbonate - Protease	<0.0001
	Protease – Protease	<0.0001
	Protease – Amylase	<0.0001
	Protease – Main wash time	<0.0001
	Amylase – Main wash time	0.002

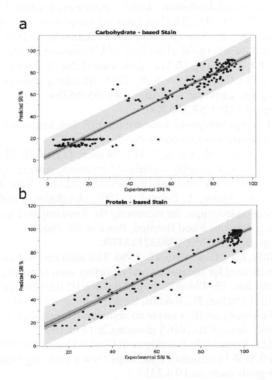

Figure 2: Validation cleaning models 95 % of confidence: (a) Carbohydrate – based stain, (b) Protein – based stain.

5. Conclusions

A three – stage methodology is employed to build a dynamic cleaning mechanistic model for carbohydrate – based and protein – based stains. Initially, a statistical analysis based on response surface methodology allowed the identification of the main factors involved on the removal of the stains. Enzymatic action was the main route in the decomposition of both stains. However, only in the protein – based stain, color removal by bleaching agents was observed. The statistical analysis allowed the identification of kinetic cleaning mechanism, allowing the formulation of a non – linear differential model for both stains. High statistical significance and adequate accuracy of the models were obtained, concluding that the use of the previous statistical study helps in the identification of mechanisms and fitting complex models for ADW cleaning process.

Acknowledgments

This work was supported by funding to José Enrique Roldán San Antonio under the call for predoctoral contracts USAL 2021, co-funded by Banco Santander.
We would like to thank the Procter & Gamble Newcastle Innovation Centre (UK) for providing the experimental data as well as software licenses required in the research.

References

Ali, A., De'Ath, D., Gibson, D., Parkin, J., Alam, Z., Ward, G., Wilson, D.I., 2015. Development of a 'millimanipulation'device to study the removal of soft solid fouling layers from solid substrates and its application to cooked lard deposits. Food Bioprod. Process. 93, 256–268. https://doi.org/10.1016/j.fbp.2014.09.001

Berkholz, P., Stamminger, R., Wnuk, G., Owens, J., Bernarde, S., 2010. Manual dishwashing habits: an empirical analysis of UK consumers. Int. J. Consum. Stud. 34, 235–242. https://doi.org/10.1111/j.1470-6431.2009.00840.x

Bianchetti, G.O., Devlin, C.L., Seddon, K.R., 2015. Bleaching systems in domestic laundry detergents: a review. Rsc Adv. 5, 65365–65384. https://doi.org/10.1039/C5RA05328E

Copley, T., 2017. Testing Detergents: Establishing Efficient Methods for Formulation, QC and Comparative Assessment. Sofw. J. 143, 38–42.

Gordon, P.W., Brooker, A.D.M., Chew, Y.M.J., Wilson, D.I., York, D.W., 2010. A scanning fluid dynamic gauging technique for probing surface layers. Meas. Sci. Technol. 21, 85103. https://doi.org/10.1088/0957-0233/21/8/085103

Liu, W., Christian, G.K., Zhang, Z., Fryer, P.J., 2002. Development and use of a micromanipulation technique for measuring the force required to disrupt and remove fouling deposits. Food Bioprod. Process. 80, 286–291. https://doi.org/10.1205/096030802321154790

Neiditch, O.W., Mills, K.L., Gladstone, G., 1980. The stain removal index (SRI): A new reflectometer method for measuring and reporting stain removal effectiveness. J. Am. Oil Chem. Soc. 57, 426–429. https://doi.org/10.1007/BF02678931

Pérez-Mohedano, R., Letzelter, N., Bakalis, S., 2017. Integrated model for the prediction of cleaning profiles inside an automatic dishwasher. J. Food Eng. 196, 101–112. https://doi.org/10.1016/j.jfoodeng.2016.09.031

United Nations, 2023. Sustainable Development Goals | United Nations Development Programme [WWW Document]. URL https://www.undp.org/sustainable-development-goals (accessed 9.4.23).

Flavio Manenti, Gintaras V. Reklaitis (Eds.), Proceedings of the 34th European Symposium on Computer Aided Process Engineering / 15th International Symposium on Process Systems Engineering (ESCAPE34/PSE24), June 2-6, 2024, Florence, Italy

From Municipal Solid Waste to Sustainable Aviation Fuel: Process Design

Mohammad Alherbawi, Ridab Khalifa, Yusuf Bicer, Tareq Al-Ansari

College of Science and Engineering, Hamad Bin Khalifa University, Qatar Foundation, Doha, Qatar.
talansari@hbku.edu.qa

Abstract

The aviation industry has rebounded post-pandemic, where carbon dioxide (CO_2) emissions escalated to 80% of pre-pandemic levels. Studies on waste-to-energy have been exploring various methods like catalytic hydro-processing, pyrolysis of waste plastic, and gasification coupled with Fischer–Tropsch processes. In this work, Aspen Plus was utilized to model a waste-based system, aiming to showcase the production of bio-jet fuel from municipal solid waste (MSW), treated wastewater and captured CO_2. The system involved steam gasification, Fischer-Tropsch synthesis, dry reforming, hydrocracking, and isomerization processes. The results demonstrated the product distribution after each stage, emphasizing the potential of producing jet fuel with the highest selectivity of 53.3% and a total production of 357,781 tonnes/year out of 3 million tonnes of MSW. Such initiatives presented a promising pathway to mitigate aviation emissions while harnessing waste as a valuable resource for energy production.

Keywords: Bio-jet fuel, Sustainable aviation, Sustainable jet fuel, waste-to-energy

1. Introduction

The aviation industry accounted for 2.5% of the global carbon dioxide (CO_2) emissions (Ritchie, 2020). As aviation's demand recovered in 2022, post-pandemic, the emissions rebounded to nearly 80% of pre-pandemic levels, reaching 800 million metric tons of CO_2 (IEA, 2023). Meanwhile, the world's rapid industrial progress boosted the economy but also led to the emission of diverse pollutants into the environment, stemming from various industries. These pollutants have seeped into the soil and groundwater, posing a threat due to improper treatment of industrial effluents and solid wastes discharged into the surroundings (Verma & Sharma, 2023). Currently, the world has been generating 2.01 billion tonnes of municipal solid waste (MSW) annually, with a minimum of 33% of waste not being managed in an environmentally safe manner. The average daily waste produced per person globally used to be 0.74 kg, but the range varied significantly, from 0.11 to 4.54 kg. By 2050, global waste is expected to grow to 3.40 billion tonnes (The World Bank, 2023). The utilization of waste for energy production holds a significant importance in meeting future global energy needs.

1.1. Sustainable aviation and alternative jet fuels from waste

Producing sustainable aviation fuel from waste materials like MSW, encompassing food waste and waste cooking oils alongside agricultural and forestry residues, utilizing existing conversion technologies has been studied recently by Emmanouilidou et al. (2023). Their systematic review revealed that the catalytic hydro-processing of waste lipid feedstocks had been the most employed method for producing bio-jet fuel (BJF).

Additionally, the catalytic pyrolysis of waste plastic and co-pyrolysis with solid biomass residues had the potential to contribute to effective policy support and enhance current technologies to reduce production costs. Moreover, the combination of gasification with Fischer–Tropsch (FT) processes emerged as an intriguing pathway for sustainable aviation fuel production. In this study, a completely waste-based processing pathway is evaluated, where all feeding streams are of a waste-nature. The system is based on gasification and Fischer-Tropsch processes, where the key feedstock is MSW, while treated sewage effluent is used as a gasifying agent, and captured CO_2 is utilized within the process to run dry reforming of methane. The system is simulated and evaluated in Aspen Plus.

2. Methodology

Aspen Plus (V.12) ® was utilized to simulate the system for a Qatar case study, aiming to mainly generate BJF from various biomass sources, considering earlier modelling approaches (Alherbawi et al., 2023). Figure 1 illustrates the process flow of the system. The primary process involves integrating biomass gasification, followed by FT synthesis. Additionally, a dry-reforming phase employing CO_2 was implemented to increase BJF production and decrease the environmental impact of the system. The system's design was based on assumptions of an isothermal system and steady-state reactions. Thermodynamic properties were estimated using the Redlich-Kwong-Soave (RK-SOAVE) and non-random two-liquid model (NRTL).

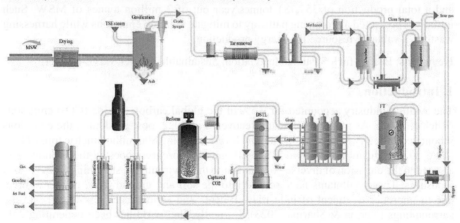

Figure 1: Flowsheet of a hybrid biorefinery.

MSW was used as a feedstock and was processed through gasification. 3 million tonnes of locally available MSW were simulated as an initial feed after the drying process, where a rotary dryer was modelled by an "RYield" block, operating at a temperature of 110°C. Table 1 shows the characteristics of the utilized feedstock. At first, Aspen Plus categorized all solid inputs as non-conventional elements based on their elemental and proximate attributes from Table 1. MSW in a dry state were fed into the gasifier. The outcomes of the drying phase were calculated using the proximate analysis of the inputs, redefining the dried remnants with a zero-moisture content.

Table 1: Attributes of MSW characteristics utilized in this study.

Moisture (%)	Fixed carbon (%)	Volatile matter (%)	Ash (%)	C (%)	H (%)	N (%)	S (%)	O (%)
7.56	24.21	57.99	17.8	48.23	5.16	1.21	0.29	27.31

2.1. Steam gasification

A steam gasifier was implemented at 1100°C with a steam-to-feed ratio of 1.2. A high steam input is considered to enhance the H_2/CO ratio for optimal FT process. Local treated sewage effluent (TSE) is used as a gasifying agent, where it is treated locally to a high purity level, however, it is mostly unutilized due to public unacceptance. Decomposition was the initial step in breaking down biomass into simpler components, fulfilling equation 1, the process was simulated employing two successive units. The first unit was represented by an "RYield" block, converting unconventional components into conventional components like char, solid sulfur, hydrogen, nitrogen, and oxygen. The second unit, using an "RGibbs" block, aimed to convert volatile carbon into potential products (e.g., methane, carbon dioxide, carbon monoxide), while nitrogen and sulfur contents were transformed into ammonia and hydrogen sulfide, respectively, due to the considerable presence of hydrogen in the reaction medium. Subsequent oxidation processes occurred due to the presence of oxygen as an intermediate product from the breakdown stage at a high operating temperature. Key reactions were simulated based on Equations 5–8 using an "REquil" unit. The final phase of the gasification reactions was modelled utilizing an "RGibbs" block, which calculates phase and chemical equilibrium by minimizing the Gibbs free energy. The resulting gases throughout these stages included H_2, CO, CO_2, H_2O, CH_4, NH_3, and H_2S. However, it was expected that nearly all char content would be volatilized by the end of the reaction, while ash was discharged as waste at the end of the process. Eq. (1) to Eq. (10) were assumed to be the only processes occurring under the specified conditions to ensure model simplicity.

A two-phase flash separator was employed to dry the wet syngas. The optimal operating conditions of the flash unit were determined through automated sensitivity analysis, running at a high pressure of 40 bar. Wax generation is assumed to be negligible due to the high processing temperature. Crude syngas is then underwent purification to remove impurities (primarily CO_2, NH_3, and H_2S) using the methanol absorption system, wherein these impurities were dissolved into chilled methanol at high pressure.

$$\text{Pyrolysis: Biomass} \rightarrow \text{Char} + \text{Tar} + NH_3 + H_2S + H_2 + CO + CO_2 \qquad (1)$$

$$\text{Complete Char Combustion: } C + O_2 \rightarrow CO_2 \qquad (2)$$

$$\text{Incomplete Char Combustion: } C + 0.5O_2 \rightarrow CO \qquad (3)$$

$$\text{Steam production: } H_2 + 0.5\, O_2 \rightarrow H_2O \qquad (4)$$

$$\text{CO oxidation: } CO + 0.5O_2 \leftrightarrow\leftrightarrow CO_2 \qquad (5)$$

$$\text{Boudouard reaction: } C + CO_2 \leftrightarrow 2\, CO \qquad (6)$$

$$\text{Methanation: } C + 2H_2 \leftrightarrow CH_4 \qquad (7)$$

Steam gasification: $C + H_2O \leftrightarrow CO + H_2$ (8)

Methane reforming: $CH_4 + H_2O \leftrightarrow CO + 3H_2$ (9)

Water gas shift reaction: $CO + H_2O \leftrightarrow CO_2 + H_2$ (10)

2.2. Fischer-Tropsch

A cobalt-based catalyst was selected due to its high stability. A slurry-phase reactor (SPR) using Co/Al$_2$O$_3$ catalyst is considered. The FT simulation began by introducing purified syngas into the FT reactor represented by an "RYield" block. The process occurred at 240 °C and 25 bar, predicting product distribution using a Fortran code based on the Anderson-Schulz-Flory (ASF) correlation (Schulz, 1999). For maximizing kerosene-range compounds, an α value of 0.85 was set, assuming an 80% CO conversion into paraffins, olefins, and alcohols (AlNouss et al., 2019). Around 50 chemical reaction equations were employed to predict hydrocarbon pathways leading to the occurrence of expected compounds.

2.3. Dry Reforming, Hydrocracking and Isomerization

In this particular model, the integration of dry reforming within the system was implemented to enhance fuel production while mitigating the carbon dioxide produced. The simulation was conducted at 800°C, with a CO$_2$/CH$_4$ ratio of 5, using a Ni/Al$_2$O$_3$ catalyst. Dry reforming, being highly endothermic, demands a substantial amount of heat to drive the reaction. On the contrary, FT is notably exothermic. Consequently, after an initial energy analysis, it was observed that the heat released by FT was approximately equivalent to the heat required by the reforming unit. Thus, to minimize the overall energy demand, heat integration was carried out between the two units. To simulate the process, an "RStoic" block was employed, defining potential reactions with a fractional conversion of 95% (Jiang et al., 2013; Lavoie, 2014).

Hydrocracking and isomerization were employed to increase the selectivity of BJF production. In this study, hydrocracking reactions occurred at 350°C and 80 bar, using a zeolite catalyst at a specified rate. Additionally, to enhance the quality of BJF, the product required isomerization. This step aimed to alter the structure of branched paraffins, as they display lower freezing points and surface tension, positively impacting the final product. Isomerization was conducted separately using a platinum-alumina catalyst system at defined conditions. The product distribution from this process was determined through an experimental study (Dhar et al., 2017). Operating at specific temperature and pressure conditions, this process was essential to control the characteristics of the resulting products.

2.4. Downstream Process

In the downstream process, the combined stream from various sub-units was passed through subsequent units to obtain different fuel categories. Initial processing involved a multi-stage flash drum to remove water and achieve gas-liquid separation. Higher hydrocarbons were then directed into a fractionation column with 18 stages to gather distinct fuel types, including BJF, gasoline, and diesel.

3. Results

The utilization of 3 million tonnes of MSW with the input of 3.6 million tonnes of TSE has led to the production of 3.96 and 0.5 million tonnes of gas and solids, respectively, per year, as per the ratio represented in Figure 2. In addition, the crude syngas composition is reflected in Figure 3. H_2/CO ratio of 2.05 is achieved, which is optimal for FT process.

At FT and reforming stage, the obtained syncrude's composition is illustrated in Figure 4. Where at the selected operating conditions, jet fuel (C8-C16) dominated the fuel mixture, reducing the need for intensive cracking.

At the final stage of cracking, isomerization and distillation, 0.37 million tonnes of jet fuel are obtained (53.3% liquid fuel selectivity), followed by fuel gas (0.26 Mtonnes/year), gasoline (0.2 Mtonnes/year) and green diesel (0.13 Mtonnes/year) as represented in Figures 5 and 6.

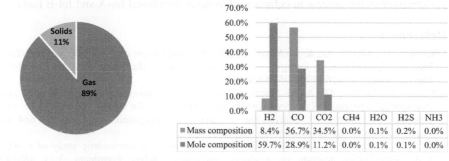

Figure 2: Gasification's products distribution.

	H2	CO	CO2	CH4	H2O	H2S	NH3
Mass composition	8.4%	56.7%	34.5%	0.0%	0.1%	0.2%	0.0%
Mole composition	59.7%	28.9%	11.2%	0.0%	0.1%	0.1%	0.0%

Figure 3: crude syngas composition.

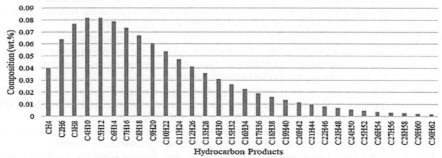

Figure 4: FT products distribution.

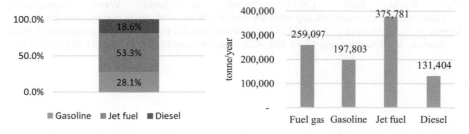

Figure 5: Final liquid products selectivity.

Figure 6: Final products total yield.

4. Conclusions

Post-pandemic aviation recovery led to a surge in emissions, emphasizing the urgency for sustainable practices. Waste-to-energy solutions, especially deriving aviation fuel from waste materials, offer promising environmental benefits. Research in waste-to-energy conversion methods like gasification highlights technical hurdles and avenues for improvement, aiming to refine technologies and reduce operational costs. The system for bio-jet fuel production from MSW illustrated here has shown the potential for generating environmentally friendly aviation fuel. The results exhibited product distributions, indicating the viability of producing substantial quantities of jet fuel from waste. Future expansion of this study may focus on evaluating the environmental and economic performance of the process in comparison to the conventional Jet-A and Jet-B fuels.

References

Alherbawi, M., McKay, G., & Al-Ansari, T. (2023). Development of a hybrid biorefinery for jet biofuel production. Energy Conversion and Management, 276, 116569. https://doi.org/10.1016/J.ENCONMAN.2022.116569

AlNouss, A., McKay, G., & Al-Ansari, T. (2019). A techno-economic-environmental study evaluating the potential of oxygen-steam biomass gasification for the generation of value-added products. Energy Conversion and Management, 196, 664–676. https://doi.org/10.1016/J.ENCONMAN.2019.06.019

Dhar, A., Vekariya, R. L., & Sharma, P. (2017). Kinetics and mechanistic study of n-alkane hydroisomerization reaction on Pt-doped γ-alumina catalyst. Petroleum, 3(4), 489–495. https://doi.org/10.1016/J.PETLM.2017.02.001

Emmanouilidou, E., Mitkidou, S., Agapiou, A., & Kokkinos, N. C. (2023). Solid waste biomass as a potential feedstock for producing sustainable aviation fuel: A systematic review. Renewable Energy, 206, 897–907. https://doi.org/10.1016/J.RENENE.2023.02.113

IEA. (2023). Aviation - IEA. https://www.iea.org/energy-system/transport/aviation

Jiang, Z., Liao, X., & Zhao, Y. (2013). Comparative study of the dry reforming of methane on fluidized aerogel and xerogel Ni/Al2O3 catalysts. Applied Petrochemical Research 2013 3:3, 3(3), 91–99. https://doi.org/10.1007/S13203-013-0035-9

Lavoie, J. M. (2014). Review on dry reforming of methane, a potentially more environmentally-friendly approach to the increasing natural gas exploitation. Frontiers in Chemistry, 2(NOV), 92076. https://doi.org/10.3389/FCHEM.2014.00081/BIBTEX

Ritchie, H. (2020). Climate change and flying: what share of global CO2 emissions come from aviation? https://ourworldindata.org/co2-emissions-from-aviation#article-citation

Schulz, H. (1999). Short history and present trends of Fischer–Tropsch synthesis. Applied Catalysis A: General, 186(1–2), 3–12. https://doi.org/10.1016/S0926-860X(99)00160-X

The World Bank. (2023). Trends in Solid Waste Management. https://datatopics.worldbank.org/what-a-waste/trends_in_solid_waste_management.html

Verma, S. K., & Sharma, P. C. (2023). Current trends in solid tannery waste management. Critical Reviews in Biotechnology, 43(5), 805–822. https://doi.org/10.1080/07388551.2022.2068996

Flavio Manenti, Gintaras V. Reklaitis (Eds.), Proceedings of the 34th European Symposium on Computer Aided Process Engineering / 15th International Symposium on Process Systems Engineering (ESCAPE34/PSE24), June 2-6, 2024, Florence, Italy

Economic resilience evaluation of wastewater resource recovery in Qatar's fertilizer market

Fatima-Zahra Lahlou, Ahmed Al-Nouss, Rajesh Govindan, Tareq Al-Ansari*

Division of Sustainable Development, College of Science and Engineering, Hamad Bin Khalifa University, Qatar foundation, Doha, Qatar
talansari@hbku.edu.qa

Abstract

Commercial fertilizers production represents a substantial contributor to greenhouse gas emissions when it comes to irrigated agriculture, and can, in some instances, hold the biggest share of emissions associated with the practice. With the intensification of climate change, energy intensive Haber-Bosh process, and declining phosphorus reserves, there is a need to adopt alternative and sustainable methods to meet farming nutrient requirements while reducing the burden on the environment, and while maintaining reduced economic costs. Using wastewater treatment plants byproducts, treated sewage effluent and sludge, can represent an effective mitigation strategy to reduce these emissions. In addition to that, sludge can be converted into soil amendment which can also substitute for conventional fertilizers. This study aims to evaluate the economic resilience of substituting commercial fertilizers with treated wastewater and sludge produced from wastewater treatment plants. The economic resilience of directing treated wastewater to the agricultural sector in the State of Qatar for fertigation purposes and converting sludge into biochar using pyrolysis modeled in Aspen Plus is compared with that of urea production. The economic resilience is assessed using the redundancy ratio as well as a price comparison under carbon taxation, thus internalizing the environmental cost of carbon emissions. Results indicate that treated sewage effluent and sludge biochar maintain stable redundancy ratios from 2017 to 2021, suggesting a dependable supply that could reduce reliance on traditional fertilizers. The introduction of a carbon tax, ranging from \$0.03 to \$0.1 per kg of CO_2, markedly shifts cost competitiveness in favor of sludge biochar, highlighting its potential as an economically and environmentally superior alternative. the substitution of conventional fertilizers with treated sewage effluent and sludge biochar in Qatar presents a viable pathway to enhance agricultural sustainability. The study emphasizes the need for policy frameworks that support the integration of these practices, ensuring that Qatar's agricultural sector remains resilient in the face of environmental and economic challenges.

Keywords: Wastewater, sludge, resilience, resource recovery, greenhouse gas emissions.

1. Introduction

As the global population continues to rise, the demand for food production intensifies, leading to a parallel increase in the need for commercial fertilizers. These fertilizers, essential for boosting crop yields and ensuring food security, have become a cornerstone of modern agriculture. However, the production of commercial fertilizers is an energy-intensive process, predominantly through the Haber-Bosch method for synthesizing ammonia, contributing significantly to the carbon footprint (CF) of the food sector (Gellings and Parmenter, 2004). With agriculture accounting for a substantial share of global greenhouse gas emissions, there is a pressing need to explore alternative resources that can sustain agricultural productivity while mitigating environmental impacts (Lahlou et al., 2023b).

The State of Qatar, with its arid climate and limited arable land, faces unique challenges in its pursuit of food security. In recent years, Qatar has made significant strides in recycling water, resulting in treated wastewater that is highly purified and enriched with nutrients, including nitrogen. This treated sewage effluent (TSE) is a byproduct of urban water use that, after treatment, could serve as a sustainable irrigation resource. Yet, the potential of TSE as a partial substitute for commercial fertilizers remains largely untapped (Lahlou et al., 2021). Additionally, Qatar's wastewater treatment plants produce a substantial amount of sludge which currently represents an underexplored resource. Through processes such as pyrolysis, sludge can be converted into biochar, a soil amendment known for its nutrient-rich content and carbon sequestration capabilities (Ghiat et al., 2020).. Biochar not only offers a way to recycle waste but also provides a means to improve soil health and fertility, presenting an opportunity to enhance the sustainability of Qatar's agricultural practices.

The transformation of TSE and sludge into valuable agricultural inputs could significantly reduce the reliance on imported commercial fertilizers, decrease the carbon footprint associated with food production, and align with the broader environmental goals of reducing greenhouse gas emissions (Lahlou et al., 2020). The integration of these resources into the agricultural sector may serve as a model for sustainable agriculture in arid regions worldwide. In this context, economic resilience becomes a critical factor. It is essential to ensure that the adoption of alternative fertilizers such as TSE and biochar derived from sludge is not only environmentally sustainable but also economically viable. Economic resilience in this framework refers to the agricultural sector's ability to absorb and adapt to economic shocks, such as fluctuations in global fertilizer prices or changes in trade policies, without significant disruption to agricultural output or profitability.

This study examines the economic resilience of TSE and sludge biochar in Qatar's fertilizer market. It evaluates whether these alternatives can provide a stable and reliable supply of nutrients, offset the need for commercial fertilizers, and withstand market and environmental uncertainties. This research provides insights into the feasibility of these alternative nutrient sources in supporting Qatar's agricultural sustainability and food security ambitions. The findings of this research aim to contribute to the development of informed policies and practices that balance economic, environmental, and regulatory considerations, paving the way for a more sustainable agricultural future.

2. Methodology

2.1. Study area and scenario description

2.1.1. Base scenario

The State of Qatar relies heavily on commercial fertilizers. For the existing fodder farms, it is estimated that a range of 382-556 kg of N is required every year for optimum fodder yield, of which 89% is supplied through commercial fertilizers while the rest comes from TSE (Lahlou et al., 2023a). For this study, it is assumed that urea produced locally is used as commercial fertilizer, and that the total capacity of N sourced from urea can be calculated using Eq. (1), such that TC^Y is the total capacity in year Y [t.y^{-1}], P_u^Y designates the urea production in year Y, retrieved from QAFCO (2021) [t.y^{-1}], E_u^Y is the urea exports in year Y retrieved from (Planning and Statistics Authority, 2021), Pop_Q^Y designates Qatar population in year Y [capita], $U_{fert-use}$ is the average urea usage per capita retrieved from (FAO, 2023) [t.capita^{-1}.y^{-1}], $\%_{N-urea}$ is the % of N in urea (QAFCO, 2021).

$$TC^Y = \left(P_u^Y - E_u^Y - Pop_Q^Y \cdot U_{fert-use}\right) \cdot \%_{N-urea} \qquad (1)$$

2.1.2. Treated sewage effluent scenario

The State of Qatar has been producing increasingly more TSE over the past few years. While some of the TSE is used for irrigation purposes, important volumes are disposed of, which could, in turn, not only reduce the reliance on the heavily abstracted groundwater resources but also supply Qatar's irrigated agricultural sector with loads of Nitrogen (N) that can offset the burden associated with commercial fertilizers. With an average N concentration of 6.76 mg.L^{-1}, TSE has the potential to meet the total N fertilization requirement of the agricultural sector in Qatar (Lahlou et al., 2023a). The total capacity of N sourced from urea can be calculated using Eq. (2), such that C_i^Y is the capacity of plant i in year Y [million $m^3.y^{-1}$], and C_i^N is the N concentration in plant i [mg.L^{-1}]

$$TC^Y = \sum_i^n C_i^Y \cdot C_i^N \qquad (2)$$

2.1.3. Sludge scenario

Important volumes of sludge are produced every year in the State of Qatar, which can be converted into biochar. When used as soil amendment under the right conditions, biochar can supply a plant's nutrient requirements (Haider et al., 2020). Aspen Plus is used to simulate the pyrolysis of sludge produced between the year 2017 and 2021 (Figure 1). the proximate and ultimate analysis are retrieved from the literature (AlNouss et al., 2019).

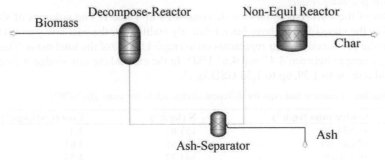

Figure 1: Pyrolysis flowsheet in Aspen Plus.

2.2. Economic resilience assessment

The redundancy ratio (RR) is used as a measure of the economic resilience of the three scenarios from 2017 to 2021 and is calculated using Eq. (3) and (4) such that EC is the excess capacity, AD is the average demand, and TC is the total capacity.

$$RR = \frac{EC}{AD} \qquad (3)$$
$$EC = TC - AD \qquad (4)$$

To assess the economic impact of environmental taxation on fertilizer choices, the costs of sludge and urea are compared, incorporating a variable carbon tax. The analysis simulates the introduction of a carbon tax, ranging from \$0.03 to \$0.1 per kg of CO2, to the price of urea. This is done to reflect the higher carbon emissions from urea production, compared to the lower emissions from sludge. The carbon tax is applied based on the emissions differential per unit of nitrogen, $\Delta CO2$, given by Eq. (5). Where E_{urea} and E_{sludge} represent the kg of $CO_{2\text{-eq}}$ emissions per kg of nitrogen from urea and sludge, respectively. The adjusted cost of urea, $C_{urea,adj}$ is then calculated using Eq. (6) such that C_{urea} is the initial cost of urea per kg of nitrogen and Tax_{CO2} is the applied carbon tax per kg of $CO_{2\text{-eq}}$. To maintain revenue neutrality for urea production, the carbon tax's financial impact is directly added to its market price.

$$\Delta CO2 = E_{urea} - E_{sludge} \qquad (5)$$

$$C_{urea,adj} = C_{urea} + \Delta CO2 . Tax_{CO2} \qquad (6)$$

The sensitivity of the cost difference to variations in sludge pricing is also analyzed, considering price changes ranging from a 50% decrease to a 50% increase. This comparison provides insights into how environmental policies, such as carbon taxes, could shift economic preferences between traditional and more sustainable nitrogen sources in agriculture. The price of urea ranges averages 1.51 USD.kg⁻¹, while that of biochar is obtained from the Aspen Plus model considering that raw sludge cost is set to 0.1 USD.t⁻¹ (AlNouss et al., 2021; Quinn, 2023).

3. Results

3.1. Pyrolysis model results

The results of the Pyrolysis model are demonstrated in Table 1. The volumes of sludge produced in the state of Qatar have been relatively stable over the past few years. The N content in the produced biochar represents on average 11.17% of the total mass. The cost of each kg_N ranges between 4.47 and 4.61 USD. In the case where raw sludge is free, the cost would reduce to 1.39, up to 1.53 USD.kg$_N$⁻¹.

Table 1: Biochar N content and cost for different sludge inlets for years 2017-2021.

Year	Sludge inlet (kg.h⁻¹)	N (kg.h⁻¹)	Cost (USD.kg$_N$⁻¹)
2017	41,554.00	153.87	4.47
2018	37,688.00	139.55	4.61
2019	39,096.00	144.77	4.56
2020	40,960.00	151.67	4.49
2021	41,349.00	153.11	4.48

3.2. Redundancy Ratio

The RR analysis reveals that TSE and sludge maintain stable surpluses, indicating robust systems against supply shocks (Figure 2). Although TSE's surplus slightly declined in 2021, its overall contribution to resilience remains strong. Conversely, urea's RR displayed volatility, surging in 2017 due to export disruptions from the regional blockade, which suggests a potential for stockpile inefficiency and vulnerability to trade dynamics. The significant RR increase in 2021 for urea might reflect strategic adjustments in response to the Russia-Ukraine conflict's impact on global demands. High RR for urea signals management inefficiencies, reflecting potential overproduction or underuse, rather than economic strength. These inefficiencies underscore the need for a supply strategy attuned to market demands. Conversely, the stable RRs for TSE and sludge indicate reliable, sustainable alternatives. Policy efforts must, therefore, balance urea's output with actual needs and foster the integration of TSE and sludge to safeguard against urea's market uncertainties.

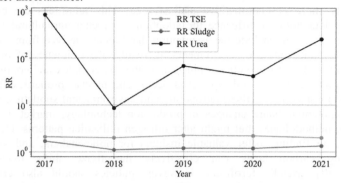

Figure 2: RR values for TSE, Sludge, and Urea from 2017 to 2021.

3.3. Price comparison under carbon taxation

The introduction of a carbon tax significantly influences the cost competitiveness between sludge and urea-based fertilizers. Figure 3 demonstrates that as the carbon tax increases from $0.03 to $0.10 per kg of CO_2, the cost advantage of sludge over urea becomes more pronounced. Each line represents the cost differential between sludge and urea at various levels of carbon taxation, holding the sludge price constant while urea's price adjusts to include the tax, maintaining revenue neutrality. The smallest tax rate, $0.03/kg CO_2, presents a modest difference in cost, but as the tax rate escalates, the gap widens consistently. At the highest carbon tax rate, $0.10/kg CO_2, the price of urea surpasses that of sludge significantly, indicating a strong economic incentive to shift towards sludge as a more cost-effective and environmentally sustainable nitrogen source. These results suggest that implementing a carbon tax could be an effective policy tool for promoting sustainable agricultural practices through the adoption of lower-emission fertilizers.

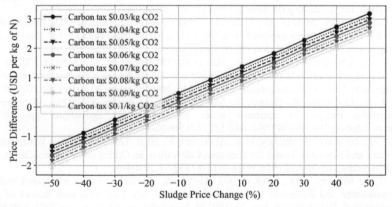

Figure 3: Price Difference Between Sludge and Urea with Varying Carbon Tax Rates.

4. Conclusions

The RR for urea displayed considerable volatility over the five-year period, with significant peaks in 2017 and 2021 due to geopolitical events that disrupted exports. In contrast, TSE and sludge demonstrated stable RRs, suggesting they are more resilient to external shocks and could serve as sustainable alternatives within Qatar's agricultural sector. The carbon tax simulation showed that as the tax rate increases, sludge becomes progressively more cost-competitive against urea. This reinforces the potential of environmental taxation to influence the adoption of greener agricultural inputs. The limitations of the study include the assumption that biochar derived from sludge could fully meet a plant's nitrogen requirements. In reality, the efficiency of biochar as a complete substitute for conventional nitrogen sources is complex and contingent upon multiple variables, such as soil type, crop needs, and biochar properties. The study's model does not account for these agronomic and environmental subtleties, nor does it incorporate potential future changes in production technology, market dynamics, or agricultural methodologies that might alter nitrogen utilization patterns. Despite these constraints, the findings offer valuable insights for policymakers, and suggest that implementing carbon taxes can effectively incentivize shifts towards more environmentally friendly fertilizers. However, policies should also consider the agronomic suitability of biochar and TSE, promoting research and development to optimize their use in Qatar's unique environmental conditions.

References

AlNouss, A., Namany, S., McKay, G., Al-Ansari, T., 2019. Applying a Sustainability Metric in Energy, Water and Food Nexus Applications; A Biomass Utilization Case Study to Improve Investment Decisions. pp. 205–210. https://doi.org/10.1016/B978-0-12-818634-3.50035-7

AlNouss, A., Parthasarathy, P., Mackey, H.R., Al-Ansari, T., McKay, G., 2021. Pyrolysis Study of Different Fruit Wastes Using an Aspen Plus Model. Front. Sustain. Food Syst. 5. https://doi.org/10.3389/fsufs.2021.604001

FAO, 2023. FAOSTAT analytical brief 68 - Inorganic fertilizers 2000–2021. https://doi.org/10.1007/978-1-4020-3995-9_293

Gellings, C., Parmenter, K., 2004. production and use. In Knowledge for Sustainable Development—An Insight into the Encyclopedia of Life Support Systems II, 419–450.

Ghiat, I., AlNouss, A., Mckay, G., Al-Ansari, T., 2020. Modelling and simulation of a biomass-based integrated gasification combined cycle with carbon capture: comparison between monoethanolamine and potassium carbonate. IOP Conf. Ser. Earth Environ. Sci. 463, 012019. https://doi.org/10.1088/1755-1315/463/1/012019

Haider, G., Joseph, S., Steffens, D., Müller, C., Taherymoosavi, S., Mitchell, D., Kammann, C.I., 2020. Mineral nitrogen captured in field-aged biochar is plant-available. Sci. Rep. 10, 13816. https://doi.org/10.1038/s41598-020-70586-x

Lahlou, F.-Z., AlNouss, A., Govindan, R., Hazrat, B., Mackey, H.R., Al-Ansari, T., 2023a. Water and sludge resource planning for sustainable agriculture: An energy-water-food-waste nexus approach. Sustain. Prod. Consum. 38, 130–148. https://doi.org/10.1016/j.spc.2023.03.027

Lahlou, F.-Z., Govindan, R., Hazrat, B., Mackey, H.R., Al-Ansari, T., 2023b. Water resource portfolio design for optimum water productivity in agriculture. pp. 569–574. https://doi.org/10.1016/B978-0-443-15274-0.50090-1

Lahlou, F.-Z., Namany, S., Mackey, H.R., Al-Ansari, T., 2020. Treated Industrial Wastewater as a Water and Nutrients Source for Tomatoes Cultivation: an Optimisation Approach. pp. 1819–1824. https://doi.org/10.1016/B978-0-12-823377-1.50304-9

Lahlou, F.Z., Mackey, H.R., McKay, G., Al-Ansari, T., 2021. Reuse of treated industrial wastewater and bio-solids from oil and gas industries: Exploring new factors of public acceptance. Water Resour. Ind. 26, 100159. https://doi.org/10.1016/j.wri.2021.100159

Planning and Statistics Authority, 2021. Agricultural Statistics. Doha.

QAFCO, 2021. Sustainability Report.

Quinn, R., 2023. DTN Retail Fertilizer Trends. DTN Progress. Farmer.

Flavio Manenti, Gintaras V. Reklaitis (Eds.), Proceedings of the 34th European Symposium on Computer Aided Process Engineering / 15th International Symposium on Process Systems Engineering (ESCAPE34/PSE24), June 2-6, 2024, Florence, Italy

Computational Molecular Dynamics in Emerging Biological Fields

Joseph Middleton (a), Joan Cordiner (a)

(a) *Chemical and Biological Engineering, University of Sheffield, S10 2TN*
j.cordiner@sheffield.ac.uk

Abstract

Due to the pandemic, mRNA-based vaccines and therapeutics are now a widely tested technology rather than just emerging. It is an efficient way of delivering genetic information to the target, making use of human cellular transcription enzymes to convert the mRNA strands into proteins that can then perform their intended functions in vivo. Traditional methods of process discovery, which require exploring the process conditions experimentally, would be prohibitively expensive meaning instead we must use modelling to construct a digital twin of the process from first principles, or *Ab Initio*. This can be achieved using Molecular Dynamics, which sits at the intersection between the fields of Quantum Chemistry and Classical Mechanics; with quantum operations being expressed directly as combinations of the quantum mechanical operator and classical (Hamiltonian) Mechanics.

Similar to classical mechanics there is a limited set of analytically solvable problems, meaning that the multi-dimensional, many-body problems must be solved numerically. Even using approximations, numerical solutions take a significant amount of time to converge. This paper reviews the current state of molecular modelling, through the practice lens of modelling the mRNA transcription process. The paper will discuss the theory behind, the mathematics and the physical computation of the process and will make suggestions as to what the next stage of computational molecular dynamics might look like and what might be required in order to make large scale biological simulation more accessible.

Keywords: mRNA, Quantum Dynamics, Computing

1. Quantum Chemistry

Quantum Dynamics, which was reformulated by Schrödinger in 1926, is based around Equation 1. A quantum evolution operator for the wave equation, Ψ, which takes the form of a standard eigen-problem (Galler et al., 2021). H, is the Hamiltonian operator, which represents the total sum of energies in the system. Similar to the classical interpretation, when describing a quantum system, T is the kinetic energy, dependent on the momentum of the particle, and V is the potential energy of the system, granted by an external field. In order to accurately represent a particle system in 3D space H, becomes Equation 2 (Arnol'd et al., 1989). In the equation, the Hamiltonian is time dependent though it does not have to be. In addition, if the time element of the Schrödinger equation is separable, then the wave function becomes stationary as it represents the probability of a particle's position in space.

$$\hat{H}\,\Psi(x,t) = i\bar{h}\frac{\delta}{\delta t}\Psi(x,t) \qquad (1)$$

$$H = \left[V(x,y,z,t) - \frac{i\bar{h}}{2m}\nabla^2\right] \qquad (2)$$

The Schrödinger formulation is not the only interpretation of how a wave function can propagate, by using the Heisenberg or Interaction formulation as seen in Eugen Merzbacher (1998), the view of the wave system can change. In the interaction formulation the wave equation propagation can be broken down into discrete time-steps, allowing for easier understanding and analysis.

Because the set of analytically solvable solutions to the Schrödinger equation is limited, numerical solutions must be used to calculate the results to non-trivial molecules. All of the particles, both electrons and nuclei, interacted within the simulation scaling the computational complexity required at least quadratically. The Time-Independent or Time-Dependent Schrödinger can be used depending on the type of simulation required. Time-Dependent models allow for bonding and other dynamic processes to be simulated, whereas using the Time-Independent equation produces a stationary wave-function that varies only with position, Dick (2020, pp. 265–310).

In order to make simulations more accessible and allow for larger, more complex molecules, the computational complexity must be reduced. The most well-known method is the Born-Oppenheimer approximation (Born & Oppenheimer, 1927), which uses the momentum basis of the Schrödinger Equation to separate the nuclei and electrons into two wave-functions, which can then be propagated whilst the other remains fixed. Since the electrons contribute most significantly to the simulation complexity, solving the electronic wave-function through approximation will decrease computation times and increase the maximum simulation size possible.

The nature of Equation 1 is that both the wave-function and constants are undetermined; before attempting to solve a problem the wave-function for the quantum system must be found. Generating an approximate wave-function that correlates well with the analytical solution can be found through perturbation or variational methods (Nayfeh, 2011). Achieved by either perturbing one of the analytically solvable problems, in the case of an electronic wave-function this will be a single electron Schrödinger, or by combinations of basis sets as set out by Roothaan (1951). This and the use of a Self-Consistent Field, Hartree-Fock, approach with a Slater determinant (Slater, 1929) allows for iterative convergence for better approximate answers. Now, in the age of fast computing, the possibility of utilising Post Hartree-Fock methods such as Density Field Theory (DFT) and valence band theory can be fulfilled, these methods can be tuned to a variety of tasks, from full-scale ab-initio quantum dynamics to stationary bond energy predictions.

2. Molecular Dynamics & Mechanics

Full molecular modelling can be used when attempting to study the makeup of molecules and their behaviour in different environments. This takes a molecular potential field generated and uses it to propagate the atoms whilst being constrained by the molecule's bonds and other interactive forces.

Methods to generate molecular potential fields generally take three forms: Molecular Dynamics, being Quantum Chemistry extended over molecules and their interactions; Molecular Mechanics using a macro-mechanical interpretation of bond mechanics, energies and inter-molecular forces to generate potentials; and Semi-Empirical methods, which only utilise full quantum chemistry in certain parts of the molecule, including just the outer valence electrons or only certain physical locations of interest.

Though computationally cheaper molecular mechanics methods have significant drawbacks, including requiring extensive parametrisation and are restricted to ground-state simulations Levine (2014, pp. 634–654). Generally, all the parameters of the models need to be found empirically, requiring models to interpolate or extrapolate to the conditions of the simulation. Quantum chemistry, or *full ab-initio* methods, which take a variety of forms, are more computationally complex. Which becomes problematic when the potential of the molecule must be calculated each timestep. For this cost, they generate accurate approximate solutions, including information on the electronic properties of the material, meaning that these methods are capable of modelling covalent bond reactions. Semi-empirical simulations present a good balance of accuracy and speed. In these parts of the molecular structure which require complex reaction modelling use Quantum Chemistry methods while the rest can be treated with less complex methods, such as Molecular Mechanics (Warshel & Levitt, 1976).

All of these methods can be improved by implementing adaptive resolution, in which parts of the molecular structure that influence the molecule as a group can be summed into a single large pseudo-atom to reduce the number of interactions, possibly improving calculation speed (Kmiecik et al., 2016). During long simulations the size of the time-step contributes significantly to the computation required with larger time-steps reducing the number of costly potential calculations. In order to ensure this does not cause simulation decoherence, fast vibrations of light nuclei can be damped (Tuckerman et al., 1991) whilst still potentially allowing them to quantum tunnel.

As these simulations do not take place in a vacuum, the effects of the solvent must be taken into consideration in order to accurately study how large molecules behave. In continuum solvent models, the solvent potential fields are averaged and applied to the molecule of interest through the external potential field, Equation 2. Like the molecule being studied, the method of deriving the external potential can be of a quantum or classical nature, with classical solutions being tuned for specific solvents. In contrast, the quantum chemical methods are more generally applicable (Levine, 2014, pp. 634–654). Alternative models are able to predict how the physical properties of the solvent affect the molecule directly.

3. mRNA and Biological Applications

In eukaryotic cells, mRNA, or messenger RNA, is the encoded form of DNA transcribed by a polymerase in the nucleus of a cell where it has its on-coding regions spliced, is capped and has a poly(A) tail added before being transported into the cells' cytoplasm 'for reading by a ribosome (Cooper & Adams, 2023, pp. 263–278). mRNA-based vaccines and therapeutics exploit this process in order to confer immunity or immune responses to various threats, either by delivering mRNA directly into the cells cytoplasm or by using viral vectors directly into the nucleus of the cell. Beyond both vectors utilising human cellular transcription and translation to provoke an immune response, they are not similar. Viral vectors may cause host-genome integration and have provoked adverse reactions in recipients; it is by this process that the conditions of an active infection can be simulated, provoking a strong immune response (Ura et al., 2014). In order to facilitate direct mRNA delivery, lipid nanoparticles can be used, which are taken up most readily by dendritic cells, which then translate the mRNA stored within into antigens and communicate them with T-cells, generating an immune response (Kliesch et al., 2022).

The presence and general understanding of mRNA-based therapies has increased dramatically due to the pandemic in which the technology developed vaccines at unprecedented speed, and the widespread deployment of those vaccines has seen the technology tested on a scale not often seen (Wherry et al., 2021). The enormous interest and large quantity of funding available for mRNA-based products has further spurred development in wider applications such as disease mitigation or cancer treatments.

Unlike viral based products, which can be produced in cell-based reactors (Ura et al., 2014), products that are transcribed raw within the reactor setup require external control to ensure the purity, and therefore safety, of the product. In order to model the transcription process, the rate kinetics, side products and promoters of the reaction must be understood. Given the significant cost of experimenting with mRNA in a research environment, the challenge of observing products, and the large number of experiments required to construct an accurate digital twin of the process, finding an alternative way of understanding and predicting process properties is desirable. In understanding the initial rate kinetics of the polymerase, typically T7rna, the development of any reactor would be significantly sped up. Future processing steps, such as lipid encapsulation, could also be modelled depending on the modelling methods. Correctly understanding and modelling these processes will increase production efficiency and decrease potential time to market.

4. Hardware and Software Implementations

When assessing the calculation speed for a problem there exists no standard metric. For example, for programs that do not rely on parallelisation, the maximum operating frequency matters considerably more than the number of simultaneous cores present in the processor, with the opposite also being true. The instruction set architecture, generally register-memory in modern processors (Flynn, 1998), largely determines calculation speed by the efficiency of its physical instructions and how long it takes to access and return the data to registers, the best case, or memory required in the calculation, the worst case. In-between these memory layers exist processor caches, which have fast access times due to their physical architecture and proximity to the processor, reducing the amount of power and time wasted waiting for memory.

Ensuring as much required program data is in the cache can lead to performance increases by reducing memory fetch time. Generally, this can be achieved by storing data that is likely to be used together sequentially in arrays the size of the target cache. The desired precision of the calculations can have a significant impact on performance, as performing more precise floating-point logic takes longer. This can be addressed by choosing an acceptable level of accuracy for the entire program or specifically for time-consuming or repetitive calculations.

Though the individual time savings from each optimisation are very small, on the scale of single nanoseconds, the overall impact after calculating each particle-particle interaction over a representative large molecule is significant. For example, within T7rna a simple lab transcription polymerase, there are 6802 non-hydrogen atoms (Yin & Steitz, 2002), that must be accounted for in at most in each timestep. Running calculations on a single processor, even with optimisations, is still limiting; calculation throughput can be dramatically increased by parallelising the workflow through partitioning, though at the cost of complexity. This can be partially achieved within the processor or within processor groups, such as graphics processing units, by using SIMD instructions in which single instructions act on multiple pieces of data. Whereas fully horizontal parallelisation requires inter-processor networking and careful partitioning to avoid simulation decoherence.

Moving away from general purpose computing hardware towards specialised machines such ANTON, an application-specific integrated circuit (ASIC) based MD machine built by Shaw et al. (2008) capable of simulating 17 μs of time per day presents a way forward. Instead of utilising ASICs, field programmable gate arrays (FPGA) present a cheaper and more dynamic approach to solving MD simulations by supporting field reprogramming whilst allowing simulation logic to be integrated directly into the hardware. Because of this flexibility, FPGAs generally perform logic operations slower than ASICs, though the gap can be closed by utilising embedded hardware units. However, both are significantly more power efficient than performing the same calculations on general-purpose hardware (Betkaoui et al., 2010). Making this dedicated hardware approach more accessible would be a novel step to creating more performant molecular dynamics simulations.

5. Conclusions

Developing new methods of modelling large biological molecules accurately, benefits not only the manufacturing of mRNA therapies, but also the modelling of new designs and understanding the why and how of potential adverse reactions. Rather than performing calculations on more general-purpose compute hardware, time and money can be saved by using a more specialised hardware. This is by no means a new approach; existing MD systems already exist, albeit not in a widely available manner, and were not developed recently. Dedicated hardware is already in use in processes like digital signal processing, which are required to keep up with the high speed networking that our computational infrastructure requires to operate. The potential computational and cost benefits are real and can be used to enable reaction modelling and deepen our understanding of large pharmaceutical molecules which are becoming important in our global health.

References

Betkaoui, B., Thomas, D. B., & Luk, W. (2010). *Comparing performance and energy efficiency of FPGAs and GPUs for high productivity computing*. 94–101. https://doi.org/10.1109/FPT.2010.5681761

Born, M., & Oppenheimer, R. (1927). Zur quantentheorie der molekeln. *Annalen Der Physik, 389*, 457–484. https://doi.org/10.1002/andp.19273892002

C, S. J. (1929). The theory of complex spectra. *Physical Review, 34*, 1293–1322. https://doi.org/10.1103/physrev.34.1293

Cooper, G. M., & Adams, K. W. (2023). *The Cell: A Molecular Approach* (9th ed., pp. 263–278). Oxford University Press.

Dick, R. (2020). Time-dependent perturbations in quantum mechanics. In *Graduate texts in physics* (pp. 265–310). Springer International Publishing. https://doi.org/10.1007/978-3-030-57870-1_13

Eugen Merzbacher. (1998). *Quantum mechanics*. Wiley.

Flynn, M. J. (1998). *Computer architecture pipelined and parallel processor design*. Jones and Bartlett Publishers Inc.

Galler, A., Canfield, J., & Freericks, J. K. (2021). Schrodinger's original quantum-mechanical solution for hydrogen. *European Journal of Physics*, *42*, 035405. https://doi.org/10.1088/1361-6404/abb9ff

I, A. V., Vogtmann, K., & Weinstein, A. (1989). *Mathematical methods of classical mechanics*. Springer.

Kliesch, L., Delandre, S., Gabelmann, A., Koch, M., Schulze, K., Guzmán, C. A., Loretz, B., & Lehr, C.-M. (2022). Lipid–Polymer Hybrid Nanoparticles for mRNA Delivery to Dendritic Cells: Impact of Lipid Composition on Performance in Different Media. *Pharmaceutics*, *14*(12), 2675–2675. https://doi.org/10.3390/pharmaceutics14122675

Kmiecik, S., Gront, D., Kolinski, M., Wieteska, L., Dawid, A. E., & Kolinski, A. (2016). Coarse-Grained Protein Models and Their Applications. *Chemical Reviews*, *116*(14), 7898–7936. https://doi.org/10.1021/acs.chemrev.6b00163

Levine, I. N. (2014). *Quantum chemistry* (pp. 634–654). Pearson.

Nayfeh, A. H. (2011). *Introduction to perturbation techniques*. John Wiley & Sons.

Roothaan, C. C. J. (1951). New developments in molecular orbital theory. *Reviews of Modern Physics*, *23*, 69–89. https://doi.org/10.1103/revmodphys.23.69

Shaw, D. E., Deneroff, M. M., Dror, R. O., Kuskin, J. S., Larson, R. H., Salmon, J. K., Young, C., Batson, B., Bowers, K. J., Chao, J. C., Eastwood, M. P., Gagliardo, J., Grossman, J. P., Ho, C. R., Ierardi, D. J., Kolossváry, I., Klepeis, J. L., Layman, T., McLeavey, C., & Moraes, M. A. (2008). Anton, a special-purpose machine for molecular dynamics simulation. *Communications of the ACM*, *51*(7), 91–97. https://doi.org/10.1145/1364782.1364802

Thakkar, A. J. (2014). *Quantum chemistry : a concise introduction for students of physics, chemistry, biochemistry, and materials science*. Morgan & Claypool Publishers.

Tuckerman, M. E., Berne, B. J., & Martyna, G. J. (1991). Molecular dynamics algorithm for multiple time scales: Systems with long range forces. *The Journal of Chemical Physics*, *94*(10), 6811–6815. https://doi.org/10.1063/1.460259

Ura, T., Okuda, K., & Shimada, M. (2014). Developments in Viral Vector-Based Vaccines. *Vaccines*, *2*(3), 624–641. https://doi.org/10.3390/vaccines2030624

Warshel, A., & Levitt, M. (1976). Theoretical studies of enzymic reactions: Dielectric, electrostatic and steric stabilization of the carbonium ion in the reaction of lysozyme. *Journal of Molecular Biology*, *103*(2), 227–249. https://doi.org/10.1016/0022-2836(76)90311-9

Wherry, E. J., Jaffee, E. M., Warren, N., D'Souza, G., & Ribas, A. (2021). How did we Get a COVID-19 Vaccine in Less Than 1 Year? *Clinical Cancer Research*, *27*(8). https://doi.org/10.1158/1078-0432.ccr-21-0079

Yin, Y. W., & Steitz, T. A. (2002). Structural Basis for the Transition from Initiation to Elongation Transcription in T7 RNA Polymerase. *Science*, *298*(5597), 1387–1395. https://doi.org/10.1126/science.1077464

Flavio Manenti, Gintaras V. Reklaitis (Eds.), Proceedings of the 34th European Symposium on Computer Aided Process Engineering / 15th International Symposium on Process Systems Engineering (ESCAPE34/PSE24), June 2-6, 2024, Florence, Italy

Scoping and Identifying Data-Driven Optimization Prospects in the Danish Processing Industry

Adem R.N. Aouichaoui,[a] Brynjolf B. Ernstsson, [a] Peter Jul-Rasmussen,[a] Nicklas H. Iversen, [b] Laurent Vermue, [c] Jakob K. Huusom[a*]

[a] *Dept. of Chemical and Biochemical Engineering, Technical University of Denmark, Søltofts Plads 228A, 2800 Kgs. Lyngby, Denmark*
[b] *Viegand Maagøe A/S, Nørre Søgade 35, 1370 København K, Denmark.*
[c] *BioLean ApS, Søndre Jernbanevej 32, 3400 Hillerød, Denmark.*

[*]jkh@kt.dtu.dk

Abstract

Increased investment in various technological upgrades (sensors and IT infrastructures) in the processing industry has resulted in large data lakes over the last decade. These can be leveraged for various purposes such as gaining further process insight and optimization through advanced analytics and data-driven modelling. Nevertheless, this potential is rarely fulfilled due to the required wide range of expertise and time constraints. This work presents an extended framework towards bridging the gap between data availability and data utilization for the process and energy optimization within the Danish processing industry. The overall project aims to produce a real-world implementation of data-driven methods on-site at five different industries to highlight the potential, opportunities, and barriers of such endeavours. The multifaceted nature of the project requires multi-disciplinary teams with various expertise and a revised systematic framework to cover the complete life cycle of a data-driven model, from scoping the problem and model development to the on-site deployment and contentious maintenance. The core phases of such a framework are proposed herein.

Keywords: Optimization, Modelling, Framework, Digitalization

1. Introduction

The increased technological development and digitalization witnessed over the last decade has heralded the fourth industrial revolution also referred to as Industry 4.0 (Cannavacciuolo et al., 2023). One important catalyzer is the increased data availability and quality enabled through improved sensor technologies and IT infrastructures (A. Udugama et al., 2021). These developments are expected to play an important role in reducing the current industrial carbon footprint and producing more sustainable processes through the efficient use of various resources. As such, there is an increased interest from industry, policymakers, and governments to implement measures to accelerate and realize the full potential of Industry 4.0. This should be seen in light of the ever-increasing concerns about resource consumption and the negative effects of anthropological emissions. Efforts to mitigate these effects differ across countries and industries. In Denmark, the overshoot day in the year 2023 was around the 28th of March, making it among the top 15 fastest countries to reach overshoot day (Wackernagel and Lin, 2023). Despite reducing greenhouse gas (GHG) emissions by almost 23% (from 69.3 to 51.9 million t CO2e) between 2005 to 2018, it is mainly driven by the energy sector.

Meanwhile, the Danish processing industry accounts for almost 15% of the total CO_2e in GHG emissions (Danish Agency of Energy, 2023). Consequently, efforts have been made to publicly fund endeavours aiming to improve industrial energy efficiency and reduce GHG emissions. Yet, Despite this, a survey revealed few companies reported using data-driven approaches in plant operations (A. Udugama et al., 2021). Such projects require a wide range of expertise across disciplines ranging from project management, process engineering, operation, data science, and IT expertise. An important factor is that data-driven solutions must be integrated with on-site systems to allow for data input from users and systems (automatically and manually). It should also provide a means to display information and insights generated by the solution. This results in additional considerations including organizational impact on existing procedures, which requires the development of a framework encapsulating the entire process of problem scoping to model implementation and evaluation.

The development and use of systematic frameworks are not foreign in the field of process systems engineering (PSE), with various frameworks developed for process design and product design (Cignitti et al., 2018). However, most of these frameworks only focus on the analysis, synthesis, and validation steps. As such, few available frameworks focus on data-driven modelling. Hwangbo et al., (2020) proposed a framework that covers four steps: data pipeline (including data collection augmentation and reconstruction), model development (including data splitting and model training), model validation (model testing and optimal model selection) and global sensitivity analysis to provide an interpretable aspect to the data-driven model. InfoQ is another framework developed for assessing the quality of information in data-driven applications for Industry 4.0 (Reis, 2018). This framework is made of four components: analysis goal, data management, empirical analysis, and utility function evaluation. Common for both methods is that they are both model-centric rather than user-centric. Therefore, they omit important elements such as problem scoping, model deployment and maintenance. Model deployment is a crucial step and is often overlooked when developing a model whether it is based on mechanistic knowledge or machine learning (ML). A survey among data scientists revealed that only 13% see their models get deployed (Davenport and Malone, 2020). Another survey has reported that the deployment phase takes between 8 to 90 days with over 18% stating that longer deployment time is rather common (Paleyes et al., 2023). Therefore, the inclusion of a deployment step is vital for any systematic framework intended for developing models for production.

2. Machine Learning for Energy and Process Optimization (MLEEP)

2.1. Introduction

"<u>M</u>achine <u>L</u>earning for <u>E</u>nergy and <u>P</u>rocess Optimization (MLEEP)" is a publicly funded cross-industry project with the main objective of identifying the potential, opportunities, and barriers to leveraging ML for optimization purposes across various industries in the Danish processing ecosystem. This is done through the direct development and implementation of such tools at the production sites of a series of companies from different industries. The selected companies cover a wide range of industries including the food and beverage industry in the form of malt and whey protein production, the oil industry in the form of a refinery of used lubrication oil, a utility company providing district heating as well as the paint and coating industry. This variety is intentional and aims to broaden the applicability of the solution approach and findings. The project involves four stakeholders: a consultancy company (Viegand Maagøe A/S) specialising in project management and providing services within energy and utility optimization, an academic research centre in process systems engineering (DTU Chemical Engineering) at the Technical University of Denmark (DTU), an IT service and solution provider within data-driven modelling and digitalization (BioLean ApS) and the selected companies. The schematic of the central actors in the project is shown in Figure 1. Such a project involves many steps and in order to handle the complexity characterizing this project, a systematic framework was developed and is described in the following.

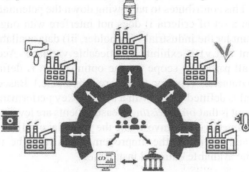

Figure 1: MLEEP involved stakeholders; inner circle (core) representing Viegand Maagøe A/S, DTU, and BioLean ApS, outer circle representing selected companies.
.

2.2. Systematic framework for model development and production deployment
The following presents the various steps and the activities associated with each phase of the framework. Figure 2 provides an overview of these steps and activities.

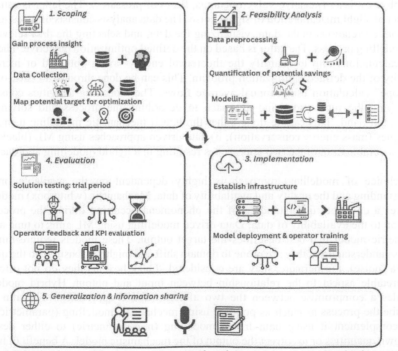

Figure 2: Main phases of the proposed framework and the major activities involved.

2.2.1. Scoping

The first phase of the framework is the *scoping* process which involves the core project partners. This step consists of gaining as much insight as possible into the industrial processes present in the company. This knowledge covers available measurements (amount, frequency, quality, format etc.), the various units (PIDs, technical drawings, etc.), the structure of the control system (manipulated variables, controlled variables, any manual intervention etc.) as well as operator experience concerning challenging aspects of running the plant. This contributes to narrowing down the potential case studies, which are selected based on a set of criteria i) does not interfere with ongoing projects at the company, ii) is relevant for the industrial stakeholder, iii) data availability, iv) the process is operator dependent and v) it exhibits a noticeable variance. Accordingly, and with agreement between all parties, a scope for the optimization is defined covering which process and which parameters (energy, time, throughput, etc.). Based on this, a concrete optimization potential is defined along with a suitable key performance indicator (KPIs). An important constraint is that *only existing* measurements are leveraged for this purpose. This is an important prerequisite to investigate the potential of ML to build upon existing optimization efforts and provide further improvements as such, the aim is to produce the best solution given the available data and installation.

2.2.2. Feasibility Analysis

The second phase consists of a *feasibility analysis*. This phase covers i) data preprocessing, ii) data analysis, iii) base case calculation, and iv) modelling activities. The data preprocessing aims to unify the various measurements provided by storing them in a single source enabling easy access to them. Further processing includes aligning timestamps, filling gaps in the data, denoising the data, and rationalizing measurements with physical laws (negative valve openings, flows or pressures). The latter is usually a product of slight miscalibration or signal noise. The data analysis consists of investigating potential correlations in the data, visualizing the data, and selecting the data relevant to the modelling process. The latter is based on the defined optimization objective. The base case calculations aim to quantify the theoretical energy-saving potential or increased capacity of the defined optimization problem. This can be done through a "back-of-the-envelope" calculation using annual average flows. The modelling activities consist of identifying the most promising approach to describing the process and solving the problem at hand. This can be done either through i) mechanistic modelling using first principles (mass/energy conservation), ii) data-driven approaches using ML (black box) or iii) a combination of the two approaches resulting in a hybrid modelling approach.

The choice of modelling approach is highly dependent on the available process understanding and the quality and availability of data. Mechanistic (white-box) modelling requires a complete understanding of the phenomena occurring during the process in addition to the availability of data. Data-driven modelling using ML aims to tune a non-parametric model that best describes the target output. These models require minimal process understanding, they are prone to domain shift and might be sensitive to the quality of data (noise). Furthermore, they are considered "black box" as they do not offer any interpretable aspect to the relationship between input and output. Hybrid modelling provides a compromise between the two aforementioned approaches. The aim is to describe the process as much as possible using mechanistic modelling (parametric) and then complement it using data-driven modelling (non-parametric) to either describe unknown quantities or to correct the output of the mechanistic model. A benefit of hybrid modelling is the added aspect of interpretability and explainability compared to purely data-driven modelling which contributes to the broader acceptance and adaptation of data-driven approaches, especially in fields that historically have been relying on mechanistic modelling or semi-empirical correlations. It is important to note that multiple approaches can be used to solve a given problem, this provides the opportunity to benchmark the various approaches. The selected models are then trained on available historical data and validated using a hold-out portion of the data. All modelling approaches are considered initially, and a final model is selected considering various criteria such as accuracy, simplicity, speed and required data input.

2.2.3. Implementation

When a model has achieved satisfactory accuracy on historical data, it is implemented at the plant. An important aspect of this implementation is that the solution is not directly implemented into the currently existing set-up but rather considered as an add-on where the operators are provided with suggestions on how to change the process conditions in real-time through a dashboard. This solution includes establishing the necessary infrastructure on-site to collect the data, deploy the model and provide real-time suggestions for the operators. As part of the implementation, the operators are instructed on how to use the tools they have at their disposal to ensure wide usage and acceptance.

2.2.4. Evaluation

After successful implementation of the model on-site, a trial period commences. During this, operators are encouraged to use the implemented solution and report their experience with it as a part of qualitative assessment. This aims to i) unravel aspects such as user-friendliness, ii) compare model output with operator experience, iii) evaluate the defined steps of the framework and revise any missing elements and iv) evaluate the communication between stakeholders. A quantitative assessment of the solution is done by revisiting the selected KPIs.

The continuous development and maintenance of the implemented solution is an important step in the life cycle of a data-driven model in production. The established infrastructure enables online retraining of the model whenever a deterioration of the performance is detected, a phenomenon also known as temporal quality degradation, which is expected to occur when operator response changes (Vela et al., 2022). Furthermore, the existence of several case studies allows for revisiting the proposed framework and developing further on it as required.

2.2.5. Generalization

An important aspect of the project is to disseminate the findings and experiences gained throughout the project. This includes whether the framework adopted was sufficient and any recommendation on how to extend and revise it and how to adapt it to cover the different elements to overcome the challenges faced in each industry. The aim of this step is also to highlight the successes and failures and more importantly, the elements that contributed to each. With this step, the aim is also to motivate and encourage the processing industry to adopt measures to become more energy efficient by presenting real-life applications.

3. Conclusion

Leveraging the increasing availability of data in the process industry requires a multidisciplinary effort that ranges from project management, process engineering, data scientists and software architects. Existing systematic frameworks within the field of process systems engineering mainly focus on model development and testing and do not include surrounding activities. In this work, we proposed a framework that addresses missing elements and consists of five phases: problem scoping, feasibility analysis, implementation, evaluation, and generalization. An added value of the framework is the further focus on the on-site implementation of data-driven models throughout its life cycle (development, training, validation, deployment, and retraining). Future work will aim to incorporate empirical validation of the proposed framework and a comparative analysis between the various case studies.

Acknowledgements

This study was financially supported by the Danish Energy Agency. ELF221-496435: Machine Learning til Energi- og Proces-optimering (MLEEP).

References

A. Udugama, I., Öner, M., Lopez, P.C., Beenfeldt, C., Bayer, C., Huusom, J.K., Gernaey, K.V., Sin, G., 2021. Towards Digitalization in Bio-Manufacturing Operations: A Survey on Application of Big Data and Digital Twin Concepts in Denmark. Frontiers in Chemical Engineering 3.

Cannavacciuolo, L., Ferraro, G., Ponsiglione, C., Primario, S., Quinto, I., 2023. Technological innovation-enabling industry 4.0 paradigm: A systematic literature review. Technovation 124, 102733.

Cignitti, S., Mansouri, S.S., Woodley, J.M., Abildskov, J., 2018. Systematic Optimization-Based Integrated Chemical Product–Process Design Framework. Ind. Eng. Chem. Res. 57, 677–688.

Danish Agency of Energy, 2023. Klimastatus og -fremskrivning 2023.

Davenport, T., Malone, K., 2020. Deployment as a Critical Business Data Science Discipline. Harvard Data Science Review 3.

Hwangbo, S., Al, R., Sin, G., 2020. An integrated framework for plant data-driven process modeling using deep-learning with Monte-Carlo simulations. Computers & Chemical Engineering 143, 107071.

Paleyes, A., Urma, R.-G., Lawrence, N.D., 2023. Challenges in Deploying Machine Learning: A Survey of Case Studies. ACM Comput. Surv. 55, 1–29.

Reis, M.S., 2018. A Systematic Framework for Assessing the Quality of Information in Data-Driven Applications for the Industry 4.0. IFAC-PapersOnLine 51, 43–48.

Vela, D., Sharp, A., Zhang, R., Nguyen, T., Hoang, A., Pianykh, O.S., 2022. Temporal quality degradation in AI models. Sci Rep 12, 11654.

Wackernagel, M., Lin, D., 2023. Earth Overshoot Day, in: Wallenhorst, N., Wulf, C. (Eds.), Handbook of the Anthropocene. Springer International Publishing, Cham, pp. 569–572.

Flavio Manenti, Gintaras V. Reklaitis (Eds.), Proceedings of the 34th European Symposium on Computer Aided Process Engineering / 15th International Symposium on Process Systems Engineering (ESCAPE34/PSE24), June 2-6, 2024, Florence, Italy

GCN-based Soft Sensor Utilizing Process Flow Diagram

Hiroki Horiuchi, Yoshiyuki Yamashita*

Tokyo University of Agriculture and Technology, 2-24-16 Naka-cho, Koganei, Tokyo 184-8588, Japan
yama_pse@cc.tuat.ac.jp

Abstract

Ensuring the secure and steady operation of chemical plants, online and real-time measurement of process variables is inevitable. However, certain important variables remain unmeasured due to prohibitive costs or inherent complexities. Soft sensors offer a pivotal solution to estimate unmeasured variables from the available data. Prevailing methods employ either physical models or data-driven models, such as PLS, LSTM, and CNN, to construct the estimation model. This study proposes the integration of piping and instrumentation diagram (P&ID) knowledge into data-driven soft sensor modeling, deploying a graph convolutional network (GCN) to capture spatial relationships. The proposed methodology has been validated through its application to concentration estimation in the Tennessee Eastman Benchmark Simulation, demonstrating its effectiveness and robustness to the accurate estimation.

Keywords: Graph Convolutional Networks, P&ID, soft sensor

1. Introduction

For safe and stable operation of chemical plants, it is essential to monitor plant conditions and properly control. However, there are many process variables where frequent measurement is difficult or costly. It is difficult to control such process variables in real time, resulting in control lag. Therefore, soft sensors are used as a method to estimate the values of process variables that are difficult to measure in real time. In recent years, data-driven methods such as Deep Neural Network (DNN) and Convolutional Neural Network (CNN) have been studied and their prediction accuracy has been improved. However, these methods are susceptible to the influence of training data and are weak in the extrapolation domain, making it difficult to cope with faults and changes in operation (D. Wu et al., 2021). When used for quality control and operation monitoring, soft sensors must have the versatility to make stable predictions under a variety of process conditions.

In this study, we develop a machine learning method for soft sensors that utilizes process knowledge. Piping & Instrumentation Diagrams (P&IDs), which are the blueprints of chemical plants, contain a variety of information about the plant. By utilizing this information for soft sensors, we aim to develop a method that has high prediction accuracy and is not easily affected by operational changes or faults in the plant.

2. Method

2.1. Graph Convolutional Network (GCN)

The concept of CNN, originally proposed by Y. LeCun et al. (1998), has been extensively used for Euclidean data such as images and texts. However, there kinds of direct embedding methods lack generalization capabilities for dynamic graphs (H. Cai et al., 2018). GCN aggregate information from graph structure and embedding (M. Xu et al., 2020). They introduce spectral convolutions, where a signal vector is multiplied with a filter parameterized in the Fourier domain. However, this approach has high computational complexity due to the multiplication with the Fourier basis. To address this problem, a graph convolutional network was proposed by T. N. Kipf and M. Welling (2017).

A multilayer GCN model is generalized as follows:

$$H^{(l+1)} = \sigma(\tilde{D}^{-\frac{1}{2}}\tilde{A}\tilde{D}^{-\frac{1}{2}}H^{(l)}W^{(l)}) \qquad (1)$$

where $H^{(l+1)}$ and $H^{(l)}$ are the output and input of layer l, $\theta^{(l)}$ is the parameter of layer l and $\sigma(\cdot)$ is the activation function used for nonlinear modelling. Generally, a two-layer GCN model is generalized as follows:

$$f(X, A) = \sigma(\hat{A} \, ReLU(\hat{A} \, XW^{(0)})W^{(1)}) \qquad (2)$$

where $W^{(0)} \in \mathbb{R}^{C \times H}$ is the weight matrix from the input layer to the hidden unit layer, C is the length of time, and H is the number of hidden units. $W^{(1)} \in \mathbb{R}^{H \times F}$ is the weight matrix from the hidden layer to the output layer, $f(X, A) \in \mathbb{R}^{N \times F}$ denotes the output with a forecasting length of F, and $ReLU(\cdot)$ is a common nonlinear activation function.

2.2. Directed graph

2.2.1. Definition of graph

A graph is represented as $G = (V, E)$ where V is the set of nodes and E is the set of edges. Let $v_i \in V$ to denote a node and $e_{ij} = (v_i, v_j) \in E$ to denote an edge pointing from v_j to v_i. The neighborhood of a node v is defined as $N(v) = \{u \in V \mid (v, u) \in E\}$. The adjacency matrix A is a $n \times n$ matrix with $A_{ij} = 1$ if $e_{ij} \in E$ and $A_{ij} = 0$ if $e_{ij} \notin E$. A graph may have node attributes X, where $X \in \mathbb{R}^{n \times d}$ is a node feature matrix with $x_v \in \mathbb{R}^d$ representing the feature vector of a node v. Meanwhile, a graph may have edge attributes X^e, where $X^e = \mathbb{R}^{m \times c}$ is an edge feature matrix with $x_{(v,u)} \in \mathbb{R}^c$ representing the feature vector of an edge (v, u).

A directed graph is a graph with all edges directed from one node to another. An undirected graph is considered as a special case of directed graphs where there is a pair of edges with inverse directions if two nodes are connected. A graph is undirected if and only if the adjacency matrix is symmetric.

2.2.2. k-NN graph

The k-NN graph is a representation of the relationships between data points in each dataset. It operates on the principle that similar data points in a feature space tend to exhibit similar characteristics. This graph is constructed by connecting each data point to its k-nearest neighbors, forming a network of interconnected nodes. To build the k-NN graph, one must first define a distance metric to quantify the similarity between data points. Common distance metrics include Euclidean distance, Manhattan distance, or cosine similarity. For each data point, the algorithm identifies its k-nearest neighbors based on the chosen metric and establishes edges between them in the graph (Y. Zhang, Jianbo Yu, 2022).

2.2.3. Proposed method

In chemical processes, unit operations and streams are physically connected with pipes. P&ID contains a large amount of knowledge about the relationships between different variables. To utilize the information of P&ID, it should be transformed into a graph. The graph works as loose constraints for the relationships between variables. Wu et al. showed a method to create a graph and to process the node data (D. Wu et al., 2021). Detailed steps to construct graphs from chemical processes are listed below.

Step 1: Create a stream node and a unit operation node based on the P&ID. Set edges between the stream node and the unit operation node from upstream to downstream.
Step 2: Correspondence between the node created in step 1 and the measurement node.
Step 3: Set edges from the stream and unit operation node created in step 1 toward the measurement node based on the P&ID.
Step 4: Add a manipulated node to the graph in step3 based on the P&ID. Set edges from the manipulated node to the control node and from the control node to the measurement node.
Step 5: Remove unnecessary nodes and directly connect existing nodes.
Step 6: Create self-loop edges.

In this study, lagged explanatory variables were also used to predict the objective variable. The total number of nodes never changed before or after adding the lag, because each lagged explanatory variable was stacked in a row of one measurement node. After P&ID was transformed into a graph, graph convolution is introduced by GCN, which is utilized to extract information from a graph.

3. Case study

3.1. Setup and dataset

The proposed method is evaluated on Tennessee Eastman Process (TEP) which is one of most popular benchmarks in chemical process. This process contains five main unit operations: a reactor, a condenser, a product stripper, a vapor-liquid separator, and a recycle compressor. This process flow is shown in Figure 1. The process produces two products from four reactants, including an inert and a byproduct. The process has 41 measurements and 12 manipulated variables in (Downs and Vogel, 1993).

In our study, we used 10,000 data measured every five minutes (Rieth et al., 2017). However, concentrations were measured every 10 or 15 minutes. Therefore, the most recent data was used for missing data. As a pre-processing step, data standardization was performed. In addition, data were forecasted using data up to three previous time points. The objective variable was set to the concentration of product component H.

Furthermore, 20 different faults were generated in the TEP simulator to check the prediction accuracy in the extrapolation region. We obtained 960 pieces of data for each fault and used 8,000 pieces of normal data for training and all data after the occurrence of the fault as test data for validation.

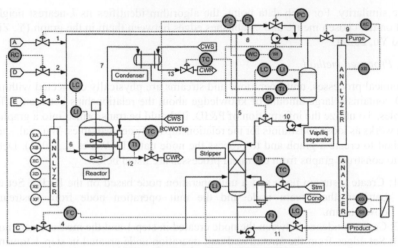

Figure 1. Process flow of the TEP

Table 1. RMSE and MAE for each method

	PLS	MLP	2D-CNN	*k*-NN +GCN	Proposed method
RMSE	0.3578	0.4184	0.3492	0.3286	**0.3198**
MAE	0.2527	0.3143	0.2512	0.2285	**0.2194**

3.2. Result and Discussion

3.2.1. Prediction for normal data

The proposed method was applied to predict the concentration of product component H. The prediction model was based on the training dataset for normal run (IDV00). The prediction accuracy was evaluated on the test dataset for normal run. The GCN model consists of an input layer, two hidden layers, and an output layer. The input layer has 136 units, while the hidden layers have 64 and 16 units, respectively. The output layer has 1 unit. ReLU activation functions were employed in the hidden layer, and MSE was used for the loss function. In addition, Adam was used to optimize the model. For comparison, PLS, MLP, 2D-CNN, *k*-NN+GCN were also used to predict the concentration H. Here, the MLP model consists of an input layer, two hidden layers, and an output layer. The hidden layers have 64 and 16 units, respectively. ReLU activation functions were employed in the hidden layer, and MSE was used for the loss function. The *k*-NN +GCN is a GCN model with *k*-NN graph. This method is proposed by Y. Zhang and J. Yu (2022). The architecture of each of these four methods is optimized by grid search. First, we tested whether incorporating P&ID into the model would improve the accuracy of the model. Prediction errors are summarized in Table 1 by MSE, RMSE and MAE.

Table 1 shows that the GCN model using the adjacency matrix created based on the P&ID has the highest accuracy. This shows the effectiveness of utilizing domain knowledge in a graphical form. In addition, it can be said that incorporating P&ID into GCN as graphical data can leverage domain knowledge and improve accuracy. An example of component concentration prediction is shown in Fig. 2.

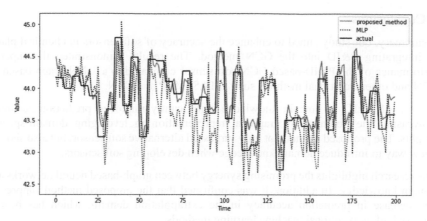

Figure 2. Prediction results for normal data

Table 2. RMSE for each fault type

Fault number	3	9	11	15	20
PLS	0.3027	0.3023	0.3039	0.3024	0.3142
MLP	0.3284	0.3285	0.3305	0.3281	0.3418
2D-CNN	0.3025	0.3021	0.3018	0.3011	0.3291
k-NN+GCN	0.3233	0.3240	0.3251	0.3227	0.3423
Proposed method	**0.2932**	**0.2930**	**0.2949**	**0.2951**	**0.3031**

3.2.2. Prediction for fault data

The proposed method was applied to the prediction of the concentration of product component H. The prediction model was based on the training dataset for normal run. The prediction accuracy was evaluated on the test dataset for faults run. PLS, MLP, 2D-CNN, k-NN+GCN were also used to predict the concentration of component H. The results of applying the model trained on normal data to all fault data showed that the proposed method improved RMSE for most faults. Some excerpts of the prediction errors are summarised in Table 2 by RMSE.

Table 2 confirms that the proposed method has the best results for five faults data. Since fault data was not used in the training data, the prediction of fault data is an example of prediction in the extrapolation domain. The high prediction accuracy of the proposed method suggests that the prediction in the extrapolation region was enhanced by utilizing the P&ID information. It is thought that the use of P&ID enabled more appropriate setting of learning parameters, resulting in higher prediction accuracy.

4. Conclusions

In summary, our study aimed to enhance the accuracy of soft sensors in chemical plants by integrating P&ID into the GCN method. The results demonstrated the superior performance of the P&ID-based GCN soft sensor compared to a *k*-NN graph based on GCN and other traditional methods such as PLS, MLP and CNN.

The use of P&ID as domain knowledge significantly improved prediction accuracy, showcasing the potential of graphical representation in leveraging domain-specific insights. The proposed method not only surpassed alternative soft sensor, but also showed a new way to introduce domain knowledge when developing soft sensors.

This research highlights the promising synergy between graph-based neural networks and domain knowledge. In addition, it was confirmed that the proposed method suppresses the decrease in prediction accuracy in the extrapolation domain, which has been a drawback of conventional machine learning methods.

Soft sensors are expected to be used for quality control and operation monitoring. When used for quality control purposes, they are required to have high prediction accuracy and versatility to make stable predictions under various process conditions. The proposed method was found to improve the results of these two elements.

References

H. Cai, W. Vincent, W. Zheng, K. C. Chang, 2018, A Comprehensive Survey of Graph Embedding: Problems, Techniques and Applications, IEEE Transactions on Knowledge & Data Engineering, vol. 30, no. 9, pp. 1616-1637

J. J. Downs, E. F. Vogel, 1993, A plant-wide industorial process control problem, Comput. Chem. Eng., vol. 17, pp. 245-255

L. Feng, C. Zhao, Y. Li, M. Zhou, H. Qiao, C. Fu, 2021, Multichannel Diffusion Graph Convolutional Network for the Prediction of Endpoint Composition in the Converter Steelmaking Process, IEEE Trans. Instrument. and Meas., vol. 70, 3000413

M. Jia, J. Hu, Y. Liu, Z. Gao, Y. Yao, 2023, Topology-Guided Graph Learning for Process Fault Diagnosis, Industrial & Engineering Chemistry Research, vol. 62, pp. 3238-3248

T. N. Kipf, M. Welling, 2017, Semi-Supervised Classification with Graph Convolutional networks, International Conference on Learning Representation (ICLR), Toulon, France

G. D. Molina, D. A. R. Zumoffen, M. S. Basualdo, 2011, Plant-wide control strategy applied to the Tennessee Eastman process at two operating points, Comput. Chem. Eng., vol. 35, pp. 2081-2097

C. Reinartz, M. Kulahci, O Ravn, 2021, An extended Tennessee Eastman simulation dataset for fault-detection and decision support systems, Comput. Chem. Eng., vol. 149, 107281

Rieth, A. Cory, Amsel, D. Ben, Tran, Randy, Cook, B. Maia, 2017, Additional Tennessee Eastman Process Simulation Data for Anomaly Detection Evaluation, Harvard Dataverse, V1

F. A. A. Souza et al., 2016, Review of soft sensor methods for regression applications, Chemometrics and Intelligent Laboratory Systems, vol. 152, pp. 69–79

D. Wu, J. Zhao, 2021, Process topology convolutional network model for chemical process fault diagnosis, Process Safety and Environmental Protection, vol. 150, pp. 93-109

T. Wu, I. Carreno, A. Scaglione, D. Arnold, 2023, Spartio -Temporal Graph Convolutional Neural Networks for Physics-Aware Grid Learning Algorithms, IEEE Trans. Smart Grid, vol. 14, pp. 4086-4099

M. Xu, 2020, Understanding graph embedding methods and their applications, SIAM, vol. 63, no. 4, pp. 825-853

Y. Zhang, Jianbo Yu, 2022, Pruning graph convolutional network-based future learning for fault diagnosis of industrial processes, Journal of Process Control, vol. 113, pp. 101-113

Flavio Manenti, Gintaras V. Reklaitis (Eds.), Proceedings of the 34th European Symposium on Computer Aided Process Engineering / 15th International Symposium on Process Systems Engineering (ESCAPE34/PSE24), June 2-6, 2024, Florence, Italy

Hybrid Model-based Design Space Determination for an Active Pharmaceutical Ingredient Flow Synthesis using Grignard Reaction

Junu Kim[a], Yusuke Hayashi[a], Sara Badr[a], Kazuya Okamoto[b], Toshikazu Hakogi[b], Satoshi Yoshikawa[c], Hayao Nakanishi[c], Hirokazu Sugiyama[a*]

[a]*Department of Chemical System Engineering, The University of Tokyo, 7-3-1, Hongo, Bunkyo-ku, Tokyo, 113-8656, JAPAN*
[b]*Technology Development Department, Pharmira Co., Ltd. Hyogo, 660-0813, JAPAN*
[c]*Production Technology Department, Shionogi Pharma Co., Ltd. Hyogo, 660-0813, JAPAN*
sugiyama@chemsys.t.u-tokyo.ac.jp

Abstract

This work presents a hybrid model-based design space determination for active pharmaceutical ingredient flow synthesis using Grignard reaction. A set of flow experiments was conducted to gather kinetic data for model development. One of the significant challenges encountered during this modelling process was the incorporation of impurity generation, specifically halohydrin, the mechanism of which remains elusive. To tackle this challenge, a hybrid modelling approach was adopted, integrating a mechanistic component to capture the main product and a data-driven component to account for the impurity. The resulting hybrid model was then employed to simulate various evaluation indices such as yield and temperature change towards the identification of the design space. Furthermore, the analysis of potential disturbances was conducted by leveraging this model, which can aid in the sensor monitoring and the management of disturbances.

Keywords: Flow chemistry; Drug substance; Machine learning; Continuous manufacturing; Random forest regression model

1. Introduction

Flow synthesis has recently attracted significant attention especially in the pharmaceutical industry. This growing interest can be attributed to the advantages offered by compact reactors in flow synthesis, including efficient heat and mass transfer, improved yields, cost savings, reduced ecological impact of manufacturing plants, and minimized waste and energy consumption. As such, extensive research has been conducted on the flow synthesis of active pharmaceutical ingredients (APIs) (Nqeketo, *et al.* 2023, Xu, *et al.* 2023, Burange, *et al.* 2022). These advancements encompass not only experimental work but also extend to the application of modelling and simulation techniques towards optimising pharmaceutical processes (Kim, *et al.* 2023a, Diab, *et al.* 2021).

The utilization of modelling and simulation techniques in API synthesis holds the potential for alignment with regulatory frameworks like the design space (ICH Q8) and state of control (ICH Q13). Design space is a concept within the framework of Quality by Design (QbD), referring to the multidimensional combination of input parameters that ensures product quality. Meanwhile, the state of control is a crucial notion, particularly in flow synthesis, signifying a condition that ensures ongoing process performance and product quality assurance. Creating a design space and assessing the state of control involves analysing various parameters, which requires a substantial volume of experimentation. The digital techniques can offer a promising avenue to reduce reliance on extensive experimentation, as exemplified by the works of García-Muñoz, *et al.* (2015) and Sagmeister, *et al.* (2023). Nonetheless, model-based investigations within this domain are still in their early stages of development, highlighting the need for ongoing efforts to seamlessly integrate new methodologies, such as digitalization and flow synthesis, into established regulatory frameworks. Process systems engineering can serve as a powerful tool for addressing this aspect (Gani, *et al.* 2022).

This work presents the model-based design space determination for an active pharmaceutical ingredient flow synthesis using Grignard reaction. The Grignard reaction is fast and highly exothermic reaction, making it difficult to collect kinetic data in batch synthesis. Thus, a set of flow experiments was conducted to collect kinetic data for model development. One of the key challenges encountered during this modelling process was incorporating the generation of impurities, such as halohydrin, the mechanism of which remains unknown. To address this challenge, a hybrid modelling approach was adopted, integrating a mechanistic component to capture the main product and a data-driven component to account for the impurity. The design space was then determined based on simulations using the developed hybrid model. Furthermore, the study conducted an assessment of process robustness in the presence of pulse disturbances.

2. Materials and Methods

2.1. Target Reaction and Flow Experiment

Figure 1 shows the target Grignard reaction and the overview of flow experiment (modified from Kim et al. (2023b)). In this reaction, (2R)-2-(Phenoxymethyl)oxirane (SM, starting material) undergoes a reaction with allylmagnesium chloride (RMgCl, Grignard reagent). The reaction is subsequently quenched with methanol, resulting in (2R)-1-phenoxy-5-hexen-2-ol (DP, desired product). As a by-product of this reaction, (2R)-1-chloro-3-phenoxy-2-propanol (BP, by-product) is also produced.

In the flow experiments, the inner diameter, temperature, concentration of RMgCl, and residence time were varied in different runs. High performance liquid chromatography (HPLC) was used to quantify the concentrations of SM, DP, and BP in the samples. The heat of reaction was measured by reaction calorimeter, and it was -258 kJ mol^{-1}.

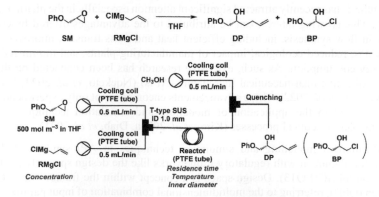

Figure 1. Target Grignard reaction and flow experiment

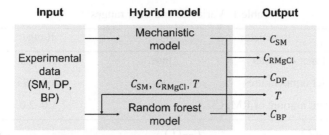

Figure 2. Overview of the hybrid model

2.2. Hybrid Model Development

In this study, the hybrid model was developed to quantify both DP and BP. **Figure 2** shows the overview of the model (modified from Kim et al. (2023b)). The mechanistic component of this model involves the mass and energy balance models, while the data-driven part of the model incorporates the application of a random forest regression model to estimate the trace amount of produced by-proudct. Although the miniscule amount of BP produced during the reaction is problematic from a product quality point of view, the effect on the overall mass and energy balances is negligible. Thus, the mechanistic component of the hybrid model considered only the main reaction to give DP. The mechanistic model is presented in Eqs. (1)–(4).

$$\frac{\partial C_{SM}}{\partial t} = -u\frac{\partial C_{SM}}{\partial l} + D_{SM}\frac{\partial^2 C_{SM}}{\partial l^2} - r \tag{1}$$

$$\frac{\partial C_{RMgCl}}{\partial t} = -u\frac{\partial C_{RMgCl}}{\partial l} + D_{RMgCl}\frac{\partial^2 C_{RMgCl}}{\partial l^2} - r \tag{2}$$

$$\frac{\partial C_{DP}}{\partial t} = -u\frac{\partial C_{DP}}{\partial l} + D_{DP}\frac{\partial^2 C_{DP}}{\partial l^2} + r \tag{3}$$

$$\frac{\partial T}{\partial t} = -u\frac{\partial T}{\partial l} + \alpha\frac{\partial^2 T}{\partial l^2} + \frac{r(-\Delta H)}{\rho C_p} - \frac{2(T - T_c)}{\rho C_p r_{in}^2 \left(\frac{1}{hr_{in}} + \frac{1}{k_{thermal}}ln\left(\frac{r_{out}}{r_{in}}\right)\right)} \tag{4}$$

Here, C_i [mol m^{-3}] is the concentrations of i (SM, RMgCl, and DP), u [m s^{-1}] is the flow velocity, D_i [m^2 s^{-1}] is the diffusion coefficients, r [mol m^{-3} s^{-1}] is the reaction rate, T [K] is the temperature, α [m^2 s^{-1}] is the thermal diffusivity, ΔH [J mol^{-1}] is the heat of reaction, ρ [kg m^{-3}] is the density of the solvent, C_p [J kg^{-1}, K^{-1}] is the specific heat capacity, r_{in} and r_{out} [m] are the inner and outer radius, h [J s^{-1} m^{-2} K^{-1}] is the heat transfer coefficient, $k_{thermal}$ [J s^{-1} m^{-1} K^{-1}] is the thermal conductivity of the PTFE tube.

For BP, it was not feasible to develop a mechanistic model due to the unresolved nature of the underlying mechanism. Instead, a random forest regression model was constructed in this work. To evaluate the reliability of the model, a Leave-One-Out (LOO) cross-validation process was conducted.

2.3. Evaluation Model Development and Conditions Settings for Design Space Analysis

To assess the design space, several evaluation indices were determined in this work. Evaluation models were developed for yields (DP and BP) and maximum temperature change as shown in Eqs. (5)–(7). The constraints for each evaluation index are defined as follows: $y_{DP} \geq 0.90$, $y_{BP} \leq 0.005$, and $\Delta T_{max} \leq 3.0$.

Table 1. Variables and their ranges

Parameter		Range
Inner diameter	mm	1.0–2.4
Set temperature	K	263–293
Equivalent amount of RMgCl	eq.	1.0–2.0

$$y_{DP} = \frac{C_{DP}(L)}{C_{SM}(0)} \quad (5) \qquad y_{BP} = \frac{C_{BP}(L)}{C_{SM}(0)} \quad (6) \qquad \Delta T_{max} = T_{max} - T(0) \quad (7)$$

Here, y_{DP} and y_{BP} [–] are the yields of DP and BP, ΔT_{max} [K] is the maximum temperature change, $C_{SM}(0)$ [mol m^{-3}] is the initial concentration of SM at the inlet, $C_{DP}(L)$ and $C_{BP}(L)$ [mol m^{-3}] are the concentrations of DP and BP at the outlet, T_{max} [K] is the maximum temperature in the reactor, and $T(0)$ [K] is the initial temperature.

The design space was investigated for a steady-state process, as outlined in Eq. (8). The process variables and their ranges are summarised in **Table 1**. Furthermore, the process was assessed in the presence of pulse disturbances, as defined in Eq. (9). These disturbances, characterised by their intensity and duration, were applied to the steady-state process. This study conducted two case studies involving variations in inlet and coolant temperature, considering the highly exothermic nature of the Grignard reaction.

$$DS = \left\{ x | x \in X, \prod_{k=1}^{n} z_k = 1, z \in Z(x) \right\} \qquad (8)$$

$$DS = \left\{ x_{dis} | x_{dis} \in X_{dis}, \prod_{t=t_{steady1}}^{t_{steady2}} \prod_{k=1}^{n} z_{t,k} = 1, z \in Z(x_{dis}) \right\} \qquad (9)$$

Here, DS is the design space, x is the element of the set of variables X, z is the element of the set Z which represents whether the results of each evaluation index fulfil the constraints, x_{dis} is element of the set of disturbance variables X_{dis}, $t_{steady1}$ is the time point when the system reaches the steady state, and $t_{steady2}$ is the time point when the system reaches the steady state after the occurrence of the disturbance. When z equals to 1, it indicates compliance with the constraint, while a value of 0 indicates non-compliance.

3. Results and Discussion

3.1. Kinetic Analysis and Regression Results
The kinetic analysis and regression results are shown in **Figure 3** (modified from Kim et al. (2023b)). The average coefficient of determination (R^2) of SM and DP for all experimental conditions was 0.96. For the random forest regression model, the R^2 was calculated using the mean values of prediction results by the LOO cross validation, and the resulting value was 0.96.

3.2. Assessment of Design Space
The simulation was conducted using the predefined ranges of variables outlined in **Table 1**, with the aim of investigating the design space for the Grignard reaction. **Figure 4** presents the results of (a) yield of DP, (b) yield of BP, (c) maximum temperature change, and (d) design space.

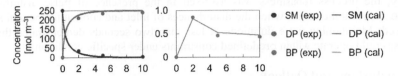

Figure 3. Prediction results of the hybrid model

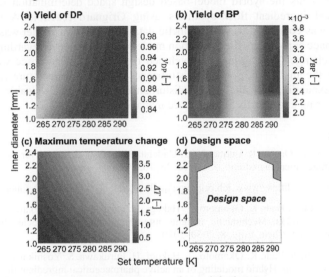

Figure 4. Individual evaluation results and design space

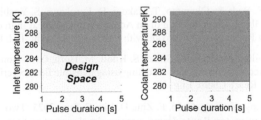

Figure 5. Design space considering pulse disturbances, with the x-axis representing pulse duration and the y-axis representing intensity

In general, the simulation results show that a smaller inner diameter and higher set temperatures significantly enhance the yield of DP, albeit concurrently escalating the maximum temperature change. Furthermore, as illustrated in **Figure 4(b)**, elevated temperatures lead to increased production of BP. A notable increase in BP production is observed when the inner diameter falls within the intermediate range of 1.4 to 1.8 mm. This trend aligned with the experimental observations, and the data-driven model successfully reflected the trends identified in the experimental data.

Figure 4 (d) shows the design space which fulfils the predefined constraints ($y_{DP} \geq 0.90$, $y_{BP} \leq 0.005$, and $\Delta T_{max} \leq 3.0$), indicated by the white area. Within the domain of larger inner diameters and lower temperatures, the process violated the DP yield constraint. In the case of larger inner diameters and elevated temperatures, the process could not satisfy the maximum temperature constraint, surpassing the predefined value.

Finally, the process robustness was assessed in the presence of pulse disturbances. Figure 5 shows the results when the disturbances of inlet and coolant temperature were occurred. Maintaining pulse durations of less than two seconds demonstrated that the process could operate within predefined constraints under specific conditions.

4. Conclusions and Outlook

This work presents the hybrid model-based design space determination for an active pharmaceutical ingredient flow synthesis using Grignard reaction. A set of flow experiments was conducted to gather kinetic data, and a hybrid model was developed. For design space exploration, the evaluation models were developed regarding yields and temperature change. Leveraging the developed model, the design space was assessed. Furthermore, the analysis of potential disturbances related to temperature variables was conducted. Expanding this analysis of disturbances can provide valuable insights for sensor monitoring and effective disturbance management.

References

A. S. Burange, S. M. Osman, R. Luque, 2022. Understanding flow chemistry for the production of active pharmaceutical ingredients. iScience, 25, 103892.

ICH Official website, https://www.ich.org/page/quality-guidelines (acessed October 12th, 2023)

J. Kim, Y. Hayashi, S. Badr, K. Okamoto, T. Hakogi, H. Furukawa, S. Yoshikawa, H. Nakanishi, H. Sugiyama, 2023a. Mechanistic insights into the amination via nucleophilic aromatic substitution. React. Chem. Eng., 8, 2060–2070

J. Kim, Y. Hayashi, S. Badr, K. Okamoto, T. Hakogi, H. Furukawa, S. Yoshikawa, H. Nakanishi, H. Sugiyama, 2023b. Hybrid modeling of an active pharmaceutical ingredient flow synthesis in a ring-opening reaction of an epoxide with a Grignard reagent. Ind. Eng. Chem. Eng.

J. Kim, Y. Hayashi, S. Badr, K. Okamoto, T. Hakogi, H. Furukawa, S. Yoshikawa, H. Nakanishi, H. Sugiyama, 2023b. Hybrid Modeling of an Active Pharmaceutical Ingredient Flow Synthesis in a Ring-Opening Reaction of an Epoxide with a Grignard Reagent. Ind. Eng. Chem. Res.

P. Sagmeister, C. Schiller, P. Weiss, K. Silber, S. Knoll, M. Horn, C. A. Hone, J. D. Williams, C. O. Kappe, 2023. Accelerating reaction modeling using dynamic flow experiments, part 1: design space exploration. React. Chem. Eng.

Q. Xu, J. Chen, Z. Wang, Y. Zang, G. Li, F. Zhu, D. Liu, C. Sun, 2023. Two‐step flow synthesis of Olaparib in microreactor: Route design, process development and kinetics research. Chem. Eng. J., 471, 144304.

R. Gani, X. Chen, M. R. Eden, S. S. Mansouri, M. Martin, I. M. Mujtaba, O. Padungwatanaroj, K. Roh, L. Ricardez-Sandoval, H. Sugiyama, J. Zhao, E. Zondervan, 2022. Challenges and opportunities for process systems engineering in a changed world. Comput. Aided Chem. Eng., 49, 7–20.

S. Diab, M. Raiyat, D. I. Gerogiorgis, 2021. Flow Synthesis Kinetics for Lomustine, an Anti-Cancer Active Pharmaceutical Ingredient. React. Chem. Eng., 6, 1819–1828.

S. García-Muñoz, C. V. Luciani, S. Vaidyaraman, K. D. Seibert, 2015. Definition of Design Spaces Using Mechanistic Models and Geometric Projections of Probability Maps. Org. Process Res. Dev., 19, 1012-1023.

S. Nqeketo, P. Watts, 2023. Synthesis of Dolutegravir Exploiting Continuous Flow Chemistry. J. Org. Chem., 88, 12024–12040.

Flavio Manenti, Gintaras V. Reklaitis (Eds.), Proceedings of the 34th European Symposium on Computer Aided Process Engineering / 15th International Symposium on Process Systems Engineering (ESCAPE34/PSE24), June 2-6, 2024, Florence, Italy

Development of a Strategy for the Analysis of the Fluid Dynamic Behavior of Sieve Trays for Distillation

Jose Alfredo Paredes-Ortiz [a] , Mario Alberto Rodríguez-Angeles [b] , Vineet Vishwakarma [c] , Markus Schubert [d,e] , Uwe Hampel [e,f] , Fernando Israel Gómez-Castro [a*] , Mayra Margarita May-Vázquez [g] .

[a] *Departamento de Ingeniería Química, División de Ciencias Naturales y Exactas, Campus Guanajuato, Universidad de Guanajuato, Noria Alta S/N Col. Noria Alta, Guanajuato, Guanajuato, Mexico 36050.*

[b] *Universidad Politécnica de Juventino Rosas, Departamento de Metalúrgica, Hidalgo No. 102, Comunidad de Valencia, Juventino Rosas, Guanajuato, Mexico 38253.*

[c] *University of Michigan, Department of Nuclear Engineering & Radiological Sciences, 2355 Bonisteel Boulevard, Ann Arbor, Michigan 48109, USA.*

[d] *Chair of Chemical Process Engineering, Technische Universität Dresden, 01062 Dresden, Germanye.*

[e] *Institute of Fluid Dynamics, Helmholtz-Zentrum DresdenRossendorf, Bautzner Landstraße 400, 01328 Dresden, Germany*

[f] *Chair of Imaging Techniques in Energy and Process Engineering, Technische Universit¨at Dresden, 01062 Dresden, Germany*

[g] *Unidad Académica de Educación Virtual, Dirección General de Desarrollo Académico, Universidad Autónoma de Yucatán, Benito Juárez 421 Cd. Industrial, Mérida, Yucatán, Mexico 97288.*

fgomez@ugto.mx

Abstract

An Eulerian-Eulerian model was applied to simulate the two-phase flow behaviour in a sieve tray column, i.e., typically used for distillation. Following grid independence, the phase fraction was simulated, which exhibit good agreement with experimental data, especially in the liquid-dominant zones of the two-phase dispersion. In the upper gas-dominant dispersion regions, a slight deviation of less than 10% was observed in the gas fraction data. The results demonstrate that the proposed modeling strategy can adequately predict the fluid dynamic behaviour in tray-equipped distillation columns.

Keywords: Sieve tray, CFD, Eulerian-Eulerian model, Two-phase flow, Hydrodynamics.

1. Introduction

Tray distillation columns are the cornerstone of the chemical industry, and they are employed extensively for the separation of fluid mixtures. The intricate fluid dynamics of the gas-liquid interactions in these columns, however, pose a significant challenge in their numerical modeling (Yang Kong et al., 2022). Two-phase gas-liquid flows in tray columns have complex interfacial dynamics due to the interplay of phenomena such as coalescence, breakup, turbulence, and so forth.

Despite recent advancements in Computation Fluid Dynamics (CFD), an accurate prediction of the fluid dynamics in distillation columns has not been fully achieved. This holds even for the sieve trays, which are the most common column internals. Sieve trays are nothing but perforated plates that promote interfacial contact between gas and liquid phases thereby improving the efficiency of the separation process (Vishwakarma et al., 2021). However, the fluid dynamics related to these trays are influenced by a multitude of factors, including tray design, operating conditions, and physical properties of the fluids, to name but a few. The interplay of these factors leads to a range of flow regimes and conditions, which makes the accurate prediction of the fluid behaviour in tray columns incredibly challenging. The inadequate selection of design variables as the tray diameter, the width and length of downcomers, among others, may lead to operational as flooding or weeping. Therefore, a robust modelling and simulation strategy is desired to address this challenge, which is the scope of the present proposal. This work emphasizes on validating the applied CFD model using high-fidelity experimental data from a sieve tray column (recently reported by Vishwakarma et al. (2021)). The work aims to provide valuable insights of the fluid dynamics in a sieve tray column and encourages CFD application for the design and analysis of high-performance trays in the future.

2. Modeling methodology

The Euler-Euler approach is popular in CFD for modeling multiphase flows such as for gas-liquid interactions in distillation columns (Drumm et al., 2010). This approach treats each phase as a separate interpenetrating continuum with its own set of conservation equations for mass and momentum (Alzyod et al., 2018).

The conservation equation for the mass of phase i is given by Eq. (1)

$$\frac{\partial(\alpha_i \rho_i)}{\partial t} + \nabla \cdot (\alpha_i \rho_i \mathbf{u}_i) = 0, \tag{1}$$

where α_i is the phase fraction, ρ_i is the density, and $\mathbf{u}_i$ is the velocity of the phase i. The sum of the fractions for all phases is unity, which is Eq. (2) as

$$\sum \alpha_i = 1. \tag{2}$$

The momentum equation for the phase i is given as

$$\frac{\partial(\alpha_i \rho_i \mathbf{u}_i)}{\partial t} + \nabla \cdot (\alpha_i \rho_i \mathbf{u}_i \mathbf{u}_i) = -\alpha_i \nabla p + \nabla \cdot \mathbf{T}_i + \alpha_i \rho_i \mathbf{g} + \sum \mathbf{R}_{ij}, \tag{3}$$

where p is the pressure, $\mathbf{T}_i$ is the stress tensor for phase i, $\mathbf{g}$ is the gravitational acceleration, and $\mathbf{R}_{ij}$ the interphase momentum transfer from phase j to phase i. Solving these conservation equations lead to the prediction of the phase distribution and velocity profiles that are directly related to the column performance.

3. Column geometry and boundary conditions

In this work, the Euler-Euler approach was implemented in FLUENT 23R1 to solve the Eqs. (1-3). The geometry used in this work corresponds to a sieve tray column, directly derived from the experimental work of Vishwakarma et al. (2021). The dimensions of the tray column facility are summarized in Table 1, and its CAD drawing is shown in Figure 1.

Table 1. Specifications of the tray column derived from Vishwakarma et al. (2021).

Particulars	Dimensions
Internal column diameter	800 mm
Inlet weir (L × W × H)	532 mm × 2 mm × 35 mm
Outlet weir (L × W × H)	465 mm × 2 mm × 20 mm
Flow path length	620 mm
Active tray area	0.44 m²
Hole specifications	3052 × 5 mm Ø, pitch: Δ × 12 mm
Fractional free area	13.55%
Downcomer clearance	20 mm
Tray spacing	365 mm
Tray thickness	15 mm
Calming zone	36 mm (inlet), 30 mm (outlet)

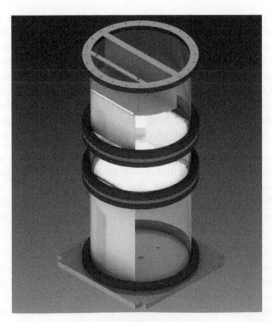

Figure 1. CAD drawing of the tray distillation column.

For the shown experimental setup, Vishwakarma et al. (2021) recently reported the distributions of effective froth height, phase fraction and liquid residence time at several weir loadings and F-factors. These data were obtained in their work based on the application of a multiprobe conductivity sensor and novel data processing algorithms (Vishwakarma et al., 2020, 2021). The availability of the aforesaid high-resolution data along with other information pertaining to setup, operating conditions, and physical properties make them ideal for CFD model validation. From their work, the data corresponding to 1.77 Pa$^{0.5}$ F-factor and 2.15 m^{3}m^{-1}h^{-1} and 4.30 m^{3}m^{-1}h^{-1} weir loading are considered for CFD simulation.

To solve the proposed equations and systems, a specific solution method was employed. The turbulence modeling was carried out using the k-ε model with the RNG sub-model, which allows for a more precise modeling of the turbulent behavior with high vorticity (Rodríguez-Ángeles et al., 2015). Furthermore, air and water were the two phases in the simulation (just like in the reported experimental campaign). Based on the selected operating conditions, a coupled resolution approach was used due to the nonlinearity of the model. Likewise, the resolution parameters were configured to be second order. The phase solution parameter was Modified HRIC due to its affinity in multiphase resolution models as well as due to greater precision at the water-air interface, which reduces its probability of divergence.

To achieve stability in the models, an F-Cycle approach was used, supported by the BCGSTAB stabilization method. This was due to the high nonlinearity of system's model, which generates floating points in situations where turbulent viscosity reaches very high values.

4. Model validation and assessment

Figure 2 compares the average liquid phase fraction (α) obtained experimentally at different heights above the tray with the CFD model predictions for the considered loadings. This Figure also shows the CFD-based liquid phase distribution at given heights above the tray for the given weir loadings. Good agreement in the holdup values from CFD and experiments in the Figure validates the modeling approach used here. The analysis of the CFD data suggests an increase in the average liquid phase with height until the gas jets have enough momentum to keep the liquid suspended. Afterwards, the liquid phase tends to reduce and rightfully becomes nil above the two-phase dispersion. The observations from the CFD data analysis comply with those from the experimental data.

The simulation results reveal that higher loadings lead to an increase in the liquid phase downstream, particularly around 50 mm above the tray. In contrast, under lower loading conditions, the average value is much higher, occurring around 40 mm above the distillation tray. This suggests that the liquid phase distribution is significantly influenced by the loading conditions, with higher loadings promoting a more downstream liquid phase distribution.

Interestingly, very similar behavior was observed between the average liquid phase values from the experimental data and the simulations at most heights above the tray. However, between 40-60 mm elevation above the deck, a higher deviation in the average liquid phase was observed between the two. Further, at heights greater than 60 mm, the CFD predicted the average liquid phase again similar to the experimental data.

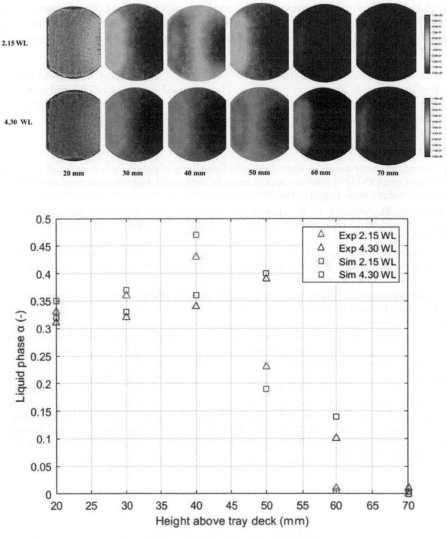

Figure 2. Comparison of experimental and simulated average liquid phase versus height above tray deck for the loadings (lower), and CFD-led phase fraction distribution at different heights above the deck (upper).

5. Conclusions

The computational fluid dynamics (CFD) model used in this study has demonstrated its effectiveness in accurately predicting the liquid phase distribution in a sieve tray column under two loading conditions ($2.15 \ m^3m^{-1}h^{-1}$ and $4.30 \ m^3m^{-1}h^{-1}$). The model predictions align very well with the experimental data except at few heights above the tray deck indicating the scope for further model refinement. Particularly, as unidirectional liquid velocities based on tracer-based experimental data are reported only for one height, the

CFD-based velocity data requires post processing to make it suitable for comparison. The study also revealed that the loading conditions significantly influence the liquid phase distribution on a sieve tray. Higher weir loading was found to promote elevation-specific liquid phase in the tray downstream.

These findings validate the use of the CFD model and highlight its potential for optimizing trays in distillation processes. The proposed analysis strategy will contribute to the development of more efficient and effective distillation processes, and ultimately leading to significant energy and cost savings in the chemical industry.

References

S. Alzyod, M. Attarakih, H.-J. Bart, 2018. CFD modelling of pulsed sieve plate liquid extraction columns using OPOSPM as a reduced population balance model: hydrodynamics and mass transfer. Comput. Aided Chem. Eng., 43, 451–456.

C. Drumm, M. Attarakih, M. W. Hlawitschka, H.-J. Bart, 2010, One-group reduced population balance model for CFD simulation of a pilot-plant extraction column, Ind. Eng. Chem. Res., 49, 3442–3451.

M.A. Rodríguez-Ángeles, F.I. Gómez-Castro, J.G. Segovia-Hernández, A.R. Uribe-Ramírez, 2015. Mechanical design and hydrodynamic analysis of sieve trays in a dividing wall column for a hydrocarbon mixture, Chem. Eng. Process., 97, 55-65.

V. Vishwakarma, S.A Haq, E. Schleicher, M. Schubert, U. Hampel, 2021. Experimental analysis of the hydrodynamic performance of an industrial scale cross-flow sieve tray, Chem. Eng. Res. Des., 174, 294-306.

V. Vishwakarma, E. Schleicher, A. Bieberle, M. Schubert, U. Hampel, 2021, Advanced flow profiler for two-phase flow imaging on distillation trays, Chem. Eng. Sci., 231, 116280.

V. Vishwakarma, P. Wiedemann, E. Schleicher, M. Schubert, U. Hampel, U., 2021, A new approach for estimating the effective froth height on column trays, Chem. Eng. Sci., 231, 116304.

V. Vishwakarma, E. Schleicher, M. Schubert, M. Tschofen, M. Löschau, 2020, Inventors; Helmholtz Zentrum Dresden Rossendorf eV, assignee. Sensor zur Vermessung von Strömungsprofilen in großen Kolonnen und Apparaten. German patent DE1020181245012020.

Z. Yang Kong, H. Yeh Lee, H., J. Sunarso, 2022, The evolution of process design and control for ternary azeotropic separation: Recent advances in distillation and future directions, Sep. Purif. Technol., 284, 120292.

Flavio Manenti, Gintaras V. Reklaitis (Eds.), Proceedings of the 34th European Symposium on Computer Aided Process Engineering / 15th International Symposium on Process Systems Engineering (ESCAPE34/PSE24), June 2-6, 2024, Florence, Italy

Dynamic Modelling of a Milk Triple Effect Falling Film Evaporator with Mechanical Vapor Recompression

Hattachai Aeowjaroenlap[a], Jakob Strohm[b], Isuru Udugama[a], Wei Yu[a], Brent Young[a*]

[a]*The University of Auckland, Auckland 1010, New Zealand*
[b]*Technical University of Munich, Munich 80333, Germany*
b.young@auckland.ac.nz

Abstract

The dairy industry plays a crucial role in the global food supply chain, with milk powder processing plants being integral components of this industry. The efficient operation of the plant is essential to ensure high-quality milk powder production while minimizing resource consumption and operational costs. Conventional steady-state modelling is not enough to handle the complex dynamic scenarios in the plant, as well as the demonstration of process control. In this work, we developed a novel dynamic model with integrated control loops for a Triple Effect Falling Film Evaporator with Mechanical Vapor Recompression (MVR) using the industrial process simulator, Aspen HYSYS. To represent the complexity of real compounds, we simulated milk by employing hypothetical components based on literature data and regressions. We then investigated the effects of changes in milk composition, production throughput, and product quality versus energy consumption on the industrial standard control-based control scheme with both feed-forward and feedback elements. Finally, model limitations and future possibilities were also examined.

Keywords: Milk Evaporator, Dynamic Modelling, Process Control

1. Introduction

The significance of food processing in modern society is continually growing, and in New Zealand, dairy products play a vital role in contributing to the economy (Ballingall and Pambudi, 2017). One of the most important parts of the dairy industry is the large-scale production of milk powder, a process characterized by the removal of water under strictly hygienic conditions, all executed with a focus on minimizing costs (Packer *et al.*, 1998). Evaporation is the second most energy-intensive stage in milk powder production, behind the drying process (Ruan *et al.*, 2015). Some researchers have suggested energy-saving opportunities, including evaporating milk at temperatures exceeding 70°C using multi-effect evaporators and employing mechanical vapor recompression (MVR) in conjunction with falling-film evaporators (Bylund, 2003; Pisecky *et al.*, 2012).

For modelling and simulation studies, there is a limited amount of literature addressing the modeling of falling-film evaporators for milk powder application using industrial process simulators (Zhang *et al.*, 2018). Previously, researchers have done extensive studies on developing artificial milk components in commercial process simulation software (Zhang *et al.*, 2014; Munir *et al.*, 2016). Subsequently, a steady-state model of a commercial multiple effects Falling Film Evaporator (FFE) was proposed (Zhang *et al.*,

2018). In this study, we extend these prior efforts by developing a dynamic model of a triple-effect falling film evaporator tailored for the milk powder production process and then integrating it with a conventional process control scheme currently used in the actual plant.

2. Process Description

Generally, the water content in raw milk varies, typically ranging from 87 wt% to 91 wt%. To transform milk into milk powder, the removal of this water is essential, achieved through the evaporation of water in milk. While cost-effectiveness is a priority, hygiene standards must be maintained throughout the process. Moreover, it is also important to preserve desirable natural properties such as nutritional value, flavour, colour, and solubility. To meet these requirements, the process is commonly conducted under reduced pressure, lowering the boiling point of water, and therefore avoiding excessively high temperatures (Packer *et al.*, 1998).

2.1. Multi-effect Evaporator

For the evaporator, the water content in pasteurised and standardised feed milk is decreased from an initial 87 % to a final concentration of around 50 % by boiling water from the milk. Heat is a critical factor in this operation, but due to the sensitivity of certain milk components to heat, evaporation is also normally carried out under reduced pressure.

In the process, milk circulates within vertical tubes, while steam provides direct heating from the shell for evaporation. A lower pressure is maintained inside of the tubes compared to the shell side. Gravity facilitates the downward flow of liquid, creating a film on the inner tube wall. On the shell side of the FFE, the heat source, often in the form of saturated steam, provides the necessary heat for evaporation. This steam flows between the tubes, condenses, and forms a falling film on the exterior of the tubes. The heat transferred during this exchange raises the temperature of the liquid inside the tubes. The reduced pressure on the tube side results in a lower boiling point for the water within the liquid compared to the shell side steam, leading to water evaporation inside the tubes. The simplicity of this design is a key reason why falling film evaporation is favored in dairy processes. The process offers advantages such as a short residence time and a substantial heat transfer area, making it well-suited for milk powder production (Zhang *et al.*, 2018). Due to the significant energy intensity of the evaporation process, Falling Film Evaporators (FFEs) are often configured as multi-effect evaporators to minimize steam consumption (Bylund, 2003).

2.2. Mechanical Vapor Recompression (MVR)

The concept of vapor recompression involves the recycling of steam generated from the product to enhance heat recovery. Through the compression of vapor, its temperature is elevated, allowing the recompressed steam to be used in heating a subsequent effect of the Falling Film Evaporator. Mechanical Vapor Recompression (MVR) is particularly suitable for compressing larger quantities of steam, enabling a higher degree of heat recovery. In the case of MVR, electric power is employed instead of thermal energy. This introduces the opportunity to significantly decrease the reliance on thermal energy.

2.3. Process Flow Diagram and Process Control

To study the dynamic behaviors of a triple-effect FFE with MVR in a milk powder processing plant, the process flow diagram, along with the control scheme and a simulation baseline, is shown in Figure 1.

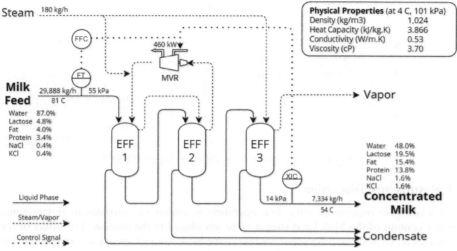

Figure 1 Process Flow Diagram (PFD)

Two conventional control loops were employed in this model. Firstly, the feed-forward control adjusted the electric power of the MVR according to the inlet flow rate of the evaporator. This scheme was designed to handle the fluctuation or the disturbance in the inlet flow. The second scheme considered the final water content in the milk product stream, which was controlled by a feedback control loop with the electric power supplied to MVR in order to ensure product quality.

3. Model Development in Aspen HYSYS

3.1. Pseudo-milk Component

In previous research (Zhang *et al.*, 2014; Munir *et al.*, 2016), a simulated milk stream incorporating pseudo-milk components was developed for industrial process simulators. This involved "hypothetically" generating key components of milk, including fat, proteins, lactose, and minerals, by populating key properties in the simulator. Thermodynamic and physical properties such as heat capacity, thermal conductivity, density, and viscosity of these pseudo-milk components were developed and validated before use in the model.

3.2. Flowsheet Model

Although Aspen HYSYS is one of the most rigorous commercial process simulation software with up-to-date databases, there is no model of an evaporator included in the model library. Therefore, a combination of shell and tube heat exchanger and separator models was utilized to represent the behaviors of the triple-effect FFE. The entire simulation flowsheet is shown in Figure 2.

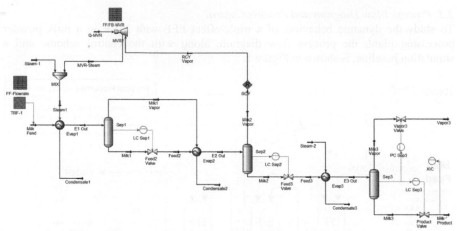

Figure 2 Simulation Flowsheet in Aspen HYSYS

4. Results and Discussion

To investigate process stability, two operating scenarios i.e., changes in flowrate and water content of the Milk Feed stream were introduced to the process. The effect of the changes on Q-MVR and the water content in the Milk Product stream was monitored and described below.

4.1. Changes in Feed Flowrate

In Figure 3, the results showed variations in feed flowrate at -5% and +5%, respectively. In terms of process stability, the control loops effectively returned the system to its steady state, maintaining the water content in the product at 48% despite the introduced changes. Notably, the -5% case demonstrated a significantly quicker return to a steady state compared to the other scenarios. In the standard scenario for Q-MVR, the process required 460 kW. Interestingly, when the feed was increased by +5%, the Q-MVR surged up to 1220 kW, while at -5% feed flowrate, it required only 145 kW.

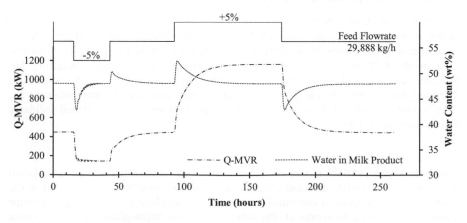

Figure 3 Variations of Feed Flowrate on Process Stability

4.2. Changes of Water Content in Feed

In Figure 4, the outcomes of variations in the water content of the Milk Feed stream, specifically +2% and -2% were presented. The control loops still maintained both the setpoint of water content in the product and process stability across the variations. Similarly, the -2% case exhibited a quicker return to a new steady state compared to the other scenarios. Moreover, the +2% case demanded a substantial Q-MVR of 1170 kW, whereas a low 140 kW was required for the -2% case.

The surges in Q-MVR observed in the higher range cases, namely +5% and +2%, could be attributed to the characteristics and efficiency of the compressor model utilized in this simulation, which required relatively higher power at higher throughput.

However, it should be noted that the Autotuning feature in HYSYS applied in this model resulted in the tuning parameters for the controllers. The result of this tuning, which is not necessarily optimal, could be seen from the observed non-linear behaviour and non-stable regions of Q-MVR and setpoints, particularly in the lower range cases, such as -5% and -2%. This research gap could be filled by the implementation of the advanced controller model and tuning strategy, which is included in our future research plans.

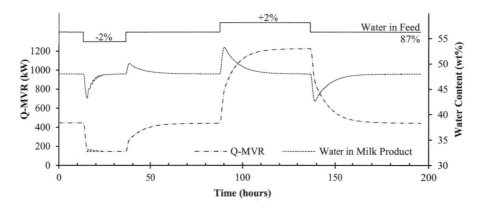

Figure 4 Variations of Water Content in Feed on Process Stability

5. Conclusions

This study described a novel dynamic model with integrated control loops for a triple-effect FFE with MVR tailored for a milk processing plant. The two control loops, aligned with the control scheme of the actual plant, allowed us to investigate the impact of variations in feed flowrate and water content on the process stability. The dynamic model could be used as a foundation for future work in controller tuning optimisation, process optimisation, and improvement in process control. The knowledge obtained from this study could also provide a stepping stone for refining operational strategies and enhancing the efficiency of milk processing plants.

References

J. Ballingall and D. Pambudi, 2017, Dairy trade's economic contribution to New Zealand, New Zealand Institute of Economic Research

G. Bylund, 2003, Dairy Processing Handbook, Tetra Pak Processing Systems AB

M.T. Munir, Y. Zhang, W. Yu, D.I. Wilson, B.R. Young, Virtual milk for modelling and simulation of dairy processes, Journal of Dairy Science, Volume 99, Issue 5, Pages 3380-3395

J. E. Packer, J. Robertson, H. Wansbrough, 1998, Chemical Processes in New Zealand. 2nd ed. 1. New Zealand Institute of Chemistry

I.J. Pisecky, 2012, Handbook of Milk Powder Manufacture, GEA Process Engineering A.

Q. Ruan, H. Jiang, M. Nian, Z. Yan, 2015, Mathematical modeling and simulation of countercurrent multiple effect evaporation for fruit juice concentration, Journal of Food Engineering, Volume 146, Pages 243-251

Y. Zhang, M.T. Munir, W. Yu, B.R. Young, 2014, Development of hypothetical components for milk process simulation using a commercial process simulator, Journal of Food Engineering, Volume 121, Pages 87-93

Y. Zhang, M.T. Munir, I. Udugama, W. Yu, B.R. Young, 2018, Modelling of a milk powder falling film evaporator for predicting process trends and comparison of energy consumption, Journal of Food Engineering, Volume 225, Pages 26-33

Flavio Manenti, Gintaras V. Reklaitis (Eds.), Proceedings of the 34th European Symposium on Computer Aided Process Engineering / 15th International Symposium on Process Systems Engineering (ESCAPE34/PSE24), June 2-6, 2024, Florence, Italy

A Pre-train and Fine-tune Paradigm of Fault Detection and Fault Prognosis for Chemical Process

Yiming Bai,[a] Jinsong Zhao[a,b]

[a]*Department of Chemical Engineering, Tsinghua University, Beijing 100084, China*
[b]*Beijing Key Laboratory of Industrial Big Data System and Application, Tsinghua University, Beijing, 100084, China*
jinsongzhao@mail.tsinghua.edu.cn

Abstract

With the transformation of industrial production digitization and automation, process monitoring has been an indispensable technical method to realize the safe and efficient production of chemical process. Deep learning methods had a significant impact in chemical process fault detection and fault prognosis. But for each process fault or process variable, a specific deep learning model needs to be trained to solve the problem, which would consume a lot of computing resources and time. Inspired by the pre-trained models in natural language processing and computer vision, we proposed a pre-train and fine-tune paradigm of fault detection and fault prognosis for chemical process to improve the robustness of deep learning model for various process states. Specially, we took a deep attention model pretrained on the raw datasets and fine-tuned models based on fault detection and fault prognosis. The Tennessee Eastman process (TEP) was used to demonstrate the validity of the method. Based on the test results, we discussed the relevant issues in applying the pre-train and fine-tune paradigm to real chemical process data.

Keywords: Pre-training, Fine-tuning, Fault detection, Fault prognosis, Chemical process.

1. Introduction

In recent decades, the digitization and automation reforms in industrial production have resulted in the generation of vast amounts of data across various industrial processes. This surge in data has necessitated an unprecedented level of computing power in machines for comprehensive processing and application (Rehman et al. 2019). In the contemporary chemical industry, the advancements in artificial intelligence technology have significantly contributed to industrial intelligence. However, the simultaneous drive for automation and intelligence has led modern chemical plants to evolve into larger, highly complex entities compared to their historical counterparts.

The escalating complexity inherent in contemporary chemical engineering systems has presented formidable challenges to ensuring process safety within the domain. Instances of faults occurring within these chemical processes can result in profound economic ramifications and human casualties, posing a substantial threat to overall chemical safety (Bi et al. 2022). The process monitoring stands as an indispensable technical methodology critical to the realization of safe and efficient chemical production. Its significant theoretical framework within the chemical industry was visually represented in Fig. 1 (Bai and Zhao 2023a). Within this context, fault detection pertains to the identification of the current system state, while fault prognosis predominantly aims to anticipate potential and forthcoming system faults.

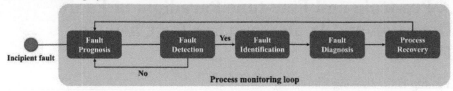

Fig. 1. The process monitoring loop. (Bai and Zhao 2023a)

The prevailing methodologies for process monitoring can be broadly categorized into three main types: model-based methods, knowledge-based methods, and data-based methods. Data-based methods ascertain the system's condition within a specific timeframe by scrutinizing previously gathered data. The progression of automation and digitization has significantly elevated the prominence of data-based models within the process industry (Bai and Zhao 2023b).

Conventional fault detection in the industrial domain primarily relies on established statistical techniques like principal component analysis, partial least squares, and independent component analysis. The advent of machine learning has ushered in a rapid proliferation of sophisticated models such as deep belief networks, convolutional neural networks, and variational recurrent autoencoders. These advanced models demonstrate superior efficacy in handling intricate problems with heightened accuracy (Bi and Zhao 2021). The evolution of fault prognosis methodologies has been intricately tied to the evolving characteristics of industrial data. As data complexity increases, many data-driven methods are flexibly applied in nonlinear scenarios, such as artificial neural networks, autoencoder, support vector machine, and radial basis function network. recurrent neural network, long short-term memory and Transformer model have shown advantages in learning nonlinear features of sequences and extracting long-dependent information (Wen et al. 2022).

However, it typically necessitated the training of distinct deep learning models for effective fault detection and prognosis of each process fault or variable. This process consumed substantial computing resources and time. A prevalent approach involves the training of expansive models on unsupervised or weakly supervised objectives initially, followed by fine-tuning or assessing zero-shot generalization for downstream tasks (Brown et al. 2020). This practice finds widespread application, notably in domains such as natural language processing and computer vision, which inspired the pre-trained models for time series data (Ma et al. 2023).

In this paper, we proposed a pre-train and fine-tune paradigm of fault detection and fault prognosis for chemical process, inspired by the pre-trained models in natural language processing and computer vision. We took a deep masked attention model pretrained on the raw datasets and fine-tuned models based on fault detection and fault prognosis. TEP was used to demonstrate the validity of the method. The results show that the proposed method obtain the strong performance on the chemical process fault detection and prognosis task. Additionally, we discussed the relevant issues in applying the pre-train and fine-tune paradigm to real chemical process data.

The remainder of this paper is organized as follows. Section 2 reviews the basic knowledge of the self-attention structure and Transformer model. Section 3 includes the descriptions of the proposed pre-train and fine-tune paradigm. The applications and discussions are revealed in Section 4. Section 5 gives summary.

2. Methodology

2.1. Self-attention

The attention mechanism is the most important structure in Transformer model and could overcome the difficulty in modeling long sequential data by learning the weights of input information at different times. We show a version of a commonly used attention in Fig. 2, which was proposed as self-attention (Vaswani et al. 2017).

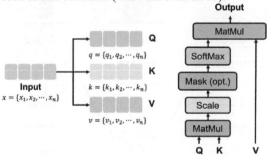

Fig. 2. A schematic diagram of self-attention. (Bai and Zhao 2023a)

Firstly, given an input sequence $x = \{x_1, x_2, \cdots, x_n\}, x_i \in R^k$, query, key and value vectors can be obtained by mapped with linear transformations and $q_i, k_i, v_i \in R^{d_k}$ are examples to show the calculation process of each element.

$$q_i = W^q x_i, \qquad k_i = W^k x_i, \qquad v_i = W^v x_i, \qquad i = 1, 2, \cdots, n \qquad (1)$$

Where $W^q, W^k, W^v \in R^{d_k \times k}$ are learnable matrices.

Then, the similarity score of q_i and k_j, which are chosen as examples to show calculation, is calculated through a scaled dot-product function, and the SoftMax function is then used to normalize the scores and obtain the weights.

$$\alpha_{i,j} = \frac{q_i \cdot k_j}{\sqrt{d_k}} \qquad (2)$$

$$\hat{\alpha}_{i,j} = \frac{\exp(\alpha_{i,j})}{\sum_{t=1}^{n} \exp(\alpha_{i,t})} \qquad (3)$$

Finally, the output is calculated as the weighted sum of values.

$$z_i = \sum_{j=1}^{n} \hat{\alpha}_{i,j} v_j \qquad (4)$$

3. The proposed method

The proposed pre-train and fine-tune paradigm is shown in the Fig. 3. The offline training stage included model pretraining and model finetuning. During the offline modeling stage, the historical data were spilt into large unlabeled dataset and two small labeled datasets for fault detection and fault prognosis. Z-score normalization is utilized for data pre-processing. Then a multilayer Transformer model was designed according to chemical process system. We got the pretrained model on large unlabeled dataset by training model to generate process variables at the next time point. In the model finetuning step, specific detection and prognosis heads were integrated into the pretrained model architecture, and then the modified models were finetuned on the labeled datasets. The online testing included fault detection and fault prognosis. Online data were collected and normalized by the z-score method. Then the data were transformed into the same time window as the offline model training step. For fault detection, we calculated the abnormal score through the finetuned model and compared the abnormal score with the threshold to detect if a fault occurred. For fault prognosis, we predicted multi-step process variables and used the key variables to prognose whether an abnormality occurred.

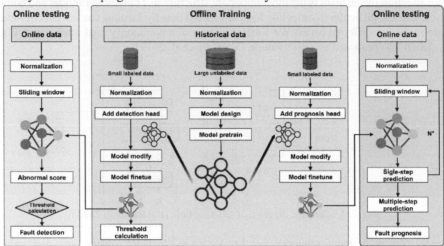

Fig. 3. The diagram of the proposed pre-train and fine-tune paradigm.

4. Applications

TEP is a classical model benchmark for simulating chemical processes, which has been widely used as a study case for FDD and fault prognosis. It is mainly composed of 5 unit operations, including a reactor, a condenser, a recycle compressor, a vapor-liquid separator, and a stripper. The initial version of TEP simulation involves 12 manipulated variables, 41 measured variables and 20 different types of faults. In 2015, Bathelt revised the original TE model and used MATLAB's Simulink to build a new one, adding 8 measurement variables, 24 component variables, and 8 new process disturbances (Bathelt, Ricker, and Jelali 2015). In our work, the applications are based on the revised version of Bathelt, which is shown in the Fig. 4.

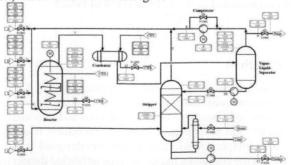

Fig. 4. P&ID of the revised TEP model (Bathelt, Ricker, and Jelali 2015).

Since the sampling frequency of component analysis measurement is much lower, we only choose the process manipulated variables and continuous process measurements for dataset building. Each set of test data included 480 normal samples followed by 600 fault samples, corresponding to 8 h of normal data and 10 h of fault data.

4.1. Fault detection

We selected and applied 5 distinct step type faults, 5 random variation type faults, and 1 slow drift type fault for our analysis. The results will be compared with LSTM-VAE (Bi and Zhao 2021), an advanced and effective method for fault detection in chemical processes, which originated from published process monitoring study and was evaluated for performance on the same dataset we generated. To assess the efficacy in fault detection, we utilized fault detection rates (FDR). Additionally, we employed fault detection time (FDT) as a metric to gauge the method's timeliness in fault identification. The comparative results between LSTM-VAE and the proposed method are presented in Table. 1.

Table. 1. Detection results of proposed method and LSTM-VAE.

Fault Number	FDR of LSTM-VAE	FDT of LSTM-VAE (min)	FDR	FDT (min)
1	98.8%	7	99.4%	3
2	96.3%	28	97.3%	**13**
3	99.8%	1	99.8%	1
4	99.8%	1	99.8%	1
5	99.8%	1	99.8%	1
8	93.3%	37	95.2%	**23**
9	92.5%	44	95.9%	**20**
10	91.8%	49	85.7%	46
11	99.5%	3	96.1%	15
12	99.2%	5	94.6%	24
13	72.5%	165	81.4%	**46**

Table 1 reveals that both LSTM-VAE and the proposed method exhibited similar FDR, with both achieving rates approximately exceeding 90%. Notably, in the case of the slow drift-type fault, Fault 13, the proposed method demonstrated a significantly higher FDR compared to LSTM-VAE. Regarding FDT, the proposed method generally outperformed LSTM-VAE in early fault detection, except for Fault 11 and Fault 12. Remarkably, the proposed method showcased a notable advancement of over 100 minutes in detecting Fault 13 compared to LSTM-VAE.

4.2. Fault prognosis

The accurate prediction of future changes in process variables stands as the fundamental challenge in fault prognosis. We selected faults with long FDT for fault prognosis analysis, which were Fault 8, 9, 10, 13. To comprehensively evaluate the proposed method's performance across distinct time scales, we examined varied forecast times. Our approach finetuned the pretrained model solely with a batch of simulation data. Employing the autoregression concept, we iteratively predicted process variables at forecast times of 1, 4, 8, 16, 32, and 64 minutes across the four fault types. The accuracy of the forecasts was evaluated using Mean Absolute Error (MAE), with the results detailed in Table 2. The results demonstrated the robust predictive capabilities of the finetuned model across the four fault types. The MAE for short term forecasts is about less than 0.1, while for long term forecasts is around 0.5.

Table. 2. MAE of various faults in different forecast times.

Fault Number	1	4	8	16	32	64
8	0.014	0.052	0.163	0.412	0.256	0.505
9	0.006	0.027	0.122	0.458	0.260	0.461
10	0.007	0.025	0.068	0.279	0.258	0.461
13	0.009	0.028	0.076	0.331	0.230	0.533

5. Summary

In the article, we proposed a pre-train and fine-tune paradigm of fault detection and fault prognosis for chemical process. This proposed paradigm exhibited validity and efficacy in both fault detection and fault prognosis across the TEP. Additionally, the capacity to finetune the same pretrained model for different process monitoring tasks related to various fault types suggests considerable potential for conserving computational resources and time. In the future, the widespread application of pretrained models within process monitoring systems and their utilization across real industrial datasets presents an open field replete with several unresolved challenges demanding attention.

References

Bai, Yiming, and Jinsong Zhao. 2023a. "A Novel Transformer-Based Multi-Variable Multi-Step Prediction Method for Chemical Process Fault Prognosis." *Process Safety and Environmental Protection* 169 (January): 937–947.

Bai, Yiming, and Jinsong Zhao. 2023b. "A Process Data Prediction Method for Chemical Process Based on the Frozen Pretrained Transformer Model." In *Computer Aided Chemical Engineering*, edited by Antonios C. Kokossis, Michael C. Georgiadis, and Efstratios Pistikopoulos, 52:1717–1723. 33 European Symposium on Computer Aided Process Engineering. Elsevier.

Bathelt, Andreas, N. Lawrence Ricker, and Mohieddine Jelali. 2015. "Revision of the Tennessee Eastman Process Model." *IFAC-PapersOnLine*, 9th IFAC Symposium on Advanced Control of Chemical Processes ADCHEM 2015, 48 (8): 309–314.

Bi, Xiaotian, Ruoshi Qin, Deyang Wu, Shaodong Zheng, and Jinsong Zhao. 2022. "One Step Forward for Smart Chemical Process Fault Detection and Diagnosis." *Computers & Chemical Engineering* 164 (August): 107884.

Bi, Xiaotian, and Jinsong Zhao. 2021. "A Novel Orthogonal Self-Attentive Variational Autoencoder Method for Interpretable Chemical Process Fault Detection and Identification." *Process Safety and Environmental Protection* 156 (December): 581–597.

Brown, Tom, Benjamin Mann, Nick Ryder, Melanie Subbiah, Jared D. Kaplan, Prafulla Dhariwal, Arvind Neelakantan, et al. 2020. "Language Models Are Few-Shot Learners." *Advances in Neural Information Processing Systems* 33: 1877–1901.

Ma, Qianli, Zhen Liu, Zhenjing Zheng, Ziyang Huang, Siying Zhu, Zhongzhong Yu, and James T. Kwok. 2023. "A Survey on Time-Series Pre-Trained Models." arXiv.

Rehman, Muhammad Habib ur, Ibrar Yaqoob, Khaled Salah, Muhammad Imran, Prem Prakash Jayaraman, and Charith Perera. 2019. "The Role of Big Data Analytics in Industrial Internet of Things." arXiv..

Vaswani, Ashish, Noam Shazeer, Niki Parmar, Jakob Uszkoreit, Llion Jones, Aidan N Gomez, Łukasz Kaiser, and Illia Polosukhin. 2017. "Attention Is All You Need." In *Advances in Neural Information Processing Systems*. Vol. 30. Curran Associates, Inc.

Wen, Qingsong, Tian Zhou, Chaoli Zhang, Weiqi Chen, Ziqing Ma, Junchi Yan, and Liang Sun. 2022. "Transformers in Time Series: A Survey." arXiv.

Flavio Manenti, Gintaras V. Reklaitis (Eds.), Proceedings of the 34th European Symposium on Computer Aided Process Engineering / 15th International Symposium on Process Systems Engineering (ESCAPE34/PSE24), June 2-6, 2024, Florence, Italy

A Multi-Criteria Based Approach for Large-Scale Deployment of CO_2 Capture, Utilization and Storage

Thuy Thi Hong Nguyen*, Satoshi Taniguchi, Takehiro Yamaki, Nobuo Hara, Sho Kataoka

National Institute of Advanced Industrial Science and Technology
1-1-1 Higashi, Tsukuba, Ibaraki 305-8565, Japan
Corresponding author email: nguyen.thuy@aist.go.jp

Abstract

In this study, CCU was coupled with CCS to totally abate the target CO_2 emission sources in an economically viable manner. Herein, CO_2 was captured from a target source, and separated into two different streams, one of which was prepared for storage. The remaining CO_2 stream, together with H_2, was utilized for methane production. Rigorous process design was performed to provide necessary data for constructing a multi-criteria based evaluation model, based on which the total system is optimized for maximizing both total economic benefit and CO_2 abatement. The results show that coupling CCU with CCS can help reduce the incurred cost and achieve high CO_2 reduction when a low carbon tax rate (i.e., \$130/ton) is applied and when a low-cost H_2 source (i.e., below \$1.3/kg) is available. CCU is not necessary to be coupled with CCS if a higher carbon tax rate is applied or when a large supply of low cost green H_2 is available.

Keywords: Carbon capture and utilization/storage, rigorous process design, multi-criteria based evaluation model, economic benefit, CO_2 abatement

1. Introduction

Greenhouse gases (GHG), mainly due to CO_2 emissions, have become a worldwide concern. CCS and CCU are key technologies for mitigation of CO_2 emissions. In Japan, utilizing CO_2 for methane synthesis and storing CO_2 at available injection well sites have been considered as important strategies to fulfil the domestic energy demand and to achieve the target GHG emissions reduction. Large-scale implementation of either CCS or CCU is limited due to some technical challenges (Dziejarski et al., 2023) and economic drawbacks (Chen at al., 2022). CCS incurs additional cost for CO_2 capture, liquefaction, transport and injection (Storrs et al., 2023). CCU can lead to much lower CO_2 emissions than CCS while creating new added value products. Thus, coupling CCU and CCS is expected to be a wise strategy for abating large-scale CO_2 emissions economically. However, most of CCU processes need large supplies of heating utilities and raw materials (Hao et al., 2022). In addition, allocating suitable amounts of CO_2 for storage and utilization is the key to sustaining economic and environmental viability of an integrated CCUS system. Therefore, optimizing the entire system is strongly needed.

In this study, an integrated CCUS system producing e-fuel as the main product is designed and optimized. Herein, CO_2 is captured from a target source, and separated into two different streams, one of which is liquefied, transported to an injection site, and injected while the other CO_2 stream, together with H_2, is utilized for methane production. A multi-criteria evaluation model is built. It is used to evaluate and optimize the entire system by maximizing total economic benefit and CO_2 emission reduction. The impacts of external factors such as carbon tax rate and H_2 cost and sources (i.e., grey and green sources) on total process performances are also investigated.

2. Method of study

Figure 1 shows the boundary of the coupling CCUS system considered in this study. The flue gas containing CO_2 is led to the capture process, in which CO_2 is captured by monoethanolamine solvent and purified. A fraction amount of the CO_2 product stream is led to the liquefaction process, producing liquid CO_2 which is then transported to the injection site and stored underground. The rest amount of CO_2 is mixed with hydrogen and it is used as input for the methane production process, in which methane is synthesized following reaction (1) and purified. The product methane can be sold and used as a replacement to the conventional natural gas fuel.

$$CO_2 + 4 H_2 \rightleftharpoons CH_4 + 2 H_2O \qquad \Delta H_{298K} = -165 \text{ kJ/mol} \qquad (1)$$

In this study, the following assumptions are made for process design and optimization:

- CO_2 is captured from a target emission source at a rate of 208.4 t/h
- Amine-based capture process, CO_2 liquefaction and methane production processes are located nearby the CO_2 emission source.
- The liquid CO_2 is transported by ship for 1,000 km.
- Different sources of H_2, both grey and green sources, are available and their amounts are unlimited.
- The produced methane can be used for many purposes; thus, its demand is unlimited.

The methane production and CO_2 liquefaction processes are designed using process simulator Pro/II (developed by AVEVA, 2022). The structure and operating conditions (i.e., temperature and pressure) of each process are also optimized to reduce its energy consumption and material loss. However, separately optimizing each process is not

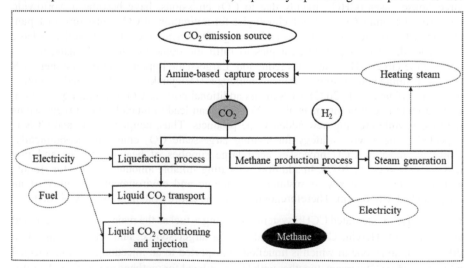

Figure 1. Boundary of process design and evaluation

necessary a guarantee of an optimal coupling CCUS system which would yield high economic benefit and low CO_2 emissions. Therefore, to optimize the target of the integrated CCUS system, a multi-criteria based evaluation model is constructed. It consists of two evaluation indicators, the net present value (NPV) and the potential CO_2 reduction (PCR), which are used for maximizing economic benefit and CO_2 abatement amount. The model is operated using some fundamental data such as mass and energy balance data obtained from the above process design and simulations, life cycle assessment database, and market price data.

Here, NPV ($) accounts for all kinds of investment and operating costs (i.e., process plant construction, injection well, CO_2 transport, energy and raw material consumption costs, etc.) and benefits gained from selling the product produced by the system (Biegler et al.,1997). Basic parameters used for evaluating NPV are shown in Table 1. PCR (t CO_2/year) is the sum of all direct and indirect CO_2 emissions due to energy and raw material consumptions minus the amount of CO_2 avoided due to substitution of conventional natural gas. The scales of CO_2 utilization and storage, which depend on the fractions of CO_2 utilized and stored, respectively, directly impact the amounts of salable product (i.e., methane), the size of process equipment, and the amounts of raw material and energy consumed. As a result, they directly impact the evaluation indicators NPV and PCR. Therefore, the optimization problem is defined as following:

$$\max NPV = f(w_1, w_2) \tag{2}$$
$$\max PCR = g(w_1, w_2) \tag{3}$$
$$\text{subject to: } w_1 + w_2 = 1 \tag{4}$$

Here, w_1 and w_2 are fractions of CO_2 utilized for methane production and stored, respectively.

Table 1. Basic parameters used for evaluating NPV.

Operating hours per year	8400 hours
Useful plant life	15 years
Depreciation life (straight line method)	12 years (straight line method)
Contingency and fee costs	20% bare module cost
Auxiliary facilities costs	20% bare module cost
Working capital	19.4% fixed capital investment cost
Labor	10% operating cost
Other cost (maintenance, supplies, etc.)	11% process capital cost
Tax rate	35%
After tax rate of return	10%

The optimization of each objective is performed using an optimization solver, namely What's Best (an add-in to Excel) (Lindo Systems Inc., 2023). To support different stakeholders to make decision based on different targets of CO_2 abatement, different target PCR values, calculated based on ε (varying in the range of 0 and 1) and optimal PCR result, are assigned as constraints (equation (5)) for optimizing the NPV.

$$\text{Target PCR} = \varepsilon * PCR_{max} \tag{5}$$

To illustrate evaluation results, varying in wide ranges, the following equations are used to normalize the NPV and PCR results:

$$NPV_{Norm} = exp\left(NPV/NPV_{max}\right) \tag{6}$$

$$PCR_{Norm} = exp\left(PCR/PCR_{max}\right) \tag{7}$$

Thus, when NPV_{Norm} is higher than 1, economic benefit is gained, and vice versa. When PCR_{Norm} is higher than 1, CO_2 emission can be reduced, and vice versa.

3. Evaluation result

To examine the impact of the external factors on the evaluation objective NPV and PCR, different scenarios of carbon tax rates and H_2 sources and prices (S1 – S6) were considered. As Table 2 clearly shows, the optimal results of NPV and PCR markedly change when these external factors change. Figure 2 shows the changing tendencies of these objectives as functions of fraction amount of CO_2 utilized for methane production.

Table 2. Different scenarios considered in this study.

Scenario ID	S1	S2	S3	S4	S5	S6
Input conditions						
H_2 type	Grey	Grey	Green	Green	Green	Green
H_2 price [a] ($/kg H_2)	1.26	1.26	2.0	2.0	1.3	1.3
H_2's CO_2 factor[a] (kg CO_2/kg H_2)	12.4	12.4	1.99	1.99	1.99	1.99
CH_4 price[b] ($/kg CH_4)	0.88	0.88	0.88	0.88	0.88	0.88
Carbon tax rate ($/ton CO_2)	0	130	0	130	0	130
Optimization results						
max NPV (x $100MM)	5.90	4.75	8.59	1.14	5.11	14.8
max PCR (MMt CO_2/year)	0.96	0.96	1.58	1.58	1.58	1.58

a) data are obtained from Parkinson et al., (2019)
b) average value in Japan as of 2022 (New Power Net, 2022)

As the results clearly show, S1 can gain some economic benefit when CO_2 in the target source is utilized for methane production. However, because grey H_2 has extremely high CO_2 emission factor, the total amount of CO_2 emitted is much higher than the amount of CO_2 utilized and avoided. As a result, PCR_{Norm} result is lower than 1 when $w1$ increases. Thus, with the conditions of S1, CO_2 emissions cannot be reduced. S2, using the H_2 source similar to S1, has much worse economic performance, as the carbon tax rate increases from 0 to $130/ton.

Scenarios S3-S6 use green H_2 source, thus, their PCR_{Norm} results are always higher than 1, indicating high potential of CO_2 abatement. However, their economic viability heavily depends on the H_2 market price and on the carbon tax rate. When the H_2 price increases beyond \$1.5/kg, a carbon tax rate of at least \$130/ton CO_2 is required to compensate for the H_2 cost expense and to make the target system economically viable (scenario 4). A H_2 price below \$1.3/kg (scenarios S5 and S6) will make the target system economically feasible without need of applying carbon tax.

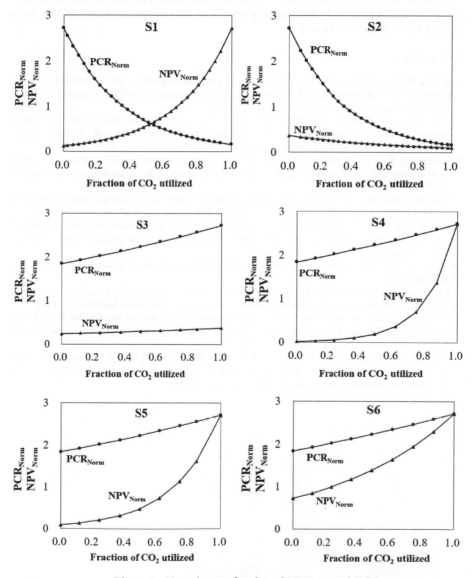

Figure 2. Changing tendencies of NPV_{Norm} and PCR_{Norm} towards CO_2 utilization fraction

4. Conclusion

In this study, a coupling CCUS system was designed for abating the target fixed CO_2 emission source. Here, methane is considered as the main product. Based on a constructed multi-criteria evaluation model, the system was optimized, considering different scenarios of carbon tax rate and H_2 sources. The optimization results show that when grey H_2 is used and low carbon tax rate is applied, coupling CCU and CCS can help reduce the additional cost incurred by CCS and increase CO_2 emission reduction. Nevertheless, when large supply of green H_2 is available and high carbon tax rate is considered, it is more economically and environmentally beneficial for the whole amount of CO_2 to be utilized for methane production.

References

AVEVA, PRO/II simulation v2022. https://www.aveva.com/en/products/pro-ii-simulation/.

L.T. Biegler, I.E. Grossmann, A.W. Westerberg, 1997, Systematic methods of chemical process design. Prentice Hall PTR, Upper Saddle River, N.J.

S. Chen, J. Liu, Q. Zhang, F. Teng, B.C. McLellan, 2022, A critical review on deployment planning and risk analysis of carbon capture, utilization, and storage (CCUS) toward carbon neutrality, Renewable and Sustainable Energy Reviews 167, 112537.

B. Dziejarski, R. Krzyżyńska , K. Andersson, 2023, Current status of carbon capture, utilization, and storage technologies in the global economy: A survey of technical assessment, Fuel 342 127776.

Z. Hao, M.H. Barecka, A.A. Lapkin, 2022, Accelerating net zero from the perspective of optimizing a carbon capture and utilization system. Energy & Environmental Science 15, 2139-2153.

Lindo Systems Inc, 2023, What'sBest! 18 - Excel add-in for linear, nonlinear, and integer modeling and optimization. Available at: https://www.lindo.com/index.php/products/what-sbest-and-excel-optimization.

New Power Net, 2022, Statistical data of comodity. Available at: https://pps-net.org/statistics/gas. (in Japanese)

B. Parkinson, P. Balcombe, J.F. Speirs, A.D. Hawkes, K. Hellgardta, 2019, Levelized cost of CO_2 mitigation from hydrogen production routes, Energy & Environmental Science, 12, 19-40.

K. Storrs, I. Lyhne, R. Drustrup, 2023, A comprehensive framework for feasibility of CCUS deployment: A meta-review of literature on factors impacting CCUS deployment, International Journal of Greenhouse Gas Control, 125, 103878.

J.D. Tapia, J.Y. Lee, R.E.H. Ooi, D.C.Y. Foo, R.R. Tan, 2018, A review of optimization and decision-making models for the planning of CO_2 capture, utilization and storage (CCUS) systems, Sustainable production and consumption 13, 1-15.

Flavio Manenti, Gintaras V. Reklaitis (Eds.), Proceedings of the 34th European Symposium on
Computer Aided Process Engineering / 15th International Symposium on Process Systems
Engineering (ESCAPE34/PSE24), June 2-6, 2024, Florence, Italy

Accelerating Unsteady Fluid Dynamic Simulations for Taylor-Couette Crystallizer using Snapshot POD and Recurrent Neural Networks

Osamu Tonomura, Tsuyoshi Maruyama, Atsuto Nishihata, Ken-ichiro Sotowa

Dept. of Chem. Eng., Kyoto University, Nishikyo, Kyoto 615-8510, Japan
tonomura@cheme.kyoto-u.ac.jp (O. Tonomura)

Abstract

Digital twins are expected to be applied in the chemical process industry as important
technology for improving productivity and profits. While research and development is
being conducted on elemental technologies related to digital twins such as sensing,
communication, data management and utilization, and optimization, detailed flow models
are necessary to reproduce flow inside equipment used in chemical processes.
Computational fluid dynamics (CFD) is essential in equipment design because it can
provide high-fidelity field data to better understand flow physics. CFD has been used to
analyse flow and mass transfer in various equipment. However, CFD simulations,
especially for unsteady flows, are computationally expensive because they must march
the flow solution with a small step in time. Introducing CFD into digital twins, there is a
need to reduce the above cost and make unsteady CFD more affordable. In this study, the
reduced order modelling of CFD using snapshot proper orthogonal decomposition
(snapshot POD) and deep neural networks was applied to a Taylor-Couette crystallizer,
which has a double cylindrical structure with an inner cylinder and an outer cylinder and
is characterized by periodic vortices (called Taylor-Couette flow) generated by shear
when the inner cylinder is rotated. Previous research reported that Taylor-Couette flow is
effective in promoting nucleation and crystal growth. This study showed that the reduced
order model (ROM) developed in this study is 1000 times faster than CFD and can
accurately predict the flow that changes depending on the rotation speed of the inner
cylinder.

Keywords: Digital twin, Modelling and simulation, Computational fluid dynamics
(CFD), Reduced order model (ROM), Taylor-Couette crystallizer.

1. Introduction

Numerical analysis using computers based on the finite element method or finite volume
method, namely computer aided engineering (CAE) simulation, has long been introduced
in the fields of product design and research and development (R&D). In recent years, new
technologies such as the Internet of Things (IoT) and artificial intelligence (AI) are being
introduced to product design and manufacturing sites. Against this background, attempts
are being made to integrate conventional CAE technology with AI and IoT technology.
In order to advance such attempts, this research aims to develop a virtual sensor for
measuring the flow conditions that affect mass transfer and crystal growth inside the
Taylor-Couette crystallizer, and the final goal is to apply the developed virtual sensor to
a digital twin system. If the digital twin of the crystallization process is realized, it will
be possible to numerically visualize what is happening in the operating equipment,
monitor the internal state of the equipment that cannot be detected with sensors, predict

the occurrence of abnormalities, and enable timely control and adjustment to optimal conditions, which is expected to lead to cost reductions and quality improvements.

A Taylor-Couette crystallizer has a double cylindrical structure with an inner cylinder and an outer cylinder and is characterized by periodic vortices (called Taylor-Couette flow) generated by shear when the inner cylinder is rotated. Taylor-Couette flow is known to be effective in promoting nucleation and crystal growth. Therefore, the Taylor-Couette crystallizer is said to be superior to the conventional stirred tank crystallizer in terms of shortening crystal precipitation time, improving yield, and narrowing the crystal size distribution. As research on detailed analysis and quantitative evaluation of the flow, Sha et al. (2016) used computational fluid dynamics (CFD) to study the relationship between the flow in the Taylor-Couette crystallizer and a dimensionless number called the Taylor number (Ta), which is defined as a function of the radius and angular velocity of the inner cylinder, the gap distance between the inner and outer cylinders, and the kinematic viscosity of the fluid. As a result, they suggested that Taylor-Couette flow was generated when Ta was greater than 700 and that CFD was a useful flow analysis tool. However, CFD, especially for unsteady flows, is computationally expensive because it must march the flow solution with a small step in time. Introducing CFD into digital twins, there is a need to reduce the above cost and make unsteady CFD more affordable.

Achieving real-time CFD simulations is the focus of our study, and one of the keys is the reduced order modelling of CFD. The reduced order modelling methods have been developed for many fields like electronics or structural mechanics. This study focuses on transient fluid simulation applications. According to related past research, projection-based reduced order modelling is the most popular method and is divided into two: 1) posteriori methods such as the proper orthogonal decomposition (POD), which build a reduced order model (ROM) from a large set of simulations called "snapshots" and require a computationally intensive offline phase, and 2) priori methods such as the proper generalized decomposition, which reduce the model during the problem solving process itself (Calka et al., 2021). To our knowledge, none of these reduced order modelling methods have been applied to Taylor-Couette crystallization processes. A constraint on crystallizer modeling is the need to solve the governing equations of flow over time. For this reason, the posteriori methods described above are considered desirable.

In this paper, first, a flow analysis of a three-dimensional Taylor-Couette crystallizer is performed using Ansys Fluent, a commercial software for the general-purpose finite volume method. After that, in order to speed up this calculation for building a virtual sensor, a snapshot POD and deep neural networks based ROM is derived using flow field data obtained through ANSYS Fluent and Dynamic ROM Builder.

2. CFD simulation for Taylor-Couette crystallizer

In this study, CFD simulations were employed to analyze the fluid flow in a Taylor-Couette crystallizer, which has a double cylindrical structure with an inner cylinder and an outer cylinder, as shown in Fig. 1. Mathematical model of the crystallizer and simulation settings are as follows: the equations used to describe the system are the continuity and Navier-Stokes (pressure and velocity) equations. The liquid-gas interface is resolved using the volume-of-fluid (VOF) model. In all cases studied, the flow is

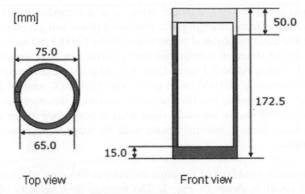

Fig. 1. A schematic diagram of crystallizer. Dark gray and light gray regions in the equipment indicate liquid and gas phases, respectively.

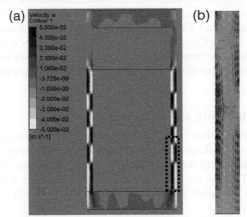

Fig. 2. CFD simulation result (v_r = 400 rpm).

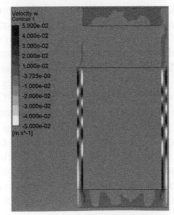

Fig. 3. CFD simulation result (v_r = 250 rpm).

turbulent and simulated by k-ε model, while adiabatic conditions are applied at domain boundaries. Slip effects are negligible. The liquid phase and gas phase in the crystallizer are assumed to have the physical properties of water (293K) and air (293K), respectively. Unless otherwise stated all simulations are performed in three dimensions. Simulations are performed using ANSYS Fluent. Unstructured grids are used and the total number of cells is approximately 700,000 in all cases. The SIMPLEC method is implemented for pressure-velocity coupling and the spatial discretization was performed using the second order upwind scheme. An unsteady simulation is performed under a time step of 0.01 seconds starting from a stationary state until 10 seconds have passed after the inner cylinder rotates.

When the rotation speed (v_r) of the inner cylinder is 400 rpm, which corresponds to Ta = 2660, CFD result is shown in Fig. 2. The average flow velocity of the fluid flowing through the gap between the inner and outer cylinders was 0.07 m/s. Fig. 2(a) shows the velocity in the direction of gravity on the vertical cross section of the equipment, with black indicating upward direction and white indicating downward direction. Fig. 2(b) is an enlarged vector representation of the dotted line frame in Fig. 2(a). Periodic vortices were confirmed, indicating that Taylor-Couette flow occurs. Fig. 3 shows the result when the rotation speed was changed to 250 rpm. The average flow velocity of the fluid flowing through the gap between the inner and outer cylinders was 0.045 m/s. Taylor-Couette flow was confirmed, but its behavior was different from that shown in Fig. 2.

3. Reduced order modeling and its results

It took four hours of calculation time to perform a 10-second unsteady CFD simulation of the Taylor-Couette crystallizer mentioned above. Reduced order modelling methods can be used for problems requiring real-time results or numerous simulations. In this study, ROM of the Taylor-Couette crystallizer was developed using flow field data obtained through ANSYS Fluent and Dynamic ROM Builder (DRB), which is accessible in the ANSYS Twin Builder.

The changes in Taylor-Couette flow behaviour, which are described with time varying flow velocity at spatial coordinates of the crystallizer, called outputs, are induced by the time variations of the rotation speed of the inner cylinder, called inputs. After this input-output relationship was computed with a full-order transient solver in ANSYS Fluent, the learning and validation of a ROM is performed from a limited number of the computed data sets, called scenarios, in DRB. The DRB computes one or more outputs from one or more dynamic inputs, based on recurrent neural networks (RNN). Here, the dimensionality of the output vector is reduced using the snapshot POD. Namely, as shown in Fig. 4, the spatial distribution of velocity is expressed as a superposition of modes representing the characteristics of the original data. By learning the behaviour of the superposition coefficients, the spatial distribution of velocity can be approximately reconstructed. The learning process aims at finding the nonlinear function that minimizes the error. It is implemented as a three-layer recurrent neural network with the same number of variables in the hidden and the output layers. The activation function used in the hidden and output layers is a sigmoid. The gradient descent optimization methods are used in the optimization.

In this study, DRB was tested in its ability to explain the dynamical Taylor-Couette flow behavior of the crystallizer in response to the rotation speed of the inner cylinder. Firstly, the flow field data when the rotation speed of the inner cylinder was varied by a sine function, as shown in Fig. 5, was computed with a full-order transient solver in ANSYS Fluent. The details of the computation are as in the previous section. The input and output variables sampled every 0.01 seconds were used as snapshots necessary for developing a ROM by DRB. Fig. 6 shows the developed ROM, and its input and output are the rotation speed of the inner cylinder and the flow velocity in the equipment, respectively. As a result, it was shown that the Taylor-Couette flow was reproduced using the developed ROM with an average relative error of 4.5%. In addition, it took four hours of calculation time to perform a 10-second unsteady CFD simulation of the Taylor-Couette crystallizer, but using the developed ROM, the calculation time was reduced to 6 seconds.

$$\Phi(x,t) \approx \sum_{k=1}^{r} \alpha_k(t)\varphi_k(x)$$

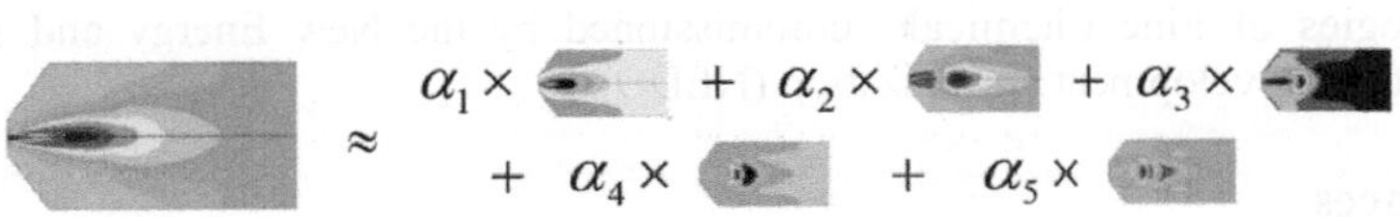

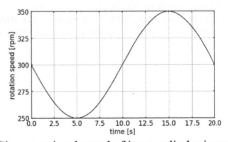

Fig. 4. Overview of POD for spatial distribution of velocity

Fig. 5. Given rotational speed of inner cylinder in case study.

Ansys Fluent   Ansys Twin Builder

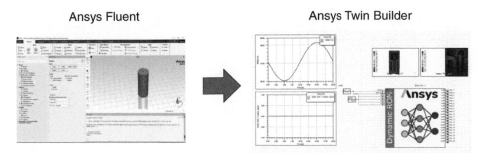

Fig. 6. ROM developed with ANSYS Twin builder.

4. Conclusions

This study presented the results of proper orthogonal decomposition (POD) and recurrent neural network (RNN) based reduced-order modeling method applied to a complex 3D fluid dynamic model of a Taylor-Couette crystallizer. It was demonstrated that if some training simulations were performed using CFD simulators, time-series flow field prediction in response to rotational speed of inner cylinder could be instantaneously performed using Dynamic ROM Builder (DRB), which is based on model estimation using deep learning. The reduced order model (ROM) obtained by DRB was comparable in accuracy to 3D simulation, and the calculation time can be significantly reduced, making it suitable for use as a virtual sensor. In the next study, the developed ROM is linked with experimental equipment and the usefulness of digital twin-based monitoring and control is verified.

Acknowledgement

This work was partially supported by the Grant-in-Aid for Scientific Research (B) (No. 23H01754) and a project, Development of Continuous Production and Process Technologies of Fine Chemicals, commissioned by the New Energy and Industrial Technology Development Organization (NEDO).

References

ANSYS Inc., 2022, Dynamic ROM components, Twin Builder 2022 Online Help.

M. Calka P. Perrier, J. Ohayon, C. Grivot-Boichon, M. Rochette and Y. Payan, 2021, Machine-Learning based model order reduction of a biomechanical model of the human tongue, Computer Methods and Programs in Biomedicine, 198, 105786

Y. Sha, J. Fritz, T. Klemm and S. Ripperger, 2016, Untersuchung der Strömung im Tylor-Couette-System bei geringen Spaltbreiten", Chemie Ingenieur Technik, 88(5), 640-647.

Flavio Manenti, Gintaras V. Reklaitis (Eds.), Proceedings of the 34th European Symposium on Computer Aided Process Engineering / 15th International Symposium on Process Systems Engineering (ESCAPE34/PSE24), June 2-6, 2024, Florence, Italy

Process simulation of effects of ammonia co-firing on the thermal performances of the supercritical circulating fluidized bed boiler

Seong-il[a] Kim and Won Yang[a]*

aCarbon Neutral Technology R&D Department; Research Institute of Clean Manufacturing System; Korea Institute of Industrial Technology, Cheon-an, 31056, Republic of Korea
Email_of_the_Corresponding_yangwon@kitech.re.kr

Abstract

The use of ammonia co-firing technology for carbon neutrality has emerged in the power generation industry. It is necessary to determine the optimal operating conditions by investigating boiler performance during ammonia co-firing. In this study, a process simulation was performed to determine the effect of ammonia co-firing on the thermal performances of the 550 MWe supercritical circulating fluidized bed (CFB) boiler. The process simulations were conducted using co-firing ratios of 5, 10, 15, 20, and 30%. Although the amount of CO_2 reduction during ammonia co-firing was confirmed, the radiation and convective heat transfer rates decreased due to changes in the adiabatic flame temperature, flue gas flow rate, and increased moisture heat loss. Accordingly, the main and reheat steam temperatures decrease. In conclusion, the boiler thermal efficiency decreases under the condition of ammonia co-firing. These findings can be used to establish optimal operating conditions for CO_2 reduction and improve plant efficiency, which are operational trade-offs during ammonia co-firing.

Keywords: ammonia co-firing, CFB boiler, process simulation, heat transfer

1. Introduction

With the global declaration of carbon neutrality, the reduction of greenhouse gases in the power generation sector must be accompanied. According to Korea's Nationally Determined Contribution, greenhouse gas emissions from the power generation sector were adjusted downward from 192.7 million tons to 145.9 million tons by 2030. According to Korea's power mix, the power generations of a coal-fired power plant and a combined cycle power plant account for 32.8% and 68.7%, respectively. Therefore, for carbon neutrality in the power generation field, the transition from fossil fuel-based to carbon-free fuel is essential. Under this power supply situation, Korea announced the 10th basic plan for electricity supply and demand this year as shown in Table 1. In 2030, electricity production from coal-fired plants by utilizing ammonia will begin. Based on the power supply and demand plan, Korea started an ammonia co-firing demonstration project in 2030. The target co-firing rate is 20%, and it is planned to be demonstrated in pulverized coal (PC) boilers and circulating fluidized bed (CFB) boilers in 2027. In particular, the world's first ammonia co-firing project in a circulating fluidized bed boiler was started in Korea. The target power plants were two PC boilers and two CFB boilers in Korea.

Table 1 10th Basic plant for electricity supply and demand in KOREA, 2023

Year	Division	Nuclear	Coal	LNG	Renewable	H$_2$ & NH$_3$	Etc.	Total
2030	Generation [TWh]	201.7	122.5	142.4	134.1	**13.0**	8.1	621.8
	Ratio	32.4%	19.7%	22.9%	21.6%	**2.1%**	1.3%	100%
2036	Generation [TWh]	230.7	95.9	62.3	204.4	**47.4**	26.6	667.3
	Ratio	34.6%	14.4%	9.3%	30.6%	**7.1%**	4.0%	100%

Table 2 shows the fuel characteristics of ammonia compared to other hydrocarbon fuels. Ammonia combustion only generates water and nitrogen. Therefore, ammonia can be considered as a carbon-free fuel. Ammonia is easily transportable and storable due to high boiling temperature compared to other fuels. However, ammonia has lower combustibility than other fuels, and fuel-NOx may be emitted, when ammonia is combusted due to the N component in ammonia. Therefore, it is important to minimize NO$_x$ emissions, while ensuring sufficient ammonia combustibility.

Previous studies have mainly been conducted to identify ammonia combustion and NO$_x$ emission characteristics [Kobayashi et al.]. However, under the ammonia co-firing condition, the flue gas properties change due to the change in adiabatic flame temperature, required air flow rate, flue gas composition, and flue gas mass flow rate. These changes affect the radiation and convection heat transfer rate of each heat exchanger in a boiler. Accordingly, main and reheat steam temperatures change, which means a change in plant efficiency. Therefore, for ammonia co-firing, it is necessary to establish optimal operating conditions by analyzing the heat transfer characteristics in the boiler in terms of power operation.

The aim of this study is to predict main and reheat steam temperature by investigating the variation of heat transfer rate through process simulation techniques. The process simulation model was developed by In-house code to analyze the change in heat transfer mechanism. The target plant was selected as the 550 MW$_e$ CFB boiler, which is the subject of the demonstration project. The simulation results were compared with the 870 MW$_e$ PC boiler. Based on the simulation data, it is possible to understand the relationship between CO$_2$ reduction and boiler thermal efficiency according to the change in heat transfer rate due to ammonia co-firing.

2. Model descriptions

2.1. Target boiler system and CFB loop model

Fig. 1 shows a schematic diagram of the target CFB boiler and the model diagram of the CFB loop. Unlike pulverized coal boilers, the CFB boiler has a solid return part due to the bed material. The CFB loop includes a cyclone, a loop-seal, and external heat exchanger (EXHE) or internal heat exchanger (INTREX). EHEX and INTREX are a type of heat exchanger that utilizes the sensible heat of circulating solids. The bed material is circulated in this CFB loop. Due to these bed materials, the range of the furnace temperature of the CFB boiler is about 800~900 °C, which is lower than that of the PC boiler. Due to these characteristics, thermal NOx emissions are lower than PC boilers and low-grade coal can be used. The solid return part of the target boiler has 8 cyclones, 8 loop seals, and 8 INTREXs. The INTREXs of the target boiler system are used as the

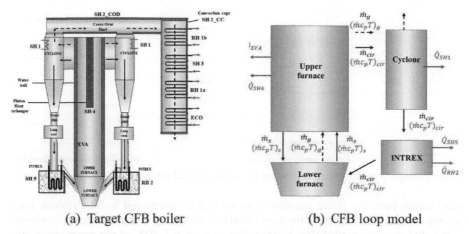

(a) Target CFB boiler   (b) CFB loop model

Figure 1 Schematic diagram of the target CFB boiler (a) and model diagram of the CFB loop (b) [Kim et. al.]

final superheater (SH) and reheater (RH). The heat exchangers of the target boiler system include economizer (ECO), evaporator (EVA), SH1~5, and RH1~2. The CFB loop includes EVA and SH4 in the combustion chamber, SH1 in the cyclone, and SH5 and RH2 in ITNREX. Since the heat load of the heat exchangers of the CFB loop accounts for 68% of the total boiler heat load, it is important to evaluate the heat transfer characteristics within the CFB loop. Due to the bed material, the following model consideration has to be applied. Ascending and descending solids in the furnace and circulating solids must be considered. Therefore, the CFB furnace was divided into lower furnace (LF) and upper furnace (UF) by considering the up-down flows of solids between LF and UF. Additionally, because circulating solids affect the heat transfer rate and temperature of the CFB loop, the solid circulation rate must be derived. In this study, the temperature of the CFB loop was adjusted to match the design data by adjusting the solid circulation rate. Therefore, a fitting equation of a second order of the flue gas velocity was presented. In addition, to derive the heat transfer coefficient (HTC) of INTREX, a correlation equation is developed based on the solid circulation rate and inlet solid temperature of INTREX by using the design data of HTC of INTREX.

2.2. Process simulation model

The model approach is the heat exchanger block simulation model. Fig. 2 shows a schematic diagram of the process simulation model and each heat exchanger model. In this model approach, the entire boiler system is divided into finite heat exchangers. Each heat exchanger is composed of a gas-solid side model and a water-steam side model, in which a lumped parameter model was applied. The gas-solid side and water-steam side models are connected through heat transfer. Therefore, the mass and energy balances of the solid-gas and water-steam blocks are solved. In this model, the heat transfer rate is calculated by overall HTC, heat transfer area, and logarithmic mean temperature difference. Overall HTC is calculated using the external HTC of the gas-solid side and the internal HTC of the water-steam side.

The applied correlations for the calculation of HTC of the CFB boiler are as follows. The calculation of outer HTC is divided into the CFB furnace, solid return part, and back pass. In the CFB furnace, the cluster-renewal model is applied, which can calculate the radiation and convection heat transfer of bed material and flue gas [Basu, 2006]. In the

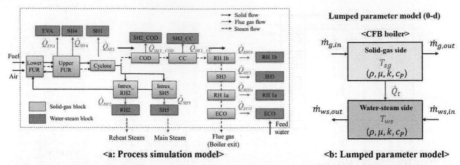

Figure 2 Diagram of process simulation model (a) and lumped parameter model (b) [Kim. et. al.]

solid return part, the fitting equation was applied as mentioned in the previous page [Zhang et al.]. In the back pass, the correlation for calculating the radiation and convection of the flue gas was applied [Basu, 2000; Incropera et al.]. HTC of the water-steam side was applied according to subcritical and supercritical pressures [Incropera et al.; Sallevelt et al.]. This process simulation model was validated with the design data. The error rate of solid-gas temperature, the water-steam temperature, and the heat transfer rate are below 5%. Therefore, model reliability was confirmed.

3. Process simulation results

The simulation conditions are Boiler Maximum Continuous Rating with the design coal. The ammonia co-firing ratio is 0, 5, 10, 20, and 30% based on high heating value (HHV). As shown in Table 2, the CFB boiler utilizes low-grade coal compared to the PC boiler. Table 3 shows the mass balance results and adiabatic flame temperature (AFT) according to ammonia co-firing. The total flue gas flow rate and AFT decrease according to the increase in ammonia co-firing ratio. Solid circulation rate decreases due to the decrease in air flow rate. At 30% co-firing condition, the solid circulation rate decreases by approximately 150 kg/s. It is assessed that the reduction in solid circulation does not significantly affect heat transfer. Apparently, as the ammonia co-firing ratio increases, the mole fraction of CO_2 decreases, and the mole fraction of H_2O increases.

Table 2 Design coal of the CFB boiler and PC boiler

	C (%)	H (%)	O (%)	N (%)	S (%)	Ash (%)	Mois. (%)	HHV (kcal/kg)
CFB	47.80	3.67	22.5	0.75	0.22	4.5	20.5	4,350
PC	57.42	3.32	11.7	1.15	0.31	6.0	20.0	5,300

Table 3 Mass balance results and adiabatic flame temperature according to ammonia co-firing ratio.

Design coal_BMCR load condition		Unit	Ammonia co-firing ratio				
			0%	5%	10%	20%	30%
Coal flow rate		kg/s	76.20	72.39	68.58	60.96	53.34
Ammonia flow rate		kg/s	0	3.08	6.16	12.32	18.47
Total air flow rate		kg/s	532.32	528.31	524.31	516.29	508.27
Total flue gas flow rate		**kg/s**	**604.83**	**600.28**	**595.73**	**586.63**	**577.53**
Solid circulation rate		**kg/s**	**3,093**	**3,069**	**3,046**	**2,999**	**2,953**
Flue gas composition	O_2	%	3.38	3.37	3.36	3.35	3.33
	N_2	%	66.75	67.17	67.60	68.47	69.36

(Mass fraction)	CO_2	%	21.93	20.99	20.04	18.09	16.08
	H_2O	%	7.94	8.46	9.00	10.1	11.23
Adiabatic flame temperature		°C	1,668	1,665	1,662	1,657	1,651

Fig. 3 shows the simulation results of the heat transfer rate of each heat exchanger according to the ammonia co-firing ratio (a) and the comparison with the PC boiler (b). The process simulation model and results of the PC boiler can be referred to the reference [Kim et. al.]. Although flame emissivity slightly increases due to the increase in H_2O mole fraction, the radiation heat transfer rate decreases due to the decrease in AFR, circulating solids, and mass fraction of coal particles. In addition, the convective heat transfer rate decreases due to the decrease in the flue gas mass flow rate. However, as shown in Fig. 3(b), the decrease in heat transfer rate of the CFB boiler is lower than in the PC boiler. This is because the CFB boiler uses low-grade coal (low HHV and large content of moisture). Therefore, the reduction in AFT and heat loss of moisture of the CFB boiler is lower than that of the PC boiler. In addition, due to the heat transfer mechanism of the circulating solids, the reduction in radiative heat transfer in CFB boilers is lower compared to PC boilers, where only a flue gas heat transfer mechanism exists. Accordingly, the main and reheat steam temperature decreases due to the decrease in the radiation and convection heat transfer rate under the ammonia co-firing condition as shown in Fig. 4. Because there is a difference in the heat transfer rate of both boilers, the reduction of main and reheat steam temperature of the CFB boiler is lower than that of the PC boiler. At ammonia co-firing of 30%, the main and reheat steam temperatures of the CFB boiler decrease by 18.5°C and 6.5°C, respectively.

Fig. 5 shows the reduction of CO_2 and boiler thermal efficiency. As shown in Fig. 5, As the ammonia co-firing rate increases, the reduction in CO_2 emissions increases, but the boiler thermal efficiency decreases. This shows the trade-off relationship between CO_2 reduction and boiler thermal efficiency. The boiler thermal efficiency of the CFB boiler decreases by 1.9%, which is lower than that of the PC boiler (2.3%). Therefore, it is important to establish optimal operation conditions with ammonia-co-firing conditions.

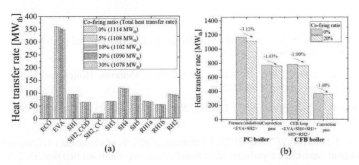

(a)   (b)

Figure 3 Simulation results of heat transfer rate of each heat exchanger according to ammonia co-firing ratio (a) and the comparison with PC boiler (b)

4. Conclusions

We performed a process simulation of a CFB boiler utilizing ammonia as a carbon-free fuel for the evaluation of thermal performance. We demonstrated the reduction of radiation and convective heat transfer rate with ammonia co-firing conditions. This means a reduction in plant efficiency. However, we also confirmed the reduction of the heat transfer rate of the CFB boiler is lower than that of the PC boiler due to the use of low-grade coal and the heat transfer mechanism of circulating solids. This study can be used to establish optimal operating conditions for CO_2 emissions and the improvement of plant efficiency during ammonia co-firing in terms of power plant operation.

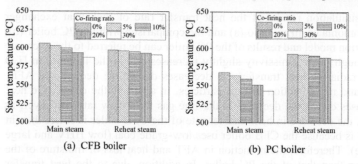

Figure 4 Simulation results of main and reheat steam temperatures of the CFB boiler and the PC boiler

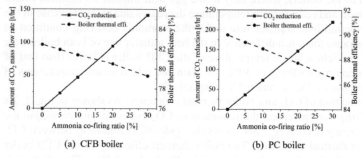

Figure 5 Relationship between the reduction of CO2 and boiler thermal efficiency of the CFB boiler and PC boiler.

References

P. Basu, 2006, Chap. Heat Transfer in Circulating Fluidized Beds, Combustion and gasification in fluidized beds. New York: Taylor & Francis, pp. 178.

P. Basu, C. Kefa, L. Jestin, 2000, Chap. Boiler Furnace Design Methods, Boiler and burners (Design and Theory). New York, Springer, pp. 128

FP. Incropera, DP. Dewitt, TL. Bergman, AS. Lavine, 2008, Fundamentals of Heat and Mass Transfer, New York, Wiley.

S. Kim, M. Lim, Y. Lee, J. Lee, and W Yang, 2023, Evaluation of effects of ammonia co-firing on the thermal performances of supercritical pulverized coal and circulating fluidized bed boilers, 276, 116528.

H. Kobayashi, A. Hayakawa, K. Somarathne, E. Okafor. Science and technology of ammonia combustion. 2019, Proc Combust Inst 37, 109–133. https://doi.org/10.1016/j.proci.2018.09.029.

J. Sallevelt, J. Withag, E. Bramer, D. Brilman, G. Brem, 2012, One-dimensional model for heat transfer to a supercritical water flow in a tube, J Supercrit Fluids, 68, 1–12. https://doi.org/10.1016/j.supflu.2012.04.003.

M. Zhang, H. Wu, Q. Lu, Y. Sun, and G. Song, 2012, Heat transfer characteristics of fluidized bed heat exchanger in a 300MW CFB boiler, Powder Technol, 222, 1–7.

Flavio Manenti, Gintaras V. Reklaitis (Eds.), Proceedings of the 34th European Symposium on Computer Aided Process Engineering / 15th International Symposium on Process Systems Engineering (ESCAPE34/PSE24), June 2-6, 2024, Florence, Italy

Modelling of PEM Electrolyzer Dynamics for Green Hydrogen Production

Dhana Lakshmi Gosu[a]*, Anuradha Durvasula[a], Jagan Annamalai[b], Abhisek Roy Chowdhary[a], Guillermo Sanchez Freire[c],

aCognizant Technology Solutions, Hyderabad, India,
bAveva Group Limited, Houston, USA
cSchneider Electric Limited, Madrid, Spain
dhanalakshmi.gosu@cognizant.com

Abstract

In the current paradigm of sustainable chemical production, Green Hydrogen production using electrolysis of water is gaining popularity. This necessitates the development of rigorous first principles-based electrolyzer simulation models that can be leveraged during design and for training of operators in optimal and safe operation of the equipment. PEM electrolysis is one of the efficient routes for Green Hydrogen generation. Given the complexity of the operation, unique challenges are faced in modelling the dynamic behavior. The study here involves developing a high-fidelity dynamic model of PEM Electrolyzer leveraging AVEVA Dynamic Simulation™. Modelling of electrolyzers is challenging, as standard reactor models cannot be used for electrolysis since electrolysis has a strong dependency on electrical current flowing in the cell. The model should be capable of predicting correct pressure and temperature responses for any changes in feed water flow and/or electrical load. The model should be also capable of accurately modelling the heat generation in the system due to low material inventory at cathode and should have provision for the heat extraction. In addition, the model should also have provision to handle gas leakage through membrane as well as water permeation due to electro-osmatic drag and diffusion through the membrane. The model developed here was used to study the cell voltage, ohmic, activation and mass transfer losses along with the typical relationship between voltage and current density in the electrolyzer cell. The use case of Hydrogen leakage based on electrical load was also simulated using user specified diffusion coefficient. Membrane rupture which leads to direct flow from cathode to anode is another use case that was studied. These together provided the ability to test the boundary case of explosion limit. Multiple electrolyzer arrangements, connected in series and in parallel, were also simulated and studied. The results generated from the studies using the model were in close match with the expected dynamic response and results. In summary, the PEM Electrolyzer model developed using AVEVA Dynamic Simulation™ has the rigor to train engineers in operating this complex piece of equipment.

Keywords: Sustainability, green hydrogen production, proton exchange membrane (PEM) electrolyzer, water permeation, electro-osmatic drag, membrane hydrogen leakage.

1. Introduction

Hydrogen is an important chemical that is widely used in the Oil & Gas, Refining and Chemical industries to produce other chemicals. Hydrogen demand is also increasing globally as a clean source of energy, as it emits only water on combustion. Hydrogen has been traditionally manufactured using hydrocarbon feedstock via steam reforming, coal gasification or methane pyrolysis process. However, of late, the focus is on green hydrogen, where hydrogen is produced via water electrolysis, using electricity generated from a renewable source like solar or wind energy. The process gives high purity hydrogen, with oxygen as a by-product, and has minimal impact on the environment. The commercial electrolyzers used for hydrogen production are Proton Exchange Membrane (PEM) and Alkaline Electrolyzers. The study presented here simulates PEM electrolyzer using AVEVA Dynamic Simulation™ (ADS), which is a commercial high fidelity dynamic process simulator used for building Operator Training Simulators (OTS).

2. Description

In PEM electrolyzer, water is electrochemically split into hydrogen and oxygen at the cathode and anode respectively. Pure water, used as the electrolyte solution, is pumped to the anode section, where it splits into oxygen (O2), protons (H+) and electrons (e−). These protons migrate via the polymer-based proton conducting membrane towards cathode. At the cathode, migrated protons re-combine with electrons to produce hydrogen gas.

2.1 Design

Anode and Cathode are considered as holdups in ADS where the fluids can be accumulated for some time before moving forward. These holdups have their own pressure dynamics and mass/energy balance. The Hydrogen and Oxygen formation is based on the electrochemical reaction. The flow at cathode and anode sides are based on the differential pressure and mass balance equation. The temperature is determined based on the energy balance equation.

2.2 Voltage calculations

In the electrochemical model, the cell voltage is calculated by adding the open circuit voltage, the maximum voltage, and various losses (activation and ohmic losses) based on Abdol et al., 2015 [1] as shown below:

$$V_{cell} = V_{ocv} + V_{act} + V_{\Omega}$$

The open-circuit voltage is calculated based on Saebea et al., 2017 [2].

$$V_{ocv} = V_{rev} + \frac{R * T_m}{2F} \ln\left(\frac{pp_{H_2}}{P_{std}} * \sqrt{\frac{pp_{O_2}}{P_{std}}}\right)$$

where V_{rev} is the reversible voltage i.e., the minimum voltage required for each reaction at the electrodes, F is Faraday constant i.e., 96485 C/mol, R is gas constant, pp_{H2} and pp_{O2} are partial pressures of Hydrogen and Oxygen respectively.

Activation over-potential at each electrode is defined based on Abdol et al., 2015 [1]:

$$V_{act,x} = \frac{R * T_x}{2\alpha_x F} \ln\left(\frac{J}{2i_{ax}} + \sqrt[2]{\left(\frac{J}{2i_{ax}}\right)^2 + 1}\right)$$

where x is either anode or cathode, α_x is charge transfer coefficient for the electrode which is user-defined and dimensionless, J is current density on the PEM stack electrodes and i_{ax} is current density at each electrode.

Ohmic overpotential is caused by the membrane resistance to the flow of ions. This can be expressed as (Abdol et al., 2015 [1]):

$$V_{ohm} = R_{mem} * I$$

R_{mem} is an ionic resistance as a function of membrane thickness conductivity (σ_{mem}, S/m), membrane height (ϕ, m), I is electrolyzer current to each electrolyzer.

2.3 Material Balance

Simplified form of equations based on Paolo et al., 2017 [3], are used in the simulation model. The hydrogen production rate is calculated using the equation:

$$FH_{2,prod} * 1000 = \frac{etaF * Ncell * I}{2 * F}$$

where *Ncell* is Number of electrolyzer cells, I is Electrolyzer total current (Amp), and *etaF* is Faraday efficiency (1.0).

The permeation rate can be represented as being proportional to the diffusivity in the membrane and the concentration gradient between two flow channels, which is described as:

$$FH_{2permeated} = D_{H2} \frac{(C_{H2,c} - C_{H2,a}) N_{Cell} A_e}{t_m}$$

where t_m is the thickness of the membrane and $C_{H2\,c}$ and $C_{H2,a}$ are hydrogen concentration for the cathode and anode surface of the membrane, respectively.

Diffusivity of hydrogen in membrane (m²/s),

$$D_{H_2} = DiffCoeff_{H_2} * e^{(-\frac{16510}{RT_m})}$$

where *DiffCoeff*$_{H2}$ is diffusion coefficient for hydrogen which is user-defined, and the default is 4.9E-5, R is Gas Constant and is equal to 8.314 kJ/kg-mol/K and T_m is electrolyzer temperature.

The large fraction of permeated hydrogen from the membrane decomposes at the anode. The current understanding is that 80% of the permeated hydrogen decomposes, while the remaining 20% mixes with oxygen at the anode.

$$FH_{2anode} = 0.2 * FH_{2permeated}$$

Oxygen production rate is half of FH_2. Oxygen undergoes the same membrane migration phenomenon in the opposite direction as hydrogen permeation.

$$FO_{2cathode} = DiffCoeff_{O2} \frac{(C_{O2,a} - C_{O2,c}) N_{Cell} A_e}{t_m}$$

where $C_{O2,c}$ and $C_{O2,a}$ are oxygen concentration for the cathode and anode surface of the membrane, respectively. *DiffCoeff*$_{O2}$ is diffusion coefficient for oxygen which is user-defined, and default is 6.5E-7.

Water permeates through membrane due to electro-osmatic drag and diffusion which are functions of the water content of the membrane and are based on Paolo et al., 2017 [3]. Water permeation due to electro-osmatic drag,

$$FH_2O_{eod} = EODragCoeff * \frac{M_{H2O} n_d J N_{cell} A_e}{1000 * F}$$

where A_e is the area of the electrolyzer cell, J is current density, λ is degree of humidity in membrane and n_d is the electro-osmotic drag coefficient calculated based on degree of humidity as follows:

$$n_d = 0.0029\lambda^2 + 0.05\lambda - 3.4 * 10^{-19}$$

EODragCoeff is user-defined factor to tune electro-osmatic drag flow rate.

Water diffusion through the membrane is given from Fick's first law of diffusion as follows:

$$FH_2O_{Diff} = D_w \frac{(C_{wa} - C_{wc}) NCell A_e}{t_m}$$

where $C_{w,c}$ and $C_{w,a}$ are water concentration for the cathode and anode surface of the membrane, respectively. The water diffusion coefficient is computed as:

$$D_w = 1.25 * 10^{-10} DiffCoeff e^{2416(\frac{1}{303} - \frac{1}{T_{cat}})}$$

DiffCoeff is user-defined factor to tune water flow rate due to diffusion, T_{cat} is cathode side temperature.

2.4 Energy Balance

Overall energy balance includes the heat of reaction as well as the electrical power including heat loss from fluid to metal (Paolo et al., 2017 [3]).

$$M_m Cp_m \frac{dT_m}{dt} = (Q_{loss} - Qf_{Ano} - Qf_{Cat} - HRxn + Power)$$

where Qf_{Ano} and Qf_{Cat} are heat losses from fluid to metal at the corresponding electrodes, H_{rxn} is the heat of reaction and Q_{loss} is the heat loss from metal to ambient.

3. Simulation model details and results of case study

3.1 Simulation model

In ADS, the PEM electrolyzer model consists of an anode chamber with one inlet and one outlet stream and a cathode chamber with two process outlet streams. An electrical source is connected to the cell using electrical stream. A fan-based cooler is used to remove excess heat generated in the electrolyzer. Water enters the anode chamber, where it is split into hydrogen and oxygen using electrical potential. Hydrogen then flows to cathode, along with permeated water. The mixture is separated in a gas-liquid separator and hydrogen is sent to storage. Oxygen flows out of anode along with the unused water and is separated in a gas-liquid separator. Oxygen is taken to storage, while water is recycled back to electrolyzer. Diffusion coefficients are used to simulate cross-over of hydrogen and oxygen through the membrane. The schematic representation of the simulation model is depicted in Figure.1.

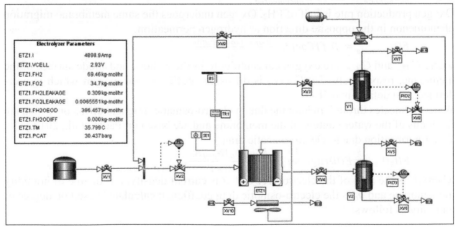

Electrolyzer Parameters	
ETZ1.I	4898.9Amp
ETZ1.VCELL	2.93V
ETZ1.FH2	69.46kg-mol/hr
ETZ1.FO2	34.7kg-mol/hr
ETZ1.FH2LEAKAGE	0.309kg-mol/hr
ETZ1.FO2LEAKAGE	0.0065551kg-mol/hr
ETZ1.FH2OEOD	396.457kg-mol/hr
ETZ1.FH2ODIFF	0.000kg-mol/hr
ETZ1.TM	35.799C
ETZ1.PCAT	30.437barg

Figure.1. Schematic representation of electrolyzer process and corresponding ADS simulation model.

3.2 Impact of membrane thickness

Current density is a key parameter that determines the operating efficiency of PEM Electrolyzer. The effect of membrane thickness on performance was modeled using the ADS simulation model (Figure.2) and the results match with literature (Saebea et al., 2017 [2]). It is found that at high pressure operation thinner membranes provide higher performance because when membrane gets thicker, ohmic resistance increases, thereby necessitating higher voltages.

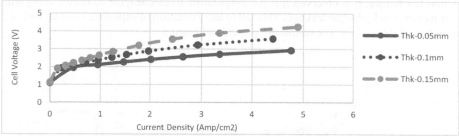

Figure.2. Effect of Current Density and membrane thickness on Cell Voltage

3.3 Impact of cathode pressure

When hydrogen leak accumulates over time or if the leakage increases to such an extent that it is near the explosion limit i.e., 4% of H_2 in O_2, it poses safety risk. So, the detection of hydrogen leakage is very important. The effect of pressure on hydrogen permeation towards anode is modelled using the PEM model (Figure.3) and the results match with literature (Saebea et al., 2017 [2]). The rise in pressure increases hydrogen permeation towards anode, i.e., hydrogen leakage.

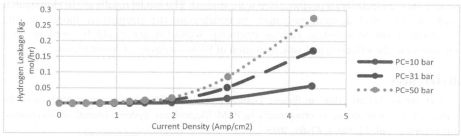

Figure.3. Effect of Current Density and Pressure on Hydrogen leakage

3.4 Impact of membrane degradation

Membrane degradation takes a long time to observe in actual operation. Its effect on electrolyzer can be simulated at an accelerated rate using the ADS model. For this purpose, a malfunction scenario is created where in membrane degradation is simulated. Membrane degradation causes a decrease in the performance of electrolyzer; thereby, a decrease in hydrogen production and increase in hydrogen leakage. The results are plotted in Figure.4.

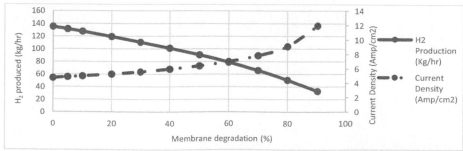

Figure.4. Effect of Membrane Degradation on Hydrogen Production and Current Density

3.5 Impact of external cooling source

Excess heat produced is removed by fan to maintain the temperatures at around 35^0 C. Upon fan failure, it is observed that the temperature of electrolyzer increases along with the increase in hydrogen leakage. This led to the formation of explosive mixture within 2 minutes of fan cut off. The results are plotted in Figure.5.

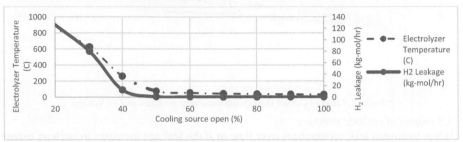

Figure.5. Cooling source cut off vs Hydrogen Leakage and Electrolyzer Temperature

Conclusions

PEM electrolyzer was successfully simulated using AVEVA Dynamic Simulation™. The impact of various parameters like membrane thickness, current density, and cooling source on the electrolyzer performance were studied. The model results are in close match with the expected dynamic response. In conclusion, the electrolyzer model has the necessary rigor to simulate various operating scenarios and could be used to train engineers in the effective and safe operation of PEM electrolyzer based plants.

References

1. A.H. Abdol Rahim, Alhassan Salami Tijani, Farah Hanun Shukri, Simulation Analysis of the Effect of Temperature on Overpotentials in PEM Electrolyzer System, Journal of Mechanical Engineering, Vol. 12, No. 1, 47-65, 2015
2. Dang Saebea, Yaneeporn Patcharavorachot, Viktor Hacker, Sutthichai Assabumrungrat, Amornchai Arpornwichanop, Suthida Authayanun, Analysis of Unbalanced Pressure PEM Electrolyzer for High Pressure Hydrogen Production, Chemical Engineering Transactions, VOL. 57, 2017, A publication of The Italian Association of Chemical Engineering https://doi.org/10.3303/CET1757270
3. Paolo Colbertaldo, Sonia Laura Gómez Aláez , Stefano Campanari, Zero-dimensional dynamic modeling of PEM electrolyzers, Energy Procedia 142 (2017) 1468–1473 https://doi.org/10.1016/j.egypro.2017.12.594

Flavio Manenti, Gintaras V. Reklaitis (Eds.), Proceedings of the 34th European Symposium on Computer Aided Process Engineering / 15th International Symposium on Process Systems Engineering (ESCAPE34/PSE24), June 2-6, 2024, Florence, Italy

Discovering zeolite adsorption isotherms: a hybrid AI modeling approach

Arijit Chakraborty[a], Akhilesh Gandhi[b], M. M. Faruque Hasan[b] and

Venkat Venkatasubramanian[a]*

[a]*Complex Resilient Intelligent Systems Laboratory, Department of Chemical Engineering, Columbia University, New York, NY 10027, USA*
[b]*Artie McFerrin Department of Chemical Engineering, Texas A&M University, College Station, TX 77843, USA*
*venkat@columbia.edu

Abstract

Zeolites renowned for their porous structure, and adsorption capabilities, play a pivotal role in various applications across energy, health, and the environment. Despite advances in experimental techniques, the problem of predicting the adsorption isotherm of a zeolite – given its structural properties – remains inherently challenging. A key aspect of such a problem is that it is inherently a "precious data" problem, due to the limitation of having a small dataset. In this work, we propose a hybrid AI modeling approach for obtaining predictive models for zeolite adsorption isotherms. Given the structural properties of a zeolite, we are able to predict the adsorption property reasonably well, in contrast to conventional machine learning techniques which – despite their relatively complex parameterizations – are unable to provide reliable results. We rely on a combination of symbolic AI, and intelligent feature engineering, resulting in concise and interpretable adsorption isotherm models for different temperatures. The benefits of the same are that the models generalize well on zeolites that have not been encountered during the training phase, while being relatively simple in its form. This enables post-hoc analyses for gaining insights on the descriptive capabilities of the structural properties used.

Keywords: Zeolites, Adsorption isotherm, Artificial intelligence, Machine Learning, Hybrid AI.

1. Introduction

The advent of enhanced computational power along with vast amounts of data, has expanded the applicability of artificial intelligence (AI) – specifically, machine learning (ML) – across a wide range of domains. These include speech recognition, entertainment recommender systems, chat-bots like ChatGPT (Brown et al., 2020), to name a few. Unlike sciences and engineering, these domains have colossal amounts of data at their disposal, with more being acquired every day. We refer to the sciences and engineering as "precious" data domains as opposed to "big data", since the cost of obtaining a new datapoint for analysis is prohibitively expensive. Accordingly, conventional AI/ML techniques are not feasible for performing analyses on such limited datasets. Further, unlike the sciences and engineering, these domains are not governed by underlying fundamental laws and mathematical relations, nor do they have underlying expert knowledge that can be used to guide one towards a better prediction.

Thus, in order to tackle such problems, we suggest the incorporation of relevant first-principles knowledge, along with modifications in conventional AI/ML techniques to enhance the predictive capabilities of data-driven models. In this work, we tackle such a problem for the prediction of the adsorption capacity of carbon-dioxide (CO_2) gas on various zeolites at 2 temperatures – 323 K and 373 K. The goal is to predict the adsorption isotherm, given the structural properties of a zeolite. We are limited by the size of the dataset (here, 188 zeolites). Accordingly, this is a *precious* data problem. Relying on the inclusion of first-principles' based feature engineering, and evolutionary-algorithmic search for mathematical models, we successfully obtain reasonably accurate models using such a restrictive dataset. We highlight the efficacy of our approach, and how it results in more interpretable models that enable a domain-expert to gain additional insights.

2. Background

Zeolites and other porous structures such as metal-organic frameworks (MOFs) are known for their adsorption properties, and thus, have applications in the most crucial chemical industries. Towards the goals for sustainability, these structures play a pivotal role in gas adsorption such as applications in carbon-capture for lowering global CO_2 levels, hydrogen storage for energy transition, oxygen concentrators (which helped save lives during the pandemic), and in catalysis. Prediction of adsorption properties for these materials is paramount, and has thus far been heavily reliant on experimental methods, which were time and resource intensive. With the advent of computational modeling, recent methods such as Monte Carlo (MC) simulations were developed to address the highly resource-intensive experiments. However, with the advent of computational methods, came the possibility of design of new structures and thus, several million pure-Si zeolite frameworks have been hypothesized. The computational requirement of the Grand Canonical Monte Carlo (GCMC) models are substantially high, when considering millions of such hypothetical designs.

Several works have tried to address the adsorption prediction for these zeolites due to their key importance in the materials design space (Raji et al., 2022; Okello et al., 2023; Alizadeh et al., 2022). However, most of the current literature is focused on building models to predict properties of known zeolites, and to enhance the predictions of molecular simulations. Some approaches also focus on the high throughput screening of zeolites (Moliner et al., 2019). However, property prediction for unknown zeolites adsorption data for which is not readily available – has remained an open challenge for exploration.

Towards property prediction of hypothetical structures, one of the key limitations is the representation for these crystal structures. These are cumbersome, in contrast to molecular representations which are studied in detail such as SMILES, SELFIES, molecular fingerprints etc. This limitation was overcome recently in a work by Gandhi et al. (Gandhi and Hasan, 2021), where a graph-theoretic representation for crystal structures such as zeolites, was developed. Specifically for zeolites, the architecture of the process for inverse design and property prediction has been laid out in literature (Gandhi and Hasan, 2022). A key limitation however in the case of adsorption is the lack of data to build autonomous ML models. Towards this end, we propose a hybrid AI model for the prediction of adsorption in zeolites based on structural features that can be computed easily, and can be modeled with the physical knowledge of the system to overcome the data limitations.

In this section, we have introduced the importance of adsorption isotherm modeling, and enumerated some of the pertinent challenges faced by researchers in this domain. We emphasize the need for an alternate modeling approach – one that combines the benefits of first-principles knowledge, with a data-driven approach – resulting in a hybrid AI (Chakraborty et al., 2022) model. In the next section, we discuss the approach used for modeling the adsorption isotherm of zeolites for CO_2 gas. We subsequently present the results of our approach, and highlight the efficacy of domain-knowledge-driven feature engineering on the limited dataset available (188 zeolites). Finally, we conclude this manuscript by summarizing our work, and provide future directions that are currently being undertaken, and for guiding researchers in this field.

3. Methods

In this section, we discuss the data used for the problem at hand, the approach used to obtain an interpretable data-driven model for the prediction of CO_2 adsorption isotherm of zeolites based on their structural properties.

3.1. Data

Adsorption of CO2 on pure-silica zeolites is a resource-intensive computation for different pressures and temperatures. Also, experimental methods are limited by synthesis and availability of these zeolite structures limiting the scope to only computational molecular simulations. In this case, we used the GCMC simulations to determine the adsorption capacity of each of the zeolite structures at different P-T conditions (Iyer and Hasan, 2019). Furthermore, the data validation only makes sense for observed zeolites where we can validate the GCMC parameters. Thus, expanding the dataset over 188 zeolites is extremely resource-intensive and challenging to validate.

The descriptors in the dataset used in this work capture the physical structure of the zeolites. Some of these descriptors, including density, Si-O-Si bond angle (average, harmonic mean, min, max, std, etc), statistics on the bond distance, and details on accessible and non-accessible surface area, among others, are computable from the location of the Si and O atoms. Some of these descriptors are already pre-computed in the IZA database and have been used in other studies regarding ML-based modeling of mechanical properties of zeolites. The data used in our work is taken from a previous study (Evans and Coudert, 2017) where the data is freely available. Accordingly, we proceed with 16 descriptor variables for each zeolite, which are further transformed by the modelling tool, as discussed in the next section.

In our work, we split the dataset into a train and test set with 80% zeolites (150 zeolites) categorized into the train set, with the remaining 20% (38 zeolites) categorized into the test set.

3.2. Modeling tool – AI-DARWIN

Relying on the formulation of obtaining a mathematical model as a search problem, we build on the genetic algorithmic-based model discovery engine AI-DARWIN (Chakraborty et al., 2021), which is a more generalized extension of a linear model discovery engine titled GFEST (Chakraborty et al., 2020). Previously, this approach has yielded success in the prediction of a bubble column aeration process (Jul-Rasmussen et al., 2023), which relied on real and noisy data from a pilot plant. It relies on the use of genetic algorithm (GA), coupled with a function library where user-specified function transformations act upon the input variables, to generate features for subsequent regression. For a detailed description of the algorithm, the reader is directed to the original research articles discussing this method, and its efficacy on a variety of case-studies.

3.3. First-principles' based feature engineering

Upon utilizing this approach without any modifications and permitting all function transformations, we obtained unsatisfactory results. A common result from these inaccurate models was the prevalence of pressure as a major contributing factor to the prediction of the adsorption loading, which overshadowed the effect of the structural properties of the zeolite. Such a model relied entirely on the pressure of adsorption process, and failed to capture the variability caused due to a different zeolite structure, and by extension, its properties. Accordingly, it would be incorrect to apply such an approach for different zeolites.

Empirical models of adsorption include the Langmuir adsorption model, given as

$$\theta = \frac{K_{eq}P}{1 + K_{eq}P} \tag{1}$$

Here, θ is the fraction of the adsorption sites occupied, K_{eq} is the equilibrium constant, and P is the pressure at which adsorption is taking place. Here, the effect of pressure is accounted for as a direct effect. Inspired by the form of the Langmuir adsorption model, we choose to separate the effects of the structural properties from pressure, and account for both, such that the effect of pressure does not overwhelm the predictive capability of the zeolite's physical properties. Our proposed model takes the following form:

$$\theta_{AI-DARWIN} = \sum_i^{14} \frac{\beta_i f_i P}{1 + \beta_i f_i P} \tag{2}$$

Here, f_i represents the features generated by the AI-DARWIN model, and β_i is the parameters obtained upon performing parameter estimation on the model. It must be noted that the model form does not have an intercept. It was formulated as such, to impose the physical constraint that at zero pressure, the absolute loading (adsorption capacity) be zero.

4. Results and Discussion

The results are depicted for 323 K in Figure 1, and 373 K in Figure 2. The model for 323 K (R^2 of 0.65 on training set, and 0.87 on test set) was less accurate as compared to the model for 373 K (R^2 of 0.87 on training set, and 0.94 on test set), likely due to the training data for numerous zeolites in the dataset being limited to low and medium pressures (up to approximately 4 MPa). Thus, there was not sufficient variability in the training data to yield more accurate predictions.

We see that by virtue of having chosen a model without an intercept, we can enforce that the absolute loading would be zero, at a pressure of zero Pa. The model is able to fit accurately ($R^2 > 0.9$) for most zeolites. There are some errors where the model predicts adsorption isotherms which are vastly different from the true isotherms in some cases, such as for zeolites SBS at 323 K and GOO at 373 K (depicted in Figure 1 (a) and Figure 2 (a) respectively). This is likely due to reduced variability in the training data, for the model to recognize salient features effectively. In future research, this can be addressed by using alternate parametrizations, and/or enforcing constraints which account for the smoothness of the curves.

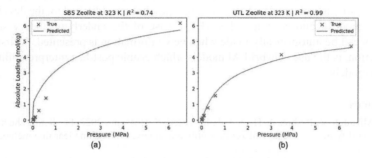

Figure 1: Adsorption isotherms at 323 K for (a) SBS zeolite, and (b) UTL zeolite.

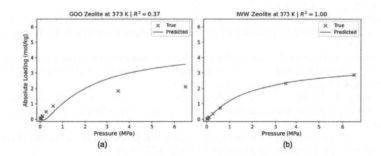

Figure 2: Adsorption isotherms at 373 K for (a) GOO zeolite, and (b) IWW zeolite

We obtained high test R^2 values for both the temperatures, which is a success of our hybrid AI approach. While it might seem that these scores could be a result of the small size of the test dataset (38 zeolites), we tested the model multiple times with different randomized splits of zeolites, with 80% used for training, and the remaining 20% used for testing, such that our testing metrics are not biased. Accordingly, we obtained similar performance with negligible variability across the different random splits of the dataset.

A key advantage of our method is its inherent interpretability since we know which feature combinations (and their weights) affect the adsorption capacity. These can be used to estimate the differentiating factor for adsorption at a higher temperature (373 K), when compared to one at a lower temperature (323 K). Such analysis is not possible when using a purely black-box modeling technique such as a neural network.

5. Conclusions and future work

Having obtained a reasonably accurate predictive model for adsorption isotherm from a dataset with less than 200 zeolites, it is important for future research to explore the explainability of such models. Particularly, researchers must probe into how the variables and their interactions captured in the model, relate to the adsorption capacity. Such directions have the potential to provide novel insights, previously unknown by the respective subject matter experts. While quantitative estimates of these interactions and their resultant effects are possible, for e.g., using Shapley values (Shapley, 1953)), they do not provide us any insight into the fundamental process(es) that take place. Hence, these explanations will need to be guided by first-principles, since there is ample theory on the topic of adsorption in zeolites. This includes, but is not limited to, the development of causal models, and creating a knowledge base of the underlying theory such that inferences can be automatically made when new information is presented. A first step in this direction, is the use of hybrid AI models which enable post-hoc interpretability of the obtained models.

References

S. M. S. Alizadeh, Z. Parhizi, A. H. Alibak, B. Vaferi, S. Hosseini, 2022. Predicting the hydrogen uptake ability of a wide range of zeolites utilizing supervised machine learning methods. International Journal of Hydrogen Energy 47 (51), 21782–21793.

T. Brown, B. Mann, N. Ryder, M. Subbiah, J. D. Kaplan, P. Dhariwal, A. Neelakantan, P. Shyam, G. Sastry, A. Askell, et al., 2020. Language models are few-shot learners. Advances in neural information processing systems 33, 1877–1901.

A. Chakraborty, S. Serneels, H. Claussen, V. Venkatasubramanian, 2022. Hybrid ai models in chemical engineering–a purpose-driven perspective. Computer Aided Chemical Engineering 51, 1507–1512.

A. Chakraborty, A. Sivaram, L. Samavedham, V. Venkatasubramanian, 2020. Mechanism discovery and model identification using genetic feature extraction and statistical testing. Computers & Chemical Engineering 140, 106900.

A. Chakraborty, A. Sivaram, V. Venkatasubramanian, 2021. Ai-darwin: A first principles-based model discovery engine using machine learning. Computers & Chemical Engineering 154, 107470.

J. D. Evans, F.-X. Coudert, 2017. Predicting the mechanical properties of zeolite frameworks by machine learning. Chemistry of Materials 29 (18), 7833–7839.

A. Gandhi, M. F. Hasan, 2021. A graph theoretic representation and analysis of zeolite frameworks. Computers & Chemical Engineering 155, 107548.

A. Gandhi, M. F. Hasan, 2022. Machine learning for the design and discovery of zeolites and porous crystalline materials. Current Opinion in Chemical Engineering 35, 100739.

S. S. Iyer, M. F. Hasan, 2019. Mapping the material-property space for feasible process operation: application to combined natural-gas separation and storage. Industrial & Engineering Chemistry Research 58 (24), 10455–10465.

P. Jul-Rasmussen, A. Chakraborty, V. Venkatasubramanian, X. Liang, J. K. Huusom, 2023. Identifying first-principles models for bubble column aeration using machine learning. In: Computer Aided Chemical Engineering. Vol. 52. Elsevier, pp. 1089–1094.

M. Moliner, Y. Roman-Leshkov, A. Corma, 2019. Machine learning applied to zeolite synthesis: the missing link for realizing high-throughput discovery. Accounts of chemical research 52 (10), 2971–2980.

F. O. Okello, T. T. Fidelis, J. Agumba, T. Manda, L. Ochilo, A. Mahmood, A. Pembere, 2023. Towards estimation and mechanism of co2 adsorption on zeolite adsorbents using molecular simulations and machine learning. Materials Today Communications 36, 106594.

M. Raji, A. Dashti, M. S. Alivand, M. Asghari, 2022. Novel prosperous computational estimations for greenhouse gas adsorptive control by zeolites using machine learning methods. Journal of Environmental Management 307, 114478.

L. S. Shapley, 1953. A value for n-person games 2, 307–317.

Flavio Manenti, Gintaras V. Reklaitis (Eds.), Proceedings of the 34th European Symposium on Computer Aided Process Engineering / 15th International Symposium on Process Systems Engineering (ESCAPE34/PSE24), June 2-6, 2024, Florence, Italy

Simulation and 3E assessment of pre-combustion CO_2 capture process using novel Ionic liquids for blue H_2 production

Sadah Mohammed[a], Fadwa Eljack[a], Saad Al-Sobhi[a], Monzure-Khoda Kazi [b]

[a]*Qatar University, Department of Chemical Engineering, College of Engineering, P.O. Box 2713, Doha, Qatar*
[b]*Texas A&M Energy Institute, Texas A&M University, College Station, TX 77843-3372, United States*
Fadwa.Eljack@qu.edu.qa

Abstract

Incorporating pre-combustion CO_2 capture with H_2 production is expected to be a crucial technology that significantly contributes to CO_2 emissions reduction from the current H_2 production methods that rely on natural gas. This work evaluates the exergy, energy, and economic aspects of a CO_2 capture process using new ionic liquid (IL) [NMIM][DCN] for H_2 production, and compares it with the established IL [HMIM][DCN]. The package "PySCF" in Python was used for geometry optimization, and a COSMO-based/Aspen approach was employed for component definition in Aspen Plus. The pre-combustion CO_2 capture was simulated using Aspen Plus V12®. Through the simulation, it was evident that IL [NMIM][DCN] exhibited promising results with a CO_2 capacity of 92.84% and 99.7% H_2 recovery, in contrast to [HMIM][DCN] (CO_2 capacity: 80% and H_2 recovery: 94.6%). Additionally, [NMIM][DCN] demonstrated lower energy consumption, utility cost and CAPEX compared to [HMIM][DCN] due to its low energy requirement for CO_2 separation. However, it is noteworthy that exergy and OPEX values were significantly high for [HMIM][DCN], and this suggests further analysis to optimize the [HMIM][DCN]-based process.

Keywords: CO_2 Capture, Ionic Liquids, Aspen Plus, COSMO-SAC, Exergy

1. Introduction

With the imperative shift toward a low-carbon economy and the objective of reaching net-zero emissions by 2025, the global demand for hydrogen production is expected to rise. Hydrogen, a non-carbonaceous fuel, offers a low-emission alternative to traditional fossil fuels and can be blended with natural gas to meet diverse energy system demands. Around 95% of the global hydrogen demand is fulfilled via the reforming of fossil fuels, of which steam methane reforming (SMR) constitutes 50% of the overall production (Oh et al., 2022). However, the SMR process, being energy intensive, leads to high CO_2 emissions (89.1 gCO_2/MJ) (Birol, 2019). To address this, blue hydrogen production offers a viable solution, where CO_2 is removed from hydrogen effluent gases in fossil fuel processes using CO_2 capture technology, resulting in lower carbon intensity (22.4 gCO_2/MJ) (Regulator, 2020). Therefore, it is imperative to develop a highly efficient process for capturing CO_2, aimed at producing H_2 with a low-carbon footprint. Achieving this involves adopting the pre-combustion CO_2 capture approach, due to its high efficiency in removing high CO_2 concentration. Normally, in the pre-combustion process, CO_2 removal is accomplished by the absorption process. Physical absorption application has become a standard practice for developing versatile, energy-efficient, and economically viable pre-combustion CO_2 capture systems. The latter emphasizes the energy and solvent requirements as vital aspects during the design stage. In this regard, Ionic liquids (ILs) have gained popularity due to their desirable CO_2 capture properties in pre-combustion applications like low vapor pressure. This study focuses on improving efficiency while minimizing capital and operating costs in pre-combustion CO_2 capture for the production of blue H_2 using novel ILs. The study scrutinizes the system engineering aspects, including energy, exergy, and economics (3E), comparing them to established IL.

2. Methodology

2.1 Geometry optimization and property method

The novel IL introduced in this work, namely 1-nonyl-3-methylimidazolium dicyanamide [NMIM][DCN] (see Figure 1), was designed using a predictive deep learning model (DL), which was developed in our previous work (Mohammed et al., 2023).

Figure 1: Molecular structure of the novel IL

Based on the DL results, the novel IL exhibited the highest CO_2 solubility ($x_{CO2} = 0.425$) and low viscosity ($\gamma_{IL} = 0.061$ Pa.s) at 1.5 MPa and 283.15 K, compared to other designed ILs. Following the acquisition of the IL's molecular structure, molecular geometry optimization was conducted following a procedure outlined by (Wang & Song, 2016). This method explicitly incorporates collective translations and rotations, allowing for effective traversal of many-dimensional potential energy surfaces. The "PySCF" package available in the programming language "Python" was employed for all the density functional theory (DFT) calculations. The "B3LYP" functional coupled with the "STO-3G" basis set, was selected. Subsequently, a file ".XYZ" that contains the optimized geometry was generated via a Python code based on the above-mentioned information. Following that, the ".XYZ" file was integrated into COSMO-RS software to acquire the necessary parameters, including sigma profile (SGPRF1 to SGPRF5), activity coefficients, molecular volume "CSACVL", and Henry's constant to define [NMIM][DCN] in simulation software "Aspen Plus" following a COSMO-based/Aspen approach described in previous work (Ferro et al., 2018). A similar approach was implemented to an established IL "1-hexyl-3-methylimidazolium dicyanamide, [HMIM][DCN]". Both results of the studied ILs were compared and analyzed.

2.2 Process modeling of IL-based CO_2 capture

Figure 2 shows the IL-based pre-combustion CO_2 capture process, following a similar process configuration proposed by (Zhai & Rubin, 2018). The process was developed in simulation software "Aspen Plus V12®". In this configuration, the process comprises a "RADFRAC" packed absorber column (C-101) where the CO_2 in the simplified inlet syngas stream (CO_2/H_2: 0.37/0.63) is absorbed by the IL. Pure hydrogen stream (S-03) exists the overhead of the absorber tower. Subsequently, the CO_2-rich stream (S-04) leaving the absorber is entered flash drums (F-101 & F-102) that are arranged in series for regeneration. The two CO_2 streams (S-05 & S-07) from the two flash drums are mixed using a mixer (M-101), and the stream (S-08), which mainly contains IL exits the second flash drum, then cooled (E-101) and circulated back to the absorber.

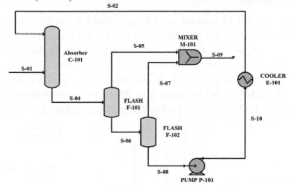

Figure 2: IL-based pre-combustion CO₂ physical absorption process scheme.

2.3 Energy and exergy analyses

To assess the total energy consumption associated with equipment duty, an energy analysis was conducted. Consequently, the proposed process was optimized via sensitivity analysis to identify the optimal operating parameters that would result in lower energy requirements. Exergy analysis accounts for the maximum amount of efficient work accomplished from a process through the identification of the irreversibility within the process. Exergy assessment in any process involves the combination of two components: physical exergy and chemical exergy. In this study, only physical exergy is considered. Exergy destruction takes place within processing equipment, identifying the generation of entropy that causes deviation in the process and introduces inefficiencies. Exergy analysis for calculating exergy destruction (equations (1)-(4)) and exergy efficiency (equations (5)-(7)) of the processing equipment are listed below (Kazmi et al., 2021):

$$I_{absorber,min} = \sum(\dot{m}.\dot{e})_{in} - \sum(\dot{m}.\dot{e})_{out} \tag{1}$$

$$I_{pump} = \sum(\dot{m}.\dot{e})_{in} - \sum(\dot{m}.\dot{e})_{out} + W \tag{2}$$

$$I_{heat\ exchanger} = \sum(\dot{m}.\dot{e})_{in} - \sum(\dot{m}.\dot{e})_{out} \tag{3}$$

$$I_{separator} = \sum(\dot{m}.\dot{e})_{in} - \sum(\dot{m}.\dot{e})_{out} \tag{4}$$

$$eff_{pump} = \frac{\sum(\dot{m}.\dot{e})_{out} - \sum(\dot{m}.\dot{e})_{in}}{W} \tag{5}$$

$$eff_{heat\ exchanger} = 1 - \left[\left\{\frac{\sum_1^n m\Delta e}{\sum_1^n m\Delta h}\right\}_{hot} - \left\{\frac{\sum_1^n m\Delta e}{\sum_1^n m\Delta h}\right\}_{cold} \right] \tag{6}$$

$$eff_{separator} = \frac{\sum(\dot{m}.\dot{e})_{out}}{\sum(\dot{m}.\dot{e})_{in}} \tag{7}$$

2.4 Economic analysis

In this section, economic analysis was implemented based on the energy and exergy analysis. The overall cost of the proposed pre-combustion CO₂ capture process was assessed using a built-in feature called Aspen Process Economic Analysis (APEA). This feature uses the Aspen plus simulation run to evaluate the total process costs encompassing operational expenditure (OPEX) and capital expenditure (CAPEX). APEA employs a bottom-up methodology, breaking down the process to assess each component through the appropriate cost model.

3. Results and discussion

3.1 Sensitivity analysis

To examine the impact of crucial operating variables on CO_2 removal, a sensitivity analysis was carried out on the pre-combustion CO_2 capture process. To attain high CO_2 solubility and pure H_2 stream, the inlet flow rate of the IL to the absorber as well as the operating pressure were varied to explore the effect on the CO_2 capture. Figure 3 shows the optimal IL flow rate for [NMIM][DCN] and [HMIM][DCN]. It is conspicuous from the figure that [NMIM][DCN] requires less flow rate ($\approx 490\ kmol/h$) to achieve high CO_2 capture, unlike [HMIM][DCN] which requires a higher flow rate ($\approx 660\ kmol/h$).

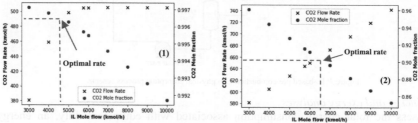

Figure 3: Impact of IL flow rate (1) [NMIM][DCN], and (2) [HMIM][DCN] on purity and yield of CO_2

Figure 4 depicts the CO_2 and H_2 flow rate (kmol/h) in the rich-solvent stream (stream S-04, see Figure 2). It can be seen from the results that [NMIM][DCN] absorbs less hydrogen even at a higher pressure compared to [HMIM][DCN].

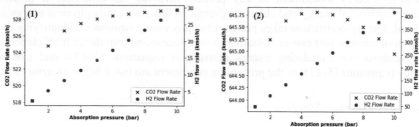

Figure 4: Impact of absorber operating pressure on CO_2 and H_2 solubility in (1) [NMIM][DCN], and (2) [HMIM][DCN].

The separation process is a crucial step in the CO_2 removal process. In this step, two flash columns were required to completely separate the CO_2 from the IL. In column F-101, a pressure and temperature-swing was implemented as seen in Figure 5-1 and Figure 5-2, and only temperature-swing is applied in column F-102 as shown in Figure 5-3. More CO_2 is separated when IL [NMIM][DCN] is used at low temperatures and pressure. The findings from this analysis prove that increasing the alkyl chain in ILs improves CO_2 solubility and enhances separation processes. Based on the above analysis and optimizing the operating parameters, remarkable H_2 and CO_2 recovery (%) results were obtained. Specifically, [NMIM][DCN] exhibited an excellent recovery value of 99.7% for H_2 and 92.84% for CO_2. In comparison, [HMIM][DCN] demonstrated slightly lower but still significant recovery rates, with 94.6% for H_2 and 80% for CO_2.

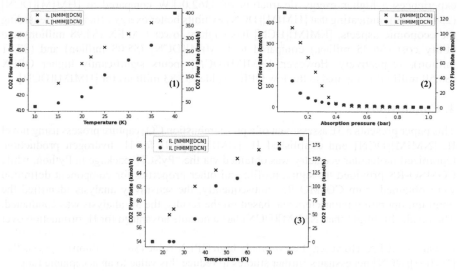

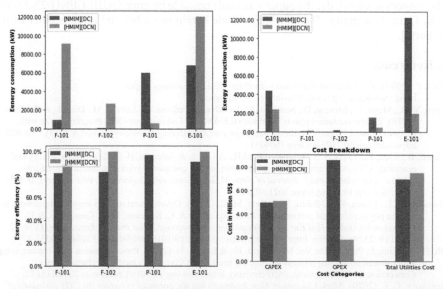

Figure 5: Effect of temperature (K) and pressure (bar) on CO₂ purity in (1) Column F-101, and (2) Column F-102

3.2 3E assessment

Figure 6: 3E assessment of [NMIM][DCN] and [HMIM][DCN] IL-based CO₂ capture process

Figure 6 illustrates the 3E analysis for the proposed process under the optimized operating parameters. Notably, when considering energy consumption, IL [NMIM][DCN] exhibited low energy consumption due to the CO₂ separation occurring at low temperatures and pressure as outlined in section 3.1. Specifically, in column F-101, [NMIM][DCN] shows significantly low energy consumption (900 kW) compared to [HMIM][DCN] (9137 kW). A similar trend is noticed in column F-102, where [NMIM][DCN] consumes lower energy than [HMIM][DCN]. Furthermore, in terms of exergy destruction, a notable difference is highlighted. In column C-101, [NMIM][DCN] experiences a higher exergy destruction of 4362.0 kW compared to [HMIM][DCN] (2347.5 kW), indicating that [HMIM][DCN] exhibits better exergy efficiency. Regarding the economic aspects, [NMIM][DCN] has slightly lower CAPEX ($4.98 million) and utility cost ($6.95 million) compared to [HMIM][DCN] ($5.08 million) and ($7.51 million), respectively. However, [NMIM][DCN] incurs significantly higher OPEX ($8.59 million) compared to the low OPEX value ($1.83 million) for [HMIM][DCN].

4. Conclusions

This paper presents a 3E assessment of a pre-combustion CO_2 capture process using novel IL [NMIM][DCN] and established IL [HMIM][DCN] for H_2 hydrogen production. Optimized molecular geometry was obtained via the "PySCF" package in Python, while COMSO-RS provided the sigma profile and other properties for component definition were obtained from COSMO-RS. Subsequently, the sensitivity analysis identified the optimum operating parameters and based on the results, the 3E analysis was conducted. The results highlight that [NMIM][DCN] has a notable advantage for H_2 production over [HMIM][DCN] due to its high CO_2 capacity (92.84%), low energy consumption, utility cost and CAPEX. However, the extremely high OPEX value ($8.59 million) suggests that [NMIM][DCN] necessitates further studies to reduce this value to an acceptable range.

5. Acknowledgment

The authors acknowledge the paper was made possible by grant QUHI-CENG-22-23-465 from Qatar University. The statements made herein are solely the responsibility of the author[s].

6. Reference

Birol, F. (2019). *The Future of Hydrogen – Analysis - IEA*. Technology Report. https://www.iea.org/reports/the-future-of-hydrogen

Ferro, V. R., Moya, C., Moreno, D., Santiago, R., De Riva, J., Pedrosa, G., Larriba, M., Diaz, I., & Palomar, J. (2018). Enterprise Ionic Liquids Database (ILUAM) for Use in Aspen ONE Programs Suite with COSMO-Based Property Methods. *Industrial and Engineering Chemistry Research*, 57(3), 980–989. https://doi.org/10.1021/acs.iecr.7b04031

Kazmi, B., Raza, F., Taqvi, S. A. A., Awan, Z. ul H., Ali, S. I., & Suleman, H. (2021). Energy, exergy and economic (3E) evaluation of CO2 capture from natural gas using pyridinium functionalized ionic liquids: A simulation study. *Journal of Natural Gas Science and Engineering*, 90(February), 103951. https://doi.org/10.1016/j.jngse.2021.103951

Mohammed, S., Eljack, F., Al-Sobhi, S., & Kazi, M.-K. (2023). Development of deep learning framework to predict physicochemical properties for Ionic liquids. In A. C. Kokossis, M. C. Georgiadis, & E. Pistikopoulos (Eds.), *33rd European Symposium on Computer Aided Process Engineering* (Vol. 52, pp. 2119–2124). Elsevier. https://doi.org/https://doi.org/10.1016/B978-0-443-15274-0.50337-1

Oh, H. T., Kum, J., Park, J., Dat Vo, N., Kang, J. H., & Lee, C. H. (2022). Pre-combustion CO2 capture using amine-based absorption process for blue H2 production from steam methane reformer. *Energy Conversion and Management*, 262(February). https://doi.org/10.1016/j.enconman.2022.115632

Regulator, C. E. (2020). *Market Snapshot: How hydrogen has the potential to reduce the CO2 emissions of natural gas*. Canada Energy Regulator. https://www.cer-rec.gc.ca/en/data-analysis/energy-markets/market-snapshots/2020/market-snapshot-hydrogen-potential.html

Wang, L. P., & Song, C. (2016). Geometry optimization made simple with translation and rotation coordinates. *Journal of Chemical Physics*, 144(21). https://doi.org/10.1063/1.4952956

Yu, M., Wang, K., & Vredenburg, H. (2021). Insights into low-carbon hydrogen production methods: Green, blue and aqua hydrogen. *International Journal of Hydrogen Energy*, 46(41), 21261–21273. https://doi.org/10.1016/j.ijhydene.2021.04.016

Zhai, H., & Rubin, E. S. (2018). Systems Analysis of Physical Absorption of CO2 in Ionic Liquids for Pre-Combustion Carbon Capture. *Environmental Science and Technology*, 52(8), 4996–5004. https://doi.org/10.1021/acs.est.8b00411

Flavio Manenti, Gintaras V. Reklaitis (Eds.), Proceedings of the 34th European Symposium on Computer Aided Process Engineering / 15th International Symposium on Process Systems Engineering (ESCAPE34/PSE24), June 2-6, 2024, Florence, Italy

Novel closed-loop precursor re-synthesis assisted by roasting and wastewater electrolysis: Industrial scaled design and simulation

Sunwoo Kim[a], Jeongdong Kim[a], Junghwan Kim[a,*]

[a] *Department of Chemical and Biomolecular Engineering, Yonsei University, Seoul 03722, Republic of Korea*
Email of the corresponding author: kjh24@yonsei.ac.kr

Abstract

In recent spent lithium-ion batteries (LIBs) recycling studies, the concept of direct re-synthesis of NMC ($Ni_xMn_yCo_zO_2$) precursor from spent LIBs is experimentally tested on a laboratory scale. However, beyond the laboratory scale, excessive usage of chemicals and wastewater disposal are the main issue in industrial scale application due to the accompanying operating cost and environmental degradation. Herein, to resolve the challenges, this study proposes a novel integrated system of closed-loop precursor re-synthesis assisted by roasting and wastewater electrolysis for sustainable LIB recycling. The proposed system consists of three major processes: hydrogen roasting for Li extraction, NMC precursor crystallization, and wastewater electrolysis for chemical regeneration. First, the spent LIB is decomposed into metal oxide by hydrogen roasting, and water-soluble lithium oxide is firstly separated from the metal oxide mixture. The recovered LiOH slurry undergoes further crystallization for high purity recovery. Second, the remaining metal oxide mixture undergoes acid leaching process, and pH adjustment via NaOH injection is conducted for satisfying minimum pH condition for precursor crystallization. Finally, the wastewater produced following the crystallization is fed to anion-exchange membrane-based electrolysis to regenerate sulfuric acid and NaOH in cathode and anode, separately. Recycling regenerated chemicals to leaching and pH adjustment step accomplished the closed-loop recycling system. This study firstly proposed conceptual design of closed-loop precursor re-synthesis assisted by roasting and electrolysis. The results of industrial scaled simulation showed the recovery of 98.95 % for Li and 79.00 % for NMC precursor. These results revealed that the proposed process could be a feasible solution for sustainable closed-loop battery recycling.

Keywords: Lithium-ion batteries, Battery recycling, Hydrogen roasting, Precursor crystallization, Wastewater electrolysis, Process design and evaluation

1. Introduction

Lithium-ion batteries (LIBs) occupy a dominant position in the energy storage field due to their advantages such as high energy capacity, and excellent cycling performance, leading to a rapidly growing market size. Among the various types of cathode materials, NMC ($Ni_xMn_yCo_zO_2$)-type cathodes are expected to account for more than 80% of production by 2023. Due to the high share of the NMC market and limited reserves of valuable metals (e.g., Li, Ni, Mn, and Co), recycling spent LIBs is essential for a sustainable battery recycling system.

In the field of the batteries recycling technology, a method of direct re-synthesis of NMC precursor from spent LIB has recently been studied on a laboratory scale. NMC precursor is synthesized by adding a chelating agent and chemicals for pH adjustment to a leached solution of spent NMC cathode. Fang et al. experimentally achieved the recovery of NMC precursor and Li_2CO_3 through coprecipitation by adding Na_2CO_3 and $NH_3 \cdot H_2O$ to leached solution. The synthesized precursor showed reliable electrochemical performance with the highest discharge specific capacity. Chen et al. synthesized the NMC precursor by simultaneously reacting spent cathode leached solution with Na_2CO_3 and $NH_3 \cdot H_2O$, and found that it reached the electrochemical properties of commercial cathode on a laboratory scale.

Despite various efforts in the direct resynthesis of NMC precursor to improve performance, application in industrial scale recycling requires to resolve following challenges. First, excessive usage of chemicals to achieve high metal recovery leads to increased operating costs. Second, excessive usage of chemicals results in large amounts of wastewater, causing environmental pollution. At industrial scale, operating costs and environmental impact are major issues. Thus, strategy to regenerating the chemicals via electrochemical treatment of wastewater may reduce the operating cost and environmental impacts.

In this study, a novel closed-loop precursor re-synthesis process design and simulation at industrial scale assisted by hydrogen roasting and wastewater electrolysis was conducted. By roasting, lithium can be first extracted as LiOH form by water leaching, which can reduce the usage of chemicals. Furthermore, the recovery of the NMC precursor was derived through a population balance equation (PBE) approach. Finally, the regeneration of sulfuric acid and NaOH was achieved through wastewater electrolysis.

2. Methodology

2.1. Process description

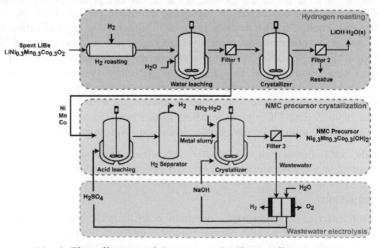

Fig. 1. Flow diagram of the proposed LIB recycling system.

Fig. 1 shows the flow diagram of proposed battery recycling system. The proposed system consists of three major processes: hydrogen roasting, NMC precursor crystallization, and wastewater electrolysis. First, the NMC cathode materials are decomposed into metal oxide by hydrogen roasting, and water-soluble lithium oxide firstly separated from the

metal oxide mixture by water leaching. Then, the lithium oxide is recovered as
$LiOH \cdot H_2O(s)$ by crystallization. Second, the remaining metal oxide mixture undergoes
acid leaching process, followed by chelating agent injection and pH adjustment through
NaOH injection to meet minimum pH conditions for precursor crystallization. Finally,
the wastewater produced following the crystallization is fed to anion-exchange
membrane-based electrolysis to regenerate H_2SO_4 and NaOH, separately.

2.2. Process model

2.2.1. Hydrogen roasting

The overall hydrogen roasting involves two main reactions: irreversible thermal
decomposition to metal oxides and hydrogen roasting of metal oxides. In the thermal
decomposition step shown in Eq. (1), the interlayer crystal structure of NMC is destructed
and converted into a mixture of metal oxides. In the roasting step, the reduction of metal
oxide with hydrogen is described as Eq. (2-4).

$$LiNi_{1/3}Mn_{1/3}Co_{1/3}O_2 \rightarrow \frac{1}{2}Li_2O + \frac{1}{3}NiO + \frac{1}{3}MnO_2 + \frac{1}{3}CoO + \frac{1}{12}O_2 \tag{1}$$

$$NiO + H_2 \rightarrow Ni + H_2O \tag{2}$$

$$MnO + H_2 \rightarrow Mn + H_2O \tag{3}$$

$$CoO + H_2 \rightarrow Co + H_2O \tag{4}$$

The generated Li_2O is soluble in water, and water leaching is performed for selective Li
extraction as shown in Eq. (5-6). Following, the LiOH slurry is converted into $LiOH \cdot$
$H_2O(s)$ in the crystallization process as shown in Eq. (7).

$$Li_2O(s) + H_2O(l) \rightarrow 2LiOH(s) \tag{5}$$

$$LiOH(s) \rightarrow Li^+(aq) + OH^-(aq) \tag{6}$$

$$Li^+(aq) + OH^-(aq) + H_2O(l) \rightarrow LiOH \cdot H_2O(s) \tag{7}$$

2.2.2. NMC precursor crystallization

In NMC precursor crystallization, mathematical modeling through population balance
equation (PBE) approach is applied. In this approach, the evolution of the crystal size
distribution is calculated by the particle growth rate and nucleation rate as shown in Eq.
(8). Then, mass balance and energy balance equation are integrated with the PBE and
solved iteratively in time steps.

$$\frac{\partial Vf(L,t)}{\partial t} + V\frac{\partial[G(L)f(L,t)]}{\partial L} = VB\delta(L - L_0) \tag{8}$$

The particle growth rate is described as Eq. (9), and nucleation rate is expressed by Eq.
(10).

$$G(L) = k_0 \exp\left(-E_g/RT\right)(1 + k_1L)^{k_2}(S - 1)^\alpha \tag{9}$$

$$B = k_b(S - 1)^\beta \tag{10}$$

Detailed parameter definitions and ranges of Eq. (8-10) can be referred to in the study by Hu, Q et al.

2.2.3. Wastewater electrolysis

After NMC precursor crystallization, the wastewater mainly composed of sodium sulfate is fed into the anion-exchange membrane electrolysis. Deionized water and wastewater are introduced into the anode and cathode compartments, respectively. Applying voltage, the water splitting and anion diffusion through membranes leads to the production of sodium hydroxide and hydrogen at cathode compartment via Eq. (11), and sulfuric acid and oxygen at anode compartment via Eq. (12).

$$Cathode: 2H_2O + 2e^- \rightarrow H_2 + 2OH^- \tag{11}$$

$$Anode: 2H_2O \rightarrow O_2 + 4H^+ + 4e^- \tag{12}$$

The electrochemical model is applied for wastewater electrolysis. The total cell voltage is defined in Eq. (13) as the sum of the reversible voltage, activation overpotential, and ohmic overpotential.

$$V = V_{rev} + \eta_{act} + \eta_{Ohm} \tag{13}$$

The activation overpotential is related to the electrochemical reaction at the electrode and is expressed by the Butler-Volmer equation in Eq. (14).

$$\eta_{act} = \frac{RT_a}{2\alpha_a F} \text{arcsinh}\left(\frac{i}{2i_{0,a}}\right) + \frac{RT_c}{2\alpha_c F} \text{arcsinh}\left(\frac{i}{2i_{0,c}}\right) \tag{14}$$

Ohmic overpotential is a term related to resistance when current flows and is expressed by ohmic's law in Eq. (15).

$$\eta_{ohm} = l_a \frac{i}{\sigma_a} + l_m \frac{i}{\sigma_m} + l_c \frac{i}{\sigma_c} \tag{15}$$

Detailed parameter definitions and ranges of Eq. (10-12) can be referred to in the study by Federica Mosca. The ionic conductivity of anolyte and catholyte varies with the concentration of OH⁻ and H⁺ ions produced over time. Therefore, the current density over time is calculated at constant voltage, and the experimental values are used for parameter fitting using a genetic algorithm.

2.3. Process simulation

The overall system simulation is conducted using the Python environment and a process simulator. Fig. 2 shows the simulation framework of the proposed overall system. Steps 1 to 3, roasting, lithium recovery, and acid leaching, are performed in Aspen plus. The stream properties after acid leaching are linked with Python to calculate the NMC precursor crystallization process. The concentration of each ion in the wastewater generated after the crystallization process is taken to calculate the current, time, and flow rate of regenerated chemicals in the electrolysis process. The flow rate of the regenerated chemicals is then fed back into the input streams of the unit processes.

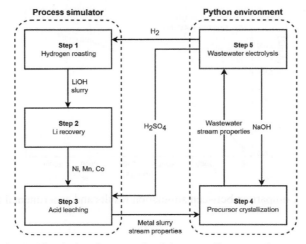

Fig. 2. Simulation framework of the overall proposed system.

3. Results

Table 1 shows the simulation results of the recovery efficiency of lithium and NMC precursor. Lithium is recovered in the form of LiOH·H$_2$O(s) with a recovery efficiency of about 98.95 % and NMC precursor with a recovery efficiency of about 79.00 %.

Table 1. Simulation result of the proposed process.

	Li crystallization	NMC precursor crystallization
Temperature (°C)	80	90
pH value	7.0	10.6
Recovery efficiency (%)	98.95	79.00

Fig. 3 shows the validation of the electrochemical model of the electrolysis cell with experimental data at voltage of 4.75 V. To improve the accuracy, parameter fitting is performed using a genetic algorithm.

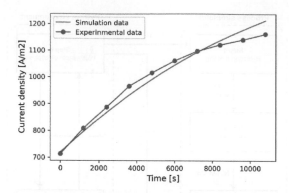

Fig. 3. Comparison between simulation results and experimental data.

4. Conclusions

This study firstly proposed conceptual design of closed-loop precursor re-synthesis assisted by hydrogen roasting and wastewater electrolysis. Furthermore, the results of industrial scaled simulation showed the recovery of 98.95 % for Li and 79.00 % for NMC precursor. As future work, techno-economic analysis will be conducted to reveal that the proposed process could be a feasible solution for economic and environmentally sustainable battery recycling.

References

C. Curry, 2017, Lithium-ion battery costs and market. Bloomberg New Energy Finance.

Jiahu Fang, 2022, Green recycling and regeneration of LiNi0.5Co0.2Mn0.3O2 from spent Lithium-ion batteries assisted by sodium sulfate electrolysis, Chemical Engineering Journal.

Xiaoqing Chen, 2022, Co-precipitation preparation of Ni-Co-Mn ternary cathode materials by using the sources extracting directly from spent lithium-ion batteries, Journal of Alloys and Compounds.

Jucai Wei, 2021, A zero-emission method for recycling phosphogypsum using Na2SO4 electrolysis: Preliminary study, Separation and Purification Technology.

Hyejeong Lee, Novel lithium production process using desalination wastewater and waste heat from natural gas combined cycle, Energy Conversion and Management.

F. Liu, 2021, Selective lithium recovery and integrated preparation of high-purity lithium hydroxide products from spent lithium-ion batteries, Separation and Purification Technology.

Jeongdong Kim, 2023, Process design and economic analysis of hydrogen roasting integrated with CCU for a carbon-free spent LIB recycling process, Chemical Engineering Journal.

Federica Mosca, 2023, Modelling and simulation of an Offshore Wind-Hydrogen combined system for hydrogen production using an anion exchange membrane electrolyzer, NTNU.

Hu, Q., 2004, Nonlinear kinetic parameter estimation for batch cooling seeded crystallization, AIChE Journal.

Jeongdong Kim, 2023, Integration of wastewater electro-electrodialysis and CO2 capture for sustainable LIB recycling: Process design and economic analyses, Journal of Cleaner Production.

Jeongdong Kim, 2023, Sequential flue gas utilization for sustainable leaching and metal precipitation of spent lithium-ion battery cathode material: Process design and techno-economic analysis, Journal of Cleaner Production.

Flavio Manenti, Gintaras V. Reklaitis (Eds.), Proceedings of the 34th European Symposium on Computer Aided Process Engineering / 15th International Symposium on Process Systems Engineering (ESCAPE34/PSE24), June 2-6, 2024, Florence, Italy

Use of Image Data in Kinetic Model Development for the Design of Mesenchymal Stem Cell Cultivation Processes

Keita Hirono,[a] Yusuke Hayashi,[a] Isuru A. Udugama,[a,#] Yuto Takemoto,[b] Ryuji Kato,[b] Masahiro Kino-oka,[c] Hirokazu Sugiyama[a,*]

[a]*Department of Chemical System Engineering, The University of Tokyo, 7-3-1, Hongo Bunkyo-ku, Tokyo 113-8656, Japan*
[b]*Department of Basic Medicinal Sciences, Graduate School of Pharmaceutical Sciences, Nagoya University, Furocho, Chikusa-ku, Aichi 464-8601, Japan*
[c]*Department of Biotechnology, Osaka University, 1-2, Yamadaoka, Suita-shi, Osaka 565-0871, Japan*
*sugiyama@chemsys.t.u-tokyo.ac.jp
#Current affiliation
Department of Chemical and Materials Engineering, University of Auckland, Private Bag 92019, Auckland, New Zealand

Abstract

This work presents the use of image data in kinetic model development for designing mesenchymal stem cell (MSC) cultivation processes. To incorporate the initial spatial distribution of seeded cells, seeding bias in static MSC cultivation was investigated with phase contrast microscopic image acquisition on Day 1. Subsequently, a parameter was newly defined from the image analysis result calculating the standard deviation of cell number fraction among numerical 64 tiles on a squared culture space. Finally, the parameter was integrated with our previous kinetic model to consider spatial growth limitation. The model was then applied to simulate static MSC cultivation processes. Based on ordinary differential equations (ODE), Monte-Carlo simulations were conducted to find cell harvesting time when cultivation ensured given specifications for cell number and confluence degree. Feasible range of harvesting time was illustrated subject to seeding heterogeneity and density. The presented image-based ODE model could help incorporate spatial heterogeneity into the MSC cultivation process design.

Keywords: Cell therapy, stem cell, spatial heterogeneity, hybrid model, image analysis

1. Introduction

There is an increased demand for mesenchymal stem cells (MSCs) for cell therapy applications due to self-renewability, immunosuppression, and the lack of ethical concerns (Levy et al., 2020). MSCs have been studied for a wide range of therapy including acute graft vs. host diseases (Najima and Ohashi, 2017). In anticipation of the demand growth of MSCs, quality cell manufacturing by process design has been required (Lipsitz et al., 2016). To this end, ordinary differential equations (ODEs) have been developed to describe cell growth dynamics in static MSC cultivation, e.g., Jossen et al. (2020) and Hirono et al. (2022). ODEs benefit from fast simulation compared to spatially high-resolution models, which demand long calculation time. Besides, ODEs can be smoothly integrated with stochastic technology including the Monte-Carlo method to predict process variations. For example, Hirono et al. (2022) conducted an ODE-based stochastic simulation incorporating system dynamics and variabilities in MSC cultivation processes.

Primary assumptions in ODEs are regarding the system as homogeneous. For example, specific cell growth rate in cultivation is typically formulated based on the Monod kinetics (Monod, 1949) as a function of average concentration of limiting substrates and bulk cell density. However, in actual static cultivation, heterogeneous initial cell distribution can happen, which potentially causes significant growth delay due to cell–cell contact inhibition (Kagawa and Kino-oka, 2016). To model such spatial growth limitation by compensating for the ODE limitation, imaging technology can be a promising tool. In the PSE field, image-based technology has been developed for process monitoring and control applications. For bioprocess development, Oh et al. (2009) developed an automated vision system to identify human embryonic stem cell differentiation in teratoma section tissues. In static MSC cultivation studies, phase contrast microscopic images were analyzed to predict osteogenic potentials (Matsuoka et al., 2013) and to detect quality decay through passaging (Takemoto et al., 2021). However, integration and application of image information for designing MSC cultivation processes are still in infancy.

In this work, we utilized image data in kinetic modeling for the design of MSC cultivation processes. To incorporate the initial spatial distribution, our previous ODE model (Hirono et al., 2022) was extended by defining a new parameter from phase contrast microscopic images. Seeding bias in static MSC cultivation was investigated with the image acquisition on Day 1. Subsequently, the observed spatial distribution was parameterized using the image analysis results. Finally, the parameter was integrated with the kinetic model to predict spatial growth limitation due to seeding heterogeneity.

2. Model development

2.1. MSC cultivation

Figure 1 shows the experimental setup for in vitro static cultivation of bone marrow MSCs at passage 6 (19TL281098, Lonza Group Ltd., Basel, Switzerland) with MSCGM (Lonza Group Ltd) culture medium. A restricted culture area (17×17 mm^2) was designed at the center of polydimethylsiloxane in each well of 6-well plates (353046, FALCON, NY, USA). MSCs were initially seeded in a cloning ring (20140514, AGC Inc., Tokyo, Japan) to produce seeding bias by changing hold time, t_r, specifically 0 (numerically), 15, and 60 min. In the case of $t_r = 0$, the colony ring was not used. The cultivation was started with seeding densities, X_{seed}, of 1000, 2000, and 3000 cells cm^{-2} under 37°C, 5% CO_2 in the incubator until Day 8. Partial medium change was conducted on Days 2, 4, and 6 when half of the working medium was replaced with fresh medium. Duplicate samples were prepared for each culture condition, resulting in a total of 18 samples.

2.2. Image acquisition

Phase contrast microscopic images were automatically acquired every 6 h during the cultivation using BioStation CT (Nikon Corporation, Tokyo, Japan) at 4× magnification (8×8 tiling per well, covering 15.3×15.3 mm^2; 1000 pixels^2/image). The number of evaluation time points was 32 from 6 h to 192 h after removing the ring. All images were quantified by original Python code implementing the image processing pipeline based on Matsuoka et al. (2013) and Sasaki et al. (2014). Using the image analysis results, cell numbers at each tile were measured. Specifically, the analysis result on Day 1 was used for the following parameterization.

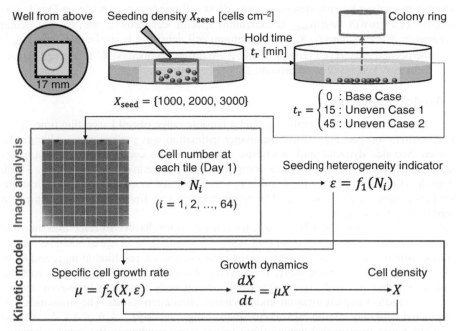

Figure 1. Use of image data in kinetic model development for MSC cultivation.

2.3. Parameterization

The cell number at each tile, N_i ($i = 1, 2, ..., 64$), was counted using the images acquired from the cultivation on Day 1. Dividing N_i by the total cell number on Day 1, the cell number fraction at each tile was obtained. Subsequently, calculating the standard deviation for the fraction among the 64 samples, seeding heterogeneity was quantified as a new parameter, ε (see Eq. (1)).

2.4. Integrated kinetic model

To incorporate the image-driven ε into our previous kinetic model (Hirono et al., 2022), spatial growth limitation was formulated as a function of ε, which modified the Monod equation (see Eq. (2)). The overall integrated kinetic model was developed as follows:

$$\varepsilon = \left\{ \frac{1}{64} \sum_{i}^{64} \left(\frac{N_i}{\sum_j^{64} N_j} - \frac{1}{64} \right)^2 \right\}^{\frac{1}{2}} \tag{1}$$

$$\mu = \mu_{\max}(1 - \varepsilon K_S)\left(1 - \frac{X}{X_{\max}}\right) \tag{2}$$

$$\frac{dX}{dt} = \mu X \qquad \left(t \geq t_{\text{lag}}\right) \tag{3}$$

$$t_{\text{lag}} = \beta \cdot \ln\left(\frac{\alpha X_{\text{seed}}}{X^*}\right) + \gamma \tag{4}$$

where ε is seeding heterogeneity indicator; N_i is cell number at each tile on Day 1; μ is specific cell growth rate; μ_{max} is maximum specific cell growth rate; K_S is a fitting parameter; X is cell density; X_{max} is maximum cell density; t is cultivation time; t_{lag} is lag time needed to start cell growth; β and γ are fitting parameters; X^* is unit cell density ($= 1$ cells cm^{-2}); α is adhesion ratio on Day 1.

3. Model application

The developed model was applied to find a feasible range of cell harvesting time, t_h, in MSC cultivation processes. First, parameter estimation was performed to calibrate the model. Second, dynamic and stochastic simulation was conducted with parameters sampled from discrete distributions composed of the experimental observation. Third, from the simulation outputs, the feasible t_h was calculated to satisfy given specifications for cell number and confluence. The feasible range was finally visualized subject to seeding heterogeneity and density.

Figure 2 illustrates the parameter estimation results. As an error indicator, the normalized root-mean-square error (NRMSE) was used. The average of NRMSE for cell number among the 32 measuring time points was minimized, resulting in μ_{max} and K_S of 24.7 and 3.03×10^{-2}, respectively. The developed model showed better fitness with the observation than the model ignoring seeding heterogeneity, especially in uneven cases.

Figure 3 depicts dynamic and stochastic simulation results. The simulation was conducted until Day 10 with 50 vol% medium changes on Days 2, 4, 6, and 8. As an input variable, ε was investigated among three values corresponding to the mean calculated from Base Case, Uneven Case 1, and Uneven Case 2, respectively. Besides, X_{seed} was

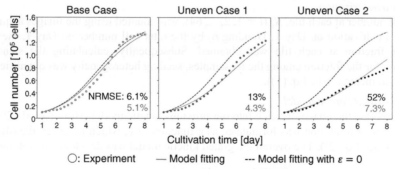

Figure 2. Parameter estimation results ($X_{seed} = 2000$ cells cm^{-2}).

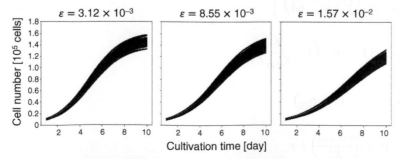

Figure 3. Dynamic and stochastic simulation results ($X_{seed} = 2000$ cells cm^{-2}).

varied from 1000 to 3000 cells cm^{-2} based on the experimental conditions with an interval of 500 cells cm^{-2}. Regarding uncertain parameters, α and X_{max} were randomly sampled from independent discrete distributions consisting of their experimental observations. The stochastic simulation with the sampling was conducted 10000 times to capture the resulting variations in the cultivation.

Using the simulation outputs, a feasible range of t_h was obtained subject to X_{seed} and ε such that MSC cultivation ensured given specifications. The feasible range, FR, was mathematically defined as follows:

$$FR = \{t_h | f(\boldsymbol{x}, \boldsymbol{\theta}, t_h) \in A\} \tag{5}$$

$$A = \left\{5.0 \times 10^4 \leq SX \wedge \frac{X}{X_{max}} < 0.80\right\} \tag{6}$$

where f is the developed model; $\boldsymbol{x}$ is a set of input variables (i.e., ε and X_{seed}); $\boldsymbol{\theta}$ is a set of uncertain parameters (i.e., α and X_{max}); A is quality specifications; S is surface area. Here, A was based on the reason that cell quantity was directly related to medical needs, while confluence level would decrease stem cell quality (Jossen et al., 2020).

Figure 4 visualizes the calculated FR depending on both ε and X_{seed} as a box-whisker plot with the whiskers within 1.5 times the interquartile range. Given ε of 3.12 $\times$ 10^{-2} and X_{seed} of 2000 cell cm^{-2}, for example, Days 4–7 were feasible, while harvesting before Day 4 or after Day 7 was unfeasible. Areas under and over the feasible range were out of the specifications due to insufficient cell number and exceeded confluence level, respectively. Interestingly, wider feasible ranges were obtained in higher ε (i.e., uneven seeding) because of lower growth rate (see Eq. (2)) even though long-term cultivation was needed. Regardless of ε, higher X_{seed} achieved the feasible range earlier due to larger initial cell density and shorter lag time (see Eqs. (3)–(4)). Figure 4 can serve as a basis for quantitative design of MSC cultivation processes given seeding heterogeneity and input cell resources. The default *scipy.integrate.solve_ivp* was used in Python 3.9 for solving the ODEs. The total CPU time for the feasible range calculation was ca. 10 min using an Intel Xeon Gold 6142 CPU @ 2.60 GHz with 128 GB RAM.

4. Conclusions and outlook

This work presented the use of image data in kinetic model incorporating the effects of initial spatial distribution into the design of MSC cultivation processes. Utilizing image

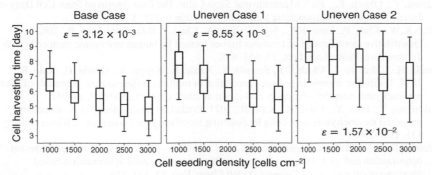

Figure 4. Calculated feasible range of cell harvesting time.

analysis result with phase contrast microscopy, a new parameter was defined representing seeding heterogeneity and integrated with our previous kinetic model. As a model application, a feasible range of harvesting time was calculated subject to seeding heterogeneity and density. Future work would include more experimental and numerical investigations to achieve practical probability distributions of seeding heterogeneity towards robust process conditions for MSC cultivation. Recently, computer-aided studies on cell and gene therapy have gathered attentions in the PSE field (Hayashi et al., 2023; Triantafyllou et al., 2023). This work would contribute further model-based investigation in the relevant area.

Acknowledgements

This research was supported by the Japan Agency for Medical Research and Development (AMED) under Grant Number JP20be0704001. HK is grateful for the financial support of Grant-in-Aid for JSPS Fellows [Grant Number 23KJ0375].

References

Hayashi, Y., Scholz, B.X., Sugiyama, H., 2023. A CFD-model-based approach to continuous freezing process design for human induced pluripotent stem cells. Compt. Aided Chem. Eng. 52, 65–70.

Hirono, K., A. Udugama, I., Hayashi, Y., Kino-oka, M., Sugiyama, H., 2022. A Dynamic and Probabilistic Design Space Determination Method for Mesenchymal Stem Cell Cultivation Processes. Ind. Eng. Chem. Res. 61, 7009–7019.

Jossen, V., Muoio, F., Panella, S., Harder, Y., Tallone, T., Eibl, R., 2020. An Approach towards a GMP Compliant In-Vitro Expansion of Human Adipose Stem Cells for Autologous Therapies. Bioengineering 7, 77–99.

Kagawa, Y., Kino-oka, M., 2016. An in silico prediction tool for the expansion culture of human skeletal muscle myoblasts. R Soc Open Sci 3, 160500.

Levy, O., Kuai, R., Siren, E.M.J., Bhere, D., Milton, Y., Nissar, N., De Biasio, M., Heinelt, M., Reeve, B., Abdi, R., Alturki, M., Fallatah, M., Almalik, A., Alhasan, A.H., Shah, K., Karp, J.M., 2020. Shattering barriers toward clinically meaningful MSC therapies. Sci Adv 6, eaba6884.

Lipsitz, Y.Y., Timmins, N.E., Zandstra, P.W., 2016. Quality cell therapy manufacturing by design. Nat. Biotechnol. 34, 393–400.

Matsuoka, F., Takeuchi, I., Agata, H., Kagami, H., Shiono, H., Kiyota, Y., Honda, H., Kato, R., 2013. Morphology-based prediction of osteogenic differentiation potential of human mesenchymal stem cells. PLoS One 8, e55082.

Monod, J., 1949. THE GROWTH OF BACTERIAL CULTURES. Annu. Rev. Microbiol. 3, 371–394.

Najima, Y., Ohashi, K., 2017. Mesenchymal Stem Cells: The First Approved Stem Cell Drug in Japan. Journal of Hematopoietic Cell Transplantation 6, 125–132.

Oh, S.K.W., Chua, P., Foon, K.L., Ng, E., Chin, A., Choo, A.B.H., Srinivasan, R., 2009. Quantitative identification of teratoma tissues formed by human embryonic stem cells with TeratomEye. Biotechnol. Lett. 31, 653–658.

Sasaki, H., Takeuchi, I., Okada, M., Sawada, R., Kanie, K., Kiyota, Y., Honda, H., Kato, R., 2014. Label-free morphology-based prediction of multiple differentiation potentials of human mesenchymal stem cells for early evaluation of intact cells. PLoS One 9, e93952.

Takemoto, Y., Imai, Y., Kanie, K., Kato, R., 2021. Predicting quality decay in continuously passaged mesenchymal stem cells by detecting morphological anomalies. J. Biosci. Bioeng. 131, 198–206.

Triantafyllou, N., Shah, N., Papathanasiou, M.M., Kontoravdi, C., 2023. Combined Bayesian optimization and global sensitivity analysis for the optimization of simulation-based pharmaceutical processes, Comput. Aided Chem. Eng. 52, 381–386.

Flavio Manenti, Gintaras V. Reklaitis (Eds.), Proceedings of the 34th European Symposium on Computer Aided Process Engineering / 15th International Symposium on Process Systems Engineering (ESCAPE34/PSE24), June 2-6, 2024, Florence, Italy

Ab Initio Prediction of Surface Tension from Fundamental Equations of State using Density Gradient Theory (DGT)

Anna Šmídová, Lukáš Šatura, Alexandr Zubov*

Department of Chemical Engineering, University of Chemistry and Technology Prague, Technická 5, Praha 6 – Dejvice, 166 28, Czech Republic
Corresponding author: alexandr.zubov@vscht.cz

Abstract

The density gradient theory (DGT), is used to predict interfacial and surface tension, along with other thermodynamic parameters, for various chemical systems. DGT can be formulated based purely on the equations of state (EoS) describing thermodynamic equilibrium at the interface between studied phases without need of any parameter estimation. To ensure better numerical stability, we introduce a time-dependent Cahn-Hilliard equation into DGT formulation for the calculation of the density equilibrium profile. In this contribution we focus on using the van der Waals EoS and its modification, van der Waals-711, for one-component vapor-liquid systems, and the use of Flory-Huggins equation is demonstrated for one polymer-solvent system. The obtained results are compared with available experimental data and indicate that this *ab initio* approach is capable of predicting the surface/interfacial tension in reasonable agreement with experimental data.

Keywords: density gradient theory, surface tension, vapor-liquid equilibria, Cahn-Hilliard model, van der Waals equation of state.

1. Introduction

Knowledge of surface/interfacial tension is crucial in many branches of science and engineering. Motivation for this work stems from mathematical modelling of morphology evolution (typically done using Cahn-Hilliard approach) in complex polymer mixtures undergoing phase separation, such as during production of hollow fiber membranes, high-impact polymers or porous polymeric materials with well-defined porous microstructure. Models describing these processes contain the so-called interfacial parameter usually denoted as κ, which has significant impact on the resulting multi-phase morphology, is connected to interfacial tension between co-existing and evolving phases, and has usually highly uncertain value (Nistor *et al.*, 2017).

A thermodynamically consistent approaches, such as the density gradient theory (DGT) and related classical density functional theory (cDFT), are used to predict interfacial and surface tension, along with other thermodynamic parameters, for various chemical systems. DGT can be formulated based purely on the equations of state (EoS) describing thermodynamic equilibrium at the interface between studied phases without need of any parameter estimation. In this formulation, the total free energy of the system as a function of the density at the boundary between two phases consists of the bulk contribution and weakly non-local terms, which are dependent on the first gradients of density (Landau and Lifshitz, 1980).

DGT is usually formulated as a boundary value problem (BVP), in which the pre-calculated equilibrium bulk densities of both phases form boundaries of the computational domain, while the interface is treated as a continuous spatial profile of varying density, having typical sigmoidal shape. From the estimated spatial density profile, surface (interfacial) tension can be predicted. However, BVP formulation of DGT often suffers from numerical issues, as large systems of nonlinear equations need to be solved to obtain equilibrium density profile at the interface (Liang and Michelsen, 2017).

    To ensure better numerical stability, we follow the work of Šatura *et al.* (2022) and introduce a time-dependent Cahn-Hilliard equation into DGT formulation for the calculation of the density equilibrium profile. In this contribution we focus on using the van der Waals EoS and its modification, van der Waals-711, for one-component vapor-liquid systems (series of saturated liquid hydrocarbons), and the use of Flory-Huggins equation is demonstrated for one polymer-solvent system (polystyrene-cyclohexane). We compare performance of the extended vdW-711 EoS with the original vdW EoS in calculation of vapor-liquid equilibria (equilibrium densities and saturation pressures) and demonstrate ability of the presented approach to estimate surface and interfacial tension with decent accuracy without a need for model parameter estimation, i.e. *ab initio*.

2. Mathematical Model

2.1. Cahn-Hilliard Equation

Nonstationary, spatially one-dimensional molar balance of species in a one-component system with molar density ρ as a state variable, including non-local term describing surface energy of the interface between co-existing phases, is described by Cahn-Hilliard equation

$$\frac{\partial \rho}{\partial t} = \frac{D}{RT} \frac{\partial}{\partial x} \left[\rho \frac{\partial}{\partial x} \left(\frac{\partial f}{\partial \rho} - \kappa \frac{\partial^2 \rho}{\partial x^2} \right) \right] \tag{1}$$

where R denotes universal gas constant, T is temperature, f represents Helmholtz free energy density, κ is the interfacial (influence) parameter, and t, x denote temporal and spatial coordinate, resp. Mobility/diffusion coefficient D in our approach does not represent the actual diffusion coefficient of the species, but rather reflects rate at which the density profile at the interface evolves in time from the initial condition to the equilibrium shape between two co-existing phases. Term $\partial f / \partial \rho$ represents chemical potential of diffusing species and takes specific form depending on the actual form of the Helmholtz free energy density f, which is derived from the equation of state describing the system (see Subsections 2.2 and 2.3 below). The interfacial parameter κ can be estimated from properties of pure components as described in Section 2.4.

    Initial condition for molar density is in our case represented by linear profile between two fixed boundary points (boundary conditions), which are found by solving conditions of thermodynamic equilibrium for given system. In the case of vapor-liquid systems, the conditions are represented by equality of chemical potential and pressure in both phases (Šatura *et al.*, 2022), while for polymer-solvent system the equilibrium (boundary) points are found as the so-called binodal points in the energy vs. composition phase diagram, i.e., by finding common tangent line to these points, where the slope of the line represents equal value of chemical potential of polymer species in polymer-rich and polymer-lean phase (Nistor *et al.*, 2017). Eq. (1) is integrated numerically, using explicit Euler method.

2.2. Van der Waals Equation of State

As the predictive power of the original vdW EoS is rather limited, more advanced cubic EoS have been developed and commonly used in chemical/process engineering practice, with Peng-Robinson (PR) and Soave-Redlich-Kwong (SRK) being probably the most popular of them. However, there exists also less known advanced version of vdW EoS called vdW-711, which in some cases even outperforms PR and SRK EoS (Tassios, 1992). In addition to already existing parameters *a* and *b* in the original formulation, new parameter, *t* (being function of temperature and acentric factor), is added to the equation, and parameter *a* is newly made temperature-dependent. Final form of vdW-711 EoS and its corresponding Helmholtz free energy density *f*, which is needed for evaluation of chemical potential used in Eq. (1) and can be derived following the methodology of Mauri (2013), is then

$$p = \frac{RT}{V_m + t - b} - \frac{a}{(V_m + t)^2} \tag{2}$$

$$f(\rho) = \frac{a\rho}{t(\rho t + 1)} + \rho RT \ln\left[\frac{\rho(t - b)}{\rho(b - t) - 1}\right] \tag{3}$$

The interfacial parameter κ appearing in Eq. (1) is estimated as

$$\kappa = \frac{9\pi T_c}{4T} N_a d^5 \tag{4}$$

where T_c is critical temperature, N_a is Avogadro number and *d* denotes the hard-sphere (hard-core) diameter of the molecule.

2.3. Flory-Huggins Equation of State

The Helmholtz free energy density of a polymer-solvent mixture is, according to the Flory-Huggins lattice theory,

$$f(\phi) = \frac{\phi}{N_A}\ln\phi + \frac{1 - \phi}{N_B}\ln(1 - \phi) + \chi\phi(1 - \phi) \tag{5}$$

In Eq. (5) symbol ϕ represents volume fraction of polymer in the mixture, N_A and N_B denote the number of lattice sites occupied by each molecule (A = polymer, B = solvent), and χ is Flory-Huggins interaction parameter, which can be estimated, for instance, using Hansen solubility parameters.

The interfacial parameter κ in this case can be estimated as

$$\kappa = \frac{R_g^2}{3}\left(\chi + \frac{1}{N_A\phi} + \frac{1}{1 - \phi}\right) \tag{6}$$

where R_g represents polymer chain radius of gyration.

2.4. Evaluation of surface/interfacial tension

From the equilibrium density profile at the vapor-liquid interface calculated via solution of Eq. (1), the surface tension γ is estimated by integration of density spatial derivative square over the interface thickness L in the following way

$$\gamma = \kappa RT \int_0^L \left(\frac{d\rho}{dx}\right)^2 dx \tag{7}$$

In the case of polymer-solvent system, the state variable in the Cahn-Hilliard equation (1) is polymer volume fraction ϕ rather than molar density ρ, and evaluation of interfacial tension at polymer-solvent interface then must be slightly modified and involves molar density of polymer ρ_m

$$\gamma = \kappa RT \rho_m \int_0^L \left(\frac{d\phi}{dx}\right)^2 dx \tag{8}$$

3. Results and Discussion

Figure 1 presents comparison of saturation pressures and liquid densities estimated using vdW-711 EoS (continuous curves) and their experimental values (discrete points) for series of hydrocarbons, namely *n*-pentane, *n*-hexane, *n*-heptane, *n*-octane, *n*-nonane, *n*-decane and cyclohexane at different temperatures. From the plots it is evident that the agreement between experimental data and predictions by vdW-711 is excellent.

Figure 2 demonstrates similar results for a single compound – cyclohexane – but this time the comparison is made between the original vdW EoS and its extended form, vdW-711. The prediction error of the original vdW is relatively high, especially in the case of liquid densities, where it deviates from the experimental values almost by 50%. Figure 3 demonstrates how molar density and chemical potential evolve throughout the simulation during integration of Eq. (1) from the initial linear density profile into the final equilibrium profile of sigmoidal shape. The observed flat (constant) profile of chemical potential at the end of the simulation is a proof of reached thermodynamic equilibrium.

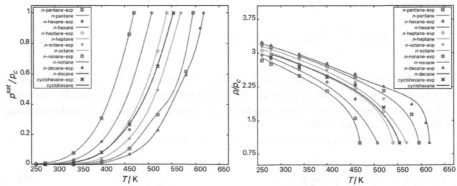

Figure 1. Comparison of predicted (continuous curves) and experimental (discrete points) temperature dependencies of (i) saturation pressure scaled by critical pressure (left), and (ii) liquid phase densities scaled by critical densities (right). The predictions were performed for *n*-pentane, *n*-hexane, *n*-heptane, *n*-octane, *n*-nonane, *n*-decane, cyclohexane using vdW-711 EoS. The experimental data are taken from the NIST Webbook of Chemistry.

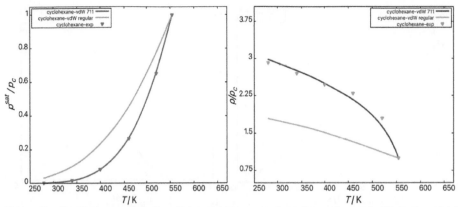

Figure 2. Comparison of predicted (continuous curves) and experimental (discrete points) temperature dependencies of (i) saturation pressure scaled by critical pressure (left), and (ii) liquid phase densities scaled by critical densities (right) for cyclohexane. Light grey curves represent predictions using classical vdW EoS, dark curves are calculated using extended vdW-711 EoS. The experimental data are taken from the NIST Webbook of Chemistry.

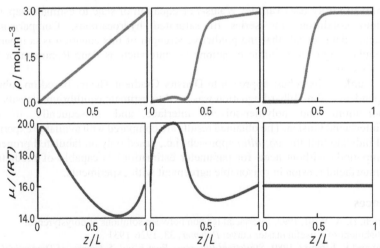

Figure 3. Dynamic evolution of spatial profiles of cyclohexane molar density (top graphs) and normalized chemical potential (bottom graphs) as predicted by Cahn-Hilliard model from initial condition (left) to the final equilibrium profiles (right). Spatial coordinate z is normalized by the length of the computational domain $L = 10$ nm. In this simulation, system temperature was set to $T = 298.15$ K.

Figure 4 shows comparison of predicted and experimental dependency of n-heptane surface tension on temperature and pressure. Considering that no parameters have been estimated from the experimental data for this prediction, the agreement is surprisingly good, rendering not only qualitative trends, but also being quantitatively close to the measured values.

Also for the polymer-solvent system, the interfacial tension can be estimated with reasonable accuracy – using Flory-Huggins equation, we predicted value of 0.0022 mN·m⁻¹ for system polystyrene-cyclohexane at 285 K, where the value measured at these conditions by Heinrich and Wolf (1992) was found to be 0.0015 mN·m⁻¹.

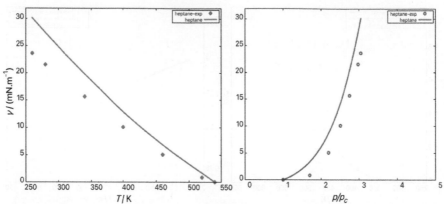

Figure 4. Comparison of predicted (using vdW-711 EoS) and experimental values of surface tension of *n*-heptane in dependence on temperature (left) and pressure (right). The experimental data are taken from the NIST Webbook of Chemistry.

4. Conclusions

We employed rarely used van der Waals-711 equation of state to estimate vapor-liquid equilibrium conditions for series of saturated hydrocarbons. Comparison with experimental data reveals that the predictive strength of this equation is comparable to more popular engineering cubic equations of state such as Peng-Robinson or Soave-Redlich-Kwong.

Numerically robust approach to Density Gradient Theory based on solution of time-dependent Cahn-Hilliard equation was used to estimate equilibrium density profile of vapor-liquid and polymer-solvent interface and subsequently to predict surface/interfacial tension. The obtained results are compared with available experimental data and indicate that this *ab initio* approach (i.e., based only on tabulated properties of pure compounds without need for parameter estimation) is capable of predicting the surface/interfacial tension in reasonable agreement with experiments.

References

M. Heinrich, B. A. Wolf, 1992, Interfacial tension between solutions of polystyrenes: establishment of a useful master curve, *Polymer*, 33, 1926-1931.

L. Landau and E. Lifshitz, 1980, *Statistical Physics – Part 1, vol. 5, Course of Theoretical Physics*, Butterworth-Heinemann.

X. Liang and M. L. Michelsen, 2017, General approach for solving the density gradient theory in the interfacial tension calculations, *Fluid Phase Equilibria*, 451, 79–90.

R. Mauri, 2013, *Non-Equilibrium Thermodynamics in Multiphase Flows*, Springer.

NIST Webbook of Chemistry, National Institute of Standards and Technology, available at: https://webbook.nist.gov/

A. Nistor, M. Vonka, A. Rygl, M. Voclová, M. Minichová, J. Kosek, (2017), Polystyrene Microstructured Foams Formed by Thermally Induced Phase Separation from Cyclohexanol Solution, *Macromolecular Reaction Engineering*, 11, 1600007.

L. Šatura, M. Minichová, M. Pavelka, J. Kosek, A. Zubov, 2022, A Robust Physics-Based Calculation of Evolving Gas-Liquid Interfaces, *Journal of Non-Equilibrium Thermodynamics*, 47, 143-154.

D. P. Tassios, 1993, *Applied Chemical Engineering Thermodynamics*, Springer.

Flavio Manenti, Gintaras V. Reklaitis (Eds.), Proceedings of the 34th European Symposium on Computer Aided Process Engineering / 15th International Symposium on Process Systems Engineering (ESCAPE34/PSE24), June 2-6, 2024, Florence, Italy

Model-Based Assessment of Electrochemical Cells and Systems for Atmospheric CO_2 Removal

Guokun Liu, Aidong Yang[*]

Department of Engineering Science, University of Oxford, Oxford, OX13PJ, UK
aidong.yang@eng.ox.ac.uk

Abstract

Electrochemical systems utilizing alkali solutions and the pH-swing method demonstrate promising potential for large-scale direct air capture (DAC) applications, addressing the imperative need to remove CO_2 from the atmosphere alongside emissions reduction to meet climate goals. However, a consistent and systematic comparison between different electrochemical schemes is still lacking. This research focuses on numerically modelling and simulating electrolysis and electrodialysis cells and systems, with the aim of providing guidance for further system optimization. Considering hydrogen production with a referred efficiency of 80%, the minimum DAC energy of the electrolysis system is obtained as 587 kJ mol^{-1}, while the electrodialysis system consumes a minimum energy of 273 kJ mol^{-1}. The DAC energy of the two systems can be further reduced to 290 kJ mol^{-1} and 170 kJ mol^{-1} as the defined CO_2 absorption reaches 1. The results highlight the importance of further optimization in CO_2 absorber design and system operating parameters.

Keywords: CO_2 capture; Electrochemical systems; Hydrogen production; Ion exchange membrane, Numerical modelling.

1. Introduction

Disturbing reports indicate that the world is poised to deplete the remaining carbon budget allocated to limit global temperature rise to 1.5 °C by 2030 (Liu et al., 2023). Responding to this pressing scenario, increasing attention is given to artificial CO_2 removal methods as a direct intervention to mitigate atmospheric CO_2 levels. Among these methods, direct air capture (DAC) stands out due to its merits of requiring less land, fewer location constraints, and simpler quantification of removal (Hanna et al., 2021), positioning it as a focal point in recent research and development.

Currently, DAC technologies fall into two primary categories: solid adsorption and liquid absorption (Sanz-Perez et al., 2016). The prevalent solid adsorption technique employs amine sorbents through a chemical-thermal process, demanding substantial thermal energy for CO_2 liberation and sorbent regeneration (Zhu et al., 2022). Alternatively, wet scrubbing with aqueous alkali hydroxide has emerged as a more energy-efficient alternative with higher productivity (Sabatino et al., 2021). The wet-scrubbing-based DAC plant by Carbon Engineering Ltd, achieving an annual productivity of 1 Mt CO_2 (Keith et al., 2018), serves as a testament to the feasibility of the liquid absorption pathway. However, challenges persist, particularly in the high-temperature heating process required for CO_2 recovery, introducing complexity and energy demands. In response, an alternative pH-swing concept has been suggested, employing an electrochemical system powered by electricity and conducive to integration with renewable energy sources.

In the realm of electrochemical systems, various methods, including electrolysis, bipolar membrane (BPM) electrodialysis, capacitive deionization, and proton-coupled electron transfer agents, have been explored for DAC purposes (Sharifian et al., 2021). BPM electrodialysis has garnered significant attention due to its capacity to directly dissociate water into H^+ and OH^- ions, resulting in lower energy consumption for generating pH differences (Eisaman et al., 2011). In contrast, electrolysis systems, generally deemed more energy-intensive, have demonstrated promise with innovative designs and approaches such as hydrogen recycling (Zhu et al., 2023).

Despite the potential demonstrated by both electrolysis and electrodialysis systems for energy-efficient DAC, a comparable evaluation under standardized conditions is lacking. This study aims to bridge this gap through numerical modeling and simulation, considering actual cation exchange membrane (CEM) ion selectivity. By scrutinizing and comparing the impacts of applied current, CO_2 absorption ratio, and CO_2 recovery ratio, the study endeavors to offer valuable insights into the performance of both systems. The results hold the potential to steer further technological advancements in the pursuit of effective and sustainable DAC solutions.

2. Systems and Models

Figure 1 presents the electrolysis and electrodialysis DAC systems modelled and simulated in this study. The base solute utilized is K_2CO_3, and both systems follow a four-step cycle to achieve solution pH-swing and DAC. Firstly, in the alkali generation process $(4 \rightarrow 1)$, OH^- ions are continuously generated in the alkaline compartment through the cathodic semi-reaction, leading to the alkalization of the solution and the attainment of a high CO_2 absorption capacity. Subsequently, in the CO_2 absorption process $(1 \rightarrow 2)$, the strongly basic solution is directed through the membrane contactor to absorb CO_2 from the air, transforming into a CO_2-rich solution. Following this, in the acidification process $(2 \rightarrow 3)$, the CO_2-rich solution undergoes acidification in the acidic compartment, where K^+ ions migrate from the acidic to the alkaline compartment through the CEM, replacing K^+ ions with H^+ ions. Finally, in the CO_2 release process $(3 \rightarrow 4)$, the acidification leads to a reduction in CO_2 solubility, prompting the release of CO_2. The purified CO_2 gas is separated and collected in the separator, while the solution reverts to its initial state.

While both the electrolysis and electrodialysis systems share similar CO_2 absorption processes, the electrochemical reactions responsible for solution alkalization and acidification differ between the two. In electrolysis, the reactions involve hydrogen generation, serving the dual purpose of hydrogen production and partially mitigating the energy consumption of DAC. On the other hand, the electrodialysis system incorporates a crucial component known as the BPM, which includes an anion exchange layer and a cation exchange layer. The BPM facilitates the direct dissociation of water into OH^- and H^+ ions, creating a pH difference between the two compartments. Unlike water electrolysis, the water splitting reaction in electrodialysis does not necessitate energy for hydrogen and oxygen generation, resulting in a reduced overall energy demand.

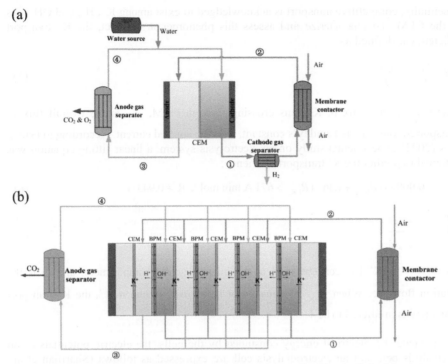

Figure 1: System layouts: (a) Electrolysis systems; (b) Electrodialysis system

The solution's dissolved inorganic carbon (DIC) and total alkalinity (TA) are defined as follows to facilitate numerical analysis:

$$DIC = c_{CO_3^{2-}} + c_{HCO_3^-} + c_{CO_2\,(aq)} \tag{1}$$

$$TA = c_{K^+} = 2c_{CO_3^{2-}} + c_{HCO_3^-} + c_{OH^-} - c_{H^+} \tag{2}$$

Achieving full absorption or recovery of CO₂ in the membrane contactor and anode gas separator, reaching their respective equilibrium limits, can pose challenges under actual conditions. To quantify the scope of these limitations, the CO₂ absorption and recovery ratios are delineated as follows:

$$R_a = \frac{DIC_2}{DIC_{eq,2}} \tag{3}$$

$$R_r = \frac{DIC_2 - DIC_4}{DIC_2 - DIC_{eq,4}} \tag{4}$$

where the subscript numbers correspond to the node numbers in Figure 1.

The ideal selectivity of the CEM for K⁺ ions is typically not achieved, as there could be the crossovers of H⁺ and OH⁻ ions. However, the crossovers of CO₃²⁻ and HCO₃⁻ ions are disregarded due to their significantly lower concentrations compared to OH⁻ ions.

Essentially, competitive transport is acknowledged to exist among K^+, H^+, and OH^- ions in the CEM. To characterize and assess this phenomenon's impact, the K^+ transport efficiency is defined as:

$$\beta = \frac{n_{K^+}}{N_+} = \frac{n_{K^+} F}{I} \tag{5}$$

where n_{K^+} is the flux of K^+ ions crossing through CEM, N_+ is the overall flux of transported ions, F is Faraday's constant, I is the applied current. According to Gao et al.'s (2022) experimental study on an electrolysis system, a linear fitting equation was obtained to predict the K^+ transport efficiency:

$$\beta = -0.000537 R_{I/K_+} + 1.36 \quad (R_{I/K_+} > 671 \text{ A min mol}^{-1}, \text{ R}^2 = 0.9437) \tag{6}$$

$$R_{I/K_+} = \frac{i \cdot A}{c_{K^+,a} Q} \tag{7}$$

where $c_{K^+,a}$ is K^+ ion concentration at the inlet of acidic compartment, and Q is the solution flowrate. When R_{I/K_+} is equal to or below 671 A min mol^{-1}, the K^+ transport efficiency is predicted to be 1.

To calculate the electrical energy consumed by the cells, the electric potentials of an electrolysis cell and an electrodialysis cell are expressed as follows (Sharifian et al., 2021):

$$E_{\text{electrolysis}} = E_{\text{rev}} + E_{\text{a}} + E_{\text{c}} + E_{\text{ohm}} \tag{8}$$

$$E_{\text{electrodialysis}} = 0.15 + \frac{0.059T}{T_0} \Delta pH + E_{\text{ohm}} \tag{9}$$

where E_{rev}, E_{a}, E_{c} and E_{ohm} are the reversible, anode activation, cathode activation and ohmic potentials, and ΔpH is the solution pH difference between the two compartments.

The foundational operational parameters are provided in Table 1. The systems are simulated and iterated, and the results are obtained when the systems reach a steady state.

Table 1: Parameters of electrolysis and electrodialysis systems

Parameter	Electrolysis system (Gao et al., 2022)	Electrodialysis system (Iizuka et al., 2012)
K^+ concentration	1.54 M	1.54 M
Solution flow rate (single cell)	1 ml min^{-1}	1ml min^{-1}
Single cell effective area	16 cm^2	210 cm^2
Cell compartment thickness	0.1 cm	0.075 cm
CO_2 absorption ratio	0.9	0.9
CO_2 recovery ratio	0.9	0.9

3. Results

Conducting simulations across a range of applied current densities facilitates the determination of optimal conditions and enables the comparison of overall performance between electrolysis and electrodialysis systems. Figure 2 provides an overview of energy consumption trends (per mol of CO$_2$ capture) for both systems. In the case of the electrolysis system, the energy consumption of hydrogen production is deducted by considering a reference hydrogen production efficiency (HPE). The results show that the reference HPE significantly impacts DAC energy consumption evaluation in the electrolysis system. At low applied currents and with a lower HPE (70%), the DAC energy consumption of the electrolysis system can be as low as 121 kJ mol^{-1}. However, assuming an HPE of 80%, the minimum DAC energy consumption of the electrolysis system significantly increases to 587 kJ mol^{-1} at a current of 2.2 A. On the other hand, the electrodialysis system achieves a minimum DAC energy consumption of 273 kJ mol^{-1}. These findings suggest that the water electrolysis system could be competitive in terms of DAC energy demand when considering the impact of hydrogen production, particularly at a low HPE. Nevertheless, a notable drawback of electrolysis is the mixing of recovered CO$_2$ gas with O$_2$, rather than producing a pure product.

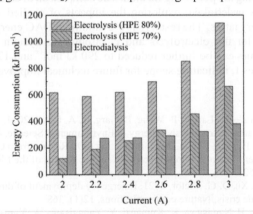

Figure 2: System energy consumption under various single-cell applied current

Figure 3 compares the DAC energy consumption of the two systems under different CO$_2$ absorption and recovery ratios. In Figure 3(a), the DAC energy for both systems can be significantly reduced to 290 kJ mol^{-1} (electrolysis, 80% HPE) and 170 kJ mol^{-1} (electrodialysis) as the CO$_2$ absorption ratio increases to 1. This highlights the importance of improving CO$_2$ absorber design and enhancing CO$_2$ absorption performance. Given the low CO$_2$ concentration in the ambient air, achieving this improvement could be challenging and warrants in-depth exploration. In Figure 3(b), an increasing CO$_2$ recovery ratio has a positive effect on saving the DAC energy of the electrolysis system. However, an opposite effect is observed for the electrodialysis system. As the CO$_2$ recovery ratio increases, the pH difference between the compartments rises, elevating the BPM cell potential. Compared with the electrolysis cell, the overall electrical potential of the electrodialysis cells is much lower, and the negative effect of the increasing potential is more significant than the positive effect brought about by the increasing CO$_2$ recovery amount. These results show the significance and necessity of further comprehensive optimization of operating parameters.

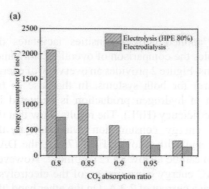
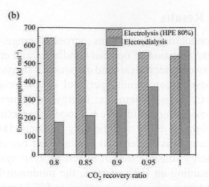

Figure 3: System DAC energy under various (a) CO_2 absorption and (b) CO_2 recovery ratios

4. Conclusions

This study conducts a modeling and parametric analysis of electrolysis and electrodialysis DAC systems employing the pH-swing method. The comparison of DAC energy demand for both systems was undertaken, exploring the impacts of applied current, and CO_2 absorption and recovery ratios. The results show a minimum DAC energy of 587 kJ mol^{-1} and 273 kJ mol^{-1} for the electrolysis and electrodialysis system under the given conditions. These values can be further reduced to 290 kJ mol^{-1} and 170 kJ mol^{-1} as the CO_2 absorption reaches 1, indicating scope for future technical improvement.

References

M. D. Eisaman, L. Alvarado, D. Larner, P. Wang, B. Garg, K. A. Littau, 2011, CO2 separation using bipolar membrane electrodialysis, Energy & Environmental Science, 4(4), 1319-1328.

X. Gao, A. Omosebi, R. Perrone, K. Liu, 2022, Promoting CO2 Release from CO3 2−-Containing Solvents during Water Electrolysis for Direct Air Capture, Journal of the Electrochemical Society, 169(4), 044527.

R. Hanna, A. Abdull, Y. Xu, D. G. Victor, 2021, Emergency deployment of direct air capture as a response to the climate crisis, Nature communications, 12(1), 368.

A. Iizuka, K. Hashimoto, H. Nagasawa, K. Kumagai, Y. Yanagisawa, A. Yamasaki, 2012, Carbon dioxide recovery from carbonate solutions using bipolar membrane electrodialysis, Separation and purification technology, 101, 49-59.

D. W. Keith, G. Holmes, D. S. Angelo, K. Heidel, 2018, A process for capturing CO2 from the atmosphere, Joule, 2(8), 1573-1594.

Z. Liu, Z. Deng, S. Davis, P. Ciais, 2023, Monitoring global carbon emissions in 2022, Nature Reviews Earth & Environment, 4(4), 205-206.

F. Sabatino, A. Grimm, F. Gallucci, M. van Sint Annaland, G, J. Kramer, M. Gazzani, 2021, A comparative energy and costs assessment and optimization for direct air capture technologies, Joule, 5(8), 2047-2076.

E. S. Sanz-Pérez, C. R. Murdock, S. A. Didas, C. W. Jones, 2016, Direct capture of CO2 from ambient air, Chemical reviews, 116(19), 11840-11876.

R. Sharifian, R. M. Wagterveld, I. A. Digdaya, C. Xiang, D. A. Vermaas, 2021, Electrochemical carbon dioxide capture to close the carbon cycle, Energy & Environmental Science, 14(2), 781-814.

P. Zhu, Z. Y. Wu, A. Elgazzar, C. Dong, T. U. Wi, F. Y. Chen, Y. Xia, Y. Feng, M. Shakouri, J. Y. Kim, Z. Fang, T. A. Hatton, H. Wang, 2023, Continuous carbon capture in an electrochemical solid-electrolyte reactor, Nature, 618(7967), 959-966.

X. Zhu, W. Xie, J. Wu, Y. Miao, C. Xiang, C. Chen, B. Ge, Z. Gan, F. Yang, M. Zhang, D. O'Hare, J. Li, T. Ge, R. Wang, 2022, Recent advances in direct air capture by adsorption, Chemical Society Reviews.

Flavio Manenti, Gintaras V. Reklaitis (Eds.), Proceedings of the 34th European Symposium on Computer Aided Process Engineering / 15th International Symposium on Process Systems Engineering (ESCAPE34/PSE24), June 2-6, 2024, Florence, Italy

Enhancing Unit Operation Design: Leveraging Neural Networks to Enforce Physical Hard Constraints

Jana Mousa,* Stéphane Negny, Rachid Ouaret, Alessandro Di Pretoro, Ludovic Montastruc.

Laboratoire de Génie Chimique, Université de Toulouse, CNRS, INPT, UPS, Toulouse, France.

Correspondence : jana.mousa@toulouse-inp.fr

Abstract

Neural networks have emerged as promising computational tools in machine learning for modelling complex systems due to their ability to learn intricate patterns from data. However, they come with inherent limitations as they often disregard the fundamental principles of the processes modelled making them incapable to incorporate physical laws and constraints into the learning and training processes. Tackling these limitations, in this work physics-informed neural networks and data reconciliation were the main approaches tested to embed such constraints within the neural network architecture and training process, aiming to bridge the gap between data-driven modelling and physics-based comprehension of systems. This innovative framework exposes the potentiality of hybrid modelling by forcing a neural network to integrate and harmonize with physical laws, paving the way for a robust fusion between fundamental laws and AI for engineering applications. In this study we propose and compare data reconciliation methods with standard neural networks and physics-informed neural networks and test their capabilities on the design of a flash drum for the separation of a binary mixture.

Keywords: neural networks, physics-informed neural networks, data reconciliation, unit operations.

1. Introduction

Neural networks have become an integral part of process engineering, serving as powerful computational tools for modelling complex systems. Their power lies in their ability to decipher intricate patterns within data sets, making them valuable tools for predictive modelling, optimization and control in a variety of industrial processes. However, despite their capabilities, neural networks often operate as black boxes, neglecting the fundamental physical principles that underpin the processes they seek to model.

In the realm of process engineering, where precision and reliability are paramount, the inability of neural networks to incorporate physical laws and constraints is a significant limitation. The black-box nature of these networks renders their predictions uninterpretable, making it difficult to understand the underlying mechanisms controlling a system. This is particularly problematic in industries where compliance with specific physical laws and constraints is critical for safe and efficient operation (Solle et al., 2017).

The importance of incorporating physical meaning into neural networks becomes apparent when considering their applications in fields such as chemical engineering, where processes adhere to fundamental principles of mass and energy conservation. Neural networks that ignore these principles can make inaccurate predictions and fail to capture the nuanced behavior of systems (Hussain et al., 2021). To address these drawbacks, recent research in process engineering has focused on the integration of neural networks with physics-based models, which we call hybrid modelling. This integration aims to exploit the strengths of both approaches, combining the data-driven capabilities of neural networks with the physical insight provided by first-principles models (Bhutani et al., 2006). This makes the neural network a more reliable tool for tasks such as model calibration, design optimization and control in industrial processes.

In conclusion, while neural networks offer remarkable capabilities in modeling complex systems, their limitations in incorporating physical significance can hinder their effectiveness in process engineering (Cavalcanti et al., 2021). The integration of physics-based principles into neural networks, represents a promising solution to overcome these limitations, fostering a more robust and interpretable framework for applications in various engineering domains. To achieve this, the aim of this study is to propose and test a new approach that couples fundamental principles and neural networks and to test and compare it with other approaches and methods

2. Case Study and Methodologies

This work focuses on optimizing the design of a flash drum—a vital component in chemical engineering for liquid-vapor separation. The flash drum's efficiency hinges on strict adherence to mass balances (Eq. (1)) which are the constraints to be respected. To achieve this, we employed two approaches, the first one being physics-informed neural networks (PINNs) and the second being a data reconciliation approach merged with the neural network.

$$m_{i,in} = \sum m_{i,out} \tag{1}$$

2.1. Physics-Informed Neural Networks

Physics-informed neural networks (PINNs) represent an eye-catching approach that integrates physical knowledge into the architecture of neural networks. These networks are designed to incorporate underlying physical principles and constraints directly into the learning process, making them highly desirable for modeling systems where disregarding the fundamentality is detrimental for efficient operation. The primary motivation behind utilizing PINNs lies in their ability to fuse the strengths of traditional physics-based models with the data-driven capabilities of neural networks (Raissi et al., 2019). In our approach, the physics-informed neural network was done by creating a custom loss function that aims to enforce the physical constraints of the system by minimizing an objective function during training. As seen in Figure 1, the custom loss function is meant to be embedded inside the neural network's training process so that it learns from it. This function is composed of two components: a data-driven loss term and a physics informed constraint term.

The data-driven component (Eq. (2)) is responsible for minimizing the difference between the model's predictions and the actual observed data (mean squared error). While the second term (Eq. (3)) ensures that the neural network respects the mass balance equality by introducing a penalty term α to effectively guide the network to generate predictions that align with the underlying physics of the problem.

$$Data\text{-}driven\ loss = MSE(y_{true} - y_{pred}) \quad (F1) \tag{2}$$

$$Physics\text{-}informed\ loss = \alpha * mass\ balance\ equality\ constraint \quad (F2) \tag{3}$$

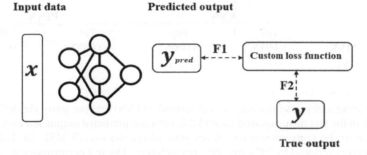

Figure 1. Visualization of a physics-informed neural network with a custom loss function.

2.2. Data-Reconciliation

Data reconciliation is a process used in a variety of fields, including engineering, process industries and data management, to ensure the consistency and accuracy of data collected from different sources or obtained through different measurements(Crowe, 1996). The primary goal of data reconciliation is to identify and correct discrepancies or errors in the data, resulting in a more accurate and reliable dataset. In our case (Figure 2), data reconciliation techniques were applied by generating multiple initializations of the gradient descent used in the learning step.

Through iterative adjustments and corrections guided by reconciliation methods, we sought to reconcile any discrepancies in the network's predicted outputs, thereby refining the model's predictions to align with the fundamental principles of mass conservation.

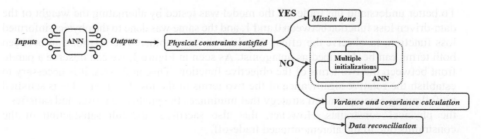

Figure 2 Steps to constrained outputs via data reconciliation

3. Results

The flash drum is dealing with a binary mixture of Toluene (Tol) and Biphenyl (Bip) where the inlet flow rate, temperature and pressure, along with the top and bottom outlet flowrates for both components are acquired as simulated data of 1594 points. An artificial neural network was studied for the prediction of the output flowrates of both components as well as a physics-informed neural network that was specifically equipped with physical constraints embedded in a custom loss function to uphold the mass balance equality, distinguishing it from the standard neural network (the train-test data split ratio was 80:20). Results are shown in Table 1.

Table 1. Performance of ANN and PINN.

	ANN		PINN	
	Tol	Bip	Tol	Bip
Train MSE	0.0025	0.00052	0.0022	0.0004
Test MSE	0.01	0.0016	0.007	0.00078

Incorporating a physics-informed neural network (PINN) into our study yielded a notable decrease in the test mean squared error (MSE between predicted output and expected one) compared to the conventional neural network where the overall MSE for Toluene and Biphenyl was reduced by 70% and 49% respectively. The test performance of the model is improved when the physics term is introduced to the loss function. By enforcing these constraints, the PINN enhances the model's capability to capture the underlying physics of the system, resulting in more accurate and meaningful predictions during the testing phase.

Moreover, when working with a typical neural network, the mass balance equality was never satisfied which is an expected observation knowing that a neural network operates as a black box thereby disregarding the physical aspects of the process. However, when applying the physical term in the loss function, we notice a decrease of 58% in the mass balance values for both chemical components, which is considered as an improvement. Yet, the values indeed don't hit 0, so on the broader aspect; the mass balance equality was not 100% satisfied even with a physics-informed neural network.

Accordingly, one can conclude that while the physics-informed neural network demonstrated improved results compared to the standard neural network, it is essential to acknowledge the inherent trade-off between predictive accuracy and the fulfilment of physical constraints.

To better understand this trade-off the model was tested by alternating the weight of the data-driven loss function between 0 and 1, and the same was done to the physics-informed loss function, to observe the effect they had on one another. The relationship between both terms can be described as antagonist. As seen in Figure 3, we can observe a pareto front between the two terms of the objective function. This, in PINNs, it is necessary to establish the relative significance of the two terms of the loss function. This is satisfied by following a compromised strategy that minimizes the predictive power and satisfies the physical constraints. However, this also sacrifices the full satisfaction of the constraints due to the aforementioned trade-off.

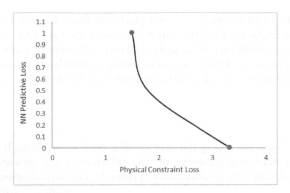

Figure 3. Observations on the loss terms upon varying the penalty term.

This antagonistic interplay reflects the inherent challenge of striking a balance between achieving accurate predictions and upholding the physical constraints when working with physics-informed neural networks.

As previously discussed, the usage of a neural network on is own did not respect properly the mass balance constraint on the chemical components. Accordingly, a data reconciliation methodology has been proposed to address this issue. This procedure is grounded in the concept that a neural network is treated as a soft sensor generating "prediction" errors that is consequently rectified using classical data validation techniques where the linear optimization problem (Eq. (4)) and its relative solution (Eq. (5)) are shown below:

$$\begin{cases} \min\limits_{\tilde{x}_{ann}} \frac{1}{2} (\tilde{x}_{ann} - x_{ann})^t V^{-1} (\tilde{x}_{ann} - x_{ann}) \\ M\tilde{x}_{ann} = B \end{cases} \tag{4}$$

And the corresponding analytical solution is the following:

$$\tilde{x}_{ann} = x_{ann} - VM^t(MVM^t)^{-1}(M\tilde{x}_{ann} - B) \tag{5}$$

The matrix M encodes the linear structure of the constraints as seen below:

$$\begin{pmatrix} 1 & 0 & 0 & 0 & 0 & 0 \\ 0 & 0 & 0 & 1 & 0 & 0 \\ 1 & -1 & -1 & 0 & 0 & 0 \\ 0 & 0 & 0 & 1 & -1 & -1 \end{pmatrix} \tilde{x}_{ann} = \begin{pmatrix} Tol\ inlet\ flow \\ Bip\ inlet\ flow \\ 0 \\ 0 \end{pmatrix} \tag{6}$$

Where V is the variance-covariance matrix calculate by running the neural network on multiple initializations for the same input. Also, $\tilde{x}_{ann}$ is a vector of validated data of the predicted output of the neural network and x_{ann} is that of the desired expected output.

Therefore, the data to be reconciled is that predicted from the neural network. This methodology completely satisfied the physical constraints but at the cost of having slightly less accurate predictions of the individual output flowrates. Accordingly, further investigations are ongoing to achieve a balance between adhering to physical constraints and the performance level associated with a machine learning model.

4. Conclusion

This study delved into enhancing the design of a flash drum through three distinct approaches: traditional neural networks, physics-informed neural networks, and neural networks integrated with data reconciliation (Figure 4). Our investigation revealed that while each method demonstrated its merits, a recurring theme emerged, that is the inherent trade-off between satisfying stringent physical constraints and achieving robust predictive power. Traditional neural networks showcased predictive powers but struggled to align with the physical significance. In

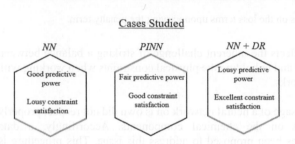

Figure 4. Insights from the three studied approaches

contrast, physics-informed neural networks demonstrated a more nuanced understanding of underlying principles but to a limit to not lose the predictive power. The incorporation of data reconciliation into neural network models provided a promising avenue for enforcing physical constraints, yet this enhancement was matched with a subpar accuracy. Lastly, our findings underscore the importance of weighing the trade-offs inherent in each methodology according to one's needs and the importance of further future investigations.

References

Bhutani, N., Rangaiah, G.P., Ray, A.K., 2006. First-Principles, Data-Based, and Hybrid Modeling and Optimization of an Industrial Hydrocracking Unit. Ind. Eng. Chem. Res. 45, 7807–

Cavalcanti, F.M., Kozonoe, C.E., Pacheco, K.A., Alves, R.M. de B., 2021. Application of Artificial Neural Networks to Chemical and Process Engineering, in: Mazzeo, P.L., Spagnolo, P. (Eds.), Deep Learning Applications. IntechOpen, Rijeka.

Crowe, C.M., 1996. Data reconciliation — Progress and challenges. Journal of Process Control, Process Systems Engineering 6, 89–98. https://doi.org/10.1016/0959-1524(96)00012-1

Hussain, M.M., Bari, S.H., Mahmud, I., Siddiquee, M.I.H., 2021. Chapter 5 - Application of different artificial neural network for streamflow forecasting, in: Sharma, P., Machiwal, D. (Eds.), Advances in Streamflow Forecasting. Elsevier, pp. 149–170.

Raissi, M., Perdikaris, P., Karniadakis, G.E., 2019. Physics-informed neural networks: A deep learning framework for solving forward and inverse problems involving nonlinear partial differential equations. Journal of Computational Physics 378, 686–707.

Solle, D., Hitzmann, B., Herwig, C., Pereira Remelhe, M., Ulonska, S., Wuerth, L., Prata, A., Steckenreiter, T., 2017. Between the Poles of Data-Driven and Mechanistic Modeling for Process Operation. Chemie Ingenieur Technik 89, 542–561.

Flavio Manenti, Gintaras V. Reklaitis (Eds.), Proceedings of the 34th European Symposium on Computer Aided Process Engineering / 15th International Symposium on Process Systems Engineering (ESCAPE34/PSE24), June 2-6, 2024, Florence, Italy

Design and Energy Evaluation of a Multi-Stage CO₂ Separation Process Using Surrogate Model

Yota Fujii,[a] Keigo Matsuda [a*]

[a]*Department of Complex Systems Science, Nagoya University, Furo-cho, Chikusa-ku, Nagoya, Aichi 464-8601, Japan*
Corresponding Author's E-mail: matsuda@i.nagoya-u.ac.jp

Abstract

In recent years, CCS (Carbon dioxide Capture and Storage) has been promoted worldwide, and among CCS processes, the adsorption and membrane separation processes require lower energy than gas absorption using amine but have lower separation performance. The combination of adsorption and membrane is one way to improve the separation performance, but studies on the combination and configuration of multi-stage processes are still insufficient. In addition, membrane separation is operated in a steady state and adsorption separation is operated in an unsteady state, which makes the design of a hybrid process complex. Therefore, to design CO_2 capture processes for both adsorption and membrane separation, we applied statistical model with machine learning. For each process, parameters such as compressor and vacuum pump operating pressures, CO_2 composition in the feed gas, and flow rate were set within predefined ranges, and data were collected from these ranges. Using this data, a statistical model was developed using a neural network, and multi-objective optimization was performed using NSGA-II (Non-dominated Sorting Genetic Algorithms II). The results showed that the required performance of the CCS could not be achieved with a single stage separation. The Pareto solution obtained by optimization indicated that a high compression ratio and low vacuum pressure were required. The prediction accuracy of the physical and statistical models in the Pareto solution was about 0.99 for both processes. Based on these results, a multi-stage separation process was synthesized using surrogate models. Multi-objective optimization was performed on this process, and it was found that the combined membrane and adsorption process not only met the CCS requirements, but also reduced the operating pressures of the compressor and vacuum pump compared to the single-stage process. These results highlight the effectiveness of using statistical models for process optimization.

Keywords: CO₂ separation, Machine learning, Process design

1. Introduction

The Industrial Revolution occurred between the 18th and 19th centuries and brought revolutionary changes to human life and the economy. In recent years, there has been a consensus that greenhouse gas emissions from human activities are the main cause of global warming (L. Al-Ghussain, 2019). As a solution to the current situation, CCS technologies have gained much attention. Three main processes are considered for CO_2 separation technology: amine absorption, membrane separation and gas adsorption separation. Among the three processes, amine absorption is currently the most widely used CO_2 separation process. However, the reaction heat between the absorbed solvent and CO_2 is high, and the CO_2 capture energy is high (4 - 6 MJ/kg-CO_2 (M. Mondal et al., 2012)), because more energy than the reaction heat is required for CO_2 stripping. On the other hand, the energy consumption of adsorption and membrane separation processes, which are alternative technologies to the amine absorption process, are reported to be 2 - 3 MJ/kg-CO_2 and 0.5 - 6 MJ/kg-CO_2, respectively (M. Mondal et al., 2012). However, adsorption and membrane separation processes have the problem of low separation efficiency (Y. Fujii et al., 2023). One solution is to combine these processes into a multi-stage process. In the design of multi-stage processes, membrane separation is operated in a steady state and adsorption separation is operated in an unsteady state, which makes the calculation complex to design processes that combine these processes. Recently, it has been reported that this complexity can be reduced by using a surrogate model based on machine learning (M. Rahimi et al., 2021). In this study, we perform efficient membrane-adsorption multi-stage process synthesis using machine learning.

2. Modelling

Figure 1 shows a schematic of the trained adsorption and membrane separation process. First, a physical model of the adsorption and membrane separation processes was developed. The mass balance and energy balance equations for component i in this process model are expressed by Eq. (1) and (2) respectively. The mass transfer to the adsorbent is represented by the Linear Driving Force (LDF) model in Eq. (3). The adsorption process assumes pressure swing adsorption; initially, the inlet gas is pressurized by a compressor and fed into the column for adsorption. After adsorption is completed, the column is depressurized by a vacuum pump for desorption. The solution was obtained by a dynamic simulation in which these steps are repeated alternately.

$$-\varepsilon_b D_{ax,i}\frac{\partial^2 c_i}{\partial z^2} + \frac{\partial(v_g c_i)}{\partial z} + \left(\varepsilon_b + (1-\varepsilon_b)\varepsilon_p\right)\frac{\partial c_i}{\partial t} + \rho_s(1-\varepsilon_b)\frac{\partial q_i}{\partial t} = 0 \tag{1}$$

$$-k_g\varepsilon_b\frac{\partial^2 T_g}{\partial z^2} + C_{vg}v_g\rho_g\frac{\partial T_g}{\partial z} + \varepsilon_b C_{vg}\rho_g\frac{\partial T_g}{\partial t} + P\frac{\partial v_g}{\partial z} + h_f\alpha_p(T_g - T_s)$$
$$+ \frac{4h_w}{D_b}(T_g - T_0) = 0 \tag{2}$$

$$\frac{\partial q_i}{\partial t} = k_{LDF,i}(q_i^* - q_i) \tag{3}$$

The membrane separation is based on the cross-plug flow model expressed in Eq. (4), and the pressure difference caused by pressurizing the inlet gas and depressurizing the drop side was used as the driving force for the separation in the simulation.

$$\frac{dF_i}{dA} = -Q_i(p_h x - p_l y) \tag{4}$$

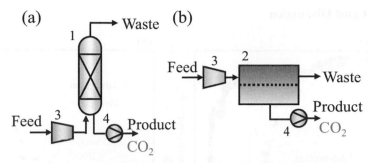

1 : Adsorption bed, 2 : Membrane module
3 : Compressor, 4 : Vacuum pump

Figure 1 Schematic diagram of (a) adsorption and (b) membrane process

The material of the two processes was fixed, assuming Zeolite13x (N. Jiang et al., 2020) as adsorbent and polymer membrane (B. Ghalei et al., 2017) as membrane material, 1000 GPU (=3.3×10-10 [mol/(m^2 s Pa)]), CO$_2$/N$_2$ selectivity 50. The sampling ranges for the training data are shown in **Table 1**. The neural network was trained with about 1000 data samples. The data were collected randomly in these ranges. As input parameters, the inlet flow information, adsorption time for adsorption, and membrane area for membrane were trained. The range of adsorption time was 250 s - 2500 s, and the range of membrane area was 0.5 - 30 m^2. Inlet gas is assumed to be two components, CO$_2$ and N$_2$. The information of outlet flow and energy consumption were trained as output parameters. The neural network was tuned for the number of hidden layers, the number of neurons in each layer, and the number of epochs, and the most accurate model was used. These learned statistical models were used as alternative models for optimization. A genetic algorithm, NSGA-II, was used for multi-objective optimization. Surrogate models were combined when constructing a multi-stage process, and multi-objective optimization was performed for the entire system.

Table 1 Operating range

Parameter		unit
Inlet CO$_2$ mole fraction	15 - 99	mol%
Inlet gas flowrate	0.01 – 0.20	mol/s
Compressor	200 - 500	kPa
Vacuum pump	5 - 20	kPa

3. Result and Discussion

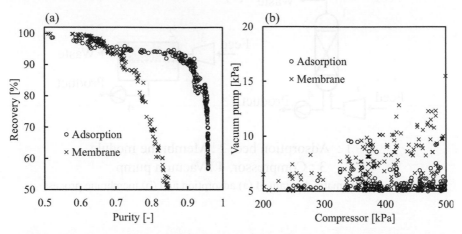

Figure 2 Pareto solutions of multi-objective optimization in single-stage process

To optimize product composition and recovery in a single-stage process, a surrogate model trained on adsorption and membrane models was used. **Figure 2**(a) shows the Pareto solutions obtained from the multi-objective optimization. At 90% recovery, the composition is 0.92 for adsorption and 0.74 for membrane, which does not meet CCS requirements (product composition 0.95 [-], recovery 90%). In addition, it was shown that the membrane gives a higher recovery if the product composition is less than 0.70. Figure 2(b) shows the pressure conditions for the compressor and vacuum pump in the Pareto solutions. It is clearly seen that both adsorption and membrane separation processes operate with a high compression ratio for the compressor and vacuum pressure. Adsorption requires a significantly low vacuum pump operating pressure, around 5 kPa, despite its separation performance close to CCS requirements. This indicates that while adsorption provides superior separation performance compared to membrane separation, it also requires a higher operating pressure. The coefficients of determination between the physical and statistical models at this point showed high accuracy, with product composition at 0.993 [-] and 0.998 [-] for adsorption and membrane, respectively, and recovery at 0.978 [-] and 0.997 [-].

Next, **Figure 3(a)** shows an example of a multi-stage process constructed in this study, the membrane adsorption process. This process was combined with a surrogate model to create a system. The optimization was performed. **Figure 3(b)** shows the Pareto solutions obtained through multi-objective optimization of the membrane adsorption process. This result maximizes product composition and recovery and minimizes energy consumption. The CCS requirement (product composition 0.95 [-], recovery 90%) was achieved by the multi-stage process; the minimum energy to achieve the CCS requirement was 1.15 MJ/kg-CO_2. The operating pressures of the first stage compressor and vacuum pump were 221 kPa and 9.8 kPa, respectively, and those of the second stage compressor and vacuum pump were 336 kPa and 19.2 kPa, respectively. Thus, the use of a multi-stage process avoids higher compressor and vacuum pump operating pressures compared to a single-stage process. Other similar two-stage processes considered in this study were the membrane-membrane, adsorption-adsorption, and adsorption-membrane processes, and the minimum energy consumption after achieving the respective CCS requirements was 1.26, 1.37, and 1.41 MJ/kg-CO_2 for the membrane-adsorption process, respectively.

In two-stage processes such as those considered in this study, the flow entering the compressor or vacuum pump in the first stage is lower than that in the second stage. Therefore, increasing the CO_2 concentration in the membrane process, which has a low energy consumption, followed by separation by the adsorption process with high separation efficiency, is considered optimal, as concluded from the optimization of the single-stage process.

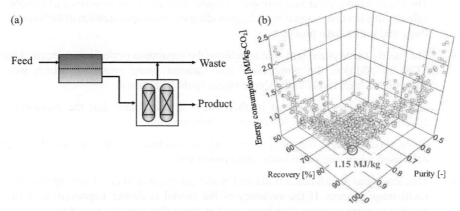

Figure 3 (a)Schematic diagram of membrane-adsorption process and (b) Result of multi-objective optimization of membrane-adsorption process

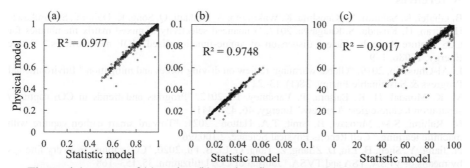

Figure 4 Accuracy of (a) purity, (b) product flow rate, and (c) recovery in Pareto solution of membrane-adsorption process

Figure 4 shows the model accuracy of the two-stage process in the Pareto solutions of the membrane adsorption process, which was a best case in this study. The coefficients of determination for product composition and product flow show a high accuracy of 0.97, while the accuracy for recovery is relatively low at 0.90. This is probably because recovery was not trained directly, but calculated from the trained product composition and product flow. Therefore, due to some error in the recovery, it is expected that there may be some error in the optimization results of this study. However, if models with improved accuracy can be developed, they could be applied to more complex explorations, such as the search for complicated multi-step process structures. This could make it a powerful method for process synthesis.

4. Conclusions

In this study, a surrogate model for adsorption and membrane processes was created using machine learning. Using this model, the design of single and multi-stage processes was performed, leading to the following conclusions:

- The use of machine learning surrogate models allowed for the avoidance of complex calculations in conventional models, providing efficient optimization of both single-stage and multi-stage processes.

- The multi-stage system was able to achieve the separation performance required for CCS. In addition, the compressor and vacuum pump pressures, which were operating at high levels in the single stage, could be reduced.

- Among the cases considered in this study, it was found that the membrane-adsorption process had the lowest energy consumption.

- Since recovery was not directly trained, the accuracy of the statistical model decreased when applied to multi-stage processes.

- This study has shown that a statistical model can be used to efficiently synthesize a multi-stage process. If the accuracy of the model is further improved, it can be applied to more complex structures, such as those that consider recycling.

References

B. Ghalei, K. Sakurai, Y. Kinoshita, K. Wakimoto, A. Isfahani, Q. Song, K. Doitomi, S. Furukawa, H. Hirao, H. Kusuda, S. Kitagawa, 2017, "Enhanced selectivity in mixed matrix membranes for CO_2 capture through efficient dispersion of amine-functionalized MOF nanoparticles." Nature Energy, 17086, 1–9

L. Al-Ghussain, 2019, "Global warming: review on driving forces and mitigation." Environmental Progress & Sustainable Energy, 38(1), 13–21

M. K. Mondal, H. K. Balsora, P. Varshney, P., 2012, "Progress and trends in CO_2 capture / separation technologies: A review." Energy, 46, 431–441.

M. Rahimi, S.M. Moosavi, B. Smit, T.A. Hatton., 2021, "Toward smart carbon capture with machine learning." Cell Reports Physical Science, 2,100396

N. Jiang, Y. Shen, B. Liu, D. Zhang, Z. Tang, G. Li, B. Fu, 2020, "CO_2 capture from dry flue gas by means of VPSA, TSA and TVSA." Journal of CO_2 Utilization, 35, 153–168

Y. Fujii, A. Yamamoto, K. Matsuda, 2023, "Synergistic Effects of Adding a Rinse Step and Investigating Adsorbents in PVSA Systems for Enhanced CO2 Separation Performance in Postcombustion" Industrial & Engineering Chemistry Reserch, 62, 19764-19772

Flavio Manenti, Gintaras V. Reklaitis (Eds.), Proceedings of the 34th European Symposium on Computer Aided Process Engineering / 15th International Symposium on Process Systems Engineering (ESCAPE34/PSE24), June 2-6, 2024, Florence, Italy

Advancing Low-Carbon LNG within Qatar: Integrating CCUS and Carbon Tax Strategies

Razan Sawaly, Mohammad Alherbawi, Ahmad Abushaikha, Tareq Al-Ansari[*]

College of Science and Engineering, Hamad Bin Khalifa University, Doha, Qatar
Corresponding email: talansari@hbku.edu.qa

Abstract

As global efforts intensify to reduce CO_2 emissions and combat climate change, this study explores the integration of Carbon Capture, Utilisation, and Storage (CCUS) and Carbon Tax (CT) for producing low-carbon Liquefied Natural Gas (LNG) in Qatar. The research develops a model employing Post Combustion Capture (PCC) to capture and strategically allocate CO_2 from a single major source to various sinks. Utilising a Linear Programming (LP) approach, the model aims to balance profitability with emission reduction, considering factors such as operational costs, carbon tax implications, and the market price of CO_2. The study examines three scenarios: maximising profit, minimising emissions, and a hybrid of both, to assess the impact of different CO_2 allocation strategies on economic and environmental outcomes of the source. Results indicate that while specific CO_2 allocation decisions subtly affect profitability and emissions, the overall economic and environmental performance remains consistent across scenarios. This research highlights the potential of CCUS and CT in enhancing sustainable energy production and advancing a circular economy in carbon management.

Keywords: Liquified Natural Gas, CO_2 allocation, Sustainability, Carbon capture

1. Introduction

In the quest to mitigate climate change, the production of low or zero-carbon products has become increasingly important, especially in industries traditionally associated with high CO_2 emissions. Carbon Capture, Utilisation, and Storage (CCUS) plays a pivotal roles in this transformation (Zhu, 2019). This technology is not just about reducing emissions; but also about reimagining production processes to create environmentally sustainable products. Moreover, the utilisation aspect of CCUS opens up opportunities for converting captured CO_2 into valuable products, thereby adhering to the principles of a circular economy (Zeng et al., 2017). This could include the transformation of CO_2 into building materials, chemicals, or even as a feedstock for synthetic fuels, creating a market for what was once considered waste (Alper & Yuksel Orhan, 2017). Yang et al. (2022) introduces a CO_2 allocation approach in China, aiming to achieve carbon peak and carbon neutrality by integrating carbon sinks with sources. It demonstrates that this method results in a fairer regional distribution, particularly benefiting carbon-rich areas like Southwest China. One of the key areas where this transformation is critical is in the production of energy-intensive commodities. In this evolving landscape, the emphasis on producing low-carbon products extends beyond specific sectors, encompassing a wide range of industries. This shift towards sustainable production methods, driven by CCUS and similar technologies, marks a significant stride in redefining how industries operate. By transforming CO_2 from a byproduct into a resource, these technologies not only reduce emissions but also foster the creation of diverse low-carbon products. Examples include construction materials where CO_2 is used in creating more sustainable concrete, and in the chemical industry for producing greener polymers and plastics. Additionally, the development of synthetic fuels from captured CO_2 illustrates the potential for innovation in energy sources. This holistic approach aligns with the principles of a circular economy, setting a new standard for environmental responsibility across all sectors. Kätelhön et al., (2019) explore the use of carbon capture and utilisation (CCU) in the chemical industry to potentially reduce greenhouse gas emissions by up to 3.5 Gt CO_2-equivalent in 2030, requiring over 18.1 PWh of low-carbon electricity. Biermann et al., (2020)'s study aims to produce low-carbon steel, ethanol, and electricity by investigating carbon allocation methods in facilities that co-process biogenic and fossil feedstocks with carbon capture utilisation and storage technology. It evaluates two allocation schemes to determine how they impact the emission intensities of these products, highlighting the importance of such schemes in enabling the production of low-carbon products in various industries. In this context, the production of low-carbon liquefied natural gas (LNG) emerges as a crucial area of focus. Through the integration of CCUS in the LNG production process, it is possible to capture a substantial portion of the CO_2 emissions generated. This captured CO_2 can then be either sequestered or utilised in other industrial processes, thereby reducing the overall carbon footprint of the LNG produced. Building on the foundational work by the author (Sawaly et al., 2023), which established a comprehensive framework for CO_2 allocation, this paper delves into the production of low-carbon LNG in Qatar, where the focus is narrowed to QatarEnergy LNG, analysing the environmental and economic impacts from the source's perspective. The aim is to balance profit maximisation with CO_2 emission reduction, thereby addressing the critical need for sustainable LNG production practices in the region.

2. Methodology

2.1. Model Description

The proposed model aims to produce low-carbon LNG from one of the leading global LNG Producers in Qatar (QatarEnergy-LNG) through CCU within an industrial park setting. The datasets and foundational assumptions employed in this study are detailed in the previous work (Sawaly et al., 2023). Figure 1 graphically represents the framework of the allocation model, detailing the source, sinks, and product of each sink.

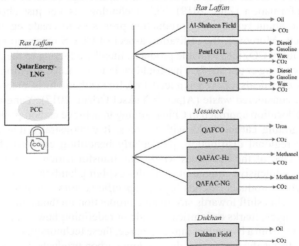

Figure 1: CO_2 Allocation Framework.

2.2. Model Formulation

The model focuses on two main objectives: first, to minimise emissions relative to annual LNG production as per Equation (1), and second, to optimise net profit from carbon sales after accounting for operational costs based on Equation (2). Foundational definitions and relationships are laid out in Equations (3-8). Essential constraints are set to ensure that CO_2 allocation does not exceed the captured amount (Equation 9), individual sink capacities are not surpassed (Equation 10), and the CO_2 allocation remains economically viable and environmentally positive (Equations 11-12).

$$\text{Min} \sum_i^7 \sum_j^5 \frac{IE + DE - Q + CCE_{ij} + x_{ij} \cdot T_{ij}}{P_{LNG}} \tag{1}$$

$$\text{Max} \sum_i^7 \sum_j^5 \frac{CS_{ij} + SCT - CCC_{ij}}{P_{LNG}} \tag{2}$$

Where;
$$CRF = \frac{Rate}{1 - (1 + Rate)^{-life}} \tag{3}$$

$$EAC_{Transport} = (CRF \times Capex) + \left(x_{ij} \times Opex\right) \tag{4}$$

$$\underline{T_{ij}} = x_{ij} \times Q \times D_i \times (2 \times 10^6) \text{ (tonne } CO_2/km) \tag{5}$$

$$CS_{ij} = \sum (x_{ij} \times Q \times R)^{DF} \tag{6}$$

$$C_{transport} = Qx_{ij} \times \left(\frac{D_i}{100}\right) \times 1.5 \tag{7}$$

$$CCE = Q \times CCE_{Rate} \tag{8}$$

Subject to
$$\sum_{i=1}^7 x_{ij} = 1 \tag{9}$$

$$\sum_{i=1}^7 \sum_{j=1}^5 x_{ij} \cdot Q \leq C_{Pij} \tag{10}$$

$$\sum_i^7 \sum_j^5 \frac{CCC_{ij} - CS_{ij} + SCT}{P_{LNG}} > 0 \tag{11}$$

$$\sum_i^7 \sum_j^5 \frac{IE + DE - Q + CCE + x_{ij} \cdot T_{ij}}{P_{LNG}} > 0 \tag{12}$$

Parameters:

- IE: Indirect emissions
- DE: Direct emissions
- Q: Quantity of CO_2 captured
- CCE: Capturing related emissions
- T_{ij}: Transportation emissions to sink
- TS_{IJ}: Emission saving through local transportation of CO_2
- P_{LNG}: Annual LNG production
- CS_{ij}: Carbon Sales ($/year)
- CCC_{ij}: Carbon Capture Cost ($/year)

- SCT: Save Carbon Tax ($/year)
- EAC: Equivalent Annual Cost ($/year)
- CRF: Capital Recovery Factor
- Capex: Capital Expenses (capture and transportation)
- Opex: Operating Expenses (capture and transportation)
- R: Price Rate of CO_2 ($/ton)
- DF: Bulk Discount Factor

3. Results and Discussion

In this case study, the system boundary is defined by the source. The primary aim is to enhance sources' profitability while minimising its environmental footprint, particularly

CO_2 emissions. Emissions and costs taken into account are those related to carbon capture and transportation, along with carbon tax.

3.1. Allocation percentages comparison across the 3 scenarios

a)   Objective 1- Maximise Profit

The results show that from 2016-2020, every sink met their CO_2 needs 100%, except for the Dukhan field (see
Figure 4). The model focuses heavily on transportation costs in its cost structure, aiming to maximise source profit, primarily influenced by these costs. This is seen as sinks closer to the source, like Oryx GTL and Pearl GTL (5 and 3 km away, respectively), are prioritised. Following them, the AlShaheen field, 80 km distant, gets its CO_2 share. Interestingly, the model favors smaller-capacity sinks like QAFAC H_2, QAFAC NG, and QAFCO, about 100 km from the source, over the larger Dukhan field. This choice is driven by Equation 8's discount factor; allocating more to Dukhan would lower CO_2 costs and thus profit. The model first meets the demands of these smaller sinks to maintain a high CO_2 selling price before attending to Dukhan. Furthermore, the consistency of carbon capture costs, which remain unaffected by the CO_2 transportation destination, ensures they do not influence the allocation decisions as the source consistently captures the same amount of CO_2 each year, irrespective of its eventual transportation or allocation destination. Instead, transportation stands out as the primary variable cost.

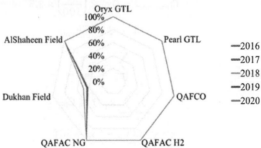

Figure 2: Percent CO_2 allocation from source to sinks (Objective 1).

b)   Objective 2 - Minimise CO_2 emissions

In addressing the objective of minimising the source's emissions, the allocation patterns exhibit distinct characteristics. With QatarEnergy-LNG as the designated system boundary, transportation-related emissions emerge as the primary factor influencing CO_2 allocation. This places the distances between the source and the sinks at the forefront of the allocation strategy, echoing trends seen in the prior case. Figure 3 visually maps out the allocation designed to mitigate sources' emissions. Owing to their closeness to the source, Oryx GTL and Pearl GTL are the initial beneficiaries, each receiving a full allocation of 1,280,000 ton CO_2/year. Following this, the AlShaheen field, 80km from the source, is allocated its CO_2 share. The remaining sinks, all roughly 100km from the source, share what is left of the captured CO_2. More importantly, the allocation to the EOR sinks, which, due to their expansive capacities, secure a larger CO_2 share. A pivotal observation is the potential for numerous allocation solutions. The model's flexibility is evident: as long as the nearby sinks—Oryx, Pearl, and AlShaheen—are the primary recipients, the residual CO_2 can be variably distributed among the other four sinks. Their equal distance from the source means the allocation order does not affect emissions. It is evident that Oryx, Pearl, and AlShaheen consistently receive full capacity allocations. In contrast, the other sinks experience varied allocations year on year, reinforcing the notion of multiple viable solutions, provided the three primary sinks are fully allocated first.

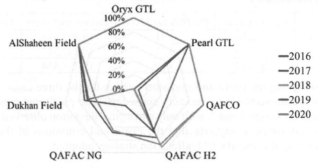

Figure 3: Percent CO_2 allocation from source to sinks (Objective 2).

c) Multi-Objective (Maximise Profit + Minimise Emissions)

To achieve the dual objectives, Matlab was employed to determine the optimal allocation solution, aiming for the highest profit and the lowest emissions. The outcomes of this scenario is depicted in Figure 4; where the amount of CO_2 allocated from the source to each sink is illustrated. The allocation trends across the years 2016-2019 exhibit remarkable consistency for most sinks. Specifically, Oryx, Pearl, QAFCO, QAFAC H_2, and AlShaheen consistently receive around 80% sink capacity satisfaction. This uniformity underscores the model's balanced approach in addressing the CO_2 needs of these sinks over these years. In contrast, Dukhan and QAFAC NG consistently receive a slightly lower satisfaction percentage, hovering around 70%, marking them as the sinks with the lowest percentage satisfaction compared to their counterparts during this period. However, 2020 presents a deviation from this trend. The allocation percentages for QAFAC NG, Pearl, and Oryx surge, with each achieving a capacity satisfaction of over 90%. This spike suggests a shift in the model's priorities or external factors influencing the allocation for that year. Dukhan field, on the other hand, consistently emerges as the sink with the lowest capacity satisfaction across all years, reinforcing its position as a lower priority in the allocation hierarchy. It is worth noting that this scenario is designed to harmonise both objectives: maximising profit and minimising emissions. The allocation percentages across all sinks hover within a relatively close range, indicating the model's endeavour to strike a balance between the two objectives while ensuring equitable CO_2 distribution among the sinks.

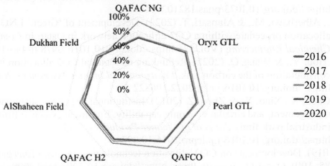

Figure 4: Percent CO_2 allocation from source to sinks.

3.2. Profit and emissions comparison for the 3 Scenarios

Table 1: Comparative analysis of the total profit and total emissions across all cases (for years 2016-2020).

	Profit ($/year)	Profit($/ton LNG)	Emissions(t/year)	Emissions(t CO_2/t LNG)
Case 1	2.55×10^8	3.31	19,292,315	0.251
Case 2	2.55×10^8	3.31	19,292,315	0.251
Case 3	2.53×10^8	3.29	19,292,434	0.251

Upon analysing the profit and emission metrics for the three cases, it is evident that the sources' performance is remarkably consistent. The profit and emission values remain consistent for Cases 1 and 2, with only a marginal deviation observed in Case 3 (See Table 1). This consistency suggests that the profit and emissions of the source are largely unaffected by the specific CO_2 allocation strategy adopted.

4. Conclusion

In the ongoing global effort to reduce CO_2 emissions and combat climate change, the integration of Carbon Capture, Utilisation, and Storage (CCUS) and Carbon Tax (CT) holds great promise. This study explored the application of these technologies to produce low-carbon Liquefied Natural Gas (LNG) in Qatar, with a focus on QatarEnergy-LNG as a significant CO_2 source. The research developed a model employing Post Combustion Capture (PCC) to allocate CO_2 strategically to various industries, using Linear Programming (LP) to balance profitability with emission reduction. The study assessed three scenarios, aiming to maximise profit, minimise emissions, and a hybrid approach. The results indicate that while CO_2 allocation strategies subtly affect profitability and emissions, the overall economic and environmental performance remains consistent across scenarios.

References

Alper, E., & Yuksel Orhan, O. (2017). CO2 utilization: Developments in conversion processes. *Petroleum*, *3*(1), 109–126. https://doi.org/10.1016/J.PETLM.2016.11.003

Biermann, M., Montañés, R. M., Normann, F., & Johnsson, F. (2020). Carbon Allocation in Multi-Product Steel Mills That Co-process Biogenic and Fossil Feedstocks and Adopt Carbon Capture Utilization and Storage Technologies. *Frontiers in Chemical Engineering*, *2*. https://doi.org/10.3389/fceng.2020.596279

Kätelhön, A., Meys, R., Deutz, S., Suh, S., & Bardow, A. (2019). Climate change mitigation potential of carbon capture and utilization in the chemical industry. *Proceedings of the National Academy of Sciences of the United States of America*, *166*(23). https://doi.org/10.1073/pnas.1821029116

Sawaly, R., Alherbawi, M., & Alansari, T. (2023). Development of 'Green' LNG through a CO2 allocation procedure within a CO2 utilisation network in Qatar. In *Computer Aided Chemical Engineering* (Vol. 52). https://doi.org/10.1016/B978-0-443-15274-0.50347-4

Yang, M., Hou, Y., & Wang, Q. (2022). Rethinking on regional CO2 allocation in China: A consideration of the carbon sink. *Environmental Impact Assessment Review*, *96*. https://doi.org/10.1016/j.eiar.2022.106822

Zeng, H., Chen, X., Xiao, X., & Zhou, Z. (2017). Institutional pressures, sustainable supply chain management, and circular economy capability: Empirical evidence from Chinese eco-industrial park firms. *Journal of Cleaner Production*, *155*. https://doi.org/10.1016/j.jclepro.2016.10.093

Zhu, Q. (2019). Developments on CO2-utilization technologies. In *Clean Energy* (Vol. 3, Issue 2, pp. 85–100). Oxford University Press. https://doi.org/10.1093/ce/zkz008

Flavio Manenti, Gintaras V. Reklaitis (Eds.), Proceedings of the 34th European Symposium on Computer Aided Process Engineering / 15th International Symposium on Process Systems Engineering (ESCAPE34/PSE24), June 2-6, 2024, Florence, Italy

Process efficiency enhancement of integrated hydrogen enrichment and liquefaction

Muhammad Islam[a], Ahmad Naquash[a], Ali Rehman[b], Moonyong Lee[a*]

[a] *School of Chemical Engineering, Yeungnam University, Gyeongsan 38541, Republic of Korea*
[b] *Ningbo Institute of Dalian University of Technology, Ningbo 315200, Zhejiang, PR China*
mynlee@yu.ac.kr

Abstract

Hydrogen has gained a remarkable position in the global market as a complete package for a cleaner fuel with carbon neutrality. Increasing hydrogen demand as a clean fuel sheds light on hydrogen transportation and large-scale storage. Liquid hydrogen is the viable way to go with the purest form of hydrogen. The purity of hydrogen is the key to avoiding impairments in hydrogen liquefaction. Primarily, hydrogen is produced from fossil-originated fuel, which requires purification to obtain pure hydrogen. Pressure swing adsorption is the commonly used technique for this purpose, which offers high purity (~99.99 %) and low recovery (~95 %). The emerging technique of cryogenic separation of hydrogen still has limitations: low recovery and purity on top of high energy consumption. However, cryogenic separation can have a dual purpose: the pre-cooling of hydrogen as well as the solidification of CO_2. The present work explores this exploitation towards lower energy requirements and process intensification via simulation in Aspen HYSYS® V14. The de-sublimation process conditions were based on the pure CO_2 and H_2 /CO_2 mixture phase behaviour. De-sublimation was operated in an equipped refrigeration chamber designed to accommodate the refrigeration cycle. After hydrogen enrichment and pre-cooling with 99.99 % purity and 99.99 % recovery, hydrogen was liquefied utilizing the same refrigeration cycles, making the process simpler. It subsequently provides energy benefits. There is a 7.5 % reduction in overall specific energy consumption (8.90 kWh/kg) from the base case (9.62 kWh/kg). The current study will be a building block in developing the hydrogen supply chain.

Keywords: Hydrogen liquefaction, CO_2 solidification, Integrated process, Process simulation, Cryogenic separation

1. Introduction

Hydrogen is acclaimed as a potential game-changer due to its clean energy characteristics, but its long-distance transportation presents a substantial challenge owing to its low energy density (0.01 MJ/L) in gas-phase, which can be enriched to 8.5 MJ/L by liquefying H_2, as mentioned by Valenti, 2016. As the demand for hydrogen continues to rise, driven by its clean attributes, there is a foreseeable potential for hydrogen to replace conventional fuels soon (IEA, 2022). To facilitate this transition, the crucial aspects of intercontinental or international transportation of hydrogen become vital components of the global energy mix. Long-term hydrogen storage is equally as important as transportation. Liquefying hydrogen at extremely low temperatures (−253 °C) increases its density, enabling the storage of larger quantities in a given space. This is vital for applications such as long-distance transportation, where the volume efficiency of hydrogen becomes paramount.

Owing to such issues in hand, development in the hydrogen liquefaction process is inevitable. In the commercial liquefaction of hydrogen quoted by Naquash et al., 2022, a standardized three-step process is employed: down to −193 °C at pre-cooling, down to −243 °C at cooling, and down to −253 °C at liquefaction step. In the initial pre-cooling phase, liquid nitrogen (N_2) serves as the refrigerant. At the same time, an integrated sequence of Joule Thomson valves (or expanders) and hydrogen is utilized in the adjoining cooling and liquefaction sections. Recent advancements have introduced mixed refrigerants (MR) as a substitute for pure liquid N_2. A study conducted by Qyyum et al., 2021 has highlighted the existence of numerous investigations proposing varied compositions incorporating low-boiling hydrocarbons and N_2, in MR. Neon (Ne) and Helium (He) are also considered along with H_2 in the cooling and liquefaction sections. These efforts aim to minimize the Specific Energy Consumption (SEC), striving to align it with the theoretical ideal case as closely as possible (Bi et al., 2022). Industrial processes typically exhibit SEC falling within the 12-15 kWh/kg$_{LH2}$ range, having efficiency levels as minimal as 20-30 % (Krasae-in et al., 2010).

Moreover, the conditions during hydrogen liquefaction (around −253 °C) lead to the solidification of impurities, necessitating effective removal (Voldsund et al., 2016). The purification process must be highly efficient to prevent impurities from compromising the quality of the liquid hydrogen. Among the well-established technologies ensuring the production of high-purity hydrogen are pressure swing adsorption (PSA), cryogenic purification, and selective permeation gas membrane separation, which are well known. PSA stands out for its capacity to achieve remarkable purity, potentially attaining up to 99.999 % pure H_2. However, hydrogen recovery is typically 80 - 95%, depending on the specific process scheme and the adsorbent's capacity (Bernardo et al., 2020).

On the other hand, the cryogenic purification process offers an alternative route, providing up to 98 % pure H_2 with an impressive recovery rate of approximately 95% (Aasadnia et al., 2021). A potential study by (Naquash et al., 2022a) has proposed cryogenic separation of impurities (as solidified CO_2) and hydrogen liquefaction integrated with mixed refrigerant cycles. The study presented remarkable results with 99.9999 % purity and ~99 % recovery. The current research builds upon and enhances this work by (Naquash et al., 2022a), utilizing Aspen HYSYS® V14 to simulate an intensified process configuration. A comprehensive energy and exergy analysis was conducted to assess the enhancement in the proposed process.

2. Hydrogen Enrichment and Liquefaction

2.1. Process description

The proposed configuration integrates a cryogenic process employing two mixed refrigeration (MR) cycles for hydrogen enrichment and liquefaction (Figure 1). This cryogenic technique serves the dual purpose of pre-cooling hydrogen and solidifying carbon dioxide in the hydrogen enrichment process. Unlike the previous approach (Naquash et al., 2022a), both MR cycles are integrated at the first multi-stream heat exchanger (HX1). Following pre-cooling, a specially designed chamber facilitates de-sublimation, extracting hydrogen in vapor form from the top and solidifying CO_2 at the bottom. CO_2 de-sublimation conditions were set at $-61°C$ and 5 bar based on its phase diagram (Naquash et al., 2022a).

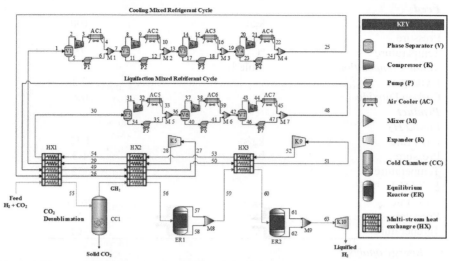

Figure 1. Proposed integrated process schematic diagram

Subsequently, the process achieves hydrogen enrichment and pre-cooling with high purity (99.9999%) and recovery (99.9999%). Pure hydrogen exhibits a temperature-dependent ortho/para composition (Qyyum et al., 2021), managed by two equilibrium reactors (ER1 and ER2)(Harkness and Deming, 1932). The liquefaction process involves cooling and liquefaction through HX2 and HX3, incorporating mixed refrigeration cycles containing various components. The resulting hydrogen stream undergoes further cooling and conversion to 99.7% para-H_2 in ER2.

The cooling and liquefaction of H_2 occurred by passing through HX2 and HX3. CMR cycle contain C1, C2, C3, n-C4, n-C5, N_2 and LMR cycle contain H_2 and He. Table 1 outlines the key design parameters that govern the proposed process, providing a comprehensive overview of the specified conditions and criteria employed in the simulation.

At the outlets of HX2 and ER1, the H_2 stream (59), at a temperature of -160 °C and comprising 34.7 % para-H_2, undergoes additional cooling in HX3. Upon leaving HX3, the temperature of H_2 is further decreased to the liquefaction temperature of -252 °C. Following this, the H_2 converts to 99.7% para-H_2 as it passes through ER2 and attains a pressure of 1.3 bar after passing through K10 (expander). The thermodynamic properties of the refrigeration cycles were calculated using the PR equation of state (Peng and Robinson, 1929). For the H_2 streams, the Modified Benedict-Webb-Rubin (MBWR) equation of state (Eckroll et al., 2017) was explicitly chosen to accurately utilize the relevant thermodynamic behaviours.

Table 1. Key parameters of the proposed H$_2$ separation and liquefaction process

Design Parameters	Unit	Values (Xu et al., 2012)
Feed		
Flow rate	kg/s	100
Pressure	bar	5.0
Temperature	°C	35
Molar Composition	mole %	
Hydrogen		0.2
Carbon Dioxide		0.8
Product		
Liquid H$_2$		
Flow rate	kg/s	1.13
Pressure	bar	1.3
Temperature	°C	−252.2
Solid CO$_2$		
Flow rate	kg/s	98.87
Pressure	bar	5
Temperature	°C	−61

2.2. Energy analysis

The energy analysis in this study is performed with a focus on SEC. Different design variables, as specified in Table 2, including refrigerant mass flow rates, suction pressure, and discharge pressure of the refrigeration cycles, are methodically modified to observe and analyse their influence on SEC.

2.3. Exergy analysis

Exergy analysis is a valuable technique for assessing process inefficiencies by considering the principles of the second law of thermodynamics. Physical and chemical exertion are the two main categories of energy, and they are computed as:

$$ex_{ph,i} = (h_i - h_o) - T_o(s_i - s_o) \tag{1}$$

$$ex_{ch,i} = \sum x_i e_i^{CH} + RT_o \sum x_i \ln x_i \tag{2}$$

For physical and chemical exergy, $ex_{ph,i}$ and $ex_{ch,i}$ are used as symbols, the mole fraction of the ith component in a stream is represented by x_i and e_i^{CH} is referred to as standard chemical exergy of ith component obtained from (Szargut, 1989). While chemical exergy values are determined using Eq. 2, physical exergy values are derived from Aspen HYSYS® stream properties. The exergy destruction of every piece of equipment is calculated to gain complete insight into the exergy analysis. The reason behind the deviation of equipment performance from an ideal scenario can be attributed to energy destruction. The process performance decreases as the energy destruction increases. The formulas used to determine the equipment's exergy destruction are provided in (Naquash et al., 2022b).

3. Results and discussion

3.1. Energy analysis

The study analysed design variables concerning SEC, including suction/discharge pressures and refrigerant flow rates, as shown in Table 2. The proposed case demonstrates approximately 7.5% higher energy efficiency than the base case. The study streamlined operations by reducing two previously employed mixed refrigerant (MR) cycles - one for de-sublimation and the other for pre-cooling enriched hydrogen. This simplification noticeably benefits energy consumption. The load on the cold chamber increased to accommodate CO_2 de-sublimation, emerging as the most energy-consuming unit at 14552.14 kW. Additionally, omitting previously assisting cycles led to increased refrigerant load on the cooling MR cycle. However, discharge pressure adjustments were made to align with the energy consumption requirements.

Table 2. Process design variables and specific energy consumption (SEC)

Pre-cooling MR cycle	Units	Base case (Naquash et al., 2022a)	Proposed study
Refrigerants flowrate	kg/s	20.26	-
Suction / Discharge pressure	bar	1.10 / 36.0	-
Cooling MR cycle			
Refrigerants flowrate	kg/s	18.85	44.02
Suction / Discharge pressure	bar	1.70 /59.0	2.04 /32.48
Liquefication MR cycle			
Refrigerants flowrate	kg/s	6.0	10.78
Suction / Discharge pressure	bar	1.32 /65.0	1.12 /18.13
SEC	**(kWh/kg)**	**9.62**	**8.90**

Similar adjustments were applied in the Liquefaction MR cycle, contributing to the overall enhanced energy efficiency observed in the proposed case compared to the base case.

3.2. Exergy analysis

Exergy destruction is provided in Table 3, indicating the equipment-wise exergy destruction analysis. Comparing the current study with the base case is not feasible due to the change in process configurations. In the proposed study, multi-stream exchangers emerge as focal points, contributing approximately 26% to total exergy destruction, emphasizing their potential for efficiency refinement. Air coolers account for around 20%, functioning as heat sinks with implications for energy recovery.

Table 3. Exergy destruction of each unit operation for the proposed study

Equipment	Exergy destruction (kW)
Compressors (K)	6780.69
Pumps (P)	30.53
Coolers (AC)	7803.87
Separators (V)	−6.74
Equilibrium reactors (ER)	6543.91
Expanders (K)	6965.30
Cold Chamber (CC)	−2405.42
Multi-stream exchangers (HX)	10019.40
Total	**35731.54**

Compressors, expanders, and equilibrium reactors collectively share the third-highest exergy destruction, each contributing around 17 to 18%. Intriguingly, phase separators and cold chamber exhibit negative exergy destruction values, warranting scrutiny and potential exploration into advanced exergy analyses to unveil distinct system intricacies.

4. Conclusions

The study emphasizes the significance of hydrogen liquefaction, detailing challenges in impurity removal and proposing an innovative cryogenic method. By integrating a de-sublimation-based cryogenic process, the research achieves high-purity hydrogen with energy and exergy destruction considerations. The proposed method showcases promising energy efficiency, paving the way for advancements in hydrogen production processes. The comprehensive analysis contributes insights into optimizing hydrogen liquefaction for a sustainable future with lower energy impact.

References

Aasadnia, M., Mehrpooya, M., Ghorbani, B., 2021. A novel integrated structure for hydrogen purification using the cryogenic method. J Clean Prod 278, 123872.

Bernardo, G., Araújo, T., da Silva Lopes, T., Sousa, J., Mendes, A., 2020. Recent advances in membrane technologies for hydrogen purification. Int J Hydrogen Energy 45, 7313–7338.

Bi, Y., Yin, L., He, T., Ju, Y., 2022. Optimization and analysis of a novel hydrogen liquefaction process for circulating hydrogen refrigeration. Int J Hydrogen Energy 47, 348–364.

Eckroll, J., Berstad, D., Wilhelmsen, Ø., Ept, S./, 2017. Concepts for Large Scale Hydrogen Liquefaction Plants.

Harkness, R.W., Deming, W.E., 1932. The Equilibrium of Para and Ortho Hydrogen. J Am Chem Soc 54, 2850–2852.

IEA, 2022. Global Hydrogen Review. Global Hydrogen Review 2022.

Krasae-in, S., Stang, J.H., Neksa, P., 2010. Development of large-scale hydrogen liquefaction processes from 1898 to 2009. Int J Hydrogen Energy 35, 4524–4533.

Naquash, A., Haider, J., Qyyum, M.A., Islam, M., Min, S., Lee, S., Lim, H., Lee, M., 2022a. Hydrogen enrichment by CO2 anti-sublimation integrated with triple mixed refrigerant-based liquid hydrogen production process. J Clean Prod 341, 130745.

Naquash, A., Qyyum, M.A., Islam, M., Sial, N.R., Min, S., Lee, S., Lee, M., 2022b. Performance enhancement of hydrogen liquefaction process via absorption refrigeration and organic Rankine cycle-assisted liquid air energy system. Energy Convers Manag 254, 115200.

Naquash, A., Riaz, A., Lee, H., Qyyum, M.A., Lee, S., Lam, S.S., Lee, M., 2022c. Hydrofluoroolefin-based mixed refrigerant for enhanced performance of hydrogen liquefaction process. Int J Hydrogen Energy 47, 41648–41662.

Peng, D., Robinson, D.B., 1929. A New Two-Constant Equation of State, Int. J. Heat Mass Transfer.

Qyyum, M.A., Riaz, A., Naquash, A., Haider, J., Qadeer, K., Nawaz, A., Lee, H., Lee, M., 2021. 100% saturated liquid hydrogen production: Mixed-refrigerant cascaded process with two-stage ortho-to-para hydrogen conversion. Energy Convers Manag 246, 114659.

Szargut, J., 1989. Chemical exergies of the elements. Appl Energy 32, 269–286.

Valenti, G., 2016. Hydrogen liquefaction and liquid hydrogen storage, Compendium of Hydrogen Energy. Elsevier Ltd.

Voldsund, M., Jordal, K., Anantharaman, R., 2016. Hydrogen production with CO2 capture. Int J Hydrogen Energy 41, 4969–4992.

Xu, G., Li, L., Yang, Y., Tian, L., Liu, T., Zhang, K., 2012. A novel CO2 cryogenic liquefaction and separation system. Energy 42, 522–529.

Flavio Manenti, Gintaras V. Reklaitis (Eds.), Proceedings of the 34th European Symposium on Computer Aided Process Engineering / 15th International Symposium on Process Systems Engineering (ESCAPE34/PSE24), June 2-6, 2024, Florence, Italy

Physics-informed neural networks and time-series transformer for modeling of chemical reactors

Giacomo Lastrucci[a], Maximilian F. Theisen[a], and Artur M. Schweidtmann[a,*]

[a] *Process Intelligence Research Group, Department of Chemical Engineering, Delft University of Technology, Van der Maasweg 9, Delft 2629 HZ, The Netherlands*
Corresponding author. Email: a.schweidtmann@tudelft.nl

Abstract

Multiscale modeling of catalytical chemical reactors typically results in solving a system of partial differential equations (PDEs) or ordinary differential equations (ODEs). Despite significant progress, the numerical solution of such PDE or ODE systems is still a computational bottleneck. In the past, deep learning techniques have gained attention for developing surrogate models in chemical engineering. Also, hybrid models and physics-informed neural networks (PINNs) have been developed to integrate physical knowledge and data-driven approaches. However, it is often unclear how such modeling approaches compare for specific case studies. In this study, we investigate and compare state-of-the-art surrogate and hybrid models for the spatial evolution of the state variables in a packet-bed reactor for methanol production. Firstly, we develop a tailored hybrid model based on PINNs, thereby seamlessly integrating physical knowledge and data. Secondly, we investigate a recently-developed time-series transformer model to learn the spatial evolution of the state variables. As a benchmark model, we train a traditional multilayer perceptron (MLP) and compare the models to a standard numerical integration technique. We achieve orders of magnitude in speedup using MLPs and PINNs when compared to classical ODE solvers, while maintaining high levels of accuracy in modeling the underlying system.

Keywords: Reactor modeling, hybrid modeling, physics-informed machine learning, time-series transformer

1. Introduction

Multiscale modeling of catalytical reactors is a well-known discipline and commonly results in partial differential equation (PDE) or ordinary differential equation (ODE) systems describing the spatial and temporal evolution of the state variables involved (e.g., temperature, pressure, and composition). Despite significant progress, the numerical solution of such PDE or ODE systems is still a computational bottleneck when integrated into large, plant-wide process simulation and optimization studies.

The selection of suitable surrogate modeling approaches for chemical reactor systems is complex and remains to be decided circumstantially. Deep learning techniques have recently gained attention in their function as surrogate models in chemical engineering. The most popular deep learning technique, multilayer perceptrons (MLPs), consists of layers of nonlinear transformations that act as universal approximators. MLPs have been successfully applied to model many reactor systems. As an alternative, foundation models have recently yielded promising results in fields such as computer vision or natural language processing (Kolides, et al., 2023). In the context of chemical engineering, time-series transformers (TSTs) specifically allow for modeling of complex temporal dynamics and relationships within sequential data, enhancing the accuracy and efficiency of predictions. They have been applied to crystallization systems before (Sitapure and Kwon, 2023). However, the black-box nature of such deep learning models poses issues when applied outside training boundaries or when insufficient data is available (Schweidtmann, et al., 2021). Moreover, despite sufficient overall accuracy, physical consistency may not be ensured. Considering those challenges, hybrid models (von Stosch, et al., 2014) and physics-informed neural networks (PINNs) deep learning architectures (Raissi, et al., 2019) have been proposed to integrate physical knowledge and data-driven approaches. While many promising data-driven model architectures exist, the comparison and selection of suitable hybrid model approaches for chemical reactor surrogation is complex and remains an open research question.

In this study, we investigate and compare three state-of-the-art surrogate and hybrid models for describing the spatial evolution of the state variables in a packet-bed reactor for methanol production (Vanden Bussche and Froment, 1996). (1) We develop and train a tailored hybrid model based on PINNs, seamlessly integrating physical knowledge and data. We employ PINNs, commonly used in scientific machine learning, for solving Partial Differential Equations (PDEs) under a specific initial and boundary condition. We expand the PINN framework to enable predictions for a broad range of initial conditions. (2) We deploy a TST to model the sequential state evolution along the reactor length. (3) As a baseline surrogate model, we train a traditional multilayer perceptron. Finally, we systematically compare the developed surrogate and hybrid models with standard ODE solvers in terms of accuracy and runtime for potential application in complex plant-wide simulation and optimization.

2. Methodology

2.1. Physics-informed neural networks

Physics-informed neural networks are deep learning models aimed to bridge the gap between data-driven machine learning models and rigorous scientific computing (Lawal, et al., 2022). PINNs are neural networks trained in a supervised manner, while also minimizing errors on a given equation, e.g., physical law. They allow to approximate the solution of ODE and PDE systems, without performing numerical integration, thus potentially speeding up the computation.

We extend the original PINNs framework (Raissi et al., 2019) by considering the solution of an ODE system across a variety of initial conditions, specifically targeting relevant operating inlet variables. The original PINN formulation was developed to address forward PDE problems provided with fixed initial and boundary conditions. Our approach involves leveraging existing data to gain insights, while ensuring physical accuracy and interpretability through the enforcement of fundamental physical laws, such as material, energy, and momentum balances. Given an autonomous initial value problem (IVP):

$$\frac{ds}{dz} = f(s), \quad s(z = 0) = s_0 \tag{1}$$

Based on (Eq. 1), a *residual* function $r(z, s) := \frac{ds}{dz} - f(s)$, is defined. An MLP can be trained to predict the solution s while also considering physical laws by optimizing the following multi-task loss function $L = L_{SD} + \sum_k L_{R_k}$. The first term ($L_{SD}$), known as - *sensors data loss*, represents the data-driven contribution, computed as the Mean Squared Error (MSE) between the available ground-truth in the domain and the model prediction. The second term ($\sum_k L_{R_k}$), known as *residual loss*, measures the physical discrepancy represented by the difference (residual) between the right-hand side f of the ODE system, and the spatial derivative of the state variables, computed through automatic differentiation:

$$L_{R_k} = \frac{1}{N_C} \sum_{i=1}^{N_C} r_{ik}^2 = \frac{1}{N_C} \sum_{i=1}^{N_C} \left(\frac{d\hat{s}_k}{dz} \Big|_{z_i} - \hat{f}_k(z_i, s_i^0, \theta) \right)^2 \tag{2}$$

where N_C is the number of *collocation points*, which can be arbitrary distributed in the domain of interest, and k is the kth equation in the IVP (Eq. 1). Note that the residual loss is entirely dependent on model prediction, parameters, and initial conditions, ensuring that spatial derivative and predictions comply with given differential equations, i.e., the physical laws. Beyond physical explainability, PINNs potentially facilitate extrapolation and generalization outside training boundaries, which can be achieved by placing collocation points where training data are unavailable.

2.2. Tailored training procedure for physics-informed neural networks
The network parameters are optimized with respect to the multitask loss through a gradient descent algorithm, but the PINN training is often challenging. As observed by S. Wang et al. (2020) the multiscale nature of both variables involved and loss terms can lead to preferential optimization, favoring the minimization of certain terms at expense of others. Also, it has been demonstrated that imbalanced contributions can lead to a "*stiff*" loss surface characterized by large eigenvalues of the hessian, ultimately resulting in gradient descent failure and training divergence. This has also been observed in our study.

To improve the PINN training, we implement a series of mitigation actions following the recommendations by Wang et al. (2023). Specifically, *(i)* we perform system nondimensionalization and *(ii)* we implement an adaptive loss balancing algorithm. The scaling is a best-practice for solving stability issues in gradient-based optimizers. Balancing the loss terms is crucial to avoid task dominance and "stiff" loss surface and several solutions have been proposed in the literature (McClenny and Brada-Neto, 2022). We develop a simple adaptive loss balancing routine based on losses magnitude, by defining the following weights:

$$\lambda_{SD}^* = \frac{L_{SD}^t + \sum_k L_{R_k}^t}{L_{SD}} \tag{3}$$

$$\lambda_{R_k}^* = \frac{L_{SD}^t + \sum_k L_{R_k}^t}{L_{R_k}^t} \qquad k = 1, \dots, N_{eq} \tag{4}$$

We update the loss weights every j iterations, through a moving average $\lambda^{t+1} = \alpha \lambda^t + (1 - \alpha)\lambda^*$, where, j and α are tunable hyperparameters. We initialize the weights to be equal to 1 in the first iteration. Finally, we define the weighted loss function as $L = \lambda_{SD} L_{SD} + \lambda_R^T L_R$.

To comply with the physical system (i.e., all the variables must be greater than zero), we add the term $\|\max(0, -s)\|_1$ to the physical loss function to penalize negative predictions. Moreover, the residual loss is excluded in case output variables are below zero to avoid numerical overflow due to physical inconsistency.

For the PINNs model, we uniformly distributed the collocation points along the reactor length to enforce physical consistency. In order to use mini-batch optimization, the number of collocation points should coincide with the batch size, which is set to 200. We trained the three model on the same generated *sensor* dataset (Sect. 3) using Adam optimizer and a learning rate of 10^{-5}.

2.3. Time-series transformers
Time-series transformers, a subset of transformers, have recently been developed to handle multivariate problems along a temporal dimension (Wen, et al., 2022).

Transformers are neural networks designed for autoregressive sequence prediction based on inputs and past outputs. They operate by embedding and processing initial input through an encoder. A decoder then uses this encoded information to predict outcomes one time step at a time. In TSTs, each prediction step corresponds to a system state, represented by a single vector for each timestep. We have adapted these transformers for spatial sequence modeling, where each step corresponds to a discretized spatial unit instead.

The foundation of transformer architectures is attention (Vaswani, et al., 2017). Attention measures the relatedness of tokens to each other in a computationally efficient way by calculating a dot product based similarity of the mapped inputs. For a detailed explanation of attention and its application to TSTs, we refer to Wen, et al. (2022).

3. Case study

3.1. Case study description

We investigate and compare the methodologies by modeling a tubular packet-bed reactor for methanol synthesis. For simplicity, we assume pseudo-homogeneous conditions and plug flow hydrodynamics with negligible axial dispersion. The model consists of an ODE system representing the material (Eq. 5), energy (Eq. 6) and momentum balance (Eq. 7) along the reactor axial coordinate.

$$\frac{d\dot{n}_i}{dz} = A\rho_C(1-\epsilon)\sum_j^{NR} \nu_{ij}r_j \qquad i = CO, CO_2, CH_3OH, H_2, H_2O, CH_4, N_2 \tag{5}$$

$$\frac{dT}{dz} = \frac{A}{\dot{m}_{tot}\tilde{c}_{P,mix}}\left(US_V(T_C - T) - \sum_j^{NR} \Delta H_{R_j}^0 r_j \rho_C(1-\epsilon)\right) \tag{6}$$

$$\frac{dP}{dz} = -\left(150\frac{(1-\epsilon)^2}{d_{eq}^2\epsilon^3}\mu u_s + 1.75\frac{1-\epsilon}{d_{eq}\epsilon^3}\rho u_s^2\right) \tag{7}$$

In Eq.5, $\dot{n}$ is the molar flowrate, A is the cross sectional area of the tube, ρ_C is the catalyst density, ϵ is the bed void fraction, $\nu_{ij}r_j$ indicate the product between the reaction rate of the reaction j and the correspondent stoichiometric coefficient of species i. In the energy balance (Eq.6), $\dot{m}_{tot}$ is the constant total mass flowrate, S_V is the exchange area *per* unit of volume and T_C is the temperature of the coolant medium.

The reactive system comprises 7 species and 2 reactions, kinetically modelled by Vanden Bussche and Froment (1996). The reactor is cooled through boiling water at 38 bar and Antoine equation is considered. Specific heat capacity ($\tilde{c}_{P,mix}$), reaction enthalpy (ΔH_R^0), overall heat transfer coefficient (U), void fraction (ϵ), viscosity (μ) are assumed to be constant along the reactor. The pressure drop follows the Ergun relation (Eq.7), where d_{eq} is the equivalent diameter of the cylindrical catalyst pellet, ρ is the gas density and u_s is the superficial velocity.

3.2. Data generation

We generate the training dataset by uniformly and randomly perturbating a main set of operating conditions (Table 1) and solving the system accordingly. The training dataset is composed of 5,000 distinct initial condition combinations, each varying within a +/- 20% range from a central set of industrial relevant conditions. Each sample describes the spatial profile of the 9 state variables, discretized in 800 points along the domain. We solve the system in Python, using `scipy.integrate.solve_ivp` function, setting the absolute and relative tolerance to 10^{-12} and 10^{-8}, respectively. The dataset, comprising 4 million data points, is used to train the models, enabling them to forecast the values of state variables at a specific longitudinal coordinate based on a given set of initial conditions.

Table 1: Main set of initial conditions: the training dataset is generated within a range that spans +/- 20% of the standard operating conditions. The molar flowrate is expressed in mol/s *per* tube.

T_0 [°C]	P_0 [bar]	$\dot{n}_{CO,0}$	$\dot{n}_{CO_2,0}$	$\dot{n}_{CH_4,0}$	$\dot{n}_{CH_3OH,0}$	$\dot{n}_{H_2,0}$	$\dot{n}_{H_2O,0}$	$\dot{n}_{N_2,0}$
245	65	0.0409	0.0764	0.1786	0.0032	0.4316	0.0007	0.0261

For training and test, we nondimensionalize the system, defining the nondimensional state variables T^*, P^*, and $\dot{n}_i^*$ and longitudinal coordinate z^*, by scaling the original values using the characteristic dimensions $< T >$, $< P >$, $< \dot{n}_i >$, and $< z >$, specifically taken as the average of each feature in the dataset. For a fair comparison, we trained and tested all the models on the scaled, nondimensional, dataset.

4. Results and discussion

We evaluate and compare the accuracy and runtime of the proposed data-driven and hybrid modeling methods with a conventional Python ODE solver. We build an MLP and PINN with 3 hidden layer with 512 neurons each. The TST consists of 5 encoder and decoder layers respectively.

4.1. Prediction runtime and accuracy

We compare the performance of a conventional ODE solver in Python (`scipy.integrate.solve_ivp`) with the presented alternatives in terms of accuracy and runtime. The test set, which includes 500 varied scenarios distributed within the same range as the training set, utilizes an ODE solver with very small tolerances, considered to be 100% accurate, and serves as the baseline for comparison. We assessed the solver's performance by adjusting the tolerance to quicken computation, albeit at the expense of reduced accuracy. Ultimately, we evaluated the outcome of MLP, PINNs and TST when applied to the unseen test dataset. We measure the accuracy as the complement of the Mean Absolute Percentage Error (MAPE).

Table 2 summarizes the main results. We evaluate the mean performance across 500 simulations using varying solver tolerances, MLP, PINN, and TST methodologies. As expected, the runtime of the ODE solver decreases significantly when the tolerances are reduced. Nonetheless, the MLP and PINNs offer a significantly shorter computational time, being 35 times faster than the least accurate version of the ODE solver, and still maintain a higher accuracy of approximately 99.5%. This outcome indicates a potential for utilizing MLP and PINNs models within larger optimization or simulation studies. For instance, in superstructure optimization problems, the reactor model often needs to be evaluated thousands of times at every iteration, depending on the plant complexity and assessed configurations. Thus, our work can potentially shift the overall computational time of such problems from several hours to few minutes. Considering that this preliminary analysis excludes hyperparameter tuning, it is anticipated that incorporating extended optimization analysis would likely improve accuracy of the models further. The computational time of the TST is higher due to its large architectural complexity. The accuracy of the TST prediction is simultaneously the lowest, suggesting the need for further investigation, for instance into more data-efficient modeling techniques.

Table 2: Runtime and accuracy comparison of the proposed methods. The ODE solver is tested when different relative (r) and absolute (a) tolerances are required. In the table, the tolerance numbers indicate the decimal digits (e.g., r10 stands for "relative tolerance = 10^{-10}"). The runtime is referring to CPU computation (11th Gen Intel Core i7).

Method	Runtime [s]	Accuracy [%]
ODESolver - r10a12 (baseline)	0.063	100.00
ODESolver - r3a3	0.013	99.97
ODESolver - r1a1	0.0056	98.55
MLP	0.00016	99.54
PINNs	0.00016	99.51
TST	12.06	96.87

5. Conclusions

We demonstrate that MLPs and PINNs can provide an alternative to classical modeling techniques given their high accuracy and low runtime. These considerations are especially important in the context of highly exhaustive modeling, for instance plant-wide optimization and simulation, where the reactor model needs to be solved repetitively. Notably, the MLP and PINNs demonstrate a 35-fold decrease in computation time compared to the least accurate version of the ODE Solver, while maintaining higher accuracy. Further investigation is needed to evaluate the ability of PINNs model to overcome the limitations in the application scope commonly associated with black-box models by extending the solution beyond the training limits. Finally, we envision the opportunity to embed physical knowledge into a simplified transformer architecture for steady-state and dynamic modeling of chemical systems.

Acknowledgements

This research is supported by Shell plc, for which we express sincere gratitude.

References

K. M. V. Bussche and G. F. Froment, 1996, A Steady-State Kinetic Model for Methanol Synthesis and the Water Gas Shift Reaction on a Commercial Cu/ZnO/Al2O3 Catalyst, Journal of Catalysis, Volume 161, June, p. 1–10.

A. Kolides, A. Nawaz, A. Rathor, D. Beeman, M. Hashmi, S.Fatima, D. Berdik, M. Al-Ayyoub, Y. Jararweh, 2023, Artificial intelligence foundation and pre-trained models: Fundamentals, applications, opportunities, and social impacts, Simulation Modelling Practice and Theory, Volume 126, July, p. 102754

L. McClenny and U. Braga-Neto, 2022, Self-Adaptive Physics-Informed Neural Networks using a Soft Attention Mechanism, ArXiv preprint at arXiv:2009.04544

M. Raissi, P. Perdikaris and G.E. Karniadakis, 2019, Physics-informed neural networks: A deep learning framework for solving forward and inverse problems involving nonlinear partial differential equations, Journal of Computational Physics, Volume 378, February, p. 686–707.

A.M., Schweidtmann, E. Esche, A. Fischer, M. Kloft, J. Repke, S. Sager, A. Mitsos, 2021, Machine Learning in Chemical Engineering: A Perspective, Chemie Ingenieur Technik, Volume 93, October, p. 2029–2039.

N. Sitapure and J.S. Kwon, 2023, Exploring the Potential of Time-Series Transformers for Process Modeling and Control in Chemical Systems: An Inevitable Paradigm Shift?, Chemical Engineering Research and Design, Volume 194, June, p. 461-477

A. Vaswani, N. Shazeer, N. Parmar, J. Uszkoreit, L. Jones, A.N. Gomez, L. Kaiser, I. Polosukhin, 2017, Attention Is All You Need, 31st Conference on Neural Information Processing Systems (NIPS 2017)

M. von Stosch, R. Oliveira, J. Peres, and S.F de Azevedo, 2014, Hybrid semi-parametric modeling in process systems engineering: Past, present and future, Computers & Chemical Engineering, Volume 60, January, Pages 86–101.

S. Wang, S. Sankaran, H. Wang, and P. Perdikaris, 2023, An Expert's Guide to Training Physics-informed Neural Networks, Arxiv preprint at arXiv:2308.08468

S. Wang, Y. Teng, and P. Perdikaris, 2020, Understanding and mitigating gradient pathologies in physics-informed neural networks, SIAM Journal on Scientific Computing, Volume 43, 5, p. A3055-A3081

Q. Wen, T. Zhou, C. Zhang, W. Chen, Z. Ma, J. Yan, and L. Sun, 2022, Transformers in Time Series: A Survey, Arxiv preprint at arXiv:2202.07125

Flavio Manenti, Gintaras V. Reklaitis (Eds.), Proceedings of the 34th European Symposium on
Computer Aided Process Engineering / 15th International Symposium on Process Systems
Engineering (ESCAPE34/PSE24), June 2-6, 2024, Florence, Italy

Techno-Economic Analysis of H$_2$ Extraction from Natural Gas Transmission Grids

Homa Hamedi,[a,b]* Torsten Brinkmann[b]

[a] *TU Dresden, Process Systems Engineering Group, Helmholtzstraße 10, 01069 Dresden, Germany*

[b] *Helmholtz-Zentrum Hereon, Institute of Membrane Research, Max-Planck-Straße 1, 21502 Geesthacht, Germany*

homa.hamedimastanabad@tu-dresden.de

Abstract

The storage and transportation of renewable H$_2$ using natural gas (NG) pipelines is regarded as a promising solution to overcome the challenges with renewable energy storage and distribution. However, among other concerns regarding safety and reliability, further investigation is still required to unveil all the aspects associated with the techno-economics of the technology. Levelized H$_2$ separation cost (LHSC) is a crucial metric to evaluate and compare the cost-effectiveness of H$_2$ blending technology as an alternative to other incubators for H$_2$ transportation such as compressed or liquefied H$_2$ tube trailers. H$_2$ extraction from the NG distribution system (NGDS), where the operating pressure is moderate values and H$_2$ contents can be up to 30-50%, have been already studied in the literature using a standalone pressure swing adsorption (PSA) process. However, utilizing the PSA technique for mixtures with low H$_2$ contents of 5-10% as allowable in the NG transmission system (NGTS) requires much larger PSA equipment with high capital expenses and additional significant costs needed for the recompression to reinject the NG into the gas pipelines. To customize the PSA separation technology for such scenarios, a hybrid process scheme has been proposed in the literature. This configuration comprises one membrane stage followed by a PSA purification step. This study presents a rigorous techno-economic model for the aforementioned process scheme using an integrated platform developed based on Aspen Adsorption V11 and Aspen Costume Modeler. The system is composed of one stage hollow fiber matrimid membrane module followed by a 4-bed adsorption process operating on 8 sequential steps. to fulfil H$_2$ market purity requirements with an acceptable recovery rate for H$_2$. To reduce the computational time, we used a cyclic steady-state approach to solve the governing equations of the PSA. The system is studied for different scenarios relevant to NGTS operating conditions. The results show that the viability of hydrogen blending technology from technoeconomic standpoint becomes questionable, particularly at lower grid pressures when allowable hydrogen content is less than 10%. However, for industries requiring lower hydrogen purity levels (below 90-99%), this transportation method seems promising.

Keywords: Pressure swing adsorption optimization, H$_2$ purification, Aspen Adsorption, Cyclic steady-state simulation approach, Hybrid Process

1. Literature review

The amount of hydrogen that can be blended with NG depends on several factors, such as the pipeline infrastructure and the end-use sectors. Numerous economic, logistical, and safety challenges are still to be addressed before possibly implementing this strategy to gradually decarbonize NG contents. Therefore, at this stage, the lack of inconclusive regulations and guidelines causes a level of indetermination to report resolute codes and values for technical and economic aspects of H_2 blending technology. According to a primary technical report by Bard et al., house installations can endure up to 5-10% for NG grids operating at pressures higher than 16 bar. However, using H_2 as an NG substitution for heating purposes reduces emissions only by 6%-7% for a blend of 20% H_2 (Bard et al., 2022). Moreover, due to today's high price of green H_2, the blending significantly increases NG price. In comparison, up to 50% reductions in emissions are achievable by utilizing the same amount of the unblended H_2 in mobility sectors and industrial applications (Bard et al., 2022). The last two statements suggest that H_2 use in the heating sector is not an economically and environmentally effective solution for the NG system decarbonization. These may however imply that the downstream deblending of H_2 can increase the economic viability of H_2 storage and transportation via the existing pipelines. Techno-economic perspectives of H_2 extraction from the NGDS, where the operating pressure is moderate and H_2 contents can be up to 30-50%, have been already studied in the literature (Hamedi et al., 2023). However, utilizing the PSA technique for mixtures with low H_2 contents of 5-10% as allowable in the NG transmission grids requires much larger PSA equipment with high capital expenses and additional significant costs needed for the recompression to reinject the NG into the gas pipelines. To customize this separation technology for such scenarios, a hybrid process scheme has been proposed in the literature (Liemberger et al., 2017; Purrucker and Balster, 2021). This configuration comprises one membrane stage followed by a PSA purification step. In this scheme, a bulk amount of NG can be separated in the membrane module at almost the same pressure as that of the grid before the H_2-rich permeate stream is sent to the PSA columns. This combined arrangement is expected to outperform a standalone PSA due to lower power demand for recompression. This approach is only possible where sufficient pressure is available for the permeation separation process. Figure 1 presents the detail of this process scheme.

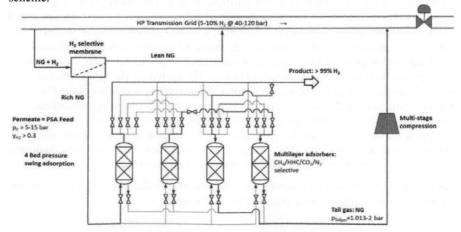

Figure 1 Process flow diagram for a hybrid process composed of one-stage membrane separation and a 4-bed PSA filled with activated carbon

2. Mathematical modeling

2.1. PSA process description and simulation

A 4-bed 8-step PSA process was chosen in this study to feed the system with a continuous flow and increase H_2 recovery compared to 2-bed PSA configuration which is commonly used in the literature. The columns are packed with activated carbon, which is widely used in the literature for H_2 extraction from other gas mixtures like coal gas, biogas, and refinery gasses. Figure 2 demonstrates the periodic operation of the PSA process, illustrating pressure changes within one periodic cycle and the material interaction among the 4 bed columns.

2.1.1. Cyclic steady-state approach

PSA systems are inherently dynamic and are regarded as highly nonlinear systems. A common technique in the literature to solve the respective set of nonlinear partial differential equations is the time integration method. However, the ultimate operation of a PSA process can be regarded as a cyclic steady-state performance. Approaching the problem from the latter angle, we can uniformly discretize the time domain over one single adsorption period and make the states of the system equal at the beginning and the end of the cycle, ensuring periodic steady-state performance. In other words, instead of numerical integration along time and initial conditions, temporal discretization and temporal periodic conditions are required. Fig.3 highlights the underlying difference between the approaches. Though the CSS technique is memory-intensive, it offers significantly less computational time by avoiding computations for a considerable number of dynamic cycles preceding the ultimate cyclic steady-state status. This method is highly advantageous for optimization, due to not only faster objective function evaluations but also easier handling of constraints on the product or tail gas specifications. The latter is carried out by averaging the respective specifications over one steady-state cycle, which can be readily evaluated in the CSS approach. Moreover, the CSS approach facilitates the integration of a PSA system with other components in a process flowsheet. In this study, the spatial derivatives of the 1-D bed model were evaluated based on the second-order finite difference scheme, and the temporal discretization was approximated using the first-order backward finite difference method.

2.2. Membrane process description and simulation

As earlier mentioned, a membrane module at the upstream of the PSA system serves to separate a bulk amount of NG while remaining its pressure at that of the grid. A hollow fiber membrane module is modeled in Aspen Custom Modeler. Some details of the model can be found in our previous publications (Hamedi and Brinkmann., 2021).

3. Integration of process simulations

To integrate the two modules inside one single flowsheet, membrane model was exported from Aspen Custom Modeler to the Aspen Adsorption environment. To solve the problem, two sets of initial guesses should be generated for each module separately and then combined together to solve the overall flowsheet.

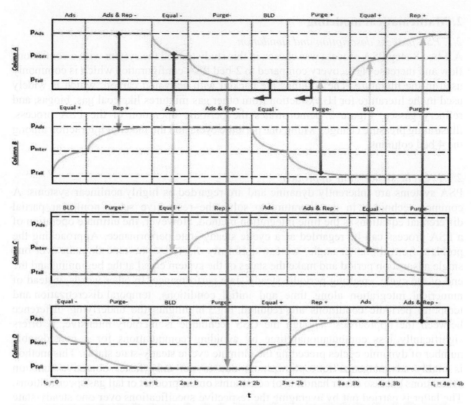

Figure 2 The schedule diagram for a 4-bed 8-step PSA. Green arrows indicate material interaction among columns and the yellow graph depicts pressure changes (Hamedi et al. 2022)

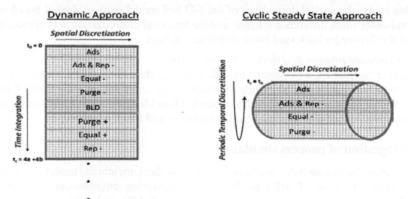

Figure 3 Dynamic versus cyclic steady-state approaches for solving periodic adsorption processes (Hamedi et al. 2022)

4. Design data and specifications

Table 1 presents the specifications that form the basis for the result section. The Langmuir adsorption isotherm parameters and mass transfer coefficients are adopted from the literature (Hamedi et al., 2022).

Table 1 Design data and specifications

Parameters	Value
H_2 feed content, mol%	10.00
Feed temperature, K	298.15
Feed flowrate, kmol/hr	250
Total adsorbent height (column height), m	5.00
PSA column's L/D ratio	10.00
Velocity in retentate and permeate sides, m/s	< 4
Project lifetime, year	25
Adsorbent lifetime, year	2
Matrimid membrane lifetime, year	5
Discount rate	0.10
Adsorbent price, $/kg adsorbent	2.00
Matrimid membrane price, $/m²	36

5. Results and discussions

Tables 2 and 3 show the simulation results for different relevant scenarios for two levels of NG grid pressures. The H_2 recovery rates and LHSC are presented in Table 1 when the grid pressure is at 60 bar. As shown in Table 2, at this pressure, H_2 purity of 99.97%, which is the requirement of the mobility sector, was not achievable. However, it is possible to achieve 99.00% H_2 purity, LSHC falls in the range of $1.626-3.251 per kg H_2. The difference of overall H_2 recoveries for Scenarios 1 and 3 are slight but the LHSC can be reduced by about 13%. As the product pressure (which slightly varies from the permeate pressure or adsorption pressure due to the limited pressure drop in the column) increases, LSHC rises to $3.251 per kg H_2. However, it should be noted that the H2 product pressure is higher in Scenario 5. At the grid pressure of 120 bar, the 99.97% purity is achievable and LSHC varies between $1.137-3.602 per kg H_2.

LSHC should be a reasonable portion of the competitive H_2 cost goal for fuel cell vehicle markets, which is reported at $2–$4/kg-H_2 (Melaina et al., 2013). Considering the latter and the high LSHC for purity values larger than 99.97%, the H_2 blending technology

Table 2 Results for 10% H_2 content at 60 bar NG grid pressure

	Unit	Scenario 1	Scenario 2	Scenario 3	Scenario 4	Scenario 5	Scenario 6
H_2 Product Pressure	bar	8	8	10	10	15	15
Product Purity	%	99.00	99.97	99.00	99.97	99.00	99.97
LHSC	USD/kg H_2	1.626	NA	1.870	NA	3.251	NA
Membrane H_2 Recovery	mol %	0.34	NA	0.345	NA	0.29	NA
PSA H_2 Recovery	mol %	0.733	NA	0.727	NA	0.585	NA
Overall H_2 Recovery	mol %	0.249	NA	0.251	NA	0.169	NA

Table 3 Results for 10% H_2 content at 120 bar NG grid pressure

	Unit	Scenario 1	Scenario 2	Scenario 3	Scenario 4	Scenario 5	Scenario 6
H_2 Product Pressure	bar	8	8	10	10	15	15
Product Purity	%	99.00	99.97	99.00	99.97	99.00	99.97
LHSC	USD/kg H_2	1.137	2.363	1.175	2.216	1.596	3.602
Membrane H_2 Recovery	mol %	0.48	0.285	0.55	0.35	0.54	0.36
PSA H_2 Recovery	mol %	0.763	0.511	0.78	0.523	0.745	0.38
Overall H_2 Recovery	mol %	0.366	0.146	0.429	0.183	0.402	0.137

cannot be economically (even technically for lower grid pressures) viable when the allowable H_2 content is less than 10%. However, for the industry which operates with lower H_2 purity (<90-99%), this transportation mode seems a promising option.

6. Conclusions

We presented an integrated simulation framework for a hybrid process separation flowsheet to separate H_2 from NG. The cyclic steady state approach was used to solve the PSA system. This facilitates the combination of the membrane module at steady state condition with the PSA system, which operates inherently dynamically. The results show that at 60 bar grid pressure, LSHC is $1.626-3.251 per kg H_2. Moreover, in this scenario, the purity of 99.97% was not achieved based on simulation. The cost rises to $1.137-3.602 per kg H_2 when the grid pressure is at 120 bar; however, in this case 99.97% purity is achievable. LSHC needs to constitute a substantial component of the competitive cost target for hydrogen fuel cell vehicles, as stated in the range of $2–$4 per kg H_2 (Melaina et al., 2013). Taking into account this cost range and the elevated LSHC associated with purity levels exceeding 99.97%, the economic feasibility of hydrogen blending technology becomes questionable, especially under lower grid pressures, when the permissible hydrogen content is below 10%. Nevertheless, for industries operating with lower hydrogen purity levels (below 90-99%), this mode of transportation appears to be a promising alternative.

References

[1] M.W. Melaina, O. Antonia, M. Penev, 2013, Blending hydrogen into natural gas pipeline networks: a review of key issues, National Renewable Energy Laboratory.
[2] H. Hamedi, T. Brinkmann, T. Wolf, 2023, Techno-economic assessment of H_2 extraction from natural gas distribution grids: A novel simulation-based optimization framework for pressure swing adsorption processes, Chemical Engineering Journal Advances, 16, 100541.
[3] L. Dehdari, I. Burgers, P. Xiao, K.G. Li, R. Singh, P.A. Webley, 2022, Purification of hydrogen from natural gas/hydrogen pipeline mixtures, Separation and Purification Technology, 282, 120094.
[4] J. Bard, N. Gerhardt, P. Selzam, M. Beil, M. Wiemer, M. Buddensiek, 2022, The limitations of hydrogen blending in the European gas grid, Fraunhofer Institute for Energy Economics and Energy System Technology.
[5] W. Liemberger, M. Groß, M. Miltner, M. Harasek, 2017, Experimental analysis of membrane and pressure swing adsorption (PSA) for the hydrogen separation from natural gas, Journal of Cleaner Production 167, 896-907.
[6] O. Purrucker, J. Balster, 2021, Hydrogen on tap: Supporting decarbonization by pipelining H2 to the point of use, Linde Group.
[7] H. Hamedi, T. Brinkmann, 2021, Rigorous and customizable 1d simulation framework for membrane reactors to, in principle, enhance synthetic methanol production.

Flavio Manenti, Gintaras V. Reklaitis (Eds.), Proceedings of the 34th European Symposium on Computer Aided Process Engineering / 15th International Symposium on Process Systems Engineering (ESCAPE34/PSE24), June 2-6, 2024, Florence, Italy

Aiding Modular Plant Design by Linking Capability and Transformation Models

Amy Koch[*], Florian Kunkel, Louise Theophile, Veronica De la Vega Hernandez, Jonathan Mädler, Leon Urbas

Chair of Process Control Systems & Process Systems Engineering Group, TU Dresden, Dresden 01062, Germany;
amy.koch@tu-dresden.de

Abstract

The specialty chemical and pharmaceutical industries require faster time-to-market, and more flexible production. Modular Plants built from Process Equipment Assemblies (PEAs) and designed according to VDI 2776 are suitable to address these challenges. Since PEAs are designed independently of the products they produce, this creates a problem of exchange of information between the Owner/Operator (O/O) and the PEA vendors. In this paper, an information model that links information about the products from the O/O (transformation model) to the information about the PEA from the PEA vendors (capability model) is introduced. Although, the necessary information from these two parties is contained and distributed amongst existing standards for information models such as DEXPI and OntoCAPE, these standards must be connected and extended to support modular plants. The application of this information model is illustrated for connecting a neutralization process with a reactor PEA.

Keywords: digital twins, information model, modular plants, knowledge distribution

1. Motivation

Modular Plants (MP) built from Process Equipment Assemblies (PEA) can be used to address the challenges of faster time-to-process and increased flexibility facing the specialty chemical and pharmaceutical industries. Additional benefit can be achieved from the integration of a Digital Twin (DT) of these PEAs to facilitate with process planning, scale-up, and validation (Mädler et al. 2022; Koch et al. 2023). Compared to conventional production plants, Modular Plants require adapted engineering workflows as information must be exchanged across different domains throughout different lifecycles and lifecycle phases. This highlights the increasing importance of information management and seamless data exchange between software tools. Here, information models can be used to support the digital data exchange and semantic connections as part of a Digital Twin Concept. For modular plants, significant challenge stems in matching a product specification and process, described in a transformation model, to the necessary PEA assets described by capability models. In this work we intend to propose a new information model to match simulation-based transformation models with equipment datasheet-based capability models based on existing standards of OntoCAPE and DEXPI[1]. This paper is structured as follows: first the challenges of knowledge distribution in modular plants are presented. Next, the relevant submodels of the Digital Twin are aligned with the OntoCAPE ontology and the resulting architecture of the information

---

[1] DEXPI = Data Exchange in the Process Industry.

model based on the Digital Twin workflow is presented. Finally, the use case of a neutralization process is used to evaluate the results followed by a critical discussion and conclusion.

2. Background

2.1. Knowledge Distribution and Digital Twin Workflows for Modular Plants

To address matching Owner/Operator (O/O) knowledge with PEA vendor knowledge, first the relationship between the partial models of the Digital Twin to the relevant workflows and lifecycle phases must be understood. Bamberg et al. (2021) describe the Digital Twin as a Product-Process-Resource model. Mädler et al. (2022) applied the model of Bamberg et al. (2021) to modular plants and illustrated how the knowledge distribution in a Digital Twin is divided between the PEA manufacturers and the Owners. For this purpose, the **capability model** is a product independent description of a PEA containing structural, operational and behavioral information about the PEA, independent of a concrete process. The **transformation model** is a plant independent and product specific description of the process and is used to derive information e.g., residence times. To leverage this knowledge distribution in modular plants, information regarding the product and process has to be compared and matched with equipment specific capabilities. Currently, this process is carried out using vendor specific toolsets or by exchange of non-machine-readable documentations between both parties and all participating domains. The transfer of documents between different parties increases the risk of potential loss of important information. This highlights the need for seamless data exchange using information models. Supporting this with suitable digital representation would not only make this exchange more efficient, but also creates the potential to automate the interaction of different parties and enable an automated selection of PEAs.

2.2. Vendor-Independent Standards for Information Models

To support the semantic connection between submodels in a Digital Twin, existing vendor-independent standards for information models and ontologies are needed. Koch et al. (2023) evaluated the suitability of OntoCAPE, DEXPI, among others, across the lifecycle phases relevant to the modular plant and PEA assets; here, DEXPI and OntoCAPE were demonstrated to be most suitable to address describing the lifecycle phases for a Digital Twin, where information from the transformation model must be matched to the capability model. DEXPI is a vendor-independent standard which can be used to describe the data in Piping & Instrumentation Diagrams (P&ID) for exchange between Computer Aided Engineering (CAE) systems such as COMOS (Wiedau et al., 2019). As of 2023, the DEXPI P&ID has been supplemented with the DEXPI Process specification to describe the functional requirements phase in the asset lifecycle of Wiedau et al. (2019) and describe the information covered in block flow diagram (BFD) and process flow diagram (PFD) as outlined by the ISO 10628 standard. However, for the remainder of this paper DEXPI refers to the DEXPI P&ID standard. OntoCAPE is an ontology used for chemical process and the engineering of chemical plants (Marquart et al. 2010). The suitability of these standards for the domain of modular plants is well validated in literature: Mädler et al. (2022) introduce a linked data information model for Digital Twins where DEXPI is used to describe structural models. Rahm et al. (2021) has illustrated how different data models, specifically DEXPI and VDI/VDE/NAMUR 2658, can be synchronized.

3. Methodology

3.1. Requirements

The goal of the information model is to match the information documented in the transformation model (product specification + gPROMS flowsheet) with the comparative information in the capability model (COMOS). The seamless transfer of the information is a crucial aspect in supporting Digital Twin workflows for modular plants. To develop the information model the OntoCAPE ontology (Marquart et al. 2010), which describes the domain of chemical engineering and has broad applicability, was chosen due to the suitability of OntoCAPE to describe both the transformation model as well as the capability model (Koch et al., 2023). In addition to OntoCAPE, DEXPI (Wiedau et al., 2019) is used to expand some the design attributes of the equipment class, as this information model is already used for the P&ID domain. However, the aforementioned attributes might not necessarily be sufficient to describe all of the technical criteria and might potentially need to be further extended with custom attributes to describe all relevant technical criteria for each equipment class.

3.2. OntoCAPE

Athough direct equivants of the transformation model and capability model are not directly described with OntoCAPE, a releationship between these models and aspects of systems of the OntoCAPE model *technical system module* can be identified (see Figure 1). Here, aspects of the **transformation model** are described by the *system requirements* and *system function*. The **capability model** can be related to the *system realization* and *systems behavior*. In OntoCAPE the *system realization* constrains the *system behavior* which is true of the capability model as well. For the transformation model, literature such as Koch et al. (2023) and Mädler et al. (2022) have indicated

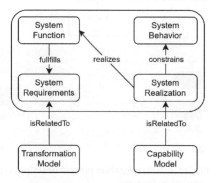

Figure 1. Transformation models and capability models in OntoCAPE.

that a direct result of the transformation model is a list of capability requirements, which the capability model of the PEA asset must fulfil. By understanding this relationship, a connection between the transformation model and capability model can be made using the OntoCAPE ontology.

3.3. Information Model

As a first step in modular plant engineering, it is necessary to match the product specific transformation model with a PEA asset specific capability model. For this purpose, we suggest to use linked data to describe the connections (cf. Mädler et al. 2022; Rahm et al. 2021). We propose that the transformation model and capability model be connected using the relationship *realizes*. This is an approach which is consistent with examples shown in OntoCAPE (Marquardt et al., 2010). The transformation model is modelled exclusively using the OntoCAPE terminology. For the capability model we propose that the description of the PEA be extended with the corresponding DEXPI class, enabling a clear connection to the descriptive design data provided in the DEXPI specification such as the upper and lower design limits for temperature and pressure as well as aspects relevant to substance class such as material of construction.

4. Use Case

In order to explain the information exchanged between the transformation model and the capability model, we propose the use case of a neutralization reaction which must be matched to a set of PEAs. Here, the considered reaction is:

$$NaOH\ (aq) + HCl(aq) \rightarrow NaCl\ (aq) + H_2O\ (aq)$$

This specific reaction is chosen due to its simplicity and suitability to explain certain technical criteria from the product model which must be transferred. A block flow diagram with the relevant process parameters is shown in Figure 2.

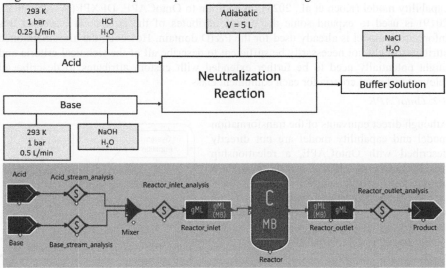

Figure 2. Neutralization process description as a block flow diagram (top) and gPROMS flowsheet simulation (bottom).

From this example, a list of system requirements from the transformation model can be derived which must be matched to a set of capabilities described in the capability model of the reactor. Therefore, in addition to the block flow diagram in Figure 2 and the reaction equation, an example PEA is also required. For this purpose, a stirred tank reactor, whose capabilities are listing in Table 1, is provided. A selected list capability requirements and capabilities is presented in Table 1.

Table 1: List of capability requirements derived from the neutralization process transformation model and the corresponding capabilities described by the capability model of the reactor PEA.

Capability Requirements from the Transformation model	Reactor Module Capabilities from the Capability model
required reaction volume of 5 L	> 5 L available reaction volume
293K reaction temperature	upper design temperature > 293 K > lower design temperature
1 bar reaction pressure	upper design pressure > 1 bar > lower design pressure
material of construction suitable for corrosive materials	material of construction suitable for corrosive materials

4.1. Linked Data Representation

To illustrate this connection between the transformation model and capability model, a linked data representation based on the neutralization process and reactor PEA is provided in Figure 3. For the neutralization use case, the connection between transformation model and capability model is demonstrated between a Reactor (*PlantItem*) which realizes a Chemical Reaction (*ProcessStep*). The transformation model is modelled exclusively using the OntoCAPE terminology. Here, the key information is related to the chemical reaction class. Then, corresponding reactants and products are described with the substance class, which is important information for connecting material of construction suitability. Additionally, the thermodynamics properties of the chemical reaction are provided such as temperature, pressure, and reaction volume using the *hasProperty* connector. This information is shown in an exemplary way and is not an exhaustive list of all information which would be required to link transformation and capability model. To describe the capability model, the class plant item is selected. OntoCAPE suggests that additional information describing the real equipment can be implemented using the *hasDesignProperty* descriptor. Here, the capability requirement of reaction temperature is connected to the upper and lower temperature design limits of the reactor PEA. Additionally, we suggest that the material of construction be constrained by substance class. This enables a direct connection between the transformation and capability models.

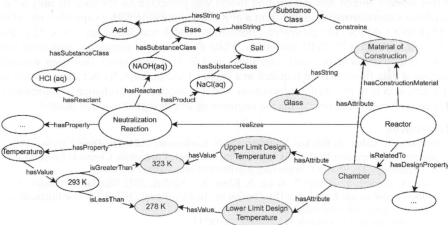

Figure 3. Connection of the transformation and capability models using linked data.

4.2. Implementation

To test the developed information model, a manual implementation is carried out for the neutralization reaction use case. Information from the gPROMS flowsheet isas extracted and combined with additional technical criteria from the product specification. To simplify the scope, only basic information from gPROMS (temperature, pressure, volumetric flowrate, density, and viscosity) and the most relevant information product specification (pH, substance) are considered. However, challenge came when trying to match this information between the relevant models in the different software tools. In gPROMS, only information regarding the material stream and process parameters is available from this kind of simulation model. However, some information regarding the unit operation is missing. This is potentially problematic when trying to match between the gPROMS and COMOS domains. Here, mathematical relationships are needed to compare aspects such as residence time from the simulation model to required reactor volume in the capability model. This suggests that some of the information from the gPROMS simulation file must be adapted to connect it to a capability model in COMOS.

5. Discussion

The information model presented in this work and exemplified for the neutralization reaction and reactor module use case illustrates that it is possible to directly connect the transformation model and the capability model. This enables a direct connection between the information contained in the Digital Twin submodels needed to enable a semantic connection. Furthermore, the implementation in gPROMS and COMOS highlights that certain information requires adaption or was missing from gPROMS to directly connect it to the reactor model in COMOS. Although the results presented in the work are a significant first step, the information model and case study are limited in scope and do not consider all aspects of the modular plant domain. In particular, aspects such as suitability of services and safety aspects must also be considered. Additionally, other types of data models might be advantageous over linked data.

6. Conclusion

In this work, a linked data information model was presented for the specific purpose of matching a transformation model from a gPROMS simulation with a capability model of a PEA in COMOS. The presented information model was based on OntoCAPE and extended with DEXPI P&ID attributes. While the presented results are indicative that connecting transformation and capability models is a significant first step towards integrated data flows that are required in a Digital Twin concept, further work is needed. The next step is to extend the information model to support the exchange of information at specific handover points during the engineering and design of modular plants.

References

A. Bamberg, L. Urbas, S. Bröcker, M. Bortz, and N. Kockmann, 2021. The Digital Twin – Your Ingenious Companion for Process Engineering and Smart Production. Chemical Engineering & Technology 44, 954–961. https://doi.org/10.1002/ceat.202000562

A. Koch, N. Hamedi, L. Furtner, T. Kock, A. Klose, & J. Mädler, 2023. Standards for Information Models Considering Knowledge Distribution in Modular Plants. 2023 IEEE 21st International Conference on Industrial Informatics (INDIN), 1–7. https://doi.org/10.1109/INDIN51400.2023.10218218

J. Mädler, J. Rahm, I. Viedt, and L. Urbas, 2022, A digital twin-concept for smart process equipment assemblies supporting process validation in modular plants. In L. Montastruc & S. Negny (Eds.), Computer Aided Chemical Engineering (Vol. 51, pp. 1435–1440). Elsevier. https://doi.org/10.1016/B978-0-323-95879-0.50240-X

W. Marquardt, J. Morbach, A. Wiesner, and A. Yang, 2010. OntoCAPE: A re-usable ontology for chemical process engineering. Springer.2010. doi: 10.1007/978-3-642-04655-1

J. Rahm, M. Hanselmann, and L. Urbas, 2021, Synchronization network of data models in the process industry. 2021 26th IEEE International Conference on Emerging Technologies and Factory Automation (ETFA), 1–8. https://doi.org/10.1109/ETFA45728.2021.9613647.

VDI/VDE/NAMUR, 2019, VDI/VDE/NAMUR 2658-Automation engineering of modular systems in the process industry-General concept and interfaces – Part 1, Beuth Verlag GmbH.

VDI 2776-1:2020, "Process engineering plants - Modular plants - Fundamentals and planning modular plants - Part 1 (VDI 2776:2020-11)." Beuth Verlag, Nov. 2020.

M. Wiedau, L. von Wedel, H. Temmen, R. Welke and N. Papkonstantinous, 2019. ENPRO Data Integration: Extending DEXPI Towards the Asset Lifecycle. Chemie Ingenieur Technik, 91(3), 240–255. https://doi.org/10.1002/cite.201800112

Flavio Manenti, Gintaras V. Reklaitis (Eds.), Proceedings of the 34th European Symposium on Computer Aided Process Engineering / 15th International Symposium on Process Systems Engineering (ESCAPE34/PSE24), June 2-6, 2024, Florence, Italy

Optimization-based design of distillation columns using surrogate models

Marc Caballero, Anton A. Kiss, Ana Somoza-Tornos*

Delft University of Technology, Van der Maasweg 9, 2629 HZ, Delft, The Netherlands
A.SomozaTornos@tudelft.nl

Abstract

Continuous process re-design and optimization are required to assess and implement emerging sustainable technologies. Being distillation one of the most energy intensive operations in the process industry, the design and optimization of such is crucial. This paper presents a novel approach to distillation column optimized design using a neural network surrogate model embedded in particle swarm optimization (PSO). The surrogate model resembles the input/output structure of shortcut models, yet it is trained with data from rigorous calculations carried out in Aspen Plus. The surrogate model is integrated into a PSO algorithm, enabling to obtain global optimum points while simultaneously including simplified constraints through cost function penalty. A case study illustrates how this novel methodology combines the simplicity of a shortcut method while offering design parameters closer to an already optimized rigorous model.

Keywords: distillation design, shortcut distillation, surrogate modelling, particle swarm optimization.

1. Introduction

The energy transition introduces new challenges for process design, from the shift to alternative feedstocks to the electrification of thermal demands (*Mallapragada et al., 2023; Lopez et al., 2023*). Continuous efforts are dedicated to exploring novel reactions, enhancing yields and selectivities, and transitioning to alternative fuels. Translating these efforts into fully-operable processes raises the critical need for constant process re-design and optimization, which poses a challenge for commonly used optimization approaches.

Traditional distillation column design often relies on heuristics and experience. Rigorous models, despite providing detailed insights, are computationally expensive and prone to convergence issues, challenging their use in deterministic optimization algorithms, and resulting in non-integer nonlinear problems. Meanwhile, simplified models lack the detail necessary for robust column design. Surrogate modeling, lying between these two alternatives, yields results close to rigorous simulations within the surrogate's training space, avoiding convergence issues and significantly reducing computational time.

Recent literature displays the use of surrogate modelling embedded in distillation optimization. It has been used for specific distillation systems, such as crude oil columns (*Ibrahim et al., 2018*) or vacuum distillation (*H'ng et al, 2021*). More generic approaches to this problem involve substituting the entire optimization for a surrogate model (*Quirante et al., 2015 and Keßler et al., 2018*).

Therefore, a novel optimization-based approach for designing distillation columns, using surrogate models, is proposed.

2. Methodology

The proposed methodology is depicted in Figure 1, formed by three main parts. First, gathering and pre-processing the data, followed by using it to train a surrogate model for distillation. Lastly, embedding the trained surrogate model into an optimization framework. This section provides a general overview of each part.

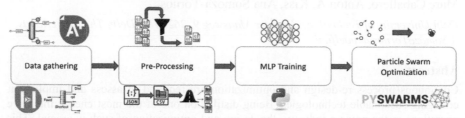

Figure 1. General workflow for the proposed methodology

2.1. Data Gathering

The data collection approach for training the surrogate model involves several steps. First, the feed to the distillation column and the desired separation must be defined. Once the separation is chosen, the variability in the column feed is calculated to enable the model to handle a spectrum of compositions and operating conditions. This involves heuristically defined ranges and simplified simulations of the upstream.

After defining the ranges for the feed stream, the bounds for the column's design parameters can be defined. The proposed method for that task is to run enough calculations with a distillation shortcut method (Fenske-Underwood-Gilliland-Kirkbride method) within the different feed conditions established.

To generate the data for training the surrogate model, Latin hypercube sampling is employed to generate input data for rigorous distillation calculations, within the bounds obtained previously. These calculations involve solving the MESH equations (Material balance, Phase Equilibrium, Mole fraction Summation, and Heat) using Aspen Plus. Both simulation results and corresponding inputs are stored in both .csv files with raw data and .json files with the object structure developed for distillation calculations.

As part of the data gathering process, raw data undergoes pre-processing. Non-converged simulations are filtered out, and normalization is applied to enhance machine learning model performance and prevent training issues. An essential pre-processing step is defining the input and output structure of the model. In this methodology, the model's inputs include feed compositions, pressure, temperature, and distillate flows, while the outputs encompass column design variables (number of trays, number of feed trays, reflux ratio, distillate to feed ratio), reboiler and condenser temperatures and duties, and the annualized fixed cost of production (FCOP) of the column. This input/output (I/O) structure resembles to the shortcut model, compensating for its lack of accuracy by training the surrogate model with data from rigorous simulations.

Finally, the data used for surrogate model training is divided into 3 subsets: a training set, a validation set, and a test set. These splits are standard in machine learning training procedures, with the training set used for model training, the validation set for adjusting hyperparameters, and the test set for evaluating model performance with "unseen data."

2.2. Model Training

Distillation, inherently nonlinear due to phase equilibrium between liquid and vapor in the column, requires nonlinear machine learning models. Once the machine learning model is chosen, a dedicated implementation is performed to unlock hyperparameter tuning and architecture modification capabilities.

Regarding the performance, the r^2 factor is the metric chosen over other regression metrics since its scale ranges from 0 to 1. The learning curves are useful for assessing the model's performance, as well as for hyperparameter tuning. Learning curves are valuable for assessing model performance and hyperparameter tuning. The divergence between the training and validation curves suggests overfitting, while a significant gap between the curves indicates the model may not be effectively learning the relationships between inputs and outputs.

2.3. Optimization framework embedding

Given the complexity of implementing gradient-based methods with nonlinear machine learning models, a stochastic (derivative-free) optimization method is recommended. In this work, the chosen method is particle swarm optimization (PSO).

3. Case Study

The general case study (regarding compositions and conditions for streams) is derived from *Somoza et al. 2020*. The target product is ethylene, with an alternative sustainable upstream process based on polyethylene pyrolysis; hence the aim to train the surrogate model toward the C2 fractionator column, which is the main product of the case study.

To capture compositions and design parameters resembling the C2 column, the distillation train feed stream from the case study is taken and fed into a cascade of two shortcut models in Aspen HYSYS V12. The first models the C1 column upstream, while the second estimates the design parameters for the C2 column; number of trays, feed tray, and reflux ratio. The first shortcut specifies molar fractions of light and heavy key components in bottoms and distillate, uniformly distributed from 0.1 to 0.001. Both shortcuts are set with a reflux ratio spec of 1.5 times the minimum reflux.

Rigorous simulations in Aspen Plus V12 are carried out using a RadFrac model for column simulations, with bounds from the shortcut simulations. Compositions are derived from the shortcut simulations, and the remaining parameters have their ranges defined as shown in Table 1. Pressure, feed temperature, and distillate to feed ratio (DFR) values are within commonly seen ranges in literature, such as in the work of *Spallina et al., 2017*. Combining 1,000 normally distributed values from Table 1 with 100 different compositions from the shortcut simulations results in 100,000 rigorous simulations. These simulations yield distillate flows and parameters necessary to calculate the total annualized cost (TAC) of the column, following the equations proposed by Sinnot & Towler, 2020.

Table 1. Bounds for design parameters in data gathering rigorous simulations

	Pressure [kPa]	Feed Temp. [°C]	DFR [-]	Reflux Ratio [-]	Number of trays [-]	Feed Tray [-]
Upper Bound	3200	92	0.7	4.54	40	16
Lower Bound	1600	25	0.3	0.72	16	6

Following the proposed methodology, a multilayer perceptron (MLP) is trained using the Keras API of Tensorflow 2.14; with the following "shortcut-like" input/output structure described in section 2.2. The model architecture is comprised of an input layer matching the input dimension, 8 hidden layers with 256 neurons each, and an output layer with the same size as the output dimension; all densely connected. The training was conducted for 10,000 epochs.

A PSO library from L.J.V Miranda, 2018, *Pyswarms 1.3,* is used to optimize a column with a fixed feed flow (by specifying the input bounds). The optimization problem is:

$$\min TAC \tag{1}$$

$$TAC = FCOP + Ut \tag{2}$$

$$Ut = Q_r \left(CEPCI \cdot \left[7 \cdot 10^{-7} \cdot Q_r^{-0.9} \cdot T_r^{0.5} \right] + C_{S,f} \left[6 \cdot 10^{-8} \cdot T_r^{0.5} \right] \right)$$
$$+ Q_c \left(CEPCI \cdot \left[0.6 \cdot Q_c^{-0.9} \cdot T_c^{-3} \right] + C_{S,f} \left[1.1 \cdot 10^{-6} \cdot T_c^{-5} \right] \right) \tag{3}$$

$$s.t. \, C2_{recov} \geq r_c \tag{4}$$

Where *"FCOP"* in Eq. (2), corresponds to the (annualized) Fixed Cost of Production; Q_r, T_r, Q_c and T_c are the reboiler and condenser duties and temperatures, which are taken as an output of the neural network. The cost for the utilities (*Ut*) in Eq. (3) comes from the work of *Ulrich and Vasudevan, 2006*, as well as the $C_{S,f}$ parameter, used to relate energy with fuel cost. The *CEPCI* corresponds to the Chemical Engineering Plant Cost Index, which value is the one corresponding to 2022 from *ChemEngOnline.com*. The recovery constraint is implemented as a penalty constraint. The optimizing algorithm used was the global optimizer, with cognitive, social and inertia parameters being respectively 1, 1 and 0.0001. The number of particles used in the swarm is 500.

4. Results and discussion

The results of the MLP training are shown in Figure 2, Figure 3 and Table 2. Figure 2 displays the learning curves for the training, showing consistent trends with minor differences. This indicates that the MLP model effectively learns the nonlinear relationships between inputs and outputs without overfitting the data. While the majority of learning is achieved within 2,000 epochs, extending the process to 10,000 epochs enhances model performance. The parity plot in Figure 3 illustrates the correlation between the data and the outputs generated from the MLP model for the FCOP, the variable with the lowest r^2 factor, yet it can be appreciated how the deviations resemble more outliers rather than a deviation trend.

In addition, the calculated R^2 factors with the test dataset, shown in Table 2, reflect the performance of the MLP model in predicting the output values. Moreover, the MLP model outperforms in terms of computational time the rigorous models, in single simulations by a factor of x17, and in performing multiple simulations, thanks to the parallelization capabilities of the neural network, by a factor of x1500.

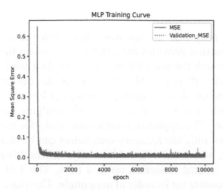

Figure 2. Learning curves of MLP

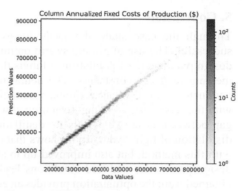

Figure 3. FCOP parity plot

This computational time advantage plays a crucial role in the optimization step, where multiple function evaluations are performed. For the optimization, values in Table 3 are used as a case study, retrieved from the test data slice, with the lowest cost that meets the recovery constraint $r_c \geq 0.95$.

The data from Table 3 is processed by the surrogate-based optimization methodology proposed, as well as in the "traditional" workflow for column design (first use a shortcut model to obtain the design variables, and use those results in a rigorous model). For a more meaningful comparison, the values are processed in a different instance with the pressure and temperature values from the optimization. Table 4 lists the design parameters and results. This reveals that using a rigorous simulation based on the parameters from a shortcut model yields a case with low cost, but does not meet the C2 recovery constraint. Using the design parameters from the shortcut model and the parameters obtained from the optimization, the constraint is still not met (despite better recovery), in addition to a significant cost increase. From the optimization output, the cost is much lower and the recovery constraint is met. However, when using the feed conditions and design parameters in a rigorous simulation, although the C2 recovery is increased significantly, the cost suffered more than a twofold increase. Nevertheless, this case meets the constraints and has less cost than the case where the shortcut design parameters and optimized pressure and temperature for the feed were used.

Table 2. R^2 factors of the trained MLP

	Number of trays	Feed Tray	Reflux Ratio	DFR	Reb. Temp	Cond. Temp.	Reb. Duty	Cond Duty	FCOP
R^2	0.978	0.981	0.988	0.992	0.999	0.998	0.993	0.994	0.939

Table 3. Feed stream summary

	Methane	Ethylene	Propylene	Propyne	1-Butene	1,3-Butadiene	Benzene	Temp. [°C]	Pres. [kPa]
Flow [kg/s]	0.00778	2.31663	0.91118	0.05499	0.09434	0.58862	0.69578	49.89	1603

Table 4. Parameters and results of different approaches

Case	Temp. [°C]	Pres. [kPa]	NT [-]	FT [-]	RR [-]	DFR [-]	C2 Recov. [-]	TAC [$]
Shortcut	49.89	1603	13.58	10.49	1.17	0.4845	0.971	N/A
Shortcut + Rigorous	49.89	1603	14	11	1.17	0.4845	0.882	1,557,575
Shortcut + Rig. + Opt. P,T	45.13	2230	14	11	1.17	0.4845	0.929	3,587,728
Optimization	45.13	2230	32.14	11.71	2.84	0.6380	0.974	1,051,084
Optimization + Rigorous	45.13	2230	32	12	2.84	0.6380	0.999	2,227,929

5. Conclusion

Through the case study, the methodology proposed in this work has been proven successful. The use of particle swarm optimization provides the capability to obtain the design parameters of distillation columns through the use of surrogate modelling while meeting recovery constraints. The neural network is capable of providing the design parameters for a distillation column, resembling the shortcut model input/output structure, with the reliability of having been trained with data from rigorous models, bridging the gap between these two models for distillation. The application on a case study for the distillation of light hydrocarbon shows that not only the recovery constraints can be met with this method, but also improved up to an additional 4%. Additionally, the resulting design parameters and feed conditions lead to cost reduction up to a 62%. The TAC obtained from the optimization provides an estimate for its order of magnitude. Therefore, this methodology holds great promise for speeding up the design process of distillation columns compared to the traditional workflow, while meeting constraints.

References

D. S. Mallapragada, Y. Dvokin, M. A. Modestino, D. V. Esposito, W. A. Smith, B. Hodge, M. P. Harold, V. M. Donnelly, A. Nuz, C. Bloomquist, K. Baker, L. C. Grabow, Y. Yan, N. N. Rajput, R. L. Hartman, E. J. Biddinger, W. S. Aydil, A. D. Taylor, 2023, Decarbonization of the chemical industry through electrification: Barriers and opportunities, Joule, 7, 1, 23-41.

G. Lopez, D. Keiner, M. Fasihi, T. Koiranen, C. Breyer, 2023, From fossil to green chemicals: sustainable pathways and new carbon feedstocks for the global chemical industry, Energy & Environ. Sci., 26, 2879.

N. Quirante, J. Javaloyes, J. A. Caballero, 2015, Rigorous Design of Distillation Columns Using Surrogate Models Based on Kriging Interpolation, AIChE Journal, 61, 7, 2169 - 2187.

D. Ibrahim, M. Jobson, J. Li, G. Guillén-Gosálbez, 2018, Optimization-based design of crude oil distillation units using surrogate column models and a support vector machine, Chemical Engineering Research and Design, 134, 212-225.

S. X. H'ng, L. Y. Ng, D. K. S. Ng, V. Andiappan, 2021, Optimising of vacuum distillation units using surrogate models, IOP Conf, Ser.: Mater. Sci. Eng., 1195, 012050.

T. Keßler, C. Kunde, N. Mertens, D. Michaels, A. Kienle, 2018, Global optimization of distillation columns using surrogate models, SN Appl. Sci. 1, 11(2019).

A. Somoza-Tornos, A. Gonzalez-Garay, C. Pozo, M. Graells, A. Espuña, G. Guillén-Gosálbez, 2020, Realizing the Potential High Benefits of Circular Economy in the Chemical Industry: Ethylene Monomer Recovery via Polyethylene Pyrolysis, ACS Sustainable Chem. Eng., 8,9, 3561 – 3572.

V. Spallina, I. Campos Velarde, J.A. Medrano Jimenez, H. Reza Godini, F. Gallucci, M. van Sint Annaland, 2017, Techno-economic assessment of different routes for olefins production through the oxidative coupling of methane (OCM): Advances in benchmark technologies, Energy Conservation and Management, 154, 244 – 261.

R. Sinnot and G. Towler, Chemical Engineering Design (Sixth Edition), 2020.

G. D. Ulrich, P. T. Vasudevan, How to Estimate Utility Costs, 2006, Chemical Engineering, 113, 66-69.

L. J. V. Miranda, 2018, PySwarms: a research toolkit for Particle Swarm Optimization, The Journal of Open Source Software, 3(21) 433.

The Chemical Engineering Plant Cost Index. Chemical Engineering, http://www.chemengonline.com

Flavio Manenti, Gintaras V. Reklaitis (Eds.), Proceedings of the 34th European Symposium on Computer Aided Process Engineering / 15th International Symposium on Process Systems Engineering (ESCAPE34/PSE24), June 2-6, 2024, Florence, Italy

Towards PEA Matching from Simulation as Part of a Digital Twin Concept for Scale-Up in Modular Plants

Amy Koch[a*], Jonathan Mädler[a], Andreas Bamberg[b], Leon Urbas[a]

[a]*Chair of Process Control Systems & Process Systems Engineering Group,*
TU Dresden, Dresden 01062, Germany;
[b]*Merck Electronics KGaA, Frankfurter Str. 250, Darmstadt 64293, Germany,*
amy.koch@tu-dresden.de

Abstract

Digital artefacts stored in different engineering tools must be semantically linked as part of a Digital Twin (Rosen et al., 2019). However, in the case of Digital Twins for modular plants made up of process equipment assemblies (PEAs) according to VDI 2776-1, different domains interact with different phases across a PEA asset lifecycle and a modular plant lifecycle, challenging existing vendor independent standards for information models (Koch et al., 2023b). To address this challenge, we present a new information model based on OntoCAPE and DEXPI P&ID to support selecting a PEAs from a PEA pool based on queries. To illustrate the functionality of the presented information model, an implementation is presented for matching a Buchwald-Hartwig reaction to a stirred tank reactor PEA using the Siemens engineering toolchain of gPROMS and COMOS. An example query is also provided.

Keywords: Digital Twin, modular plants, PEA pool, PEA match, information model

1. Motivation

The utilization of Modular Plants is crucial to maintain a competitive edge in the dynamic chemical and pharmaceutical markets by reducing time-to-process. Further benefit can be leveraged by combination of a Digital Twin with predesigned Process Equipment Assemblies (PEA) according to VDI 2776-1, a standard outlining modular plants. A Digital Twin is a semantically linked collection of all digital artefacts, including design and engineering data as well as operational data and behavior descriptions. Here, this semantic linkage between digital artefacts stored in different engineering tools is an essential part of a Digital Twin (Rosen et al., 2019). This semantic linkage is supported by descriptive models. However, in the case of Digital Twins for modular plants made up of process equipment assemblies (PEAs) according to VDI 2776-1, different domains interact with different phases across a PEA asset lifecycle and a modular plant lifecycle, challenging existing vendor independent standards for information models (Koch et al., 2023b). In this paper we present a new information model to support selecting a PEA from a PEA pool based on querying technical criteria (Schindel et al., 2021). The presented information model extends the OntoCAPE with the DEXPI P&ID standard to support PEA matching as part of a Digital Twin concept for scale-up in modular plants. This paper is structured as follows: in section 2, relevant workflows and lifecycles for Digital Twins are presented and aligned. Additionally, relevant standards for information models are presented, requirements for information exchanged during PEA matching is outlined, and the distinction between PEA types and PEA instances is discussed. Section 3 introduces the case study, information model architecture with the Siemens toolchain of gPROMS and COMOS is presented, and an example query is provided. We conclude with a discussion of the results and an outlook toward future work.

2. Methodology

2.1. Digital Twin Workflows in COMOS and gPROMS

Engineering of modular plants requires adapted workflows compared to conventional plants. As indicated by the VDI 2776-1, a significant portion of the design phase is dedicated to the matching of PEAs instead of the detailed engineering phase for conventional plants. Additional benefit can be taken from incorporating a Digital Twin combing structural, behavioral, and descriptive models. Koch et al. (2023a) has presented a Digital Twin workflow in COMOS and gPROMS based on stakeholder requirements. To support this Digital Twin workflow, descriptive models based on vendor-independent standards for information models are required. For this purpose, the Digital Twin workflows from Koch et al. (2023a) refined with the procedure model for process validation in modular plants from Mädler (2023) (see Figure 1). Additionally, this workflow is aligned to the Modular Plant Lifecycle (Mädler et al., 2022a) and the PEA Lifecycle (Menschner et al., 2018; Wiedau et al., 2019) presented in Koch et al. (2023b) to evaluate the suitability of existing vendor independent standards to describe the required information in each lifecycle phase. The workflow is described as follows: (1) product specification is provided from product development, (2) development of process steps in gPROMS, (3) block flow diagram (BFD) with technical criteria in COMOS, (4) query of PEA types in COMOS, (5) query of PEA instances in COMOS, followed by (6) additional steps in gPROMS. For the focus on this paper, emphasis is placed on the information used in steps 4 and 5 to describe PEAs for the PEA matching process.

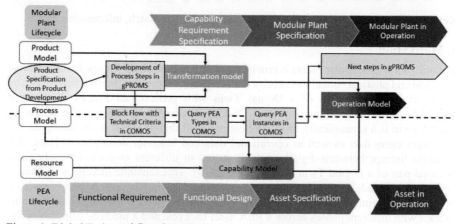

Figure 1: Digital Twin workflow from Koch et al. (2023a) derived from Mädler (2023) aligned to the PEA Lifecycle and Modular Plant Lifecycles illustrated in Koch et al. (2023b).

2.2. Methodology for PEA Matching

The selection of PEAs is well described in literature. The VDI 2776-1 outlines the PEA selection process generally, describing a database with modular planning documentation to compare process requirements and PEA characteristics. In literature, Harding et al. (2021) have also proposed using technical criteria along with a Master Block Flowchart describing required process functions to evaluate different potential technologies while Schindel et al. (2021) suggest matching PEAs based on technical and evaluate criteria using matching matrices. First, technical criteria are used to knock out unsuitable technologies. The suitable technologies are ranked based on evaluative criteria.

2.3. Selection of PEA Types and Instances

In the procedure workflow of Mädler (2023), the querying of PEA types serves to match the capability requirements with the capabilities of the PEA using technical and evaluative criteria of Schindel at al. (2021), process functions from Harding et al. (2021), and the Module Type Package (MTP) (VDI/VDE/NAMUR 2658). Once non-suitable technologies are knocked out, a list of potential PEA types is generated. In the subsequent step, PEA instances are queried and again a list of potential PEA instances is propagated and ranked based on the relevant technical and evaluative criteria (Mädler, 2023). Klose et al. (2022) also distinguish between type-based and instance-based engineering workflows. Here, type-based engineering is part of the functional design phase while instance-based engineering is part of the asset specification phase. Analogue to the DT workflow in Figure 1, the process begins with a general recipe (IEC 61512-1), from which a block flow diagram (ISO 10628), functional plant topology, process-based safety requirements, and process-based service requirements are derived. For instance-based engineering, the MTP (VDI/VDE/NAMUR 2658) is used to define the software instances, from which the plant topology is derived and used for recipe design, design of control and interlocks, as well as safety aspects such as HAZOP.

2.4. Vendor-Independent Standards

Koch et al. (2023b) illustrated the suitability of DEXPI P&ID, OntoCAPE, IEC 61512-1, and VDI/VDE/NAMUR 2658 to describe the different phases of the PEA and Modular Plant Lifecycles. For the functional design phase, OntoCAPE and DEXPI P&ID were classified as potentially suitable to describe the information exchanged during this phase, which DEXPI P&ID, OntoCAPE, IEC 61512-1 and VDI/VDE/NAMUR 2658 were all considered to be potentially suitable for description of the asset specification phase. DEXPI P&ID is an industry wide standard for the exchange of P&ID data between CAE tools, OntoCAPE is an ontology describing the domain of chemical engineering, VDI/VDE/NAMUR 2658 describes the MTP which describe the automation perspective of the PEA, and IEC 61512-1 is a widely adopted information model for describing Batch Control and is used in the VDI/VDE/NAMUR 2658 for the description of services.

3. Case Study

3.1. Example System

The owner/operator (O/O) of a PEA is often from the pharmaceutical and speciality chemical industry where rapid time to market and short product lifecycles are a consistent reality and challenge. Thus, the Buchwald-Hartwig coupling, a palladium catalyzed reaction mechanism for the synthesis of aryl amines from amines and aryl halides, was selected as a case study due to its relevance in the production of pharmaceutical products. The overall schema for the Buchwald-Hartwig reaction mechanism is shown in (1):

$$\tag{1}$$

In the reaction schema, X is Cl or Br, R_1 is an Alkyl, CN, or a CON group, and R_2 is an Alkyl or Aryl group. In addition to the generalized reaction schema, relevant technical criteria, cf. Mädler (2023) for the Buchwald-Hartwig reaction, an STR PEA type description, and a PEA instance of an STR "M02" are provided in Table 1.

Table 1: Technical Criteria for selection of reactor type with technical criteria for the Buchwald-Hartwig reaction mechanism, an STR PEA type, and a PEA instance (c.f. Mädler, 2023).

Technical Criteria	Buchwald-Hartwig Reaction	PEA Type STR	PEA Instance M02
Reaction Type	Liquid-Liquid	Liquid-Liquid	Liquid-Liquid
Pressure	Atmospheric	Mild	Mild
Temperature	25 – 100°C	30 – 150 °C	25 – 100°C
Reaction Time	6 to 24 hours	Broad	Broad

3.2. OntoCAPE and DEXPI P&ID

The goal of the information model is to describe the capability models of the PEA type and instance. As this is the first step towards PEA matching, the information model only considers the technical criteria as described by Schindel et al. (2021). As this information in the PEA pool must be searchable, linked data is used due to enable searchability with SPARQL or SQL query languages. As a first step, OntoCAPE (Marquardt et al., 2010), is used due to its suitability to describe both the process and PEA resources domains (Koch et al., 2023b). Also, DEXPI P&ID (Wiedau et al., 2019) is used to extend the descriptions of the PEA instances. Although for the information model in this paper, OntoCAPE would be entirely suitable, DEXPI P&ID has large support in the process industry, meaning that the P&ID data of documented PEA instances might often already exist in DEXPI P&ID or is easily done so using a plant engineering tool such as COMOS. The information model architecture using OntoCAPE is as follows: to describe the relationship between capability requirements for PEA matching with the PEA capabilities, the *System Module* and *Technical System Modules* in the Upper Layer of OntoCAPE are used. In the conceptual design phase, *system requirements* (functional requirements) are transformed into system functions (PEA type) during the basic design (functional design phase), which are detailed in the *system realization* (PEA instance). Ultimately it is the capabilities of PEA instance which constrain the *system behavior* during the asset specification and operation phases. This is in good agreement with the proposed Digital Twin workflow and the alignment to the PEA lifecycle in Figure 1.

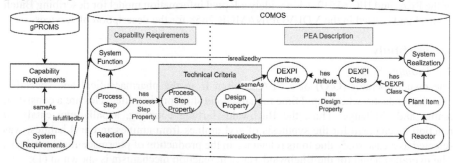

Figure 2: Linked-data based information model architecture for querying PEAs in COMOS.

Figure 2 illustrates the architecture of the information model with respect to the data sources of gPROMS and COMOS. Here, the PEA description is extended with the DEXPI P&ID classes and attributes as DEXPI P&ID is widely used in the process industry to document the equipment capabilities. The information model architecture in Figure 2 shows how the relevant technical criteria for PEA matching for both PEA types and instances can be linked to the capability requirements propagated by gPROMS and transferred to COMOS via BFD. In order to query the PEA instances and types, capability models must be catalogued in a PEA Pool database. The technical criteria of a PEA are described using the *ProcessStep* and capability requirements are described with *ProcessStepProperty*. For the PEA description, the *Plant Item* is used and the *DesignProperty* describes the technical capabilities. Here we suggest to extend the PEA description with DEXPI P&ID, due to its acceptance in the process industry to describe the equipment topography and design capabilities. The distinction is that PEA type has a more generalized description and the PEA instance has a more concrete constraint of the system behavior as the description considers additional aspects beyond the technical criteria. Next, we use the architecture in Figure 2 to describe the capability requirements of the Buchwald – Hartwig reaction with the PEA type "STR", and the PEA instance M02. These connections are modelled in Figure 3. Here, a distinction is made between the PEA type "STR" and the PEA instance "M02" via the use of DEXPI P&ID classes and attributes for the description of the technical criteria of the M02 PEA instance.

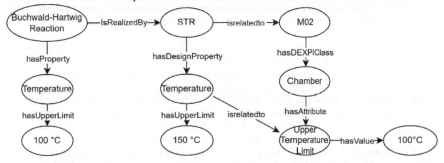

Figure 3: Connection of the Buchwald-Hartwig reaction with the STR PEA type and with the PEA Instance M02 for the technical criteria *upper temperature limit* using OntoCAPE and DEXPI P&ID.

3.3. Formulation of Queries

The goal of the queries is to generate lists of suitable PEA types and instances, with the intention of ranking the suitability based on technical and evaluative criteria (Mädler, 2023). We propose the use of the SPARQL query language to search the capability models of both PEA types and instances in the PEA Pool. In the example provided in Table 2, all PEA types are queried for the ProcessStep of reaction, and the ?Properties, which are the technical criteria, are displayed for two example reactor types (STR and PFR). We propose, that these technical criteria-based queries be used to propagate the lists of suitable PEA types and PEA instances. However, to evaluate these criteria, the methods from Schindel et al. (2021) or Harding et al. (2021) should be employed.

Table 2: Example result from a SPARQL query for ProcessStep = Reaction from a PEA databank.

Query	Result		
	?Property	STR	PFR
?Reaction?Properties	?ReactionType	Liquid-liquid	Gas-liquid; liquid-liquid
	?ReactionTime	Broad	Narrow
	?Temperature	30 – 150 °C	80 – 200 °C
	?Pressure	Mild	Up to 20 bar

4. Discussion and Conclusion

This paper presented an information model describing capability models of both PEA types and instances, with emphasis on the technical criteria. Using this information model, an example query and result was illustrated for PEA types for the process step "reaction". As a use case, the technical criteria from the Buchwald-Hartwig reaction linked to a stirred tank reactor type, and a specific PEA instance of M02. Although the information model presented in this paper is a reasonable step towards PEA matching, the problem is in reality much more complex. Here, aspects such as service matching and recipes must be considered with respect to the IEC 61512-1 and VDI/VDI/NAMUR 2658 standard for the MTP. Additionally, the product specification in the DT workflow is considered to be a recipe therefore alignment and integration of the IEC 61512-1 is a key next step to expand upon the presented information model.

References

A. Bamberg, L. Urbas, S. Bröcker, M. Bortz, and N. Kockmann, 2021. The Digital Twin – Your Ingenious Companion for Process Engineering and Smart Production. Chemical Engineering & Technology 44, 954–961. https://doi.org/10.1002/ceat.202000562

D. Harding, M. Polyakova, D. Nowara, S. Rech, M. Grünewald, & C. Bramsiepe, 2021. New process function-based selection and configuration methodology for Process Equipment Assemblies (PEAs) exemplified on the unit operation distillation. Chemical Engineering and Processing - Process Intensification, 168, 108531. https://doi.org/10.1016/j.cep.2021.108531

A. Klose, J. Lorenz, L. Bittorf, K. Stark, M. Hoernicke, A. Stutz, H. Weinhold, N. Krink, W. Welscher, M. Eckert, S. Unland, A. Menschner, P. Da Silva Santos, N. Kockmann, L. Urbas, 2022. Orchestration of modular plants: Procedure and application for orchestration engineering. atp 63, 68–77. https://doi.org/10.17560/atp.v63i9.2599

A. Koch, J. Mädler, A. Bamberg, and L. Urbas, Digital Twins for Scale-Up in Modular Plants: Requirements, Concept, and Roadmap. In A. C. Kokossis, M. C. Georgiadis, & E. Pistikopoulos (Eds.), Computer Aided Chemical Engineering (Vol. 52, pp. 2063–2068). Elsevier. https://doi.org/10.1016/B978-0-443-15274-0.50328-0.

A. Koch, N. Hamedi, L. Furtner, T. Kock, A. Klose, & J. Mädler, (2023). Standards for Information Models Considering Knowledge Distribution in Modular Plants. 2023 IEEE 21st International Conference on Industrial Informatics (INDIN), 1–7. https://doi.org/10.1109/INDIN51400.2023.10218218

IEC 61512-1:1997, "Batch control - Part 1: Models and terminology." IEC, Aug. 1997.

W. Marquardt, J. Morbach, A. Wiesner, and A. Yang, OntoCAPE. in RWTHedition. Berlin, Heidelberg: Springer Berlin Heidelberg, 2010. doi: 10.1007/978-3-642-04655-1

J. Mädler, 2023. Smarte Process Equipment Assemblies zur Unterstützung der Prozessvalidierung in modularen Anlagen.

J. Mädler, J. Rahm, I. Viedt, and L. Urbas, 2022, A digital twin-concept for smart process equipment assemblies supporting process validation in modular plants. In L. Montastruc & S. Negny (Eds.), Computer Aided Chemical Engineering (Vol. 51, pp. 1435–1440). Elsevier. https://doi.org/10.1016/B978-0-323-95879-0.50240-X

A. Menschner, S. Hensel, L. Urbas, T. Holm, C. Schäfer, T. Scherwietes, A. Stutz, M. Maumaier, and M. Hoernicke, 2018."Module Lifecycle Model for Modular Plants," Chem. Ing. Tech., 90(9), 1220-1220, doi: 10.1002/cite.201855192.

R. Rosen, J. Fischer, & S. Boschert, 2019, Next generation digital twin: An ecosystem for mechatronic systems?, IFAC-Paper, 52, 15, 265-270. https://doi.org/10.1016/j.ifacol.2019.11.685

A.-L. Schindel, M. Polyakova, D. Harding, H. Weinhold, F. Stenger, M. Grünewald, and C. Bramsiepe, 2021, General approach for technology and Process Equipment Assembly (PEA) selection in process design, CEP:PI, 159, 108223. https://doi.org/10.1016/j.cep.2020.108223

VDI/VDE/NAMUR, 2019, VDI/VDE/NAMUR 2658-Automation engineering of modular systems in the process industry-General concept and interfaces – Part 1, Beuth Verlag GmbH.

VDI 2776-1:2020, "Process engineering plants - Modular plants - Fundamentals and planning modular plants - Part 1 (VDI 2776:2020-11)." Beuth Verlag, Nov. 2020.

M. Wiedau, L. von Wedel, H. Temmen, R. Welke and N. Papkonstantinous, 2019. ENPRO Data Integration: Extending DEXPI Towards the Asset Lifecycle. Chemie Ingenieur Technik, 91(3), 240–255. https://doi.org/10.1002/cite.201800112

Flavio Manenti, Gintaras V. Reklaitis (Eds.), Proceedings of the 34th European Symposium on Computer Aided Process Engineering / 15th International Symposium on Process Systems Engineering (ESCAPE34/PSE24), June 2-6, 2024, Florence, Italy

Comparative Study on Hydrothermal Gasification and Thermal Gasification via Hydrothermal Carbonization of Digestate Residues from Anaerobic Digestion

Fadilla Noor Rahma[a]*, Khanh-Quang Tran[b], Roger Khalil[c]

[a]*Norwegian University of Science and Technology, Høgskoleringen 1, Trondheim 7034, Norway*
[b]*SINTEF Energy Research, Sem Sælands vei 11, Trondheim 7034, Norway*
fadilla.n.rahma@ntnu.no

Abstract

Anaerobic digestion (AD), one of the key technologies for achieving the climate-neutral target, is still facing various digestate management issues. The digestate's high organic matter and energy content make it a potential feedstock for further processing into value-added products, i.e., through thermochemical conversion technologies. However, its high moisture content poses additional challenges related to extensive energy requirement for feedstock drying. This paper investigates two technologies for efficient processing of digestate, i.e., 1) hydrothermal gasification (HTG) and 2) hydrothermal carbonization coupled with thermal gasification of hydrochar (HTC-TG). Both technologies can convert digestate into producer gas while avoiding the energy-extensive drying step. Aspen Plus software is used to compare the performance of the two processes. The process performance is comparatively assessed based on the producer gas generation and overall system efficiency (OSE).

Keywords: Hydrothermal gasification, hydrothermal carbonization, thermal gasification, anaerobic digestion, Aspen Plus.

1. Introduction

Anaerobic digestion (AD) for biogas production from low-grade biomass is one of the essential technologies in the transition towards climate-neutral energy production. Despite its potentials, the implementation of AD technology is currently facing significant challenges related to digestate management (Nkoa 2014). The digestate stream is rich in organic matters and energy content, making it suitable for generating value-added products, heat, and power. Valorization of digestate through gasification technology is particularly promising since it mainly generates producer gas, a valuable source for alternative fuels or chemicals. However, digestate's high moisture content leads to substantial energy demand during the initial drying step in gasification (Parmar and Ross 2019).

This paper investigates two potential gasification strategies to avoid the energy-intensive drying step, i.e., 1) hydrothermal gasification (HTG) and 2) thermal gasification through hydrothermal carbonization of digestate (HTC-TG). HTG is performed in supercritical water, where water acts as solvent, reactant, and catalyst to assist the chemical reactions (Tekin, Karagöz et al. 2014). The feedstock drying process is therefore not necessary during HTG, making it an interesting option for processing of high-moisture biomass such as digestate. Another strategy is to employ hydrothermal carbonization (HTC) for digestate pre-processing before thermal gasification (TG). HTC, which takes place in subcritical water environment, can assist mechanical dewatering of high-moisture biomass (Salaudeen, Acharya et al. 2021), thus reducing the feedstock drying requirement. Additionally, hydrochar from HTC is known to have improved fuel properties, leading to better efficiency of the TG process (Gai, Guo et al. 2016).

In comparison to TG, HTG is carried out at lower temperature without the energy-intensive drying step (Kumar, Oyedun et al. 2018). However, the technology requires considerable energy for the high-pressure reactor pumping system (Macrì, Catizzone et al. 2020). On the other hand, HTC-TG involves relatively lower pressure, with lower drying energy than stand-alone TG. Nevertheless, drying is still required to some extent, and the gasifier temperature is higher than HTG. These operational aspects can significantly influence the energy performance of both systems. To the best of the authors' knowledge, research comparing the energy performance of HTG and HTC-TG has not been performed. Therefore, the present study aims to comparatively evaluate the energy performance of HTG and HTC-TG of biogas digestate.

2. Methods

2.1. Process Simulation

Aspen Plus models of HTG and HTC-TG are employed to comparatively assess the performance of both systems. The HTG and HTC-TG models are presented in Figure 1 and Figure 2, respectively. The digestate feedstock is regarded as non-conventional component specified by its ultimate composition, as displayed in Table 1.

Both the HTG and TG models are based on thermodynamic equilibrium using Gibbs free energy minimization principle. This approach is widely used for evaluating the performance of gasification systems. Previous works employing the same model have been proven to accurately represent the experimental data for both HTG system (Hantoko, Kanchanatip et al. 2019, Okolie, Nanda et al. 2020) and TG system (Salaudeen, Acharya et al. 2021).

2.1.1. Hydrothermal Gasification (HTG)

The digestate and water mixture enter the system as stream FEED. The PUMP and HEATER increase the temperature and pressure of the feedstock mixture to the reactor condition, as specified in Table 3. The HTG reactor is theoretically represented with two reactor blocks in Aspen Plus, i.e., RYIELD (HTGYIELD) and RGIBBS (HTGGIBBS). The HTGYIELD breaks down the non-conventional digestate component into its element (C, H, O, N, S, and ash) according to its ultimate composition. To perform this operation, a Fortran statement is written on a calculator block embedded in the HTGYIELD reactor. The decomposed feedstock from HTGYIELD is reacted in HTGGIBBS, which performs the Gibbs free energy minimization. The resulting products are flown through COOLER and VALVE to reduce the temperature and pressure before being separated in SEP.

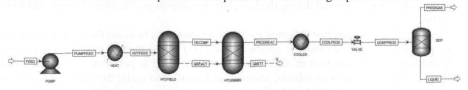

Figure 1. Process Flow Diagram of HTG in Aspen Plus

2.1.2. Hydrothermal Carbonization Integrated with Thermal Gasification (HTC-TG)
The stream FEED, consisting of digestate and water mixture, is flown through PUMP 1
and HEAT1 before being fed to the HTC unit. Due to the complexity of HTC reaction
kinetic and mechanism, the data for HTC unit in the simulation is taken from experimental
results available in literature (Ghavami, Özdenkçi et al. 2022). The reference data
included the mass balance of the HTC system and the ultimate composition of hydrochar,
as displayed in Table 2. In addition, energy requirements for the dewatering and drying
steps are approximated according to experimental filtration (Aragon-Briceño, Pożarlik et
al. 2022) and thermal drying of hydrochar (Zhao, Shen et al. 2014).

The dried hydrochar in stream TGFEED enters the gasification reactor system. Similar
with the HTG process, the TG system is also theoretically modelled with RYIELD and
RGIBBS reactor blocks, denoted by TGYIELD and TGGIBBS, respectively. TGYIELD
decomposes the non-conventional hydrochar component into its elements with the aid of
a calculator block, whereas TGGIBBS carries out the Gibbs free energy minimization.
The product flows through SOLIDSEP for solid removal and COOLER and SEP for
separation of condensate.

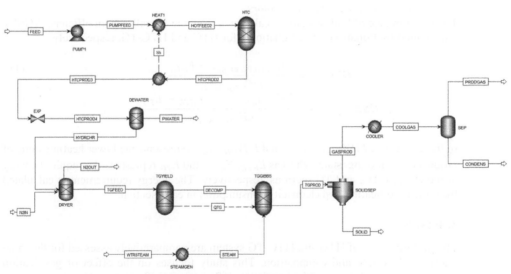

Figure 2. Process Flow Diagram of HTC-AD in Aspen Plus

Table 1. Input Data from Characterization of Digestate Feedstock

Parameters	Value	Units
C	44.10	%
H	5.10	%
O	31.30	%
N	3.20	%
S	0.30	%
Ash	16.00	%
HHV	17.80	MJ/kg

Table 2. Input Data from Characterization of Hydrochar from HTC

Parameters	Value	Units
C	51.20	%
H	6.10	%
O	22.80	%
N	3.50	%
S	0.00	%
Ash	16.30	%
HHV	20.90	MJ/kg

Table 3. Parameters for Simulation

Parameters	HTG	TG
Temperature	400-600 °C	600-1000 °C
Pressure	250 bar	1 bar
Digestate wt%	30%	30%

2.2. Performance Assessment Indicators

The performance of both systems is calculated from the overall system efficiency (OSE), represented by Equation 1 and Equation 2 for HTG and HTC-TG, respectively.

$$OSE_{HTG} = \frac{LHV_{Producer\ gas} - E_{HTG}}{LHV_{Digestate}} \tag{1}$$

$$OSE_{HTC-TG} = \frac{LHV_{Producer\ gas} - E_{HTC} - E_{TG}}{LHV_{Digestate}} \tag{2}$$

In the equations, $LHV_{Producer\ gas}$ and $LHV_{Digestate}$ represents the lower heating value of producer gas and digestate; whereas E_{HTG}, E_{HTC}, and E_{TG} represents the required energy in the HTG, HTC, and TG sections, respectively. The energy requirement is calculated from the total heating and electricity consumption in each section.

3. Results

The performances of HTG and HTC-TG system are comparatively assessed for the same digestate flow rate and composition. This study focuses on the effect of gasification temperature, which is one of the key factors influencing gasification systems performance (Sanaye, Alizadeh et al. 2022). The other parameters for simulation are summarized in Table 3.

(a) 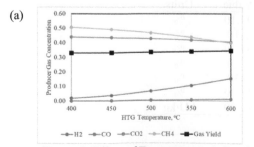   (b)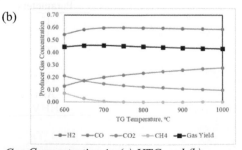

Figure 3. Effect of Temperature on Producer Gas Concentration in (a) HTG and (b) HTC-TG Systems

Figure 3 (a) and (b) display the comparison of gas yield and composition produced from HTG and HTC-TG systems, respectively. The results suggest that temperature plays a significant role influencing producer gas composition in both systems. In general, higher temperature results in increasing H_2 and CO concentration, whereas CH_4 and CO_2 production decrease. The effect of temperature can be explained by the basic law of chemical reactions: exothermic reactions are favored by lower temperature, whereas endothermic reactions are enhanced by higher temperature (Salaudeen, Acharya et al. 2021). At the particular range of operating conditions observed in this study, HTC-TG is more suitable for hydrogen production, whereas HTG produces higher methane concentration. However, other influencing factors such as feedstock-to-steam ratio and feedstock concentration can significantly change the process behavior.

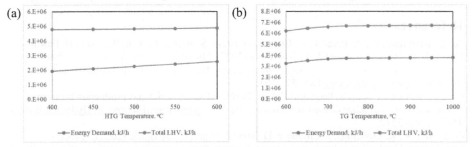

Figure 4. Effect of Temperature on Total Energy Demand and Producer Gas LHV in (a) HTG and (b) HTC-TG Systems

Figure 4 (a) and (b) display the system total energy demand and the product heating value, which is calculated by multiplying the producer gas flow rate with the producer gas LHV. The result shows that for both HTG and HTC-TG systems, total energy demand and heating value increase with higher temperature. The increase in energy demand is mostly contributed by the higher heating requirement in the gasification system, with regards to the higher temperature. The increasing temperature also enhances the product heating value in both systems. However, the increase in product heating value is not as significant as the increase in heating requirement, causing the overall system efficiency (OSE) to decrease, as displayed in Figure 5. It is possible to reduce energy demand of the system through better heat integration, which may result in better OSE. Furthermore, it was found that within the operating conditions observed in this study, HTC-TG system performs with higher OSE compared to HTG.

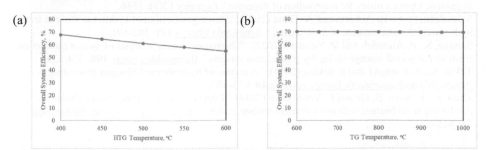

Figure 5. Effect of Temperature on Overall System Efficiency (OSE) in (a) HTG and (b) HTC-TG Systems

4. Conclusions

Two technologies for efficient processing of digestate, i.e., 1) hydrothermal gasification (HTG) and 2) hydrothermal carbonization coupled with thermal gasification of hydrochar (HTC-TG) are comparatively assessed using Aspen Plus software. The effect of temperature is observed, and it was found that higher temperature has a positive influence on H_2 and CO concentration, whereas CH_4 and CO_2 production are negatively affected. At the operating conditions observed in this study, HTC-TG is more suitable for hydrogen production, while HTG produces more methane. The higher temperature also increases both energy demand and product heating value in both systems. However, the OSE of the two systems decreases with increasing temperature. Within the operating conditions observed in this study, HTC-TG system performs with higher OSE compared to HTG.

References

Aragon-Briceño, C., A. Pożarlik, E. Bramer, G. Brem, S. Wang, Y. Wen, W. Yang, H. Pawlak-Kruczek, Ł. Niedźwiecki and A. Urbanowska (2022). "Integration of hydrothermal carbonization treatment for water and energy recovery from organic fraction of municipal solid waste digestate." Renewable energy **184**: 577-591.

Gai, C., Y. Guo, T. Liu, N. Peng and Z. Liu (2016). "Hydrogen-rich gas production by steam gasification of hydrochar derived from sewage sludge." International Journal of Hydrogen Energy **41**(5): 3363-3372.

Ghavami, N., K. Özdenkçi, S. Chianese, D. Musmarra and C. De Blasio (2022). "Process simulation of hydrothermal carbonization of digestate from energetic perspectives in Aspen Plus." Energy Conversion and Management **270**: 116215.

Hantoko, D., E. Kanchanatip, M. Yan, Z. Weng, Z. Gao and Y. Zhong (2019). "Assessment of sewage sludge gasification in supercritical water for H2-rich syngas production." Process Safety and Environmental Protection **131**: 63-72.

Kumar, M., A. O. Oyedun and A. Kumar (2018). "A review on the current status of various hydrothermal technologies on biomass feedstock." Renewable and Sustainable Energy Reviews **81**: 1742-1770.

Macrì, D., E. Catizzone, A. Molino and M. Migliori (2020). "Supercritical water gasification of biomass and agro-food residues: Energy assessment from modelling approach." Renewable Energy **150**: 624-636.

Nkoa, R. (2014). "Agricultural benefits and environmental risks of soil fertilization with anaerobic digestates: a review." Agronomy for Sustainable Development **34**: 473-492.

Okolie, J. A., S. Nanda, A. K. Dalai and J. A. Kozinski (2020). "Hydrothermal gasification of soybean straw and flax straw for hydrogen-rich syngas production: Experimental and thermodynamic modeling." Energy Conversion and Management **208**: 112545.

Parmar, K. R. and A. B. Ross (2019). "Integration of hydrothermal carbonisation with anaerobic digestion; Opportunities for valorisation of digestate." Energies **12**(9): 1586.

Salaudeen, S. A., B. Acharya and A. Dutta (2021). "Steam gasification of hydrochar derived from hydrothermal carbonization of fruit wastes." Renewable Energy **171**: 582-591.

Sanaye, S., P. Alizadeh and M. Yazdani (2022). "Thermo-economic analysis of syngas production from wet digested sewage sludge by gasification process." Renewable Energy **190**: 524-539.

Tekin, K., S. Karagöz and S. Bektaş (2014). "A review of hydrothermal biomass processing." Renewable and sustainable Energy reviews **40**: 673-687.

Zhao, P., Y. Shen, S. Ge and K. Yoshikawa (2014). "Energy recycling from sewage sludge by producing solid biofuel with hydrothermal carbonization." Energy conversion and management **78**: 815-821.

Flavio Manenti, Gintaras V. Reklaitis (Eds.), Proceedings of the 34th European Symposium on Computer Aided Process Engineering / 15th International Symposium on Process Systems Engineering (ESCAPE34/PSE24), June 2-6, 2024, Florence, Italy

Strategic Decision-Making in Duopolistic Energy and Water Markets: Examining Competition versus Cooperation for Farm Siting Optimization

Sarah Namany, Maryam Haji, Mohammad Alherbawi, Tareq Al-Ansari

College of Science and Engineering, Hamad Bin Khalifa University, Qatar Foundation, Doha, Qatar

*talansari@hbku.edu.qa

Abstract

The purpose of this paper is to delve into the intricate dynamics of duopolistic energy and water markets, wherein the main players are represented by a set of energy technologies inclusive of combined-cycle gas turbines power plants competing with photovoltaics plants. Considering the water sector, competition is happening amongst desalination plants and wastewater treatment plants. Both sectors are feeding an agricultural farm producing a diversified food basket. The first stage of the proposed framework aims to investigate the complex interactions between technologies by means of game theoretic models that simulate scenarios of both competition and cooperation amongst energy technologies and water technologies. Insights are unveiled on how these strategies impact the minimum profitable supply of each technology. This information is then integrated in the second stage of the study where a multi-objective optimisation framework is formulated to determine the optimal farm sites based on soil quality indicators and supply data. The optimisation results are represented by suitability maps that offer a comprehensive view of the interplay between resource supply, environmental conditions, and technology dynamics, aiding in the identification of optimal locations for agricultural farms. By combining economic and environmental considerations, this research paves the way for more sustainable farming practices. Ultimately, this study contributes to a deeper understanding of the strategic decision-making process in duopolistic energy and water markets. It offers a framework for stakeholders to navigate the complexities of these markets, leading to more efficient decisions with regards to farm siting and resource allocation.

Keywords: Duopolistic Markets, Game Theory, Cooperation, Competition, Farm Siting Optimization.

1. Introduction

Ensuring a continuous and sufficient supply of food products is one of the goals that all countries are determined to sustain and improve. However, the ever-increasing internal food system's challenges and the external pressures imposed by the surrounding environment such as political instabilities, market turmoil and climate change are rendering the process of growing nutritious and sufficient food quantities very strenuous (Namany et al., 2022). These challenges do not impact the food system exclusively, yet the influence is also affecting the water and energy sectors which represent the major enablers of the food provision system. In fact, the reliance on these two resources exposes the food system to the risks impacting these sectors as well which further hampers the food security target. Competition and cooperation are some of the behaviors that influence the performance of sectors and their productivity. In the context of the energy and water resources, the market structure is constantly changing due to the frequent emergence of new technologies and volatilities of resources prices such as oil and gas for the energy sector and precipitation for the water system, leading to uneven supply patterns which pose multiple pressures on the provision of food products (Gielen et al., 2019). In order to account for the influence of competitive and cooperative behaviors on food systems performance, game theory approaches are deployed to determine the payoffs and strategies of water and energy industries contributing to food production. In this regard, Yan *et al.* (2023) investigated the oligopoly of the soybean market in China and the influence of monopolistic behaviors on trade activities. Dianat *et al.* (2022) analysed the impact of competitiveness of the renewable energy market on the energy sector's performance based on game theory, agent-based modeling and system dynamics. In this paper, cooperative and non-cooperative games are used to determine the minimum quantities that a set of energy and water technologies should supply for the provision of a diversified perishable food basket such that their economic profits are maximised. The equilibrium payoff amounts generated by the games is then used as capacity constraints deployed in a farm sitting optimisation model. The framework, based on geographic information system (GIS) determines the most appropriate farm locations given land suitability, energy and water technologies dynamics that drive their supply patterns. This paper is structured such that the following section 2 describes the data and methods used to develop the suggested framework while section 3 presents the major findings of the study and finally section 4 concludes and paves the way for future research.

2. Data and Methods

The methodology developed in this paper aims to design a framework that assists decision-makers in the food sector in determining the optimal locations of crop-producing farms given the competitive and cooperative behavior of multiple energy and water technologies. Two different scenarios are developed to represent these behaviors wherein the first scenario assumes a competitive set up between energy technologies and amongst water sources while the second scenario implies their cooperation. For the energy sector, a mix of renewable and non-renewable plants are considered, counting combined-cycle gas turbine (CCGT) and photovoltaics (PV), as for water, the contributing technologies consist of reverse osmosis (RO) and multi-stage flash (MSF) desalination plants and treated sewage effluent (TSE) as a water treatment plant. Tables 1 and 2 summarises the set of technologies utilised in this study along with their geospatial locations. Results of the games are used to locate the optimal farms that supply a range of perishable food products counting vegetables and fruits. The following section 2.1 describes the games adopted to determine the payoff of every contributing technology.

2.1. Game theory to model competitive and cooperative markets

Industries or sectors can adopt different strategies as means to maximise their profits. Profit maximisation is generally enabled by either raising the quantities produced or increasing the selling prices. In order to model competition, the Cournot non-cooperative game is adopted wherein industries that produce to the same product compete and act independently and simultaneously to produce their targeted quantities then the market price is set.

Table 1. Locations of electricity power plants.

		Latitude	Longitude
	Messaid (E1)	24.989	51.577
	Rass Laffan B (E2)	25.924	51.548
CCGT Power Plants	**Rass Laffan C (E3)**	25.935	51.526
PV Power Plant	**Al kharsaah (E4)**	25.22305	51.0182

Table 2. Locations of water treatment and desalination plants.

		Coordinates	
		Latitude	Longitude
MSF Desalination Plants	**Ras Abu Fontas A (W1)**	25.207	51.61646
	QEWC A1 (W2)	25.210	51.61613
	Ras Laffan A (RLPC) (W3)	25.889	51.438
	Umm al-Houl (W4)	25.112	51.612
RO Desalination Plants	**Ras Abu Fontas A3 (W5)**	25.213	51.586
	Ras Abu Aboud (W6)	25.287	51.561
TSE	**TSE Pumping station (W7)**	25.233	51.523
	TSE Pumping station Markhiya (W8)	25.021	51.158
	Lusail Sewage Treatment (W9)	25.408	51.473

In this case, the water technologies and the energy power plants compete within the water and energy markets, respectively, and their best response function can be described by the following equation (Eq.1) (Namany et al., 2023), wherein q_i^* is the Nash equilibrium quantity wherein i is the set of water or energy technologies, a and b are derived from the inverse market demand equation while c is the unit production cost.

$$q_i^* = \frac{a-c_i}{2b} - \frac{\sum_{i \neq j} q_j}{2} \tag{1}$$

The second scenario describes a case of cooperation between water technologies and energy production systems. This collaboration is represented by a Cartel game which implies an agreement to collude given a set of predetermined quotas that restrict the quantities to be produced. In this case, technologies form a water monopoly and an energy monopoly wherein the contribution of each plant is in accordance with the agreed upon percentages that are influenced by the current market presence. The optimal Cartel quantity can be depicted by the following equation (Eq. 2), such that p_i is the predetermined Cartel quota of the technology studied and p_j is the quotas of the remaining involved technologies:

$$q_i^* = p_i \frac{\frac{a-c_i}{2b}}{p_i - \frac{\sum p_j}{b}} \tag{2}$$

2.2. Farm Siting Using Geospatial Information Technology

Initially, the soil types of Qatar were mapped in ArcGIS. Where four types of soil have been identified, including lithosol, sandy, sabkha and rawdha soils. Meanwhile, restricted regions were also mapped including natural reserves, built-up areas, and Qatar Petroleum reserved areas. These restricted areas are excluded from the farms siting process. Then, nodes network layer with (3x3 km) inter-nodes distance is created in ArcGIS, covering

3

all the Qatar's map. The nodes where then shortlisted based on their locations on the maps. Where only nodes they overlay the most fertile soil (rawdha) and away from restricted areas were selected as candidate farm sites.

The concluding step in the optimization process for site selection is determined through the application of ArcGIS's location-allocation function. This involves assessing the identified candidate sites in relation to existing energy and water supply locations, considering the associated expenses associated with water transportation and power transmission to these candidate sites. The purpose of the location-allocation function within the network analyst module is to identify a site capable of efficiently handling supply/demand points. The algorithm systematically assesses various candidate sites to determine one or more locations that optimize the specified objective. In this particular scenario, the goal is to minimize the total travel distance and, consequently, transportation/transmission costs. Thus, the problem aligns with the p-Median Problem (PMP), defined by the mathematical model presented in Eq (3-7):

$$\text{Min } \sum_i \sum_j (Q_i * D_{ij} * C_{ij} * X_{ij}) \tag{3}$$

Subject to:

$$X_{ij} \leq Y_j \quad \forall i, j \quad \text{(No allocation occurs unless a farm is selected)} \tag{4}$$

$$\sum_J Y_J = 5 \quad \text{(Only 5 farm sites to be selected)} \tag{5}$$

$$X_{ij} \leq \{0,1\} \quad \forall i, j \quad \text{(Integer requirement)} \tag{6}$$

$$Y_j \leq \{0,1\} \quad \forall j \quad \text{(Integer requirement)} \tag{7}$$

Decision variables:

$$\text{Yj} = \begin{cases} 1, \text{if candidate farm site "j" is selected} \\ 0, \text{otherwise} \end{cases}$$

$$\text{Xij} = \begin{cases} 1, \text{if utility supply "i" is alloocated for farm site "j"} \\ 0, \text{otherwise} \end{cases}$$

Where,

i: is the set of initial farming candidate sites.

j: is the set of utilities supply sites.

Q_i: is the supplied quantity of water (m^3) or energy (MWh) at site "i".

D_{ij}: is the distance between the supplying sites "i" to the farming candidate site "j".

C_{ij}: cost of utility delivering from site "i" to the farming candidate site "j".

An allocation model is developed and solved by Excel Solver to allocate 2 Mm^3 and 300 MWh for each of the selected 5 farming sites, based on Eq (3), and the upper and lower bounds created through game theory in the previous section. Every technology is allowed in the contributing mix if the contribution amount outweighs its payoff quantity.

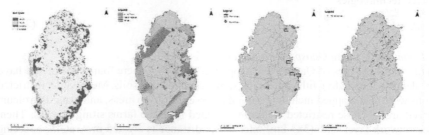

Figure 1. From left to right: soil map, restricted areas, energy and water technologies distribution and preferred farm locations.

Table 3. Energy and water mix under the Cartel case.

	E1	E2	E3	E4	W1	W2	W3	W4	W5	W6	W7	W8	W9
Site1	46%	33%	8%	13%	0%	0%	73%	0%	0%	0%	0%	0%	27%
Site2	33%	33%	33%	0%	0%	0%	100%	0%	0%	0%	0%	0%	0%
Site3	41%	24%	13%	21%	0%	0%	100%	0%	0%	0%	0%	0%	0%
Site4	45%	27%	27%	0%	22%	0%	4%	25%	4%	0%	28%	18%	0%
Site5	54%	24%	23%	0%	57%	19%	0%	0%	0%	25%	0%	0%	0%

3. Results and discussion

After modelling the competition and cooperation between the different energy power plants and water resources, two different scenarios were run to determine the impact of these behaviors on the distribution of resources to the 5 selected farms. Two maps were generated illustrating the allocation networks, wherein figure 2 presents the Cournot case and is also representative of the formed Cartels. Under both scenarios, the energy supply is granted by all energy power plants, with a significant monopoly of CCGT Mssaid under both scenarios. These results can be explained by the proximity of the plant in addition to its large capacity and relatively low cost. PVs contribution to the mix is relatively limited in the Cartel compared to Cournot as it is not economically profitable for it to supply all farms. Considering water technologies, all water sources are supplying the farms with the required amount of water, with a large contribution of desalination in the Cartel case and a more diversified mix in the Cournot scenario. This can be explained by the flexibility of the production and supply under the Cournot competition wherein quantities are not restricted in advance by market leaders (Table 3-4).

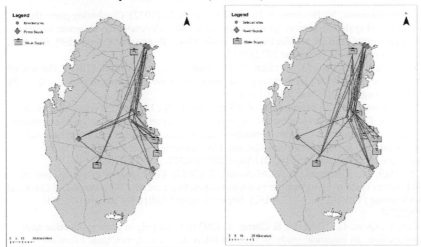

Figure 2. Left to right: Cournot and Cartel network of energy and water technologies.

Table 4. Energy and water mix under the Cournot case.

	E1	E2	E3	E4	W1	W2	W3	W4	W5	W6	W7	W8	W9
Site1	78%	8%	9%	6%	0%	0%	0%	0%	0%	0%	0%	100%	0%
Site2	0%	64%	36%	0%	0%	0%	0%	71%	14%	0%	15%	0%	0%
Site3	24%	31%	22%	23%	67%	6%	0%	27%	0%	0%	0%	0%	0%
Site4	0%	30%	52%	18%	0%	10%	0%	0%	0%	13%	0%	0%	77%
Site5	44%	26%	0%	31%	0%	0%	82%	0%	0%	0%	0%	0%	18%

4. Conclusion

Optimally locating crop producing farms is a multifaceted problem especially in areas with limited land and scarce water resources. Several factors such as the soil quality and the proximity of the water and energy sources are factors amongst others that influence the selection of suitable farms 'sites. Market dynamics is also a critical risk that needs consideration while addressing farms siting due to the variation of water and energy resources supplies. In this paper, the interactions between resources technologies are factored into the selection of optimal farms locations through considering the competition and cooperation of energy and water systems. Result of the games are integrated as constraints in a geospatial optimisation combining soil quality, proximity, and cost. Findings of the study asserts that competitive environments allow more diverse contribution from both the energy and water sectors while collaboration gives a preferential supply to market leader due to the restriction of quantities produced.

5. Acknowledgement

The research is funded by the Qatar National Research Fund (MME01-0922-190049).

6. References

Dianat, F., Khodakarami, V., Hosseini, S. H., & Shakouri G, H. (2022). Combining game theory concepts and system dynamics for evaluating renewable electricity development in fossil-fuel-rich countries in the Middle East and North Africa. Renewable Energy, 190, 805–821. https://doi.org/10.1016/j.renene.2022.03.153

Gielen, D., Boshell, F., Saygin, D., Bazilian, M. D., Wagner, N., & Gorini, R. (2019). The role of renewable energy in the global energy transformation. *Energy Strategy Reviews*, *24*, 38–50. https://doi.org/10.1016/j.esr.2019.01.006

Namany, S., Govindan, R., Di Martino, M., Pistikopoulos, E. N., Linke, P., Avraamidou, S., & Al-Ansari, T. (2022). Developing intelligence in food security: An agent-based modelling approach of Qatar's food system interactions under socio-economic and environmental considerations. *Sustainable Production and Consumption*, *32*, 669–689. https://doi.org/https://doi.org/10.1016/j.spc.2022.05.017

Namany, S., Haji, M., Alherbawi, M., & Al-Ansari, T. (2023). Food security in an oligopolistic EWF nexus system: a cooperative vs a non-cooperative case. In *Computer Aided Chemical Engineering* (Vol. 52, pp. 1427–1432). https://doi.org/10.1016/B978-0-443-15274-0.50227-4

Yan, J., Xue, Y., Quan, C., Wang, B., & Zhang, Y. (2023). Oligopoly in grain production and consumption: an empirical study on soybean international trade in China. *Economic Research-Ekonomska Istrazivanja* , *36*(2). https://doi.org/10.1080/1331677X.2022.2142818

Flavio Manenti, Gintaras V. Reklaitis (Eds.), Proceedings of the 34th European Symposium on Computer Aided Process Engineering / 15th International Symposium on Process Systems Engineering (ESCAPE34/PSE24), June 2-6, 2024, Florence, Italy

Dynamic Modeling of Precipitation in Electrolyte Systems

Niklas Kemmerling,[a] Sergio Lucia[a]

aTU Dortmund, Emil-Figge Str. 70, Dortmund 44227, Germany
niklas.kemmerling@tu-dortmund.de

Abstract

This study presents a dynamic modeling approach for precipitation in electrolyte systems, focusing on the crystallization of an aromatic amine through continuous processes. A novel model, integrating equilibrium and crystallization kinetics, is formulated and applied to a continuous oscillatory baffled reactor. The approach assumes rapid equilibrium establishment and is formulated as a set of differential algebraic equations. Key features include a population balance equation model to describe the particle size distribution and the modeling of dynamically changing equilibria. The predictions of the dynamic model show good agreement with the available experimental measurements. The model is aimed at aiding the transition from a batch process to continuous process by forming the basis for numerical optimization and advanced control.

Keywords: dynamic modeling, reaction model, continuous crystallization, Continuous Oscillatory Baffled Reactor, reactive crystalization.

1. Introduction

Precipitation processes in electrolytic solutions are integral to various parts of the chemical industry, including water treatment and pharmaceutical production. These processes are characterized by dynamic changes in reactions and equilibria, necessitating sophisticated modeling to enhance production efficiency and quality. In these processes, changes in variables like compound concentration, temperature, or pressure shift electrolytic equilibria, leading to the precipitation of insoluble compounds due to a supersaturation of the solution. The dynamics of supersaturation play a crucial role in defining the nature and characteristics of the resulting crystals.

This work builds upon the dynamic modeling approach for systems with various equilibria, as initially established by Moe et al. (1995). The applicability of this methodology is demonstrated through its adoption by Kakhu et al. (2003), who incorporated phase appearance and disappearance using state-transition networks, and by Bremen et al. (2022), who integrated this concept into an open-source Modelica library. The approach presented here combines Moe et al.'s method with a population balance model.

This work's main objective is to develop a dynamic model that will be the basis for the optimization and model-based control for a large-scale production of an aromatic amine in a continuous oscillatory baffled reactor (COBR).

2. Model

A key complexity in modeling electrolyte systems lies in determining the complete set of concentrations of all species present in the system. The concentrations in these systems are primarily determined by equilibrium reactions, commonly assumed as rapid and reversible processes. In aqueous systems, the autoprotolysis of water must be accounted for, represented by the equilibrium: $2 \cdot H_2O \rightleftharpoons H_3O^+ + OH^-$. Commonly, for simplification, this reaction is exchanged by assuming the formation of free protons (H^+) instead of the hydronium ion. The dissociation of acids and bases then follows the equilibrium reactions: $HA \rightleftharpoons H^+ + A^-$; $OHB \rightleftharpoons OH^- + B^+$. The species involved in these equilibrium reactions conform to the equilibria according to the formula:

$$ln(K_j) = \sum_{i \in \mathcal{R}} v_{ij} \, ln(a_i), \tag{1}$$

where K_j is the equilibrium constant for equilibrium j, with $\mathcal{R}$ representing the set of reactants, a_i is the activity of species i, which is usually modelled using activity coefficient models, and v_{ij} is the stoichiometric coefficient of species i in reaction j.

The second key aspect in describing precipitation in electrolyte solutions involves modeling the particulate system of crystals and their characteristics. This is achieved using a population balance equation (PBE) model, which describes the evolution of the particle size distribution. The PBE is expressed as:

$$\frac{\partial f}{\partial t} + \frac{\partial G f}{\partial L} = B. \tag{2}$$

In this equation, f represents the number density function of the particles, L is the characteristic length of the crystals, G is the crystal growth rate, and B is the nucleation rate for crystals of infinitesimal size. Employing a PBE model, the rate at which the crystallizing compound precipitates out of the solution can be determined as (Jha et al. (2017)):

$$r_c = 3 \frac{\rho_c k_v}{M_c} \int_0^\infty G f L^2 dL. \tag{3}$$

Here, ρ_c denotes the density of the crystals, k_v is the volumetric shape factor, and M_c is the molar mass of the crystallizing compound. The combination of both the electrolyte and particulate systems is achieved by considering the component balances of the electrolytic species present as:

$$\frac{d\bar{n}}{dt} = \dot{\bar{n}}_{in} - \dot{\bar{n}}_{out} + V\left(\bar{v}_{eq}^T \bar{r}_{eq} + \bar{v}_r^T \bar{r}_r\right). \tag{4}$$

In this balance, $\bar{n}$ is the vector of moles of components in the systems and $\dot{\bar{n}}_{in}$ and $\dot{\bar{n}}_{out}$ represent the rates of flow entering and exiting the system, respectively. The second term on the right-hand side accounts for the equilibrium reactions (denoted by the subscript eq) and rate terms for slow processes (denoted by the subscript r), such as crystallization, where $\bar{v}$ are matricies of stoiciometric coefficients and $\bar{r}$ are the vectors of volume specific reactions rates. This balance is lumped over the entire reaction volume V.

2.1. Index reduction

Addressing the high index of the set of differential algebraic equations derived from the equilibrium equations Eq. 1 and the component balance equations Eq. 4, an index reduction strategy is employed to formulate a semi-explicit system with index one. This approach introduces reaction invariants related to the equilibrium reactions, initially presented by Asbjørnsen and Field 1970 and later applied by Moe et al. 1995. In the first step, it involves determining the basis vectors for the nullspace, λ_k, corresponding to the stoichiometric coefficient matrix associated with these equilibrium reactions:

$$\bar{v}_{eq}\lambda_k = 0 \quad for \ k = 1, \dots, N_{comp} - N_{eq}, \tag{5}$$

where N_{comp} represents the total number of species in the equilibria, and N_{eq} is the count of equilibria considered. The equilibrium reaction invariant states, denoted as $\tilde{n}$, are subsequently defined through:

$$\tilde{n}_k = \lambda_k^T \tilde{n} \quad for \ k = 1, \dots, N_{comp} - N_{eq}. \tag{6}$$

The dynamics governing these invariant states are captured by differential equations:

$$\frac{d\tilde{n}_k}{dt} = \lambda_k^T(\dot{n}_{in} - \dot{n}_{out} + \bar{v}_r^T \bar{r}_r V) \quad for \ k = 1, \dots, N_{comp} - N_{eq}. \tag{7}$$

By maintaining Eq. 1, alongside Eq. 6, and Eq. 7 where $\tilde{n}$ are considered as algebraic variables, the desired index reduction is achieved, while maintaining the component balances and equilibria and disregarding the dynamics of the equilibrium reactions.

2.2. Method of moments

Considering the partial differential PBE to describe the particulate system often becomes computationally intractable. Even in its discretized form, the complete model is likely not feasible for real-time application, particularly in a distributed reaction system like a COBR. This is the reason for employing the method of moments, which introduces moments of order p, μ_p, as differential states, defined as:

$$\mu_p = \int_0^\infty L^p f dL. \tag{8}$$

Assuming that the growth rate, G, is independent of particle size, and that there is no particle agglomeration or breakage, the resulting moment balances are:

$$\frac{d\mu_0}{dt} = B; \ \frac{d\mu_p}{dt} = pG\mu_{p-1} \quad for \ p \in \mathbb{Z}^+ \backslash \{0\}. \tag{9}$$

To model the dependence of the growth rate G and nucleation rate B on supersaturation, the following exponential relationships are introduced:

$$G = k_g S^{\sigma_g}; \ B = k_b S^{\sigma_b}. \tag{10}$$

The supersaturation, S, is defined as:

$$S = max\left(\frac{C - C_{sol}}{C_{sol}}, 0\right). \tag{11}$$

Here, C is the concentration of the precipitating solute, and C_{sol} is its concentration at equilibrium, which is a function of the system state. While the use of the maximum-function in the definition of supersaturation is important for parameter estimation to avoid negative values, it introduces non-smoothness in the solution of the model. To address this, the function is approximated as follows:

$$max(x, y) \approx \frac{x + y + \sqrt{(x-y)^2 + 2k^2}}{2}. \tag{12}$$

In this approximation, k acts as a smoothing parameter, such that as $k \rightarrow 0$, the maximum-function is accurately approximated.

2.3. Continuous Oscillatory Baffled Reactor Model

In modeling precipitation within a COBR, a plug flow reactor model is assumed, based on the COBR's effective mixing characteristics (Stonestreet and Van Der Veeken, 1999), which lead to the assumption of negligible backmixing. The model is discretized using a backward difference scheme, effectively transforming it into a sequence of small; in-series connected continuously stirred tank reactors. Side-injections are modelled as feed streams into single reactor elements. A schematic view of the modelled COBR is depicted in Figure 1.

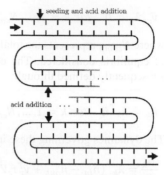

Figure 1: COBR with acid side injections.

3. Case Study

The case study explores the industrial crystallization of an aromatic amine. Initially, the amine is dissolved in a basic solution. The addition of acid neutralizes this solution, leading to the protonation of the amine, which then precipitates due to its low solubility in its protonated form. The primary objective is to enhance manufacturing efficiency by shifting from batch processing to continuous production methods. This change aims to achieve greater space efficiency, environmental sustainability, and cost reduction. A crucial part of this study is the optimization of the crystallization process. Key factors influencing this include the acid addition, which causes solution supersaturation, and the control of process variables like temperature.

For simplification, it is assumed that strong acids and bases fully dissociate and salts completely dissolve in the solution, predominantly forming ionic species. Thus, adding hydrochloric acid is treated as directly producing hydronium and chloride ions, enabling the disregard of equilibrium equations for these reactions. Additionally, due to current unavailability of thermodynamic data, the activities of the components are assumed to be represented by their concentrations. The equilibrium constant of the acidic and basic components of the system are derived from known pKa and pKb values.

The model is implemented in the open-source Python library *Pyomo*, utilizing its modules, *pyomo.dae* (Nicholson et. al, 2018) and *pyomo.paramest* (Klise et. al, 2019).

Parameter estimation for the model is conducted using data from isothermal semi-batch experiments where hydrochloric acid is continuously added to the process solution. A solubility function specific to the neutralized form of the amine is used to calculate the supersaturation. The model is discretized using orthogonal collocation on finite elements. The parameter estimation problem is structured using a least squares objective function, aiming to minimize the difference between the model's predicted and the experimentally measured time-dependent concentration of the amine in the solution. The parameters to be determined include the equilibrium constant for the amine's protonation and the parameters that describe the crystallization kinetics, as indicated in Eq. 10. Given the highly non-linear nature of the optimization problem and the lack of a solver with global guarantees, the parameters are initially set to random values within a predefined range, and the optimization process is repeated 5000 times to ensure a good fit.

4. Results

4.1. Parameter Estimation

The parameter estimation process results in a model that closely matches the experimental data, with an average relative error of 8%. Figure 2 shows the normalized results for one of the ten experiments conducted. The lower section of the figure illustrates the concentration of free protons, highlighting the pH change during the acidification and precipitation.

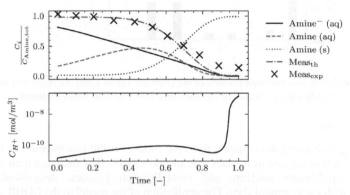

Figure 2: Relative Concentration profiles of the aromatic amine in the process solution (top) and the free protons (bottom).

While the model effectively captures the overall trend of the measurements, it deviates towards the end of the trajectory, predicting lower amine concentrations than observed. This discrepancy could be attributed to a potential inaccuracy in the solubility function, which may estimate a lower solubility than what actually occurs in the experiments.

4.2. Continuous Oscillatory Baffled Reactor Simulation

In the COBR simulation, six hydrochloric acid addition flows are chosen to maintain consistent supersaturation and ensure complete conversion of the aromatic amine. The simulation results are illustrated in Figure 3. This figure is generated from a simulation where the COBR operates until it reaches a steady state, where most of the aromatic amine is precipitated from the solution.

The profile for adding hydrochloric acid is designed to initially introduce approximately 40% of the total acid flow, to achieve a desired level of supersaturation, followed by a period of reduced addition, to maintain the desired level. Towards the end of the process, the acid feed rate is significantly increased. This change corresponds to the increasing rate of precipitation, which is influenced by the growing surface area of the forming and expanding crystals. The crystallization process inherently leads to sharp peaks in supersaturation. These peaks can cause scaling on certain sections of the reactor walls, a challenge that has also been observed in laboratory experiments. This aspect is a critical consideration in the design of the acid feed profile for the reactor.

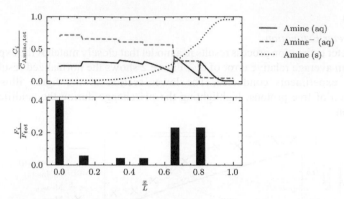

Figure 3: Relative concentration profiles of the aromatic amine along the reactor of length L at steady state (top) and the hydrochloric acid side feed flows (bottom).

5. Conclusion

This study successfully developed a dynamic model for the precipitation in electrolyte systems, which is applied in a batch reactor and COBR. The model, which effectively integrates equilibrium reactions and crystallization kinetics, demonstrates strong agreement with experimental data. The application of the model in the COBR simulation highlights its potential for optimizing and controlling a continuous production process. The insights gained, particularly in controlling the acid addition and addressing scaling issues, provide valuable guidance for enhancing manufacturing efficiency and environmental sustainability in large-scale industrial applications.

Acknowledgement

The project leading to this publication has received funding from the European Union's Horizon research and innovation programme under grant agreement No 101058279.

References

Asbjørnsen O. A., Field M., 1970, Response modes of continuous stirred tank reactors. Chemical Engineering Science, *25*(11), 1627-1636.

Bremen, A.M., Ebeling, K.M., Schulte, V., Pavšek, J., Mitsos, A., 2022, Dynamic modeling of aqueous electrolyte systems in Modelica, Computers & Chemical Engineering 166.

Jha S.K., Karthika S., Radhakrishnan T.K., 2017. Modelling and control of crystallization process, Resource-Efficient Technologies 3, 94–100.

Klise K.A., Nicholson B.L., Staid A., Woodruff D.L., 2019, Parmest: Parameter Estimation Via Pyomo, Computer Aided Chemical Engineering. Elsevier, pp. 41–46.

Kakhu A.I., Pantelides, C.C., 2003. Dynamic modelling of aqueous electrolyte systems, Computers & Chemical Engineering 27, 869–882.

Moe H.I., Hauan S., Lien K.M., Hertzberg T., 1995, Dynamic model of a system with phase- and reaction equilibrium, European Symposium on Computer Aided Process Engineering.

Nicholson B., Siirola J.D., Watson J.-P., Zavala V.M., Biegler L.T., 2018, pyomo.dae: a modeling and automatic discretization framework for optimization with differential and algebraic equations, Math. Prog. Comp. 10, 187–223.

Ramkrishna D., 2000, Population balances: theory and applications to particulate systems in engineering, Academic Press, San Diego, CA.

Stonestreet P., Van Der Veeken P.M.J., 1999, The Effects of Oscillatory Flow and Bulk Flow Components on Residence Time Distribution in Baffled Tube Reactors, Chemical Engineering Research and Design 77, 671–684.

Flavio Manenti, Gintaras V. Reklaitis (Eds.), Proceedings of the 34th European Symposium on Computer Aided Process Engineering / 15th International Symposium on Process Systems Engineering (ESCAPE34/PSE24), June 2-6, 2024, Florence, Italy

Assessing Climate Variability in Tomato Supply Chain Optimization: A Multi-Objective Approach for Qatar's Tomato Imports

Bashar Hassna, Sarah Namany, Mohammad Alherbawi, Adel Elomri, Tareq Al-Ansari*

College of Science and Engineering, Hamad Bin Khalifa University, Doha, Qatar
talansari@hbku.edu.qa

Abstract

In an era of escalating climate variability, characterized by unpredictable monthly trends in temperature and precipitation, the stability of global food supply chains faces unprecedented challenges. Nowhere is this more evident than in countries like Qatar, which heavily depend on imports to meet their food demands. This research investigates the influence of climate variability on determining the optimal network of trade partners to supply tomatoes to Qatar. The suggested framework employs a stochastic multi-objective optimization framework, simultaneously addressing critical objectives: the minimization of transportation costs, the reduction of water consumption, and the tracking of virtual water trade in tomato imports, while accounting for climate uncertainty, such as rainfall, modeled as stochastic parameters following a normal distribution. Findings of this study shed light on the intricate dynamics within Qatar's tomato sourcing network. A notable observation relates to the identification of discernible seasonal variability in the percentage contribution of certain countries, with a visible aggregation of supply sources during the summer months. This observation underscores the susceptibility of tomato supply chains to the influence of climatic factors, particularly during periods characterised by heightened fluctuations in temperature and low precipitations. Furthermore, results discern explicit correlations between climate variability and trade networks selection. These findings not only furnish invaluable insights for Qatar's policymaking in the field of imports but also contribute to a broader comprehension of the strategies employed by nations reliant on food imports as they endeavor to strengthen the resilience and sustainability of their supply chains in the midst of prevailing climatic uncertainties.

Keywords: optimisation, precipitation, supply chain, virtual water, climate variability.

1. Introduction

The foundation of food security lies in food production. Availability, defined as the physical presence of food supplies in appropriate quantities, hinges on factors such as net domestic production, net imports, and stock drawdown, adjusted for feed, seed, and waste. Physical availability in specific areas necessitates market integration, robust storage, and transportation infrastructure. Yet, the cornerstone of food availability remains food production. As highlighted by Swaminathan, the global challenge is to sustainably increase production amid decreasing per capita arable land, irrigation water supplies, and mounting abiotic and biotic pressures, intensified by a projected global population exceeding 8.5 billion by 2030 (Swaminathan MS, 2013).

Given the significant impact of weather on food production, the focus is on remediation strategies across various domains. These encompass food production and quality yields, irrigation water requirements, technological advancements in crops and livestock, erosion-induced loss of arable land, fish production needs, and emerging risks affecting food security. Adverse weather events, marked by unusually high or low temperatures and erratic rainfall patterns, negatively impact cattle and agriculture performance. Although contemporary technologies can mitigate these effects (Gomez-Zavaglia, A., 2020), the interplay between temperature, crop land, water quality, and irrigation significantly influences the environment.

Irrigation, a key factor in altering hydrological conditions, poses environmental concerns, especially as regions transition from rain-fed to irrigated agriculture. Climate variability, including temperature extremes during critical phenophases, affects crop growth, development, and yields. This climate uncertainty, coupled with concerns about water availability, highlights the importance of integrated water management. Water scarcity is a potential constraint for crop production and food security, with studies emphasizing the critical role of efficient water use (Fujihara et al., Kang et al.). Assessing water availability is essential for maintaining biodiversity, human life, and the environment, especially considering the strains on the hydrological cycle due to pollution, population growth, land use changes, and climate change.

The paper's objective is to propose a framework utilising multi-objective optimisation that addresses critical objectives such as minimising transportation costs, reducing water consumption, and tracking virtual water trade in tomato imports, the framework accounts for climate uncertainty, modelled as stochastic parameters with a normal distribution. The findings unveil the intricate dynamics of Qatar's tomato sourcing network, highlighting discernible seasonal variability in certain countries' contributions and an aggregation of supply sources during summer months.

2. Data and Methods

The methodology suggested in this paper aims to determine the optimal trade network that satisfies the demand for tomatoes in the state of Qatar while considering environmental and economic aspects. The purpose is to quantify the impact of the variability of weather conditions of trade partners on their contributions in supplying the overall demand. In this model, the influence of weather fluctuations is expressed by means of the crop water requirement (CWR). It is defined as the optimal amount of water required for the crop to grow adequately. Due to the limited access to data of trade partners, the CWR computation was assumed based on the CWR of Qatar which is estimated based on the following equation (Eq.1), such that w is the crop water requirement, A is the area cultivated and ET_C is the evapotranspiration.

$$w = \frac{A \times ET_C}{1000} \qquad (1)$$

Linear regression is used to link the water to temperature and forecast the values for all participating countries (Namany *et al.*, 2020). The dataset was subjected to a fitting process utilizing a normal distribution model. Subsequently, the parameters of the normal distribution were estimated to best align with the observed data, thereby characterizing the underlying statistical distribution of the dataset as approximately normal. The crop water data along with cost environmental data were used to in a multi-objective optimisation model that minimises economic costs (Eq 2) and GHG emissions (Eq 3) along with global water consumption (Eq 4). Table 1 summarises the set of data used in the model. The mathematical formulation of the optimisation framework is described in the following section:

Objective functions:

$$min \ \sum_{i=1}^{7} Dc_i x_i \tag{2}$$
$$min \ \sum_{i=1}^{7} De_i x_i \tag{3}$$
$$min \ \sum_{i=1}^{7} Dw_i x_i \tag{4}$$

Subject to:

$$\sum_{i=1}^{7} x_i = 100\%$$
$$Dx_i < Ex_i$$

Such that,

x_i is the decision variable that represents the share of each trade partner i. D is total demanded quantity, c_i is the unit cost of crop production, e_i is the unit GWP emissions as for w_i , it represents the unit crop water requirement. Considering Ex_i, it refers to the allowed exportable quantity of every exporting country.

Table 1. Economic and environmental data used for the optimisation model.

Objective Function	USA	Iran	Lebanon	India	Turkey	Morocco	Nether-lands
CWR (m³/kg)	μ=0.0152 σ=0.0024	μ=0.0147 σ = 0.0092	μ=0.0187 σ = 0.0046	μ=0.0238 σ = 0.00314	μ=0.0103 σ = 0.0074	μ=0.0151 σ = 0.0033	μ=0.0085 σ = 0.0048
Cost ($/kg)	5.4007	0.7541	3.5796	1.1799	3.9110	2.8374	4.1137
GWP(kg of CO₂eq)	0.0002	0.0002	0.0002	0.0003	0.0001	0.0002	0.0001

3. Results

The developed multi-objective optimization model is solved in MATLAB using the genetic algorithm. Results of the framework are summarised in the 3-D Pareto front illustrated in Figure 1. Figure 2 displays the percentage contributions of each trade partner to the overall demand considering 200 data points. Findings display some noticeable variations in supply from certain countries such as India, Morocco and Lebanon. This can be explained by the variations in their weather conditions. The US on the other hand has a low and fixed contribution. This is due to its high economic costs and emissions associated with transportation. In order to further investigate the results and understand the distributions, the optimal graphical solution is generated and illustrated in figure 3. The major contributions are from Netherlands, Iran then India thanks to their low crop water requirement, and relatively cheap cost that is due to their proximity to Qatar. Generally, the cost and proximity which is a major influence of the environmental performance are the main driver for the selection of trade partner. The weather conditions come to play when the variability is significant between the hot and cold season.

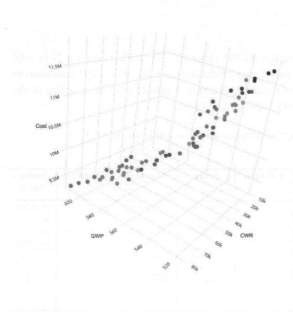

Figure 1. The Pareto Front.

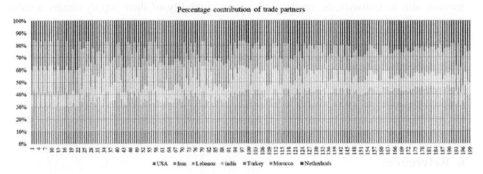

Figure 1. The contribution of each trade partner.

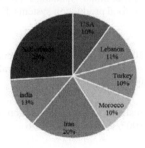

Figure 3. Average optimal solution.

4. Conclusions

This research examines how climate variability impacts the determination of the most effective network of trade partners for supplying tomatoes to Qatar. The proposed framework utilizes a stochastic multi-objective optimization approach that simultaneously addresses key objectives, including minimizing transportation costs, reducing water consumption, and monitoring virtual water trade in tomato imports. This framework also considers climate uncertainty, such as rainfall, modeled as stochastic parameters that follow a normal distribution. The study's findings provide insights into the complex dynamics within Qatar's tomato sourcing network. Notably, the research identifies distinct seasonal variations in the percentage contribution of specific countries, with a noticeable concentration of supply sources during the summer months. This observation highlights the vulnerability of tomato supply chains to climatic factors, especially during periods characterized by significant temperature fluctuations and low precipitation. Additionally, the results reveal clear correlations between climate variability and the selection of trade networks. These findings not only offer valuable insights for Qatar's import-related policymaking but also contribute to a broader understanding of the strategies employed by nations reliant on food imports. These nations aim to enhance the resilience and sustainability of their supply chains amidst ongoing climatic uncertainties.

5. Acknowledgement

This research was made possible by an Award (GSRA7-1-0521-20080) and supported by proposal number MME01-0922-190049 from Qatar National Research Fund (a member of Qatar Foundation). The contents herein are solely the responsibility of the authors[s].

6. References

Gomez-Zavaglia, A., Mejuto, J. C., & Simal-Gandara, J. (2020). Mitigation of emerging implications of climate change on food production systems. Food Research International, 134, 109256. https://doi.org/10.1016/j.foodres.2020.109256

Kang, Y., Khan, S., & Ma, X. (2009). Climate change impacts on crop yield, crop water productivity and food security – A review. Progress in Natural Science, 19(12), 1665–1674. https://doi.org/10.1016/j.pnsc.2009.08.001

Luo, Q. (2011). Temperature thresholds and crop production: a review. Climatic Change, 109(3-4), 583–598. https://doi.org/10.1007/s10584-011-0028-6

Namany, S. *et al.* (2020) 'Sustainable food security decision-making: an agent-based modelling approach.', *Journal of Cleaner Production*. Elsevier, p. 120296. doi: 10.1016/J.JCLEPRO.2020.120296.

Swaminathan MS, Bhavani RV. Food production & availability--essential prerequisites for sustainable food security. Indian J Med Res. 2013 Sep;138(3):383-91. PMID: 24135188; PMCID: PMC3818607.

Flavio Manenti, Gintaras V. Reklaitis (Eds.), Proceedings of the 34th European Symposium on Computer Aided Process Engineering / 15th International Symposium on Process Systems Engineering (ESCAPE34/PSE24), June 2-6, 2024, Florence, Italy

Developing the final product attribute prediction model in a continuous direct compression process

Yuki Kobayashi[a], Sanghong Kim[b], Takuya Nagato[c], Takuya Oishi[b,c], Manabu Kano[a,*]

[a]*Department of Systems Science, Kyoto University, Yoshida-honmachi, Sakyo-ku, Kyoto, 6068501, Japan*
[b]*Department of Applied Chemistry, Tokyo University of Agriculture and Technology, 2-24-16 Naka-cho, Koganei, 1840012, Japan*
[c]*Research and Development Division, Powrex Corporation, 5-5-5 Kitagawara, Itami, 6640837, Japan*
manabu@human.sys.i.kyoto-u.ac.jp

Abstract

To analyze and design pharmaceutical manufacturing processes, models that can predict critical quality attributes (CQAs) of final products from process parameters (PPs) and material attributes (MAs) have been investigated. Such a model depends on a type of equipment; thus, the entire model needs to be rebuilt when equipment is changed. To reduce the effort of model rebuilding, we propose a series model, in which an equipment-dependent model predicts inputs of the equipment-independent model that predicts CQAs. This study focuses on the equipment-independent model of a continuous direct compression (CDC) process. In experiments by changing nine PPs, MAs of intermediate products and CQAs of final products were measured. CQA prediction models were constructed by using different sets of input variables. As a result, equipment-independent models accurately predicted the disintegration time of tablets. Moreover, ten MAs and the main compression force were identified as important variables for accurate prediction.

Keywords: Continuous manufacturing, continuous direct compression, variable selection, process model, intermediate product material attribute

1. Introduction

Quality assurance methods in the pharmaceutical industry are shifting from Quality by Testing (QbT), in which the quality is ensured by product testing, to Quality by Design (QbD), in which the quality is ensured by the good design of the products and manufacturing processes. To realize QbD, mathematical models that express the relationship between CQAs of final products and PPs and MAs play a significant role. This study focuses on a continuous direct compression (CDC) process. CDC is one of the methods of continuous manufacturing of tablets and its process consists of feeders, mixers, and a tablet press.

For integrated CDC processes, Kreiser et al. (2022) developed multiple linear regression models that predict final product CQAs from the PPs of the mixer. Bekaert et al. (2022) developed a partial least squares (PLS) model to predict final product CQAs from the PPs and the blend properties. These models use the PPs of the mixers, however, the types of mixers used are different and the mixers have different PPs. As shown in Figure 1 (top), previous studies have predicted CQAs from equipment-dependent PPs, equipment-independent PPs, and MAs. Because such models depend on equipment, the entire CQA prediction model needs to be rebuilt when equipment is changed. On the other hand, as shown in Figure 1 (bottom), this study aims at a series model that predicts MAs independent of equipment from variables including equipment-dependent PPs and then predicts CQAs from equipment-independent PPs and MAs. Such a model has the advantage of requiring less effort when equipment is changed since only the equipment-dependent model needs to be rebuilt.

In this study, aiming for the concurrent use of equipment-independent models that predict CQAs and equipment-dependent models that predict MAs in a CDC process, an attempt was made to construct equipment-independent models. Moreover, important variables for accurate prediction were identified.

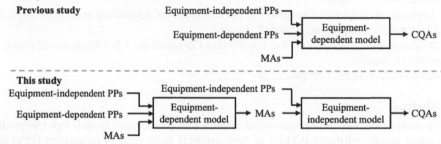

Figure 1: Comparison of prediction models in the previous (top) and this study (bottom).

2. Materials and Methods

2.1. Process

Figure 2 shows the process flow diagram investigated in this study. Acetaminophen (APAP) (Spera Nexus, Japan) was used as an active pharmaceutical ingredient, SuperTab 11SD (DFE Pharma, Germany) was used as an excipient, and magnesium stearate (MgSt) (Taihei Chemical Industrial, Japan) was used as a lubricant. APAP and SuperTab were fed into mixer 1 (MG100, Powrex, Japan) by different feeders (LIW-300-P, Ishida, Japan) and mixed in mixer 1. Intermediate product 1 coming out from mixer 1 and MgSt were fed into mixer 2 and mixed. Intermediate product 2 coming out from mixer 2 was fed into the tablet press (FETTE 102i, Fette Compacting, Germany) and compressed and made into tablets, the final product.

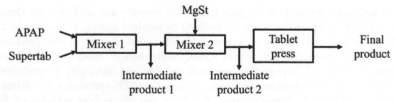

Figure 2: Process flow diagram.

2.2. Experiments

A definitive screening design (DSD) with 19 runs (Jones et al., 2011) was adopted to determine the combinations of levels of nine variables listed in Table 1 because DSDs require a small number of experiments and accommodate second-order effects. The nine variables were grouped into four groups (X_{1-4}). The measured MAs of the intermediate products 1 and 2 are listed in Table 2 and they were denoted as groups Z_1 and Z_2. The measured CQA was the disintegration time of tablets.

Table 1: Variables and their levels in the experiments.

Variable group	Variable	Level
X_1	Mass fraction of APAP	5, 10, 15 %
	Production rate	15, 20, 25 kg/h
X_2	Scraper blade rotation speed of mixer 1	30, 50, 70 rpm
	Center blade rotation speed of mixer 1	500, 1500, 2500 rpm
X_3	Scraper blade rotation speed of mixer 2	30, 50, 70 rpm
	Center blade rotation speed of mixer 2	100, 550, 1000 rpm
X_4	Force feeder blade rotation speed	10, 30, 50 rpm
	Ratio of pre-compression force to main compression force	20, 40, 60 %
	Main compression force	5, 15, 20 kN

Table 2: Intermediate product material attributes.

Variable	Unit
Measured mass fraction of APAP	-
Standard deviation (SD) of measured mass fraction of APAP	%
Relative standard deviation (RSD) of measured mass fraction of APAP	%
Aerated bulk density	g/cm^3
Tapped bulk density	g/cm^3
Compressibility	-
Angle of repose	Degree
Angle of rupture	Degree
Angle of difference	Degree
Angle of spatula	Degree
Degree of agglomeration	%
Degree of dispersion	%
Particle size distribution, D_{10}, D_{50}, D_{90}	μm
Basic flowability energy (BFE)	mJ
Flow rate index (FRI)	-
Specific energy (SE)	mJ/g
Cohesion	kPa
Unconfined yield strength (UYS)	kPa
Major principal stress (MPS)	kPa
Angle of internal friction (AIF)	Degree
Flow function coefficient (FF)	-

2.3. Modeling

The evaluation criteria of the model accuracy are root mean squared error (RMSE) and Q^2.

$$RMSE = \sqrt{\frac{1}{N} \sum_{n=1}^{N} (y_n - \hat{y}_n)^2} \tag{1}$$

$$Q^2 = 1 - \frac{\sum_{n=1}^{N}(y_n - \hat{y}_n)^2}{\sum_{n=1}^{N}(y_n - \bar{y})^2} \qquad (2)$$

, where N is the number of the samples, y_n is the nth actual value, $\hat{y}_n$ is the nth predicted value, and $\bar{y}$ is the mean of the output variable.

The modeling procedure is as follows:
(1) Set the output variable of the model to be the disintegration time.
(2) Select the model from PLS (Geladi et al., 1986) or random forest (RF) (Breiman, 2001).
(3) Set $p = 0$.
(4) Set $m = 0$.
(5) Put the variables in the input variable set p and in the nominal variable set $\acute{S}_p$ in Table 3 into the input variable set S_m.
(6) Let one sample be test data, and the other be training data.
(7) If PLS is selected in step 2, construct a PLS model using cross-validation. In the cross-validation, tune the number of latent variables. If RF is selected in step 2, construct a RF model using cross-validation. In the cross-validation, tune three parameters: the number of trees, the maximum depth of trees, and the number of features to consider when looking for the best split.
(8) Calculate RMSE with test data.
(9) Calculate each variable's permutation importance (PI) (Altmann et al., 2010).
(10) Perform steps 6 through 9 using each sample as test data once.
(11) Calculate the mean of RMSEs, referred to as RMSECV.
(12) Calculate the mean of PIs for each variable, referred to as PICV.
(13) Exclude the input variable with the smallest PICV among those not included in the nominal variable set $\acute{S}_p$ from S_m.
(14) Update $m = m + 1$
(15) Perform steps 6 through 14 until S_m has the same number of input variables as that in $\acute{S}_p$.
(16) Select S_m with the smallest RMSECV and let the selected variable set be denoted $\tilde{S}_p$.
(17) Update $p = p + 1$, and perform steps 4 through 17.
(18) Perform steps 2 through 17 for PLS and RF.

Table 3: Variable groups containing input variable candidates.

Number of the input variable set	X_1	X_2	Z_1	X_3	Z_2	X_4	Nominal variable set $\acute{S}_p$
1	✓					✓	Empty set
2					✓		$\tilde{S}_1$
3				✓			$\tilde{S}_2$
4			✓				$\tilde{S}_2$
5		✓					$\tilde{S}_2$

3. Results and Discussions

3.1. Experiments

Figure 3 shows the relationship between disintegration time and both the APAP mass fraction setpoint (left) and the main compression force (right). The disintegration time varied even when the APAP mass fraction setpoint was the same, thus, it is affected not only by the formulation but also by PPs such as the main compression force.

3.2. Modeling

Table 4 shows the prediction results of disintegration time. The PLS models achieved better prediction accuracy than the RF models, therefore, the results of the PLS models are focused on in the following paragraph.

The comparison of RMSECV and Q^2 between the input variable sets 1 and 2 in Table 4 shows that using MAs of intermediate product 2 significantly improves prediction accuracy. The comparison of RMSECV and Q^2 among the input variable sets 2, 3, 4, and 5 in Table 4 shows that variables in X_2, X_3, and Z_1 do not significantly improve the prediction accuracy. The variables selected from the input variable set 2 are enough to accurately predict the disintegration time. This means that the equipment-independent model can be constructed for disintegration time prediction. The variables selected from the input variable set 2 are shown in Table 5 and the actual and predicted values are shown in Figure 4.

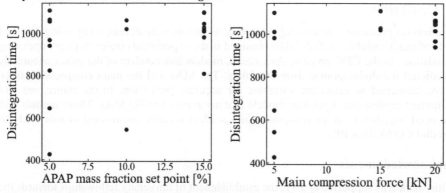

Figure 3: Relationship between disintegration time and APAP mass faction setpoint (left) and main compression force (right).

Table 4: Prediction results of disintegration time

| Number of input | RMSECV | | Q^2 | |
variable set	PLS	RF	PLS	RF
1	122.2	99.8	0.18	0.50
2	48.2	99.7	0.86	0.36
3	48.2	99.7	0.86	0.36
4	48.2	99.7	0.86	0.36
5	45.5	99.7	0.85	0.36

Table 5: The variables selected from the input variable set 2 in Table 3.

Z_2	X_4
RSD of measured mass fraction of APAP, MPS, SE, SI, UYS, angle of spatula, degree of agglomeration, compressibility, angle of difference, aerated bulk density	Main compression force

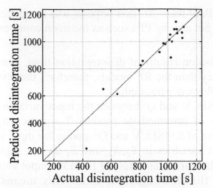

Figure 4: Prediction results of disintegration time by the PLS model with variables in Table 5.

4. Conclusions

Models for predicting the disintegration time of tablets were developed by using different sets of input variables with the data obtained in the experiments under different operating conditions in the CDC process. As a result, models independent of the mixer accurately predicted the disintegration time of tablets. Ten MAs and the main compression force were identified as important variables for accurate prediction. In the future, we will construct equipment-dependent models that accurately predict MAs. These models will be used together with the equipment-independent models constructed in this study to predict CQAs from PPs.

Acknowledgments

This work was supported by JST, the establishment of university fellowships towards the creation of science technology innovation, Grant Number JPMJFS2123, and by JSPS KAKENHI Grant Number JP21H01704.

References

A. Altmann, L. Tološi, O. Sander, T. Lengauer, 2010, Permutation importance: a corrected feature importance measure. *Bioinformatics*. **26**, 1340–1347.

B. Bekaert, B. Van Snick, K. Pandelaere, J. Dhondt, G. Di Pretoro, T. De Beer, C. Vervaet, V. Vanhoorne, 2022, Continuous direct compression: Development of an empirical predictive model and challenges regarding PAT implementation. *International Journal of Pharmaceutics: X*. **4**, 100110.

L. Breiman, 2001, Random Forests. *Mach. Learn.* **45**, 5–32.

P. Geladi, B. R. Kowalski, 1986, Partial least-squares regression: a tutorial. *Anal. Chim. Acta* **185**, 1–17.

B. Jones, C. J. Nachtsheim, 2011, A Class of Three-Level Designs for Definitive Screening in the Presence of Second-Order Effects. *J. Commod. Sci. Technol. Qual.* **43**, 1–15.

M. J. Kreiser, C. Wabel, K. G. Wagner, 2022, Impact of Vertical Blender Unit Parameters on Subsequent Process Parameters and Tablet Properties in a Continuous Direct Compression Line. *Pharmaceutics*. **14**, 278.

Flavio Manenti, Gintaras V. Reklaitis (Eds.), Proceedings of the 34th European Symposium on Computer Aided Process Engineering / 15th International Symposium on Process Systems Engineering (ESCAPE34/PSE24), June 2-6, 2024, Florence, Italy

A Genetic Algorithm-Based Design for Hydrogen Pipeline Infrastructure with Real Geographical Constraints

Joseph Hammond,[a] Manou Rosenberg,[b] Solomon Brown,[a*]

[a]*Department of Chemical & Biological Engineering, University of Sheffield, Sheffield, S1 3JD, England, UK*
[b]*Curtin Centre for Optimisation and Decision Science, Curtin University, Perth, Australia*
*s.f.brown@sheffield.ac.uk

Abstract

As industries grapple with net-zero constraints, increasing attention is directed at new hydrogen pipeline infrastructures that benefit from economies of scale and shared-use between producers and consumers. However, uncertainties in the design and cost of new infrastructure deters investment and, in turn, obstructs the progress of carbon-intensive sectors achieving net-zero targets (DESNZ, 2023).

Accurate pipeline routing is a key stage when determining optimised infrastructure in network design models, such as AENI (Ejeh, 2023a). Routing needs to handle costly-to-cross areas, significantly impacting the capital and operational costs. This typically requires the traversal of complex environments, which can be represented as obstacles of two types: 1) hard obstacles that cannot be traversed, such as protected areas; and 2) soft obstacles that can be traversed with a higher financial cost, such as densely populated areas and difficult terrains. Previous attempts at modelling the routing of pipeline infrastructures have evaded hard obstacles (Heijnen et al., 2013), but none yet demonstrate a capability to design a path with considerations of penetrable but resistant (soft) obstacles.

In this paper, we propose a two-step pipeline design methodology for a hydrogen network within an industrial cluster. A graph-based genetic algorithm (Rosenberg et al., 2021) defines the topological arrangement of new pipeline infrastructure in the presence of geographical constraints without using predefined corridors. Then, we apply a state-of-the-art MINLP model to define the dimensions based on commercially available sizes, with the objective of minimising the network's capital and operational costs.

This paper's novelty is defined by its approach to design the topology of a spatially explicit pipeline infrastructure with considerations of hard and soft obstacles that represent real geographical constraints.

The assessment looks at the regional case study of the Humber cluster in the UK, based upon infrastructural development plans for large-scale hydrogen production and industrial consumers (National Grid, 2022). We demonstrate the method's ability to consider geographical constraints to design practical pipeline routes.

Keywords: Network optimisation with obstacles, Genetic Algorithm, MINLP, Pipeline design.

1. Background

In the UK, forthcoming hydrogen pipeline infrastructures are planned to connect large industrial sites over vast regions (National Grid, 2022). Historically, gas pipeline networks evolved incrementally, growing node-by-node with increasing demand, leading to sub-optimal designs. A push for net-zero emissions has incentivised designing new infrastructures, such as hydrogen networks, from the ground up with the advantage of leveraging advanced computational methods to optimise the design and test configurations before the practical implementation.

Pipeline networks are commonly modelled using graphs, where the producers and consumers serve as nodes and the pipeline routes as capacitated arcs (Heijnen et al., 2013). Given that both network topology and pipeline sizing are computationally challenging problems, comprehensive design often bifurcates into separate models for each. Uncapacitated Minimum Spanning Tree (MST) algorithms, such as Kruskal's, have been widely used in pipeline design to minimise the network's length as a proxy for cost (André et al., 2013). Yet, real pipeline routes are characterised by complex geometries that consider real-world constraints, such as areas that cannot be traversed by pipelines or where traversal incurs a higher cost or risk. In graphs, these can be represented by hard (non-traversable) and soft (traversable) obstacles. Considering this, some studies avoid optimising the network structure by leveraging existing infrastructure, such as roads or gas networks, to predefine candidate routes (Parolin et al., 2022). This provides a convenient method to evade obstacles by nature of the route's existence. However, it does not provide a method to define a novel route around the obstacles nor indicates its applicability to new infrastructure, thereby limiting topological flexibility and potential for improvement.

Graph-based methods have utilised Steiner points - nodes with no supply or demand – which can further reduce the network length and enhance topological flexibility. Heijnen et al. (2013) explored this concept, offering a geometric heuristic method for designing a single-source capacitated network around hard obstacles using Steiner points. However, this approach proved limiting, displaying slow procedural times for case studies featuring few hard obstacles and terminals. Furthermore, the method is not easily extendable to soft obstacles. More recently, Rosenberg et al. (2021) demonstrated the feasibility of designing networks around hard and soft obstacles using Steiner points. Rosenberg et al. developed a Genetic Algorithm (GA) for the Steiner tree problem with soft and hard obstacles with small errors. To the author's knowledge, a topological method of comparable fidelity as Rosenberg et al. has not been applied to a gas pipeline infrastructure problem.

While the network topology aims to overcome spatial challenges, the pipeline's physical dimensions have their own set of constraints, requiring another layer of optimisation. In gas pipeline design, the capacity is derived from the relationships between the flow rate, pipe diameter, and pressure gradient, resulting in a non-linear optimisation space. To alleviate computational intensity, some studies, such as Baufumé et al. (2013) and Welder et al. (2018), fix the gas velocity, linearising the optimisation space. However, Reuß et al. (2019) highlighted the limitations of this approach in hydrogen network design, noting that the cost discrepancy between the linear and non-linear approaches increased with larger flows. Additional considerations in pipeline sizing have been put forth by Robinius et al. (2019) and Ejeh et al. (2023b). Robinius et al. described the optimisation formulation with discrete arc sizing, making an allowance for commercially available sizes. More recently, Ejeh et al.'s Mixed-Integer Non-Linear Programming (MINLP) formulation for CO_2 pipeline design presented a highly encompassing model with hydraulic, structural, commercial, and velocity constraints.

In the UK, hydrogen networks planned to decarbonise industrial clusters are expected to have high flowrates and complex, cross-region topologies. Therefore, a need arises for high-fidelity techniques that consider the complexities of pipeline design. Candidate routing methods do not provide the flexibility and generality desired from design models, while graph-based topological methods have sparsely tackled obstacle handling within pipeline design, particularly at large spatial scales and with soft obstacles. In the proposed approach, a Genetic Algorithm (GA) defines the topology around user-defined soft polygonal obstacles, and a MINLP model defines the pipeline dimensions according to commercially available sizes with structural and velocity constraints.

The computational methods are described in Section 2 and then applied to a case study described in Section 3. Section 4 discusses the findings and results, and Section 5 concludes this work.

2. Methodology

The proposed method finds optimised infrastructures for hydrogen pipeline networks that connect a given set of demand and supply points while considering real-world constrained areas, where pipelines cannot be built or incur a higher cost. The problem is represented as a graph with constrained areas represented as simple polygonal obstacles in the plane. The obstacles are classified as soft or hard obstacles, depending on whether they can be traversed. This is determined by their attributed crossing weight which is used as a multiplying factor to the length of the pipeline connection that crosses the obstacle. In case of a hard obstacle, this crossing weight is set to infinity.

2.1. Topology

Separating the topology and sizing requires an implicit assumption that minimising the total length will lead to the most cost-efficient network infrastructure. To find a minimal-length network connecting supply and demand points considering constrained areas we apply the Euclidean Steiner tree problem with soft and solid obstacles (ESTPO). This generalises the well-known computationally difficult Steiner tree problem in the plane.

To solve this problem for the given set of hydrogen supply and demand points as terminals and case-specific constrained regions as obstacles, we apply the genetic algorithm for Steiner trees with obstacles (StObGA) developed by Rosenberg et al. (2021). This GA is based on the following chromosome structure that is comprised of two parts: the first part represents the number and x-y-coordinates of Steiner points and is of variable length, and the second part is a fixed-length binary list determining which of the obstacle corner points are used in the network. For example, the chromosome

$$\{(x_1, y_1), (x_2, y_2), (x_3, y_3), \dots, (x_s, y_s)\} \\ + [\, 1 \mid 0 \mid 1 \mid 0 \mid 0 \mid 0 \mid 0 \mid 1 \mid 1 \mid 0 \mid 0\,] \tag{1}$$

includes s Steiner point locations and uses four obstacle corner nodes. The final network connections are then calculated using Kruskal's MST algorithm to connect all terminals, Steiner points, and obstacle nodes using edges weighted based on their length and whether they cross any obstacles. For further details, refer to Rosenberg et al. (2021).

To finalise the topology, the nodes and edges found by the GA are input into an optimisation that directs the edges as appropriate for the given production and consumption. It is assumed that the edges are unidirectional.

2.2 Sizing

The sizing of pipelines is designed using a MINLP optimisation with hydraulic, structural, and commercial constraints. The objective function, equation 2, minimises the total cost C of a set of pipelines, arcs A. The total cost is the addition of the capital costs C^c, given in equation 3 (NREL, 2018), and operating costs, assumed to equal 4% of the capital cost. The costs are a function of a pipeline's length L and outer diameters D^{po}, and consider the material, labour, right-of-way, and other miscellaneous costs.

$$C = (1 + 0.04) \sum_{i|(i,j) \in A} C_{ij}^C \tag{2}$$

$$C_{ij}^C = 43079 L_{ij} \times e^{0.0697 D_{ij}^{po}} + 251.7(D_{ij}^{po})^2 + 60848 D_{ij}^{po} + 303657 \tag{3}$$

The optimisation is constrained by a material balance, hydraulic behaviour, discrete sizing, and the erosion velocity, as described by Ejeh et al. (2023b).

3. Case Study

The methodology of Section 2 was applied to the case study in Figure 1, which displays the pipeline route planned by the National Grid (2022) as validation. The case study depicts a high-capacity pipeline infrastructure with two producers and four consumers across the geographically complex Humber region. Additional pipelines are known to be planned for storage points in Easington and Aldbrough, but high-resolution route plans have not been disclosed, so those pipelines are out-of-scope. We assume compressor stations exist at the terminal sites and junctions (nodes with more than two connections).

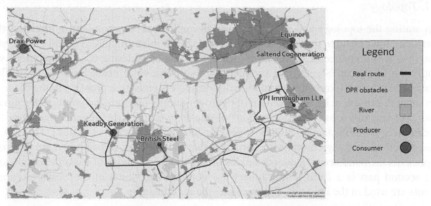

Figure 1: Humber region with planned pipeline route outlined by the National Grid (2022). Hydrogen sites are sized proportionally according to their production/consumption.

Soft obstacles are used to represent densely populated regions (DPRs) and the river Humber, which is the most significant geographical obstacle. The obstacles are represented by polygons, with a total of 406 corners. The weightings for the soft obstacles are literature-based terrain factors; however, due to a lack of literature on the terrain factors concerning hydrogen, we assume CO_2 infrastructure values will be satisfactory. The spread of the terrain factor values is broad, so the assumption is not presumptuous. We assume that DPRs and rivers require two times and four times the standard investment, respectively (van den Broek et al., 2013).

4. Results and Discussion

The topological results are presented in Figure 2. The GA-generated route is compared to the real pipeline route. The GA's route reflects a pathway previously considered by National Grid (Geels et al., 2023). The routes share the same interconnections between terminals but differ in structure. The most significant deviation is around Scunthorpe; the GA's route takes a shorter route between Keadby and British Steel. The real network's manoeuvre around the bottom of Scunthorpe could be due to obstacles not considered here, such as the crossing of significant roads, terrain, land permissions, etc. or further network considerations such as anticipated connections. Table 1 presents the total network lengths, total costs (capital + operating), and the difference in cost from the real pipeline route's values, sized using the same technique. An MST configuration is similarly costed without consideration of obstacles for comparison to an existing method.

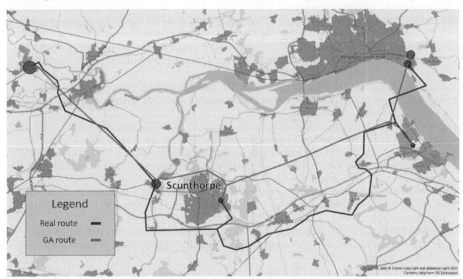

Figure 2: Humber region with the GA-generated route and real pipeline route.

Table 1: Infrastructure summary with calculated costs based on the derived pipeline lengths and diameters.

Scenario	Total length (km)	Total cost (£million)	Cost difference from real (%)
Real	101	79.1	-
GA	72	58.5	26
MST	69	38.8*	51*

*Calculated without consideration of obstacles.

5. Conclusion

In this work, we tackled hydrogen pipeline design, focused mostly on improving the topological arrangement. The consideration of soft obstacles allows the model to operate more flexibly and compromise the length on behalf of reducing costs. We have shown that this technique leads to more realistic route proposals and lengths that coincide with real networks. Furthermore, it opens the possibility of backwards analysis to understand the heuristic judgements that experts make on the costs and risks of traversing areas of interest. Future work should expand on the obstacles considered here, increase the interconnectedness between the design stages to avoid sub-optimal decisions and increase the temporality in the short- and long-term to understand how the design stages are altered with the prospect of network evolution.

References

André, J., Auray, S., Brac, J., Wolf, D., Maisonnier, G., Ould-Sidi, M., Simonnet, A., 2013. Design and dimensioning of hydrogen transmission pipeline networks. European Journal of Operational Research, Vol. 229, No. 1, p. 239-251.

Baufumé, S., Grüger, F., Grube, T., Krieg, D., Linssen, J., Weber, M., Hake, J.-F., Stolten, D., 2013. GIS-based scenario calculations for a nationwide German hydrogen pipeline infrastructure. International Journal of Hydrogen Energy, Vol. 38, No. 10, p. 3813-3829.

DESNZ, Department for Energy Security and Net Zero, 2023. Hydrogen transport and storage infrastructure: government response. UK.

Ejeh, J. O., Blazejewski, T., Martynov, S., Brown, S., 2023a. AENI: Agent-based energy network infrastructure design and operation. AIChe Annual Meeting 2023.

Ejeh, J. O., Marynov, S., Brown, S., 2023b. An MINLP model for the optimal design of CO_2 transportation infrastructure in industrial clusters. 33rd European Symposium on Computer Aided Process Engineering (ESCAPE33).

Geels, F., Sovacool, B., Iskandarova, M., 2023. The socio-technical dynamics of net-zero industrial megaprojects: Outside-in and inside-out analyses of the Humber industrial cluster. Energy Research & Social Science.

Heijnen, P. W., Ligtvoet, A., Stikkelman, R. M., Herder, P. M., 2013. Maximising the worth of nascent networks. Networks and Spatial Economics, Vol. 14, No. 1, p. 27-46.

National Grid, 2022. Humber low carbon pipelines project: Section maps of the proposed route for consultation. Available at: https://www.nationalgrid.com/our-businesses/national-grid-ventures/humber-low-carbon-pipelines/documents (Accessed: 13 September 2023).

NREL, National Renewable Energy Laboratory, 2018. H2A: Hydrogen Analysis Production Models Version 3. Available at: https : / / www . nrel . gov / hydrogen / h2a - production-models.html.

Parolin, F., Colbertaldo, P., Campanari, S., 2022. Development of a multi-modality hydrogen delivery infrastructure: An optimization model for design and operation. Energy Conversion and Management, Vol. 266.

Reuß, M., Welder, L., Thürauf, J., Linßen, J., Grube, T., Schewe, L., Schmidt, M., Stolten, D., Robinius, M., 2019. Modeling hydrogen networks for future energy systems: A comparison of linear and nonlinear approaches. International Journal of Hydrogen Energy, Vol. 44, No. 60, p. 32136-32150.

Robinius, M., Schewe, L., Schmidt, M., Stolten, D., Thürauf, J., Welder, L., 2019. Robust optimal discrete arc sizing for tree-shaped potential networks. Computational Optimization and Applications, Vol. 73, No. 3, p. 791-819.

Rosenberg, M., French, T., Reynolds, M., While, L., 2021. A genetic algorithm approach for the Euclidean Steiner tree problem with soft obstacles. Genetic and Evolutionary Computation Conference (GECCO'21), p. 618-626.

van den Broek, M., Mesquita, P., Carneiro, J., Silva, J. R., Berghout, N., Ramírez, A., Gouveia, J. P., Seixas, J., Cabal, H., Martinez, R., Rimi, A., Zarhloule, Y., Sardinha, M., Boavida, D., Tosato, G. C., 2013. Region specific challenges of a CO2 pipeline infrastructure in the West Mediterranean sea. Energy Procedia, Vol. 37, p. 3137-3146.

Welder, L., Ryberg, D. S., Kotzur, L., Grube, T., Robinius, M., Stolten, D., 2018. Spatio-temporal optimization of a future energy system for power-to-hydrogen applications in Germany. Energy, Vol. 158, p. 1130-1149.

Flavio Manenti, Gintaras V. Reklaitis (Eds.), Proceedings of the 34th European Symposium on
Computer Aided Process Engineering / 15th International Symposium on Process Systems
Engineering (ESCAPE34/PSE24), June 2-6, 2024, Florence, Italy

An Attainable Region Approach to Chemical Reaction Equilibrium

Diane Hildebrandt,[a*] James Fox,[b] Neil Stacey,[c] Celestin Sempuga [d]

[a]*Department of Chemical and Biochemical Engineering, Rutgers, The State University
of New Jersey, 98 Brett Road, Piscataway, NJ,08854, USA*
[b]*Helica Energy, 20 Eastbourne Terrace, Paddington, London, W2 6LG, UK*
[c]*School of Chemical and Metallurgical Engineering, University of the Witwatersrand, P
Bag X3, PO Wits, 2050, South Africa*
[d]*University of South Africa, Corner Christiaan de Wet Road & Pioneer Avenue,
Florida, 1709, South Africa*
Diane.Hildebrandt@Rutgers.edu

Abstract

Consider a process with a defined feed(s) and a set of potential products. The Material
Balance Limited Attainable Region (AR^{MB}) encompasses all feasible combinations of
process product streams consistent with the feed(s). The dimension of the AR^{MB} is equal
to the number of independent material balances (IMBs) that define the relationship
between the process species. We can transform AR^{MB} into a 2D region in the space of
Gibbs Free Energy (G) and Enthalpy (H), designated as the Thermodynamic Limited
Attainable Region (AR^T). At a temperature of T°=25°C and ambient pressure, the
AR^T(T^0) represents the required process work and heat flows needed to transform a given
feed into a specified product. However, the AR^T(T) describes the driving forces governing
chemical transformations within reactors under different temperatures T. In particular,
the minimum G on the boundary of the AR^T(T) is related to thermodynamic equilibrium.
We describe how the AR^T is a simple but powerful representation of all possible reactions
consistent with the given atomic inputs and how it can facilitate the identification of shifts
in equilibrium species with temperature and assess the anticipated sensitivity of process
outcomes to varying conditions. We demonstrate this conceptual framework across
diverse processes, encompassing biological and emerging eco-friendly feeds.

Keywords: Material Balance Limited Attainable Region, Thermodynamic Attainable
Region, Reaction Equilibrium.

1. Introduction

The early research by Shinnar (1998) demonstrated how thermodynamic constraints can
be leveraged to design chemical reactors and improve thermal efficiency. He asserted that
the limitations on efficiency are imposed by technology and catalysts rather than
thermodynamics. Shinnar defined the reachable or accessible 'composition space' for a
reacting system, bounded by overall specified reactions, where $\Delta G \leq 0$. He computed
reachable compositions for systems with multiple reactions demarcated by iso-G lines.
This approach considered both individual and combinations of reactions, offering insights
into potential reaction routes to overcome thermodynamic limitations and facilitate the
development of novel catalysts. However, this visualization is constrained by the
dimensionality of the problem, limited by the number of species involved.

Understanding thermodynamic constraints, namely enthalpy (H) and Gibbs Free energy (G), is crucial for designing more efficient chemical and biochemical processes that meet emission targets. Efficient utilization of raw materials and optimized heat and workflows are critical considerations (Shinnar, 1988). This understanding enhances process synthesis techniques for determining process structures, represented as flowsheets and reactions, based on specified inputs and outputs (Patel et al., 2007).

The most efficient, productive and sustainable processes, must first be identified through the governing laws of thermodynamics. This is evidenced through the work of Tula et al. (2019), who described the software tool ProCAPD which used a hybrid approach to evaluate all feasible alternatives with a search space, Monjur et.al. (2021) describing the tool SPICE, which evaluates combinations of process building blocks rather than unit operations. SPICE also uses the minimisation of Gibbs Free Energy to determine the theoretical attainable limits for any reaction-separation system and, more recently Pazmiño-Mayorga et.al. (2022) developed a decision-making approach for the synthesis of reactive distillation systems using basic thermodynamic properties and kinetic data.

The development of flowsheets is governed by thermodynamics, where the change in G across a process, flow sheet, or operating unit is related to workflows or irreversibility. Similarly, the change in H corresponds to minimum or net heat flows in the processes. The Attainable Region (AR) approach, introduced by Horn (1964) and Glasser et al. (1987), provides a systematic method for developing and improving process flowsheets. In this paper, we extend the AR approach to determine both material and thermodynamically limited attainable regions.

2. Theoretical Development

This section follows the development and nomenclature used by Shinnar and Feng (1985). Consider a set of reactions involving N species, where a species j is denoted by A_j and $1 \leq j \leq N$. We specify that *j* must either be a feed to the process or reaction or alternatively be a product of the process or reaction. We can represent the set of species as vector $\underline{A}^T = \{A_1, A_2 ..., A_N\}$ where $\underline{A}^T$ represents the transpose of vector $\underline{A}$. $\underline{A}$ must contain the species in the feed as well as all the species identified as possible products. Let n_j be the number of moles of species A_j. We can represent the composition of a mixture as vector $\underline{n}^T = \{n_1, n_2, ..., n_N\}$.

Aris and Mah (1963) state that a set of chemical reactions is linearly independent if any of the reactions in the set cannot be written as a linear combination of the other reactions. Yin (1989) points out that the independent reactions may not correspond to actual reactions occurring in the system and are merely stoichiometric relationships between the chemical species in the system. In this paper, we have adopted the term Independent Material Balances (IMBs) to emphasise that individual relationships may not correspond to actual reactions in a process or reactor. We have denoted the number of IMBs that describe the relationship between the elements of $\underline{A}$ as S. The Gibbs stoichiometric rule gives an upper bound for S; different procedures to determine the IMBs have been reported, and we use the method of Yin (1989) in this paper.

We can define the stoichiometric matrix $\boldsymbol{v}$, with elements v_{ij} where $1 \leq i \leq N$ and $1 \leq j \leq S$ such that the S IMBs can be written as $\boldsymbol{v}\underline{A} = \underline{0}$. We define the extent vector ε of the S IMBs as $\underline{\varepsilon}^T = \{\varepsilon_1, \varepsilon_2, ..., \varepsilon_s\}$. The relationship between the molar composition of the reaction mixture $\boldsymbol{n}$, the initial or feed composition $\underline{\boldsymbol{n}}^o$ and the stoichiometric matrix $\boldsymbol{v}$ is given by:

$$\underline{\boldsymbol{n}} = \underline{\boldsymbol{n}}^0 + \boldsymbol{v}^T \underline{\boldsymbol{\varepsilon}}, \quad \text{where the range of } \underline{\boldsymbol{\varepsilon}} \text{ is such that all } n_i \geq 0 \qquad (1)$$

2.1. Definition of the Material Balance Limited AR^{MB}

The AR^{MB} is the set of all possible outputs $\underline{\boldsymbol{n}}$ for a given $\underline{\boldsymbol{n}}^o$ that is consistent with the material balance constraints (that is, not taking other constraints such as kinetics, equilibrium, energy, etc. into account). This region contains all possible outputs from all possible processes or reactors that have a feed(s) of given composition $\underline{\boldsymbol{n}}^o$ and possible products of composition $\underline{\boldsymbol{n}}$. Thus, for a given feed $\underline{\boldsymbol{n}}^o$:

$$\underline{\boldsymbol{n}} \in AR^{MB} \text{ iff } n_j = n_j^o + \sum_{i=1}^{S} v_{ij}\varepsilon_i \geq 0 \text{ for } j = 1,2,\dots N \qquad (2)$$

2.1.1. Properties of the AR^{MB}

Some of the properties of the AR^{MB} that will be used in this paper are:

- Consider two systems, one with a feed of $\underline{\boldsymbol{n}}^{o1}$ and the second with a feed $\underline{\boldsymbol{n}}^{o2}$, where $\underline{\boldsymbol{n}}^{o2}$ and $\underline{\boldsymbol{n}}^{o1} \in AR^{MB}$. This implies that, in principle, we can achieve $\underline{\boldsymbol{n}}^{o2}$ from feed $\underline{\boldsymbol{n}}^{o1}$ and conversely, we can achieve $\underline{\boldsymbol{n}}^{o1}$ from feed $\underline{\boldsymbol{n}}^{o2}$. The vectors $(\underline{\boldsymbol{n}}^{o1} - \underline{\boldsymbol{n}}^{o2})$ and $(\underline{\boldsymbol{n}}^{o2} - \underline{\boldsymbol{n}}^{o1})$ are contained in the AR^{MB}.
- We denote an extreme point e of the AR^{MB} as $\underline{\boldsymbol{E}}_e(AR^{MB})$. We denote the set of all $\underline{\boldsymbol{E}}(AR^{MB})$ as $\{\underline{\boldsymbol{E}}(AR^{MB})\}$. The AR^{MB} is the convex hull of $\{\underline{\boldsymbol{E}}(AR^{MB})\}$, and is an S dimensional polytope.

2.2. Definition of the Thermodynamic Limited Attainable Region $AR^T(\underline{\boldsymbol{n}}^o,T,p)$

We begin by defining various properties that are useful in subsequent development. The specific enthalpy of component A_i in a mixture at temperature T and pressure p is denoted $\hat{H}_i(T,p)$. The specific enthalpies of $\underline{A}$ can be combined into vector $\underline{\hat{\boldsymbol{H}}}(T,p)^T = \{\hat{H}_1(T,p), \hat{H}_2(T,p), \dots \hat{H}_N(T,p)\}$. Similarly, the specific Gibbs Free energy vector of $\underline{A}$ in a mixture at T and p is defined as $\underline{\hat{\boldsymbol{G}}}(T,p)^T = \{\hat{G}_1(T,p), \hat{G}_2(T,p), \dots, \hat{G}_N(T,p)\}$, where $\hat{G}_i(T,p)$ is the specific Gibbs Free energy of component A_i at T and p.

We define the Thermodynamic Limited Attainable Region with respect to feed, $AR^T(\underline{\boldsymbol{n}}^o,T,p)$, as the set of $\{\Delta H, \Delta G\}$ of all the products $\boldsymbol{n}$ that can be achieved in a process or reaction from a given feed $\underline{\boldsymbol{n}}^o$, where the feed and product enter and leave at T and p. Mathematically we can define the region as follows:

$$AR^T(\underline{\boldsymbol{n}}^0,T,p) = \{\Delta H(T,p), \Delta G(T,p)\} = (\underline{\boldsymbol{n}} - \underline{\boldsymbol{n}}^0)^T \cdot \{\underline{\hat{\boldsymbol{H}}}(T,p), \underline{\hat{\boldsymbol{G}}}(T,p)\} = \{\Delta H_{rxn}(T,p), \Delta G_{rxn}(T,p)\} \qquad (3)$$

and $\{\Delta H(T,p), \Delta G(T,p)\} \in AR^T(\underline{\boldsymbol{n}}^o,T,p) \; \forall \; \underline{\boldsymbol{n}} \in AR^{MB}$.

2.2.1. Properties of the AR^{MB}

The following can be deduced from this definition:

- AR^T is a linear transformation of the elements of AR^{MB}. It follows that AR^T is a surjective mapping of the elements of AR^{MB}, or equivalently that there is a one-to-many relationship between elements of AR^{MB} and elements of AR^T. Consequently, $\underline{E}(AR^{MB})$ are linearly transformed and are elements of AR^T ($\underline{n}^o$,T,p) but may not be $\underline{E}(AR^T$ ($\underline{n}^o$,T,p)). However, all $\underline{E}(AR^T$ ($\underline{n}^o$,T,p)) correspond to $\underline{E}(AR^{MB})$.
- We define the set of $\underline{E}(AR^T$ ($\underline{n}^o$,T,p)) as $\{\underline{E}(AR^T$ ($\underline{n}^o$,T,p))\}$. The $\{\underline{E}(AR^{MB})\}$ do not depend on either T or p. As T and/or p are changed, the $\{\underline{E}(AR^T$ ($\underline{n}^o$,T,p))\}$ will be linear transformations of $\{\underline{E}(AR^{MB})\}$, however, the $\{\underline{E}(AR^T$ ($\underline{n}^o$,T,p))\}$ will depend on T and p. The implications of this will be discussed later.
- The AR^T is the convex hull of $\{\underline{E}\,(AR^T)\}$; thus the AR^T is a convex, connected set.

3. Example: All possible material balances for conversion of biomass (glucose) into syngas, biogas and/or methanol

Consider a feed of 1 mole of glucose ($C_6H_{12}O_6$) and 6 moles of water; we consider possible products CO, CO_2, H_2, CH_3OH and CH_4. We are thus considering the anaerobic conversion of glucose to the specified products. There are 4 IMB's and the $\underline{E}(AR^{MB})$ are given in Table 1. Thus, for example, E_1 corresponds to the formation of CH_4 and CO_2, or biogas, E_3 corresponds to a 1:2 mixture of CO_2:H_2 and E_7 corresponds to a syngas mixture of CO:H_2 of 1:1. Any point in the AR^{MB} can be considered a feed point and thus the line between E_7 and E_1 corresponds to converting CO_2 and H_2 to glucose and water.

Table 1 Vertices of AR^{MB} for a feed of 1 mole $C_6H_{12}O_6$ and 6 moles H_2O

IMB's		E	ε_1	ε_2	ε_3	ε_4	Composition at Vertex
$C_6H_{12}O_6 \rightarrow 6CO+6H_2$	ε_1						
		1	0	0	0	1	$6H_2O+3CH_4+3CO_2$
$C_6H_{12}O_6+6H_2O \rightarrow 6CO_2+12H_2$	ε_2	2	0	$1/3$	$2/3$	0	$4H_2O+4CH_3OH+2CO_2$
		3	0	1	0	0	$12H_2+6CO_2$
$C_6H_{12}O_6+6H_2 \rightarrow 6CO_3OH$	ε_3	4	$2/3$	$-1/3$	0	$2/3$	$4CO+8H_2O+2CH_4$
		5	$1/2$	0	$1/2$	0	$3CO+6H_2O+3CH_3OH$
$C_6H_{12}O_6 \rightarrow 6CH4+3CO_2$	ε_4	6	0	0	0	0	$C_6H_{12}O_6+6H_2O$
		7	1	0	0	0	$6CO+6H_2+6H_2O$

The next step is to transform the $\underline{E}(AR^{MB})$ to the {H,G} space and then find $\{\underline{E}(AR^T)\}$ by finding the convex hull of the transformed points, as shown in Fig 1. Ambient pressure is assumed in all 4 scenarios in Fig.1 and in (a) T=298K and H_2O and CH_3OH are liquid; in the three other figures, H_2O and CH_3OH are in the gas phase and (b) T= 548 K, (c) T=950K and (d) T=1300 K. The enthalpy of reaction (ΔH_{rxn}) was assumed constant, and the Gibbs-Helmholtz relationship was used to calculate the temperature dependence of G.

We see that the boundary of the AR^T at 25 ^{0}C (Fig. 1(a)) with H_2O and CH_3OH as liquids is defined by 4 extreme points, namely E_6,E_1,E_3, and E_7. Changing the H_2O and CH_3OH to gas results in E_4 moving the boundary of the region and furthermore, E_1, E_3, E_4 and E_7 being approximately linear in Fig. 1(b)..(d). Under the assumption of constant ΔH_{rxn}, extreme points in AR^T remain in the boundary as the temperature varies, and only their relative positions change. Conversely, interior points do not move into the boundary of

AR^T as temperature changes. Thus, G of the extreme points change with temperature, although H stay constant, resulting in the relative positions of E_6 and the line between E_1 and E_7 changing with temperature.

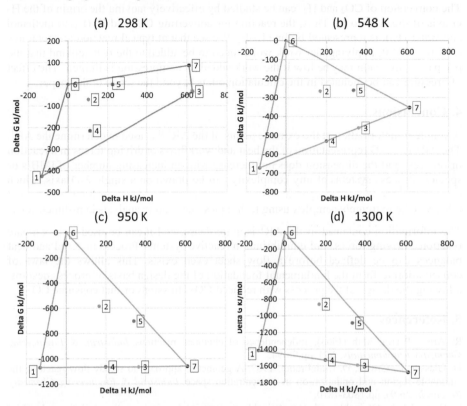

Figure 1 $\{\underline{E}(AR^T)\}$ in the {H,G} space at different temperatures. Extreme points of the AR^{MB} are shown; the numbers in rectangles correspond to the extreme points E in Table 1.

In Fig 1(a), G of E_7 >0, implying that work needs to be added to convert glucose to syngas. G <0 of the other vertices, which allows us to conclude that these reactions could proceed for example, biologically. The anaerobic production of biogas is represented by the line between E_6-E_1. The production of hydrogen (E_6-E_3) releases less G, implying that there is less work available for the bacteria to use for metabolism and reproduction than those who produce CH_4. G ≤0 for all E(AR) in Fig 1 (b) to Fig (d), which implies that thermo-catalytic reactions could potentially achieve the reactions between the feed and all possible products.

Minimum G corresponds to the thermodynamic equilibrium of the system. We can see that the overall system equilibrium changes from E_1 (CH_4 and CO_2) at 298 K (Fig 1(a)) to E_7 (syngas) at 1300 K (Fig 1(d)). At 950 K, the lower boundary of the AR^T is approximately a horizontal line containing E_1, E_4, E_3 and E_7. This means that the equilibrium conversion is determined by the mixing rather than the reaction term. Thus, practically, as the reaction proceeds to equilibrium, the composition will be path-dependent, and the resulting composition could look very different in different pieces of equipment, depending on how the reaction approaches equilibrium. This is indeed what is found in gasification and reforming reactions.

The conversion of CO_2 and H_2 can be studied by effectively moving the origin of the H-G axis of the AR^T to E_3. Thus, the reaction for converting CO_2 and H_2 (E_3) to methanol and water (E_2), is represented by line E_3-E_2. We see that at typical reaction temperatures (Fig 1(b)) G>0, implying that work would need to be added to the reaction and that the per-pass conversion may be low if the work added by compression is limited. The effect of pressure can be included in the calculation but is beyond the scope of this paper.

4. Conclusions

A simple transformation of the extreme points of the AR^{MB} is used to determine the AR^T. The AR^T is a 2-D representation of the heat and workflows or driving forces in a reaction or process, and the dimension does not increase with an increasing number of IMB's or species. Process systems of any complexity can be drawn on a single 2-D axis, which offers considerable advantages in visualization, analysis and optimization that would otherwise be extremely complex using techniques that require analysis in n-dimensions.

The information contained in the AR^T is quite dense, and it can be used to investigate reactions, driving forces, and equilibrium in reactive systems. Process material and heat balances can be defined before a flow sheet even exists. This allows the laws of conservation to form the fundamental foundation of the design basis of process designs, allowing the design of the process with reduced CO_2 emissions or that consume CO_2.

5. References

R. Aris, R.H.S. Mah (1963), Independence of chemical reactions. *Industrial & Engineering Chemistry Fundamentals*, 2(2), pp.90-94.

D. Glasser, C. Crowe, D. Hildebrandt, 1987. A geometric approach to steady flow reactors: the attainable region and optimization in concentration space. *Industrial & Engineering Chemistry Research*, 26(9), pp.1803-1810.

F. Horn, 1964, Attainable and non-attainable regions in chemical reaction technique. In *Third European Symposium on Chemical Reaction Engineering* (pp. 1-10). London: Pergamon Press.

A.S. Monjur, S..E. Demirel, J. Li, M.M.F.Hasan, 2021, A Computer-Aided Platform for Simultaneous Process Synthesis and Intensification, Computer Aided Chemical Engineering, (50). pp. 287-293 doi.org/10.1016/B978-0-323-88506-5.50046-2.

B. Patel, D. Hildebrandt, D.Glasser, B. Hausberger, 2007, Synthesis and Integration of Chemical Processes from a Mass, Energy, and Entropy Perspective,Industrial & Engineering Chemistry Research 46(25), pp. 8756-8766. DOI: 10.1021/ie061554z.

I. Pazmiño-Mayorga, A.A. Kiss, M. Jobson, (2022), Synthesis of Advanced Reactive Distillation Technologies: Early-Stage Assessment Based on Thermodynamic Properties and Kinetics, Computer Aided Chemical Engineering, 49, pp. 643-648, doi.org/10.1016/B978-0-323-85159-6.50107-X.

R. Shinnar, C. A. Feng (1985), Structure of complex catalytic reactions: thermodynamic constraints in kinetic modeling and catalyst evaluation, *Industrial & Engineering Chemistry Fundamentals*, 24 (2), pp. 153-170.

R. Shinnar, 1988. Thermodynamic analysis in chemical process and reactor design. Chem.Eng. Sci. 43(8), pp.2303-2318.

A.K. Tula, M R. Eden, R Gani, 2019, ProCAFD: Computer-aided Tool for Sustainable Process Synthesis, Intensification and Hybrid solutions, Computer Aided Chemical Engineering, 46, pp. 481-486. doi.org/10.1016/B978-0-12-818634-3.50081-3.

F. Yin, 1989, A simpler method for finding independent reactions, Chemical Engineering Communications, 83 (1), pp. 117-127, doi: 10.1080/00986448908940657.

Flavio Manenti, Gintaras V. Reklaitis (Eds.), Proceedings of the 34th European Symposium on Computer Aided Process Engineering / 15th International Symposium on Process Systems Engineering (ESCAPE34/PSE24), June 2-6, 2024, Florence, Italy

A Nash equilibrium approach to supply chain design of oligopoly markets under uncertainty

Asimina Marousi[a], Karthik Thyagarajan[b], Jose M. Pinto[b], Lazaros G. Papageorgiou[a], Vassilis M. Charitopoulos[a*]

[a]*Department of Chemical Engineering, Centre for Process Systems Engineering, University College London, Torrington Place, London WC1E 7JE, UK*
[b]*Linde Digital Americas, 10 Riverview Drive, Danbury CT 06810, United States*
v.charitopoulos@ucl.ac.uk

Abstract

An increased interest has been observed by the process systems community for the integration of game theoretic principles to the design and operation problems, especially of supply chains. This integration can achieve a better insight to the problem at hand and at the same time build resilience. In this work we aim to further increase the stability of a supply chain design problem by taking into account uncertainty. An oligopoly Nash bargaining game is proposed under a scenario-based approach in order to fairly allocate the profit among the members of a duopoly industrial gas market. The nonlinearity stemming from the Nash objective is linearised by a piecewise SOS2 linear approximation resulting in an MILP class. The design of the scenario-based approach is compared with that of the deterministic case for a number of sampled uncertainty realisations. Results suggest that the scenario-based approach can better leverage the capacity potential of the duopoly while guaranteeing higher profits and maintaining the market share of the initial market.

Keywords: game theory, Nash bargaining, supply chain optimisation, uncertainty, scenario- based optimisation.

1. Introduction

The study of supply chains under a game theoretic framework has met an increasing interest in the process systems engineering community (Marousi and Charitopoulos (2023)). Depending on the structure of the game and the timing of the decision making among the players, supply chains can be represented either as competitive or cooperative games. Competitive games are commonly modeled as Stackelberg games which result in bi-level or tri-level programming problems (Florensa et al. (2017)). On the other hand, when the total payoff of a game is to be allocated fairly among the players of the game, various egalitarian or utilitarian schemes can be employed. The Nash bargaining fairness scheme is commonly utilised for the profit allocation in supply chains since the relevant market share of the stakeholders in the status quo market is maintained. However, the main drawback of this method is that it results in a nonlinear objective function. Researchers in the past decades have approximated the Nash bargaining objective via separable programming and piecewise-linearisation in various applications (Ortiz-Gutiérrez et al. (2015), Charitopoulos et al. (2020)). Game theory has been insightful in cases where the supply chain stakeholders have conflicting objectives. Recently, Marousi et al. (2023a) have introduced a cooperative approach where manufacturers and customers of industrial gas oligopoly markets are players of the game. The former aim to maximise their profit while the latter maximise their savings by signing various contracts in a multi-periods deterministic model.

Even though supply chain design problems are inherently uncertain, especially in the case of multi-period considerations, most of the research in the field focuses on deterministic approaches (Barbosa-Póvoa et al. (2018)). Zamarripa et al. (2013) studied the problem of supply chain strategic planning as multi-objective game where the different strategies of the stakeholders were represented by different scenarios. While Zamarripa and co-authors study both a cooperative and competitive case, Hjaila et al. (2017) focused on a Stackelberg game when examining the operation of multi-echelon supply chains under uncertain competition. Different scenarios for the expected payoff were generated using the Monte Carlo sampling method. Following the Stackelberg approach, Gao and You (2019) developed a two-stage stochastic mixed-integer bi-level programming (MIBP) model in search of the optimal design and operation of multi-enterprise supply chains.

2. Problem statement

We consider the problem of a two-echelon oligopoly supply chain in multi-period framework. The problem was initially proposed as a static game in Charitopoulos et al. (2020) and extended in a multi-period framework under contractual agreements in Marousi et al. (2023b). Here, we investigate the impact of exogenous uncertainty on the fair profit allocation in an industrial gas market. The firms of the market are assumed to be acting rationally and have an estimate of the other firms' information. Customers can be assigned to firms via contractual agreements and can be re-allocated at the end of a contract's duration. Previous results of the authors have suggested that employing the Nash bargaining fairness scheme can provide significant profit increase for the oligopoly firms compared to the status quo, while at the same time maintaining the initial market structure (Charitopoulos et al. (2020), Marousi et al. (2023a,b)).

2.1. Model formulation outline

The overall optimisation problem is formulated as a MILP with key features including: (i) contract formulation, (ii) plant production shortcut model, (iii) inter-firm swap agreements, (iv) inventory levels, (v) electricity service cost. Given the planning horizon of P time periods, the number of firms F, producing I products at known plant capacities, the allocation of existing and new customers C with deterministic demand and delivery costs is to be optimized. We consider two sources of uncertainty in the model, one affecting the pricing of the contracts K and the other the expected electricity price. In this work we are following a risk-neutral two-stage scenario-based approach (Li and Grossmann (2021)).

2.2. Two-stage scenario-based approach

In the first-stage stochastic problem the allocation of the customers is decided along with the planning of the supply of the supply chain in terms amount of production, swaps between firms, outsourcing of production to spot market and inventory levels, which correspond to the here and now decision variables. The second stage decisions account for the operation of the Air Separation Unit (ASU) of the oligopoly firms. Let π_{pfs} be the profit of a firm f in a time period p and a scenario s is given by Eq.(1), where Rev_{fps} corresponds to the revenue, EC_{fps} to the electricity cost and $SC_{fp}, IC_{fp}, AFC_{fp}$ the service, inventory and acquisition/forfeit costs respectively. It can be observed that only the first two terms of the profit calculation in Eq. (1) are scenario dependent.

$$\pi_{fps} = Rev_{fps} - EC_{fps} - SC_{fp} - IC_{fp} - AFC_{fp}, \ \forall f, p, s \tag{1}$$

In order to clarify how uncertainty affects the examined supply chain we need to give a more detailed description of the terms Rev_{fps} and EC_{fps} in Eq. (2) and (3). The revenue of the firm is dictated by the selling price P_{icfpks} and the selected contract W_{cfkp}, the former being the uncertain parameter and the later a binary here and now variable for the customer allocation. In this study we consider deterministic customer demands D_{icfkp}. For a detailed description of the product price with respect to the different contracts we refer the reader to Marousi et al. (2023b).

$$Rev_{fps} = \sum_{ick} P_{icfpks} W_{cfkp} D_{icfkp}, \ \forall f, p, s \tag{2}$$

Since ASU are energy intensive industries converting atmospheric air to gas and liquid products, electricity consumption dominates the operating cost of the process (Mitra et al. (2014)). To achieve more favourable electricity prices, industrial gas firms are often part of demand side response schemes. A fixed electricity price EP_{ps} is agreed with the energy system operator within pre-specified electricity consumption tiers. In the case of over or under-consumption of energy outside those limits a penalty a is imposed on the electricity price. The electricity consumption is a wait and see decision and is derived using an ASU short-cut model, it is divided into the energy consumption within the specified consumption tiers θ_{fps}, over-consumption above the tiers δ_{fps}^{+} and under-consumption δ_{fps}^{-}. Parameter OT in Eq. (3) corresponds to the operating time of the ASU unit for the decided time period duration.

$$EC_{fps} = EP_{ps} OT (\theta_{fps} + (1 + a)(\delta_{fps}^{+} + \delta_{fps}^{-})), \ \forall f, p, s \tag{3}$$

For the scenario-based model we account for 5 equiprobable scenarios for different perturbations (low-medium-high) of the nominal product price and electricity price as displayed in Table 1.

Table 1 Perturbation of uncertain parameters in examined scenarios.

Uncertain parameter	Scenario 1	Scenario 2	Scenario 3	Scenario 4	Scenario 5
P_{icfpks}	Medium	High	High	Low	Low
EP_{ps}	Medium	Low	High	Low	High

2.3. Fair objective function

For the fair profit allocation between the oligopoly firms the Nash bargaining fairness scheme is selected and represented for each scenario in Eq. (4). However, Eq. (4) corresponds to a non-convex polynomial equation whose order depends on the number of firms. A separable programming approach is well studied in the literature for the Nash bargaining scheme following a logarithmic transformation resulting in Eq. (5) (Marousi and Charitopoulos (2023)).

$$\Psi_s = \prod_f (\pi_{fs} - \pi_{fs}^{sq}) \ \forall s \tag{4}$$

$$\Phi_s = \ln \Psi_s = \sum_f \ln(\pi_{fs} - \pi_{fs}^{sq}) \ \forall s \tag{5}$$

Given the probability of a scenario realization p_s, the scenario-based Nash bargaining objective can be formulated in Eq. (6).

$$\Phi' = \sum_s p_s \Phi_s = \sum_s p_s \left(\sum_f \ln\left(\pi_{fs} - \pi_{fs}^{sq}\right) \right) \tag{6}$$

Despite the logarithmic transformation the overall problem class remains as MINLP which can be computationally challenging in large scale multi-period problems. To this end, a SOS2 piecewise linearisation for n breakpoints is performed resulting in Eq. (7) and the additional constraints in Eq. (8), (9).

$$\widetilde{\Phi} = \sum_{fs} p_s \left(\sum_{fn} \ln\left(\tilde{\pi}_{fsn} - \pi_{fs}^{sq}\right)\lambda_{fsn} \right) \tag{7}$$

$$\sum_n \lambda_{fsn} = 1 \qquad \forall f, s \tag{8}$$

$$\pi_{fs} = \sum_n \tilde{\pi}_{fsn}\lambda_{fsn} \qquad \forall f \tag{9}$$

The objective function for the deterministic case can be similarly derived as suggested by the authors in Marousi et al. (2023b).

3. Case study

For the purpose of this paper, a duopoly case study is examined from an industrial liquid market. The market is comprised of 2 firms and 97 customers. At the status quo, Firm A serves 44 customers and Firm B 37, while there are 16 free customers that allow a market share growth. The selected time horizon is 8 years discretised in 32 quarterly periods. We examine three contracts of 1 year, 2 years and 3 years duration that result in different pricing schemes. Note that the contracts are binding and cannot be terminated before the predefined duration. The computational experiments were carried in an Intel®Core™i9-10900K CPU @ 3.70GHZ machine using GAMS Studio v.44.4 with Gurobi v10.0 with 20 threads. The optimality gap was set to 1%. The number of grid points for both the deterministic and the scenario-based model was selected as n=500. The original MINLP problem for both models was solved using BARON (Sahinidis (1996)), however no convergence was achieved for 5 days running time.

The aspects that are of interest in this study are how the introduction of scenarios affects the here and now decisions and if the scenario-based model can result in a more favorable market allocation for different uncertainty realisations.

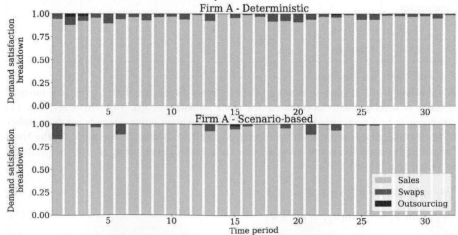

Figure 1 Demand satisfaction breakdown for Firm A for the Deterministic and Scenario-based models.

The first aspect is evaluated in Figure 1 where the demand satisfaction breakdown is displayed for Firm A and the different models. It can be observed that in the solution of the deterministic model Firm A is acquiring product from Firm B via swaps in all the examined time horizon ranging from 3 to 10% of the total demand to be covered. For three time periods Firm A refers to the

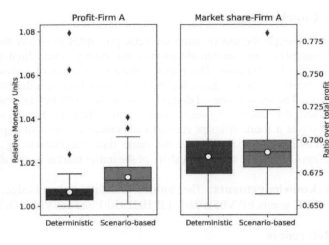

Figure 2 Profit (left) and market share (right) of Firm A for the Deterministic and Scenario-based models for Monte Carlo samples.

spot market to cover the allocated demand at a significantly higher cost. One of the modeling assumptions is that swap amount between firms should be balanced at predefined time periods in the examined time horizon (Marousi et al. (2023b)). Hence the use of the spot market does not necessarily imply that the capacity of the duopoly is not sufficient, but rather that under the deterministic customer allocation the full potential of the capacity is not fully captured. On the contrary, the scenario-based solution results in a customer allocation in which Firm A relies in fewer periods to swaps and almost no outsourcing is taking place.

To further evaluate the acquired solutions for the two models, we have generated 100 uncertainty samples using Monte Carlo simulation by allowing a perturbation of 20% of the nominal prices. For fixed customer and contract allocation, the statistical results for the profit in terms of Relative Monetary Units (RMU) and market share of Firm A are presented in Figure 2. The deterministic customer allocation results in tighter profit distribution with a median of 1.005 RMU, however for 3 samples the profit peaked at 1.02, 1.06 and 1.08 RMU. For the scenario-based allocation there is a wider interquartile range around the median of 1.01 RMU which suggest a higher flexibility of the model. Apart from the outlier values, the profit of Firm A is higher for the scenario-based customer allocation. Even for a small number of selected scenarios the extra flexibility introduced in the model is beneficial for the firm. Since the aim is to maintain a fair profit allocation, it is important to evaluate the market share allocation. In the status quo market, Firm A hold 80% of the market in terms of profit. The Nash bargaining solution results in a decreased market share for Firm A for both models. Nevertheless, Firm A maintains the dominance over the market since for the deterministic model the median market share is 68.6% and 69.1% for the scenario-based model. There is only a single uncertainty realisation at which the scenario-based model can reach a 72.5% market share. Despite the outliers detected in the profit for the deterministic model, there are no outliers for the market share, suggesting that the profit of Firm B was analogously high in those samples.

4. Conclusions

Even though the use of game theoretic principles is widely used for the supply chain design problems, the consideration of uncertainty is very limited in the process systems engineering literature. This paper examines the effect of a scenario-based approach on the Nash bargaining game for profit maximisation in a duopoly industrial gas supply chain. The scenario-based design is juxtaposed with the deterministic one for a number of sampled uncertainty realisations. Results indicate that the scenario-based approach proposes a more efficient customer allocation in terms of capacity planning and thus achieving higher profits. At the same time the fairness of the profit allocation is safeguarded since the market share of the initial market is overall maintained.

Acknowledgements: The authors gratefully acknowledge financial support from EPSRC grants EP/V051008/1, EP/T022930/1 and EP/V050168/1.

References

A. P. Barbosa-Póvoa, C. da Silva, A. Carvalho, 2018. Opportunities and challenges in sustainable supply chain: An operations research perspective. Eur. J. Oper. Res. 268, 399-431.

V. M. Charitopoulos, V. M. Dua, J. M Pinto, and L. G. Papageorgiou, 2020. A game-theoretic optimisation approach to fair customer allocation in oligopolies. Opt. Eng. 21, 1459–1486.

C. Florensa, P. Garcia-Herreros, P. Misra, E. Arslan, S. Mehta, I. E. Grossmann, 2017. Capacity planning with competitive decision-makers: Trilevel MILP formulation, degeneracy, and solution approaches. Eur. J. Oper. Res. 262(2).

J. Gao, and F. You, 2019. A stochastic game theoretic framework for decentralized optimization of multi-stakeholder supply chains under uncertainty. Comput. Chem. Eng. 122, 31–46.

K.Hjaila, L. Puigjaner, , J. M. Laínezand and A. Espuña, 2017. Integrated game-theory modelling for multi enterprise-wide coordination and collaboration under uncertain competitive environment. Comput. Chem. Eng. 98, 209–235.

M. N. Koleva, A.J Calderón, D. Zhang, C.A Styan, L. G. Papageorgiou, 2018. Integration of environmental aspects in modelling and optimisation of water supply chains. Sci. Total Environ. 636, 314–338.

C. Li and I.E. Grossmann 2021. A review of stochastic programming methods for optimisation of process systems under uncertainty. Front. Chem. Eng. 2, 34.

A. Marousi, V. M. Charitopoulos, 2023. Game theoretic optimisation in process and energy systems engineering: A review.   Front. Chem. Eng., 5, 1-13.

A. Marousi, J. M. Pinto, L. G. Papageorgiou, V. M. Charitopoulos, 2023a. Multi-period optimisation of oligopolies with contracts: A cooperative approach to customer fairness. Comput. Aided Chem. Eng. 52, 1603-1608.

A. Marousi, K.Thyagarajan, J. M. Pinto, L. G. Papageorgiou, V. M. Charitopoulos, 2023b. A multi-period game-theoretic approach to market fairness in oligopolies. arXiv:2311.10655 [math.OC].

S. Mitra, J. M. Pinto, I. E. Grossmann, 2014.Optimal multi-scale capacity planning for power-intensive continuous processes under time-sensitive electricity prices and demand uncertainty. Part I: Modeling, Comput. Chem. Eng. 65,89-101.

R. A., Ortiz-Gutiérrez, S. Giarola, N. Shah, F. Bezzo, 2015. An approach to optimize multi-enterprise biofuel supply chains including nash equilibrium models. Comput. Aided Chem. Eng. 37, 2255–2260.

N. V Sahinidis., BARON: A general purpose global optimization software package, J. Glob Opt., 8:201-205, 1996.

M. A. Zamarripa, A. M. Aguirre, C. A. Méndez, and A. Espuña,, 2013. Mathematical programming and game theory optimization-based tool for supply chain planning in cooperative/ competitive environments. Chem. Eng. Res. Des. 91, 1588–1600.

Flavio Manenti, Gintaras V. Reklaitis (Eds.), Proceedings of the 34th European Symposium on Computer Aided Process Engineering / 15th International Symposium on Process Systems Engineering (ESCAPE34/PSE24), June 2-6, 2024, Florence, Italy

Modelling of PCL Production: Multi-Scale Approach and Parameter Estimation

Jakub Staś,[a]*Alexandr Zubov[a]

[a]*Department of Chemical Engineering, University of Chemistry and Technology Prague, Technická 5 166 28 Prague 6 - Dejvice, Czech Republic*
stask@vscht.cz

Abstract

The rheological properties of polycaprolactone (PCL) are useful for handling and manufacturing of final products. In this work, we present the modification of a model developed by Zubov and Sin (2018) for polylactic acid (PLA) production. This model consists of three sub-models (macro-, micro- and meso-scale). Using experimental data from literature, we have found the values of kinetic parameters for proposed reaction mechanism of caprolactone ring-opening polymerization. To speed up the kinetic parameter estimation process, we decomposed the parametric space into two parts: (i) parameters associated with reactions that affect only monomer conversion, and (ii) parameters that influence the polymer mean molecular masses M_n and M_w. The agreement between predicted and measured data is discussed in this paper.

Keywords: polycaprolactone, ring opening polymerization, kinetics, melt rheology, Monte Carlo simulation

1. Introduction

Nowadays there are trends around the world to substitute conventional polymers with biodegradable ones, such as polycaprolactone. This polymer is fully compostable and biocompatible, has a relatively low production cost and maintains structural rigidity in a physiological environment. Thanks to these properties it can be used in medicine as a drug carrier, to produce stitches or as a tissue scaffold (Behtaj *et al.* 2021).

The information about polymerization kinetics and its relation to polymer chain length distribution is useful for production optimization to obtain material of desired properties. One of the most popular methods of industrial PCL production is ring-opening polymerization (ROP) of ε-caprolactone catalysed by stannous octoate with alcohol as a co-catalyst. (Wu *et al.* 2017). There have already been published some simple kinetic studies of PCL production by, e.g., Rafler and Dahlmann (1992), Punyodom *et al.* (2021) or Punyodom *et al.* (2022), which discussed the polymerization mechanism and focused on estimation propagation kinetics. Similarly to PCL, polylactic acid (PLA) is also produced by the ROP with stannous octoate as the catalyst – this analogy has been used by Rosa *et al.* (2018), who used the same kinetic model including the same kinetic rate coefficients as Yu *et al.* (2011) did previously for PLA.

The rheological properties of polymer melt are useful for handling and manufacturing of final products. These properties are dependent on many factors, including the type of monomeric unit(s) and chain architecture. One of the state-of-the-art tools in the prediction of polymer melt rheological properties is the publicly available Branch-on-

Branch (BoB) software published by Das *et al.* (2006). Combining this software with a model which simulates evolution of polymer chain architecture enables one to predict the rheology of the product melt during the polymerization reaction. This approach was already applied by Zubov and Sin (2018) for the ROP of PLA.

In this work, we are presenting the modification of a model by Zubov and Sin (2018) for PCL. This model consists of three sub-models (macro-, micro- and meso-scale) and aims to predict the rheological properties of polymer melt using reaction conditions. We decided to use the same kinetic scheme as was proposed by Yu *et al.* (2011) for the ROP of PLA. For estimation of the elementary reactions' kinetic constants, we used the experimental data published by Wu *et al.* (2017).

2. Methodology

Thanks to multi-scale modelling we can combine different types of information and strengths of different approaches. As it was mentioned before, our model is based on the work of Zubov and Sin (2018) and consists of 3 sub-models (macro-, micro- and meso-scale). The macro-scale and micro-scale models have been implemented in FORTRAN programming language. The last part, meso-scale model, is publicly available as C++ code and was developed by Das *et al.* (2006).

2.1. Macro-scale model

The first part is a simulation of the batch polymerisation reactor. Inputs to this sub-model are the initial composition of the reaction mixture (concentrations of monomer, catalyst, and co-catalyst), temperature and duration of the reaction. By solving the population and material balances (set of ODEs) are obtained temporal evolutions of concentrations, reaction rates, monomer conversion and number- and weight-average molecule masses of polymer, M_n and M_w, respectively.

Because of similarities between the two processes, we assume that the polymerization mechanism of PCL should follow that of PLA, with the only difference between them being the number of monomer units in one monomer molecule. For the ROP of PLA, L, L-lactide is used as a monomer molecule which contains two monomer (lactoyl) units. In the case of PCL, ε-caprolactone consisting of only one monomer unit is used. The modified kinetic scheme is presented in **Figure 1**. Reaction scheme published by Yu *et al.* (2011) modified for PCL ROP from ε-caprolactone. In **Figure 1** *M* stands for monomer unit (ε-caprolactone), *C* for catalyst (tin (II)octoate) and *A* for octanoic acid. Symbols R_n, D_n and G_n are used for the active (growing), dormant and terminated chains of length *n*, respectively. The co-catalyst (typically alcohol) molecule can be denoted as D_0. Apart from reversible catalyst activation and active chain propagation, the kinetic scheme involves also reversible chain transfer, intermolecular transesterification and nonradical chain scission.

The kinetic parameters of propagation, chain transfer, transesterification and chain scission follow the Arrhenius dependency on temperature. Therefore, we need to estimate the preexponential factors k_{p0}, k_{s0}, k_{te0}, k_{de0} and activation energies $E_{a,p}$, $E_{a,s}$, $E_{a,te}$, $E_{a,de}$. In contrast to Zubov and Sin (2018), we did not assume temperature-independent kinetics of chain transfer. It is assumed that catalyst activation is much faster than other reactions, so the kinetic constant of this reaction is set to $k_{a1} = 10^6$ L/mol/h. For calculation of the kinetic constant of catalyst deactivation, the equilibrium condition, $k_{a2} = \frac{k_{a1}}{K_{eq}}$, is used and it is assumed that equilibrium constant K_{eq} is temperature dependent in the following way:

$$K_{eq} = \exp\left(-\frac{A}{T} + B\right) \tag{1}$$

The value of the depropagation rate coefficient k_d is calculated as $k_d = k_p[M_{eq}]$, where monomer equilibrium concentration $[M_{eq}]$ is assumed dependent on temperature via reaction enthalpy and entropy according to

$$[M_{eq}] = \exp\left(\frac{\Delta H_{lc}}{RT} - \frac{\Delta S_{lc}}{R}\right) \tag{2}$$

The model further details on the model can be found in paper by Zubov and Sin (2018).

1. Catalyst (de)activation: $\quad C + D_n \xleftrightarrow{k_{a1}/k_{a2}} R_n + A$

2. Reversible propagation: $\quad R_n + M \xleftrightarrow{k_p/k_d} R_{n+1}$

3. Reversible chain transfer: $\quad R_n + D_i \xleftrightarrow{k_s} R_i + D_n$

4. Intermolecular transesterification: $\quad R_i + R_j \xleftrightarrow{k_{te}} R_{i+j-n} + R_n$

$\qquad\qquad\qquad\qquad\qquad\quad R_i + D_j \xleftrightarrow{k_{te}} R_{i+j-n} + D_n$

$\qquad\qquad\qquad\qquad\qquad\quad R_i + G_j \xleftrightarrow{k_{te}} R_{i+j-n} + G_n$

5. Nonradical random chain scission: $\quad R_i \xrightarrow{k_{de}} R_{i-k} + G_k$

$\qquad\qquad\qquad\qquad\qquad\quad D_i \xrightarrow{k_{de}} D_{i-k} + G_k$

$\qquad\qquad\qquad\qquad\qquad\quad G_k \xrightarrow{k_{de}} G_{k-j} + G_j$

Figure 1. Reaction scheme published by Yu *et al.* (2011) modified for PCL ROP from ε-caprolactone.

2.2. Micro-scale model

The Monte Carlo model is based on a stochastic simulation algorithm for dynamic modelling of coupled reactions which was pioneered by Gillespie (1976). We are following the hybrid Monte Carlo approach, as described in Zubov and Sin (2018), which uses the temporal evolutions of concentrations and reaction rates pre-calculated by the macro-scale model and simulates only the growth of polymeric chains step-by-step. This results in prediction of polymer chain architecture and its evolution in time. The hybrid approach offers a significant increase of computational efficiency of the Monte Carlo (micro-scale) simulation when compared to the original formulation by Gillespie (1976).

2.3. Meso-scale model

The last part of our multi-scale modelling framework is prediction of the rheological properties (loss and storage moduli, complex viscosity) using the distribution of molecular masses generated by the Monte Carlo model. The employed model of chain reptation is based on tube theory and was developed by Das *et al.* (2006). Besides the polymer chain architecture or mean molecular masses (in the case of linear polymers) this model also needs as input the temperature at which the rheological properties should be calculated, the molecular mass of monomer, density, mean chain entanglement length and characteristic (mean) entanglement time.

2.4. Estimation of kinetic parameters

We have re-implemented the macro-scale model in MATLAB environment for kinetic parameters estimation. MATLAB was chosen because of built-in global optimisers that are already parallelised.

For the reaction mechanism described in previous subsections, we needed to estimate 12 parameters. We have decided to decompose the problem into two parts: (i) estimation of parameters of reactions that affect only the monomer conversion, and (ii) estimation of parameters of reactions that change the polymer mean molecular masses M_n and M_w. For this task, we have used experimental data from the literature which were published by Wu *et al.* (2017), who conducted 6 experiments. Four of them were performed with the same concentration of monomer $[\varepsilon\text{-CL}]_0 = 3$ mol/L at different temperatures (60, 70, 80, and 90°C). The other two experiments were carried out at 80°C and the initial monomer concentration varied (1.5 mol/L and 0.5 mol/L). The initial concentrations of monomer, catalyst (tin octoate) and co-catalyst (*n*-butanol) were in all experiments set to the same ratio, $[\varepsilon\text{-CL}]_0:[\text{Sn(Oct)}_2]_0:[n\text{-butanol}]_0 = 300:1:1$.

The experimental data consisted of values of M_n, M_w and caprolactone conversion measured in time. We used the whole available dataset (from all six experiments) for estimating the first six parameters. However, to estimate the rest of the six parameters (affecting the change in mean molecular weights), we used only the experimental data at 70, 80 and 90°C (all with $[\varepsilon\text{-CL}]_0 = 3$ mol/L), as the course of the measured average molecular masses of polymer in the omitted experiments did not follow the expected trends, could not be rendered by the model predictions not even qualitatively, and was probably burdened with experimental error. MATLAB genetic algorithm from the Global Optimization Toolbox has been employed to estimate the desired kinetic parameters.

3. Results

As stated above, first we have found the parameters affecting the monomer conversion. Values of estimated parameters can be found in **Table 1**. The comparison of model predictions compared to experimental results – in good agreement between both – can be found in **Figure 2**.

Table 1. Values of parameters affecting monomer conversion.

Parameter	Value	Unit
A	-40506.8	K
B	127.588	-
k_{p0}	$4.53873 \cdot 10^{11}$	L/mol/h
$E_{a,p}$	73.4463	kJ/mol
ΔH_{lc}	-9.56587	kJ/mol
ΔS_{lc}	-6.80662	kJ/mol/K

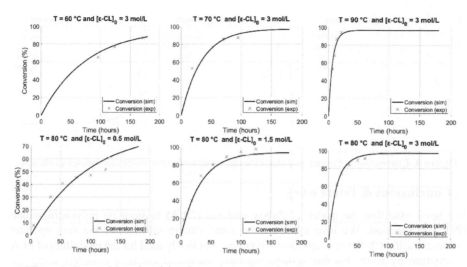

Figure 2. Comparison of model and experimental results of monomer conversion evolution.

Using the values of the first six parameters we found values of the remaining six ones, which are describing the kinetics of reactions between polymeric chains without the monomer influence. Values of these parameters are summarized in

Table 2. Simulation results compared with experimental data can be found in **Figure 3**. The agreement with experimental data in this case is significantly worse than for the monomer conversion. Qualitatively, the observed trends are described well by the model predictions, including the initial increase of the molecular weights due to fast propagation and subsequent decay due to polymeric chains re-shuffling. This suggests that the kinetic scheme used previously for PLA production can be used for PCL modelling as well. The quantitative discrepancy between model predictions and measured data can be caused either by small data set or lower quality of experimental data.

Table 2. Values of parameters related to reactions involving only polymeric species.

Parameter	Value	Unit
k_{s0}	$4.173 \cdot 10^{19}$	L/mol/h
$E_{a,s}$	186.27	kJ/mol
k_{te0}	$5.014 \cdot 10^{19}$	L/mol/h
$E_{a,te}$	158.40	kJ/mol
k_{de0}	$9.045 \cdot 10^{18}$	1/h
$E_{a,de}$	148.35	kJ/mol

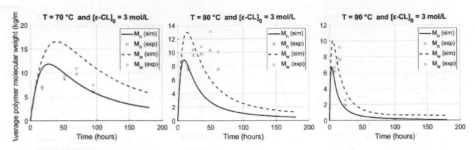

Figure 3. Comparison of model and experimental results of mean molecular masses evolution.

4. Conclusions & future work

We have modified the model of Zubov and Sin (2018) to simulate polycaprolactone (PCL) production. We have assumed the same kinetic scheme for the ring opening polymerization (ROP) of ε-caprolactone as was previously used for the modelling of PLA production by ROP. For this scheme we have estimated kinetic parameters using the experimental data published by Wu *et al.* (2017). We have achieved good agreement between simulated and experimental values of monomer conversion. Although qualitatively polymer mean molecular masses can be predicted by the presented model, the quantitative agreement is weaker. Because of that we are actively seeking for another experimental datasets (especially those carried out at higher polymerization temperatures), so that we can re-estimate kinetic parameters of our model. In the future we are planning to perform global sensitivity analysis of our model. We are also planning to modify our model, so that it will be able to simulate production of branched polymers.

References

S. Behtaj, F. Karamali, E. Masaeli, Y. G. Anissimov and M. Rybachuk (2021). "Electrospun PGS/PCL, PLLA/PCL, PLGA/PCL and pure PCL scaffolds for retinal progenitor cell cultivation." Biochemical Engineering Journal **166**: 107846.

C. Das, N. J. Inkson, D. J. Read, M. A. Kelmanson and T. C. B. McLeish (2006). "Computational linear rheology of general branch-on-branch polymers." Journal of Rheology **50**(2): 207-234.

D. T. Gillespie (1976). "A general method for numerically simulating the stochastic time evolution of coupled chemical reactions." Journal of Computational Physics **22**(4): 403-434.

W. Punyodom, W. Limwanich, P. Meepowpan and B. Thapsukhon (2021). "Ring-opening polymerization of ε-caprolactone initiated by tin(II) octoate/n-hexanol: DSC isoconversional kinetics analysis and polymer synthesis." Designed Monomers and Polymers **24**(1): 89-97.

W. Punyodom, P. Meepowpan and W. Limwanich (2022). "Determination of the activation parameters for the ring-opening polymerization of ε-caprolactone initiated by Sn(II) and Zn(II) chlorides using the fast technique of DSC." Thermochimica Acta **710**: 179160.

G. Rafler and J. Dahlmann (1992). "Biodegradable polymers. 6th comm. Polymerization of ϵ-caprolactone." Acta Polymerica **43**(2): 91-95.

R. P. Rosa, F. V. Ferreira, A. P. K. Saravia, S. A. Rocco, M. L. Sforca, R. F. Gouveia and L. M. F. Lona (2018). "A Combined Computational and Experimental Study on the Polymerization of epsilon-Caprolactone." Industrial & Engineering Chemistry Research **57**(40): 13387-13395.

D. Wu, Y. Lv, R. Guo, J. H. Li, A. Habadati, B. W. Lu, H. Y. Wang and Z. Wei (2017). "Kinetics of Sn(Oct)(2)-catalyzed ring opening polymerization of epsilon-caprolactone." Macromolecular Research **25**(11): 1070-1075.

Y. C. Yu, G. Storti and M. Morbidelli (2011). "Kinetics of Ring-Opening Polymerization of L,L-Lactide." Industrial & Engineering Chemistry Research **50**(13): 7927-7940.

A. Zubov and G. Sin (2018). "Multiscale modeling of poly(lactic acid) production: From reaction conditions to rheology of polymer melt." Chemical Engineering Journal **336**: 361-375.

Flavio Manenti, Gintaras V. Reklaitis (Eds.), Proceedings of the 34th European Symposium on Computer Aided Process Engineering / 15th International Symposium on Process Systems Engineering (ESCAPE34/PSE24), June 2-6, 2024, Florence, Italy

Improving food security through water management and allocation: A geospatial optimization approach

Maryam Haji, Mohammad Alherbawi, Sarah Namany, Tareq Al-Ansari

College of Science and Engineering, Hamad Bin Khalifa University, Qatar Foundation, Doha, Qatar

*talansari@hbku.edu.qa

Abstract

In response to the growing population and rising demand for agricultural production, the need for efficient and sustainable allocation of water resources becomes highly crucial. In the State of Qatar, these imperatives are particularly pronounced due to its arid desert climate and limited freshwater resources. Consequently, Qatar has been striving to enhance its food security. This research focuses on optimising the spatial distribution of water resources, encompassing various sources such as desalinated water (DW), groundwater (GW) and treated sewage effluent (TSE), to supply a variety of agricultural and fodder farms. By leveraging geospatial data within ArcGIS, the study deploys advanced optimisation techniques to create an optimal allocation network map that enhances resource utilization while concurrently minimising the costs associated with supplying water from various water sources to these farms. This research takes a comprehensive approach to tackle the multifaceted challenges of water allocation in agriculture, which includes integrating alternative water sources, prioritizing cost-effectiveness, and promoting sustainability. Furthermore, this study incorporates factors that will guide the geospatial distribution of different water resources. This includes groundwater depth, pH, salinity, and recharge rate. In addition to socio-environmental factors, such as public acceptance. The outcomes of this study offer the potential to provide valuable insights for policymakers along with agricultural and water resources stakeholders. By optimizing the allocation of water resources across various ranges of farms, the aim is to contribute to the long-term sustainability and resilience of agricultural practices, even in the face of evolving environmental and economic constraints. This research stands as a promising initiative to ensure a sustainable future of agriculture while addressing the serious global challenge of resource scarcity

Keywords: EWF nexus; food security; geospatial optimisation; water resource allocation

1. Introduction

The issue of limited water availability has garnered significant global attention due to the escalating population growth and ongoing socioeconomic progress observed in the 21st century (Hogeboom, 2020). Studies indicate that nearly 92% of global freshwater usage is linked to agricultural activities (Hoekstra & Mekonnen, 2012). Specifically, irrigation which is a crucial component of agricultural practices, dominates freshwater usage constituting roughly 70% of the total in 2021 (FAO, 2021). Anticipations point toward a continual surge in irrigation requirements over the forthcoming decades (FAO, 2021). By 2030, it is predicted that the worldwide demand for freshwater will soar to 160% of the current available quantities if the current rate of increase persists, a trend driven by the imperative to secure food resources (Ricart & Rico, 2019). This is particularly critical in arid regions such as the State of Qatar, where the harsh desert climate and scarcity of freshwater resources pose significant challenges to agricultural sustainability. Qatar's drive to bolster its food security and reduce its dependence on imported agricultural products has placed a premium on optimizing the utilization of its water resources. Consequently, there is an urgent need to devise effective strategies for accounting and managing water resources to harmonize the escalating demand for agricultural freshwater with the imperative of sustainable water supply. Against this backdrop, this study will adopt a holistic EWF nexus approach, to address the intricate challenges of water allocation in agriculture through the integration of alternative water sources, with a focus on cost-effectiveness, and a commitment to sustainable practices. Thus, this research embarks on a comprehensive investigation to optimize the spatial distribution of various water sources, including desalinated water (DW), groundwater (GW), and treated sewage effluent (TSE), across diverse agricultural and fodder farms. Hence, by employing an advanced spatial optimization technique within the ArcGIS platform, the study endeavors to devise an optimal allocation network map, aimed at maximizing resource efficiency and minimizing the costs associated with water supply from multiple sources to these farms. Moreover, specific risk factors that are critical to the geospatial distribution of water resources were meticulously incorporated into the analysis, this includes groundwater characteristics (such as depth, pH, salinity, and recharge rate) and socio-environmental considerations (such as public acceptance). The research's outcomes offer crucial guidance for decision-makers, farmers, and those overseeing water resources, setting the foundation for a resilient and sustainable farming industry in the State of Qatar. By optimizing the water resource allocation among various farms, the study contributes meaningfully to the long-term sustainability and resilience of agricultural practices, even in the face of evolving environmental and economic hurdles. Highlighting the urgent worldwide issue of limited resources, this research exemplifies Qatar's forward-thinking and creative strategies for a sustainable future in agriculture.

2. Methodology

2.1. Data Collection and Mapping

The geospatial data of the two different food industries were selected and located on the map, including 11 agriculture farms, and 3 fodder farms. Besides this, two water industries (including 22 treated wastewater stations and 32 desalinated reservoirs) and critical groundwater characteristics impacting food production industries; such as depth and salinity were re-mapped in ArcGIS (10.7.1) using the Universal Transverse Mercator (UTM) coordinates system, as illustrated in Figure 1 and Figure 2.

The water capacity for every treated wastewater plant, desalinated reservoir, and groundwater basin was assumed to be 3,180,000, 3,000,000, and 3,700,000 m³/year respectively. In which the water requirement to be fulfilled for each food industry was assumed to be 1,000,000 m³. Furthermore, the distances between the 54 water supply sites and the 14 food industries were calculated in ArcGIS along with the corresponding transportation costs.

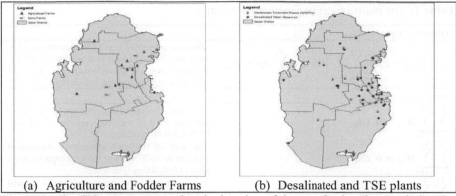

(a) Agriculture and Fodder Farms   (b) Desalinated and TSE plants

Figure 1: The GIS maps representing the location of the two food and water industries (a) agriculture and fodder farms and (b) desalinated and wastewater treatment plants.

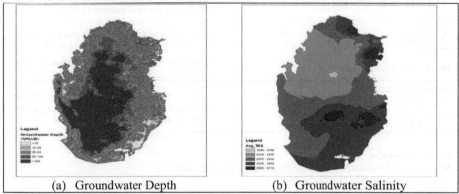

(a) Groundwater Depth   (b) Groundwater Salinity

Figure 2: The GIS maps representing geospatial groundwater characteristics including (a) depth and (b) salinity.

2.2. Optimisation Model Development

The multi-objective optimisation model aims at minimising the total transportation cost from different water industries to food production sites. A summary of the mathematical model is presented in Table 1. Five constraints were introduced to ensure the optimal contribution of groundwater sources and minimal transportation cost of the other two water sources to the food production industries (including agriculture and fodder farms). The first constraint is to allow only 10% (5,180,000 m³/year) utilisation of total groundwater capacity for food production industries. The same applied for the second and third constraints, allowing only 20% (19,200,000 m³/year) and 30% (20,988,000 m³/year) utilisation of the total capacity of desalinated water and treated wastewater respectively. The fourth constraint represents the public acceptance aspect, where people in the State of Qatar have concerns about the potential health risks associated with the use of treated wastewater for irrigation and the safety of the crops irrigated with treated wastewater. The third constraint is to restrict groundwater to participate in supplying groundwater to fodder farms, as it's considered the scarcest water resource in the country studied. Thus, Excel Solver was used to run and solve the optimization problem.

Table 1: The multi-objective function formulation.

Objective Function:	
$$Min\ Cost = C_t \sum_{i=1}^{14} \sum_{j=1}^{54} D_{ij} \frac{Q_j}{L} W_{ij} + \sum_{j=55}^{68} Dp_j\, C_p W_{ij}$$	Minimising the total cost (Transportation cost of both desalinated water and treated wastewater + Pumping cost of groundwater) for the food industries (agriculture and fodder farms).
Subject to:	
$$\sum_{i=1}^{11} \sum_{j=55}^{68} W_{ij} = 0.1A_g$$	The sum of total utilisation from groundwater must be 10% of the total capacity of 14 sites (3,700,000 m³/year/site).
$$\sum_{i=1}^{14} \sum_{j=23}^{54} W_{ij} = 0.2A_d$$	The sum of total utilisation from desalinated water must be 20% out of the total capacity of 32 sites (3,000,000 m³/year/site).
$$\sum_{i=12}^{14} \sum_{j=1}^{22} W_{ij} = 0.3A_w$$	The sum of total utilisation from treated wastewater must be 30% of the total capacity of 22 sites (3,180,000 m³/year/site).
$$\sum_{i=1}^{11} \sum_{j=1}^{22} W_{ij} = 0$$	Treated wastewater from sites (1 to 22) is not suitable for the food industry (1 to 11).
$$\sum_{i=12}^{14} \sum_{j=55}^{68} W_{ij} = 0$$	Restricts the utilisation of groundwater from sites (55 to 68) for the food industry (12 to 14).
$$W_{ij}, X_i, Y_i, X_j, Y_j \geq 0$$	It implies that all decision variables must be strictly positive.
Distance Formula:	
$$D_{ij} = \sum_{i=1}^{14} \sum_{j=1}^{54} \sqrt{(X_i - X_j)^2 + (Y_i - Y_j)^2}$$	Distance between water production site (j) to food industry (i) based on UTM coordinates.

Decision variables:
W_{ij}: *Allocated water (t/y) from water supply site (j) to food industry (i).*
(X_i, Y_i): *Coordinates of the food industry (i).*
(X_j, Y_j): *Coordinates of water production site (j).*

Parameters:
C_t: *Cost of transportation ($/km).*
C_p: *Cost of pumping groundwater ($/km).*
A_g: *Total capacity of groundwater (51,800,000 m³/year).*
A_d: *Total capacity of desalinated water (96,000,000 m³/year).*
A_w: *Total capacity of treated wastewater (69,960,000 m³/year).*
D_{ij}: *Distance (km) between water production site (j) to food industry site (i).*
Q_j: *Total quantity (t) of water production at a site (j).*
L: *Freight load (t/y).*

Where;

i=1, 2, … ,11 is representing agriculture farm
i=12,13,14 represent fodder farm
j=1, 2, … ,22 is representing treated wastewater stations
j=23, 24, … ,54 is representing desalinated water reservoirs
j=54, 55, … ,68 is representing groundwater sites

3. Result and Discussion

The Excel solver yielded an optimal water allocation from 17 sites (out of a total of 68 sites) as shown in Table 2 and Figure 3. The 17 selected sites will satisfy the total water requirements for each farm (equivalent to 1,000,000 m³). As such, the water requirement for the three fodder farms (no. 12, 13, and 14) was fulfilled using only treated wastewater from sites 2, 6, and 20. Whereas desalinated water and groundwater were allocated for agriculture farms (no. 1-11).

Besides this, Table 3 summarises the net and average transportation costs from different water sources to the 14 food industries. Hence the optimal water allocation yielded a total transportation cost of $4,231,185 per year, with an average cost equivalent to $0.30 per m^3.

Figure 3: Representation of optimal water allocation over food industries.

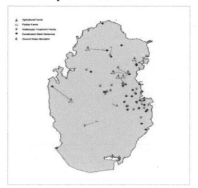

Table 2: The optimal allocation of water from 3 different sources over 11 agriculture and 3 fodder farms.

Water sources	Water allocation amount from different water sources to the food industries (m^3)													
	F1	F2	F3	F4	F5	F6	F7	F8	F9	F10	F11	F12	F13	F14
WW #2	-	-	-	-	-	-	-	-	-	-	-	-	-	1,000,000
WW #6	-	-	-	-	-	-	-	-	-	-	-	1,000,000	-	-
WW #20	-	-	-	-	-	-	-	-	-	-	-	-	1,000,000	-
DW #3	-	-	-	-	-	-	-	1,000,000	1,000,000	-	-	-	-	-
DW #4	-	-	-	320,000	-	-	-	-	-	-	-	-	-	-
DW #5	1,000,000	-	250,000	-	-	-	250,000	-	-	-	-	-	-	-
DW #6	-	-	-	-	250,000	-	-	-	-	250,000	-	-	-	-
DW #26	-	1,000,000	-	-	-	-	-	-	-	-	-	-	-	-
DW #28	-	-	-	-	-	-	-	-	-	-	250,000	-	-	-
DW #31	-	-	-	-	250,000	-	-	-	-	-	-	-	-	-
GW #3	-	-	750,000	-	-	-	-	-	-	-	-	-	-	-
GW #4	-	-	-	680,000	-	-	-	-	-	-	-	-	-	-
GW #5	-	-	-	-	750,000	-	-	-	-	-	-	-	-	-
GW #6	-	-	-	-	-	750,000	-	-	-	-	-	-	-	-
GW #7	-	-	-	-	-	-	750,000	-	-	-	-	-	-	-
GW #10	-	-	-	-	-	-	-	-	-	750,000	-	-	-	-
GW #11	-	-	-	-	-	-	-	-	-	-	750,000	-	-	-
Total Water Requirement (m3)	1,000,000	1,000,000	1,000,000	1,000,000	1,000,000	1,000,000	1,000,000	1,000,000	1,000,000	1,000,000	1,000,000	1,000,000	1,000,000	1,000,000

Table 3: The net cost of water from 3 different sources over 11 agriculture and 3 fodder farms.

Water sources	The total cost of transporting different water sources to the food industries ($/year)													
	F1	F2	F3	F4	F5	F6	F7	F8	F9	F10	F11	F12	F13	F14
WW #2	-	-	-	-	-	-	-	-	-	-	-	-	-	160,781
WW #6	-	-	-	-	-	-	-	-	-	-	-	181,805	-	-
WW #20	-	-	-	-	-	-	-	-	-	-	-	-	95,504	-

DW #3	-	-	-	-	-	-	-	625,406	625,406	-	-	-	-	-
DW #4	-	-	-	209,877	-	-	-	-	-	-	-	-	-	-
DW #5	678,838	-	173,017	-	-	-	151,330	-	-	-	-	-	-	-
DW #6	-	-	-	-	168,370	-	-	-	-	145,072	-	-	-	-
DW #26	-	601,921	-	-	-	-	-	-	-	-	-	-	-	-
DW #28	-	-	-	-	-	-	-	-	-	-	192,752	-	-	-
DW #31	-	-	-	-	-	178,632	-	-	-	-	-	-	-	-
GW #3	-	-	4,557	-	-	-	-	-	-	-	-	-	-	-
GW #4	-	-	-	9,251	-	-	-	-	-	-	-	-	-	-
GW #5	-	-	-	-	3,383	-	-	-	-	-	-	-	-	-
GW #6	-	-	-	-	-	7,691	-	-	-	-	-	-	-	-
GW #7	-	-	-	-	-	-	6,822	-	-	-	-	-	-	-
GW #10	-	-	-	-	-	-	-	-	-	6,281	-	-	-	-
GW #11	-	-	-	-	-	-	-	-	-	-	4,487	-	-	-
Net Cost ($/y)	678,838	601,921	177,575	219,129	171,753	186,323	158,152	625,406	625,406	151,353	197,239	181,805	95,504	160,781
Avg Cost ($/m³)	0.68	0.60	0.18	0.22	0.17	0.19	0.16	0.63	0.63	0.15	0.20	0.18	0.10	0.16

4. Conclusions

This study presents a novel EWF Nexus framework designed to optimally allocate water resources from three district sources (groundwater, desalinated water, and treated wastewater) over food production industries (agriculture and fodder farms). Leveraging ArcGIS and Excel tools, the framework integrates geospatial, quantitative, and qualitative considerations for both water-supplying sites and receiving food industries. The model aimed at selecting optimal water resources while minimising overall transportation costs. Among the 68 water sites evaluated, 17 were strategically chosen to supply water for the 14 food industries, resulting in an annual transportation cost of $4,231,185, which is equivalent to an average cost equivalent to $0.30 per m³. Placing our findings within the broader field of research, the existing water resource allocation models, such as WEAP, SWAT, and AQUATOOL provides comprehensive insights into specific aspects of water management. However, our model excels in seamlessly integrating geospatial considerations and cost optimization, thereby offering a unique and holistic approach to address the complex challenges within the broader field of sustainable water resource allocation. Thereby, this model provides insights on possible means to optimally utilise different water resources at lower costs while ensuring long-term sustainability and resilience of agricultural practices.

5. Acknowledgments

This research was made possible by an Award (GSRA7-1-0407-20014) and supported by proposal number NPRP11S-0107-180216 from Qatar National Research Fund (MME01-0922-190049). The contents herein are solely the responsibility of the authors[s].

6. References

FAO. (2021). Monitoring Water Use in Agriculture through Satellite Remote Sensing. Retrieved from https://www.fao.org/aquastat/en/

Hoekstra, A. Y., & Mekonnen, M. M. (2012). The water footprint of humanity. *P. Natl. Acad.Sci. USA.*, (109), 3232–3237. https://doi.org/https://doi.org/10.1073/pnas.1109936109.

Hogeboom, R. J. (2020). The water footprint concept and water's grand environmental challenges. In *One Earth 2*. https://doi.org/https://doi.org/10.1016/j.oneear.2020.02.010.

Ricart, S., & Rico, A. M. (2019). Assessing technical and social driving factors of water reuse in agriculture: a review on risks, regulation and the yuck factor. *Agric. Water Manag.*, (217), 426–439. https://doi.org/https://doi.org/10.1016/j.agwat.2019.03.017

Flavio Manenti, Gintaras V. Reklaitis (Eds.), Proceedings of the 34th European Symposium on Computer Aided Process Engineering / 15th International Symposium on Process Systems Engineering (ESCAPE34/PSE24), June 2-6, 2024, Florence, Italy

Technical Analysis of Ammonia Converter Catalyst Installation in Ammonia 1B PT Pupuk Kujang Plant

Zulkarnaen Arsadduddin

PT Pupuk Kujang, Jalan Jenderal Ahmad Yani 39, Karawang 41373, Indonesia
arzadd@gmail.com

Abstract

Selecting suitable catalyst for ammonia converter is crucial, as it can significantly affect the ammonia production as well as enthalpy generated from the reaction. A study case regarding technical selection of catalyst needs to be done in PT Pupuk Kujang. Throughout 17 years of operation, the ammonia reactor catalyst of Kujang 1B plant has never been replaced. Within that time interval, the performance of the ammonia reactor was declining. Operational data shows that the average volume fraction of NH3 products in 2008 was 15.18%, while the average for the last three years was 13.44%.

There are three types of catalysts that are commonly used in the time being, namely iron-based catalysts in the form of magnetite (Fe-M) and wustite (Fe-W), as well as ruthenium supported by graphite carbon (Ru/C). In this research, the author simulated ammonia synthesis according to the scheme of PT Pupuk Kujang Cikampek Ammonia Unit IB with a production capacity of 1000 tons/day. During 17 years of operation, Kujang 1B ammonia plant has been using the Fe-M catalyst in the ammonia reactor. Simulations were conducted using Aspen Plus© by reviewing catalyst variations, reactor inlet temperatures, cold shot opening, and optimization of operating conditions based on the Response Surface Methodology (RSM).

The simulation results conducted on PT Pupuk Kujang Cikampek Unit Ammonia IB scheme and operating conditions show that the wustite catalyst generates more ammonia products than Ru/C and magnetite. In the optimization, the reactor inlet temperature conditions for Ru/C, magnetite, and wustite catalysts are 360.4°C; 386.1°C; and 377.7°C with ammonia products of 1140.98 tons/day, 1099.00 tons/day and 1151.94 tons/day, as well as a mass enthalpy of 42.44 kcal/kg, 61.61 kcal/kg, and 52.35 kcal/kg. Variation of cold shot opening has no significant effect on the ammonia product, but there is an optimum point at 48% opening.

Keywords: ammonia converter, catalyst, magnetite, ruthenium, wustite.

1. Introduction

Ammonia is a raw material in the manufacture of fertilizers, refrigerants, and some derived chemical compounds. In Indonesia, ammonia is mostly produced by fertilizer companies, one of which is PT Pupuk Kujang Cikampek. This plant has two ammonia production units with a capacity of 330,000 tons/year each. The newest production unit, the Kujang 1B ammonia plant, was built in 2005 under Kellogg Brown & Root license. The ammonia synthesis process takes place in three packed bed type reactors with one intercooler. In addition, heat utilization is also available in the form of feed heating in the annulus and cold shot to control the temperature in the reactor.

Throughout 17 years of operation, the ammonia reactor catalyst has never been replaced. Within that time interval, the performance of the ammonia reactor was declining. Operational data shows that the average volume fraction of NH3 products in 2008 was 15.18%, while the average for the last three years was 13.44%. In addition, the cold shot valve is no longer opened to the inlet of reactor bed 1 because the bed temperature is pretty low. In 2008, the cold shot valve opened at 16.61%, while in 2022 the flow was closed (0.00% opening). If a cold shot is flowing, the reactor temperature will drop and reduce the amount of reaction in the ammonia reactor. Operational data also shows a decrease in energy at the reactor output flow rate. Data for 2008 indicates that the outlet must be cooled with a boiler feed water (BFW) flow rate of 89 tons/hour, while in 2022 it only requires 69 tons/hour.

There are three types of catalysts that are commonly used in the time being, namely iron-based catalysts in the form of magnetite (Fe-M) and wustite (Fe-W), as well as ruthenium supported by graphite carbon (Ru/C). Throughout 17 years of operation, Kujang 1B ammonia plant has been using the Fe-M catalyst in the ammonia reactor. The synthesis reaction of ammonia with Fe-M was conducted at operating conditions of 350-525°C and a pressure of 100-300 bar (Liu, 2014). This catalyst is quite active and inexpensive, but the reaction is inhibited by NH3.

In this research, the author simulated ammonia synthesis according to the scheme of PT Pupuk Kujang Cikampek Ammonia Unit IB with a production capacity of 1000 tons/day. Simulations were conducted using Aspen Plus© by reviewing catalyst variations, reactor inlet temperatures, cold shot opening, and optimization of operating conditions based on the Response Surface Methodology (RSM).

2. Research Methods

The type of this research is dry research using software. The research was begun by simulating the process of the ammonia 1B plant using the Aspen Plus© software. Furthermore, the model that has been made is validated and analyzed to observe the operating conditions using several types of catalysts.

2.1. Research Tools and Materials

This research used Aspen Plus © V11.0 (UGM academic license) software and Minitab 19.

2.2. Research Procedure

The research was conducted in a process simulation by limiting the fixed variables and independent variables. Kinetic base, field data, and assumptions were used to facilitate calculations. The response surface methodology method in Minitab 19 was utilized to optimize operating conditions.

2.2.1. Description of Process

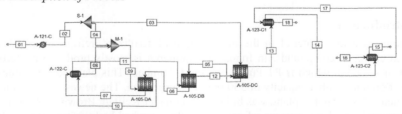

Figure 2.1. The Simulation of Ammonia Synthesis with PT Pupuk Kujang Unit IB Scheme with Aspen Plus©

Figure 2.1 shows the simulation of ammonia synthesis based on the scheme and operating conditions at PT Pupuk Kujang Cikampek Ammonia Unit IB. The reactor consists of a pressure shell and basket with three fixed bed reactors. Kinetically, there are only two beds because there is no intercooler between beds 2A (A-105-DB) and 2B (A-105-DC). The inlet gas flow A-105-D is divided into two, namely the feeds flow that enters through the annulus and the cold shot flow which serves to control the temperature of the reactor bed 1 (A-105-DA). The feed will go through the annulus on the three reactor beds and be heated in the intercooler (A-122-C) on the side of the tube. The cold shot is combined with the tube side A-122-C output current to the A-105-DA. The A-105-DA outlet moves towards the A-122-C shell side for cooling before entering A-105-DB. The product outlet from A-105-DC is cooled through the Ammonia Converter Effluent (A-123-C1/C2) which also serves to generate steam at the ammonia plant 1B.

2.2.2. Field Data and Assumptions

The thermodynamic model chosen is Peng-Robinson because it is used to calculate non-ideal gas mixtures (Tripodi et al., 2018). Incoming feed conditions are adjusted based on the latest data at PT Pupuk Kujang Cikampek Unit Ammonia IB. Feed enters the Ammonia Feed Reactor (A-121-C) at a temperature of 56°C and a pressure of 144.6 kg/cm2G. The feed flow rate was 4.2 m3/hour with a composition of 22.93% N2, 66.83% H2, 1.25% NH3, 5.23% CH4, and 3.76% Ar. The A-121-C outlet is set at 240°C. This is because there is still feed heating in the annulus and intercooler; hence the temperature entering the reactor is in the reaction temperature range of 350-525°C for Fe-based catalysts and 325-450°C for Ru/C (Liu, 2014). The product flow contains high energy; hence it is utilized for steam generation in A-123-C1 and A-123-C2. In this research, the BFW inlet temperature to A123-C2 used was 128°C with a pressure of 130 kg/cm2G. The length of the A-105-DA reactor is 5.78 meters, while the width is 2.55 meters. The A-105-DB reactor is 5.75 meters long and 2.55 meters in diameter. The A-105-DC reactor has a length of 5.80 meters and a diameter of 2.55 meters. The heat transfer coefficient for the three reactors is 5 Btu/hour.ft.°F according to the rule of thumbs for gas reactions (Couper et al., 2010). The approach temperature specification for the heat exchanger for fluids is at least 10°C (Turton, 2013); hence the approach temperature is chosen at 10°C. The ammonia synthesis cycle has a broad scope and variables, hence assumptions must be employed. Assumptions employed in this research are as follows:

a. Cycle based on steady state.
b. Pressure drop in the heat exchanger, mixer, splitter, and reactor is extremely small, hence it can be neglected.

2.2.3. Kinetic Models

The reaction for the formation of ammonia is shown in Equation (1) (Brown et al., 2014). The ammonia formation reaction is a complex heterogeneous catalytic reaction, in which N_2, H_2 and NH_3 adsorption or desorption occurs on the surface of the catalyst.

$$N_2 + 3H_2 \leftrightarrow 2NH_3 \quad \Delta H° = -46 \ kJmol^{-1} \tag{1}$$

The kinetic modeling of the reaction for NH_3 formation was conducted using the Langmuir-Hinshelwood-Hougen Watson (LHHW) approach. The driving force of this reaction is the fugacity of each component raised to the power of the exponent for an alternating reaction. The LHHW approach is shown in Equation (2) (Tripodi et al., 2018).

$$r = \frac{\left(k_0 \ e^{-\frac{Ea}{RT}}\right)\left(K_{for}\Pi_r \quad f_r^{vr} - K_{rev}\Pi_p \quad f_p^{vp}\right)}{\left(\Sigma_i \quad K_i\Pi_i \quad f_i^{vi}\right)^e} \tag{2}$$

Many previous researches have represented the reaction of ammonia formation with Fe-based catalysts, namely magnetite and wustite with the Temkin equation shown in Equation (3) (Dyson and Simon, 1968). The gas phase reaction the activity of the component a_i is expressed as a fugacity function (f_i).

$$\frac{d\eta}{d\tau} = k_{for}\left[(K_a)^2 a_{N2}\left(\frac{a_{H2}^{2,25}}{a_{NH3}^{1,5}}\right) - \left(\frac{a_{NH3}^{0,5}}{a_{NH3}^{0,75}}\right)\right] \tag{3}$$

$$a_i = \frac{f_{ip}}{P^\theta} \tag{4}$$

$$f_i = \phi_i y_i P \tag{5}$$

Equation (3) cannot represent the reaction kinetics of the Ru/C catalyst. This is due to differences in inhibitors; hence a modification of the Temkin equation is necessary so that it can be applied to Ru/C catalysts. Modification of the Temkin equation for the Ru/C catalyst with the LHHW approach is shown in Equation (6) (Rossetti et al., 2006).

$$\frac{d\eta}{d\tau} = k_{for}\frac{a_{N2}^{0,5}\left[\left(\frac{a_{H2}^{0,375}}{a_{NH3}^{0,25}}\right) - \frac{1}{K_a}\left(\frac{a_{NH3}^{0,5}}{a_{NH3}^{0,75}}\right)\right]}{1 + K_{H2}(a_{H2})^{0,3} + K_{NH3}(a_{NH3})^{0,2}} \tag{6}$$

Kinetic data in the form of Ea values based on experimental results at temperatures of 400-460°C with Fe-M and Fe-W catalysts of 47.5 and 44.9 kcal/mol and k_0 values of 3.23 x 10^9 and 7.47 x 10^8 kmol s^{-1} kg$_{cat}$$^{-1}$ (Pernicone et al., 2003). Ru/C catalyst kinetics data has been validated in a simulation conducted by Tripodi et al., (2018) with an Ea value of 23 kcal/mol and k_0 of 426 kmol s^{-1} kg$_{cat}$$^{-1}$. To implement the data in Aspen Plus©, the pre-exponential factor was corrected by considering the density of the catalyst being equal to 0.59 g cm^{-3} (Ru/C), 2.8–2.9 g cm^{-3} (magnetite) and 3.25 g cm^{-3} (wustite) (Tripodi et al., 2018).

2.2.4. Optimization Method with RSM

Response Surface Methodology (RSM) is a statistical calculation tool used for modeling and analyzing problems. The response it resulted is affected by parameters that can be controlled. The objective is to optimize by developing a model with the response surface method. RSM uses a sequential procedure where the first step is to decide the boundaries of the experimental area by identifying the range for each parameter. Next, the tool will design and plan the experiment using the appropriate DOE RSM approach.

3. Results and Discussions

3.1. Validation

This research compares Fe-M, Fe-W, and Ru/C catalysts. The scheme used is in accordance with PT Pupuk Kujang Cikampek Unit Ammonia IB which uses Fe-M catalyst. For validation purposes, the simulations with Fe-M catalyst were compared with the field data. The simulation results indicate a difference of 0.08% - 3.94%, hence the simulation model already represents the conditions in the field and can be used for further evaluation.

3.2. The Comparison of Simulation Results on Three Catalyst

The operating conditions in the simulation are in accordance with the PT Pupuk Kujang Cikampek Unit IB scheme, namely the inlet pressure at A121-C is 144 kg/cm²G, the temperature is 56°C and the flow rate is 4.2 m³/hour. Figure 3.1. shows the volume fraction profile of ammonia in each bed for the three types of catalysts. Based on Figure 3.1., ammonia conversion with Fe-based catalyst in bed 1 increases significantly. This indicates that the reaction is running rapidly. Meanwhile, the conversion in bed 2 and bed 3 shows a slower rate of formation. This is because the increasing volume fraction of ammonia will increase the inhibitory activity of Fe catalysts by ammonia. Meanwhile, the formation rate with the Ru/C catalyst in bed 1 is slower. This is due to the inhibition of the Ru/C catalyst by H_2 where the volume fraction was high at the beginning of the reaction, thus affecting the formation rate. According to Figure 3.1, the higher the temperature, the higher the ammonia volume fraction. However, ammonia formation does not occur at an overly high temperature because the reaction goes beyond the reaction temperature range. At outlet bed 3, the highest percentage of ammonia volume fraction was produced by the Fe-W catalyst of 15.86%. This result indicates that Fe-W has better catalytic activity compared to Fe-M and Ru/C in the PT Kujang Fertilizer Ammonia Unit 1B scheme.

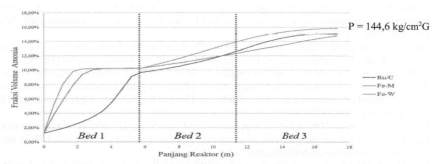

Figure 3.1 Changes in Ammonia Concentration toward Reactor Length with Catalyst Variations

3.3. The Analysis of Inlet Temperature Sensitivity toward Volume Fraction, Ammonia Product, and Mass Enthalpy Product

Response surface analysis (RSM) method is used to determine the temperature of incoming feed reactor which produces the optimum volume fraction, ammonia product, and mass enthalpy. Based on the optimization conducted with Minitab 19, the results showed that the Fe-W catalyst produced the most product of 1151.91 tons/day and a mass enthalpy of 52.35 kcal/kg. These results are obtained at a current 09 temperature of 377.7°C.

3.4. The Analysis of Wustite Cold Shot Opening Variations toward Ammonia Product

The simulation was tried on the catalyst that gave the most ammonia conversion and products, namely the Wustite-type Fe-based catalyst. The operating conditions used were adjusted to the operational conditions with A-121-C exit temperature at 140°C with 144.6

kg/cm₂G pressure. The higher the fraction that enters the annulus bed 3, the less the opening in the cold shot. The chart shows that opening a cold shot has no significant effect on the product. However, there is one optimum point that generates the maximum product, which is 1176.19 tons/day at the inlet fraction to the annulus bed 3 of 52% or the opening of a cold shot of 48%. After reaching the optimum condition, the product will decrease if the opening of the cold shot is increased.

4. Conclusion

The simulation results conducted on PT Pupuk Kujang Cikampek Unit Ammonia IB scheme and operating conditions show that the wustite catalyst generates more ammonia products than Ru/C and magnetite. In the optimization, the reactor inlet temperature conditions for Ru/C, magnetite, and wustite catalysts are 360.4°C; 386.1°C; and 377.7°C with ammonia products of 1140.98 tons/day, 1099.00 tons/day and 1151.94 tons/day, as well as a mass enthalpy of 42.44 kcal/kg, 61.61 kcal/kg, and 52.35 kcal/kg. Variation of cold shot opening has no significant effect on the ammonia product, but there is an optimum point at 48% opening.

References

Brown, D. E. et al. (2014) 'The genesis and development of the commercial BP doubly promoted catalyst for ammonia synthesis', Catalysis Letters, 144(4), pp. 545–552. doi: 10.1007/s10562-014-1226-4.

Couper, J. R. et al. (2010) Chemical Process Equipment. 2nd edn, Process Technology. 2nd edn. Burlington, United States of America: Butterworth-Heinemann. doi: 10.1515/9783110712445-007.

Dyson, D. C. and Simon, J. M. (1968) 'A kinetic expression with diffusion correction for ammonia synthesis on industrial catalyst', Industrial and Engineering Chemistry Fundamentals, 7(4), pp. 605–610. doi: 10.1021/i160028a013.

Hagen, S. et al. (2003) 'Ammonia synthesis with barium-promoted iron-cobalt alloys supported on carbon', Journal of Catalysis, 214(2), pp. 327–335. doi: 10.1016/S0021-9517(02)00182-3.

Kalaivanan, R., Ganesh, N. V. and Al-mdallal, Q. M. (2020) 'Case Studies in Thermal Engineering An investigation on Arrhenius activation energy of second grade nanofluid flow with active and passive control of nanomaterials', Case Studies in Thermal Engineering, 22(November), p. 100774. doi: 10.1016/j.csite.2020.100774.

Kowalczyk, Z. et al. (1998) 'Effect of potassium and barium on the stability of a carbon-supported ruthenium catalyst for the synthesis of ammonia', Applied Catalysis A: General, 173(2), pp. 153–160. doi: 10.1016/S0926-860X(98)00175-6.

Liu, H. (2014) 'Ammonia synthesis catalyst 100 years: Practice, enlightenment and challenge', Cuihua Xuebao/Chinese Journal of Catalysis, 35(10), pp. 1619–1640. doi: 10.1016/S1872-2067(14)60118-2.

Liu, H. et al. (2020) 'Development and application of wüstite-based ammonia synthesis catalysts', Catalysis Today, 355, pp. 110–127. doi: 10.1016/j.cattod.2019.10.031.

Pernicone, N. et al. (2003) 'Wustite as a new precursor of industrial ammonia synthesis catalysts', Applied Catalysis A: General, 251(1), pp. 121–129. doi: 10.1016/S0926-860X(03)00313-2.

Smith, J. M., Van Ness, H. C. and Abbott, M. M. (2005) Introduction to Chemical Engineering Thermodynamics. 7th edn. Boston: McGraw-Hill.

Tripodi, A. et al. (2018) 'Process simulation of ammonia synthesis over optimized Ru/C catalyst and multibed Fe + Ru configurations', Journal of Industrial and Engineering Chemistry, 66, pp. 176–186. doi: 10.1016/j.jiec.2018.05.027.

Turton, R. (2013) Analysis, Synthesis, and Design of Chemical Processes Fourth Edition, Journal of Chemical Information and Modeling.

Yoshida, M. et al. (2021) 'Economies of scale in ammonia synthesis loops embedded with iron- and ruthenium-based catalysts', International Journal of Hydrogen Energy, 46(57), pp. 28840–28854. doi: 10.1016/j.ijhydene.2020.12.081.

Flavio Manenti, Gintaras V. Reklaitis (Eds.), Proceedings of the 34th European Symposium on Computer Aided Process Engineering / 15th International Symposium on Process Systems Engineering (ESCAPE34/PSE24), June 2-6, 2024, Florence, Italy

Model-Based Design of Experiments for Isotherm Model Identification in Preparative HPLC

Konstantinos Katsoulas, Maximillian Besenhard, Federico Galvanin, Luca Mazzei, Eva Sorensen*

Department of Chemical Engineering, Sargent Centre for Process Systems Engineering, University College London, Torrington Place, London WC1E 7JE, UK
*e.sorensen@ucl.ac.uk

Abstract

Chromatography is a pivotal purification process widely used in the pharmaceutical and fine chemical industries. Design of such processes are increasingly based on the use of mathematical models, within which the isotherm model equation is crucial. Traditionally, the values of the parameters present in the isotherm model are obtained through extensive experimentation such as Frontal Analysis (FA). This study focuses on identifying suitable isotherm models, and accurate values for their parameters, using Model-Based Design of Experiments (MBDoE) based on curve-fitting rather than FA. The methodology involves screening potential isotherm models, assessing model identifiability, selecting the model, and refining the precision of the parameter values. We implemented the methodology in a case study, benchmarking the MBDoE refinement against the traditional approach of factorial designs.

Keywords: Chromatography, Isotherm identification, Model-Based Design of Experiments, Parameter estimation.

1. Introduction

In the pharmaceutical industry, mathematical models for chromatography are an integral part of the concept of Quality-by-Design (QbD). Accurate models and model parameters, particularly related to adsorption isotherms, are essential for precise design. Nevertheless, obtaining the values of these parameters is challenging, often requiring costly and time-consuming experimental methods such as Frontal Analysis (FA). A common alternative is curve-fitting, also known as Inverse Method (IM), where one obtains the values of the parameters by fitting simulated chromatograms to experimental chromatograms based on a given isotherm model. IM requires less material, but users rarely consider the statistical accuracy of the parameter values obtained in terms of their precision and model structure (i.e. set of model equations), with most researchers focusing only on the goodness of the fit (Andrzejewska et al., 2009; Gétaz et al., 2013).

In selecting isotherm models, researchers often rely on intuition or experience, but methods exist for systematically identifying suitable model structures (Waldron et al., 2019). Also, experiments are rarely designed optimally, thus unnecessarily increasing the experimental effort. Model-Based Design of Experiments (MBDoE) is a powerful method for obtaining precise parameter estimates for a given model structure based on fewer experiments (Franceschini & Macchietto, 2008).

2. Theoretical background

2.1. Parameter estimation and MBDoE theoretical background

Curve fitting consists in matching the results of the simulation with the corresponding experimental measurements. Fitting can be achieved by varying the parameter values to be estimated in order to maximise or minimise an objective function. This work focuses on maximising the log-likelihood objective function, $\Phi(\theta)$ (Bard, 1974):

$$\max_{\theta}\big(\Phi(\theta)\big) = \max_{\theta}\left(\sum_{i=1}^{Nexp}\sum_{j=1}^{Nm}\left(-\frac{1}{2}\ln(2\pi) - \frac{1}{2}\ln(\sigma_{ij}^2) - \frac{1}{2}\left(\frac{\rho_{ij}}{\sigma_{ij}}\right)^2\right)\right) \tag{1}$$

where θ is a vector of the model parameters, σ_{ij} is the normally distributed error of the jth measurement of each ith experiment, and ρ_{ij} are the residuals of the jth measurement of the ith experiment that represent the difference between the ij-th model prediction $\widehat{y_{ij}}$ and the corresponding experimental measurement y_{ij}:

$$\rho_{ij} = y_{ij} - \hat{y}_{ij} \tag{2}$$

The goodness of fit of the simulated measurements can be evaluated through the x^2 test (Bard, 1974; Quaglio, 2020):

$$x^2 = \sum_{i=1}^{Nexp}\sum_{j=1}^{Nm}\left(\frac{\rho_{ij(\hat{\theta})}}{\sigma_{ij}}\right)^2 \tag{3}$$

where $\hat{\theta}$ is the vector of optimal parameter estimates. The Fisher Information Matrix (FIM) is a $N_\theta \times N_\theta$ semi-definite matrix whose elements are estimated according to (Franceschini & Macchietto, 2008):

$$[H_\theta]_{kl} = \sum_{i=1}^{Nexp}\sum_{j=1}^{Nm}\frac{1}{\sigma_{ij}}\left(\frac{\partial\hat{y}_{ij}}{\partial\theta_k}\frac{\partial\hat{y}_{ij}}{\partial\theta_l}\right) \text{ for } k = 1\ldots N_\theta \text{ and } l = 1\ldots N_\theta \tag{4}$$

where $\frac{\partial\hat{y}_{ij}}{\partial\theta_k}$ is the sensitivity of the ith measurement of the jth experiment with respect to the parameters θ. From the FIM, we can acquire the variance-covariance matrix, V_θ, as the inverse of the former (Bard, 1974):

$$V_\theta \approx H_\theta^{-1} + H_\theta^{0^{-1}} \tag{5}$$

where H_θ^0 is the initial FIM. The square root of the diagonal elements of the variance-covariance matrix, $\sqrt{V_{\theta,ii}}$, yields the standard deviation of each parameter θ_i. The product between the standard deviation and the $a\%$ t_{ref}-value gives the $a\%$ confidence interval (according to the available Degrees of Freedom) of the parameters. For instance, the 95% confidence interval is given by (Bard, 1974):

$$95\% \ confidence \ interval_i = \sqrt{V_{\theta,ii}} \times t(95\%, DoF) \text{ for } i = 1\ldots N_\theta \tag{6}$$

The 95% confidence interval of each parameter may, in turn, be utilised in the estimation of the t-value of the parameters:

$$t_i = \frac{\hat{\theta}_i}{95\% \ confidence \ interval_i} \text{ for } i = 1\ldots N_\theta \tag{7}$$

The t-test is utilised to evaluate the relative precision of the parameters (Bard, 1974; Quaglio, 2020). Note that if the elements of the FIM are linearly independent, that is the rank of the matrix is equal to the number of parameters to be estimated, then the model and its parameters are considered to be *identifiable*. Naturally, the FIM is a local matrix; i.e., it depends on the collected measurements of the performed experiments:

$$\hat{H}_\theta'(\hat{\theta}, \varphi) = \sum_{l=1}^{Nd}\hat{H}_\theta(\hat{\theta}, \varphi) + H_\theta^0 \tag{8}$$

where $\hat{H}_\theta'$ is the updated FIM based on the observed FIM, $\hat{H}_\theta$ of Nd experiments. Thus, although an experiment or a combination of experiments do not yield an identifiable model, there is the possibility that a combination of experiments might exist which renders a model and its parameters identifiable.

If the model is found to be identifiable, one can proceed with refining the model parameters using MBDoE. MBDoE aims to maximise properties of the FIM, or minimise properties of the variance-covariance matrix, to design the most informative experiments (Franceschini & Macchietto, 2008):

$$\varphi = \arg\max \psi_{criteria}\{[\widehat{H}'_\theta(\hat{\theta}, \varphi)]\} = \arg\min \psi_{criteria}\{[\widehat{V}_\theta(\hat{\theta}, \varphi)]\} \tag{9}$$

The most widely used criteria for MBDoE are the D-, E-, and A- optimal criteria that, in terms of the FIM, aim to maximise the determinant, the smallest eigenvalue, and the trace of the matrix, respectively.

2.2. Mathematical model

The Equilibrium Dispersive Model (EDM) has been widely used in chromatography. It accounts for mass transfer resistances and dispersion by lumping these together in an apparent dispersion coefficient, D_a^t. The mass balance of the model is then (Schmidt-Traub et al., 2020):

$$\frac{\partial C_m}{\partial t} + F\frac{\partial q}{\partial t} + u_{int}^{hypo}\frac{\partial C_m}{\partial x} = D_a^t\frac{\partial^2 C}{\partial x^2}, \quad 0 < x < L \tag{10}$$

where C_m and q are the concentrations in the mobile and stationary phases, respectively, u_{int}^{hypo} is the hypothetical interstitial velocity given by $u_{int}^{hypo} = \frac{Q}{S\varepsilon_t}$, where Q, S, and ε_t are the volumetric flow rate, cross-section of the column, and total porosity, respectively. The apparent dispersion coefficient depends on the velocity and efficiency of the column, that is $D_a^t = \frac{u_{int}^{hypo}L}{2N}$, where L is the column length and N is the number of theoretical plates.

A Danckwerts boundary condition is considered at the inlet of the column (Katsoulas et al., 2023):

$$C_{in} = C_m(x = 0, t) - \frac{D_a^t}{u_{int}^{hypo}}\frac{\partial C_m(x=0,t)}{\partial x} \tag{11}$$

as well as at the outlet:

$$\frac{\partial C_m(x=L,t)}{\partial x} = 0 \tag{12}$$

Initially, the concentration is zero everywhere within the column:

$$C_m(t = 0) = 0, \quad 0 < x < L \tag{13}$$

Assuming that the concentration in the stationary phase is always in equilibrium with the concentration in the mobile phase, then the stationary concentration is a function of the mobile concentration, $q = f(C_m)$. In preparative conditions, however, non-linear isotherms, for instance of the Langmuir-type, are required. Assuming that the number of potential types of binding sites can vary, we proceeded with the following potential isotherms models (Carta, 2020):

$$q_n = \sum_{i=1}^{n}\frac{K_i C_m}{1+\frac{K_i}{q_{sat,i}}C_m}, \quad n = 1 \dots 3 \tag{14}$$

Regardless of the number of sites, we have found it necessary to implement a reparameterization for the first type of sites to prevent high correlation between model parameters:

$$\theta_1 = \ln q_{sat,1} \tag{15}$$

Note that in diluted conditions, the isotherm is represented by a linear function:

$$q = HC_m \tag{16}$$

where $H_n = \sum_{i=1}^{n} K_i$ for $n = 1 \dots 3$.

3. Methodology

The identification of potential isotherm models considered in this work follows the methodology outlined in Fig. 1 (adapted from Waldron et al., 2019). In Step 0, the Henry coefficient is estimated by performing experiments under diluted conditions, and these values are used as initial guesses for the parameter estimation later. Step 1 involves the design of factorial experiments that are used for screening purposes and are evaluated via the x^2 test. If the model structure passes the test, it is a candidate model for the system and it proceeds to the model identifiability test. The identifiability test is a local test that is conducted under a specified number of experiments, determined by the user. Using the preliminary parameter estimates acquired from the previous step, several factorial experiments are designed and performed *in-silico*. If the model passes the t-test after parameter estimation, it is considered practically identifiable. In case there are more than one identifiable model, the methodology also involves a step of MBDoE for model discrimination. Further experimentation will eventually identify the most appropriate model, and we then proceed to improve the precision of its parameters, terminating the procedure when the effect of the parameter uncertainty on the model output, quantified by uncertainty analyses, is believed to be negligible, albeit this is judged differently in each application.

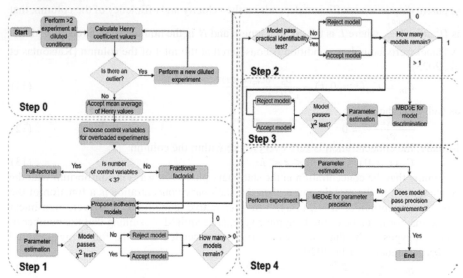

Figure 1: Flow diagram showing the methodology for isotherm model identification and parameter estimation.

4. Case Studies

The methodology presented above is illustrated by a case study that investigates the isotherm model identification and estimation of its isotherm parameters for the separation of oligopeptide of tri-Leucine (LLL) in a Reversed Phase Liquid Chromatography (RPLC) column. For this case study, computer simulations are used to create what would normally be real experimental data. This is done so that the correct model and parameters are known for comparison. The bi-Langmuir isotherm was considered, and a Gaussian distributed error of 2% was introduced in the concentration measurements to simulate real-world applications. Plant parameters, as well as isotherm parameters, were based on the work of Andrzejewska et al. (2009) as follows: column length $L = 15$ cm, column diameter $D = 4.6$ mm, total porosity $\varepsilon_t = 0.6355$, and number of theoretical plates $N = 11,860$. The reader is referred to Table 1 for the real values of the isotherm parameters. The time-invariant controlled factors considered for the 'experiments' were the inlet concentration, the sample volume, and the volumetric flow rate. Simulations were performed in gPROMS® ModelBuilder, discretising the grid in 800 elements using the method of Orthogonal Collocations. The procedure began with Step 0, the estimation of Henry coefficient initial guesses by performing two experiments in diluted conditions. The next step, Step 1, involves designing factorial experiments to initiate the screening of the proposed models. Here, two different strategies were considered:

- Strategy A - 2^3 factorial using all three factors,
- Strategy B - 2^2 factorial using only the factors of inlet concentration and sample volume.

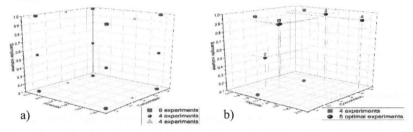

a)   b)

Figure 2: Illustration of the experimental conditions of the experiments performed by a) Strategy A and b) Strategy B.

Three models were proposed, the simple, bi-, and tri- Langmuir isotherms. (Note that the 'real' isotherm is the bi-Langmuir isotherm). From these, only the bi- and tri- Langmuir isotherms passed to the next step, and for both strategies. In Step 2, 27 total experiments were disbursed on a 27-factorial for each of the two remaining models. As expected, only the bi-Langmuir isotherm passes the identifiability test, and for both strategies, and thus the discrimination step is skipped. In the final step, the precision of the parameters was refined either by:

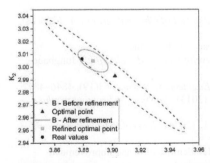

Figure 3: The 95% confidence ellipsoid between K_1 and K_2 in Strategy B before

- Strategy A – two rounds of 2^2 factorial, or,
- Strategy B – four D-Optimal and one E-Optimal experiments, sequentially.

Note that decisions on the usage of the different MBDoE criteria were taken based on statistics evaluation (i.e., correlation statistics). Fig. 2 (a) and (b) depict the total number of experiments in Strategy A and B, respectively. Note that the experiments designed by the 27-factorial and used for the identifiability test are not used in the parameter estimation procedure since they were experiments simulated using the preliminary estimates of each model and not the real values of the parameters. Table 1 summarises the estimated values of the parameters after improving their precision with Strategies A and B. Note that t_{ref} is 1.65 in both strategies. Although the optimal estimates in Strategy A are closer to the real values, we can only judge the quality of the estimates based on their relative precision. Strategy B employed fewer preliminary experiments than Strategy A while employing only 5 rounds of sequential MBDoE's totalling 9 experiments. On the other hand, Strategy A required a total of 16 experiments designed by full factorials, yet the precision of the parameter estimates of the two strategies is equivalent. Fig. 3 shows the improvement of the precision of parameters K_1 and K_2 illustrated in the 95% confidence ellipsoids before and after refinement by MBDoE.

Table 1: The optimal parameter estimates and statistics after improving the precision.

	x^2/x_c^2	K_1 $\pm 95\%$	$\theta_1 \pm 95\%$	K_2 $\pm 95\%$	$q_{sat,1}$ $\pm 95\%$	Total No. Experiments
Strategy A	411.9/475.1	3.88±0.008	6.06±0.029	3.01±0.006	2.35 ± 0.018	16 No MBDoE
t-value		504	211	485	134	
Strategy B	119.3/199.2	3.89±0.008	6.04±0.02	3.01±0.007	2.33±0.024	9
t-value		495	301	453	94	
Real values		3.88	6.06	3.01	2.35	

5. Concluding remarks

This study presents a step-by-step methodology for isotherm model identification in preparative chromatography. For the selected case study, we estimated the values of the isotherm parameters through curve-fitting following two distinct identification strategies. Refining parameter precision by MBDoE was benchmarked against factorial designs. MBDoE reduced the required number of experiments while maintaining precision. This work highlighted the benefits of MBDoE in making informed decisions in process development based on limited information.

Acknowledgement

The authors wish to acknowledge the financial support given to this research project by Eli Lilly and Company, and the Engineering and Physical Sciences Research Council (EPSRC), grant code EP/T005556/1.

References

Andrzejewska, A., Gritti, F., & Guiochon, G. (2009). *Journal of Chromatography A, 1216*(18), 3992–4004.

Bard, Y. (1974). *Nonlinear Parameter Estimation*. Academic Press Inc.

Carta, G. (2020). *Protein chromatography: process development and scale-up* (A. Jungbauer, Ed.). Wiley-VCH.

Franceschini, G., & Macchietto, S. (2008). *Chemical Engineering Science, 63*(19), 4846–4872.

Gétaz, D., Stroehlein, G., Butté, A., & Morbidelli, M. (2013). *Journal of Chromatography A, 1284*, 69–79.

Katsoulas, K., Tirapelle, M., Sorensen, E., & Mazzei, L. (2023). *Journal of Chromatography A, 1708, 464345*.

Quaglio, M. (2020). Novel techniques for kinetic model identification and improvement. In *Doctoral thesis, UCL (University College London)*.

Schmidt-Traub, H., Schulte, M., & Seidel-Morgenstern, A. (2020). *Preparative Chromatography*. John Wiley & Sons, Incorporated.

Waldron, C., Pankajakshan, A., Quaglio, M., Cao, E., Galvanin, F., & Gavriilidis, A. (2019). *Industrial and Engineering Chemistry Research, 58*(49), 22165–22177.

Flavio Manenti, Gintaras V. Reklaitis (Eds.), Proceedings of the 34th European Symposium on Computer Aided Process Engineering / 15th International Symposium on Process Systems Engineering (ESCAPE34/PSE24), June 2-6, 2024, Florence, Italy

Modeling of Carbon Offsetting Industries for Performance Prediction of Emissions Trading Systems

Soo Hyoung Choi

Division of Chemical Engineering, Clean Energy Research Center, Jeonbuk National University, 567 Baekje-daero Deokjin-gu, Jeonju, 54896, S. Korea
soochoi@jbnu.ac.kr

Abstract

A mathematical programming approach is suggested to predict the environmental impact of carbon offsetting by emissions trading systems under various circumstances. The proposed method is to simulate industries that seek to minimize the cost of reduction and/or offsetting of carbon emissions minus the value of retained carbon credits. Case study indicates that reduction of net emissions can be promoted by emissions trading systems if rightly operated. The necessary conditions are industrial abilities to reduce emissions and/or capacities for extension of sequestration, correct issuance of carbon credits, proper policies on allowances, and rational rules on carryovers.

Keywords: carbon offset, carbon credit, emissions trading system

1. Introduction

Carbon offsetting means to compensate for carbon emissions by earning or purchasing carbon credits, which are issued by governmental organizations as reward for carrying out or investing in projects that contribute to reducing atmospheric carbon dioxide (Wikipedia, 2023a). Typical carbon offset projects include forestry, renewable energy, energy conservation, and conversion of waste to energy (Jennifer L, 2023). If carbon credits are correctly evaluated for these projects, and the total amount of carbon credits is limited to the total capacity of carbon sequestration, net zero emissions can be attained. However, carbon neutrality defined as such only guarantees a steady state of carbon cycle, at which the concentration of atmospheric carbon dioxide can still be high, if emissions and sequestration are equally large. Therefore, rigorous carbon cycle impact assessment is necessary in order to correctly evaluate carbon credits for various carbon offset projects (Choi, 2023). Carbon credits basically represent allowances for emissions, and can be sold and bought in carbon markets (Wikipedia, 2023b). Speculation is that, as the government reduces allowances, their price increases, until the industry is forced to reduce emissions rather than to offset them. However, carbon markets are geared towards offsetting (Frank, 2023). In this work, the effectiveness of carbon markets is investigated in terms of carbon cycle dynamics (Choi and Manousiouthakis, 2022).

The emissions trading system (ETS) facilitates the process industry's achievement of carbon neutrality. The root reason is that companies can offset their emissions by purchasing carbon credits even if they emit more than the allowances allocated by the government. The only obstacle seems to be the future price of carbon credit, if available, which is expected in general to increase as carbon neutrality is approached (Tiseo, 2023). This prediction is reasonable because, after carbon neutrality is achieved, the government can no longer allow offsets more than the capacity of sequestration attained up to that time. The ETS differs from country to country, but generally, the total allowances are reduced every year, and a fixed amount of allowance is additionally placed on the market if the price of carbon credit increases over a certain preset value. If the price of carbon credit is significantly higher than the cost of carbon capture, companies will prefer reduction to offset. Besides, surplus credits, if any, can be sold for profits. However, most companies prefer carryover to the next year in order to prepare for an uncertain future. Carryover rules also differ from country to country. If not allowed, excessive supplies will lower the price, weakening the driving force for reduction. If unlimited, accumulated allowances will lead to insecure carbon neutrality. The effects of carryover rules are also evaluated in this work.

In order to get ETS to properly work, carbon credits should be evaluated to represent actual decrease in net emissions. Evaluation of direct reduction is straightforward, but assessment of indirect reduction is apt to cause duplicate credits. Actually, over-crediting is known to be happening (Gurgel, 2022). Credits should be correctly issued for increase in sequestration also. Forestry is essential because fundamental sequestration is being performed by forests (Norman, 2023). However, the problem is that credits are issued not only for afforestation, but also for stopping deforestation (Greenfield, 2023). As an alternative to forestry, direct air capture (DAC) is in business (Twidale et al., 2023). However, from a technical point of view, unlike its name, it indirectly captures carbon after released into the air. Therefore, DAC should confront low efficiency, hence strong skepticism (Shelton-Thomas and DiFelice, 2023). Nonetheless, in this work, it is assumed that carbon credit is equivalent to decrease in net emissions.

2. Proposed Method

A model is proposed in order to simulate emissions trading that affects carbon cycle as shown in Fig. 1. The emitter represents the entire group of carbon positive industries, and the absorber represents the entire group of carbon negative industries. Therefore, the emitter should reduce emissions and/or buy credits from the absorber in order to satisfy the requirements from the government. For more simplicity, it is assumed that the government buys all surplus credits at market price, i.e., $d = u + w + z - y$. Then, the emitter's behavior for a given period can be modeled as a mathematical program, which minimizes the total expenses minus the final value of retained carbon credit. The proposed formulation is as follows:

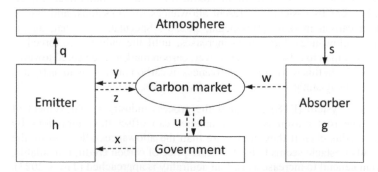

Figure 1. Carbon (solid) and credit (dashed) flow model.

$$\min \sum_{i=1}^{n} [c_i(q_0 - q_i) + p_i(y_i - z_i)] - p_n h_n \tag{1}$$

subject to

$$x_i = r_x x_{i-1}, x_0 = q_0 \tag{2}$$

$$h_i = h_{i-1} + x_i + y_i - z_i - q_i, h_0 = 0 \tag{3}$$

$$g_i = g_{i-1} + s_i - w_i, g_0 = 0 \tag{4}$$

$$p_i(x_i + u_i) = c_i q_i \tag{5}$$

$$q_i \geq r_q q_{i-1} \tag{6}$$

$$s_i - s_{i-1} \leq b_s, s_0 = 0 \tag{7}$$

$$h_i - h_{i-1} \leq b_h \tag{8}$$

where

c_i = average cost for reduction of emissions in the i-th year
p_i = average price of carbon credit in the i-th year
q_i = amount of emission in the i-th year
s_i = amount of absorption in the i-th year
u_i = carbon credit supplied by the government in the i-th year
w_i = carbon credit sold by the absorber in the i-th year
x_i = carbon credit allotted to the emitter for the i-th year
y_i = carbon credit bought by the emitter in the i-th year
z_i = carbon credit sold by the emitter in the i-th year

The proposed optimization problem is a nonconvex nonlinear program, because p_i's are unknowns, while c_i's, r_x, and r_q are constants. All variables and parameters are nonnegative, and can be normalized so that $q_0 = 1$ can be used. For maximum simplicity, it is assumed in this work that the emitter buys the absorber's credit, and the government buys the emitter's credit, i.e., $w_i = y_i$ in (4). Equation (5) represents a carbon price model which is obtained from the assumption that the reference price is equivalent to the cost of reduction, and the change factor is proportional to the industrial demand, q_i, and inversely proportional to the governmental supply, $x_i + u_i$, i.e., $p_i \approx c_i q_i/(x_i + u_i)$, where u_i is a manipulated variable for the government to control the market price, which is omitted in this work, i.e., $u_i = 0$. Net emissions can be calculated by $q_i - s_i$.

3. Case Study

Consider a governmental plan for reduction of emissions by 30% in 5 years. The allowances are to be reduced by 7% every year. The cost of carbon capture is assumed to be constant so that $c_i = 1$ can be used. The suggested parameter values are listed in Table 1. Case 1 is designed to use reduction only, and case 2 to use offset only. Case 3 uses both. In cases 1–3, carryover is unlimited. In case 4, it is assumed that carryover is limited to the same amount as sales. In case 5, carryover is not allowed. The proposed mathematical program is solved by MATLAB's fmincon. For numerical stability, 10^{-8} is used for 0, and 10^8 for ∞ as parameter values in Table 1.

Table 1. Parameters for case study

Case	r_x	r_q	b_s	b_h
1	0.93	0.90	0	∞
2	0.93	1	0.10	∞
3	0.93	0.95	0.05	∞
4	0.93	0.95	0.05	z_i
5	0.93	0.95	0.05	0

The results of simulation are shown in Figs. 2 and 3. Case 1 indicates that excessive allowance causes decrease in the carbon price, and ends up failing to achieve target reduction of emissions. Case 2 indicates that deficient reduction causes rapid increase in the carbon price. Cases 1–3 indicate that if carryover is unlimited, net emissions may suddenly increase at last. Case 4 shows the best performance, which suggests that it is necessary to properly limit carryovers in order to ultimately minimize net emissions. Case 5 indicates that it is not desirable to completely prohibit carryovers.

4. Conclusions

Case study indicates that the effectiveness of ETS depends on governmental policies and industrial capabilities. Excessive allowances would worsen the carbon cycle, and deficient allowances would threaten the industrial economy. Unlimited carryovers would endanger carbon neutrality, and extreme limitation might delay its achievement. These issues also need optimization, and extension of the proposed method is suggested as future work.

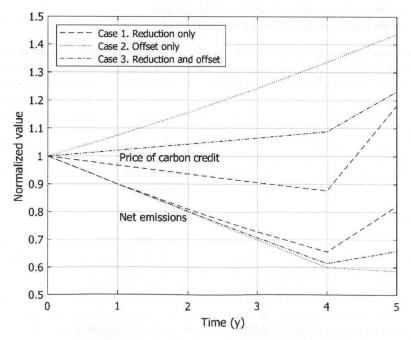

Figure 2. Predicted carbon price and net emissions for reduction and/or offset cases.

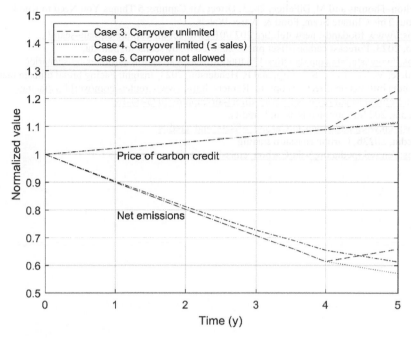

Figure 3. Predicted carbon price and net emissions for different carryover rules.

Acknowledgement

This work was supported by National Research Foundation of Korea (NRF) funded by Ministry of Science and ICT (MSIT) (Grant No. 2022R1F1A1062901).

References

S. H. Choi, 2023, Modeling of carbon offset networks for process systems to achieve net zero emissions, Computer Aided Chemical Engineering, 52, 1083-1088, https://doi.org/10.1016/B978-0-443-15274-0.50173-6

S. H. Choi and V. I. Manousiouthakis, 2022, Modeling the Carbon Cycle Dynamics and the Greenhouse Effect, IFAC-PapersOnLine, 55-7, 424-428, https://doi.org/10.1016/j.ifacol.2022.07.480

Jennifer L, 2023, How Do Carbon Offsetting Projects Work? CarbonCredits.Com, https://carboncredits.com/how-do-carbon-offsetting-projects-work/

P. Greenfield, 2023, Revealed: more than 90% of rainforest carbon offsets by biggest certifier are worthless, analysis shows, The Guardian, https://www.theguardian.com/environment/2023/jan/18/revealed-forest-carbon-offsets-biggest-provider-worthless-verra-aoe

S. Frank, 2023, Does carbon offsetting do more harm than good? Carbon Market Watch, https://carbonmarketwatch.org/2023/07/06/does-carbon-offsetting-do-more-harm-than-good/

A. Gurgel, 2022, Carbon Offsets, MIT Climate Portal, https://climate.mit.edu/explainers/carbon-offsets

C. Norman, 2023, Kreye, M. How Forests Store Carbon, Penn State Extension, https://extension.psu.edu/how-forests-store-carbon

O. Shelton-Thomas and M. DiFelice, 2023, Direct Air Capture: 5 Things You Need to Know About This Climate Scam, Food & Water Watch, https://www.foodandwaterwatch.org/2023/01/19/direct-air-capture-climate-scam/

I. Tiseo, 2023, Forecast carbon offset prices 2030-2050, by scenario, Statista, https://www.statista.com/statistics/1284060/forecast-carbon-offset-prices-by-scenario/

S. Twidale, V. Volcovici, S. Jessop, and P. Henderson, 2023, Insight: Facing brutal climate math, US bets billions on direct air capture, Reuters, https://www.reuters.com/world/us/facing-brutal-climate-math-us-bets-billions-direct-air-capture-2023-04-18/

Wikipedia, 2023a, Carbon offsets and credits, https://en.wikipedia.org/wiki/Carbon_offsets_and_credits

Wikipedia, 2023b, Carbon emission trading, https://en.wikipedia.org/wiki/Carbon_emission_trading

Flavio Manenti, Gintaras V. Reklaitis (Eds.), Proceedings of the 34th European Symposium on Computer Aided Process Engineering / 15th International Symposium on Process Systems Engineering (ESCAPE34/PSE24), June 2-6, 2024, Florence, Italy

Multi-objective optimization for pharmaceutical supply chain design: application to COVID-19 vaccine distribution network

Jonathan J. Cuevas-Lopez[a], Catherine Azzaro-Pantel[a]* and Sofia De-Leon Almaraz[b]

[a]*Laboratoire de Génie Chimique, Université de Toulouse, CNRS, INPT, UPS, Toulouse, France*
[b]*Corvinus University of Budapest. Department of Supply Chain Management. 8 Fővam tér. 1093 Budapest, Hungary*

jonathanjair.cuevaslopez@toulouse-inp.fr,

Abstract

In response to the exigencies of the COVID-19 pandemic, this study proposes a novel Pharmaceutical Supply Chain (PSC) model for vaccination. Our model encompasses a five-echelon structure, integrating manufacturing plants, fill-finish facilities, distribution centres, and administration points focusing on a stratified deterministic demand for COVID-19 vaccines. A distinctive feature of our work is the incorporation of a multi-objective optimization approach within the General Algebraic Modeling System (GAMS) environment using Mixed Integer Linear Programming (MILP). This approach is designed to optimize cost, CO_2 emissions, and backlog over a 40-week vaccination campaign. Utilizing an augmented epsilon-constraint method, our model facilitates the identification of trade-offs between these objectives, thus enabling informed decision-making in PSCs. The results underscore the inherent trade-offs among the objectives, reflecting the complexity of the supply chain management for vaccination as shown in the Pareto fronts.

Keywords: Pharmaceutical Supply Chain, Mixed Integer Linear Programming, multi-objective optimization, COVID-19 vaccines, epsilon-constraint.

1. Introduction

The COVID-19 pandemic has highlighted the essential role of Pharmaceutical Supply Chains (PSCs) in maintaining global health. Unlike other supply chains, the pharmaceutical supply chain is unique, particularly in its crucial role during a pandemic for vaccine distribution. Disruptions in this supply chain can lead to severe consequences. Inadequate management strategies in healthcare industries can result in substantial financial losses and significantly impact patient care and outcomes (Uthayakumar, 2013).

To address the optimization of the supply chain, approaches can be either single-objective, focusing mainly on cost, or multi-objective where traditional models, primarily focused on cost and service efficiency (Papageorgiou, 2009), often overlook environmental considerations (Alnaji, 2013). This oversight, coupled with concerns about social equity in vaccine distribution, (Sazvar, 2021) underscores the need for a more comprehensive approach to PSC management. To bridge this gap, our study introduces a new PSC model for the distribution of COVID-19 vaccines. In this work, we are particularly interested in understanding the impact of integrating various sustainability objectives when designing a specific PSC. This model incorporates a multi-echelon structure, reflecting insights from studies like (Papageorgiou, 2009). It employs a multi-objective optimization strategy using an augmented epsilon-constraint method (Mavrotas, 2009), enabling the identification of balanced solutions between cost, backlog, and environmental impacts. Our objective is to address the existing gaps in PSC management and offer a model that seamlessly integrates economic, service-level, and environmental objectives. This model serves as a valuable tool for stakeholders and modelers engaged in multi-criteria decision-making within the pharmaceutical supply domain.

2. Formulation of the PSC Mathematical Model

Our model for COVID-19 vaccine distribution integrates various stages of the supply chain, from manufacturing plants to administration points. Figure 1 showcases the vaccine's journey through the supply chain. The primary manufacturers echelon involves vaccine synthesis at import locations (i) or manufacturing plants (m), followed by fill-finish plants (f) in the secondary manufacturers echelon. Subsequently, vaccines are stored and distributed through the main and local storage echelons, represented by

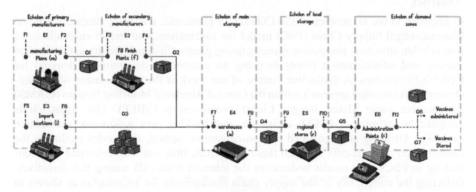

Figure 1. PSC scheme

warehouses (w) and regional stores (r), respectively. The final demand zone echelon includes administration points (c), where the public receives vaccines. Likewise, the variable "F" indicates the flow of vaccines entering and leaving each entity, which are denoted as variables with the letter "E", while the "Q" is interpreted as the quantity of vaccines transported between the different entities.

Table 1 briefly encapsulates the Pharmaceutical Supply Chain (PSC) model's components, illustrating the intricate framework designed to optimize the distribution of vaccines. Parameters bring real-world context into the model, grounding it with empirical data such as costs and capacities. Central to the model are the objective functions, which guide the system towards cost efficiency, backlog minimization, and reduced environmental impact. Optimization variables dictate the supply chain's functionality, while sets categorize the key elements, from manufacturing entities to distribution factors, within a structured matrix. This integration is key to optimizing the three above mentioned objectives. The use of continuous integer and binary variables in linear constraints results in an MILP model and a strategy of multi-objective optimization through an enhanced epsilon-constraint approach is implemented, as outlined by Mavrotas (2009) within the GAMS environment.

Table 1. Summary of the PSC model framework

Inputs	Optimization	Outputs
Sets:	*Constraints:*	*Optimization Variables:*
- Manufacturing plants	- Supply flow conservation	- Vaccine quantities
- Fill-finish facilities	- Capacity limitations	- Associated costs
- Distribution centers	- Inventory bounds	- Emission metrics
- Administration points	- Transportation logistics	
- Time periods (weeks)	- Service level requirements	*Network Configuration:*
		- Supply chain structure
Parameters:	*Objective Functions:*	*Performance Metrics:*
- Cost factors	- Minimizing cost	- Total cost
- Capacity constraints	- Minimizing backlog	- Environmental impact
- Distance matrices	- Minimizing CO_2 emissions	- Backlog
- Demand forecasts		
- CO_2 emissions		

3. Case study

In this work, insights from the framework proposed by (Ibrahim, 2021) have been used to validate our methodology. Within the four echelons used in the COVID-19 vaccine supply chain, the case study utilizes a single import location since the vaccine comes from Puurs, Belgium. Regarding warehouse locations, there are four major cities in the UK, including the capital. Additionally, there are 12 potential sites for regional stores and 12 administration points..

Our focus is exclusively on the Pfizer vaccine, selected due to the availability of detailed data and its widespread use during the study period. This choice allowed for a more precise and data-driven analysis, crucial for the model's accuracy. The vaccination campaign was planned for a duration of 40 weeks, aiming for full vaccination (two doses per person). It is noteworthy that the demand is entirely met by imported vaccines, as domestic production of Pfizer vaccines was not available.

For cost parameters, including fabrication, storage, and application, we drew insights from studies such as those by (Martonosi, 2021). These studies provided a comprehensive understanding of various cost factors in the vaccine supply chain. Logistics parameters, particularly those associated with vaccine transportation, cold chain requirements, and packaging, were informed by research including (Holm, 2021) and (Organization, 2022). Finally, the environmental impact, focusing on CO_2 emissions related to refrigeration, transportation, and waste, was a critical aspect. The ecological footprint of the vaccine supply chain is assessed similarly to the approach taken by (Kurzweil, 2021), ensuring that our model also incorporates sustainability.

4. Results

The statistics of the MILP model are presented in Table 2. The computations were carried out on a system running Microsoft Windows 10 as the operating system. The core processing tasks were managed by an 11th Gen Intel(R) Core(TM) i7-11850H processor, operating at a base frequency of 2.50GHz and equipped with 8 cores. The system was supported by 16 GB of RAM.

Table 2. Typical features of the model instance.

Typical features of the instance	Value
Blocks of Equations.	143
Single Equations.	237,306
Blocks of Variables	92
Single Variables	111,090
Non-Zero Elements	591,480
Discrete Variables	24,376
Generation Time (seconds)	1.382

The model consists of 143 blocks of equations, further detailed into 237,306 single equations, highlighting the detailed level of modelling. The variable blocks are fewer, at 92, but expand into 111,090 individual variables, of which 24,376 are discrete, likely representing binary decisions within the supply chain. The large number of non-zero elements, 591,480, reflects the density and connectivity of the model constraints. The model was generated and solved in a relatively swift 1.382 seconds, demonstrating the efficiency of the solver in handling such an extensive model.

Table 3 presents some typical results of a single-objective optimization for the PSC model for COVID-19 vaccines. The objectives were individually minimized, showcasing the minimal results for cost, backlog (BL), and CO_2 emissions.

Table 3. Payoff table with single-objective results

	Min Cost (billion $)	Min BL (Million vaccine)	Min CO_2 (Million Kg CO_2-eq)
Cost	5.693	321.534	50.514
BL	6.326	0	85.989
CO_2	5.693	321.534	33.426

When minimizing cost, the model achieved a balance with a cost of 5.693 billion dollars, a backlog of 321.534 million vaccines, and CO_2 emissions of 50.514 million kg. Focusing solely on minimizing the backlog resulted in zero backlog but increased cost and CO_2 emissions, indicating the trade-offs involved. Lastly, targeting CO_2 reduction led to lower emissions but at the cost of higher expenses and backlog. These results underscore the inherent compromises in supply chain optimization, emphasizing the need for a balanced approach for practical implementation.

Table 4 encapsulates the outcomes of a multi-objective optimization using the epsilon constraint method, which simultaneously considers cost, backlog (BL), and CO_2 emissions.

Table 4. Payoff table with multi-objective optimization results

	Cost (billion $)	BL (Million vaccine)	CO_2 (Million Kg CO_2-eq)
Multi-objective Optimization	6.131	32.153	37.763

This trade-off solution showcases the efficacy of the epsilon constraint method in deriving a balanced solution that does not excessively favour one objective over the others. With a cost of 6.131 billion dollars, a backlog of 32.153 million vaccines, and CO_2 emissions of 37.763 million kg, a feasible and sustainable approach to managing the vaccine supply chain is proposed, underlining the method's capacity to find a middle ground that could be acceptable on multiple fronts. The outcomes of the multi-objective optimization, conducted through the augmented epsilon constraint method, are graphically depicted in Figures 2, 3, and 4. These Pareto fronts collectively encapsulate the trade-offs within the vaccine supply chain optimization process. Figure 2 presents the trade-offs between the cost (USD) and backlog (Vaccines) in vaccine distribution. Lower costs are associated with higher backlogs, demonstrating the complexities inherent in optimizing for cost-effectiveness while maintaining high levels of service.

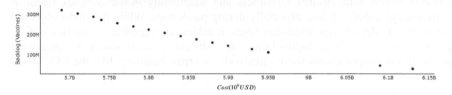

Figure 2. Pareto front with cost and backlog

In Figure 3, the interplay between operational costs (USD) and CO_2 emissions (kg) is explored. It becomes evident that more cost-effective operations tend to result in increased emissions, positing a challenge for balancing financial and environmental considerations in supply chain management.

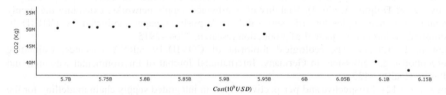

Figure 3. Pareto front with cost and CO_2 emissions

Figure 4 demonstrates the correlation between backlog (Vaccines) and CO_2 (kg) emissions. It is observed that scenarios with reduced backlogs are associated with higher emissions, underscoring the environmental cost of enhanced service levels in vaccine distribution.

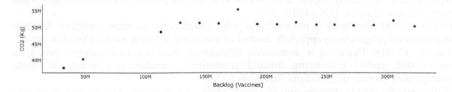

Figure 2. Pareto front with backlog and CO_2 emissions.

Together, these Pareto fronts offer a multifaceted view of the decision-making landscape, where the goals of minimizing costs and backlogs must be weighed against the imperative of reducing carbon footprint. The analyses provide strategic insights into achieving an optimal compromise in the complex domain of global vaccine distribution logistics.

5. Conclusions

In this study, a novel Pharmaceutical Supply Chain (PSC) model for the effective distribution of COVID-19 vaccines has been developed. This model showcases a significant advancement in supply chain management by integrating a multi-objective optimization framework using MILP within the GAMS environment. Its unique capability lies in balancing critical objectives like minimizing costs, reducing CO_2 emissions, and optimizing backlog during a 40-week vaccination campaign. Employing the augmented epsilon-constraint method, the model effectively navigates trade-offs between these objectives, aiding informed PSC management decisions. Its comprehensive approach, drawing from diverse data sources including academic research and public health data, ensures robustness and adaptability in the complex realm of pharmaceutical supply chains, especially during pandemics. While the multi-objective nature of the model impacts resolution times, it achieves a significant reduction in CO_2 emissions, at least 38%, highlighting its substantial contribution to managing pharmaceutical supply chains more effectively in crisis situations like the COVID-19 pandemic.

References

Alnaji, L. (2013). The role of Supply Chain Applications in Jordanian Pharmacies: A case study on Pharmacies in the capital city Amman. Industrial Engineering Letters, 65--71.

Holm, M. (2021). Critical aspects of packaging, storage, preparation, and administration of mRNA and adenovirus-vectored COVID-19 vaccines for optimal efficacy. Vaccine, 457.

Ibrahim, D. (2021). Model-based planning and delivery of mass vaccination campaigns against infectious disease: Application to the COVID-19 pandemic in the UK. Vaccines, 430--445.

Ivanov, D., & Dolgui, A. (2020). Viability of intertwined supply networks: extending the supply chain resilience angles towards survivability. A position paper motivated by COVID-19 outbreak. International journal of production research, 2904--2915.

Kurzweil, P. (2021). The ecological footprint of COVID-19 mRNA vaccines: estimating greenhouse gas emissions in Germany. International Journal of Environmental Research and Public Health, 7425.

Lainez, J. (2012). Prospective and perspective review in integrated supply chain modelling for the chemical process industry. Current Opinion in Chemical Engineering, 430--445.

Martonosi, S. (2021). Pricing the COVID-19 vaccine: A mathematical approach. Omega, 102451.

Mavrotas, G. (2009). Effective implementation of the ε-constraint method in multi-objective mathematical programming problems},. Applied mathematics and computation, 455--465.

Organization, W. H. (2022). Operational guidance on establishing an ultra-cold chain system in support of the Pfizer-BioNTech COVID-19 vaccine rollout, 1 February 2022.

Papageorgiou, L. G. (2009). Supply chain optimisation for the process industries: Advances and opportunities. Elsevier, 1931--1938.

Sadjadi, S. (2019). The design of the vaccine supply network under uncertain condition: A robust mathematical programming approach. Journal of Modelling in Management, 841--871.

Sazvar, Z. (2021). Designing a sustainable closed-loop pharmaceutical supply chain in a competitive market considering demand uncertainty, manufacturer's brand and waste management. Annals of Operations Research, 1--32.

Uthayakumar, R. (2013). Pharmaceutical supply chain and inventory management strategies: Optimization for a pharmaceutical company and a hospital. Operations Research for Health Care, 52--64.

Zahiri, B. (2018). Design of a pharmaceutical supply chain network under uncertainty considering perishability and substitutability of products. Information sciences, 257--283.

Flavio Manenti, Gintaras V. Reklaitis (Eds.), Proceedings of the 34th European Symposium on Computer Aided Process Engineering / 15th International Symposium on Process Systems Engineering (ESCAPE34/PSE24), June 2-6, 2024, Florence, Italy

The Reverse Water-Gas Shift Reaction as an Intermediate Step for Synthetic Jet Fuel Production: A Reactor Sizing Study at Two Different Scales

Antoine Rouxhet,[a]* Grégoire Léonard[a]

[a]Chemical Engineering, University of Liège, Allée du 6 Août 13, 4000 Liège, Belgium
antoine.rouxhet@uliege.be

Abstract

This paper presents a reactor model for the reverse water-gas shift reaction (rWGS) implemented in the framework of captured CO_2 conversion. Kinetics are included in the model and validated with experimental data from the literature. The model is used to size a reactor at two scales: a small pilot (inlet H_2 of 1.5 Nm³/h) and a mature plant (inlet H_2 of 1,500 Nm³/h). The designs at both scales differ by the heating configuration; it is assumed that the small-scale unit is isothermal while the industrial-scale unit is adiabatic. For the small-scale unit, it is shown that the equilibrium conversion (65.6 %) can easily be reached within 30 cm at 1 bar. However, this reactor is not optimal for a 20-bar operation as the maximum conversion (65.2 %) is reached in the first centimetres before decreasing to 62.1 %, as methanation occurs, leading to an outlet CH_4 selectivity of 17.3 %. In the large-scale adiabatic unit, both operating pressures lead to a sudden temperature drop due to the endothermic reaction followed by a temperature increase, but this latter is more important at high pressure due to methanation accentuation. This difference in the temperature profile results in a CO_2 conversion of 64.8 % at 20 bar against 51.1 % at 1 bar. In summary, the equilibrium conversion in an isothermal unit is slightly higher at 1 bar, even in a reactor adequately sized for each pressure. In an adiabatic unit, the equilibrium conversion is reached within the same length for both pressures and is significantly higher at 20 bar, at the extent of an accentuated methanation.

Keywords: reverse water-gas shift (rWGS), modelling, kinetics, process design

1. Introduction

The energy sector is the largest contributor to global greenhouse gas emissions, with transportation accounting for around 25 % of energy-related emissions (Ritchie, 2020). Thus, extensive efforts are underway to mitigate this issue. One part of the solution lies in increasing the share of electrified vehicles on the market and developing hydrogen-based transport. Although these solutions might be promising for road transportation, they seem limited options for long-freight ships and aircraft because of too low energy density. Hence, it is crucial to find an alternative, such as high-density fuels with low carbon footprint. In this perspective, Power-to-X processes seem to be an appealing option. The general principle is to combine captured CO_2 with H_2 produced by water electrolysis powered by renewable energies to yield hydrocarbon chains, which can be further upgraded to fuels. When CO_2 is directly captured from the atmosphere with a low-carbon energy, the process becomes circular, as waste CO_2 production is avoided by its reuse as the process feedstock. It leads to a potential net-zero emissions way to synthesise transportation fuels. In this paper, the transition from CO_2 to the hydrocarbon chains is considered to happen in two separate reactors. In the first reactor, the highly stable CO_2 molecule is activated and converted to CO through the rWGS reaction:

$$CO_2 + H_2 \rightleftharpoons CO + H_2O \qquad \Delta H°_{298.15\,K} = 41.2 \frac{kJ}{mol} \qquad (1)$$

The reactor design dedicated to this reaction is the core of this paper. The resulting syngas (H_2/CO mix) is sent to the second unit to produce the hydrocarbon chains through the Fischer-Tropsch synthesis. The description of this unit is beyond the scope of this work, but it was optimised to yield a mix of hydrocarbons suitable to be upgraded to kerosene, as this latter is the value-added product targeted at ULiège (Morales and Leonard, 2022).

2. Reactor Model Description

The simulation model of the rWGS unit was developed using Aspen Custom Modeler (ACM), a modelling package of AspenTech. This tool was chosen as it enables a complete reactor model construction, including material, heat, and momentum balances while taking advantage of the available physical properties databases. On top of the rWGS reaction (see Eq. (1)), the developed model considers two side reactions, which decrease the CO selectivity, as both yield CH_4:

$$CO_2 + 4\,H_2 \rightleftharpoons CH_4 + 2\,H_2O \qquad \Delta H°_{298.15\,K} = -165.0 \frac{kJ}{mol} \qquad (2)$$

$$CO + 3\,H_2 \rightleftharpoons CH_4 + H_2O \qquad \Delta H°_{298.15\,K} = -206.2 \frac{kJ}{mol} \qquad (3)$$

These exothermic methanation reactions are not the only side reactions that can appear alongside the rWGS reaction. Coking could also manifest as carbon-containing species, i.e. CO_2, CO and CH_4, could decompose into solid carbon. However, it turns out that coking reactions are exothermic and can generally be neglected when modelling an rWGS reactor. Adelung et al. (2021) showed that, from a thermodynamic equilibrium point of view, carbon formation is suppressed above 600 °C under rWGS conditions, which we verified (Rouxhet et al., 2024). Furthermore, Wolf et al. (2016) stated that these reactions are not likely to be observed due to slow kinetics. The kinetics implemented in this model are based on a Langmuir-Hinshelwood-Hougen-Watson (LHHW) model and were developed by Vidal Vázquez et al. (2017) based on a 2 wt-% Ni/Al$_2$O$_3$ catalyst. They selected a mechanistic model from the literature (Xu and Froment, 1989) and regressed the kinetic parameters based on their own experiments. These kinetics have been selected for different reasons. First, the original model developed by Xu and Froment (1989) has already been successfully used in numerous works (Bisotti et al., 2023). Besides, the experiments conducted by Vidal Vázquez et al. (2017) were conducted in a wide range of temperatures (between 550 and 800 °C) and pressures (1 and 30 bar). It is generally not the case for other rWGS experiments, which are more often carried out at atmospheric pressure and lower temperatures (Daza and Kuhn 2016). Finally, these kinetics have presumably been used to model a pilot installation approximately the same size as the one to be installed at ULiège (Vidal Vázquez et al., 2018).

3. Model Validation

The proper implementation of the kinetic model and the material balance is validated with experimental data produced by Vidal Vázquez et al. (2017). For this purpose, the CO_2 conversion at the reactor outlet is calculated between 450 and 850 °C for six operating conditions sets, corresponding to the ones investigated by the authors in their article. The validation is demonstrated here for two sets in Figures 1a and 1b. The curves generated in these figures differ only in the operating pressure while other conditions are kept constant, i.e. an initial H_2/CO_2 ratio of 2, a total catalyst mass of 0.25 g and a total gas flow rate of 2.087 NL/min, including 42.5 % of N_2. It appears that the developed kinetic model can reproduce the 33 experimental points generated by Vidal Vázquez et al. (2017) with satisfying accuracy, as their squared correlation coefficient (R^2) equals 92.5 %.

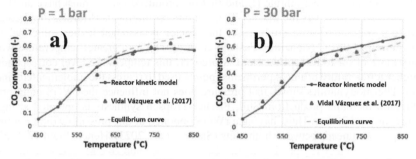

Figure 1 - CO_2 conversion at different temperatures, for an inlet $H_2/CO_2 = 2$ and a catalyst mass of 0.25 g, at 1 bar (Fig. 1a) and 30 bar (Fig. 1b).

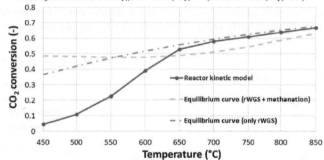

Figure 2 - Kinetic model comparison with two different equilibria
(30 bar, inlet $H_2/CO_2 = 2$, catalyst mass = 0.25 g).

The slight drop in conversion observed in Figure 1a at high temperature is an issue inherent to the model, as it seems that the experimental points are tending towards equilibrium. The authors explain this phenomenon by a lack of experiments conducted at these temperatures. Interestingly, it can also be seen in Figure 1b that at high temperatures, CO_2 conversion is beyond equilibrium. The way this equilibrium curve has been generated must be detailed to understand this trend. The equilibrium conversion is calculated through Gibbs' free energy minimisation, thus assuming that each chemical reaction reaches its equilibrium, i.e. both the rWGS and the methanation reactions in the present case. Following Eq. (2) and (3), methanation appears favoured at high pressure, inducing a more pronounced parasitic effect at 30 bar. As these reactions require more H_2 molecules to convert one molecule of CO_2, the resulting CO_2 conversion is slightly smaller when methanation is favoured compared to the case where only the rWGS reaction is observed. Now, from a kinetic point of view, if it is true that the rWGS reaction is close to equilibrium at high temperature, it is not the case for the methanation reactions, given their exothermic nature. Thus, the kinetic curve tends towards an equilibrium where only the rWGS reaction would exist, as can be justified with Figure 2 in which the equilibrium curve for this reaction alone is added for the data set at 30 bar.

4. Reactor Design at Two Different Scales

4.1. Small Pilot Scale

The global objective of this project is to build a small-scale Power-to-kerosene pilot plant. Its production rate is based on the electrolysis capacity available at ULiège, which is a maximum of 1.5 Nm³/h of H_2. It is assumed that the rWGS reactor is a tubular fixed bed, which consumes all the available H_2 and that the produced syngas must have an H_2/CO ratio close to 2.1 to satisfy the Fischer-Tropsch unit requirements (Morales and Leonard, 2022). For this small-scale study case, the reactor is assumed to be isothermal at a temperature of 800 °C (González-Castaño et al., 2021), as a greater temperature would require more effort to be kept constant. Moreover, this temperature level corresponds to the higher temperature of any experimental points used to regress the implemented kinetic model. Thermodynamics shows that pressure does not influence the rWGS equilibrium but favours methane production. Still, Fischer-Tropsch synthesis is operated at higher pressure, around 20 bar. Thus, having a rWGS operated at this pressure would eliminate the compression needs between both units (Santos et al., 2023). Hence, Figure 3 displays conversion and selectivity profiles inside the reactor at 1 and 20 bar. Note that this paper presents only the first insights into the unit design. Thus, as a first approach, the constraint imposed on the selected length is that the equilibrium conversion is reached within this length for both pressures. In the future, the optimal outlet conversion should be refined, as it might require unnecessarily more effort to reach the equilibrium conversion.

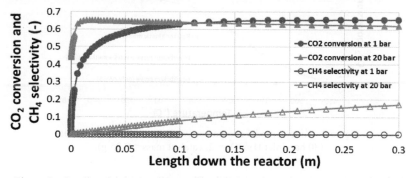

Figure 3 - Small-scale reactor sizing at 1 and 20 bar and 800 °C based on an H_2 inlet flow rate of 1.5 Nm³/h.

As observable in Figure 3, it turns out that the equilibrium can easily be reached within 30 cm at 1 bar, where the CO_2 conversion equals 65.6 % with only negligible formation of methane. When the pressure is increased to 20 bar, a maximum CO_2 conversion is quickly reached (65.2 %) but then decays to its final value of 62.1 %. This decrease should be seen in conjunction with the increase in methane selectivity up to 17.1 %. The inlet H_2/CO_2 ratio must be close to 2 at both pressures to ensure a 2.1 H_2/CO ratio for the following Fischer-Tropsch synthesis. However, given the higher methane production at 20 bar, 20 % less syngas is produced than at 1 bar. Figure 3 illustrates that the reactor length should be limited at high pressure to prevent a too large methanation extent. Indeed, simulating a 5 cm long reactor at 20 bar yields an equilibrium conversion of 64.8 % and a CH_4 selectivity of only 4.6 %.

4.2. Industrial Scale

A single tubular reactor is still assumed for the industrial scale. The main difference with the small-scale reactor lies in the unit heating. While it is reasonable to have an isothermal pilot-scale reactor, it demands more effort to maintain an industrial unit at such high temperatures. An alternative could be an adiabatic reactor, which only requires gas preheating. The industrial size is determined to be 1,000 times larger than the small-scale unit in terms of inlet H_2 flow rate, which is in line with the current biggest electrolysis units on the market (Shiva Kumar and Lim, 2022). In this adiabatic configuration, with an inlet H_2 flow rate of 1,500 Nm³/h and an inlet temperature of 800 °C, the equilibrium CO_2 conversion is reached at 1 bar and 20 bar roughly after the same reactor length (4 m). At 1 bar, the conversion equals 51.1 % and 64.8 % at 20 bar. The CH_4 selectivity is no longer negligible at 1 bar as it equals 2.2 %, while it is 16.9 % at 20 bar. Figure 4 depicts the temperature profiles in the reactor, which help to explain these results.

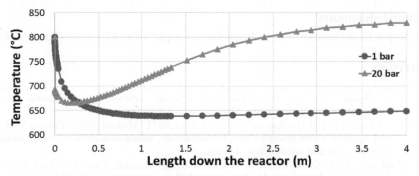

Figure 4 - Temperature profiles along the reactor length at 1 and 20 bar for an inlet H_2 flow rate of 1,500 Nm³/h and an inlet temperature of 800 °C.

At 1 bar, the temperature rapidly decreases to 640 °C before stabilising and increasing slowly to 650 °C. This temperature decay is attributed to the endothermicity of the rWGS reaction, which is favoured at the beginning. The temperature in the reactor being globally lower than 800 °C explains the lower CO_2 conversion and the higher CH_4 selectivity, as conditions are more favourable for methanation. At 20 bar, the same phenomenon occurs, but the initial temperature decay is milder due to the higher pressure favouring the methanation. Furthermore, as both the temperature and pressure are now in favour of methanation, the heat released by the side reactions is sufficient to significantly increase the temperature in the reactor, returning to conditions optimal for the rWGS reaction and explaining the high CO_2 conversion at the expense of a higher CH_4 selectivity.

5. Conclusions

This paper treats the development of an rWGS reactor in Aspen Custom Modeler in the framework of captured CO_2 reutilisation. It is shown that Vidal Vázquez et al. (2017)'s experimental results are reproduced with satisfying accuracy ($R^2 = 92.5$ %). Then, the model is tested isothermally for an inlet H_2 of 1.5 Nm³/h at 1 and 20 bar and leads to the conclusion that the equilibrium conversion is reached in a reactor six times shorter at 20 bar. Thus, if the reaction aims to be tested at both pressures in the same facility, the reactor would be oversized as pressure increases. This oversizing results in a significant propensity for methanation, increasing the outlet CH_4 selectivity. Consequently, a trade-off should be made between sufficient CO_2 conversion at low pressure and sufficiently low methanation at high pressure. The model is also tested in an adiabatic configuration for an inlet H_2 flow of 1,500 Nm³/h. In this case, the same length (4 m) is required to reach the equilibrium conversion irrespective of the operating pressure. It is shown that at both pressures, the temperature drops due to the preponderance of the endothermic rWGS reaction before increasing again. The increase at 20 bar is much more significant because of the temperature and pressure combined effect, favouring methanation reactions. This larger temperature increase at 20 bar enables an outlet conversion greater by more than 25 % compared to the value obtained at 1 bar but at the expense of a larger CH_4 production. This analysis shows that, in adiabatic configuration, the inlet temperature optimisation will be a decisive design factor. In the future, these results will be refined to account for the pressure drop effects on the presented profiles. Besides, the possibility of operating a large-scale unit isothermally will also be investigated.

Acknowledgements

This papper is supported by the Walloon Region as part of a FRIA grant (F.R.S.-FNRS).

References

S. Adelung, S. Maier, and R-U. Dietrich, 2021, 'Impact of the Reverse Water-Gas Shift Operating Conditions on the Power-to-Liquid Process Efficiency', *Sustainable Energy Technologies and Assessments,* 43 (February): 100897.

F. Bisotti, M. Fedeli, P. P. S. Quirino, K. Valverde Pontes, and F. Manenti, 2023, 'Process Modeling and Apparatus Simulation for Syngas Production', *Advances in Synthesis Gas : Methods, Technologies and Applications*, 43–101.

Y. A. Daza, and J. N. Kuhn., 2016, 'CO_2 Conversion by Reverse Water Gas Shift Catalysis: Comparison of Catalysts, Mechanisms and Their Consequences for CO_2 Conversion to Liquid Fuels', *RSC Advances,* 6 (55): 49675–91.

M. González-Castaño, B. Dorneanu, and H. Arellano-García, 2021, 'The Reverse Water Gas Shift Reaction: A Process Systems Engineering Perspective', *Reaction Chemistry & Engineering,* 6 (6): 954–76.

A. Morales, and G. Leonard, 2022, 'Simulation of a Fischer-Tropsch Reactor for Jet Fuel Production Using Aspen Custom Modeler', *Computer Aided Chemical Engineering*, 51:301–6.

H. Ritchie, 2020, 'Sector by Sector: Where Do Global Greenhouse Gas Emissions Come From?', OurWorldInData.Org, 2020, https://ourworldindata.org/ghg-emissions-by-sector.

A. Rouxhet, and G. Léonard, 2024, 'Using H_2 for CO_2 activation on the way towards synthetic fuels', Under submission for the EPHyC conference 2024, Ghent, Belgium.

M. F. Santos, A. E. Bresciani, N. L. Ferreira, G. S. Bassani, and R. M. B. Alves, 2023, 'Carbon Dioxide Conversion via Reverse Water-Gas Shift Reaction: Reactor Design', *Journal of Environmental Management,* 345 (November): 118822.

S. Shiva Kumar, and H. Lim, 2022, 'An Overview of Water Electrolysis Technologies for Green Hydrogen Production', *Energy Reports,* 8 (November): 13793–813.

F. Vidal Vázquez, P. Pfeifer, J. Lehtonen, P. Piermartini, P. Simell, and V. Alopaeus, 2017, 'Catalyst Screening and Kinetic Modeling for CO Production by High Pressure and Temperature Reverse Water Gas Shift for Fischer–Tropsch Applications', *Industrial & Engineering Chemistry Research,* 56 (45): 13262–72.

F. Vidal Vázquez, J. Koponen, V. Ruuskanen, C. Bajamundi, A. Kosonen, P. Simell, J. Ahola, C. Frilund, J. Elfving, M. Reinikainen, N. Heikkinen, J. Kauppinen, and P. Piermartini, 2018, 'Power-to-X Technology Using Renewable Electricity and Carbon Dioxide from Ambient Air: SOLETAIR Proof-of-Concept and Improved Process Concept', *Journal of CO2 Utilization,* 28 (December): 235–46.

A. Wolf, A. Jess, and C. Kern, 2016, 'Syngas Production via Reverse Water-Gas Shift Reaction over a Ni-Al2O3 Catalyst: Catalyst Stability, Reaction Kinetics, and Modeling', *Chemical Engineering & Technology,* 39 (6): 1040–48.

J. Xu, and G. F. Froment, 1989, 'Methane Steam Reforming, Methanation and Water-Gas Shift: I. Intrinsic Kinetics', *AIChE Journal,* 35 (1): 88–96.

Flavio Manenti, Gintaras V. Reklaitis (Eds.), Proceedings of the 34th European Symposium on Computer Aided Process Engineering / 15th International Symposium on Process Systems Engineering (ESCAPE34/PSE24), June 2-6, 2024, Florence, Italy

Modelling, simulation, exergy and economic analyses of thermal cracking of propane using CO_2 and steam as diluents

Yao Zhang [a], Hui Yan [a,b], Daotong Chong [b], Joan Cordiner [a], *Meihong Wang [a]

[a]*Department of Chemical and Biological Engineering, The University of Sheffield, Sheffield S1 3JD, UK*

[b] *State Key Laboratory of Multiphase Flow in Power Engineering, Xi'an Jiaotong University, Xi'an, Shaanxi, China*

Corresponding author: meihong.wang@sheffield.ac

Abstract

Introduction of diluents can increase the yield of valuable products and reduce the coking rate during thermal cracking of propane. Use of captured CO_2 as diluent for the ethylene manufacturing is preferred to avoid high cost of CO_2 transport and storage. A 1-dimentional pseudo-dynamic model of plug flow reactor (PFR) was developed and implemented in gPROMS ModelBuilder®. The model was validated with industrial data. Economic and exergy analyses of the PFR using steam or CO_2 as diluent were then carried out. The results indicate that using CO_2 as diluent can increase the run length of PFR by 13.0% and annual profit by 10.20%. When operating at the ratio achieving highest annual profit, using CO_2 as diluent can reduce exergy destruction by 20.53%. The key findings from this study indicate that using CO_2 as alternative diluent has high potential to increase the profit and reduce energy consumption for ethylene manufacturing. Further study will focus on the effects of diluent-to-propane ratio using different diluents and the potential of using mixed diluents.

Keywords: *First principles modelling, process simulation, thermal cracking furnace, ethylene manufacturing, economic analysis, exergy analysis*

1. Introduction

As one of the most important products in the petro-chemical industry, ethylene has a rapidly increasing demand among the world. Therefore, it is urgent to improve the yearly production of thermal cracking furnace, which is the heart of ethylene manufacturing. However, in the background of global warming, energy consumption and CO_2 emission of thermal cracking furnace are two aspects that have to be focused on.

Several studies aimed to improve the economic benefits of thermal cracking furnace for the ethylene manufacturing. Berreni and Wang (2011a) developed a first principle model of plug flow reactor (PFR) in thermal cracking furnace and carried out dynamic optimization to maximize the annual operating profit. Higher operating profit can be achieved by dynamic optimization but the computation demand also improved a lot. Caballero et al. (2015) carried out an optimization to find out the optimal heat flux profiles along the PFR to improve the ethylene yield. Jarullah et al.(2015) optimized the flow rates of different hydrocarbons feeds to obtain the maximum profit.

Reducing the energy consumption and CO_2 emission of ethylene manufacturing are equally important to improve economic benefit. Yuan et al.(2019) developed a steady state model of whole thermal cracking furnace including radiation section, convection section and quench system. Energy and exergy analyses of the thermal cracking furnace were carried out based on the steady state model to find out the energy saving potential. Zheng et al. (2023) proposed a low-carbon ethylene production system, which can achieve 57.5% CO_2 reduction but 15.92 annual cost increase.

In this paper, using captured CO_2 as alternative diluent for thermal cracking furnace was compared with using steam based on a 1-D pseudo-dynamic model considering both energy consumption and economic benefits.

2. Mathematic modelling and model validation

2.1 Mathematic model

Since the coking rate is extremely slow compared with the reaction rates of thermal cracking reactions, only the coke thickness change with time in this pseudo-dynamic model.

$$\frac{\partial F_i(z)}{\partial z} = \sum_{j=1}^{NR} s_{ij} \times r_j(z) \times M_{wi} \times \frac{\pi \times D(z)^2}{4} \tag{1}$$

The reaction rates and coking rate can be calculated by Eq. (2) and Eq. (3) according to the Arrhenius law. The reaction scheme used is from Sundaram and Froment (1979).

$$r_j(z) = A_j \times e^{\frac{-Ea_j}{RT(z)}} \times \prod_{i=1}^{NC} C_i^{nij}(z) \tag{2}$$

$$r_c(z) = A_c \times e^{\frac{-E_c}{RT_c(z)}} \times C_{C_3H_6}(z) \tag{3}$$

The concentration of each component can be calculated by Eq. (4) based on ideal gas law.

$$C_i(z) = \left(\frac{\frac{F_i(z)}{MW_i}}{N_{diluent} + \sum_{i=1}^{NC}\left(\frac{F_i(z)}{MW_i}\right)}\right) \times \frac{P(z)}{R \times T(z)} \times \frac{1}{1000} \tag{4}$$

The coke thickness and PFR internal diameter reduction caused by coke formation can be obtained by Eq. (5) and Eq. (6).

$$\frac{d\delta(z)}{dt} = \frac{r_c(z)}{\varphi c} \tag{5}$$

$$D(z) = Di - 2\delta(z) \tag{6}$$

Energy balance and momentum balance are shown in Eq. (7) and Eq. (8).

$$(\sum_{i=1}^{NC} F_i(z) \times Cpm_i(z) + F_d(z) \times Cpm_d(z)) \times \frac{\partial T(z)}{\partial z} = \frac{\dot{Q}(z)}{L} + \frac{\pi \times D(z)^2}{4} \times \sum_{j=1}^{NR} r_j(z) \times \left(-\Delta H_{r,j}(z)\right) \tag{7}$$

$$\frac{\partial P(z)}{\partial z} = \frac{\partial\left(\frac{1}{Mm(z)}\right)/\partial z + \frac{1}{Mm(z)} \times \left(\frac{1}{T(z)} \times \frac{\partial T(z)}{\partial z} + Fr(z)\right)}{\frac{1}{Mm(z)P(z)} - \frac{P(z)}{G(z)^2 RT(z)}} \tag{8}$$

2.2 Model validation

This 1-D pseudo-dynamic model was implemented in gPROMS ModelBuilder®. The predication of physical properties used in this model is based on Peng-Robinson equation of state and can be obtained from Multiflash®.

Modelling, simulation, exergy and economic analyses of thermal cracking of
propane using CO_2 and steam as diluents
693

Table 1 Simulation results of propane conversion and main products yields compared with industrial data (clean tube condition).

	Simulation results	Industrial data	Relative error
C_2H_4	35.80%	34.50%	3.16%
C_3H_6	14.77%	14.70%	0.50%
CH_4	25.31%	24.00%	5.46%

Table 2 Simulation results of PFR outlet coke thickness compared with industrial data.

Run length	Simulation results	Industrial data	Relative error
100 hours	0.13 cm	0.14 cm	-7.1%
300 hours	0.41 cm	0.44 cm	-6.8%
700 hours	0.94 cm	0.98 cm	-4.08%

Reactor parameters and operating parameters from Sundaram and Froment (1979) are used. As shown in Table 1 and Table 2, the simulation results show good agreement with industrial data from Sundaram and Froment (1979).

3. Exergy and economic analyses

3.1 Methodology
This section aims to compare two kinds of diluents (CO_2 vs steam) at different diluent-to propane ratios. The total inlet mass flow rate of process gas (propane + diluent) is fixed at 1.0689 kg/s and inlet pressure is fixed at 3 bars, which are the same as the base case used to validate the model. The diluent-to-propane ratio varies from 0.2-1.0, which is the operating range reported by Sundaram and Froment (1979).

3.1.1 Economic analysis
Annual production of valuable products and annual profit are two important indicators to compare the economic benefits of using two diffetent kinds of diluents. Annual production of valuable products can be calcuated using Eq.(9) and annual profit can be calculated using Eq.(10-12).

$$F_{i\,annual} = 8160 \times \frac{F_{i\,total}}{t_c + t_d} \tag{9}$$

$$P_{annual} = INC_{annual} - COS_{annual} \tag{10}$$

$$INC_{annual} = \sum F_{i\,annual} \times COS_i \tag{11}$$

$$COS_{annual} = \frac{t_c * 8160}{t_c + t_d} \left(F_{C3H6} \times COS_{C3H6} + F_d \times COS_D + \dot{Q}_{annual} \times COS_H + \frac{DCC}{t_c} \right) \tag{12}$$

The cost of CO_2 as diluent is assumed to be 0 since use of captured CO_2 can avoid the extremely high cost of CO_2 transport and storage. Other chemical price factors and costs for decoking can be found in Berreni and Wang (2011b).

3.1.2 Exergy analysis

Exergy analysis can be used to evaluate the energy consumption in quality of using two kinds of diluents.The exergy balance of PFR can be writtern as Eq. (13).

$$\dot{E}_Q + \dot{E}_{f,in} = \dot{E}_{f,out} + \dot{E}_D \tag{13}$$

$\dot{E}_Q$ can be determined by Eq. (14) which is the exergy of required heat and $\dot{E}_{f,in}$ and $\dot{E}_{f,out}$ can be determined by Eq.(15) which are the exergy of inlet and outlet process gas.

$$\dot{E}_Q = \sum_{Z=0}^{L} \frac{Q(z)}{L} \times \left(1 - \frac{T_{we}(z)}{T}\right) \tag{14}$$

$$\dot{E}_f = \dot{E}^{ph} + \dot{E}^{ch} \tag{15}$$

The rational exergy efficiency is used in this study to analyse the exergy efficiency, which can be calculated by Eq.(16). Since the exergy of required heat is much smaller than the exergy of stream, the conventional exergy efficiency is close to 1.

$$\varepsilon = \frac{\dot{E}_{f,out} - \dot{E}_{f,in}}{\dot{E}_Q} \tag{16}$$

Considering the run length and shut down of PFR, the annual exergy of required heat ($\dot{E}_{Q,annual}$), annual exergy destruction ($\dot{E}_{D,annual}$) and annual average exergy efficiency (ε_{annual}) can be used to compare the PFR performance for the whole year.

$$\dot{E}_{Q,annual} = 8160 \times \frac{\dot{E}_{Q,total}}{t_c + t_d} \tag{17}$$

$$\dot{E}_{D,annual} = 8160 \times \frac{\dot{E}_{D,total}}{t_c + t_d} \tag{18}$$

$$\varepsilon_{annual} = 1 - \frac{\dot{E}_{D,annual}}{\dot{E}_{Q,annual}} \tag{19}$$

3.2 Economic analysis

To compare the economic benefits of using two different kinds of diluents, the yield of valuable products under clean tube condition operating at different diluent-to-propane ratios are shown in Fig.1.From Fig.1 (a) and (b), using CO_2 has a higher ethylene yield but lower propylene yield. Compared with steam, CO_2 has a higher molecular weight. Therefore, at the same diluent-to-propane ratio, the molar flowrate of CO_2 is smaller. It leads to a higher partial pressure of propylene and more propylene participates in the reaction to produce ethylene and methane. Using CO_2 as diluent also has a longer run length, which is shown in Fig.1(c). This can be explained by that using CO_2 as diluent will lead to a lower coking rate because of lower coke surface temperature.

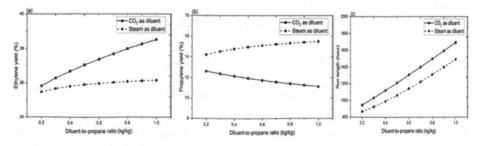

Fig. 1. Comparison of (a) Ethylene (b) Propylene yield under clean tube condition and (c) Run length using CO_2 and steam

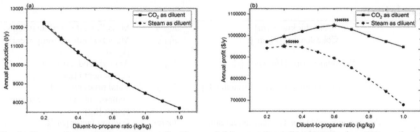

Fig. 2. Comparison of (a) annual production and (b) annual profit of PFR using CO_2 and steam

The total annual production and annual profit using CO_2 and steam are compared to evaluate the economic benefits of using two different kinds of diluents. As shown in Fig.2 (a), the annual production of valuable products are quite close using CO_2 and steam. This is because using CO_2 has a higher ethylene yield but lower propylene yield, which are both considered as valuable products. As shown in Fig.2 (b),using CO_2 has a much higher annual profit since longer run length, lower energy consumption and lower cost for diluent. Using CO_2 can increase the highest annual profit from 950,590 $/y (operating at steam-to-propane ratio 0.3) to 1,046,555 $/y (operating at diluent-to-propane ratio 0.6).

3.3 Exergy analysis

When operating at the diluent-to-propane ratio achieve the highest annual profit, the annual exergy of required heat, annual exergy destruction and annual average exergy efficiency were compared using CO_2 as diluents.

As shown in table 3, when achieving the highest annual profit, using CO_2 as diluent can reduce annual required exergy by 16.16% , reduce annual exergy destruction by 20.53% and increase annual average exergy efficiency by 1.5%. This is because using CO_2 can reduce the temperature difference between tube outer wall of PFR and process gas compared with using steam.

4. Conclusion

A 1-D pseudo-dynamic model for PFR in thermal cracking furnace was developed to compare using CO_2 and steam as diluents. The model was validated with industrial data from literature. The results indicated that using CO_2 as diluent can increase the run length of PFR and increase annual profit by 10.10%. When operating at the ratio achieving highest annual profit, using CO_2 as diluent can reduce exergy destruction by 20.53%. The key findings of this study demonstrate that using CO_2 as alternative diluent has high potential to increase the profit and reduce energy consumption in ethylene manufacturing.

Table 3 Comparison of annual exergy utilization when PFR operating at the diluent-to-propane ratio with highest annual profit using CO_2 and steam

	Run length (h)	Annual exergy of required heat (GJ)	Annual exergy destruction (GJ)	Annual average exergy efficiency (%)
Steam as diluent	926	49896	10872	78.18
CO_2 as diluent	1307	41832	8640	79.36
Absolute difference	381	-8064	-2232	1.18
Relative difference	29.15%	-16.16%	-20.53%	1.50%

Nomenclature

$\delta(z)$	Coke thickness [m]	$F_d(z)$	Mass flowrate of diluent [kg/s]
φ_c	Coke density [kg/m³]	$F_i(z)$	Mass flowrate of component i [kg/s]
ε	Rational exergy efficiency [-]	$F_{i\,annual}$	Annual production of component i [kg/y]
ε_{annual}	Annual average rational exergy efficiency [-]	$F_{i\,total}$	Total production of component i in a cycle of run length [kg]
A_c	Pre-exponential factor for coking reaction [kg m³/(kmol m² s)]	$Fr(z)$	Friction factor [m⁻¹]
A_j	Pre-exponential factor for reaction j [s⁻¹] or [m³/(kmol s)]	$G(z)$	Total mass flow rate [kg/m² s]
$C_i(z)$	Molar concentration of component i [kmol/m³]	INC_{annual}	Annual income [\$/y]
COS	Annual cost [\$/y]	L	Reactor length [m]
COS_D	Price factor of diluent [\$/kg]	MW_i	Molecular weight of component i [kg/kmol]
COS_H	Price factor of heat [\$/kg]	$N_{diluent}$	Molar flowrate of diluent [kmol/s]
COS_i	Price factor of component i [\$/kg]	NC	Number of components [-]
Cpm_d	Specific heat capacity of diluent [kJ/(kg K)]	NR	Number of reactions [-]
$Cpm_i(z)$	Specific heat capacity of component i [kJ/(kg K)]	n_{ij}	Reaction order [-]
$D(z)$	Internal diameter of PFR with coke formation [m]	$P(z)$	Pressure [Pa]
DCC	Decoking cost per cycle [\$]	P_{annual}	Annual profit [\$/y]
Di	Internal diameter of PFR under clean tube condition [m]	$\dot{Q}_{annual}$	Annual energy consumption [GJ/y]
Ea_j	Activation energy of reaction j [kJ/kmol]	R	Ideal gas constant [J/ (K mol)]
E_c	Activation energy of coking reaction [kJ/kmol]	$r_c(z)$	Reaction rate of coking [g /(m² s)]
$\dot{E}^{ch}$	Chemical exergy [kj/s]	$r_j(z)$	Reaction rate of reaction j [kmol/(s m³)]
$\dot{E}^{ph}$	Physical exergy [kj/s]	s_{ij}	Stoichiometry coefficient [-]
$\dot{E}_D$	Exergy destruction of PFR [kJ/s]	$T(z)$	Process gas temperature [K]
$\dot{E}_{f,in}$	Exergy of PFR inlet stream [kJ/s]	t_c	Run length [h]
$\dot{E}_{f,out}$	Exergy of PFR outlet stream [kJ/s]	t_d	Decoking time per cycle [h]

References

Berreni, M. and Wang, M. (2011a) ,Modelling and dynamic optimisation for optimal operation of industrial tubular reactor for propane cracking, *Computer Aided Chemical Engineering*, pp. 955–959.

Berreni, M. and Wang, M. (2011b) Modelling and dynamic optimization of thermal cracking of propane for ethylene manufacturing, *Computers & Chemical Engineering*, 35(12), pp. 2876–2885.

Caballero, D.Y., Biegler, L.T. and Guirardello, R. (2015) Simulation and optimization of the ethane cracking process to produce ethylene, *12th International Symposium on Process Systems Engineering and 25th European Symposium on Computer Aided Process Engineering*, pp. 917–922.

Sundaram, K.M. and Froment, G.F. (1979) Kinetics of coke deposition in the thermal cracking of propane, *Chemical Engineering Science*, 34(5), pp. 635–644.

Jarullah, A.T. *et al.* (2015) Optimal operation of a pyrolysis reactor, *12th International Symposium on Process Systems Engineering and 25th European Symposium on Computer Aided Process Engineering*, pp. 827–832.

Yuan, B. *et al.* (2019) Assessment of energy saving potential of an industrial ethylene cracking furnace using advanced exergy analysis, *Applied Energy*, 254, p. 113583.

Zheng, C., Wu, X. and Chen, X. (2023) Low-carbon transformation of ethylene production system through deployment of carbon capture, utilization, storage and Renewable Energy Technologies, *Journal of Cleaner Production*, 413, p. 137475.

Flavio Manenti, Gintaras V. Reklaitis (Eds.), Proceedings of the 34th European Symposium on Computer Aided Process Engineering / 15th International Symposium on Process Systems Engineering (ESCAPE34/PSE24), June 2-6, 2024, Florence, Italy

Moisture content prediction model of a wet granulator – fluid bed drier system based on LSTM networks

Kai Liu[a], Mahdi Mahfouf[b], James Litster[a] and Daniel Coca[b,c]

[a] *Department of Chemical and Biological Engineering, University of Sheffield, UK*

[b] *Department of Automatic Control and Systems Engineering, University of Sheffield, UK*

[c] *School of Engineering, Newcastle University, UK*

kai.liu@sheffield.ac.uk

Abstract

In recent years, continuous manufacturing has garnered considerable attention within the pharmaceutical industry, owing to the advantages it offers over traditional batch manufacturing. In the production of solid dosage forms, wet granulation is a widely used approach because of advantages such as improved flowability and compressibility. In comparison to the other wet granulation methods, twin screw granulation (TSG) exhibits greater suitability for continuous processing. The process flowsheet encompasses various units including feeders, blender, twin screw granulation (TSG), fluidized bed dryer (FBD), milling and tableting. While conventional mechanistic models primarily concentrate on individual unit operations, data-driven models can be employed to simulate the integration of these units and account for their collective effects. In this research, LSTM networks are employed to model the complex relationships between process parameters in TSG/FBD and the moisture content of granules after drying. Input parameters of LSTM model consist of liquid flowrate and solid flowrate in TSG, meanwhile, drying air temperature and drying air flowrate of FBD unit. The model is trained and validated by well-designed pseudorandom binary sequence experiment data from the Diamond Pilot Plant (DiPP) which is located in the University of Sheffield (UK). In order to showcase the efficacy of the LSTM model, a comparative investigation is conducted against conventional RNNs and FNNs. The results demonstrate that the proposed approach exhibits superior modelling performance in comparison to the alternative modelling methods.

Keywords: Wet granulation, twin screw granulation, fluidized bed dryer, dynamic modelling, LSTM

1. Introduction

Wet granulation is a widely employed pharmaceutical manufacturing process crucial for the production of high-quality solid dosage forms. It involves the agglomeration of fine powder particles through the addition of liquid binders, creating granules that exhibit improved flowability, compressibility, and content uniformity. The process typically comprises four key stages: mixing of the active pharmaceutical ingredient (API) and excipients, wet massing to form granules, drying to remove excess moisture, and sizing to achieve the desired particle size distribution. It enhances the compressibility of the formulation and facilitates uniform drug distribution.

The success of wet granulation depends on several critical factors, with moisture content playing a pivotal role in determining the quality and efficacy of the final tablets. Accurate and timely prediction of moisture content during the wet granulation process is crucial for ensuring product consistency, optimizing manufacturing efficiency. To reach this goal, first principle models that consider multiple factors have been proposed. These models rely on fundamental principles and conservation laws. While these models are crucial for understanding drying processes, they have limitations when it comes to making accurate predictions. As a result, they are not commonly used for real-time control and optimization in industrial drying. One big reason for this is that these models are computationally taxing, adding to the challenge of applying them in real-time settings. Another reason is that these mechanistic models are primarily designed for individual operating units. When applied to combined multiple operating units, the exercise to integrate different mechanistic models into a cohesive framework becomes challenging.

Understanding and modelling all these relationships is crucial in drying technology in order to meet the requirements of consumption or further processing. Heuristic computational methods such as artificial neural networks offer a promising alternative for handling nonlinearity and complexity, particularly in scenarios involving ambiguously defined processes and system behaviours, even when the underlying mechanisms and principles remain incompletely elucidated. For the representation of nonlinear processes in sequential data, Recurrent Neural Network (RNN) finds extensive application. RNN is a type of neural network with short-term memory capabilities. The hidden layers consist of multiple recurrent structures, and the output at a particular time step is not only dependent on the current input but also on the output from the previous time step. The primary limitation of RNNs lies in their inability to handle long-term dependencies, meaning that information with significant temporal gaps in time sequences cannot be effectively learned and produced by RNNs, this issue is commonly referred to as the gradient vanishing problem. LSTM neural networks represent a specific type of RNN that differs from standard RNNs, which rely solely on a simple activation function in their internal structure. In LSTM neural networks, gated units are introduced within the hidden layers to determine whether to remember or forget information. The incorporation of these gated units allows LSTM networks to selectively retain critical information while optionally discarding previous memory information.

In this research, LSTM is used to predict the moisture content of dried granules in the integrated TSG-FBD process, LSTM prediction accuracy is compared with those of conventional RNN and FNN. Moreover, the prediction difference of moisture content between integrated TSG-FBD model and single FBD model is address ed. The rest of this paper is organized as follows: Section 2 provides the methodology, including the architecture of LSTM network and experiment design. Section 3 presents the LSTM-model prediction results compared to the different models. Finally, Section 4 concludes the paper with a summary of key findings.

2. Methodology

2.1 *Long Short-Term Memory Network*

The essence of LSTM lies in its cell state (Hochreiter and Schmidhuber, 1997), which is designed to address the challenges of long-term dependencies in sequence data. In the Figure 1, f_t represents the forget gate, where the preceding time step's hidden layer state h_{t-1} and the current time step's input x_t undergo a series of vector outputs ranging between 0 and 1 via the sigmoid activation function. This determines how much information from the previous time step's internal state c_t needs to be forgotten. Correspondingly, the outputs of the new memory gate and input gate, along with the output of the forget gate, collectively decide the retention of the new memory information, as depicted by the fundamental principles in Eq. (1) to Eq. (6):

$$i_t = \sigma_g(W_i \cdot x_t + U_i \cdot h_{t-1} + b_i) \tag{1}$$

$$f_t = \sigma_g(W_f \cdot x_t + U_f \cdot h_{t-1} + b_f) \tag{2}$$

$$\tilde{c}_t = tanh\ tanh\ (W_c \cdot x_t + U_c \cdot h_{t-1} + b_c) \tag{3}$$

$$c_t = f_t \odot c_{t-1} + i_t \odot \tilde{c}_t \tag{4}$$

$$o_t = \sigma_g(W_o \cdot x_t + U_o \cdot h_{t-1} + b_o) \tag{5}$$

$$h_t = o_t \odot tanh\ tanh\ (c_t) \tag{6}$$

where the activation functions are $\sigma_g(x) = \frac{1}{(1+e^{-x})}$ $\sigma_g(x) = \frac{1}{(1+e^{-x})}$ and $tanh\ tanh\ x = \frac{e^x - e^{-x}}{e^x + e^{-x}}$, W_i **W** and U_i **U** represent weight parameters and **b**b_i denotes bias, $\odot \odot$ signifies the pointwise multiplication.

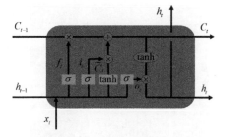

Figure 1. diagram of the long short-term memory network

2.2 *Experiment design*

Despite the existence of a distinctive continuous production line encompassing the entire journey from powder to tablet in the ConsigmaTM-25 of DiPP, this study concentrates specifically on the integrated TSG and FBD, as depicted in Figure 2. A hopper and a feeder deposit and transport the blended powder to the TSG. . The blended powder formulation consisted of 72% lactose (DFE Pharma, Germany), 24% microcrystalline cellulose (MCC), and 4% polyvinylpyrrolidone (PVP) (Harke Pharma GmbH, Germany). The conveying elements further transport the blended powder, which undergoes mixing at a designated port where a liquid binder is injected for nucleation purposes. Subsequently, the wet granules are gravimetrically transported to the FBD, which consists of six segments, sequentially receiving wet granules for a loading time. Upon completion of the loading time, the powders transition to subsequent cell, while the preceding cell continues the drying process. A Near Infrared (NIR) spectrometer (Fibre Optic FP710e, NDC Technology, Essex, UK) is installed in a specific section of the fluid bed dryer to enable real-time monitoring of the moisture content of the granules (Lige et al., 2022).

The manipulated parameters in the experiment are mixed powder flowrate, liquid flowrate, drying air temperature and drying air flowrate. To increase the diversity of the data, a Pseudo-Random Binary Sequence (PRBS) type of data was chosen, signifying that these parameters undergo staged changes during the TSG and FBD processes. For TSG, the intervals at which mixed powder flowrate and liquid flowrate change are 60 seconds, while in FBD, the intervals for the variations of drying air temperature and drying air flowrate are 330 seconds. The ranges of variations for these four parameters are detailed in Table 1, ensuring that, irrespective of the changes in mixed powder flowrate and liquid flowrate, the Liquid-to-Solid (L/S) ratio is maintained between 0.05 and 0.38 in order to prevent blockages in the TSG that may result from excessively high L/S ratios. The TSG process was executed at a constant screw speed of 500 RPM. Each cell undergoes a filling time of 240 s, followed by a drying stage lasting 760 s, resulting in a cumulative duration of 1000 s for a complete drying cycle. Totally, 9 batch experiment data were collected, an example of manipulated parameters in the experiment is illustrated in Figure 3. The TSG ran for 240s in total, however, the residence time of FBD was 1000s so data was collected from the FBD up to 1000s.

Table 1. Input range of manipulated parameters.

Mixed Powder Flowrate (kg/h)	5-20
Liquid Flowrate (kg/h)	0.9-4
Drying Air Temperature (°C)	50-70
Drying Air Flowrate (m³/h)	180-360

Figure 2. Integrated TSG and FBD in DiPP (a. Hopper b. Twin screw granulation c. Six-segment Fluid bed dryer).

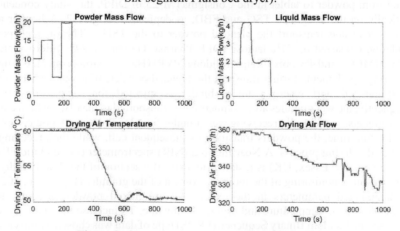

Figure 3. An example of the manipulated inputs.

2.3 *The proposed modelling strategy*

The inputs to the LSTM model consist of liquid flowrate and mixed powder flowrate in TSG, meanwhile, drying air temperature and drying air flowrate of FBD unit. The output of the model is the moisture content after drying process. A total of 9 batches of

experimental data were collected, with 8 batches designated as training data and the
remaining batch used as test data.

The loss function employed in this study is the root mean squared error (RMSE), defined
as follows:

$$RMSE = \sqrt{\sum_{i=1}^{N} \frac{(y_i - \hat{y}_i)^2}{N}})$$

(7)

where y_i are the actual value and $\hat{y}_i$ are the prediction results, N is the number of data
points.

3. Results and discussions

To validate the predictive accuracy of the LSTM model, its moisture content forecasts
were compared with those from conventional RNN and FNN models. For all three
models, the number of hidden units was fixed at 30, and the maximum training epochs
were limited to 100. The Root Mean Square Error (RMSE) values for the validation set
are presented in Table 2. The comparison of these values demonstrates that, with this
particular type of input, both RNN and FNN models exhibit significantly lower accuracy.
While the conventional RNN displays limited long-term memory capacity, and the FNN
lacks the functionality for making sequential predictions, the LSTM model can accurately
predict the process. To further validate the accuracy of the LSTM-based integrated TSG-
FBD model, results are juxtaposed with those of the single FBD model.

Table 2. RMSE comparison of LSTM, RNN and FNN.

Network	RMSE
LSTM	0.345
Conventional RNN	2.798
FNN	10.379

For the integrated TSG-FBD model, its inputs comprise the TSG flowrates (mixed
powder flowrate, liquid flowrate) and FBD drying air conditions (drying air temperature,
drying air flowrate) or the single FBD model, the inputs are the FBD drying conditions.
The prediction outcomes are illustrated in Figure 4, and the corresponding prediction
errors are depicted in Figure 5. Figure 4 clearly shows that during the loading phase (0-
240s), the integrated TSG-FBD model closely matches the actual values. In the drying
phase after 240s,the difference in predictions between the two models becomes less
marked, with the integrated TSG-FBD model demonstrating an overall better
performance. Analysis of Figure 5 reveals that the discrepancies in errors primarily occur
in the 0-240s region. Beyond 240s, in the drying phase, the two error curves fluctuate
around the 0 value. Physically, during the loading phase, the moisture content is largely
influenced by the moisture content of granules post-TSG. At this stage, the moisture
content is affected by both TSG flowrates and FBD drying conditions, whereas in the
single FBD model only the drying conditions influence the drying curve. Thus, the TSG
inputs significantly impact the final moisture content of the integrated TSG-FBD model,
underscoring the long-term memory capabilities of the LSTM model.

In conclusion, the LSTM model consistently outperforms both RNN and FNN in
predicting moisture content, regardless of whether it is for the single FBD model or the
integrated TSG-FBD model. Furthermore, the integrated TSG-FBD model with its four
inputs yields more accurate predictions compared to the single FBD model.

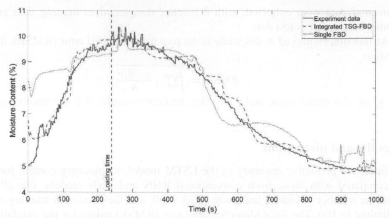

Figure 4. Moisture content prediction of the integrated TSG-FBD and single FBD.

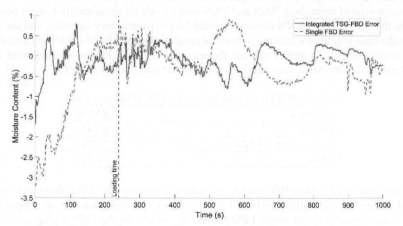

Figure 5. Prediction error of integrated TSG-FBD and single FBD.

4. Conclusion

This paper, proposed an integrated TSG-FBD model based on an LSTM network for wet granulation. The LSTM model demonstrates superior accuracy in predicting moisture content compared to conventional RNN and FNN models. The initial findings indicate that the integrated model which incorporates four inputs (TSG inlet flowrates and FBD drying conditions) achieves higher accuracy than the individual FBD model. The developed model shows promise for future process optimization and control applications.

References

Hochreiter, S., & Schmidhuber, J., 1997. Long short-term memory. Neural Computation, 9(8), 1735-1780.

Wang, L.G., Omar, C., Litster, J., Slade, D., Li, J., Salman, A., Bellinghausen, S., Barrasso, D. & Mitchell, N., 2022. Model driven design for integrated twin screw granulator and fluid bed dryer via flowsheet modelling. International Journal of Pharmaceutics, 628, 122186.

Flavio Manenti, Gintaras V. Reklaitis (Eds.), Proceedings of the 34th European Symposium on
Computer Aided Process Engineering / 15th International Symposium on Process Systems
Engineering (ESCAPE34/PSE24), June 2-6, 2024, Florence, Italy

Simulation study of a hybrid cryogenic and membrane separation system for SF_6 recovery from aged gas mixture in electrical power apparatus

Huong Trang Vo [a], Lujia Chen [a], Patricia Gorgojo [a,b]

[a,*]*Department of Electrical and Electronic Engineering, The University of Manchester, Manchester M13 9PL, United Kingdom*
[b] *Department of Chemical Engineering, University of Zaragoza, Zaragoza 50005, Spain*
Email: lujia.chen@manchester.ac.uk

Abstract

Despite its wide use in the electrical industry, sulphur hexafluoride (SF_6) is a potent greenhouse gas with high global warming potential (GWP) and atmospheric lifetime. Therefore, there is a growing interest in recycling, reconditioning, and reusing aged SF_6. In addition, the upcoming F-gas regulation with increasing restrictions imposed on SF_6 could result in banning the production of virgin SF_6, consequently putting pressure on end-users to recycle aged SF_6 for further reuse. Currently, the cryogenic separation process is often utilised to recondition aged SF_6 in electrical power apparatus. However, this method is not suitable to treat mixtures with SF_6 content lower than 40 mol.% and produces a waste gas containing up to 15 mol.% of SF_6. This study aims to tackle the aforementioned limitations of the existing SF_6 reconditioning process by implementing membrane units. For this purpose, a mathematical model of the membrane gas separation process is developed and integrated into an Aspen Plus simulation via CAPE-OPEN interface. A set of cost model equations collected from the open literature is also used to study the cost of the SF_6 recovery process. The simulation for a feed mixture of 20 mol.% SF_6 and 80 mol.% N_2 with a flow rate of 10,000 kg/day demonstrated the system's capability to recover SF_6 efficiently, achieving the product purity of 98.3 mol.% and SF_6 recovery of 97.5 %. Further optimisation reduced the specific recovery cost and specific energy consumption by 3.1 % and 10.5 %, respectively. The SF_6 content in the waste gases was also reduced to less than 1 mol.%.

Keywords: sulphur hexafluoride; gas separation; membrane; simulation; Aspen Plus.

1. Introduction

Sulphur hexafluoride (SF_6) is highly stable due to its chemical inertness that is comparable to nitrogen (N_2) gas and incombustible nature. Two prominent factors for the use of SF_6 in the electrical industry are its excellent dielectric property, almost 2.5-3 times higher than air, under similar test conditions, and its low dissociation temperature and high dissociation energy, which makes it an excellent insulating and arc-quenching gas medium. However, SF_6 is a potent greenhouse gas with a global warming potential (GWP) 24,300 times higher than carbon dioxide (CO_2) and an atmospheric lifetime of 3,200 years.

Despite these facts, demand for SF_6 in medium- and high-voltage switchgear continues to rise, driven by increasing electricity demand and electrification of transportation. The installed base of medium-voltage switchgear units is expected to increase between 40 to 90 per cent by 2050 (Zhai et al., 2021). In the UK, National Grid is the largest user of SF_6 with an inventory of 920 tonnes with SF_6 presently accounting for 92 % of their Scope 1 emissions (National Grid, 2022). The continued phase-out of SF_6 is paramount to ensure the wider electricity industry in the UK and Europe as a whole can realise the ambition of Net Zero by 2050. Despite decades of research, there is no perfect like-for-like replacement for SF$_6$ across different power equipment. For example, some SF$_6$ alternatives, such as fluoronitrile and perfluoroketone, despite having been commercialised and successfully operational in gas-insulated equipment (Hyrenbach et al., 2017; Kieffel and Biquez, 2014), belong to the group of per- and polyfluoroalkyl substances (PFAS), which will potentially be banned according to the EU PFAS Restriction Proposal (Zeinoun and Reisser, 2023). This leaves non-greenhouse gases including N$_2$ and air to be the only viable option to substitute SF$_6$. However, due to their low dielectric strength, it would dramatically increase the equipment size and/or the filling pressure of the device by a factor of at least 2.5 (Kieffel et al., 2016). This would not only impact enclosure design, safety and cost but also go against the need to develop more compact electrical equipment in metropolitan areas.

The new proposed F-gas regulation draft prohibits the sale of new SF$_6$ switchgear starting in 2026 (European Commission, 2022). However, it does not ban the use of existing SF$_6$ switchgear which has a typical asset lifespan of 20 to 30 years for medium-voltage and >40 years for high-voltage equipment. Hence, there is a growing need to recondition aged SF$_6$ for further use. The conventional method of reconditioning used SF$_6$ is a commercialised cryogenic process to recover SF$_6$ from aged mixtures of SF$_6$ with other components mainly including N$_2$ or air (DILO, 2022). The recovered product contains 99 mol.% SF$_6$ and approximately 85 % recovery of SF$_6$ is achieved. As shown in Figure 1, the process consists of two separation columns: one operates at relatively lower pressures of approximately 8-14 bara, and the other operates at a higher pressure of up to 40 bara. The refrigeration units are used to cool the gas mixture to the optimal temperature between -40 °C to -30 °C. Note that the operating pressure and temperature of the two separation columns are dependent on the composition of the gas mixture. The main limitation of this process is that it is only suitable for treating insulating gas mixtures with a high content of SF$_6$ (>60 mol.%), while the SF$_6$ content in a binary mixture with a buffer gas such as N$_2$ typically ranges from 5 to 20 mol.% (Hama et al., 2018). In addition, a certain amount of waste gas is produced and cannot be directly released into the atmosphere, which requires a further stage of incineration that can be energy intensive.

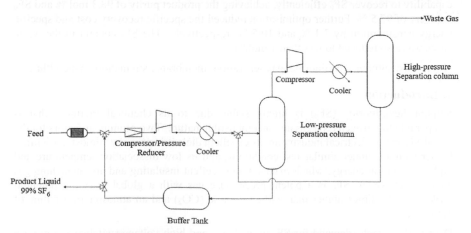

Figure 1: Conventional SF$_6$ cryogenic process (simplified from (DILO, 2022))

In gas separation applications, there is a growing interest in membrane-based hybrid systems due to their inherent benefits. These include straightforward installation and operation, minimal energy requirements, continuous operation with the option of partial or complete recycling, and the potential for integration with other separation units. This study aims to address the current shortcomings of the SF$_6$ reconditioning process by introducing a design of a membrane-cryogenic hybrid system.

2. Methodology

2.1. Aspen Plus simulation

Aspen Plus was used to model a plant capable of treating approximately 10,000 kg/day of SF_6 mixture, a representative scale for a small gas treatment plant. The SF_6 content in the feed mixture was set at 20 mol.% with the remaining gas being N_2. Two membrane units were incorporated: the first one was used to enhance the SF_6 content in the feed for the cryogenic cycle, while the second unit was employed to minimise the SF_6 concentration in the residual gas. A process flow diagram of the process is included in Figure 2.

Since a membrane module is not available in Aspen Plus and in its Model Library, the membrane model was solved numerically in MATLAB and integrated into Aspen Plus as a CAPE-OPEN COM unit operation with a license provided by AmsterCHEM. Additionally, the Peng-Robinson property package was chosen for the simulation.

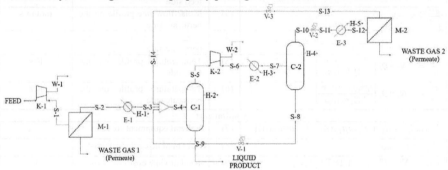

Figure 2: Membrane-cryogenic process

2.2. Model assumption

The model was assumed to operate for 358 days per year (7 maintenance days) for a period of 5 years. The membrane is assumed to be in constant operation during this period with no decrease in productivity. The process operated at steady state, which means no replacements or adjustments to the production rate were considered and any replacements during the 5-year period were accounted for in the contingency and fee costs (15 % and 3 % of the bare module cost, respectively) (Turton et al., 2018). For simplification, other costs including operational labour, waste treatment and administration were neglected.

The membrane module was simulated as a hollow fibre module using a mathematical model that describes the counter-current flow pattern with bore-side feed, based on a number of assumptions. The most important ones are as follows:

   i)   The permeation rate obeys Fick's law;
   ii)   Permeance of gas components through the membrane is independent of pressure on both sides;
   iii)   Plug flow on the feed side with changes in velocity and pressure are assumed to be negligible on the feed side;
   iv)   Laminar flow on the permeate side.
   v)   Gas properties including viscosity and specific heat capacity are independent of pressure and temperature.

Before being utilised to examine the plant performance under various operating conditions, the membrane model was validated using experimental data of SF_6/N_2 separation using a commercial membrane from Yamamoto et al. (2002). The analysis revealed that the membrane model aligned well with experiment data for high feed-side pressures of 4 and 5 bara but deviated significantly at a lower pressure of 3 bara. To address this, a corrective factor of 1.12 was applied to the membrane area variable, reducing the model's average percentage error to 8 %. Turton et al. (2018) reported the correlations of cost estimation in USD back in 2001. Hence, all the costs were updated to 2023 price using cost indexes and converted to GBP adopting the current exchange rate of 1 USD to 0.8 GBP.

2.3. Simulation study of a hybrid SF₆ recovery system

The simulation of the process was carried out to estimate the key performance indicators including the specific SF_6 recovery cost and energy consumption. Table 1 presents a set of important equations of the membrane and economic models used in this study.

Table 1: Membrane and economic models and equations for key performance indicators

Equation		Description	Unit
Membrane model			
$A_m = N_T \pi D_i L$	(1)	Effective membrane area	m²
$\frac{dN_t x_i}{dz} = -N_T \pi D_i J_i$	(2)	Molar flow rate profile on the feed side	mol/m·s
$\frac{dN_s y_i}{dz} = -N_T \pi D_i J_i$	(3)	Molar flow rate profile on the permeate side	mol/m·s
$\frac{dp_s}{dz} = \frac{192\mu R T_s N D_0 (D+ND_0)}{\pi(D^2 - ND_0^2)^3 P_s} \sum_{i=1}^{nc} N_s y_i$	(4)	Pressure profile on the permeate side	Pa/m
$\frac{dT_t}{dz} = -N_T \pi D_i \left[\frac{\sum_{i=1}^{nc} J_i c_{pi}(T_t - T_s)}{\sum_{i=1}^{nc} N_t x_i c_{pi}} \right]$	(5)	Temperature profile on the feed side	K/s
$\frac{dT_s}{dz} = -N_T \pi D_i \left[\frac{\sum_{i=1}^{nc} J_i c_{pi}(T_t - T_s)}{\sum_{i=1}^{nc} N_s y_i c_{pi}} \right]$	(6)	Temperature profile on the permeate side	K/s
Capital cost			
$log_{10} C_P^0 = K_1 + K_2 log_{10}(A) + K_3 [log_{10}(A)]^2$	(7)	General equipment cost	£
$C_{BM} = C_P^0 F_{BM}$	(8)	Bare module equipment cost	£
$C_{TM} = \sum_{i=1}^{n} C_{TM,i} = 1.18 \sum_{i=1}^{n} C_{BM,i}$	(9)	Total module cost	£
$C_M = 1.12 A_m C_m$	(10)	Membrane cost	£
$C_{TM} = 0.2(C_{TM} + C_M)$	(11)	Annual capital related cost	£/y
Operational cost			
$C_{Elec} = C_E \cdot P_C$	(12)	Annual electricity cost	£/y
$C_{UT} = C_R \cdot Q_C + C_S \cdot Q_H$	(13)	Annual utilities cost	£/y
$C_{OPEX} = (C_{Elec} + C_{UT}) \cdot \tau$	(14)	Annual operating cost	£/y
Performance indicator			
$(C_{TM} + C_{OPEX})/F_{SF_6}$	(15)	Specific recovery cost	£/kg SF₆
$(P_C + Q_C + Q_H)/F_{SF_6}$	(16)	Specific energy consumption	kWh/kg SF₆

2.4. Optimisation of a hybrid SF₆ recovery system

The operation of the SF_6 recovery system was optimised to ensure the purity of the liquid product is at 99 mol.% SF_6. The feed conditions were fixed at 10,000 kg/day at 25 °C and 4 bara. The nonlinear single-objective optimisation framework can be described as:

Objective function: Minimise specific recovery cost (Equation 15)
Constraint: Minimum purity of 99 mol.% SF_6 in the liquid product
Decision variables: The initial values, lower and upper bounds for the decision variable in the optimisation problem are listed in Table 2

Table 2: Initial values, lower and upper bounds for the decision variables in the optimisation problem

Decision variable	Unit	Initial	Minimum	Maximum
C-1 operating pressure	bara	9	7	14
C-1 operating temperature	°C	-30	-40	-30
C-2 operating pressure.	bara	38	30	40
C-2 operating temperature	°C	-38	-40	-30
M-1 area	m²	145	140	160
M-2 area	m²	45	30	60

3. Results and discussion

The simulation was first carried out using initial operating conditions as listed in Table 2 and the results obtained are shown in Table 3. Results from the initial simulation show that the hybrid process is capable of treating feed gas with a low concentration of SF_6 of 20 mol.%, while achieving a high SF_6 recovery of 97.5 % and product purity of 98.3 mol.%. The SF_6 content in the waste gas streams was 0.555 and 0.844 mol.%, respectively, which are significantly lower in comparison to that of the conventional cryogenic process.

Table 3: Simulation results using initial operating conditions

SF_6 recovery productivity	Product purity	SF_6 content in Waste Gas 1	SF_6 content in Waste Gas 2	SF_6 recovery	Total capital cost	Total operational cost	SF_6 recovery cost	Specific energy consumption
kg/h	mol.%	mol.%	mol.%	%	10^6£/y	10^6£/y	£/kg	kWh/kg
230	98.3	0.555	0.844	97.5	1.058	0.649	0.580	0.171

Table 4 shows the optimisation results of several performance indicators. In general, there was no significant change in the product purity and SF_6 recovery in comparison to the initial simulation. Notably, the purity of the liquid product was only influenced by the operational parameters of the low-pressure separation column C-1, with a purity of 99 mol.% attainable at operating pressures below 8 bara. The SF_6 content in the waste gas streams was kept below 1 mol.% at 0.539 and 0.549 mol.%, respectively. This could potentially eliminate the requirement of the further destruction of the waste gas. Furthermore, increasing the membrane area in both stages led to a reduction in the total energy consumption, indicating a trade-off between capital and operational costs. Nevertheless, the associated increase in capital costs was negligible, resulting in a lower specific recovery cost by 3.1 %.

Table 4: Optimisation results

Optimised decision variable			Optimised results		
Parameter	Unit	Value	Parameter	Unit	Value
C-1 operating pressure	bara	8	SF_6 recovery productivity	kg/h	230
C-1 operating temperature	°C	-33	Product purity	mol.%	99.0
C-2 operating pressure	bara	32	SF_6 content in Waste Gas 1	mol.%	0.539
C-2 operating temperature	°C	-40	SF_6 content in Waste Gas 2	mol.%	0.549
M-1 area	m^2	150	SF_6 recovery	%	97.8
M-2 area	m^2	50	Total capital cost	10^6£/y	1.058
			Total operational cost	10^6£/y	0.531
			SF_6 recovery cost	£/kg	0.562
			Specific energy consumption	kWh/kg	0.153

4. Conclusion

The combination of Aspen Plus simulations with a techno-economic analysis provided an overview of the feasibility of a hybrid membrane-cryogenic plant to recover SF_6 from the aged gas mixture. Firstly, the initial simulation showed success in addressing the two main drawbacks of the current SF_6 reconditioning process: i) capable of treating SF_6/N_2 insulation mixture with a low SF_6 content below 20 mol.%, and ii) achieve high SF_6 recovery of 97.5 % and low SF_6 content of less than 1 mol.% in the waste gas streams. This degree of recovery was significantly higher, as opposed to the 85 % recovery of SF_6 achieved by the conventional cryogenic process. The content of SF_6 in the waste gas streams produced of less than 1 mol.% could potentially obviate the need for additional disposal process of the waste gas. The cost analysis reveals that the compressor cost constitutes the primary component of the overall capital expenditure, whereas the operational cost is predominantly influenced by the combined expenses of electricity required for the compressors and the cost of refrigerants used in the refrigeration unit.

Therefore, the incorporation of membrane units does not substantially affect the overall process cost. The determination and optimisation of various process variables has attained the reduction in both SF_6 recovery cost and SF_6 content in waste gas. The specific SF_6 cost, and energy consumption were reduced by 3.1 % and 10.5 % respectively.

Acknowledgement

Authors acknowledge funding support by the School of Engineering of the University of Manchester for this PhD studentship.

References

Hama, H., Buhler, R., Girodet, A., Juhre, K., Kindersberger, J., Lopez-Roldan, J., Neumann, C., Riechert, U., Schichler, U., Yasuoka, T., Okabe, S., Gautschi, D., Hering, M., Kieffel, Y., Koltunowicz, W., Neuhold, S., Pietsch, R., Rokunohe, T., Steyn, D., 2018. Dry air, N2, CO2 and N2-SF6 mixtures for gas-insulated systems. CIGRÉ.

Hyrenbach, M., Paul, T.A., Owens, J., 2017. Environmental and safety aspects of AirPlus insulated GIS. In: CIRED - Open Access Proceedings Journal. Institution of Engineering and Technology, pp. 132–135.

Kieffel, Y., Biquez, F., 2014. SF6 alternative development for high voltage switchgears. In: 33rd Electrical Insulation Conference, EIC 2015. Institute of Electrical and Electronics Engineers Inc., pp. 379–383.

Kieffel, Y., Irwin, T., Ponchon, P., Owens, J., 2016. Green Gas to Replace SF6 in Electrical Grids. IEEE Power and Energy Magazine 14, 32–39.

National Grid, 2022. Re-opener Report MSIP-SF6 Asset Intervention [WWW Document]. URL https://www.nationalgrid.com/electricity-transmission/document/140901/download (accessed 11.29.23).

European Commision, 2022. Proposal for a Regulation of the European Parliament and of the Council on fluorinated greenhouse gases, amending Directive (EU) 2019/1937 and repealing Regulation (EU) No 517/2014. Strasbourg.

DILO, 2022. Separating unit - Translation of original operating manual [WWW Document]. DILO. URL www.dilo-gmbh.com

Turton, R., Shaeiwitz, J.A., Bhattacharyya, D., Whting, W.B., 2018. Analysis, Synthesis, and Design of Chemical Processes, 5th ed. Prentice Hall.

Yamamoto, O., Takuma, T., Kinouchi, M., 2002. Recovery of SF6 from N2/SF6 gas mixtures by using a polymer membrane. IEEE Electrical Insulation Magazine.

Zeinoun, E., Reisser, C., 2023. The EU's PFAS Restriction Proposal Explained [WWW Document]. APCO worldwide. URL https://apcoworldwide.com/blog/the-eus-pfas-restriction-proposal-explained/#:~:text=The%20proposal%20will%20effectively%20ban%20all%20forms%20of,for%20which%20there%20is%20no%20currently%20known%20alternative. (accessed 9.6.23).

Zhai, P., Pirani, A., Connors, S.L., Péan, C., Berger, S., Caud, N., Chen, Y., Goldfarb, L., Gomis, M.I., Huang, M., Leitzell, K., Lonnoy, E., Matthews, J.B.R., Maycock, T.K., Waterfield, T., Yelekçi, O., Yu, R., Zhou, B., 2021. Climate Change 2021: The Physical Science Basis. Contribution of Working Group I, IPCC.

Flavio Manenti, Gintaras V. Reklaitis (Eds.), Proceedings of the 34th European Symposium on Computer Aided Process Engineering / 15th International Symposium on Process Systems Engineering (ESCAPE34/PSE24), June 2-6, 2024, Florence, Italy

Investigating Fluid Flow Dynamics in Triply Periodic Minimal Surfaces (TPMS) Structures Using CFD Simulation

Kasimhussen Vhora,[a,b] Tanya Neeraj,[b] Dominique Thévenin,[b] Gábor Janiga,[b] Kai Sundmacher [a,b,*]

[a]*Max Planck Institute for Dynamics of Complex Technical Systems, Sandtorstr. 1, 39106 Magdeburg, Germany*
[b]*Otto von Guericke University Magdeburg, Universitätsplatz 2, 39106 Magdeburg, Germany*
kai.sundmacher@ovgu.de

Abstract

Efficient absorption processes require optimized packed bed column structures, which affect gas-liquid contact, flow distribution, and pressure drop. An optimal setup ensures efficient mass transfer with high surface area while keeping down the pressure drop, which leads to energy savings and better absorption. TPMS structures such as the Gyroid, Schwarz-P, and Schwarz-D were investigated in this study, with a focus on balancing porosity and surface area to achieve reduced pressure drops and optimal phase contact. Single-phase flow simulations were conducted using the commercial software STAR-CCM+, compared to the lattice Boltzmann method (LBM) to provide an alternative perspective on fluid dynamics. Validation, analysis of the results and identification of possible improvements were achieved through these comparisons. According to the results, the Schwarz-D structure with 70% porosity and 2 mm unit cell leads to the best performance, exhibiting a pressure drop of 655 Pa m^{-1} and a specific surface area of 1776 m^2 m^{-3} when analysed with STAR-CCM+. The predicted pressure drop was successfully confirmed using LBM simulations, adding robustness to the findings.

Keywords: Computational Fluid Dynamic, TPMS Structure, Pressure Drop, LBM.

1. Introduction

Optimizing packed bed column structures is vital for various chemical engineering and industrial separation processes. Structured packings play a key role in enhancing mass transfer, with success dependent on factors like separation efficiency, minimal pressure drop, and maximal capacity (Lange and Fieg, 2022). Innovations in packing design, such as 3D-printed structured packings made from materials like clear resin and polyamide allow for fine-tuning of geometric parameters to improve performance (Kawas et al., 2021).

TPMS (Triply Periodic Minimal Surface) structures offer low-pressure drops, contributing to energy savings, and possess considerable structural strength for durability. The modern manufacturing method, 3D printing, provides the ability to easily adjust designs, enabling the creation of specialized solutions that meet the particular requirements of industrial separation processes. This makes TPMS a forward-thinking and efficient option for such applications. In the context of packed columns, the optimization variables of high specific surface area and low-pressure drop are intrinsically linked to the performance efficiency of the columns (Rix and Olujic, 2008). Efficiency in packed columns involves a trade-off between high-surface-area packing for efficiency but limited capacity, or low surface area for high capacity but lower efficiency (Jaya and Kolmetz, 2020).

Hawken et al. conducted experimental studies on 3D-printed Schwarz-D TPMS structure, focusing on their impact on pressure drop in chemical engineering applications. Their research measured pressure drop per unit length over two different lengths (46 mm and 89 mm) of Schwarz-D packing with a unit cell size of 3.14 mm and various velocities within the Reynolds number range of 1-1000 (Hawken et al., 2023). Building on this foundation, our study introduces a single-phase Computational Fluid Dynamics (CFD) model for accurately predicting pressure drops in TPMS structures. Successful validation against experimental values enables its later applicability to predict pressure drops and surface areas for Gyroid, Schwarz-D, and Schwarz-P TPMS structures (Hawken et al., 2023). The investigation explores porosity variations within these structures, aiming to identify an optimized structure balancing low-pressure drop and high specific surface area.

The lattice Boltzmann method (LBM) was additionally used in our study to validate the pressure drop of the optimized structure. The documented ability of LBM to handle complex fluid flows in porous media adds robustness to our analysis, ensuring consistent and reliable comparisons.

2. CFD Digital Twin Setup

Figure 1 depicts two selected CFD configurations, each showcasing a column with an 18 mm diameter and the Schwarz-D structural pattern. In CFD Setup 1, the structure spans 89 mm, with designated points (P1 to P4) strategically placed for monitoring pressure gradients. Inlet and outlet boundary conditions are illustrated to aid in simulating airflow through the Schwarz-D structure. CFD Setup 2 presents a more compact structure, measuring 46 mm in length, and incorporates points P5 and P6 for analogous pressure drop measurements.

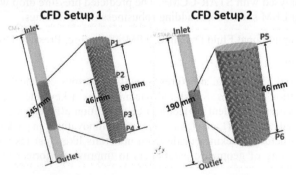

Figure 1: CFD Digital twin setup1 & setup2.

2.1. CFD Simulation setup

The present study first utilized the commercial CFD package STAR-CCM+ for solving the governing equations. The geometry was created using Autodesk Fusion360, a computer-aided design (CAD) software. The generated mesh consists of unstructured polyhedral grids, with a total of 4.58 million cells (for CFD setup2). For the gas phase, air was selected with a constant density (ρ) of 1.18 kg m^{-3}. Typical computational run time for the simulations were around 45 minutes on a system equipped with an 11th Gen Intel(R) Core(TM) i7-11700 processor at 2.50 GHz, 64.0 GB RAM, and running on a 64-bit operating system. For the column inlet and outlet, the boundary conditions were set to velocity inlet and pressure outlet, respectively. The temporal resolution was managed through an implicit unsteady model with an initial time step of 0.001 s, for a total physical simulation time of 1 s. In the post-processing of the simulation, the working pressure is measured by the surface mean average on the 2D plane. This measurement occurs at different points, ranging from P1 to P6, in both CFD setups as shown in Figure 1.

2.2. Lattice Boltzmann Method

2.2.1. Enhanced Central Hermite Multiple Relaxation Time Lattice Boltzmann Solver

The lattice Boltzmann method, introduced for instance by Krüger et al., 2017, stands as a robust solver for the Boltzmann equation, particularly in the hydrodynamic regime. Specifically designed to reconstruct the Navier–Stokes equations, this method initiates with the discretization of the Boltzmann equation in phase space, representing particle degrees of freedom. Utilizing the projection onto Hermite polynomials and Gauss-Hermite quadrature, this discretization results in a system of interconnected hyperbolic equations. Integration along characteristics yields the widely-recognized "stream-collide" equation:

$$f_i(r + c_i\delta t, t + \delta t) - f_i(r, t) = \Omega_i \qquad \text{(i)}$$

Here, f_i represents discrete distribution functions, c_i denotes discrete particle velocities, Ω_i is the collision operator, and δt signifies the time-step.

2.2.2. Advanced Boundary Conditions

Solid boundaries in our study are effectively modeled using the half-way bounce-back scheme as described in Hosseini, 2020. The determination of missing distribution functions after collision-streaming steps is succinctly expressed as:

$$f_i(x, t + \Delta t) = f_{\underline{i}}^*(x, t) \qquad \text{(ii)}$$

where f^* represents the post-collision population, and $\underline{i}$ is the index of the particle velocity opposite i. At inlets and outlets , we implement constant velocity and/or constant pressure conditions using the non-equilibrium extrapolation approach (Hosseini, 2020). A notable advantage of the lattice Boltzmann method is its capability to handle solid boundaries with a complex geometry without the need for complex grid adaptation in packed bed structures. Simple, regular, equidistant grids suffice for all simulations.

3. Validation of the CFD Simulations

The pressure drop profiles for the CFD setups 1 and 2 are depicted in Figure 2a. In CFD setup 1, the pressure drop measurements over lengths of 89 mm (P1-P4) and 46 mm (P2-P3) in the Schwarz-D structure yielded similar values of pressure drop per unit length (1571 -1578 Pa m^{-1}). This indicates that a longer structure length is unnecessary for CFD simulations to measure pressure drop, given the periodic nature of the Schwarz-D structure; the impact of inlet and outlet boundary conditions is already negligible for the shorter configuration. This is confirmed by CFD Setup 2, in which the measured pressure drop (P5-P6) is 1588 Pa m^{-1}, nearly identical to that of CFD setup 1 (less than 1% relative difference). Consequently, CFD setup 2 has been chosen for all subsequent simulations and various case studies exploring different levels of porosity and unit cell sizes.

The subsequent CFD simulation maintained the configuration of CFD setup 2, varying only the inlet velocity to measure the resulting pressure drop. Figure 2b presents a graph comparing the pressure drop against inlet velocity for the CFD simulations, alongside experimental data from (Hawken et al., 2023). This comparison underscores a robust correlation between computational predictions and experimental findings. Notably, the CFD pressure drop aligns with the experimentally measured pressure drop from the literature for the same configurations at different velocities. Utilizing this validated CFD model, further simulations were conducted to measure the pressure drop of the Gyroid, Schwarz-P, and Schwarz-D structures.

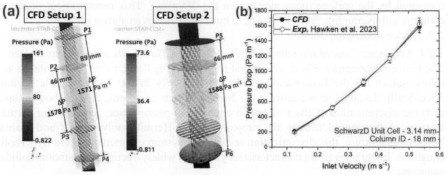

Figure 2: (a) Pressure drop calculation of CFD setup 1 & 2. (b) Pressure drop comparison with experimental literature data at different velocities.

4. Results and Discussion

Figure 3a illustrates the specific surface area and pressure drop characteristics of various porous structures, Gyroid, Schwarz Primitive and Diamond within a column with an internal diameter (ID) of 18 mm and a structure unit cell size of 3.14 mm. The plot reveals that as porosity increases, the specific surface area for each structure initially rises and then decreases, while the pressure drop consistently decreases. The Gyroid and Schwarz-D structures exhibits the highest specific surface area and also a low pressure drop at 80% porosity. For a porosity between 30 and 60%, we observed that the specific surface area remains relatively unchanged for each structure. As expected, low-porosity structures exhibit a higher pressure drop when compared to those with higher porosity. Consequently, for further investigations concentrating on low pressure drops, structures with porosities of 50%, 60%, and 70% were selected. Additionally, the unit cell size within these structures was varied, with sizes of 2.5 mm and 2 mm being examined.

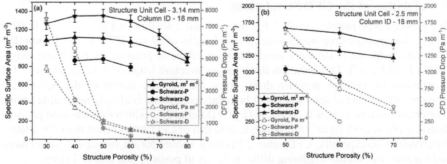

Figure 3: Specific surface area and CFD pressure drop comparison across Gyroid, Schwarz-P, and Schwarz-D structures, with porosity: (a) 30-80%, unit cell size 2 mm; (b) 50-70%, unit cell size 2.5 mm.

Figures 3b and 4a present a clear picture of how structural porosity impacts the properties of different structures. As the porosity increases from 50% to 70%, there is a notable trend: the specific surface area for all structure types tends to decrease slightly. In contrast, the pressure drop demonstrates an inverse relationship with porosity. This indicates that higher porosity correlates with less resistance to fluid flow within the structure. A particularly interesting observation is that both the Gyroid and Schwarz-D structures exhibit nearly identical pressure drops. However, the specific surface area is significantly higher in the Schwarz-D structure compared to the Gyroid.

Therefore, when optimizing for both pressure drop and surface area, the Schwarz-D structure emerges as the superior choice. With 70% porosity and a unit cell size of 2 mm, it achieves an optimal balance, characterized by a pressure drop of 655 Pa m^{-1} and a specific surface area of 1776 m^2 m^{-3}. These findings are crucial in the optimization of porous media designs across various engineering applications. They highlight the necessity of striking a delicate balance between maximizing surface area and minimizing fluid resistance, which is essential for efficient design and operation. Furthermore, the feasibility of 3D printing the Schwarz-D structure, featuring a 2 mm unit cell size, was assessed through its practical implementation via in-house 3D printing using clear resin as shown in Figure 4b.

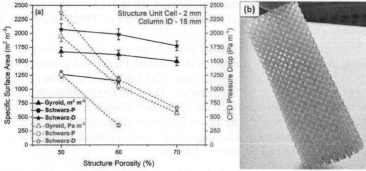

Figure 4: (a) Comparison of surface area and pressure drop in Gyroid, Schwarz-P, and Schwarz-D (50-70% porosity, 2 mm unit cell size). (b) 3D printed Schwarz-D, 70% porosity, 2 mm unit cell size.

4.1 Comparison between LBM and STAR-CCM+ simulation

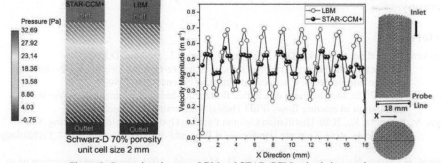

Figure 5: Comparison between LBM and STAR-CCM+ simulation results.

Finally, an LBM simulation was carried out. Figure 5 on the left illustrates the comparison between LBM and STAR-CCM+ pressure drop profiles in the Schwarz-D structure. It reveals minimal discrepancies in the pressure drop, with values of 31.4 Pa for LBM and 30.2 Pa for STAR-CCM+ (less than 4% difference). This good agreement in the pressure drop values underscores the consistency and accuracy of both computational approaches in capturing the macroscopic behaviour of fluid flow through the porous medium of this structure. Further analysis involved studying the velocity profiles along the streamwise direction, as shown in the Figure 5 on the right side. The velocity profile in STAR-CCM+ appears averaged over the domain due to its reliance on the Reynolds-Averaged Navier-

Stokes (RANS) model and to a coarser grid. In contrast, LBM utilizes the Direct Numerical Simulation (DNS) approach on a considerably finer, regular grid, revealing large velocity changes.

This fundamental difference in modeling techniques results in distinctive characteristics in the velocity profiles generated by each method. However, LBM comes with a noticeably larger numerical cost. Overall, the close agreement in pressure drop (the target quantity of this study) attests to the robustness of both LBM and STAR-CCM+ as effective tools for simulating fluid flow through porous structures, providing researchers and engineers with reliable insights into the complex dynamics of such systems.

5. Conclusions

The study concludes that the use of TPMS in the design of packed bed columns presents a significant advancement in the optimization of the absorption process. This is established by examining the fluid dynamic behaviours of TPMS structures, including Gyroid, Schwarz-P, and Schwarz-D, through single-phase flow simulations. Notably, the Schwarz-D structure, with 70% porosity and a 2 mm unit cell, demonstrates promising performance, characterized by a pressure drop of 655 Pa m^{-1} and a specific surface area of 1776 m^2 m^{-3}. The pressure drop comparison of STAR-CCM+ with the LBM simulation provides a good agreement regarding pressure. These findings indicate that TPMS-based structures could significantly enhance the design and efficiency of packed bed columns. The study highlights a research gap in the long-term stability and scalability of TPMS-based structures for industrial use, suggesting future research should focus on their performance, scalability, and economic viability.

Acknowledgment

This research was supported by the International Max Planck Research School for Advanced Methods in Process and Systems Engineering (IMPRS ProEng), Magdeburg, Germany.

References

Hawken, M.B., Reid, S., Clarke, D.A., Watson, M., Fee, C.J., Holland, D.J., 2023. Characterization of pressure drop through Schwarz-Diamond triply periodic minimal surface porous media. Chemical Engineering Science 280, 119039.
Hosseini, S.A., 2020. Development of a lattice Boltzmann-based numerical method for the simulation of reacting flows (PhD Thesis). University of Magdeburg.
Jaya, A., Kolmetz, K., 2020. Distillation Column Packing Hydraulics Selection, Sizing And Troubleshooting, Kolmetz Handbook Of Process Equipment Design. KLM Technology Group.
Kawas, B., Mizzi, B., Dejean, B., Rouzineau, D., Meyer, M., 2021. Design and conception of an innovative packing for separation column – Part II: Design and characterization of a wire based packing. Chemical Engineering Research and Design 169, 189–203.
Krüger, T., Kusumaatmaja, H., Kuzmin, A., Shardt, O., Silva, G., Viggen, E.M., 2017. The lattice Boltzmann method. Springer International Publishing 10, 4–15.
Lange, A., Fieg, G., 2022. Designing Novel Structured Packings by Topology Optimization and Additive Manufacturing, in: Yamashita, Y., Kano, M. (Eds.), 14th International Symposium on Process Systems Engineering, Computer Aided Chemical Engineering. Elsevier, pp. 1291–1296.
Rix, A., Olujic, Z., 2008. Pressure drop of internals for packed columns. Chemical Engineering and Processing: Process Intensification 47, 1520–1529.

Flavio Manenti, Gintaras V. Reklaitis (Eds.), Proceedings of the 34th European Symposium on Computer Aided Process Engineering / 15th International Symposium on Process Systems Engineering (ESCAPE34/PSE24), June 2-6, 2024, Florence, Italy

A deep learning-based energy and force prediction framework for high-throughput quantum chemistry calculations

Guoxin Wu, Qilei Liu*, Jian Du, Qingwei Meng, Lei Zhang*

State Key Laboratory of Fine Chemicals, Frontiers Science Center for Smart Materials Oriented Chemical Engineering, Institute of Chemical Process Systems Engineering, School of Chemical Engineering, Dalian University of Technology, Dalian 116024, China

Qilei Liu (liuqilei@dlut.edu.cn), Lei Zhang (keleiz@dlut.edu.cn)

Abstract

Quantum chemistry (QC) calculations rely on solving the Schrödinger equation and have emerged as a significant computational tool in solvent design and drug discovery. However, the computational cost of conventional QC calculation approaches increases dramatically with the complexity of molecular systems, hindering the large-scale screening of solvents and drugs. In this work, a deep learning-based energy and force prediction (DeepEF) framework is established for high-throughput QC calculations. In this framework, a machine learning potential model is first built by utilizing the atomic self-attention mechanism, and then integrated with a geometry optimization algorithm, GeomeTRIC, for efficient molecular geometry optimizations. Finally, an external test set containing 125 solvents is used to test the prediction accuracy and computational speed of the DeepEF framework.

Keywords: quantum chemistry, deep learning, attention mechanism, machine learning potential.

1. Introduction

Quantum chemistry (QC) calculations use the principles and methodologies of quantum mechanics to precisely predict electronic structures, energy characteristics, and other attributes of chemical substances, thus yielding physicochemical properties. QC has become an indispensable tool in fields such as solvent design and drug discovery. For example, QC calculations can be used to predict the solubilities and the absorption rates of CO_2 in different solvents, which helps to reduce the blindness of experiments and shorten the research and development cycle to some extent. The commonly used QC calculation methods are the density functional theory (DFT) method (Parr, 1980) and the Hartree-Fock method (Matthew & Sutcliffe, 1978). It is often hard to adopt these rigorous methods to perform high-throughput calculations on a large number of molecules. Therefore, the efficient and precise execution of QC calculations has emerged as a worldwide challenge.

One possible solution to this challenge is developing the machine learning potential (MLP) model, which uses machine learning models to directly represent the functional relationship between atomic coordinates and molecular energies, thus avoiding solving the explicit Schrödinger equation. In this way, the potential energy surface (PES) can be constructed in a fast manner. According to the Born-Oppenheimer approximation (Born & Oppenheimer, 1927), the nucleus mass is much larger than the electron mass. Thus, the molecular energy of the fundamental electronic state can be regarded as a function of nuclear coordinates, which is the theoretical basis of the MLP model. The first MLP for high-dimensional systems is the high-dimensional neural network potential (BPNN) proposed by Behler and Parrinello (2007), which borrows from the atomic contribution method of classical molecular mechanics, that is, the system energy is obtained by adding up the atomic energy output of each atomic network. However, the input of BPNN is an atom-centric symmetric function, where all parameters need to be manually tuned, which is often a painstaking and tedious task. Han et al. (2018) have developed a simple but versatile end-to-end MLP model naming Deep Potential for molecular and atomic systems. The time cost of the MLP model is comparable to that of the empirical force field, but its accuracy is comparable to that of QC calculations. Unke et al. (2021) have added the charge and spin multiplicity to the traditional MLP model, so that the MLP model is able to predict the long-range interaction and the PES of different electronic states. Although MLP models have demonstrated their ability to accurately predict energy and forces for a variety of molecular systems, most of the MLP models are difficult to couple with downstream applications. This limits the users to the secondary development and the extended application of MLP models. The ASE library (Hjorth Larsen et al., 2017) has the capability to use the MLP models for geometry optimizations. However, its current geometry optimization algorithm is characterized by a single convergence condition, requiring that the maximum atomic force value be below a specified threshold. This criterion has the limitation to find the accurate equilibrium geometries. Therefore, it is still a challenge to achieve the seamless integration of the MLP model with external molecular geometry optimization algorithms that have rigorous convergence criteria.

In this work, a deep learning-based energy and force prediction (DeepEF) framework is established for high-throughput QC calculations. In Section 2, a MLP model is developed using an atomic self-attentive neural network for accurate and rapid predictions of molecular energy and forces. Then, in Section 3, the MLP model is coupled with the GeomeTRIC algorithm (Wang & Song, 2016) for molecular geometry optimizations.

2. DeepEF framework for QC calculations

2.1. Construction of the MLP model

In this work, the QM7-X dataset (Hoja et al., 2021) is used to construct the MLP model. The QM7-X dataset was generated starting from ~7,000 molecular graphs with up to five non-hydrogen atoms (C, N, O, S, Cl) drawn from the GDB13 chemical universe (Blum & Reymond, 2009). After sampling and optimizing the structural and constitutional (stereo) isomers of each graph, 42,000 equilibrium structures are obtained. Each of them was then perturbed in order to obtain 99 additional non-equilibrium structures, leading to a total of about 4.2 million samples. To rigorously evaluate the generalizability of the MLP model, 25 molecular graphs and their conformations are selected from the original 7,000 molecular graphs as the test dataset, and the remaining data are used to train the model in this work.

The feature matrix size varies for different molecules due to differences in the number of atoms in each molecule. The self-attention mechanism in deep learning is used to deal with the case of matrices with different sizes (Vaswani et al., 2023). The architecture of the MLP model is shown in **Figure 1**. The input to the model is the atomic cartesian coordinate matrix $R = [\vec{r_1} \quad \vec{r_2} \quad \cdots \quad \vec{r_N}]$ and the corresponding nuclear charge number vector $Z = [Z_1 \quad Z_2 \quad \cdots \quad Z_N]$, where N is the number of atoms. The nuclear charge number vector Z is mapped through the embedding layer to the elemental information $X = [X_1 \quad X_2 \quad \cdots \quad X_N]$. The structural information G can be obtained from the cartesian coordinate matrix R via the coordinate vector difference $R_{ij} = R_j - R_i$ and the radial basis function (Gasteiger et al., 2022) to ensure the molecular rotation and translation invariances. Elemental information X and structural information G are collected in three interaction layers (**Figure 1(b)**) dominated by a multi-head self-attention network (**Figure 1(c)**) and a residual block (**Figure 1(d)**). Finally, the molecular energy and forces are obtained by integrating the information using the output layer (**Figure 1(e)**). The MLP model developed in this work has achieved a high prediction accuracy in estimating the energies and forces of the molecules in the test set with the determination coefficients of 1.00 and 0.98, as well as the mean absolute errors of 0.048 eV and 0.044 eV/Å, respectively.

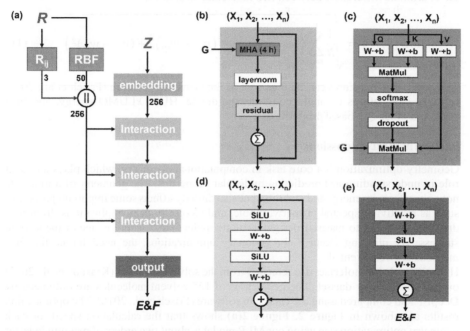

Figure 1. The machine learning potential model. **a** Atom self-attentional neural network. **b** Interaction layer. **c** Multi-head self-attention network. **d** Residual block. **e** Output layer.

2.2. Integration of the MLP model and the geometry optimization algorithm

Accurate calculations of QC-based properties are essential for the study and design of new materials, drugs, and so on. Common QC tasks include the optimizations of minimum energy structures and excited states, the calculations of spectral characteristics, thermodynamic properties of reactions, and so on. All of these properties can be inferred from QC calculations of electronic structures by solving the Schrödinger equation. However, as the complexity of the system increases, the computation time can range from hours to days, and the need for efficient computing cannot be satisfied.

The DeepEF framework is able to optimize molecular structures by integrating the developed MLP model and the GeomeTRIC algorithm (Wang & Song, 2016). The convergence criteria of the GeomeTRIC algorithm are:

(1) The root mean square deviation (*RMSD*) from the previous step is less than 1.2×10^{-3} Å.

(2) The maximum atomic displacement from the previous step is less than 1.8×10^{-3} Å.

(3) The root mean square of the nuclear gradient is less than 3.0×10^{-4} a.u.

(4) The maximum nuclear gradient for any atom is less than 4.5×10^{-4} a.u.

(5) The energy change from the previous step is less than 1.0×10^{-6} a.u.

Compared with the Gaussian software, the convergence criteria of GeomeTRIC has a large probability to find the molecular equilibrium geometry. The equilibrium geometry optimized by DeepEF and the equilibrium geometry optimized by DFT (B3LYP/6-31G*) are compared by *RMSD* (as shown in **Eq. (1)**), which represents the root mean square of the coordinate differences between two molecular structures.

$$RMSD(V, W) = \sqrt{\frac{1}{N}\sum_{i=1}^{N}\left((v_{ix} - w_{ix})^2 + \left(v_{iy} - w_{iy}\right)^2 + (v_{iz} - w_{iz})^2\right)} \qquad (1)$$

The optimized structures can be then sent to the Gaussian software (Frisch et al., 2016) to make calculations of single point energy, dipole, HUMO/LUMO energy, chemical shift, and other QC-based properties.

3. Results and discussions

Geometry optimization is a core task in computational chemistry, which plays a crucial role in understanding and predicting molecular properties. The geometry of a molecule not only affects its energy and stability, but also directly affects some important properties such as reactivity, spectral property, dipole, and so on. In many applications, from new drug development to material design, accurate molecular structure is one of the keys to success. By utilizing DeepEF for geometry optimizations, the need for costly DFT methods is circumvented.

Here, 125 solvent molecules are screened from the solvent database (Krasnov et al., 2022) outside the training dataset. The geometries of 125 solvent molecules are optimized by DeepEF and compared using the Gaussian software (Frisch et al., 2016). The optimization results are shown in **Figure 2**. **Figure 2(a)** shows that the calculation speed for each geometry optimization step using our MLP model is about two orders of magnitude faster than that using the DFT method in the Gaussian software. As the number of atoms increases, the optimization time of our MLP model hardly changes at each step, exhibiting a linear relationship between the optimization time and the number of atoms, while the optimization time of the DFT method increases exponentially with the increase of the atom number. **Figure 2(b)** shows that the *RMSD*s of 70% molecules are less than 0.1 Å, indicating that our MLP model has a high prediction accuracy in predicting molecular energy and forces, and our DeepEF framework enables the geometry optimizations in a rapid and accurate manner. After optimizing the molecular structures, some QC tasks are performed by the Gaussian software to obtain the molecular QC-based properties. A script has been provided by DeepEF to convert molecular ".xyz" files into the Gaussian input files. The highest occupied molecular orbital and lowest unoccupied molecular orbital (HOMO/LUMO) of 125 solvent molecules are computed. The calculation results are shown in **Figure 3**, and the HOMO/LUMO properties calculated from the DeepEF optimized structures are closely consistent with those calculated from the DFT optimized structure.

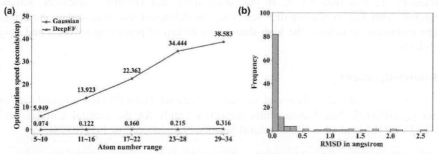

Figure 2. The geometry optimization results of 125 molecules using the DeepEF framework and the gaussian software. **a** The calculation speed for each geometry optimization step using the MLP model in DeepEF and the DFT (B3LYP/6-31G*) method in Gaussian. **b** The *RMSD* between the DeepEF optimized molecular geometries and the Gaussian optimized molecular geometries.

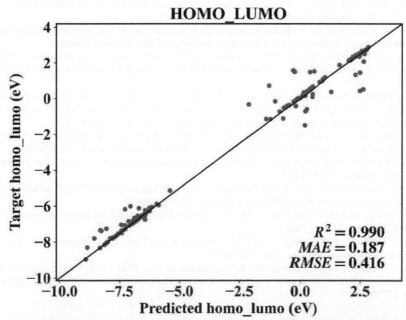

Figure 3. The HOMO/LUMO calculation results of 125 molecules using the MLP model in DeepEF and the DFT (B3LYP/6-31G*) method in Gaussian.

4. Conclusions

In this work, a DeepEF framework is proposed for efficient QC calculations. With the developed MLP model, the DeepEF framework has a powerful ability to fast optimize the structures of various small organic molecules at the cost of minor prediction error. The computational speed of our MLP model is two orders of magnitude faster than that of the traditional DFT method (B3LYP/6-31G*), and the *RMSD*s of 70% optimized molecular geometries are less than 0.1 Å. In the future work, our DeepEF framework will be integrated with the conductor-like screening model-based computer-aided molecular design technique to achieve the high-throughput design of promising solvents and drug candidates.

Acknowledgement

The authors are grateful for financial supports of National Natural Science Foundation of China [22208042], "the Fundamental Research Funds for the Central Universities" [DUT22YG218], and China Postdoctoral Science Foundation [2022M710578].

References

R. G. Parr, 1980. Density functional theory of atoms and molecules, Dordrecht.

O. T. Unke, S. Chmiela, M. Gastegger, K. T. Schütt, H. E. Sauceda, and K.-R. Müller, 2021. SpookyNet: Learning force fields with electronic degrees of freedom and nonlocal effects. Nature Communications, 12(1), 7273.

M. J. Frisch, G. W. Trucks, H. B. Schlegel, et al., 2016. Gaussian 16 Rev. C.01. Wallingford, CT.

M. Born, and R. Oppenheimer, 1927. Zur Quantentheorie der Molekeln. 389(20), 457-484.

L.-P. Wang, and C. Song, 2016. Geometry optimization made simple with translation and rotation coordinates. The Journal of Chemical Physics, 144, 214108.

L. Krasnov, S. Mikhaylov, M. V. Fedorov, and S. Sosnin, 2022. BigSolDB: Solubility dataset of compounds in organic solvents and water in a wide range of temperatures. Zenodo.

L. C. Blum, and J.-L. Reymond, 2009. 970 million druglike small molecules for virtual screening in the chemical universe database GDB-13. Journal of the American Chemical Society, 131(25), 8732-8733.

J. Hoja, L. Medrano Sandonas, B. G. Ernst, A. Vazquez-Mayagoitia, R. A. DiStasio Jr, and A. Tkatchenko, 2021. QM7-X, a comprehensive dataset of quantum-mechanical properties spanning the chemical space of small organic molecules. Scientific Data, 8(1), 1-11.

J. Han, L. Zhang, R. Car, and E. Weinan, 2018. Deep Potential: A general representation of a many-body potential energy surface. Communications in Computational Physics, 23(3), 629-639.

J. Gasteiger, J. Groß, and S. Günnemann, 2020. Directional message passing for molecular graphs. arXiv preprint arXiv:2003.03123.

J. Behler, and M. Parrinello, 2007. Generalized neural-network representation of high-dimensional potential-energy surfaces. Physical Review Letters, 98(14), 146401.

J. A. D. Matthew, and B. T. Sutcliffe, 1978. The Hartree–Fock method for atoms: A numerical approach. Physics Bulletin, 29(4), 178.

A. Vaswani, N. Shazeer, N. Parmar, et al., 2017. Attention is all you need. Advances in Neural Information Processing Systems, 30, 5998-6008.

A. Hjorth Larsen, J. Jørgen Mortensen, J. Blomqvist, et al., 2017. The atomic simulation environment—a Python library for working with atoms. Journal of Physics: Condensed Matter, 29(27), 273002.

Flavio Manenti, Gintaras V. Reklaitis (Eds.), Proceedings of the 34th European Symposium on Computer Aided Process Engineering / 15th International Symposium on Process Systems Engineering (ESCAPE34/PSE24), June 2-6, 2024, Florence, Italy

Mathematical Modeling of an Enzyme Catalyzed Transamination Reaction with Integrated Product Removal

Jessica Behrens [a,*], Sven Tiedemann [a], Jan Von Langermann [a], Achim Kienle [a,b]

[a] Otto-von-Guericke-University, Universitätsplatz 2, D-39106 Magdeburg, Germany
[b] Max-Planck-Institute for Dynamics of Complex Technical Systems, Sandtorstrasse 1, D-39106 Magdeburg, Germany
* jessica.behrens@ovgu.de

Abstract

Enzyme catalyzed reactions have a big potential for the targeted synthesis of fine chemicals and pharmaceuticals. To enable the possibility of time- and resource-saving evaluation of process operation strategies computer-based process simulations can be done. This requires process models being able to reliably predict the process. In this contribution a mathematical model of the enzyme catalyzed transamination reaction for the production of (S)-(3-methoxyphenyl)ethylamine with integrated product removal is proposed and compared to preliminary experimental data. A kinetic model for the trans-aminase-catalyzed reaction is combined with a method-of-moments model for the product crystallization. The model shows great potential for representing the process and builds an important basis for future computer aided design and control strategies.

Keywords: Transaminase-catalyzed reactions, Integrated product removal, Complex mathematical model, King-Altman method, Method of moments

1. Introduction

Transaminase (TA)-catalyzed reactions combine high selectivities with mild reaction conditions and environmentally friendly solvents, and can be deployed for the synthesis of enantiomerically pure amines. However, the product conversion in these biocatalytic reactions are often constrained by an unfavorable chemical equilibrium (Gundersen et al., 2015). To enhance the process performance a variety of strategies are utilized, which range from protein engineering for higher catalytic substrate specificity, to process operation strategies like integrated product and co-product removal. An overview about the different strategies was given by Guo and Berglund (2017) and references therein. The combination of different process strategies renders a high number of degrees of freedom, which makes it time and resource consuming to find optimal operation conditions with common empirical approaches. Process models offer the possibility to study the process in a simulation environment to conduct feasibility studies, examine important process parameters and optimize the process performance. Two examples for simulation studies of process feasibility are the work of Tufvesson et al. (2014), who developed a mathematical model and investigated critical process parameters for co-product removal to estimate the substrate loss and product yields and the work of Esparza-Isunza et al. (2015), who investigated the most important process parameters for a membrane reactor system. Furthermore, process models can be used for online process observation, optimization andcontrol.

In our work we focus on the batch process operation for the production of (S)-(3-methoxyphenyl)ethylamine (3MPEA), a valuable intermediate in the synthesis of Rivastigmine, an important drug for the treatment of Parkinson's and Alzheimer's disease. We aim to develop a process model that provides a basis for future model-based design and control. To our knowledge this is the first report of a process model for the production of 3MPEA. The present paper is based on the experimental results in Neuburger et al. (2021). They combined the ω-TA catalyzed reaction with integrated evaporation of the co-product and integrated crystallization of the main product 3MPEA to shift the reaction equilibrium and increase the product yield. In the following sections we will present a process model for the batch production of 3MPEA with focus on the reaction kinetics, the crystallization, donor salt dissolution, miscibility limitations of compounds and the co-product evaporation.

2. Process Model

The mechanisms of the TA catalyzed reaction with integrated product removal is schematically shown in Figure 1. There are 5 interacting thermodynamic phases in the reactor namely two solid phases, two liquid phases and a gas phase. Note, that for reasons of readability the phases are shown separately in the scheme, however, the reactor is continuously stirred and the solid and liquid phases are mixed. The enzyme (E) catalyzed reaction of isopropylamine (IPA) and 3-methoxyacetophenone (3MAP) to the co-product acetone (Ac) and the product (S)-(3-methoxyphenyl)ethylamine (3MPEA) takes place in the aqueous liquid phase. The reaction is shown in the right half of Figure 1. For more information on the biocatalyst and the chemicals, the interested reader is referred to Neuburger et al. (2021). The carbonyl compound 3MAP has a limited miscibility in the polar phase and separates into a second non-polar liquid phase. The amine donor IPA is injected into the system by the donor salt isopropylammonium 3,3-diphenylpropionate (IPA-3DPPA), which forms the first solid phase. It continuously dissolves in the polar liquid phase together with 3,3-diphenylpropionate (3DPPA). The latter acts as counter-ion for the product 3MPEA and both form the product salt (S)-(3-methoxyphenyl)-ethylamine 3,3-diphenylpropionate (3MPEA-3DPPA) by continuous crystallization. The product salt resembles the second solid phase. The co-product Ac is a volatile compound and can be removed from the process by continuous evaporation. Therefore, it leaves the process via the gas phase. The interactions of the phases with one another are depicted by material fluxes $\dot{N}(t)$ in terms of molar amounts. To discriminate between the different phases in the process model a short nomenclature is introduced.

Liquid Phase(s)
The liquid phases are denoted by the superscript $'$. The polar liquid phase is further

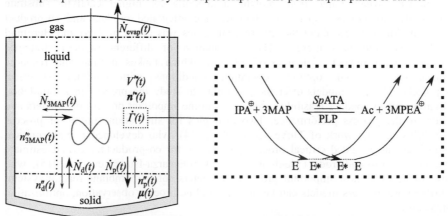

Figure 1: Process scheme. The left scheme shows the batch reactor and the right scheme the chemical reaction.

denoted by the superscript $\cdot$ and the non-polar liquid phase by the superscript $\circ$.
The molar amount vector of the compounds in the polar liquid phase is given by
$\boldsymbol{n}'^{\cdot}(t) = [n'_{\text{IPA}}(t) \quad n'_{\text{3MPEA}}(t) \quad n'_{\text{Ac}}(t) \quad n'_{\text{3MAP}}(t) \quad n'_{\text{3DPPA}}(t) \quad n'_{\text{E}}(t)]$. The non-polar liquid
phase is resembled by the molar amount of excess 3MAP $n'^{\circ}(t) = n'^{\circ}_{\text{3MAP}}(t)$. The vector
containing all molar amounts of the liquid phase is given by $\boldsymbol{n}'(t) = [\boldsymbol{n}'^{\cdot}(t) \quad n'^{\circ}(t)]$.

Solid Phase(s)
The solid phases are denoted by the superscript $''$. The molar amount of the donor salt
IPA-3DPPA is denoted by $n''_{\text{d}}(t) = n_{\text{IPA-3DPPA}}(t)$. The molar amount of the product salt
3MPEA-3DPPA is denoted by $n''_{\text{p}}(t) = n_{\text{3MPEA-3DPPA}}(t)$. The vector containing all molar
amounts of the solid phase is given by $\boldsymbol{n}''(t) = [n''_{\text{d}}(t) \quad n''_{\text{p}}(t)]$. The moments of the
product salt number density distribution are denoted by
$\boldsymbol{\mu}_{\text{k}}(t) = [\mu_0(t) \quad \mu_1(t) \quad \mu_2(t) \quad \mu_3(t)]^{\text{T}}$.
All molar amounts are contained in $\boldsymbol{n}(t) = [\boldsymbol{n}'(t) \quad \boldsymbol{n}''(t)]^{\text{T}}$.

2.1 Reaction kinetic
The reaction catalyzed by a ω-transaminase is known to follow a "ping-pong bi-bi"
mechanism (Walsh, 1998). The steady state reaction rate derivation is based on the
schematic method of King-Altmann and Cleland. For in depth information the reader is
referred to the book of Segel (1993). The formation of nonproductive complexes is
neglected in this work. The reaction rate $\dot{\Gamma}(t)$ is given by

$$\dot{\Gamma}_{\text{Nom}}(t) = V_{\text{f}}V_{\text{r}}\left(c'_{\text{IPA}}(t)c'_{\text{3MAP}}(t) - \frac{c'_{\text{Ac}}(t)c'_{\text{3MPEA}}(t)}{K_{\text{eq}}}\right) \tag{1}$$

$$\dot{\Gamma}_{\text{Denom}}(t) = V_{\text{r}}K_{\text{m,IPA}}c'_{\text{3MAP}}(t) + V_{\text{r}}K_{\text{m,3MAP}}c'_{\text{IPA}}(t) + V_{\text{f}}\frac{K_{\text{m,3MPEA}}}{K_{\text{eq}}}c'_{\text{Ac}}(t)$$

$$+V_{\text{f}}\frac{K_{\text{m,Ac}}}{K_{\text{eq}}}c'_{\text{3MPEA}}(t) + V_{\text{r}}c'_{\text{IPA}}(t)c'_{\text{3MAP}}(t) + V_{\text{f}}\frac{K_{\text{m,3MAP}}}{K_{\text{i,IPA}}K_{\text{eq}}}c'_{\text{IPA}}(t)c'_{\text{Ac}}(t) \tag{2}$$

$$+V_{\text{r}}\frac{K_{\text{m,IPA}}}{K_{\text{i,3MPEA}}}c'_{\text{3MPEA}}(t)c'_{\text{3MAP}}(t) + \frac{V_{\text{f}}}{K_{\text{eq}}}c'_{\text{3MPEA}}(t)c'_{\text{Ac}}(t)$$

$$\dot{\Gamma}(t) = \frac{\dot{\Gamma}_{\text{Nom}}(t)}{\dot{\Gamma}_{\text{Denom}}(t)}V''(t) \tag{3}$$

with the forward (f) and reverse (r) rate

$$V_{\text{f}} = K_{\text{f}}c''_{\text{E}}(t) \tag{4}$$

$$V_{\text{r}} = K_{\text{r}}c''_{\text{E}}(t) \tag{5}$$

and the equilibrium constant

$$K_{\text{eq}} = \left(\frac{K_{\text{f}}}{K_{\text{r}}}\right)^2\frac{K_{\text{m,3MPEA}}K_{\text{m,Ac}}}{K_{\text{m,3MAP}}K_{\text{m,IPA}}}. \tag{6}$$

The Michaelis constants are denoted by K_{m} and the inhibition constants by K_{i} for the
respective compounds. The forward and reverse rate constant are denoted by K_{f} and K_{r},
respectively. The reaction rate is expressed in terms of molar concentrations
$\boldsymbol{c}''(t) = \frac{n''(t)}{V''(t)}$, where $V''(t)$ denotes the volume of the polar liquid phase.

2.2 Crystallization – Moment model
The crystallization of the product salt 3MPEA-3DPPA is modeled by a moment model
approach (Hulbert and Katz, 1964). The crystallization kinetics are restricted to crystal
growth only. Nucleation is neglected due to seeding. The temporal evolution of the
kth-moment $\mu_{\text{k}}(t)$ of the number density distribution of crystals is given by

$$\frac{d\mu_{\text{k}}(t)}{dt} = kG(t)\mu_{\text{k-1}}(t) \tag{7}$$

where the crystal growth rate $G(t)$ is given by

$$G(t) = \begin{cases} k_G(S_p(t) - 1) & \text{if } S_p(t) > 1 \\ 0 & \text{if } S_p(t) \leq 1 \end{cases}, \quad S_p(t) = \frac{\min(n'_{3MPEA}(t), n'_{3DPPA}(t))/V'(t)}{c_{sat,p}}. \quad (8)$$

Crystallization of the product salt is driven by oversaturation $S_p(t)$ of the solutes 3MPEA and 3DPPA. The saturation concentration of the product salt is denoted by $c_{sat,p}$ (≈ 4.7 mM). Oversaturation ($S_p(t) > 1$) is reached, when both substance concentrations in the polar liquid phase are higher than the saturation concentration. Otherwise ($S_p(t) \leq 1$) no crystal growth occurs. The growth rate constant is denoted by k_G. The flux of material for crystal growth from the liquid to the solid phase is given by

$$\dot{N}_d(t) = \frac{\rho_p}{M_{w,p}} \Phi 3G(t)\mu_2(t) \quad (9)$$

and is proportional to the crystal surface, which is expressed by the second moment $\mu_2(t)$. The shape of the crystals is assumed to resemble needles, which is considered by the shape factor Φ. The molar weight and density of 3MPEA-3DPPA are denoted by $M_{w,p}$ and ρ_p, respectively.

2.3 Donor Salt Dissolution and 3MAP Miscibility

In the experiments of Neuburger et al. (2021) the donor salt IPA-3DPPA and the ketone 3MAP are injected into the process. The salt, as well as the ketone show limited solubility in the liquid polar phase and therefore form additional phases. We model those as material storages and assume a continuous exchange of material from the storages to the polar liquid phase. Both mechanisms are modeled in the same way. The flux from the aqueous phase to the solid phase in the case of donor salt dissolution, and the non-polar phase in the case of 3MAP miscibility, is given by

$$\dot{N}_X(t) = k_X(1 - S_X(t))n_X \quad \text{with} \quad n_X = \begin{cases} \min(n'_X(t)) & \text{if } S_X > 1 \\ n^u_X(t) & \text{if } S_X \leq 1 \end{cases} \quad (10)$$

with the flux constants k_X, subscript $X \in \{d, 3MAP\}$ and superscript $u \in \{'', '\circ\}$. Driving force for the material flux is the saturation $S_X(t)$ of the compounds in the polar liquid phase

$$S_X(t) = \frac{\min(n'_X(t))/V'(t)}{c_{sat,X}} \quad (11)$$

The saturation concentration is denoted by $c_{sat,X}$. The saturation concentration of 3MAP is given by the miscibility limit of 3MAP in the polar liquid phase (≈ 25 mM). The saturation concentration of IPA-3DPPA is the solubility concentration of the salt in the aqueous phase (≈ 55 mM). In case of donor salt storage ($S_d(t) > 1$) the flux $\dot{N}_d(t)$ is proportional to the minimal amount of the donor salt solutes in the polar liquid phase $\min(n'_d(t)) = \min(n'_{IPA}(t), n'_{3DPPA}(t))$. Otherwise ($S_d(t) \leq 1$) it is proportional to the molar amount of the donor salt in the solid phase. In case of an oversaturated polar liquid phase in terms of 3MAP ($S_{3MAP}(t) > 1$) the material flux $\dot{N}_{3MAP}(t)$ is proportional to the molar amount in the polar liquid phase $\min(n'_{3MAP}(t)) = n'_{3MAP}(t)$. Otherwise ($S_{3MAP}(t) \leq 1$) the flux is proportional to the molar amount of 3MAP in the non-polar phase.

2.4 Evaporation

The co-product Ac is a volatile compound and can be removed from the process by evaporation. In the experiments of Neuburger et al. (2021) this is achieved by pressure reduction. Based on their data they conclude that low pressures of about 300 mbar are enough to remove the co-product Ac and record their data under reduced pressure conditions. Therefore, it is assumed that the produced Ac is entirely and instantaneously removed from the aqueous phase

$$\dot{N}_{\text{evap}}(t) = \dot{\Gamma}(t). \tag{12}$$

2.5 Reactor Model

This concludes our model of the process, which is given by

$$\frac{dn(t)}{dt} = \begin{bmatrix} -1 & 1 & 0 & 0 & 0 \\ 1 & 0 & 0 & -1 & 0 \\ 1 & 0 & 0 & 0 & -1 \\ -1 & 0 & 1 & 0 & 0 \\ 0 & 1 & 0 & -1 & 0 \\ 0 & 0 & 0 & 0 & 0 \\ 0 & 0 & -1 & 0 & 0 \\ 0 & -1 & 0 & 0 & 0 \\ 0 & 0 & 0 & 1 & 0 \end{bmatrix} \begin{bmatrix} \dot{\Gamma}(t) \\ \dot{N}_{\text{d}}(t) \\ \dot{N}_{\text{3MAP}}(t) \\ \dot{N}_{\text{p}}(t) \\ \dot{N}_{\text{evap}}(t) \end{bmatrix}, \qquad \frac{d\mu_{\text{k}}(t)}{dt} = kG(t)\mu_{k-1} \tag{13}$$

3. Results and Discussion

The process model is fitted to two data sets presented in Neuburger et al. (2021). We use their first data set of the time evolution of the product salt concentration for different biocatalyst loadings (see Figure 2 in Neuburger et al., 2021). However, only measurements of the product salt concentration are not sufficient to reliably parameterize the full process model. In a first approximation we use literature values to parameterize the reaction kinetic from a very similar biocatalytic reaction reported in Al-Haque et al. (2012). They propose a robust methodology for the parameter estimation for a biocatalytic reaction kinetic. Their final results are used in this study and given in Table 1. However, Neuburger et al. (2021) use a different biocatalyst for what reason K_{f} is estimated from their experimental data. Note that K_{r} is linked to K_{f} via K_{eq} and thus differs from Al-Haque et al. (2012) as well. Additionally, the constants k_{G}, k_{d} and k_{3MAP} are fitted to the experimental data. The model is fitted to the data set of 40 U/ml and 60 U/ml, where the parameters were estimated such that the quadratic error between the data points and simulation is minimized. The third set with 80 U/ml is used for validation. The data fit is shown in Figure 2 with the mean estimation error $\bar{E}$. Additionally, the time evolution of the concentrations and molar amounts of the different phases are given for the simulation result with 80 U/ml. Note, that the illustration of Ac is omitted. The simulation of the time evolution of the product salt concentration is in good agreement with the experimental data. The time evolution of the polar liquid phase shows that the concentration of 3MAP is decreasing even though the amount of 3MAP in the non-polar phase is nonzero and thus constantly mixing into the polar phase. A similar dynamic can be observed for the donor salt and its solutes. This is caused by the small estimated diffusion rate k_{3MAP} and k_{d}, which might be a hint for a lowered 3MAP miscibility or mixing effects and an interesting phenomenon to be analyzed in future experiments. However, the predictive capabilities of the model are limited and the results should be cautiously interpreted. Further individual investigation on the reaction, crystallization and evaporation dynamics need to be done to guarantee a well parameterized process model.

Table 1: Model parameter. The reaction kinetic parameters are taken from Al-Haque et al. (2012). The highlighted values are fitted to experimental data (Neuburger et al., 2021).

param.	1/s	param.	mM	param.	mM	param.	
K_{f}	1.05e-2	$K_{m,\text{Ac}}$	148.99	$K_{m,\text{IPA}}$	101.28	k_G	0.2175 $\frac{mm}{s}$
K_{r}	1.78e-2	$K_{m,\text{3MAP}}$	1.85	$K_{i,\text{IPA}}$	4.281	k_{3MAP}	3.132e-5 $\frac{1}{s}$
k_d	8.8e-7	$K_{m,\text{3MPEA}}$	0.12	$K_{i,\text{3MPEA}}$	1e5	K_{eq}	0.033

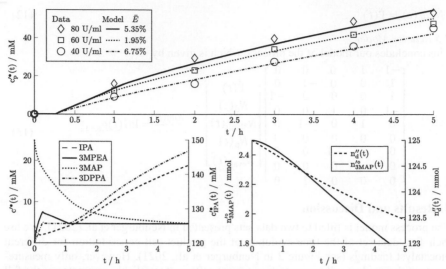

Figure 2: Simulation results. The data is taken from Neuburger et al. (2021). For the simulation with 80 U/ml the time evolution of the concentrations and molar amounts is shown in the lower figures.

4. Conclusion

In this contribution a mathematical model for the batch production of (S)-(3-methoxy-phenyl)ethylamine is presented. The model links the transaminase catalyzed reaction with product crystallization and co-product evaporation. The model shows good agreement with experimental data and builds a basis for further model refinement. In future work we aim to fully parameterize the model and suggest optimal operation strategies for the production of 3MPEA. Furthermore, we aim to use the model for online process observation and to develop process control strategies.

Acknowledgment

We would like to thank Deutsche Forschungsgemeinschaft (DFG, German Research Foundation) for funding our research with the project ID 501735683.

References

N. Al-Haque, P. A. Santacoloma, W. Neto, P. Tufvesson, R. Gani, J. M. Woodley, 2012, Biotechnol. Prog., 28 (5), 1186-1196.

T. Esparza-Isunza, M. González-Brambila, R. Gani, J. M. Woodley, F. López-Isunza, 2015, Chemical Engineering Journal, 259, 221-231.

M. T. Gundersen, R. Abu, M. Schürmann, J. M. Woodley, 2015, Tetrahedron: Asymmetry, 26, 10-11, 567-570.

F. Guo, P. Berglund, 2017, Green Chem., 19, 333-360.

H. M. Hulburt, S. Katz, 1964, Chem. Eng. Sci., 19, 555-574.

J. Neuburger, F. Helmholz, S. Tiedemann, P. Lehmann, P. Süss, U. Menyes, J. v. Langermann, 2021, Chemical Engineering and Processing - Process Intensification, 168, 108578.

I. H. Segel, 1993, Enzyme kinetics: Behavior and Analysis of Rapid Equilibrium and Steady-State Enzyme Systems, Wiley Classics Library, ISBN: 978-0-471-30309-1.

P. Tufvesson, C. Bach, J. M. Woodley, 2014, Biotechnol. Bioeng., 111, 2, 309-319.

C. Walsh, 1998, Enzymatic Reaction Mechanisms, W. H. Freeman & Co, ISBN: 0716700700.

Flavio Manenti, Gintaras V. Reklaitis (Eds.), Proceedings of the 34th European Symposium on Computer Aided Process Engineering / 15th International Symposium on Process Systems Engineering (ESCAPE34/PSE24), June 2-6, 2024, Florence, Italy
© 2024 Elsevier B.V. All rights reserved. http://dx.doi.org/10.1016/B978-0-443-28824-1.50122-8

Process Integration of Urea Production with Methane Pyrolysis for Reduced Carbon Emissions

Muhamad Reda Galih Pangestu,[a] Umer Zahid [a,b*]

[a]*Department of Chemical Engineering, King Fahd University of Petroleum and Minerals, Dhahran 31261, Saudi Arabia*
[b]*Interdisciplinary Research Center for Membranes & Water Security, King Fahd University of Petroleum & Minerals, Dhahran 31261, Saudi Arabia*
uzahid@kfupm.edu.sa

Abstract

The increasing global population is expected to drive a surge in fertilizer demand in the future. However, traditional ammonia-urea production methods, such as steam methane reforming (SMR), contribute significantly to greenhouse gas (GHG) emissions. The purpose of this study was to mitigate CO_2 emissions during an unbalanced plant load of ammonia and urea, which leads to a CO_2 vent. The introduction of methane pyrolysis (MP) enabled the continuous production of ammonia and urea while concurrently generating marketable solid carbon black as a co-product.

A detailed SMR-ammonia production plant was simulated using Aspen Plus v11 software. The MP unit was then integrated into SMR by splitting the natural gas feed after the sulfur removal unit. Several schemes were conducted to find the most preferable scenario for the industries. The schemes vary in the splitting ratio of natural gas feed to SMR and MP. Techno-economic analysis from this study demonstrates favorable results when incorporating methane pyrolysis into current SMR-ammonia-urea plants. This includes a significant reduction in CO_2 emissions by 71.70% compared to the base case. In a scenario with a carbon black revenue of $0.5/kg and a carbon tax of $20/ton-$CO_2$e, diverting 15% of the natural gas feed into a methane pyrolysis integration unit could yield an extra annual profit of up to $1.6 million.

Keywords: Ammonia, CO_2 emission, Methane pyrolysis, SMR, Urea.

1. Introduction

Urea is the most widely used nitrogen fertilizer globally. The increasing global population is expected to drive a surge in fertilizer demand in the future. The International Fertilizer Association (IFA) predicts that in 2027, nitrogen-based fertilizer demand will reach 115 million tons or increase by around 9% from 2022 (International Fertilizer Association, 2023). However, traditional ammonia-urea production methods, such as steam methane reforming (SMR) are known as one of the main contributors to greenhouse gas (GHG) emissions. SMR is a highly energy-intensive process that uses natural gas as a raw material to produce hydrogen, which is then used to synthesize ammonia (Pruvost et al., 2022). Ammonia then could be converted to urea.

In recent years, there has been growing interest in developing more sustainable urea production methods. One promising approach is to integrate methane pyrolysis (MP) into the existing SMR-Ammonia-Urea process plant. Methane pyrolysis is a thermochemical process that converts methane into hydrogen and solid carbon in the temperature range of $600 – 1,400\ ^{\circ}C$. The hydrogen produced from methane pyrolysis can synthesize ammonia, while the solid carbon can be sold as a valuable byproduct (Fromm, 2023). The integration of methane pyrolysis with SMR offers several potential benefits, such as (i) reduced CO_2 emissions: unlike SMR, MP doesn't have CO_2 as a byproduct; (ii) increased production flexibility: The integrated system can continuously produce ammonia and urea, while also generating marketable solid carbon; (iii) improved economic performance. The sale of solid carbon can offset the cost of methane pyrolysis, making the integrated system more economically viable.

Numerous recent studies indicate that the MP reaction can occur both with and without a catalyst. When employing catalysts, the fluidized bed reactor (FBR) emerges as the optimal option for solid catalysts, typically those based on Nickel, while liquid metal bubble column (LMBC) reactors find application for liquid catalysts, including metals like Ni-Bi or combinations with salts such as KBr (Fan et al., 2021). In the case of non-catalytic reactions, Monolith's plasma reactor has successfully reached an early commercial scale. It is crucial to note that variations in temperature and catalyst choice lead to distinct types of solid carbon co-products (Korányi et al., 2022).

Methane pyrolysis (MP) is a developing technology that has nearly reached a large-scale commercial production plant. However, integrating SMR with MP in an ammonia-urea plant is a new research area. This study will focus on developing the process design for the integration, whereas there are numerous possibilities regarding the splitting of natural gas feed, utilizing the gas product of the MP reactor, including the heat integration. A detailed simulation model of the integrated system is developed using Aspen Plus software. The simulation results are analyzed to assess the environmental and economic performance of the system.

2. Process description

2.1. Conventional SMR-Ammonia-Urea unit (base case)
Natural gas undergoes pretreatment to eliminate sulfur impurities. The purified gas is then compressed and combined with steam in a certain ratio before being injected into the primary reformer. The steam methane reforming (SMR) process typically employs a nickel catalyst and operates within a temperature range of $800 – 1,000\ °C$ (Katebah et al., 2022). The SMR reaction is followed by the water-gas shift (WGS) reaction, as shown in Reactions (1) and (2).

$$CH_4 + H_2O \rightarrow 3H_2 + CO \tag{1}$$

$$CO + H_2O \rightarrow H_2 + CO_2 \tag{2}$$

$$3H_2 + N_2 \rightarrow 2NH_3 \tag{3}$$

$$2NH_3 + CO_2 \rightarrow CO(NH_2)_2 + H_2O \tag{4}$$

Reaction (1) + (2) yields hydrogen and carbon dioxide, collectively termed synthesis gas or syngas. The syngas, along with any remaining unreacted gases are subsequently mixed with air to produce ammonia through ammonia synthesis as shown in the Reaction (3). The ammonia synthesis reaction is also endothermic and requires the presence of a catalyst. This process is widely recognized as Haber-Bosch (Ojelade et al., 2023).

Approximately 98% of the produced ammonia and almost all the carbon dioxide co-product are diverted to the urea plant for urea production via an exothermic reaction (4). Urea synthesis demands a high-pressure reactor (150-200 bar) and a temperature range of $180 - 200$ °C. Figure 1 provides an overall block diagram of this production process.

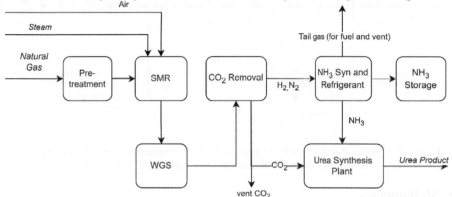

Figure 1. SMR-Ammonia-Urea production overview.

Typically, industrial ammonia and urea plants are designed to operate at full capacity, ensuring a well-balanced production process that prevents the release of CO_2 from the ammonia plant into the atmosphere. These plants are engineered to fully convert the CO_2 co-product from the ammonia plant into urea in the urea plant, with only a minimal amount being vented for safety purposes. However, operational challenges often hinder the achievement of perfect balance within these plants.

Moreover, it is common for companies with ammonia and urea plants also to sell liquid ammonia to their customers. Given that excess ammonia is stored in ammonia storage, retaining it without sales is impractical. During periods of heightened ammonia sales and dwindling storage levels, it becomes necessary to increase the ammonia plant's load or maintain it at full capacity, while simultaneously decreasing the load on the urea plant to replenish ammonia storage levels. Consequently, $15 - 25\%$ of CO_2 can be released into the atmosphere during operational adjustments (SABIC Agri-Nutrients Company, 2022).

2.2. Methane Pyrolysis and SMR Integration (proposed design)

Methane pyrolysis technology could be a promising solution for balancing ammonia and urea production without worrying about carbon dioxide venting. Methane pyrolysis converts methane into hydrogen gas and solid carbon with or without the help of a catalyst (Fromm, 2023). The methane pyrolysis reaction is shown in reaction (5).

$$CH_4 \rightarrow 2H_2 + C(s) \tag{5}$$

Integrating methane pyrolysis (MP) with steam methane reforming (SMR) process would result in the controlling of hydrogen and carbon dioxide products. Additionally, methane pyrolysis gives valuable solid carbon co-products that can be sold to reduce the overall ammonia and urea production cost. The integration process block diagram is shown in Figure 2.

Techno-economic analysis was conducted for this integration. Detailed methodology and results are discussed in the section below.

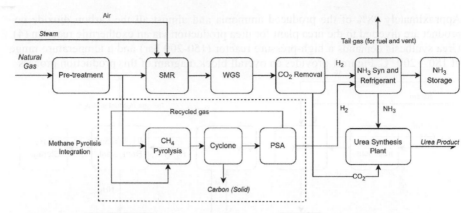

Figure 2. Methane Pyrolysis (MP) integration in SMR-Ammonia-Urea production. Splitting natural gas feed to SMR and MP will control the CO_2 emission.

3. Methodology

The techno-economic analysis was conducted through simulation using Aspen Plus software v11 and the Aspen Process Economic Analyzer (APEA) to evaluate the feasibility and economic viability of the proposed methane pyrolysis integration in the SMR-ammonia-urea production process.

3.1. Model Basis and Assumptions
The SMR-ammonia plant model was built with a capacity design of 1,000 t/d ammonia production from 21,540 kg/h natural gas feed then the methane pyrolysis unit was integrated into the same flowsheet. Several assumptions are taken into consideration:

- The model is under steady-state conditions.
- The properties method used is RKS-BM for ammonia and methane pyrolysis simulation while for the urea plant using SR-POLAR properties.
- Natural gas (NG) feed pressure is 11 bar and temperature 32 °C with composition in dry mol basis: $CH_4 = 91\%$; $C_2H_6 = 2.88\%$; $N_2 = 2.54\%$; $CO_2 = 1.83\%$ and a small amount of C3, C4, and sulfur.
- Air composition was fixed at 78% mol of N_2, 21% mol O_2 and 1% mol argon.
- The 1,725 t/d urea plant was modeled employing Aspen Plus.
- The reference plant was assumed to only utilize 85% of CO_2 from the ammonia plant to be converted into urea product (SABIC Agri-Nutrients Company, 2022).
- Solid carbon coproduct from methane pyrolysis is carbon black.

3.2. Economic Analysis
Capital cost was annualized with operational cost (including raw material and utility costs) in every scheme. It was assumed that the natural gas price was $ 6.00/MMBTU; ammonia and urea selling prices were $550/ton and $390/ton respectively. Profit estimation was conducted using a carbon black revenue assumption of $0.5/kg (market value range of 0.4 – 2.0 $/kg) and a carbon tax policy of $20/ton-$CO_2$e. It also assumed that the operation days in a year is 330 days.

4. Results and Discussion

The integrated simulation of methane pyrolysis and the SMR-ammonia-urea plant was successfully conducted and converged in Aspen Plus v11. Production and emission comparisons from several schemes are shown in Figure 3.

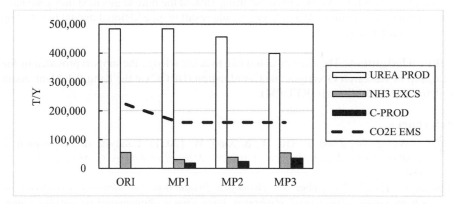

Figure 3. Production and emission outcomes among various schemes: the baseline plant without integration (ORI), MP1 means 15% integration, MP2 is 20%, and MP3 is 30% integration.

All the methane pyrolysis (MP) integration schemes (MP1 to MP3) could reduce the CO_2 emission to around 71.70% (160 kt-CO_2/y) compared to the base case (ORI) emission (223 kt-CO_2/y). The reduced annual CO_2 (160 kt/y) only came from flue gas emissions. MP1 with 15% integration or 15% of natural gas feed directed to MP will result in similar urea production to the base case with additional solid carbon black product (C-PROD) but less ammonia excess. MP2 and MP3 with a higher integration scheme (20% and 30%) will give less urea and less ammonia compared to the base case, but they will give more carbon black products.

Profit estimation and comparison for each scheme are shown in Table 1. The base case shows a good profit revenue with a CO_2 tax policy of $20/ton. However, the MP1 scheme shows the highest profit. The MP1 scheme with 15% natural gas splitting promises an additional profit of 1.6 million USD annually than the base case (ORI). A higher integration scheme (e.g., MP3) will produce more carbon black coproduct but produce less urea. This scheme gained the lowest profit for the assumed product prices scenario but could become the highest profit scheme if the carbon black price increases.

Table 1 Economic comparison for different integration schemes with a solid carbon revenue of $0.5/kg and a CO_2 tax assumption of $20/ton

	AN.OP.COST	**UREA PROD**	**NH3 EXCS**	**CO2E EMS**	**C-PROD**	**PROFIT EST**
	$/Y	*T/Y*	*T/Y*	*T/Y*	*T/Y*	*$/Y*
ORI	107,988,109	483,863	55,166	223,171	0	90,880,419
MP1	110,506,513	483,863	30,416	159,984	17,820	**92,536,470**
MP2	109,122,880	455,400	38,333	159,984	23,760	88,134,654
MP3	106,360,780	398,475	54,166	159,984	35,640	79,474,222

The results also reveal that the annualized operational cost for integration schemes MP1 and MP2 was slightly higher than ORI due to the additional utilities and maintenance costs for the methane pyrolysis unit. However, in the MP3, the operational cost is lower due to the lower load of the urea plant (less energy consumption).

5. Conclusions

Based on the techno-economic analysis conducted, methane pyrolysis integration in existing SMR-ammonia-urea plants shows promising results for lowering CO_2 emissions to 71.70% compared to the existing plant. For a carbon black revenue of $0.5/kg and a carbon tax of $20/ton-$CO_2$e scenario, splitting 15% of the natural gas feed into a methane pyrolysis integration unit (MP1 scheme) would result in an additional annual profit of up to $1.6 million per year.

Acknowledgment: The authors would like to acknowledge the support provided by the Deanship of Research Oversight and Coordination (DROC) at the King Fahd University of Petroleum and Minerals (KFUPM).

References

Fan, Z., Weng, W., Zhou, J., Gu, D., & Xiao, W. (2021). Catalytic decomposition of methane to produce hydrogen: A review. *Journal of Energy Chemistry*, *58*, 415–430.

Fromm, C. (2023, November 1). *Hydrogen Production via Methane Pyrolysis: An Overview of "Turquoise" Hydrogen*. https://www.chemengonline.com/hydrogen-production-via-methane-pyrolysis-an-overview-of-turquoise-h2/?printmode=1

International Fertilizer Association. (2023). *Public Summary Medium-Term Fertilizer Outlook 2023-2027*. https://api.ifastat.org/reports/download/13912

Katebah, M., Al-Rawashdeh, M., & Linke, P. (2022). Analysis of hydrogen production costs in Steam-Methane Reforming considering integration with electrolysis and CO2 capture. *Cleaner Engineering and Technology*, *10*, 100552.

Korányi, T. I., Németh, M., Beck, A., & Horváth, A. (2022). Recent Advances in Methane Pyrolysis: Turquoise Hydrogen with Solid Carbon Production. In *Energies* (Vol. 15, Issue 17). MDPI.

Ojelade, O. A., Zaman, S. F., & Ni, B. J. (2023). Green ammonia production technologies: A review of practical progress. In *Journal of Environmental Management* (Vol. 342). Academic Press.

Pruvost, F., Cloete, S., Arnaiz del Pozo, C., & Zaabout, A. (2022). Blue, green, and turquoise pathways for minimizing hydrogen production costs from steam methane reforming with CO2 capture. *Energy Conversion and Management*, *274*, 116458.

SABIC Agri-Nutrients Company. (2022). *Performance and Activities of the Company for the Fiscal Year 2021*. https://www.sabic.com/en/investors/performance-financial-highlights/annual-report

Flavio Manenti, Gintaras V. Reklaitis (Eds.), Proceedings of the 34th European Symposium on Computer Aided Process Engineering / 15th International Symposium on Process Systems Engineering (ESCAPE34/PSE24), June 2-6, 2024, Florence, Italy

Modelling study of CO₂ sequestration by mineral waste carbonation process

Natalia Vidal de la Peña[a], Dominique Toye[a], Grégoire Léonard[a],

aChemical Engineering, University of Liège, B6a Sart-Tilman, 4000 Liège, Belgium
nvidal@uliege.be

Abstract

The objective of this work is to develop a COMSOL Multiphysics numerical model to represent the physico-chemical phenomena associated to CO_2 sequestration by mineral waste carbonation process. In the model, the bed of particles is described as a dual-scale porosity medium. The two scales considered are the intragranular and intergranular scales. The model is formulated to predict phenomena occurring in the dual-scale system, considering the diffusion of CO_2 between particles and within them. This involves the characterization of crucial parameters such as the porosity and the amount of water present both inside and outside the solid particles. The proposed model is qualitatively validated against existing literature and quantitatively validated against experimental results. This article presents the results obtained for the $Ca(OH)_2$ carbonation during three days, emphasizing the significance of the COMSOL software in comprehending the process thoroughly.

Keywords: Mineral carbonation, multi-scale modelling, CO_2 capture, CO_2 sequestration

1. Introduction

Nowadays, the continuous pursuit of human well-being and the improvement of quality-of-life lead to the increasingly high consumption of natural resources, reaching depletion and generating waste and pollutants that are difficult to manage. This is evident in the construction industry, as it is one of the main contributors to CO_2 emissions, producing around 10 Gt of CO_2 in 2022, as presented by the United Nations Environment Program (2022). In addition to CO_2, this industry significantly contributes to waste streams, accounting for 37.5% of the total European waste in 2020, as reported by Eurostat (2020). Taking in consideration the increasing trend of the presence of CO_2 in the atmosphere, the EU established a Green Deal to achieve carbon neutrality by 2050 and a reduction of 55% in the CO_2 concentration by 2030. To this end, the EU is focusing on improving energy efficiency, renewable energy and the CCUS (Carbon Capture, Utilization and Storage) technologies. Regarding this goal, it is necessary to find solutions that can help the reduction of emissions in the construction sector.

As a response to this situation, the capture of CO_2 is proposed through the carbonation of hydrated lime, the symbol of which is "$Ca(OH)_2$" in the chemical engineering sector and "CH" in the construction one. The hydrated lime has been chosen due to its presence in the recycled concrete aggregates (RCA) which are nowadays getting more importance to mitigate the consumption of natural resources of the concrete industry (Vidal de la Peña et al., 2023). In this case, the residues are collected, treated and then, carbonated so the carbonation of construction waste is made ex-situ. The amount of carbonatable components will define the carbonation potential of the material. In case of need, the efficiency of the reaction can be increased by applying some pre-treatments to the material, such as the decreasing of the particle size distribution or the optimization of the relative humidity in the system. During this process, the highest energy consumption phases are the obtention of the carbonatable material by separation processes, if needed, and the preparation of the material by crushing, among others (Vidal de la Peña et al., 2023).

Ex-situ mineral carbonation can be conducted both directly and indirectly. This article focuses on the gas-solid direct carbonation which consists of the reaction between carbonatable materials with CO_2 in a single stage, avoiding the several stages in the indirect carbonation processes (Vidal de la Peña et al., 2023). As the process allows the simultaneous recycling and valorization of two waste streams, i.e., CO_2 and $Ca(OH)_2$, it is of direct interest for the construction sector to strengthen its contribution to a more sustainable circular economy. In this work, a COMSOL Multiphysics numerical model is proposed to explain and optimize a direct carbonation process of construction materials and wastes.

2. Mathematical description of the model

The carbonation process has been simulated in a 2D axisymmetric cylindrical reactor (length = 70 mm – diameter 70 mm), filled with spherical $Ca(OH)_2$ particles with 1 mm of particle diameter, arranged in linear rows configuration (Figure 1a). On the reactor conceptual design implemented in COMSOL (Figure 1b), CO_2 is introduced into the reactor from the bottom (x=0), with the maximum concentration of CO_2 at the reactor inlet.

Besides this bottom entry flux, the flux conditions on the other boundaries are "no flux". The reactor top side and lateral wall (right side of the scheme) are hermetic, so there is no outgoing CO_2 flux. The left boundary on the scheme corresponds to the central axis of the cylindrical bed, where the symmetry condition imposes a null flux. Furthermore, the initial concentration of CO_2 at the bottom of the reactor is 4 mol/m^3 during the whole process, equivalent to a 10% of molar-volume fraction. The initial conditions in the reactor are uniform: zero concentration of CO_2. The $Ca(OH)_2$ concentration in the particles equals to 4000 mol/m^3, which leads to $Ca(OH)_2$ concentration of 2400 mol/m^3 at the reactor scale if the bed porosity (ϕ_{bed} = 0.4) is considered.

Carbonation is a complex process that takes place in a heterogeneous (gas-solid) system with two levels of porosity, the results of which depend not only on the chemical reaction between $Ca(OH)_2$ and CO_2, but also on the coupling of this reaction with various physical and chemical phenomena occurring at different scales.

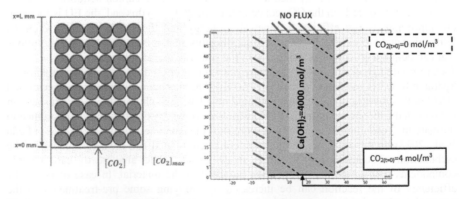

Figure 1: a. Scheme of the principle of the granular bed configuration. b. Axisymmetric reactor and conditions modelled in COMSOL.

Basically, the system is described by the following parameters: the porosity of the bed, which is defined by the bulk density of the material and the particle density, the particle porosity that is calculated by the bulk density and the porous volume, the water content present along the granular bed and the water content present in the particles. These parameters influence diffusion phenomena both along the bed and inside the particles. Higher porosity (both bed and particle) will enhance the diffusion phenomenon, while higher water content (both in the bed and in the particles) will worsen the diffusion process. In contrast to the diffusion process, chemical reactions occur only inside the solid particles, and they are influenced by the water-to-solid mass ratio inside them. The chemical reaction considered in this model is represented by the following equation, the rate of which is determined using a second-order mass action law.

$$Ca(OH)_2(s) + CO_2(g) => CaCO_3(s) + H_2O(l) \tag{1}$$

From the reaction, water is released, being a critical aspect due to its substantial impact on the system. Initially, the material must contain a minimum amount of water to initiate the carbonation reaction; without sufficient humidity, no reaction occurs.

On the contrary, an excess of water in the system can lead to material agglomerations, pore clogging, impeding the entry of CO₂ and thereby reducing system efficiency. The quantity of water significantly influences both the reaction kinetics and the diffusion of CO₂ inside the particles and into the granular bed, i.e., between the particles. Given its significance, characterizing this parameter is crucial. So, its influence is incorporated into the calculation of the kinetic constant in our model as well as in the diffusion coefficient(D_p) in the particles. Therefore, the influence of water quantity is introduced in the rate law as expressed in the following equation.

$$R = k_0 \cdot [CO_2] \cdot [Ca(OH)_2] \cdot sl \tag{2}$$

Where k_0 represents the kinetic constant with a value equal to 0.00025 mol/m³s (L.Reich et al., 2014), $[CO_2]$ denotes the concentration of carbon dioxide in mol/m³, $[Ca(OH)_2]$ is the concentration of Ca(OH)₂ in mol/m³, and sl is the water-to-solid mass ratio in the particle. As mentioned earlier, this parameter promotes the reaction as it is necessary to initiate it, but its effect becomes negative after reaching a certain maximum value. To incorporate this influence into the reaction rate, specific conditions have been imposed. If sl is less than 0.05, its value is corrected to 0. If it is between 0.05 and 0.9, it is taken as it is. And if sl is equal or greater than 0.9, its value is equal to 1. In this work, the initial value of sl considered is 0.1.

The diffusion of CO₂ in the granular bed (between particles) will be affected by the bed porosity (ϕ_{bed}) and by the bed water-to-solid mass ratio (S_l). The equation to determine the bed diffusion ($\mathbb{D}$) is based on the Belin et al.(2013) work and is described as follows.

$$\mathbb{D} = D_{CO_2}^0 \cdot \phi_{bed}^{4/3} \cdot (1 - S_l)^{10/3} \tag{3}$$

Where $D_{CO_2}^0$ is the specific diffusion coefficient of CO₂ at a pressure of 0.1 MPa and a temperature of 273 K, with a value of 1.6E-5 m²·s⁻¹. The correction term used to account for porosity and water content effects is calculated using the Millington's correlation equation (Belin et al., 2013). The bed porosity ϕ_{bed} is taken equal to 0.4 and remains constant. The water saturation around the particles, S_l, is assumed to be zero, as the presence of a water film around the particles can be neglected due to their small size (less than 1 mm) (Belin et al., 2013).

On the other hand, the diffusion coefficient inside the particles is calculated as follows.

$$D_p = D_{CO_2}^0 \cdot \Phi_p^{2.7} \cdot (1 - sl)^{4.2} \tag{4}$$

The correction factor for porosity and water content is calculated using the expression of Thiery (2005). Φ_p represents the particle porosity. sl denotes the water-to-solid mass ratio. As the carbonation process progresses, the $Ca(OH)_2$ is progressively converted to $CaCO_3$, the molar volume of which (calculated as $\frac{MM_i}{\rho_i}$) is larger. If a constant total solid volume is considered, the carbonation process leads to a reduction of particle porosity. The porosity evolution may be related to the degree of carbonation of the particles (η_p), using the following expression.

$$\Phi_p(t) = \Phi_{p,t=0} \cdot \left[(1 - \eta_p) + \eta_p \cdot \frac{MM_{Ca(OH)_2}}{MM_{CaCO_3}} \cdot \frac{\rho_{CaCO_3}}{\rho_{Ca(OH)_2}} \right] \tag{5}$$

Where $\Phi_{p_{t=0}}$ is the initial porosity of the particles, $MM_{Ca(OH)_2}$ and MM_{CaCO_3} are the molar masses of $Ca(OH)_2$ and $CaCO_3$ equal to 74 g/mol and 100 g/mol, respectively, ρ_{CaCO_3} and $\rho_{Ca(OH)_2}$ are the densities of $Ca(OH)_2$ and $CaCO_3$, equal to 2.21 g/cm^3 and 2.71 g/cm^3. η_p is the degree of carbonation of the particles, calculated as follows.

$$\eta_p = \frac{[CaCO_3]_p}{[Ca(OH)_2]_{p,t=0}} \tag{6}$$

A complete carbonation ($\eta_p=1$) corresponds to a 10% relative variation in porosity. As computed in equation (6), η_p compares the molar concentration of $CaCO_3$ inside particles ($[CaCO_3]_p$) to the maximum concentration of $CaCO_3$ that can theoretically be reached, equal to the initial molar concentration of $Ca(OH)_2$ inside particles, $[Ca(OH)_2]_{p,t=0}$. η_p values were computed from concentration profiles simulated in COMSOL. They were then compared to experimental data.

3. Results and discussion

After introducing the equations and correlations in the model, concentration profiles were simulated for different reaction times. The temperature and pressure were kept constant at 273 K and atmospheric pressure. The objective was to get a better understanding of the carbonation process as a whole and to analyze how the particle porosity and water-to-solid mass ratio evolve and impact the process.

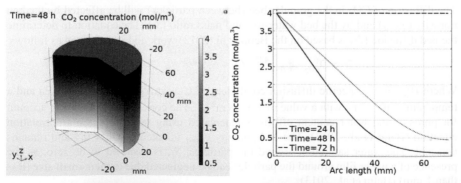

Figure 2: a. CO$_2$ gradient concentration along the 3D reactor. b. CO$_2$ concentration along the reactor after 24 h, 48 h and 72 h of carbonation.

Figure 2a shows the 3D distribution of CO_2 concentration in the reactor after 48 h. Figure 2b presents the CO_2 concentration vertical profiles along the granular bed (from bottom to top) at different times. Figure 2a shows that CO_2 enters at the bottom of the reactor and penetrates gradually. After 48 hours, the reactor is not yet filled with CO_2. Figure 2b shows the CO_2 concentration in different reactor cross-sections after one, two and three days. Interestingly, after 72 hours, the reactor is completely filled with CO_2.

Complementary to Figure 2, Figure 3 shows the $CaCO_3$ concentration distribution in the reactor. As above, Figure 3a shows the 3D distribution at a given time (48 hours) and Figure 3b the 2D vertical profiles at different times (24, 48, and 72 hours). Logically, the bottom of the reactor exhibits a higher concentration of carbonated material compared to the top. These results confirm those presented in Figure 2.

Regarding the maximal concentration of $CaCO_3$ at the bed scale, it logically equals the initial bed scale concentration of $Ca(OH)_2$ equal to 2400 mol/m³. If these concentrations are related to the solid particle volume, one obtains 4000 mol/m³ solid for both. After 72 h, the carbonation process is completed: the maximal carbonate calcium concentration is achieved in all the reactor and the latter is filled with CO_2 in excess.

The evolution of the water-to-solid mass ratio and of the particle porosity throughout the carbonation process are also accounted for in the model. The initial value of the water-to-solid mass ratio considered is equal to 0.1 and the initial particle porosity is 0.49.

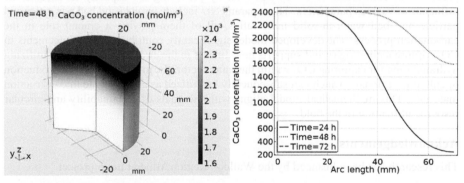

Figure 3: a. $CaCO_3$ gradient concentration along the 3D reactor. b. $CaCO_3$ concentration along the reactor after 24 h, 48 h and 72 h of carbonation.

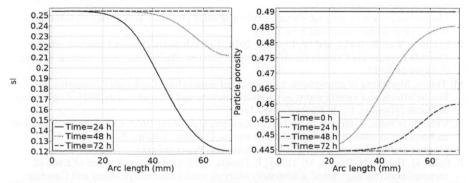

Figure 4: a. Water-to-solid mass ratio variation along the reactor during the carbonation process. b. Particle porosity variation along the reactor during the carbonation process

Figures 4a and 4b show the evolution, during the carbonation process, of the water-to-solid ratio (sl) and of the particle porosity (Φ_p) profiles, respectively. Concerning the water content, Figure 4a shows that it increases at the same time as $Ca(OH)_2$ is consumed and $CaCO_3$ is produced (Figures 2 and 3), due to the generation of H_2O by the carbonation reaction. This parameter reaches a maximum value of 0.25.

Concerning the particle porosity, Figure 4b shows that it decreases as the solid is carbonated. The porosity decreases from 49% to 44.5%, which corresponds to the 10% relative variation predicted by Equation (5). The model enables the observation of how this parameter varies across different layers of the reactor, reaching its minimum value throughout the entire reactor after 3 days of carbonation.

Finally, these simulation results were compared with experimental results obtained by the company Centre Terre et Pierre (CTP) within the framework of the Mineral Loop project financed by the Walloon Region. The agreement is quite satisfying as the gap between simulations and experiments is smaller than 20%, if one uses the total mass carbon ratio as comparison criterion.

4. Conclusions

In conclusion, this study facilitates a comprehensive understanding of the carbonation process and contributes to its optimisation by providing the capacity to analyse the influence of various parameters. These parameters include particle and bed porosities and particle and bed water-to-solid mass ratios, all of them playing a crucial role in the carbonation process. The developed model significantly contributes to advancements in CO_2 capture technologies and supports investments in the circular economy. By utilizing hydrated lime for capturing CO_2 within the same sector, the model enables the production of calcium carbonate, which can subsequently be utilized as an aggregate in construction materials. This mathematical model aligns with the goals of sustainability and circular resource utilization in the field.

Acknowledgements

This research has been financed by the Walloon Region Mineral Loop project.

References

Eurostat, Waste Statistics for EU.
    <https://ec.europa.eu/eurostat/statisticsexplained/index.php?title=Waste_statistics#Hazardous_waste_generation>, (accessed 1/12/2023).
L. Reich, L. Yue, R. Bader, and W. Lipiński, 2014, Towards Solar Thermochemical Carbon Dioxide Capture via Calcium Oxide Looping: A Review, Aerosol Air Qual, Volume 4, Issue 2, pages 500-514.
M. Thiery, Modelling of atmospheric carbonation of cement based materials considering the kinetic effects and modifications of the microstructure and the hydric state,2005, Engineering Sciences [physics], Ecole des Ponts ParisTech
N. Vidal de le Peña, S. Grigolleto, D.Toye, L.Courard, G.Léonard, 2023, CO_2 Capture by mineral carbonation of construction and industrial wastes, Circular Economy Processes for CO_2 Capture and Utilization: Strategies and case studies, Pages 163-185
P. Belin, G. Habert, N. Rousse, M. Thiery, P. Dangla, 2013, Carbonation kinetics of a bed of recycled concrete aggregates: a laboratory study on model materials, Cement and Concrete Research,Volume 46, Pages 50-65
United Nations Environment Programme, 2022, Global Status Report for Buildings and Construction: Towards a Zero-emission, Efficient and Resilient Buildings and Construction Sector, Nairobi

Flavio Manenti, Gintaras V. Reklaitis (Eds.), Proceedings of the 34th European Symposium on Computer Aided Process Engineering / 15th International Symposium on Process Systems Engineering (ESCAPE34/PSE24), June 2-6, 2024, Florence, Italy

Evaluation of Two-Dimensional Pseudo-Homogeneous and Heterogeneous Modeling Approaches in Steam-Methane Reforming Reactors

Chengtian Cui [a], Yufei Zhao [a], Cornelius Masuku [a, *]

[a] Davidson School of Chemical Engineering, Purdue University, West Lafayette, IN 47907, United States
cmasuku@purdue.edu

Abstract

Steam-methane reforming (SMR) serves as a key method for hydrogen production, using natural gas as the primary feedstock. Traditionally, the simulation of this process has relied on one-dimensional (1D) models, which come in two forms: pseudo-homogeneous or heterogeneous. These models, while useful, neglect the radial gradients in temperature and concentration by assuming perfect radial mixing within the reactor. In contrast, two-dimensional (2D) models offer a detailed view of the reactor radial behavior by incorporating terms for mass dispersion and thermal conduction. When modeling SMR packed-bed reactor, one can opt for a pseudo-homogeneous approach, which employs an effectiveness factor to consider diffusion resistances, or a heterogeneous approach, which calculates diffusion within the catalyst particles by adding a catalyst domain. This work presents a case study of an SMR reactor to evaluate these modeling approaches. Steady-state analysis is given to highlight the distinct outcomes produced by each method.

Keywords: Steam Reforming, Hydrogen Production, Dynamic Modeling, Reaction Engineering.

1. Introduction

Hydrogen is a crucial raw material for numerous industrial sectors, including petroleum refinery, methanol, and ammonia production. Currently, more than 40% of hydrogen is produced via steam-methane reforming (SMR). The SMR reaction proceeds in the following three reaction steps: (I) and (II) show methane reforming, while (III) is the water gas shift reaction.

$$CH_4 + H_2O = CO + 3H_2 \quad \Delta H_{298} = 206.1 \; kJ/mol \tag{I}$$

$$CH_4 + 2H_2O = CO_2 + 4H_2 \quad \Delta H_{298} = 164.9 \; kJ/mol \tag{II}$$

$$CO + H_2O = CO_2 + H_2 \quad \Delta H_{298} = -41.15 \; kJ/mol \tag{III}$$

These methane reforming reactions are notably endothermic and equilibrium limited. Predominantly, SMR is operated at high-pressure and high-temperature tubular reactors packed with catalyst active for reforming reactions (Kuncharam and Dixon, 2020). The modeling of SMR packed-bed reactors varies in complexity. For the entire reformer flowsheet design, often a simplified one-dimensional (1D) representation of the reactor tubes is used (Rout and Jakobsen, 2015). However, due to the considerable radial temperature gradients, which can surpass 80°C as a result of intense heat transfer through the tube walls (Wesenberg and Svendsen, 2007), it becomes essential to employ two-dimensional (2D) models to accurately depict the radial temperature profile.

Regarding modeling approaches, packed-bed reactor models for the SMR process are typically divided into pseudo-homogeneous and heterogeneous models. The pseudo-homogeneous models disregard any limitations caused by intraparticle diffusion as well as resistances to interparticle mass and heat transfer. If diffusion limitations are significant, they are accounted for by incorporating a constant effectiveness factor (Lao et al., 2016). On the other hand, heterogeneous models take a detailed approach by including the physical presence of solid catalyst and directly assess diffusion limitations within these catalysts, thus eliminating the need for an effectiveness factor (Cruz et al., 2017).

In this study, we examine both 2D pseudo-homogeneous and 2D heterogeneous models in the context of a standard SMR process. We present steady-state results to underscore the unique results obtained from each modeling technique.

2. SMR models

2.1. 2D pseudo-homogeneous model

Xu and Froment (1989) derived the intrinsic kinetics of the endothermic SMR reactions based on Langmuir-Hinshelwood approach. The intrinsic kinetics are well-known and not presented here. The 2D pseudo-homogeneous model includes mass, energy, momentum balances as well as reaction kinetics (Xing et al., 2021). The main governing equations are listed below while the reaction kinetics are given in the next section:

$$\frac{\partial (C_i U_z)}{\partial z} = \varepsilon_b D_{i,z}^e \frac{\partial^2 C_i}{\partial z^2} + \varepsilon_b D_{i,r}^e \left(\frac{\partial^2 C_i}{\partial r^2} + \frac{1}{r} \frac{\partial C_i}{\partial r} \right) + (1 - \varepsilon_b) \rho_p \sum \eta_j R_j \nu_{ij} \qquad (1)$$

$$\rho_g C_{p,g} U_z \frac{\partial T}{\partial z} = \lambda_z \frac{\partial^2 T}{\partial z^2} + \lambda_r \left(\frac{\partial^2 T}{\partial r^2} + \frac{1}{r} \frac{\partial T}{\partial r} \right) + (1 - \varepsilon_b) \rho_p \cdot \sum (-\Delta H_j) \eta_j R_j \qquad (2)$$

$$\frac{\partial P}{\partial z} = -\frac{G}{\rho_g d_p} \left(\frac{1 - \varepsilon_b}{\varepsilon_b^3} \right) \left[\frac{150(1 - \varepsilon_b) \mu_g}{d_p} + 1.75G \right] \qquad (3)$$

in which C_i is the concentration of component i, U_z the superficial velocity, ε_b the bed porosity, D_i^e the effective mass dispersion coefficient of component i, η the effectiveness factor, ρ the density, R_j the reaction rate of reaction j, ν_{ij} the stoichiometric coefficient of component i in reaction j, T the temperature, P the pressure, d_p the catalyst diameter, μ_g the viscosity of gas, and G the superficial mass velocity. It is considered that the SMR reactor has a length of L, tube radius of r_t, and wall temperature of T_w. The boundary conditions are:

$$C_i(0, r) = C_{i,0}; \quad \frac{\partial C_i(L, r)}{\partial z} = 0; \quad \frac{\partial C_i(z, 0)}{\partial r} = 0; \quad \frac{\partial C_i(z, r_t)}{\partial r} = 0 \qquad (4)$$

$$T(0, r) = T_0; \quad \frac{\partial T(L, r)}{\partial z} = 0; \quad \frac{\partial T(z, 0)}{\partial r} = 0; \quad \lambda_r \frac{\partial T(z, r_t)}{\partial r} = U(T_w - T) \qquad (5)$$

$$P(0, r) = P_0; \quad \frac{\partial P(L, r)}{\partial z} = 0; \quad \frac{\partial P(z, 0)}{\partial r} = 0; \quad \frac{\partial P(z, r_t)}{\partial r} = 0 \qquad (6)$$

2.2. 2D heterogeneous model

2D heterogeneous model includes separate mass and energy transport equations for the reactor domain and the catalyst domain. These two domains are then coupled together by incorporating mass and energy boundary conditions (Ghouse and Adams, 2013). The pressure drop equation is the same to the pseudo-homogeneous model, while the reactor domain mass and energy equations are listed as follows:

$$\frac{\partial(C_i U_z)}{\partial z} = \varepsilon_b D_{i,z}^e \frac{\partial^2 C_i}{\partial z^2} + \varepsilon_b D_{i,r}^e \left(\frac{\partial^2 C_i}{\partial r^2} + \frac{1}{r}\frac{\partial C_i}{\partial r}\right) + a_v k_i (C_{i,p} - C_i) \tag{7}$$

$$\rho_g C_{p,g} U_z \frac{\partial T}{\partial z} = \lambda_z \frac{\partial^2 T}{\partial z^2} + \lambda_r \left(\frac{\partial^2 T}{\partial r^2} + \frac{1}{r}\frac{\partial T}{\partial r}\right) + a_v h(T_p - T) \tag{8}$$

in which a_v is external surface area per unit volume of catalyst bed, k_i the mass transfer coefficient between fluid and solid, and h the heat transfer coefficient between fluid and solid. The boundary conditions of the heterogenous model in reactor domain are same to that of the pseudo-homogeneous model. For the catalyst domain, we consider a spherical particle. So, the mass and energy balances are:

$$\varepsilon_p D_{i,\xi}^e \left(\frac{\partial^2 C_{i,p}}{\partial \xi^2} + \frac{2}{\xi}\frac{\partial C_{i,p}}{\partial \xi}\right) + (1 - \varepsilon_p)\rho_p \sum R_j v_{ij} = 0 \tag{9}$$

$$\lambda_\xi \left(\frac{\partial^2 T_p}{\partial \xi^2} + \frac{2}{\xi}\frac{\partial T_p}{\partial \xi}\right) + (1 - \varepsilon_p)\rho_p \cdot \sum (-\Delta H_j) R_j = 0 \tag{10}$$

in which ε_p is catalyst porosity, $C_{i,p}$ component concentration of i in catalyst particle, and T_p the temperature in catalyst particle.

$$C_{i,p}(0,r,\xi) = C_{i,0}; \quad \frac{\partial C_{i,p}(L,r,\xi)}{\partial z} = 0; \quad \frac{\partial C_{i,p}(z,0,\xi)}{\partial r} = 0; \quad \frac{\partial C_{i,p}(z,r_t,\xi)}{\partial r} = 0 \tag{11}$$

$$\frac{\partial C_{i,p}(z,r,0)}{\partial \xi} = 0; \quad -\varepsilon_p D_{i,\xi}^e \frac{\partial C_{i,p}(z,r,r_p)}{\partial \xi} = k_i [C_{i,p}(z,r,r_p) - C_i(z,r)] \tag{12}$$

$$T_p(0,r,\xi) = T_0; \quad \frac{\partial T_p(L,r,\xi)}{\partial z} = 0; \quad T_p(z,0,\xi) = T(z,0); \quad T_p(z,r_t,\xi) = T(z,r_t) \tag{13}$$

$$\frac{\partial T_p(z,r,0)}{\partial \xi} = 0; \quad -\lambda_\xi \frac{\partial T_p(z,r,r_p)}{\partial \xi} = h[T_p(z,r,r_p) - T(z,r)] \tag{14}$$

2.3. Primary setting

In this works, some of primary assumptions have been made: (1) The reformer tubes are assumed to be homogeneous within the reformer, which means the conditions of any one tube are sufficient to represent all other tubes (Vo et al., 2019). (2) Ideal gas model is used to descript the fluid phase in the SMR reactor. (3) CH_4, CO, CO_2, H_2, H_2O, and N_2 are the components considered in this model. Their initial feed pressures are 5.46 bar, 0 bar, 0.31 bar, 0.68 bar, 18.34 bar, and 0.9 bar, respectively. The length of reactor is 12 m with a tube diameter of 0.1 m. (4) Carbon deposition was not considered in this work, and catalyst partials are considered as spherical form. The diameter of catalyst is 0.01 m. (5) The feed temperature is at 793.15 K, and the tube wall temperature is at 1000 K.

3. Results and discussions

For the selection of effectiveness factor in the 2D pseudo-homogeneous model, Pantoleontos et al. (2012) listed many works with different value, ranging from 0 to 1. Most of works considered 1 to ignore diffusional limitations. In this work, we select the value of 0.01 as the constant effectiveness factor. In the following, the comparison of the pseudo-homogeneous and heterogeneous models is given:

3.1. Pressure and temperature profiles

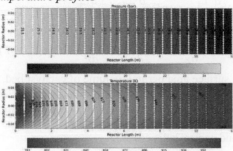

Figure 1. The pressure and temperature profiles of the pseudo-homogeneous model.

Figures 1 and **2** show the pressure and temperature profiles of pseudo-homogeneous and heterogeneous models, respectively. The results demonstrate the pressure profiles of two models are generally flat. However, the temperature profiles are not the case. In the pseudo-homogeneous model, the maximal radial temperature gradient is about 50°C, while this temperature is around 45°C in heterogeneous model.

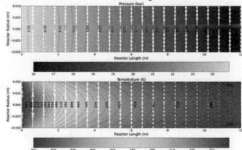

Figure 2. The pressure and temperature profiles of the heterogeneous model.

3.2. Concentration profiles

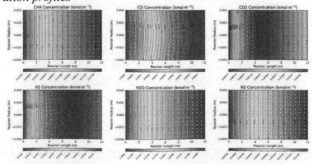

Figure 3. The concentration of each component in the pseudo-homogeneous model.

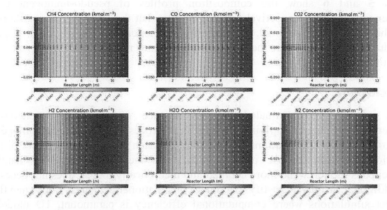

Figure 4. The concentration of each component in the heterogeneous model.

Figures 3 and **4** show the concentration profiles of pseudo-homogeneous and heterogeneous models, respectively. Taking the concentration of hydrogen as an example, in pseudo-homogeneous model, the maximum concentration is appearing in the around 4 – 10 m in the reactor, but it appears in around 6 – 12 m in the heterogeneous model.

3.3. Composition profiles

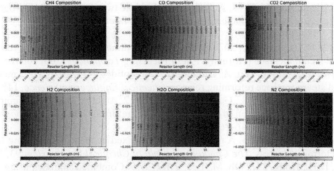

Figure 5. The composition of each component in the pseudo-homogeneous model.

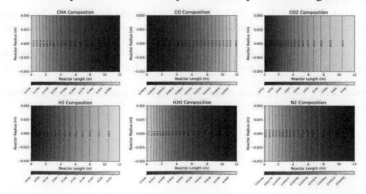

Figure 6. The composition of each component in the heterogeneous model.

Figures 5 and **6** show the composition profiles of pseudo-homogeneous and heterogeneous models, respectively. It can be seen that the composition profile is very flat in the heterogeneous model. This indicate 1D model might be sufficient to present the product distribution, while the pseudo-homogeneous model has slight curved product distributions.

4. Conclusions

In this research, we evaluated the performance of pseudo-homogeneous and heterogeneous models by applying them to SMR processes. The pseudo-homogeneous model offers a less computationally intensive approach as it bypasses direct diffusion calculations through the use of an effectiveness factor. On the other hand, the heterogeneous model, while more detailed and theoretically robust, does not necessarily yield additional insights to justify its higher computational demands. Consequently, for flowsheet simulations where computational efficiency is paramount, 1D models are recommended. However, 2D models, which provide more detailed radial information, should be considered when the specificity of the application demands it.

References

Cruz, F.E.d., Karagoz, S., Manousiouthakis, V.I., 2017. Parametric studies of steam methane reforming using a multiscale reactor model. *Industrial & Engineering Chemistry Research*. 56, 14123-14139.

Ghouse, J.H., Adams II, T.A., 2013. A multi-scale dynamic two-dimensional heterogeneous model for catalytic steam methane reforming reactors. *International Journal of Hydrogen Energy*. 38, 9984-9999.

Kuncharam, B.V.R., Dixon, A.G., 2020. Multi-scale two-dimensional packed bed reactor model for indutrial steam methane reforming. *Fuel Processing Technology*. 200, 106314.

Lao, L., Aguirre, A., Tran, A., Wu, Z., Durand, H., Christofides, P.D., 2016. CFD modeling and control of a steam methane reforming reactor. *Chemical Engineering Science*. 148, 78-92.

Pantoleontos, G., Kikkinides, E.S., Georgiadis, M.C., 2012. A heterogeneous dynamic model for the simulation and optimisation of the steam methane reforming reactor. *International Journal of Hydrogen Energy*. 37, 16346-16358.

Rout, K.R., Jakobsen, H.A., 2015. A numerical study of fixed bed reactor modelling for steam methane reforming process. *The Canadian Journal of Chemical Engineering*. 93, 1222-1238.

Vo, N.D., Oh, D.H., Hong, S.H., Oh, M., Lee, C.H., 2019. Combined approach using mathematical modelling and artificial neural network for chemical industries: Steam methane reformer. *Applied Energy*. 255, 113809.

Wesenberg, M.H., Svendsen, H.F., 2007. Mass and heat transfer limitation in a heterogeneous model of a gas-heated steam reformer. *Industrial & Engineering Chemistry Research*. 46, 667-676.

Xing, Y., Masuku, C.M., Biegler, L,T., 2021. Modular gas-to-liquids process with membrane steam-methane reformer and Fischer–Tropsch reactive distillation. *AIChE Journal*. 67, e17467.

Xu, J., Froment, G.F., 1989. Methane steam reforming, methanation and water-gas shift: I. Intrinsic kinetics. *AIChE Journal*. 35, 88-96.

Flavio Manenti, Gintaras V. Reklaitis (Eds.), Proceedings of the 34th European Symposium on Computer Aided Process Engineering / 15th International Symposium on Process Systems Engineering (ESCAPE34/PSE24), June 2-6, 2024, Florence, Italy

Implementing a Model Library for Wastewater Treatment Plants in an Equation-Oriented Simulator: Numerics and Architecture

Jochen Steimel[a,*] , Sahand Iman Shayan[b]

[a]AVEVA, Mainzer Landstraße 178-190, 60327 Frankfurt, Germany
[b]AVEVA, 920 Memorial City Way, Suite 1200, Houston, TX 77024, USA
jochen.steimel@aveva.com

Abstract

This contribution discusses the lessons we have learned during the development of a library of models for municipal wastewater treatment plants in AVEVA Process Simulation, an equation-oriented next-generation process simulator, capable of both steady-state and dynamic simulation.

While available models in the literature for activated sludge processes and anaerobic digestion commonly used in wastewater treatment plants are structurally straight-forward, they exhibit characteristics that make solving them reliably a challenge. Especially the existence of multiple steady states can be hard to overcome for an equation-solving system.

Two approaches have been identified as helpful in determining a valid and desired steady-state operating point for these kinds of processes.

1. Repeated solution of embedded steady-state problems (or homotopy continuation) from a known operating point to the actual operating point.

2. Seamless switching to dynamic simulation mode, using the integrator to drive the system closer to the good steady state, and then directly solving the equations to the desired operating point.

This article will discuss the challenges in solving these models, give concrete examples generated with the software AVEVA Process Simulation, and provide recommendations to handle these problems in general.

Keywords: Wastewater Treatment, Simulation, Numerical Methods, Modeling

1. Introduction

Municipal wastewater treatment plants have been the subject of modeling for a long time. In the last 30 years, efforts have been made to standardize the nomenclature and model formulations by publishing reference models such as for activated sludge (ASM1, ASM2, ASM3, Henze 2000) or anaerobic digestion (ADM1, Batstone 2002) through the International Water Association. If possible, these models are expressed as explicit ordinary differential equations, and algebraic loops are avoided. One example is the ADM1 model (Batstone 2002), for which 2 formulations are provided, one as an ordinary differential equation system and one as a differential algebraic equation system.

Implementers can use these models with suitable equipment models (such as continuously stirred tank reactors) to describe entire treatment plants. The models describe the mass and energy balance of the treatment plant and can be used for answering design and operability questions.

Many of the reaction models exhibit aspects that might provide a challenge for solving these models:

1. *Multiple solutions:* Many of the reaction rate expression are proportional to both the concentration of a substrate and of a biomass species that catalyses the reaction. For reactions of these kind, two solutions are mathematically valid, the desired productive state and the wash-out state. When calculating a steady-state solution from an arbitrary initial point, equation-based solvers can get stuck in the trivial solution.

2. *Very slow reactions and fast hydraulics:* Many reactions involved in biological treatment are very slow and take place over several days (or even weeks). In contrast, the hydraulic residence time is smaller (often in the range of minutes to hours), so the integration time step needs to be fine enough to correctly model the mass flows and accumulation terms, but also be fast enough to simulate the entire duration.

3. *High demand for customizability: Even* though considerable research went into the development of the ASM type models, practitioners still must tune and calibrate a model to their plant. This often involves tuning stoichiometric factors and rate-expressions or adding new conversion processes for species (i.e., adding completely new reactions).

Any simulation environment for wastewater treatment plants should be aware of these challenges and take measures to address them, so that the resulting plant simulation models provide results quickly, reliably, and correctly. In the following we will discuss the measures taken in the development of the Wastewater library for AVEVA Process Simulation to address the challenges discussed above.

2. Open-equation modeling for extending activated sludge models

It appears that two major approaches for modeling wastewater treatment plants are used in the literature. On the one hand, academic research often focuses on modeling and simulation, and uses programming languages like MATLAB or Python, to directly encode the differential equations. For example, the benchmark simulation models library WWT (Gernaey 2014) is implemented in C and used as an external function in MATLAB.

On the other hand, industry-focused contributions are often implemented in commercial, special purpose simulation programs. These tools are often graphical, flowsheet-based, interactive and implement the reaction models natively in their source-code. These tools offer extensibility via Petersen/Gujer matrix editors which allow users to modify the stoichiometric coefficients and rate-expressions of the involved processes. While this approach offers a considerable amount of flexibility, users are still confined to the limitations of the matrix formalism and often do not have access to the actual equations as they would have in for example a MATLAB-based implementation.

In AVEVA Process Simulation, every model is represented in a tabular modeling language, that allows the user to see and modify each variable, parameter, and equation. The entire Wastewater library is visible to the user, and it is easy to copy and modify a library model to represent a specific process or phenomenon. A screenshot of the model editor showing some variables and equations of the ASM3 model are shown in Fig. 1.

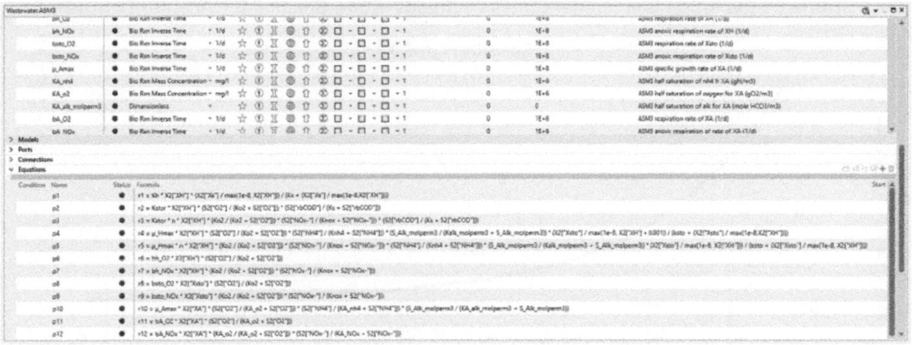

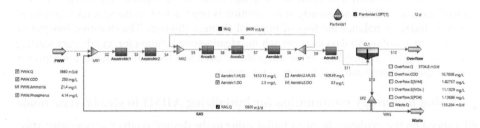

Figure 1: Selected variables and equations of the ASM3 model

Later, a simulation builder can put these models onto a flowsheet and assemble larger simulations representing entire plants from the building blocks. An example of such a flowsheet can be seen in Fig. 2. AVEVA Process Simulation combines the expressivity of modeling languages and the ease of assembling larger system of flowsheeting tools.

Figure 2: Screenshot of a wastewater treatment plant as modelled in AVEVA Process Simulation

3. Transient initialization of the anaerobic digester model

The problem of finding the desired steady-state solution out of all the possible solutions (Bornhöft 2013) of the anaerobic digester is caused by the formulation of the rate expressions. For example, we will consider the process of update of acetate (Rosen 2005):

$$\rho_{11} = k_{m,ac} \frac{S_{ac}}{K_{s,ac} + S_{ac}} X_{ac} I_{11} \tag{1}$$

Where S_{ac} is the concentration of acetate, X_{ac} is the concentration of acetate degraders (micro-organisms), I_{11} is an inhibition term and $k_{m,ac}$ and $K_{s,ac}$ are rate constants.

Even when acetate is present in the feed stream, the resulting reaction rate might still be zero when there are no degraders ($X_{ac}=0$) in the initial state. As the growth of degraders is also linked to this reaction, there is a strong attraction to the trivial (wash-out) steady-state solution if initial values are not chosen carefully and a Newton-type method is used for solving the equation set.

One approach to overcome the challenge of finding a steady-state solution from a trivial starting point is to use transient (or pseudo-transient) models to provide a continuous solver path (Pattison 2014). We can apply this technique to solve the steady-state version of the anaerobic digester model.

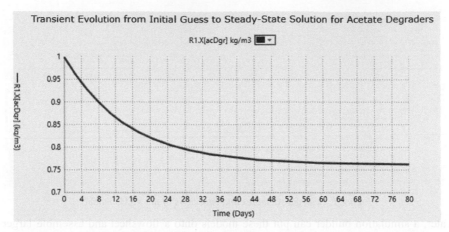

Figure 3: Transient evolution of acetate degraders towards the steady-state solution

In Fig. 3 a transient is shown that illustrates the development of the concentration of a particulate species (X_{ac}, which is labeled X[acDgr] in the implementation) over time, starting from an initial seed value to the actual steady-state solution. By starting from an initial seed, a well-defined steady-state solution is approached, as the species is produced by the sludge to balance the loss of material through dilution. The dynamic integration requires to simulate 80 days to approach the steady-state solution for an arbitrary starting point of 1 kg/m³. Depending on the integrator chosen and the initial values, this approach (while quite robust) can be very time consuming.

4. Using homotopy continuation for initializing ADM1 in steady-state mode

To speed up the evolution from an initial value to the desired productive operating state, completely inside a steady-state solver, homotopy continuation can be applied. In eq. 2, the modified mass-balance equation for the particulate species is shown:

$$\frac{dX}{dt} = \zeta_H \left(\frac{1}{HRT} \cdot (X_{in} - X) + R \right) + (1 - \zeta_H) \cdot (X - X_{seed}) \qquad (2)$$

where HRT is the hydraulic residence time, X is the concentration of particulate species inside the digester, X_{in} is the concentration in the feed, R is the reaction rate, ζ_H is the homotopy factor and X_{seed} is the initial concentration.

In addition to the usual terms (in/out flows and reaction), a second term was added that forces the steady-state value of X to be equal to X_{seed} when $\zeta_H = 0$. For $\zeta_H = 1$ the differential equation is used, and the initial values are ignored. Once the anaerobic digester has been solved for fixed values of X_{seed}, the homotopy factor can be moved towards 1, providing a smooth convergence path for X even when X_{in} is equal to zero.

Using this approach, the tedious integration over days or weeks of simulation time can be avoided and convergence to the steady-state solution is typically achieved in a handful of homotopy steps. Fig. 4 shows a typical homotopy evolution for a few selected concentrations. For illustrative purposes, the maximum variable move in each homotopy iteration was limited to 5%. Over 20 iterations, the solution transitioned from the trivial solution to the desired solution.

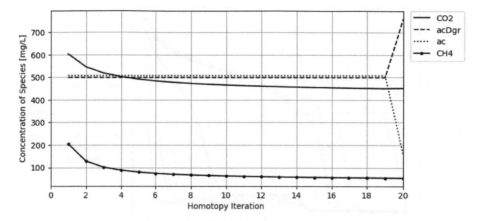

Figure 4: Illustrative homotopy path for selected variables of the ADM1 model

5. Solving for the real steady-state solution of the discretized settler

The last example is concerning the discretized settler, as described in (Takacs 1991). For any given feed concentration and flow split a steady-state profile of solids along the layers of the discretized model will manifest. In (Verdickt 2002) numerical analysis of this model was performed and the results were shown in Fig. 6 in that article. When we tried to reproduce these results, we found that for Xf=5300 mg/L different solutions could be obtained depending on the solution method. Our results are shown in Fig. 5.

The line labelled *Xf=5300* represents the steady-state solution found by direct solution of the equation system with a Newton-type method, whereas the line labelled *Xf=5300 (t=500h)* represents the apparent steady-state solution after running a dynamic integration for 500h. The latter line is also matching the results of (Verdickt 2002). When we let the integration run for longer times, we observe profiles that are slowly approaching the steady-state profile. As details for the calculation method are not given in the reference article, we assume that they terminated the integration too early and assumed a steady-state, as the longest time reported for a profile evolution is given as 200 hours (in their Fig.7). Note that we were not able to reach the steady-state profile even after 2000 hours.

These results illustrate the need for both dynamic integration and direct solution for simulations of Wastewater treatment plants. Due to the very slow processes involved, using only a dynamic integrator can be misleading or be very time consuming.

6. Conclusions

In this article, we have highlighted and discussed three major challenges inherent in the simulation of Wastewater plants and proposed methods to overcome them, giving examples on their implementation and shown numerical results to ground our claims.

The three tools and methods that we recommend are:

1. Model extensibility on the equation level to be able to calibrate the model to a real plant and extend reaction schemes for new/unknown species.

2. Use of transient initialization and / or homotopy continuation to solve challenging multiple-solution problems.

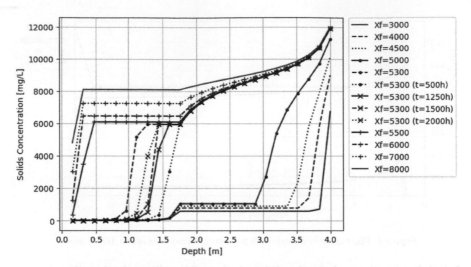

Figure 5: Solid concentration profiles of a settler for different feed concentrations. Solutions are calculated from steady-state, except those marked (t=...h), which are obtained by integration.

3. Use of both dynamic integration and direct solve to steady-state to be able to capture both dynamic transients and find the true steady-state of extremely slow processes.

While we did not invent any of the methods discussed here, we combined them and applied them to create a new library of models that were validated against literature data. The library is made available in AVEVA Process Simulation 2024, and the user can combine all the platform features such as Sensitivity Analysis, Scenarios and Optimization with Wastewater simulations. In summary, we can recommend any practitioner to consider using the discussed methods and tools in their simulations of Wastewater treatment plants.

References

Henze, M., Gujer, W., Mino, T., van Loosdrecht, M. C. M., 2000, Activated sludge models ASM1, ASM2, ASM2d and ASM3, reprint ed., IWA Publishing Company

Batstone, Damien J., Keller, J. Angelidaki, I., Kalyuzhnyi, S.V. Pavlostathis, S.G., Rozzi, A., Sanders, W.T.M., Siegrist, H. Vavilin, V.A., 2002, The IWA anaerobic digestion model no 1 (ADM1), Water Science and technology 45.10, 65-73.

Bornhöft, A., Hanke-Rauschenbach, R. & Sundmacher, K., 2013, Steady-state analysis of the Anaerobic Digestion Model No. 1 (ADM1). Nonlinear Dyn 73, 535–549

Gernaey, K. V., Jeppsson, U., Vanrolleghem, P. A., & Copp, J. B. (Eds.), 2014, Benchmarking of Control Strategies for Wastewater Treatment Plants. Scientific and Technical Report No. 23, IWA Publishing

Pattison R.C., Baldea, M., 2002, Equation-oriented flowsheet simulation and optimization using pseudo-transient models, AIChe Journal 60, 4104-4123

Rosén, C., Jeppsson, U., 2005, Aspects on ADM1 Implementation within the BSM2 Framework. (TEIE; Vol. 7224). Department of Industrial Electrical Engineering and Automation, Lund Institute of Technology. http://www.iea.lth.se/publications/Reports/LTH-IEA-7224.pdf

Takács, I., Patry G.G., Nolasco, D., 1991, A dynamic model of the clarification-thickening process. Water Research, 25, 1263-1271.

Verdickt L. B., Van Impe, J. F., 2002, Simulation Analysis of a one-dimensional sedimentation model, IFAC Proceedings, 35, 1, 473-478

Flavio Manenti, Gintaras V. Reklaitis (Eds.), Proceedings of the 34th European Symposium on Computer Aided Process Engineering / 15th International Symposium on Process Systems Engineering (ESCAPE34/PSE24), June 2-6, 2024, Florence, Italy

Waste Heat Recovery from PEM Electrolyzer for Desalinated Water Production in Polygeneration Systems

Artur S. Bispo,[a] Leonardo O.S. Santana,[a] Gustavo S. dos Santos,[a] Chrislaine B. Marinho,[a] Daniela Hartmann,[b] Fernando L.P. Pessoa.[a,c]

[a]*SENAI CIMATEC, Av. Orlando Gomes, 1845 - Piatã, Salvador - BA, 41650-010, Brazil.*
[b]*CNODC Brasil Petróleo e Gás, Botafogo, Rio de Janeiro – RJ, 22250-040, Brazil.*
[c]*UFRJ, Av. Athos da Silveira Ramos, 149 - RJ, 21941-909, Brazil.*
Artur.bispo@outlook.com.br

Abstract

In the upcoming future hydrogen-based economy, water electrolysis appears as a good alternative for producing clean hydrogen without carbon emissions, however, the rising water scarcity call for the wider adaptation of desalination techniques. One of the most promising desalination techniques is pervaporation (PV), an emerging technology for desalination of water that works by temperature. A mathematical model of the PEM electrolyser heat generation is used and experimental data from literature is used to simulate the pervaporator. The proposed plant led to a polygeneration system that consists of interdependent processes that seek opportunities to increase the thermodynamic efficiency of the system.

Keywords: Water Electrolysis, Water desalination, Emerging membrane technologies, Polygeneration system, process integration.

1. Introduction

In the current global industrial/energy scenario, there has been a growing interest in sustainable energy matrices, driven by the need to reduce dependency on fossil sources and mitigate the impacts arising from GGEs (Greenhouse Gas Emissions). In this context, GH_2 (green hydrogen) emerges as one of the main actors in decarbonization, enabling the efficient storage and transport of energy generated from renewable sources, such as wind and solar power (Atilhan *et al.*, 2021).

GH_2 refers to hydrogen produced specifically through the electrolysis of water using renewable energy sources. Furthermore, among the various electrolysis technologies, PEM (Proton Exchange Membrane) stands out for its advantages in aspects such as energy efficiency, design, H_2 purity, operational conditions, and rapid response time. However, to make this energy vector competitive, several bottlenecks need to be overcome, mostly involving the high costs associated with both the implementation – noble metals in equipment structure and catalysts – and operation – energy and water treatment – of these electrolytic systems (Kumar e Himabindu, 2019).

1.1. PEM Electrolysis

The structure known as a stack, according to Mehta and Cooper (2003), is composed by a set of electrolytic cells, electrically connected in series, consisting of bipolar plates surrounding the MEA (Membrane-Electrode Assembly). The MEA, in turn, is made up of a polymer membrane, electrocatalysts, and the GDL (Gas Diffusion Layer).

Authors such as Kumar and Himabindu (2019), Rozain *et al.* (2016), Garcia Valverde *et al.* (2011) and Wang *et al.* (2021) explains, in summary, that PEM process begins with water being supplied to the electrolyzer through the anodic channel, where it is split (Eq. 1), with the aid of metallic oxide catalysts, following mechanisms such as AEM and LOM. After crossing the polymer membrane, the protons are reduced to gaseous hydrogen (Eq. 2) at the cathode through Volmer, Heyrovsky, and Tafel mechanisms. Meanwhile, gaseous oxygen molecules (O_2) are generated at the anode, diffusing through the GDL, and being removed as a process byproduct.

$$H_2O \rightarrow 2H^+ + \frac{1}{2}O_2 + 2e^- \tag{1}$$

$$2H^+ + 2e^- \rightarrow H_2 \tag{2}$$

As for the typical PEM electrolyzers make up materials, specific catalysts are employed at both anode and cathode. At the anode, metallic oxides like RuO_2 and IrO_2 are preferred for their high metallic conductivity properties and favorable molecular characteristics. At the cathode, platinum is the most efficient catalyst, although palladium is also used due to its lower cost and greater abundance. Moreover, Nafion®/PFSA polymer membranes are widely used as solid electrolyte, for their effectiveness in proton conduction and long durability (Kumar and Himabindu, 2019).

1.2. Pervaporation

Pervaporation (PV) is an emerging membrane desalination technology that uses thermal energy as an external gradient and phase change as physicochemical process (Saavedra, *et al.* 2021). This technology is widely used in poly and cogeneration systems as it increases the exergy efficiency of energy resources (Kumar *et al.*, 2023).

Therefore, and due to the increase in the world population and urbanization, water scarcity is today one of the most difficult challenges faced by society. Pervaporation is probably the membrane operation with the lowest permeation rate, but providing high salt rejection, so the use of this technology allows drinking water to be obtained more efficiently and with reuse of waste heat (Castro-Muñoz, 2020) such as solar thermal energy. There are several solar-powered desalination systems around the world with internal heat recovery. The system configuration can be seen in Figure 1, heat is required to heat the feed (preheated) to the hot inlet, i.e., from about 72 ∘C to about 80 ∘C (the numbers are given as an example of standard module operation) (Cipollina *et al.*, 2012).

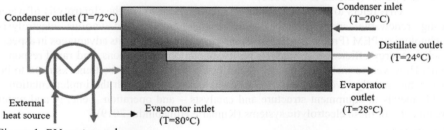

Figure 1. PV system scheme

Considering this entire scenario, the present work aims to assess the energy integration of the GH_2 production by PEM electrolysis, directing the stack waste heat produced during the electrolysis to the feed water treatment process. This approach not only makes feasible the use of the heat generated by the production process itself but also opens up the possibility of using nontraditional water sources, such as seawater, in a polygeneration system that produces potable water, hydrogen and oxygen.

2. PEM modeling

To quantitatively determine the heat generated by a PEM electrolyzer, it is necessary to model the entire process, considering the occurring phenomena and inherent losses. Thus, Eq. (3) works as the modeling basis, determining the stack operating voltage (V_{stack}) as a function of phenomenology and non-ideality (losses) of the process (Garcia Valverde *et al.*, 2011).

$$V_{stack} = n_{cell}\left(U_{cell} + \eta_{act,C} + \eta_{act,A} + \eta_{diff} + \eta_{ohm}\right) \tag{3}$$

Where n_{cell} is the number of cells, U_{cell} is the open circuit potential, $\eta_{act,C}$ and $\eta_{act,A}$ are the activation overpotentials at the electrodes, η_{diff} is the diffusion or mass transfer overpotential, and η_{ohm} is the ohmic overpotential. For convenience, this work will make use of a more simplified version of this model, disregarding the activation overpotential at the cathode – as it is very small when compared anode side – and the diffusion overpotential – since it can be neglected at operating current densities below 1.6 A/cm² (Agbli *et al.*, 2010).

For the open circuit potential, the empirical Eq. (4) – also used by Garcia Valverde *et al.* (2011) and Agbli *et al.* (2011) – was considered, relating it to the process operating temperature.

$$U_{cell} = 1,5184 - 1,5421 * 10^{-3}T + 9,523 * 10^{-5}T \ln(T) + 9,84 * 10^{-8}\,T^2 \tag{4}$$

The anode activation overpotential can be determined from Eqs. (5) and (6), which describes it, based on the Butler-Volmer equation, as a function of operating temperature and current density (i) (Garcia Valverde *et al.*, 2011).

$$\eta_{act,a} = \frac{RT}{\alpha z F} \ln\left(\frac{i}{i_0}\right) \tag{5}$$

$$i_0 = i_{0,ref}\, e^{\frac{-E_{exc}}{R}\left(\frac{1}{T}-\frac{1}{T_{ref}}\right)} \tag{6}$$

Regarding the ohmic overpotential, it can be described as the ratio of the membrane thickness to its conductivity (Eq. 7). Therefore, according to Bispo *et al.* (2023), the conductivity can be determined as a function of the temperature and the membrane water content (λ_m), based on Arrhenius equation, as seen in Eq. (8).

$$\eta_{ohm} = \frac{t_m}{\sigma_{H^+}} \tag{7}$$

$$\sigma_{H^+} = (0,046459 \ln \lambda_m - 0,051556) * e^{\left[1268\left(\frac{1}{298,15}-\frac{1}{T}\right)\right]} \tag{8}$$

Finally, it is possible to determine the stack heat generation ($\dot{Q}_{gen}$) from Eq. (9), (Garcia Valverde *et al.*, 2011). The thermoneutral potential (U_{tn}), in turn, is calculated using Eqs. (10 – 12), and the experimental parameters from a to d – to a component j – can be found in Marangio *et al.* (2009).

$$\dot{Q}_{gen} = (V_{stack} - [U_{tn} * n_{cell}]) * (i * A) \tag{9}$$

$$U_{tn} = \frac{\Delta H}{nF} \tag{10}$$

$$\Delta H = H_{H_2}(T) + \frac{1}{2} H_{O_2}(T) - H_{H_2O}(T) \tag{11}$$

$$H_j(T) = a_j T + 4b_j \frac{T^{5/4}}{5} + 2c_j \frac{T^{3/2}}{3} + 4d_j \frac{T^{7/4}}{7} \tag{12}$$

3. Process Simulation

The simulation of the integrated seawater purification and GH$_2$ production process was carried out using two software, simultaneously: Excel VBA and Aspen Plus V14. Given the absence of a PEM electrolyzer simulation block in Aspen Plus V14, all the modeling developed and explained in previous section was implemented in Excel VBA, to determine the water consumption and heat generation of a given electrolyzer. The simulated electrolyzer specifications can be seen in Table 1.

Table 1. Electrolyzer Modeling Specs.

Parameter	Symbol	Value	Unit
Temperature	T	353.15	K
Current density	i	1.35	A/cm^2
Number of cells	n_{cell}	20	-
Membrane thickness	t_m	0.03306	cm
Water content	λ_m	22	-
Area	A	100	cm^2
Power	Pot	1	MW
Reference exchange current density	$i_{0,ref}$	10^{-7}	A/cm^2
Reference temperature	T_{ref}	353.15	K
Activation energy	E_{exc}	53.99	kJ/mol
Transferred electrons	z	2	-
Charge transfer coefficient	α	0.5	-

Furthermore, Aspen Plus also lacks a simulation block for defining membrane distillation. Therefore, for simulation purposes and to determine the distillation heat duty, a "Hierarchy" block was used, and the flowsheet shown in Figure 2 was structured.

As explained by Cipollina *et al.* (2012), brine distillation occurs through the increase of its temperature and, consequently, the evaporation of the solvent. For the simulation, a *HeatX* was used to simulate this heat transfer and a *Heater* to describe the fluid heating and the heat duty. Moreover, to 'emulate' the membrane behavior and based on the parameters presented by Khalifa *et al.* (2017), a *Flash2* was used, setting the separation temperature value - 80 °C - and the molar vapor fraction - 4.8 %.

Besides, in order to determine the brine flow to simulated membrane distillation, its output was considered to be equal to the feed of purified water in the electrolyzer described by Table 1. Pressure and temperature conditions where also based on Khalifa *et al.* (2017) and Cipollina *et al.* (2012). Finally, the two simulated processes were integrated, with the pervaporator distillate being fed into the stack, and the energy generated by the stack transferred to heat the brine.

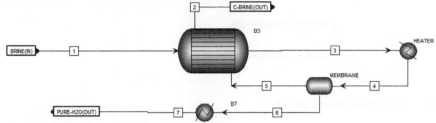

Figure 2. Membrane distillation flowsheet.

4. Results

Figure 3 shows the integrated process flowsheet. Additionally, Table 2 displays the input and output values for each process. As can be observed, the simulation shows that to produce 174.32 kg/h of distilled water – electrolyzer demand, the pervaporator requires 24.56 kW of energy and 4791.81 kg/h of brine (35% NaCl). Furthermore, the stack specified in Table 1 exhibits an overall excess heat of 98.20 kW, therefore, approximately 25 % of this heat needs to be utilized to meet the water purification system demand.

Thus, these results present a promising alternative in the energy integration of the process (considering a conservative scenario in terms of heat recovery), enabling the use of brine in PEM electrolysis systems without the need for an external energy demand for its purification.

Table 2. Simulation results.

	Pervaporator			PEM Electrolyzer		
Stream	Composition		Temperature	Voltage	35,40	Volts
Brine	H₂O	3114.68 kg/h	25 °C	H₂O feed	174.32	kg/h
	NaCl	1677.13 kg/h				
Pure – H₂O	H₂O	174.32 kg/h	24 °C	H₂ Produced	19.37	kg/h
	NaCl	≅ 0				
C-Brine	H₂O	2940.36 kg/h	28 °C	O₂ Produced	154.95	kg/h
	NaCl	1677.13 kg/h				
Heat Duty/ Excess	24.56 kW			98.20 kW		

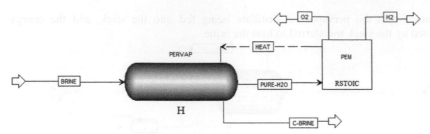

Figure 3. Integrated process flowsheet.

5. Conclusions

Finally, this study demonstrates a novel approach to integrating PEM electrolyzer with a membrane desalination process, effectively utilizing waste heat (with an efficiency of 25%) for desalinated water production in polygeneration systems. The results reveal that the recovered waste heat from the PEM stack significantly enhances the pervaporation process efficiency, contributing to a more energy-efficient water desalination method. In conclusion, this study not only contributes to the field of renewable energy and environmental engineering but also aligns with the global efforts to develop innovative solutions for energy efficiency and water resource management.

References

A. Bispo, C. Marinho, F. Pessoa & J. Almeida (2023). An ajusted model of proton conductivity in Nafion® membranes. Journal of Bioengineering and Technology Applied to Helth.

A. Saavedra, H. Valdés, A. Mahn & O. Acosta (2021). Comparative analysis of conventional and emerging technologies for seawater desalination: Northern Chile as a case study. Membranes, 11(3), 180.

C. Rozain, E. Mayousse, N. Guillet, & P. Millet (2016). Influence of iridium oxide loadings on the performance of PEM water electrolysis cells: Part I–Pure IrO2-based anodes. Applied Catalysis B: Environmental, 182, 153-160.

F. Marangio, M. Santarelli & M. Calì (2009). Theoretical model and experimental analysis of a high pressure PEM water electrolyser for hydrogen production. International journal of hydrogen energy, 34(3), 1143-1158.

G. Kumar, D. Ayou, C. Narendran, R. Saravanan, M. Maiya & A. Coronas (2023). Renewable heat powered polygeneration system based on an advanced absorption cycle for rural communities. Energy, 262, 125300.

K. Agbli, M. Péra, D. Hissel, O. Rallières, C. Turpin & I. Doumbia (2011). Multiphysics simulation of a PEM electrolyser: Energetic Macroscopic Representation approach. International journal of hydrogen energy, 36(2), 1382-1398.

R. Castro-Muñoz (2020). Breakthroughs on tailoring pervaporation membranes for water desalination: A review. Water research, 187, 116428.

R. García-Valverde, N. Espinosa & A. Urbina (2012). Simple PEM water electrolyser model and experimental validation. International journal of hydrogen energy, 37(2), 1927-1938.

S. Atilhan, S. Park, M. El-Halwagi, M. Atilhan, M. Moore, & R. Nielsen (2021). Green hydrogen as an alternative fuel for the shipping industry. Current Opinion in Chemical Engineering, 31, 100668.

S. Kumar, & V. Himabindu (2019). Hydrogen production by PEM water electrolysis–A review. Materials Science for Energy Technologies, 2(3), 442-454.

S. Wang, A. Lu, & C. Zhong (2021). Hydrogen production from water electrolysis: role of catalysts. Nano Convergence, 8, 1-23.

V. Mehta, & J. Cooper (2003). Review and analysis of PEM fuel cell design and manufacturing. Journal of power sources, 114(1), 32-53.

Flavio Manenti, Gintaras V. Reklaitis (Eds.), Proceedings of the 34th European Symposium on Computer Aided Process Engineering / 15th International Symposium on Process Systems Engineering (ESCAPE34/PSE24), June 2-6, 2024, Florence, Italy

Modeling and Optimizing Sugarcane-Livestock Integration Systems in Brazil

Igor L. R. Dias[a*], Terezinha F. Cardoso[b], Ana C. M. Jimenez[b], João G. O. Marques[c], Luís G. Barioni[d], Flávia Barbosa[e], Adriano P. Mariano[a], Marcelo P. Cunha[f], Antonio Bonomi[b]

[a]*Faculdade de Engenharia Química (FEQ), Universidade Estadual de Campinas (Unicamp), Av. Albert Einstein, 500, CEP 13083-852, Campinas - SP, Brasil.*

[b] *Laboratório Nacional de Biorrenováveis (LNBR), Centro Nacional de Pesquisa em Energia e Materiais (CNPEM), CEP 13083-970, Campinas - SP, Brasil.*

[c]*Parrakat, Edinburgh, United Kingdom.*

[d]*Embrapa Agricultura Digital - Embrapa, Av. Dr. André Tosello, 209 - Cidade Universitária, 13083-886, Campinas – SP, Brasil.*

[e] *INESC TEC, Faculdade de Engenharia, Universidade do Porto, Porto, Portugal.*

[f]*Instituto de Economia (IE), Universidade Estadual de Campinas (Unicamp), R. Pitágoras, 353 - Cidade Universitária, 13083-857, Campinas - SP, Brasil.*

i261185@dac.unicamp.br

Abstract

The expansion of ethanol production in Brazil sparks several sustainability concerns, including debates on "food versus fuel", the environmental impacts of monocultures, and indirect land-use change. Since livestock farming occupies a significantly greater area than sugarcane for ethanol production in Brazil and has a large yield gap, sugarcane-livestock integration can be a promising alternative. This integrated system considers crop production systems, biorefinery processing and meat production in both intensive and extensive livestock farming. Optimizing this system for both economic and environmental aspects can be challenging to implement and computationally expensive as this system's complexity arises from nonlinear subsystems and their intertwining input-output flows. For these reasons, this paper develops metamodels from detailed models to: (i) Optimize the extensive livestock farming, (ii) Optimize the confined animal feeding, and (iii) Optimize the integrated system. The main objective is to maximize the Net Present Value relative to investment. This study contributes to the literature by developing innovative models for ethanol-beef integrated production systems and methods for optimizing such systems to avoid negative externalities on food security and environmental impacts.

Keywords: Biorefineries, Beef cattle production, Experimental Design, Optimization metamodels, Food versus biofuels

1. Introduction

The production of biofuels, especially ethanol, has been encouraged in Brazil for a long time. Currently, Brazil has a National Biofuels Policy, RenovaBio, whose main objective is to increase the production of biofuels seeking environmental, economic, and social sustainability (UNICA, 2020; ANP, 2020). Under this program, the country's first official carbon credit, CBIO, has been established. The expansion of ethanol production has raised concerns regarding land change, such as competition with food crops and impacts on deforestation (ACHINAS et al., 2019; ALLEN et al., 2019).

Sugarcane-livestock integration systems could improve the sustainability of ethanol production expansion by minimizing negative impacts and improving yields for both sugarcane plantations and livestock (SOUZA et al., 2019). The complexity of this system is evident, including nonlinear models dependent on numerous variables, such as the amount of processed sugarcane, the nutritional quality of feed ingredients, and nitrogen doses for forage production, among others. Therefore, optimizing this system, considering economic and environmental aspects, requires the application of advanced methods, which can be challenging to implement and properly adjust.

This study designs the integrated system by developing metamodels and solving sub-optimization problems. Therefore, this work aims to answer the research question: "Under what conditions is it feasible to expand the area dedicated to biofuel production without negatively impacting food or livestock production while avoiding expansion into natural vegetation areas?".

2. Methodology

2.1. Description of the system

In the integrated system analyzed (Figure 1), sugarcane is produced and transported to the biorefinery (1G autonomous) for processing. Soybean grains produced in the sugarcane reform area are sold on the market, and their revenues are added to the system. The sugarcane straw and bagasse are used to produce bioelectricity. Ethanol and the surplus of bioelectricity produced in the biorefinery are sold, while part of the excess bagasse is used for animal feed or sold. Livestock farming, in turn, considers only beef cattle fattening.

2.2. Development of metamodels

A crucial point to consider in optimizing the defined integrated system is that detailed models would render a highly complex or even unsolvable optimization problem, given that these models exhibit varying degrees of mathematical complexity and convergence issues. Furthermore, creating distinct data flow strategies among models built on various platforms would be necessary. To address these challenges, an optimization strategy similar to the one presented by Bressanin et al. (2021) was adopted. This involves creating simplified metamodels derived from more detailed models. To achieve this, several data sampling strategies from the detailed models were employed for each system.

The Virtual Biorefinery (VB) tool was employed as a detailed base for obtaining metamodels of the agricultural and industrial phases of the biorefinery (BONOMI et al., 2016). In the VB, CanaSoft is used to estimate the amount and costs of sugarcane, soybeans production, and sugarcane straw recovery (BONOMI et al., 2016), while the Aspen Plus® commercial process simulator (JUNQUEIRA et al., 2017) is used to simulate the industrial phase. The metamodels for this ethanol production chain were developed using the Factorial Design of Experiment 2^k with a Central Composite Circumscribed Design (CCC) based on the sampling and simulation of detailed models from the VB. Seven factors were considered in the crop system to obtain the system outputs (Figure 1). In total, 143 points were collected for each response without repetition at the central point. Likewise, four factors were considered for the industrial system. In total, 25 points were collected for each response without repetition at the central point. Statistical metamodels were generated by second-degree polynomial regression from the datasets (inputs and outputs) obtained from detailed models according to the experimental design matrix. In this fitting process, the main and interaction effects of two factors were tested, adopting a confidence level of 95%. The quality of the fit of the statistical metamodels was assessed using the coefficient of determination (R^2) and Analysis of Variance (ANOVA).

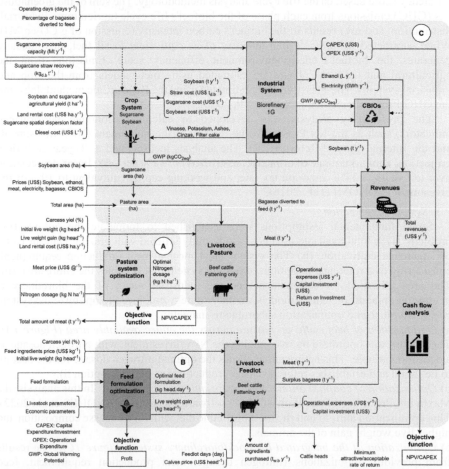

Figure 1. (A) Frontiers of optimizing pasture livestock farming. (B) Frontiers of optimizing feed for confined animals. (C) Frontiers of optimizing the integrated system of the sugarcane and beef cattle production chain

Regarding pasture, detailed models of beef cattle production developed by the Brazilian Agricultural Research Corporation (Embrapa) were used to obtain metamodels. In this analysis, a prototypical farm in São Paulo state (Brazil) was considered. To generate the metamodels, a second-degree polynomial statistical regression was conducted among different scenarios with varying levels of nitrogen fertilization for pasture maintenance. The response variables considered were the cattle stocking rate and the system's operational costs.

In the context of feedlot operations, the total cost per head per day was calculated, considering the costs of animal acquisition and feed, which represent the major portion of daily costs. Other operational costs, such as labor and sanitary management, as well investment costs, were estimated based on Sartorello et al. (2018). Thus, it was possible to estimate the total cost per cycle by providing information on the number of days in the confinement cycle, the number of confined animals, and daily costs.

Estimation of CBIOs generated depends on the biofuel's energy-environment efficiency score based on the life cycle analysis methodology. The sum of the greenhouse gas (GHG) emissions from each phase of the biofuel's life cycle (agricultural, industrial, distribution, and use) results in the biofuel's carbon intensity (emission of g CO_2eq/MJ). When subtracted from the carbon intensity of its equivalent fossil fuel, this intensity generates the biofuel's energy-environment efficiency score. Emissions from the agricultural and industrial phases are estimated using metamodels from the crop and industrial systems. The emissions were estimated using the RenovaCalc methodology and the GHG Protocol emission factors (ANP, 2020).

The profitability analysis was developed to calculate economic performance indicators, such as the Net Present Value relative to the investment (NPV/Investment) and the Internal Rate of Return (IRR). The analysis incorporated a 30-year cash flow projection, a minimum rate of attractiveness (MRA) of 12% per year, linear depreciation over ten years, a 34% corporate tax rate, and working capital equivalent to 10% of the investment (BONOMI et al., 2016).

2.3. System optimization

Some fundamental assumptions were established to formulate the optimization problem: (i) A total available area needs to be fully utilized with grazing land for cattle or sugarcane cultivation. (ii) The system must meet a meat production requirement, considering as a reference an optimized pasture system that occupies the entire area. (iii) The area allocated for confinement and the industrial sector of the biorefinery is negligible. In addition to metamodels, another strategy employed to optimize the system was to divide it into optimization subproblems, namely:

a) *Optimization of beef cattle on pasture across the entire available area (Figure 1A):* The pasture was optimized by maximizing the NPV/Investment, with nitrogen dosage as a decision variable. This optimization allows the optimal nitrogen dosage for the pasture to be found and establishes reference meat production (as per assumption ii).

b) *Optimization of confined animal feeding formulation (Figure 1B):* In this analysis, the Maximum Profit Feed Formulation (RLM) software was used (MARQUES et al., 2022). This optimization aims to find the most profitable diet with high bagasse composition and higher live weight gain of the animals.

c) *Optimization of the integrated livestock/biorefinery system (Figure 1C):* The results from previous optimizations (nitrogen dosage, meat production requirement, feed formulation, and confinement weight gain) are input parameters for the integrated system optimization. In this context, decision variables include the production capacity of the biorefinery and the straw recovery, aiming to maximize NPV/Investment.

Optimization algorithms were implemented in Python. To solve the optimization problem, the PYOMO library and the IPOPT ("Interior Point Optimizer") solver, which is suitable for nonlinear programming, were utilized. To run the optimization, the input parameter values were considered for São Paulo state (Table 1).

Table 1. Input parameters considered in the optimization for the state of São Paulo (Brazil) – valid for 2019

Variable	Value	Variable	Value
Total area (10^3 ha)	100	Carcass yield (%)	54
Sugarcane agricultural yield (t ha^{-1})	80	Feedlot days (day)	100
Land rental cost (US\$ ha.y^{-1})	240	Initial feedlot live weight (kg head^{-1})	380
Sugarcane spatial dispersion factor (%)	24	Calves price (US\$ head^{-1})	500
Diesel cost (US\$ L^{-1})	0.70	Ethanol price (US\$ L^{-1})	0.39
Soybean agricultural yield (t ha^{-1})	3.2	Electricity price (US\$ MWh^{-1})	42.2
Operating days (days y^{-1})	200	Soybean price (US\$ (60kg)$^{-1}$)	17.6
Bagasse diverted to feed (%)	10	Meat price (US\$ (15kg)$^{-1}$)	51.6
Initial pasture live weight (kg head^{-1})	322	Bagasse price (US\$/kg)	0.03
Pasture live weight gain (kg head^{-1})	150	CBIOS price (US\$/un)	10

3. Results and discussion

The metamodels showed statistical significance, with an R^2 between 0.95-0.99 for the biorefinery fits and 0.82-0.88 for the pasture-based livestock system.

For the input parameters, the subproblem optimization results were: (a) The optimal nitrogen dose equals 0 kg ha^{-1}, meaning it is economically better not to fertilize the pasture. This system produces 21.5×10^3 t y^{-1} of meat, with a maximum NPV/Investment equal to -0.75. (b) The optimal feed consists of 18.9% bagasse, 46.5% corn, 30% DDGS, 1.3% vegetable oil, 0.5% potassium chloride, 1% calcite, 0.8% urea, and 1% mineral salt. This feed results in a weight gain of 170 kg head^{-1} y^{-1} (c) The optimal capacity of the biorefinery is 6.33 Mt y^{-1}, and the optimal recovered straw is 6.74 kg$_{db}$ t^{-1}. Under these conditions, sugarcane cultivation occupies the entire available area, and all cattle are confined. The maximum NPV/Investment obtained was 0.81 (IRR = 21.7%).

In the optimization, a sensitivity analysis was performed for different values of meat price and land rent (Figure 2). In Figure 2A, the upper level indicates that all cattle are confined, while the lower level represents the majority kept on pasture, considering the integrated system. With the same meat price, an increase in land rent leads to a shift of cattle from pasture to confinement. For the same land rent, pasture is more valued at higher meat prices. As the meat price increases, it is possible to maintain pasture even with higher rental prices. The NPV/Investment decreases with increasing land rent and decreasing meat price (Figure 2B), and this change in the slope of the curve coincides with the shift in levels in Figure 2A. This curve can be characterized by the junction of two lines (Figure 2C): one where the biorefinery operates at its minimum capacity (2 Mt y^{-1}) and another at its maximum capacity (6.33 Mt y^{-1}), considering the integrated system with livestock. Up to the point of slope change, the 2 Mt y^{-1} line shows a higher NPV/Investment, making it more advantageous to keep pasture. From this point, the 6.33 Mt y^{-1} line exhibits a higher NPV/Investment, indicating that it is more advantageous to keep all cattle confined.

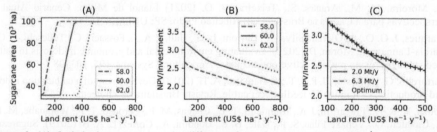

Figure 2. (A) Optimized sugarcane area for different values of meat price (US\$ 15kg^{-1}). (B) Optimized NPV/Investment for different values of meat price (US\$ 15kg^{-1}). (C) NPV/Investment for an integrated biorefinery of 2.0 and 6.3 Mt y^{-1}, considering 62 US\$ 15kg^{-1} of meat.

4. Conclusion

The methodology presented in this study allows evaluation under which conditions it is feasible to integrate biofuel production with beef cattle farming. In situations with high land rental costs and low meat prices, it is economically advantageous to confine cattle and release the area for the biorefinery. This analysis considers the expansion of sugarcane cultivation without negatively impacting food or livestock production while avoiding expansion into natural vegetation areas. Therefore, this study contributes by developing useful metamodels for ethanol-beef integrated production systems and methods for optimizing such systems, providing insights into the decision-making process. These tools also allow other scenarios to be analyzed to find other economically attractive solutions for the expansion of ethanol production in Brazil, utilizing pasture areas.

Acknowledgments

The authors gratefully acknowledge the financial support received from FAPESP (Grant No: 2017/11523-5) and CNPq (Grant No: 140029/2022-2). Additionally, we extend our appreciation to Embrapa Pecuária Sudeste for their invaluable contributions and support in successfully executing this research project.

References

Achinas, S., Horjus, J., Achinas, V., Euverink, G. J. W. (2019). A PESTLE Analysis of Biofuels Energy Industry in Europe. Sustainability, 11(21), 5981.

Allen, P. E., & Hammond, G. P. (2019). Bioenergy utilization for a low carbon future in the UK: the evaluation of some alternative scenarios and projections. BMC Energy, 1(1), 3.

ANP Agência Nacional de Petróleo, Gás Natural e Biocombustíveis: RenovaBio (2020). URL: https://www.gov.br/anp/pt-br/assuntos/renovabio. (accessed 1.12.23).

Bonomi, A., Cavalett, O., Cunha, M.P., Lima, M.A.P. (2016). Virtual Biorefinery – An Optimization Strategy for Renewable Carbon Valorization. Springer International Publishing, Switzerland.

Bressanin, J. M., Geraldo, V. C., Gomes, F. de A. M., Klein, B. C., Chagas, M. F., Watanabe, M. D. B., Bonomi, A., Morais, E. R. de, Cavalett, O. (2021). Multiobjective optimization of economic and environmental performance of Fischer-Tropsch biofuels production integrated to sugarcane biorefineries. Industrial Crops and Products, 170.

Junqueira, T. L., Chagas, M. F., Gouveia, V. L. R., Rezende, M. C. A. F., Watanabe, M. D. B., Jesus, C. D. F., Cavalett, O., Milanez, A. Y., Bonomi, A. (2017). Techno-economic analysis and climate change impacts of sugarcane biorefineries considering different time horizons. Biotechnology for Biofuels 10(1).

Neves, M. F., Milan, M., Valerio, F. R., Marques, V. N., Delsin, F. G., Cambauva, V., Martinez, L. F., Moreira, M. M., Arantes, S., Teixeira, G. O. (2021) Etanol de Milho: Cenário Atual e Perspectivas para a Cadeia no Brasil. (1.ed) Ribeirão Preto, SP: UNEM. v.1. 116p.

Marques, J. G. O., de Oliveira Silva, R., Barioni, L. G., Hall, J. A. J., Fossaert, C., Tedeschi, L. O., Garcia-Launay, F., Moran, D. (2022). Evaluating environmental and economic trade-offs in cattle feed strategies using multiobjective optimization. Agricultural Systems, 195, 103308.

Sartorello, G. L., Bastos, J. P. S. T., & Gameiro, A. H. (2018). Development of a calculation model and production cost index for feedlot beef cattle. Revista Brasileira de Zootecnia, v. 47.

Souza, N. R. D., Fracarolli, J. A., Junqueira, T. L., Chagas, M. F., Cardoso, T. F., Watanabe, M. D. B., Cavalett, O., Venzke Filho, S. P., Dale, B. E., Bonomi, A., Cortez, L. A. B. (2019). Sugarcane ethanol and beef cattle integration in Brazil. Biomass and Bioenergy, 120, 448–457.

UNICA – União da Indústria de Cana-de-açúcar. RenovaBio (2020). URL: https://unica.com.br/iniciativas/renovabio/. (accessed 1.11.23).

Flavio Manenti, Gintaras V. Reklaitis (Eds.), Proceedings of the 34th European Symposium on Computer Aided Process Engineering / 15th International Symposium on Process Systems Engineering (ESCAPE34/PSE24), June 2-6, 2024, Florence, Italy

Waste-to-X: An enviro-economic assessment of circular chemical production via municipal solid waste gasification

Ben Lyons,[a] Saxon Stanley,[a] Andrea Bernardi,[a] Benoît Chachuat[a,*]

[a] The Sargent Centre for Process Systems Engineering, Department of Chemical Engineering, Imperial College London, SW7 2AZ, UK

*b.chachuat@imperial.ac.uk

Abstract

As the generation of municipal solid waste continues to increase, its inadequate disposal is becoming a meaningful threat to the health of both humanity and the planet. Most waste is either landfilled or incinerated, which exacerbates climate change, incidence of respiratory diseases, and damage to marine environments. As the chemical industry moves towards a circular economy model, the gasification of municipal solid waste has garnered increased attention as it offers both an alternative waste management solution and the exploitation of an abundant renewable feedstock. In this work, three distinct municipal solid waste gasification pathways are investigated, with each leading to different chemical products. The routes considered are: i) methanol synthesis; ii) methanol-to-olefins; and iii) ethanol synthesis via dimethyl ether hydrocarbonylation. The results suggest that whilst the methanol route was competitive with the incineration base case in 2017-18 due to elevated methanol prices, all three routes would require higher gate fees on average (up to 3 times higher) to break-even across 2017-22 market conditions. The ethanol route displays the lowest required gate fees on average due to a lower H_2 requirement and lower capital costs. However, all routes achieve a 40-80% reduction in scope 1 and 2 CO_2 emissions when compared to incineration.

Keywords: municipal solid waste, gasification, circularity, economics, CO_2 emissions

1. Introduction

As the world population grows and standards of living improve, the excess generation and inadequate disposal of waste is increasingly becoming a fundamental issue. In 2016, it was estimated that two billion tonnes of waste were generated globally, approximately two thirds of which were either landfilled or openly dumped. This represented a significant contribution to global CO_2 emissions, with 1.6 billion tonnes emitted and is predicted to rise to 2.6 billion tonnes by 2050. Not only does it intensify climate change, but the improper burning and dumping of waste leads to the release of toxins and particulate matter, resulting in respiratory diseases. Additionally, marine ecosystems are damaged as waste runoff infiltrates rivers (Kaza et al., 2018). It is therefore clear that proper waste management is crucial to a lasting and secure future for the planet.

To address this environmental challenge, municipal solid waste (MSW) gasification has garnered increased attention as a more sustainable waste disposal option than landfilling or incineration (Ouedraogo et al., 2021). Historically, both research and implementation of MSW gasification has focused on either waste-to-energy via the combustion of the produced syngas, or fuel production, where the syngas is catalytically converted into fuels

(Niziolek et al., 2015; Kumar et al., 2017). However, as the chemical industry is beginning to transition towards a circular economy model, there is a growing research focus on chemical production via MSW gasification. The reported benefits of this route are three-fold: i) it utilizes an abundant waste resource; ii) it offers a more sustainable waste management option as previously mentioned; and iii) it provides an attractive economic element because a gate or tipping fee is typically received for processing the waste. However, syngas is a versatile chemical precursor, and it poses the question of what the optimal use of this syngas is when generated from MSW gasification. In this paper, three alternative chemical production routes via MSW gasification are compared. These routes are: i) methanol synthesis; ii) methanol-to-olefins (MTO); and iii) ethanol synthesis via dimethyl ether (DME) hydrocarbonylation (HC). The main aim of this work is to conduct a comparative assessment between these three routes by considering both their economic performance and their scope 1 and 2 emissions.

2. Methodology

All three processes were modelled using Aspen HYSYS and considered a 360 kt/y feed of MSW. A block flow diagram summarizing the routes is shown in Figure 1.

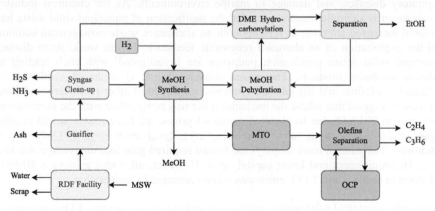

Figure 1: Block flow diagrams for the three routes (blue = methanol; yellow = ethanol; red = olefins)

2.1. Process Modelling

Feed Preparation: The MSW is first processed in a refuse-derived fuel (RDF) facility. Preprocessing the MSW into RDF reduces the heterogeneity of the feedstock via three main mechanisms: classification i.e., the removal of inert or incombustible materials (e.g., glass, scrap metal etc.), size reduction and drying. First the unwanted material is removed, which is assumed to be 20 wt% of the MSW feed. The remaining waste is then dried to a moisture content of 12 wt%, where the duty for this task is provided by cooling the syngas exiting the gasifier. It is assumed that the RDF facility requires 33 kWh of electricity per tonne of MSW for size reduction and grinding (Nuss et al., 2013). The resulting RDF is modelled as a hypothetical solid in HYSYS, where the elemental composition and lower heating value are taken from Jones et al. (2009). This fixed composition is an important assumption, as in reality it would still vary considerably even with RDF processing.

Gasification: The RDF is fed to a fluidized bed gasifier operating at 750°C and atmospheric pressure (Griggs et al., 2023). Since gasification is a complex process with many potential simultaneous reactions, it is modelled with a conversion reactor and a Gibbs reactor. The conversion reactor splits the RDF into its elemental components:

carbon, hydrogen, oxygen, nitrogen, sulphur, and ash, the latter of which is assumed to be SiO_2, CaO, Al_2O_3 and Fe_2O_3. The components are then fed to the Gibbs reactor alongside O_2 and steam as gasifying agents, the flowrates of which were adjusted until the desired outlet temperature and composition was achieved. Since gasification is an autothermal process, the duties of the conversion and Gibbs reactors must sum to zero. The resulting contaminated syngas leaving the Gibbs reactor contains CO, CO_2, H_2, H_2S, NH_3, CH_4 and steam. Tar formation was not modelled in this work, but the cost of a catalytic tar converter is included in the costing of the gasification unit, along with cyclones and particulate matter filters to remove any remaining ash and solids.

The syngas is then cooled (providing the RDF drying duty) and increased in pressure to 20 bar via a multi-stage compressor train. The Rectisol process with refrigerated methanol is then used to reduce H_2S and NH_3 levels to <0.1 ppmv, which is crucial for preventing downstream catalyst poisoning. This process was modelled as a component splitter, but the cost and performance of such a unit is considered (Kreutz et al., 2008). Rectisol also removes CO_2, but it can be recovered separately and re-introduced to the syngas stream to maintain a high carbon efficiency. The clean syngas contains CO_2, CO, H_2 with some CH_4 and can be fed to the subsequent chemical processes.

Methanol Synthesis: The methanol route is based on work by Van-Dal et al. (2013), where the clean syngas is further compressed to 78 bar. Additional H_2 is added to achieve the desired stoichiometric ratio for methanol production. This mixed stream is then fed to a packed-bed isothermal reactor filled with a copper-based catalyst at 250°C, where the reverse-water gas shift reaction takes place alongside methanol synthesis:

$$CO_2 + 3H_2 \rightarrow CH_3OH + H_2O \qquad (1)$$
$$CO + H_2O \rightarrow CO_2 + H_2 \qquad (2)$$

Unreacted syngas is separated in a flash vessel and recycled to the reactor feed. The methanol product is then separated from water in a distillation column.

Methanol-to-Olefins: This route is based on the UOP/Norsk Hydro process where methanol is fed to a circulating fluidized bed reactor with a SAPO-34 catalyst. Here, the methanol undergoes various dehydration reactions to form ethylene and propylene as the desired products. Additional side products include higher olefins, alkanes, coke, H_2, CO, CO_2 and water. This product stream is then sent to an olefins separation section, which includes a water quench, multi-stage compression, CO_2 removal, and a distillation train to separate the olefins and alkanes. An olefin cracking process (OCP) is included to crack the higher olefins into additional ethylene and propylene (Hannula et al., 2015).

Ethanol Synthesis via DME Hydrocarbonylation: In this route, only a small fraction of the clean syngas leaving the Rectisol process is fed to the methanol synthesis section; the rest is diverted to the HC reactor further downstream. The produced methanol is pumped to 12 bar and fed to an adiabatic packed bed reactor to form DME at 250°C in the following dehydration reaction (Ng et al., 1999):

$$2CH_3OH \rightarrow CH_3OCH_3 + H_2O \qquad (3)$$

The DME is separated from unreacted methanol and water (which are sent back to the methanol-water distillation column) and fed to the HC reactor at 220°C and 15 bar. This reactor combines both the carbonylation of DME to form methyl acetate (MA), and the hydrogenation of MA to form ethanol and methanol (Zhang et al., 2010):

$$CH_3OCH_3 + CO \rightarrow CH_3COOCH_3 \qquad (4)$$
$$CH_3COOCH_3 + 2H_2 \rightarrow C_2H_5OH + CH_3OH \qquad (5)$$

The required CO and H_2 is provided by the diverted syngas, which is also topped up with additional H_2 to achieve the desired stoichiometric ratio for DME HC. Note that DME carbonylation requires a 10:1 CO:DME ratio, hence why most of the syngas is diverted away from the methanol synthesis section. The product stream contains a mixture of unreacted syngas, DME, MA, methanol, ethanol and CO_2. The CO_2 is removed first before a flash vessel separates the syngas and most of the DME in the gaseous stream, which is recycled to the HC reactor. The liquid stream consisting of methanol, ethanol, MA and DME is further separated in a sequence of two distillation columns and a dimethyl sulfoxide extractive distillation process to break the MA-methanol azeotrope. DME and MA are recycled to the HC reactor, the methanol is recycled to the DME synthesis reactor, and the ethanol is obtained as the final product.

2.2. Route Assessment

Economics: The factorial method with purchased equipment costs was used to obtain the total capital investment for each route, where a 20-year lifetime and 10% interest rate was chosen. The cost of the RDF facility, gasifier (with tar conversion and solids separation), and Rectisol process are obtained from Niziolek et al. (2015). The metric used to compare each route is the MSW gate-fee that is required to break-even across 2017-22 market conditions i.e., the break-even gate fee (BEGF). UK wholesale data was used for the natural gas and electricity prices and it is assumed that there is sufficient renewable energy for the generation of green H_2 at 3 USD/kg (IEA, 2019). All costs and prices were adjusted to a USD 2022 basis. The routes were also compared to the most common gate fee range used in Energy-from-Waste (EfW) facilities in the UK (WRAP, 2017-22).

CO_2 Emissions: The scope 1 and 2 emissions for each route were used to compare environmental performance, which accounts for direct emissions from the process and emissions from energy use. The carbon intensity of electricity was taken as the 2022 grid average in the UK (257 gCO_2/kWh) and natural gas was the intensity of heating from a CCGT plant (394 gCO_2/kWh; Staffel, 2017). An EfW plant acts as the base case, where it is assumed that all the carbon in the RDF is converted into CO_2 and the net heat and electricity generation efficiencies are 5% and 22.5% respectively (Hogg, 2023).

3. Results and Discussion

Break-Even Gate Fees: Figure 2(a) shows the calculated BEGF for each route across varying market conditions from 2017-22. All three routes require a higher BEGF than a typical EfW plant in the UK in most of the market conditions considered. The MTO route is the most expensive, with a BEGF ranging from 0.23-0.39 USD/kg$_{MSW}$ or 1.6-2.8 times greater than the EfW base case. This is predominantly caused by the higher capital investment for an additional compression train and refrigeration cycle in the olefins separation section. Olefin prices also dropped from 2019-20 leading to a steady increase in the BEGF. The further increases in BEGF in 2021-22 were both caused by sharp increases in the cost of electricity and natural gas in the UK. The results for the methanol route demonstrate a better economic performance than MTO, with BEGFs ranging from 0.12-0.28 USD/kg$_{MSW}$. It is also the only route that manages to be both competitive with and cheaper than EfW in 2017-18, when higher methanol prices were coupled with low energy prices. The ethanol route shows the lowest variability in BEGF at 0.15-0.25 USD/kg$_{MSW}$ or 1.2-1.8 times greater than EfW. This route has similar capital costs to the methanol route despite having a more complex process. This is caused by two main factors: i) most of the syngas is diverted away from the high-pressure compressor train for methanol synthesis, so compression costs are substantially reduced; and ii) the route has a lower H_2 requirement as the stoichiometric ratio of $H_2:CO$ for the global hydrocarbonylation reaction is lower than that of methanol synthesis. Coupled with a higher value product, these points lead to a lower BEGF compared to the other two routes.

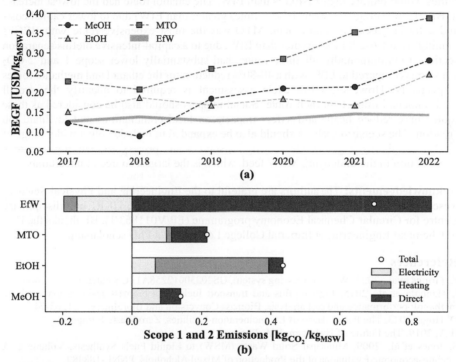

Figure 2: (a) Break-even gate fees and (b) Scope 1 & 2 emissions for each Waste-to-X route.

CO₂ Emissions: Figure 2(b) shows the levelized scope 1 and 2 CO_2 emissions, where each route achieves a substantial reduction in emissions compared to EfW, even after considering the credits for generated heat and electricity. The ethanol route has the greatest emissions of the three routes at 0.43 kg_{CO2}/kg_{MSW}, which still represents a ~40% reduction compared to EfW at 0.70 kg_{CO2}/kg_{MSW}. Most emissions from the ethanol process are due to its high heating requirements, which totals to ~3 MJ/kg_{MSW} with heat integration. The main heating sources are the reboiler duties in the methanol-ethanol and methanol-water distillation columns, for which there is limited potential for integration. A less carbon-intensive heating source would be the main improvement to reduce emissions further in this route. The MTO route has the next highest levelized emissions at 0.22 kg_{CO2}/kg_{MSW}. In contrast to the ethanol route, the MTO option shares most of its emissions equally between direct emissions and electricity. The large direct emissions are due to the combustion of multiple side product streams, including alkanes and coke, whereas the larger electricity requirement comes from the second compression train and refrigeration cycle. Finally, the methanol route has the lowest emissions overall at 0.14 kg_{CO2}/kg_{MSW}, an 80% reduction over EfW. After heat integration, this route has no heating requirement, but it has a higher electricity requirement than the ethanol route since all of the syngas is compressed to 78 bar for methanol synthesis.

4. Conclusions

This paper determined the economic and environmental performance of producing three different chemicals (methanol, ethanol, and light olefins) via municipal solid waste gasification and compared them to an incineration plant (EfW). The economic results revealed that whilst methanol was competitive with EfW in 2017-18, on average all three routes would require higher BEGFs than EfW. The ethanol route had the lowest increase in BEGF on average (between 1.2-1.8 times greater than EfW) due to lower capital costs and a lower green H_2 requirement. MTO was the most expensive route with BEGFs ranging from 1.6-2.8 times greater than EfW, due to a capital-intensive olefins separation section. Environmentally, all three routes had substantially lower scope 1 and 2 CO_2 emissions compared to EfW, with a 40-80% reduction for the ethanol and methanol routes respectively. However, a life cycle assessment is required to quantify the overall environmental impacts of each route. Social impacts should also be considered as these routes may reduce health and safety impacts on local communities and facilitate job creation. The scenario analysis should also be expanded to include both a wider range of market conditions (e.g., include varying H_2 sources and prices), and varying feedstock composition in the underlying MSW feed, which is the largest source of uncertainty.

Acknowledgements: The authors are grateful to the Engineering and Physical Sciences Research Council (EPSRC) for funding the research under the UKRI Interdisciplinary Centre for Circular Chemical Economy programme (EP/V011863/1). BL thanks the Dpt. of Chemical Engineering at Imperial College London for a PhD scholarship.

References

K. Griggs et al., 2023, Waste processing system, US20230012258A1 (US Patent)
I. Hannula et al., 2015, Light olefins and transport fuels from biomass residues via synthetic methanol: performance and cost analysis, Biomass Conversion and Biorefinery, 5, 1, 63-74
D. Hogg, 2023, The Performance of EU Incineration Facilities, Zero Waste Europe
IEA, 2019, The Future of Hydrogen, IEA, Paris
S. Jones et al., 2009, Municipal Solid Waste (MSW) to Liquid Fuels Synthesis, Volume 2: A Techno-economic Evaluation of the Production of Mixed Alchohols, PNNL-188482
S. Kaza et al., 2018, What a Waste 2.0: A Global Snapshot of Solid Waste Management to 2050, Urban Development Series. Washington, DC: World Bank
T. Kreutz et al., 2008, Fischer-Tropsch Fuels from Coal and Biomass, 25th Annual International Pittsburgh Coal Conference
A. Kumar, S. R. Samadder, 2017, A review on technological options of waste to energy for effective management of municipal solid waste, Waste Management, 69, 407-422
K. Ng et al., 1999, Kinetics and modelling of dimethyl ether synthesis from synthesis gas, Chemical Engineering Science, 54, 15-16, 3587-3592
A. Niziolek et al., 2015, Municipal solid waste to liquid transportation fuels – Part II: Process synthesis and global optimization strategies, Computers & Chemical Engineering, 74, 184-203
P. Nuss et al., 2013, Environmental Implications and Costs of Municipal Solid Waste-Derived Ethylene, Journal of Industrial Ecology, 17, 6, 912-925
A. Ouedraogo et al., 2021, Comparative Life Cycle Assessment of Gasification and Landfilling for Disposal of Municipal Solid Wastes, Energies, 14, 21, 7032
I. Staffel, 2017, Measuring the progress and impacts of decarbonising British electricity, Energy Policy, 102, 463-475
E. Van-Dal et al., 2013, Design and simulation of a methanol production plant from CO_2 hydrogenation, Journal of Cleaner Production, 57, 38-45
WRAP, 2017-22, Comparing the costs of alternative waste treatment options, Gate Fees Report
Y. Zhang et al., 2010, Novel Ethanol Synthesis Method via C1 Chemicals without Any Agriculture Feedstocks, Indsutrial & Engineering Chemistry Research, 49, 11, 5485-5488

Flavio Manenti, Gintaras V. Reklaitis (Eds.), Proceedings of the 34th European Symposium on Computer Aided Process Engineering / 15th International Symposium on Process Systems Engineering (ESCAPE34/PSE24), June 2-6, 2024, Florence, Italy

Vulnerability of Microbial Communities to Dishonest Signals: A biology driven complexity reduced model.

Jian Wang, Ihab Hashem, Satyajeet S. Bhonsale, Jan F.M. Van Impe*

BioTeC+ - Chemical and Biochemical Process Technology and Control, KU Leuven Ghent, Gebroeders de Smetstraat 1, 9000 Ghent, Belgium
Email: jan.vanimpe@kuleuven.be

Abstract

In multispecies communities, bacteria are capable of coordinating population-level behaviour through a signalling mechanism termed quorum sensing (QS). However, these signals can be either 'honest,' mutually beneficial, or 'dishonest,' benefiting only the sender. This study performs the biology-driven complexity reduction on the system with two species to evaluate the vulnerability and adaptability of microbial communities to dishonest signalling. We corroborate the finding that sensitive bacterial species S can gain advantages from early toxin production by a specific QS species P, even without detecting the signal. Adopting a dynamic analysis framework, we examine the potential dynamics of a bacterial community in relation to various physiological factors, focusing on the stability of different equilibria as a function of factors. Our analysis identifies the key QS-regulatory factors, such as toxin thresholds and signalling production costs, which critically determine the vulnerability of bacterial communities to dishonest signalling. By incorporating the effects of nutrient diffusion and stochastic elements, we verified the population level findings with individual based models. These obtained insights could extend our ecological understanding of microbial signalling and offer avenues for the development of quorum-quenching antibacterial and probiotic strategies.

Keywords: Quorum sensing; Biology-driven complexity reduction; Dynamic analysis; Stability and structure; Antibiotic strategy

1. Background

The original model consists of five state variables describing the dynamics of two bacterial strains, P and S, and their interactions with nutrients N, toxin T, and another signalling molecule Q. The system captures the complex interactions between different bacterial populations, toxin, nutrient, and quorum-sensing molecules; Quorum sensing enables species P to alter its behavior based on population density, influencing toxin production; The toxin T acts as a control variable that affects the growth of sensitive bacteria S but is also produced by P. The equations are given by (Bucci et al., 2011):

$$\frac{dP}{dt} = (1 - f\,H(Q - G_{th}) - qp)\,\mu_{max}\left(\frac{N}{N + K_n}\right)P$$

$$\frac{dS}{dt} = (1 - qs)\left(\mu_{max}\left(\frac{N}{N + K_n}\right) - K_t T\right)S$$

$$\frac{dT}{dt} = \alpha f\,H(Q - G_{th})\,\mu_{max}\left(\frac{N}{N + K_n}\right)P - \beta_T T$$

$$\frac{dN}{dt} = -\frac{1}{Y}\,\mu_{max}\left(\frac{N}{N + K_n}\right)(P + S)$$

$$\frac{dQ}{dt} = qp\,\mu_{max}\left(\frac{N}{N+K_n}\right)P + qs\mu_{max}\left(\frac{N}{N+K_n}\right)S$$

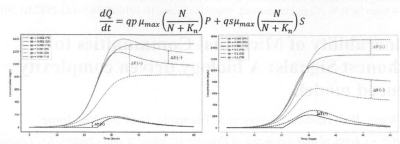

Figure 1: Growth dynamics of species P and S in response to variations in quorum sensing parameters q_p and q_s

With f the fraction of toxin produced, $H(Q - G_{th})$ the step function, μ_{max} the maximum specific growth rate, K_n the nutrient half-saturation constant, K_t toxin killing rate for sensitive bacteria, α the toxin production rate, β_T the toxin decay rate, Y the yield coefficient, q_p the quorum sensing molecule production by toxin-producing species, q_s the quorum sensing molecule production by sensitive bacteria, G_{th} the quorum sensing threshold.

2. Complexity Reduction

2.1. Assumption

The quasi-steady assumption allows for a simplification of this system. This assumption is based on the idea that certain variables reach a quasi-steady state much faster than others. In the simplified model, we assume that the toxin concentration T and the quorum sensing signal Q reach a quasi-steady state quickly compared to the bacterial populations P and S. This enables us to eliminate T and Q from the differential equations, reducing the system to a more tractable form for both mathematical analysis and computational simulations. Biologically, this simplification is valid under certain conditions:

- Rapid equilibration of metabolic processes compared to cellular division and death rates.
- Short time scales for nutrient uptake and release of metabolic by-products.
- Negligible time delays in regulatory responses within the cell.

2.2. Reduced models

Using the quasi-steady assumption, the equations for describing the dynamics of P and S simplified as:

$$\frac{dP}{dt} = \left(1 - \frac{f}{1 + \exp(-k(P - G_{th}))} - qp\right)\mu_{max}\left(\frac{N_0}{N_0 + K_n}\right)P + aPS$$

$$\frac{dS}{dt} = (1 - qs)\left(\mu_{max}\left(\frac{N_0}{N_0 + K_n}\right) - K_t\frac{\alpha.f}{1 + \exp(-k(P - G_{th}))}\frac{\mu_{max}}{\beta_T}P\right)S - bPS$$

In this above model, toxin T, quorum sensing signal Q and N nutrient is explicitly incorporated into the P and S dynamics, the limiting effects of toxin producing strain to sensitive strain are captured by Lotka-Volterra competitive structure. Comparisons of parametric plots from both the original and reduced models, within a similar value range, demonstrate parallel evolutionary dynamics

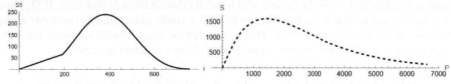

Figure 2: Comparative dynamics of species P and S in the reduced model

3. Model Analysis

3.1. Dynamic analysis of the models

To investigate the stability of the system, we begin by finding the equilibrium points, where $(\frac{dP}{dt} = 0)$ and $(\frac{dS}{dt} = 0)$, *with certain parameters setting* . The Jacobian matrix J of the system is given by:

$$\begin{bmatrix} \dfrac{\partial\left(\frac{dP}{dt}\right)}{\partial P} & \dfrac{\partial\left(\frac{dP}{dt}\right)}{\partial S} \\[2em] \dfrac{\partial\left(\frac{dS}{dt}\right)}{\partial P} & \dfrac{\partial\left(\frac{dS}{dt}\right)}{\partial S} \end{bmatrix}$$

$$\lambda_1 = \frac{\mu_{max} \times N_0 \times (1 - q_s)}{Kn + N_0}$$

$$\lambda_2 = \frac{\mu_{max} \times N_0 \times \left(-f - q_p \times e^{Gth \times k} - q_p + e^{Gth \times k} + 1\right)}{K_n \times e^{Gth \times k} + K_n + N_0 \times e^{Gth \times k} + N_0}$$

Equilibrium points 1:

$$(P[t], S[t]) = (0,0)$$

In the biological realistic parameters settings , λ_1 is always positive, making this equilibrium point an unstable node regardless of the value of k. This implies that any small perturbation in the bacterial populations P and S will result in the system moving away from this equilibrium point. In the biological context, this indicates that a state with zero populations for both P and S strains is not sustainable and will naturally evolve to a state with non-zero populations.

Equilibrium points 2:

$$(P[t], S[t]) = \left(\frac{\ln\left(\dfrac{(1 - q_p)\exp(Gth \times k)}{f + q_p - 1}\right)}{k}, 0 \right)$$

$$\lambda_1 = \frac{0.0009999 \times (0.1k - 1000.1 \times \ln(-1.1 \times e^{1.4k}))}{k}$$

$$\lambda_2 = 0.0009 \times \ln(-1.1 \times e^{1.4k})$$

If both (λ_1) and (λ_2) are positive, the equilibrium point is an unstable node, meaning the system will move away from this state. If both (λ_1) and (λ_2) are negative, the equilibrium point is a stable node, and the system will naturally evolve towards this state. If (λ_1) and (λ_2) have opposite signs, the equilibrium point is a saddle point, which is unstable along one axis and stable along the other. This means that the system could approach this state from certain initial conditions but depart from it along a different trajectory.

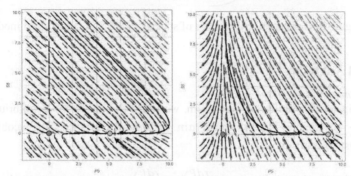

Figure 3: Phase plane analysis of the system's equilibrium points by varying the value of q_p and q_s.

3.2 Parameter sensitivity analysis of the reduced model

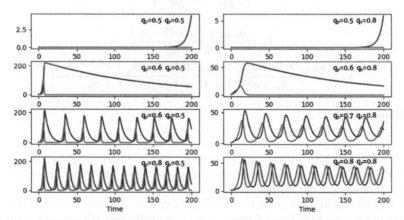

Figure 4: Sensitivity analysis of microbial communities to parameter variations. By increasing the q_p and q_s value, the system is gradually transitioning to the co-existence.

3.2. Design of Individual based model (IbM)

This IbM includes two types of entities: microbial cells and their environment. Cells, represented as disks in the graphical interface, are defined by their species, size, mass, spatial coordinates, and nutrient uptake rates, with their community structure indicated by the color intensity. The environment is a 40 mm by 20 mm section of 4 mm thick gel, divided into 10 mm square units, each with specific nutrient and toxin concentrations. The model encompasses various processes: intracellular activities like growth, maintenance, and division, managed in dense populations through cell shoving;

intercellular and environmental interactions, including glucose uptake and toxin secretion affecting neighboring cells; and diffusion processes following Fick's second law, which dictate nutrient and toxin distribution. The simulation uses two time resolutions: a smaller time step ($\Delta t_1 = 0.005min$) for rapid dynamics like diffusion, nutrient uptake, cell growth, and toxin secretion, and a larger time step ($\Delta t_2 = 0.1min$) for slower processes such as cell reproduction and shoving (Tack et al. 2015).

3.2.1. Physical and chemical sub-models

- *Growth and Maintenance:* cells uptake nutrients from their local environment and metabolize them for cell maintenance. When the maintenance requirements are fulfilled, the excess of assimilated glucose is directed toward increasing their mass and volume. The cellular glucose uptake from the environment is assumed to follow Monod kinetics:

$$\frac{dS_{(i,j)}^k}{dt_1} = v_{(i,j)}^k = q_{S,max}^k \cdot \frac{C_{S,(i,j)}}{K_S + C_{S,(i,j)}}$$

where cell k located at the at coordinates (i,j); $C_{S,(i,j)}$ is the glucose concentration of the environmental unit.

- *Shoving:* The high density of cells in a colony result in spatial overlap of neighboring cells, which causes pressure between the cells. This pressure can be alleviated through a mechanism called cell shoving.

$$r_o = s.r_k + r_n - d$$

where r_k and r_n define the radius of cell k and its neighbor n, d is the distance between the centers of these two cells

- *Nutrient Diffusion:* The second law of Fick can be used to describe the movements of nutrients through 2D isotropic media. Simulated cells can sense the nutrient concentration of their immediate environment and adjust their nutrient uptake accordingly based on this local availability.

$$\frac{\partial C_i}{\partial t} = D_i \cdot \left(\frac{\partial^2 C_i}{\partial x^2} + \frac{\partial^2 C_i}{\partial y^2} \right)$$

where C_i represents the concentration of substance i, D_i is the diffusion coefficient of substance i

- *Toxin Secretion:* Each species secretes toxins affecting neighbouring cells of other genres, modulating their growth rates.

$$\mu_s = \mu(1 - q_p) - TK_p \times T_p$$

where TK_p represents the toxin killing rate of species P, μ_s is the growth rate of species.

Figure 5: Patterns of two species competing by varying the toxin production costs from 0.15, 0.30 to 0.50.

4. Conclusions

This study presents a comprehensive examination of the interactions between toxin-producing P and sensitive bacterial strains S. The primary aim is to understand the conditions that lead to either dominance of coexistence of these strains in microbial communities. To achieve this, we employ the biology-driven quasi-steady assumption to perform a nuanced equilibrium analysis. This analysis specifically identifies the biological and environmental factors that dictate whether P strains dominate or coexist with S strains in the microbial community. To capture the complexities not addressed by the deterministic model, we introduce Individual-Based Models (IbMs). These models allow us to incorporate stochastic elements and spatial heterogeneity, thereby offering a more comprehensive understanding of bacterial interactions. Our findings reveal that key parameters, such as toxin production costs, play a significant role in determining the stability and structure of bacterial communities. This has important practical implications, offering guidelines for the development of more effective antibacterial strategies and probiotic formulations.

Acknowledgement

This work was supported by the E-MUSE project funded by the European Union's Horizon 2020 Research and Innovation Programme [MSCA grant 956126]

References

Bucci, V., Nadell, C. D., & Xavier, J. B. (2011). The evolution of bacteriocin production in bacterial biofilms. The American Naturalist, 178(6), E162-E173.

Strogatz, S. H. (2018). Nonlinear dynamics and chaos with student solutions manual: With applications to physics, biology, chemistry, and engineering. CRC press.

Hashem, I., & Van Impe, J. F. (2022). Dishonest signaling in microbial conflicts. Frontiers in microbiology, 13, 468.

Tack, I. L., Logist, F., Fernández, E. N., & Van Impe, J. F. (2015). An individual-based modeling approach to simulate the effects of cellular nutrient competition on Escherichia coli K-12 MG1655 colony behavior and interactions in aerobic structured food systems. Food Microbiology, 45, 179-188.

Flavio Manenti, Gintaras V. Reklaitis (Eds.), Proceedings of the 34th European Symposium on Computer Aided Process Engineering / 15th International Symposium on Process Systems Engineering (ESCAPE34/PSE24), June 2-6, 2024, Florence, Italy

Accelerating Steam Cracking Simulations with Surrogate-Assisted Parameter Estimation

Qiming Zhao,[a,b] Dong Qiu,[a,b] Kexin Bi,[c] Tong Qiu[a,b*]

[a]Department of Chemical Engineering, Tsinghua University, Beijing 100084, China
[b]Beijing Key Laboratory of Industrial Big Data System and Application, Tsinghua University, Beijing, China
[c]School of Chemical Engineering, Sichuan University, Sichuan 610065, China
ilmaple@163.com

Abstract

The paper presents a method that accelerates steam cracking simulations by employing machine learning surrogates for parameter estimation. This approach utilizes Latin hypercube sampling to create initial simulation datasets. It applies Bayesian ridge regression to kernel approximations of the data, enabling efficient prediction of simulation parameters with measurable uncertainty. Compared to traditional iterative techniques, this method speeds up simulations by reducing the average number of iterations by 4.33. Data-driven models using pre-computed simulation data have been shown to enhance the efficiency of simulators based on first principles significantly.

Keywords: Steam Cracking, Parameter Estimation, Surrogate Modeling

1. Introduction

Steam cracking is the primary process for producing light olefins. Research in this field depends on process simulation models, including rigorous and data-driven ones, to enable yield optimization by adjusting operating conditions. Rigorous models use equations based on first principles, allowing for broad applicability. On the other hand, data-driven models operate as black boxes capable of approximating input-output relationships but have constrained predictive power during extrapolation (Zhao et al., 2023). It is expected to use data-driven techniques to enhance rigorous simulations, which can speed up calculations or improve model fidelity.

When solving complex simulations numerically, iterative methods help manage the challenge of measuring specific parameters. Sequential iterations can be time-consuming, but choosing initial values wisely can accelerate convergence. Due to the nonlinear and dynamic nature of real-world processes, algorithms generally rely on heuristics for parameter initialization, which are customized for particular operating conditions and model frameworks. An alternative approach is using machine learning to provide better initial values of parameters. This can be done by constructing a surrogate model using a dataset of offline simulations that were pre-computed in parallel by the rigorous model.

This paper presents a surrogate-assisted approach to parameter estimation, which uses a regression model to guide the selection of parameters. A simulation database is built from parallel pre-computation of simulations. Using a Bayesian ridge regression model allows for estimating parameters with quantifiable uncertainties, thereby reducing the iterations required to attain similar levels of precision. Measuring uncertainty enables users to set simulation search limits informed by probabilistic understanding.

2. Methods

Steam cracking process efficiency is impacted by feedstock quality and furnace performance. Critical parameters like feedstock and steam flow rates, temperature, and pressure must be monitored to evaluate industrial production efficiency (Fakhroleslam et al., 2020). We developed a first-principles simulator for the radiation section of an industrial steam cracker that processes light hydrocarbons. The simulator only considers axial changes in temperature and pressure, resulting in simplified one-dimensional profiles. Once the reaction network and furnace settings are established, the simulator takes input such as feedstock compositions and operating conditions.

Integral calculations are made during simulation to achieve the desired COT and COP. Two parameters, dTg and VPR, are adjusted at each iteration. dTg measures fuel adjustments needed within furnace control systems to meet COT targets, while VPR reflects pressure adjustments for desired COP. When started from fixed parameters, a simulation needs multiple sequential iterations to reach an acceptable error threshold.

The proposed surrogate-assisted parameter estimation approach is applied to the simulator mentioned above, with feedstocks assumed to contain only ethane, propane, and n-butane. It aims to approximate the relationship between input variables, estimated parameters, and convergence loss. The approach consists of two phases: surrogate construction and surrogate-assisted simulation, as shown in Figure 1.

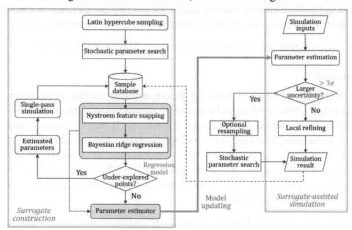

Figure 1. Flowchart of the proposed framework

2.1. Surrogate construction

This phase involves sampling the input space, conducting simulations on these samples to create a database, and developing a surrogate model for parameter estimation.

Variable ranges of concern in industrial production are identified using production datasets and furnace design specifications. Given the vastness of the resulting input space, simple random or grid sampling may introduce bias and patterns. To address this, Latin hypercube sampling (LHS), a traditional technique for space-filling experimental designs, is used for sampling. LHS ensures adequate coverage of both the input space and subspaces (Liu et al., 2018), producing representative samples. This leads to improved performance in industrial settings where few variables change simultaneously. The method is also sample-efficient, making it ideal for use with costly simulators.

Figure 2 illustrates the process of collecting and modeling simulated data.

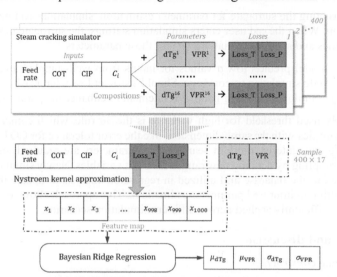

Figure 2. A schematic diagram of the simulator and the regression model

Initial simulations begin using random parameters. The diagram above displays the scenario with 400 samples undergoing 16 parallel stochastic parameter searches. These results are the basis of the simulation database. To ensure the database captures scenarios near convergence, an extra simulation is conducted for each sample using parameters derived from a simple linear interpolation of losses from the previous searches.

The regression model uses the Nystroem method for kernel map approximation combined with Bayesian ridge regression to balance accuracy and efficiency. Since linear models are inadequate for this complex system, kernel tricks are used to capture the nonlinear traits of the data, where the idea is to map data into a high-dimensional feature space to make it easier to perform linear classification or regression. However, working in such high-dimensional spaces can be computationally intensive, especially for large datasets. Gaussian processes (GPs), which use covariance functions as kernels, do not scale well to large datasets. Training GPs also require hyperparameter tuning, which complicates real-time model updates.

The Nystroem method produces a low-rank approximation of the original kernel matrix using a strategically selected subset of the training data, typically using a radial basis function (RBF) kernel (Martín et al., 2019). It is essential to tune the number of components in the feature map. Meanwhile, Bayesian ridge regression enhances the conventional ridge regression by integrating it into a Bayesian framework and introducing Gaussian priors on the coefficients for improved regularization. This method provides point estimates similar to standard linear regression and measures uncertainty about these estimates via the posterior distributions of random variables (Shi et al., 2016).

The surrogate model estimates parameters based on simulation inputs and convergence losses. It is trained using data that includes actual loss values, but it treats convergence losses as zero during prediction. This approach enables model training with data from simulations that have not fully converged.

2.2. Surrogate-assisted simulation

After constructing the surrogate for parameter estimation, simulations can be assisted by its prediction. When the end user enters simulation settings, the surrogate model predicts average values and standard deviations (σ) for those parameters.

A surrogate model predicts low σ values for inputs from well-sampled areas, typically requiring a single simulation to converge. Should a stricter loss tolerance be required, the software may employ iterative or parallel refinement to enhance the parameter further.

A commonly used threshold for high variance is the 3σ rule, which compares 3σ with physical quantities with the same dimension, like the error tolerance for COT. When high σ values are encountered, random searches for parameters are required, or users may broaden resampling by defining new ranges for input variables. Newly simulated samples are appended to the database and utilized in regression analysis to update the surrogate model, resulting in improved parameter estimations. This approach allows computational efforts to be efficiently applied across different simulation scenarios.

3. Results and discussion

3.1. Surrogate construction

Input variables were sampled within the bounds specified in Table 1.

Table 1. Sampled input variables

Variable	Description	Lower bound	Upper bound
Propane	Propane mass fraction	0.1	0.2
n-Butane	n-Butane mass fraction	0.1	0.2
(Ethane)	Ethane mass fraction	0.6	0.8
FFR	Feed flow rate (t/h)	9	13
COT	Coil output temperature (°C)	860	875
CIP	Coil input pressure (MPa)	0.23	0.37

The decisions and variations in industrial production determined the boundaries shown above. Ethane mass fractions were calculated from the remaining feed compositions. The LHS method generated 400 input samples, and each sample led to 16 random parameter searches in parallel, where dTg ranges from -25 K to 25 K, and VPR values are between 0.55 and 0.85. Subsequently, an additional 400 simulations were conducted using linearly interpolated parameter values. The training set gathered 6,800 instances in total.

The training data were first normalized to reduce numerical artifacts. Subsequently, the Nystroem method was used to project the inputs to 1,000 components. The regression model then correlated these features to two targets, dTg and VPR. Table 2 displays the model metrics, including the root-mean-square error (RMSE) and coefficient of determination (R^2) for both the training and test sets.

Table 2. Regression model test metrics

Target	RMSE (train)	R^2 (train)	RMSE (test)	R^2 (test)
dTg (K)	0.291	0.9997	0.521	0.9990
VPR	0.00142	0.9998	0.00280	0.9992

The model fits the simulated data well without obvious overfitting. Considering the simulator's convergence tolerance, the estimated parameters are satisfactory, with predicted temperature deviations near 0.5K and pressure deviations under 1 kPa.

3.2. Surrogate-assisted simulation

An additional 100 samples were LHS-generated as the test set, mimicking end-user inputs in the same region of furnace operation. The surrogate model predicted parameters corresponding to the converged simulations for these samples, and simulations are carried out using these parameters. Meanwhile, stochastic parameter searches were also tested for comparison. Figure 3 illustrates the predictive capabilities of the surrogate-assisted simulation using two evaluation methods: 1) reduced loss compared with random searches and 2) reduced iterations compared with a sequential iterative algorithm.

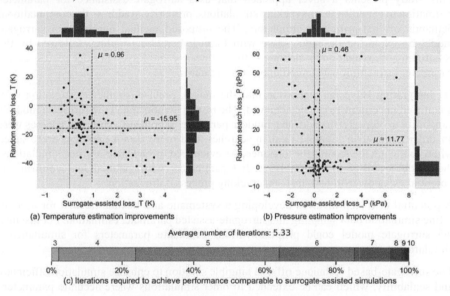

(a) Temperature estimation improvements   (b) Pressure estimation improvements

(c) Iterations required to achieve performance comparable to surrogate-assisted simulations

Figure 3. Surrogate model evaluation

Figures 3(a) and 3(b) illustrate the improvements made in parameter estimation of temperature and pressure. The surrogate model generated predictions for each sample, verified using the original simulator (shown on the x-axis). The results were compared to a random parameter search (shown on the y-axis), where the best outcome (minimum loss) was selected for each sample. Random search losses demonstrated a mean shift and high variance, and pressure data also took a very skewed distribution. On the contrary, surrogate-assisted simulations contributed to average losses that were an order of magnitude lower than the best outcomes achievable through random searches. The results were still slightly skewed, probably due to biased training datasets generated by stochastic searches. Increasing the sample size may reduce bias and the long tail in distribution. The surrogate model consistently outperformed the random search by reducing computational resource usage while maintaining accuracy.

Regarding uncertainties, the average simulated COT loss was consistent with the model's predicted standard deviations of dTg. In the 100 surrogate-assisted simulations, the mean COT loss ($\mu_{\text{loss\_T}} = 0.96$) was almost identical to the mean predicted standard deviation

($\bar{\sigma}_{dTg} = 0.97$). The rationale remains the same for both parameters, but VPR must be multiplied by CIP to restore the unit during pressure variance calculations.

Figure 3(c) demonstrates the number of iterations required to reach the same or smaller absolute loss for dTg and VPR as the simulator starts from a fixed parameter set (dTg=0 and VPR=0.7). Most samples needed four or more iterations, and some even reached 8~10. The mean iteration times were 5.33, indicating that the surrogate model can reduce simulation time by more than 4 iterations on average. This is valuable when a single iteration takes several minutes.

4. Conclusions

This study presents a novel approach that uses surrogate assistance for parameter estimation to expedite steam cracking simulations, primarily to address the computational demands of first-principles simulators. The proposed method constructs a surrogate model by exploring the input space with Latin hypercube sampling, followed by the Nystroem approximation for dealing with nonlinearity and Bayesian ridge regression for parameter estimation with uncertainty measurements.

By integrating Bayesian regression into the model, precise parameter estimation with quantified uncertainties can be achieved. The results demonstrated the surrogate model's capability to outperform traditional random parameter searches. Convergence losses were an order lower than multiple random searches, and more than 4 iterations were saved when achieving the same error tolerance. As evidenced in surrogate-assisted simulations, the model's ability to accurately predict parameters with minimal loss confirms its potential as a powerful tool in steam cracking process optimization.

A potential next step involves developing a systematic and flexible method for users to refine simulation data sampling and surrogate-assisted parameter estimation. For instance, the surrogate model could propose various candidate parameters for simultaneous simulations.

The surrogate-based technique offers a tangible solution to enhance simulation efficiency and scalability, which can be extended to other simulations where accurate parameters are crucial for calculation efficiency and outcome reliability.

References

M. Fakhroleslam, & S. M. Sadrameli, 2020, Thermal Cracking of Hydrocarbons for the Production of Light Olefins; A Review on Optimal Process Design, Operation, and Control, Industrial & Engineering Chemistry Research, 59, 27, 12288–12303.

H. Liu, Y. Ong, & J. Cai, 2018, A survey of adaptive sampling for global metamodeling in support of simulation-based complex engineering design, Structural and Multidisciplinary Optimization, 57, 1, 393–416.

M.L. Martín, B. Carro, A.J. ánchez-Esguevillas, & J. Lloret, 2019, Shallow neural network with kernel approximation for prediction problems in highly demanding data networks, Expert Systems with Applications, 124, 196-208.

Q. Shi, M.A. Abdel-Aty, & J. Lee, 2016, A Bayesian ridge regression analysis of congestion's impact on urban expressway safety, Accident Analysis & Prevention, 88, 124–137.

Q. Zhao, K. Bi, & T. Qiu, 2023, Data-driven intelligent modeling framework for the steam cracking process, Chinese Journal of Chemical Engineering, 61, 237–247.

Flavio Manenti, Gintaras V. Reklaitis (Eds.), Proceedings of the 34th European Symposium on Computer Aided Process Engineering / 15th International Symposium on Process Systems Engineering (ESCAPE34/PSE24), June 2-6, 2024, Florence, Italy

The effect of location and inter-cluster networks on the optimal decarbonization of ammonia production

Julia L. Tiggeloven[a], Charidimos Makrakis, André P.C. Faaij[a,b], Gert Jan Kramer[a], Matteo Gazzani[a,c]

[a] Copernicus Institute of Sustainable Development, Utrecht University, The Netherlands
[b] TNO Energy Transition, the Netherlands
[c] TU Eindhoven, Department of Chemistry and Chemical Engineering, The Netherlands

j.l.tiggeloven@uu.nl

Abstract

Ammonia is a key platform chemical, especially for the synthesis of fertilizers, and an important energy vector in net-zero energy systems. Yet, it faces environmental challenges due to its energy-intensive production and substantial greenhouse gas emissions. As such, prioritizing the decarbonization of the ammonia production process is crucial. In this study, we employ mixed integer linear programming optimization to assess different decarbonization strategies for low-carbon ammonia production. As a case study, we consider two existing production sites in the Netherlands, which allows us to investigate the influence of site-specific factors and the advantages of inter-cluster networks. Results show, that retrofitting plants with carbon capture achieves an 83-97.3 % cost-effective emission reduction. We also show, that achieving an early net-zero ammonia industry via H_2O electrolysis presents challenges due to the availability of green electricity and the CO_2 intensity of the electricity grid.

Keywords: Ammonia production, decarbonization, MILP, location, inter-cluster networks

1. Introduction

Ammonia is a vital platform chemical used in various products, primarily serving as a key component in 70 % of the world's nitrogen fertilizer production, which is essential for supporting global food production. Traditionally relying on energy-intensive grey hydrogen derived from natural gas-based steam reforming, the ammonia industry currently accounts for 2 % of global energy consumpsion and contributes 1.3 % of greenhouse gas emissions (IEA, 2021; IRENA and IEA, 2022). As such, prioritizing the decarbonization of the ammonia production process is crucial.

More sustainable, alternative routes for ammonia production include hydrogen production with carbon capture, electrified steam methane reforming, and electrolysis. These rely on the availability of low-carbon electricity and/or long-term CO_2 storage. The accessibility of these resources across production site locations might vary, adding significant complexity to identifying the most efficient decarbonization route.

Here, we employ a mixed-integer linear program (MILP) model to determine the optimal decarbonization configurations – considering both sizing and operation – for ammonia production at two existing locations in the Netherlands. The analysis considers three alternative production routes and the impact of inter-cluster connections involving electricity, CO_2, and hydrogen. A full-year, hourly resolution is adopted to incorporate seasonal and hourly energy variations in energy prices and renewables availability.

2. Method

2.1. MILP modeling framework

The decision variables that are optimized within the MILP model include design variables (i.e. selection and size of technologies) and operational variables (i.e. energy and material flows and storage levels). The framework uses hourly resolved input data on weather conditions, prices, and demand data together with a set of available technologies and the corresponding cost and performance coefficients. The objective function of the problem is to minimize the total annualized system cost, J, that is the sum of the technology and infrastructure cost, J_c, and the operating cost, J_o. The annualized cost is defined as

$$J_c = \sum_{i \in \mathcal{M}} (1 + \psi_i) \lambda_i S_i a_i \tag{1}$$

where λ_i is the size-dependent cost parameter of technology/infrastructure i and the annuity factor a_i is used to compute the annualized capital costs for each $i \in \mathcal{M}$. The maintenance cost is included as fraction of the annual capital costs ψ_i. The operating cost of the system is determined by the annual amount electricity and methane import, and the annual amount of CO_2 that is exported and stored, which is expressed as

$$J_o = \sum_{t=1}^{T} \left(u_{j,t} U_{j,t} + v_{CO_2} V_{CO_2,t} \right) \tag{2}$$

where $U_{j,t}$ is the import and $u_{j,t}$ the hourly price at hour t of each imported carrier $j \in \mathcal{N}$ (only methane and electricity in this work) and $V_{CO_2,t}$ is the export of CO_2 to an offshore storage with the costs v_{CO_2}. The total emissions include indirect emissions from the import of electricity and direct emissions from the ammonia/hydrogen production processes, and are calculated as

$$e_{CO_2} = \sum_{t=1}^{T} \left[\varepsilon_{elec,t} U_{elec,t} + \sum_{i \in \mathcal{M}} \epsilon_i F_{input,i,t} \right] \tag{3}$$

where $\varepsilon_{elec,t}$ is the carbon intensity of the electricity grid, ϵ_i is the emission factor of the technology per unit of main input carrier. It is assumed that all pure CO_2 that is not supplied to the demand or transported and stored is emitted. There are no CO_2 emission costs included in this work.

The balance for each material and energy carrier is formulated as:

$$\sum_{i \in \mathcal{M}} \left(U_{j,i,t} + P_{j,i,t} - V_{j,i,t} - F_{j,i,t} \right) - D_{j,t} = 0 \tag{4}$$

where $U_{j,i,t}$ is import, $P_{j,i,t}$ is the production, $V_{j,i,t}$ is export, $F_{j,i,t}$ is consumption and $D_{j,t}$ is the demand of each carrier $j \in \mathcal{N}$ at each hour t. The conversion performances of the technologies are described as

$$P_{j',t} = \alpha_{j'} F_{input,t} \tag{5}$$

and

$$F_{j'',t} = \beta_{j''} F_{input,t} \tag{6}$$

where $\alpha_{j'}$ and $\beta_{j''}$ represent the conversion efficiencies per unit of main input carrier of the outputs $j' \in J' \subset J$ and the ratio between inputs $j'' \in J'' \subset J$, respectively. The operation range of the technologies is expressed as

$$\gamma_i S_i \leq F_{input,t} \leq S_i \tag{7}$$

where γ_i is the minimum feasible operating point as a fraction of the installed capacity. Finally, the ramping rate (RR) is enforced by the following inequality constraints:

$$- \text{RR} \leq F_{\text{input},t} - F_{\text{input},t-1} \leq \text{RR} \qquad (8)$$

For a detailed description of the modeling of storage, the reader is referred to Gabrielli et al. (2018).

2.2. Site data and boundary conditions

Our case study includes two ammonia production sites in the Netherlands, placed at the Chemelot industrial cluster and Sluiskil, respectively. The data for the respective sites is shown in Table 1. An hourly ammonia demand needs to be produced, which is based on the annual ammonia production capacity in 2017 (1819 kt for Sluiskil and 1184 kt for Chemelot) (Batool & Wetzels, 2019). As both sites also produce urea downstream of the ammonia plant, we include an hourly CO_2 demand based on the annual urea production capacity (1300 kt/yr at Sluiskil, and 525 kt/yr at Chemelot) and a CO_2 consumption of 0.75 t CO_2/t Urea (Batool & Wetzels, 2019). The additional CO_2 is either emitted or, in the case of the Sluiskil site, it can be transported and stored offshore for 54.56 €/t CO_2 (IEA, 2022). Moreover, hydrogen and CO_2 can be transported between the production sites, for which we assume the required hydrogen infrastructure exists. The overall length of the pipelines is assumed to be 155 km, including the distance between the two sites and a 10 % extra accounting for terrain factors (Roussanaly et al., 2013); the investment costs for the CO_2 pipeline is 3.4 m€/km (Mikunda et al, 2011). Finally, we use a methane price of 35 €/MWh and electricity can be bought from the grid at both production sites for an hourly fluctuating price and CO_2 intensity that reflects 2030 as the base year. For more information on the modeling of electricity prices and CO_2 rates, the reader is referred to (Koirala et al., 2021).

2.3. Ammonia production processes

The traditional ammonia production process relies on the energy-intensive steam reforming process to produce hydrogen from fossil fuels, mainly methane, which is used with nitrogen in the Haber-Bosch process. One of the most common production processes is the Kellogg, Brown and Root (KBR) process, which we consider as the baseline process. The following alternative hydrogen production technologies are evaluated: KBR with carbon capture (CC) from flue gas, electric steam methane reforming (eSMR), and alkaline electrolyzers (AEC). The technology cost and performance are shown in Table 2 and 3. In contrast to the KBR, where stoichiometric nitrogen for the Haber-Bosch process is added via the secondary autothermal reformer (ATR), the eSMR and the electrolyzers require a complementary ASU to produce the required nitrogen. Furthermore, we assume that the CO2 from the syngas in the KBR is always separated. Although conventional production is typically steady-state, the electric technologies allow for a more flexible production responding to the availability of renewables (Wismann et al., 2021), which is why ammonia and CO2 buffer storage is included in the analysis to deal with fluctuations in production.

Table 1: Site-specific input data for the Sluiskil and Chemelot ammonia production sites.

Node	Hourly ammonia demand [t/h]	Hourly CO_2 demand [t/h]	Offshore CO_2 storage cost [€/t CO_2]	Average electricity price [€/MWh]	Average electricity CO_2 intensity [t/MWh]
Sluiskil	208	111	54.56	65.89	0.0636
Chemelot	135	45	-	91.02	0.0636

Table 2: Technology performance data for the conventional and alternative hydrogen production technologies, ammonia synthesis process and storage.

TECHNOLOGY	Inputs	Input ratio $(\beta_{j''})$	Outputs	Output ratio $(\alpha_{j'})$	Emission factor (ϵ_i) [t/input]	Minimum load (γ_i) [%]	Ramping rate [unit/h]
KBR	{methane, electricity}	[1, 0.001]	{hydrogen, CO_2, nitrogen}	[0.807, 0.171, 0.137]	0.035	0.7	7
KBR with CC	{methane, electricity}	[1, 0.015]	{hydrogen, CO_2, nitrogen}	[0.807, 0.203, 0.137]	0.003	0.7	7
eSMR	{electricity, methane}	[1, 3.676]	{hydrogen, CO_2}	[3.408, 0.739]	0.01	0.3	70
Alkaline electrolyzer	{electricity}	[1]	{hydrogen}	[0.665]	0	0.2	-
Haber Bosch process	{hydrogen, nitrogen}	[1, 0.139]	{ammonia}	[0.168]	0	0.3	70
Air separation unit	{electricity}	[1]	{nitrogen}	[4.000]	0	0	-
Ammonia storage	{ammonia, electricity}	[1, 0.01]	{ammonia}	[1]	0	-	-
CO_2 buffer storage	{CO_2}	[1]	{CO_2}	[1]	0	-	-

Table 3: Technology cost data for the conventional and alternative hydrogen production technologies, the ammonia synthesis process and storage.

TECHNOLOGY	CAPEX [€/unit]	Unit	OPEX [%]	Lifetime [y]	Discount rate
KBR	800902	MW_{th}	0.04	25	0.1
KBR with CC	949004	MW_{th}	0.04	25	0.1
eSMR	2484888	MW_{el}	0.02	25	0.1
Alkaline electrolyzer	1400000	MW_{el}	0.02	30	0.1
Haber bosch process	460650	MW_{th}	0	25	0.1
Air separation unit	3906200	MW_{el}	0	20	0.1
Ammonia storage	1034	tonne ammonia	0	25	0.1
CO_2 buffer storage	551	tonne CO_2	0.06	25	0.1

3. Results

To assess the impact of geographical location and inter-cluster networks on the decarbonization of the ammonia industry, we employed a two-step optimization approach: first an individual site optimization followed by the optimization of both sites simultaneously with interconnecting networks. Several optimizations were carried out minimizing costs for different (decreasing) emission limits, thus resulting in the Pareto fronts shown in Figure 1. Notably, our findings show that, across all three case studies, a net-zero ammonia process is not feasible considering the technologies portfolio adopted here. The reasons lie in two main factors; firstly, the demand for pure CO_2 requires on-site production, often via technologies with less than 100 % capture rate, leading to direct emissions. Secondly, the carbon intensity associated with electricity consumption in the eSMR and AEC processes contributes to indirect emissions. The KBRCC is the most cost-effective technology due to the limited changes and the high CO_2 capture ratio; electric technologies are more expensive and have higher indirect emissions from the electricity grid. In fact, 83.0 %, 97.1 %, and 97.3 % reductions can be achieved by installing only the KBRCC in the Chemelot, Sluiskil, and the combined case, respectively.

Our analysis shows how site-specific factors affect the optimal decarbonization strategy, as the Sluiskil site outperforms the Chemelot case by achieving lower emissions more cost-effectively. This advantage is attributed to several factors, including the accessibility of offshore CO_2 transport and storage, and a higher on-site CO_2 consumption. Moreover, our study underscores the potential synergies achievable through collaborative efforts among industrial clusters. The combined Chemelot and Sluiskil case exhibits similar minimum emissions compared to the stand-alone Sluiskil scenario, thus highlighting the benefits of inter-site cooperation in the pursuit of decarbonizing production processes within industrial clusters.

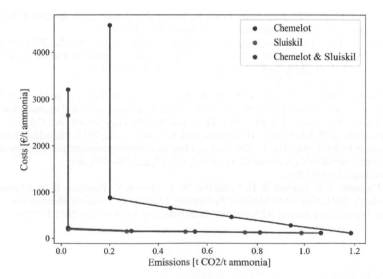

Figure 1: Pareto fronts for the two individual Chemelot and Sluiskil site optimizations and the simultaneous optimization of both sites.

4. Conclusions

In this study, we assessed how (i) different ammonia production processes, (ii) the location of production sites, and (iii) the integration of inter-cluster networks impact the emissions reduction strategies. Using a detailed MILP model with a full year, hourly resolution, we optimized the selection and size of technologies and their operation, while considering also fluctuations in electricity prices across seasons and hours. Our analysis shows that cost-effective CO_2 reduction (up to 97.3 %) can be achieved by retrofitting traditional ammonia production plants with carbon capture. The results highlight the effect of site-specific factors, such as varying electricity prices and CO_2 transport and storage costs, on the feasibility and cost-effectiveness of emission reduction efforts. Moreover, our findings demonstrating that similar specific costs and emissions can be achieved when both sites are optimized simultaneously, emphasizing the benefits of inter-cluster cooperation. To expand our understanding, we propose extending this approach beyond the ammonia industry to explore potential synergies across different production processes. Finally, we conclude that achieving a net-zero ammonia industry by 2030 presents significant challenges, as deep decarbonization must rely on a fully decarbonized electricity grid.

References

M. Batool & W. Wetzels, 2019, Decarbonisation Options for the Dutch Fertiliser Industry, https://www.pbl.nl/sites/default/files/downloads/pbl-2019-decarbonisation-options-for-the-dutch-fertiliser-industry_3657.pdf.

P. Gabrielli, M. Gazzani, E. Martelli, & M. Mazzotti, 2018, Optimal design of multi-energy systems with seasonal storage. Applied Energy, 219, 408–424. https://doi.org/10.1016/J.APENERGY.2017.07.142

IEA, 2021, Ammonia Technology Roadmap, IEA, Paris https://www.iea.org/reports/ammonia-technology-roadmap, License: CC BY 4.0

IEA, 2022, Low-Carbon Hydrogen from Natural Gas: Global Roadmap. International Energy Agency

IRENA and AEA, 2022, Innovation Outlook: Renewable Ammonia, International Renewable Energy Agency, Abu Dhabi, Ammonia Energy Association, Brooklyn

B. Koirala, S. Hers, G. Morales-España, Ö. Özdemir, J. Sijm, and M. Weeda, 2021, Integrated electricity, hydrogen and methane system modelling framework: Application to the Dutch Infrastructure Outlook 2050, Applied Energy, vol. 289, p. 116713, doi: 10.1016/j.apenergy.2021.116713.

T. Mikunda, J. Van Deurzen, A. Seebregts, K. Kerssemakers, M. Tetteroo, and L. Buit, 2011, Towards a CO_2 infrastructure in North-Western Europe: Legalities, costs and organizational aspects, Energy Procedia, vol. 4, pp. 2409–2416, doi: 10.1016/j.egypro.2011.02.134.

S. Roussanaly, J. P. Jakobsen, E. H. Hognes, and A. L. Brunsvold, 2013, "Benchmarking of CO_2 transport technologies: Part I—Onshore pipeline and shipping between two onshore areas," International Journal of Greenhouse Gas Control, vol. 19, pp. 584–594, doi: 10.1016/j.ijggc.2013.05.031.

S. T. Wismann, J. S. Engbæk, S. B. Vendelbo, W. L. Eriksen, C. Frandsen, P. M. Mortensen, I. Chorkendorff, 2021, Electrified Methane Reforming: Elucidating Transient Phenomena. Chemical Engineering Journal 425, 131509. https://doi.org/10.1016/j.cej.2021.131509.

Flavio Manenti, Gintaras V. Reklaitis (Eds.), Proceedings of the 34th European Symposium on Computer Aided Process Engineering / 15th International Symposium on Process Systems Engineering (ESCAPE34/PSE24), June 2-6, 2024, Florence, Italy

Applicability of Energy Storage for Mitigating Variability of Renewable Electricity Considering Life cycle Impacts

Ayumi Yamaki [a], Shoma Fujii [a], Yuichiro Kanematsu [b], Yasunori Kikuchi*[a,b,c]

[a] *Institute for Future Initiatives, The University of Tokyo, 7-3-1 Hongo, Bunkyo-ku, Tokyo 113-8654, Japan*
[b] *Presidential Endowed Chair for "Platinum Society", Organization for Interdisciplinary Research Project, The University of Tokyo, 7-3-1 Hongo, Bunkyo-ku, Tokyo 113-8656, Japan*
[c] *Department of Chemical System Engineering, The University of Tokyo, 7-3-1 Hongo, Bunkyo-ku, Tokyo 113-8656, Japan*
ykikuchi@ifi.u-tokyo.ac.jp

Abstract

To achieve carbon neutrality, power generation from variable renewable energy (VRE) has been accelerated. VRE variability can be mitigated by installing energy storage. This study aims to evaluate life cycle impacts of an energy storage system with battery, H_2 storage, or thermal energy storage. A model of the energy storage system with VRE was constructed and the annual energy flow was simulated. The energy storage system targeted in this study assumed that all the energy derived from VRE was stored in the energy storage and supplied to consumers. The amount of power to-be-sold from the energy storage system based on the VRE and the energy storage installation were calculated. A life cycle assessment was performed to evaluate the greenhouse gas (GHG) emissions, abiotic resource depletion (ARD), and intensity of GHG and ARD as life cycle impacts. The smallest life cycle impacts varied depending on the type and amount of energy storage. A quantitative evaluation of the energy supply capacity and the environmental impacts of the energy storage system could assist in designing an energy storage system with VRE, considering regional energy supply systems.

Keywords: battery, hydrogen storage, thermal energy storage, life cycle greenhouse gas emissions, abiotic resource depletion

1. Introduction

The installation of variable renewable energy (VRE) has expanded around the world toward the achievement of a sustainable society, although ensuring a stable power supply by power generation from VRE is a challenge because of lack of stability. To accelerate VRE penetration, there are projects underway to install energy storage, such as batteries, hydrogen (H_2) storage, and thermal energy storage (TES). Batteries store electricity, whereas H_2 storage and TES should be converted into electricity before use. Therefore, batteries can supply electricity more effectively than H_2 storage or TES. However, battery production has a relatively high environmental impact (Peters et al., 2017) and batteries require relatively high investment costs for long-term storage (Battke et al., 2013). Life cycle assessments (LCAs) of other energy storage methods have been conducted (Strazza et al., 2015, Oró et al., 2012). Because the environmental impacts of energy storage differ according to the characteristics of the region where the energy storage is installed, even for installing the same type and amount of energy storage (Yamaki et al., 2023), a model is needed to compare energy storage considering energy demand and the VRE characteristic of the region.

This study constructed a model of energy storage systems assumed to be required for batteries, H_2 storage, or TES; the systems were compared by LCA. An energy storage system is installed as a new energy supply system to mitigate the variability of VREs and connect to the grid. An energy storage system with the battery, H_2 storage, and TES is referred to as a battery system, H_2 system, and TES system, respectively. The model of the energy storage system was constructed to simulate the energy flows for a one-year period. We calculated the amount of power that would be sold from the energy storage system and analyzed life cycle impacts. For life cycle impacts, life cycle greenhouse gas (LC-GHG), abiotic resource depletion (ARD), and greenhouse gas (GHG) and ARD intensities were evaluated. We discuss the applicability of energy storage systems.

2. Materials and methods

2.1. Energy and material flows of the energy storage system

The energy and material flow of the energy storage systems targeted in this study is shown in Figure 1. In energy storage systems, VRE is assumed to be stored and supplied as electricity. In the battery system, electricity is assumed to be generated by a VRE, stored, and supplied. In the H_2 system, H_2 is assumed to be produced by electrolyzers using VRE electricity, stored as liquefied H_2 in a tank, and converted into power in fuel cells. In the TES system, heat is generated directly from VRE, stored in molten salts, and converted into steam in a heat exchanger for power generation in steam turbines.

An LCA was conducted for the cradle-to-grave energy storage and other facilities, assumed to have a new energy storage system installed. The LCA covered the facilities of power/heat generation from VRE, the battery in the battery system, the H_2 tank, electrolyzer, and fuel cell in the H_2 system and the molten salt, its tanks, heat exchanger, and steam turbine in the TES system. The functional unit was the annual VRE instability. The foreground data for the LCA was obtained through energy flow simulations. The background data, such as materials for the energy systems, were extracted from the Japanese LCA database, the inventory database for environmental analysis version 3.2 (IDEA) (AIST, 2019). LC-GHG and ARD were calculated by adding the GHG emission and ARD from production to disposal of each facility extracted from the data, respectively. The GHG and ARD intensities were expressed in LC-GHG and ARD per power to be sold. Although GHG emissions are reduced by VRE installation, ARD would increase due to rare metal utilization in the production of energy storage. Therefore, it is important to evaluate both GHG emissions and ARD.

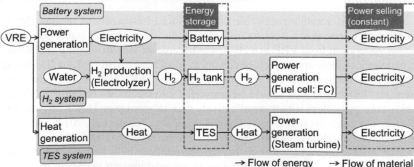

Figure 1 Energy and material flows of the energy storage systems.

2.2. Model of the energy storage systems

The model constructed in this study assumed that all VRE-derived energy was stored in energy storage and supplied electricity to consumers. The simulations were performed in Microsoft Excel. The energy flow was calculated hourly from 0:00 on January 1st to 24:00 on December 31st based on the model made by Yamaki et al. (2020). Life cycle impacts were calculated from the amount of energy storage and wind installation.

The amounts of energy storage and VRE installation were set, and then the maximum amount of power to be sold from the energy storage system was estimated. In the simulation, the stored energy was calculated hourly from the charge of VRE-derived power/heat and the discharge of power to be sold using Eq. (1).

$$Q_{storage}(t) = Q_{storage}(t - \Delta t) + Q_{in}(t) - Q_{out}(t) - Q_{leakage}(t) \tag{1}$$

Here, $Q_{storage}(t)$ is stored energy in the targeted energy storage at time t. $Q_{in}(t)$ is input energy supplied from VRE, and $Q_{out}(t)$ is output energy from the energy storage, which is calculated from the amount of power sales. $Q_{leakage}(t)$ is energy leakage from energy storage, i.e., battery self-discharge, H_2 leakage, and heat leakage. The amount of power sales is adjusted to occur in periods of zero $Q_{storage}$, and the power sales value represents the maximum that the targeted energy storage system could sell. This study presents the value of power to be sold as a result of power sales. To supply electricity for more than one year, it was assumed that the amount of stored energy, i.e., the amounts of electricity in the battery, H_2 in the H_2 tank, and the temperature and amount of molten salt in the TES tank, were the same at 0:00 on January 1st and at 24:00 on December 31st.

2.3. Analysis settings

In energy flow simulations, wind energy was assumed to be delivered as a type of VRE. Facilities for power/heat generation from wind energy are assumed to be installed as wind turbines (WTs) or wind-thermal energy converters (WEC_th). A WEC_th directly converts wind-derived rotational energy into thermal energy (Okazaki et al., 2015).

In the battery system, the number of battery cells was assumed to be two. Cells were assumed to be incapable of charging and discharging simultaneously. In the H_2 system, the capacity of the electrolyzer was determined based on the capacities of the WT and the H_2 tank. When sufficient WTs and an H_2 tanks are installed, the electrolyzer capacity is assumed to be equal to the WT capacity. When a small H_2 tank is installed, it might not be able to store all the H_2 generated from WTs, so the electrolyzer capacity is limited to a scale that could produce the amount of H_2 that could be stored in the small H_2 tank. To avoid low-load operation, the electrolyzer was assumed to produce H_2 when power generation exceeded 50% of the electrolyzer capacity. For the installation of the 100.8 MW-WT, H_2 was assumed to be produced when generating power over 50.4 MW because the electrolyzer was 100.8 MW. If large volumes of H_2 are to be produced, several electrolyzers are assumed to be installed, configuring the total capacity of the electrolyzers.

The energy conversion rate was assumed to be fast enough to operate the energy storage system. In this study, all the wind energy input for one hour was assumed to be converted into the target energy, such as H_2 and heat, in one hour. Also, H_2 and heat were assumed to be converted into power depending on the amount of power sales.

3. Results and discussions

3.1. Power sold from the energy storage systems

The energy flows of the energy storage systems were simulated, and the power to be sold was calculated for various amounts of WT/WEC_th and energy storage. The maximum amount of power to be sold from the energy storage system was calculated when the amount of energy storage and WT/WEC$_{th}$ installation were set.

The amount of power to be sold was assumed to be constant throughout the year. Regardless of the type of energy storage, the amount of power to be sold increased as the amount of WT/WEC$_{th}$ and energy storage installed increased (Figure 2). However, if the large energy storage facilities were installed, the power sales increased slightly or decreased. In the case of the TES system in particular, the molten salt stored in the cold tank also releases heat due to its temperature of 290 °C. If an excessive TES was installed, the excess molten salt stored in the cold tank released heat, resulting in a decrease in the amount of power to be sold. Therefore, to use stored energy effectively, the amount of energy storage should be installed in proportion to the amount of WT/WEC$_{th}$ installation.

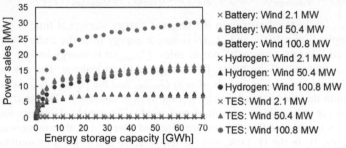

Figure 2 Power sold from the energy storage system.

3.2. Life cycle impacts of the energy storage systems

Life cycle impacts were evaluated for various amounts of WT/WEC$_{th}$ and energy storage. The GHG and ARD intensities are shown in Figure 3. LC-GHG and ARD were considered from production to disposal of WT/WEC$_{th}$, energy storage, and other utilities. Other utilities included an electrolyzer and fuel cell in the H$_2$ system and a steam turbine and heat exchanger in the TES system. The GHG and ARD intensities were calculated from LC-GHG and ARD and expressed per power sales.

GHG intensities were compared in Figure 3 (a). In all systems, GHG intensities were high with small energy storages due to the large impact of WT/WEC$_{th}$ production. When a few large energy storage facilities were installed, GHG intensities decreased because the amount of power sold increased. When excess energy storage facilities were installed, GHG intensities were high due to high GHG emissions from the production of energy storage facilities. For the H$_2$ system, in particular, although it showed the lowest GHG intensity with large energy storage in the three systems, it showed the highest GHG intensity with small energy storage. This is because GHG emissions from the production of the electrolyzer, which increased in scale with the WT installation, increased in order to use the wind-derived electricity for H$_2$ production. Therefore, even with smaller H$_2$ tanks, the GHG intensity was higher when a large WT was installed. In addition, when H$_2$ production increased due to a large H$_2$ tank, GHG emissions from liquefied H$_2$ were high. When WT/WEC$_{th}$ was installed at 100.8 MW, the minimum GHG intensities were 0.42 kg-CO$_2$eq/kWh for a battery of 1.1 GWh, 0.34 kg-CO$_2$eq/kWh for 53 GWh-H$_2$ storage, and 0.17 kg-CO$_2$eq/kWh for 2.9 GWh-TES.

The ARD intensities of the energy systems are shown in Figure 3 (b). Like the GHG intensities, ARD intensities were high with small or excess large energy storage, but they were low with appropriate capacity of energy storage. However, ARD from the operation, such as H$_2$ liquefied, was low, so ARD from the production of the electrolyzer in the H$_2$ system, WT/WEC$_{th}$, and energy storages had a high impact. When WT/WEC$_{th}$ was installed at 100.8 MW, the minimum ARD intensities were 0.011 g-Sbeq/kWh for 2.9 GWh-battery, 0.027 g-Sbeq/kWh for 53 GWh-H_2 storage, and 0.0066 g-Sbeq/kWh for 2.9 GWh-TES.

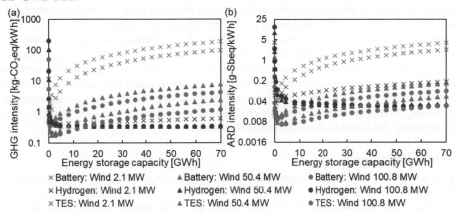

Figure 3 Life cycle impacts of the energy storage system. (a) GHG intensity for power sales. (b) ARD intensity for power sales.

3.3. Applicability of the energy storage systems

When a large WT/WEC$_{th}$ and energy storage was installed, the H_2 or TES systems seemed to be suitable because of their smaller life cycle impacts than batteries. There was a trade-off between the H_2 and TES systems because the H_2 system had a lower GHG intensity, whereas the TES system had a lower ARD intensity. To decide whether to install an H_2 storage or TES, it is important to consider the demand for H_2 and heat in the region where the energy storage system is to be installed. H_2 is used as fuel for vehicles and materials for factories, and heat is used as an energy source for factories and houses. This study enabled a quantitative evaluation of the environmental performance of the energy storage system, including its operation, and detailed energy storage system installation plan.

In areas where the energy storage system is installed, life cycle impacts can be reduced and prevent from blackouts of conventional power plants due to natural disasters. Furthermore, installing new equipment, such as energy storage, would create employment opportunities to operate the energy storage systems, stimulate the local economy, and promote regional revitalization. Therefore, energy storage systems have many types of applicability, such as contributing to carbon neutrality and revitalizing the region.

In this study, the installed energy storage systems covered only one type of energy storage and the amount of power sales were constant. In addition, all facilities were assumed to be disposed of without recycling. The energy storage system has the possibility of improvement through recycling of the energy storage (Kikuchi et al., 2021) and operation change. Although optimizations are not enough to operate an energy storage system, in this study, low environmental impacts and efficient energy storage systems were expected to be designed by applying the simulation conducted here.

4. Conclusions

In this study, we developed a model of energy storage system installing VRE and conducted annual energy flow simulations and LCA. The amount of power sold and life cycle impacts of the energy storage systems were evaluated, and the smallest ones varied with the type and amount of energy storage. When WT/WEC$_{th}$ was installed at 100.8

MW, the minimum GHG intensity was 0.17 kg-CO_2eq/kWh for 2.9 GWh-TES, and the minimum ARD intensity was 0.0066 g-Sbeq/kWh for 2.9 GWh-TES. The energy storage installed in a region should be selected based on the amount of power to be sold, i.e., the energy demand in the region, because the amount of power sold and the minimum GHG and ARD intensities differ depending on the type of energy storage.

Improvements in energy storage technology and its operation to accommodate the unstable nature of VRE could lead to more effective use of energy storage systems. Energy storage systems are expected to have many types of applicability, such as contributing to carbon neutrality and revitalizing the region. This study was able to quantitatively evaluate the environmental performance of energy storage systems, including their operation. Appropriate energy storage systems are expected to be designed by applying the simulation conducted in this study.

Acknowledgement

This study was financially supported by the New Energy and Industrial Technology Development Organization (NEDO) (Grant number, JPNP21026). This research contained the achievements of the JST-MIRAI Program Grant Number JPMJMI19C7, JST COI-NEXT (JPMJPF2003), JST PRESTO (Grant Number JPMJPR2278), JSPS KAKENHI (Grant Number JP21H03660), JSPS KAKENHI Grant-in-Aid for Early-Career Scientists (Grant Number JP22K18061), and the Environment Research and Technology Development Fund (Grant Number JPMEERF20213R01) of the Environmental Restoration and Conservation Agency of Japan. Activities of the Presidential Endowed Chair for "Platinum Society" at the University of Tokyo are supported by Mitsui Fudosan Corporation, Sekisui House, Ltd., the East Japan Railway Company, and Toyota Tsusho Corporation.

References

AIST (National Institute of Advanced Industrial Science and Technology), 2019, Inventory Database for environmental analysis version 2.3

B. Battke, T. S. Schmidt, D. Grosspietsch,V. H. Hoffmann, 2013, A review and probabilistic model of lifecycle costs of stationary batteries in multiple applications, Renew Sustain Energy Rev, 25, 240–250

Y. Kikuchi, I. Suwa, A. Heiho, Y. Dou, S. Lim, T. Namihira, K. Mochizuki, T. Koita, C. Tokoro, 2021, Separation of cathode particles and aluminum current foil in lithium-ion battery by high-voltage pulsed discharge Part II: Prospective life cycle assessment based on experimental data, Waste Manag, 132, 86–95

T. Okazaki, Y. Shirai, T. Nakamura, 2015, Concept study of wind power utilizing direct thermal energy conversion and thermal energy storage, Renew Energy, 83, 332–338

E. Oró, A. Gil, A. de Gracia, D. Boer, L. F. Cabeza, 2012, Comparative life cycle assessment of thermal energy storage systems for solar power plants, Renew Energy, 44, 166–173

J. F. Peters, M. Baumann, B. Zimmermann, J. Braun, M. Weil, 2017, The environmental impact of Li-Ion batteries and the role of key parameters – A review, Renew Sustain Energy Rev, 67, 491–506

C. Strazza, A. Del Borghi, P. Costamagna, M. Gallo, E. Brignole, P. Girdinio, 2015, Life Cycle Assessment and Life Cycle Costing of a SOFC system for distributed power generation, Energy Convers Manag, 100, 64–77

A. Yamaki, S. Fujii, Y. Kanematsu,Y. Kikuchi, 2023, Life cycle greenhouse gas emissions of cogeneration energy hubs at Japanese paper mills with thermal energy storage, Energy, 270, 126886

A. Yamaki, Y. Kanematsu,Y. Kikuchi, 2020, Lifecycle greenhouse gas emissions of thermal energy storage implemented in a paper mill for wind energy utilization, Energy, 205, 118056

Flavio Manenti, Gintaras V. Reklaitis (Eds.), Proceedings of the 34th European Symposium on Computer Aided Process Engineering / 15th International Symposium on Process Systems Engineering (ESCAPE34/PSE24), June 2-6, 2024, Florence, Italy
© 2024 Elsevier B.V. All rights reserved. http://dx.doi.org/10.1016/B978-0-443-28824-1.50133-2

Pinch curves computation using differential continuation algorithm

Nataliya Shcherbakova[a*], Ivonne Rodriguez-Donis[b], Vincent Gerbaud[a]

[a] *Laboratoire de Génie Chimique, Université de Toulouse, CNRS, INP-ENSIACET, UPS Toulouse, France*
[b]*Laboratoire de Chimie Agro-industrielle, LCA, Université de Toulouse, INRA, Toulouse, France*
*nataliya.shcherbakova@ensiacet.fr

Abstract

The paper presents a novel algorithm for numerical computation of pinch curves in homogeneous ternary mixtures under assumption of constant molar flow in an infinitely long column. The method is based on the analysis of topological structures of the surfaces associated to the vapor-liquid equilibrium and pinch conditions in the 3D composition - temperature space. The computational problem is formulated in terms of a system of ordinary differential equations. The efficiency of this approach to compute the pinch curves and to detect the bifurcations of their structure is demonstrated with four examples of real mixtures covering ternary diagram classes 1.0-1a, 1.0-2 and 2.0-2b.

Keywords: pinch curves, bifurcation, homogenous extractive distillation, differential continuation

1. Introduction

Pinch curves are an important concept in the design of extractive distillation process since they are related to limiting operating conditions, which are essential for assessing process feasibility. They represent the set of compositions that remain constant along the increasing number of stages in the column, so that the separation becomes impossible once these compositions are reached. The location of pinch curves delimits the operation domain for a given set of operation parameters (reflux ratio and the entrainer flow rate) allowing to evaluate the performance of the entrainer using the Infinitely Sharp Split (ISS) method as described by Petlyuk et al. (2015) and Rodriguez-Donis et al. (2023).

This paper aims to present a new numerical algorithm for pinch curves computation based on the differential continuation method, which reduces the computation to the integration of a system of ordinary differential equations (ODE) instead of solving the systems of algebraic equations by an iterative procedure, as it is usually done. It represents the further development of the new algorithmic approach that was successfully applied to compute univolatility curves (Shcherbakova et al., 2017, Cots et al. 2021) and binodal curves (Shcherbakova et al., 2023) of ternary diagrams.

Starting from the code of Poellmann and Blass (1994), there were several attempts to construct the pinch curves via integration of ODE systems (Feldbab, 2012) or differential-algebraic systems of equations (DAE) (Skiborowski at al. 2016). The novelty of our approach comparing to the other authors consist in exploration of the topology and the mutual arrangement of the pair of surfaces associated to the vapor-liquid equilibrium (VLE) and pinch conditions in a complete 3D composition-temperature state space, under the standard assumptions of constant molar flow rates and the infinite height of the column. We show that in this setting the pinch curves are just projections on the 2D composition space of the intersection of these surfaces, and their singularities results from the common tangent points of these surfaces.

The described geometric model can be formalized in a set of ODE that can be solved by any conventional ODE solver. This reduces the computation time gaining in the numerical accuracy and flexibility of the code.

This paper is organized as follows. In Section 2 we present the detailed geometrical model of pinch curves. In particular, we derive the bifurcation condition for the pinch curves splitting caused by the change of operation parameters. In Section 3 four examples of different configurations of pinch curves of real ternary mixtures are presented. In Conclusion we discuss the possible practical issued of the presented algorithm.

2. Computational method

2.1. Pinch points and pinch curves in ternary extractive distillation

Assume that the extractive distillation of a ternary homogeneous mixture takes place in an infinitely high column with constant liquid and molar flow rates L and V. It is also assumed that the ratio of the entrained molar feed rate E and molar flow rate of the distillate D is fixed.

Pinch points of the distillation diagram associated to this process represent the critical state inside the column corresponding to the VLE between the vapor and the liquid phases at the same stage of the column, so that their further separation becomes impossible. The geometrical loci of such points form the pinch branches or pinch curves of the diagram. Pinch curves play the central role in the infinitely sharp split (ISS) method to identify the possible splits in a given mixture. For the detailed description the pinch points and the ISS method we refer the reader to papers cited above and references therein. In this paper we focus on the purely computational aspects, and to this end we first need to recall the main mathematical formulae used in the computations.

In what follows T is the temperature of the mixture, $x_i, y_i \in [0,1]$, $i=1,2,3$ denote the molar fractions of the three component mixtures in liquid and vapor phases so that $y_i = K_i(x_1, x_2, x_3, T)x_i$, where the functions K_i represent the distribution coefficients of the component i. Since $\sum_{i=1}^3 x_i = 1$, only two of molar fraction are independent, so that below x_3 will be replaced by $x_3 = 1 - x_1 - x_2$. The VLE condition then takes the form

$$x_1 K_1(x_1, x_2, T) + x_2 K_2(x_1, x_2, T) + (1 - x_1 - x_2)K_3(x_1, x_2, T) = 1 \qquad (1)$$

According to Petlyuk et al. (2015), the pinch point condition associated to the first component that should be separated from the second one (using the third one as an entrainer) can be written in the form

$$x^\Delta = \frac{K_1(x_1, x_2, T) - K_2(x_1, x_2, T)}{1 - K_2(x_1, x_2, T)} x_1 \qquad (2)$$

Here $x^\Delta \notin [0,1]$ denoted the difference or delta point, it lies outside of the concentration triangle and represents a virtual concentration corresponding to the composition of the difference between the distillate and the entrainer feed of the column. Once the ratio E/D is fixed, the delta point is defined by the relation

$$\frac{E}{D} = \frac{x^\Delta - 1}{x^\Delta} \qquad (3)$$

The set of points on the composition space verifying conditions Eqs. (1, 2) is called the pinch curve associated to the operating conditions expressed by Eq. (3). These curves are bounded by the binary sides of the composition triangle and by the univolatility curves

$\alpha_{ij} = 1$ that represent the points where $K_i(x_1, x_2, T) = K_j(x_1, x_2, T)$. The pinch and univolatility curves can intersect only at azeotropic points of the diagram.

2.2 Geometrical model of pinch curves and their singularities

Eqs.(1-3) admit a very clear geometric interpretation. Indeed, denote by $\Omega = \{x_i \in [0,1], x_1 + x_2 \ll 1, i = 1,2\}$ the composition triangle associated to the ternary mixture and consider a 3D cartesian space $\Sigma = \{z = (x_1, x_2, T): (x_1, x_2) \in \Omega\}$ over Ω endowed with the coordinates x_1, x_2 and T. Define two functions

$$\Phi(x_1, x_2, T) = \sum_{i=1}^{3} x_i K_i(x_1, x_2, T) - 1 \tag{4}$$
$$\Psi(x_1, x_2, T) = x_1^{\Delta}(1 - K_2(x_1, x_2, T)) - x_1(K_1(x_1, x_2, T) - K_2(x_1, x_2, T))$$

and consider the pair of surfaces associated to zero levels of these functions: $W_{\Phi} = \{z \in \Sigma: \Phi(z) = 0\}$ and $W_{\Psi} = \{z \in \Sigma: \Psi(z) = 0\}$. The intersection of these surfaces defines a smooth curve Γ in Σ. Comparing Eqs. (1, 2) with Eq. (4) one can easily see that Γ projects on a pinch curve on Ω. Indeed, since $\Gamma = W_{\Phi} \cap W_{\Psi}$, it is generated by some vector field which is orthogonal to both normal vectors $N_{W_{\Phi}} = \nabla\Phi(z)$ and $N_{W_{\Psi}} = \nabla\Psi(z)$. In other word, at each point $z \in \Gamma$, the vector

$$U(z) = \nabla\Phi(z) \times \nabla\Psi(z) \tag{5}$$

belongs to the tangent space Γ. Therefore, knowing one pinch point $z_0 \in \Gamma$ would be enough to construct the whole pinch curve by solving the following system of ODE

$$\dot{x}_1 = U_1(x_1, x_2, T), \qquad \dot{x}_2 = U_2(x_1, x_2, T), \qquad \dot{T} = U_3(x_1, x_2, T) \tag{6}$$

where, according to Eq.(5),

$$U_1 = \frac{\partial\Phi}{\partial x_2}\frac{\partial\Psi}{\partial T} - \frac{\partial\Phi}{\partial T}\frac{\partial\Psi}{\partial x_2}, \qquad U_2 = \frac{\partial\Phi}{\partial T}\frac{\partial\Psi}{\partial x_1} - \frac{\partial\Phi}{\partial x_1}\frac{\partial\Psi}{\partial T}, \qquad U_3 = \frac{\partial\Phi}{\partial x_1}\frac{\partial\Psi}{\partial x_2} - \frac{\partial\Phi}{\partial x_2}\frac{\partial\Psi}{\partial x_1} \tag{7}$$

Observe, that if W_{Φ} and W_{Ψ} have a common tangent plane at z, then $U(z) = 0$ and such a point is a singular point of Γ. Moreover, due to Eq.(4), the gradient of the at any point of W_{Φ} verifies $\frac{\partial T}{\partial x_i} = -\frac{\partial\Phi}{\partial x_i}/\frac{\partial\Phi}{\partial T}$, $i = 1,2$. Combining this expression with Eq. (7) yields

$$U_3 = U_1\frac{\partial T}{\partial x_1} + U_2\frac{\partial T}{\partial x_2} \tag{8}$$

Hence the singular points of Γ are in one-to-one correspondence with the singularities of the underlying pinch curve. Such isolated singular points can be of elliptic of hyperbolic type. The first case corresponds to a pinch curve shrinking into an isolated pinch point under certain operation conditions, whereas in the hyperbolic case there are four pinch branches meeting at the singular point. Examples of such configurations will be given in Section 3.

The topology of the pinch diagram is entirely determined by the value of E/F (and hence by x^{Δ}) and by the pressure in the column. The modification of one of these parameters will cause the transformation of the shape and of the mutual arrangement of the surfaces W_{Φ} and W_{Ψ}, leading to the transformation of the topological structure of the underlying pinch diagram. In particular, it may cause its bifurcation. According to the geometrical model described above, the bifurcation occurs when the curve Γ has a critical point. In view of Eq. (8), the corresponding pinch point and operating conditions can be found by solving the following system of four equations

$$\Phi(x_1, x_2, T) = 0, \Psi(x_1, x_2, T) = 0, U_1(x_1, x_2, T) = 0, U_2(x_1, x_2, T) = 0 \tag{9}$$

with respect to x_1, x_2, T, P or x_1, x_2, T, x^Δ, according to the type of bifurcation.

2.3 Numerical computation of pinch curves

Using Eq. (6) the whole pinch curve can be computed numerically using the standard Runge-Kutta schemes for ODE integration. Initial points for such integration can be found by solving Eqs. (1,2) on the binary sides of the composition triangle, by an iteration procedure. The dot symbol in the left-had side of Eq. (6) has the meaning of a derivative with respect to some scalar parameter. In practical computations, is would be convenient to normalize the vector field U and thus rewrite Eq. (6) with suspect to arc length s. This will also avoid an eventual stiffness problem along the integration.

The same computation can be performed with respect to another component, yielding other types of pinch curves. To this end, in Eq. (2) x_1 should be replaced by x_2 or x_3, modifying the indexes of distribution coefficients accordingly. Taking into account that a pinch curve may be composed of two branches, the complete pinch diagram of a given mixture can be obtained by the following steps:

- build a list of all pinch points on the boundary of Ω.
- for each binary pinch point from in the list compute the pinch curve issued from this point with an appropriate choice of the initial direction. The numerical integration should be continued until the boundary of Ω is attained.
- exclude both initial and final points of the already computed pinch curve and continue until the list of binary pinch points is emptied.

The described algorithm allows to compute all pinch branches of the given ternary mixture that start from the binary pinch points. However, in some rare cases the mixture may have closed pinch curves entirely lying in the interior part of Ω. An example of such a situation is presented in the next section. In this case the pinch curve still can be computed by solving Eq. (6), but it will require to detect a pinch point in the inner part of Ω. The clear indicator of such a configuration is the existence of a bifurcation point of elliptic type inside Ω for some value of E/D or pressure.

A numerical implementation of Eq. (7) requires the access to the derivatives of the expressions defining the thermodynamical model of the mixture: activity coefficients, vapor-temperature equation, etc. The complexity of such a computation is the main reason for which the differential continuation algorithms are very poorly used in Process Engineering compared to the other fields. In fact, the use of difference formulae for the derivatives do not satisfy the necessary accuracy requirements to guarantee the numerical stability and accuracy of computation. On the other hand, the automatic differentiation technology (Hasco and Pascual, 2012), already implemented in many other applications, can successfully build these expressions. A working example of a code based on the coupling of the differential continuation with automatic differentiation of the thermodynamic model is described in Cots et al. (2020) for the univolatility curves computation. For academic use, any package of symbolic computation can be employed. The results of pinch curves computation presented in the next section were done with Mathematica.

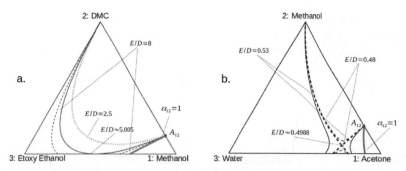

Figure 1. Serafimov's class 1.01-a. Examples of possible evolution of the pinch diagrams with the variation of E/D parameter. Case a: regular splitting of the pinch curve. Case b : splitting caused by a saddle-type biffurcation.

3. Case studies

All the pinch diagrams presented below were computed using Mathematica 9 package. We present here 4 examples corresponding to Serafimov's classes 1.01-a (Fig. 1), 1.02 and 2.02-b (Fig. 2). The DIPPR equation and database were used for the vapor pressure computation. For the mixtures shown in Fig. 1 and Fig. 2 a the non-random two-liquid (NRTL) thermodynamic model was used, the VLE binary coefficients of mixtures were taken from the Simulis Thermodynamics (Prosim, 2023). For the case shown in Fig. 2b the binary coefficients were estimated using the UNIFAC modified Dortmund 1993 model (Gmehling et al., 1993). The pressure is assumed constant and equal to 1 atm.

Fig1. shows two examples of ternary mixtures of Serafimov's class 1.01-a. (Kiva et al., 2003). In the first case methanol is separated from dimethyl carbonate (DMC) using the ethoxy ethanol as an entrainer. The pinch curve is bounded from the right by the univolatility curve α_{12} (thick grey curve) issued from the binary azeotrope A_{12}. A unique pinch curve exists for E/D smaller than 5.005. For greater values a part of the pinch curve lies behind of the composition triangle, causing the regular splitting of the unique pinch curve into two disjoin branches. The situation shown in Fig.1b is different. This example describes the separation of the acetone from methanol using water as the entrainer. Near the value 0.48 of E/D, the diagram has 2 pinch branches that meet each other when $E/D=0.4988$ in the interior of Ω, and then they split again in the other direction. This is the saddle-type splitting caused by the presence of the bifurcation point of hyperbolic type verifying Eq. (9).

Fig. 2a provides an example of a mixture of class 2.0-2b with two binary azeotropes. It shows the evolution of the pinch curve associated to the separation of component 2 from component 3 using component 1 as the entrainer. The pinch curves are limited from the left and from the right by two univolatility curves α_{13} and α_{23}. Fig. 2b shows an example of class 1.0-2 exhibiting closed type pinch curves in the zone comprised between two branches of the same univolatility curve α_{13}. Here toluene is separated from ethanol using acetone as an entrainer. The elliptic type bifurcation at the critical value $E/D=0.017$ causes the vanishment of the pinch curve for the greater values of E/D.

4. Conclusion

The new computational method presented in this paper assures an easy computation of pinch curves of ternary mixtures with high accuracy. It can be naturally generalized to compute the curves of more than three component mixtures. It is based on a geometrical

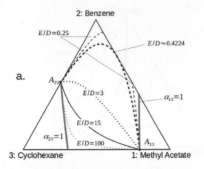
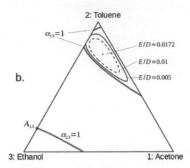

Figure. 2. Case a: Serafimov's class 2.0 – 2b. Pinch curves associated to the separatiion of benzene from cyclohexane using methyl acetate as the entrainer. Case b : Serafimov's class 1.0-2, pinch separtion of touluene from acetone using ethanol as an entrainer. Ellyptic-type biffurcation: closed pinch curves shrink into a single pinch point before disappearing.

model that allows to predict the bifurcation in the topological structure of pinch diagrams related to the change of operational conditions. Accurate computation of pinch branches facilitates the synthesis and design of an extractive distillation column since they give access to limiting values of reflux ratio and entrainer flow rate that help select entraincr and assess process feasibility.

References

O.Cots, N. Shcherbakova, J. Gergaud, 2021. SMITH: Differential homotopy and automatic differentiation for computing thermodynamic diagrams of complex mixtures. Comput. Aided Chem. Eng., 50, 1081–1086

N. Felbab, 2012. An Efficient Method of Constructing Pinch Point Curves and Locating Azeotropes in Nonideal Distillation Systems. Ind. Eng. Chem. Res, vol 51, pp. 7035-7055.

J. Gmehling, J. Li, M.A. Schiller, M.A., 1993. Modified UNIFAC Model 2.Present parameter matrix and results for different thermodynamic properties. Ind. Eng. Chem. Res. 32, 178–193.

L. Hasco and V. Pascual, 2012. The tapenade automatic differentiation tool: principles, model, and specification. Research report R-7957, INRIA, http://tapenade.inria.fr:8080/tapenade/

V.N. Kiva, E.K. Hilmen, S.Skogestad, 2003. Azeotropic phase equilibrium diagrams: a survey. Chem. Eng. Science, 58, 1903-1953.

Mathematica 9. URL http://www.wolfram.com/mathematica/.

F. Petlyuk, R. Danilov, J. Burger, 2015. A novel method for the search and identification of feasible splits pf extractive distillations in ternary mixtures, Chem. Eng.Res. Des., 99, 132 – 148

P. Poellmann, E. Blass, 1994. Best products of homogeneous azeotropic distillations. Gas Separation & Purification, 8, 194-228

I.Rodriguez-Donis, N. Shcherbakova, J. Abildskov, V.Gerbaud, 2024. Entrainer selection using the Infinitely Sharp Split method for separating binary minimum boiling azeotrope by extractive distillation (in preparation)

Simulis Thermodynamics; ProSim: Labege Cedex, France, 2023. https://www.prosim.net/

Skiborowski, M., Bausa, J., Marquardt W., 2016. A Unifying Approach for the Calculation of Azeotropes and Pinch Points in Homogeneous and Heterogeneous Mixtures. Ind. Eng. Chem. Res., 55, 6815–6834.

Flavio Manenti, Gintaras V. Reklaitis (Eds.), Proceedings of the 34th European Symposium on Computer Aided Process Engineering / 15th International Symposium on Process Systems Engineering (ESCAPE34/PSE24), June 2-6, 2024, Florence, Italy

Assessment of Plastic Recycling Technologies Based on Carbon Resource Circularity Considering Feedstock and Energy Use

Takuma Nakamura,[a] Shoma Fujii,[a,b] Aya Heiho,[c,d] Heng Yi Teah,[c]
Yuichiro Kanematsu,[c] Yasunori Kikuchi,[a,b,c*]

[a]Department of Chemical System Engineering, The University of Tokyo, 7-3-1 Hongo, Bunkyo-ku, Tokyo 113-8656, Japan
[b]Institute for Future Initiatives, The University of Tokyo, 7-3-1 Hongo, Bunkyo-ku, Tokyo 113-8654, Japan
[c]Presidential Endowed Chair for "Platinum Society", The University of Tokyo, 7-3-1 Hongo, Bunkyo-ku, Tokyo 113-8656, Japan
[d]Fuculty of Environmental Studies, Tokyo City University, 3-3-1 Ushikubo-Nishi, Tsuzuki-Ku,Yokohama, Kanagawa 224-8551 ,Japan
ykikuchi@ifi.u-tokyo.ac.jp

Abstract

In this study, we are tackling the impact assessment of introducing plastic recycling technologies based on resource circularity for the design of a circulating plastic recycling system. For a sustainable circulation of carbon resources in the chemical industry, CO_2 emitted from refineries needs to be converted back into resources and the remaining carbon needs to be substituted with bio-derived carbon and recycle-derived carbon. This study assessed the "circulating" plastic recycling technologies that convert waste plastics into raw materials that can be fed into existing facilities. The impact of plastic recycling has been evaluated in terms of greenhouse gas emissions, with a few cases evaluated from the perspective of the degree of resource circularity; however, an appropriate environmental indicator for recycling has not been constructed to date. To evaluate the impact of recycling technologies, we developed an indicator for the degree of carbon resource circulation based on the model consisting of process flows of refineries and multiple recycles. The results showed that catalytic cracking and material recycling contribute to carbon resource circulation more than other recycles.

Keywords: circularity indicator, carbon resource circulation, life cycle assessment, process modeling

1. Introduction

Regarding the circulation of carbon resources in the chemical industry, the consumption of fossil-derived resources needs to be reduced for carbon neutrality. The CO_2 emitted from refineries needs to be converted back into resources and the remainder of the carbon needs to be substituted with bio-derived and recycle-derived carbon (Meng et al., 2023). In the context of carbon resource circulation, "circulating" recycles, including material recycling (MR), catalytic cracking, and monomerization, are important for converting used plastic products into chemical feedstocks and plastic resins that can be fed into existing processes such as refineries. Particularly, it was reported that the performance of catalytic cracking could be linked with improvements in environmental performance

(Kikuchi et al., 2023). These technologies allow carbon resources to be used circularly and reduce the consumption of crude oil. By contrast, waste to energy mainly contributes to energetic circulation. Its main objective is to reduce greenhouse gas (GHG) emissions. Therefore, the impact of introducing recycling technologies must be evaluated in terms of both GHG emissions and contribution to carbon resource circulation. Evaluation by these two indicators will lead to an appropriate impact assessment framework for plastic recycling technologies.

In this study, we are tackling the impact assessment of introducing plastic recycling technologies based on resource circularity for the design of the circulating plastic recycling system. To date, the impact of introducing recycling has been evaluated mainly by GHG emissions, with a few cases evaluated from the perspective of the degree of resource circularity in previous studies (Lase et al., 2023). The study also reported results that an increase in material circularity generally decreases the environmental load regarding MR of plastic packaging (Vadoudi et al., 2022). However, evaluation based on the circularity of waste plastics is not sufficient. In other words, to date, an appropriate evaluation environment for plastic recycling has not been constructed. In this study, we made progress by setting the following: system boundaries, process modeling, developing an indicator for carbon resource circulation, inventory analysis, and environmental impact assessment. The system boundary includes oil refining, petrochemical processes, and several recycling processes. The unit configuration, capacity, energy consumption, and component yields of each unit refineries were set by the literature. In addition, the yields of products and energy consumption for each recycling were also set by the literature. From the above, GHG emissions were calculated by multiplying the energy consumption of each process by the CO_2 emissions intensity taken from life cycle assessment (LCA) database. As for the calculation of resource circulation, the value was calculated by combining the mass of crude oil and waste plastics processed by incineration, landfill, and waste to energy.

2. Materials and methods

2.1. System boundary

The production flows of petroleum products and basic chemicals derived from oil refining and petrochemical processes to produce plastics are shown in Figure 1. At the end-of-life, plastic wastes are treated for resource recycling (i.e., MR, mechanical recycling, monomerization, or catalytic cracking) or other purposes (energy recovery (ER), gasification, blast furnace reduction, or coke oven reduction). The manufacturing of plastic products and the use of the products were excluded from the system boundary of this study. The oil refining converts crude oil into naphtha and other products through separation and refining. The petrochemical process uses the naphtha to produce basic chemicals such as olefins and benzene, toluene, and xylene (BTX). The catalytic cracking converts waste plastics into hydrocarbons, which are similar in composition to conventional oil refineries. The steps include mixing with solvents, cracking reaction, and distillation. The cracking products were separated into four groups based on their carbon chains. Product A (C1 to C4) was similar to gas products used as raw materials at cryogenic separation. Product B (C5 to C7) and product C (C8 to C9) were fed as naphtha substitutes into naphtha cracking and catalytic reforming, respectively. Product D (>C10) contained the heavy content that must be fed to fluid catalytic cracking (FCC) for further cracking. Product distribution depends on the composition and decomposition conditions of the waste plastics. MR produces recycled pellets by washing, crushing, sorting, and pelletization. The mechanical recycling produces recycled pellets with improved quality by sorting and removal of impurities and solid-phase polymerization. The monomerization produces recycled pellets by depolymerization and repolymerization. The ER generates electricity from incinerating waste plastics. Other processes, including ER, produce energy sources.

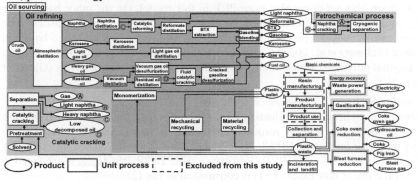

Figure 1 System boundary including the plastic recycling system and refineries

2.2. Process modeling and parameter setting

The model was created for the system boundary diagram shown in Figure 1 to analyze the impact of each unit process on the overall process, the relationship between unit processes, the contribution of technology performance at refineries and recycles, and the flow of resources when the system boundary is considered as a series of resource loops. For this model, the unit process configuration of the oil refining and petrochemical process, in addition to the capacity, the component yields, and the energy consumption per throughput of the unit process, were determined based on the reports on an existing refinery (Yoshitome et al., 2022). The production of basic chemicals and petroleum products was calculated by tracking the production of each unit process. We were also able to confirm the flow of fossil-derived feedstocks and waste plastics-derived feedstocks and calculate the throughput for each unit process capacity. Regarding

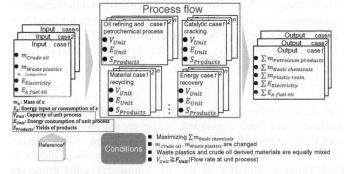

Figure 2 Process modeling

catalytic cracking, energy consumption per amount of waste plastics to be processed and the yields of cracking products were estimated by process simulation using Aspen HYSYS™. The product yields, which depend on the decomposition temperature and the composition of waste plastics, were collected from literature values (Keller et al., 2022). Energy consumption per amount of waste plastics to be processed and yield of products at MR and ER were established based on reports (JaIME, 2022). The composition of waste plastics can be arbitrarily changed. Basically, it was assumed that polypropylene (PP), polyethylene (PE), and polystyrene (PS) account for at least 60% of the total, and polyethylene terephthalate (PET), polyvinyl chloride (PVC), and other types account for the rest based on the current composition of Japan. Additionally, catalytic cracking was set to process mixed PP, PE, and PS, and MR was set to process pure PP or PE, and monomerization was set to process pure PET in this model. On the other hand, other recycling including ER, were assumed to process all types of waste plastics. The substitution rate of recycled resins for petroleum-derived resins was set to 0.5 based on the literature (Schwarz et al., 2021) for those by MR, and 1.0 for those by monomerization considering that polymerization process is included with reference to the study (Ghosh et al., 2023).

2.3. The indicator of carbon resource circulation

In this study, the indicator for evaluating the impact of a recycling process introduction based on the degree of carbon resource circulation is named the recycling system circularity indicator for plastics (RSCI_Plastics). The RSCI_Plastics was developed by modifying the product circularity indicator (PCI) (Bracquené et al., 2020). RSCI_Plastics is presented in Eq. (1), and the linear flow index (LFI) is constructed as Eq. (2).

$$RSCI_{Plastics} = 1 - LFI \tag{1}$$

$$LFI = \frac{M_\alpha + M_\beta + M_\gamma}{M_{\alpha,liner} + M_{\beta,liner} + M_{\gamma,liner}} \tag{2}$$

RSCI_Plastics is indicated as dimensionless numbers from 0 to 1, where 1 means complete carbon resource circulation. In this study, the complete carbon resource circulation is defined as "the state in which all waste plastics are processed by circulating recycles in resource loop and are used as raw materials for basic chemicals and plastic resins, not used for fuels". This indicator was mainly calculated by LFI, which indicates the degree of resource use in a linear economy. LFI was composed of mass of crude oil input (M_α), mass of waste plastics that are incinerated or landfilled (M_β), and mass of waste plastics that are processed by waste to energy (M_γ). Also, LFI was calculated as a percentage in Eq. (2), with the denominator being the value obtained when the resources are used in a completely linear economy with the numerator being the value obtained when recycling processes are introduced. To calculate M_α, M_β, and M_γ, the distribution ratio of waste plastics to each recycling process, yields of products, and others are also used as parameters.

A previous study (Matos et al., 2023) showed that indicators for circularity that focus on specific materials and products have been developed. However, the indicators do not adapt to system boundaries that cover multiple recycling processes and include design variables of recycling processes. RSCI_Plastics covered the system boundary shown in Figure 1, which can include existing refineries, as well as multiple recycling processes for waste plastics. This indicator makes it possible to evaluate each recycling process per processing of 1 kg of waste plastics comparatively and evaluate recycling systems per processing the total amount of waste plastics. Plastic resins are considered to be final products, and the resource loop starts at the stage after the collection of waste plastics within the system boundary. As described above, RSCI_Plastics is characterized by the fact that it does not take into account the contribution of the recycling process to energetic circulation, but makes possible an evaluation focused on material-to-material.

3. Results and discussion

Figure 3 shows the evaluation results of recycling technologies based on two indicators, $RSCI_{Plastics}$ and GHG emissions reduction. The upper right corner of the graph indicates a greater GHG reduction and a higher $RSCI_{Plastics}$, which represents a better plastic recycling path. The graph also identifies any potential burden shift between GHG reduction and carbon resource circulation. We show that although circulating recycles are inferior to waste to energy in terms of GHG emission reduction, they have an advantage in contributing to carbon resource circulation. This result can be attributed to the fact that catalytic cracking and MR can convert waste plastics into chemical feedstocks and plastic resins on a material-to-material basis when compared to other recycling technologies. When we consider scenarios that move the plot to the upper right, the results are calculated under the assumption that energy sources for all recycles are fossil-derived, so if renewable energy resources can replace these, the plot can move to the right. In other words, there are other ways to increase GHG emission reductions besides implementing the recycling process itself. In addition, to move the plot upward, it is important to introduce recycling technologies like catalytic cracking and MR. Therefore, in the context of carbon resource circulation, it is essential to promote the implementation and introduction of such recycling technologies. Another important aspect is that the arrangement of these plots also changes depending on the composition of the waste plastics. In this figure, the composition of waste plastics was determined based on the current composition in Japan, and the results were calculated under the condition that all technologies process waste plastics with this composition. This was done to standardize the processing conditions for comparison of multiple recycling technologies. In the future, when we consider the strict introduction of these recycling processes, it would be useful to analyze what kind of condition or composition of waste plastics can be collected and sorted for a more precise evaluation.

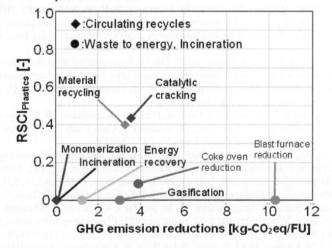

Figure 3 Results based on carbon resource circularity and GHG emission reductions

4. Conclusions

In this study, the model that linked the process flow based on existing refineries with the process flow of multiple recycling processes was constructed. In addition, $RSCI_{Plastics}$, an indicator for evaluating the impact of introducing a recycling technology from the perspective of carbon resource circulation, was developed. As a result, we confirmed the impact of introducing recycling in terms of not only GHG emission reduction, but also its contribution to resource circularity. It shows the importance of increasing the amount of waste plastics treated by catalytic cracking and MR in promoting carbon resource circulation. However, because the evaluation results are highly dependent on the state and composition of waste plastics, it is also essential to set up multiple scenarios that take these factors into account in detail. In the future, we need to clarify the supply and composition of waste plastics by material flow analysis and proceed to set up multiple scenarios. Also, RSCI$_{Plastics}$ does not take into consideration the quality of plastic resins. Therefore, in future studies it would be preferable to identify the destination products of plastic resins and incorporate parameters that indicate changes into physical properties.

Acknowledgement

This work contains the results of projects supported by New Energy and Industrial Technology Development Organization (NEDO, Grant number JPNP20012), JST COI-NEXT JPMJPF2003. Activities of the Presidential Endowed Chair for "Platinum Society" at the University of Tokyo are supported by Mitsui Fudosan Corporation, Sekisui House, Ltd., the East Japan Railway Company, and Toyota Tsusho Corporation.

References

Japan Initiative for Marine Environment (JaIME), 2022, Life cycle assessment of industrial waste plastics, https://www.nikkakyo.org/system/files/産業系廃プラスチックの環境負荷評価 (LCA) _0.pdf

T. Ghosh, G. Avery, A. Bhatt, T. Uekert, J. Walzberg, A. Carpenter, 2023, Towards a circular economy for PET bottle resin using a system dynamics inspired material flow model, J. Clean. Prod., 383, 135208

H. Jeswani, C. Krüger, M. Russ, M. Horlacher, F. Antony, S. Hann, A. Azapagic, 2021, Life cycle environmental impacts of chemical recycling via pyrolysis of mixed plastic waste in comparison with mechanical recycling and energy recovery, Science of the Total Envir., 769, 144483

F. Keller, R.L. Voss, R.P. Lee, B. Meyer, 2022, Life cycle assessment of global warming potential of feedstock recycling technologies: Case study of waste gasification and pyrolysis in an integrated inventory model for waste treatment and chemical production in Germany, Resour. Conserv. Recycl. 179, 106106

Y. Kikuchi, Y. Nomura, T. Nakamura, S. Fujii, A. Heiho, Y. Kanematsu, 2023, Application of CAPE Tools into Prospective Life Cycle Assessment: A Case Study in Feedstock Recycling of Waste Plastics, Comput. Aided Chem. Eng., 52, 2477-2482

I. S. Lase, D. Tonini, D. Caro, P. F. Albizzati, J. Cristobal, M. Roosen, M. Kusenberg, K. Ragaert, K. M. Van Geem, J. Dewulf, S. De Meester, Resour., Conserv. Recycl., 192, 106916

J. Matos, C. Martins, C. L. Simoes, R. Simoes, 2023, Susta. Prod. and Cons., 39, 521–533

F. Meng, A. Wagner, A. B. Kremer, D. Kanazawa, J. J. Leung, P. Goult, M. Guan, S. Herrmann, E. Speelman, P. Sauter, S. Lingeswaran, M. M. Stuchtey, K. Hansen, E. Masanet, A. C. Serrenho, N. Ishii, Y. Kikuchi, and J. M. Cullena, 2023, Planet-compatible pathways for transitioning the chemical industry, Proc. Natl. Acad. Sci., 120(8), e2218294120

A.E. Schwarz, T.N. Ligthart, D. G. Bizarro, P. De Wild, B. Vreugdenhil, T. van Harmelen, 2021, Plastic recycling in a circular economy; determining environmental performance through an LCA matrix model approach, Waste Manage., 121, 331–342

K. Vadoudi, P. Deckers, C. Demuytere, H. Askanian, V. Verney, 2022, Comparing a material circularity indicator to life cycle assessment: The case of a three-layer plastic packaging, Susta. Prod. and Cons., 33, 820–830

T. Yoshitome, K. Saito, K. Tamura, 2022, Modeling of refinery equipment configuration and CO$_2$ emissions, PETROTECH, 45 (1), 21-28

Flavio Manenti, Gintaras V. Reklaitis (Eds.), Proceedings of the 34th European Symposium on Computer Aided Process Engineering / 15th International Symposium on Process Systems Engineering (ESCAPE34/PSE24), June 2-6, 2024, Florence, Italy

Optimal Experimental Design for (Semi-)Batch Crystallization Processes

Gustavo L. Quilló,[a,b] Wannes Mores,[a] Satyajeet S. Bhonsale,[a] Alain Collas,[b] Christos Xiouras,[b*] Jan F. M. Van Impe[a*]

[a]BioTeC+, Chemical and Biochemical Process Technology and Control, KU Leuven Campus Ghent, Gebroeders De Smetstraat 1, 9000 Ghent, Belgium
[b]Chemical & Pharmaceutical Development and Supply, Janssen R&D, Turnhoutseweg 30, 2340 Beerse, Belgium
cxiouras@its.jnj.com , jan.vanimpe@kuleuven.be

Abstract

Knowledge gain in crystallization R&D is a stepwise process where conditions are systematically explored to develop a working process, and model-based approaches to facilitate tech transfer are gaining momentum. Despite being useful to reduce overall experimental effort, these techniques are often laden with statistical issues in the context of crystallization kinetics determination. This work demonstrates how to valorize a developed model through Optimal Experimental Design in a case-study of industrial semi-batch crystallization, achieving as much as 3.71 times more efficient data collection from a D-optimal experiment versus an experiment that uses linear cooling and dosing profiles while showcasing modelling nuances faced in the field.
Keywords: crystallization, optimal experimental design, population balance modelling.

1. Introduction

Crystallization from solution is present in approximately 80% of pharmaceutical development pipelines due to its capability of impurity rejection. The interplay between hydrodynamics, thermodynamics, and kinetics in (semi-)batch processing dictates the produced polymorph, process yield, and particle size distribution (PSD), direct impacting in the efficiency of filtration, washing, drying, mixing, tableting, and flow properties.

Population Balance Modeling (PBM) is the state-of-the-art mathematical framework to simulate crystallization, and it classically requires auxiliary kinetic expressions whose parameters need to be determined prior to process design and control. However, these are often ill-determined due to a combination of (1) high parameter correlation due to (nonlinear) model structure or data collinearity, causing degeneracy during regression, (2) insufficient system understanding, leading to poor model choice, (3) insufficient exploration of the feasibility domain, (4) experimental inaccessibility of some states, (5) the applied control policy. The advent of Process Analytical Technologies (PATs) mitigated issues, enabling frequent sampling of concentration through spectral techniques, and indirect assessment of the PSD through chord length distributions. Despite the capability of the setups, crystallization model development is almost always built (only) upon series of experiments with constant control policies because these designs offer more interpretability and easier troubleshooting. It is common to gather data through a factorial design, where isothermal desupersaturation decays are evaluated at various solvent composition levels, impurity content, seed loads, and stirring rates. Next, nonlinear regression problems are posed looping through each valid combination of the kinetic expressions from the model repository, out of which the best performing, physically relevant overall model defines the structure employed in all subsequent endeavors (Orosz et al., 2023). Optimal experimental design (OED) shines after this screening stage, where there is rudimentary knowledge about the system, but there is room for model improvement. Then the model candidate is used to plan experiments in a constrained experimental space reflecting the equipment capabilities and system complexity, avoiding polymorph conversions, phase-separation, excessive breakage, or agglomeration, dead growth zones, etc. (Quilló et al., 2023a). Practically, *in silico* experiments are sequentially generated and intercalated with data collection until the regressed parameters are within the confidence interval (CI) of the previous iteration.

Optimal experimental design (OED) is a procedure that maximally exploit experimental conditions based on the model sensitivity to the parameter values or to output prediction variance, depending on the optimized scalar statistical criterion. The most popular OED criteria for parameter estimation are the D-, A- and ME-criterion, which respectively embody the volume, outer hyperbox volume and conditioning number of the parameter joint confidence interval hyper ellipsoid. These (scalar) metrics are computed via Fisher Information Matrix (FIM), sigma-points, or other methods (Bhonsale et al., 2022).

When formulated for systems in differential form such as in crystallization, an open-loop control problem is created in which the discretized control policy, i.e., the temperature or dosing trajectory, provide (some of) the degrees of freedom to improve the objective function (Nagy et al., 2008; Quilló et al., 2023a). Likewise, the optimization can encompass the batch time, the initial condition and sampling time stamps for expensive or laborious offline data collection, e.g., PSD measurements. Moreover, the OED framework can be applied in the context of algebraic mixture models for solubility prediction to support kinetics investigations (Quilló et al., 2023b).

Although recent advances extended its application to continuous systems (Kilari et al., 2023), literature on OED for crystallization kinetics remains scarce and does not explore more sophisticated situations adjusting both temperature and solvent composition under activity-dependent supersaturation nor emphasize limitations or pitfalls of the technique. This work applies one iteration of the OED approach on the PBM to obtain the most informative piecewise linear control policy for a combined cooling-dosing (semi-)batch seeded crystallization through optimization of the D-criterion using the FIM approach.

2. Model-based Experimental Design

Let a system of ordinary differential equations (ODEs) be composed of nonlinear functions f of the state vector x, the controls u, and parameters p, and time t. Consider there are n_x states and n_p parameters. Further, assume the outputs y are linear functions of the states x. Then, the sensitivity of the i-th output with respect to the j-th parameter can be retrieved with the i-th row A_i of coefficient matrices A, as in Eq. (1).

$$\frac{dx_i}{dt} = f_i(x, u, p, t) \qquad y = Ax \Rightarrow \frac{\partial y_i}{\partial p_j} = A_i \frac{\partial x}{\partial p_j} \tag{1}$$

The sensitivity matrix element $s_{i,j}$ is computed as in Eq. (2). The derivatives of the states with respect to the parameters at each time point are obtained by solving the augmented the original ODE system with $n_x \times n_p$ coupled sensitivity equations.

$$\frac{\partial}{\partial p_j}\left(\frac{dx_i}{dt}\right) = \frac{d}{dt}\left(\frac{\partial x_i}{\partial p_j}\right) \Rightarrow \frac{d}{dt}\left(\frac{\partial x_i}{\partial p_j}\right) = \frac{ds_{i,j}}{dt} = \sum_{i=1}^{n_x} \frac{\partial f_i}{\partial x_i}\frac{\partial x_i}{\partial p_j} + \frac{\partial f_i}{\partial p_j}, s_{i,j}(t=0) = 0 \; \forall i,j \tag{2}$$

The FIM approach is rooted on the local linearization of the model by truncating a Taylor series expansion around nominal parameter estimates $\hat{\boldsymbol{p}}$ and its link with the variance-covariance matrix of the outputs $\boldsymbol{\theta}$ through the Cramér-Rao bound (Bhonsale et al., 2022). Assuming additive white measurement noise and n_y outputs each sampled at n_t regular time intervals, the **FIM** is given by Eq. (3). The model output Jacobian $\boldsymbol{J}$ is represented by Eq. (4). Note the matrix $\boldsymbol{\theta}$ can be omitted for optimization purposes.

$$\text{FIM} \cong \boldsymbol{J}^T \boldsymbol{\theta}^{-1} \boldsymbol{J} \tag{3}$$

$$\boldsymbol{J} = \begin{bmatrix} \frac{\partial y_1}{\partial p_1}(\boldsymbol{x}, \boldsymbol{u}, \boldsymbol{p}, t_1) & \cdots & \frac{\partial y_1}{\partial p_{n_p}}(\boldsymbol{x}, \boldsymbol{u}, \boldsymbol{p}, t_1) \\ \vdots & \ddots & \vdots \\ \frac{\partial y_1}{\partial p_1}(\boldsymbol{x}, \boldsymbol{u}, \boldsymbol{p}, t_{n_t}) & \cdots & \frac{\partial y_1}{\partial p_{n_p}}(\boldsymbol{x}, \boldsymbol{u}, \boldsymbol{p}, t_{n_t}) \\ \vdots & \cdots & \vdots \\ \frac{\partial y_{n_y}}{\partial p_1}(\boldsymbol{x}, \boldsymbol{u}, \boldsymbol{p}, t_1) & \cdots & \frac{\partial y_{n_y}}{\partial p_{n_p}}(\boldsymbol{x}, \boldsymbol{u}, \boldsymbol{p}, t_1) \\ \vdots & \ddots & \vdots \\ \frac{\partial y_{n_y}}{\partial p_1}(\boldsymbol{x}, \boldsymbol{u}, \boldsymbol{p}, t_{n_t}) & \cdots & \frac{\partial y_{n_y}}{\partial p_{n_p}}(\boldsymbol{x}, \boldsymbol{u}, \boldsymbol{p}, t_{n_t}) \end{bmatrix}_{p=\hat{p}} \tag{4}$$

3. Population Balance Modelling via Standard Method of Moments

The 1-D population balance for a cooling-dosing crystallization in a well-mixed semi-batch reactor under negligible breakage and agglomeration is given by Eq. (5), where $n(L, t)$ is the number density [#/(kg-neat solvent.m)], t is time, $\bar{G}$ is the growth rate [m/s], L is the characteristic particle size and M is the solvent mixture mass [kg-neat solvent].

$$\frac{\partial \tilde{n}(L, t)}{\partial t} + \frac{\partial [\bar{G}(L, t) \tilde{n}(L, t)]}{\partial L} = 0 \qquad \text{where } \tilde{n}(L, t) = n(L, t)M \tag{5}$$

The method of moments (SMOM) uses the moment transformation to obtain a set of ODEs from the original equation. The i-th moment $\tilde{\mu}_i$ of $\tilde{n}(L, t)$ is given by Eq. (6).

$$\tilde{\mu}_j = \int_0^\infty L^j \tilde{n}(L, t) \, \mathrm{d}L \tag{6}$$

The resulting ODEs are given by Eq. (7), where the concentration of the API in the liquid phase [g-solid API/g-neat solvent] is the fourth state and only output, as in Eq. (8).

$$\frac{\mathrm{d}\tilde{\mu}_j}{\mathrm{d}t} = j\bar{G}\tilde{\mu}_{j-1} + BML_0^j \qquad j = 0,1,2 \tag{7}$$

$$\frac{\mathrm{d}C}{\mathrm{d}t} = -\frac{3}{M}\rho_c k_v \bar{G}\tilde{\mu}_2 - \rho_c k_v BL_0^3 - \frac{C}{M}Q_{\text{in}} \tag{8}$$

The particles are assumed to nucleate at rate B [#/(kg-neat solvent.s)] with size L_0 [m], volume shape factor k_v [-], and true crystal density ρ_c [kg/m³], whereas Q_{in} is the neat (anti)solvent stream mixture with mass fraction composition w_{in}. For surface-integration controlled crystallization, both $\bar{G}$ and B are nonlinear functions of supersaturation (Eq. 10 and 11). Lastly, the solvent mixture volume V_r is estimated by $M/[1/\sum(w_i/\rho_i)]$, where M_i, w_i, and ρ_i are the mass, mass fraction and density of solvent component i.

4. Case study: OED for parameter estimation of crystallization kinetics

The model is used to obtain the temperature and dosing trajectories that optimize ψ under constraints of similar complexity to an industrial situation as in Eq. (9). This case maximizes the determinant of **FIM** (i.e., the D-criterion: $\psi(\mathbf{FIM}) = -\det(\mathbf{FIM})$.

$$\min_{RM_{\text{seed}},\,C(t_0),\,T_k,\,M_{\text{in,k}}} \quad \psi(\mathbf{FIM}) \qquad \text{for } k = 1 \dots n_{\text{nodes}}$$

Subject to:

$$
\begin{aligned}
&RM_{\text{seed,min}} \leq RM_{\text{seed}} \leq RM_{\text{seed,max}} && C_{t_0,\text{min}} \leq C(t_0) \leq C_{t_0,\text{max}} \\
&M_{\text{in,k=1}} = 0 \leq M_{\text{in,k}} \leq M_{\text{in,k+1}} && T_{\text{min,k}} \leq T_{k+1} \leq T_k \leq T_{\text{max,k}} \\
&R_{\text{min}} \leq (T_{k+1} - T_k)/(t_{k+1} - t_k) \leq 0 && V_{\text{r}}(t_{\text{f}}) \leq V_{\text{max}} \\
&0 \leq \left(M_{\text{in,k+1}} - M_{\text{in,k}}\right)/(t_{k+1} - t_k) \leq Q_{\text{in,max}} && 1 < \sigma_{\text{API}}(t_{\text{f}}) \leq \sigma_{\text{API,max}} \\
&0.9\,\max(\text{Yield}_{\text{theoretical}}) \leq \text{Yield}_{\text{process}}(t_{\text{f}}) && 0 \leq C(t_{\text{f}}) \leq C_{t_{\text{f}},\text{max}}
\end{aligned}
\tag{9}
$$

The specified nodes $T(k), M_{\text{in}}(k)$ are uniformly spaced in time, where k is the node index. The controls were discretized in 10 nodes defining piecewise linear profiles. The initial concentration $C(t_0)$ is optimized along with the control trajectories to leverage data on metastable zone at initial condition. Similarly, the relative seed mass RM_{seed} is optimized given the initial moments scale linearly with seed mass. The solution is obtained by single shooting in MATLAB® 2022b. The augmented ODE system is solved through ode23s, and optimization is compiled in FORTRAN (called internally by Intel oneAPI 2022) through a combination of multi-start optimization using the COBYLA (Powell, 1994) and SIMPLEX algorithms. Levenberg-Marquardt, genetic algorithm, particle-swarm etc., were applied but COBYLA was faster and achieved lower minima. Growth and secondary nucleation are given by Eq. (10), with parameters $k_{\text{g}}, g, k_{\text{g}}, b$.

$$
\left.
\begin{aligned}
\bar{G} &= 10^{k_{\text{g}}} \left[\ln(\sigma_{\text{API}})\right]^g \\
B &= 10^{k_{\text{b}}} \left[\ln(\sigma_{\text{API}})\right]^b
\end{aligned}
\right\} \text{ for } \sigma_{\text{API}} > 1, \; \bar{G} = B = 0 \text{ otherwise}
\tag{10}
$$

The supersaturation σ_{API} is computed by Eq. (11), where the denominator refers to the saturated state, and γ and x are the API activity coefficient and molar fraction. The procedure to obtain the activity coefficients is detailed elsewhere (Quilló et al., 2021).

$$\ln(\sigma_{\text{API}}) = \ln\left[\gamma_{\text{API}} x_{\text{API}} / \left(\gamma_{\text{API}}^{\text{sat}} x_{\text{API}}^{\text{sat}}\right)\right] \tag{11}$$

The API solubility is given by the Van't Hoff Jouyban-Acree model (Eq. 12), with T as temperature [K], w_{p} as the mass fraction of solvent p and ψ as parameters, as in **Table 1**.

$$
\ln(x_{\text{API}}^{\text{sat}}) = \sum_{p \in \{1,2,3\}} \left(\psi_{0,p} w_{\text{p}} + \psi_{1,p} \frac{w_{\text{p}}}{T}\right) \\
+ \sum_{p,q \in \{1,2,3\};\, p \neq q} \frac{w_{\text{p}} w_{\text{q}}}{T} \left[\psi_{2,p,q} + \psi_{3,p,q}\left(w_{\text{p}} - w_{\text{q}}\right) + \psi_{4,p,q}\left(w_{\text{p}} - w_{\text{q}}\right)^2\right]
\tag{12}
$$

Table 1. Van't Hoff Jouyban-Acree solubility parameters for the case study.

$\psi_{0,1} = -8.77$	$\psi_{1,1} = 9.43 \times 10^6$	$\psi_{2,1,2} = 7.80 \times 10^6$	$\psi_{3,1,2} = -3.40 \times 10^7$	$\psi_{4,1,2} = -3.47 \times 10^7$
$\psi_{0,2} = 17.2$	$\psi_{1,2} = -1.39 \times 10^7$	$\psi_{2,1,3} = -1.55 \times 10^7$	$\psi_{3,1,3} = -6.93 \times 10^6$	$\psi_{4,1,3} = -8.03 \times 10^5$
$\psi_{0,3} = 3.36$	$\psi_{1,3} = -3.06 \times 10^3$	$\psi_{2,2,3} = 2.07 \times 10^7$	$\psi_{3,2,3} = 7.94 \times 10^6$	$\psi_{4,2,3} = 1.15 \times 10^6$

Analytical derivatives were used for the sensitivity equations in the Jacobian to improve accuracy and reduce the computational burden. The value of the concentration variance σ_C is approximated by the model mean squared error (assumed to be 8.74×10^{-7}). The sampling interval $(t_{i+1} - t_i)$ for the states is assumed to be 5 min, in an experiment of $t_f = 16$h. The initial solvent composition is $w_1 = 0.503$, $w_2 = 0.313$ and $w_3 = 0.184$, and pure solvent 2 is dosed. Other information for the case-study is supplied in **Table 2**.

Table 2. Summary of model parameters used for OED and process parameters for the case study.

$k_b = 14.78$	$b = 7.18$	$k_g = -5.13$	$g = 2.03$
$\rho_c = 1300$ kg/m³	$k_v = 0.1$	$C_{t_f,\max} = 0.21$ kg/kg	$M_{t=0} = 0.303$ kg
$RM_{\text{seed,min}} = 0.1\%$	$RM_{\text{seed,max}} = 1.0\%$	$C_{t_0,\min} = 0.303$ kg/kg	$C_{t_0,\max} = 0.333$ kg/kg
$\tilde{\mu}_{0,t_0} = 2.05 \times 10^7$ #	$\tilde{\mu}_{1,t_0} = 2.83 \times 10^2$ m	$\tilde{\mu}_{2,t_0} = 9.67 \times 10^{-3}$ m²	$V_{\max} = 400$ mL
$T_{\min} = 33$ °C	$T_{\max} = 60$ °C	$R_{\min} = -1$ °C/min	Yield$_{\text{theoretical}} \approx 1$
$Q_{\text{in,max}} = 1.2$ mL/min	$\sigma_{\text{API,max}} = 1.01$		

Figure 1 shows it is D-optimal to produce supersaturation peaks at the start of the experiment (in an attempt) to distinguish growth and birth rate in time. This is followed by a nearly constant supersaturation period and secondary peaks caused by the dosing at the end to satisfy constraints. The sensitivities behave similarly, with a small offset due to the rate of increase of $\tilde{\mu}_2$ during the simulation. Moreover, (re)estimating k_g and g is easier than k_b and b during subsequent data collection due to their larger sensitivities.

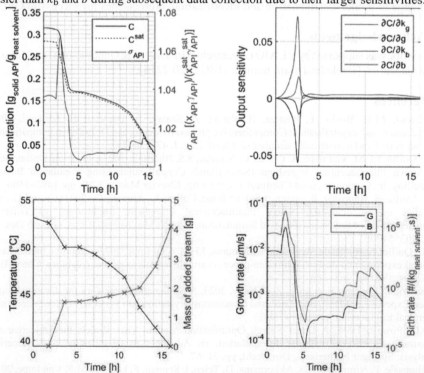

Figure 1. D-optimal *in silico* experiment, with $c_{\text{API},t=0} = 0.314$ kg/kg and 0.187% seed load. (top-left) Evolution of the concentration, solubility, and supersaturation, (top-right) output sensitivities, (bottom-left) control policy for temperature and dosing, (bottom-right) growth and nucleation rates.

5. Conclusions

While the possibility of more precise parameters estimates is a tempting route for data collection, OED limitations are evidenced by the multitude of combinations of kinetic rates models. OED epitomizes a test for model quality, which is not always satisfied due to poor statistical properties of the overall model structure or the absence of a model output relating to the PSD to partially decorrelate parameters from different phenomena. The D-efficiency of the obtained design (with $\mathbf{FIM_1}$) with respect to an analogous experiment with one-step linear cooling and dosing to the values of the final nodes of the D-optimal experiment (feasible, with $\mathbf{FIM_2}$) with is defined as $[\det(\mathbf{FIM_1})/\det(\mathbf{FIM_2})]^{1/n_p}$ and was computed as 3.71, meaning the analogous experiment would have to be replicated that many times for a similar reduction in parameter CI volume (i.e., assuming the parameters do not change). However, the low values of the determinants (2.33×10^{-6} for the OED) indicate some level of parameter unidentifiability. The same behavior is observed in other mechanistic functions for growth such as Burton-Cabrera-Frank and birth-and-spread, or expressions for nucleation.

OED is a useful addition to the strained R&D timelines, reducing the experimental effort through *in silico* designs. However, the screening experiments need to generate enough data for an imprecise model and comprehensively define a feasibility domain. Moreover, the correlated kinetic expressions used in PBMs, likely mismatch between a true follow-up experiment and the *in silico* experiment, along the experimental hurdles, pose a challenge to model discrimination and identification, prerequisites for successful OED.

6. Acknowledgements

This work was supported by VLAIO (Agentschap Innoveren & Ondernemen) and Janssen Pharmaceutica [Baekeland grant number HBC.2020.2214].

References

Á. Orosz, M.H. Bosits, É. Pusztai, H. Pataki, Z. Szalay, Á. Demeter, B. Szilágyi, 2023. Diastereomer salt crystallization: Comprehensive process modeling and DoE-driven comparison of custom-coded and user-friendly simulators. Chem. Eng. J. 473, 145257.

G.L. Quilló, J.F.M. Van Impe, A. Collas, C. Xiouras, S.S. Bhonsale, 2023a. Dynamic Optimization of Active Pharmaceutical Ingredient (Semi-)Batch Crystallization using Population Balance Modelling, in: Computer Aided Chemical Engineering. Elsevier Masson SAS, pp. 1495–1500.

G.L. Quilló, S. Bhonsale, B. Gielen, J.F. Van Impe, Collas, A., Xiouras, C., 2021. Crystal Growth Kinetics of an Industrial Active Pharmaceutical Ingredient: Implications of Different Representations of Supersaturation and Simultaneous Growth Mechanisms. Cryst. Growth Des. 21, pp. 5403–5420.

G. L. Quilló, S. Bhonsale, A. Collas, C. Xiouras, J.F.M. Van Impe, 2023b. Iterative model-based optimal experimental design for mixture-process variable models to predict solubility. Chem. Eng. Res. Des. 189, pp. 768–780.

H. Kilari, Y. Barhate, Y.-S Kang, Z.K. Nagy, 2023. A Systematic Framework for Iterative Model-Based Experimental Design of Batch and Continuous Crystallization Systems, in: Computer Aided Chemical Engineering. pp. 1501–1506.

M.J.D. Powell, 1994. A Direct Search Optimization Method That Models the Objective and Constraint Functions by Linear Interpolation, in: Advances in Optimization and Numerical Analysis. Springer Netherlands, Dordrecht, pp. 51–67.

S. Bhonsale, P. Nimmegeers, S. Akkermans, D. Telen, I. Stamati, F. Logist, J.M.F. Van Impe, 2022. Optimal experiment design for dynamic processes, in: Simulation and Optimization in Process Engineering. Elsevier, pp. 243–271.

Z.K.Nagy, M. Fujiwara, R.D.Braatz, 2008. Modelling and control of combined cooling and antisolvent crystallization processes. J. Process Control 18, pp. 856–864.

Flavio Manenti, Gintaras V. Reklaitis (Eds.), Proceedings of the 34th European Symposium on
Computer Aided Process Engineering / 15th International Symposium on Process Systems
Engineering (ESCAPE34/PSE24), June 2-6, 2024, Florence, Italy

Novel Process for Bio-jetfuel Production Through the Furan Pathway. Techno-Economic, Environmental and Safety Assessment

Melissa Pamela Hinojosa-Esquivel[a], Cesar Ramirez-Marquez[b], Julio Cesar Velazquez-Altamirano[a], Juan José Quiroz-Ramirez [c], Juan Gabriel Segovia-Hernandez[d], Gabriel Contreras-Zarazua[c,d]

[a]Escuela Superior de Ingeniería Química e Industrias Extractivas Instituto Politécnico Nacional s/n Edificio 7, de, Av Instituto Politécnico Nacional, Lindavista, Ciudad de México, CDMX, 07738, México.
[b]Departamento de Ingeniería Química, Universidad Michoacana de San Nicol´as de Hidalgo, Morelia, Mich 58060, Mexico.
[c]CONACYT-CIATEC, A.C. Centro de Innovación Aplicada en Tecnologías Competitivas. Omega 201, Industrial Delta, 37454 León, GTO, México.
[d]Departamento de Ingeniería Química, Universidad de Guanajuato, Noria Alta s/n, Guanajuato, Gto., 36050, México.
g.contreras@ugto.mx

Abstract

This study proposes a new biofuel production process using furans derived from lignocellulosic residues to convert them into gasoline, bio-jet fuel, or diesel. The process consists of three fundamental stages: aldol condensation, hydrodeoxygenation (HDO), and separation. Different catalysts and technologies were evaluated using as metrics the total annual cost (TAC), the eco-indicator 99 (EI99), and the individual risk (IR) to determine the economic, environmental and safety aspects of the process, respectively. This metrics were chosen in order to evaluate the sustainability of each technology. The results indicate that a process using DBU as a catalyst is the option with the best trade-off between environmental impacts, costs, and safety. This process shows costs and environmental impact reductions of up to 75% compared to other typical processes such as ATJ.

Keywords: Furans to fuels, Furans to jet, FTJ, Biofuels

1. Introduction

Currently, there is a clear need to replace fossil fuel energy sources with renewable energy sources to address climate change accelerated by the accumulation of greenhouse gases and the increasing costs of oil extraction. Data from the Energy Information Administration (IAE) indicate that the transportation sector is highly dependent on fossil fuels, which accounted for 96.8% of all the energy required by this sector in 2018 (EIA, 2022). In the United States alone, the transportation sector annually consumes the equivalent of 37% of the generated energy, representing the emission around 2000 million tons of carbon dioxide, surpassing emissions from other sectors such as industrial or residential (EIA, 2022). To reduce the transportation sector's dependence on fossil fuels and decrease its greenhouse gas emissions, a process of electrifying the transportation sector has been initiated, where the automotive sector is the main driving force. However, electrification presents some disadvantages that have made its implementation impossible in some sectors of transportation, such as aviation. Some of these limitations are the power-to-weight ratio, the capacity for storage, and the power generated by batteries, as well as the lack of infrastructure for a fast implementation of electrification. Therefore, the use of alternative energy sources, such as biofuels, emerges as a viable alternative to achieve a carbon-neutral or negative transportation sector (IRENA, 2021).

Despite the environmental and economic benefits of biofuels, their production remains expensive compared to fossil fuels. Although many studies have been conducted to improve biofuel production processes, most research efforts have focused on reducing production costs and finding new raw materials, rather than exploring new production routes (Ko et al., 2020). An alternative that has not been explored is the production of biofuels through the furan route, known as FTF (Furans to Fuels). The production of furans like furfural or HMF from lignocellulosic waste is widely known, and these bioproducts are already produced on an industrial scale, making them ideal raw materials for biofuel production. In addition, this route is thermochemical, which means that biochemical routes are not required, thereby reducing in this way resident time and increasing the yields to fuels. Despite the benefits of this route, up to now, there has been not reported a process for producing biofuels, such as bio-jetfuel, using this route. Based on aforementioned, in this work is proposed the synthesis, design, and simulation of a novel biofuel production process from these furans. This process consists of three stages: aldol condensation, hydrodeoxygenation (HDO), and hydrocarbon separation. In each stage, different catalysts and technologies are analyzed. The total annual cost, Eco-Indicator 99, and individual risk were considered as metrics to evaluate the economic, environmental, and safety aspects of the different process options. These indicators were selected according to the 12 principles of sustainability proposed by Jimenez Gonzalez and Constable (2011).

2. Methodology

In this study, furfural was considered as the main furan produced from lignocellulosic residues: therefore, it was used as the main raw material for this novel process. It is important to note that this work does not consider the production of furfural from biomass, since this production has been extensively explored in previous studies. The process for producing biofuels from furfural consists of three stages: aldol condensation, hydrodeoxygenation (HDO), and a separation stage. This process was designed and simulated using the Aspen Plus software. The design was performed in order to minimize the energy requirement, which is directly related to lower operating costs. Due to the difference temperature and pressure conditions of each stage, as well as the compounds involved, the Carlson's algorithm was used to determine the most suitable thermodynamic model for each stage (Carlson, 1996).

During the aldol condensation stage, furfural and a ketone compound (commonly acetone) react in the presence of basic catalysts and solvents, generating higher molecular weight oxygenated compounds called adducts. These reactions can be carried out at conditions of 1 atm and temperatures between 25 °C and 58 °C to guarantee that the ketone remains in the liquid phase. In this stage three different catalyst were studied: dolomite, sodium hydroxide solutions, and 1,8-Diazabicyclo[5.4.0]undec-7-ene (DBU). A CSTR reactor was considered to carry out the reactions in this stage. The kinetic model for all the catalysts fits to the Arrhenius equation behavior and were taken from O'Neill et al., 2014 and Jiang et al., 2018. After producing the adducts, it's necessary to eliminate the oxygen in these compounds. Additionally, the double bonds must be removed to generate saturated compounds. This is essential to achieve the specific physicochemical properties needed for certain biofuels, like biojet -fuel, to meet ASTM standards. This oxygen and double bond removal is performed by means of the hydrodeoxygenation stage, where hydrogen is added. The oxygen reacts with the hydrogen to form water. This reactive stage can be performed in a fixed bed reactor over a Nickel- Platinum supported on alumina catalyst, at conditions of 10 to 50 atm and 200 to 400 °C. The kinetic parameters used to design and simulate this equipment were taken from Faba et al, (2016) whereas the missing kinetic parameters were estimated using the methodology proposed by Sánchez-Ramírez et al., (2022). Owing to the high pressures and temperatures required in this stage, the thermodynamic model used to simulate it was the Soave-Redlich-Kwong model.

During the HDO stage, water, unreacted hydrogen, and hydrocarbons(mainly octanes and tridecanes) are produced. The unreacted hydrogen is separated from the mixture using flash distillation. On the other hand, the water can be separated from the organic mixture using an azeotropic distillation column, due to the presence of azeotropes and two liquid phases. The azeotropic distillation process was proposed and designed according to the previous work reported by Contreras-Zarazua et al. (2018) in order to obtain anhydrous hydrocarbons. Once the water is removed, the hydrocarbons can be separated into different cuts using conventional distillation columns. The azeotropic distillation system was simulated using the NRTL-RK thermodynamic model due to the presence of organic compounds and water. On the other hand, the flash tank and the conventional distillation sequence were simulated using the Peng-Robinson method. Both thermodynamic models were chosen according to Carlson's algorithm. Figure 1 shows a flowsheet of the novel proposed process.

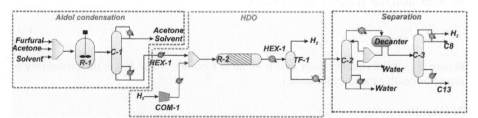

Figure 1. Process flowsheet for producing biofuels from furfural

As aforementioned, the Total Annual Cost (TAC), Eco-Indicator 99, and Individual Risks were the metrics selected to evaluate the economic, environmental, and safety aspects of the different process options. The Total Annual Cost is as shown below:

$$TAC = \frac{Capital\,cost}{Payback\,period} + Operating\,cost \qquad (1)$$

The TAC was calculated using Guthrie's method, considering a payback period of 10 years. Stainless steel was considered as the construction material. The costs were upgraded to the year 2023 using the chemical engineering plant cost index (CEPCI) of 816. The base costs corresponds to the year 2003 (CEPCI= 382). The necessary equations for estimating costs were taken from Turton et al., (2018). The operating costs were calculated considering 8500 hours of plant operation per year. Additionally, the following costs for services were considered: Cooling water ($0.355/USD/GJ), electricity ($16.8 USD/kWh), and Fired heat ($20.92 USD/GJ) Turton et al., (2018).

The Eco-Indicator 99 was the metric used to assess environmental impact, which is a life cycle methodology. The calculation for the eco-indicator was carried out according to what was reported by Contreras-Zarazua et al. (2020) The mathematical formula used to calculate the Eco-Indicator 99 is as follows:

$$EI99 = \sum_b \sum_d \sum_{k \in K} \delta_d \omega_d \beta_b \alpha_{b,k} \tag{2}$$

Where, β_b represents the total amount of chemical b released per unit of reference flow due to direct emissions, $\alpha_{b,k}$ is the damage caused in category k per unit of chemical b released into the environment, ω_d is the weighting factor for damage in category d, and δ_d is the normalization factor for damage in category d, respectively. The values of the weights and each impact category were taken from Contreras-Zarazúa et al., (2020)

Finally, the Individual Risk Index (IR) was used to quantify the safety of the process. This index is calculated through a quantitative risk analysis methodology and aims to measure the probability of harm caused to a person located at a certain distance from the epicenter of the accident. To identify potential accidents that could occur in the process, the HAZOP method was used, and seven possible accidents were found, which were classified into two categories: instantaneous accidents (BLEVE, UVCE, flash fire, and toxic release) and continuous release accidents (fire dart, flash fire, and toxic release). The complete set of equations for each incident and for the calculation of the IR was presented by Medina-Herrera et al. in 2014. The mathematical expression that describes the individual risk is expressed as follows:

$$IR = \sum f_i P_{x,y} \tag{3}$$

Where fi is the frequency with which incident i occurs, while $P_{x,y}$ is the probability of injury or death caused by incident i.

3. Results

In this section, the results obtained during the sensitivity analysis for the different processes are shown. The results obtained with the sensitivity analysis are presented in Table 1. Design specification for different the processes.

Design specifications	NaOH	Dolomite	DBU
Reactor volume R-1 (m³)	20	10	1
Temperature reactor R-1 (°C)	40	40	55
Pressure reactor R-1 (atm)	1	1	1
Feed mass flowrate acetone (kg/h)	600	500	3600
Solvent used	Methanol-Water (40-60%wt)	-------	------
Feed mass flowrate Solvent (kg/h)	9000	0	0
Number of Stages Column C-1	13	14	60
Reactor volume R-2 (m³)	11.7	26.9	19.79
Temperature reactor R-1 (°C)	273	359	192
Pressure reactor R-2 (atm)	45	25	45

Energy COM-1 (kW)	22529	16600	13081
Number of Stages Column s C-2	50	40	50
Number of Stages Column C-3	50	50	5
Mass flow C8 (kg/h)	2.46	411.33	10.3
Mass flow C13 (kg/h)	851.39	585.49	912.2
Total energy of the process (kW)	26231	10769	7012

Additionally, the results, such as cost, environmental impact, safety, as well as the sensitivity analysis for Reactor 1 of the NaOH process, which was selected as a representative case, are shown in Figure 2.

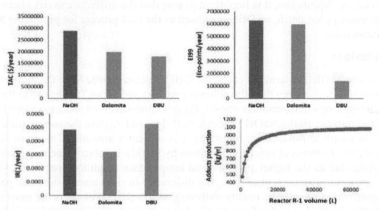

Figure 2. Results for TAC, EI99 for the process and sensitivity study for reactor R-1

As can be seen, of the three options, the process catalyzed with sodium hydroxide turned out to be the most expensive and with the highest environmental impact. This is mainly due to the fact that this process requires the use of a large amount of solvent (9000 kg/hr), which is composed of water and methanol. This causes not only an increase in cost and environmental impact due to the heating of a larger amount of solvent, but also an increase in the cost of water separation in the separation zone. Therefore, it is not surprising that this process is the one that requires the most energy, as can be seen in Table 1. It is important to note that this process is also very poorly selective to gasoline, with its main product being bio-jet fuel. Note also that this process is one of the worst in terms of safety, due to the need for higher pressure and temperature conditions to carry out the HDO reactions, this due to the dilution of the products with water.

In contrast, the process involving dolomite showed a lower total cost compared to the NaOH process. This is largely due to two important factors. Firstly, this process does not require the use of water, which significantly reduces the costs associated with heating and separating it. Secondly, this process is selective in the production of gasoline, as can be observed in Table 1. The hydrogenation of the gasolines is carried out under more moderate pressure and temperature conditions, which lowers the costs and significantly improves the safety of the process. It should be noted that, in this case, the reduction in costs does not correlate proportionally with the decrease in environmental impact. This is mainly because a greater amount of medium and low-pressure steam is required compared to the other processes, resulting in a lower environmental impact in economic terms, but not necessarily in environmental terms.

Finally, the DBU process proved to be the best in terms of cost and environmental impact. It should be noted that this process, like the hydroxide process, is not selective to gasoline.

This is mainly because the process does not use solvents for the aldol condensation; instead, it increases the amount of acetone. This results in significant cost reductions in heating during the HDO stage and reduced operational costs in separation. This can be clearly seen in Table 1, where this process consumes 30% less energy than the dolomite process and about 75% less than the NaOH process. This cost and environmental impact reduction is contrasted with safety, which has drastically worsened. This is because a larger amount of acetone and higher temperatures are required to carry it out, in contrast to the other processes. In addition, it is important to note that acetone volatilizes easily, which has severe safety consequences. It is for this reason that the DBU presents greater risks. However, despite this, it is important to note that the difference in risks between the three processes is not much, so DBU is chosen as the best process for producing biofuels via the furan route.

4. Conclusions

It was determined that in terms of costs, the DBU process proved to be the most efficient, consuming 30% less energy than the dolomite process and about 75% less than the NaOH process. Regarding environmental impact, the dolomite-involved process showed a lower total cost compared to the NaOH process, as it does not require the use of water and is selective in gasoline production, but it requires a greater amount of medium and low-pressure steam. In terms of safety, the sodium hydroxide-involved process presented the highest risks due to the higher pressure and temperature conditions required. Although the DBU process presents higher risks, the difference in risks among the three processes is not significant. Overall, the results indicate that the DBU process is the most suitable for producing biofuels via the furan route, due to its cost efficiency and environmental impact. However, it is important to take appropriate measures to ensure safety in its application. To validate the feasibility of this process and to compare its costs with those of conventional fossil fuels, a supply chain study is suggested for future research."

References

Contreras-Zarazúa, G., Sánchez-Ramírez, E., Vazquez-Castillo, J. A., Ponce-Ortega, J. M., Errico, M., Kiss, A. A., & Segovia-Hernández, J. G. (2018). Inherently safer design and optimization of intensified separation processes for furfural production. Industrial & Engineering Chemistry Research, 58(15), 6105-6120.

E. Sánchez-Ramírez, B. Huerta-Rosas, J.J. Quiroz-Ramírez, V.A. Suárez-Toriello, G. Contreras-Zarazua, J.G. Segovia-Hernández, Optimization-based framework for modeling and kinetic parameter estimation, Chem. Eng. Res. Des. 186 (2022)

Faba, L., Díaz, E., Vega, A., & Ordóñez, S. (2016). Hydrodeoxygenation of furfural-acetone condensation adducts to tridecane over platinum catalysts. Catalysis Today, 269, 132-139.

Jiménez-González, C., & Constable, D. J. (2011). Green chemistry and engineering: a practical design approach. John Wiley & Sons.

Ko, J. K., Lee, J. H., Jung, J. H., & Lee, S. M. (2020). Recent advances and future directions in plant and yeast engineering to improve lignocellulosic biofuel production. Renewable and Sustainable Energy Reviews, 134, 110390.

O'Neill, R. E., Vanoye, L., De Bellefon, C., & Aiouache, F. (2014). Aldol-condensation of furfural by activated dolomite catalyst. Applied Catalysis B: Environmental, 144, 46-56.

Turton, R., Bailie, R. C., Whiting, W. B., & Shaeiwitz, J. A. (2008). Analysis, synthesis and design of chemical processes. Pearson Education

U.S. Energy Information Administration (EIA), U.S. energy facts explained, (2022). https://www.eia.gov/energyexplained/us-energy-facts/.

I.R.E.A. (IRENA), Reaching zero with renewables biojet fuels, 2021. https://www. irena.org/-/media/Files/IRENA/Agency/Publication/2021/Jul/IRENA_Reachi ng_Zero_Biojet_Fuels_2021.pdf.

Flavio Manenti, Gintaras V. Reklaitis (Eds.), Proceedings of the 34th European Symposium on Computer Aided Process Engineering / 15th International Symposium on Process Systems Engineering (ESCAPE34/PSE24), June 2-6, 2024, Florence, Italy

Embedding Physics into Neural ODEs to learn Kinetics from Integral Reactors

Tim Kircher[a], Felix A. Döppel[a], Martin Votsmeier[a,b,*]

[a]*Technische Universität Darmstadt, Peter-Grünberg-Straße 8, 64287 Darmstadt, Germany*
[b]*Umicore AG & Co. KG, Rodenbacher Chaussee 4, 63457 Hanau, Germany*
martin.votsmeier@tu-darmstadt.de

Abstract

While the digitalization of chemical research and industry is vastly increasing the amount of data for developing kinetic models, model parametrization is not keeping up. To take advantage of the full potential of this data, machine learning tools are required that autonomously learn kinetics from reactor data. Previously, we introduced Global Reaction Neural Networks with embedded stoichiometry and thermodynamics for kinetic modelling. When trained as a neural ordinary differential equation (neural ODE), they discovered kinetics from integral reactor measurements of an equilibrium limited steam reforming reactor whereas conventional neural ODEs failed. We now extend their application to another industrially relevant case of reactors operating at full conversion. Using the preferential oxidation of CO in H_2 rich streams as an example, we show that the physics-embedded neural network discovers kinetics from stiff systems containing cases of both full conversion and equilibrium limitation using integral reactor data.

Keywords: Neural ODE, Digitalization, Kinetic Modelling

1. Introduction

The design and control of chemical reactors requires accurate kinetic models. While the wealth of data for parameterizing kinetic models is increasing due to the emergence of big data frameworks (Wulf et. al, 2021), model parametrization is not keeping up. Machine learning models that autonomously learn kinetics from reactor data are required to tackle this challenge. Neural ordinary differential equations (neural ODEs) emerged as the state-of-the-art for learning system dynamics from time-series data, such as integral reactor measurements, with neural networks (Chen et al., 2018). Still, neural ODEs face several challenges in the case of kinetic modelling. The limited data availability in a laboratory setting requires neural network architectures that generalize with little training data. Also, chemical source terms frequently vary over many orders of magnitude and conventional neural networks struggle to model such data distributions (Döppel and Votsmeier, 2022). Recently, Kircher et al. (2023) proposed a Global Reaction Neural Network for efficiently modelling chemical source terms. This physics-embedded neural network was used together with neural ODEs and accurately learned kinetics from reactor data of a steam reforming reactor operating close to the equilibrium. In this paper we extend the application of the physics-embedded neural networks to stiff chemical systems operating close to or at full conversion. We use the preferential oxidation of CO in H_2 rich streams as an example, which contains both equilibrium limitations (water-gas shift equilibrium) and full conversion (oxidation reactions) and concentrations and source terms ranging over many orders of magnitude.

1.1. Global Reaction Neural Networks

Global Reaction Neural Networks with embedded stoichiometry and thermodynamics (GRNN) use prior knowledge of the reaction system to map reaction conditions to the corresponding chemical source terms (Kircher et. al, 2023). When coupled with reactor physics, they can be trained as neural ODEs on integral reactor data (Figure 1).

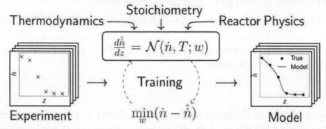

Figure 1: Training a GRNN coupled with reactor physics on integral reactor data.

The neural network layers of the GRNN $\mathcal{N}$ with trainable weights w map the molar flows $\dot{n}$ and temperature T to latent representations of the forward reaction rates $\vec{r}$:

$$\vec{r} = \mathcal{N}(\dot{n}, T; w) \tag{1}$$

The De Donder relation (Dumesic, 1999) is used to compute thermodynamically consistent net reaction rates r^{net} from the forward rates using tabulated thermochemistry. These net rates approach zero if reactants are depleted and predict the correct equilibrium:

$$r_j^{net} = \vec{r}_j \cdot \left(1 - \frac{\prod_i x_i^{\nu_{i,j}}}{K}\right) \tag{2}$$

With the equilibrium constant K, mole fraction x, stoichiometric coefficient ν, species index i and global reaction index j. Then, chemical source terms $\dot{s}$ are computed using the embedded stoichiometry. These source terms strictly conserve the atom balance:

$$\dot{s}_i = \sum_j r_j^{net} \cdot \nu_{i,j} \tag{3}$$

The embedded balance equations are then used to calculate the change in mole flow.

2. Numerical Experiments and Simulations

2.1. Microkinetic Model

We consider the same reaction mechanism as used in our previous work for modeling the detailed surface kinetics of the preferential oxidation of CO (Döppel and Votsmeier, 2022). The mechanism describes the CO oxidation, H_2 oxidation, water-gas shift reaction as well as the preferential oxidation of CO and the promoting role of H_2O on CO oxidation on platinum. The mechanism is described by a microkinetic model with 36 reactions and 9 surface species. We use the kinetic parameters provided by Hauptmann et. al (2010). Elementary reaction rates are parametrized by Arrhenius kinetics. Deviating from the original implementation, we enforce thermodynamic consistency by calculating the rate constants of the backward reactions from the kinetic constant of the forward reaction and the equilibrium constant. A detailed description of this procedure as well as all kinetic and thermodynamic parameters are provided by Kircher et. al (2023).

2.2. Generating Synthetic Experimental Data

The input range was chosen to cover typical operating conditions met in a reactor for the removal of CO impurities from high concentration H_2 streams by preferential oxidation of CO with small amounts of added O_2. Table 1 defines ranges of temperatures and mole fractions that capture this process (Manasilp & Gulari, 2002). Inlet conditions with a total molar fraction over 1 are discarded and resampled. Instead of sampling the O_2 mole fractions, a factor f_{O_2} is sampled. The inlet O_2 mole fraction is calculated by multiplying this factor with the inlet CO mole fraction:

$$x_{O_2} = x_{CO} \cdot f_{O_2} \tag{4}$$

Table 1: Range of sampled inlet temperature, mole fractions and oxygen factor.

Inlet Condition	Minimum	Maximum
T / K	400	440
x_{H_2}	0.400	0.750
x_{H_2O}	0.050	0.150
x_{CO}	0.005	0.020
x_{CO_2}	0.100	0.200
f_{O_2}	0.5	1.5

The reactor is modeled as an isothermal, isobaric steady state 1-D plug flow reactor:

$$d\dot{n}/_{dz} = A_R \cdot c_{kat} \cdot \dot{s} \tag{5}$$

with cross-section area A_R (1 m²), molar flow $\dot{n}$ (mol s⁻¹), axial reactor coordinate z (m), catalyst concentration c_{kat} (26.3 mol m⁻³) and chemical source term $\dot{s}$ (s⁻¹). The reactor has a length of 1 m, and the flow velocity is 1 m s⁻¹. For a given inlet condition, axial reactor trajectories are calculated by numerical integration of the balance equation by ode15s in MATLABR2021a. Synthetic reactor experiments use molar flows measurements from six evenly distributed locations along the reactor length, the first at the inlet and the last at the outlet, and the temperature. The molar flows $\dot{n}_{true}$ are perturbed by multiplicative (ϵ) and additive (η) gaussian noise with a mean of 0 and a standard deviation of 1. The minimum molar flow is defined as 10^{-6} mol s⁻¹, and all molar flows below are set to this value.

$$\dot{n}_{training} = \dot{n}_{true} \cdot (1 + 0.1 \cdot \epsilon) + 10^{-5} \cdot \eta \tag{6}$$

2.3. Neural Network Hyperparameters

We train both GRNN and conventional neural networks. The models take the thermo-chemical state consisting of the temperature and molar flow of reactants as inputs and map them to the chemical source terms of all reactants. The inputs are transformed by mapping the temperature to the inverse temperature and the mole flows to their logarithm. Subsequently, the features are normalized by min-max scaling to a range of -1 and 1. The transformed and normalized features are input to the single hidden layer of the neural network with 50 nodes. We use hyperbolic tangent activation functions in the hidden layer. The weights are initialized by Xavier initialization. For the conventional neural network, the weights and biases of the last hidden layer are initialized as zeros such that the model is initially an identity mapping to stabilize training and improve performance.

2.4. Neural ODE Training

Training is performed in full batch using the torch.optim.LBFGS quasi-newton optimizer implementation with strong wolfe line search, a learning rate of 1, absolute and relative tolerance of 10^{-50} and a history of size 100. Backpropagation is performed through the numerical solution of the differentiable ODE. Ten models with different seeds were trained. For training, we use a two-phase training method. First, the models are trained by a single step euler solver for 2500 epochs. During this phase, each reactor experiment is segmented into input-output pairs of adjacent measurements, such that the models approximate the gradient between molar flow measurements. For each seed, the model with the lowest validation error during training was retained. Second, the solver of the retained models is changed to RK4 with 1000 steps, and they are trained for 25 epochs. During this phase the models are given inlet molar flows of each reactor experiment, are integrated for the full length of the reactor, and the solution is evaluated where measurement data is available. For each seed, the model with the lowest validation error during training is saved. Of the ten saved models, the model with the lowest validation error is chosen for evaluation. L_2 regularization is used with a regularization parameter of 10^{-4}. The models are trained to minimize the root mean squared error loss of predicted logarithmic molar flows $\hat{n}_{i,k,l}$ and noisy molar flows $\dot{n}_{i,k,l}$ of $n=5$ species over the $m=5$ measurement positions along the reactor and N experiments:

$$L = \sqrt{\frac{1}{n \cdot m \cdot N} \left[\sum_{l=1}^{N} \sum_{k=1}^{m} \sum_{i=1}^{n} \left(\ln(\dot{n}_{i,k,l}) - \ln(\hat{n}_{i,k,l}) \right)^2 \right]} \qquad (7)$$

Training and inference of neural networks is performed using pytorch 1.12.1. Differentiable ODE-solvers by torchdiffeq 0.2.3 are used for training the neural ODEs. The datasets were created in MATLABR2021a. Training is performed on an Intel® Xeon® Platinum 9242 with 96 cores and 2x Intel® AVX-512 units per core.

3. Results and Discussion

We demonstrate that our Global Reaction Neural Network with embedded stoichiometry and thermodynamics (GRNN) autonomously learns kinetics from multiple noisy integral reactor measurements. As a first step, we defined feed conditions where we required our machine learned kinetic model to work (Table 1). Note that the machine learned kinetic models need to cover a wider range, since molar flows change along the reactor. In a second step we created artificial reactor data by randomly sampling inlet conditions from the input range and running the reactor model for these conditions. In this manner we generated a training- (80 experiments), a validation- (20 experiments) and a test dataset (100 experiments). In a third step, the reactor experiments were discretized to five measurements locations per trajectory to match the limited data obtainable in a laboratory reactor experiment. Finally, the data was perturbed by gaussian noise. This yielded data that is typically used for parametrizing kinetic models. A neural ODE was then set up and trained on this data. Figure 2 shows an example of the ground truth (dots), the noisy training data (diamonds), and the predictions of the trained GRNN (lines) for the O_2 and CO molar flows (left) and the O_2 source terms (right). The global reactions governing the system are also shown on the right. The model is only trained on the noisy training data and does not have access to information on the ground truth molar flows or source terms.

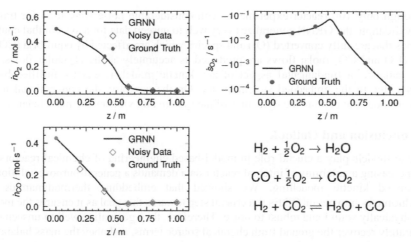

Figure 2: Profile of O_2 and CO molar flow (left) and O_2 source terms (right) predicted by the trained GRNN (line), the noisy training data (diamonds) and the ground truth (circles).

The GRNN recovers the true molar flows with a higher accuracy than the perturbed data used for training it (Figure 2, left). The relative error of all model predictions to the ground truth mole flows, ranging over 8 orders of magnitude, is 5.9 %. This is lower than the relative deviation of the noisy training data towards the ground truth of 16.5 %. While this might seem unexpected, it is important to note that the neural ODE is highly biased towards the true solution due to the embedded information on stoichiometry and equilibria. Additionally, the model is improved by training on multiple experiments. We also show that the GRNN accurately recovers the non-monotonic dynamics ranging over multiple orders of magnitude (Figure 2, right). We then evaluated the GRNN on the previously unseen test dataset. We also trained and evaluated a conventional neural network in a same manner using the same data. Figure 3 shows parity plots of the selected ground truth molar flows and total mass flow and predictions by the trained GRNN (left) and conventional neural network (right).

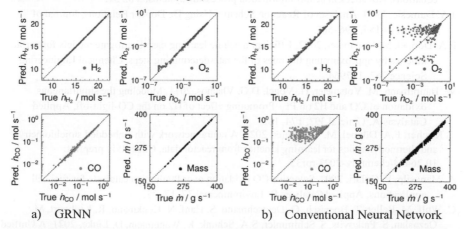

a)   GRNN   b)   Conventional Neural Network

Figure 3: Parity plots of selected test set ground truth molar flows as well as the total mass flow and the predictions of the trained a) GRNN and b) conventional neural network.

Although only 100 reactor experiments with considerable noise were used for training and validation, the GRNN generalizes very well to unseen data for a system that contains species that are fully converted (O_2) and species which are affected by equilibria (CO and H_2). H_2O and CO_2 molar flows are predicted as accurately as the H_2 molar flows. The mass balance, a fundamental aspect of any kinetic model, is exactly fulfilled by the GRNN due to the embedded stoichiometry. On the contrary, the conventional neural network does not capture the dynamics of the system and violates the mass balance.

4. Conclusion and Outlook

Kinetic models play a crucial role in model-based engineering of chemical reactors and the increasing availability of integral reactor data demands a general-purpose solution for automated kinetic modelling. We showed that embedding thermodynamics and stoichiometry into neural ODEs is a crucial step towards this goal as it ensures the models are physically sound and robust to noise. Therefore, they generalize well to unseen data, accurately recover the ground truth chemical source terms, and obey the mass balance by design. As this approach has proven successful for the two most important edge cases of industrial applications, namely equilibrium limitation and high conversion, we anticipate that it will be applicable to a wide range of systems and thus contribute to the acceleration of kinetic model development.

5. Acknowledgements

The authors gratefully acknowledge the computing time provided on the high-performance computer Lichtenberg at the NHR Centers NHR4CES at TU Darmstadt. This is funded by the Federal Ministry of Education and Research, and the state governments participating on the basis of the resolutions of the GWK for national high-performance computing at universities.

References

R.T.Q. Chen, Y. Rubanova, J. Bettencourt, D.K Duvenaud, 2018, Neural ordinary differential equations, Advances in neural information processing systems, Vol. 31.

J.A. Dumesic, 1999, Analyses of Reaction Schemes Using De Donder Relations, Journal of Catalysis 185, 496–505.

F.A. Döppel, M. Votsmeier, 2022, Efficient machine learning based surrogate models for surface kinetics by approximating the rates of the rate-determining steps, Chemical Engineering Science 262, 117964.

W. Hauptmann, M. Votsmeier, H. Vogel, D.G. Vlachos, 2011, Modeling the simultaneous oxidation of CO and H2 on Pt – Promoting effect of H2 on the CO-light-off, Applied Catalysis A: General 397, 174–182.

T. Kircher, F.A. Döppel, M. Votsmeier, 2023, A neural network with embedded stoichiometry and thermodynamics for learning kinetics from reactor data, ChemRxiv preprint: 10.26434/chemrxiv-2023-rpr35.

A. Manasilp, E. Gulari, 2002, Selective CO oxidation over Pt/alumina catalysts for fuel cell applications, Applied Catalysis B: Environmental 37, 17–25.

C. Wulf, M. Beller, T. Boenisch, O. Deutschmann, S. Hanf, N. Kockmann, R. Krähnert, M. Oezaslan, S. Palkovits, S. Schimmler, S.A. Schunk, K. Wagemann, D. Linke, 2021, A Unified Research Data Infrastructure for Catalysis Research – Challenges and Concepts, ChemCatChem 13, 3223–3236.

Flavio Manenti, Gintaras V. Reklaitis (Eds.), Proceedings of the 34th European Symposium on Computer Aided Process Engineering / 15th International Symposium on Process Systems Engineering (ESCAPE34/PSE24), June 2-6, 2024, Florence, Italy
© 2024 Elsevier B.V. All rights reserved. http://dx.doi.org/10.1016/B978-0-443-28824-1.50138-1

Sustainable Acetaldehyde Synthesis from Renewable Ethanol: Analysis of Reaction and Separation Processes

Jean Felipe Leal Silva*, Rubens Maciel Filho

School of Chemical Engineering, University of Campinas, Av. Albert Einstein, 500, Campinas, São Paulo, 13083-852, Brazil
jeanf@unicamp.br

Abstract

Acetaldehyde can be industrially manufactured by means of ethanol oxidation, ethylene oxidation, and hydration of acetylene. The preferred technology since the 1960s has been the direct oxidation of ethylene via the Wacker process. However, demand for the development of greener processes because of the aggravation of the climate crisis demands process engineers to develop greener processes and reduce reliance on non-renewable feedstocks. Ethanol costs and environmental impact have decreased substantially since the development of the Wacker process in 1957–1959, and alternative catalysts for ethanol oxidation have been evaluated as well. However, the oxidation of ethanol in the gas phase is a cumbersome process because of the dilution of products in inert gas (nitrogen). Therefore, this work evaluates the energy demand of acetaldehyde production via ethanol oxidation. Four synthesis conditions were considered, with air-to-ethanol molar ratios of 3.2, 6.1, 9.0, and 21.6, based on experimental data. Acetaldehyde was recovered using compression, gas absorption, and distillation, using different operating pressures in the gas absorption step. Dilution of reactants and recycling of unconverted ethanol in the synthesis loop has a great impact on the energy demand of the process, and results indicate that a molar ratio of 9.0 of O_2 to ethanol is desirable. Despite the high energy demand, this process could be integrated into a sugarcane ethanol biorefinery to integrate the use of renewable feedstock and renewable energy, thus reducing the depletion of fossil resources.

Keywords: process simulation, optimization, oxidation, absorption, green chemistry

1. Introduction

Acetaldehyde, a key chemical intermediate with versatile applications in the chemical industry, is crucial for synthesizing various compounds, including acetic acid, perfumes, plastics, and pharmaceuticals (Hagemeyer, 2000). The increasing global demand for acetaldehyde has fueled extensive research into optimizing production processes. Traditionally, acetaldehyde is produced through the oxidation of ethylene via the Wacker process, using metal-based catalysts (Jira, 2009). Despite reliability, these methods have drawbacks such as high energy consumption and environmental concerns regarding the use of fossil feedstock (ethylene). Sustainable alternatives are imperative considering the current state of climate change and the use of non-renewable resources.

An alternative for the oxidation of ethylene is the oxidation of ethanol. Ethanol can be produced from renewable sources (*e.g.*, sugarcane, corn), with a considerably low carbon footprint (Cantarella *et al.*, 2023), making it a perfect feedstock to replace ethylene. The catalytic process has been studied in the past (Maciel Filho *et al.*, 1996), considering the use of the Fe and Mo catalysts. However, despite the good conversion performance, recovering the acetaldehyde from the resulting product stream was a cumbersome and energy-intensive task.

Considering examples of studies on biorefinery processes indicating the attractiveness of integrating chemical processes into sugarcane mills (Leal Silva *et al.*, 2022), this work has the goal of revisiting the conversion of ethanol to acetaldehyde. Process design, modeling, and simulation were employed to explore potential operating conditions and their impact on the energy demand of the process.

2. Methodology

Ethanol oxidation and acetaldehyde recovery processes were modeled on Aspen Plus 8.6 using the Redlich-Kwong equation of state for the vapor phase and the Non-Random, Two-Liquid activity coefficient model for the liquid phase. Binary interaction parameters for all components were retrieved from the PURE32 database of Aspen Plus. Air was assumed to be composed of N_2 and O_2 at a molar ratio of 79:21, with both gases considered as supercritical components whose Henry's constant in water was based on data retrieved from the PURE32 database. The process is shown in Figure 1.

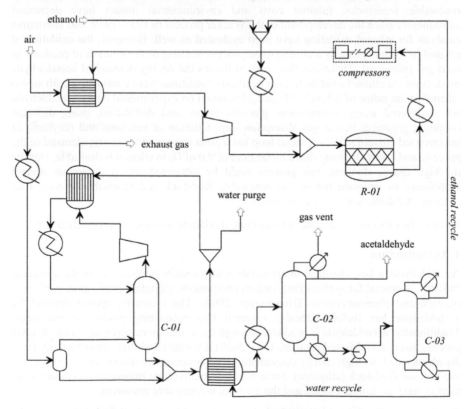

Figure 1. Block flow diagram of the proposed process for oxidation of ethanol to acetaldehyde.

In the process, ethanol (virgin and recycled) was vaporized and mixed with a stream of preheated air containing the desired proportion of O_2 to ethanol (3.2, 6.1, 9.0, or 21.6). The oxidation of acetaldehyde, represented in Eq. (1), was indicated on a stoichiometric reactor (*RStoic* model) that operates with a pressure drop of 60 mbar and outlet temperature of 240 °C. The mixture reacted in the presence of a catalyst, and conversion for each process condition was based on experimental data (Maciel Filho *et al.*, 1996). Heat was removed from the reactor as 2 bar saturated steam was produced.

$$CH_3CH_2OH + 0.5\ O_2 \rightarrow CH_3CHO + H_2O \tag{1}$$

Acetaldehyde was recovered via absorption in water. To reduce the volume of water required to carry out this operation, different operating pressures for the absorption process were tested. The total energy consumption includes the energy demand for the compression of the product stream and the recycling of the water used in the absorption step. Before compression, the reactor outlet stream was cooled to 150 °C, and then it was fed to a series of compressors whose number depends on the final desired pressure: two for 3.1 bar, three for 5.0 bar, four for 8.1 bar, and five for 13 bar. After compression, the product stream was cooled by heat exchange with the air feed stream, available at 25 °C, and then cooled to 5 °C using chilled brine. After cooling, part of the water, acetaldehyde, and ethanol was recovered in a knockout pot, and the gas stream was fed to the bottom of the absorber.

Water was fed at the top of the absorber (*C-01*), with a flow rate required to reduce volatile organic compounds of the leaving gas to a concentration of 20 mg$_C$ Nm^{-3}, compatible with European legislation (European Union, 1999). The temperature of the water was a design variable to be analyzed. The gas leaving the top of the absorber was expanded on a turbine with an isentropic efficiency of 83% to drive an electric generator with a mechanical efficiency of 98%. The water containing the acetaldehyde and unreacted ethanol was then heated to its boiling point and fed to a first distillation column (*C-02*) that operates at near atmospheric pressure to recover acetaldehyde at the top. This column included a gas vent in the condenser to remove trace amounts of nitrogen and oxygen that were dissolved in water. Then, in a second column (*C-03*), hydrated ethanol was recovered at the top, yielding water at the bottom to be recycled to the absorber. The water stream was first cooled by heat exchange with the cold stream leaving the bottom of the absorber and, after a purge, it was further cooled to the absorber operating temperature. The purge fraction varies based on the water produced during the oxidation reaction (Eq. 1) and the water required in the absorber; thus, it is calculated using a design specification. Both distillation columns have reflux ratio and number of stages calculated using shortcut methods, and these results were used to model a rigorous column based on the *RadFrac* model of Aspen Plus. In the first distillation column, the reflux ratio was varied to achieve 99.5% recovery of acetaldehyde on the top with 99% purity (the balance is ethanol). In the second column, the reflux ratio was varied so that the mass fraction of water at the bottom was 99.9%.

Process options (different O_2 to ethanol molar ratios and different operating pressures for the absorber) were compared based on utility demand: saturated steam at 2 bar for process heaters and reboilers (after the steam credit from the steam produced in the reactor), cooling water and chilled brine for condensers and coolers, and electricity for compression (after the electricity credit from the expansion of the exhaust gas). The chosen plant capacity, 256 kt/y of acetaldehyde, is compatible with the production of anhydrous ethanol in a sugarcane mill processing 4 Mt/y of sugarcane.

3. Results and discussion

The first process variable to define is the temperature of the water used in the absorption step. Figure 2 shows the required flow rate of water to achieve the emission standards required for the exhaust gas and the combined reboiler duties of the subsequent distillation columns, C-02 and C-03. This was evaluated only for the case of O_2 to ethanol ratio of 21.6, which is the most dilute system with the highest proportion of acetaldehyde to ethanol (the latter having more affinity to water than the former). It was observed that the flow rate of water increased substantially when increasing the temperature of the water. Naturally, as acetaldehyde has a low boiling point, removing it from the gas phase requires a cold solvent to increase its solubility. The increase in water demand is also reflected in the increased reboiler duty of the following distillation columns (C-02 and C-03), with a steep increase observed when working at lower pressures. This happens because, as the pressure decreases, the water demand steeply increases as the temperature rises because the recycled solvent (water) contains 0.1% ethanol. Therefore, it becomes impossible to achieve emission standards for the exhaust gas independently of flow rate.

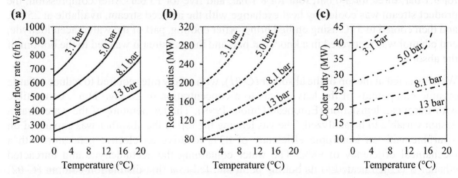

Figure 2. a) water flow rate, b) combined reboiler duties of columns C-02 and C-03, and c) cooler duty for the recycled water stream feeding the absorber for different operating temperatures of the absorption tower (assuming O_2 to ethanol ratio of 21.6 – most dilute system).

A counterintuitive result is the demand for cooling the stream of recycled water fed at the top of the absorption tower: cooling the water to lower temperatures reduces the total demand for chilled brine. This happens because the reduced demand for water as the temperature decreases compensates for the increased duty to cool it to a lower temperature. Based on these results, the lowest temperature possible is desired. This conclusion considers that any operating temperature below 20 °C would require chilled brine to adjust the final desired temperature of the feeding streams. Bearing in mind the freezing point of water, the temperature of 5 °C was chosen to ensure operability.

Figure 3 presents the results of the use of process utilities by the different process configurations (different pressures in the absorption step and different O_2 to ethanol ratios in the reactor). Generally, the demand for heating and cooling utilities decreases considerably as the pressure of the absorption step is increased: when shifting from 3.1 to 5.0 bar, the steam demand decreases 5–17% depending on the O_2 to ethanol ratio. However, this comes at the cost of higher spending with compression power, as observed in Figure 3e. It was observed that the steam demand (Figure 3c) for ratios of 21.6 and 9.0 was almost identical and the lowest for the whole range of evaluated absorption pressure, and it is 9–20% lower than the steam demand of the next best case (ratio of 6.1).

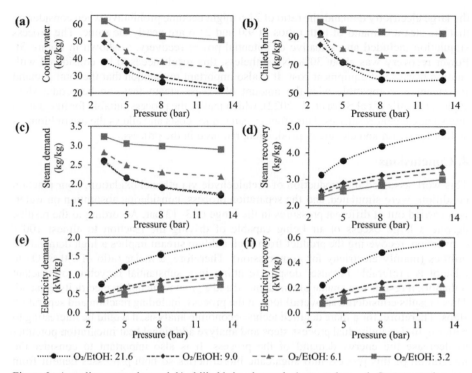

Figure 3. a) cooling water demand, b) chilled brine demand, c) steam demand, d) recovered steam, e) electricity demand, and f) recovered electricity for different pressures of the absorption step and different O_2 to ethanol ratios (O_2/EtOH in the legend); values are reported per kg of acetaldehyde produced in the process.

Steam demand increased considerably for lower O_2 to ethanol ratios because of the need to recycle ethanol. When using a ratio of 21.6, the reaction is almost driven to completion, reaching 96.2%. However, when using a ratio of 3.2, single-pass conversion is 48.2%. A ratio of 9.0 yields a conversion of 79.69% of ethanol. Despite the increased ethanol recycling compared to the ratio of 21.6, the considerably lower excess of air demands less solvent recycling to the absorption tower, which also reduces the demand for steam.

According to Figure 3b, chilled brine demand decreases considerably for the O_2 to ethanol ratio of 21.6 as the pressure increases. Because of the large excess of air and the use of low pressure in the absorption tower, the water demand increases substantially (Figure 2a), and the recycling of this water demands chilled brine. For pressures higher than 8 bar, the demand for chilled brine levels off because, despite the decrease in demand to adjust the temperature of the water stream of the recycling loop, there is also an increased demand in column C-02 because the separation between components demands greater reflux (whose cooling is carried out using this utility because of the boiling point of acetaldehyde). However, the chilled brine demand for the O_2 to ethanol ratio of 21.6 is only about 10% lower than the demand for the ratio of 9.0 for pressures greater than 8.1 bar. These savings are small compared to the increased electricity demand, which can reach 80% for 13.1 bar when comparing the O_2 to ethanol ratio of 21.6 and 9.0.

The proper comparison of process options demands an economic assessment (Leal Silva and Maciel Filho, 2024). Nevertheless, bearing in mind the high cost of compressors and the large electricity demand, the ratio of 21.6 might become prohibitive when considering that the steam demand of both ratios of 9.0 and 21.6 are nearly the same. The process simulation included an estimative of potential power recovery, as shown in Figure 3f. Power recovery was about 30%. Nevertheless, this result needs to be considered with caution because of equipment cost. It is also important to mention that the total demand for steam is compatible with the amount of excess steam that can be produced in sugarcane mills (Leal Silva *et al.*, 2022), which makes the process suitable for integration into sugarcane biorefineries. This strategy can reduce costs and allows the use of biomass as fuel for steam and electricity production, required in the process.

4. Conclusions

This work assessed the production of acetaldehyde via ethanol oxidation. Four reaction conditions were simulated, and the separation process, considering absorption on water, was carried out at different pressures in the range of 3–13 bar. According to the results, despite a large excess of air being capable of driving the reaction to almost 100% conversion, recovering the product from such a dilute stream implies a high demand for utilities (mainly electricity for compression). Therefore, a molar ratio of 9.0 of O_2 to ethanol is preferable because, despite resulting in a substantial recycling of reactant (ethanol), it results in a similar steam demand and a great reduction in electricity demand. These results considered an overall look at the process, including reaction and separation steps. Therefore, for a more detailed techno-economic analysis, it would be interesting to better optimize individual process steps and analyze additional heat integration potential to decrease the energy demand of the process. It is also important to consider the integration of this process into sugarcane biorefineries so both processes benefit from utility integration.

Acknowledgments

The authors acknowledge the support from FAPESP (the Sao Paulo Research Foundation, #2022/16379-8 and #2015/20630-4) and the Coordenação de Aperfeiçoamento de Pessoal de Nível Superior - Brasil (CAPES) - Finance Code 001.

References

H. Cantarella, J.F. Leal Silva, L.A.H. Nogueira, R. Maciel Filho, R. Rossetto, T. Ekbom, G.M. Souza, F. Mueller-Langer. 2023. Biofuel Technologies: Lessons Learned and Pathways to Decarbonization. GCB Bioenergy 15 (10), 1190–1203.

European Union. 1999. Council Directive 1999/13/EC, on the Limitation of Emissions of Volatile Organic Compounds Due to the Use of Organic Solvents in Certain Activities and Installations. Brussels.

H.J. Hagemeyer, 2000, Acetaldehyde. In: Kirk-Othmer Encyclopedia of Chemical Technology. John Wiley & Sons, Ltd.

R. Jira, 2009. Acetaldehyde from Ethylene—A Retrospective on the Discovery of the Wacker Process. Angewandte Chemie International Edition 48 (48), 9034–37.

J.F. Leal Silva, R. Maciel Filho. 2024. Techno-Economic Analysis and Life Cycle Assessment of Renewable Acetaldehyde from Sugarcane Ethanol. Chemical Engineering Transactions.

J.F. Leal Silva, P. Y. S. Nakasu, A. C. da Costa, R. Maciel Filho, S. C. Rabelo. 2022. Techno-Economic Analysis of the Production of 2G Ethanol and Technical Lignin via a Protic Ionic Liquid Pretreatment of Sugarcane Bagasse. Industrial Crops and Products, 189, 115788.

R. Maciel Filho, M. R. Wolf Maciel, A. Domingues, A. O. Stinghen. 1996. Computer Aided Design of Acetaldehyde Plant with Zero Avoidable Pollution. Computers & Chemical Engineering 20, (SUPPL.2), S1389–93.

Flavio Manenti, Gintaras V. Reklaitis (Eds.), Proceedings of the 34th European Symposium on Computer Aided Process Engineering / 15th International Symposium on Process Systems Engineering (ESCAPE34/PSE24), June 2-6, 2024, Florence, Italy

Operational condition optimization of *Schinus terebinthifolius* supercritical extraction using machine learning models

Ana L. N. Santos,[a,c] Leonardo O. S. Santana,[a] Victor L. S. Dias,[a] Ana L. B. Souza,[a,b] Fernando L. P. Pessoa[a,b]

[a] *SENAI CIMATEC University Center, Avenida Orlando Gomes, Salvador 41650-010, Brazil*

[b] *Federal University of Rio de Janeiro (UFRJ), Av. Athos da Silveira Ramos, Rio de Janeiro 21941-909, Brazil.*

[c] *Federal University of Bahia (UFBA), R. Prof. Aristídes Novis, Salvador 40210-630, Brazil*

laise.nascimento.sant@gmail.com

Abstract

Schinus terebinthifolius, or pink pepper, a native Brazilian species, is globally significant for its phytothcrapeutic, value in essential oil. This oil, rich in bioactive compounds, offers nutraceutical benefits and has diverse industrial applications, including in food, pharmaceuticals, and pesticides. Supercritical fluid extraction (SFE) using carbon dioxide is a promising alternative to traditional solvent extraction, producing residue-free products that comply with health regulations. Optimizing extraction conditions is crucial for maximizing yield. However, traditional experimentation for optimization is costly and extensive. This study employs machine learning to predict the yield of pink pepper essential oil from supercritical extraction. Data was generated through experimental methods, using a 2^3 factorial design with central point in triplicate, further analysed by bootstrap method (200 resampling's) and Monte Carlo simulation (1000 random data points). Six Machine Learning Regression Methods (Support Vector, K-Nearest Neighbour, Random Forest, Ridge, Lars, and Lasso regression) were employed. The Support Vector Regression with Standard Scaler demonstrated superior accuracy (MAE = 0.1668, MSE = 0.0653, RMSE = 0.2555, NMSE = 0.0374, Pearson = 0.9812), closely mirroring experimental results. This approach enables precise optimization for extracting high-value compounds from pink pepper essential oil in the best cxtraction conditions to maximize the yield of some compounds of interest. The yield is maximized with lower temperature and higher-pressure conditions, reaching higher yield values when the temperature is around 40 °C and the pressure is around 200 bar.

Keywords: *Schinus terebinthifolius*, Machine Learning, Super Critical Fluid Extraction, Support Vector Regression

1. Introduction

Schinus terebinthifolius, commonly referred to as pink pepper, is a plant of significant medicinal and industrial value, renowned for its bioactive compounds with antioxidant and antimicrobial properties. The extraction of these compounds, particularly through supercritical CO_2 extraction methods, has become increasingly relevant due to its environmentally friendly nature and ability to yield high-quality extracts. (Araújo dos Santos et al., 2020)

Despite its advantages, the supercritical extraction process presents several experimental challenges. Firstly, the complexity of the extraction process, influenced by various factors such as pressure, temperature, and the nature of the plant material, makes the standardization of extraction conditions difficult. High pressures and temperatures, essential for achieving supercritical states, pose operational challenges and safety concerns, necessitating specialized equipment and expertise. Moreover, Brazil's rich biodiversity, including plants like *Schinus terebinthifolius*, adds another layer of complexity due to the varied chemical compositions of these plants, which can significantly influence extraction outcomes. (De Souza et al.,2023)

Considering these challenges, there is a growing interest in applying machine learning models to predict and optimize extraction yields. Machine learning offers a promising solution to overcome the limitations of traditional experimental approaches. By analyzing complex datasets and identifying patterns, machine learning models can predict the yields of essential oils under various conditions, thereby reducing the need for extensive and costly experimentation. This approach is particularly advantageous given the difficulty in obtaining experimental data and the complexity of controlling and replicating extraction conditions in a laboratory setting.

The training of machine learning models requires a large amount of data. Therefore, statistical methods can be used when the amount of experimental data is small (less than 10 experimental points). Thus, data resampling using the bootstrap method appears as an alternative to increase the amount of data. Furthermore, increasing the amount of data with the bootstrap method can estimate the uncertainty and probability distribution of the data. With the distribution of probability and uncertainty, it is possible to generate an even greater amount of data using Monte Carlo simulation. (Thebelt et al., 2022)

Lastly, this study aims to employ machine learning techniques to predict the yield of *Schinus terebinthifolius* extract from supercritical CO_2 extraction and determine the better extraction conditions. By integrating advanced data analysis tools with experimental insights, this research endeavors to enhance the efficiency and effectiveness of the extraction process, contributing to the sustainable and economic production of essential oils.

2. Methodology

2.1. Dataset

The experimental data was obtained from the experiment done in the work of De Souza et al., (2023). An extensive range of data was generated from this experimental data using a combination of the bootstrap method and Monte Carlo simulation in MS Excel, using Microsoft Excel's programming language VBA, and the generation of random data tool. The original experimental data was obtained using a 2^3 factorial design with the central point (50 °C, 145 bar) in triplicate, and then the bootstrap method was performed (with

200 resamples) using this central point data. From these resampling data, it was assumed that the uncertainty was the same for the other P-T conditions, displayed in Table (1) below.

Table 1. Experimental P-T conditions.

Temperature (°C)	36	40	40	50	50	50	60	60	64
Pressure (bar)	145	90	200	67	145	223	90	200	145

To achieve this, pseudo-experimental data were generated from the bootstrap using a sample of 200 data for each P-T condition. A normal distribution was obtained, and the uncertainty of the central point was generated for the other P-T conditions. then, a Monte Carlo simulation was performed using the standard deviation and mean obtained in the bootstrap, with 1000 randomly generated data.

For the data preprocessing the dataset generated underwent an exploratory analysis, correlation analysis, and distribution assessment. The data was divided into 75% for training and the remaining 25% as test data, never seen by the models during the training. Within the training set, a split of 95% of the data for training and 5% for validation during the training process was made.

2.2. Implementation of regression Machine Learning models

In this study, regression methods such as support vector regression (SVR) (Drucker, H., et al., 1997), k-nearest neighbors regressor (KNN) (Cover, T., and Hart, P., 1967), random forest regressor (RFR) (Breiman, L., 2001), ridge regression (RDG) (Hoerl, A. E. and Kennard, R. W., 1970), least angle regression (LARS) (Efron, B., et al., 2004), and lasso regression (LASSO) (Tibshirani, R., 1996) were used. Machine learning models typically perform better with standardized data (BISHOP, Christopher M., 2006), hence the implementation of the MinMaxScaler, StandardScaler, and RobustScaler methods as well.

These regression and standardization methods were developed using the scikit-learn library version 1.3.2, in the Python programming language. A total of 24 models were developed, combining the six models with four methods of normalization.

The models underwent an exploratory search to identify the best set of hyperparameters using GridSearchCV, available in the Keras Tuner library. Cross-validation (CV) with 20 "folds" was conducted to compare performance and enhance the models' generalization capability.

2.3. Evaluation metrics

Initially, the Mean Squared Error (MSE), Eq. (1) below, was used to evaluate the performance of all possible configurations during GridSearchCV.

$$MSE = \frac{1}{n}\sum_{i=1}^{n}(y_i - y_i')^2 \tag{1}$$

In this metric, n represents the number of samples, y and y' are the actual and predicted data, respectively, and i indicates the index of the sample.

To assess the best models from each configuration, specifically those with the lowest MSE, additional evaluations were conducted using the Mean Absolute Error (MAE), Eq. (2), Root Mean Squared Error (RMSE), Eq. (3), Normalized Mean Square Error (NMSE), Eq. (4), and Pearson Correlation Coefficient (ρ), Eq. (5).

$$MAE = \frac{1}{n}\sum_{i=1}^{n}|y_i - y_i'| \tag{2}$$

$$RMSE = \sqrt{\frac{1}{n}\sum_{i=1}^{n}(y_i - y_i')^2} \tag{3}$$

$$NMSE = \frac{\frac{1}{n}\sum_{i=1}^{n}(y_i - y_i')^2}{var(y)} \tag{4}$$

$$\rho = \frac{\sum_{i=1}^{n}(y_i - \overline{y})(y_i' - \overline{y'})}{\sqrt{var(y)var(y')}} \tag{5}$$

For the above statistical metrics, values close to 0 are more suitable for MAE, MSE, RMSE, and NMSE, while values close to 1 are preferred for ρ.

3. Results and Discussion

3.1. Data Analysis and generation

The dataset generated from the bootstrap and Monte Carlo simulation had a total of 9011 datapoints, for better physical suitability the negative values were removed from the dataset resulting in a total of 8570 datapoints. To analyze the correlation of the target variable which is the extraction yield with the pressure and temperature conditions Pearson and Spearman correlation analysis, that said, for the Pearson correlation the pressure variable showed a positive correlation of 0.8 and for the temperature variable a negative correlation of -0.4. For the Spearman correlation the pressure also presented a positive correlation slightly higher of 0.9 and for the temperature also a negative correlation was observed of -0.3. That means in practical terms that the extraction yield is increased with higher pressures and decreased with higher temperatures.

3.2. Performance evaluation of the proposed machine learning models

Twenty-four prediction models were created and trained, the metrics of these configurations (model + standardization method) for the yield prediction are displayed in figure 1 bellow.

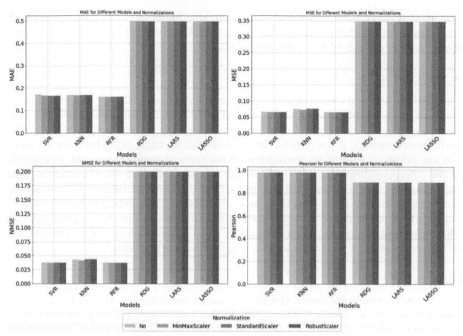

Figure 1. Performance metrics for every model trained.

Figure 1 shows that in the concern of the performance metrics and Pearson correlation the different methods of normalization did not affect the model performance resulting in little variation of the performance. Regarding the different models evaluated SVR + standard scaler normalization showed better overall performance with the lowest errors and higher correlation (MAE = 0.1668, MSE = 0.0653, RMSE = 0.2555, NMSE = 0.0374, and Pearson = 0.9812), this can be explained by the nature of the SVR model that has an advantage predicting non-linear values because of the introduction of a kernel function that project the input data into high dimensional linearly separable space (Ji et al., 2022) To visualize the prediction results a surface graph was plotted as it can be seen in figure 2 bellow.

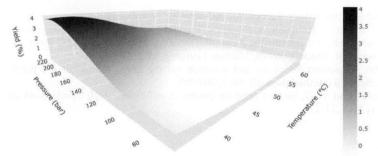

Figure 2. Surface response graph for the SVR model

The yield is maximized with lower temperature and higher-pressure conditions, which corresponds to the Pearson correlation values reaching higher yields values when the temperature is around 40°C and the pressure is around 200 bar.

4. Conclusion

The machine learning models demonstrated varying degrees of efficacy in optimizing the yield of essential oils. Among them, certain models, particularly SVR, showed high accuracy across the range of P-T conditions, showing higher Pearson coefficient e higher performance metrics, it was possible to conclude that the temperature hurts the oil yield, this can occur because of the degradation of the material compromising the final result, on the other hand, the increase of pressure has a positive effect on the oil extraction. This work was efficient in showing the best extraction conditions for supercritical extraction, providing a relevant result with three base experiments. This indicates the potential of machine learning models in precisely optimizing extraction conditions to maximize yield. The use of different normalization techniques in the models highlighted the stability and effectiveness of these methods under varying data scaling scenarios.

References

A. C. Araújo dos Santos, G.A.Batista Nascimento, N.Barbosa da Silva,V. Laurent, Y. Valdez,A. Barbosa, E. Calixto, F. Pessoa,2020, Mathematical modeling of the Extraction Process essential oils *Schinus terebinthifolius* Raddi Using Supercritical Fluids, Journal of bioengineering and technology applied to health,2,130-135

A. de Souza, G. Santos, A. Bispo, K. Hodel, B. Machado, D.Chaves, M.Mendes, F. Pessoa, 2023,Technical and Economic Evaluation of Bioactive Compounds from *Schinus terebinthifolius* Using Supercritical Carbon Dioxide, Applied Sciences, 13,21,11897

A. E. Hoerl, R. W. Kennard, 1970, Ridge regression: Biased estimation for nonorthogonal problems, Technometrics, 12, 1, 55-67.

A.Thebelt, J.Wiebe, J.Kronqvist, C.Tsay, R. Misener,2022, Maximizing information from chemical engineering data sets: Applications to machine learning, Chemical Engineering Science,252,117469

B. Efron, T. Hastie, I. Johnstone, R. Tibshirani, 2004, Least angle regression, The Annals of Statistics, 32, 2, 407-499.

C. Ji, F. Ma, J. Wang, W. Sun,2022, Early Identification of Abnormal Deviations in Nonstationary Processes by Removing Non-Stationarity, Computer Aided Chemical Engineering, 49, 1393-1398

C. M. Bishop, 2006, Pattern Recognition and Machine Learning, Springer.

H. Drucker, C. J. C. Burges, L. Kaufman, A. Smola, V. Vapnik, 1997, Support Vector Regression Machines, In Advances in Neural Information Processing Systems, 9, pp. 155-161, MIT Press.

L. Breiman, 2001, Random forests, Machine Learning, 45, 1, 5-32.

R. Tibshirani, 1996, Regression shrinkage and selection via the lasso, Journal of the Royal Statistical Society: Series B (Methodological), 58, 1, 267-288.

T. Cover, P. Hart, 1967, Nearest neighbor pattern classification, IEEE Transactions on Information Theory, 13, 1, 21-27.

Flavio Manenti, Gintaras V. Reklaitis (Eds.), Proceedings of the 34th European Symposium on Computer Aided Process Engineering / 15th International Symposium on Process Systems Engineering (ESCAPE34/PSE24), June 2-6, 2024, Florence, Italy

Optimizing circular economy levers to achieve global sustainability

Shubham Sonkusare[a], Yogendra Shastri[a*]

a Department of Chemical Engineering, Indian Institute of Technology Bombay, Mumbai 400076, India
* yshastri@iitb.ac.in

Abstract

Circular economy (CE) adoption can contribute towards the sustainable development goal of responsible consumption and production. Understanding the broader implications of CE adoption and optimizing different levers of CE such as reduce, reuse, and recycle is required. A global integrated planetary model has been used in this work to address these needs. This model represents the global ecosystem in a compartmental form, resembling a complex food web. The industrial sector within this model incorporates the crucial elements of reuse and recycling associated with the CE concept. In this work, optimization theory is used to develop time-dependent policies for implementing CE to achieve long term sustainability. Sustainability is quantified here using Fisher information (FI), and the objective function is minimization of FI variation along a desired trajectory. Result shows that if product reuse time can be increased by five times, sustainability is achieved even with "business as usual" scenario. Furthermore, as reuse time increases, there is less variation in decision variable profiles, indicating a policy that is easier to implement for sustainable system. Overall, the research provides valuable insights into the adoption, strategies, and limitations in the implementation of CE.

Keywords: Sustainability, Circular Economy, Reuse, Recycle, Reduce, Optimal control

1. Introduction

The consumption pattern and population growth in today's world are matters of concern. There are various examples of consumption increase in various fields. Since 1950, the production of plastic has increased by 200-fold reaching 381 million tonnes in 2015. The rate of fossil fuel consumption has doubled since 1980 (Sonkusare et al. (2023)). There are many such examples of consumption increases that exploit the ecosystem. The reason for this exploitation is a linear model of resource consumption. The solution to one-way linear resource consumption is the implementation of a circular economy (CE). CE has gained importance in policy formulation, advocacy, consultancy, and natural sciences in the past decade. In the CE model, waste becomes valuable resources through recovery using reuse and recycling. CE implementation can balance between the economy, environment, and society (Sehnem et al. (2019)). The study by the Ellen MacArthur Foundation and Mckinsey for Business and Environment (2015) shows that CE implementation could increase the total annual productivity by 3% by 2030, turning to a total annual benefit of 1.8 trillion at the EU level, which could increase GDP by 7% (Rizos et al. (2017)). To effectively address CE challenges and benefit from the implementation of CE at the global level, the imperative lies in adopting a comprehensive and cooperative strategy that encompasses the principles CE. To understand CE, a systemic global model is needed. To this end, Hanumante et al. (2019) studied the implementation of circular economy in a global planetary model. Their work, however, did not consider reuse aspect of CE. Moreover, optimal combinations of different CE options were not studied. This work addresses those limitations. First, reuse of industrial goods is modeled in an integrated planetary model. Second, combination of reduce-recycle-reuse has been optimized to develop a strategy for the implementation of a circular economy at a global level.

The paper is arranged as follows. Section 2 briefly describes the selected model and details how CE aspects are incorporated—section 3 deals with the Fisher Information index and optimization problem formulation, followed by scenario planning. Section 4 looks into the results and discussion related to optimization.

2. Integrated planetary model

The Integrated planetary model has some features, such as being closed to mass and energy, incorporating a legal foundation, and coupling economic and ecological sectors via price setting model. There have been various studies on a model to study global sustainability. Hanumante et al. (2019) worked on implementing a circular economy, particularly, incorporating the recycle and reduce aspect of circular economy.

2.1 Model Description

The integrated planetary model (Figure 1) is a predator-prey model with multiple trophic levels and mass distributed to various compartments. P1, P2, and P3 represent agriculture, open grassland, and forest, while H1 represents livestock. H2 and H3 represent feral herbivores, and C1 and C2 are feral carnivores. HH denotes human households, IS stands for the industrial sector, and EP represents energy production. RP and IRP refer to the resource pool and Inaccessible resource pool, respectively.

In the model, the higher trophic level provides nutrients to the lower trophic level. Primary producers P1, P2, and P3 take nutrients from the resource pool (RP) and make resources available to the rest of the food web. The model includes five value creation compartments, namely P1, H1, TI, CLI, and EP. TI produces valuable industrial goods with the resources available from P1, RP, and EP. These industrial goods are consumed by human households and are discarded to an inaccessible resource pool (IRP). This is linear economy. However, with the implementation of the recycling aspect of circular economy, post-consumption, industrial goods are recycled through the circulation industry in the purple loop. The circular economy has been incorporated in the circulation industry (CLI), which recirculates post-consumption waste back to human households.

The Integrated planetary model in Figure 1 shows the model with the recycle aspect of CE. This work deal with modeling of the reuse aspect along with recycle where goods will be used for 'k' time step. It is assumed here that the reuse of goods happens through a change in the behavior of consumers. Thus, a fraction of the population (consumers) who are more concerned about sustainability challenges change their behavior patterns and decide to use goods for multiple time steps. This is incorporated for modeling by splitting the human compartment into two parts: HH-SU (SU-single use) and HH-RU (RU-Reuse). Consumers categorized as HH-SU continue to use goods for one-time step only. In contrast, consumers in HH-RU use these goods for k time steps. It may be noted that the goods produced by traditional industry for both HH-RU and HH-SU are the same. The rest of the model does not change.

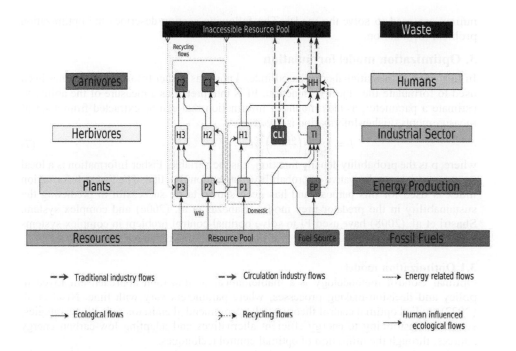

Figure 1: Integrated planetary model. Ecological trophic levels are shown on the left-hand side. Compartments corresponding to that trophic level are shown in the central part.

Implementing behavioral change in human household is coupled with the recycling industry. Post consumption the goods will be recycled in circulation industry. Here, τ gives total amount of goods required for a particular timestep (n) by the whole population. Q represents different flows in the model. The subscript to flow represents the name of the source followed by the destination. For example, flow from traditional industry (TI) to reuse compartment (RU) is represented by (Q_{TI-RU}). Reuse option is modeled using Eq. (1-6).

$$Q_{TI-RU}(n) = \tau(n) \times f \tag{1}$$

$$Q_{CLI-RU}(n) = \tau(n) \times f - RU_{stock}(n) \tag{2}$$

$$Q_{TI-SU}(n) = \tau(n) \times (1 - f) \tag{3}$$

$$Q_{CLI-SU}(n) = \tau(n) \times (1 - f) \tag{4}$$

$$Q_{HH-IRP}(n) = Q_{TI-SU}(n) + Q_{CLI-SU}(n) + Q_{TI-RU}(n - k) + Q_{CLI-RU}(n - k) \tag{5}$$

$$RU_{stock} = RU_{stock}(n) + Q_{TI-RU}(n) + Q_{CLI-RU}(n) - Q_{TI-RU}(n - k) - Q_{CLI-RU}(n - k) \tag{6}$$

Here, 'n' and '$n + 1$' represents two consecutive time steps. $Q_{HH-IRP}(n)$ represents the flow discarded from both human compartments. 'k' represents the number of time-steps the good is reused in reuse compartment. RU_{stock} represents the stock of goods in the HH-RU compartment that can be used in the next time step. The model is coded in Python programming community version 2023.2. Libraries such as mathplotlib, pandas, and numpy are used to solve the model. The following section describes the optimization problem formulation.

3. Optimization model formulation

In this work, information theory-based index known as Fisher Information (FI) has been used to formulate the objective function. FI is interpreted as a measure of the ability to estimate a parameter, as the amount of information that can be extracted from a set of measurements. Fisher Information is given as

$$I = \int \frac{1}{p(x)} \left(\frac{dp(x)}{dx} \right)^2 dx \tag{7}$$

where, p is the probability density function, x is the variable. Fisher Information is a local property as it has a derivative of probability distribution function. The Fisher Information index is used for this purpose and has been shown to be successful in predicting the sustainability in the predator-prey model (Cabezas et al. (2006) and complex system. Shastri et al. (2008) have used FI to solve optimal control problem in complex systems where minimization of FI variance over time as a possible objective function.

3.1 Optimization model

Optimal Control methodology is a mathematical optimization approach employed in policy and decision-making processes, where parameters vary with time. Nisal et al. (2022) studied optimal control theory, study recommended emission reduction strategies, such as transitioning to energy-efficient alternatives and adopting low-carbon energy sources, through the utilization of optimal control techniques.

In this work, we combined the FI with the integrated planetary model to identify the optimal combination of CE options to achieve sustainability. Based on one of the sustainability hypothesis, minimization of Fisher Information variation around a target trajectory has been used as the objective function. The objective function is defined as:

$$J = min \int_0^T (I(t) - Ic(t))^2 dt \tag{8}$$

where, $I(t)$ is the current FI profile, $Ic(t)$ is the targeted FI profile for a stable system which is sustainable, and T is the total time horizon under consideration. The model offers various potential decision variables that can be directly or indirectly adjusted. The parameters associated for decision variables are linked to CE elements: reuse, recycle, and reduce. They encompass metrics like circulation fraction (CF), human resource consumption rate (C), population reuse fraction (f), and reuse time (s). Policies like regulations, subsidies, and taxes can more feasibly influence by CF and f. Therefore, we opt for these as the time-dependent decision variables. The next section deals with the scenario planned to solve the optimization problem.

3.2 Scenario planning

The objective here is to optimize the problem with circulation fraction and fraction of population as controlled variable and get dynamically stable system. For that the following scenarios are formulated:

- Business-as-usual uncontrolled case: This scenario represents the "business as usual" approach, focusing solely on the recycling of industrial goods. Currently, the recycling rate for industrial goods stands at approximately 8%, as per the circularity gap report.
- Static and dynamic circulation fraction and fraction of population reusing: In this scenario, both recycling and reuse are considered, with different reuse time ranging from 1 to 5. The options with static decision variable implies that the decision

variables 'CF' and 'f' remain fixed throughout time step. In dynamic scenarios, 'CF' and 'f' can change over the course of the time horizon. The results presented here are for static scenario; outcomes from dynamic scenarios are not included.

All simulations maintain a consistent consumption level of 6, which increases industrial goods' consumption over a total time horizon for 200 years by six times, with each time step representing one week. The next section reports the results and discussion.

4. Results and discussion

Figure 2 represents a comparison between 'business-as-usual' and constant circulation fraction and fraction of population reusing industrial goods. In 'business-as-usual' the uncontrolled scenario, system failure occurs at the 111-year, characterized by the decrease in mass of the agricultural sector (P1) to zero. The result represents five different cases, each corresponding to a distinct reuse time ranging from 1 to 5. In evaluating all reuse time options, a consistent pattern emerges: each shows a delayed system collapse in comparison to the uncontrolled case. Specifically, for a reuse time of 1, the agricultural sector (P1) survives until the 191 year. For reuse time extending from 2 to 5 years, the system remains both sustainable and stable. Among these, the reuse time of 5 time step is the optimal solution as it gives minima when compared to all other scenarios. The uncontrolled case leads to a decline in human population due to system collapse, while it does not fall in controlled scenarios. For controlled cases with reuse times of 1 and 2, the human population not only stabilizes but also experiences growth.

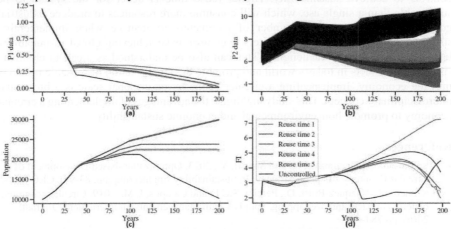

Figure 2: Controlled system compared for different reuse time for a constant circulation fraction and fraction of population reusing.

4.1 Discussion

The trend of decision variables for the above optimized problem is shown in Figure 3. It is found for static CF, f gives an optimal solution for reuse time of five time steps when compared to reuse time ranging from 1 to 5. This suggests that the higher the reuse time, the better the system from a sustainability viewpoint, which can be seen from Figure 2d, a stable Fisher information profile. The circulation fraction and fraction of the population reusing goods keeps reducing as the reuse time increases (Figure 3). By doubling the reuse time, both fractions can be reduced to half. Interestingly, increasing the reuse time to five is sufficient to achieve system sustainability with a current circulation rate of 8%.

It is worth noting that there is an inverse relationship between reuse time and circulation and reuse fraction.

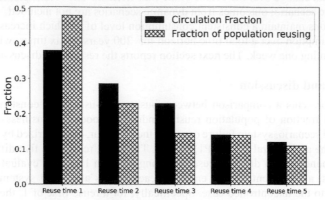

Figure 3: Optimal CF and fraction of population reusing for different reuse time

5. Conclusions

Longer reuse times result in fewer fluctuations and reduce the need for time-based controls to achieve sustainability. Higher reuse time is better for the system to be sustainable than a single use, which will consume more resources to produce industrial goods. This minimizes the number of parameters to manage when implementing sustainability policies. More parameters to control make achieving global sustainability less efficient and more challenging. This can also be understood as a call to eliminate single-use products in today's world and promote the incorporation of reusable products into the economy. Implementing a circular economy on a global scale is a broad and interdisciplinary concept. This study highlights the advantages of implementing a circular economy to promote both environmental and economic sustainability

References

- Sonkusare, S., Hanumante, N. and Shastri, Y., 2023. Optimization-based Development of a Circular Economy Adoption Strategy. In Sustainability Engineering (pp. 83-99). CRC Press.
- Sehnem, S., Vazquez-Brust, D., Pereira, S.C.F. and Campos, L.M., 2019. Circular economy: benefits, impacts and overlapping. Supply Chain Management: An International Journal, 24(6), pp.784-804.
- Rizos, V., Tuokko, K. and Behrens, A., 2017. The Circular Economy: A review of definitions, processes and impacts. CEPS Papers, (12440).
- Hanumante, N.C., Shastri, Y. and Hoadley, A., 2019. Assessment of circular economy for global sustainability using an integrated model. Resources, conservation and recycling, 151, p.104460.
- Cabezas, H. and Fath, B.D., 2002. Towards a theory of sustainable systems. Fluid phase equilibria, 194, pp.3-14.
- Shastri, Y., Diwekar, U., Cabezas, H. and Williamson, J., 2008. Is sustainability achievable? Exploring the limits of sustainability with model systems. Environmental science & technology, 42(17), pp.6710-6716.
- Nisal, A., Diwekar, U., Hanumante, N., Shastri, Y. and Cabezas, H., 2022. Integrated model for food-energy-water (FEW) nexus to study global sustainability: The main generalized global sustainability model (GGSM). PloS one, 17(5), p.e0267403.

Flavio Manenti, Gintaras V. Reklaitis (Eds.), Proceedings of the 34th European Symposium on Computer Aided Process Engineering / 15th International Symposium on Process Systems Engineering (ESCAPE34/PSE24), June 2-6, 2024, Florence, Italy
© 2024 Elsevier B.V. All rights reserved. http://dx.doi.org/10.1016/B978-0-443-28824-1.50141-1

Climate Change Adaptation Measures for Water-stressed Thermal Power Plants in India

Reshma Shinde[a], Yogendra Shastri[a,b*], Anand B. Rao[a,c]

[a] IDP in Climate Studies, Indian Institute of Technology Bombay, Mumbai 400076, India
[b] Department of Chemical Engineering, Indian Institute of Technology Bombay, Mumbai 400076, India
[c] Centre for Technology Alternatives for Rural Areas, Indian Institute of Technology Bombay, Mumbai 400076, India
Corresponding author*: yshastri@iitb.ac.in

Abstract

Thermal power plants situated in water-stressed regions of India are highly vulnerable to the impacts of climate change. Reduced availability and increased temperature of cooling water can cause shutdowns or under-performance of these plants, and these issues are highly seasonal. By considering four different water-stressed regions of India, it was shown that water withdrawal increased during the summer months by nearly 24% compared to winter months. The quantification for future climate scenarios of SSP2-RCP 4.5 and SSP5-RCP 8.5 showed that there could be severe water stress for the months of December-June in the region of Rajasthan. The objective of this work is to quantify the effect of retrofits with the environmental control technologies such as flue gas desulphurization (FGD) and carbon capture and sequestration (CCS) on the water withdrawals for a representative coal-fired power plant in the region of Rajasthan. It was found that water withdrawals further increased by $300 - 3000$ l/MWh. Adaptation of thermal power plants to climate change is, therefore, crucial. The choice of boiler technology, type of condenser for cooling, and type of fuel used in a thermal power plant can bring down the water withdrawals for electricity generation substantially. Hybrid cooling technology for water-stressed regions like Rajasthan and Punjab in India can reduce water withdrawal by 2000-2500 l/MWh. The ultra-supercritical boiler reduces water withdrawals by $500 - 700$ l/MWh compared to the sub-critical boiler. However, capital investments are considerably higher for these technologies. Therefore, a cost-benefit analysis of the available options for the adaptation along with the provision of optimal solutions for the water-stressed power plants regionally is extremely important.

Keywords: water stress, thermal power plants, environmental control technologies, cooling technologies, adaptation measures

1. Introduction

India is the third-largest producer of electricity in the world with an annual production of 1714.8 TWh after China and the United States (Statista, 2023). Coal-fired power plants in India alone were responsible for emission of 1104 Mt CO_2 which was the highest among all the sectors (IEA, 2022). These emissions are responsible for the climate change impacts such as rise in temperature, decreasing summer monsoon precipitation, an increase in the duration, intensity, and coverage of pre-monsoon season heatwaves, tropical cyclones, and droughts. Though power sector is majorly responsible for climate change, the sector itself is vulnerable to its impacts. These impacts can be infrastructural and/or operational. India particularly is vulnerable to the problem of water stress in power plants. Historically, there have been many instances of shutdowns and underperformance of thermal units due to cooling water shortages in India. Consequently, the water-power-climate nexus studies are extremely important in the states such as, Maharashtra, Rajasthan, Karnataka where most of the plant shutdowns and under-performances were reported. Taking into account such instances, uninterrupted thermal power generation with sufficient water availability is going to be very vital for future periods. Moreover, suitable adaptation measures such as hybrid cooling, efficient boiler choice; on a regional basis will be paramount.

In the previous work, we evaluated the intra-annual water withdrawal variations at the plant as well as the regional level (Shinde et al., 2023). The work focused on the states that are considered as high or extremely high water stressed. Furthermore, the work estimated the overall surface water availability for current and alternate future scenarios of a chosen region based on simple conceptual framework – Budyko. By considering this, additional water requirements because of implementation of environmental control technologies such as Flue Gas Desulphurisation (FGD), Selective Catalytic Reduction (SCR), and Carbon Capture and Sequestration (CCS) will further increase the water withdrawals. The outcome of the study will help in understanding the intra-annual water requirements for thermal power generation and increments in the water withdrawals due to the control technologies.

This manuscript is arranged as follows. Section 2 deals with materials and methods. Section 3 reports the results and key insights. The article ends with a conclusion in section 4.

2. Materials and methods

Considering the historical shutdowns in the region, high water stress, and the type of cooling technology used; four different states - Rajasthan (RJ), Maharashtra (MH), Uttar Pradesh (UP), and Punjab (PN) in India are chosen for the study. The cooling technology dominant in these regions is re-circulating cooling. The weather parameters and other plant-level data are then obtained for the selected regions and Integrated Environmental Control Model (IECM) is used to calculate monthly water withdrawal and consumption intensities (website: https://tcktcktck.org/). Further, the water withdrawal intensities are calculated for additional environmental control technologies and how the reduction in water requirements occurs with the boiler choice and cooling technologies is shown.

2.1 Integrated Environmental Control Model (IECM)

IECM allows the user to configure the plant to be modelled for a variety of pollutant control technologies and types of power plants. IECM also includes a set of major cooling technologies, including once-through cooling, wet towers, and air-cooled condensers (ACCs) for dry cooling. The water models in the IECM are based on the mass and energy balances to estimate water use, energy penalties, and costs of cooling systems (Zhai et al., 2011). The *IECM 11.5 version* is used as it allows the selection of India as the location for analysis. The plant-level data and India-specific fuel data are obtained from the literature and the Indian coal-data directory is created in IECM (Nagarkatti & Kolar,

2021). The weather parameters such as ambient air temperature (°C), mean ambient air pressure (MPa), mean relative humidity (%), and mean annual precipitation (cm/year) are also the inputs to the model and are discussed in the subsequent section.

2.2 Climate projections

For future projections, the temperature and precipitation data are obtained from the World Bank's open access climate model - 'Climate Change Knowledge Portal' (CCKP) specific to the chosen region for the period 2020 - 2039 (The World Bank Group, 2021). The data are based on CMIP-6 (Coupled Model Inter-comparison Project phase-6) viz. a standard experimental framework for studying the output of coupled atmosphere-ocean general circulation models, administered by the World Climate Research Program (WCRP).

The monthly relative humidity (Rh) for the current scenario is calculated using the ambient temperature and dew point temperature data for all the regions between the period 2011 and 2020 (Shinde et al., 2023). The climate models only estimate future ambient temperature. The intra-annual variation of Rh in the future is unknown. Hence, regression models are developed correlating T and Rh for their simplicity. Two different regression models are considered for monsoon (June-July-August-September) and non-monsoon months (Shinde et al., 2023).

2.3 Estimation of overall surface water availability: Current and Future

Budyko framework is used to calculate current as well as future water availability for the Shared Socio-economic Pathway (SSP) - Representative Concentration Pathway (RCP) of SSP2-RCP 4.5 and SSP5-RCP 8.5 climate scenarios. SSP2-RCP 4.5 represents 'middle of the road scenario' which spans the warming of 4.5 W/m^2; caused by the activities under certain assumptions of population, economic growth, and other factors such as fossil fuel usage, energy demands etc. On the other hand, SSP5-RCP8.5 is referred as 'high emission scenario' or 'pessimistic scenario' which may cause global temperature rise of 3 to 5 °C (https://www.carbonbrief.org/). Further, Budyko's hypothesis states that the ratio of mean annual transpiration to mean annual precipitation is controlled by the ratio of mean annual potential evapotranspiration to mean annual precipitation as stated by Eq. 1.

$$\frac{E}{P} = f\left(\frac{E_P}{P}\right) \tag{1}$$

Fu's functional equation is used to enhance the Budyko model for concerned local region and to account for different surface types; where a single parameter ω can be calibrated against local data as per Eq. 2 (Zhang et al., 2001; Arora, 2002).

$$\frac{E}{P} = 1 + \frac{E_P}{P} - \left[1 + \left(\frac{E_P}{P}\right)^\omega\right]^{\left(\frac{1}{\omega}\right)} \tag{2}$$

where E is actual evapotranspiration, E_P is potential evapotranspiration, P is precipitation, and ω is the single value parameter.

2.4 Water withdrawal intensities of environmental control and other cooling technologies

Based on the feasibility studies, coal power plants in India are required to install FGD systems to reduce SOx emissions. Various elements of the FGD system require water such as the wash system of the mist eliminator, the absorber system, the limestone grinding and slurry preparation system, and the gypsum de-watering system. This means additional water would be required in the power plants. The additional water requirements will increase substantially with the implementation of CCS to curb the CO_2 emissions and limit the global warming. The anthropogenic water footprints are anticipated to be doubled with the deployment of CCS technology (Brandl et al., 2017). Hence, water withdrawal for typical Indian coal plant for different control technologies, the boiler types, and other cooling technologies such as hybrid cooling are evaluated for a 650 MW subcritical coal plant with an average ambient temperature of 29 °C and average Rh of 50%. The mean annual precipitation is considered as 50 cm/year. These conditions are representative of RJ region of India. Rest of the parameters are kept at default in the IECM.

3. Results and Discussion

In this section, we will discuss the results of water withdrawal and consumption based on IECM and increase water withdrawals of control technologies. Further, water savings with other cooling technologies will be highlighted.

3.1 Water withdrawal and consumption

Water withdrawal necessarily represents water abstracted from nearby resources such as lakes, rivers to the power plants. On the other hand, water consumption represents evaporation losses of water during the cooling process. We found that intra-annual water withdrawal lied in the range of 2500 - 4600 l/MWh for the current scenario whereas future scenarios withdrawal intensity ranged between 2500 – 5400 l/MWh for all the regions. In the summer months, the withdrawal intensity increased by as much as 24% compared to winter months. This change was particularly observed in the state of Punjab. Water consumption range was found between 1700 and 2200 l/MWh for all the regions in current and future scenarios. The water withdrawal intensities of current and future scenarios were then used to evaluate average monthly gross water withdrawal for overall thermal power generation on sub-annual basis. It ranged between 5 Million m³ (MCM) to 56 MCM (Shinde et al., 2023).

It was shown by using Budyko framework how the overall surface water availability for the state of RJ decreased by almost 49% compared to current one which caused the intra-annual deficiencies in the months of Dec-June for the future scenarios of SSP2-RCP 4.5 and SSP5-RCP 8.5 (Shinde et al., 2023). Moreover, the water availability in the state of MH is decreased by 39.14% in SSP2-RCP 4.5 and 41.35% in SSP5- RCP 8.5 compared to current one.

3.2 Water withdrawals by control technologies

Some of the retrofitting technologies commonly used in a coal thermal power plants are – Electrostatic Precipitator (ESP), Selective Catalytic Reduction (SCR), FGD and CCS. The SCR is used to reduce the NOx emissions from the coal power plants. As we moved from the base plant to the CCS system retrofits; the water withdrawals went on increasing. Table 1. shows the water withdrawal increments with these technologies. It could be observed that as we went on adding the environmental control technologies from SCR to CCS, the water withdrawal rose by 17 – 2700 l/MWh. However, as we moved from sub-critical to super-critical and ultra-supercritical boiler technologies; the overall plant operation water withdrawals reduced by 700 – 1500 l/MWh. Nevertheless, apart from climate change-driven increased water withdrawals; these additional quantities must also be satisfied in order to facilitate the working of the power plant. Wet FGDs are considered to be more efficient historically. Dry FGDs require lower capital costs as there is no need of water treatment as no liquid waste is generated. However, operating costs are higher due to reagent viz. expensive. For the cooling requirements of FGD, de-mineralized water is required.

Hybrid cooling essentially does cooling by water when it's hot and by air when it's not. It can also use the combination simultaneously. When hybrid cooling method was chosen in IECM for the same parameters above, it was found that the water withdrawal dropped to 1047 l/MWh whereas the water consumption was 609.2 l/MWh. These numbers were without the implementation of environmental control technologies. Still, it could be seen that significant water can be saved as compared to the wet cooling tower. However, it comes with the higher costs and significant penalty in the plant efficiency.

The uninterrupted power supply is crucial for country's social, economic, and industrial development. The output, efficiency, and financial viability of power generation could be affected because of climate change. As of now, there is no policy to phase out old coal power plants in India. Moreover, it is stated by the government that no coal power plant will be phased out before 2030 given the expected rise in the electricity demand (Ministry of Power, 2023). Therefore, water withdrawal is only going to increase in the near future. Retrofitting or usage of less water-intensive cooling technologies can play a major role in significant water savings. Further, choice of boiler technology can lead to improved efficiency and water savings.

4. Conclusion

The water withdrawals for thermal power generation in the water stressed regions of India increased substantially compared to current situation as a result of climate change. The greatest intra-annual rise was observed in the state of Punjab where withdrawal in the month of June was nearly 24% higher than that of winter month of January. Overall withdrawal rates of UP were greatest because of actual higher withdrawal rates of non-compliant plants. The withdrawals increase with the addition of control technologies such as FGD, CCS by 17 – 2700 l/MWh. The hybrid cooling without consideration of environmental control technologies can reduce the water withdrawals significantly to 1047 l/MWh.

At a broader level, this work strongly emphasized the need to perform detailed region specific intra-annual assessment to ensure sustainability of the thermal power sector. Further, the regional adaptation measures such as alternative cooling technologies and water resources could be proven to be a solution to reduce the water withdrawals. Future work can include evaluation of long-term water needs and availability along with economic analyses of control technology options.

Table 1. Water withdrawals for different environmental control technologies (l/MWh)

Sub-critical coal plant	Base plant with ESP	ESP + Hot side SCR	ESP + Hot side SCR + FGD	all + CCS (amine based)
Water withdrawal (l/MWh)	2774	2791	3082	5432
Water consumption (l/MWh)	1892	1902	2178	3842
Super-critical coal plant	**Base plant with ESP**	**ESP + Hot side SCR**	**ESP + Hot side SCR + FGD**	**all + CCS (amine based)**
Water withdrawal (l/MWh)	2510	2526	2796	4739
Water consumption (l/MWh)	1698	1706	1963	3336
Ultra supercritical coal plant	**Base plant with ESP**	**ESP + Hot side SCR**	**ESP + Hot side SCR + FGD**	**all + CCS (amine based)**
Water withdrawal (l/MWh)	2066	2078	2316	3861
Water consumption (l/MWh)	1410	1417	1645	2741

References

Arora, V. K. (2002). The use of the aridity index to assess climate change effect on annual runoff. *Journal of Hydrology*, *265*(1–4), 164–177. https://doi.org/10.1016/S0022-1694(02)00101-4

Brandl, P., Soltani, S. M., Fennell, P. S., & Dowell, N. Mac. (2017). Evaluation of cooling requirements of post-combustion CO 2 capture applied to coal-fired power plants. *Chemical Engineering Research and Design*, *122*, 1–10. https://doi.org/10.1016/j.cherd.2017.04.001

IEA 2022. (2022). World Energy Outlook. IEA Report, Paris

Ministry of Power. (2023). *Phasing out of coal-based thermal power plants and adoption of super-critical technologies in thermal power plants.* https://Pib.Gov.In/.

Nagarkatti, A., & Kolar, A. K. (2021). Advanced Coal Technologies for sustainable power sector in India. *Electricity Journal*, *34*(6). https://doi.org/10.1016/j.tej.2021.106970

Shinde, R., Shivansh, Shastri, Y., Rao, A. B., & Mondal, A. (2023). Quantification of climate change-driven water stress on thermal power plants in India. *Computers & Chemical Engineering*, *179*, 108454. https://doi.org/10.1016/j.compchemeng.2023.108454

Statista, 2023. (2023). *Leading countries in electricity generation worldwide in 2021.* Leading Countries in Electricity Generation Worldwide in 2021.

The World Bank Group. (2021). *Climate Change Knowledge Portal.* The World Bank Washington DC.

Zhai, H., Rubin, E. S., & Versteeg, P. L. (2011). Water Use at Pulverized Coal Power Plants with Postcombustion Carbon Capture and Storage. *Environmental Science & Technology*, *45*(6), 2479–2485. https://doi.org/10.1021/es1034443

Zhang, L., Dawes, W. R., & Walker, G. R. (2001). Response of mean annual evapotranspiration to vegetation changes at catchment scale. *Water Resources Research*, *37*(3), 701–708. https://doi.org/10.1029/2000WR900325

Flavio Manenti, Gintaras V. Reklaitis (Eds.), Proceedings of the 34th European Symposium on Computer Aided Process Engineering / 15th International Symposium on Process Systems Engineering (ESCAPE34/PSE24), June 2-6, 2024, Florence, Italy

Techno-economic impacts of using alternative carbon-based feedstocks for the production of methanol

James Tonny Manalal,[a],* Mar Pérez-Fortes [a] and Andrea Ramírez Ramírez [b]

[a]*Department of Engineering, Systems and Services, Faculty of Technology Policy and Management, Delft University of Technology, Jaffalaan 5, 2628BX, Delft, Netherlands.*
[b]*Department of Chemical Engineering, Faculty of Applied Sciences, Delft University of Technology, Van der Maasweg 9, 2629 HZ, Delft, Netherlands.*
E-mail: j.t.manalal@tudelft.nl

Abstract

Different alternative carbon sources like CO_2, biomass and plastic waste, can be used to replace fossil carbon as feedstock in the production of methanol. Based on current literature, the plastic-based methanol route is the most competitive one among the three based on price indicator, but there is still a lack of comprehensive understanding of the techno-economic differences between alternative feedstock technologies. In this study, three technologies from each alternative feedstock were assessed to evaluate the techno-economic trade-offs between them. The research shows that even though currently the plastic-based route is comparatively cost competitive with the conventional route of producing methanol, the CO_2-based methanol route can also be competitive with green hydrogen prices in the range of 1400-1100 EUR/t. While the biomass-based route showed superior energy performance compared to the other two.

Keywords: ex-ante technology assessment, alternative carbon-based methanol, comparative process assessment, techno-economic impacts.

1. Introduction

Methanol is a widely used chemical solvent and a chemical building block in the petrochemical industry. It is currently produced at industrial scale using the reforming process of natural gas (methane), at a minimum selling price (MSP) of 300-550 EUR/t. To defossilize the methanol production process, different alternative carbon sources (ACS) like CO_2, biomass and plastic waste are being considered. However, the production processes from ACS are significantly different compared with the conventional process. There have been several techno-economic studies to understand the feasibility of using ACS for the production of methanol. For instance, Sollai et al., (2023) conducted a techno-economic assessment (TEA) for the production of green methanol using captured CO_2 from flue gas and hydrogen from a proton exchange membrane (PEM) water electrolyser (WE). The study showed that a MSP of 960 EUR/t is required to achieve break-even in 20 years for an internal rate of return (IRR) of 8%. Almost 97% of the variable cost and 52% of the bare equipment cost was due to the PEM electrolyser. In the case of biomass-based methanol the price varied according to the technology. For example, de Fournas & Wei, (2022) compared the production of methanol using oxygen gasification of biomass with CO_2 storage and PEM-WE for varying feedstock carbon utilisation. For the biomass based methanol, they reported that based on the feedstock carbon utilisation, a MSP between 700-1000 EUR/t is required to achieve break-even in 25 years for an IRR of 8%. Afzal et al., (2023) conducted similar TEA on plastic waste to methanol by steam gasification and syngas to methanol process. The research showed that at a MSP of 700 EUR/t, the process achieves a 10% IRR. It also showed that about 58% of the plastic-based methanol MSP was due to the feedstock cost. If we compare these studies, it seems that plastic-based methanol is the most cost competitive. However, because of varying assumptions between such studies, such simple comparisons cannot help to comprehensively assess them against each other. Hence, in this work, we conducted a comprehensive study including technology screening, ex-ante modelling and TEA to compare the techno-economic performance of different ACS feedstocks. The comparison also provides insights into the main hotspots for each route.

2. Methodology

Thirty-two different process routes were identified in literature to produce methanol from ACS feedstocks. To select the most promising technology from each feedstock for ex-ante modelling and TEA, a screening methodology based on the stage-gate concept (section 2.1) was used. The three selected process routes were modelled in Aspen Plus v.12 (section 2.2) and their TEA at process level were calculated using harmonized conditions. The indicators used for the TEA are detailed in section 2.3.

2.1. Technology screening

The screening methodology by Manalal et al., (2023) was applied in this study. It uses different constraints (or gates) to select technologies at each stage; these are technology readiness level (TRL), number of reaction steps, theoretical energy need, carbon utilization efficiency (CUE) and economic constraint (EC) (Manalal et al., 2023). In stage-1, technologies with TRL> 3 were selected to the next stage. In stage-2, the ideal stoichiometric reactions of each process route were used and technologies with less than 4 process steps (i.e., stoichiometric reactions) were selected. In stage-3, theoretical heat and electricity requirements were estimated through the standard enthalpy change (ΔH^0) and Gibbs energy change (ΔG^0). At this stage, using the ideal stoichiometric reactions, the CUE of each process route was also calculated using Eq. (1). Based on the needed energy and CUE, the different process routes were ranked.

$$Carbon\ utilisation\ efficiency\ (CUE) = \frac{Moles\ of\ carbon\ atom\ in\ product}{Moles\ of\ carbon\ atom\ in\ feedstock} \qquad (1)$$

Processes with an electricity need below 750 kJ/mol methanol and a CUE =100% were selected from each feedstock to the next stage. In stage-4, EC was calculated using Eq. (2) and chemical prices adjusted to 2018 base year.

$$Economic\ constraint\ (EC)$$
$$= \frac{\Sigma_{reactants}\ mass\ flow * Component\ price + (\Delta H^0 or\ \Delta G^0)_{endergonic} * Utility\ price}{\Sigma_{products}\ mass\ flow * Component\ price} \qquad (2)$$

Technologies with EC < 1 were selected in stage-4. In stage-5, the technologies were ranked based on TRL, number of process steps and EC; only one process route for each feedstock was selected.

2.2. Ex-ante modelling

The methanol production capacity of each of the three selected technologies was 400 kt/y, as it corresponds to the total methanol demand in the port of Rotterdam, with a purity >

99 wt%. The three selected technologies modelled in Aspen Plus were: CO_2 direct hydrogenation to methanol (C2M), biomass steam gasification (BSG) with syngas to methanol (S2M) and, plastic steam gasification (PSG) with S2M. Green hydrogen from WE was considered as hydrogen feedstock source in all three cases. The C2M process was modelled following the work by E. Lücking & de Jong, (2017), using an RPlug reactor with kinetic input data. The BSG process was modelled as per the GoBiGas Project in Sweden (Larsson et al., 2018), with the gasification section modelled as a combination of an RStoic (gasification & char formation section) and RGibbs reactors (tar formation section). After gasification, a Rectisol process and cryogenic distillation were used for syngas cleaning. The P2G process was modelled as per Quevedo et al. (Quevedo et al., 2021), combining an RStoic and an RGibbs reactor. The methanol synthesis section (S2M) in both, biomass and plastic-based routes, was modelled based on the work by E. Lücking & de Jong, (2017).

Four different steam levels were used as heating utilities, namely: low low pressure (LLP at 3.9 bar), low pressure (LP at 5.5 bar), medium pressure (MP at 21 bar) & high pressure (HP at 51 bar) steams. To supply heating needs at a temperature above 250 °C when the process could not provide them, combustion-based heat was provided and natural gas (81.4 wt% CH_4, 14.4 wt% N_2, 3 wt% C_2H_6, 1 wt% CO_2, 0.2 wt% C_3H_8) was used as utility. Cooling water (25 °C to 40 °C), chilled water (propylene glycol mixture) and refrigerants (R134a, R1150 & R50) were used as cooling utilities. Electricity was the only non-thermal utility defined in the simulation models. The models considered process heat integration, maximising the internal use of heat.

2.3. Techno-economic assessment

The indicators used to compare the three processes were: CUE of the process (Eq. (1)), energy requirements, MSP, capital expenditure (CAPEX) and operational expenditure (OPEX). A sensitivity analysis by varying the electricity price was conducted to understand the impact of hydrogen price on MSP. The CAPEX was calculated based on working capital, inside battery limit (ISBL), outside battery limit (OSBL), engineering (EN) and contingency (CN) costs with estimates from Towler & Sinnott, (2013), as shown in Table 1. The bare equipment (BE) cost for ISBL cost calculations were obtained from Aspen Economic Analyzer. The OPEX calculation was based on variable and fixed OPEX with estimates also from Towler & Sinnott, (2013), as shown in Table 1.

Table 1: Assumptions for economic calculations (Towler & Sinnott, 2013).

CAPEX		Fixed OPEX	
ISBL cost	3.3*BE	Maintenance cost	3%*ISBL
OSBL cost	40%*SBL	Capital charges & royalties	4%*(ISBL+ OSBL+EN)
Engineering costs	10%* (ISBL+OSBL)	Labour, supervision & overhead cost	1.875*Labour estimates
Contingency	10%* (ISBL+OSBL)	Corporate income tax (CIT)	25%*Operating income
Working capital	5%*(ISBL+OSBL+ EN+ CN)	Land & building rents, insurance, property taxes & environmental charges	5%*(ISBL+ OSBL)

Variable OPEX such as raw materials, utilities and wastewater treatment costs were calculated based on mass flows, obtained from Aspen, and prices. The revenue was calculated from products and excess utilities, and using a payback period of 12 years, the MSP of methanol was calculated. In this study, it was assumed that only the methanol price varied and the other product prices remained unchanged, among the lifetime of the plant (i.e., 25 years). The economic calculations were based on 2018 as the base year and Netherlands as the plant location.

3. Results and discussion

3.1. Technology screening

From the literature, 11, 15 and 6 possible 3-step routes were identified to synthesise methanol from CO_2, biomass and plastic waste, respectively. Regarding H_2 feedstock, electrochemical-based technologies were crucial for CO_2 routes, while due to the inherent hydrogen content in biomass and plastic molecules; electrochemical, biochemical and thermal routes were possible. The three selected technologies for further assessment were:

Table 2: Selected technologies for each ACS and their corresponding screening values.

Technology	Reaction steps	Electricity	Heat	CUE	EC
WE + C2M	2	712 kJ/mol	-93 kJ/mol	100%	0.89
BSG + WE + S2M	3	237 kJ/mol	-19 kJ/mol	100%	0.39
PSG + S2M	2	0 kJ/mol	112 kJ/mol	100%	0.34

Note: WE- Water electrolysis, C2M- CO_2 to methanol, BSG- Biomass steam gasification, S2M- Syngas to methanol, PSG- Plastic steam gasification.

3.2. Ex-ante modelling and techno-economic assessment

A comparative assessment of mass flows for the selected technologies shows that the CO_2 and plastic-based methanol routes have a higher CUE. Moreover, the CO_2-based route showed least technical complexity in terms of number of process streams and temperature range. The biomass-based route showed the lowest CUE and the highest process complexity. This is due to a higher range of waste and by-product streams, as shown in Table 3.

Table 3: Mass flow comparison of the selected technologies for each ACS vs base-case natural gas for the synthesis of methanol.

Parameter		Base case fossil methanol	CO₂ to methanol	Biomass to methanol	Plastic waste to methanol
Feedstock	Carbon feedstock	330 kt (natural gas)	800 kt (CO₂)	860 kt (biomass)	310 kt (plastic waste)
	Water	899 kt	981 kt	881 kt	220 kt
Products	Methanol	405 kt	406 kt	411 kt	408 kt
	Oxygen	-	871 kt	334 kt	-
	Methane	-	-	44 kt	-
Waste	Wastewater	614 kt	240 kt	670 kt	5 kt
	Purge	198 kt	248 kt	70 kt	34 kt
	Off gases	12 kt	16 kt	48 kt	8 kt
	Char & ash	-	-	143 kt	13 kt
	Tar	-	-	21 kt	62 kt
Process CUE		80%	69%	49%	68%

Two differences can be observed between CO_2 and biomass-based routes. First is the significant amount of oxygen produced compared to other routes. The second difference regards to the hydrogen requirement. As the biomass-based route produces hydrogen during steam gasification, only part of the total amount of hydrogen required for methanol production was provided by WE, while in the CO_2-based route the hydrogen is fully provided by WE. For the plastic-based route, no external hydrogen was needed as all the required hydrogen was produced during the plastic steam gasification process.

Table 4: Energy flow comparison of the selected technologies for each ACS vs base-case natural gas for the synthesis of methanol. (Note: *A negative sign means generation.*)

Main utilities	Base case fossil methanol	CO₂ to methanol	Biomass to methanol	Plastic waste to methanol
Electricity	198 GWh/y	7168 GWh/y	3553 GWh/y	220 GWh/y
LLP steam	-385 TJ/y	-76 TJ/y	-89 TJ/y	-
LP steam	259 TJ/y	-1895 TJ/y	-1899 TJ/y	1792 TJ/y
MP steam	-1085 TJ/y	-269 TJ/y	-1127 TJ/y	-1389 TJ/y
HP steam	-36 TJ/y	-	-3203 TJ/y	-615 TJ/y
Natural gas	2635 TJ/y	-	-	2509 TJ/y
Cooling water	3894 TJ/y	6209 TJ/y	10543 TJ/y	2565 TJ/y

Table 4 highlights that the need and/or production of utilities are significantly different between the different ACS routes. Thus, changing from conventional methanol production to ACS routes, can affect other interlinked utility providers or users in an existing industrial cluster thereby leading to potential cascading impacts. In terms of steam generation, the biomass-based routes outperformed the other two.

Table 5: Economic comparison of the selected technologies for each ACS vs base-case natural gas for the synthesis of methanol.

Economic indicators	Base case fossil methanol	CO₂ to methanol	Biomass to methanol	Plastic waste to methanol
Methanol MSP (EUR/t)	520	1950	1390	860
CAPEX* (MEUR)	358	432	957	403
OPEX (MEUR/y)	171	746	495	309

*H_2 is considered as raw material & cost is included in OPEX

The economic assessment showed that none of the alternative routes is competitive at the moment with the fossil-based route. The plastic-based methanol could be produced at the lowest MSP (860 EUR/t) compared to biomass (1390 EUR/t) and CO_2 (1950 EUR/t) based methanol. For the biomass-based route, the higher price was due to the higher capital expenditure on the biomass gasification and syngas cleaning steps. For the CO_2-based route, the higher price was due to the high operational expenditure affected by the high price of green hydrogen (4000-6000 EUR/t). A sensitivity analysis to understand the impact of the price of green hydrogen on these technologies was conducted by varying the electricity price used for green hydrogen production and it is shown in Figure 1.

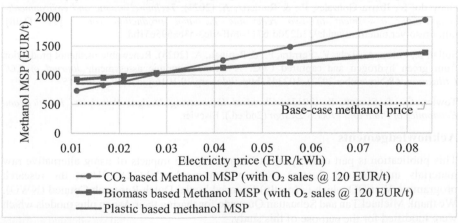

Figure 1: Impact of electricity price on CO_2, biomass & plastic based methanol prices.

The analysis shows that at a green hydrogen price of 1400-1100 EUR/t, the CO_2 route can be cost-competitive. For instance, at a green hydrogen price of 1350 EUR/t, CO_2 route had an MSP of 735 EUR/t, compared to the biomass route (930 EUR/t) and the plastic route (860 EUR/t).

4. Conclusions

This paper indicates that plastic waste gasification to produce methanol currently performs technically and economically better than CO_2 and biomass-based synthesis routes. The work also shows that the CO_2 hydrogenation-based route can potentially become competitive in a scenario with renewable electricity prices below 0.02 EUR/kWh, outperforming biomass & plastic gasification-based methanol. The bio-based route can be competitive when demand of heat/steam along with methanol is critical.

References

Afzal, S., Singh, A., Nicholson, S. R., Uekert, T., DesVeaux, J. S., Tan, E. C. D., Dutta, A., Carpenter, A. C., Baldwin, R. M., & Beckham, G. T. (2023). Techno-economic analysis and life cycle assessment of mixed plastic waste gasification for production of methanol and hydrogen. *Green Chemistry*, *25*(13), 5068–5085. https://doi.org/10.1039/D3GC00679D

de Fournas, N., & Wei, M. (2022). Techno-economic assessment of renewable methanol from biomass gasification and PEM electrolysis for decarbonization of the maritime sector in California. *Energy Conversion and Management*, *257*(December 2021), 115440. https://doi.org/10.1016/j.enconman.2022.115440

E. Lücking, L., & de Jong, W. (2017). *Methanol Production from Syngas: Process modelling and design utilising biomass gasification and integrating hydrogen supply.* https://repository.tudelft.nl/islandora/object/uuid%3Ac0c5ebd2-c336-4f2d-85d1-014dae9fdf24

Larsson, A., Gunnasrsson, I., & Tengberg, F. (2018). The GoBiGas Project-Demonstration of the Production of Biomethane from Biomass via Gasification. In *Goteborg Energi AB, Gothenburg, Sweden.*

Manalal, J. T., Pérez-Fortes, M., Gonzalez, P. I., & Ramirez, A. R. (2023). Evaluation of alternative carbon based ethylene production in a petrochemical cluster: Technology screening & value chain impact assessment. In *Computer Aided Chemical Engineering* (Vol. 52, pp. 2453–2458). Elsevier Masson SAS. https://doi.org/10.1016/B978-0-443-15274-0.50390-5

Quevedo, S., Ibarra Gonzales, P., & Rameriz, A. (2021). *Techno-economic and environmental comparative assessment of two renewable methanol production routes* [TU Delft]. http://resolver.tudelft.nl/uuid:923d22dd-b711-46ff-9fc3-a3ba695e31bd

Sollai, S., Porcu, A., Tola, V., Ferrara, F., & Pettinau, A. (2023). Renewable methanol production from green hydrogen and captured CO2: A techno-economic assessment. *Journal of CO2 Utilization*, *68*(November 2022), 102345. https://doi.org/10.1016/j.jcou.2022.102345

Towler, G., & Sinnott, R. (2013). *Chemical Engineering Design: Principles, Practice and Economics of Plant and Process Design* (2nd ed.). Elsevier.

Acknowledgements

This publication is part of the project Unravelling the impacts of using alternative raw materials in industrial clusters (with project number VI.C.183.010 of the research programme Vici DO which is (partly) financed by the Dutch Research Council (NWO). We thank Michael Tan and Sebastian Quevedo for sharing their Aspen plus models which were modified for the purpose of this study.

Flavio Manenti, Gintaras V. Reklaitis (Eds.), Proceedings of the 34th European Symposium on Computer Aided Process Engineering / 15th International Symposium on Process Systems Engineering (ESCAPE34/PSE24), June 2-6, 2024, Florence, Italy

An Application of a New Nonlinearity Measure to a Rotary Dryer Model

Jan M. Schaßberger[a], Anton Ponomarev[a], Veit Hagenmeyer[a], Lutz Gröll[a]

[a]*Karlsruhe Institute of Technology, Hermann-von-Helmholtz-Platz 1, 76344 Eggenstein-Leopoldshafen, Germany*
jan.schassberger@kit.edu

Abstract

Nonlinearity measures can be a helpful tool when selecting a control concept, especially if flexible plant operation over a large operating range is of importance. In this paper, the application of a new nonlinearity measure to a distributed-parameter rotary dryer model is presented. Besides assessing the model's nonlinearity, special emphasis is put on the measure's application and its interpretation based on the model. The analysis shows that the nonlinear model can be represented by a linear approximation in a large operating range, if operating conditions that lead to complete drying of the particles are avoided.

Keywords: Nonlinearity analysis, Rotary dryer, Partial differential-algebraic equation

1. Introduction

Demand-side flexibility is seen as a key factor in the transition of the energy grid towards renewable energy sources (Heffron et al., 2020). The provision of flexibility goes hand in hand with changes in process variables, for instance due to the variation of the material throughput. In order to continuously meet the product requirements despite flexible operation, a precise process control is required. An important factor in the design of a control concept is the nonlinearity of the system, e.g. whether sufficient performance can be expected from a linear control concept when applied to nonlinear system dynamics. While a linear controller is often sufficient for the current predominantly stationary operation, this question is more difficult to answer when considering controller design for demand-side flexibility. One approach that can be helpful in assessing this question is the use of nonlinearity measures. As one application, these measures allow the systematic investigation of the nonlinearity of a model by comparing it with its linear approximation. In the following, we describe the application of a new dynamic nonlinearity measure to a distributed-parameter rotary dryer model. On the one hand, the contribution focuses on the interpretation of the measure based on the model and the assessment of its nonlinearity. On the other hand, the article addresses challenges that arise in the measure's application to models of industrial processes instead of the usually considered rather academic concentrated-parameter examples. We first describe the considered dryer model in Section 2, before analysing the model's nonlinearity in Section 3.

2. Rotary Dryer Model

As a detailed description of the considered rotary dryer model would go beyond the scope of this article, we will limit ourselves to the core ideas and the presentation of the model structure. For a more detailed description, the reader is referred to Schaßberger et al. (2023). The dryer model is based on the idea of two interacting plug flows exchanging energy and material while traveling through the dryer. Each phase is modeled by one equation for the mass density of dry matter, for water/vapor and the temperature. Furthermore, the model covers the dryer shell and heat losses to the environment. Different from most existing models, time- and location-dependent gas and particle velocities are regarded. For modeling heat and mass transfer, common approaches are used that assume nonlinear transfer coefficients. The obtained model is given by a system of quasilinear partial differential-algebraic equations (PDAEs) in conservative form

$$E\partial_t x(z,t) + \partial_z F_1(x(z,t)) = F_0(x(z,t), d(t)), \tag{1}$$
$$x(z,0) = x_0(z),$$
$$x(0,t) = G(u(t)),$$
$$y(t) = H(x(1,t)),$$

where $t \in \mathbb{R}^+$ and $z \in [0,1] \subset \mathbb{R}$. The state vector $x(z,t) \in \{x \in \mathbb{R}^8 : x_i \geq 0, i = 1,2,\ldots,8\}$ consists of the mass densities of the phases, their temperatures, the temperature of the shell and the gas velocity. From the input streams $u(t) \in \{\mathbb{R}^6 : u_i \geq 0, i = 1,2,\ldots,6\}$, the boundary conditions of the PDEs are calculated using the function $G(\cdot)$. In addition to all elements of the state vector belonging to the particle and gas phase, also the mass shares of vapor and water at $z = 1$ are considered as output variables $y(t) \in \{y \in \mathbb{R}^9 : y_i \geq 0, i = 1,2,\ldots,9\}$ determined by means of $H(\cdot)$. The variable $d(t) \in \mathbb{R}$ represents the disturbances corresponding to the ambient temperature. The remaining system parts are the singular diagonal matrix E as well as the nonlinear functions $F_1(\cdot)$ and $F_0(\cdot)$. Linearization of (1) using the Gateaux derivative w.r.t. an equilibrium leads to a system of linear PDAEs

$$E\partial_t \tilde{x}(z,t) + A_1(z)\partial_z \tilde{x}(z,t) = A_0(z)\tilde{x}(z,t) + b_d(z)\tilde{d}(t), \tag{2}$$
$$\tilde{x}(z,0) = \tilde{x}_0(z),$$
$$\tilde{x}(0,t) = B\tilde{u}(t),$$
$$\tilde{y}(t) = C\tilde{x}(1,t),$$

where $A_1(z)$, $A_0(z)$ are matrices and $b_d(z)$ a vector with spatially varying coefficients. The matrices B and C have constant entries. The variables $\tilde{u}(t)$, $\tilde{x}(z,t)$, $\tilde{y}(t)$ and $\tilde{d}(t)$ refer to that of the nonlinear PDEs, but now to be understood as deviations from their values at the considered equilibrium.

3. Nonlinearity Analysis

Many approaches for assessing the nonlinearity of models exist in literature (Reyero-Santiago et al., 2020). Most approaches are based on the comparison of the nonlinear $N(\cdot): \mathfrak{U} \to \mathfrak{Y}$ and the linear $L(\cdot): \mathfrak{U} \to \mathfrak{Y}$ input-output map of a system, where $\mathfrak{U}$ is the set of admissible input signals and $\mathfrak{Y}$ the set of admissible output signals. Hence, the explicit realisations of the respective systems are disregarded in these approaches. However, all existing approaches of this type have certain disadvantages. First, in general also the initial values of the nonlinear and the linear system need to be taken into account. This not only increases the calculation effort but poses difficulties when different system classes like an infinite- and a finite-dimensional system are to be compared. Second, the existing measures typically consider arbitrary input signals, which has little relevance for many practical applications. For this reason, we propose a new measure based on the classical idea (Kelemen, 1986) that the output y of a (nonlinear) system

$$y = N(u; x_0), \tag{3}$$

moves close to a sequence of points in the equilibrium set $\mathfrak{Y}_{eq}$ during the transition between two operating points y_0, $y_1 \in \mathfrak{Y}_{eq}$ for a sufficiently long transition time T. The variable x_0 denotes the initial values of the system corresponding to y_0. This motivates the consideration of $N(\cdot)$ as a sum of a static input-output map $y_{eq}(u)$ and a dynamic part quantifying the deviation from equilibrium corresponding to the current input u. Thus, the approach makes full use of the typically approximately known static input-output map $y_{eq}(u)$. The linear dynamical approximation y_{lin} of y can be defined as

$$y_{lin} = y_{eq}(u) + L(u - u_0), \quad \text{where} \quad L(0) = 0, \ \lim_{t \to \infty} L(u)(t) = 0, \tag{4}$$

where u_0 is the input corresponding to y_0. Due to the additional requirements, the linear map $L(\cdot)$ is a system with a static gain equal to zero. This approach has the advantage that, due to the latter property, zero initial values of the linear system can be considered in case of transitions between equilibria. With these system descriptions, the difference between the nonlinear output and its approximation denoted by $\rho(\cdot)$ can be written as

$$\rho(N, L; u, y_0, y_1) = N(u; x_0) - y_{eq}(u) - L(u - u_0), \tag{5}$$

with which the following nonlinearity measure $v(N)$ can be defined

$$v(N) = \inf_{L \in \mathfrak{L}} \sup_{\substack{y_0, y_1 \in \mathfrak{Y}_{eq} \\ u \in \mathfrak{U}_{fwd}(y_0, y_1)}} \|\rho(N, L; u, y_0, y_1)\|. \tag{6}$$

This means that $v(\cdot)$ is the lowest maximal value of $\rho(\cdot)$ for all linear maps $L(\cdot)$ within a permissible set $\mathfrak{L}$, outputs y_0 and y_1 in the set of permissible equilibria $\mathfrak{Y}_{eq}$ and permissible inputs $\mathfrak{U}_{fwd}$ steering the system from y_0 to y_1. In order to make this measure applicable to the purpose of the paper, the sets considered in (6) must be narrowed down. While the set $\mathfrak{L}$ can possibly cover all linearizations of the nonlinear system about $y \in \mathfrak{Y}_{eq}$, we limit ourselves in the present contribution to the linearization with respect to the nominal operating point. Thus, the infimum in (6) can be omitted. On the one hand, this is a natural choice from a control engineering point of view, but on the other hand, it in general leads to an overestimation of the model's nonlinearity. In addition, for a system with a large number of inputs and disturbances, as in the case under consideration, it is necessary to severely restrict the sets $\mathfrak{Y}_{eq}$ and $\mathfrak{U}_{fwd}$ in order to limit the number of simulations to be carried out. Instead of considering all changes between operating point y_0, $y_1 \in \mathfrak{Y}_{eq}$, in the following we examine ramp-shaped changes starting from y_0, so that the starting point is fixed to the nominal operating point. Of course, the output of a nonlinear system being steered from an operating point y_1 back to y_0 generally behaves differently than the application of a similar change of the input starting from y_0. However, for a reasonable choice of the operating range that is not in close proximity to a critical point, this should not be too much of a simplification for most technical systems. Nevertheless, in order to exclude the existence of such points, like e.g. hysteresis, we also consider the case of steering the system from all possible y_1 back to y_0 for one particular input. By fixing the starting point y_0, the input signal is parameterized by y_1 and a transition time $T \in \mathfrak{T}$. The input signal ranges examined in the following comprise a symmetrical interval of at least 20 % around the nominal operating point as standard. The intervals are shortened if large deviations between the linear and the nonlinear model already occur for lower variations of the respective input or if the input loses its physical relevance, e.g. humidity below zero. Within the present contribution, (6) is evaluated separately for each output, where $L^\infty(\mathbb{R}, \mathbb{R})$ is considered. Thus, the nonlinearity measure simplifies to

$$v(N) = \sup_{\substack{y_1 \in \mathfrak{Y}_{eq} \\ u \in \mathfrak{U}_{fwd}(y_1, T; y_0)}} \|\rho(N, L; y_1, T)\|_{L^\infty}. \tag{7}$$

Whereas the nonlinear system model (1) can be directly used to represent $N(\cdot)$, the linear system model (2) cannot be applied directly for $L(\cdot)$ because the requirement of zero static gains is not met. However, by using the static characteristic curve $y_{lin} - y_0 = K \cdot (u - u_0)$ of (2), a linear map in the sense of the norm can be defined by

$$L(u - u_0) = \tilde{L}(u - u_0) - K \cdot (u - u_0), \tag{8}$$

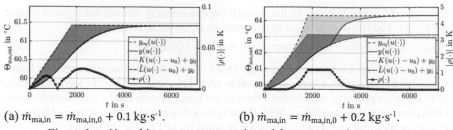

(a) $\dot{m}_{\mathrm{ma,in}} = \dot{m}_{\mathrm{ma,in,0}} + 0.1\ \mathrm{kg\cdot s^{-1}}$.   (b) $\dot{m}_{\mathrm{ma,in}} = \dot{m}_{\mathrm{ma,in,0}} + 0.2\ \mathrm{kg\cdot s^{-1}}$.

Figure 1: $\rho(\cdot)$ and its components evaluated for $\Theta_{\mathrm{ma,out}}$ and $T = 1800$ s.

where $\bar{L}(\cdot)$ is realised by (2). As the expression for $\rho(\cdot)$ is rather abstract, it is shown in Figure 1 together with all the components used in its calculation. The variables of the linear system are shown shifted to the nominal operating point y_0 for display reasons. The ramp-shaped curves in Figure 1 correspond to the static input-output maps applied to the input signal. Whereas the static input-output map of the linear system is obviously only a scaling of the input signal, a distortion due to the nonlinearity of $y_{\mathrm{eq}}(u)$ can be observed in Figure 1b. The remaining components of $\rho(\cdot)$ are the gas temperatures calculated by means of $N(\cdot)$ and $\bar{L}(\cdot)$. After inserting (8) in (5), $\rho(\cdot)$ can be interpreted as the difference between the areas spanned by the static input-output map and the transient system response of the linear and nonlinear system, respectively. This results in specific properties of the $\rho(\cdot)$ and $v(\cdot)$. First, $\rho(\cdot)$ becomes zero for $t \to \infty$ so that integral norms can be applied. Second, $v(\cdot)$ decreases for increasing T. Third, the deviation in the static input-output relation $y_{\mathrm{eq}}(u)$ and $K \cdot (u - u_0)$, become visible in the measure $v(\cdot)$ for small T. The latter two properties are particularly interesting with regard to linear control, as they directly account for the static differences between the original nonlinear model and the linear approximation as a function T. These differences can be typically well compensated by an integrator component in the controller for large values of T, whereas they can be problematic for small ones. Since the steady-state deviation between the models takes an important role in the measure and to simplify its interpretation, we will first look at the static deviations in Section 3.1 and consider afterwards the dynamic nonlinearity measure $v(\cdot)$ in Section 3.2. Due to the large number of possible input-output combinations, only a very small number of results can be presented within the paper. The results are shown exemplarily using the gas outlet temperature $\Theta_{\mathrm{ma,out}}$ for a varying mass flow of gas $\dot{m}_{\mathrm{ma,in}}$ to the dryer and its vapor share $w_{\mathrm{v,in}}$.

3.1. Steady-State Nonlinearity

The deviations between $y_{\mathrm{eq}}(u)$ and $K \cdot (u - u_0)$ are predominantly caused by the nonlinearity of $y_{\mathrm{eq}}(u)$ visualized in Figure 2. The Figures show the transient response of $\Theta_{\mathrm{ma,out}}$ for several values of $\dot{m}_{\mathrm{ma,in}}$, where their stationary values correspond to $y_{\mathrm{eq}}(u)$. From Figure 2, the existence of discontinuous nonlinearities, like hysteresis, can be

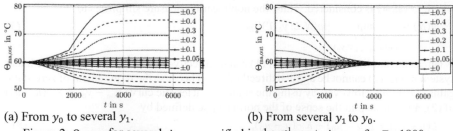

(a) From y_0 to several y_1.   (b) From several y_1 to y_0.

Figure 2: $\Theta_{\mathrm{ma,out}}$ for several $\dot{m}_{\mathrm{ma,in}}$ specified in kg·s⁻¹ w.r.t. $\dot{m}_{\mathrm{ma,in,0}}$ for $T = 1800$ s.

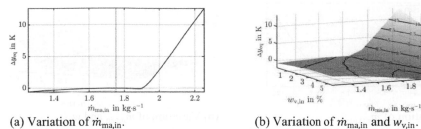

(a) Variation of $\dot{m}_{\mathrm{ma,in}}$. (b) Variation of $\dot{m}_{\mathrm{ma,in}}$ and $w_{\mathrm{v,in}}$.

Figure 3: Steady-state nonlinearities $\Delta y_{\mathrm{eq}} = y_{\mathrm{eq}}(u(\cdot)) - (y_0 + K \cdot (u(\cdot) - u_0))$ for $\Theta_{\mathrm{ma,out}}$.

excluded as all curves starting from y_0 in Figure 2a return from y_1 to y_0 in Figure 2b. Moreover, an asymmetry above a certain deviation of $\dot{m}_{\mathrm{ma,in}}$ from its nominal value $\dot{m}_{\mathrm{ma,in,0}}$ can be observed. This becomes particularly apparent, when looking at the difference between the static input-output maps depicted in Figure 3a. From this picture, it can be seen that the linear map is in good approximation a tangent to the nonlinear one at the nominal operating point, indicated by the dashed line. Moreover, one can observe that there are in principle two types of nonlinearities. The first one slowly increases away from the operating point as can be seen for $\dot{m}_{\mathrm{ma,in}} < 1.7\,\mathrm{kg\cdot s^{-1}}$ in Figure 3a. This type may be caused by nonlinear heat and mass transfer coefficients, which can only be represented in the linear model approximately. The second type are strongly increasing nonlinearities that are active above $\dot{m}_{\mathrm{ma,in}} = 1.9\,\mathrm{kg\cdot s^{-1}}$. This abrupt change in the nonlinear system's behaviour is caused by the almost complete drying of the particles due to the increased energy input. As a result, the energy supplied to the particle phase by convection is no longer consumed by the evaporation of moisture, causing the particles to heat up. This qualitative change cannot be represented by the linear system. In addition to the linearity of the outputs with respect to individual inputs, it is also of particular interest whether the superposition principle is approximately fulfilled. This can be investigated by looking at the simultaneous change in several input variables, such as the $\dot{m}_{\mathrm{ma,in}}$ and $w_{\mathrm{v,in}}$ in Figure 3b. From the Figure, two observations can be made. First, as long as no saturation in the nonlinear model occurs the superposition principle is fulfilled in good approximation. Second, the range of linearity of different inputs influence each other. In Figure 3b one can see that the maximal nonlinearity decreased with increased vapor share in the gas stream, as the latter reduces the evaporation rate and thus delays the appearance of completely dried particles. The observations made can be transferred to all outputs for all pairwise combinations of the input variables and disturbances.

3.2. Dynamic Nonlinearity

The dependence of the measure on the transition time in presence of static deviations can be investigated from Figure 4. It shows $\rho(\cdot)$ for the same inputs as in Figure 1 for three different transition times. One can conclude from Figure 4a that as long as the saturation

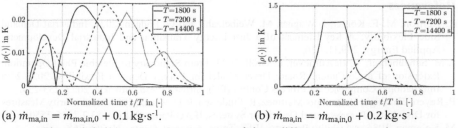

(a) $\dot{m}_{\mathrm{ma,in}} = \dot{m}_{\mathrm{ma,in,0}} + 0.1\,\mathrm{kg\cdot s^{-1}}$. (b) $\dot{m}_{\mathrm{ma,in}} = \dot{m}_{\mathrm{ma,in,0}} + 0.2\,\mathrm{kg\cdot s^{-1}}$.

Figure 4: $|\rho(\cdot)|$ evaluated for $\Theta_{\mathrm{ma,out}}$ and three different transitions times.

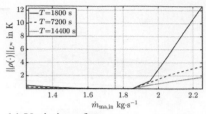

(a) Variation of $\dot{m}_{\text{ma,in}}$.

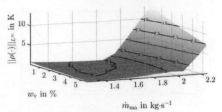

(b) Variation of $\dot{m}_{\text{ma,in}}$ and $w_{\text{v,in}}$.

Figure 5: $\|\rho(\cdot)\|_{L^\infty}$ evaluated for $\Theta_{\text{ma,out}}$, 5b shown for $T = 1800$ s.

is avoided, T has hardly any effect on $\rho(\cdot)$. Otherwise, $\rho(\cdot)$ strongly depends on T as shown in Figure 4b. The observed decrease of $\rho(\cdot)$ for increasing T can be attributed to the fact that, due to the longer transition time, the system states approach the sequence of equilibrium points between y_0 and y_1. In that case, the steady-state deviation is largely compensated by the static input-output maps. It can also be seen that the maximum value of the curve for $T = 1800$ s in Figure 4b has approximately the same value as the corresponding static deviation shown in Figure 3a. By applying the supremum norm to the signals in Figure 4 as well as to all other analysed changes of $\dot{m}_{\text{ma,in}}$, one obtains Figure 5a. When evaluating the measure for a certain fixed T, $\nu(\cdot)$ can be readily determined by taking the maximal value of the individual curve. If a range of T is examined, $\nu(\cdot)$ is approximately the maximal value of all curves corresponding to the investigated samples $T \in \mathfrak{T}$. Of course, it is not sufficient to consider only the variation of individual inputs because, as in the case of static nonlinearities, interactions can occur between the individual inputs, as shown in Figure 5b for $T = 1800$ s. The results presented on the basis of the gas temperature can be transferred to the other output variables for all pairwise combinations of inputs and disturbances.

4. Conclusion

The contribution presents the application of a new dynamic nonlinearity measure to a rotary dryer model. The usage of the measure is shown in detail and the physical relevance of the measure's components is explained using the underlying process. The analysis shows comparatively small differences between the nonlinear and the linear model both w.r.t. static as well as the dynamic nonlinearities, as long as operating conditions that lead to complete drying of the particles are avoided. Furthermore, it is illustrated that for such operating points the superposition principle is approximately fulfilled. Otherwise, strong static nonlinearities can occur causing high values of control-relevant nonlinearity for fast operating point changes. It is also found that interactions between the inputs can influence the permissible ranges of the individual inputs needed to avoid the highly nonlinear operating conditions.

References

R. Heffron, M.-F. Körner, J. Wagner, M. Weibelzahl, and G. Fridgen, 2020, Industrial Demand Side Flexibility: A Key Element of a Just Energy Transition and Industrial Development, Applied Energy, 269,115026.

J. M. Schaßberger, V. Hagenmeyer, L. Gröll, 2023, From Stationary to Flexible Plant Operation: Extension of a Co-Current Rotary Dryer Model for Energy Demand Flexibility, 2023 24th International Conference on Process Control (PC), Strbske Pleso, Slovakia, 2023, 233-239.

P. Reyero-Santiago, C. Ocampo-Martinez, R. Findeisen, R.D. Braatz, 2020, Nonlinearity Measures for Distributed Parameter and Descriptor Systems, IFAC PapersOnLine, 53,2,7545-7550.

M. Kelemen, 1986, A Stability Property, IEEE Trans. Automat. Contr., 31, 766-768.

Flavio Manenti, Gintaras V. Reklaitis (Eds.), Proceedings of the 34th European Symposium on Computer Aided Process Engineering / 15th International Symposium on Process Systems Engineering (ESCAPE34/PSE24), June 2-6, 2024, Florence, Italy

Reduction of an aerated fermenter CFD model using proper orthogonal decomposition

Pedro M. Pereira,[a]* Rui C. Martins,[a] Bruno S. Ferreira,[b] Fernando P. Bernardo,[a]

[a]*CIEPQPF, Department of Chemical Engineering, University of Coimbra, R. Sílvio Lima, 3030-790 Coimbra, Portugal.*
[b]*Biotrend SA, Biocant Park Núcleo 04, Lote 2, 3060-197 Cantanhede, Portugal*
pedro.mesquita.pereira@tecnico.ulisboa.pt

Abstract

The development of CFD + reaction models has been accepted as an important tool to solve the problems of fermentation scale-up, however the computational cost of multiphase CFD makes simultaneous solution unfeasible. In this work a method for model reduction of CFD models for aerated stirred tanks is proposed as a fast and reliable model for the fluid dynamics. The CFD model took one week to simulate 11 seconds on 20 cpu cores. The reduced model can be reconstructed in less than one second with an error of 10% on the velocity of the water and 20 % on the velocity of the air. With the least well described areas of the tank being also the least relevant to the application. These results show great promise for the use of POD + regression reduced models for aerated stirred tank CFD.

Keywords: CFD, Proper orthogonal decomposition, Fermentation, Aerated stirred tank.

1. Introduction

The scale-up of fermentation processes is an active area of research in bioprocess engineering. As in other applications, the increase in scale of fermentation from bench to pilot and then to commercial scale introduces inefficiency in the mixing process, namely undesired and hard to predict heterogeneities in critical parameters such as dissolved oxygen concentration. Detailed process models are thus required to resolve the spatial gradients in large scale bioreactors, and for that purpose the most useful tool is computational fluid dynamics (CFD) coupled with transport equations for the most relevant species and kinetics of the biochemical reactions.

The major barriers to the development of such CFD + kinetic models are the different time scales of the phenomena taking place and the high computational cost of CFD simulations. The characteristic time for liquid circulation in the reactor and gas-liquid mass transfer is in the order of 10 s while the characteristic time for the biochemical reactions may be as large as 10^4 s (Amanullah et al 2003). With such separate time scales, the two model components can be decoupled (Yamamoto, 2013) by first calculating a representative set of stationary fields obtained from CFD simulations and then solving on top of these the biochemical kinetics.

The alternative here presented is the creation of a computationally light, reduced order model, able to capture the main fluid dynamics aspects of the system. This reduced model will then allow future extensions to a parametric reduced model, with parameters being fed-batch operation variables and also design variables, and as such the model can be used to support scale-up decisions. Such parametric reduced CFD models have been described for equally complex systems, for example in aerospace applications (Chen et al, 2019).

Proper orthogonal decomposition (POD) is a well-established method to construct reduced versions of large-scale CFD models (Holmes et al, 2012), and therefore is the method used in this work. In the context of CFD, POD is traditionally used to identify, study and model the dynamics of spatial structures of the turbulent velocity field. POD creates the key spatial ingredients, called modes, that compose the dynamics of the system. With the time-dependent combination of modes, the reduced model can dynamically recreate the coherent structures and by extension the fluid flow (Holmes et al, 2012). Equation 1 represents the form of a POD decomposition, with $\vec{U}(x,t)$ being the velocity field calculated from the higher order CFD model, which is then approximated by a linear combination of p orthogonal functions named eigenvectors (or modes) $\vec{\phi}_k$. The combination is weighted by a_k time varying coefficients that capture the system dynamics, while vectors $\vec{\phi}_k$ retain the spatial distribution of state variables, in this case the fluid velocity.

$$\vec{U}(x,t) \approx \vec{U}_{POD}(x,t,p) = \sum_{k=1}^{p} a_k(t)\vec{\phi}_k(x) \tag{1}$$

Multiple authors have applied POD in the context of stirred tank fluid dynamics. One example of this is the work of Lammote et al (2018), that performed POD to CFD and experimental results of stirred tanks for identification of the dominant spatial and time characteristic of flow dynamics in stirred tanks, including the simulation of the free-surface. The authors observed some limitations of k-ε RANS simulations to predict the energy-containing higher modes. Due to its computational efficiency this method is still employed in this work. Other authors like Jin and Fan (2020) also applied POD to experimental data of fluid flow in stirred tanks. Mendonza et al (2018) used experimental and CFD results with POD to study the transitional flow regime in stirred tanks. The work of Janiga (2019) can be seen as a landmark work on POD in stirred tanks due to its use of higher fidelity Large eddy simulations with highly detailed description of coherent structures in the flow at varying Reynolds numbers. The more recent work of Arosema and Solsvik (2023) presents a similar approach with investigation of the effect of bubbly flows and the modelled shape of the bubbles. Tabib and Joshi (2008) also studied the flow of multiple pieces of equipment with experimental data, LES and POD. Mikhaylov et al (2023) propose an innovative use of POD for reconstruction of the instantaneous 3D velocity field from sparse pressure measurements at the impeller blades.

The work of Mayorga et al (2022) is one of the first to propose what is also adopted in this work, the reconstruction of the model and not just the use of POD as an analytical tool for stirred tanks, this work only included single phase stirred tanks while in the present work aerated flows are studied.

The reconstruction of the model can be based on projection methods like the Galerkin projection or interpolative/regression methods. The Galerkin projection requires a great degree of manipulation of the higher order CFD model. For a greatly complex multiphase CFD model this was not deemed practical and for this reason a regression-based method was used to recreate the $a_k(t)$ function. Due to the existence of fluctuating stationary fields $a_k(t)$ are expected to be periodic functions as observed by Mayorga et al (2022) for single phase stirred tanks. Fourier series are well established to approximate periodic functions and as such are here used to approximate $a_k(t)$.

2. Methods

2.1. CFD methods

The CFD model used is an Euler-Euler model with a mass conservation and momentum conservation equation for each phase as seen in Eq. 2,3 and 4. The two phases share the same pressure field and exchange momentum with each other. The interphase drag is modelled by the Schiller and Naumann model (1933) and the virtual mass forces are modelled by a constant coefficient model.

$$\frac{\partial}{\partial t}(\alpha_k \rho_k) + \nabla \cdot (\alpha_k \rho_k U_k) = 0 \tag{2}$$

$$\frac{\partial}{\partial t}(\alpha_k \rho_k U_k) + \nabla \cdot (\alpha_k \rho_k U_k U_k) = -\alpha_k \nabla p + \nabla \cdot (\alpha_k \tau_k) + \alpha_k \rho_k g + M_{jk} \tag{3}$$

$$\frac{\partial}{\partial t}(\rho_m k_m) + \nabla \cdot (\rho_m \tilde{U}_m k_m) = \nabla \cdot \mu_{mt} \sigma_m \nabla k_m + P_k^m - \rho_m \varepsilon_m + S k_m \tag{4}$$

The turbulence model is Reynolds Averaged Navier-Stokes with mixture k-ε closure (Behzadi et al 2004). The rotation of the impellers is modelled by a multiple reference frame (MRF) model. The geometry of the bioreactor here studied was replicated from the description of the equipment used by Warmoeskerken and Smith (1985). The mesh was produced with the openFOAM® tools blockMesh and snappyHexMesh. The mesh was refined to achieve recommended values of y^+ for the used wall functions and composed of 517868 elements. Predicting the power spent by the impeller was done by a custom openFOAM® function. The operating conditions modelled are: rotation speed of the impeller constant at 3 rotations per second and the gas flow-rate at 2×10^{-4} m³s⁻¹ corresponding to 0.2 vvm on a 7L filled volume tank.

2.2. POD and regression methods

The construction of the reduced basis with POD started with the construction of the $3L \times N$ sized matrix of snapshots $\underline{M}$. The reduced basis was constructed with the snapshot method. This method starts with the construction of the correlation matrix $\underline{R}$ as seen in Eq. 5. This was followed by the solution of the eigenvalue problem in Eq. 6. With Y being the matrix whose columns are the eigen vectors of the correlation matrix and $\vec{E}$ being a vector of the eigenvalues of the correlation matrix. The coefficient array T that contain the values of $a_k(t)$ for each time t_n for each snapshot was calculated by equation 7. Finally the matrix Φ containing the modes $\bar{\phi}_k(x)$ as columns was constructed with equation 8.

$$\underline{R} = \underline{M}^T \underline{M} \tag{5}$$

$$\underline{R}Y = Y\vec{E} \tag{6}$$

$$T = Y\vec{E}^{\frac{1}{2}} \tag{7}$$

$$\Phi = \underline{M}T \tag{8}$$

$$a_k(t) = a_{0,k} + a_{1,k}\cos(w_k t) + a_{2,k}\cos(2w_k t) + a_{3,k}\cos(3w_k t) \\ + b_{1,k}\sin(w_k t) + b_{2,k}\sin(2w_k t) + b_{3,k}\sin(3w_k t) \tag{9}$$

The python Library "modred" was used to perform these steps and a small python script was created to construct the matrix of snapshots. To reconstruct the 3D fields a continuous function for each $a_k(t)$ is necessary. From the POD results the array T was obtained which contains the values of $a_k(t)$ for the times corresponding to the snapshots. To construct a continuous function from this data a regression method was used. The data was fitted to a third order Fourier Series approximation, with the form presented in equation 9 using the python library "symfit".

3. Results and Discussion

The use of POD in the literature is mostly applied to find coherent structures in pseudo-stationary or fully developed turbulent flow. Works including this type of analyses include those of Lamote et al (2018) and Mayorga et al (2022). Another use of POD is for analysis of dynamics of CFD results. In figure 1 are presented the results of instantaneous power developed by the impeller and *a(t)* for the 4 principal modes obtained from POD of the full domain of time up to 11.5 s including start-up. Power expended in agitation is a well established indicator of the operating conditions of an aerated stirred tank (Paul et al 2003). As such the stabilization of the expended power is a good indicator that the system arrived at a pseudo-stationary state. For the two most important modes 0 and 1, *a(t)* stabilizes at 6 s of aerated flow as the aerated power also stabilizes. Further POD analyses were done from 6 s of simulation for these reasons. The *a(t)* profiles present a step perturbation with a plateau from 2 s to 3.6 s, this coincides with the transition of the impeller from operating in the regime with 6 equally sized trailing air cavities to 3 larger and 3 smaller sized trailing air cavities.

POD was performed based on the snapshots from 6 s to 11.5 s, the sampling rate was 10 s^{-1} which corresponds to 55 snapshots. The vector fields of velocity of air $(\vec{u}_{air})$ and velocity of water $(\vec{u}_{water})$ were decomposed. For each of the velocity fields 11 modes were necessary to effectively represent the CFD data. These 11 modes represent more than 99% of the eigen values of both correlation matrices, the advised threshold (Chen et al, 2019).

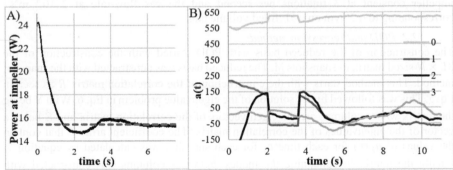

Figure 1 A) Power spent on agitation calculated online with CFD simulation B) a(t) of the main 4 modes of the 0 s to 11 s POD.

The values of a(t) of the most important modes in both vector fields represent a constant profile independent of time. For $\vec{u}_{water}$ α(t) of the first mode is 630 and for $\vec{u}_{air}$ it is 766. The other 3 most important modes are presented in figure 2. These modes present an oscillatory behaviour as is expected, but the profile of the curves is not well conforming to simple periodic functions, as has been observed for non-aerated flows (Mayorga et al 2022). Reconstruction of the vector fields using equation 8 was done to investigate the error associated with POD alone. Error was defined as seen in equation 10. The average error over the space and time domains was 7% for u_{air} 5% for u_{water}.

$$e_{rel} = \frac{\|U_{POD} - U_{CFD}\|}{\|U_{CFD}\|} \tag{10}$$

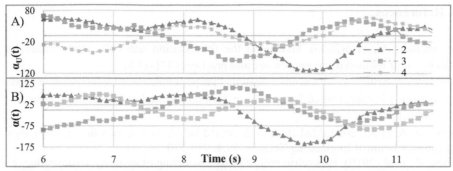

Figure 2 Evolution of a(t) of the 2nd, 3rd and 4th most important modes for: A) $\vec{u}_{water}$ and B) $\vec{u}_{air}$.

The regression of the discrete values of a(t) seen as points in figure 2 was done for all 11 modes for $\vec{u}_{air}$ and $\vec{u}_{water}$. In general good fits were obtained with r^2 greater than 0.9. The relative error for $\vec{u}_{water}$ was of 10 % and 20% for $\vec{u}_{air}$ averaged over space and time. To verify the spatial distribution of the relative error a time average and spatial discretized map of relative error is presented in figure 3. From this image it's observed that the regions of greater relative error in the prediction are located in the air headspace of the tank. This is a good indication as this area is not directly relevant for modelling of a reaction occurring in the liquid phase. If the headspace is not accounted the error goes to 13% for $\vec{u}_{air}$ and 8 % for $\vec{u}_{water}$. The errors in estimation of velocity imply error in estimation of local residence time for bubbles, fluid and biomass. This influences the estimation of gas-liquid mass transfer and estimation of concentration gradients for the fermentation model. The simulation is speed-up from taking one week to simulate the 11s of the full order model to less than one second to reconstruct the POD + regression model.

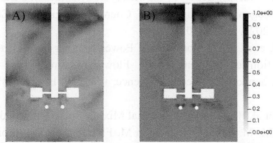

Figure 3 Map of the relative error on a cut of the stirred tank: A) of $\vec{u}_{water}$ B) of $\vec{u}_{air}$.

4. Conclusion

The results here presented show that application of POD to aerated stirred tanks is feasible. In particular the use of POD-Regression methods appears as a viable option for model reduction of multiphase CFD results. This also opens the possibility of application on reaction modelling of fermentation with fine spatial discretization. This spatially fine discretization compared to other surrogate models such as compartment models allows the modelling of concentration gradients deemed as central for the effects of scale-up. Ultimately, also parametric models should be possible. The data-set here presented is somewhat limited, as such, further investigation is necessary on the effects of the reactor operating regime and the construction of models sufficiently robust to allow for extrapolation.

5. References

Amanullah, A., et a(2003). Mixing in the fermentation and cell culture industries. Handbook of industrial mixing: science and practice, 1071-1170.

Arosemena, Arturo A., and Jannike Solsvik. "Proper Orthogonal Decomposition Modal Analysis in a Baffled Stirred Tank: A Base Tool for the Study of Structures." Flow, vol. 3, Cambridge UP (CUP), 2023.

Behzadi, A. et al., 2004, Modelling of dispersed bubble and droplet flow at high phase fractions. Chemical Engineering Science, 59.4: 759-770

Chen, Z., et al (2019). Parametric reduced-order modeling of unsteady aerodynamics for hypersonic vehicles. Aerospace Science and Technology, 87, 1-14.

Holmes, Philip. Turbulence, Coherent Structures, Dynamical Systems and Symmetry. Cambridge UP, 2012.

Janiga, Gábor. "Large-eddy Simulation and 3D Proper Orthogonal Decomposition of the Hydrodynamics in a Stirred Tank." Chemical Engineering Science, vol. 201, Elsevier BV, June 2019, pp. 132–44.

Jin, Jie, and Ying Fan. "PIV Experimental Study on Flow Structure and Dynamics of Square Stirred Tank Using Modal Decomposition." Korean Journal of Chemical Engineering, vol. 37, no. 5, Springer Science and Business Media LLC, Apr. 2020, pp. 755–65

Lamotte, Anne de, et al. "Identifying Dominant Spatial and Time Characteristics of Flow Dynamics Within Free-surface Baffled Stirred-tanks From CFD Simulations." Chemical Engineering Science, vol. 192, Elsevier BV, Dec. 2018, pp. 128–42

Mayorga, C., et al. "Reconstruction of the 3D Hydrodynamics in a Baffled Stirred Tank Using Proper Orthogonal Decomposition." Chemical Engineering Science, vol. 248, Elsevier BV, Feb. 2022, p. 117220.

Mikhaylov, Kirill, et al. "Decomposition of Power Number in a Stirred Tank and Real Time Reconstruction of 3D Large-scale Flow Structures From Sparse Pressure Measurements." Chemical Engineering Science, vol. 279, Elsevier BV, Sept. 2023, p. 118881

Paul, Edward L., et al. Handbook of Industrial Mixing. John Wiley and Sons, 2004

Warmoeskerken, M. M. C. G.; Smith, John M. Flooding of disc turbines in gas-liquid dispersions: A new description of the phenomenon. Chemical Engineering Science, 1985, 40.11: 2063-2071

Schiller, Links. A drag coefficient correlation. Zeit. Ver. Deutsch. Ing., 1933, 77: 318-320

Tabib, Mandar V., and Jyeshtharaj B. Joshi. "Analysis of Dominant Flow Structures and Their Flow Dynamics in Chemical Process Equipment Using Snapshot Proper Orthogonal Decomposition Technique." Chemical Engineering Science, vol. 63, no. 14, Elsevier BV, July 2008, pp. 3695–715.

Yamamoto, M. (2013). Multi-physics CFD simulations in engineering. Journal of Thermal Science, 22(4), 287-293.

Flavio Manenti, Gintaras V. Reklaitis (Eds.), Proceedings of the 34th European Symposium on Computer Aided Process Engineering / 15th International Symposium on Process Systems Engineering (ESCAPE34/PSE24), June 2-6, 2024, Florence, Italy

A Techno- Economic Evaluation of Potential Routes for Industrial Biosolids Conversion

Hesan Elfaki[a*] and Dhabia M. Al-Mohannadi[b]

aTexas A&M University, Artie McFerrin Department of Chemical Engineering, College Station, Texas, United States

bTexas A&M University at Qatar, Chemical Engineering Program, Education City, Doha, Qatar

h.elfaki@tamu.edu

Abstract

Renewable sources of energy have been widely investigated to reduce the reliance on fossil fuel energy sources that are depleting and have their adverse impacts on the climate and environment. Biosolid, a form of biomass holds the potential of a renewable energy source due to their carbon content. These solids are removed as waste residues from municipal and Industrial wastewater treatment plants. The most dominant biosolids management method has been landfilling for decades, a practice that can be seen as a waste of a valuable resource. In addition, recent research findings proved that this practice has unfavourable impacts on the food chain. This has been an additional motivation to research the possibilities of unlocking the potential of these solids. Published research work investigated different types of biomass waste but limited literature is available on biosolids in general and on industrial biosolids in specific.

This paper evaluates the potential of converting industrial biosolids into value added products that can be used as feedstocks in chemical production or can be utilized for heat and power co-generation. The study focuses on well-established thermal conversion technologies mainly pyrolysis, gasification anaerobic digestion and the developing hydrothermal technology. A MILP model is developed to provide a techno economic assessment of the treatment routes using these technologies. The model assists in deciding on the optimal treatment route with the objective of minimum cost. The model was executed using a case study of industrial biosolids stream with a flowrate of 50 ton/d and water moisture of 30% The results show that for the given biosolid stream anaerobic digestion treatment route will provide the optimal minimum cost route. The model demonstrates that the selection of the suitable route highly depends on technology limitation and biosolids stream characteristics.

Keywords: Biosolids, Renewable Energy, Utilization.

1. Introduction

A significant waste stream at industrial facilities consists of biosolids, which are the waste leftovers generated by the wastewater treatment plants that are connected to these facilities. These wastes require careful attention due to the worldwide emphasis on energy preservation, waste reduction, and the transition to clean energy objectives. Landfilling and direct land application have been the primary methods for managing biosolids for many years (Winchell et al., 2022). In response to limited land availability and the negative effects on food chains, efforts were focused on finding alternatives that promote the circularity of biosolids. These alternatives would employ biosolids as a feed stream and convert it into value-added products, energy sources, and feedstocks, including biogas, biooils, and biochar (Egan, 2013). The high energy and nutrient content of biosolids makes them highly suitable for reuse applications, rather than being disposed of as waste materials, which necessitates investment. The potential biosolids management applications to alternate land applications encompass anaerobic digestion, gasification, pyrolysis, and hydrothermal treatment. In their study, (Zhao et al., 2019) performed comparative research on the management of biosolids using co-anaerobic digestion, specifically focusing on the treatment plants in the United States and China. Current research endeavours have been concentrated on the process of pyrolysis applied to organic matter derived from biological sources. In their publication, (Paz-Ferreiro et al., 2018) conducted a comprehensive analysis of the utilization of pyrolysis biochar in land-based activities.

The variability in the properties of biosolids streams resulting from their source of generation and the specific technological constraints on their reuse applications are the primary factors determining the potential compatibility between waste and application. Furthermore, economic issues and environmental repercussions play a pivotal role in the process of decision making. Although there is a wealth of literature on the progress made in biosolids conversion technologies, most of the existing studies focus on testing a particular type of waste stream and conducting an economic evaluation for a predetermined technology. There is a need for a comprehensive selection process and evaluation method to determine the suitability and desirability of one technology over another, taking into consideration the economic and environmental consequences of the decision. The objective of this project is to develop a practical decision-making tool that can be used to assess the viability of matching biosolid waste streams with potential reuse applications. An adaptable optimization model is being created to choose the appropriate treatment pathway based on a certain objective that can vary between minimizing costs, reducing emissions, or optimizing certain production or energy generation.

2. Methodology

A techno-economic model is constructed to assess the cost of biosolids treatment applications. The model serves as a decision-making tool for determining the most optimal application to be implemented for a continuous flow of biosolids. The calculating model is a mixed integer linear programming paradigm, MILP. The model inputs consist of the flowrate and composition of the biosolids stream, as well as the cost parameters. Binary variables are established. The model's restrictions are set by applying concepts of mass and energy conservation, while also considering the limitations of the utilization technologies included in the model. The cost minimization function is the primary objective function.
The model demonstrates the capacity to modify the objective function as necessary to attain alternative goals, such as maximizing revenues or minimizing emissions.

3. Mathematical Model

3.1 Model Formulation

The model assumes P to be the set of treatment processes to convert biosolids waste stream into a value-added product i.e., sinks to biosolids streams.

$$P = \{P_i\} \tag{1}$$
$$i = \{1,2,3,4\} \tag{2}$$

$$P = \{P_1, P_2, P_3, P_4\} \tag{3}$$

Where P_1, P_2, P_3 and P_4 represent pyrolysis, hydrothermal liquefaction, gasification, and anaerobic digestion respectively. Each of these processes requires certain feed flowrate, F_{pi}, and is limited to a certain composition of water moisture in the feed, X_{wpi} and a certain composition of biosolids content in the feed, X_{spi}. The variables F_{pi}, X_{wpi}, and X_{spi} are bounded by lower and upper limits. Thus, they must satisfy the following process requirements:

$$F_{pi}^{min} \leq F_{pi} \leq xF_{pi}^{max} \tag{4}$$
$$x_{wpi}^{min} \leq x_{wpi} \leq x_{wpi}^{max} \tag{5}$$
$$x_{spi}^{min} \leq x_{spi} \leq x_{spi}^{max} \tag{6}$$

Each process produces a certain effluent stream, G_{pi} that contains a variety of components. Each component has a composition, G_{epi}. Where, e in the composition subscript stands for the set of possible products in the effluent stream, G_{pi}.

$$e = \{g, o, r, c, w, x\} \tag{7}$$

The elements of this set are biogas, g, biooil, o, biochar r, compost, c, water, w, and extra residues denoted as x.
The model assumes S to be the be the set of sources that produce biosolids. These are the individual industrial processes that generates biosolid stream as waste.

$$S = \{S_j\} \tag{8}$$
$$j = \{1,2,3\} \tag{9}$$
$$S = \{S_1, S_2, S_3\} \tag{10}$$

Each source has a flowrate F_{sj}, a defined composition of water moisture X_{wsj} and a defined composition of biosolids in the source stream X_{ssj}. In the case study illustrated in the following section, the problem exhibits a single biosolids source stream. Thus, S= $\{S_1\}$.

Service streams with defined flowrates are also included in the model. These are simply extra streams that are required to aid the biosolids conversion processes. The aiding streams can possibly include a stream of steam with flowrate F_{v1}, a stream of air with flowrate F_{v2} and a stream of nitrogen with a flowrate of F_{v3}.

The model assumes that the biosolids source stream, represented by the flowrate F_{s1}, undergoes pretreatment to reduce its metal concentration, aligning the stream with the heavy metal composition standards set by the EPA for agricultural purposes, landfills, and land applications. Metals are incorporated into a biochar product during the biosolids conversion process, which is then used in agricultural applications. Post the pretreatment, the presence of heavy metal compositions in the source stream is negligible compared to the amount of biosolids and water content. Therefore, these compositions are not considered in the mass balance representation of this model. The pre-treatment process for the removal of metals is not within the scope of this article. The technology flowrate restrictions are determined by the capabilities of commercially operated plants in the United States.

3.2 Model Equations

The model objective function is an annual cost minimization equation where y_n represents the binary variables, C_{Pi} represents the total cost of the reutilization process and C_{vk} is the cost of the aiding agent used in gasification process:

$$min \ Cost = \sum y_1 C_{P1} + y_2 C_{p2} + y_3 (C_{p3} + C_{vk} F_{vk}) + y_4 C_{p4} \tag{11}$$

$$y_1 = \begin{cases} 1, \text{ if Pyrolysis is in operation} \\ 0, \text{ otherwise} \end{cases}$$

$$y_2 = \begin{cases} 1, \text{ if Hydrothermalis in operation} \\ 0, \text{ otherwise} \end{cases}$$

$$y_3 = \begin{cases} 1, \text{ if Gasification in operation} \\ 0, \text{ otherwise} \end{cases}$$

$$y_4 = \begin{cases} 1, \text{ if Anaerobi Digestion in operation} \\ 0, \text{ otherwise} \end{cases}$$

The total cost of the process is given by:

$$C_{pi} = C_{pi}^{fixed \ cost} + C_{pi}^{operation \ cost} \tag{12}$$

Material balances around source allocation point:

$$\sum F_{si} = \sum F_{pi} \tag{13}$$
$$x_{wsi} F_{si} = \sum x_{wpi} F_{pi} \tag{14}$$
$$x_{ssi} F_{si} = \sum x_{spi} F_{pi} \tag{15}$$

The objective function is subjected to the following constraints:

$$y_2 + y_3 + y_4 + y_5 \leq 1 \tag{16}$$

$$F_{pi}^{min} . y_n \leq F_{pi} \leq F_{pi}^{max} . y_n \tag{17}$$

$$0 \leq x_{wpi} \leq x_{wpi}^{max} \tag{18}$$

$$0 \leq x_{spi} \leq x_{spi}^{max} \tag{19}$$

$$F_{s1} \geq 0 \tag{20}$$

$$F_{vk} \geq 0 \tag{21}$$

4. Case Study

4.1 Problem Statement

The wastewater treatment facility releases a continuous flow of industrial biosolids waste, with a rate of 50 tons per day and a moisture content of 30%. The stream underwent pretreatment to reduce its metal level. The purpose of deploying the developed model is to determine the most viable treatment technique for biosolids consumption, while minimizing the annual cost. This will be achieved through the implementation of Python code. The yearly expense is determined by using a plant's lifespan of 10 years. Table 1 provides a concise overview of the inputs used in the model. The flowrate limits for the treatment procedures are determined by the capacity of commercial-scale plants in the United States. The energy requirements and cost components in the table are derived from pertinent sources. (Linville et al., 2015; Perkins et al., 2019; Shahbaz et al., 2021).

Table 1 Case Study Inputs

Notation	Process	F_{pi}^{min} (ton/d)	F_{pi}^{max} (ton/d)	x_{wpi}^{max}	Energy Consumption (kWh/ton)	OPEX Factor ($/kWh)	CAPEX Factor ($/kWh)
P_1	Pyrolysis	2	110	0.8	800	0.07	0.33
P_2	Hydrothermal. L	0.3	6	0.9	10	0.07	0.23
P_3	Gasification	6	90	0.85	1000	0.07	0.44
P_4	Anaerobic D.	8	285	0.5	35	0.07	0.18

4.2 Results and Discussion

The case study inputs were integrated into the model framework utilizing an optimization algorithm implemented in Python 3.8 programming language. The code was executed to achieve the ideal answer within a few of seconds. The answer proposed the utilization of anaerobic digestion for the treatment of the specified biosolids stream, as it is linked to the most economical annual treatment cost of $154,000. The model analyses the biosolids stream parameters, including flowrate and compositions, and compares them to the constraints of the four reutilization processes included in the model. The flowrate limits, moisture content, and water composition are all evaluated simultaneously in comparison to the feed criteria. The biosolids feed, with a flowrate of 50 ton/d, might potentially be distributed to three technology choices. At this stage of the analysis, the hydrothermal treatment method is ruled out as an option since the feed flowrate exceeds the maximum permitted flowrate specified in Table 1 for this application. The composition limits of pyrolysis, gasification, and anaerobic digestion processes permit the inclusion of the biosolids stream for conversion using any of these methods. After conducting the initial screening, the model proceeds to calculate the precise quantity of biosolids that requires treatment. The projected quantity is approximately 175,000 tons of biosolids throughout the operational lifespan of the treatment facility. The specified energy consumption factors in the case study are utilized to determine the entire energy requirements using this amount. The model thereafter performs economic computations to estimate the total capital and operational expenses for each treatment option. Finally, an optimization analysis is conducted with the aim of decreasing the overall annual expenses. Hence, anaerobic digestion, a widely recognized method for converting biosolids and biomass, is the most cost-effective choice in this case study. The results showcased the model's capacity to choose the appropriate approach by considering the feed specifications and the constraints of the technologies used in the model, in terms of their compositions and capacities.

5. Conclusions

The model presented in this work aids in determining the most suitable treatment method for converting biosolids waste stream, considering the stream's characteristics and the constraints of the reuse application. The model is an optimization mixed integer model that aims to minimize treatment costs, serving as an example of potential optimization objectives. The model is generic and adaptable, allowing for modifications to address the optimization problem of biosolids treatment for various objectives, such as maximizing revenues, minimizing emissions, and specific pollutant removal.

6. Acknowledgment

The authors gratefully acknowledge funding provided by Qatar Shell Research and Technology Centre (QSTRC).

7. References

M. Egan, 2013, Biosolids management strategies: An evaluation of energy production as an alternative to land application, Environmental Science and Pollution Research, Vol. 20, Issue 7, pp. 4299–4310

M. Shahbaz, A. AlNouss, I. Ghiat, G. Mckay, H. Mackey, S. Elkhalifa, T. Al-Ansari, 2021, A comprehensive review of biomass based thermochemical conversion technologies integrated with CO2 capture and utilisation within BECCS networks. Resources, Conservation and Recycling, *173*

L.J. Winchell, J.J. Ross, D.A. Brose, T.B. Pluth, X. Fonoll, J.W. Norton, K.Y. Bell, 2022, Pyrolysis and gasification at water resource recovery facilities: Status of the industry, Water Environment Research, Vol. 94, Issue 3

J.L. Linville, Y. Shen, M.M. Wu, M. Urgun-Demirtas, 2015, Current State of Anaerobic Digestion of Organic Wastes in North America, Current Sustainable/Renewable Energy Reports, Vol. 2, Issue 4, pp. 136–144

J. Paz-Ferreiro, A. Nieto, A. Méndez, M.P.J. Askeland, G. Gascó, 2018, Biochar from biosolids pyrolysis: A review. In International Journal of Environmental Research and Public Health, Vol. 15, Issue 5

G. Zhao, M. Garrido-Baserba, S. Reifsnyder, J.C. Xu, D. Rosso, 2019, Comparative energy and carbon footprint analysis of biosolids management strategies in water resource recovery facilities. Science of the Total Environment, 665, 762–773.

G. Perkins, N. Batalha, A. Kumar, T. Bhaskar, M. Konarova, 2019, Recent advances in liquefaction technologies for production of liquid hydrocarbon fuels from biomass and carbonaceous wastes, Renewable and Sustainable Energy Reviews, Vol. 115

Flavio Manenti, Gintaras V. Reklaitis (Eds.), Proceedings of the 34th European Symposium on Computer Aided Process Engineering / 15th International Symposium on Process Systems Engineering (ESCAPE34/PSE24), June 2-6, 2024, Florence, Italy

A Rigorous Integrated Approach to Model Electrochemical Regeneration of Alkaline CO$_2$ Capture Solvents

Fariborz Shaahmadi, [a] Katia Piscina, [a] Qingdian Shu, [b] Sotirios Efstathios Antonoudis, [a,c] Sara Vallejo Castaño, [b] Philip Loldrup Fosbøl, [d] Uffe Ditlev Bihlet, [d] Mijndert van der Spek [a,*]

[a]*Research Centre for Carbon Solutions (RCCS), Heriot Watt University, Edinburgh EH14 4AS, UK*
[b]*Wetsus, European Centre of Excellence for Sustainable Water Technology, Netherlands*
[c]*Centre for Research & Technology Hellas/Chemical Process and Energy Resources Institute (CERTH/CPERI), Greece*
[d]*Technical University of Denmark - Department of Chemical Engineering, Denmark*

*m.van_der_spek@hw.ac.uk

Abstract

This work develops a rigorous model for electrochemical regeneration in Aspen Custom Modeler (ACM), designed to seamlessly integrate into ASPEN Plus, allowing to model complete carbon dioxide (CO$_2$) capture – electrochemical regeneration cycles on a single modelling platform. The modelling of CO$_2$ electrochemical cells has gained significant attention in CO$_2$ capture and utilization processes. This emphasizes the importance of modelling in driving the progress of CO$_2$ electrochemical cells which combines absorption by alkaline solvents and electrochemical solvent regeneration. In such process, potassium hydroxide (KOH, or other metal hydroxides) is used as a solvent for CO$_2$ capture. This process involves a series of chemical reactions that result in the formation of potassium carbonate (K$_2$CO$_3$) and potassium bicarbonate (KHCO$_3$). After CO$_2$ is captured through absorption, the K$_2$CO$_3$/KHCO$_3$ solution is directed towards the regeneration cell where an electrochemically driven pH swing takes place facilitating the desorption of CO$_2$. The cell's primary objective is to lower the pH of the solution by generating protons at the anode, thereby moving its chemical equilibrium towards carbonic acid. Given the limited solubility of CO$_2$ in water, it desorbs once it reaches saturation. The residual solution can be reclaimed in the cathode compartment and recycled. Electrochemistry models are currently unavailable in popular simulation software like ASPEN Plus, thus making the development of integrated process models, in this case for CO2 capture, more challenging. Here, we introduced a rigorous model to be applied in ACM/ASPEN Plus software to simulate the CO$_2$ regeneration process. The model's validity was assessed against experimental measurements. Following this validation, the model was subsequently employed to design pilot plant campaigns for the Horizon 2020 project ConsenCUS.

Keywords: CO$_2$ capture, electrochemical cell, ConsenCUS, absorption, ASPEN.

1. Introduction

Human actions are a key driver of global warming and its resulting impacts, thus necessitating urgent action to address the current climate crisis. The significant increase in the global average surface temperature by more than 1°C from 1880 to 2012 unequivocally indicates the effect of anthropogenic climate change. This rise in temperature has substantial negative consequences, underscoring the need to decrease carbon emissions and seek innovative solutions. The vital importance of sustainable solutions to counteract the increase in CO_2 emissions has stimulated research into groundbreaking technologies for effective CO_2 capture, utilization, and storage. Of these, absorption in alkaline solutions coupled with electrochemical regeneration processes may prove relevant, as the leverage electricity-based regeneration while mitigating harmful emissions from, e.g., amine-based processes. Recently, the modelling of electrochemical cells is getting considerable noticed in CO_2 capture and utilization processes (Abdin et al., 2015, Ali and Kwabi, 2022, Seo and Hutton, 2023). Sabatino et al. (2020) introduced a simplified equilibrium model in MATLAB for a bipolar membrane electrodialysis regenerator for a direct air capture process. Shu et al. (2020) investigated the energetic requirements of CO_2 capture using a spent alkaline solution. This study introduces a fully developed model for electrochemical regeneration specified in Aspen Custom Modeler (ACM), integrated into an ASPEN Plus flowsheet. The primary objective was to create a unified platform that models complete CO_2 capture and electrochemical regeneration cycles, with an emphasis on integrating electrochemical unit operations into larger process models.

2. Model development

The CO_2 capture process is based on alkaline absorption and electrochemical regeneration as shown in Figure 1. Potassium hydroxide (1M KOH) is used as solvent due to its ability to react with CO_2, forming a potassium carbonate/bicarbonate solution (K_2CO_3/$KHCO_3$) through a selective chemical reaction. Aspen Plus commercial software was utilized to design the absorption column with the primary aim of constructing a model that describes the operation of pilot plants. For the system design, we used the Electrolyte-NRTL thermodynamic property method from the Aspen Plus, which was re-regressed inhouse. The equilibrium reactions' chemistry and kinetics were then modified according to Rastegar & Ghaemi's work (2022).

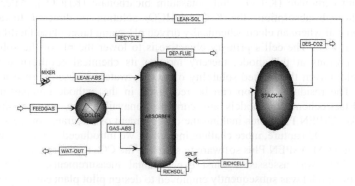

Figure 1- Schematic of the CO_2 capture process based on alkaline absorption and electrochemical regeneration. The configuration includes a recycle over the absorber to allow achieving higher solvent loading.

The absorber was simulated in ASPEN Plus with an effective packing height of 12 m and a diameter of 0.6 m, while the packing void fraction and the surface area were set equal to 0.9 and 200 m^2/m^3 respectively, in line with the design of the pilot plant absorber. Cement plant flue gas with a CO_2 concentration of 18% was used in the simulation.

The regeneration unit includes two stacks of 126 bipolar membrane electrodialysis cells, where the solution from the absorber experiences a pH swift to facilitate the desorption of dissolved CO_2. The aim is to decrease the pH of the solution by introducing protons to the solution, resulting in a change in the equilibrium state. Due to the low solubility of CO_2 in water, the CO_2 will degasify from the solution once it reaches its saturation point. Most of the CO_2 will therefore be desorbed in a flash drum and can be sent for further processing. The remaining solution can be recovered in the cells' cathode compartments and be recycled to the absorber. As this is a novel technology no suitable models for this process exist in ASPEN Plus. Therefore, we implemented a bespoke mathematical model in Aspen Custom Modeller to represent the regeneration unit.

The rich solvent from the absorber consists of water, dissolved carbon ions, potassium ions, plus protons and hydroxides. The series of homogeneous equilibrium reactions between the carbon ions and the water dissociation reaction results in a $CO_2/HCO_3^-/CO_3^{2-}$ buffer system process as follows:

$$CO_2(aq) + H_2O \overset{K_1}{\Leftrightarrow} HCO_3^- + H^+ \tag{1}$$

$$HCO_3^- \overset{K_2}{\Leftrightarrow} CO_3^{2-} + H^+ \tag{2}$$

$$H_2O \overset{K_w}{\Leftrightarrow} H^+ + OH^- \tag{3}$$

where K_1 and K_2 denote the equilibrium constants of reaction equations see above; K_w is the equilibrium constant for the water dissociation reaction. K_1, K_2 and K_w. These Equations show that the carbon ion distribution depends on the pH. The equilibrium reactions determine the system composition in the acidifying compartment of the BPMED cells as well as in the cathode compartment. Since the potassium cations will transfer through the membrane there will be a concentration gradient along the acidifying compartment with the outlet concentration being lower than the inlet concentration. This change in the longitudinal direction has not been modelled in detail yet. The potassium inlet concentration is equal to the absorber potassium concentration. Its outlet concentration depends on the applied current density.

In principle, both K^+ and H^+ cations migrate through the membrane for the sake of charge neutrality and electrical circle completion. It has been reported by several studies that H^+ transportation is more hindered, and thus the transfer rate of H^+ is much smaller compared with the anodic generation and cathodic consumption rate (Zhang et al., 2019). This means that the cations with higher concentrations (here, K^+) transfer transfer more predominantly (Cao et al., 2009) and strongly impede H^+ transportation across the membrane (Chae et al., 2008).

The Nernst-Planck equations for the flux (mol s⁻¹ m⁻²) describes the movement of ions across the membrane:

$$J_{K^+} = -D\, c_x \left(\frac{df_{K^+}}{dx} + f_{K^+} \frac{d\Phi}{dx} \right) \tag{4}$$

$$J_{H^+} = -a\, D\, c_x \left(\frac{df_{H^+}}{dx} + f_{H^+} \frac{d\Phi}{dx} \right) \tag{5}$$

where J_{H^+} is flux of protons (mol m⁻² s⁻¹); a denotes ratio between diffusivities of protons and metal ions; D is diffusivity coefficient of metal ion inside the membrane; c_x is total cation concentration inside membrane (also called fixed charge density); Φ is electric potential; x is position inside the membrane; $f_K{}^+$ is concentration fractions of K^+; and $f_H{}^+$ is concentration fraction of H^+.

Furthermore, the total current density must satisfy the ionic current density through the membrane:

$$j_c = (J_{K^+} + J_{H^+})\, F \tag{6}$$

where F is Faraday constant equals to 96485 C mol⁻¹.

The flux of CO_2 produced (i.e., the amount of CO_2 produced per unit of membrane area), or the production rate can then be derived from the following equation:

$$J_{CO_2} = \frac{Q}{A_a} C_{CO2\,out} \tag{6}$$

where Q is the rich solvent flow rate and A_a stands for the membrane area.

A set of Key Performance Indicators (KPIs) specific to the CO_2 capture process have been presented within this framework. These KPIs play a pivotal role in evaluating the operational effectiveness, efficiency, and overall performance of the ConsenCUS CO_2 capture approach. The efficiency of CO_2 capture by the absorber was determined using the following equation:

$$\eta_{CO_2\ capture}(\%) = \frac{y_{CO_2,in} - y_{CO_2,out}}{y_{CO_2,in}} \tag{8}$$

where $y_{CO_2,in}$ and $y_{CO_2,out}$ represent the mole fractions of CO_2 in the gas stream at the absorber's inlet and outlet, respectively.

In electrochemical regeneration part, the calculations were performed to determine the specific energy consumption (expressed in GJ per ton of CO_2) and CO_2 production rate.

$$SEEC = \frac{j_c A_a V_{total}}{j_{CO_2}} \tag{9}$$

where V_{total} is the stack voltage (V), j_c is the current density (A/m²), A_a is the active membrane area (m²), and j_{CO_2} is the measured CO_2 gas flow rate produced in the regeneration cell. It's worth mentioning that we employed two stacks consisting of 126 pairs of cells for each in the regeneration unit for the base case scenario (one stack (A) is shown in Figure 1). Additionally, the process incorporates a recycle stream that circulates from the rich solvent back to the lean solvent within the absorber unit (as shown in Figure

3. Results and discussion

The validity for electrochemical regeneration model was assessed against experimental measurements as shown in Figure 2. As can be seen from Figure 2, the model predictions for CO_2 production at various current densities, flow rates and rich carbon loadings are compatible with experimental data. The prediction of energy consumption is more scattered (while the trend seems to be predicted well), explained by cell inefficiencies not included in the mathematical model (especially at low current densities).

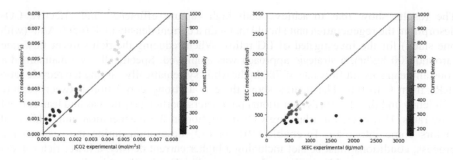

Figure 2- Predicted desorbed CO_2 (mol/m²s) and SEEC (kJ/mol) for
electrochemical regeneration of CO_2 loaded aqueous KOH versus
experimental measurements.

To evaluate the performance of the whole CO_2 capture unit, we opted to explore various parameter ranges and further refined our analysis using the insights gained from Figure 3. By narrowing down our focus within these selected ranges, we can investigate the regeneration unit impact on the overall efficiency of the CO_2 capture process. We selected a specific range for the pilot plant solvent flow rate, ranging from 1000 to 1200 kg/hr, and for the flue gas flow rate spanning from 60 to 80 m3/hr (without recycle ratio, RR=0). The results are shown in Figure 3.

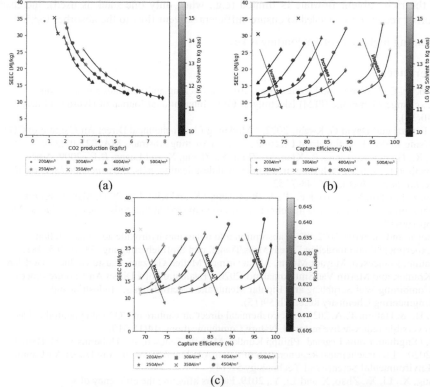

Figure 3- The relation between specific energy consumption as a function of
current density and capture efficiency where lean solvent flow rate is 1000
to 1200 kg/hr and gas flow rate is 60 to 80 m³/hr (no recycle ratio).

The results show that to achieve both high capture efficiency and effective CO_2 desorption, the regeneration unit should maintain a current density of 400-500 A/m^2 (with one stack) for the investigated of LG ratios. When reducing the rich solvent flow rate (around 700 kg/hr), a strategic approach was employed. Specifically, we maintained a constant lean solvent flow rate of 700 kg/hr while systematically varying the recycle ratio (RR) from 0 to 0.6. The figures show there is a strong correlation between capture efficiency and the absorber's operational parameters, particularly the gas and lean solvent flow rates. However, for achieving CO_2 desorption in the regeneration unit, additional factors come into play. To enable efficient CO_2 desorption during the regeneration process, conditions must be met including a higher current density, a lower rich solvent flow rate, an increased lean solvent flow rate, and high rich solvent loading.

4. Conclusions

The conclusion drawn is that there is a strong correlation between capture efficiency and the operational parameters of the absorber, particularly the gas and lean solvent flow rates. However, achieving CO_2 desorption during the regeneration process requires additional considerations. Efficient CO_2 desorption and capture efficiency are dependent on specific conditions, including:

• Higher current densities leading to more CO_2 desorbed and lower energy consumption.

• If the rich solvent flowrate is limited (e.g., when only one stack is used), increased absorber recycling is needed to ensure sufficient solvent flow to the absorber top.

• Ensuring high rich solvent loading is key.

References

Abdin, Z., C. J. Webb, and E. Maca Gray. 2015. "Modelling and Simulation of a Proton Exchange Membrane (PEM) Electrolyser Cell." International Journal of Hydrogen Energy 40(39).

Ali, Fawaz, and David G. Kwabi. 2022. " Modeling Electrochemical Direct Air Capture of CO2 Using Redox-Functionalized Electrodes ." ECS Meeting Abstracts MA2022-02(27).

Cao, X., Huang, X., Liang, P., Xiao, K., Zhou, Y., Zhang, X. and Logan, B.E., 2009. A new method for water desalination using microbial desalination cells. *Environmental science & technology*, *43*(18), pp.7148-7152.

Chae, K.J., Choi, M., Ajayi, F.F., Park, W., Chang, I.S. and Kim, I.S., 2008. Mass transport through a proton exchange membrane (Nafion) in microbial fuel cells. *Energy & Fuels*, *22*(1), pp.169-176.

Rastegar and Ghaemi, 2022, "CO2 absorption into potassium hydroxide aqueous solution: experimental and modeling," Heat Mass Transf. und Stoffuebertragung, 58(3), 365–381.

Sabatino, Francesco, Mayank Mehta, Alexa Grimm, Matteo Gazzani, Fausto Gallucci, Gert Jan Kramer, and Martin Van Sint Annaland. 2020. "Evaluation of a Direct Air Capture Process Combining Wet Scrubbing and Bipolar Membrane Electrodialysis." Industrial and Engineering Chemistry Research 59(15).

Seo, H., & Hatton, T. A. 2023. Electrochemical direct air capture of CO2 using neutral red as reversible redox-active material. Nature Communications, 14(1), 313.

Shu, Qingdian, Louis Legrand, Philipp Kuntke, Michele Tedesco, and Hubertus V. M. Hamelers. 2020. "Electrochemical Regeneration of Spent Alkaline Absorbent from Direct Air Capture." Environmental Science and Technology 54(14).

Zhang, X., Li, X., Zhao, X. and Li, Y., 2019. Factors affecting the efficiency of a bioelectrochemical system: a review. *RSC advances*, *9*(34), pp.19748-19761.

Flavio Manenti, Gintaras V. Reklaitis (Eds.), Proceedings of the 34th European Symposium on Computer Aided Process Engineering / 15th International Symposium on Process Systems Engineering (ESCAPE34/PSE24), June 2-6, 2024, Florence, Italy

Global modelling and simulation of essential oil extraction processes

Zouhour Limam[1,2], Catherine Azzaro-Pantel[1], Noureddine Hajjaji[2], Mehrez Romdhane[3], Jalloul Bouajila[1]

[1] *Laboratoire de Génie Chimique, Université de Toulouse, CNRS, INPT, UPS, Toulouse France.*

[2] *Laboratory of Catalysis and Materials for Environment and Processes (UR11ES85), National School of Engineers, University of Gabes, Tunisia*

[3] *Laboratory of Energy, Water, Environment and Processes (LR18ES35), National School of Engineers, University of Gabes, Tunisia*

Abstract

Traditional essential oil extraction methods using organic solvents such as ethanol, hexane, methanol, acetone, and even water, have raised environmental concerns due to their high energy and solvent consumption. In response, researchers are exploring alternative techniques like ultrasound-assisted, supercritical fluid, and microwave-assisted extractions to reduce the ecological impact. This study aims to develop a comprehensive modeling framework for essential oil extraction, with a specific focus on supercritical fluid and water distillation technologies. The models developed have been successfully validated against experimental data. Supercritical fluid extraction has proven to be the most energy-efficient method. Additionally, this modeling facilitates the transition from laboratory or pilot scale to industrial production. The results obtained from this study will serve as inventory data for conducting environmental Life Cycle Assessment (LCA).

Keywords: Essential oil, Simulation, Aspen plus™, *Eucalyptus intertexta, Rosemary.*

1. Introduction

Essential oils (EOs) are formed in aromatic and medicinal plants as products of secondary metabolism. Extracting them can be seen as a complex and delicate process to capture and collect the most volatile, subtle, and delicate products that the plant produces, all without compromising their quality. Various extraction methods have been developed for distilling the terpenic molecules from fragrance plants, categorized into ancient and modern technologies. Traditional extraction methods have historically relied on organic solvents such as ethanol, hexane, methanol, acetone, and even water to extract essential oils from aromatic and medicinal plants. However, these methods have raised environmental concerns due to significant energy consumption and solvent usage. In response to these issues, researchers have been actively investigating alternative extraction techniques, including ultrasound-assisted extractions, supercritical fluids, and microwave-assisted extractions, with the goal of mitigating emissions and reducing ecological impacts. According to the state-of-the-art analysis, the majority of research reported to date is focused on investigating specific extraction processes, both through experimental methods and modeling techniques to gain a deeper understanding of the physico-chemical phenomena governing the extraction process. It is noteworthy that there have been relatively few studies focused on simulating essential oil extraction processes from plants thus far. One notable contribution is the work of (Moncada et al., 2016): in this paper, a techno-economic and environmental assessment of the extraction of essential oil from *Oregano* and *Rosemary* in Colombia is thus performed with the Aspen Plus™ simulator for process modeling. Only a few investigations have also delved into exploring the essential oil supply chain (González-Aguirre et al., 2020). In this vein, the scientific objective of this study is thus to develop a comprehensive modelling framework for the extraction of essential oils that can serve as a basis for process selection. For this purpose, we will explore two different technologies, namely supercritical fluid extraction and water distillation. The extraction of essential oils from two types of leaves, namely *Eucalyptus intertexta* and *Rosemary* plants, has been studied. This paper is divided into four sections. Section 1 covers the essential oil extraction technique. Section 2 is dedicated to the modeling of essential oil extraction processes, including water distillation and supercritical fluid methods. Section 3 provides a detailed discussion and emphasizes key results. Finally, we draw conclusions and outline potential avenues for future research in Section 4.

2. Methodology

2.1. Traditional vs. emerging method for EO extraction
There are several techniques for extracting essential oils. We will focus on water distillation (WD) (Kant et al., 2022) as a traditional extraction method and on supercritical fluid extraction (SFE) (Kant et al., 2022) as a modern technology for comparison purpose.

2.2. Modeling of Essential Oil Extraction Processes
2.2.1. Modelling principles
The EO from the plant was modeled using Aspen Plus V14 software from Aspen Technology Inc., USA. The thermodynamic models employed include Unifac-Dortmund for calculating activity coefficients in the liquid phase and the Hayden-O'Connell equation of state to model the vapor phase (Moncada et al., 2016). The predictive models UNIFAC have proven their efficiency to obtain reliable results over a large range of applicability (Gmehling et al., 2012). Hayden and O'Connell's method is well-suited to ideal and non-ideal systems at low pressures. To assess the accuracy and reliability of our modeled process scheme, two approaches were considered. Firstly, the modeling of essential oil extraction from *Rosemary* (*Rosmarinus officinalis*) was conducted using the operational data provided by (Moncada et al., 2016) for both water distillation and supercritical fluid extraction for validation purposes. Given that the publication did not explicitly detail the equipment employed, we leveraged insights from other literature pertaining to biomass valorization simulations (Rosha et al., 2022). Simultaneously, we compared the results obtained from the modeled process scheme for *Eucalyptus intertexta*, with experimental data provided in (Chamali, 2020). This comparison allowed us to evaluate the consistency and agreement between simulation results and laboratory data.

2.3. Process description
- *Water Distillation Extraction (WDE)*

The essential oil extraction process through WD involves the intake of 200kg/h of the raw material (RM), which is modeled as a non-conventional compound within the Aspen software. To identify it, it is necessary to determine its composition using proximate and ultimate analyses. To address this, analytical correlations established by (Park et al., 2023) are employed and the composition is solved using MATLAB software.

Subsequently, the plant material undergoes a drying process at a temperature of 30°C in a dryer. The dried solids are then ground to achieve a particle size of less than 0.5 cm in a mill, aiming to expose the oil fraction and increase the extraction yield (due to increased interfacial tension) in line with the observations of (Moncada et al., 2016).

The reduced solid is then directed into an extraction vessel. However, the Aspen software cannot simulate a non-conventional solid in an extraction column. To overcome this limitation, an RYield reactor is utilized to convert the non-conventional compound into a conventional compound, with reference to the compositions of *Eucalyptus* (Chamali, 2020) and *Rosemary* (Moncada et al., 2016) essential oils. A splitter is also employed for the extraction phase. Steam is generated within a heat exchanger and then introduced into the extraction column. The steam temperature is kept relatively low (around 90 - 100 °C) to prevent the degradation of the essential oil compounds. Some authors have suggested using a solid/fluid ratio (kg:kg) for steam extraction. For instance, (Chamali, 2020) experimentally studied the impact of the solvent/dry plant material ratio and found the optimal ratios to be 10, 12, and 14. (Cassel et al., 2009) reported solid/vapor ratios of 2.2, 2.1, and 3.0 respectively for *Rosemary*, *Basil*, and *Lavender*, while Moncada reported a solid (dry)/vapor ratio of 5:1. For our modeling, we adopt Moncada's approach since we are working with a vapor-phase fluid. In this process, utilities included low-pressure steam (3 bars) as well as cooling water for heating and cooling respectively. After extraction, the steam containing the essential oil is rapidly cooled, resulting in two distinct liquid fractions: one rich in oil and the other rich in water. These fractions are separated in a decanter. Figure 1 illustrates the schematic of essential oil extraction through WD and Table 1 provides a description of the equipment used. Additionally, a step of waste recovery was incorporated to facilitate the generation of both heat and electricity (the section surrounded in Figure 1 integrates a waste combustion reactor and a power generation turbine.).

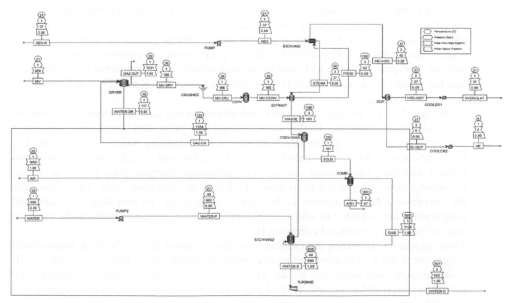

Figure 1. Flowsheet diagram for the extraction of essential oils using water distillation.

Table 1. Equipment used for WDE of essential oils

Step	Aspen reference	Description
Pump & pump2	**Pump**	To pump the water.
Dyer	**RStoic**	Drying process to reduce moisture content, temperature (30°C), and pressure (1 bar).
Crusher	**Crusher**	To reduce the size of the RM particles.
Conv & conv-was	**RYield**	To convert non-conventional compounds into conventional ones.
Extract	**Splitter**	To extract essential oils from eucalyptus using steam at 100°C and 3 bars.
Exchang & Exchang2	**HeatX**	To evaporate water using the product from other equipment
Sep	**Separator**	To separate the essential oil from water.
Cooler1	**Heater**	To cool the hydrolat to 21°C (1 bar).
Cooler2	**Heater**	To cool the essential oil to 4°C (1 bar)."
Comb	**RGibs**	To incinerate waste
Turbine	**Turbine**	To produce electricity.

- *Supercritical Fluid Extraction (SFE)*

The processes for preparing the raw material follow the same steps as explained for water distillation extraction. However, in this case, the fluid used for extraction is carbon dioxide (CO_2), a non-toxic and environmentally friendly solvent. The primary advantage of SFE lies in its selectivity and the conditions associated with this method reduce the risks of thermal decomposition of the components. Due to the volatility of SFs, it is possible to extract active compounds, which facilitates solvent recovery. SFE ensures high extraction yields in a short amount of time, without generating toxic waste. This process impresses with its solubility, selectivity, and rapid mass transfer. Currently, over 90% of SFE processes use CO_2 due to its accessibility, non-flammability, non-toxicity, and reasonable cost. The solvent is then routed to an evaporator to be transformed into the gaseous phase.

Subsequently, CO_2 undergoes a series of compressions and cooling processes to reach the required supercritical state. In this state, CO_2 exhibits properties intermediate between a gas and a liquid, making it an effective solvent for extracting volatile compounds. The supercritical solvent is then introduced into the extraction vessel where it comes into contact with the plant material. The volatile compounds from the plant are extracted into the supercritical solvent, due to its ability to diffuse through plant cells. After extraction, the solvent-compound mixture is directed to a separator. At this stage, pressure and temperature are adjusted to allow for the separation of the solvent from the mixture, leaving behind the extracted compounds.

These processes, involving the preparation if the raw material and the implementation of supercritical fluid extraction are pivotal steps in producing high-quality essential oils from aromatic and medicinal plants. Figure 2 illustrates the schematic of essential oil extraction through CO_2-sc and Table 2 provides a description of the equipment used.

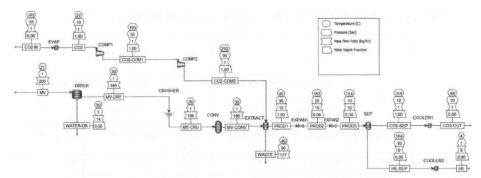

Figure 2. Flowsheet diagram for the extraction of essential oils using CO_2.

Table 2. Equipment used for the SFE of essential oils

Bloc	Aspen	Description
Evap	**heater**	To raise CO_2 from -60°C to 27°C (phase change) at 10 bars.
Comp1&2	**compressor**	To compress CO_2 from 10 bars to 50 and 90 bars successively
Expan1&2	**Valve**	To depressurize the mixture produced by the extraction tank from 90 bars to 50 and 10 bars successively.
Cooler1	**Heater**	To cool CO_2 to -60°C (10 bars).
Cooler2	**Heater**	To cool the essential oil to 4°C (1 bar).

3. Results

The primary focus of the process simulation analysis for extracting essential oil from *Rosemary* and *Eucalyptus intertexta* revolves around processing yield, defined as the quantity of essential oil obtained per ton of fresh raw material. Extraction yield is contingent upon both the employed technology and the inherent characteristics of the raw material. Notably, *Rosemary*, with a moisture content ranging between 60-70%, and *Eucalyptus*, with a moisture content of 7%, exhibit distinct yield variations. For *Rosemary*, the yields are recorded at 10.61 kg/t for SFE and 8.95 kg/t for WDE. SFE consistently demonstrates superior yields, with only a marginal impact on energy consumption. In the case of *Eucalyptus*, SFE yields 8 kg/t, whereas WDE yields 4 kg/t. To check how accurate our model is, we compared our simulation results with those obtained by (Moncada et al., 2016). Significant differences are observed, as shown in Table 3, but our results are within the range of the values obtained by the authors. These variations could be due to different factors such as assumptions, simulation conditions, input data, and calculation methods used by both approaches, regardless of whether they consider energy integration. In summary, supercritical fluid extraction consistently outperforms water distillation extraction in terms of yield for both *Rosemary* and *Eucalyptus intertexta*. Additionally, by repurposing the waste (i.e., 181 kg/h) generated after the extraction process for electricity and heat generation, the process can generate 197 kWh of electricity for water distillation, while generating 237 kWh for supercritical fluid extraction from 177 kg/h of waste. This innovative approach not only addresses waste management effectively but also makes a significant contribution to sustainable energy production.

Table 3. Comparison between the results found by Aspen and the results of (Moncada et al., 2016).

	SFE			WDE		
	ASPEN	Article of (Moncada et al., 2016).		ASPEN	Article of (Moncada et al., 2016).	
		Without Integration	With Integration		Without Integration	With Integration
	kJ/kg RM	kJ/kg RM	kJ/kg RM	kJ/kg RM	kJ/kg RM	kJ/kg RM
Heating	62,71	128,94	79,87	84,08	148,43	44,56
Cooling	37,44	136,88	31,17	107,19	148,08	44,14

4. Conclusion

In conclusion, this study has developed a comprehensive modeling framework for essential oil extraction, specifically emphasizing supercritical fluid and water distillation technologies. The models developed have been successfully validated against experimental data, showcasing their accuracy and reliability. Notably, the interest of this modeling lies in providing access to energy data that were not attainable at the laboratory scale. The findings affirm the superior energy efficiency of supercritical fluid extraction. Furthermore, the modeling approach not only contributes to a deeper understanding of the extraction process but also facilitates a seamless transition from laboratory or pilot scale to industrial production. The results obtained from this study serve as crucial inventory data for conducting environmental Life Cycle Assessment (LCA), thereby enhancing the sustainability assessment of essential oil extraction processes.

References

E. Cassel, R. M. F.Vargas, N. Martinez, D. Lorenzo D., E. Dellacassa , Steam distillation modeling for essential oil extraction process », Industrial Crops and Products, 1 Janvier 2009, vol. 29, no 1, p. 171-176.

S. Chamali, Extraction et valorisation des principes actifs d'une plante tunisienne : optimisation et modélisation, thèse, Gabes, Tunisie, 2020.

G. Jürgen, X. Zhimin, M. Tiancheng, Reply to "Comments on 'Comparison of the a Priori COSMO-RS Models and Group Contribution Methods: Original UNIFAC, Modified UNIFAC(Do), and Modified UNIFAC(Do) Consortium'", Industrial & Engineering Chemistry Research 2012 51 (41), 13541-13543.

A. González, A. José, J. Solarte-Toro A. CardonaCarlos, Supply chain and environmental assessment of the essential oil production using Calendula (Calendula Officinalis) as raw material, Heliyon, 2020. Volume 6, Issue 11, e05606, ISSN 2405-8440.

Kant, Ravi, K. Anil« Review on essential oil extraction from aromatic and medicinal plants: Techniques, performance and economic analysis ». Sustainable Chemistry and Pharmacy 30 December 2022: 100829.

J. Moncada, J. Tamayo A. Cardona Carlos, « Techno-economic and environmental assessment of essential oil extraction from Oregano (Origanum vulgare) and Rosemary (Rosmarinus officinalis) in Colombia », Journal of Cleaner Production, 20 Janvier 2016, vol. 112, p. 172-181.

S. Park , J. Kim Seok , C. Oh Kwang, H Cho La , D. Kim, « Developing a Proximate Component Prediction Model of Biomass Based on Element Analysis », Energies, Janvier 2023, vol. 16, no 1, p. 509.

P. Rosha , S. Kumar, H. Ibrahim « Sensitivity analysis of biomass pyrolysis for renewable fuel production using Aspen Plus », Energy, 15 March 2022, vol. 247, p. 123545.

Flavio Manenti, Gintaras V. Reklaitis (Eds.), Proceedings of the 34th European Symposium on Computer Aided Process Engineering / 15th International Symposium on Process Systems Engineering (ESCAPE34/PSE24), June 2-6, 2024, Florence, Italy

Mathematical issues in Bilevel Mixed-Integer Linear Programming applied to systems biology.

Guilherme De Olivera Mendes Ohira[a]*, Galo A. C. Le Roux[a]

[a]Escola Politécnica da Universidade de São Paulo (USP), Av. Prof. Luciano Gualberto, trav.3, 380, 05508-020, Brazil

guilhermeohira@gmail.com

Abstract: Gene knockouts can be seen as a tool to improve microorganism performance of an industrial objective function. Despite molecular biology own several state-of-the-art techniques to implement genetic modifications, there is a lack of strategies to guide the decision on what target genes knockouts should be performed. It is possible to tackle the problem by Fluxomic techniques and Constraint-Based Models. To secure the biological objective, while trying to enhance bioprocess performance, it is possible to formulate the problem as a Bilevel Mixed-Integer Linear Programming Problem (BMILP). The first algorithm suggested in literature that makes use of the BMILP strategy, is called OptKnock (2003). It applies the BMILP by enforcing optimality, which translates the problem into a Mathematical Program with Complementarity Constraints (MPCC). OptKnock's authors suggested to apply dual theory to circumvent complementary and keep linearity (MILP). However, the mathematical description never clear if the optimal achieved by the MPCC is in fact the same of the original BMILP. In the present study an algebraic deduction was suggested. A set of *in silico* experiments were performed in order to compare the approaches. The results show that, in spite of the that KKT conditions hold, the industrial optimal achieved by the resulting MILP may not match the original BMILP. It is possible to conclude that the approach suggested by OptKnock authors can reshape the feasible region when compared to the BMILP. Hence, it is possible to obtain values of industrial that wrongly appear to enhance the objective functions, which can mislead the binary search tree (BST), and then produce suboptimal results instead.

Keywords: Systems Biology, Fluxomic, Computational Strain Optimization Models (CSOM), Bi-level Programming Problem (BLPP), MPCC.

1. Introduction

Microorganisms can be seen as micro-factories since they can convert substrate into products. There are several products that are desired for industrial scale production, but microorganisms are biased to enhance their survivability, usually maximizing biomass production instead. This behaviour can be seen as a competition between biological objectives (i.e., maximizing survivability) and industrial objectives (i.e., human desire).

To build algorithms that optimize biochemical networks it is necessary to interpret the metabolism. One way to do this is by Fluxomic techniques, such as Flux Balance Analysis (FBA). This approach looks at the metabolic network as a huge collection of chemical reactions that are represented by their stoichiometry. Hence, it is possible to reconstruct biochemical networks that contain known metabolites, reactions and genes in genomic scale (Orth *et al.*, 2010). It is possible to build a space that describes everything that is possible in terms of the fluxes inside a microorganism. It is common to explore that space with a bias, which corresponds to the maximization of biomass growth.

FBA, when seen mathematically, is a Linear Programming problem (LP), which can have degenerate results. In Systems Biology, and when considering genomic scale, a degenerate result is very likely to take place. This means that there are infinite optimal solutions. This result depicts a key feature in metabolism, the notion of silent phenotypes (Price *et al.*, 2004).

Algorithms such as OptKnock (Burgard *et al.*, 2003), hold the biological objective function (i.e., the inner problem – the biological objective) while looking for high-throughput industrial desired phenotypes (i.e., outer problem – industrial objective). These algorithms switch off (i.e., knockout) given reactions, in order to enhance industrial objective function.

OptKnock-like algorithms formulate a Bilevel Mixed-Integer Linear Programming Problem (BMILP). When first described, authors suggested to enforce KKT conditions to describe the BMILP as a Mathematical Problem with Complementarity Constraints (MPCC), and then use lagrangian duality to finally achieve a Mixed-Integer LP (MILP) (Burgard *et al.*, 2003).

OptKnock-like algorithms account for a reasonable number of successful strains and patents. But, even after twenty years, it has not been proved that the resulting MILP has the same result as the first suggested BMILP framework. Dempe and Dutta, 2012, showed that when Slater's Condition Qualification (CQ) (in the outer problem) is fulfilled both Bilevel and MPCC have the same optimal global solution. Despite both levels are LP, Slater CQ may not be achieved in the outer level.

Our main objective is to review the mathematical framework, initially proposed by Burgard *et al.*, 2003, to verify possible issues that can be faced in further studies and highlight them. In the present work we show that KKT conditions hold in the inner problem (FBA), but a gap can be achieved when comparing both BMILP and the resulting MILP.

2. Methodology

2.1. Experimental (in silico) set-up

A process to convert a bilevel into a single level problem is described in Burgard *et al.*, 2003. Briefly, the BMILP, shown in Eq. (1), has an inner level (MILP) that can have its KKT conditions enforced.

$$
\begin{aligned}
\max_{y,v} \quad & v_{obj} = c_2^t v \\
\text{s.t.} \quad & \max_{v} \quad v_{bio} = c_1^t v \\
& \quad \text{s.t.} \quad Sv = 0 \\
& \qquad diag(y)LB \leq v \leq diag(y)UB \\
& \qquad LB \leq v \leq UB \\
& \qquad v \in \mathbb{R}^n \\
& \qquad 1^t y = n - K \\
& \qquad y \in \{0,1\}^n
\end{aligned}
\tag{1}
$$

This results in a MPCC with mixed-integer variables. This problem can be solved by lagrangian duality, which has linear dependency with KKT conditions. Hence, it is possible to replace complementarity with lagrangian strong duality (which is still non-linear for the mixed-integer case). It is possible to handle the current problem applying an additional (redundant) constraint ($LB \leq v \leq UB$). Due to duality theory and based on how the lagrangian duality works (Bazaraa *et al.*, 2003) the redundant constraint unfolds in such way that non-linear terms of strong duality can be replaced by a zero, and hence, circumventing the non-linearity (this approach was first suggested by Xu *et al.*, 2013).

Eq. (1) shows the BMILP, where v is the optimization column vector of the inner level that represents flux of every reaction (n). S is the stoichiometric matrix, which has m lines and n columns, representing every reaction rule in the metabolism. Both LB and UB stands for *lower bounds* and *upper bounds*, which are imposed limits estimated by Flux Variability Analysis (FVA). FVA is a pre-processing analysis (details described in Maranas and Zomorrodi, 2016). y is a binary optimization vector (of the outer level) that can switch-off (i.e., knockout) reactions. Due to limited computational power, number of knockouts are limited by the constraint $\mathbf{1}^t y = n - K$. In the present study, only 5 knockouts are allowed ($K = 5$). Additionally, there are reactions that cannot be knocked out since the pre-processing analysis showed that they are fundamental to viability: Biomass formation, ATP maintenance and phosphate transport (diffusion) (algorithm described in Machado *et al.*, 2016).

A genomic scale model of *Escherichia coli*, strain K-12, substrate MG1655 was used (Feist *et al.*, 2007). This model is available in BiGG database (King *et al.*, 2016). After FVA pre-processing the model has $n = 1532$ reactions and $m = 1032$ metabolites. The resulting MILP is similar to the one presented by Xu *et al.*, 2013 (I), but we assumed the stoichiometric matrix constraint as an equality one ($Sv = 0$), and our knockout number is fixed ($\mathbf{1}^t y = n - K$). The final build had 8692 continuous variables and 1532 binary variables, 2671 equalities constraints and 9192 inequalities constraints. A total of 3 industrial objective functions were chosen: Succinate production, Hydrogen production and Threonine (the difference between each industrial objective is due only to c_2^t).

2.2. Slater's CQ verification and degeneracy (outer level)

A LP always fulfil Slater's CQ if in standard form, so there is no reason to test the resulting MILP. However, it is possible for a BMILP not to fulfil these conditions. To investigate that, the inner optimal was fixed in both LB_{bio} and UB_{bio} and FBA-like optimization was solved for each reaction (primal variables) maximizing and minimizing each of them (Eq. (2)).

$$
\begin{aligned}
\min/\max_{v} \quad & v_i = c^t v \\
s.t. \quad & Sv = 0 \\
& LB \leq v \leq UB \\
& LB_{bio} \leq v_{bio} \leq UB_{bio} \\
& v \in \mathbb{R}^n
\end{aligned}
\tag{2}
$$

Once solved, it is possible to know the possible horizon of each reaction, and hence, to know the domain of variation of each reaction in the degenerate region of FBA. Then, if a given v_i is at either LB_i or UB_i bound, while there is a gap between $\bar{v}_i^{min}$ and $\bar{v}_i^{max}$, then Slater's CQ is unfulfilled.

Both results of $\bar{v}_i^{min}$ and $\bar{v}_i^{max}$ show how wide a reaction can vary in the degenerate region of FBA. That kind of analysis is very close to FVA, which shares the same objective, but with a slight difference. In this study this kind of analysis is important to verify if the degenerate region is the same for both conditions.

2.2.1. Medium, growing conditions and implementation

It was assumed that carbon source, oxygen uptake and ATP maintenance (LB_i) take the following values: -10 mmol/gDW/h (gDW stands for grams of Dry Weight biomass), -18.5 mmol/gDW/h and 8.39 mmol/gDW/h, respectively.

Optimizations were run in a desktop computer with a Ryzen 5 3600X, implemented in Matlab, solved by Gurobi (v. 10) with a fixed processing time of 90 minutes (deterministic). Due to computational limitations, a heuristic algorithm was chosen for the binary search tree while the node solver chosen was Dual Simplex.

3. Results and discussion

3.1. Algebraic deductions

3.1.1. Inner optimal

When dealing with Bilevel problems it is possible to enforce KKT conditions of the inner level problem and achieve a MPCC. In LP, complementarity can hold in both contexts: elementwise and as vector multiplication. In the last one it is possible to see that strong duality is linear dependent with KKT conditions, so is possible to circumvent the complementarity non-linearity by replacing it with strong duality (first suggested by Burgard et al., 2003). Note that complementarity is implicit in the final framework.

An important concern is that if a numerical violation eventually crosses a line to be negative (either dual variable associated to an inequality constraint or the inequality constraint itself), then the complementarity seen as a vector multiplication is not equivalent to original KKT condition (hence it is not possible to replace it by a strong duality). In such a way, complementarity can be violated, and the inner result be non-optimal.

3.1.2. Gap between MPCC and Bilevel

Theorem 2.3 and Slater's CQ showed in Dempe and Dutta (2012) study states that if the outer problem fulfils Slater's CQ then a bilevel can be replaced by an MPCC and achieve global optimal solution (and hence exhibiting no gap between them).

Slater condition is a sufficient condition for strong duality, which is mandatory to enforce KKT conditions. Slater condition, in LP, is always fulfilled if it is in its standard form. When deducing strong duality, in LP, it is possible to see that complementarity is only the term which stands between weak and strong duality.

In bilevel LP context, it is possible not to fulfil Slater CQ in the outer problem. For example, consider an inner level problem that a dual variable (associated to an inequality) is positive, and hence inequality is binding in the optimal solution. That particular inequality does not fulfil Slater CQ (regarding outer level problem) since it becomes an equality and not a strict inequality. This behaviour is usual in the context of Systems Biology.

Furthermore, also usual in Systems Biology, there is a degeneracy in dual problem, meaning that Lagrange multipliers (i.e., dual variables) are non-unique (i.e., degenerate). Note that, in the process of enforcing KKT conditions, dual variables become optimization variables as well. Then, it is possible to glimpse that some of these variables can reshape constraints related to inner level problem. For example, it is possible that a $LB_i \leq v_i \leq UB_i$ variable gets bonded to a lower bound or upper bound if the degeneracy of dual variables enhances the industrial objective function. Note that the original bilevel, Eq. (1), does not consider dual variables as optimization variables, and hence, it is not possible to observe that kind of behavior.

3.1.3. What cannot be seen only using algebra?

In section 3.1.1. there is a motivation to investigate KKT conditions of the inner level. Obviously numerical violations, below a given threshold (10^{-9}), is expected, but there is no guarantee that this value is small enough to hold KKT conditions of the inner level. If so, the matter of how small the gap between an optimal inner level solution (bilevel) and a (possible) non-optimal solution (resulting MILP), due unfulfilled complementarity, can only be evaluated in practice.

Section 3.1.2. outlines possible issues due to the gap between BMILP and the resulting MILP. The magnitude of this gap must be discussed, as the size of the gap determines the magnitude of this (possible) issue. Additionally, a greater value wrongly observed in the Bilevel Problem, can also be translated into a false degeneracy (regarding to outer problem), which can hinder further analysis on phenotypes.

In order to verify the magnitude of possible gaps and mathematical issues in the resulting MILP (mathematical simplification of a Bilevel), an experimental (*in silico*) set-up was suggested.

3.2. Experimental (in silico) results

3.2.1. Inner level optimality

In all studied industrial objective functions, there was no significant numerical violation ($>10^{-7}$). Moreover, once knockouts were identified, original bilevel was solved sequentially (inner level was solved, then $c_1^t v = v_{bio}$ becomes a constraint bond into its respective bounds). Comparing both results (resulting MILP and Bilevel) were basically the same. From these results, it is reasonable to assume that inner level optimal was achieved and the "height" (i.e., the optimal result of the inner level) of a feasible region is the same for both resulting MILP and BMILP.

3.2.2. Slater's CQ and gap

All conditions did not fulfill Slater's CQ. Succinate and Hydrogen show 2 reactions (common to both) "ATPM" (maintenance energy requirements) and glucose uptake, that could both be translated into equality constraints. Threonine showed 82 reactions that violated Slater's CQ. All of them at once cannot be translated into equalities without violating predictive feature of Eq. (1).

Succinate, Hydrogen and Threonine showed low gap between BMILP and the resulting MILP. Succinate was of the order of 10^{-9} and both the others were 10^{-6}, all had a greater result in the resulting MILP.

It was possible to verify if the FBA's degenerate region for BMILP and resulting MILP is the same, as was suggested in Eq. (2). Table 1 shows the results.

Table 1: FBA's degenerate region differences

Target	# of LB cases	# of UB cases	Order of greatest gap (LB)	Order of greatest gap (UB)
Succinate	2	1	10^{-3}	10^{-4}
Hydrogen	14	25	10^{-2}	10^{-5}
Threonine	277	1174	10^{-6}	10^{-6}

The second and third columns of Table 1 stands for the number of cases where LB and UB were different between BMILP and resulting MILP. The last two columns show the order of greatest values (in absolute) of gap between BMILP and MILP.

It is notable that the gap between objective functions is reasonably smaller than the gap between these two cases of lower bounds shown in Table 1. Succinate's lower bound greatest gap (10^{-3}) is associated with "GLYLC" (Glycine Cleavage System) reaction. While Hydrogen's lower bound greatest gap (10^{-4}) is associated with "EAR160x" a reaction related to Cell Envelope Biosynthesis. Both need coenzymes in its reactions (NAD and NADH).

4. Conclusions

Present work highlights a possible gap between a BMILP and resulting MILP in the context of OptKnock-like algorithms. It is possible to find gaps between BMILP and the resulting MILP as well as in FBA's degenerate region. That kind of problem can produce unrealistic results in nodes while an algorithm searches the binary tree (BST). Because of it is a MILP, lower node solutions usually are not searched, which can mislead the BST if one node shows a high enough unrealistic optimal. Hence, by the end of the BST, a suboptimal result can be achieved in the resulting MILP.

Additionally, the cause for these gaps remains unclear. Despite all industrial objective functions show violation of Slater's CQ, Succinate and Hydrogen can be avoided by removing inequalities from "ATMP", glucose uptake and oxygen uptake, and introducing equivalent equalities instead. On the other hand, there are gaps in FBA's degenerate region too.

Probably the non-unicity (degeneracy) of the dual variables can reshape the feasible region of the outer level (i.e., FBA's degenerate region), turning upper or lower feasible bounds tighter.

Acknowledgements

We gratefully acknowledge grant 2023/09130-6, São Paulo Research Foundation (FAPESP), and support of the RCGI – Research Centre for Gas Innovation, hosted by the University of São Paulo (USP) and sponsored by FAPESP (2014/50279-4 and 2020/15230-5) and Shell Brasil, and the strategic importance of the support given by ANP (Brazil's National Oil, Natural Gas and Biofuels Agency) through the R&D levy regulation.

References

Bazaraa, M. S., Sherali, H. D., & Shetty, C. M. (2013). *Nonlinear programming: theory and algorithms*. John wiley & sons.

Burgard, A. P., Pharkya, P., & Maranas, C. D. (2003). Optknock: a bilevel programming framework for identifying gene knockout strategies for microbial strain optimization. Biotechnology and bioengineering, 84(6), 647-657.

Dempe, S., & Dutta, J. (2012). Is bilevel programming a special case of a mathematical program with complementarity constraints?. Mathematical programming, 131, 37-48.

Feist, A. M., Henry, C. S., Reed, J. L., Krummenacker, M., Joyce, A. R., Karp, P. D., ... & Palsson, B. Ø. (2007). A genome-scale metabolic reconstruction for Escherichia coli K-12 MG1655 that accounts for 1260 ORFs and thermodynamic information. Molecular systems biology, 3(1), 121.

Gurobi Optimization, L.. (2023). Gurobi Optimizer Reference Manual

King ZA, Lu JS, Dräger A, Miller PC, Federowicz S, Lerman JA, Ebrahim A, Palsson BO, and Lewis NE. BiGG Models: A platform for integrating, standardizing, and sharing genome-scale models (2016) Nucleic Acids Research 44(D1):D515-D522. doi:10.1093/nar/gkv1049.

Machado, D., Herrgård, M. J., & Rocha, I. (2016). Stoichiometric representation of gene–protein–reaction associations leverages constraint-based analysis from reaction to gene-level phenotype prediction. PLoS computational biology, 12(10), e1005140.

Maranas, C. D., & Zomorrodi, A. R. (2016). Optimization methods in metabolic networks. John Wiley & Sons.

Orth, J. D., Thiele, I., & Palsson, B. Ø. (2010). What is flux balance analysis?. *Nature biotechnology*, 28(3), 245-248.

Price, N. D., Reed, J. L., & Palsson, B. Ø. (2004). Genome-scale models of microbial cells: evaluating the consequences of constraints. Nature Reviews Microbiology, 2(11), 886-897.

Xu, Z., Zheng, P., Sun, J., & Ma, Y. (2013). ReacKnock: identifying reaction deletion strategies for microbial strain optimization based on genome-scale metabolic network. PloS one, 8(12), e72150.

Flavio Manenti, Gintaras V. Reklaitis (Eds.), Proceedings of the 34th European Symposium on Computer Aided Process Engineering / 15th International Symposium on Process Systems Engineering (ESCAPE34/PSE24), June 2-6, 2024, Florence, Italy

Combining Energy System Planning with Carbon Emissions Trading Optimisation for Decarbonisation of Emerging Economies

Gul Hameed,[a] Purusothmn Nair S. Bhasker Nair,[b] Raymond R. Tan,[c] Dominic C. Y. Foo,[b] Michael Short [a*]

[a]*School of Chemistry and Chemical Engineering, University of Surrey, Guildford GU2 7XH, UK*
[b]*Department Chemical and Environmental Engineering, University of Nottingham Malaysia, Broga Road, Semenyih 43500, Selangor, Malaysia*
[c]*Department of Chemical Engineering, De La Salle University, 2401 Taft Avenue, Manila 0922, Philippines*
Corresponding Author: m.short@surrey.ac.uk

Abstract

This study integrates scalable carbon trading models into an energy planning framework to address carbon emissions challenge associated with increasing energy consumption in emerging economies. The models utilise mixed-integer nonlinear programming (MINLP) formulations to optimise power generation, emissions, and costs. A case study is demonstrated on an ASEAN (Association of Southeast Asian Nations) country, Malaysia. Results reveal that carbon trading enhances both financial gains and environmental sustainability, with direct optimisation (where emission rights or carbon prices are variables) approach proving more effective in maximising carbon markets' efficacy, leading to better cost to emission ratios when compared to indirect optimisation (with emission rights prices as parameters, estimated via demand-supply curve). For emissions minimisation, direct optimisation resulted in cost to emissions ratios about 1.3 to 2.1 times more than those for indirect optimisation for three out of four studied periods. The study highlights the potential of coordinated emissions trading and optimal investment decision-making, providing valuable insights for energy planning in Malaysia which is aiming for net-zero target by year 2050.

Keywords: Emissions trading; energy planning; decarbonisation

1. Introduction

Energy planning is pivotal in achieving climate targets, such as net-zero emissions and limiting global temperature rise to 1.5°C above pre-industrial levels. While developed nations are transitioning towards sustainability, the rapid growth of emerging economies, particularly in Southeast Asia, has led to escalating emissions and resource depletion. This has created a web of issues, such as deforestation, global warming, inflation, and energy crisis, inextricably linked to the energy market crisis (specifically the emissions stemming from energy sector), impacting ASEAN and other countries (Kabanov et al., 2022). These challenges can jeopardise ASEAN economies if not managed ("Open Access Government," 2021). ASEAN region is on track to become the fourth-largest global economy by 2030, with electricity as a key player in meeting energy demands

(Permatasari, 2020). However, predominant use of fossil fuels for electricity generation raises concerns about achieving emission targets while sustaining economic growth.

Among ASEAN, Malaysia's GDP (Gross Domestic Product) growth stands out; however, its significant greenhouse gas (GHG) emissions, exceeding 400 Mt in 2016, position it among top 25 emitters globally (Ertugrul et al., 2016). Malaysia's emissions will persist until at least 2030 due to current coal agreements (Mohd Salleh, 2022). Its commitment to maintain 50% of forest cover to curtail atmospheric CO_2 level, highlights its climate action goals (Tan, 2021). Reliance on forests alone is insufficient, necessitating diversification in climate change mitigation strategies to achieve carbon neutrality (Bhasker Nair et al., 2022b). National Renewable Energy Policy set targets (Sustainable Energy Development Authority, 2011), but deeper emissions cuts are required by Malaysia Renewable Energy Roadmap (MRER) (Sustainable Energy Development Authority, 2021), targeting 31% renewable by 2025, further increasing to 40% by 2035. This aligns with Malaysia's goal of a 45% reduction in GHG emissions intensity per unit GDP by 2030, projected to rise to 60% by 2035.

Malaysia's aim to achieve net-zero by year 2050 requires immediate implementation of feasible climate policies. Energy optimisation tools that may predict decarbonisation pathways are efficient in guiding decision-making towards sustainable energy, appropriate decarbonisation technologies and implementing economic instruments (e.g., carbon trading and taxation). A multi-criteria method for Malaysian energy planning outperformed alternative tools like MARKAL, ENPEP (Energy and Power Evaluation Program), and LEAP (Low Emissions Analysis Platform) in handling diverse objectives (Hussain, 2011). A similar multi-criteria approach proposed solar to be potentially the best renewable energy for Malaysia (Ahmad & Tahar, 2014). An MILP (mixed-integer linear programming) model was developed focusing on CO_2 reduction strategies in Iskandar Malaysia via fuel switching, renewable energy, and CCS (Carbon Capture and Storage) deployment (Lee & Hashim, 2014), adoption of these technologies by 2025 was predicted. Results showed a transition from coal to natural gas in existing plants and greater biomass utilisation in new facilities to meet emission targets. An MILP model was developed (Bhasker Nair et al., 2022a), which optimised the deployment of energy resources, alternative fuels (AFs), CCS and NETs (negative emission technologies) for Malaysian power sector to forecast the CO_2 neutrality by 2050 scenarios (Bhasker Nair et al., 2022b). This formulation was published as DECarbonisation Options Optimisation (DECO2) software, a tool for multiperiod energy planning and long-term decarbonisation strategy assessment (Bhasker Nair et al., 2023).

A carbon trading policy is inevitable for Malaysia to meet its environmental and economic targets (Izlawanie, 2021). However, uncertainty surrounds the strategies Malaysian government should adopt to navigate carbon trading policies effectively. Combined carbon trading and energy planning optimisation are shown to be promising for both environmental and economic sustainability (Hameed et al., 2023). While previous studies have focused on CCS, the potential of CCUS (Carbon Capture Utilisation and Storage), which not only captures and stores CO_2 but also leverages it for economic benefits, remains unexplored for Malaysia. This study bridges these gaps by pioneering Malaysian case study using modified DECO2 framework (Hameed et al., 2023) to include emissions trading, fuel balancing, electricity pricing, and decarbonisation technologies' (CCUS, NETs, AFs, Renewable Energy) deployment, to determine long-term planning scenarios, and predict carbon and electricity prices.

2. Mathematical Model

The original DECO2 framework encompassed a range of power plant types, technologies, and factors like energy prices, budget targets, and emission goals to meet energy demands (Bhasker Nair et al., 2023). The modified DECO2 (Hameed et al., 2023) is an MINLP formulation that introduces a carbon banking financial concept, which encourages the rapid adoption of costly emission reduction technologies and renewable energy solutions to achieve net-zero emissions within electricity sector. This model provides indirect incentives, allowing power plants to obtain profit from emissions trading apart from their core energy operations. When carbon prices are high, plants may opt to install emission reduction technologies to consume fewer emission rights than they have purchased and sell surplus/unused rights to generate revenue and to offset the costs of emission reduction technologies. The enhanced framework also considers electricity prices as variables and incorporates CCUS for additional revenue. Various taxes may also be applied to promote cleaner fuels, ultimately striving for both financial and environmental sustainability. Overall, various interconnected components within the model such as financial factors, energy prices, carbon prices, energy demands, CO_2 gas prices, technologies, plants, emissions, etc., will assist the analysis of long-term energy system. Emerging economies have dynamic energy systems, and hence the model can help to assess policies, uncover potential challenges and opportunities for their power sectors.

Two approaches are used for establishing emission prices and caps: a direct optimisation method that simultaneously optimises both prices and caps for each plant in each period, and an indirect optimisation approach where emission prices for each plant are treated as parameters while emission caps are optimised (Hameed et al., 2023). Four-stage carbon trading model as in Eq. (1) is implemented via the above-mentioned approaches.

$$
\begin{aligned}
&Total\ Carbon\ Trading\ Income \\
&= Per\ Unit\ Emissions\ Price \\
&\times ((Emissions\ Bought\ from\ the\ Previous\ Period\ or\ Stage\ 1) \\
&+ (Emissions\ Bought\ from\ the\ Government\ or\ Stage\ 2) \\
&- (Emissions\ Traded\ within\ the\ Plants\ or\ Stage\ 3) \\
&- (Emissions\ Left\ to\ Trade\ in\ the\ next\ Period\ or\ Stage\ 4))
\end{aligned}
\tag{1}
$$

The net income from carbon trading can either be positive (cost) or negative (revenue) depending on the carbon trading prices and the volume of emission rights traded in the final two stages of carbon trading. In the initial two stages, purchasing emission rights from the government adds to plant costs, contributing to government profits. In the third stage, a plant incurs cost when it acquires emission rights from another plant, while it generates revenue if it sells emission rights. It's worth noting that in the same period, one plant may earn revenue while another incurs costs. In the last stage, if a plant consumes more emission rights than it purchases, it pays a cost to the government. Conversely, if the plant has unused emission rights, the government will pay the plant to buy back the unused rights, resulting in revenue for the plant. These rights bought by the government will be available for trading in the next period. Consequently, some plants may generate revenue, while others may incur costs in the fourth stage.

The mathematical formulations handle two objectives (Hameed et al., 2023). In adapted combined energy planning and carbon trading framework, one objective function minimises costs to achieve emission targets, while the other aims to minimise emissions while adhering to a predetermined budget. The complete mathematical model is available on GitHub (https://github.com/gulhameed361/DECO2-2023v.git).

3. Case Study

This study's primary focus is to create a practical energy planning strategy for advancing decarbonisation initiatives utilising carbon trading in Malaysia. Data used in this study accurately represents Malaysia's existing energy landscape. The planning covers four periods, each spanning five years, encompassing 20 years from 2023 to 2043. Values are predicted for the fifth year of each period, listed as 2028, 2033, 2038 and 2043. Aligning with MRER (Sustainable Energy Development Authority, 2021), the goal is to propose a forward-looking approach to expedite Malaysia's journey towards achieving net-zero by year 2050. A total of 35 existing (15 renewables, 14 natural gas and 6 coal) and 41 planned (19 renewables and 22 natural gas) power plants are considered to meet energy demand between 133 and 166 TWh/y for the duration between years 2023 and 2043 (Bhasker Nair et al., 2022b). It is expected that 19% CO_2 emission reduction will be achieved by 2043 from current emission levels, i.e., from 116 Mt/y to 94 Mt/y (Bhasker Nair et al., 2022b).

3.1. Minimisation of Budget

Direct optimisation (with variable carbon prices) for budget minimisation anticipates an initial focus on installing renewable plants. With time, there's a shift towards natural gas, attributed to CCUS retrofit and the utilisation of AFs in plants. At the same time, rapid decommissioning of coal plants is observed. By year 2043, the optimised model projects the Malaysian power sector to emit 56.3 Mt/y of CO_2, much lower than the targeted CO_2 emissions of 94 Mt/y (Figure 1). Higher projected carbon prices lead to reasonable electricity prices for years 2028, 2033, 2038, and 2043 (70, 140, 176, and 225 million USD/TWh, respectively), resulting in positive net incomes (profits). On the other hand, budget minimisation via indirect optimisation (with carbon prices as parameters) predicts 30 Mt/y emissions by 2043. Note that, via indirect optimisation, negative net income is foreseen for the first three periods, while the last period predicts positive net income (i.e., profit) possibly because of a comparatively higher carbon and electricity prices.

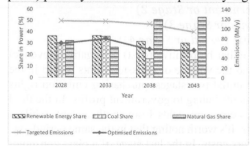

Figure 1. Minimisation of budget with direct optimisation approach

3.2. Minimisation of Emissions

Minimisation of emissions using direct optimisation (variable carbon prices) envisions an initial emphasis on renewable plants. Natural gas share remains prominent, facilitated by CCUS retrofit and AFs' usage. The coal share, though minimal, persists due to the use of AFs. In year 2043, the Malaysian power system is projected to achieve zero emissions target via direct optimisation (Figure 2). Higher predicted carbon prices lead to overall positive profits and low electricity prices (70, 125, 173, and 231 million USD/TWh, for 2028, 2033, 2038 and 2043, respectively). Through indirect optimisation model, the Malaysian power system again achieves zero Mt/y of CO_2 emissions by year 2043.

However, negative net income is projected for the initial three periods, while last period anticipates positive net income due to a relatively higher electricity and carbon prices.

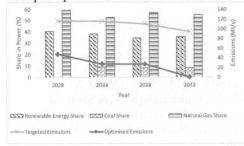

Figure 2. Minimisation of emissions with direct optimisation approach

3.3. Policy Recommendations for Malaysia

There is a crucial need for high carbon prices to ensure both environmental sustainability and economic viability. The old carbon prices forecasted by British Petroleum (British Petroleum, 2020) are deemed inadequate for the post-COVID energy dynamics. Carbon trading is essential to maintain low electricity prices in Malaysia. Immediate sizeable investments in renewables are recommended, prioritising their deployment in the initial phases of the planning horizon. Given the environmental goals, it is advisable to refrain from signing further coal agreements. Relying solely on renewables for meeting targets may not be economically viable, emphasising the importance of a diversified approach (e.g., adoption of emission mitigation technologies within existing fossil-based plants) for long-term sustainability, therefore, a revision of renewable energy targets is proposed.

4. Conclusions

This study underscores the significance of combining energy system planning with carbon emissions trading optimisation for the decarbonisation of emerging economies, with a specific focus on Malaysia. Through meticulous case studies and strategic optimisation approaches, research recommends the involvement of carbon trading policy within energy planning to meet financial and environmental goals. Direct optimisation model achieved 1.67 times lower emissions than the targets for budget minimisation and achieved zero emissions for emissions minimisation. This advocates Malaysia toward achieving its ambitious net-zero targets by year 2050. Case tests should be extended to encompass other ASEAN countries, delving into carbon trading for regional emission rights allocation and industrial emissions management.

Acknowledgement

This project is funded by the Foreign, Commonwealth & Development Office (FCDO), UK
through its call for proposals "Research and Innovation for Development in ASEAN (RIDA)"

References

Ahmad, S., & Tahar, R. M. (2014). Selection of renewable energy sources for sustainable development of electricity generation system using analytic hierarchy process: A case of Malaysia. *Renewable Energy*, *63*, 458–466. https://doi.org/https://doi.org/10.1016/j.renene.2013.10.001

Bhasker Nair, P. N. S., Foo, D. C. Y., Tan, R. R., & Short, M. (2022). A Software Framework for Optimal Multiperiod Carbon-Constrained Energy Planning. In Y.

Yamashita & M. Kano (Eds.), *Computer Aided Chemical Engineering* (Vol. 49, pp. 1177–1182). Elsevier. https://doi.org/https://doi.org/10.1016/B978-0-323-85159-6.50196-2

Bhasker Nair, P. N. S., Short, M., Foo, D. C. Y., & Tan, R. R. (2022). A Process Integration-Based Optimal Decarbonisation Policymaking Software Framework. In L. Montastruc & S. B. T.-C. A. C. E. Negny (Eds.), *32 European Symposium on Computer Aided Process Engineering* (Vol. 51, pp. 1603–1608). Elsevier. https://doi.org/https://doi.org/10.1016/B978-0-323-95879-0.50268-X

Bhasker Nair, P. N. S., Tan, R. R., Foo, D. C. Y., Gamaralalage, D., & Short, M. (2023). DECO2 - An Open-Source Energy System Decarbonisation Planning Software including Negative Emissions Technologies. In *Energies* (Vol. 16, Issue 4). https://doi.org/10.3390/en16041708

British Petroleum. (2020). Energy outlook 2020 edition. *London: BP*.

Ertugrul, H. M., Cetin, M., Seker, F., & Dogan, E. (2016). The impact of trade openness on global carbon dioxide emissions: Evidence from the top ten emitters among developing countries. *Ecological Indicators*, *67*, 543–555. https://doi.org/https://doi.org/10.1016/j.ecolind.2016.03.027

Hameed, G., Nair, P. N. S. B., Tan, R. R., Foo, D. C. Y., & Short, M. (2023). A novel mathematical model to incorporate carbon trading and other emission reduction techniques within energy planning models. *Sustainable Production and Consumption*, *40*, 571–589. https://doi.org/https://doi.org/10.1016/j.spc.2023.07.022

Hussain, N. (2011). Long Term Sustainable Energy Planning for Malaysia: A Modelling and Decision Aid Framework. *Journal of Energy and Environment*, *3*(1). https://journal.uniten.edu.my/index.php/jee/article/view/118

Izlawanie, M. (2021, November 3). *Challenges in implementing carbon pricing policy in Malaysia*. THE BARTLETT. https://www.ucl.ac.uk/bartlett/news/2021/nov/challenges-implementing-carbon-pricing-policy-malaysia

Kabanov, O. V., LyubovPudeyan, V. M., Sadunova, A. G., & Avlasevich, A. I. (2022). ASEAN ENERGY MARKET CRISIS. *Procedia Environmental Science, Engineering and Management*, *9*, 751–757.

Lee, M. Y., & Hashim, H. (2014). Modelling and optimization of CO2 abatement strategies. *Journal of Cleaner Production*, *71*, 40–47. https://doi.org/https://doi.org/10.1016/j.jclepro.2014.01.005

Mohd Salleh, N. H. (2022). *Malaysia Now*. https://www.malaysianow.com/news/2022/04/13/is-malaysia-ready-to-leave-coal-behind-in-renewable-energy-push

Open Access Government. (2021). *Environment News*. https://www.openaccessgovernment.org/asean-climate-change/123591/

Permatasari, Y. (2020). Building Indonesia through ASEAN economic community. *Journal of ASEAN Studies*, *8*(1), 81–93.

Sustainable Energy Development Authority. (2011). *NATIONAL RENEWABLE ENERGY POLICY*. https://www.seda.gov.my/policies/national-renewable-energy-policy-and-action-plan-2009/

Sustainable Energy Development Authority. (2021). *MALAYSIA RENEWABLE ENERGY ROADMAP (MYRER)*. https://www.seda.gov.my/reportal/myrer/

Tan, Y. (2021). *Climate Tracker Asia*. https://climatetracker.asia/fact-sheets/what-is-happening-to-malaysias-rainforests/

Flavio Manenti, Gintaras V. Reklaitis (Eds.), Proceedings of the 34th European Symposium on Computer Aided Process Engineering / 15th International Symposium on Process Systems Engineering (ESCAPE34/PSE24), June 2-6, 2024, Florence, Italy

Optimization of a reactive distillation process to produce propylene glycol as a high value-added glycerol derivative

Jahaziel Alberto Sánchez-Gómez[a*], Fernando Israel Gómez-Castro[a], Salvador Hernández[a]

[a]*Departmento de Ingeniería Química, División de Ciencias Naturales y Exactas, Campus Guanajuato, Universidad de Guanajuato, Noria Alta S/N, Guanajuato, Gto. 36050, Mexico. fgomez@ugto.mx*

Abstract

Glycerol is a key component that is currently obtained in significant volumes as a by-product in the production of biodiesel. Due to this increase on its production, the prize of crude glycerol has become lower. Moreover, its availability has triggered its potential to be used as a platform molecule for the generation of high value-added chemical products. In the present work, process intensification and optimization techniques are applied to the production of propylene glycol (PG) from glycerol. The design parameters of the reactive distillation scheme are adjusted by rigorous simulation and optimization using genetic algorithms through a link between Python and the Aspen Plus process simulator. The TAC value obtained for the reactive distillation scheme is 16.279×10^5 USD/y, with operation and capital costs of 8.711×10^5 USD/y and 37.838×10^5 USD, respectively.

Keywords: reactive distillation, crude glycerol valorization, process simulation, process optimization, propylene glycol production.

1. Introduction

Renewable glycerol is a by-product generated in the biodiesel production process by transesterification of triglycerides. Unlike fossil fuels, biodiesel is a renewable resource since it can be obtained from biomass. Moreover, it can reduce greenhouse gas emissions, being an environmentally friendly alternative (Rasrendra et al., 2023). The increase in global biodiesel production has led to an increase in glycerol generation. In 2020, the production of more than 4.6 billion liters of crude glycerol derived from biodiesel industry has been reported (REN 21, 2021). Due to this increase on its production, glycerol is considered a financial and environmental problem for the biodiesel industry, and the need arises to look for alternatives to take advantage of glycerol as a renewable and abundant source of carbon to obtain high value-added chemical products.

Propylene glycol (PG) is an important chemical used as antifreeze, humectant, solvent, and preservative. It is usually obtained by hydrolysis of propylene oxide, which is generated from petrochemicals. Alternatively, PG can be produced by a more environmentally friendly route by using renewable glycerol as a carbon and hydrogen source. Although cost competitiveness remains a major constraint, the conversion of glycerol to PG is a promising route. An approach to improve economic and energy indicators in the renewable PG production process involves applying process intensification (PI) strategies. The goal of PI is to achieve technological improvement in chemical processes to make them more efficient, reducing the size of equipment and/or the number of basic process operations by using multifunctional equipment (Devaraja and Kiss, 2022). Furthermore, there are few precedents in the literature on the design of intensified processes for the valorization of glycerol.

This work presents the intensification of the conventional process to produce propylene glycol from renewable glycerol. The intensified process includes the implementation of a reactive distillation system. Given the large number of decision variables involved in the reactive distillation scheme, as well as the non-linearity of the model that represents this type of system, its optimization is carried out by coupling the Aspen Plus simulation software with a genetic algorithm coded in Python, with the objective of minimizing the total annual cost.

2. Methodology

2.1. Process design

Considering the pressure conditions used in the process and the presence of organic mixtures with a large amount of water, the UNIQUAC model was selected as the thermodynamic model for the simulation of the process. This model has been previously used to describe satisfactorily the phase equilibrium in similar systems (Jimenez et al., 2020). The binary interaction parameters for the thermodynamic model were obtained from the Aspen Plus database.

The production of PG ($C_3H_8O_2$) from glycerol ($C_3H_8O_3$) has as by-products ethylene glycol ($C_2H_6O_2$), methanol (CH_3OH) and water (H_2O). The reactions involved in the generation of PG are shown in Eqs. (1) and (2). The hydrodeoxygenation of glycerol to produce PG occurs in the liquid phase, using Cu/ZrO_2 as catalyst. For this catalyst reaction, the power-law first-order kinetic model developed by Gabrysch et al. (2019) has been used. The dependence of the kinetic constant (k_I) on temperature is given by the Eq. (3), where $k_{0,I}$ is the pre-exponential factor, and $E_{A,I}$ is the activation energy. The parameters required for the calculation of reaction rates are summarized in Table 1.

$$C_3H_8O_3 + H_2 \underset{k_1}{\rightarrow} C_3H_8O_2 + H_2O \tag{1}$$
$$C_3H_8O_3 + H_2 \underset{k_2}{\rightarrow} C_2H_6O_2 + CH_3OH \tag{2}$$
$$k_I = k_{0,I} e^{-E_{A,I}/RT} \tag{3}$$

Table 1.- Pre-exponential factor and activation energy for reactions.

I	$k_{0,I}$	$E_{A,I}$ (kJ/mol)
1	2 x 10^{10} L mol^{-1} s^{-1}	106
2	9 x 10^7 L mol^{-1} s^{-1}	97

The conventional process (CS) for PG production is shown in Figure 1. The simulation of the conventional process is adapted from the work of Jimenez et al. (2020), in which the technical and economic evaluation of a PG production process using glycerol as raw material was performed. In this CS a plug flow reactor is employed for carrying out the glycerol hydrodeoxygenation reaction, followed by two distillation columns for the separation and recovery of PG with a purity of 99 mol%. In the bottom of Figure 1, the reactive distillation scheme (RDS) proposed for PG production is presented (Sánchez-Gómez et al., 2023). It consists of a reactive distillation column where the glycerol hydrodeoxygenation takes place, followed by a distillation column for the purification of PG. The aqueous glycerol and hydrogen feed streams have flowrates of 100 kmol/h and 16.4 kmol/h, respectively (1:1 by weight). In the simulation of these schemes, the RadFrac module of the commercial simulator Aspen Plus has been used. A total pressure drop of 10 psia has been assumed for each distillation column.

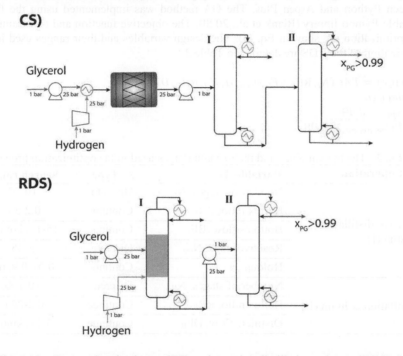

Figure 1.- Conventional scheme (CS) and reactive distillation scheme (RDS) for propylene glycol production.

2.2. Process optimization

In process optimization, it is necessary to select appropriate criteria to determine the potential of each processing scheme. Among these performance criteria, one of the most widely used is the total annual cost (TAC), which considers the capital and operating costs required in the process, Eq. (4). Operation cost (OC) includes those derived from the use of utilities, which are steam, cooling water, electricity, and catalysts. The operation time of the plant is 8600 hours per year. The unitary costs considered for these services are 12.247 USD/GJ for HP steam, 0.354 USD/GJ for cooling water, 0.0775 USD/kW for electricity and 143.87 USD/kg for catalyst (Atsbha, et al., 2021).

$$TAC = Operation\ Cost + \frac{Capital\ Cost}{Payback\ Period} \tag{4}$$

The capital cost (CC) considers the cost of distillation columns, trays, reactor, reboiler, condenser, pumps, and compressors. In the evaluation of these costs, the methodology proposed by Guthrie (Turton et al., 2003) was used, with the expressions updated to 2022. This methodology was used for coding mathematical functions in the Python programming language, which will be used later for the evaluation of the objective function in the optimization process. The design parameters of the process scheme will be adjusted by rigorous optimization using genetic algorithms (GA) and data transfer between Python and Aspen Plus. The GA method was implemented using the freely available Pymoo library (Blank et al., 2020). The objective function and constraints for the optimization are show in Eq. (5). The design variables and their ranges used in the optimization of the RDS are detailed in Table 2.

$$\min f(x) = TAC(N_I, RR_I, BF_I, N_{rea}, H, N_{II}, RR_{II}, DF_{II}) \tag{5}$$
$$subject\ to:$$
$$x_{PG} \geq 0.99$$
$$PG_{recovery} \geq 0.98$$

Table 2.- Design variables and their search ranges used in the optimization process.

Unit operation	Variable, ID	Type	Search range
Reactive distillation column (I)	Number of stages, N_I	Discrete	10-50
	Reflux ratio, RR_I	Continue	0.2-5.0
	Bottoms flow, BF_I	Continue	15-17, kmol/h
	Reactive stage, N_{rea}	Discrete	5-45
	Holdup, H	Continue	0.01-0.9, m^3
Distillation column (II)	Number of stages, N_{II}	Discrete	10-100
	Reflux ratio, RR_{II}	Continue	0.2-20.0
	Distillate flow, DF_{II}	Continue	15-17, kmol/h

3. Results and discussion

The optimized diagram of the reactive distillation scheme (RDS) is show in Figure 2. In Column I, with 29 stages, glycerol and hydrogen feed streams are introduced in stages 23 and 28, respectively. These compounds are transformed into PG in the reactive zone between stages 24 to 27. After removing the water in the top of the column, the PG produced is obtained with a mole fraction of 0.967 in the bottoms stream. Then, this PG is depressurized and sent to the Column II. In Column II, with 51 stages, the PG is fed to middle of the column (stage 25). At the bottom of the column, ethylene glycol and glycerol are found, while PG is obtained as top product with a purity of 99 mol%. Additionally, the reboiler duty of the Column I represent the highest energy consumption (1364 kW), because of the large volume of water present in the feed mixture and the pressure conditions required in the reactive distillation process.

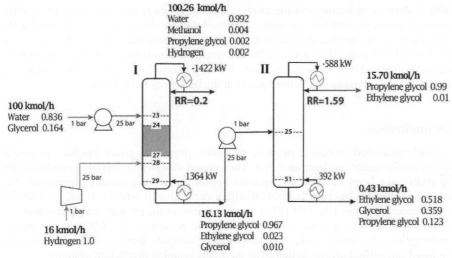

Figure 2.- Optimized RDS for propylene glycol production.

The TAC obtained for the RDS is 16.279×10^5 USD/y, with OC and CC of 8.711×10^5 USD/y and 37.838×10^5 USD, respectively. The comparation of RDS and the CS is shown in Table 2. Compared with the CS, the TAC increased by 10.19 % because of the increased operating conditions of the reactive distillation process. However, due to the combination of reaction and separation processes in a single process unit, the operating costs decreased from 11.40×10^5 USD/y for the CS to 8.71×10^5 USD/y for the RDS. In this sense, although the RDS is associated to a higher TAC, the implementation of this scheme allows a higher energy efficiency, which is also present in the reduction of 28.9 % of the energy requirements. Although it is not estimated here, this has a direct impact on the carbon dioxide emissions associated to the production process.

Table 2.- Comparison of the intensified and conventional processes for PG production.

Parameters, unit	CS	RDS
TAC, 10^5 USD/y	**14.77**	**16.27**
OC, 10^5 USD/y	**11.40**	**8.71**
Steam, 10^5 USD/y	9.35	6.65
Cooling water, 10^5 USD /y	0.3	0.22
Electricity, 10^5 USD /y	0.6	0.61
Catalyst, 10^5 USD /y	1.15	1.23
CC, 10^5 USD	**16.82**	**37.83**
Vessel	1.91	2.62
Trays	1.09	1.49
Reboiler	5.25	27.14
Condenser	5.94	3.91
Compressor	2.38	2.38
Bomb	0.25	0.25
Total Cooling Duty, kW	2789.31	2011.63
Total Heating Duty, kW	2467.69	1757.38
Energy Intensity, kW/kg of PG	**2.07**	**1.47**

Table 2 shows that heating and the catalyst account for more than 70 % and 10 % of the operating costs, respectively, in both schemes. Additionally, it can be observed that the capital cost of the heating system (reboiler) increases from 5.25×10^5 USD in the CS to 27.14×10^5 USD in the RDS because the reactive distillation column (Column I) operates at a pressure of 25 bar, which is necessary for the glycerol hydrodeoxygenation reaction, generating considerably high temperatures at the bottom of the column.

4. Conclusions

A novel intensified process to produce renewable propylene glycol has been proposed, and its performance has been analysed. The optimal design and operating conditions for the RDS have been obtained by rigorous optimization using genetic algorithms. The intensified process has been contrasted with a conventional scheme. It has been shown that an increase in the TAC (10.19 %) occurs because of the pressure conditions required in the RDS. However, this scheme allowed the reduction of 28.9% of the energy consumption. As future work, it is proposed to compare these results with the other proposed intensified process schemes as thermally coupled distillation systems.

References

T. A. Atsbha, T. Yoon, B. H. Yoo, C. J. Lee. 2021, Techno-Economic and Environmental Analysis for Direct Catalytic Conversion of CO2 to Methanol and Liquid/High-Calorie-SNG Fuels, Catalysts, 11, 687.

J. Blank, K. Deb, 2020, pymoo: Multi-Objetive Optimization in Python, IEEE Access, 8, 89497-89509.

D. Devaraja, A. A. Kiss, 2020, Novel intensified process for ethanolamines production using reactive distillation and dividing-wall column technologies, Chem. Eng. Process. Process Intensif., 179, 109073.

T. Gabrysch, M. Muhler, B. Peng, 2019, The kinetics of glycerol hydrodeoxygenation to 1,2-propanediol over Cu/ZrO2 in the aqueous phase, Appl. Catal. A Gen., 576, 47–53.

R. X. Jiménez, A. F. Young, H. L. S. Fernandes, 2020, Propylene glycol from glycerol: Process evaluation and break-even price determination, 158, 181-191.

C. B. Rasrendra , N. T. U. Culsum, A. Rafiani, G. T. M. Kadja, 2023, Glycerol valorization for the generation of acrylic acid via oxidehydration over nanoporous catalyst: Current status and the way forward. Bioresour. Technol. Rep., 23, 101533.

REN 21, 2021, Renewables 2021 Global Status Report - REN21, in https://www.ren21.net/wp-content/uploads/2019/05/GSR2021_Full_Report.pdf.

J. A. Sánchez-Gómez , F. I. Gómez-Castro, S. Hernández, 2023, Design and intensification of the production process of propylene glycol as a high value-added glycerol derivative. Comput. Aided Chem. Eng., 52, 1915-1920.

Printed and bound by CPI Group (UK) Ltd, Croydon, CR0 4YY

03/10/2024

01040328-0006